Biomathematical and Biomechanical Modeling of the Circulatory and Ventilatory Systems

Volume 5

For further volumes:
http://www.springer.com/series/10155

Marc Thiriet

Tissue Functioning and Remodeling in the Circulatory and Ventilatory Systems

Springer

Marc Thiriet
Project-team INRIA-UPMC-CNRS REO
Laboratoire Jacques-Louis Lions, CNRS UMR 7598
Université Pierre et Marie Curie
Place Jussieu 4
75252 Paris Cedex 05
France

ISSN 2193-1682 ISSN 2193-1690 (electronic)
ISBN 978-1-4614-5965-1 ISBN 978-1-4614-5966-8 (eBook)
DOI 10.1007/978-1-4614-5966-8
Springer New York Heidelberg Dordrecht London

Library of Congress Control Number: 2012952271

Printed on acid-free paper

Springer is part of Springer Science+Business Media (www.springer.com)

Preface

"πάντες ἄνθρωποι τοῦ εἰδέναι ὀρέγονται φύσει."
[All men by nature desire to know.]
(Aristotle [−384 – −322], Metaphysics, Book I)

Modeling relies on access to data accumulated on any target object (configuration, structure, controlled functioning, fate under various types of stimuli, reaction to stresses that can be experienced, etc.). Knowledge is associated with a set of conditions: truth, belief justification, reliability, and stability. Understanding does not necessarily bring certitude before validation, especially in the absence of observation data.

Volumes 1 to 4 of this series present events and their molecular participants that govern cell fate — intra-, auto-, juxta-, and paracrine — as well as possible endocrine regulations upon diverse types of stimuli, in the context of modeling and simulations. These events rely on cascades of chemical reactions that successively activate signaling mediators within the cell, at the nanoscopic scale. At the micro- and mesoscopic scale, biological phenomena comprise isolated and collective behavior of cells in interaction with their environment, either apposed cells, or the extracellular matrix. The present volume, Volume 5, treats mural tissues of airways and blood and lymph vessels — events occurring at the mesoscopic scale that can be targeted by mathematical models. Subsequent volumes focus on diseases associated with air and blood flows (Vol. 6) and mechanics of these flows (Vol. 7). Volume 8 is a set of glossaries.

Biological tissues are made up of composite material with different cell types in an extracellular matrix. Both types of tissue constituents — cells and matrix — are reinforced by filaments and fibers. In addition, the extracellular matrix can contain lamellae. Structural hierarchy relies on the multilevel structure of biological tissues. The smallest structural levels are related to biological components, i.e., macromolecules (scale $\mathcal{O}[1]$ nm), microfibrils ($\mathcal{O}[1]$ and $\mathcal{O}[10]$ nm]), fibers ($\mathcal{O}[100]$ nm), lamellae ($\mathcal{O}[1–10]$ μm). The fifth structural level usually corresponds to tissue elementary structure (scale $\mathcal{O}[100]$ μm). Biological tissues are characterized by an optimal combination of configuration and structure at a given time with appropriate chemical and physical features to achieve their functions.

Biological tissues are capable of short- and long-term adaptation (remodeling) to applied constraints that causes change in configuration and structure. These constraints include mechanical stresses. The mechanical behavior of biological tissues depends on (1) type of applied forces (tension, compression, shear, and torsion) and their possible combination; (2) loading orientation, magnitude, duration, and rate; (3) eventual periodicity (frequency); and (4) state of the surrounding environment and organism that regulate perfusion (hence, temperature, moisture, etc.).

The rheological properties of a sample of biological tissues that are related to tissue composition are usually described by a set of parameters (bulk, elasticity, and shear moduli in principal tissue directions;[1] Poisson's ratio, Lamé's parameters, bending stiffness, flexural modulus, ultimate tension, compression, and shear stress; etc.).

Even in normal conditions, biological tissues remodel to adapt to their environment. For instance, the thickness of arterial walls in a given vessel section can vary in the azimuthal direction, as it depends on the local stress field applied by the flowing blood. Tissue engineering that does not incorporate mechanical stresses fails.

Abnormal situations are also characterized by maladaptive tissue remodeling (Vol. 6). Surgical procedures such as grafting and minimally invasive, catheter-based implantation of medical devices such as stents aimed at correcting risky tissue growth can themselves trigger another type of aberrant tissue growth, such as intimal hyperplasia.

Especially when they aim at predicting organ functioning, disease evolution, and drug delivery up to cells with abnormal functioning, mathematical models not only focus on cellular events, but also on biological tissues. Major scientific developments in health technology are aimed at integrating available fragmented data and models to improve medical decision-making and actions (gestures), reduce healthcare costs, and optimize the design of implantable devices.

This strategy copes with many challenges. Among them, modeling that span multiple structural levels (from molecular to entire organ). Therefore, the three basic

[1]Viscoelastic materials exhibit a phase lag (ϕ) between stress ($\mathbf{c}$) and strain ($\mathbf{e}$):

$$\mathbf{c} = \mathbf{c_{max}} sin\{\omega t\},$$
$$\mathbf{e} = \mathbf{e_{max}} sin\{\omega t + \phi\}. \tag{0.1}$$

Complex elastic (E^*) and shear (G^*) moduli

$$E^* = E_{Re} + \imath E_{Im}, \quad G^* = G_{Re} + \imath G_{Im}, \tag{0.2}$$

are expressed from tensile ($\{E_{Re}, E_{Im}\}$) and shear ($\{G_{Re}, G_{Im}\}$) storage ($\mathbf{c}/\mathbf{e} \cdot \cos\phi$) and loss ($\mathbf{c}/\mathbf{e} \cdot \sin\phi$) moduli. Storage and loss moduli represent stored and dissipated energy of viscoelastic tissues, respectively. Moreover, viscoelastic materials can be described by a low and a high modulus at high and low temperatures, respectively, and the storage modulus by a low and a high rate modulus.

natural sciences — biology, chemistry, and physics — interact with mathematics to optimize proper descriptions of the functioning and regulation of air and blood flows.

Multiscale computational models currently being tackled include: (1) the heart pump functioning in the normal and diseased states; (2) the transport of drug nanoparticles that target a specific body region; (3) the insertion of an implantable medical device inside a blood vessel, and (4) the transfer of drugs across the arterial wall and of therapeutic aerosols within the lumen and walls of the respiratory tract. However, despite advances in modeling patient anatomy, biophysical models remain difficult to efficiently personalize.

Whereas mathematical models focus on tissue behavior from the nanoscopic to the mesoscopic scale, biomechanical modeling and simulations concentrate on macroscopic events. Pumps and conduits walls are constituted by various types of biological tissues. These living composite materials sense, transmit, and react to forces exerted by flowing physiological fluids, such as air and blood. The behavior of cells and tissues is tightly coupled to flow pattern.

Major features of flows are described by the equations of mass and momentum conservation, the so-called Navier-Stokes equations. However, because of the complicated architecture of networks of large and mid-size airways and blood vessels that are assumed to convey a homogenous fluid, the Navier-Stokes equations cannot be solved analytically and require numerical simulations.

In addition, both air and blood flow inside deformable conduits. Therefore, the fluid dynamics depend not only on the geometry of the computational domain, on the boundary conditions applied at domain inlets and outlets,[2] and on the values of flow governing parameters, but also on the constitutive law of vessel walls, as well as that of blood, when a non-Newtonian behavior must be incorporated, i.e., in the presence of stagnant flow regions.

Blood is propelled from cardiac ventricles into arterial trees during systole once the pressure difference between the ventricular and arterial sides of the ventriculo-arterial valves causes leaflet displacement and change in shape. Flow deceleration then provokes valve closing. The structure of the valved orifice can be divided into four main domains with their own geometrical and rheological properties: arterial root, valve base, leaflet, and coaptation zone (contact surfaces). Therefore, coupling cardiac contraction to blood ejection requires tackling a strong contact phenomenon.

Heart electromechanical coupling provides a good example of a feedback loop. The electrochemical wave caused by activation and deactivation of a set of ion channels, pumps, and exchangers on the surface of as well as inside nodal cells and cardiomyocytes triggers the ventricular contraction that, in turn, strains nodal cells and cardiomyocytes, thereby influencing the activity of mechanosensitive ion channels (mechanotransduction).

---

[2]Entry and exit segments should be extended in the vessel axis direction to limit the sensitivity to the boundary conditions in the region of interest and take into account up- and downstream effects of three-dimensional flows.

The respiratory epithelium that covers the luminal face of airways secretes a bilayered fluid-gel coating and surfactant in large and small ducts, respectively. Mucus in large airways clears entrapped particles from the respiratory tract. Surfactant permits a proper inflation and deflation of bronchioles and alveoli.

This set of textbooks devoted to Circulatory and Ventilatory Systems in the framework of Biomathematical and Biomechanical Modeling aims at providing basic knowledge and state of the art on the biology and the mechanics of blood and air flows. The cardiovascular and respiratory systems are tightly coupled, as their primary function is the supply of oxygen (O_2) to and removal of carbon dioxide (CO_2) from the body's cells.

The present volume comprises 14 chapters. These chapters focus on the biological tissues of the cardiovascular and ventilatory apparati (i.e., histology) and their functioning. Chapter 1 covers blood, a concentrated suspension of circulating cells in a solvent, the plasma. Blood is a peculiar type of connective tissue used for transport and body regulation. All blood cell lineages derive from hematopoietic stem cells (Chap. 2). Hematopoiesis thus aims at maintaining a steady cell density in the blood circulation (source term). The structure and function of blood cells — erythrocytes, leukocytes, and thrombocytes — are described in chapter 3. The circulatory network is connected to a specialized plasma-recycling compartment, the lymph collector. The latter conveys the lymph in lymphatic vessels (Chap. 4).

Chapters 5 to 9 cover the structure and composition of the heart and vessel walls. Chapter 5 focuses on cardiomyocyte structure and function. Chapter 6 on heart wall highlights nodal cells that constitute the cardiac natural pacemaker and conduction routes for the propagation of the electrochemical command. Blood vessel walls are studied in chapter 7 with a special emphasis on the blood–brain barrier. Chapter 8 examines the activity of the vascular and respiratory smooth myocytes that regulate the caliber of blood vessels and airways. Endothelial cell at the interface between the flowing blood and vessel wall is assigned to chapter 9. These cells play a role in clotting, extravasation of flowing cells, and transduction of mechanical stresses into chemical cues. Mechanotransduction leads to the synthesis of several substances that regulate the vasomotor tone.

Chapter 10 presents the regulated development of blood (vasculo- and angiogensis) and lymph vessels (lymphangiogensis). Chapter 11 discusses some biological processes, such as tissue growth as well as remodeling and repair, which are common to any body's tissue. The final two chapters are devoted to the coating of the respiratory tract. Chapter 12 focuses on the mucus layer that lies over a periciliary fluid in large airways as well as in medium-sized and small bronchi, while chapter 13 deals with the airway-lining surfactant in pulmonary acini.

Abbreviations and Notation

Common abbreviations such as "a.k.a." ("also known as") are used throughout the text to lighten sentences. Latin-derived shortened expressions are also widely utilized: "e.g." (exempli gratia) and "i.e." (id est) mean "for example" and "in other words", respectively. The notation mode of molecule aliases is introduced in the appendix.

A physical quantity associated with a given point in space at a given time can be: (1) a scalar uniquely defined by its magnitude; (2) a vector characterized by a magnitude, a support, and a direction represented by an oriented line segment defined by a unit vector; or (3) a tensor specified by a magnitude and a few directions. To ensure a straightforward meaning of symbols used for scalar, vectorial, and tensorial quantities, bold face upper and lower case letters **T** and **v** are used to denote a tensor and a vector, respectively, whereas both roman (plain, upright)-style upper and lower case letters designate a scalar.

Acknowledgments

These books result from lectures given at University Pierre et Marie Curie in the framework of prerequisite training of Master "Mathematical Modeling", part of Master of "Mathematics and Applications", Tbilisi State University, Centre de Recherches Mathmatiques,[3] and Taida Institute for Mathematical Sciences,[4] the latter two in the framework of agreements with the French National Institute for Research in Computer Science and Control.[5] These lectures mainly aim at introducing students in mathematics to basic knowledge of biology, medicine, rheology, and fluid mechanics in order to conceive, design, implement, and optimize appropriate models of biological systems at various length scales in normal and pathological conditions. These books may also support the elaboration of proposals following suitable calls of granting agencies. The author takes the opportunity to thank the members of ERCIM office (European Consortium of Public Research Institutes) and all of the participant teams of the working group "IM2IM" that yields a proper framework for such proposals. The author thanks especially Springer staff members S.K. Heukerott and D. Packer for their help and comments.

The author, an investigator from the French National Center for Scientific Research[6] wishes to acknowledge members of the INRIA-UPMC-CNRS team, "REO",[7] of Laboratoire Jacques-Louis Lions,[8] of CRM (Y. Bourgault, M. Delfour, A. Fortin, and A. Garon), being a staff member in these research units, as well as C.S. Lin and T.W.H. Sheu from Taida Institute for Mathematical Sciences, and R. Botchorishvili from Department of Numerical Analysis of the Vekua Institute of Applied Mathematics at the Tbilisi State University.

[3]CRM (www.crm.umontreal.ca).

[4]TIMS (www.tims.ntu.edu.tw).

[5]Institut National de la Recherche en Informatique et Automatique (INRIA; www.inria.fr).

[6]Centre National de la Recherche Scientifique (CNRS; www.cnrs.fr).

[7]www-roc.inria.fr/reo

[8]LJLL (www.ann.jussieu.fr).

The author also acknowledges the patience of his wife Anne, daughter Maud, sons Alrik and Damien, and their respective families (Julien, Jean, and Louis; Raphaëlle, Matthieu, and Alexandre; Joanna and Frédéric).

Contents

Chapter 1
Blood

"... there is a continuous and uninterrupted movement of blood from the heart through the arteries to the body as a whole, and likewise back from that body as a whole through the veins to the heart, with such flow and ebb that in such quantity and amount that it must somehow move in circle." (A letter from W. Harvey to C. Hofmann [1])

Blood is a major tissular component of the closed circulatory system, which supplies nutrients and oxygen using hemoglobin-containing red blood capsules to the body's cells and removes metabolic wastes. This specialized liquid is composed of cells suspended in a liquid, the plasma (mostly water). It flows throughout the body in blood vessels due to a pressure difference set by the heart pump.

Blood cells are involved in tissue adaptation to hypoxia via angiogenesis as well as treatment of injury via blood coagulation and healing and other types of tissue damage, such as that caused by infection, by triggering inflammation and immunity. Carried cells, like vessel wall cells, sense and respond to mechanical stresses exerted by the flowing blood.

Blood is propelled throughout the circulatory system by the heart. Blood is ejected from the left and right cardiac ventricles into the systemic and pulmonary circulation, respectively. In the heart, nodal myocytes of the natural pacemaker — the sinoatrial node — spontaneously and rhythmically generate an electrochemical wave, or the so-called action potential, which characterizes heart automatism. This action potential then propagates down to other atrial and ventricular nodal cells and cardiomyocytes. The latter also propagate the action potential and contract to propel the blood. Coupling of (1) generation and propagation of the action potential (heart electrical activity); (2) cardiac wall contraction and deformation (solid mechanics aspects of heart activity); and (3) blood filling in and ejection out of both coupled ventricular pumps associated with closing and opening of auriculoventricular and ventriculo-arterial valves and myocardium perfusion is a non-trivial task. This problem has thus been tackled by splitting and separately processing the major processes. Today, accumulated knowledge, appropriate modelings, and updated technologies allow their coupling.

M. Thiriet, *Tissue Functioning and Remodeling in the Circulatory and Ventilatory Systems*, Biomathematical and Biomechanical Modeling of the Circulatory and Ventilatory Systems 5, DOI 10.1007/978-1-4614-5966-8_1, © Springer Science+Business Media New York 2013

The cardiac pump sets up a pressure wave that impinges strongly upon walls of adjacent elastic arteries, which are then capable of ensuring a blood flow during the ventricular diastole. The blood vessel wall is a living tissue that quickly reacts to loads applied on it by the flowing blood. In any segment of the vasculature, endothelial and smooth muscle cells sense space and time variations in small-magnitude wall shear stress and large-magnitude wall stretch generated by the flowing blood. These cells respond with a short time scale (from seconds to hours) to adapt the vessel caliber according to the loading, especially when changes exceed the limits of the usual stress range. The mechanotransduction pathways determine the local vasomotor tone and subsequently the lumen bore of the reacting blood vessel. This regulatory mechanism is much faster than the nervous and hormonal control. It sets up the level of local resistance to blood flow: (1) in large arteries, which ensure the blood distribution in the body, to limit the cardiac postload; (2) in small resistive arteries, which irrigate the body's tissues, to maintain the flow rate; and (3) in arterioles, which perfuse a cell population, to adapt the blood supply to metabolic needs of active cells.

1.1 Blood Functions

The blood performs 3 major functions: (1) transport through the body; (2) regulation of bulk equilibria; and (3) body immune defense against foreign bodies.

Blood supplies oxygen, hence energy, and conveys nutrients (vitamins, mineral ions, glucose, amino acids, fatty acids, among other glucids, protids, and lipids) to the tissues and removes carbon dioxide and waste products of cell metabolism toward lungs and purification organs. Kidneys filter blood. Toxins are not only removed in urine, but also by sweating. Blood transmits metabolic factors and messengers such as hormones to target organs.

Blood volume and electrolyte concentration are regulated. Blood maintains the body temperature (36.4–37.1 °C) and acid–base equilibrium, controlling blood pH that remains in the range 7.35 to 7.45.[1] Blood circulation transports heat throughout the body, thereby contributing to the thermoregulation. Increased blood flow at the body's surface during warm weather or exercise permits higher heat loss. Conversely, when the ambient temperature decays, blood flow near the skin lowers to prevent heat loss.

Blood participates in the body's defense against infection, as it transports immunocytes and antibodies, as well as in repair after injury. Innate immunity yields

[1]The major blood buffer is composed of bicarbonate ions, from which carbonic acid and water can be obtained. Carbonic acid is unstable and decomposes into carbon dioxide and water.

the first stage of defense against invading pathogens and signals for the development of adaptive immunity.[2]

Blood limits its own loss through damaged vessel walls by blood coagulation, or clotting (Chap. 9). Blood coagulation is a component of hemostasis, i.e., cessation of blood flow through a damaged vessel wall. Clot formation covers the injury site by a platelet and fibrin-containing material to stop hemorrhage and begin repair of the damaged vessel. Platelets immediately form a plug (primary hemostasis) that is simultaneously reinforced by fibrin strands, which result from the coagulation cascade of activation of coagulation factors (or clotting factors; secondary hemostasis).

Blood contains living cells and plasma (Table 1.1). Eight to twelve hours after a meal, 100 ml of blood contains 19 to 23 g of solids and 77 to 81 g of water. Blood cells include red blood cells (RBC), or erythrocytes, white blood cells (WBC), or leukocytes, and platelets, or thrombocytes.

Leukocytes operate in immunity. Five main classes of leukocytes exist: 3 types of *granulocytes* that have about the same size — neutrophils, eosinophils, and basophils — and 2 types of *agranular leukocytes* — lymphocytes and monocytes. Leukocyte lifespan in blood is several days.

1.2 Plasma

Plasma constitutes a body fluid subcompartment. The body fluids indeed form several compartments. The 2 major compartments include intra- and extracellular fluids. The extracellular fluids comprise: (1) interstitial fluids;[3] (2) plasma; and

[2]The detection of both commensal and pathogenic microbes that trigger mutualistic or antagonistic interactions with host cells relies on host receptors and microbial sensing pathways. Two types of immunity — innate and adaptive — are used to protect the host from infections. The innate immune system is genetically programmed to detect invariant features of invading microbes. Non-specific innate responses are mainly mediated by natural killer and myelomonocytic cells. They are activated on engagement of pattern recognition receptors expressed by microorganisms and host cells (Vol. 3 – Chap. 11. Receptors of the Immune System). The initial inflammation (Chap. 11) aims at increasing the cytotoxic function of immunocytes and clearance of eliciting agents and, later, to tissue repair and immunity regression. Cells of the innate immunity — dendritic cells, macrophages, and neutrophils, among others — trigger specific adaptive immune responses. They internalize pathogens and dying cells and process them into peptidic antigens that are presented with major histocompatibility complex (MHC) molecules to conventional T lymphocytes. They integrate danger and inflammation signals and stimulate proper B- and T-cell responses. Cells of adaptive immunity and B and T lymphocytes employ antigen receptors that are not encoded in the germ line, but are generated owing to immunological memory that relies on expansion of T- and B-cell clones, yield stronger and faster responses after repeated exposure to a given eliciting antigen, after engagement of B- (BCR) and T-cell (TCR) receptors.

[3]The interstitial fluid located between cells corresponds to the interstice or interstitium.

Table 1.1. Blood composition and main characteristics in healthy adult male. The blood cells include erythrocytes (red blood cells), leukocytes (white blood cells), and platelets. Leukocytes are divided into 5 classes based on morphological and tinctorial characteristics. Neutrophils, eosinophils, and basophils are known as granulocytes due to granules in the cytoplasm. Monocytes and lymphocytes are involved in the body scavenging and defense. The blood plasma consists of water (90%), the remainder being electrolytes (sodium [Na^+], 142 mmol/l; chloride [Cl^-], 102 mmol/l; and potassium [K^+], 5 mmol/l), carbohydrates, lipids, and amino acids, etc.

Erythrocytes	$4.5–5.2 \times 10^6/mm^3$
Hematocrit	41–47%
Leukocytes	$4–10 \times 10^3/mm^3$
Neutrophils	40–70%
Eosinophils	1–2%
Basophils	0.5–1%
Lymphocytes	20–40%
Monocytes	2–10%
Platelets	$2–4 \times 10^5/mm^3$
Ions	295–310 mEq/l
Protids	70–80 g/l
Lipids	5–7 g/l
Glucids	0.8–1.1 g/l
pH	7.39–7.41
Osmotic pressure	280–300 mosm

Table 1.2. Approximative water content of body fluid compartments (l).

Intracellular space	27–30
Extracellular space	14–17
Interstitial fluid	11–13
Plasma	3–4
Total body water	41–47

(3) minor components, such as lymph,[4] cerebrospinal fluid, digestive secretions, aqueous humor, and pleural, pericardial, and synovial fluids for pleura, pericardium, and joint lubrication. About 64% of body water is found in the intracellular space, about 25% in the intertitium (∼75% of extracellular liquid, about 8% in the plasma, and about 3% in the minor compartments (Table 1.2).

[4]Because of the difference in tissue and blood pressures, blood plasma continuously leaks from the entrance segment of capillaries. Close-ended lymphatic capillaries return protein-rich lymph via lymphatic vessels and lymphaticovenous junctions back to the venous compartment of the blood systemic circulation.

1.3 Plasma Constituents

Plasma represents approximately 55% of the blood volume. The remaining is *hematocrit* (Ht), i.e., percent of packed cells[5] (Ht 38–46% in women, 42–53% in men). Plasma is mainly composed of water, a suspending fluid (or solvent) for various solutes (Table 1.3). Plasma contains 92% water, 8% proteins (~ 7 g/dl), and other substances.

1.3.1 Electrolytes

Electrolytes, or ions, contribute to the osmotic pressure (Π), which is mainly regulated by the kidneys. Major electrolytes are Na^+, K^+, Ca^{2+}, Mg^{2+} cations, and HCO_3^-, Cl^-, HPO_4^{2-}, and SO_4^{2-} anions.

Cations and anions are unevenly distributed in body fluid compartments (Table 1.4). Sodium ion (Na^+) is the major cation and chloride (Cl^-) the major anion outside the cell. Inside the cell, potassium ion (K^+) is the major cation and phosphate (HPO_4^{2-}) the major anion. At physiological pH, proteins are negatively charged.

Ion transport through the cell membrane between the body fluid compartments requires specialized plasmalemmal proteins (Fig. 1.1 and Vol. 3 – Chaps. 2. Membrane Ion Carriers to 5. Receptors of Cellular Trafficking). Ion displacements in the different compartments affect the blood volume.

Table 1.3. Plasma approximate composition (%). Plasma, the suspending fluid for peripheral blood cells, is composed of water, electrolytes, proteins (albumins 60–80%, globulins 16–36%, and fibrinogen 4%), amino acids, lipoproteins, other lipids, and glucids. Hormones, vitamins, and enzymes are conveyed by blood. The normal plasma volume is 40 to 90 ml/kg of body weight.

Water	~92
Proteins	~7.0
Electrolytes	~0.9
Lipids	~0.6
Glucids	~0.1

[5]Because erythrocytes (RBC) represent 97% of blood cell volume (Table 3.1), Ht estimates RBC concentration. The normal sex-dependent range has different limits according to different research teams: $Ht = 0.47 \pm 0.07$ and $Ht = 0.42 \pm 0.05$ in male and female subjects, respectively.

Table 1.4. Distribution of ions and proteins (mEq/l) in intra-, and extracellular spaces, and plasma.

	Cell	Intertitium	Plasma
Sodium (Na^+)	10–15	130–150	135–145
Calcium (Ca^{2+})	10^{-4}	2–5	5–10
Potassium (K^+)	135–150	4–5	3–5
Magnesium (Mg^{2+})	30–35	1–3	1–3
Cations	180	152	155
Chloride (Cl^-)	3–9	108–125	100–108
Bicarbonate (HCO_3^-)	10–12	27–30	23–28
Phosphate (HPO_4^{2-})	40–80	2–3	1–3
Sulfate (SO_4^{2-})	20	1	1
Proteins	35–55	~ 0	14–16
Organic acid	20	5	5–6
Anions	180	152	155

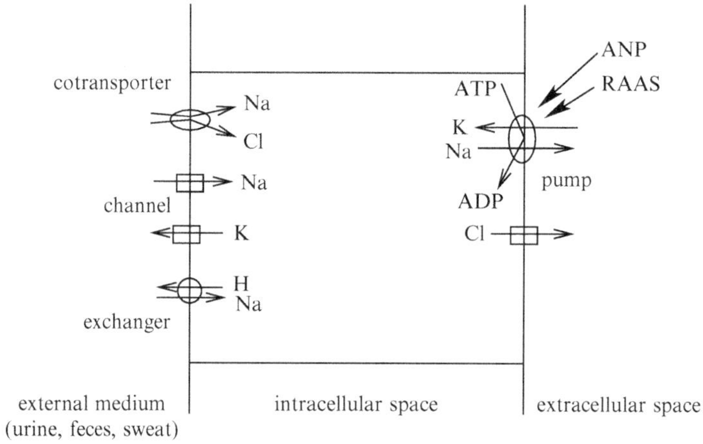

Fig. 1.1 Types of ion motions in the 3 main fluid compartments, extracellular (particularly the blood), intracellular, and external media

1.3.2 Glucids

Glucids are composed of: (1) oligosaccharides (glucose, fructose, and galactose); (2) disaccharides (saccharose, lactose, and maltose); and (3) polysaccharides (glycogen). Glucose and fructose can be used by cells. Other glucids require degradation into glucose. Glucids are important nutrients because they are energy sources.

Table 1.5. Plasma protein composition (%).

Albumin	∼0.60
α1-Globulin	∼0.04
α2-Globulin	∼0.08
β-Globulin	∼0.12
γ-Globulin	∼0.16
Fibrinogen	∼0.03

Glycemia (∼ 1 g/l) is the blood glucose concentration, which depends on the exogenous supply and degradation of hepatic glycogen.[6] Glycemia is stabilized by 2 pancreatic hormones, insulin and glucagon. Insulin decreases glucose level by cell use and storage, especially in the liver and the muscles. Glucagon increases glucose concentration. When glycemia falls below 70 mg/dl, parasympathetic nerves are stimulated and hormones are released (adrenaline, cortisol, glucagon, and growth hormone) to limit glucose uptake. On the other hand, hyperinsulinemia occurs to compensate for insulin resistance.

1.3.3 Plasma Proteins

Serum is plasma without fibrinogen (195–365 mg/dl) and other clotting factors. Fibrinogen acts on erythrocyte aggregation,[7] hence influencing blood rheology (Vol. 7), and blood coagulation (Sect. 9.8). Main non-protein nitrogens (NPN) are urea, uric acid, creatine, creatinine, ammonium salts, and amino acids.

Serum proteins are composed of albumin and globulins (Table 1.5). Plasma proteins are responsible for osmotic pressure that maintains fluid balance across capillaries (∼3.3 kPa [∼25 mmHg]; Vol. 6 – Chap. 4. Cardiovascular Physiology).

Albumin is the main plasma protein (3.0–4.5 g/dl; plasma half-life 15–19 days; ∼60% of total protein) synthesized by hepatocytes. Newly synthesized albumin is secreted into the circulation at the rate of about 15 g/d in humans. It binds many small molecules for transport in the blood and participates in blood colloidal osmotic pressure (Π), which keeps fluids within the vasculature. Plasma albumin indeed yields about 65% of osmotic pressure, whereas globulins and fibrinogen contribute according to their plasma concentrations. Molecular structure and charge of albumin

[6]The best measurement technique of glycemia targets the percentage of glycated hemoglobins (hemoglobin A1c) that reflects the blood sugar levels over the course of a few months. In healthy individuals, less than 6% of hemoglobin is bound to glucose.

[7]Fibrinogen bridges erythrocytes to form rouleaux. Erythrocyte aggregation depends on the concentration of fibrinogen and β-globulins. The *fractal dimension* of 2D finite RBC aggregates is equal to 1.59 ± 0.03 [2], which is similar to the result (1.55) of a chemically limited aggregation model [3], which gives in 3D a fractal dimension of 3.

Table 1.6. Size and lipid content (%) of lipoproteins.

	Chylomicron	VLDL	LDL	HDL
Size (nm)	80–500	∼50	∼20	∼10
Cholesterol	∼0.02	∼0.07	∼0.08	∼0.04
Cholesterol esters	∼0.03	∼0.12	∼0.42	∼0.15
Phospholipids	∼0.07	∼0.18	∼0.22	∼0.30
Triglycerides	∼0.86	∼0.55	∼0.06	∼0.04
Proteins	∼0.02	∼0.08	∼0.22	∼0.47

facilitate the cotransport of numerous hydrophobic molecules, such as enzymes and hormones, across the endothelium (Sect. 9.6). In the interstitial space, it also serves as a major interstitial osmotic agent, hence transendothelial osmotic pressure gradient. Extravasated albumin is recycled into blood circulation by lymphatic vessels. Like IgG, albumin binds the major histocompatibility complex-related Fc receptor at low pH. This interaction shields these proteins from degradation.

Like fibrinogen, globulins induce reversible RBC aggregation in stagnant blood regions.[8] Several kinds of globulins exist: (1) α-globulins that transport chemical species, such as thyroxine and retinol (vitamin-A); β-globulins such as transferrin; and γ-globulins, most of the antibodies (0.1–0.4 g/dl of α1-globulins, 0.5–1 g/dl of α2-globulins, 0.7–1.2 g/dl of β-globulins, and 0.5–1.6 g/dl of γ-globulins).

1.3.4 Plasma Lipids and Lipoproteins

The lipoprotein structure shields water-insoluble lipids (cholesteryl esters and triglycerides) from water, as it encapsulates them with polar lipids and proteins. However, core lipids can move between lipoproteins. The 4 main types of circulating lipoproteins that differ in size, density, and content include chylomicrons, and very-low-density (VLDL), low-density (LDL), and high-density (HDL) lipoproteins (Table 1.6, Fig. 1.2).

Lipoproteins convey cholesterol esters (CsE) and triglycerides (TG) in blood.[9] Triglycerides are delivered to muscles and adipose tissues for energy production and storage (blood TG concentration< 1.2 g/l at 20 years old and < 1.6 g/l at 60 years old).[10] Excess lipid and glucid intake leads to conversion into triacylglycerols in the

[8]Hemodynamic stresses oppose aggregation, limit aggregation growth, and progressively break existing erythrocyte rouleaux and possible rouleau networks.

[9]Analysis of serum lipids commonly yields concentrations of LDLs, HDLs, and TGs.

[10]Lipoprotein lipase removes triglycerides from chylomicrons and VLDLs for fat storage or energy source in adipose tissues and muscles, respectively.

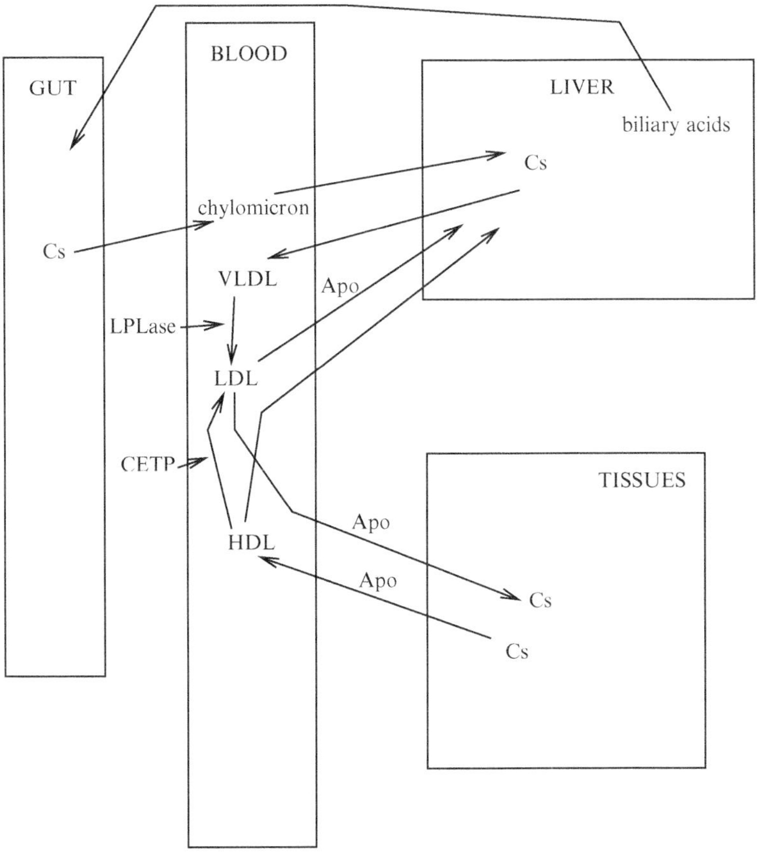

Fig. 1.2 Cholesterol turnover

liver that are packaged into VLDLs and released into the circulation. VLDL particles (size 30–80 nm) contain also cholesteryl esters and apolipoproteins (ApoB100, ApoC1–ApoC3, and ApoE).

Intermediate-density lipoproteins (IDL) are formed from VLDLs by lipoprotein lipase after triacylglycerol removal.[11] IDL particles become LDLs (size 20 nm) after losing the whole TG content. Plasma LDL contains cholesterol esters in its core and a hydrophilic coat composed of cholesterol, phospholipids, and ApoB100.

Blood lipids require many types of molecules from their synthesis, exocytosis out of producing cells, and blood transport, to cell capture (receptors) and endocytosis, and degradation (Table 1.7).

---

[11] IDL particles retain cholesteryl esters. Some IDLs are rapidly taken up by the liver, whereas others remain in the blood and are converted to LDL particles.

Table 1.7. Lipids and associated genes (Sources: [4, 5]). Investigations commonly examine high-density and low-density lipoprotein cholesterol, and total plasma triglycerides that are carried mainly in very-low-density lipoproteins and their remnants. Involved genes encode (1) apolipoproteins (ApoA–ApoC and ApoE); (2) lipoprotein receptor (LDLR) and multiligand endocytic receptor sortilin (Sort1) for lipoprotein lipase (LPase) that acts as a sorting receptor in the Golgi body and clearance receptor on the cell surface; (3) transporters of cholesterol (ABCa1) and cholesterol ester (CETP); (4) transcription factor Max-like protein X (MLX)-interacting protein-like (MLXIPL; a.k.a. class-D basic helix–loop–helix protein bHLHd14 and Williams-Beuren syndrome chromosomal region 14 protein [WBSCR14] that activates triglyceride synthesis in a glucose-dependent manner, as it binds to carbohydrate response elements of promoters of triglyceride synthesis genes; (5) enzymes of cholesterol synthesis such as mevalonate kinase (MvK) and degradation such as methylmalonic aciduria B-type protein (MMAB; a.k.a. cob(I)alamin adenosyltransferase); (6) glycosyltransferases, such as β(1,3)-galactosyltransferase 4 (β3GalLT4; a.k.a. ^{UDP}galactose β^{N}acetyl glucosamine), β(1,3)-galactosyltransferase polypeptide 4), β(1,4)-galactosyltransferase 4 (β4GalLT4), and protein-UDP ^{N}acetyl galactosaminyltransferase 2 (GalNT2; a.k.a. UDP–^{N}acetyl α^{D}galactosamine–polypeptide ^{N}acetylgalactosaminyltransferase 2); (7) lipoprotein (LPase; LPL gene), hepatic (LipC), and endothelial (LipG) lipases; and (8) inhibitor of lipase such as angiopoietin-like protein-3 (AngptL3; a.k.a. angiopoietin-5). Other targets include 3-hydroxy 3-methylglutaryl coenzyme-A reductase (HMGCR); cadherin, EGF-like, LAG-like, and seven-pass receptor CELSR2; cartilage intermediate layer protein CILP2; lysolecithin cholesterol acyltransferase (LCAT); myosin-binding protein H-like molecule (MyBPHL); neurocan (NCan; a.k.a. chondroitin sulfate proteoglycan CSPG3); preB-cell leukemia homeobox transcription factor PBx4; proline/serine-rich coiled-coil protein PSRC1; proprotein convertase subtilisin–kexin-like peptidase PCSK9 (a.k.a. neural apoptosis-regulated convertase); sterol regulatory element-binding protein (SREBP); and tribbles homolog Trib1.

Lipids	Genes
Triglycerides	Cilp2–PBX4; Mlxipl
	LPL; GALNT2; ANGPTL3; TRIB1
	APOA1–APOA4–APOA5–APOC3 cluster
	APOB, APOC–APOE cluster
HDL cholesterol	MVK–Mmab cluster
	Srebp2 (MvK and MMAB regulation)
	LCAT; CEPT; ABCA1
	APOA1–APOA4–APOA5–APOC3 cluster
	LIPC, LIPG, LPL; GALNT2
LDL cholesterol	Celsr2–Psrc1–MYBPHL–SORT1 cluster
	HMGCR; LDLR; Pcsk9; Cilp2; PBX4
	APOB, APOC1–APOC2–APOC4–APOE cluster
	SORT1; NCAN

1.3.4.1 Cholesterol

Cholesterol (Cs, Vol. 1 – Chap. 7. Plasma Membrane), from extracellular sources as well as de novo synthesis[12] is transported in blood mainly as cholesteryl esters associated with lipoproteins (blood Cs concentration < 2.2 g/l at 20 years old and < 2.6 g/l at 60 years old).

Dietary cholesteryl esters are carried from the small intestine to the liver by chylomicrons.[13] A part is excreted in biliary ducts as free cholesterol or bile acids (partial cyclic travel) and another part is conveyed in blood within LDL[14] or VLDL particles.[15] LDL particles link to specific plasmalemmal receptors. They are incorporated in vesicles, which fuse with lysosomes for degradation.[16] Endocytosis via LDLR requires ApoB100 constituent. Agent ApoE synthesized in various organs and found in high concentration in interstitial fluid, is a ligand for LDL receptors that participate in cholesterol redistribution [7].[17]

1.3.4.2 High-Density Lipoproteins

High-density lipoproteins (5–17 nm, 1.06–1.21 g/ml) are synthesized in the liver[18] and small intestine.[19] They extract cholesterol from cell membranes with ApoA1 and ApoA2 [11].[20]

[12]Endogenous cholesterol is synthesized in the liver and carried to tissues by VLDLs.

[13]Cholesterol esters and triglycerides packed into chylomicrons in the gut are conveyed in lymph and then in blood. Chylomicron is removed from blood by the liver.

[14]Endogenous and exogenous cholesterol that exceeds hepatic needs is carried in the blood circulation by LDL particles.

[15]The liver synthesizes VLDL particles. These particles are plasma carriers of cholesterol after conversion of VLDLs to LDLs by lipoprotein lipases, in particular in endothelial cells.

[16]The sequential steps of the cellular LDLR destiny include: (1) LDL receptor binding; (2) coated pit formation; (3) endocytosis; (4) lysosomal hydrolysis; and (5) metabolism via a passage into the endoplasmic reticulum and Golgi body to come back to the plasma membrane [6].

[17]ApoE mutants that cannot bind LDL receptors is associated with familial type 3 hyperlipoproteinemia, a genetic disorder with elevated plasma cholesterol levels and coronary artery disease.

[18]Synthesis of new high-density lipoproteins begins with the secretion of apolipoprotein-A1 from the liver. Apolipoprotein-A1 can bind to the ATP-binding cassette transporter-A1 (ABCa1) on macrophages in the subendothelial space. Agent ApoA1 promotes efflux of cholesterol and phospholipids from macrophages that form nascent HDLs [8]. These particles can then be modified by the lecithin-cholesterol acyltransferase into larger, less dense HDL2 and smaller, denser HDL3. Both HDL2 and HDL3 can receive additional cholesterol from macrophages before entering into blood circulation.

[19]Intestine is an important source of HDL particles. The ABCA1 gene in the intestine plays a key role in the production of high-density lipoproteins. Concentration of HDLs in mice with deleted ABCA1 gene specifically in the intestine are 30% lower than in normal mice [9]. When the ABCA1 gene is deleted in both liver and intestine, HDL concentration is lowered by more than 90%.

[20]Membrane-associated cholesterol is esterified by HDL-associated lecithin cholesterol acyltransferase (LCAT). Cholesterol efflux especially requires ApoA1-containing HDLs [10].

High-density lipoproteins are circulating stores of cholesterol, cholesteryl esters, and apolipoproteins (ApoA1, -A2, -A4, -C1 to -C4, -D to -F, -H, -J, -L1, and -M)[21] delivered to the liver, where they are endocytosed. Molecule HDL–cholesterol may be esterified in the plasma by lecithin-cholesterol acyltransferase (LCAT). Free and esterified cholesterols are removed from HDL in the liver using scavenger receptor ScaRb1, and afterward excreted into the bile. Several proteins, such as lecithin-cholesterol acyltransferase, paraoxonase Pon1, and platelet-activating factor acetylhydrolase (PAFAH),[22] are associated with HDL particles.

High-density lipoproteins contain [13]: (1) numerous proteins, such as enzymes and apolipoproteins, as well as (2) various lipids, not only free and esterified cholesterol, but also phospholipids (phosphatidylcholine, phosphatidylethanolamine, plasmalogen, lysophosphatidylcholine, glycerophospholipids), free and esterified fatty acids (mono- and triacylglycerols), and sphingolipids (ceramide, sphingomyelin, sphingosine 1-phosphate, lysosulfatide, and sphingosylphosphorylcholine).

Several HDL types differ in density, shape, size, and surface charge. Therefore, HDLs are heterogeneous in structure, physicochemical properties, metabolism, and activity. Small, lipid-poor, *nascent HDLs* are made of apolipoproteins embedded in a lipid monolayer of phospholipids and free cholesterol. They are mainly produced by the liver and intestine or can be released as fragments from triglyceride-rich lipoproteins. Large, *mature HDLs* contain a core of cholesteryl esters and triglycerides.

Nascent HDLs acquire cholesterol and phospholipids at cell membranes upon efflux via ATP-binding cassette transporter-A1. Esterification of HDL cholesterol from preβ-HDLs by lecithin cholesterol acyltransferase generates: (1) small, dense, cholesterol-poor HDL3 and (2) large, light, cholesterol-rich HDL2 that, in turn, can be converted to HDL3 (Fig. 1.2).

Metabolism of HDLs involves several processes, such as exchange of cholesteryl esters and triglycerides between HDLs and apoB-containing lipoproteins, uptake of cholesteryl esters by hepatocytes via scavenger receptor-B1, and phospholipid and triglyceride hydrolysis by hepatic lipase. Cholesterol acquired by HDLs can also be transferred to VLDLs and LDLs by cholesterol ester transfer protein (CETP).[23]

---

[21] ApoA1, ApoA2, ApoA4, ApoB, ApoC3, ApoD, ApoE, ApoH, LCAT, and CETP are detected in the ApoA4-containing lipoproteins [12].

[22] A.k.a. serine-dependent phospholipase-A2 (Vol. 4 – Chap. 2. Signaling Lipids).

[23] The glycoprotein cholesteryl ester transfer protein (CETP) acts in cholesterol distribution among plasma lipoproteins. CETP transfers not only cholesteryl esters, but also triglycerides and phospholipids between lipoproteins, including VLDLs, LDLs, HDLs, and chylomicrons. "Bad" cholesterol is mainly found in low-density lipoproteins, "good" cholesterol in high-density lipoproteins. Protein CETP mediates the transfer of cholesteryl esters from HDL to ApoB-containing lipoproteins in exchange for triglycerides. Agent ApoC1 abolishes CETP activity [14]. Lipid transfer inhibitor protein (LTIP or ApoF) regulates CETP activity in LDLs.

High-density lipoproteins are considered anti-atherogenic because they transport cholesterol to the liver for excretion and degradation.[24] Apolipoprotein-M (apoM), an HDL component, is required for the formation of preβ-HDL and cholesterol efflux to HDL, thereby hampering atherosclerosis [15]. Cholesterol diffusion and the scavenger receptor-B1 efflux pathway that allows bidirectional cholesterol exchange between cells and HDL, may also be involved in macrophage cholesterol efflux. Cholesterol efflux from arterial wall cells corresponds to the initial stage of cholesterol transport to the liver (*reverse cholesterol transport*).

Inflammation (Chap. 11) modifies lipid and lipoprotein metabolism to hinder the toxicity of various agents and enhance tissue repair by nutrient redistribution to immune cells. The inflammatory cascade decreases the concentration in cholesterol-rich HDLs to favor the synthesis and accumulation of phospholipid-rich VLDLs aimed at binding bacterial products and other toxic substances [16].[25]

Alterations in lipoproteins implicated in HDL metabolism, associated with depletion in several nuclear receptors (Vol. 3 – Chap. 6. Receptors), such as peroxisome proliferator–activated receptors (NR1c), liver X receptors (NR1h2/3), farnesoid X receptors (NR1h4/5), and retinoid X receptors (NR2b), impair the reverse cholesterol transport and increase cholesterol delivery to immunocytes [17].

Lipid efflux that depends on ABCa1 is also disturbed. During severe infection, VLDL clearance decays due to decreased lipoprotein lipase and apolipoprotein-E levels in VLDL particles. Enrichment of lipoproteins in ceramide, glucosylceramide, and sphingomyelin enhances uptake by macrophages. Changes in binding capacity of associated apolipoproteins and enzymes alter the HDL anti-oxidant capacity. HDL particles become pro-inflammatory molecules. Oxidation of LDLs and VLDLs increases.

High-density lipoproteins possess multiple anti-atherogenic activities that include reverse cholesterol transport from the arterial wall to the liver for excretion, anti-oxidative,[26] anti-inflammatory,[27] antithrombotic, and vasodilatory actions. Lipid-free apolipoproteins, primarily ApoA1, cause specific efflux of cellular

---

[24]Atherosclerosis (Vol. 6 – Chap. 8. Vascular Diseases) is a chronic inflammation of arterial wall that can be identified by increased serum levels of C-reactive protein. Wall inflammation associated with endothelial dysfunction is responsible for oxidative stresses, with possible excessive transport and intimal retention of LDL and monocyte extravasation. Oxidized LDLs induce cholesterol accumulation in macrophages. An additional source of influx is provided by vasa vasorum.

[25]The inflammation acute phase mediated by cytokines is characterized by high plasma levels of triglycerides with VLDL secretion caused by adipose tissue lipolysis, increased hepatic fatty acid synthesis, and suppression of fatty acid oxidation.

[26]Low-density lipoproteins may be the major target of HDL anti-oxidative effects. Anti-oxidative action of HDLs depends on apolipoproteins such as ApoA1, by the removal of oxidized phospholipids from LDLs, and enzymes that hydrolyze LDL-derived oxidized phospholipids, such as platelet-activating factor acetylhydrolase (or lipoprotein-associated phospholipase-A2) and lecithin cholesterol acyltransferase.

[27]Anti-inflammatory activity of HDLs can be characterized by inhibited expression of adhesion molecules, reduced neutrophil infiltration within the arterial wall and decreased generation of reactive oxygen species. HDL enzymes can hydrolyze oxidized lipids.

cholesterol and phospholipids via ATP-binding cassette transporter-A1, whereas ATP-binding cassette transporter-G1 and -G4 mediate cholesterol efflux to mature HDL particles. Mature HDLs induce a cholesterol efflux mediated by the scavenger receptor-B1, an HDL receptor. Small, lipid-poor HDL3 thus represents more efficient cholesterol acceptors than HDL2 subtype.

Small, dense HDL3 has anti-oxidative activity, transferring oxidized lipids from the tissues and blood circulation to the liver. HDL3 particles may constitute the HDL subset mostly responsible for anti-inflammatory activity under normal conditions. Alterations in HDL composition, such as: (1) decrease in cholesteryl esters, ApoA1, and enzymes bound to HDLs and (2) an increase in triglyceride content and serum amyloid-A occur during inflammation. HDL particles undergo structural and functional changes especially in atherogenic dyslipidemia [18].[28] Replacement of cholesteryl esters by triglycerides alters the conformation of ApoA1, and hence the lipid-binding capacity. The selective uptake of cholesteryl esters from high-density lipoprotein by scavenger receptor-B1 is strongly affected by dysfunctional triglyceride-rich HDLs [19]. HDL enzymes (paraoxonase-1, platelet-activating factor acetylhydrolase, and lecithin cholesterol acyltransferase) can become dysfunctional and depleted. Serum amyloid-A can replace ApoA1 and other HDL apolipoproteins [20]. Enrichment of HDL3 in serum amyloid-A[29] reduces cholesteryl ester uptake by hepatocytes, but can increase HDL binding to macrophages [21]. Sustained redirection of cholesterol from the liver to macrophages corresponds to a maladaptive response in the long term.

Bile acids interacts with triglycerides.[30] Bile acids are not only required in dietary lipid absorption in the digestive tract and cholesterol homeostasis, but also also have endocrine functions [22]. Bile acids bind G-protein-coupled bile-acid receptor GPBAR1[31] activate mitogen-activated protein kinase (MAPK) modules and nuclear receptors such as farnesoid X receptor-α (FXRα or NR1h4).[32] Farnesoid X receptor-α induces apolipoprotein-C2 expression. Agent ApoC2 is a coactivator of lipoprotein lipase that lowers serum triglyceride and VLDL levels.

---

[28] Atherogenic dyslipidemia, a major risk factor for cardiovascular diseases, is defined by higher circulating levels of pro-atherogenic, cholesterol-rich, ApoB+ lipoproteins, such as LDLs, and lower concentrations of anti-atherogenic ApoA1+ HDL particles.

[29] Lipid-free serum amyloid-A, a ligand for ScaRb1, inhibits HDL binding on ScaRb1 receptor. HDL-associated SAA has little effect on HDL binding to ScaRb1, but decreases selective cholesteryl ester uptake of HDL particles.

[30] Bile acids are synthesized in the liver from cholesterol. Bile acids secreted from hepatocytes into the bile canaliculi are stored in the gallbladder and transported down the duodenum. They are absorbed again in the terminal ileum, and convected back to the liver via the portal vein. Bile acids decrease their synthesis using a negative feedback. Main bile acids include: (1) primary bile acids, cholic and chenodeoxycholic acids, as well as their glycine and taurine conjugates and (2) secondary bile acids, deoxycholic and lithocholic acids.

[31] A.k.a. GPCR19, GPR131, and TGR5.

[32] In the liver and intestine, activated farnesoid X receptor-α protects against toxic accumulation of bile acids by transport and excretion of bile acids. Bile acids are also involved in liver regeneration by activation of farnesoid X receptors, as well as, but to a lower extent and transiently, xenobiotic receptors CAR [23].

Table 1.8. Value ranges of arterial blood gas analysis.

Parameter	Range
pH	7.35–7.45
H^+	35–45 nmol/l
p_{O_2}	9.3-13.3 kPa
p_{CO_2}	4.7–6.0 kPa
$HCO3^-$	22–26 mmol/l
$HPO4^{--}$	0.8–1.5 mmol/l

1.3.5 Blood Gas

Analysis of arterial blood gas determine the concentrations of carbon dioxide, oxygen, and bicarbonate, as well as blood pH (Table 1.8). One of the main tasks of the blood circulation is to transport dissolved gas, such as oxygen and carbon dioxide. Oxygen dissolves less and needs to bind to a suitable transporter in erythrocytes — hemoglobin. Oxygen-carrying capacity is related to the quantity of oxygen transported by hemoglobin. Oxygen transport relies on conditions and rate of its association and dissociation.

Hemoglobin loading and unloading, indeed, happen under different conditions. The higher the oxygen availability, the greater the quantity of oxygen taken up by hemoglobin. The lower the local oxygen concentration, the higher the release of oxygen by hemoglobin. The dissociation curve levels off toward the saturation line (100% of oxygenated hemoglobin). Almost full loading is ensured at usual oxygen concentration (97% of oxygenated hemoglobin).

The dissociation curve exhibits 2 main features, a sigmoid shape and a left–right position with respect to the oxygen concentration. A shift toward the left depicts an increase in oxygen availability. The higher the carbon dioxide concentration, the greater the right shift. Consequently, the higher the CO_2 concentration, the greater the oxygen dissociation and delivery.

1.4 Blood Disorders

Blood disorders include: (1) blood volume loss due to bleeding after injury and dehydration that reduces water content of the blood; (2) blood circulation dysfunctioning and resulting diminished tissue perfusion caused by abnormal cardiac output or obstacles in blood vessel lumen (thrombi and emboli); and (3) blood composition changes that can affect the heart functioning and blood vessel caliber, hence blood supply to the body's organs.

Table 1.9. Anemia grading.

Grade	Degree	Hemoglobin
0	Normal	12–16 g/dl (women) 14–18 g/ dl (men)
1	Mild	10–12/14 g/dl
2	Moderate	8–10 g/dl
3	Severe	6.5–8 g/dl
4	Life threatening	< 6.5 g/dl

1.4.1 Blood Cell Disorders

Hematological disorders[33] comprise: (1) blood cell underproduction, such as anemia and thrombopenia, and non-tumoral (e.g., polycythemia vera and essential thrombocytosis) and cancerous (e.g., leukemia) overproduction; (2) abnormal coagulation (e.g., hemophilia and thrombophilia); and (3) blood infections (parasitic, viral, or bacterial infections).

Erythrocyte diseases include: (1) anemias (iron deficiency, genetic sickle-cell disease, thalassemia, spherocytosis, autoimmune pernicious anemia, aplastic anemia due to dysfunctional erythropoiesis, aplasia, hemolysis); (2) erythrocyte infections, such as malaria (caused by Plasmodium falciparum, vivax, ovale, and malariae) and trypanosomosis (e.g., Chagas disease); (3) polycythemias or erythrocytoses with increased blood viscosity; and (4) several microangiopathic diseases. In addition, toxics can bind to hemoglobin and, in some cases, create irreversible damage. Inhaled carbon monoxide irreversibly binds to hemoglobin to form carboxyhemoglobin, thereby limiting oxygen transport in the blood circulation.

Anemia is due to acute or chronic factors. Anemias are classified according to the hemoglobin level (Table 1.9). Injuries are common examples of acute anemias by hemorrhage. Another usual cause is cancer. Tumor anemia is due to tumor bleeding, hemolysis, and deficiency in folic acid and vitamin-B12. Release of cytokines, especially interferon-γ, interleukin-1, and tumor-necrosis factor-α, impairs erythropoietin synthesis in the kidney. Activation of macrophages can lead to a shorter erythrocyte half-life and a decrease in iron use. Chemotherapy and radiotherapy further aggravate anemia. Recombinant human erythropoietins (epoetins) avoid transfusions, but thromboembolic complications have been observed [24].

Platelet and leukocyte deficiencies can also occur. Thrombocytopenia and leukopenia can result from damage to hematopoietic stem cells or precursors. Lymphocytopenia (or lymphopenia) can be also due to lack of zinc, long-term heavy alcohol consumption, and certain infections. Human immunodeficiency virus (HIV), a lentivirus of the retrovirus family, causes acquired immunodeficiency syndrome (AIDS), as it invades leukocytes.

[33] αιμα: blood, bloodshed.

1.5 Blood Therapies

1.5.1 Blood Substitutes

Blood substitutes have been studied for universal compatibilility, availabililty, and storability. Safe, effective hemoglobin substitutes such as adenosine-modified hemoglobin aim at mimicking blood's oxygen-carrying characteristic. They are especially designed to be easily transported and used in remote regions.

Hemoglobin needs to be encapsulated in erythrocytes. In the absence of its shell (RBC membrane), hemoglobin tetramers break down into potentially toxic dimers that can damage renal filter and cause vasoconstriction and inflammation. Blood substitutes must deliver oxygen to the body's organs without toxicity. There are basically 2 types of oxygen carriers based on perfluorocarbons that carry O_2 and CO_2 or hemoglobin, encapsulated or crosslinked to avoid toxicity. Perfluorocarbons are inert materials that can dissolve much more oxygen than plasma, but have an oxygen-carrying capacity much lower than that of hemoglobin. In addition, they must be combined with lipids to be conveyed into the blood stream. Perfluorocarbon emulsions are still under investigation. Hemoglobin-based oxygen carriers can be obtained by crosslinking hemoglobins. Polyhemoglobins and conjugated hemoglobins have a greater ability to release oxygen to tissues than erythrocytes. Although they can be stored during a long period, the first-generation blood substitutes have a short life in blood (¡30 h). Second-generation crosslinked hemoglobin substitutes are designed to incorporate anti-oxidant enzymes to prevent methemoglobin formation and reperfusion injury. Third-generation blood substitutes comprise encapsulated hemoglobin (nanocapsules). Another solution relies on supply of universal erythrocytes grown from umbilical cord blood. Cord blood collected asceptically from the placenta after birth of healthy new-borns has a higher hemoglobin and growth factor content than normal adult blood. An artificial oxygen transport protein similar to human neuroglobin has been designed [26].[34]

Nowadays, the production of cultured human erythrocytes using culture support that mimics bone marrow environment (cytokine supply, coculture, etc.) provides an alternative to blood transfusion. Cultured erythrocytes can also be used: (1) to model human RBC-targeted infections and (2) for gene therapy. In vitro production of mature and functional human erythrocytes from hematopoietic stem cells has been demonstrated [25].

[34]Neuroglobin transports oxygen in the nervous system. This monomeric intracellular hemoprotein synthesized in the central and peripheral nervous system is observed in the cerebrospinal fluid and endocrine glands. It reversibly binds oxygen with an affinity higher than that of hemoglobin.

1.5.2 Plasmapheresis

Plasmapheresis separates plasma and blood cells. The filtration flux per unit of plasma filter area is approximately proportional to the wall shear rate at least up to 7500/s and depends on the effective membrane length $L^{-1/3}$ [27]. Pulsatile flow with the time-mean flow rate equal to the steady flow enhances the filtration flux, with optimal modulation frequency 1 Hz and volume amplitude 1.5 ml [28]. Mathematical and numerical analysis of filtration across a fibrous medium has been performed [29], using a homogeneization technique [30–36].

Chapter 2
Hematopoiesis

Hematopoietic cells, or hemopoietic cells, represent bone marrow-derived cell types that circulate in blood (including mature cell types and their precursors). Hematopoietic cells are categorized into *myeloid cells* (basophils, eosinophils, neutrophils, erythrocytes, thrombocytes, monocytes and macrophages, and mastocytes) and *lymphoid cells* (B lymphocytes, various types of T lymphocytes, which are the only hematopoietic cells that can be generated elsewhere than in the bone marrow, and natural killer [NK] cells, that are cytotoxic lymphocytes also called large granular lymphocytes).

Stromal cells is another collective term for different cell types in a given tissue at a given time that do not pertain to functional components of this tissue, i.e., *parenchymal cells*. In particular, the bone marrow stroma yields a hematopoietic microenvironment that consists of various cell types, such as mesenchymal progenitors, fibroblasts, adipocytes, monocytes and macrophages, endothelial and smooth muscle cells, T-lymphocytes, among others. These cells can support short- and long-term hematopoiesis via membrane-bound and secreted hematopoietic factors that operate on hematopoietic stem and progenitor cells.

Hematopoietic stem cells produce by mitosis either stem cells of the same type (self-renewal) or progenitors leading to precursor, immature, and mature blood cells. In steady state, most stem cells are dormant, whereas few are active [37]. Hematopoietic stem cell self-renewal is regulated by intrinsic and extrinsic signals.

Blood-cell progenitors and mature cells of the granulocytic, monocytic, megakaryocytic, and erythroid lineages can be generated from differentiated cells. For example, human dermal fibroblasts that bypass the pluripotent stem cell state can give rise to multipotent hematopoietic progenitors. These fibroblasts synthesize predominantly octamer-binding transcription factor Oct4, or POU domain, class-5, homeodomain-containing transcription factor POU5F1, as well as PTPRc phosphatase, or panhematopoietic marker CD45, during the multipotent reprogramming [38].

M. Thiriet, *Tissue Functioning and Remodeling in the Circulatory and Ventilatory Systems*, Biomathematical and Biomechanical Modeling of the Circulatory and Ventilatory Systems 5, DOI 10.1007/978-1-4614-5966-8_2, © Springer Science+Business Media New York 2013

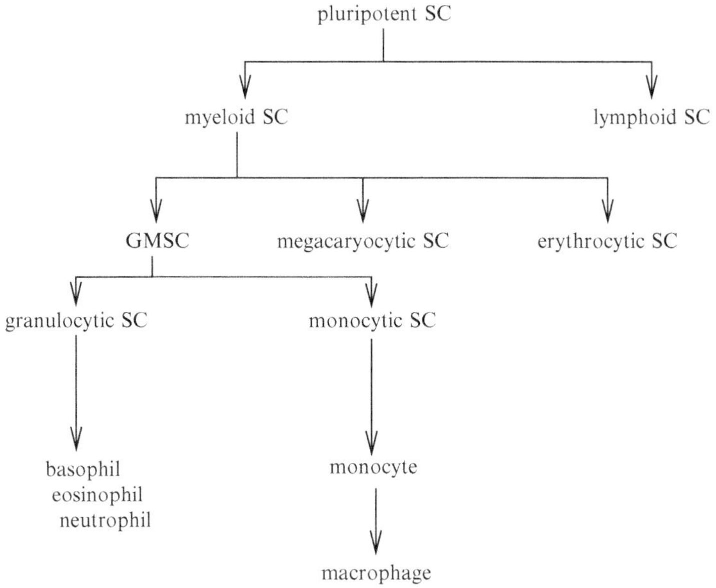

Fig. 2.1 Hematopoiesis with a focus on granulocytic and monocytic lineages (SC: stem cell, GMSC: granulocyte–monophage stem cell). After birth, hematopoiesis occurs in the bone marrow. Under the influence of stimulators, multipotent stem cells of the bone marrow differentiate into progenitors (committed stem cells). The first differentiation leads to myeloid and lymphoid stem cells. The progenitors undergo several divisions to give rise to stem cells with limited differentiation potential, the specific progenitors and precursors of a single blood cell lineage. Lymphoid stem cells (progenitors) produce B and T lymphocytes. Myeloid stem cells, or colony-forming unit **CFUgemm** (granulocyte–erythroid–macrophage–megakaryocyte) give birth to progenitors of many lineages (granulocytic, monocytic, erythrocytic, and megakaryocytic). Generator CFUgemm leads to more differentiated progenitors: (1) **CFUgm** (granulocyte–macrophage) that give rise to **CFUg** (granulocyte) and **CFUm** (macrophage); (2) **CFUeo** (eosinophil); (3) **CFUb** (basophil); (4) **CFUe** (erythroid); and (5) **CFUmeg** (megakaryocyte). Precursors have lost self-renewal capacity. They include myeloblasts, monoblasts, lymphoblasts, proerythroblasts, and megakaryoblasts. Several shared modifications characterize cell maturation: reduction in cell size and nucleocytoplasmic ratio and chromatin condensation. Precursors lead to mature blood cells after 3 to 5 divisions according to the cell lineage (one precursor can give birth to up to 32 daughter cells). Mature blood cells enter blood. Lymphocytes and monocytes can differentiate in tissues after their blood journey. Hematopoiesis is composed of several compartments: (1) multipotent stem cells such as colony-forming unit-spleen (**CFUs**), (2) progenitors, (3) precursors, and (4) mature cells (granulocytes, monoocytes, lymphocytes, erythrocytes, and thrombocytes).

The turnover of hematopoietic cells[1] is estimated to be about 10^{12} cells per day [39]. All blood cell types are produced in the bone marrow from *pluripotent stem cells* (PSC; Fig. 2.1). Hematopoietic stem cells and derived progenitor cells yield blood resupply during the entire life. Pluripotent stem cells give birth via

[1] αιμ ατικος: associated with blood, ποιεο: to do.

multi- and unipotent stem cells (USC) to the different lineages via proerythroblasts and erythroblasts, megakaryoblasts, myeloblasts and promyelocytes, and lymphoblasts.

2.1 Hematopoietic Stem Cells

Hematopoietic stem cells (HSC) generate a set of progenitors for erythroid, lymphoid, megakaryocytic, and myeloid lineages. Another type of multipotent cells resides in the bone marrow — *mesenchymal stem cells* — that can differentiate into osteoblasts, chondrocytes, and adipocytes. Most primitive hematopoietic stem cells are quiescent.

Two subpopulations of hematopoietic stem cells exist: (1) long-term quiescent (reserve) and (2) active hematopoietic stem cells [40]. In the bone marrow, endosteal regions composed of osteoblasts, osteoclasts, as well as vascular and CXCL12^{high} reticular cells, and central regions that lack osteoblasts form inhibitory and stimulatory (proliferative) zones, respectively. These 2 adjoining, but separate regions enable the coexistence of 2 different subpopulations (states) of hematopoietic stem cells.

The pool of hematopoietic stem cells comprises clonal subtypes with different lineage potential and self-renewal capacity. Aging affects the function and composition of mature blood cell compartments with a decline in lymphoid lineage potential and increased myeloid lineage commitment [41]. The gene program that specifies lymphoid fate is downregulated with age. Myeloid-biased hematopoietic stem cells produce high levels of signaling lymphocytic activation molecule family

Table 2.1. Members of the signaling lymphocytic activation molecule family (SLAMF) of immune receptors, or cluster of differentiation CD2 receptor family. They are synthesized in various subsets of immunocytes (T, B, and NK cells; BLAME: B-lymphocyte activator macrophage-expressed membrane protein; BLAST: B-lymphocyte activation surface marker; CD2F: CD2 family member; CD84H: CD84 homolog; CRACC: CD2-like receptor activating cytotoxic cells; CS: CD2 subset; Ly: lymphocyte surface antigen; NAIL: NK-cell activation-inducing ligand; NKR: NK-cell type-1 receptor; nLy: novel lymphocyte antigen; NTBA: natural killer-, T-, and B-cell antigen; SF: signaling lymphocytic activation molecule family member).

Member	Other aliases
SLAMF1	CD150, SLAM
SLAMF2	CD48, BLAST1
SLAMF3	CD229, Ly9
SLAMF4	CD244, NAIL, NKR2B4
SLAMF5	CD84, Ly9β
SLAMF6	CD352, Ly108, NTBA, SF2000
SLAMF7	CD319, nLy9, CRACC, CS1
SLAMF8	CD353, BLAME
SLAMF9	CD2F10, CD84H1, SF2001

member SLAMF1 (SLAMF1^{high} HSCs), or CD150 (Table 2.1), possess a stronger self-renewal potential, and predominate the stem cell pool in the aging body, whereas those that generate lower levels of SLAMF1 (SLAMF1^{low} HSCs) have a balanced lineage output.

Hematopoiesis depends on hematopoietic stem cell survival, self-renewal, and differentiation, hence mainly on the balance between quiescence (reserve) and cell division (regeneration of its pool and blood tissue). Lifetime hematopoiesis indeed relies on the capacity of hematopoietic stem cells to self-renew and to generate blood cells according to the body's needs. Blood homeostasis, hence hematopoiesis, requires long-term retention of hematopoietic stem cells in quiescence to maintain hematopoietic regenerative capacity. In the absence of a need of strong blood regeneration, the bone marrow, which is the site of blood cell formation from hematopoietic stem cells, maintains the circulating cell pool because mature blood cells continuously undergo senescence with a given degradation rate.

Tissue regeneration depends, at least partly, on stem cells, hence on their metabolic regulation. Hematopoietic stem cells are highly sensitive to energetic and oxidative stress; they shift between quiescence and proliferation according to the context. Kinase STK11, or LKB1, and its substrate AMPK coordinate metabolism with cell fate, thereby adapting cellular energetics with stem cell maintenance or tissue regeneration. The primary function of LKB1 in adult tissues is its inhibition of cell division, hence preventing tissue overgrowth. Enzyme LKB1 also controls cell survival and proliferation, as well as mitochondrial function and energy homeostasis in hematopoietic stem cells [42]. Inactivation of and deficiency in LKB1 in adult mice causes loss of HSC quiescence and subsequent depletion of all hematopoietic derived cells, as well as mitochondrial defects, alterations in lipid and nucleotide metabolism, and depletion of cellular ATP [42–44]. Hematopoietic stem cells relies more strongly on LKB1 for cell cycle regulation and survival than committed hematopoietic progenitor and precursor cells [42, 44]. However, LKB1 is needed for HSC maintenance via AMPK-dependent and -independent mechanisms [42,43]. Metabolic sensor LKB1 indeed controls chromosome stability in HSCs via an AMPK-independent process. The metabolic control in hematopoietic stem cells yields an additional metabolic checkpoint of the cell division cycle. The bone marrow niche controls by sending proper cues the activity of hematopoietic stem cells during regenerative hematopoiesis.

The transcriptional regulation of cell quiescence relies on cell cycle regulators (e.g., P53, retinoblastoma protein, cyclin-D, and cyclin-dependent kinase inhibitors CKI1a and CKI1b) and factors with specific functions in hematopoietic stem cells (e.g., early growth response factor EGR1, forkhead box protein FoxO, DNA sequence GATA-binding protein GATA2, growth factor-independent transcription repressor GFI1, and Runx1).[2] The transcription factor, nuclear receptor-related

[2]Cell quiescence is associated with: (1) downregulation of cyclin-D1; (2) upregulation of cyclin-dependent kinase inhibitors CKI1a and CKI2a; (3) action of T-cell-intrinsic quiescence factor FoxO1; and (4) activation of P38MAPK [45]. Factor FoxO1 maintains cell quiescence and impedes T-cell activation by self-antigens, but does not prevent polyclonal T-cell activation. Polyclonal

protein NuRR1, or NR4a2, prevents the entry into the cell cycle [46]. Factor NR4a2 may upregulate cyclin-dependent kinase inhibitor CKI2c.

Connexin-43, a constituent of gap junctions, is synthesized in hematopoietic stem and progenitor cells. However, it is downregulated during the differentiation of hematopoietic stem cells to progenitors. Connexin-43 regulates the content of reactive oxygen species inside stressed hematopoietic stem cells, as it enables ROS transfer to supporting stromal cells in the hematopoietic environment [45]. Whereas low ROS levels are needed for cell activity (Vol. 4 – Chap. 10. Other Major Signaling Mediators), a sustained increase in ROS production during stress (and aging) can prevent self-renewal and cause senescence. Hyporegenerative capacity and cell cycle arrest can result from high levels of ROS and activation of P38MAPK and FoxO1 agents.

The number of hematopoietic stem cells is assessed to be of the order of 10000 cells. However, the blood cell production rate varies with mammal size. In addition, the larger the species size, the faster the proliferation rate. The size of the hematopoietic stem cell pool can be associated with allometric scaling, which is defined by a simple power law between a given observable y and the mass m of the organism:[3]

$$y = y_0 m^p. \tag{2.1}$$

The number of active stem cells in adult mammals evaluated by measuring the number of circulating reticulocytes scales with body mass with exponent 3/4 [47].

2.1.1 LSK Hematopoietic Stem Cells

Hematopoietic stem cells belong to a bone marrow cell population that expresses high levels of stem cell antigen-1 (SCA1) and plasmalemmal stem cell factor receptor (SCFR),[4] but not plasmalemmal markers of lineage-committed hematopoietic cells (Lin−).[5] This stem cell subset is hence defined as Lin−, SCA1+, SCFR (KIT)+ (LSK) cells.

T cells diversify their activities, unlike antigen-specific T-cell clones, but generally remain restricted to certain antigens. On the other hand, polyclonal immunoglobulins (antibodies) are combination of immunoglobulins secreted against a specific antigen, each identifying a different antigen epitope (antigenic determinant) and generally obtained from different B-cell clones.

[3]Allometric scaling can also be applied to basal metabolic rate, heart rate, arterial radii, etc. Most of the scaling powers $\cdot^p$ are multiples of 1/4. Computed power 3/4 can be related to a surface-to-volume ratio.

[4]Membrane-bound stem cell factor, or cellular kinase in tyrosine (KIT or cKIT) ligand, after proteolytic cleavage, binds and activates SCFR receptor. Membrane-bound stem cell factor stimulates adhesion of hematopoietic stem and progenitor cells to stromal cells, because it activates integrins.

[5]A lineage marker negative phenotype (Lin−) corresponds, in mice, to CD2−, CD3−, CD4−, CD5−, CD8−, CD161c−, PTPRc−, Ly76−, Ly6g−. Hematopoietic stem cells can be precisely distinguished from other hematopoietic progenitors by the plasmalemmal expression of signaling lymphocytic activation molecules (SLAM), such as SLAMF1, SLAMF2, and SLAMF4 [48].

2.1.2 *Hemangioblast*

Both hematopoietic and endothelial cells arise from a mesoderm-derived VEGFR2+ common precursor, the so-called hemangioblast.[6] Hematopoietic progenitor cells arise from these bipotential precursors via a subset of early endothelial cells, i.e., differentiated endothelial cells that have a hematopoietic potential and form the *hemogenic endothelium* [49].

$TIE2^{high}$, SCFR+ hemogenic endothelium is transiently generated. Hemangioblasts isolated from differentiated mouse embryonic stem cells in culture generate tightly adherent structures (stage 1) and then non-adherent round cells that proliferate to generate a mature blast colony (stage 2) [50].[7]

During embryogenesis, SCA1+, SCFR+ (CD117+), CD41+ (Itgα_{2B}+) hematopoietic stem cells emerge directly from the aortic floor into the dorsal aortic lumen, at least in some species, following egress of Runx1+ endothelial cells from the aortic ventral wall into the subaortic space [51, 52]. Cadherin-5+ (or vascular endothelial [VE]-cadherin) endothelial precursors or hemogenic endothelial cells transiently possess the ability to give rise to multipotent hematopoietic stem and progenitor cells during vertebrate development (*endothelial–hematopoietic transition*) [53]. These cells afterward enter the blood circulation to colonize and differentiate in hematopoietic organs.

2.2 Biological Models of Hematopoiesis

Hematopoiesis relies on classical and alternative pathways. Long-term hematopoietic stem cells give rise to short-term hematopoietic stem cells that have a transient ability to self-renew and differentiate into multipotent progenitors [54].

The latter lack the capacity of self-renewal, but retain multipotency, creating a series of intermediate progenitors, such as common lymphoid and myeloid progenitors that further differentiate into lymphoid and granulocyte–macrophage progenitors, respectively.

Lymphoid-primed multipotent progenitors can develop lymphoid and myeloid progeny, but not erythroid and megakaryocytic cells.

---

[6]Marker VEGFR2 is also called fetal liver kinase FLK1 and kinase insert domain receptor (KDR). VEGFR2+, cadherin-1+ precursors can differentiate into endothelial, primitive and definitive hematopoietic progenitor cells, cardiomyocytes, and mural cells.

[7]At stage 1, a majority of cells is positive for the endothelial marker, angiopoietin-1 TIE2 receptor (a.k.a. Tyr endothelial kinase [TEK]), whereas a minority of cells is positive for α_{2B} integrin (or CD41) that defines hematopoietic engagement. Afterward, the percentage of CD41+ cells rises and most cells are rapidly CD41+, TIE2− cells. Transient $TIE2^{high}$, SCFR+ cell population that contains both CD41− and CD41+ cells hence represents a transitional population from which definitive hematopoietic progenitors originate.

Granulocyte–macrophage progenitors generate myelomonocytic and megakaryocytic–erythrocyte progenitors. However, the separation between the myeloid and megakaryocytic–erythroid cell lineages can occur before the development of common myeloid progenitors.

Myelo-erythroid progenitors include pregranulocyte–macrophage precursors that give rise to granulocyte–macrophage progenitors, whereas prepremegakaryocytic–erythroid precursors yield preCFUe (CFU erythroid), CFUe, and megakaryocytic–platelet (MKP) compartments.

2.2.1 Arborescence Models of Hematopoiesis

An arborescence model of hematopoiesis assumes branches for the differentiation of blood cells [55]. This model states that: (1) megakaryocytes as well as erythroid and myeloid cells from a *multipotent progenitor* (MPP), *lymphoid-primed multipotent progenitor* (LMPP), *early lymphoid progenitor* (ELP), or *common myeloid progenitor* (CMP); and (2) natural killer cells and B and T lymphocytes from a *common lymphoid progenitor* (CLP).

Another tree-like model supposes that: (1) a *common myeloid–lymphoid progenitor* (CMLP) gives rise to (1.1) a *myeloid–T-cell progenitor* (MTP) that differentiates into myeloid and T lymphocytes and (1.2) *myeloid–B-cell progenitor* (MBP) that gives rise to myeloid and B lymphocytes; and (2) a *common myeloid progenitor* generates myeloid cells and a *myeloid–erythroid progenitor* (MEP), a source of erythroid cells and megakaryocytes.

A third model proposes that erythroid potential is lost early from a *common myeloid progenitor* that differentiates into *granulocyte–monocyte progenitor* (GMP) with myeloid potential and *myeloid–erythroid progenitor*, whereas a *common lymphoid progenitor* leads to lymphoid cells.

Arborescence Model Drawbacks

Lymphoid progenitors can have a myeloid potential and differentiate into myeloid cells (macrophages and dendritic cells). Therefore, models of early segregation of lymphoid and myeloid differentiation axes are not validated by observations.

2.2.2 Lymphoid Progenitors

Most Lin−, SCA1+, SCFR+, FLT3+ lymphoid-primed multipotent progenitors lack erythroid and megakaryocytic potential. Several lymphoid progenitors that are able to generate B and T lymphocytes and natural killer cells include: (1) Lin−, SCA1+, SCFR+, FLT3^{high}, VCAM1− lymphoid-primed multipotent progenitors;

(2) $SCFR^{high}$ early lymphoid progenitors that express recombination-activating gene RAG1;[8] and (3) SCFR−, PTPRc+, CLP2 (type-2 common lymphoid progenitor) progenitors that express the gene encoding the preT-cell antigen receptor.[9]

Early progenitors with lymphoid and myeloid potential (EPLM), *lymphoid-primed multipotent progenitors* (LMPP), and *thymic progenitors* that have both lymphoid and myeloid potential receive many signals that govern their fate. This destiny depends on the type of growth factors as well as strength and duration of signaling.

Lymphoid-primed multipotent progenitors have both lymphoid and myeloid potential, with high granulocytic, monocytic, and lymphoid potential, but low potential for megakaryocyte or erythroid development [55]. However, an LMPP subpopulation that expresses the thrombopoietin receptor has a megakaryocytic and erthyroid potential.

The earliest thymic progenitors retain myeloid potential, hence can contribute to thymic granulocyte and macrophage populations, but not B cells. Early T-cell progenitors, or CD4−, CD8− *double-negative cells* (DN), comprise:

1. $SCFR^{high}$[10] early thymocyte progenitors (ETP);
2. $SCFR^{high}$, CD44+, CD25− DN1 cells;
3. $SCFR^{high}$, CD44+, CD25+ DN2 cells;
4. SCFR±, CD44−, CD25+ DN3 cells; and
5. SCFR−, CD44−, CD25− DN4 cells.

Notch-1 is required for T-cell development. When it is lacking in the thymus, thymocyte progenitors bifurcate toward B-cell development.

Early thymocyte progenitors (ETP) can be subdivided according to expression of stem cell protein Tyr kinase-1 (STK1) receptor[11] and chemokine CCR9 receptor. Only STK1+ or CCR9+ *thymus-settling progenitors* have B-cell potential [55]. Double-negative DN1 and DN2 cells are not committed to T-cell development, but give rise to NK and myeloid cells.

The pairwise relationship model of hematopoiesis [55] does not take into account branching patterns that determine preferred route for a given cell fate. It displays in a circle blood cell sectors (megakaryocyte, erythrocyte, basophil or mastocyte, eosinophil, neutrophil, monocyte, dendritic cell,[12] and B, NK, and T cell) and arcs

---

[8]The recombination-activating gene, or RING finger gene Rnf74, encodes enzymes that act in the rearrangement of immunoglobulin genes and T-cell receptors during VDJ recombination.

[9]Two subpopulations of common lymphoid progenitors — CLP1 and CLP2 — coexist in the bone marrow and generate lymphocytes. Type-1 common lymphoid progenitors (CLP1 cells) are IL2Rα+ (CD25), PTPRc−, SCFR− cells, whereas CLP2 cells are IL2Rα+, PTPRc+, SCFR− cells.

[10]I.e., $CD117^{high}$.

[11]A.k.a. CD135, fetal liver kinase FLK2, and FMS-like Tyr kinase receptor FLT3.

[12]Dendritic cells can have both lymphoid and myeloid origins, but have a developmental program that is independent from that of lymphoid and myeloid cells. Cells that can differentiate into dendritic cells often also have the potential to become B, T, and NK cells.

inside the disk assigned for each progenitor and precursor with overlaps (as a final cell fate can be reached via several types of intermediate progenitors) or involved transcription factors in hematopoietic stem cells.

2.3 Stem Cell Niches

The embryonic origin of adult hematopoietic stem cells (multipotent progenitors) is the aorta–gonad–mesonephros region. Hematopoietic stem cells subsequently colonize fetal and adult hematopoietic organs, i.e., fetal liver and spleen on the one hand and, in adults, the bone marrow as well as spleen and liver during hematopoietic stress on the other. Colony-forming units in the spleen reconstitute transiently multipotent progenitors.

2.3.1 Types of Bone-Marrow Niches

Different niche types for hematopoietic stem cells exist: (1) osteoblastic and (2) vascular bone marrow HSC niches. In adults, hematopoietic stem cells have their own bone marrow microenvironment,[13] close to the endosteal[14] surface of bone marrow cavities in trabecular regions of long bones, whereas more differentiated hematopoietic progenitors are mainly located in the central bone marrow region [56]. The endosteal niche is characterized by a sinusoidal endothelium. Changes in the HSC niche do not allow appropriate HSC maintenance in vivo.

In addition to osteoblastic niches, the vascular bone marrow HSC niche demarcated by bone marrow sinusoidal endothelial cells is a second specialized HSC microenvironment in the bone marrow, with a large quantity of SLAMF1+ hematopoietic stem cells attached to the fenestrated endothelium of bone marrow sinusoids. Endothelial cells of bone marrow sinusoids yield a cellular platform for the differentiation of lineage-committed progenitors such as megakaryocytic progenitor cells (CFU megakaryocyte).

Bone marrow vascular and endosteal niches cooperate to control HSC quiescence and self-renewal. A small number of hematopoietic stem cells is constantly released into the blood circulation. Endothelial cells of bone marrow sinusoids differ from endothelial cells of the microvasculature of any other organ. These endothelial cells

---

[13]Extramedullary transient hematopoiesis, in the liver or spleen, occurs after bone marrow stress; but after repair, bone marrow hematopoiesis follows.

[14]The endosteum is a bone remodeling site characterized by high concentrations of calcium ions. Extracellular Ca^{2+} concentration is assessed by calcium-sensing receptors.

express cytokines such as chemokine CXCL12,[15] like osteoblasts, and adhesion molecules for HSC mobilization, homing,[16] and lodging in endosteal bone marrow HSC niches [56]. SLAMF1+ hematopoietic stem cells close to sinusoids may monitor the concentration of blood elements.

Hematopoietic stem cell niches regulate the stem cell maintenance in cooperation with endosteal cells (osteoblasts and osteoclasts) at the interface between bone and bone marrow, as well as vascular and perivascular cells around bone marrow sinusoids. The specific microenvironment promotes stem cell survival and self-renewal and regulates cell differentiation and migration according to the body's need. Endosteal, vascular, and perivascular cells cooperate and contribute to the HSC maintenance, expansion, and location. Homeostasis of hematopoietic stem cells requires many factors (Table 2.2) [57].

Many plasmalemmal receptors control the location of hematopoietic stem cells, among which the calcium-sensing receptor [61].[17] Calcium-sensing receptors are particularly expressed by LSK cells.

2.3.2 *Cells of Bone Marrow Niches*

Osteopontin+, N-cadherin+ osteoblasts that line the endosteal surface of trabecular bone, bone marrow sinusoid endothelial cells, CXCL12-secreting stromal cells, adipocytes, and macrophages constitute functional niches of hematopoietic stem cells in adult bone marrow.

The adipocyte number is inversely correlated with the hematopoietic activity in the niche. Density of hematopoietic stem cells and progenitors decays in the adipocyte-rich bone marrow. Bone marrow adipocytes operate as inhibitors of hematopoiesis [59].

On the other hand, in vitro, undifferentiated stromal cells and preadipocytes support hematopoiesis, as they produce growth factors, such as granulocyte–macrophage (CSF2) and granulocyte (CSF3) colony-stimulating factor. Moreover,

---

[15]Chemokine CXCL12 can induce motility, chemotaxis, and adhesion of cells expressing its receptor, CXC-chemokine receptor-4, as well as secretion of matrix metalloproteinases and angiogenic factors, such as vascular endothelial growth factor. Association of CXCL12 and CXCR4 is required for retention and maintenance of adult hematopoietic stem cells.

[16]Homing corresponds to recruitment of circulating hematopoietic stem cells to the bone marrow vasculature and subsequent extravasation that requires adhesion molecules, such as selectins and integrins. Small GTPases Rac1 and Rac2 are also implicated in homing and retention of hematopoietic stem cells in the endosteal bone marrow HSC niches.

[17]The endosteal niche for hematopoietic stem cells is characterized by a high Ca^{2+} concentration (more than 20-fold the plasmatic one), due to active bone remodeling in the endosteum by osteoclasts.

Table 2.2. Factors of the maintenance of hematopoietic stem cells (Sources: [57, 58]; MPL: myeloproliferative leukemia virus proto-oncogene product). Stromal and parenchymal cells of the bone marrow (mesenchymal progenitors, osteoblasts, fibroblasts, adipocytes, and reticular and endothelial cells), endosteal cells (undifferentiated bone-lining cells, bone-forming osteoblasts, and bone-resorbing osteoclasts), and vascular and perivascular cells of bone marrow sinusoids secrete factors that promote HSC maintenance. Osteoblasts secrete angiopoietin, thrombopoietin, and CXC-chemokine ligand-12 (CXCL12). Angiopoietin (also secreted by megakaryocytes and perivascular cells in the bone marrow) and thrombopoietin (supplied by blood or produced by stromal cells) promote HSC quiescence; CXCL12 (also secreted by reticular and perivascular cells) regulates HSC migration (CXCR4: CXC-chemokine receptor-4). Alternative splicing creates many epican variants (a.k.a. CD44 and homing cell adhesion molecule [HCAM]), especially the smallest $epican_S$ (or hematopoietic CD44h) that is strongly expressed on hematopoietic cells and long $epican_L$ (CD44v). Due to diversity of extracellular domain, epican interacts with numerous ligands (hyaluronate, fibronectin, collagen-1 and -4, serglycin, osteopontin, etc.).

Factor	Receptor, partner
Angiopoietin	TIE1/2
N-cadherin	Ca^{2+}-dependent N-cadherin receptor
Calcium	Ca^{2+}-sensing receptor
CXCL12	CXCR4
Osteopontin	Epican (CD44), $\alpha_V\beta_3$-integrin
Prostaglandin-E2	Cyclooxygenase and synthase
Sonic Hedgehog	Patched
Stem cell factor	Stem cell factor receptor (KIT)
Thrombopoietin	Thrombopoietin receptor (MPL)

adipocytes secrete neuropillin-1, lipocalin-2,[18] adiponectin, and tumor-necrosis factor-α that can preclude hematopoietic cell proliferation. Although adipocytes prevent expansion of hematopoietic progenitors, they preserve hematopoietic stem cell pool.

2.3.3 Crosstalk between Niche-Resident Cells

Molecular crosstalk between hematopoietic stem cells and other cellular constituents of endosteal niches regulate the balance between HSC self-renewal and differentiation. Cellular adhesions between stem cells and stromal cells and/or the extracellular matrix anchor stem cells within the niche and regulate stem cell behavior. Whereas β_1- and β_7-integrins are not necessary to maintain retention of adult hematopoietic precursors in the bone marrow, other agents, such as membrane-

[18] A.k.a. neutrophil gelatinase-associated lipocalin (NGAL) and migration-stimulating factor inhibitor (MSFI).

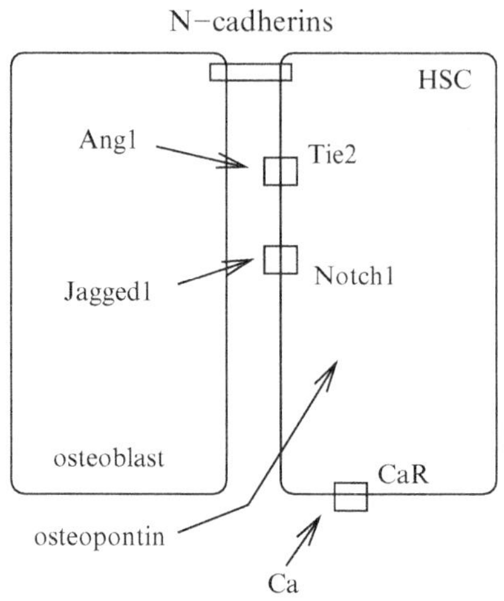

Fig. 2.2 Endosteal niche of hematopoietic stem cells. Interactions between hematopoietic stem cells and osteoblasts (Sources: [56, 60]).

bound stem cell factor[19] and its receptor SCFR, as well as CXCR4 can mediate their sequestration in the niche [62].

Osteoblasts secrete multiple cytokines that promote the proliferation of hematopoietic cells. Specialized spindle-shaped N-cadherin-expressing osteoblasts of the endosteum directly contact hematopoietic cells via N-cadherin.[20] Many stromal cell lines of endosteal bone surfaces that are involved in bone modeling, are required in HSC maintenance. The endosteal niche can contain quiescent and self-renewing hematopoietic stem cells.

Osteopontin that prevents the proliferation of hematopoietic stem cells may maintain HSC quiescence (Fig. 2.2). *Angiopoietin-1* at the osteoblast surface interacts with its receptor Tyr kinase with Ig and EGF homology domains TIE2 on stem cells to maintain stem cell quiescence in the niche.

Certain situations such as inflammation trigger the activity of osteoclasts along the stem cell niche, secretion of enzymes (cathepsin-G, elastase, MMP9, etc.) and cytokines (interleukin-8), and mobilization of progenitors from the bone marrow to the circulation [63]. Osteoclasts require TNFSF11 from osteoblasts for proliferation and bone resorption. Activating osteoclasts via either TNFSF11 or other stimuli leads to emigration of hematopoietic stem cells into the circulation. On the other hand, increased osteoblast activity via parathyroid hormone receptor or inactivation of the bone morphogenic protein receptor-1A causes a proliferation of hematopoietic stem cells.

[19]A.k.a. hematopoietic growth factor and steel factor (SLF).

[20]Receptor Tyr kinases TIE1 and TIE2 are required in HSC maintainance in HSC microenvironment. Receptor TIE2 activated by angiopoietin-1 secreted by osteoblasts upregulates N-cadherin expression in hematopoietic stem cells and maintains HSC quiescence via CKI1a cyclin-dependent kinase inhibitor.

Endothelial cells in vascular niche signal to hematopoietic stem and progenitor cells, as they release specific paracrine growth factors, the so-called *angiocrine factors*. Protein kinase-B activation in endothelial cells via TOR, but not the FoxO pathway, upregulates specific angiocrine factors that support self-renewal and expansion of CD34−, STK1−, Lin−, SCA1+, SCFR+ hematopoietic stem and progenitor cells (LSK cells), hence the long-term hematopoietic stem cell repopulation [929]. On the other hand, coactivation in endothelial cells of PKB and extracellular signal-regulated kinases ERK1 and ERK2 favors maintenance and lineage-specific differentiation of hematopoietic stem and progenitor cells. Selective activation of PKB1 in endothelial cells of adult mice increased the number of colony-forming units in the spleen and CD34−, STK1−, LSK long-term hematopoietic stem and progenitor cells in the bone marrow [929].

2.4 Regulation of Hematopoiesis

Hematopoietic stem cells express the plasmalemmal hematopoietic progenitor antigen cluster of differentiation CD34 (CD34+ cells), a glycoprotein (sialomucin protein) that acts as a cell adhesion molecule for attachment of stem cells to the extracellular matrix or stromal cells.[21] Expression of CD34 allows activated early progenitors to be distinguished from quiescent cells. CD34-related molecules — podocalyxin and podocalyxin-like molecule-2 (or endoglycan) — are also expressed on stem and early progenitor cells [65, 66] (Table 2.3).[22] Proteins of the CD34 family enhance proliferation, but repress differentiation of stem and progenitor cells, partly due to defects in adhesion and migration [66]. Both CD34 and podocalyxin improve migration of hematopoietic cells. Members of the CD34 family promote or block cell adhesion according to their expression level. They facilitate adhesion of lymphocytes to specialized vascular endothelia in lymphoid tissues such as that of high endothelial venules. On the other hand, their negatively charged, glycosylated extracellular domain can impede intercellular aggregation [66].

[21]CD34+ cells also comprise endothelial progenitor cells, vascular endothelial cells, T lymphocytes, mastocytes, and interstitial and dermic dendritic cells, as well as other tissue-specific stem cells, such as muscle satellite cells and epidermal precursors [65, 66].

[22]Podocalyxin is also known as podocalyxin-like protein PodxL1 or PCLP1, thrombomucin, and GP135; and endoglycan as podocalyxin-like protein PodxL2 or PCLP2. Podocalyxin was originally identified as a marker of kidney glomerular epithelial cells (podocytes). Podocalyxin is also expressed on vascular endothelial cells and hematopoietic stem and progenitor cells [66]. Intracellular binding partners of members of the CD34 family include scaffolds sodium–hydrogen exchanger (NHE) regulatory factors (NHERF1–NHERF2), or solute carrier family SLC8a39 regulators (SLC8a39R1 and -R2), as well as adaptor CRK-like protein (CRKL) [66]. Both NHERF1 and NHERF2 possess a C-terminal ezrin–radixin–moesin (ERM)-binding site and 2 PDZ domains. Although CD34 does not interact with NHERF proteins in hematopoietic progenitor cells, it operates with CRKL adaptor.

Numerous substances regulate hematopoiesis using a specific set of transcription factors. They include morphogens and growth factors (Table 2.4) that prime signaling cascades with given effectors down to transcription factors for gene expression. In addition, hematopoietic stem cells are mobilized by the sympathetic nervous system.

Under normal conditions, bone morphogenetic proteins, osteopontin,[23] Wnt inhibitors secreted Frizzled-related proteins sFRP1, and non-canonical Wnts are expressed in the endosteal zone to yield an inhibitory microenvironment to favor the pool of long-term quiescent hematopoietic stem cells [40].

On the other hand, fibroblast growth factors, canonical Wnt signals, and CXCL12 expressed from endothelial, megakaryocytic, and CXCL12^{high} reticular cells stimulate the active subpopulation of hematopoietic stem cells in the central zone.

2.4.1 Transcription Factors of Hematopoiesis

Transcriptional regulation controls and maintains the hematopoietic stem cell pool in the bone marrow as well as lineage commitment. Transcription factors of the E protein family (TcF3, TcF4, and TcF12), i.e., bHLHb19 to bHLHb21 (Table 2.5), and inhibitors of DNA-binding (ID), i.e., bHLHb24 to bHLHb27, which are synthesized in hematopoietic cells, regulate many developmental processes.

Progenitor fate depends on coordinated activities of multiple transcription factors, such as Activator protein AP1, CCAAT/enhancer-binding protein C/EBPα, early B-cell factor EBF1, erythroid differentiation-associated gene product (EDAG), or hemogen,[24] erythroid Krüppel-like factor (KLF1 or EKLF), members of the GATA family, zinc finger protein multitype ZFPM1,[25] and Paired box protein Pax5.

2.4.1.1 GATA Transcription Factors

The transcription factors, GATA-binding factors, interact with ZFPM1 to activate or repress transcription during tissue development. Hematopoietic factor GATA1 activates all known erythroid and megakaryocyte-specific genes, but represses GATA2, SCFR, and MyC genes.[26]

---

[23] A.k.a. bone sialoprotein BSP1, early T-lymphocyte activation ETA1, and secreted phosphoprotein SPP1.

[24] A.k.a. hemopoietic gene protein and negative differentiation regulator (NDR).

[25] A.k.a. friend of GATA protein FOG1, zinc finger domain-containing protein ZnF89a and ZnF408.

[26] The family of MYC genes that encode subclass-E basic helix–loop–helix proteins encompasses: (1) cellular (MYC or CMYC [bHLHe39]); (2) small cell lung carcinoma-derived (LMYC [bHLHe38]); and (3) neuroblastoma-derived (NMYC [bHLHe37]) genes. The MYC gene has been discovered in patients with Burkitt lymphoma.

Table 2.3. Distribution of CD34 family members (Source: [66]; +: expression; −: no expression). Cell-surface marker CD34 enables the identification of hematopoietic stem cells and progenitors. CD34-related proteins include podocalyxin and endoglycan.

Distribution	CD34	Podocalyxin	Endoglycan
Multipotent precursors	+	+	+
Eosinophils	+	−	−
Erythrocytes	−	Anemia	−
T lymphocytes	−	−	Thymocytes
B lymphocytes	−	−	TLR-activated B cells
Macrophages	−	−	−
Mastocytes	+	−	−
Thrombocytes	−	+	−
Muscle satellite cells	+	−	−
Hair follicle stem cells	+	−	−
Vascular endothelia	+	+	+
Smooth myocyte	−	−	+
Fibrocytes	+	−	−
Mesothelia	−	+	−
Neurons	Weak	+	Some
Podocytes	−	+	Weak

Table 2.4. Regulators of hematopoiesis. Canonical Wnt and Jagged-1–Notch-1 signaling promote HSC survival and proliferation particularly after damage. Insulin-like growth factor binding protein-3 (IGFBP3) favors progenitor recruitment to hypoxic sites (CSF: colony-stimulating factor; CSFR: colony-stimulating factor receptor; LRP: LDL receptor-related protein).

Factor	Receptor and partners
Erythropoietin	EpoR
Granulocyte CSF (CSF3)	CSF3R; Ras
Granulocyte–macrophage CSF (CSF2)	CSF2R; Ras
IGF1	IGF1R; IGFBP3
Interleukins	ILR
Jagged-1	Notch
Macrophage CSF (CSF1)	CSF1R; Ras
Sonic Hedgehog	Patched
Stem cell factor	SCFR
Thrombopoietin	TpoR
Tumor-necrosis factor-α	TNFR
Wnt	Frizzled, LRP; Disheveled, β-catenin
STAT	JaK

Table 2.5. Subclass-B basic helix–loop–helix proteins that belong to E/E2A protein and inhibitor of DNA binding (ID) families. Transcription factors of the E protein family can heterodimerize with subclass-A basic helix–loop–helix protein bHLHa17 (a.k.a. T-cell acute lymphocytic leukemia protein-1 [TAL1] and stem cell leukemia factor [SCL]) and bHLHa18 (a.k.a. lymphoblastic leukemia-derived sequence-1 [LyL1]).

bHLHb type	Main alias	Other aliases
	E2A transcription factors	
bHLHb19	TcF4	E2-2, TcF7L2, ITF2, SEF2
bHLHb20	TcF12	HEB, HTF4
bHLHb21	TcFE2α/TcF3	E2A, ITF1
bHLHb21^{E12}	TcFE2a	E12
bHLHb21^{E47}	TcF3	E47
	E protein partners (heterodimers)	
bHLHa17	TAL1	SCL, TCL5
bHLHa18	LyL1	
	Inhibitors of DNA binding	
bHLHb24	ID1	
bHLHb25	ID3	
bHLHb26	ID2	
bHLHb27	ID4	

The regulation of most GATA1-dependent genes requires the tethering to ZFPM1 protein that is restricted to erythroid cells and megakaryocytes. Cofactor ZFPM1 that operates as both an activator and repressor binds to the nucleosome remodeling and deacetylase (NuRD) complex that can also be detected at both inhibited and active GATA1–ZFPM1 target genes [67]. Consequently, the NuRD complex is a versatile cofactor for transcription factors that activate and repress gene transcription according to lineage-specific gene expression patterns.

2.4.1.2 Transcription Factors of Self-Renewal of Hematopoietic Stem Cells

Homeobox Gene Family

Self-renewal of hematopoietic stem cells is associated with the homeobox (HOX) gene family (Table 2.6), especially transcription factors HoxA9 and HoxB4 that cause an expansion of cultured hematopoietic stem cell populations. Homeobox gene products are homeodomain-containing proteins with tissue-specific functions that bind to DNA enhancer sequences, acting as both a repressor at given genes and activator at others. The DNA-binding homeodomain is a highly conserved 60 amino acid-long motif. Homeobox gene products regulate the proper placement of

Table 2.6. Factors of self-renewal of hematopoietic stem cells (Source: [69]; BMI1: murine leukemia viral [BMI1] Polycomb RING finger proto-oncogene product, or RING finger protein RNF51; BMP: bone morphogenetic protein; CBP: CREB-binding protein; Cdx: caudal type homeobox gene product; CKI: cyclin-dependent kinase inhibitor; ETV: ETS variant; EZH: enhancer of zeste homolog [Polycomb group; (histone) lysine N-methyltransferase]; GATA: DNA sequence GATA-binding protein; GFI1: growth factor independent-1 transcription repressor, or zinc finger protein ZnF163; Hox: homeobox gene product; MEIS: myeloid ecotropic viral integration site homeobox protein; MLL: myeloid–lymphoid or mixed-lineage leukemia proto-oncogene product [a.k.a. histone-lysine N-methyltransferase KMT2a, HRX, and trithorax homolog]; MyB: Myb myeloblastosis viral proto-oncogene homolog; MyC: Myc myelocytomatosis viral proto-oncogene homolog; PHc1: polyhomeotic homolog-1 [a.k.a. early development regulatory protein EDR1, polyhomeotic-like protein PH1, homolog of polyhomeotic HPc1, and RAE28; homeotic genes are involved in developmental patterns]; Sox: SRY (sex-determining region Y)-box; SPI1: spleen focus-forming virus [SFFV] proviral integration proto-oncogene product [or transcription factor PU.1]; ZFX: X-linked zinc finger protein, or ZnF926).

Category	Types
Hox transcription factors	HoxA7/9/10, HoxB4, CdX2/4, MEIS1
Growth factors	Notch, Wnt, BMP
Modulators	Angiopoietin-like factors, thrombopoietin, stem cell factor, retinoic acid, prostaglandin-E2
Transcription factors	CBP, ETV6, GATA2, GFI1, MyB, SPI1, Sox17, ZFX
Cell cycle regulators	CKI1a/b–CKI2a/c, MyC, P53, PTen
Chromatin-associated factors	EZH2, BMI1, MLL, PHc1

structures, such as vertebrae and limbs, in humans. Distinct Hox proteins operate in different tissues, with distinct functions, targeting different sets of genes.[27]

Expression of HOX genes is regulated by members of caudal-type (CDX), preB-cell leukemia homeobox (PBX), and myeloid ecotropic viral integration site (MEIS) families of Hox cofactors [69]. Moreover, expression of HOX genes is modulated by growth factors and morphogens, such as Hedgehog, Notch, bone morphogenetic proteins, and Wnt (Vol. 3 – Chaps. 8. Receptor Kinases and 10. Morphogen Receptors).

Chromatin Remodelers of the Polycomb Group for Epigenetic Silencing

Polycomb group proteins participate in 2 polycomb repressive complexes PRC1 and PRC2 that regulate HSC self-renewal. In addition, some substances (e.g., retinoic acid and prostaglandin-E2) modulate HSC self-renewal.

[27]HOX genes are clustered on some chromosomes. In humans, 4 clusters exist (39 HOX genes each [68]): HOXA cluster in chromosome 7; HOXB cluster in chromosome 17; HOXC cluster in chromosome 12; and HOXD cluster in chromosome 2.

2.4.1.3 E Protein Family of bHLH Transcription Factors

In mammals, the E protein family of basic helix–loop–helix (bHLH) transcription factors that are involved in B- and T-lymphocyte development includes transcription factors TcF3,[28] TcF4,[29] and TcF12.[30] Distinct proteins are formed by alternative transcription initiation, in addition to 2 splice variants TcFE2α[31] and TcF3[32] encoded by the E2A gene.[33] Ubiquitin conjugase UbC9 causes the degradation of E2A proteins by the proteasome.[34]

Inhibitors of DNA-binding proteins constitute a set of 4 members (ID1–ID4). These helix–loop–helix proteins that lack a basic region prevent E proteins from binding to DNA.

Transcription factors of the E protein family of the bHLH superclass can form: (1) homo- or heterodimers with distinct family members that acts as transcription activators; (2) heterodimers with class-2 bHLH proteins, subclass-A basic helix–loop–helix protein bHLHa17, or T-cell acute lymphocytic leukemia protein-1 (TAL1)[35] and bHLHa18, or lymphoblastic leukemia-derived sequence-1 (LyL1),[36] that function as transcription activators or repressors whether they recruit coactivator or corepressor complexes; and (3) heterodimers with any ID protein that do not bind DNA [70].[37]

---

[28] A.k.a. transcription factor E2α, E2A immunoglobulin enhancer-binding factor E12–E47, immunoglobulin transcription factor ITF1, and subclass-B basic helix–loop–helix protein bHLHb21.

[29] A.k.a. immunoglobulin transcription factor ITF2, SL3-3 enhancer factor SEF2, E2-2, and bHLHb19.

[30] A.k.a. human E-box-binding protein (HEB), human transcription factor HTF4, and bHLHb20.

[31] Or E12.

[32] Or E47.

[33] Alternative transcription start sites in genes give rise to canonical (Can) and alternative (Alt) forms of TcF4 and TcF12 proteins. Proteins of the E2A set, TcFE2α and TcF3, as well as TcF4^{Can} and TcF12^{Can}, possess 2 activation domains AD1 and AD2, whereas TcF4^{Alt} and TcF12^{Alt} have a single activation domain [70]. Subtype TcF12^{Alt} is highly expressed by CD4−, CD8− thymocytes.

[34] Carboxy terminus of heat shock protein HSP70-interacting protein (CHIP) that promotes formation of a pre-ubiquitination complex including E3-ubiquitin ligase SKP1 (S-phase-kinase-associated protein)–cullin-1–F-box-protein complex SCFSKP2 binds E2A proteins.

[35] T-cell acute lymphocytic leukemia protein-1, or stem cell leukemia factor (SCL), is involved in the generation of hematopoietic stem cells from mesoderm during embryongenesis. In adult mice, bHLHa17 is implicated in self-renewal of short-term repopulating hematopoietic stem cells and development of erythrocytes and megakaryocytes, but not for maintenance of long-term repopulating hematopoietic stem cells [70].

[36] Lymphoblastic leukemia-derived sequence factor-1 heterodimerizes with E proteins to ensure self-renewal of long-term repopulating hematopoietic stem cells [70].

[37] Homo- and heterodimeric E proteins bind to E-box sequence (preferentially CACCTG E-box sequence) of target genes and recruit transcriptional coactivators such as P300 and CBP histone acetyltransferases and RNA polymerase-2. Heterodimers composed of E proteins and class-2 bHLH factors associate with CATATG E-box sequence of target genes and recruit coactivators or corepressors, such as cyclin-D- and Runt-related transcription factor-1 translocated to protein-1 (Runx1T1), a.k.a. ETO corepressor (eight twenty one) and zinc finger MYND domain-containing

Table 2.7. E-protein-dependent genes in B and T lymphocytes, plasmacytoid dendritic cells and their multipotent progenitors (**Part 1**; Source: [70]; IGA: Igα-type of immunoglobulin heavy (IgH)-chain gene, or CD79A; IGI: preB-lymphocyte (VPREB1) Igι-chain gene, or CD179A; IGK: Igκ-type of immunoglobulin light (IgL) chains (κ) of immunoglobulin light (IgL); IGLL1: Igλ-like polypeptide-1, a.k.a. CD179B, IGL1, IGL5 [that encodes Igλ5], and VPREB2; SPI1: spleen focus-forming virus [SFFV] proviral integration proto-oncogene product; SPIb: SPI1-related transcription factor; TCRβ: T-cell receptor-β). E proteins were initially identified by their ability to bind to E-box sequences in B cells. E2A proteins control immunoglobulin gene transcription and recombination, as they bind intronic enhancers of the IGH and IGK gene for proper rearrangement of gene segments. E proteins maintain hematopoietic stem cell quiescence. They regulate the transcription of many transcription factors and proteins, such as V(D)J recombination-activating gene products RAG1 and RAG2; immunoglobulin heavy chain IgH6 [IgH of IgM]; preT-cell receptor α chain [pTCRα]; interleukin-7 receptor α chain [Il7Rα]; early B-cell factor gene [EBF], cell cycle regulators myelocytomatosis viral-related proto-oncogene NMYC (neuroblastoma-derived MYC; a.k.a. MYCN and subclass-E basic helix–loop–helix protein bHLHe37) and CMYC (or bHLHe39); messenger Notch-1 and its targets such as Hairy and enhancer of Split [HES1, or subclass-B bHLHb39] and Deltex homolog-1; as well as plasmacytoid dendritic cell-associated genes such as terminal deoxynucleotidyltransferase gene [Dntt] and CC-chemokine receptor gene Ccr9; growth factor-independent gene Gfi1 for early T-cell development). They cooperate with various transcription factors (e.g., interferon regulatory factor IRF4, Paired box protein Pax5, SPI1, and Ikaros family zinc finger-containing protein IKZF1). E2A proteins regulate cell cycle progression in hematopoietic stem cells, as they control expression of the CDKN1A and CDKN1B genes (that encode cyclin-dependent kinase inhibitors CKI1a and CKI1b), of the Gfi1 gene (growth factor-independent-1 transcription repressor), and myeloproliferative leukemia virus proto-oncogene (MPL) that encodes thrombopoietin receptor (TpoR). Transcription factors of the E2A family that target the Rag1, Dntt, and Igh6 genes regulate multipotent progenitors for the formation of common lymphoid progenitors.

Cell type	Cell-specific genes	Specification and commitment factors
B lymphocyte	IGA, IGI, IGK, IGLL1, NMYC	EBF1
T lymphocyte	deltex-1, Hes1, PTCRA, TCRB, Gfi1b	Notch-1
Plasmacytoid Dendritic cell	Irf8	SPIb

Transcription factors of the E protein family together with ID proteins regulate the transcription of many genes, as they operate in the development of many lymphoid cell types and govern the earliest restriction events in lymphopoiesis (Tables 2.7 and 2.8).

protein ZMYND2, acute myeloid leukemia AML1 (Runx1)–ETO fusion protein, and Sin3a histone deacetylase complex subunit. The most frequent chromosomal translocation in acute leukemia creates the Runx1–Runx1T1 fusion protein, which represses transcription via Runx1 and immortalizes hematopoietic progenitor cells. On the other hand, heterodimers made up of E and ID proteins fail to activate gene transcription. E-box CANNTG (N denotes any nucleotide) sequences belong to the enhancer regions of genes that encode immunoglobulin heavy (IgH) and κ chain (Igκ) of immunoglobulin light (IgL) chain. In T lymphocytes, unlike B lymphocytes, E-box binding complexes contain high amounts of both bHLHb20 and bHLHb21 factors.

Table 2.8. E-protein-dependent genes in B and T lymphocytes, plasmacytoid dendritic cells and their multipotent progenitors (**Part 2**; Source: [70]).

Cell type	Genes expressed prior to lineage restriction	
	Lymphoid-primed multipotent progenitors	Hematopoietic stem cells
B lymphocyte	Rag2, IL7RA, Ets1	Igh, Dntt
T lymphocyte	Rag2, IL7RA, Ets1	Notch-1, Dntt
Plasmacytoid Dendritic cell		Ccr9, Dntt

Transcription factors bHLHb19 to bHLHb21 participate in B-cell development before preB-cell receptor expression [70]. They also promote survival of proB cells (committed B-cell progenitors), as they prevent caspase-2 activation. They regulate the transcription of RAG1 and RAG2 recombination-activating genes. They cooperates with signal transducer and activator of transcription STAT5 to activate cell cycle regulators of the family of cellular myelocytomatosis viral-related proto-oncogene products cellular MyC and neuroblastoma-derived nMyc to support cytokine-dependent proliferation.

Transcription factors bHLHb19 to bHLHb21 operate in T-cell progenitor cells to control expression of T-cell receptor β chain (Fig. 2.3). Transcription factors $bHLHb21^{E12}$ and $bHLHb21^{E47}$ operate in the transcriptional regulation that fosters commitment to and differentiation of B- and T-cell lineages [70]. They are required for development of lymphoid-primed multipotent progenitors from hematopoietic stem cells.[38] However, they prevent the proliferation of hematopoietic stem cells and lymphoid-primed multipotent progenitors. They target lymphoid genes. They activate early B-cell factor EBF1 to promote B-cell lineage specification and cooperate with transcription factors EBF1 and Paired box protein Pax5 to activate many B-cell lineage-associated genes [70]. In addition, TcF4 intervenes in the development of mature plasmacytoid dendritic cells, in which it regulates transcription of multiple lymphoid-associated genes.

Proteins of the E2A family, $bHLHb21^{E12}$ and $bHLHb21^{E47}$, regulate Notch-1 expression in early thymic progenitors and collaborate with Notch-1 to excite Notch-1 target genes (e.g., hairy and enhancer of split HES1 (or bHLHb39), preT-cell receptor α chain [PTCRA], etc.) [70]. The Hes1 gene is needed for T-cell progenitor expansion. Conversely, Notch-1 inhibits E2A protein. In addition, transcription factor Paired box protein Pax5, a partner of TcF2α and TcF3, represses Notch-1 transcription, thereby preventing T-cell fate in proB cells.

[38]The lymphoid-primed multipotent progenitor pool is a subset of LIN−, SCFR+, SCA1+ cells that express STK1 kinase (or FMS-related Tyr kinase FLT3). They have the potential to develop into granulocytes, lymphocytes, and monocytes, but neither erythrocytes nor megakaryocytes.

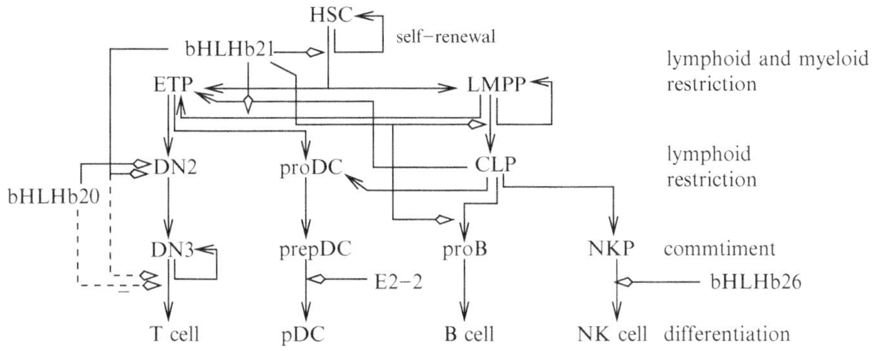

Fig. 2.3 Transcription factors of the E and ID categories (i.e., bHLHb19–bHLHb21 and bHLHb24–bHLHb27) in lymphoid lineage commitment (Source: [70]). Formation of B, T, natural killer (NK), and plasmacytoid dendritic (pDC) cells from hematopoietic stem cell (HSC), lymphoid-primed multipotent progenitor (LMPP), early thymic progenitor (ETP), common lymphoid progenitor (CLP), double-negative (DN) and NK cell progenitor (NKP). Self-renewal of stem cells and proliferation of some progenitors happens in the absence of E2A proteins. Proteins of the E2A set enforce the β-selection checkpoint during T-cell development. DN3 cells express a preT-cell receptor owing to recombination-activating RAG1 gene. In the absence of E2A protein, DN3 cells proceed past the β-selection checkpoint and become CD4+, CD8+ thymocytes. On the other hand, B-cell development happens when E2A protein level decays. However, bHLHb19 (TcF4) is required in plasmacytoid dendritic cell development. Inhibitor of DNA binding ID2 (or bHLHb26) that antagonizes E proteins is involved in the development of mature natural killer cells, as well as optimal formation of subsets of conventional dendritic cells, such as CD8−, CD11c+ dendritic cells and Langerhans cells, whereas it prevents plasmacytoid dendritic cell development.

Transcription factors bHLHb21^{E12} and bHLHb21^{E47} target recombination-activating gene Rag1, terminal deoxynucleotidyltransferase gene (Dntt), and immunoglobulin heavy chain gene Igh6 regulate multipotent progenitors for the formation of common lymphoid progenitors.

Transcription factors bHLHb21^{E12} and bHLHb21^{E47}, Early B-cell factor EBF1, and Pax5 engage hematopoietic progenitors to the B-cell lineage. Transcription factors bHLHb21^{E12} and bHLHb21^{E47} heterodimerize with bHLHa17 expressed in mesodermal cells that develop into embryonic blood cells as well as fetal and adult hematopoietic stem cells.

Activated bHLHb21^{E12} and bHLHb21^{E47} act in synergy with SPI1 and interleukin-7 receptor IL7Rα that activates signal transducer and activator of transcription STAT5 to induce EBF1 transcription that, in turn, activates Pax5 expression. Proteins bHLHb21^{E12} and bHLHb21^{E47} with EBF1 and Pax5 activate many B-cell-associated genes. In hematopoietic stem cells, bHLHb21^{E12} and bHLHb21^{E47} promote development of the entire spectrum of early hematopoietic progenitors [54]. They favor the maturation of myelolymphoid and myelo-erythroid progenitors.

2.4.1.4 Inhibitor of DNA Binding

Inhibitor of DNA binding ID2 modulates function of B- and T-cell lineages. It is required for the development of mature natural killer cells. It impedes E-protein activity. NK-cell lineage commitment can be controlled by ID2 or ID3 (bHLHb26 and bHLHb25) [70]. Protein bHLHb26 links to anaphase promoting complex, or cyclosome, and this ubiquitin ligase promotes its degradation.

2.4.1.5 bHLHa17 and Runx1 Transcription Factors

Tissue-specific basic helix–loop–helix transcription factor bHLHa17, or T-cell acute lymphocytic leukemia protein TAL1,[39] that upregulates stem cell factor receptor is needed for the establishment of hemogenic endothelium. On the other hand, transcription factor Runx1[40] is required for the formation of clusters of hematopoietic stem cells from bipotential precursors of hemogenic endothelial cells [73].[41] However, Runx1 is dispensable in cells that express Vav1 guanine nucleotide-exchange factor, one of the first pan-hematopoietic genes expressed in hematopoietic stem cells. Therefore, Runx1 requirement ceases in Vav1+ hematopoietic stem cells.

Hemodynamic stress increases the expression of Runx1 in CD41+, SCFR+ hematopoietic progenitor cells and their hematopoietic colony-forming potential [74].[42] Nitric oxide that is involved in mechanical stress signaling promotes hematopoietic potential in vivo.

---

[39]Transcription factor bHLHa17 is coexpressed with stem cell factor receptor in hematopoietic progenitors. Factor bHLHa17 constitutes a nucleation factor for a multifactorial complex with LIM-only protein-2 (LMO2; a.k.a. rhombotin-like protein-1, rhombotin-2, and T-cell translocation protein TTG2), GATA1, GATA2, bHLHb21, and LIM domain-binding protein LDB1 (a.k.a. C-terminal LIM domain-binding protein CLIM2 and nuclear LIM interactor [NLI]) that specifically enhances SCFR promoter activity via Specificity protein SP1 [71]. It also forms heterodimers with bHLHb20 and bHLHb21 factors. Activation of the TAL1 gene is the most frequent gain-of-function mutation in T-cell acute lymphoblastic leukemia. Factor bHLHa17 acts as both an activator and repressor of transcription. The Sin3a–HDAC1 corepressor complex interacts with bHLHa17 to restrict its function in erythroid differentiation [72].

[40]Runt-related transcription factor Runx1 is also termed Acute myeloid leukemia AML1.

[41]Vascular regions that generate hematopoietic stem cell in mice comprise endothelia of the dorsal aorta and vitelline and umbilical arteries, as well as yolk sac blood islands and distal allantois and chorion that are precursors of the placenta. These independent sites constitute the conceptus.

[42]Mechanical forces contribute to the regulation of the developing cardiovascular system in embryos. After initiation of the heartbeat, cells lining the dorsal aorta, placental vessels, and umbilical and vitelline arteries initiate expression of the transcription factor Runx1 [74].

2.4.1.6 SPI1 Transcription Factor

Multipotent hematopoietic stem cells have different expression of stem cell marker SCA1 with distinct transcriptomes: $SCA1^{low}$, $SCA1^{mid}$, and $SCA1^{high}$. Both $SCA1^{high}$ and $SCA1^{low}$ progenitor cells express the highest and lowest level of myeloid fate-determining transcription factor SPI1, respectively [75].[43] Transcription factor SPI1 operates in multipotent progenitors as a selector for development of B lymphocytes, monocytes, acrophages, and granulocytes. CCAAT/enhancer binding protein-α that controls myeloid specification upregulates SPI1 expression. Concentration of SPI1 is indeed higher in myeloid than lymphoid progenitors (graded expression), but transcription factors SPI1 and Ikaros are highly expressed in lymphoid-primed multipotent progenitors. Among numerous transcription factors that intervene during hematopoiesis, such as GATA1, MyB, SPI1, Ikaros, bHLHb21, EBF1, and erythroid-derived nuclear factor NFe2, GATA1 and SPI1 mutually inhibit one another from committing to the erythroid or myeloid lineage.

Transcription factor Ikaros is also required for the generation of common lymphoid progenitors, B, natural killer, and most types of T cells. Transcription factors TCF7 and GATA3 are critical for T-cell specification.

2.4.1.7 Myocyte Enhancer Factor

Transcription factor myocyte enhancer factor MEF2c that has a restricted expression (hematopoietic stem cells, myeloid progenitors, B lymphocytes, and mature monocytes) participates in lineage specification, as it promotes and suppresses lymphoid and myeloid lineage differentiation, respectively [76].

Hematopoietic stem cells give rise to multipotent progenitors that still generate all of the blood cell lineages, then progenitors with restricted potential such as lymphoid-primed multipotent progenitors that have lost the ability to form megakaryocytes and erythrocytes but retain lymphoid and myelomonocytic potential, earliest T-cell progenitors, and common lymphoid progenitors or early lymphoid progenitors that generate B, T, and NK cells. Commitment of multipotent progenitors toward the B-cell lineage requires MEF2c. Expression of MEF2c depends on SPI1 transcription factor.

---

[43]Transcription factor SPI1 is also known as PU1 factor. This tissue-specific member of the ETS family of transcription factors is predominantly expressed in myeloid cells (granulocytes, monocytes, and macrophages) and B lymphocytes.

2.4.1.8 Signal Transducers and Activators of Transcription

Signal transducers and activators of transcription intervene in hematopoiesis. Factors STAT4 and STAT6 are involved in proliferation and differentiation of T_{H1} and T_{H2} lymphocytes, respectively [77, 78]. Factor STAT3 intervenes in the differentiation of the myeloid M1 cell line into macrophages [79].

2.4.1.9 Nephroblastoma Overexpressed Protein

Nephroblastoma overexpressed protein (NOv)[44] participates in the regulation of activity of hematopoietic stem cells and their derived progenitors [80]. Protein NOv also operates in tumor suppression, wound healing, and angiogenesis.

2.4.1.10 Forkhead Box Protein FoxO3

Hematopoietic stem cells either reside in bone marrow in a dormant state or enter the cell formation. In quiescent cells, protein kinase B remains inactivated, and its downstream target FoxO3 stays in the nucleus. When used as progenitors, membrane raft clustering leads to PKB phosphorylation and FoxO3 relocalizes in the cytoplasm [81].

2.4.1.11 Nuclear Transcription Factor-Y

Nuclear transcription factor-Y (NFy, or CAAT box DNA-binding protein) induces the expression of many genes that encode regulators of HSC self-renewal and differentiation [82].[45] Regulatory and DNA-binding subunit NFyα acts in several pathways implicated in HSC self-renewal (Notch-1, lymphoid enhancer-binding factor LEF1, telomerase RNA, etc.). In addition, BCL2-associated athanogene BAG1 is necessary for survival of hematopoietic stem cells [83].

2.4.2 Hormones, Growth Factors, and Morphogens

Self-renewal, differentiation into given lineage types, and selection of lineages by the multipotential progenitors, are intrinsic properties of progenitors. Survival and proliferation are regulated by hormones, growth factors, and cytokines (Vols. 2 – Chap. 3. Growth Factors and 3 – Chaps. 8. Receptor Kinases, 10. Morphogen

[44] A.k.a. CCN3 and insulin-like growth factor-binding protein IGFBP9.

[45] Trimer NFy is composed of NFyα, NFyβ, and NFyγ subunits encoded by the NFYA, NFYB, and NFYC genes.

Table 2.9. Examples of hematopoietic growth factor receptors (Source: [85]).

Receptor type	Ligands (cytokines and chemookines)
Type-1 cytokine receptor	IL1–IL9, IL13, IL18, gmCSF and gCSF (CSF2–CSF3), Epo, Tpo
Type-2 cytokine receptor	IL10, INF
Type-3 cytokine receptor	mCSF (CSF1), SCF
G-protein-coupled chemokine receptor (CXCR1/2)	IL8 (or CXCL8)
TNFR (or TNFRSF1a)	TNFα (or TNFSF1)

Receptors, and 11. Receptors of the Immune System) [39]. Several factors promote self renewal of hematopoietic stem cells, such as Notch, Wnt3a, angiopoietin-like proteins, and prostaglandin-E2.

In addition, hematopoietic stem and progenitor cells move into and then out of the blood circulation for homing, engraftment, and retention. Their interaction with bone marrow endothelium for engraftment in bone marrow as well as mobilization into blood depends on guanine nucleotide-binding protein subunit $G\alpha s$ [84]. However, Gs signaling is not required for retention of hematopoietic stem and progenitor cells in the bone marrow.

Both growth factors and hormones bind to plasmalemmal receptors (Table 2.9). Many kinds of hematopoeitins ensure a dynamic balance between cell differentiation and proliferation. Hematopoietic stem and progenitor cells in their hematopoietic niches in bone marrow respond to their environment upon engagement of plasmalemmal receptors that leads to signal transduction.

2.4.2.1 Morphogens

Notch and Wnt signaling pathways (Vol. 3 – Chap. 10. Morphogen Receptors) interact and regulate self-renewal of hematopoietic stem cells. Self-renewal of HSCs can be regulated via inhibition of differentiation and induction of proliferation.

Signaling by Notch inhibits differentiation, thus maintaining hematopoietic stem cells in an undifferentiated state [86]. Notch signaling is required for Wnt-mediated maintenance of undifferentiated HSCs, but not for entry into the cell cycle in vitro. Notch is specifically active in vivo in the HSC population of the bone marrow and in immature thymocytes. It is reduced or absent in more mature differentiated cells.

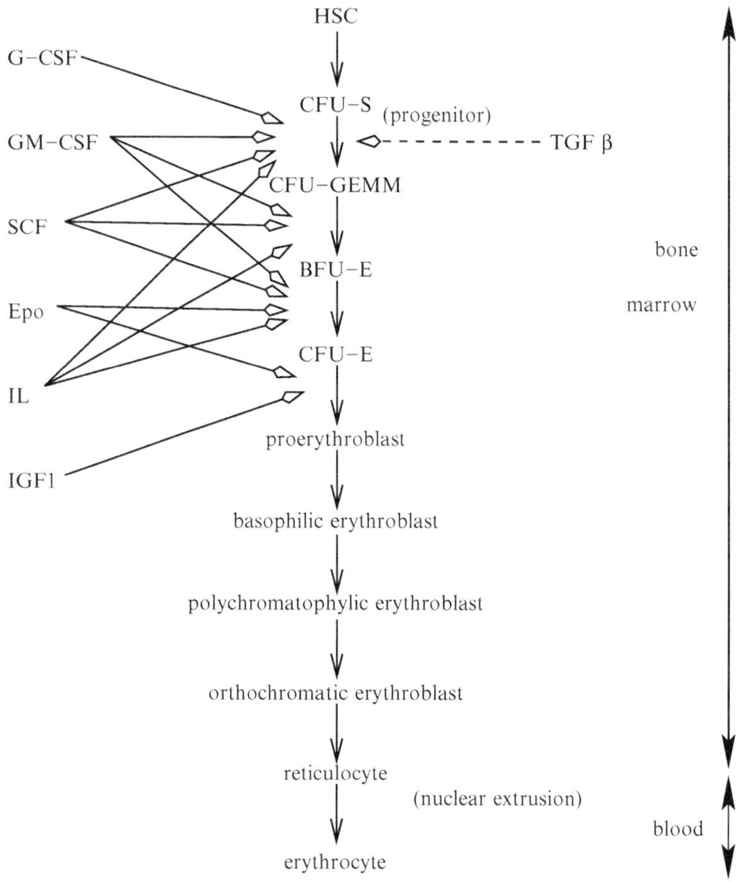

Fig. 2.4 Erythropoiesis (CFUs: colony-forming unit-spleen; CFUgemm: colony-forming unit-granulocyte–erythroid–macrophage–megakaryocyte; BFUe: burst-forming unit-erythroid; CFUe: colony-forming unit-erythoid). Many growth factors intervene, such as erythropoietin (Epo), granulocyte–monocyte colony-stimulating factor (gmCSF or CSF2), granulocyte colony-stimulating factor (gCSF or CSF3), stem cell factor (SCF), and insulin-like growth factor-1 (IGF1). Among interleukins, IL1, IL3, IL6, IL9, and IL11 are implicated in different stages of erythropoiesis. Transforming growth factor-β (TGFβ) inhibits the CFUgemm formation.

2.4.2.2 Erythropoietin

Erythropoietin (Epo) is a glycoprotein hormone synthesized mainly by the kidney cortex stimulated by hypoxemia in the renal arteries and by the liver. It activates the production of erythrocytes (Fig. 2.4).[46]

[46]Erythropoietin (Epo) and its receptor (EpoR) are also synthesized by neoplastic cells and tumor endothelial cells.

Erythropoietin acts in the bone marrow by a specific erythropoietin receptor (EpoR) located mainly on erythroid progenitors. It stimulates proliferation of erythroid progenitors and maintains their viability. More precisely, Epo induces the proliferation of burst-forming unit-erythroid (BFUe, early erythroid progenitor) and differentiation of colony-forming unit-erythoid (later erythroid precursor) [85]. It also promotes megakaryocyte differentiation, B-cell proliferation, and endothelial cell chemotaxis. Furthermore, erythropoietin prevents the destruction of viable tissue surrounding injuries, such as infarction.

2.4.2.3 Thrombopoietin

Thrombopoietin (Tpo) participates in hematopoiesis in general, and thrombopoiesis in particular (Fig. 2.5). Thrombopoietin, assisted by interleukin-11, stimulates the production of megakaryocytes via myeloid progenitors and, afterward, of platelets (megakaryocyte fragmentation produces platelets).

Thrombopoietin is manufactured by the liver, kidneys, smooth muscle, and marrow in particular. It has the longest half-life (30 h) in the circulation among hematopoietic growth factors.

2.4.2.4 Colony-Stimulating Factors

Colony-stimulating factors (CSF) are cytokines that stimulate the proliferation of stem cells of the bone marrow in adults.

Granulocyte–monocyte colony-stimulating factor (gmCSF or CSF2) induces the commitment of granulocyte–monocyte progenitors. It stimulates proliferation of multi-lineage progenitors and the growth of BFUe, granulocyte, macrophage, and eosinophil colonies [85]. It also enhances phagocyte (neutrophil, eosinophil, and macrophage) functioning. It is secreted by lymphocytes after stimulation by antibodies on the one hand and fibroblasts and endothelial cells activated by IL1 and TNF produced by macrophages and T cells on the other. Factor gmCSF can also be synthesized by macrophages and mastocytes.

Granulocyte–macrophage colony-stimulating factor receptor (gmCSFR or CSF2R) on hematopoietic cells can be activated by gmCSF, and interleukin-3 and -5. It primes the Ras–cRaf–MAPK pathway.

Granulocyte-colony stimulating factor (gCSF or CSF3), produced by endothelial cells, macrophages, and immunocytes, stimulates the proliferation and maturation of monopotent neutrophil progenitors that differentiate into neutrophils. It thus has a relatively specific activity. Further stimulated by IL5, granulocyte–monocyte progenitors develop into eosinophils. It is also produced by fibroblasts, monocytes, and macrophages stimulated by gmCSF and IL3 cytokines. Interleukin-3 participates in the differentiation of basophils. Factor gCSF enhances neutrophil functioning (chemotaxis and phagocytosis) and is involved in neutrophil extravasation.

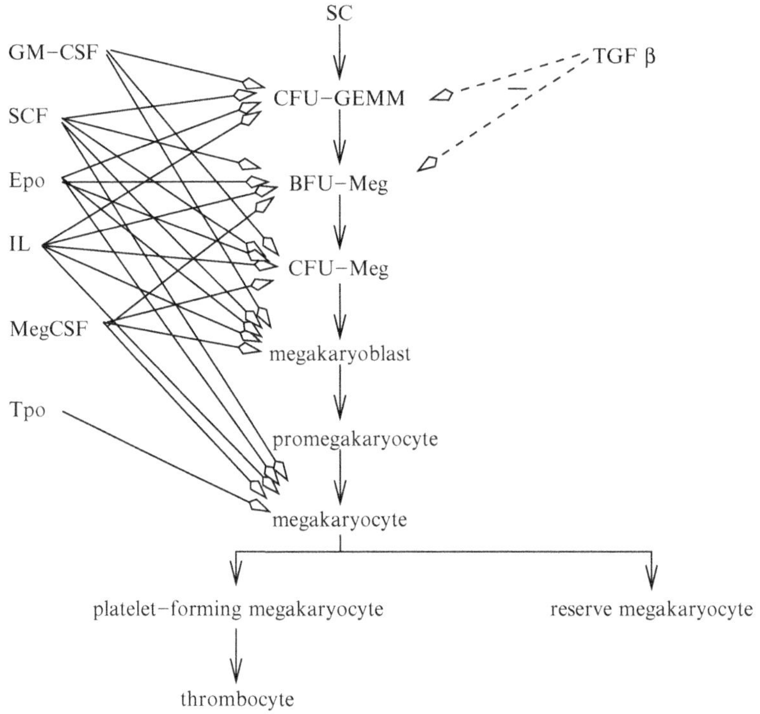

Fig. 2.5 Platelet formation in the bone marrow (SC: stem cell; CFUmeg: colony-forming unit-megakaryocyte). Many growth factors participate in the genesis of thrombocytes, such as thrombopoietin (Tpo), erythropoietin (Epo), as well as granulocyte–monocyte (gmCSF or CSF2) and megakaryocyte (megCSF) colony stimulating factor, in addition to stem cell factor (SCF). Early-acting megCSF and late-acting Tpo and other megakaryocyte potentiators stimulate maturation of megakaryocytes. Ligands that specifically binds to TpoR receptor on megakaryocytes have a potent thrombocytopoietic effect. Among interleukins, IL1, IL3, IL5 to IL7, IL9, IL11, and IL12 are implicated in different stages of thrombopoiesis. Transforming growth factor-β (TGFβ) inhibits the formation of colony-forming unit granulocyte–erythroid–macrophage–megakaryocyte (CFUgemm) and burst-forming unit megakaryocyte (BFUmeg).

Granulocyte colony-stimulating factor receptor (gCSFR) is synthesized by hematopoietic progenitors and mature hematopoietic cells, as well as bone marrow stromal cells. It targets signal transducer and activator of transcription STAT5[47] and ERK to support proliferation, survival and differentiation of neutrophilic progenitors. Suppressor of cytokine signaling proteins (SOCS) inhibits gCSFR

[47]Different STAT5 isoforms are activated during the early and late differentiation stages of granulocytes. The transcription factors Signal transducers and activators of transcription (STAT) reside in a latent form in the cytoplasm and become phosphorylated by Janus kinases (JaK). Receptor gCSFR activates neither JaK1, JaK2, TyK2, STAT1, nor STAT3 in neutrophils upon gCSF stimulation [87].

(CSF3R) signaling by recruiting ubiquitin ligase, thus controlling pathway from early to late endosomes and lysosomes [88].

Stimulated by *macrophage colony-stimulating factor* (mCSF or CSF1), granulocyte–macrophage progenitor cells differentiate into monocytes. This factor also enhances cytokine production (gCSF, IL6, and IL8) and other functions in monocytes and macrophages. Macrophage colony-stimulating factor is synthesized by fibroblasts, bone marrow stromal cells, endothelial cells, astrocytes, osteoblasts, keratinocytes, activated B and T cells, and LPS-activated macrophages. Its receptor on monocytes, macrophages, and their progenitors triggers the Ras–Raf1–ERK cascade.

2.4.2.5 Stem Cell Factor

Stem cell factor (SCF) is produced by bone marrow endothelial cells, fibroblasts, and other cells. It acts in synergy with Epo, Tpo, IL3, gmCSF, and gCSF on BFUe, CFUgm, and CFUgemm, and possibly preCFUc (precolony-forming unit in culture, i.e., granulocyte precursors), as well as on megakaryocytic progenitors [85]. Stem cell factor enhances IL7-induced proliferation of preB cells and stimulates T-lymphocyte development and maturation. With IL2, SCF promotes differentiation of CD34+ hematopoietic cells into NK cells.

Stem cell factor triggers with costimulators proliferation of hematopoietic stem cells. Many hematopoietic cell growth factors (e.g., SCF, IL3, and gmCSF), as well as other growth factors (fibroblast [FGF1 and FGF2], hepatocyte [HGF], vascular endothelial [VEGF], heparin-binding epidermal [HBEGF], platelet-derived [PDGF] growth factor, transforming growth factor-β) and cytokines (e.g., interleukins IL1, IL2, IL4, IL6–IL8, and interferon-γ), can associate with heparin and heparan sulfate proteoglycans of bone marrow stroma [89].

2.4.2.6 Pleiotrophin

Pleiotrophin, a neurite outgrowth factor[48] promotes expansion of the population of bone marrow hematopoietic stem and progenitor cells in vivo [90]. In addition to its effects as a mitogen factor for neurons and angiogenic agent, this heparin-binding, regenerative growth factor for hematopoietic stem cells, activates phosphoinositide 3-kinase.

2.4.2.7 Insulin-like Growth Factor-1

Insulin-like growth factor-1 induces erythroid colony formation.

[48] A.k.a. heparin-binding brain mitogen (HBBM), heparin-binding growth factor HBGF8, neurite growth-promoting factor NEGF1, heparin affinity regulatory peptide (HARP), and heparin-binding growth-associated molecule (HBGAM).

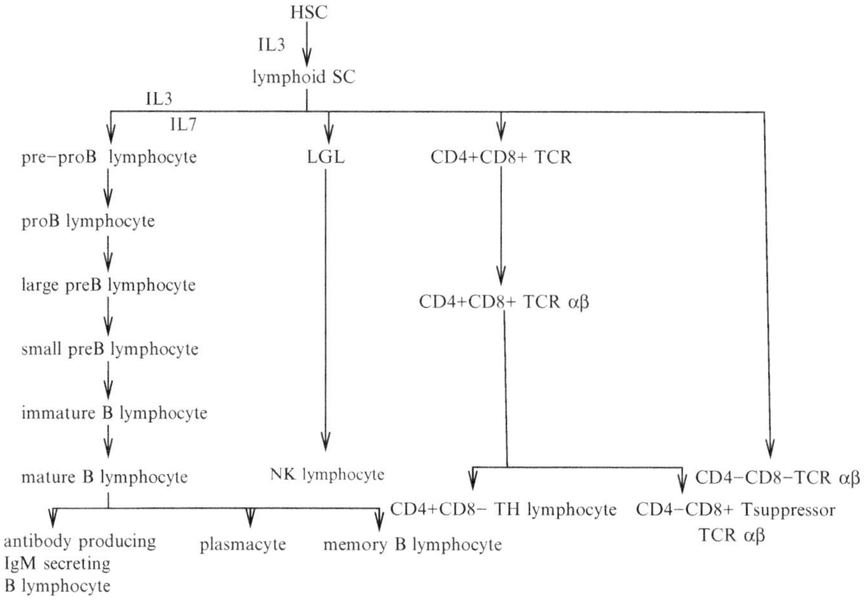

Fig. 2.6 Formation of lymphocytes. Interleukins act on lymphocyte production. T lymphocytes include: CD4+, CD8− helper (T_H) T lymphocytes; CD4−, CD8+ cytotoxic T lymphocytes; CD4−, CD8−$\alpha\gamma$ T lymphocytes; and natural killer (NK) lymphocytes. Plasmalemmal T-cell receptor (TCR) $\alpha\beta$ and $\gamma\delta$ chains characterize lymphocyte subpopulations [91]. Normal CD4+, CD8− lymphocytes have more than 99% of TCR$\alpha\beta$. Lymphocytes that express CD8 can be subdivided into distinct classes according to TCR chain expression pattern. CD4−, CD8+ lymphocytes express predominantly (99%) TCR$\alpha\beta$ receptors. CD4−, CD8− lymphocytes have the highest proportion (62%) of TCR$\gamma\delta$ receptors.

2.4.2.8 Other Cytokines

Tumor-necrosis factor-α stimulates granulopoiesis and inhibits erythropoiesis. *Interleukins* are also involved in hematopoeisis (Fig. 2.6).

Interleukin-1 increases the production of colony stimulating factors and stem cell growth factor. Expression of IL2 is induced by IL1 agent. Interleukin-3 acts on hematopoietic progenitors, in combination with other cytokines. It induces proliferation of several lineages and enhances the effects of gmCSF factor. Interleukin-4 is a T-cell growth factor and promotes the differentiation of T_{H0} precursors toward the T_{H2} lineage. Interleukin-5 influences eosinophil and basophil production and functioning. It enhances IL2-dependent differentiation and proliferation of T cells.

Interleukin-6 promotes development and functioning of lymphocytes and megakaryocyte maturation. Interleukin-7 acts on the lymphocyte production, as it stimulates stem cells to form lymphoid progenitors (Fig. 2.6). Interleukin-8 is a chemoattractant for fibroblasts, neutrophils, basophils, T lymphocytes, and endothelial cells. It also activates integrins in monocytes and eosinophils for extravasation. Interleukin-9 acts on erythroid progenitors. Interleukin-10 inhibits the synthesis of

T_{H1}-derived (IL2, Ifnγ, and gmCSF) and monocyte-derived cytokines (IL1α and IL1β, IL6, IL8, TNFα, gmCSF, and gCSF), but induces IL1RA production by macrophages. It stimulates B lymphocytes and mastocytes.

Interleukin-11 stimulates the proliferation of primitive stem cells as well as commitment and differentiation of multi-lineage progenitors. It acts synergistically to stimulate megakaryocytopoiesis and thrombopoiesis with IL3, Tpo, or SCF; erythropoiesis with IL3, SCF, or Epo; and myeloid colony formation with SCF. Interleukin-12 favors T_{H1} cell differentiation and functioning and inhibits T_{H2} cell differentiation. It synergizes with other hematopoietic factors to promote proliferation of early multipotent hematopoietic progenitors and lineage-committed precursors.

Interleukin-15 triggers the proliferation of activated B lymphocytes and their immunoglobulin production, as well as the proliferation of NK cells and activated CD4+, CD8+ T cells. Interleukin-16 is a chemokine for CD4+ cells, monocytes, and eosinophils.

Interleukin-17 exhibits indirect hematopoietic activity by enhancing the capacity of fibroblasts to sustain the proliferation of CD34+ hematopoietic progenitors and their differentiation into neutrophils. Main IL17 sources are CD4+ T cells, unconventional γδT cells, and NK and γδT cells in the gut, skin, and lungs, respectively. Interleukin-18 promotes the production of Ifnγ and TNF.

Cytokines IL34 and gCSF (CSF3) stimulate phosphorylation of extracellular signal-regulated kinases ERK1 and ERK2 (Vol. 4 – Chap. 6. Mitogen-Activated Protein Kinase Modules) in human monocytes. Interleukin-34 promotes the formation of progenitor colony-forming unit macrophage (CFUm) in human bone marrow cultures.

Proliferation and maturation of committed progenitors is controlled by late-acting factors such as Epo, mCSF, gCSF, and IL5, most of them being lineage-specific [39]. Progenitors at earlier stages of development are controlled by a group of intermediate-acting lineage-non-specific cytokines: IL3, gmCSF, and IL4. Primitive progenitors are committed by early-acting cytokines, which include IL6, IL11, IL12, and gCSF factor. Certain cytokines are either stimulatory or inhitory. However, others exhibit both effects such as IL4. Among inhibitors, lineage-non-specific interferons act at various development stages, whereas TGFβ affects early phases of hematopoiesis [92]. In addition, the synthesis of cytokines by the vascular endothelium is stress dependent (Sect. 9.10). Steady laminar shear stress increases the production of gmCSF by endothelial cells [93].

2.4.3 MicroRNAs

MicroRNA-24 is upregulated during terminal differentiation of hematopoietic cell lines (i.e., after final cell division cycle exit).[49] It inhibits histone variant H2AX

[49] When cells no longer replicate their DNA, endogenous double-stranded breaks that occur during DNA replication and compromise genomic integrity are strongly reduced or absent.

(~15% of cellular H2A) involved in double-stranded break repair [94]. Primary human, short-lived, blood cells, in which H2AX level is substantially reduced by miR24 overexpression, then become vulnerable to DNA damage.[50]

2.4.4 *Other Regulators*

2.4.4.1 Cyclooxygenases and Prostaglandins

Cyclooxygenases Cox1 and Cox2, PGE2-synthase, and prostaglandin-E2 are involved in hematopoiesis [58]. Prostaglandin-E2 augments the number of hematopoietic stem cells. Cyclooxygenase-1 serves in the formation of the hematopoietic niche, whereas Cox2 probably intervenes in HSC self-renewal and proliferation. Inhibitors of COx1 or COx2 diminish CFU activity.

2.4.4.2 Calcium-Sensing Receptor

Hematopoietic stem cells without the calcium-sensing receptor (Vol. 3 – Chap. 7. G-Protein-Coupled Receptors) can engraft in extramedullary spaces, such as the spleen or liver, and produce extramedullary hematopoiesis.

2.5 Mastocyte Formation

Mastocytes regulate physiologic and pathologic immune response, as they protect against parasites and bacteria and mediate allergy and autoimmunity.

Stem cell factor and IL3, via the PI3K, Ras–MAPK, and JaK–STAT pathways influence mastocyte number and type. Upregulation of SPI1 and GATA2 factors and downregulation of CCAAT/enhancer-binding protein-α cooperate to generate mastocytes.

Mastocytes leave the bone marrow as mastocyte progenitors that mature in target tissues. Notch signaling connects environmental cues and transcriptional control that increases mastocyte development [95]. Notch ligand Delta-1 instructs common myeloid progenitors and granulocyte–macrophage progenitors in the bone marrow toward mastocyte lineage at the expense of other granulocyte–macrophage lineages due to Notch-2. Notch-2 upregulates GATA3 (not GATA2 although GATA2 is required for in vitro mastocyte generation) and HES1 expression in

[50]During double-stranded break repair, H2AX is phosphorylated by phosphatidylinositol 3-kinase. Phosphorylated H2AX then participates in DNA repair, replication, recombination and regulation of the cell cycle, as it binds various cell cycle and DNA-repair factors.

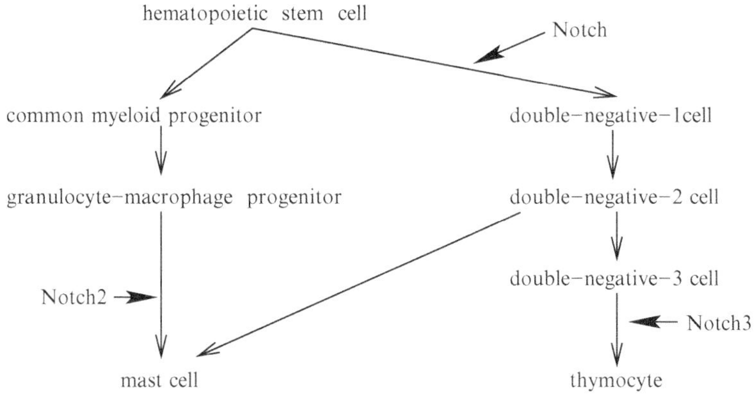

Fig. 2.7 Notch pathway in mastocyte development (Source: [95]). Mastocytes express Notch-1 and Notch-2 morphogens. Hematopoietic stem cells can differentiate into a common myeloid progenitor that differentiates into a granulocyte–monocyte progenitor. The latter is able to preferentially generate mastocytes in the presence of Delta-1 that depends on the Notch-2 targets, HES1 and GATA3 factors. Thymic progenitors that express Notch-1 and Notch-3 and have been stimulated by Notch, such as DN1 and DN2, can also develop into mastocytes. Factor GATA3 in thymocytes provokes mastocyte development.

a coordinated manner. Afterward, HES1 overexpression in myeloid progenitors represses C/EBPα, hence yielding mastocyte development. Downregulation of C/EBPα followed by GATA upregulation due to Notch-2 orchestrates mastocyte differentiation from myeloid progenitors.

Thymic progenitors that have been stimulated by Notch, such as double-negative DN1 and DN2 cells, but not those that express preT-cell receptor such as DN3, can develop into mastocytes. Factor GATA3 engages thymocytes in mastocyte generation. Commitment to mastocytes and DN3 cells requires Notch downregulation and maintenance, respectively (Fig. 2.7).

2.6 Hematopoietic Stem Cells in Angiogenesis

Vasculature repair or hypoxia-primed development is associated by migration of hematopoietic stem cells from the bone marrow into the blood circulation, which is initiated by vascular endothelial growth factor and CXCL12 chemokine. These agents also induce HSC-derived endothelial precursors to participate in angiogenesis.

Insulin-like growth factor IGF1 binds receptor IGF1R and operates via interaction with IGF-binding proteins (IGFBP1–IGFBP6). The latter serve as transporters and storage pools for IGF1 in a tissue- and developmental stage-specific manner [96]. The IGFBP activity is regulated by phosphorylation, polymerization, association to cell or matrix components, and proteolysis. The IGFBP proteins either stimulate or inhibit IGF1. Insulin-like growth factor binding protein IGFBP3 carries IGF1 and IGF2 in blood. The auto- and paracrine IGFBP3 activity influences cell adhesion, survival, motility, division, and apoptosis. Synthesis of IGFBP3 increases during hypoxia. Protein IGFBP3 enhances or prevents angiogenesis depending on the context: it favors endothelial cell migration and capillary formation, but reduces preretinal neovascularization. Hypoxia-induced IGFBP3 promotes the recruitment of progenitors into remodeling regions (at least in the retina). Agent IGFBP3 targets CD34+ endothelial precursors (not CD14+ cells involved in inflammation associated with vascular repair) that quickly differentiate into endothelial cells.

Chapter 3
Blood Cells

Leukocytes, red blood capsules, and platelets constitute the pool of circulating cells and pseudocells (Table 3.1). Red blood capsules, commonly called erythrocytes or red blood cells, are mainly devoted to blood gas transport, especially oxygen. When the demand for oxygen changes, appropriate signals are detected and transmitted so that the supply of oxygen by the cardiorespiratory system varies to match the demand. Leukocytes, or white blood cells, coordinate the immune response. Platelets, also named thrombocytes, are cell fragments that participate in blood coagulation and inflammation.

3.1 Immunophenotyping of Leukocytes

Molecules on the surface of hematopoietic cells act in the development and functioning of these cells. Leukocytes possess at their cell surface many accessory molecules that cooperate with antigen-specific receptors. They participate in cell adhesion between lymphocytes and other cell types or cell signaling. Numerous accessory molecules belong to the immunoglobulin (IGSF) superfamily.

Cell-surface antigens expressed on hematopoietic and nonhematopoietic cells include members of the cluster of differentiation (CD1–CD339) class, lymphocyte antigens (Ly1–Ly81), as well as markers that do not pertain to these molecule class.

Members of the *cluster of differentiation* class encompass G-protein-coupled, protein kinase, and cytokine receptors, ligands of the TNF superfamily, plasmalemmal carriers, cell adhesion molecules, erythrocytic antigens, etc.

Lineage markers are cell-surface antigens used for immunophenotyping cells of a particular developmental lineage. Cells that do not express given lineage markers, or express these markers at very low levels, are lineage marker negative (Lin−, i.e., in the mouse, CD2−, CD3−, CD4−, CD5−, CD8−, CD161c−, PTPRc−, Ly76−, Ly6g−).

M. Thiriet, *Tissue Functioning and Remodeling in the Circulatory and Ventilatory Systems*, Biomathematical and Biomechanical Modeling of the Circulatory and Ventilatory Systems 5, DOI 10.1007/978-1-4614-5966-8_3,

Table 3.1. Blood cell approximated geometry and relative concentration.

Blood cell	Quantity (/mm^3)	Relative proportion	Size (μm)	Cell volume percentage
RBC	5×10^6	1	~ 8	97%
Leukocyte	5×10^3	10^{-3}	~ 15	2%
Platelet	3×10^5	6×10^{-2}	~ 3	1%

Hematopoietic stem cells, early progenitor cells, and precursor cells do not yet express certain markers. The LSK (Lin−, SCA1+, KIT+) phenotype defines a primitive cell type that possesses self-renewal capacity, from which multipotential cells MMP without self-renewal capacity develop.

Immunophenotyping consists in determining expression level of cell-surface molecules on leukocytes. A molecule+ or molecule− term (plus or minus symbol) indicates whether a given molecule type is expressed or lacking on the plasma membrane of a given leukocyte type. Some leukocyte types are also defined as molecule^{high} (or molecule^{bright}), molecule^{mid} (or molecule^{int} [intermediate]), or molecule^{low} (or molecule^{dim}), according to the expression level, as well as molecule$^{\text{neg/low}}$ and molecule$^{\text{mid/low}}$ for negative-to-low and medium-to-low phenotypes.

In addition, cells can be characterized by intracellular markers such as certain transcription factors as well as plasmalemmal markers only when they are activated (molecule^{act}).

3.2 Vasculature as an Immunity Compartment

The vasculature participates in the detection and protection against pathogens. Blood vessels convey immunocytes to identify, prevent eventual dissemination of invading bacteria to different organs, and eradicate these pathogens.

Intravascular motion of immunocytes, such as neutrophils, monocytes, and invariant natural killer T cells, starts by transendothelial migration at target sites. However, many bacterial products antagonize leukocyte recruitment, such as lipopolysaccharide that can inhibit leukocyte displacement [97].

Endothelial cells contribute to innate immune responses to bacteria, as they express Toll-like receptors, lymphocyte antigen Ly96 homolog,[1] and appropriate adaptors [97]. Activated nuclear factor-κB primes synthesis of adhesion molecule on endothelial cells that promotes neutrophil infiltration.

[1] A.k.a. MD2 and ESOP1 (ES–OP9 [coculture of embryonic stem cells and OP9 stromal cell line] coculture clone-1). It is required for maximal responsiveness of TLR4 to bacterial lipopolysaccharides. It enhances TLR4-dependent activation of NFκB. It is secreted by many cell types such as monocyte-derived dendritic cells.

Neutrophils, major effector cells in innate immunity against bacteria, undergo platelet-dependent activation in the vasculature during bacteremia. They can form neutrophil extracellular traps that consist of chromatin coated with peptidases to trap and kill bacteria within the vasculature. Complement activation, specifically anaphylatoxin C5a, contributes to the recruitment of neutrophils in particular to the lung [97].

3.3 Proliferation and Mobilization of Blood Cells

Blood cell genesis, growth, proliferation in their bone marrow niches, and mobilization in the blood stream and in the body's tissues are regulated by numerous substances (Vol. 2 – Chaps. 2. Cell Growth and Proliferation and 6. Cell Motility).

Blood cells need to communicate between them and with other body's cells. They secrete *cytokines* that have a broad activity spectrum. Cytokines allow communications between immunocytes as well as between immunocytes and other cell types. They thus target manifold cell types. *Chemokines* are inducible chemotactic cytokines, which mobilize nearby responsive cells according to a chemotactic direction (Vol. 2 – Chap. 6. Cell Motility and Chaps. 10 and 11).

Three main action modes are used via suitable receptors according to the length scale: (1) *intracrine*[2] and *autocrine*[3] mechanisms act on the producing cell without or after secretion, respectively; (2) *juxtacrine*[4] and *paracrine*[5] regulation happens on apposed and neighboring cells, respectively; and (3) *endocrine*[6] activity is exerted on remote cells.

3.3.1 *Cytokines*

Cytokines are growth factors implicated in immunity and hematopoiesis (Table 3.2; Chap. 2; Vol. 2 – Chap. 3. Growth Factors)). Cytokines exhibit both pleiotropy and redundancy. Pleiotropic actions of cytokines can be explained by the presence of receptors for a given cytokine on multiple cell lineages and the activation of multiple signaling pathways contributing to different functions. Overlapping actions of different cytokines can be explained by similar cellular distributions of specific receptors for different cytokines, as well as by share of signaling pathways.

---

[2]Latin intra: within.

[3]Greek αυτο: inside.

[4]Latin juxta: immediately adjacent, adjoining.

[5]Greek παρα: close to.

[6]Greek ενδοκρινω: release inside.

Table 3.2. Cytokines in immunity and hematopoiesis. Erythropoietin, thrombopoietin, growth hormone, and prolactin have structures and signaling mechanisms similar to those of interleukins and colony-stimulating factors, referred to as type-1 cytokines. Interferons are type-2 cytokines. Many cytokines are pleiotropic. Different cytokines have overlapping activities (TGFβ: transforming growth factor-β; TNFα: tumor-necrosis factor-α).

Immunity	Hematopoiesis
Interleukins	Interleukins
TGFβ	Colony stimulating factors
Interferons	
TNFα	
Chemokines	

Moreover, cytokines can signal via several types of receptors and several cytokine types can bind to given receptors. Also, cytokine receptors have common structural components.[7] Cytokines are downregulated by cytokine antagonists.

After enzymatic cleavage of the extracellular domain of cytokine receptors, soluble cytokine receptor fragments in blood and extracellular fluid can bind to cytokines, thereby neutralizing their activity.

Most cytokine receptors, as well as growth hormone[8] and prolactin receptors of the immune and hematopoietic systems belong to the family of class-1 cytokine receptors. Most class-1 cytokine receptors are dimers, with a cytokine-specific and a signal-transducing subunit, which is usually not specific for the cytokine. Certain receptors are trimers.

Class-1 cytokine receptors are subdivided into subsets composed of receptors with identical signal-transducing subunits: (1) granulocyte–monocyte colony-stimulating factor (gmCSF) subset includes the receptors for IL3, IL5, and CSF2; (2) IL6 subset comprises the receptors for IL6, IL11, and IL12; and (3) IL2 subset encompasses the receptors for IL2, IL4, IL7, IL9, and IL15.

Ligands for class-2 cytokine receptors are interferons Ifnα, Ifnβ, and Ifnγ. The TNF receptor family includes TNF receptor-1 (TNFR1 or TNFRSF1a), TNF receptor-2 (TNFR2 or TNFRSF1b), TNFRSF5, and TNFRSF6a, among numerous

[7]Interleukins IL3 and IL5, as well as granulocyte–monocyte colony-stimulating factor (gmCSF or CSF2) share a common cytokine receptor βc chain (IL3Rβ). Receptors of IL6, IL11, leukemia inhibitory factor (LIF), oncostatin M (OSM), ciliary neurotrophic factor (CNTF), novel neurotrophin-1 (NNT1), a.k.a. B-cell-stimulating factor-3 (BSF3) and cardiotrophin-like cytokine (CLC), and cardiotrophin-1 (CT1) share the GP130 signal-transducing molecule. Interleukins IL2, IL4, IL7, IL9, IL15, and IL21 share the common cytokine receptor γ chain (IL2Rγ); IL12 and IL23, the IL12Rβ1 chain; IL4 and IL13, the type-2 IL4 receptor (IL4Rα); and IL7 and TSLP, the IL7Rα receptor subunit [98].

[8]The growth hormone is a type-1 cytokine. It acts via the JaK–STAT pathway, which mediates activation of STAT transcription factors and the Ras–MAPK pathway for phosphorylation of ELk1 transcription factor.

other members (Vol. 3 – Chap. 11. Receptors of the Immune System). Certain cytokine receptors belong to the immunoglobulin superfamily, such as receptors for IL1 and macrophage colony-stimulating factor (mCSF or CSF1).

Signal transduction via class-1 and -2 cytokine receptors begins with cytokine-induced dimerization of receptor subunits. These cytokines use the JaK–STAT pathways. Plasmalemmal Janus kinases phosphorylate tyrosine residues of signal transducers and activators of transduction. These STAT factors then activate genes via transcription factors. The JaK–STAT pathway is also used by other growth factors and hormones.

3.3.2 Chemokines

Chemokines form a family of inflammatory inducible cytokines. Chemokines have potent chemotactic activity. Most of them belong to 2 main subsets according to their architecture, i.e., the presence of 4 cysteine (C) residues with first 2 adjacent (C–C) or separated by an amino acid (C–X–C). Chemokines of the CXC category (α-chemokines) are encoded by human chromosome 4q12-21, whereas chemokines of the CC category (β-chemokines) map to human chromosome 17q11-32.

Most CXC-chemokines are chemoattractants for neutrophils, whereas CC-chemokines generally attract monocytes, lymphocytes, basophils, and eosinophils. Two additional subsets are composed of C-chemokines, more or less lymphocyte specific, and CX3C-chemokines, which target leukocytes. The chemokine receptors are composed of seven transmembrane proteins. They transduce signals via G-proteins. Chemokine receptors are classified into specific, shared, promiscuous, and viral groups.

3.4 Scaffold Proteins in Immunocytes

Signal transduction relies on location of signaling mediators inside the cell and their activation kinetics. Scaffold proteins bind to other signaling molecules to regulate signal transduction (Table 3.3), as they localize signaling effectors in specific regions within the cell to trigger and optimize signaling by acting as assembly platforms for enzymes, their regulators, and their substrates.

In immunocytes, scaffold proteins regulate calcium signaling, activation of mitogen-activated protein kinase modules, and cell polarity [99]. In T lymphocytes, calcium fluxes can oscillate repeatedly. Because scaffold proteins are able to coordinate positive and negative feedback loops, they can induce oscillatory signaling patterns.

In the immune system, scaffold proteins include immunoreceptor tyrosine-based activation motif (ITAM)-containing receptors, such as T-cell and B-cell receptors, as well as cytoplasmic scaffold proteins, such as linker for activation of T cells

Table 3.3. Scaffold proteins in immune system signaling (Source: [99]; ASC: apoptosis-associated speck-like protein containing a caspase activation and recruitment domain [CARD]; BCL: B-cell lymphoma; CARMA: CARD–MAGUK domain-containing protein; CBL: Casitas B-lineage lymphoma; CDC: cell-division cycle; CTAK: CDC25c-associated kinase; DLg: disc large homolog; ERK: extracellular signal-regulated kinase; GSK: glycogen-synthase kinase; HSP: heat shock protein; IP_3R: inositol trisphosphate receptor; IRAK: interleukin-1-receptor-associated kinase; JNK: Jun N-terminal kinase; KSR: kinase suppressor of Ras; MAGUK: membrane-associated guanylate kinase; MALT: mucosa-associated-lymphoid-tissue lymphoma-translocation gene; MAP2K: mitogen-activated protein kinase kinase; MyD: myeloid differentiation primary-response gene; NFκB: nuclear factor-κB; NIK: NFκB-inducing kinase, or MAP3K14; NLR: nucleotide-binding domain and leucine-rich-repeat-containing receptor; NLRP: NLR/pyrin-domain-containing; PLC: phospholipase-C; PKC: protein kinase-C; PP: protein phosphatase; RSG: regulator of G-protein signaling; S100: S100 calcium-binding protein; SMAD: small mothers against decapentaplegic homolog; STAT: signal transducer and activator of transcription; SuGT: suppressor of G2 allele of S-phase kinase-associated protein SKP1; TAK: TGFβ-activated kinase, or MAP3K7; TCR: T-cell receptor; TGF: transforming growth factor; TLR: Toll-like receptor; TNFR: tumor-necrosis factor receptor; TRAF: TNFR-associated factor; WASP: Wiskott-Aldrich syndrome protein; ZAP: ζ-chain-associated protein kinase).

Scaffold proteins	Binding partners
Ahnak1	Calcium channels, caveolin-1, PKC, PLCγ, S100a
BCL10	CARMA1, MALT1, TAK1 (MAP3K7), MAP2K7, JNK2
DLg1	CBL, LCK, Vav1, WASP, ZAP70, P38MAPK
Homer	IP_3R, Homer, group-1 metabolic receptors
KSR	CDC37, CTAK1, PP2a, 14-3-3, HSP70, HSP90, Raf, MAP2K1, ERK
NLRP	ASC, HSP90, SuGT1, caspase-5, BCL2, BCLxL
Pellino (Ub ligase)	BCL10, IRAK1/4, MyD88, NIK (MAP3K14), SMAD6, TAK1, TRAF6
Spinophilin ($PP1_{r9}$)	^{F}actin, PP1, RGS

(LAT), B-cell linker (BLnk),[9] and lymphocyte cytosolic protein LCP2 [99].[10] These scaffolds can be phosphorylated (hence their names "tyrosine phosphoproteins").

[9] A.k.a. B-cell adapter containing a SH2 domain protein (BASH), 65-kDa SH2 domain-containing leukocyte protein (SLP65), and lymphocyte antigen Ly57 homolog.

[10] A.k.a. 76-kDa SH2 domain-containing leukocyte protein (SLP76).

Scaffold proteins that act in ERK signaling comprise IQ motif-containing GTPase-activating protein-1 (IQGAP1), MAPK scaffold protein-1, Similar expression to FGF gene products (SEF), paxillin, β-arrestins, and kinase suppressor of Ras (KSR) [99].[11]

Scaffold B-cell lymphoma protein (BCL10) operates in the JNK pathway[12] in immunocytes. Scaffold BCL10 participates in activation of nuclear factor-κB, in addition to JNK2, in the TCR axis. It complexes with mucosa-associated lymphoid tissue lymphoma translocation peptidase MALT1 and CARMA1 to link protein kinase-Cθ with NFκB downstream from antigen receptors.

The scaffold Disc large homolog DLg1 intervenes with JNK-interacting proteins (MAPK8IP1–MAPK8IP4) in P38MAPK activation. This kinase is a sensor of pro-inflammatory stimuli primed by cytokines.

The AHNAK[13] family of PDZ-containing scaffold proteins consist of 2 members: Ahnak1, or desmoyokin, and Ahnak2. Protein Ahnak1 is strongly expressed in spleen and lymph nodes, especially in splenic CD4+ T cells, but weakly in thymus [100]. In vitro, Ahnak1 fragments bind and activate phospholipase-Cγ1 in the presence of arachidonic acid. Partners of Ahnak1 include also calcium- and zinc-binding protein S100b and the annexin-2–S100a10 complex. In cardiomyocytes, Ahnak1 associates with the β subunit of cardiac Ca_V channels at the plasma membrane.

Scaffold Ahnak in calcium signaling binds calcium channels, phospholipase-Cγ, and protein kinase-C [99]. It is phosphorylated by protein kinase-A in response to β-adrenoreceptor stimulation. Scaffold Ahnak participates in activation of nuclear factor of activated T cells via calmodulin and PP3 phosphatase.

In T lymphocytes, scaffolds Homer-2 and -3 inhibit nuclear factor of activated T cells by competing with PP3 phosphatase for NFAT binding.

Initiation of the Toll-like receptor pathway involves recruitment of the adaptor myeloid differentiation primary-response gene MyD88 to liganded TLRs. Subsequently, activated IL1 receptor-associated kinases IRAK1 and IRAK4 prime formation of a complex that contains tumor-necrosis factor receptor-associated factor TRAF6, TRAF-interacting protein with forkhead-associated domain (TIFA), MAP3K7, and its binding partner MAP3K7IP1 to stimulate P38MAPKs, JNKs, and NFκB [99].

The PELLINO family includes 3 known proteins in humans that can associate with IRAK1, TRAF6, and MAP3K7 after IL1 receptor activation.

---

[11] Scaffold KSR binds 14-3-3 proteins for cytosolic sequestration. Following cell stimulation, protein phosphatase PP2a dephosphorylates and releases KSR from 14-3-3 proteins for translocation to the plasma membrane where it can interact with activated Ras GTPase. Scaffold Connector enhancer of kinase suppressor of Ras CNKSR1 facilitates the interaction between Raf (a MAP3K) and KSR. KSR Isoforms include KSR1, which is highly expressed in brain, thymus and spleen, in addition to neutrophils and macrophages and KSR2.

[12] Kinase JNK is a stress-activated MAPK that acts in the regulation of lymphocyte activation and development.

[13] Hebrew ahnak: giant.

Members of the NLRP (NOD-like receptor [NLR] pyrin domain-containing) subfamily of the NLR family of cytoplasmic receptors of the innate immunity sense various pathogenic and chemical agents and trigger the formation of inflammasomes. These NLRP scaffolds can bind adaptor and signaling proteins, such as apoptosis-associated speck-like protein containing a CARD (ASC), HSP90, suppressor of G2 allele of S-phase kinase-associated protein SKP1 (SuGT1), and caspase-5 [99]. The procaspase-1–NLRP1–ASC complex releases active caspase-1 that cleaves precursors of IL1β and IL18 interleukins.

Membrane-associated guanylate kinases (MAGuK) constitute a family of scaffolds that include Disc large homologs (DLg1–DLg4),[14] zona occludens-1, CARMA (N-terminal caspase activation and recruitment domain [CARD]–MAGuK-like domain-containing), and MAGI1 (MAGuK-like, WW, and PDZ domain-containing with inverted domain organization protein-1) [99]. Scaffold spinophilin, more precisely $PP1_{r9}$ regulatory subunit (inhibitor) of PP1 phosphatase,[15] is involved in dendritic cell functioning, as it can act in the formation of immunological synapses.

3.5 Red Blood Capsules

In the present textbook, erythrocyte, or red blood cell (RBC), is called red blood capsule, because it defines a hemoglobin (Hb) solution[16] bounded by a thin flexible membrane (non-nucleated cell).

In its undeformed state, it has a biconcave disc shape with a greater thickness in its outer ring (diameter $7.7 \pm 0.7\,\mu$m, central and peripheral thicknesses $1.4 \pm 0.5\,\mu$m and $2.8 \pm 0.5\,\mu$m, aspect ratio ~ 0.4, surface $\sim 140\,\mu m^2$, volume $\sim 90\,\mu m^3$, membrane thickness $\sim 0.02\,\mu$m). It can then deform[17] with a negligible bending resistance, in particular with a parachute shape when moving through tiny capillaries. Its mass density is equal to about $1100\,kg/m^3$. It lives an average of 120 days. The RBC number is age- and sex-dependent ($4.2–5.5\times 10^6/mm^3$ in women, and $4.5–6.2\times 10^6/mm^3$ in men).

3.5.1 *Formation, Maturation, and Clearance*

In adults, definitive erythropoiesis occurs in the bone marrow. Erythroblasts are associated with macrophages in anatomical units (*blood islands*). Nuclei are

---

[14]Four genes in mammals encode DLg1 to DLg4 that are also called SAP97, PSD93, SAP102, and PSD95, respectively. Scaffold DLg1 that is recruited to the immunological synapse binds TCR, P38MAPK, and other scaffolds.

[15]Spinophilins are encoded by the PPP1R9A and PPP1R9B genes in humans.

[16]The kinematic viscosity (ν) of the Hb solution is greater than that of water ($\nu \sim 6$ mPa s) [101].

[17]RBC deformability can be estimated by transit time during filtration.

disconnected from reticulocytes, expelled, and then engulfed by macrophages. Phosphatidylserine, often used as an apoptosis signal, is also used for nucleus engulfment [102].

Enucleation results from asymmetrical cell division during which chromatin condensates and the nucleus surrounded by a part of the plasma membrane buds off from the immature precursor (erythroblast). MicroRNA miR191 that has minor effects on cell proliferation and differentiation is downregulated during this process [103]. In fact, the majority of microRNAs in CFUe erythroid progenitors are inhibited during terminal erythroid differentiation. In particular, miR144 and miR451, which are encoded by GATA1 target genes, are required for erythropoiesis, as they act on erythroid Krppel-like factor KLF1 and GATA2, respectively. Gene transcripts that encode erythroid-enriched RiO kinase RiOK3, which operates in chromatin condensation and enucleation, and class-C basic helix–loop–helix protein bHLHc11,[18] a MyC antagonist, are miR191 targets.

Reticulocytes are slightly immature cells that are characterized by a filament network and by granules. Reticulocytes amount to 0.5 to 2% of the total RBC count (more in anemia).

In healthy human adults, $\sim 2.5\times 10^{11}$ new red blood cells are released from the bone marrow into the blood circulation per day, and about the same number are cleared. In the few days after exit from the bone marrow, erythrocytes undergo a reduction in both volume and hemoglobin content.

A model of RBC maturation and clearance has been developed [104]. This model relies on the splitting of volume (V) and hemoglobin (Hb) dynamics of an average RBC over time into deterministic functions (ϕ) and random fluctuations (φ), using the average cell age and population means to scale time, V, and Hb quantities. The deterministic component depends on 2 parameters related to fast (β) and slow (α) changes. The effect of parameter β dominates until the hemoglobin concentration is close to the mean corpuscular hemoglobin concentration (MCHC). The random fluctuation is modeled as a Gaussian random variable with a zero mean and diffusion ($\mathcal{D}$)-dependent variance.

$$\frac{d}{dt}V(Hb) = \phi + \varphi; \tag{3.1}$$

$$\phi = \begin{cases} \alpha \exp\{\beta_V(V - Hb)\}, \\ \alpha \exp\{\beta_{Hb}(Hb - V)\}; \end{cases}$$

$$\varphi = \begin{cases} \mathcal{N}(0, 2\mathcal{D}_V), \\ \mathcal{N}(0, 2\mathcal{D}_{Hb}), \end{cases}$$

The dynamics of the complete circulating RBC population is described by an equation for time-dependent drift (φ), diffusion ($\mathcal{D}$), and birth (b) and death (d) rate of probability density of the joint volume-hemoglobin distribution ($\mathbf{P}(V, Hb, t)$)

[18] A.k.a. MAX interactor MxI1.

that is approximated by a Fokker-Planck equation, using the differential vectorial operator ∇ [104]:

$$\partial_t P = -\nabla \cdot (P\phi) + \nabla(\mathcal{D} \cdot \nabla P) + \mathrm{b}(V, Hb, t) - \mathrm{d}(V, Hb, t)P, \tag{3.2}$$

$$\mathbf{D} = \begin{pmatrix} \mathcal{D}_V & 0 \\ 0 & \mathcal{D}_{Hb} \end{pmatrix}$$

3.5.2 *Shape*

A red blood capsule experiences large reversible deformations during its 120-d lifespan. It recovers its initial shape after large deformations undergone in narrow capillaries owing to the composition of the membrane and membrane–cytoskeleton interaction. Its plasma membrane and the cell cortex, with the spectrin network tethered to the phospholipid bilayer, are the main load-bearing components. Membrane–spectrin interaction results from the phosphorylation of the interconnection protein-4.1R, or erythrocyte membrane protein band EPB4.1.

In thermal equilibrium, red blood cells, like lipid vesicles, can have various shape types according to the environmental conditions. The spectrin network enables RBCs to remain intact while deforming in blood flow through narrow capillaries. In normal conditions, RBCs adopt a biconcave disc shape with a given area (A)S and volume (V; mean radius $\mathrm{R}_m = (\mathrm{A}/[4\pi])^{1/2} \sim 3.4\,\mu\mathrm{m}$; reduced volume $\mathrm{V}/(4\pi\mathrm{R}_0^3/3) \sim 0.6$).

Membrane fluctuates over a short time scale (~ 100 ms). Membrane fluctuations depend on its bending stiffness, ATP, and integrity of the spectrin cytoskeleton connected to the membrane by actin, glycophorin-C, and protein 4.1R. Phosphorylation of 4.1R protein by protein kinase-C disassembles the EPB4.1–spectrin–actin trimer, hence spectrin–membrane connection. Inhibition and activation of this phosphorylation significantly affects tension and effective viscosity ($\sim 100\times 10^{-3}$ Pa.s) [105].

In the microcirculation, RBC deformation reduces flow resistance. Deformation also provokes ATP release from RBCs that requires cystic fibrosis transmembrane conductance regulator activated via the cAMP–PKA axis [106]. Release of ATP causes nitric oxide synthesis and augments the vascular caliber [107]. In addition, erythrocytes stimulate nitric oxide production in platelets [108].

Spectrin tetramers in the unstressed state form dimers under shear. The cell periphery undergoes fluidization beyond a shear threshold [109]. Spectrins remodel into tetramers upon unloading and the RBC membrane stiffness rises. Stretching by optical tweezers estimates the shear modulus (4–10 μN/m) [110].

The RBC shape represents an equilibrium configuration that minimizes the curvature energy of a closed surface (a vesicle) for given surface area and volume with a geometrical asymmetry.[19] Red blood cells respond to osmotic pressure (Π) changes by swelling[20] and shrinking (crenated RBC)[21] in hypotonic and hypertonic fluids, respectively. RBC Membrane has thermal agitation. These spontaneous undulations are explained by the high membrane deformability. Furthermore, the membrane is excited by permanent impacts from water molecules.

Red blood cells defend against osmotic changes by controlling their intracellular ion content, especially keeping low $[Na^+]_i$ and high $[K^+]_i$ by the activities of the ATP-driven Na^+–K^+ pumps and ion exchangers. Hereditary *stomatocytosis* and *spherocytosis* are characterized by an uncontrolled excessive membrane permeability to Na^+ and K^+ associated with amino acid substitutions in the intramembrane domain of the anion exchanger SLC4a1 (or AE1) [112].

In narrow capillaries (conduit bore slightly > particle size), the 2 essential control parameters of the collective particle flow behavior are the hematocrit and flow velocity. At high H_t values, 3 distinct phases can be predicted [113]: (1) disordered biconcave disc-shape; (2) parachute-shape, RBCs being aligned in a single file; and (3) slipper-shape, RBCs being arranged in 2 parallel interdigitated rows.

3.5.3 Plasma Membrane

The membrane is constituted by a phospholipid bilayer with protein inclusions, ion channels for chloride Cl^- and bicarbonate HCO_3^- in particular, and sialoglycopeptides (blood groups),[22] with glucids (< 10%), and with cytoskeleton meshing and mooring proteins. A peripheral skeleton, a mesh made of spectrin, actin, and other proteins (tubulin,[23] among others), resides at the inner surface of the plasma membrane. It binds to RBC membrane (ankyrin–spectrin bonds) [114]. It gives the RBC shape. Large, repeated RBC deformation associated with maintenance of structural integrity is due to the cortical membrane-associated cytoskeleton.

---

[19]The phospholipid outer layer of the RBC membrane has slightly more molecules and a greater surface area than the inner layer.

[20]In low-osmolarity solutions, RBCs becomes spherical after swelling (spherocyte). With further swelling (hemoglobin solution leakage), RBCs are reduced to the membrane (erythrocyte ghost). Hyperosmotic shock triggers the release of prostaglandin-E2 (PGe2) [111]. The latter activates Ca^{2+} channels in the RBC membrane. Subsequent Ca^{2+} entry induces ankyrin-R degradation and RBC shrinkage.

[21]Two types of spiculed cells exist: echinocyte and acanthocyte, whether spicules are uniformly distributed or not.

[22]Erythrocytes have surface antigens associated with blood groups (glycosphingolipids ABO, Rh, MN, Duffy, Lewis, and Kell).

[23]Microtubules, involved in marginal bundles of RBCs, as well as in mitotic spindles and centrioles, are composed of tubulins (Vol. 1 – Chap. 6. Cell Cytoskeleton).

3.5.3.1 Plasmalemmal Proteins

Transmembrane proteins include anion transporter SLC4a1;[24] glycophorin-A, -B, and -C; Rhesus blood group (Rh); Rh-associated glycoprotein (RhAG); Duffy; basal cell adhesion molecule (bCAM);[25] intercellular adhesion molecule ICAM4;[26] heparan sulfate proteoglycan epican (Indian blood group; a.k.a. CD44); integrin-associated protein CD47; Kell blood group metalloendopeptidase; and X-linked Kx blood group (XK).

Erythrocyte markers Rh and RhAG could be gas transporter. In addition, RhAG and Rh form the *Rh complex* with CD47, LW, and glycophorin-B. Duffy serves as a chemokine receptor. Markers BCAM, ICAM4, and epican, are involved in cell adhesion. Kell blood group metalloendopeptidase has an endothin-3-converting enzyme activity.

Transporter SLC4a1 is the preponderant transmembrane constituent of the RBC membrane that operates as an anion exchanger. It binds protein-4.2, hemoglobin, glycolytic enzymes (aldolase, phosphofructokinase, and glyceraldehyde 3′-phosphate dehydrogenase), and carbonic anhydrase-2 to form the *metabolon*. Furthermore, SLC4a1 tetramer interacts with glycophorin-A to build the *SLC4a1–ankyrin complex* that is attached by ankyrins to spectrin tetramers.

3.5.3.2 Connections between the Plasma Membrane and Cytoskeleton

The cell cortex lattice is composed of spectrin tetramers formed by self-association of heterodimers made of α- and β-spectrin. Spectrin tetramers are attached to junctional complexes that comprise short ^{F}actin protofilaments and several actin-binding proteins, such as protein-4.1R, dematin (protein-4.9), adducin, tropomyosin, and tropomodulin.

Ankyrins link the spectrin–actin cytoskeleton to the membrane. Ankyrin deficiency is also known as hereditary spherocytosis. Ankyrins bridge the spectrin–actin cytoskeleton and SLC4a1 tetramers, as well as proteins involved in ion transfer, such as Cl^-–HCO_3^- anion exchanger, Na^+–Ca^{2+} exchanger, Na^+–K^+ ATPase, IP_3 receptor, ryanodine receptor–Ca^{2+} channel, and voltage-gated Na^+ channels (Na_V) [115].

Protein-4.1R, dimeric SLC4a1, adducin, spectrin, and actin form an additional multifunctional, multiproteic complex of the RBC membrane with transmembrane glycophorin-C, palmitoylated membrane protein MPP1,[27] Rh, Duffy, Kell, and XK: the so-called *4.1R-based complex* that surrounds connections between the cortical cytoskeleton and plasma membrane [116].

---

[24] A.k.a. erythrocyte membrane protein band-3 and anion exchanger AE1.

[25] A.k.a. Lutheran blood group glycoprotein (Lu antigen) and CD239.

[26] A.k.a. Landsteiner-Wiener blood group glycoprotein (LW) and CD242.

[27] A.k.a. 55-kDa erythrocyte membrane protein P55.

Plasma membrane deformability of red blood cells results from the coupling between the phospholipid bilayer and spectrin network. Adenosine triphosphate favors dynamical fluctuations in the RBC membrane [117]. Membrane vibrations (amplitude $\mathcal{O}[10\,\text{nm}]$) that use both metabolic and thermal energies are enhanced at the scale of spectrin mesh size to maintain the biconcave shape. Phosphorylation-dependent binding of the cytoskeleton to the plasma membrane can explain the correlation between dynamical fluctuations and biconcave shape. The spectrin mesh–membrane binding creates these dynamical vibrations via local remodeling of the spectrin junctions.

3.5.3.3 Pannexon and ATP Release

Erythrocytes express pannexin-1 that forms a mechanosensitive, ATP-permeable channel (*pannexon*) in the plasma membrane [118]. These capsules release ATP in response to hypoxemia (Sect. 9.10.9) as well as shear stress.

Adenosine triphosphate is an intra- and extracellular regulator. Red blood capsules succor blood pressure regulation, as they release ATP, a paracrine messenger that functions as a vasodilator, especially in arterial constrictions. It causes vasodilation by releasing nitric oxide from endothelial cells.

The release of ATP is delayed with respect to the onset of increased shear stress [119]. Remodeling of the spectrin–actin cytoskeleton network may launch the mechanosensitive (deformation-triggered) ATP release; a shear-dependent membrane viscosity may control the rate of release.

Flow-induced RBC deformation releases ATP that enables platelet adhesion to the endothelium. Adenosine triphosphate behaves as a platelet inhibitor below 0.16 amol per platelet;[28] above this threshold, ATP is a platelet activator [120]. On the one hand, ATP stimulates the production of nitric oxide in endothelial cells via P2Y receptors and platelets; NO is vasodilator and platelet inhibitor. On the other, ATP activates platelet ATP-gated $P2X_1$ ion channel, thereby causing a rapid Ca^{2+} entry that enables platelet aggregation. In particular, mild platelet stimulation with low concentrations of collagen rapidly releases ATP that primes the $P2X_1$–Ca^{2+}–PKC–ERK2 pathway [121]. This event enhances further degranulation, thereby allowing complete platelet aggregation.

Proinsulin C-peptide increases glucose uptake by RBCs, intracellular Ca^{2+} concentration, and, subsequently, ATP release [122].[29]

---

[28] Metric prefix symbol a for atto means 10^{-18}.

[29] Preproinsulin contains an A- and B-chain, a connecting peptide (C-peptide), and a signal sequence. It is processed by peptidases to proinsulin by removal of the signal peptide. Proinsulin gives rise to insulin by elimination of the central C-peptide; both ends (A- and B-chain) remain connected by disulfide bonds. Proinsulin C-peptide produced with insulin facilitates assembly, folding, and processing of insulin in the endoplasmic reticulum. Equimolar amounts of C-peptide and insulin are then stored in secretory granules of pancreatic β cells. Both can be released to the portal circulation. β-Cell granules contain Zn^{2+} transporter proteins; in these granules, Zn^{2+}-

Red blood capsules undergo tumbling, swinging, tank-treading, and deformation in the microvasculature.[30] Tumbling-to-tank-treading transition, and not capsule deformation (i.e., stretching), contributes to the shear-thinning behavior (Vol. 7 – Chap. 5. Rheology) [126]. However, RBC deformation intervenes in ATP release, but not the transition from tumbling to tank-treading.

Both deformation-dependent and -independent ATP export occur through pannexin-1 hemichannel. Liberation of ATP through pannexin-1 mainly remains constant at shear stresses below a threshold (3 Pa), i.e., when RBC begins to deform [126]. Above this threshold, large RBC deformations occur and ATP release increases. Cystic fibrosis transmembrane conductance regulator contributes to deformation-dependent ATP release above the mechanical stress threshold. Large changes in local membrane curvature can dissociate spectrin from actin, thereby reducing the cell shear modulus and enabling actin binding to CFTR, hence its activation [127].

3.5.4 *Influence of Oxygen Content in the Local Environment*

Human RBCs circulate between lungs and tissues approximately every minute. In the lung, where O_2 partial pressure is high, RBCs are exposed to oxidative stresses that must be controlled by accelerated production of reducing equivalents derived from the pentose phosphate pathway [128]. In tissues, where O_2 partial pressure is low, deformed RBCs in capillaries undergo mechanical stresses that induce cation leaks. This leak and deoxygenation elevates glycolysis to produce ATP and restore intracellular ion balances and represses pentose phosphate pathway in RBCs.

Erythrocyte metabolism is regulated via the oxygen-dependent assembly of glycolytic enzymes into inhibitory complexes bound to the membrane anion transport SLC4a1 protein. In vivo, SLC4a1 acts as an oxygen-dependent metabolic regulator because it interacts with deoxyhemoglobin with high affinity, but not oxyhemoglobin [128]. Deoxyhemoglobin targets the same binding site on SLC4a1 as glycolytic enzymes. Therefore, upon deoxygenation, glycolytic enzymes are released and promote glycolysis. On the other hand, in oxygenated RBCs, glycolytic enzymes are inhibited, as they connect to SLC4a1 carrier.

---

bound C-peptide is involved in the breakdown of insulin hexamer. In blood, C-peptide interacts with RBC when it is linked to a metallic ion (e.g., Zn^{2+} and Fe^{2+}) [122].

[30]Tumbling is an end-over-end rigid body rotation. Tank-treading is the rotation of the cell membrane around the center of mass of the cell [123, 124]. Swinging is an oscillatory movement of the cell's inclination angle during tank-treading, with a period and an amplitude that increases with decreasing shear rates [125]. Stretching by large shear rates is an extension of the long cell axis with shape changes from an oblate to a prolate ellipsoid, without variation of the membrane surface area. Tank-treading and tumbling depend on the viscosity ratio (ratio of cytosol viscosity to suspending solution viscosity) and membrane elasticity.

3.5.5 *Oxygen Sensor*

The cardiovascular system responds to imbalances between oxygen supply and demand and adapts the blood flow. When oxygen demand changes, oxygen supply and/or extraction from circulating RBCs are adjusted to match the demand. Oxygen delivery is determined mostly by the vascular resitances of the arteriolar bed that irrigates the tissue of interest. Oxygen extraction occurs mainly in small arterioles and capillaries.

Erythrocytes have a dominant role in the supply of oxygen to tissues. Erythrocytic hemoglobin carries about 98% of blood oxygen. It releases oxygen to tissues owing to temperature, pH, and P_{CO_2} (that shift the oxygen dissociation curve).

According to the concept of metabolic control of the local blood flow, the concentration of vasodilators released from cells into the interstitial fluid that bathes resistance vessels results from the level of cell metabolism. Vasodilators, such as adenosine, nitrosothiols (e.g., ^{S}nitrosohemoglobin [Hb^{SNO}]) and nitrite, do not fill all criteria of a specific oxygen sensor [129]. Nitric oxide can modulate both the supply of oxygen via its vasodilatory action on resistance vessels as well as the demand for oxygen via its inhibitory effect on oxygen consumption via cytochrome-C oxidase in mitochondria [129].

Erythrocytes can operate as mobile sensors for local oxygenation by the oxygen-dependent release of ATP vasodilator [130]. Insulin at normal concentrations inhibits the O_2-dependent release of ATP from red blood cells, thereby impairing O_2 delivery to fit the metabolic demand.

3.5.6 *Blood Groups of the RBC Plasma Membrane*

Erythrocyte surface contains proteins, glycoproteins, and glycolipids. Some define blood types. K. Landsteiner (1868–1943) described ABO blood types in 1900 from identification of blood agglutinins. Later (1937–1940), with A. Wiener, he worked on Rhesus blood group. The ABO and Rhesus blood groups are the main blood groups considered in transfusion. Many of the blood type antigens and corresponding antibodies have been discovered from transfusion incompatibilities. There are 30 identified blood groups with multiple distinct types. Some blood group molecules contribute to substance binding at the cell surface as well as chemical import or export.[31]

[31]Duffy glycosylated membrane protein (Duffy blood group) acts as a non-specific receptor for several chemokines. In addition, Duffy binding-like receptors are used by some plasmodium parasites, such as Plasmodium vivax and falciparum.

3.5.6.1 ABO Blood Group

The ABO blood group is related to the presence on the RBC plasma membrane (as well as on the surface of endothelial and most epithelial cells) of 2 monosaccharides at the termini of oligosaccharide chains on glycoproteins and glycolipids responsible for A- and B-antigen activity.[32] Group A has anti-B antibodies that target B antigen; group B has anti-A antibodies; and group O has both as it is defined by the absence of A and B factors (i.e., group O red blood cells lack both A and B antigens). Group-O cells have at the termini of their oligosaccharide chains fucose (H antigen). Both A and B antigens are made from H antigen. The ABO blood group include subtypes A1, A1b, A2, and A2b. The *Bombay blood group* is related to rare Bombay phenotype (hh) characterized by the absence of H antigen.

3.5.6.2 Rhesus Blood Group

Five main Rhesus antigens are transmembrane proteins C, D, E, c, and e (genotype cde/cde - d means absence of D antigen - is RhD negative [RhD−], whereas the others are RhD positive [RhD+]).

3.5.6.3 Duffy Antigen–Chemokine Receptor

Duffy antigen (Dfy, GPFy, or CD234) is an acidic glycoprotein that is produced in greater amounts on reticulocytes than erythrocytes. Other isoforms can be detected in some epithelial cells (e.g., principal epithelial cells of collecting ducts in kidneys as well as type-1 epithelial cells of pulmonary alveoli [131]), Purkinje cells of the cerebellum, and endothelial cells of capillaries (e.g., thyroid and kidney such as peritubular capillaries), postcapillary venules [132], and large pulmonary venules.

Duffy antigen receptor for chemokines (DARC) is targeted by members of both CC and CXC families (e.g., CCL2, CCL5, CXCL1, CXCL2, CXCL5, CXCL7, and CXCL8). However, DARC causes neither G-protein-coupled receptor signaling nor Ca^{2+} flux. The Duffy antigen receptor for chemokines (Darc or FY) gene (located on chromosome 1) encodes the Duffy blood group determinants that consist of 5 phenotypes (FyA, FyB, FyO, FyX, and FyY). Antigens FyA and FyB differ by a single amino acid (Gly43 and Asp43, respectively).

3.5.6.4 Diego Blood Group

The Diego blood group comprises 21 rare blood factors associated with chloride–bicarbonate exchanger (counter-transporter, also known as band-3 protein) that is encoded by the SLC4A1 gene (SLC: solute carrier family) on chromosome 17.

---

[32]Monosaccharides of A and B antigens are ^{N}acetylgalactosamine and galactose, respectively.

3.5.6.5 Kell and XK Blood Groups

The KEL gene encodes transmembrane glycoprotein CD238 or Kell protein. Several alleles exist, among which K1 (Kell) and K2 (Cellano) are the most common. The Kell protein binds to Xk protein that possesses the Kx antigen. The Kell blood group precursor (Kx) gene (XK) is located on X chromosome.

3.5.6.6 Kidd Blood Group

The Kidd blood group corresponds to an antigen on membrane of red blood cells as well as renal cells. It participates in urea transport. It is encoded by the JK gene (chromosome 18). Two common JK alleles are JKa and JKb. Human subjects with 2 JkA antigens (without JkB antigens) produce antibodies against JkB antigens.

3.5.6.7 Lewis Blood Group

The Lewis blood group relies on genes on chromosome 19. Two main types of Lewis antigens include Lewis-a and -b. Lewis antigens are released by exocrine epithelial cells and subsequently adsorbed onto RBC surface. Fucosyltransferase converts Lewis-a antigen into Lewis-b.

3.5.6.8 Glycophorin-Based MNS and Gerbich Blood Groups

The MNS blood group is based on 2 genes (on chromosome 4) that encode glycophorin-A and -B that are associated with blood group MN and Ss antigens, respectively. More than 40 MNS antigens exist. The 5 most important MNS antigens are M, N, S, s, and U (U: universal, as it is extremely common).

The Gerbich (Ge) blood group is related to glycophorin-C[33] encoded by a gene on chromosome 2 and its variant glycophorin-D. These glycophorins are unrelated to other glycophorins encoded by genes on chromosome 4 (loci 4q28-q31).

Miscellaneous

The *Cartwright* (Yt) blood group corresponds to acetylcholinesterase with 2 alleles, Yta and Ytb.

The *Chido-Rodger* (Ch/Rg) blood group is represented by 2 elements of the complement system: anaphylatoxin C4a and C3-convertase subunit C4b.

[33]A.k.a. CD236, glycoprotein-β, or glycoconnectin.

The *Colton* blood group is associated with aquaporin-1 with allelic diversity (Co-a and Co-b alleles). Aquaporin-3 is a determinant for the *GIL* blood group.

The *Cromer* blood group is related to decay accelerating factor (CD55) that prevents the assembly of the C3-convertase of the alternative pathway or accelerates the disassembly of preformed convertase.

The *Dombrock* blood group that comprises many variant alleles corresponds to ^{ADP}ribosyltransferase Art4 (CD297) [133].

The *Knops* blood group squares with RBC complement receptor-1 (CR1, CD35, C3b/C4b receptor, and immune adherence receptor), a member of the the family of receptors of complement activation.

The *Indian* blood group (In) antigen tallies with glycoprotein CD44 that is involved in intercellular interactions.

The *John Milton Hagen* blood group (JMH) antigen coincides with semaphorin-7a that is expressed on RBCs and activated lymphocytes.

The *OK blood group* relies on glycoprotein CD147 that is also termed basigin and extracellular matrix metallopeptidase inducer.

The *Ii* antigen group corresponds to branched (I) or linear (i) polysaccharides, the synthesis of which depends on enzymes encoded by genes on chromosome 6. The unbranched i antigen, the precursor of I antigen, is synthesized by the sequential action of 2 glycosyl transferases, β(1,3)acetyl glucosaminyl transferase and β(1,4)galactosyl transferase. The I antigen is formed by branching initiated by a third transferase, the I β(1,6)^{N}acetylglucosamine transferase type 2. The GCNT2 gene encodes this ^{N}acetylglucosaminyltransferase[34] that converts active straight chains of carbohydrates to I-active branched chains. Three isoforms are created by alternative splicing (IGNT1–IGNT3 or IGNTa–IGNTc).

The i and I antigens are carbohydrate linear and branched repeats of ^{N}acetyl lactosamine, i.e., *n* being the number of repeating lactosamine units (6 in average for i antigen, 8–25 for I antigen) and β*i*–*j* the linkage type:

$$\left[\mathrm{Gal}\xrightarrow[\beta 1--4]{------}\mathrm{Glc^{N}Ac}\right]_n - \;-\mathrm{Gal}\xrightarrow[\beta 1--4]{------}\mathrm{Glc^{N}Ac}\xrightarrow[\beta 1--3]{------}\mathrm{Gal}-$$

and

$$\begin{array}{ccccc} \mathrm{Gal} & & \mathrm{Gal} & & \\ |\beta 1\text{–}4 & & |\beta 1\text{–}4 & & \\ \mathrm{Glc^{N}Ac} & & \mathrm{Glc^{N}Ac} & & \\ |\beta 1\text{–}6 & & |\beta 1\text{–}6 & & \\ \left[\mathrm{Gal}\xrightarrow[\beta 1--4]{------}\mathrm{Glc^{N}Ac}\right]_n - & & -\mathrm{Gal}\xrightarrow[\beta 1--4]{------}\mathrm{Glc^{N}Ac} & \xrightarrow[\beta 1--3]{------} & \mathrm{Gal}- \end{array}$$

[34]A.k.a. glucosaminyl (^{N}acetyl) transferase-2 I-branching enzyme (IGNT), ^{N}acetyl lactosaminide β(1,6)^{N}acetylglucosaminyl-transferase isoform-A, -B, and -C.

Adult human RBCs mostly express I antigens and contain only a few i antigens, whereas the latter predominates in fetal and neonatal erythrocytes. After birth, the quantity of I antigens gradually rises and the amount of i antigens falls; adult values are reached about at month 18.

The *Raph* blood group antigen dovetails with transmembrane glycoprotein GP27, a.k.a. platelet-endothelial tetraspan antigen PETA3, CD151, and tetraspanin-24.

The *Scianna* (Sc) blood group is determined by erythroblast membrane -associated protein (ERMAP) that is encoded by the Ermap gene.

The *Landsteiner-Wiener* (LW) blood group has been renamed intercellular adhesion molecule ICAM4.

The *Lutheran* (Lu) blood group with antigens LuA and LuB is based on genes on chromosome 19.

P blood group components (Pl) are encoded by genes on chromosome 22.

The *XG blood group* is encoded by the PBDX gene on X chromosome.

Other blood group antigens include globoside, a glycosphingolipid also named erythrocyte P antigen, and Rh-associated glycoprotein, a putative gas channel protein.

3.5.7 *Hemoglobin Solution in Erythrocytic Capsule*

Each RBC contains about 280×10^6 hemoglobin molecules [134]. Hb consists of four globin chains α and β. It contains four iron atom Fe^{2+}, in the center of hemes.[35] Hemoglobin carries oxygen (O_2) from lungs to tissues and help to transport carbon dioxide (CO_2) from tissues to lungs.[36] Hemoglobin is also involved in pH regulation.

---

[35]Heme is a component of several hemoproteins that is required for: (1) oxygen transport and storage (hemoglobin and myoglobin); (2) electron transfer (cytochromes); and (3) signal transduction (nitric oxide synthases).

[36]One O_2 molecule can reversibly bind to each heme to form oxyhemoglobin. The reaction is reversible because O_2 is taken when its concentration ($[O_2]$) is high and released when $[O_2]$ is low. O_2 combination to and release from Fe^{2+} is accelerated by already existing O_2–Fe^{2+} interactions. Under conditions of lower temperature, higher pH, and increased O_2 partial pressure (p_{O_2}), O_2 fixes to Hb. When temperature is higher, or when pH and O_2 partial pressure lower, the reverse reaction is promoted. With alveolar values of $p_{A_{O_2}} = 13.7$ kPa and $p_{A_{CO_2}} = 5.3$ kPa, the arterial and mixed venous blood values are $p_{a_{O_2}} = 12.7$ kPa, $p_{a_{CO_2}} = 5.1$–5.3 kPa, $p_{v_{O_2}} = 4.9$ kPa, and $p_{v_{CO_2}} = 5.9$–6.1 kPa. About 95% of CO_2 generated by cells is carried in RBCs and about 5% dissolves in plasma. A part of CO_2 is bound to hemoglobin (carbaminohemoglobin). The rest is converted by carbonic anhydrase contained in RBCs into bicarbonate ions (HCO_3^-) and hydrogen ions (H^+), which bind to the hemoglobin and do not affect pH

$$CO_2 + H_2O \Leftrightarrow H^+ + HCO_3^-.$$

Hemoglobin is the iron-containing, oxygen-transport metalloprotein in solution in the RBC cytoplasm (~97% of RBC dry content; ~35% of total content). This tetramer contains 2 copies of 2 distinct globin types (α and β; hemoglobin-A; $\alpha_2\beta_2$) and 4 binding sites, one per globin. Hemoglobin tetramer is a ellipsoidal protein (6.4×5.5×5.0 nm).

Hemoglobin carries 97 to 98% of oxygen to cells for metabolism. Hemoglobin oligomerization enables cooperativity between multiple binding sites (*allostery*). Binding of a first O_2 molecule to any heme induces a shift in the corresponding F-helix. This structural change decreases the binding energy for subsequent O_2 binding in the other 3 subunits (cooperative binding).

Oxyhemoglobin is formed when oxygen binds to Hb heme in RBCs of the pulmonary capillaries running along pulmonary alveolus wall. *Deoxyhemoglobin* corresponds to hemoglobin that has released its oxygen. Oxyhemoglobin and deoxyhemoglobin differ by their absorption spectra.[37]

3.5.7.1 Hemoglobin–Oxygen Association and Dissociation

Oxygen saturation is a relative measure of the amount of oxygen that is dissolved or carried in a given medium. Oxygen saturation in blood (S_{O_2}) yields the percentage of hemoglobin binding sites occupied by oxygen.

Functional oxygen saturation refers to the ratio of oxyhemoglobin to all types of functional hemoglobin:

$$\text{oxyhemoglobin}/(\text{oxyhemoglobin} + \text{deoxyhemoglobin}).$$

Fractional oxygen saturation is the ratio of oxyhemoglobin to the total amount of hemoglobin that is able to carry oxygen, including dyshemoglobins:

$$\text{oxyhemoglobin}/(\text{oxyhemoglobin} + \text{deoxyhemoglobin} + \text{carboxyhemoglobin} + \text{methemoglobin}).$$

Hemoglobin–oxygen association–dissociation curve has a sigmoidal shape that results from cooperative binding of oxygen to hemoglobin. At low partial pressures of oxygen (p_{O_2}), most hemoglobin is deoxygenated. When oxygen partial pressure reaches about 10 kPa, blood oxygen saturation approaches 100%.[38]

---

[37] Pulse oximeter measures functional saturation by transmitting two light wavelengths that are differentially absorbed by oxyhemoglobin and deoxyhemoglobin. Oxyhemoglobin has significantly lower absorption of 660 nm wavelength than deoxyhemoglobin, while at 940 nm its absorption is slightly higher. Functional saturation can also be invasively assessed from arterial blood gas samples. Co-oximeter uses 4 light wavelengths to differentiate between oxyhemoglobin, deoxyhemglobin, carboxyhemoglobin, and methemoglobin and to measure fractional saturation.

[38] Hypoxemia leads to an oxygen arterial saturation (S_{aO_2}) below 90%. Oxygen venous saturation (S_{vO_2}) measures the amount of oxygen consumed by the body. S_{vO_2} below 60% indicates lack of oxygen.

Hill Model

Measure of *cooperativity* of O_2 binding relies on oxygen saturation of hemoglobin that is determined by measurement of the partial pressure of oxygen in solution using Hill cooperativity coefficient n (oxygen concentration is proportional to partial pressure). Because the hemoglobin saturation curve does not satisfy the hyperbolic relationship observed for myoglobin: $S(O_2) = p_{O_2}/(p_{O_2} + p_{(50)})$ ($p_{(50)}$: oxygen partial pressure of O_2 at which half of Hb molecules are in the oxy state), Hill assumed the following equilibrium in 1913:

$$\mathrm{Hb} + n\mathrm{O}_2 \rightleftarrows \mathrm{Hb(O_2)}n.$$

The fractional saturation for the hypothetical reaction is given by:

$$S_{O_2} = \frac{[Hb(O2)n]}{[Hb(O2)n] + [Hb]} = \frac{p_{O_2}^n}{p_{O_2}^n + p_{(50)}^n}. \tag{3.3}$$

Hill plot corresponds to the trace of $\log[S_{O_2}/(1 - S_{O_2})]$ (i.e., ratio of oxyHb to desoxyHb) vs. $\log(p_{O_2}/p_{(50)})^n$. At small and large values of p_{O_2}, $n = 1$, whereas in the transition region of the relationship, $n = 2.8$.

In tissues, β chains very easily lose O_2, but α chains tightly hold oxygen. Diphosphoglycerate (DPG) of the anaerobic glycolysis pathway used by RBCs that lack mitochondria to produce enough ATP binds to Hb between 2 deoxygenated β chains. Conformational change that results promotes oxygen release from α chains. In lung capillaries, (2,3)-DPG is expelled from Hb molecules.[39]

Haldane and Bohr Effects

The *Haldane effect* states that deoxygenated Hb has a greater affinity for CO_2 than does oxyhemoglobin. Oxygen loading and unloading assist CO_2 release and pickup in lungs and tissues, respectively, and conversely.

The *Bohr effect* states that deoxygenated Hb has a greater affinity for H^+ than does oxyhemoglobin. Oxygen loading and unloading assist H^+ delivery and capture in lungs and tissues, respectively, and conversely.

The Bohr and Haldane effects are allosteric effects caused by binding of different ligands at different Hb spots that generates Hb conformational changes, thereby increasing and decreasing Hb affinity for O_2 in lung and the body's tissues, respectively.

---

[39]Life at high altitude increases (2,3)-DPG production in addition to mitochondrial concentrations and erythropoietin and RBC production.

3.5.7.2 Heme

Hemoglobin is a large protein, but its active site is a small non-proteic heme. The heme consists of a flat porphyrin ring (4 nitrogen-containing pyrrole groups bound by methyl bridges) based on building block porphyrin. Each heme group binds a central ferrous iron ion (Fe^{2+}) to become the so-called *hemin* (Fe^{3+}–heme).[40]

Heme is an iron-containing porphyrin that serves as prosthetic group in hemeproteins. These hemeproteins are involved in: (1) oxygen binding (hemo- and myoglobin); (2) oxygen metabolism (oxidases, peroxidases, catalases, and hydroxylases); (3) electron transfer (cytochromes); and (4) signal transduction (nitric oxide synthases). This signaling messenger operates in erythropoiesis, globin synthesis, microRNA processing, and circadian rhythm regulation.

Heme binds reversibly to the nuclear receptor NR1d1[41] a heme sensor that coordinates the cellular clock, glucose homeostasis, and energy metabolism [136]. Transcription factor NR1d1 inhibits the circadian core clock and hepatic gluco-neogenic gene expression.

In addition, the signaling mediator heme inactivates transcriptional repressors such as BTB and CNC homology-1, basic leucine zipper transcription factor BACH1.

Free heme is toxic. Therefore, its intracellular concentration is carefully regulated to yield a proper supply, but avoid heme toxicity. The rate of heme synthesis varies significantly among cell types. It is high in hepatocytes and erythroid cells, in which large amounts of heme are needed for cytochrome-P450 and hemoglobin, respectively. Heme is degraded by heme oxygenases to biliverdin, which is then converted to bilirubin by biliverdin reductase. Excess heme increases the transcription of heme oxygenase-1.

In addition, intracellular heme can be exported by plasmalemmal proteins, feline leukemia virus subgroup-C receptors (FLVCR1). This cytosolic heme export protein carries cytoplasmic heme as well as other cyclic planar porphyrins, such as protoporphyrin-9 and coproporphyrin, but not bilirubin.[42]

---

[40] Hemin provokes the proteasomal degradation of arginyl-transferase involved in the conjugation of arginine to aspartate, glutamate, or oxidized cysteine in the N-end rule pathway of protein degradation [135].

[41] A.k.a. reverse erythroblastoma RevErbα.

[42] Subtype FLVCR2 is a transmembrane transporter of the major facilitator superfamily that may be involved in growth regulation and calcium exchange [137]. Proliferation and motility of vascular endothelial cells during angiogenesis, which are Ca^{2+}-dependent events, require interaction of endothelial cells with the extracellular matrix and support cells such as pericytes. Mutations in the FLVCR2 gene causes proliferative vasculopathy and hydranencephaly-hydrocephaly syndrome (PVHH), or Fowler syndrome, an autosomal-recessively inherited prenatal lethal disorder characterized by hydranencephaly; diffuse clastic ischemic lesions with calcifications in the brain stem, basal ganglia, and spinal cord; glomeruloid vasculopathy of the central nervous system and retinal vessels; and a fetal akinesia deformation sequence (FADS) with muscular neurogenic atrophy [138].

The FLVCR receptor is required for terminal RBC development [139]. It is also involved in proerythroblast survival. It facilitates macrophage recycling of heme iron [140]. Moreover, FLVCR intervenes in the recycling of heme iron from senescent RBCs.

The export of heme is preferentially boosted by hemopexin [140]. Hemopexin abounds in serum (concentration 6–25 μmol; 0.4–1.5 mg/ml in humans). It accepts heme from FLVCR and carries it to the liver, thereby ensuring iron homeostasis, in addition to its scavenger role during internal bleeding and hemolysis.

Heme synthesis must be coordinated with translation of globin proteins to avoid toxic accumulation of hemoglobin precursors. Heme synthesis begins and ends in mitochondria, but several intermediate steps happen in the cytoplasm. Hemoglobin precursors are then transported across the mitochondrial membrane. The terminal step of heme synthesis in mitochondria corresponds to insertion of an imported iron atom into protoporphyrin-9. It is catalyzed by mitochondrial ferrochelatase, an integral protein of the inner mitochondrial membrane.

Transporter SLC25a37, or mitoferrin (Mfrn1), transfers ferrous iron (Fe^{2+}) across the mitochondrial membrane. The ATP-binding cassette transporter ABCb10 stabilizes the mitochondrial iron transporter SLC25a37 at least in erythroid precursors [141]. Both SLC25a37 and ABCb10 not only can homodimerize, but also complex with other partners. The SLC25a37–ABCb10 complex yields iron to ferrochelatase, thereby allowing erythroid precursors to accelerate their production of heme. Ferrochelatase, or protoheme ferrolyase, catalyses the terminal step (stage 8) of heme synthesis, converting protoporphyrin-9 into heme. Ferrochelatase is a partner for both SLC25a37 and ABCb10 transporters that forms the Fech–SLC25a37–ABCb10 complex [142]. Iron import in mitochondria is thus synergistically integrated with heme synthesis.

3.5.7.3 Functional and Dysfunctional Hemoglobin Types

Hemoglobin has many functional and dysfunctional types (Tables 3.4 and 3.5). The globin gene loci comprise mainly the α-globin locus on chromosome 16 and β-globin locus on chromosome 11. About 400 alteration types of hemoglobin genes have been detected.

Synthesis of δ chain begins late in the third trimester in utero. In the first 8 weeks, majority of the hemoglobin is embryonic hemoglobin (hemoglobin-E). In the embryo, hemoglobin variants comprise hemoglobin Gower-1 ($\zeta_2\epsilon_2$) and Gower-2 ($\alpha_2\epsilon_2$), as well as hemoglobin Portland ($\zeta_2\gamma_2$). In the fetus, fetal hemoglobin, or hemoglobin-F ($\alpha_2\gamma_2$), is produced. In newborns, hemoglobin-F is nearly completely replaced by adult hemoglobin in week 12 of postnatal life. Fetal hemoglobin binds oxygen with greater affinity than adult hemoglobin.

In adults, hemoglobin-F is restricted to a limited population of F reticulocytes and erythrocytes. Hemoglobin-A2 ($\alpha_2\delta_2$) is an adult variant (1.5–3.5%).

Table 3.4. Human hemoglobin (HB) genes and its products (Hb). The genes in the α-globin locus consist of the transiently expressed (the first 12 weeks after conception) embryonic gene HBZ followed by HBM, HBA2, HBA1, and HBQ1 genes (from 5′to 3′end). The HBQ1 gene is expressed very early in embryos erythroid cells, but not in adult erythroid or other non-erythroid cells. The θ-globin family consists of the HBQ1 gene that encodes θ1-globin and at least a truncated, processed ψθ2 pseudogene located on chromosome 22, whereas HBZP (ψζ1) and HBAP2 (ψα2) pseudogenes localize downstream from the HBZ (ζ2) gene, and HBAP1 (ψα1) pseudogene downstream from the HBM gene. The genes in the β-globin locus are also arranged sequentially (HBE1, HBG2, HBG1, HBD, and HBB from 5′to 3′end), beginning with the gene expressed during embryogenesis and ending with the gene expressed in adults. The 2 HBG genes that produce hemoglobin-F are active during fetogenesis.

Gene	Protein
	α-Globin locus on chromosome 16
HBZ	Hemoglobin-ζ, ζ-globin
HBM	Hemoglobin-μ, μ-globin
HBA2	Hemoglobin-α2, α2-globin, HbH
HBA1	Hemoglobin-α1, α1-globin, CD31, HbH
HBQ1	Hemoglobin-θ1, θ1-globin
	β-Globin locus on chromosome 11
HBE1	Hemoglobin-ϵ, ϵ-globin
HBG2	Hemoglobin-γ2(G), γ2-globin, Gγ-globin (Paulinia) methemoglobin
HBG1	Hemoglobin-γ1(A), γ2-globin, Aγ-globin
HBD	Hemoglobin-δ, δ-globin
HBB	Hemoglobin-β, β-globin

Table 3.5. Functional and dysfunctional types of hemoglobins (FHQTL: fetal hemoglobin QTL on chromosome 8).

Category	Types	Tetramer composition
	Developmental stages	
Embryo	Gower 1	$\zeta_2\epsilon_2$
	Gower 2	$\alpha_2\epsilon_2$
	Portland	$\zeta_2\gamma_2$
Fetus (≥ week 9)	Hemoglobin-F	$\alpha_2\gamma_2$
	FHQTL	
Adult	Hemoglobin-A (A1)	$\alpha_2\beta_2$
	Hemoglobin-A2	$\alpha_2\delta_2$
	Variants responsible for hemoglobinopathies	
Hemolytic anemia	Hemoglobin-C	$\alpha_2\delta_2^C$
Hemolytic anemia	Hemoglobin-E	$\alpha_2\delta_2^E$
α Thalassemia	Hemoglobin-H	β_4
α Thalassemia	Hemoglobin Barts	γ_4
Sickle cell disease	Hemoglobin-S	$\alpha_2\delta_2^S$

Table 3.6. α-Variants and deletional forms of α-thalassemia (Source: [143]).

Variant	Amino acid subtitution in α-globin
Hb Evanston	Trp14 → Arg
Hb Hasharon	Asp47 → His
HbG Philadelphia	Asn68 → Lys
HbQ Mahidol	Asp74 → His
Hb Duan	Asp75 → Ala
Hb Nigeria	Ser81 → Lys
HbJ Capetown	Arg92 → Gln
HbJ Tongariki	Ala115 → Asp

In humans, hemoglobin variants include variants of α-, β-, γ-, and δ-globin chains. An elevated proportion of hemoglobin-F and other variants, such as hemoglobin-S ($\alpha_2\beta_2^S$) and -C ($\alpha_2\beta_2^C$), correspond to hemoglobinopathies.

Thalassemia is caused by hemoglobin variants (deletional and non-deletional forms of gene mutations). Thalassemia has most often an autosomal recessive pattern of inheritance, both parents being unaffected carriers of a hemoglobinopathy trait.

α-Thalassemia results from deletions of one or both HBA1 and HBA2 genes. The $-\alpha$ determinant is associated with many different Hb variants (Table 3.6). Non-deletion mutations that are mainly associated with the HBA2 gene, affecting mRNA translation and processing, or causing post-translational instability, also generate α-thalassemia.

Hemoglobin Bart's hydrops fetalis syndrom is one of the major hemoglobinopathies responsible for α thalassemia. α Thalassemia can indeed cause hydrops fetalis when all 4 of the genetic loci for α-globin are deleted or affected by mutation. Hydrops fetalis is characterized by edema in at least 2 fetal compartments (hydrops allantois and hydrops amnion correspond to fluid accumulation in the allantoic and amniotic space respectively). Thalassemia can coexist with other hemoglobinopathies.

The most severe case of α-thalassemia is characterized by a defective or or lacking production of α-globin chains from the HBA1 and HBA2 genes, which encode hemoglobin α1 and α2 chains. *Hemoglobin Barts* (Bartholomew) consists of 4 γ chains. It has a very high affinity for oxygen, thus hampering oxygen delivery to cells.

Hemoglobin-H (HbH) is a β-chain tetramer. The excess β chains form unstable tetramers, which have abnormal oxygen dissociation curves. It has a higher affinity for oxygen than normal hemoglobin, thereby delivering a low quantity of oxygen to cells.

Inherited β-thalassemia is characterized not only by a reduced or absent synthesis of hemoglobin β chain, but also an elevated HbF concentration. In *thalassemia minor*, a single β-globin allele (among 2 alleles) bears a mutation that causes microcytic anemia; in *thalassemia major*, both alleles undergo mutations.

Sickle cell anemia, or *drepanocytosis*, is an autosomal recessive genetic disorder. The HBB gene that encodes 1 of the 2 types of polypeptidic chains of adult tetrameric hemoglobin is responsible for sickle cell anemia, characterized primarily by chronic anemia (sickled RBCs lifespan 10–20 d). Patients inherit either 2 copies of HBS variant or 1 copy of HBS and 1 copy of another variant. The substitution of amino acid in position 6 (hydrophobic amino acid valine rather than hydrophilic glutamic acid) creates a hydrophobic spot on the outside of the protein structure that links to the hydrophobic region of an adjacent hemoglobin β chain. This polymerization of HbS molecules that creates rigid structures causes the "sickling" of RBCs.[43] Several hundred HBB gene variants are known.

The hemoglobin variant — *hemoglobin-S* (HbS) — is the most common cause of sickle cell anemia, in the case of homozygosity of a mutation in the HBB gene. In heterozygous subject with a single sickle (S) gene and a normal adult hemoglobin (A) gene (HbAS; i.e., hemoglobin genotype AS rather than normal hemoglobin genotype AA), the abnormality is called *sicklemia* or *sickle cell trait*, without disease symptoms (healthy carrier subjects).

Hemoglobin-C results from a variation in the HBB gene. In homozygotes, nearly all Hb is HbC that causes a mild chronic hemolytic anemia. A heterozygous form with one sickle gene and another encoding hemoglobin-C, i.e., the gene for HbS inherited from one parent and the gene for HbC inherited from the other parent, causes hemoglobin-SC disease.

Like hemoglobin-S and -C, *hemoglobin-E* is generated by mutations of the HBB gene. This variant causes a mild chronic hemolytic anemia.

Dyshemoglobins (dysfunctional Hb) are also caused by molecules that affect binding of oxygen to Hb by competing with oxygen at heme, such as carbon monoxide (CO) that leads to *carboxyhemoglobin* (COHb; Table 3.7). Hemoglobin binding affinity for CO is 200 times greater than its affinity for oxygen. The stable complex ^{CO}Hb (blood half-life 4–6 h) does not release carbon monoxide. Therefore, hemoglobin cannot transport oxygen from lungs to the body's cells. Tobacco smoking via carbon monoxide inhalation augments the blood ^{CO}Hb concentration. However, CO, a element of the gaseous triumvirate with nitric oxide (NO) and hydrogen sulphide (H_2S), is a gasotransmitter (Vol. 4 – Chap. 10. Other Major Signaling Mediators) with anti-inflammatory, vasodilatory, and pro-angiogenic effects. Moreover, CO and NO are volatile and inhibitory coneurotransmitters [144].

Methemoglobins (MetHb) are caused by iron oxidation (Fe^{3+}) in Hb heme. Methemoglobin is unable to carry oxygen. NADH-dependent methemoglobin reductase converts methemoglobin to hemoglobin. Nitrogen dioxide (NO_2) and

[43]Polymerization of hemoglobin-S occurs when RBCs have delivered oxygen to cells and return to lungs. When hemoglobin binds oxygen, HbS molecules depolymerize or fragment. Polymerization–depolymerization cycling stiffens RBC membranes. The erythrocytic rigidity is associated with a distorted shape when RBCs do not carry oxygen; it provokes blockages of small blood vessels.

Table 3.7. Functional and dysfunctional types of hemoglobin compounds. A single hemoglobin molecule can transport 4 oxygen molecules. Each erythrocyte contains about 3×10^8 hemoglobin molecules. The affinity of hemoglobin for oxygen depends on 4 primary regulators, CO_2, hydrogen ion (H^+), chloride ion (Cl^-), and (2,3)-bisphosphoglycerate (BPG). Hemoglobin can bind, in particular, protons and CO_2 that cause a Hb conformational change and support 0_2 release. In lung capillaries, H^+ and CO_2 are liberated from hemoglobin. Higher BPG concentrations in blood enables the delivery of larger O_2 amounts to cells under smaller blood O_2 content (high altitude). Normally 1 to 2% hemoglobin is methemoglobin. NADH-dependent methemoglobin reductase converts methemoglobin to hemoglobin. Hemoglobin has a competitive binding affinity for cyanide that contains the cyano group (-CN^-), sulfur monoxide (SO), nitrogen dioxide (NO_2), and sulfide (S^{2-}), including hydrogen sulfide (H_2S), that bind to iron (Fe^{2+}) in heme and prevent O_2 binding. Cyanide is a natural, potent cellular toxin. Acute cyanide toxicity results from accidental exposure or iatrogenic toxicity. Sulfide is a sulfur dianion in its lowest oxidation state. Hydrogen sulfide is a very toxic gas and yet another natural messenger.

Type	Associated compound
	Functional hemoglobin types
Carbaminohemoglobin (^{CO_2}Hb)	Carbon dioxide Lowers hemoglobin affinity for 0_2 (Bohr effect)
$CO_2 + Hb^{NH_2} \longleftrightarrow H^+ + Hb^{NH--COO^-}$	
Oxyhemoglobin (^{O_2}Hb)	Oxygen Taut and relaxed forms have low and high oxygen affinity, respectively
Glycated hemoglobin	Glucose
Nitrosyl hemoglobin (Hb^{NO})	Nitric oxide
Nitrohemoglobin or nitrosohemoglobin (Hb^{SNO})	Nitric oxide (S-nitrosylation)
	Dysfunctional hemoglobin types
Carboxyhemoglobin (^{CO}Hb)	Carbon monoxide (inhaled or metabolic product) Stable complex
Cyanohemoglobin	CN^- anion (via Fe^{2+}) Unability to transport oxygen
Methemoglobin $^{Fe^{3+}}Hb$	Ferric rather than ferrous ion Cannot bind oxygen
Sulfhemoglobin (^{S}Hb)	Sulfur atom upon combination H_2S and Fe^{3+} Usually caused by drugs

nitrous oxide (N_2O) can convert a small fraction of hemoglobin to methemoglobin. Normally, only 1 to 2% of hemoglobin is methemoglobin.

Other molecules compete to bind hemoglobin, such as cyanide (CN^-), sulfur monoxide (SO), nitrogen dioxide (NO_2), sulfide (S^{2-}), and hydrogen sulfide (H_2S).

3.5.7.4 Nitric Oxide Carrier

Hemoglobin also carries nitric oxide that improves oxygen delivery in the peripheral tissues and contributes to the control of respiration. Nitric oxide binds reversibly to a specific cysteine residue (highly conserved Cys93 in all mammals) in globin β chain. S-nitrosylated hemoglobin influences NO-mediated regulation of vascular resistances, blood pressure, and ventilation. Nitric oxide is transported by an anion exchanger SLC4a1 out of RBCs that then dispense NO activity.

Nitric oxide undergoes oxidative and reductive reactions to produce various chemical species, such as nitrite (NO_2^-), nitrosothiols, nitrosoamines, nitrated lipids, and iron nitrosyl complexes. It is scavenged by reactions with hemoglobin in RBCs to form nitrate (NO_3^-), methemoglobin (${}^{Fe^{3+}}Hb$), and ferrous nitrosyl hemoglobin (${}^{Fe^{2+}}Hb^{NO}$) after the reduction of nitrite by deoxyhemoglobin to NO and reaction of NO with deoxyhemoglobin (*nitrite–deoxyhemoglobin reaction*) [145]:

$$NO_2^- + H^+ \rightarrow NO + {}^{Fe^{3+}}Hb + OH^-, \tag{3.4}$$

$$NO + {}^{Fe^{3+}}Hb \rightarrow {}^{Fe^{2+}}Hb^{NO}. \tag{3.5}$$

On the other hand, nitrate formation in RBCs results from the reaction of nitrite with oxyhemoglobin (${}^{O_2}Hb$; *nitrite–oxyhemoglobin reaction*) [145]:

$$4\left[NO_2^- + {}^{O_2}Hb^{2+} + H^+\right] \rightarrow 4\left[NO_3^- + 4^{Fe^{3+}}Hb + O_2 + 2H_2O. \tag{3.6}$$

In addition, nitrite can bind to methemoglobin to produce a nitrite-bound methemoglobin intermediate that can interconvert to a ferrous–nitrogen dioxide species, which reacts with NO to form ferrous hemoglobin and dinitrogen trioxide (N_2O_3) [145]:

$$NO_2^- + {}^{Fe^{3+}}Hb \rightarrow {}^{Fe^{3+}}Hb^{NO_2}, \tag{3.7}$$

$$\left({}^{Fe^{3+}}Hb^{NO_2} \longleftrightarrow {}^{Fe^{2+}}Hb^{NO_2^{\bullet}}\right) + NO \rightarrow {}^{Fe^{2+}}Hb + N_2O_3, \tag{3.8}$$

The overall nitrite–deoxyhemoglobin reaction then becomes [145]:

$$2NO_2^- + {}^{Fe^{2+}}Hb + H^+ \rightarrow {}^{Fe^{2+}}Hb + N_2O_3 + OH^-. \tag{3.9}$$

Nitrite can quickly enter RBCs (rate constant 0.25/mn) [145]. With a nitrite dose of 800 μmol, erythrocytic uptake rate is approximately 4-fold higher than that in vitro. Iron nitrosyl hemoglobin is rapidly formed. The overall consumption of ferrous nitrosyl hemoglobin is fast in comparison with elimination of methemoglobin and plasma nitrate (elimination time ~ 40, 105, and 300 mn, respectively).

In pulmonary arteries, hypoxemia normally causes vasoconstriction to match perfusion to ventilation. Nitric oxide synthases and ^{S}nitrosoglutathione reductase in RBCs determine intraerythrocytic content in nitric oxide and nitric oxide-thiol (^{S}nitrosothiol). Normal erythrocytic ratio of SNO to hemoglobin is lower than 1:1000 [146]. ^{S}nitrosohemoglobin (Hb^{SNO}) deficiency aggravates hypoxemia-mediated pulmonary hypertension [147, 148]. However, individuals who are heterozygous for a βCys93 mutation are phenotypically silent (without clinical abnormality) [149].

3.5.8 Iron Acquisition and Storage

Iron is needed by iron-dependent enzymes. Most of body's iron is found as heme in red blood cells. Iron can easily gain and lose electrons. This property is used not only by hemoglobin, but also by myoglobin, cytochromes, and various enzymes. However, it generates superoxide anions and hydroxyl radicals by electron donation to oxygen. Because reduced iron can catalyze formation of reactive oxygen species within cells, cells control the cytosolic amount of iron. Nevertheless, intracellular iron must be delivered to sites of storage and use, particularly target proteins, such as enzymes and carriers. Iron delivery is achieved by iron chaperones such as poly(rC)-binding protein-1 (PCBP1), which delivers iron to iron storage protein ferritin [150]. Ferritin can store up to 4,500 atoms of iron.

Iron balance is maintained between regulated absorption of dietary iron, iron loss via exfoliation of enterocytes and skin cells, and bleeding.[44] Iron is absorbed in the gut, predominantly in the duodenum and upper jejunum, and transported into plasma by apical importer DMT1 and basolateral exporter *ferroportin*, respectively [151].[45]

Iron in senescent RBCs that are ingested by macrophages and degraded is recycled back into plasma by ferroportin. Plasma iron is carried by glycoprotein *transferrin* that delivers iron to cells that express transferrin receptors.[46] At blood pH, apotransferrin (transferrin iron-free form) can bind 2 Fe^{3+} (dissociation constant K_d = 10–23 mol) in the presence of an anion (usually carbonate) that bridges

[44]Excess iron damages tissues, as it causes fibrosis. Iron-deficiency disorders generate anemia, which in turn lead to hypoxia. Iron-deficiency anemias have genetic or acquired (chronic inflammation) origin. Hereditary hemochromatosis or primary iron-overload disease results from inadequate hepcidin production. Type-1 hereditary hemochromatosis is associated with increased iron absorption and deposition in the liver, heart, pancreas, and skin that leads to liver cirrhosis and diabetes. It is caused by mutations in the HFE gene (hereditary hemochromatosis protein HFE1 forms a proteic complex with transferrin receptor-1). Type-2 hereditary hemochromatosis (juvenile hemochromatosis) is characterized by severe cardiac and endocrine dysfunction. It results from mutations in hepcidin (HAMP) or hemojuvelin (HFE2) gene. Type-3 hereditary hemochromatosis is caused by a mutation in TFR2 gene that encodes transferrin receptor-2. Type-4 hereditary hemochromatosis is conferred by mutations in ferroportin leading to 2 clinical phenotypes with iron accumulation in either macrophages or hepatocytes.

[45]One to 2 mg of iron is absorbed per day in the gut.

[46]Transferrin is synthesized in the liver, retina, testis and brain.

iron and transferrin. Transferrin receptor-1 is expressed on all dividing cells.[47] Erythrocyte precursors bind transferrin to incorporate iron into heme. Transferrin receptor-2 is expressed mainly in the liver and binds transferrin–iron complex with a much lower affinity than that of transferrin receptor-1.

Cellular iron is used or stored by *ferritin*.[48] According to iron status, concentrations in iron transporters, carriers, and storage proteins, it is regulated transcriptionally by cytokines and post-transcriptionally owing to iron-regulatory proteins IRP1 and IRP2.

Ferritin is an intra- and extracellular protein. Plasma ferritin has a low iron content and distinct subunit composition. Ferritin binds high-molecular-weight kininogen to inhibit its cleavage by kallikrein, hence reducing production of bradykinin and cleaved high-molecular-weight kininogen. In addition, ferritin and high-molecular-weight kininogen colocalize in inflammation sites,[49] where ferritin inhibits the cleavage of high-molecular-weight kininogen by tryptase and elastase. In addition, plasma light (L) and heavy (H) chain-enriched ferritin preferentially bind to plasmalemmal scavenger receptor ScaRa5 and T-cell immunoglobulin and mucin domain-containing protein TIMD2 that trigger endocytosis of the receptor–ferritin pair [152].

Small peptidic hormone *hepcidin* is a regulator of iron metabolism and systemic homeostasis to avoid pathological conditions of deficiency or overload. Hepcidin controls the surface expression of iron exporter *ferroportin* (Fpn or SLC40a1) on the basolateral membrane of duodenal enterocytes and on iron-recycling macrophages that process senescent erythrocytes. Interaction of hepcidin with ferroportin precludes iron export from duodenal enterocytes and macrophages. Hepcidin is secreted by the liver[50] to adapt iron absorption to the body's need. Hepcidin transcription is controlled by iron and oxygen sensors, as well as inflammation. Hepcidin level increases with iron overload. Conversely, hepcidin expression is suppressed when iron is required for erythropoiesis. In addition, hepcidin is upregulated by inflammatory cytokines. Circulating hepcidin binds to ferroportin for degradation, and hence inhibits iron export into plasma.

Hepcidin expression depends on bone morphogenetic proteins (BMP2, BMP4, and BMP9 in vitro) and the SMAD4 pathway.[51] On the hepatocyte surface,

---

[47]Transferrin receptor-1 is a transmembrane homodimer. The extracellular domain binds one transferrin per subunit. Transferrin–iron complex is then rapidly internalized via clathrin-coated pits into endosome acidified by ATP-dependent proton pump. Acidification releases iron. Endosomal Fe^{3+} is converted into Fe^{2+} by STEAP3 ferrireductase.

[48]Ferritin is a multimer of 24 subunits that include subunit-H (for heavy or heart) and -L (for light and liver). Ferritin subunits form a central core that contains up to 4,500 atoms of iron. Monomeric H-ferritin has ferroxidase activity that converts stored Fe^{2+} into Fe^{3+}.

[49]Monomeric H-ferritin is upregulated by cytokines such as TNFα.

[50]Hepcidin mRNA that is generated from gene HAMP (hepcidin antimicrobial peptide) is also detected in heart, pancreas, and hematopoietic cells.

[51]When BMP ligands bind and activate receptor Ser/Thr kinases on hepatocyte surface, these BMP receptors phosphorylate cytoplasmic effectors SMAD1, SMAD5, and SMAD8 that then form a

BMP6 forms a complex with BMP receptors (BMPR1a, BMPR1b, and BMPR2) and coreceptor hemojuvelin[52] (Hjv) to activate SMADs. Phosphorylated SMAD1, SMAD5, and SMAD8 complex with SMAD4 and translocates to the nucleus to activate HAMP gene promoter. Disruption of the Bmp6 gene in mice reduces hepcidin synthesis and rapidly causes a massive accumulation of iron in the liver,[53] exocrine pancreas (acinar cells), heart, and renal convoluted tubules [153].

Hypoxia-inducible factors HIF1 and HIF2 bind to hepcidin HAMP gene promoter and block hepcidin expression. Inflammatory cytokines such as interleukin-6 activate signal transducer and activator of transcription STAT3 that also binds to HAMP gene promoter and in the presence of SMAD4 induces hepcidin expression. Erythroid precursors secrete growth differentiation factor GDF15. This cytokine, a member of the transforming growth factor-β family, may inhibit hepcidin gene transcription.

Known genetic defects of iron transport and metabolism involve proteins that act as: (1) iron transporters and exporters such as ferroportin; (2) receptors for iron-binding proteins, such as transferrin receptor and hemochromatosis; (3) regulators of iron transporters or iron-binding protein receptors such as hepcidin; and (4) and regulators of iron metabolism such as transcription factor SpiC that controls development of red pulp macrophages that recycle senescent RBC products.

3.6 Platelets

Platelets, or thrombocytes (TC),[54] are small (size 2–4 μm) anucleate cellular fragments produced by *megakaryocytes*. The average lifetime is 10 days.[55] The usual concentration is equal to $250–500 \times 10^3/mm^3$. Platelets circulate in blood to ensure hemostasis and wound repair.

---

complex with the common mediator SMAD4. The resulting complex translocates to the nucleus, where it primes transcription of target genes.

[52]Hemojuvelin (a.k.a. RGMc) belongs to the repulsive guidance molecules (RGM) family. Members of the RGM family act as coreceptors for bone morphogenetic proteins [154]. Unlike other RGM family members (RGMa and RGMb) that are involved in axon guidance in the central nervous system, hemojuvelin is highly expressed in the liver, skeletal muscles, and heart. Hemojuvelin synthesized by hepatocytes and myocytes is cleaved from the cell surface to release a soluble form. Soluble hemojuvelin (Hjv^S) binds bone morphogenetic proteins and operates as a competitive antagonist of plasmalemmal hemojuvelin, thereby decreasing hepcidin expression.

[53]Despite iron overload, hepatocytes of Bmp6-deficient mice have low levels of phosphorylated Smad1, Smad5, and Smad8.

[54]θρομ βος: clot; κυτος: cell.

[55]The life duration of platelets is determined by the interplay between prosurvival BCLxL of the BCL2 family and the pro-apoptotic BAK, a target of BCLxL in the apoptotic pathway [155]. Platelets can survive for days. When BCLxL degrades, aged platelets are destined to die. The balance between BCLxL and BAK thus constitutes a molecular clock for platelet life, BCLxL maintaining platelet survival by restraining BAK.

Platelets are generated from megakaryocytes in bone marrow. Megakaryocytes extend proplatelet-like protrusions into microvessels. These intravascular extensions are sheared and torn from their transendothelial stems by flowing blood [156]. Proplatelets are thus released in blood, at least partly due to applied hemodynamic forces.

Platelets are involved in blood coagulation in injured vessels and thrombosis in diseased conduits,[56] as well as inflammation, immunity, and tumor progression. Platelet activation is also influenced by hemodynamic forces.

3.6.1 Platelet Cytoskeleton

Platelet cytoskeleton is involved in shape change, secretion, and clot contraction. Platelet shape rapidly changes on contact with foreign surfaces. Platelet actomyosin microfilaments (thrombostenin) organize in contractile stress fibers.

The cortex contains actin filaments and a circumferential bundle of microtubules. An intracellular tubular system is open to the cell surface via invaginations of the plasma membrane. Dense tubules contain Ca^{2+} stores and enzymes involved in arachidonic acid metabolism that are required for platelet secretion and aggregation.

3.6.2 Translational Machinery

Platelets lack nuclear DNA but retain megakaryocyte-derived mRNAs and the translational machinery for protein biosynthesis, in particular transcription factors (e.g., peroxisome proliferator-activated receptors [nuclear receptors NR1c1–NR1c3], retinoid X receptors [NR2b1–NR2b2], glucocorticoid receptor [NR3c1], and nuclear factor-κB), spliceosome components to process mRNA, and the signal-dependent protein translation machinery. Post-transcriptional and translational mechanisms regulate platelet function.

Anucleate platelets released into the blood stream from bone marrow megakaryocytes contain not only mRNAs for protein synthesis, but also microRNAs to regulate mRNA translation [157]. Platelets contain Dicer and Argonaute-2-containing ribonucleoproteic complexes to process microRNA precursors and control reporter transcripts.

3.6.3 Platelet Granules

The 2 major types of secretory granules include numerous α granules and large, dense granules (Table 3.8). Platelet *dense granule* membranes contain both granu-

[56]Platelet aggregation and fibrin formation both require thrombin, Ca^{2+}, and other clotting factors.

Table 3.8. Platelet released substances (Sources: [159, 160]; 5HT: 5-hydroxytryptamine, or serotonin; HETE: hydroxyeicosatrienoic acid; PAF: platelet-activating factor; VWF: von Willebrand factor). Peptidase nexin-2, the secreted form of β-amyloid precursor protein, is an abundant platelet protein that, once secreted, is a potent inhibitor of activated coagulation factor XI. (Peptidase nexin-1, or serpin-E2, is a potent inhibitor of the serine peptidase thrombin).

Location	Substance	Effect
α granule	IGF, PDGF, TGFβ	Chemotaxis, ECM deposition
	CXCL4, β-thromboglobulin	GF activation
	Fibronectin, fibrinogen, thrombospondin	TC aggregation
	Albumin	
	Nexin-2	Peptidase inhibition
	Tetraspanin-30	
Dense granule (δ granule)	ADP, ATP, Ca^{2+}	TC aggregation
	5HT, catecholamines	Vasoconstriction
Cytosol	vWF	TC adhesion
	Fibronectin	Coagulation
	PAF	TC activation
	TxA2, 12HETE	Vasoconstriction

lophysin and P-selectin. *α-Granule* membrane contains growth factors,[57] clotting factors,[58] adhesion molecules,[59] and chemotactic compounds.

Granules fuse by exocytosis with the plasma membrane or membrane of the tubular system for content release. Vessel wall collagen and various molecules, such as platelet-activating factor, serotonin,[60] and products of arachidonic acid metabolism, secreted by platelets and other cells, activate platelets. Activated platelets become spherical (contraction of actin filaments and microtubules), form pseudopodia for adhesion, secrete substances, and aggregate.

Activation of glycoprotein-6, a platelet-specific receptor for collagen, causes platelet-derived vesicle release and shedding in response to platelet activation [161]. These pro-inflammatory microparticles can transport molecules throughout the body, possibly in association with transmigrating leukocytes during inflammation, as they adhere to activated polymorphonuclear leukocytes and monocytes in blood circulation.

[57]Platelet growth factors target fibroblasts and smooth myocytes for wound healing (Chap. 11).

[58]Fibrinogen and von Willebrand factor (vWF) can be used as α-granule markers.

[59]Tetraspanin-29 (a.k.a. CD9, P24, cell growth-inhibiting gene product GIG2, leukocyte antigen MIC3 or 5H9, and Motility-related protein MRP1) and PECAM1 (or CD31) are glycoproteins (GP) of the platelet membrane implicated in platelet adhesion and aggregation. Tetraspanin-29 and PECAM1 also line the membrane of α granule, thus having an intracellular distribution identical to $\alpha_{2B}\beta_3$-integrin, a major aggregation platelet receptor [158].

[60]Dense granules contain serotonin (5-hydroxytryptamine) after transport from blood. Serotonin causes vasoconstriction in blood vessels with damaged endothelium and activates platelets.

3.6.4 Platelet Activation

Circulating platelets are kept in an inactive state particularly by prostacyclin (PGi2) and nitric oxide released by endothelial cells (Sect. 9.8). Endothelial cells also express plasmalemmal ectonucleotide pyrophosphatase–phosphodiesterase ENPP1 that inhibits platelet activation by adenosine diphosphate, as it converts ADP into AMP.

Platelet synthesizes and releases multiple molecules (caspases, osteonectin, ubiquitin, vimentin, tissue inhibitors of matrix metallopeptidases, etc.). Activated platelets promote blood coagulation on their surface, produce mediators of inflammation, and cause smooth myocyte proliferation.

In resting platelets, cholesterol is uniformly distributed on the cell surface and confined to multivesicular bodies, dense granules, and internal membranes of alpha granules [162]. Upon interaction with fibrinogen, cholesterol accumulates at the leading edge during platelet spreading on fibrinogen, especially at tips of filopodia. Upon stimulation with thrombin receptor-activating peptides (TRAP), which are synthetic parts of the liberated N-terminal tail of the cleaved, moderate-affinity, thrombin receptor, cholesterol also redistributes to filopodia. This process results from the concentration of Src kinase and tetraspanin-30 in these membrane nanodomains (rafts).

Thrombin, or blood coagulation factor II, is a component of the clotting cascade and strong activator of platelet aggregation. It targets peptidase-activated receptors (PAR_1–PAR_4). The serine peptidase thrombin initiates various platelet responses (shape change, pseudopod extension, eicosanoid production, granule release, adhesive receptor expression, and aggregation) via phospholipase-C, intracellular calcium flux, and activation of kinases. Another serine peptidase, trypsin, also activates platelets, stimulating their aggregation and phosphoinositide metabolism. Both thrombin and trypsin cleave the terminal region of the moderate-affinity thrombin receptor, thereby allowing the released C-terminal fragment to subsequently stimulate the processed receptor. In addition, thrombin acts by a non-proteolytic mechanism, possibly via a high-affinity receptor such as glycoprotein-1B, which is not cleaved by thrombin.

Arachidonic acid is enriched in platelet phospholipids, such as phosphatidylcholine, phosphatidylethanolamine, phosphatidylinositol, and phosphatidylserine. Upon platelet activation, arachidonic acid is released from membrane phospholipids, most likely by cytosolic phospholipase-A2, and converted by the cyclooxygenase axis to thromboxane-A2 that causes full platelet activation (positive feedback loop) [163].

At injury sites, platelets adhere to the exposed subendothelium and interact with collagen, von Willebrand factor, and fibronectin via their receptors, integrins, and glycoproteic GP1b–GP9 complex. Glycoproteins GP1b (GP1bα–GP1bβdimer), GP5, and GP9 that are members of the leucine-rich glycoprotein (LRG) family complex at the platelet surface. Glycoprotein GP1bα (encoded by the GP1BA gene) binds to von Willebrand factor and thrombin on the one hand, and platelet cytoskeleton via actin-binding proteins on the other [164].[61] Binding of the GPIb–GP9 heterodimer to von Willebrand factor exposed in the subendothelium enables platelet adhesion at sites of injury. In addition, platelet GP5 has a high-affinity binding for thrombin.

Both thrombin and ADP activate platelets, which then secrete ADP, platelet-derived growth factor, and fibrinogen from storage granules, as well as thromboxane-A2 by immediate synthesis.

Platelet $\alpha_{2B}\beta_3$-integrin[62] binds to fibrinogen, leading to platelet aggregation. With the simultaneous formation of a fibrin mesh, this process lead to the formation of a platelet thrombus.

3.6.5 Non-Genomic Function of Transcription Factors

Transcription factors in platelets can have non-genomic functions. Platelets possess ligand-regulated transcription factors peroxisome proliferator-activated receptor-β and -γ (or nuclear receptors NR1c2 and NR1c3) and their binding partner retinoid X receptor-α and -β (or nuclear receptors NR2b1 and NR2b2), as well as glucocorticoid receptor (or nuclear receptor NR3c1) [165]. The RXR ligands prevent platelet aggregation and thromboxane-A2 release on Gq activation triggered by ADP and TxA2 receptors. Nuclear receptors of the RXR category binds to Gq subunit of G protein, thereby hindering Gq-induced Rac activation and calcium release.

Factor PPARγ reduce platelet release of mediators such as tumor-necrosis factor superfamily member TNFSF5, in addition to thromboxane-A2 [166]. Moreover, collagen-stimulated platelet aggregation is attenuated by PPARγ via inhibition of the collagen receptor GP6 [167]. Ligands of PPARγ precludes phosphorylation of components of the GP6 signaling pathway. Factor PPARγ links to SYK kinase that phosphoryles Linker for activation of T cells after platelet activation. This association is impeded by PPARγ agonists.

---

[61] Density of glycoproteins on the platelet plasma membrane depends on: (1) reversible, cytoskeletal-mediated translocation to the surface-connected canalicular system (SCCS), using neutrophil cathepsin-G, thrombin receptor-activating peptide (TRAP], a synthetic molecule that corresponds to the liberated N-terminal tail of the cleaved moderate-affinity thrombin receptor, or a combination of adenosine diphosphate and adrenaline; and (2) proteolysis [164]. Glycoprotein GP1bα is processed by neutrophil cathepsin-G, neutrophil elastase, and calcium-dependent peptidase; GP9 by thrombin and cathepsin-G; GP5 by thrombin, neutrophil elastase, and calcium-dependent peptidase [164]. The Bernard-Soulier syndrome is an autosomal recessive disorder characterized by thrombocytopenia, giant platelets, and a bleeding tendency. It results from abnormalities of the GP1b–GP5–GP9 complex, especially mutations of the GP1BA, GP1BB, and Gp9 genes.

[62] A.k.a. glycoproteic GP2b–GP3a dimer.

Nuclear factor-κ light chain-enhancer of activated B cells (NFκB) exerts a negative feedback loop that limits platelet activation [168]. Factor NFκB builds a complex with protein kinase-A that contains inhibitor of NFκB IκBα. The latter is phosphorylated by inhibitor of NFκB kinase. This event releases PKA that can then phosphorylate the actin-dynamics inhibitor vasodilator-stimulated phosphoprotein. The NFκB–PKA pathway is stimulated by thrombin and collagen. It limits platelet activation and motility, hence thrombus stability, as it reduces cytoskeletal dynamics and early-stage aggregation.

Diverse elements of the NFκB family in platelets produce different associations of NFκB subunits that initiate different signalings according to the source of platelet stimulation. Fibrinogen-activated platelets cause platelet spreading via NFκB that influences fibrinogen binding to $\alpha_{2B}\beta_3$-integrin [169].

Thrombin-activated platelets, but not resting cells, synthesize B-cell lymphoma-3 protein [170]. Protein BCL3 is a member of the atypical IκB subfamily that acts not only as a transcriptional coregulator, but also as a component of the autoregulatory loop via its association with NFκB. In activated platelets, BCL3 interacts with Fyn kinase to produce a retracted fibrin clot, the clot consisting of aggregated platelets and a mesh of fibrins.

Furthermore, inhibitor of NFκB kinases have other substrates than inhibitors of NFκB, such as AMPK and MAP3K7 kinases [171]. In platelets, AMPK and MAP3K7 contribute to signaling for adhesion and aggregation of platelets as well as exocytosis of mediators.

3.6.6 Other Regulators of the Platelet Function

At least, 3 subtypes of P2 receptors are implicated in platelet activation: $P2X_1$, $P2Y_1$, and $P2Y_{12}$ (Vol. 3 – Chaps. 2. Membrane Ion Carriers and 7. G-Protein-Coupled Receptors). Ion channel $P2X_1$ activated by ATP during collagen-induced platelet activation generates a rapid influx of calcium, which leads to a change in platelet shape and platelet aggregation. Gq-coupled $P2Y_1$ receptor also modifies platelet shape and provokes aggregation via calcium influx. Adenosine diphosphate binds to the $P2Y_1$ receptor. Gi-coupled $P2Y_{12}$ receptor that binds ADP inhibits adenylate cyclase. The activation of phosphatidylinositol 3-kinase by ADP G-protein-coupled receptors triggers platelet aggregation.

The Wnt–β-catenin signaling pathway functions in human platelets [172]. Platelets possess Wnt receptors such as Frizzled-4 and -6 receptors as well as LRP5 and LRP6 coreceptors and Disheveled-2 mediator. Wnt3a ligand prevents platelet activation, dense granule secretion, adhesion, and aggregation [172]. In addition, Wnt3a hampers the thrombin-induced activation of RhoA GTPase (non-canonical Wnt signaling). Among non-genomic functions, β-catenins link adhesion receptors of the cadherin family to the actin cytoskeleton.

3.6.7 *Inflammatory Platelet*

Platelets can also be considered as inflammatory cells that synthesize proteins involved in inflammatory pathways (Chap. 11). Granules α of platelets indeed contain inflammatory modulators, such as P-selectin,[63] tumor-necrosis factor superfamily member TNFSF5,[64] thrombospondin-1, CXCL4 chemokine,[65] transforming growth factor-β, and interleukin-1 precursors.

3.7 Granulocytes

Granulocytes are innate immunocytes that contain granules involved in immune defense. They are also called polymorphonuclear leukocytes, as their nuclei are usually lobed (2–4 lobes). The names of the 3 types of granulocytes are derived from neutrally, eosinophilic, and basophilic staining of cytoplasmic granules.

3.7.1 *Basophils*

Basophils (Bφ; size 12–15 μm; 20–80 μdm^3) have numerous large basophilic granules. They represent less than 1% of all leukocytes. Basophils are the least abundant granulocytes. These short-lived cells have an expected half-life of a few days. Basophils express immunoglobulin-E receptors (FcϵR1).

Circulating basophils originate from CD34+, FcϵR1α^{high},[66] SCFR− hematopoietic progenitors in the bone marrow that express transcription factors CCAAT–enhancer-binding protein-α and GATA2 [173]. Mature basophils exit the bone marrow.

Activated T lymphocytes produce interleukin-3 and other factors to enhance basophil production in the bone marrow. Because FcϵR1 expression can change during immune response, PTPRc (or CD45) is a useful marker of basophils.[67]

---

[63] P-selectin leads to aggregates of activated platelets and leukocytes.

[64] Cytokine TNFSF5 allows platelets to bind to macrophages, T lymphocytes, and endothelial cells, as well as to other platelets via $\alpha_{2B}\beta_3$-integrin. It also activates an inflammatory cascade involving E-selectin, VCAM, and ICAM1 for leukocyte adhesion.

[65] A.k.a. platelet factor-4.

[66] Subunit of the tri- or tetrameric, high-affinity IgE receptor encoded by the FCER1A gene.

[67] Unlike other myeloid lineage cells that are $CD45_{high}$, in which high-molecular weight forms ($CD45_{hMW}$) are, after activation, gradually replaced by low-molecular weight forms ($CD45_{lMW}$), basophils are $CD45^{mid}$ [173].

Basophils are similar to mastocytes, but distinct. In normal conditions, they are not detected in the blood circulation.[68] Basophils differ from mastocytes and eosinophils, as they do not express plasmalemmal stem cell factor receptor (SCFR)[69] and chemokine receptor CCR3, respectively. Human basophils also express TLR2 and TLR4 Toll-like receptors.

The behavior of basophils in vivo can differ from that in vitro. Moreover, functioning differences exist between mammal species. For example, mouse basophils express peptidase-activated receptors, but human basophils do not. Basophils can be activated by many agents, such as cytokines, immunoglobulins, peptidases, TLR2 ligands, and parasite-associated antigens.

Committed basophils release 2 kinds of mediators: (1) preformed granule-associated mediators, such as vasodilator histamine, serotonin, bradykinin, anti-coagulant heparin, and cytokines; and (2) newly generated mediators, such as prostaglandins and leukotrienes. Both transmembrane ectonucleotide pyrophosphatase-phosphodiesterase ENPP3 (or CD203c) and Tspan30 (or CD63) are markers of activated human basophils.

Basophils release numerous potent inflammatory mediators from their cytoplasmic granules upon high temperature, mechanical stress, ionization, toxins, peptidases, cationic proteins derived from eosinophils and neutrophils, IgE-mediated processes as well as IgE-independent immune mechanisms, such as that primed by complement factors.

Low-molecular-weight mediators include histamine, serotonin, prostaglandins, leukotrienes. These mediators are released in response of allergens. They cause increased vascular permeability, smooth muscle constraction, vasodilation.

Basophils are the main source of interleukin-4 in allergen- and helminth-activated blood mononuclear cells [174]. Basophils highly express IL33 receptor (IL33R or IL1RL1) and produce IL4 and IL13 in response to IL33, a member of the IL1 family, using MyD88 adaptor [173]. Interleukin-33 provokes airway hyperresponsiveness and goblet cell hyperplasia. Activated basophils not only secrete cytokines (Tables 3.9 and 3.10) to support the development of IL4-producing CD4+ T lymphocytes and IgE-secreting B lymphocytes associated with type-2 helper T immunity, but also produce chemoattractants for eosinophils and neutrophils into inflammatory sites.[70] Ligands of TLR2 induce nuclear translocation of transcription factor NFκB and cytokine secretion. On the other hand, TLR4 ligand lipopolysaccharides remain inactive, as CD14 is not expressed by basophils [173].

---

[68] In addition to basophil progenitors, bipotent basophil–mastocyte progenitors in the mouse spleen are able to differentiate into either basophils or mast cells [173].

[69] A.k.a. cellular kinase in tyrosine cKIT or KIT and CD117.

[70] Signals from T-cell receptors are strengthened by simultaneous binding of MHC molecules to a specific coreceptor. The coreceptor corresponds to CD4 and CD8 on CD4+ and CD8+ T cells, such as helper and cytotoxic T cells, respectively. Coreceptors CD4 and CD8 tether class-2 and -1 MHC molecules, respectively.

Table 3.9. Leukocyte sources of cytokines, especially type-1 and -2 immunocytokines (**Part 1**; Sources: [175, 928]). Cytokines can be classified into type-1 cytokines (IL2, IL12, IL18, Ifnγ, TGFβ, TNFα and -β [or TNFSF2]) that promote cellular immune response and type-2 cytokines (IL4, IL5, IL6, IL9, IL10, IL13, and IL25, as well as granulocyte–macrophage colony-stimulating factor [gmCSF or CSF2]) that favor humoral (antibody-mediated) immune response. Type-2 response is defined as an attenuated cellular response, but with a strong B-cell activity such as hypergammaglobulinemia and an increase in the level of type-2 cytokines. Type-1 and -2 cytokines then refer to cytokines produced by CD8+ and CD4+ type-1 effector T cells, such as type-1 cytotoxic (T_{C1}) and helper T (T_{H1}) cells, respectively, and type-2 effector T lymphocytes (T_{C2} and T_{H2}, respectively), respectively. Some of these type-1 and -2 cytokines are cross-regulatory: interferon-γ and interleukin-12 decrease the concentrations of type-2 cytokines, whereas IL4 and IL10 reduce the levels of type-1 cytokines. Type-1 cytokines, such as interferon-γ produced in particular by CD8+ and CD4+ type-1 effector T cells, interleukin-12 by antigen-presenting cells, and IL18 by activated macrophages, are required for effective responses to intracellular pathogens such as viruses. Type-2 cytokines produced by CD8+ and CD4+ type-2 effector T cells, mediate allergic responses. On the other hand, class-1 cytokines comprise IL1$\alpha$and 1β, IL2–IL7, IL9, IL11, IL12a–IL12b, IL13, IL15, IL18, IL21, IL23, and IL33, CSF2, granulocyte colony-stimulating factor (gCSF or CSF3), B-cell stimulating factor-3, leptin, erythropoietin, growth hormone, and prolactin.

Cell	Cytokines	
	Type 1	Type 2
B cell	IL12/18, TNFβ	IL6/10
NK cell	Ifnγ, TNFβ	
CD4+ T cell	IL2/12, Ifnγ, TNFβ	IL4/5/6/10/13
CD8+ T cell	IL2, Ifnγ	IL4/5/10

Basophils represent the main antigen-binding cell population in the bone marrow, spleen, and blood after protein immunization. Their capacity to bind large amounts of antigens on their surface depends on their expression of immunoglobulin-E receptors and antigen-specific IgE produced by B cells after immunization [174]. Activated basophils express plasmalemmal receptors such as CD40L. Activated basophils also enhance humoral memory B-cell responses by producing interleukin-4 and -6 in response to specific antigens.

Basophils induce a T_{H2} phenotype in CD4+ T lymphocytes, especially during parasite infection and allergic inflammation. Basophils activate the production by CD4+ T cell of interleukins IL4, IL5, IL10, IL13, and transcription factor GATA3, but downregulate interferon-γ and IL2. Basophils synthesize both IL4 and thymic stromal lymphopoietin to elicit T_{H2} immunity. In addition, T_{H2} differentiation promoted by basophils involves intercellular contact with CD4+ T lymphocytes in draining lymph nodes [173].

Human basophils, after IgE crosslinkage, secrete IL25 (or IL17e). IL25 stimulates the production of T_{H2} cytokines IL4, IL5, and IL13 and recruitment of eosinophils. IL25 can also activate T_{H2} memory T cells that strongly express IL25 receptor.

Table 3.10. Leukocyte sources of cytokines, especially type-1 and -2 immunocytokines (**Part 2**; Sources: [175, 928]; CSF: colony-stimulating factor; gCSFR corresponds to CSF3R and CD114, gmCSFR to CSF2R and CD116; mCSFR to CSF1R and CD115; GH: growth hormone; GHR: growth hormone receptor; Ifn: interferon; IL: interleukin).

Cell	Cytokines	
	Type 1	Type 2
Basophil		IL4/5/6/13
		IL1β/3/8
		Receptors of IL1–IL6/8/13/16/17/29/31/33
		CSF2, CSF2R
		Receptors of Ifnα/β
Dendritic cell	IL12	
Eosinophil	IL2/12, Ifnγ	IL4/5/6/9/13/25
		IL1α/3/8/10–11/16
		Receptors of IL1–6/8–10/12–13/15–16/18/29/31/33
		gmCSF, CSF2R
		Receptors of Ifnα/γ
Mastocyte		IL4/5/6
Macrophage	IL12	IL6/10
Monocyte	IL12	IL6/10
Neutrophil	IL12	IL4/6/9
		IL1/8/15/16/18/20/32
		Receptors of IL1–10/12–13/15–18/29/31
		GH, GHR, CSF2R, CSF3R
		Ifnα/β
		Receptors of Ifnα/γ/ω

Basophils are rarely detected in normal tissues, but reside in sites of allergic inflammation, especially in respiratory ducts of asthmatic patients. Basophil products, such as interleukin-4 and -13, leukotriene-C4, and histamine, cause symptoms of acute and chronic allergy.

3.7.2 Eosinophils

Eosinophil (Eφ; size$\sim$15 μm) has large granules and a nucleus which often has two lobes. The granules contain digestive enzymes that are effective against parasitic larvae. These cells also phagocytize antigen–antibody complexes. These cells account for less than 5% of all leukocytes.

In addition to the release of toxic granule content and interleukin-4 (that drives lymphocytes toward a type-2 helper T phenotype and activation of Toll-like receptor-2 on dendritic cells, eosinophils defend the body against pathogens by rapidly ejecting mitochondrial DNA [177]. Expelled mitochondrial DNAs act as an extracellular immune effector, especially in the gastrointestinal tract, that generate a sticky matrix to trap bacteria and activate their death by cationic antimicrobial peptides such as cathelicidins and nuclear proteins such as histones. Ejection of mitochondrial DNAs depends on reactive oxygen species and interleukin-5 produced by T_{H2} cells.

3.7.3 *Neutrophils*

Neutrophil (Nφ; size 8–15 μm) has small plasma granules and frequently multilobed nucleus. Normally, neutrophils account for 50 to 70% of all leukocytes. Scouting neutrophils look after possible invading agents.

Neutrophils possess an arsenal of hydrolytic and oxidative substances as well as pore-forming molecules such as pannexins that can cause tissue destruction that can contribute to further ischemic injury as well as autoimmunity.

3.7.3.1 Immunostimulatory and Immunosuppressive Neutrophils

Tumors avoid the activity of the immune system, as they set up an immunosuppressive environment. Neutrophils that are major effector cells during inflammation can also control excessive inflammation, as they secrete anti-inflammatory cytokines.

The neutrophil phenotype is regulated by the acute-phase response protein serum amyloid-A1. Increased concentration of serum amyloid-A1 observed in tumors as well as blood renders neutrophils immunosuppressive [178]. Serum amyloid-A1 indeed provokes differentiation of neutrophils that release anti-inflammatory interleukin-10. However, it also fosters the interaction of invariant natural killer T cells with immunosuppressive neutrophils to limit their suppressive activity, as the production of IL10 and IL12 is reduced and enhanced, respectively. Therefore, neutrophils adopt a pro-inflammatory phenotype. Invariant NKT cells thus counteract immunosuppression and oppose cancer development [178].

3.7.3.2 Neutrophils as Scavengers in Cell Necrosis

Neutrophils are rapidly recruited to the site of inflammation, hence responding not only to microbial infection, but also sterile injury and organ necrosis to contribute to wound healing. Cell necrosis releases multiple pro-inflammatory damage-associated molecular patterns (DAMP), or host-derived indicators of cellular damage, such as proteins, nucleic acids, extracellular matrix components, and lipid mediators. These DAMPs recruit neutrophils [179].

Adenosine triphosphate generated from necrotic cells at the injury site activates the NLRP3[71] inflammasome to assist neutrophils in extravasation across the

[71]Or NALP3; NOD-like receptor (NACHT, LRR, and PYD domain-containing) family, pyrin domain-containing proteins.

vascular endothelium using integrins [179] (Vol. 3 – Chap. 11. Receptors of the Immune System).[72] The NLRP3 inflammasome promotes maturation and secretion of interleukin-1β. It can integrate signals from interleukin-1β, reactive oxygen species, and thioredoxin-interacting protein, thereby acting as a sensor for metabolic stress, such as in insulin resistance and type-2 diabetes mellitus [180]. Maturation of IL1β into an active form is actually controlled by caspase-1-activating platforms, the so-called inflammasomes. In addition, formyl peptides released from necrotic cells guide neutrophils to the injury site [179].

3.7.3.3 Neutrophils as Microphages for Pathogens

Neutrophils constitute a first line of defense against invading pathogens. These polymorphonuclear microphages are able to phagocytize foreign cells, toxins, and viruses. Neutrophils use neutrophil extracellular traps that consist of chromatin coated with peptidases, such as myeloperoxidase, elastase, and cathepsin-G, to trap and kill bacteria [97].

Neutrophils express numerous receptors for bacterial phagocytosis, such as $\alpha_M\beta_2$-integrin that binds complement component C3b on the surface of opsonized bacteria and IgG Fc receptors (FcγR) that bind IgG-coated particles and immune complexes [97]. These molecules facilitate the uptake of opsonized bacteria into phagosomes. Phagosomes then fuse with lysosomes and neutrophil granules that contain numerous proteolytic and antibacterial enzymes and peptides. NADPH oxidase of phagolysosome membranes generates reactive oxygen species to destroy bacteria.

Human neutrophil peptides are released from granules of activated leukocytes to kill pathogens. Human neutrophil peptides can also modulate inflammation and vascular tone. In endothelial cells stimulated by human neutrophil peptides, COx2 expression (but not COx1) increases in time- and dose-dependent manner [181]. However, human neutrophil peptides do not significantly enhance COx2 activity, hence PGI2 production, but provoke the release of endothelin-1.

3.7.3.4 Neutrophil Migration

Neutrophils are the first recruited cells and responders to sites of inflammation and infection by primary chemoattractants. They then attract other cells, as they release secondary (relay) chemoattractants such as leukotriene-B4 that enables neutrophil activation and migration [182].[73]

---

[72]Extracellular ATP is an NLRP3-activating DAMP that is released on cell injury or necrosis. It stimulates rapid K^+ efflux from $P2X_7$ receptor, an ATP-gated ion channel, and triggers gradual recruitment and pore formation by the pannexin-1 hemichannel [180].

[73]Leukotriene-B4 amplifies cAMP production, myosin-2 phosphorylation, ^{F}actin polymerization, and cell polarization.

Neutrophil migration relies on 2 main pathways: (1) phosphatidylinositol 3-kinase that is activated by interactions between chemokines and G-protein-coupled receptors and (2) mitogen-activated protein kinase P38MAPK pathway that dominates in chemotaxis induced by bacterial or complement chemoattractants. Lipid and protein phosphatase and tensin homolog (PTen) that counteracts PI3K acts for priorization of bacterial chemoattractants for neutrophil migration toward bacterial products via phospholipase-A2 and P38MAPK [183].

In their search for bacteria, neutrophils respond to many chemotactic substances produced by macrophages, mastocytes, or serosal cells. Near the infection site, neutrophils select chemoattractants, favoring bacterial chemoattractants (e.g., ^{N}formyl-methionyl-leucyl-phenalanine and complement component C5a) rather than chemokines and leukotrienes.

3.7.3.5 Neutrophil Extravasation

Pentraxin-3 (a.k.a. TNFα-induced protein TNFαIP5) has both inflammatory and anti-inflammatory effects.[74] On the one hand, cytokines and liganded Toll-like receptors stimulate Ptx3 synthesis in endothelial cells, macrophages, and dendritic cells, as well as Ptx3 storage in neutrophils [184]. Neutrophils secrete Ptx3 that promotes innate immunity, as it binds to pathogens and activates complement [185]. C-terminus of Ptx3 links to the complement protein C1q, among other substances such as P-selectin.

Inflammation relies on the recruitment of circulating cells. Rolling permits selectin ligands and chemokine receptors on neutrophils to prime deceleration, arrest, then extravasation. Pentraxin-3 can limit neutrophil recruitment, as it impedes rolling on P-selectin in inflamed venules. It indeed sequesters P-selectin away from its endothelial ligands. In summary, neutrophils rolling on P-selectin in venules at sites of infection or injury receive signals that release paracrine regulator Ptx3 from specific granules. Released Ptx3 binds to locally expressed P-selectin in a paracrine manner.

Therefore, pentraxin-3 is another inhibitor of leukocyte adhesion to vascular endothelia. The integrin ligand EGF-like repeats and discoidin-1 (I)-like domain-containing protein EDIL3 is a matrix glycoprotein secreted by endothelial cells.[75] EDIL3 ligand inhibits adhesion of leukocytes and endothelial progenitor cells, as it

---

[74]Pentraxins constitute a superfamily of evolutionarily conserved multimeric proteins. Classic short pentraxins include C-reactive protein and serum amyloid-P that are produced mainly in the liver. Pentraxin-3 is a long pentraxin. Unlike classic short pentraxins that vary in sequence and regulation among mammalian species, Ptx3 is highly conserved between mouse and man for sequence, ligand recognition, and regulation.

[75]Protein EDIL3, a.k.a. Developmental endothelial locus DEL1, connects to β_2- (e.g., leukocyte $\alpha_L\beta_2$-integrin), $\alpha_V\beta_3$- and $\alpha_V\beta_5$-integrins.

competes for integrin binding with ICAM1, a major β_2-integrin ligand [186,187].[76] Consequently, different proteins preclude inflammation, as they impede either selectin- or integrin-dependent leukocyte adhesion.

3.7.3.6 Neutrophils in Lungs

Alveolar type-2 cells[77] release granulocyte colony-stimulating factor (CSF3),[78] interleukin-8, and leukotriene-B4 to attract neutrophils [188]. These chemoattractants are synthesized not only in alveolar and airway epithelial cells, but also in endothelial cells and fibroblasts. Interleukin-1β, tumor-necrosis factor-α, bacterial lipopolysaccharide, and bradykinin stimulate the release of these chemoattractants, particularly CSF3.

In addition, alveolar type-2 epithelial cells secrete interleukin-6, chemokines CCL2[79] and CCL5,[80] granulocyte–macrophage colony-stimulating factor (CSF2), transforming growth factor-β, constitutively and in response to TNFα and IL1β [188].

In humans, major chemoattractants for neutrophils during lung inflammation are Glu-Leu-Arg (or ELR) motif-containing (ELR+) CXC chemokines that include CXCL1, CXCL2 and CXCL3, and CXCL8 (IL8). The receptors for these chemoattractants are CXCR1 and CXCR2 in humans. Matrix constituent-derived peptides can also behave as chemoattractants. Certain collagen fragments can provoke accumulation of pulmonary neutrophils. The tripeptide ^{N}acetyl Pro–Gly–Pro (or PGP) that arises from the breakdown of matrix constituents after exposure to lipopolysaccharides in mouse lungs and can be detected in patients with chronic obstructive pulmonary disease fosters chemotaxis of neutrophils and production of superoxide via CXCR1 and CXCR2 receptors [189].[81]

---

[76] In particular, it blocks β_2-integrin-dependent adhesion and homing of endothelial progenitor cells to ischemic tissues, hence reducing angiogenesis [187].

[77] Alveolar type-2 cell synthesizes and secretes surfactant. It controls the volume and composition of the epithelial lining fluid. It proliferates and differentiates into alveolar type-1 cells after lung injury to maintain the integrity of the alveolar wall. It contributes to the immune response in the alveolar space.

[78] Factor CSF3 is also produced in macrophages, lymphocytes, fibroblasts, and endothelial cells.

[79] A.k.a. monocyte chemoattractant protein MCP1.

[80] A.k.a. regulated on activation normal T cells expressed and secreted (RANTES).

[81] Several matrix proteins contain the PGP motif, especially collagen, but also elastin and surfactant protein-A and -D. Collagenases, such as MMP1 and MMP9, expose normally hidden PGP-containing proteic segments, thereby unmasking pro-inflammatory activity of structural constituent-derived peptides. Neutrophils or macrophages release elastases and several MMP species in response to inflammatory stimuli.

3.8 Mastocytes

Mastocytes, or mast cells,[82] are leukocytes that contain storage granules for inflammatory mediators, such as histamine, serotonin, proteolytic enzymes cleaving complement components, heparin, chondroitin sulfate, and chemotactic factors, in particular for eosinophils and neutrophils. Mastocytes contribute to innate and adaptive immunity, angiogenesis, wound healing, tissue remodeling, and immunoregulation [190].

Mastocytes are tissue-resident sentinels. Differences in concentrations of stem cell factor and other cytokines and chemokines (interleukins IL3, IL2, IL4, IL9, and IL18 as well as chemokines CCL2 and CCL5), or nerve growth factor in respective tissues, and possibly between different regions within the same tissue, can determine the number, distribution, and subtype of mastocytes at these loci. Mastocytes recruit many immunocyte types (granulocytes, monocytes, macrophages, dendritic cells, B-, T-, and NK cells; Table 3.11).

Table 3.11. Immunomodulatory mastocytes (Source: [191]; IL: interleukin; PG: prostaglandin: TNF: tumor-necrosis factor; TSLP: thymic stromal lymphopoietin). Mastocytes initiate and increase magnitude and/or duration of immune response (positive activity). They promote the initiation and augment the magnitude of inflammation and tissue remodeling, as well as tissue reaction during allergic and autoimmune disorders. Mastocytes also have anti-inflammatory and immunosuppressive (negative) activity. Mastocytes can first enhance and later limit certain innate and adaptive immune responses.

Positive functions	Negative effects
Maturation, differentiation, and recruitment of immunocytes (via chemokines, histamine, leukotriene-B4, TNF, and peptidases)	Suppression of cytokine production by T cells, monocytes, and keratinocytes (via IL10)
Antigen presentation to T cells Production of IgE by B cells (via IL4, IL13, and TNFSF5)	Enhancement of dendritic cell-mediated decrease of T-cell proliferation and cytokine production (via IL10)
Helper-2 T-cell responses (via PGd2 on maturation of dendritic cells)	
Expression of TSLP by epithelial cells (via TNF, IL4, and IL13)	
Production of chemo- and cytokines by airway smooth myocytes (via TNF, IL4 and IL13)	

[82]Mastocytes were first described with some details by P. Ehrlich in 1878.

3.8.1 Mastocyte Origin and Maturation

Mastocytes derive from hematopoietic stem cells and undergo differentiation and maturation after migration of mastocyte precursors from blood to tissues, especially those exposed to the environment (skin, conjunctiva membrane of eyes, airways, and gastrointestinal and genital tract), or serosal cavities (pericardial, peritoneal, and pleural cavities).[83]

Mastocytes derive from multipotent hematopoietic progenitor cells in the bone marrow and $Thy1^{low}$,[84] $SCFR^{high}$,[85] CD13+,[86] CD33+,[87] CD34+,[88] CD38+,[89] bone marrow-derived circulating progenitors and FcεRI− precursors in blood. These precursors generate mastocytes only in destination tissue (in which they become resident).

Stem cell factor is required to ensure mastocyte survival in the body's tissues.[90] However, the mature mastocyte type (mucosal and connective tissue mastocytes, among other types) varies according to growth factor stimuli (interleukins IL3, IL4, and IL9; transforming growth factor-β1; etc.) [191].

Mastocytes are long-lived cells that proliferate upon suitable stimulation. Mastocyte recruitment and local maturation of mastocyte progenitors contribute to the expansion of mastocyte population. Division and tissue redistribution of mastocytes, as well as mastocyte type modulation particularly occur during helper-2 T-cell responses.

Mastocytes can also derive from thymocytes at double-negative stages DN1 and DN2 (but not at DN3 stage) when transcription factor GATA3 is upregulated, although Gata3 is one of the earliest activated genes when the Notch pathway triggers T-lineage differentiation from hematopoietic precursors [192]. DN2 thymocytes can also bear respecification to the mastocyte lineage in the presence of interleukin-3 and stem cell factor and in the simulataneous absence of Notch signaling.

---

[83]Serosa is a membrane made of a thin layer of epithelial cells lying on a thin connective tissue tunica with capillaries and nerves that excrete serous fluid. In serous cavities, the cells secrete a lubricating fluid that reduces friction during motion.

[84]Thymocyte differentiation plasmalemmal glycoprotein (antigen).

[85]Mast and stem cell growth factor receptor.

[86]Myeloid plasmalemmal glycoprotein and zinc-binding metallopeptidase, a.k.a. alanyl aminopeptidase and aminopeptidase-M (APm) and -N (APn).

[87]Myeloid plasmalemmal antigen also called sialic acid-binding Ig-like lectin-3 (siglec-3).

[88]Hematopoietic progenitor cell antigen, a L-selectin ligand.

[89]$Ecto^{ADP}$ribosyl cyclase-1.

[90]Stem cell factor and interleukin-4 regulate not only the development of mastocytes from myeloid progenitors, but also the function of mature mastocytes. Stem cell factor promotes mastocyte proliferation, prevents mastocyte apoptosis, and enhances IgE-dependent mediator and cytokine release.

Table 3.12. Mastocyte functions.

Activity	Targets Procedure
Recruitment and activation of	Monocytes and macrophages
	Release of lipid mediators – icosanoids (prostaglandins, prostacyclins, thromboxanes, and leukotrienes)
	Dendritic cells
	T lymphocytes
	Basophils, eosinophils, and neutrophils
Phagocytosis	
Alterations of	Epithelial tight junctions
Degradation of	Toxins (by chymase and/or other peptidases)
	Venoms (by carboxypeptidase-A)
Bactericidal activity	Release cathelicidins and chymases
Defense against	Some parasites (innate and IgE-mediated acquired responses)
	Viruses (via CD8+ T cells)
	Fungi

3.8.2 Mastocyte Functions

Mastocytes contribute to both allergy initiation and early phase by releasing pro-inflammatory mediators (histamine, tryptase, prostaglandins, and leukotrienes) and late phase by liberating cytokines and chemokines that recruit inflammatory cells (eosinophils, basophils, neutrophils, CD4+ T cells, and macrophages).

Mastocytes also favors angiogenesis, as they produce growth factors, such as vascular endothelial growth factor, fibroblast growth factor-2, platelet-derived growth factor, transforming growth factor-β, tumor-necrosis factor-α, granulocyte–macrophage colony-stimulating factor, and interleukin-3 and -8.

Mastocytes possess features of acquired immunity cells: they phagocytize, and process and present antigens to T lymphocytes and modulate T- and B-cell responses such as lymphocyte growth, recruitment, and production of immunoglobulins (Table 3.12). Mastocytes express major histocompatibility complex class-1 molecules, and are able to produce MHC class-2 molecules under some circumstances. Mastocytes express receptor FcϵR1 for immunoglobulin-E produced by B cells. Mastocytes operate in acquired immunity against parasites and other pathogens, such as bacteria and viruses in the gut and skin. Certain mastocytes are activated by IgG via Fcγ receptors (Table 3.13).

Table 3.13. Mastocyte activating signals in innate and adaptive immunity (Source: [193]).

Molecule	Receptor
Lipopolysaccharide	Toll-like recptor TLR4
Peptidoglycan	Toll-like recptor TLR2
Complement	CR1, CR2, CR3, C3aR, C5aR
Endothelin-1	ET_A
IgE	FcεR1
IgG1	FcγR3
Parasites	FcεR1 (via IgE)
Bacteria and their products	CD48; Toll-like recptor TLR2, -4, -6, and -9
Double-stranded viral RNA	Toll-like recptor TLR3
Single-stranded viral RNA	Toll-like recptor TLR7
Venoms	G proteins

3.8.3 *Mastocyte Production and Types*

Mastocytes produce many mediators that positively or negatively regulate the fate and function of immunocytes [191]. Mastocytes promote recruitment of immunocytes via secretion of tumor-necrosis factor and histamine.

Histamine elicits helper-1 T-cell activation via H_1 receptor, but suppresses both T_{H1}- and T_{H2}-cell activation via H_2 receptor.

Mastocytes release cytokines that influence polarization and function of T-cell subsets. Interleukins IL4, IL5, IL6, and IL13 and TNFSF5 can affect B-cell development and function (e.g., IgE production). Mastocyte-derived IL10 increases the ability of dendritic cells to prevent T-cell proliferation and cytokine production and directly inhibits the production of prostanoids by neutrophils and inflammatory cytokines by macrophages.

Long-lived mastocytes experience multiple cycles of de- and regranulation. Two types of mastocytes include T and CT phenotypes according to mastocyte content: either tryptase alone (^{T}mastocyte) or chymase and tryptase (^{CT}mastocyte), respectively. Both subtypes coexist. Multiple molecules are released into the extracellular medium (Tables 3.14 to 3.16).

3.8.3.1 Mastocytes of the Genital Tract

In the genital tract, mastocytes are activated by estrogens and lutenizing hormone, as well as corticotropin-releasing hormone, endothelin-1, and substance-P [195]. Mastocytes are the main source of inducible nitric oxide synthase in the uterus.

Table 3.14. Molecules of mastocytes (Source: [194]). (**Part 1**) Granule-stored preformed mediators and de novo synthesized mediators. Mastocytes interact with blood vessels and nerves. The set of expressed cytokines varies according to the maturity state of mastocyte and tissue of residence. Histamine, eicosanoids, and other effectors are released upon crosslinking of surface-bound immunoglobulin-E by allergens or other mediators, such as T_{H2}-cell cytokines (IL3, IL5, and IL13), fibroblast growth factor-2, transforming growth factor-β, and proteoglycans. Other stimuli that provokes the release of mastocyte mediators comprise C3a to C5a, UV light, substance-P, bacterial products, α-melanocyte-stimulating hormone released by keratinocytes, IL8, etc. Stem cell factor in the skin, produced by keratinocytes, endothelial cells, fibroblasts, melanocytes, and mastocytes, stimulates melanocytes and mastocytes (autocrine regulation) via SCFR receptor. Lipid mediators (cyclooxygenase [prostaglandins] and lipoxygenase [leukotrienes] metabolites of arachidonic acid) improve innate responses by recruiting pro-inflammatory cells. Mastocytes are sources of coagulation-regulating factors. Mastocytes store several peptidases in their secretory granules (tryptase, chymase, carboxypeptidase-A, and cathepsin-G). Chymase can activate collagenase and stromelysin, destruct vitronectin and fibronectin, and elicit fibroblast proliferation.

Agent	Effects
	Granule-stored preformed mediators
Histamine	Allergy, inflammation
	Vasodilation
	Increase in blood flow and vessel wall permeability
	Upregulation of endothelium adhesion molecules
	Upregulation of keratinocyte expression of integrins
	Fibroblast proliferation
Serotonin	Vasoconstriction during injury
	Endothelial nitric oxide synthase activation
Proteoglycans (mainly heparin)	Clotting control
Peptidases	Proteolysis
	Activation of matrix metallopeptidases (MMP2 and MMP9 by chymase)
	De novo synthesized mediators
Prostaglandin-D2	Nervous regulation
Leukotriene-B4	T-cell recruitment
Leukotriene-C4	Allergen-mediated release (no bacterial priming)
Leukotriene-B4/C4/D4	Keratinocyte growth
Platelet-activating factor	Moncyte/macrophage recruitment and activation

3.8.3.2 Skin Mastocytes

Mastocytes in the skin exhibit a distribution gradient. The smaller the distance to epidermis, the denser the population. Skin mastocytes at the vanguard of the body's defense display multiple receptors and secrete numerous products (Table 3.17).

Table 3.15. Molecules of mastocytes (Source: [194]). (**Part 2**) Growth factors (HBEGF: heparin-binding EGF-like growth factor).

Stem cell factor	Melanocyte growth
Nerve growth factor	Modulation of mastocyte activity, hair growth, neuron growth and differentiation
Fibroblast growth factor-2/5/7/10	Tissue remodeling
Transforming growth factor-β	Wound healing
Vascular endothelium growth factor	Blood vessel permeability, angiogenesis
Platelet-derived growth factor	Fibroblast growth, angiogenesis
Epidermal growth factor	Keratinocyte growth and differentiation
HBEGF	Keratinocyte growth and differentiation
Granulocyte–macrophage colony-stimulating factor (CSF2)	Langerhans cell migration and maturation, angiogenesis

Table 3.16. Molecules of mastocytes (Source: [194]). (**Part 3**) Cytokines and chemokines (MCP: monocyte chemoattractant protein, MCP1 to MCP3 being CCL2, CCL8, and CCL7 chemokines, respectively; MIP: macrophage inflammatory protein, MIP1α to MIP1δ being CCL3, CCL4, CCL9, and CCL15, respectively).

TNFα	Expression of endothelial leukocyte adhesion molecule-1
	Promotion of IL8 synthesis
	Neutrophil recruitment
	T-cell recruitment
	Lymph node hypertrophy (adaptive immune response)
	Langerhans cell migration
Interleukin-1, IL3–IL6, IL8, IL10, IL11, IL13, and IL16	
IL3	Keratinocyte growth and differentiation
IL4	Langerhans cell migration and maturation
IL8	Granulocyte recruitment and activation
	Melanocyte growth
IL10	Tolerance induction
Chemokines	
CCL2 (MCP1)	Chemotaxis
CCL3 (MIPα)	Chemotaxis
Miscellaneous	
Neuropeptides	Communication with the nervous system
Endothelin-1	Vasoconstriction

Under normal conditions, skin mastocytes are preferentially located close to nerves, especially sensory nerve endings,[91] blood vessels, and hair follicles. Hair follicle cycling during periods of rest and growth, which is associated with

[91] Axon plexi can surround superficial dermal mastocytes.

Table 3.17. Skin mastocyte receptors and products (Source: [196]; FcεR: IgE Fc receptor; gmCSF: granulocyte–monocyte colony-stimulating factor; IL: interleukin; LT: leukotriene; PG: prostaglandin; TNF: tumor-necrosis factor).

Products	Receptors
Histamine	FcεR1/2
Heparin	H_2, H_4
Tryptase, chymase	PAR_2
Carboxypeptidase-A	$C_{5a}R$
Cathepsin-G	
IL1β, IL5, IL6, IL8, IL13	
TNFα	
gmCSF	
Proopiomelanocortin	
Substance-P	
PGd2, LTc4	
Endothelin-1	
Neurotrophins	

Table 3.18. Interactions between mastocytes and the nervous system (Sources: [197]). Mastocytes are associated with nerves in the skin and blood vessels in the brain, especially in meninges. Mastocytes abound around the pituitary gland, where they can control the hypothalamo–pituitary–adrenal axis. Mastocyte regulators include not only pathogens (mainly parasites and bacteria), adhesion molecules, extracellular matrix proteins, nitric oxide, and cytokines, but also neurotransmitters and neurotrophins. Substance-P released from nerve endings promotes mastocyte-dependent granulocyte infiltration. Hypothalamic mastocytes localize close to nerve endings that contain corticotropin-releasing hormone. This hormone is released not only by hypothalamus, but also dorsal root ganglia. It not only activates skin mastocytes but also acts on endothelial cells. Mastocyte proliferation can be triggered by nerve growth factor released from leukocytes and nerve endings. Neuropeptides released from nerves can induce liberation of chymase and production of E-selectin in adjacent microvascular endothelia similarly to substance-P release from axons with subsequent priming of cytokine-mediated interaction between mast and endothelial cells. However, calcitonin gene-related peptide reduces antigen-presenting capability of mastocytes by causing degranulation and release of IL10 and TNFα. Endothelin-1 is a major activator that causes mastocyte degranulation.

Mastocyte	Nerve
Histamine	Substance-P
Serotonin	Endothelin-1
Peptidases	Urocortin
TNFα	Neurotensin
IL1/6	Nerve growth factor
	Corticotropin-releasing factor

proteolysis, angiogenesis, and nerve rearrangement, is impaired in the absence of mastocytes. Mastocytes actually participate in tissue repair and remodeling. Mastocytes are associated with sensory nerves. Mastocyte clusters can be innervated directly (Table 3.18).

3.8.3.3 Arterial Wall Mastocytes

In normal arterial intima, only a few mastocytes are present (average density ~ 1 cell/mm^2, i.e., much smaller density than that in skin ~ 50 cell/mm^2) [198]. However, in atheroma, mastocytes, together with macrophages and T lymphocytes invade the intima.

3.8.3.4 Cardiac Mastocytes

Maturation of resident cardiac mastocytes is associated with formation of chymase and histamine as well as the loss of division capacity. Four stages of mastocyte maturation can be defined [200]: stage 1 in which immature cells have the smallest size, 2 in which still immature cells are able to divide and possess granules, 3 in which non-dividing cells are endowed with abundant granules, and 4 characterized by fully differentiated, mature cells.

Cardiac mastocytes that synthesize, store, and release active renin trigger a (non-kidney-derived) cardiac renin–angiotensin system. Local generation of cardiac angiotensin-2 from mastocyte-derived renin causes noradrenaline secretion from isolated sympathetic nerve terminals via AT_1 receptors [199].

Cardiac mastocytes synthesize various growth factors, cytokines, chemokines, nitric oxide and other vasoactive molecules, peptidases, fatty acid metabolites, and other mediators involved in tissue remodeling. Mastocytes store and release several agents that can activate matrix metallopeptidases that degrade the collagen matrix [200].

Mastocyte peptidases, such as chymase and tryptase, can activate zinc-dependent interstitial collagenase-1 (MMP1), gelatinases (MMP2), and stromelysin (MMP3). Matrix metallopeptidases are produced and released as inactive zymogens; their propeptide domain is removed to obtain an active enzyme. Chymase activates proMMP1; tryptase cleaves proMMP3. Peptidase MMP3 then activates other proMMPs, in particular, activating MMP1 enzyme. In addition, mastocytes can synthesize and secrete MMP1 peptidase. Moreover, chymase and tryptase degrade fibronectin and vitronectin. Chymase is also capable of cleaving inactive angiotensin-1 to active angiotensin-2. In addition, mastocytes produce TNFα that activate MMP9 synthesized from themselves and adjacent macrophages.

Identified secretagogues encompass endothelin-1, reactive oxygen species, interleukin-33, complement factor-5A, and several neuropeptides [200]. Endothelin-1 operates via its ET_A receptor and activates cardiac mastocytes. Reactive oxygen species (oxidative stress) can induce mastocyte degranulation. Interleukin-33, a member of the IL1 family of cytokines, regulates mastocyte function.

The complement molecule, C5a, a proteic fragment released from complement component C5, is a chemoattractant for mastocytes, an anaphylatoxin that causes histamine release from mastocytes, and a potent inflammatory mediator. It binds to a Gi-coupled receptor (C5aR or CD88) of target cells such as granulocytes.

Among neuropeptides, substance-P, which is released predominately from afferent fibers of the sympathetic nervous system, provokes degranulation of cardiac mastocytes. Neurotensin, which is produced in nerve fibers of the intracardiac ganglia and coronary vasculature, as well as those that innervate cardiomyocytes and nodal cells, also causes a calcium-dependent degranulation of cardiac mastocytes, in addition to its positive inotropic effects mediated by noradrenaline release.

Cardiac mastocyte density rises in the left ventricle maladaptive hypertrophy (e.g., upon mitral regurgitation [mitral insufficiency or incompetence]) that provokes congestive heart failure. The myocardial MMP2 activity elevates with subsequent collagen degradation [200].

Cardiac mastocytes are also involved in fibrosis developed in maladaptive hypertrophy caused by arterial hypertension. Mastocyte density is correlated to stem cell factor and SCFR receptor levels. Mastocytes can prevent macrophage recruitment [200]. Mastocytes regulate fibrosis that results from pressure overload via inflammatory cell recruitment and direct action on fibroblasts.

Mastocytes participate with monocytes, macrophages, and lymphocytes in inflammation during atherosclerosis. They can provoke plaque destabilization with increased risk of rupture, in addition to their role in the development of aortic abdominal aneurysms.

Cardiac mastocyte density also augments during myocardial infarction, in particular during the healing phase [200]. Recruited macrophages and neutrophils gather near degranulated mastocytes. Mastocytes release renin and chymase that form myocardial angiotensin-2. The latter elevates noradrenaline levels, hence risk of arrhythmias. Renin and noradrenaline release indeed prime ventricular fibrillation during ischemia–reperfusion events. Mastocytes also contribute to the regulation of myofibroblast activity.

3.8.3.5 Airway Wall Mastocytes

Human lung mastocytes that reside predominantly in the submucosal connective tissue of normal human bronchi synthesize renin that triggers angiotensin formation after degranulation in airways [201]. Local generation of angiotensin-2 from mastocyte renin primes contraction of bronchial smooth myocytes via AT_1 receptors.[92] The airway renin–angiotensin system thus complements the traditional, circulating renin–angiotensin axis, in which renin is released into the blood circulation from the kidneys in response to decreased renal perfusion pressure, reduced delivery of NaCl at the macula densa, and/or increased renal sympathetic nerve activity. The pulmonary endothelium is the primary source of angiotensin-converting enzyme in the body.

In allergic asthma, mastocytes release renin, histamine, β-hexosaminidase, tryptase, leukotrienes LTc4 and LTd4, adenosine, cytokines, and kinins that elicit acute and chronic airway constriction and inflammation.

[92]Receptors AT_1 (or AT1R) and AT_2 (or AT2R) reside primarily on bronchial smooth myocytes and bronchial epithelial cell brush border, respectively.

3.9 Lymphocytes

Lymphocytes (Lφ; size 8–15 μm) have a very large nucleus. Lymphocyte is much smaller than the 3 types of granulocytes. They account for 25–35% of leukocytes. They regulate adaptive immunity. The major production sites of T and B cells are the thymus and bone marrow, respectively. Both B and T lymphocytes reside in lymph nodes and in the spleen, where they mature into fully functional cells.

Cell history can lead to tissue-specific cells according to the residential type of lymphoid organs, but this property can be reverted. Effector cell behavior is influenced by cytokines involved in helper T_{H1} and T_{H2} responses. T-cell exit from lymphoid and peripheral tissues requires sphingosine 1-phosphate and chemokine CC-motif receptor CCR7, respectively.

B lymphocytes as well as some types of T lymphocytes, such as CD4+, CD8+, α/β, γ/δ, and NK (natural killer) T cells, constitute distinct lineages that do not interconvert. On the other hand, other T-cell lineages can remodel in other T-cell types, such as helper and inducible T_{Reg} lymphocytes.

Both CD4 and CD8 that bind to major histocompatibility complex molecules lodge on immature, double-positive thymocytes. Double-positive thymocytes with a T-cell receptor that interacts weakly with self-peptides bound to MHC molecules undergo a *positive selection* to become single-positive CD4+ or CD8+ T lymphocytes. Double-positive thymocytes with a TCR that links strongly to self-peptide bound to MHC molecules undergo a *negative selection* and apoptosis, thereby preventing the development of self-reactive T lymphocytes. Positive selection relies on a moderate, sustained Ca^{2+} import and ERK activation [202]. On the other hand, high-affinity TCR ligands — negative selectors — cause high magnitude, short duration Ca^{2+} import and ERK activation. Increase in cytosolic Ca^{2+} concentration induced by TCRs that activates T cells results from the stimulation of the PLCγ1–IP_3 axis.[93] In response to weak TCR ligands, Na_V1.5 bound to the regulatory Na_Vβ4 subunit favors a sustained calcium entry [202].[94]

3.9.1 Selection between CD4 and CD8 Coreceptors

Coreceptors CD4 and CD8 intervene in the recognition of major histocompatibility complex class-2 and -1 ligands, respectively, as well as the recruitment of the leukocyte-specific non-receptor Tyr kinase LCK to the TCR complex. Coreceptors CD4 and CD8 have different capacity to recruit LCK to the TCR, thus directing lineage decision [203]. In addition, differentiation into the CD8+ lineage requires

[93]In peripheral T cells, store-operated Ca^{2+} entry through plasmalemmal Orai proteins is the main route for Ca^{2+} entry.

[94]Channel Na_V1.5 is not expressed in mature CD4+ T cells.

the TCR-induced increased density of the 70-kDa TCR ζ-chain-associated protein (ZAP70). In the thymus, CD4+ and CD8+ cells arise from distinct subsets of double-positive thymocytes that possess different kinetics and sensitivities to stimulation of T-cell receptors. ZAP70 kinase determines the cell sensitivity to TCR stimulation, thereby enabling the selection of CD4+ and CD8+ fate [204].

3.9.2 Sensing of Endo- and Exogenous Molecules

Immunocyte receptors are able to sense microbial components (danger signals) as well as endogenous molecules generated during injury (damage signals) such as ribonucleoprotein complexes released from dead cells. Self nucleic acids must not be mistaken for microbial products to avoid autoimmunity. Endogenous agonists of Toll-like receptors TLR2 and TLR4 probably trigger signals distinct from those of microbial ligands, hence may not activate adaptive immune responses due to the engagement of different coreceptors. On the other hand, tumor cells produce small-molecular-weight fragments of hyaluronic acid, biglycan, and versican that can stimulate TLR2 and/or TLR4 receptors [205].

T cells can recognize 3 types of substances [206]: (1) ligands of MHC-related and -unrelated γδ T-cell receptors, such as MHC class-1-like CD1c plasmalemmal glycoprotein and phosphoantigens; (2) various types of cell-surface molecules, such as retinoic acid early transcript RAE1 and MHC class-1 polypeptide-related sequence MICA that engage natural killer receptors (NKR; e.g., NK group-2 member-D [NKG2d] that is also called killer cell lectin-like receptor KLRk1); and (3) danger- (DAMP) and pathogen (PAMP)-associated molecular patterns sensed by pattern recognition receptors, such as Toll-like receptors and dectin-1.[95]

Pathogen-specific innate immune recognition activates antigen-specific adaptive immune responses. Pattern recognition receptors (Vol. 3 – Chap. 11. Receptors of the Immune System) of the innate immune system detect relatively invariant molecular patterns found in most microorganisms of a given class, the so-called pathogen-associated molecular patterns.

Microbial pattern recognition receptors can be categorized into [205]: (1) secreted (collectins, ficolins, and pentraxins); (2) transmembrane (Toll-like receptors and C-type lectins); and (3) cytosolic types (retinoic acid-inducible gene RIG-1-like [RLR] and nucleotide-binding domain and leucine-rich repeat-containing [NLR] receptors).

Cytosolic pattern recognition receptors — RLRs — detect viral material such as RNA, recruit adaptor Mitochondria antiviral signaling protein (MAVS), and activate nuclear factor-κB and interferon regulatory factor IRF3 [205].

Intracellular sensors — NLRs — recognize degradation products of peptidoglycans and microbial products, among others.

[95] I.e., dendritic cell-associated C-type lectin; a.k.a. C-type lectin domain family-7 member CLec7a and β-glucan receptor.

Secreted pattern-recognition receptors bind to microbe surfaces, activate classical and lectin pathways of the complement system, and opsonize pathogens for phagocytosis by macrophages and neutrophils.

Transmembrane pattern recognition receptors reside either in the plasma membrane or on endosomes and lysosomes. Plasmalemmal TLRs recognize conserved microbial patterns accessible on the cell surface, such as lipopolysaccharides of Gram− bacteria (TLR4), lipoteichoic acids of Gram+ bacteria, bacterial lipoproteins (TLR1–TLR2 and TLR2–TLR6), and flagellin (TLR5) [205]. Endosomal TLRs detect microbial nucleic acids, such as double-stranded RNA (TLR3), single-stranded RNA (TLR7), and dsDNA (TLR9). C-type lectin domain-containing proteins CLec7a and CLec6a (dectin-1 and -2) sense β-glucans and mannan on fungal walls, respectively [205].

Most, if not all, pattern recognition receptors activate the transcription factors nuclear factor-κB, interferon regulatory factors, or nuclear factor of activated T cells to induce both T- (e.g., T_{H1}, T_{H17}, and CD8+ T-cell responses) and B-cell responses (IgM, IgG, and IgA immunoglobulin–antibody responses) [205]. Secreted and some endocytic pattern recognition receptors, such as scavenger and mannose receptors, cannot induce adaptive immunity by themselves.

Three signals are required for cross-priming that permits dendritic cells to present exogenous antigens to responding antigen-specific CD8+ T cells.[96] Signal 1 (MHC–AG–TCR) constitutes the complex made of peptidic antigen and major histocompatibility complex class-1 molecule that targets the relevant TCR. Signal 2 (costimulator) refers to costimulatory molecules such as B7-family members that ligate T-cell-specific surface glycoprotein CD28. Signal 3 (inflammatory cytokine) corresponds to triggering soluble factors, such as interleukin-12 and interferon-α and-β. However, this process needs help that is provided by CD4+ T cells, plasmacytoid dendritic cells, and natural killer T cells that stimulate chemokine release from cognate dendritic cells (signal 0: chemokine gradient), i.e., upstream from signal 1, to attract CD8+ T cells [207].

3.9.3 Resistance to Oxidizing Environment

Lymphocytes are exposed to oxidizing environments during inflammation. Helper T lymphocytes become progressively less sensitive to an oxidizing environment after differentiation into effector T cells, which affects their ability to proliferate, differentiate, and secrete cytokines.

In human helper T lymphocytes, calcium influx through Orai family member-based channels is involved in the cell activation, proliferation, and differentiation.

---

[96]Cross-priming refers to the stimulation of naive CD8+ cytotoxic T cells by cross-presentation, i.e., when antigen-presenting cells take up, process, and present extracellular antigens with MHC class-1 molecules.

Orai calcium release-activated calcium channel modulator-1, but not Orai3, is inhibited by oxidation by reactive oxygen species [208]. The decreased redox sensitivity of effector T_H cells is correlated with increased amounts of Orai3 and several cytosolic anti-oxidants.

3.9.4 *PIP_3 and IP_4 Signals in Lymphocytes*

Membrane lipid phosphatidylinositol (3,4,5)-trisphosphate (PIP_3) regulates membrane receptor functioning, in particular immunoreceptor signaling. Soluble inositol (1,3,4,5)-tetrakisphosphate (IP_4) participates in the development and activity of leukocytes, such as T and B cells and neutrophils, as it controls phosphoinositide 3-kinases, phosphatase and tensin homolog (PTen), SH2-domain-containing inositol 5-phosphatases SHIP1 and SHIP2 that target PIP_3 in lymphocytes [209]. Agent IP_4 regulates the membrane recruitment of proteins, as it promotes or impedes their interactions with PIP_3 Lipid.

In mouse thymocytes, membrane recruitment of inducible T-cell kinase (ITK) by IP_4 achieves a positive feedback loop on phospholipase-Cγ1 downstream from T-cell receptors to produce sufficient amount of the second messenger diacylglycerol to trigger positive selection and generation of a mature T-cell directory [209]. In humans, inositol (1,4,5)-trisphosphate 3-kinase ITPKc limits T-cell activation level. Its defect contributes to Kawasaki disease in children. Inositol trisphosphate 3-kinase ITPKb is required for B-cell development and functional competence, as it restrains store-operated Ca^{2+} entry as well as PIP_3-mediated membrane recruitment of protein kinase-B in neutrophils [209].

3.9.5 *Immunological Memory*

Lymphocytes provide antigen-specific acquired immunity, hence immunological memory. Immature T- and B-cell activity is restricted mainly to lymphoid tissues. Activated T cells migrate more efficiently than antigen-unexperienced T cells. Phenotype and certain features allow to distinguish long-lived memory B and T cells from short-lived effectors. Antigen-experienced T cells include: (1) central memory cells that circulate through lymphoid tissues and (2) effector memory cells that do not move.

Immunological memory confers long-lasting protection. Naive cells of the immune system that have not been previously exposed to a given antigen actually learn from the first experience of encountering a pathogen to adapt and react quickly to future invasion. Cells of the adaptive immune system undergo a secondary expansion and trigger a robust protective response when they re-encounter the same pathogen. The adaptive immune system is thus characterized by an antigen-specific memory response.

Adaptive immune system associated with antigen-specific B- and T-cell receptors on B and T lymphocytes relies on immunological memory, i.e., immune repertoire of BCRs and TCRs that depends on cell history vis-à-vis antigens.

When B- and T-cell receptors recognize and bind a given antigen, clones of B and T lymphocytes that produce these receptors expand and become effector cells (*expansion phase*) [210]. Pathogen-specific immunocytes become activated and proliferate over 1 wk to generate thousands of daughter effector cells. B lymphocytes secrete antibodies. T lymphocytes kill antigen-containing cells.

Once the immune response is finished, only a small fraction of B and T lymphocytes survives (*contraction phase*). Remaining long-lived memory cells maintain long-term immunity after infection or vaccination. During the third *memory maintenance phase*, stable populations of long-lived memory cells reside in lymphoid and non-lymphoid tissues, where they search for eventual previously encountered pathogens. The fourth and last phase (*recall response*) occurs when memory cells re-encounter their cognate antigen.

Secondary lymphoid organs are loci where most immune responses are initiated. However, many types of memory cells lodge preferentially in the bone marrow, such as CD4+ and CD8+ T lymphocytes and long-lived plasmocytes. In particular, CD4+ memory T cells, which give rise to different functional subsets (effector memory [T_{EM}], central memory [T_{CM}], and follicular helper [T_{FH}]) T cells with given functional capacities and sites of action, assist in the generation of high-affinity memory B cells and long-lived plasmocytes and in the maintenance and secondary expansion of memory CD8+ T cells [212]. CD4+ memory T cells promote the production of high-affinity antibodies and the generation of bone marrow, long-lived plasmocytes in the late phase of immune responses. Bone marrow-resident CD4+ T lymphocytes (memory helper T lymphocytes) have a particular plasmalemmal molecule expression pattern characterized by bone marrow-homing receptors α_2-integrin and C-type lectin CLec2c [212].[97] Early activation marker CD69 in CD4+ memory T cells enables their accumulation in the bone marrow in dedicated memory niches, where they link to laminin+ stromal and epithelial cells and can receive necessary survival signals [213].

Like cytotoxic T lymphocytes of the adaptive immunity, natural killer cells of the innate immunity (Sect. 3.11.5) possess an immunological memory against cytomegalovirus owing to killer cell lectin-like receptors KLRa8[98] and KLRb1c [211].[99] Natural killer cells that proliferate 1,000-fold in spleen and liver after the initial infection remain in the immune system for months and confer viral resistance. After the contraction phase, a subset of virus-experienced natural killer cells survives and maintains the capacity of becoming effector cells.

---

[97] A.k.a. type-2 membrane, homodimeric activation marker CD69. Trafficking molecule CD69 connects to $S1P_1$ receptor and inhibits S1P–$S1P_1$-mediated cell egress from the bone marrow [212]. CD69+ T cells are able to enter the bone marrow. CD69+, CD4+ memory helper T cells of the bone marrow contact stromal cells, but not CD69−, CD4+ cells.

[98] A.k.a. lymphocyte surface glycoprotein Ly49h.

[99] A.k.a. natural killer cell-surface protein NK1.1.

A subset of hepatic CXCR6+ NK cells acquires antigen-specific memory to haptens (small molecules that react with protein and can then elicit an immune response) and viruses [214]. NK cells from the liver, but not spleen, sensitized to a given hapten retain memory for this hapten and can mediate hapten-induced contact hypersensitivity response. Hepatic NK cells also develop specific memory of vaccines that contain antigens from influenza, vesicular stomatitis, or type-1 human immunodeficiency virus. The persistence of memory NK cells, but not antigen recognition, depends on CXCR6 chemokine receptor, instead of pattern recognition receptors.[100]

3.10 T Lymphocytes

The human hematopoietic system gives rise to all blood cell types, among which thymic (T) lymphocytes, ot T cells. An exclusion step permits T-cell specification from lymphoid-committed progenitor cells that themselves result from blockade of erythroid and myeloid lineage at an early developmental stage, as it supresses alternative development toward B and NK (natural killer) cells.

Naive T lymphocytes differentiate into mature effector cells that form distinct T-cell populations and produce a given set of cytokines. The synthesized cytokine set does not necessarily define the T-cell lineage because the T cell program of cytokine production can be modified according to circumstances.

Each T lymphocyte is unique, as it is shaped by the environmental factors encountered during its life. However, T lymphocytes can be grouped into various subsets based on their effector function and molecular phenotype. Distinct T-lymphocyte subsets and differentiation states that can be identified based on expressed plasmalemmal markers and produced effector molecules promote different types of immune response. Moreover, T lymphocytes can interconvert from one subset phenotype to another, although signaling can cause a long-term fixation of cytokine memory. The phenotype of effector T lymphocytes can actually be remodeled.

Conventional αβ T lymphocytes recognize antigens presented in the groove of classical major histocompatibility molecules via their T-cell receptors. In addition, classical MHC class-1a and -1-like molecules interact with many other receptors to control T-cell activation.

3.10.1 T-Lymphocyte Development

T lymphocytes evolve in the thymus through several developmental stages. Early thymus-settling progenitors indeed mature through double-negative DN1, DN2, and

[100]Whereas 35 to 55% of hepatic NK cells express CXCR6, only 3 to 5% of splenic NK cells produce this chemokine receptor.

Table 3.19. T-cell specification from thymus-settling progenitors (TSP; Source: [215]; ↑: increase; BCL: B-cell lymphoma (leukemia) protein). T-cell progenitors proceed through double-negative CD4−, CD8− stages DN1 to DN4. In thymic niches, Notch-1 activated by Delta-like ligand DLL4 synthesized by apposed thymic epithelial cells is necessary, but not sufficient for the T-lineage differentiation program, as alone it does not prevent potential to become a myeloid cell type. Double-negative-2 cells (DN2) proliferate in response to interleukin-7 also produced by thymic epithelial cells. Once thymocytes leave the niche, loss of IL7 causes an increase in BCL11b expression that suppresses alternate lineage development, extinguishes stem cell-like gene expression, and specifies T-cell fate.

Site	Cell type	Features
Thymus	TSP	Migration from bone marrow to thymus Proliferation as thymocytes
	DN1	Stem cell-like
	DN2	Multipotent; generation of dendritic and natural killer cells, macrophages, and T cells
Thymic niche	DN2	Interaction with thymic epithelial cells that produce IL7 and DLL4 (DLL4–Notch1 juxtacrine interaction)
Thymus	DN2	BCL11b expression ↑
	DN3	T-cell specification

DN3 cells in the thymus (Table 3.19). Differentiation of conventional αβ T cells from thymic precursors involves a succession of developmental checkpoints that allow efficient generation of lymphocytes expressing non-autoreactive, self-MHC-restricted TCRs.

The first β checkpoint allows CD4−, CD8− double-negative thymic precursors that have produced functional TCR β chains to proceed along αβ differentiation and mature into CD4+, CD8+ double-positive thymocytes (DP). These cells then undergo selection owing to interaction between αβ TCRs and self-MHC–peptide complexes expressed by epithelial or hematopoietic thymic cells. Thymic TCR selection permits production of non-harmful, indispensable T cells that can interact with antigens encountered in the periphery.

The transcription factor B-cell lymphoma BCL11b, or zinc finger protein ZNF856b, belongs to the earliest regulators for the commitment to the T-cell lineage [216–218]. Loss of BCL11b generates natural killer-like cells [216].

3.10.1.1 *αβ* and *γδ* T cells

αβ and γδ T cells originate from common thymic precursors. Whereas β selection engages DN precursors toward the αβ lineage, expression of a functional γδ TCR initiates their differentiation into mature γδ T cells. These selections are analogous processes, as both induce thymocyte proliferation and similar up- or downregulation of several developmental markers and transcription factors as well as rely on epigenetic events.

3.10.1.2 Galectins

T-lymphocyte differentiation and maturation are governed by received cues as well as the tissue site (thymus, gut, lung, skin, etc.). Maturation, activation, and migration of T lymphocytes are influenced by galectins, members of the lectin family. The activity of galectins depends on glycans and glycoproteic receptors on the T-cell surface, different galectin types targeting different species of these molecules [219].[101]

3.10.2 Antigen Presentation

The first step of T-cell activation is engagement of a T-cell receptor with a peptide–MHC component complex on the surface of an antigen-presenting cell.[102] Adaptive immunity relies on discrimination between self and non-self antigens. Secondary lymphoid organs include the spleen, lymph nodes, tonsils in the pharynx, and Peyer's patches in the intestine. The spleen does not receive lymphatic flow, but filters blood [220]. Cells of Peyer's patches and tonsils sample antigens from the mucosa. Lymph nodes process antigen from interstitial fluid that flows through tissue-draining lymphatics.

In secondary lymphoid organs that are sites of immune surveillance organized into specialized stromal niches, antigenic contents of blood, lymph, and mucosa brought by migrating and resident antigen-presenting cells are checked by leukocytes of the adaptive immune system. Migratory antigen-presenting cells that carry antigens are directed to B- and T-cell zones when they enter into secondary lymphoid organs. Antigens that are freely delivered by lymph are processed and presented by resident antigen-presenting cells [220].

---

[101] Both galectin-1 and -9 trigger death of different types of thymocytes and T cells using different pathways.

[102] In general, MHC class-1 molecules connect to endogenous peptides derived from cytosolic proteins and interact with TCRs of CD8+ T lymphocytes, whereas MHC class-2 molecules tether to exogenous antigens and bind TCRs of CD4+ T lymphocytes [220]. However, specialized subsets of dendritic cells can present peptides derived from exogenous proteins associated with MHC class-1 molecules. In addition, endogenous antigens can be loaded on MHC class-2 molecules for autophagy.

3.10.2.1 Antigen-Presenting Cells

T lymphocytes encounter their cognate antigens in specialized compartments of secondary lymphoid organs, where antigen-presenting cells, such as dendritic cells, B lymphocytes, basophils, and stromal cells present self and non-self antigens to T cells. Dendritic cell subpopulations can be characterized by distinct capacity for antigen capture, processing, and presentation.

In the thymus, epithelial cells that are specialized for immunotolerance produce a large number of peripheral tissue-restricted antigens, mainly under the control of the autoimmune transcriptional regulator (AIRe).

Lymph node stromal cells present tissue-restricted antigens to naive T cells that become self-reactive T lymphocytes for immune tolerance. Fibroblastic reticular cells, lymphatic and blood vessel endothelial cells as well as double-negative (podoplanin− and PECAM1−) stromal cells and extrathymic AIRe+ cells possess self-antigens.[103] Fibroblastic reticular cells and lymphatic and blood endothelial cells synthesize the transcriptional regulator Deformed epidermal autoregulatory factor DEAF1,[104] as the transcriptional regulator AIRe does not control the expression of peripheral tissue-restricted antigens by these cells [220]. Hematopoietic antigen-presenting cells either permanently reside in secondary lymphoid organs, such as dendritic and stromal cells, or migrate from blood (B lymphocytes and basophils) or lymph (dendritic cells) [220].

Lymph node-resident dendritic cells acquire lymph-borne antigens either directly from subcapsular and medullary sinuses or indirectly via migrating dendritic cells in perfusing and draining lymphatics [220]. Lymph nodes filter the lymph in search of microorganisms, cellular and tissular debris, and apoptotic cells. Once in lymph nodes, lymph is controlled by macrophages that line the subcapsular and medullary sinuses and phagocytose most lymph-borne antigens. Most antigens are degraded by lysosomal proteolysis, but a small amount of antigen can be transferred to B- and follicular dendritic cells. B-cell follicles localize to the outer cortex of lymph nodes. Follicular dendritic cells organize B-cell follicles. T lymphocytes lodge mainly in the the inner cortex (or paracortex).

The spleen is responsible for the immunosurveillance of blood. The spleen is composed of 2 compartments: (1) the white pulp that contains lymphocytes and (2) red pulp, where blood is filtered through venous sinuses past cords of phagocytes (Vol. 5 – Chap. 5. Lymphatic System). The *marginal zone* that separates the white and red pulp is an antigen-capture locus, where antigens in the marginal

---

[103] Fibroblastic reticular cells lodge in T-cell zones in the paracortex of lymph nodes. They produce various extracellular matrix proteins that form sieve-like fibers. They ensheath these fibers to form a conduit network [220]. They synthesize chemokines CCL19 and CCL21 that recruit chemokine receptor CCR7+ T- and dendritic cells. In addition, they secrete interleukin-7 to foster T-cell survival. Resident dendritic cells extend protrusions into the fibroblastic reticular cell-lined conduits to capture soluble antigens delivered by the lymph.

[104] A.k.a. suppressin and zinc finger MYND domain-containing protein-5 (ZMYND5).

sinuses encounter marginal zone and metallophilic macrophages as well as dendritic cells [220]. Marginal zone dendritic cells then migrate to the T-cell region — the *peri-arteriolar lymphatic sheath* — of the white pulp as well as a B-cell follicle encapsulated by the peri-arteriolar lymphatic sheath. T lymphocytes enter the spleen from blood at the marginal zone, then migrate in channels formed by fibroblastic reticular cells to reach the peri-arteriolar lymphatic sheath. Spleen-resident dendritic cells acquire blood-borne antigens either directly from marginal sinuses or indirectly from macrophages of the marginal sinuses. The marginal zone also contains a population of non-circulating B lymphocytes that transport antigens from marginal sinuses to follicular dendritic cells in the B-cell follicle. Activated lymphocytes leave the spleen and enter the blood stream.

Human dendritic cells in blood and spleen can be classified into 3 known subpopulations [220]: (1) $CD11c^{high}$,[105] HLADR+,[106] $IL3R\alpha^{neg/low}$ (CD123) cells that are blood dendritic cell antigen BDCA2+ (CD1c+) or BDCA3+ (CD141+);[107] (2) $\alpha_X Itg^{neg/low}$, HLADR+, $IL3R\alpha^{high}$, BDCA2+ cells; and (3) BDCA1+ dendritic cells.

Antigen-presenting cells foster tolerance or immunity according to the availability of danger signals. Antigen presentation by dendritic cells indeed provokes naive T lymphocytes to differentiate into effector and memory T lymphocytes or leads to different forms of tolerance according to the functional status of these dendritic cells [220].

Stromal cells of these secondary lymphoid organs coordinate the interaction between T lymphocytes and dendritic cells, as they recruit dendritic cells to T-lymphocyte zones [220]. Resident stromal cells indeed guide migrating dendritic cells within secondary lymphoid organs.

3.10.2.2 Immunological Synapse

T lymphocytes operate in immune responses either via secreted soluble mediators or via cell contact. T lymphocytes release soluble agents for intercellular communication and target killing. The adaptive immune response requires the formation of immunological synapses between mature antigen-presenting cells such as dendritic cells and CD4+ T cells in the lymph node.[108]

---

[105] I.e., α_X-integrin.

[106] The HLADR molecules are aimed at presenting peptidic antigens to the immune system to trigger or suppress helper T-cell responses. These cell-surface receptor are HLADRα–HLADRβ heterodimer. The HLADR molecules are encoded by several loci; each locusis composed of several genes.

[107] Expression of MHC (HLA) class-2 molecules (DR subunits) allows identification of an effector cell subset (DR+).

[108] Activated T cells leave the lymph node and return to the blood circulation, whereas dendritic cells die in the lymph node.

The *immunological synapse* is constituted by molecular complexes between plasmalemmal and cytoplasmic substances on both sides, i.e., in the regions of T and antigen-presenting cells. Between-cell communication is achieved first via T-cell receptor that mediates recognition of complexes between foreign peptides and major histocompatibility complex exhibited by antigen-presenting cells that primes the transformation from quiescent to active T cells. Afterward, the microtubule-organizing center polarizes to a position beneath the immunological synapse for efficient protein secretion and a burst of actin polymerization happens at the contact site with subsequent activation of integrins for cell adhesion [221].

The immunological synapse is a cellular junction composed of a central cluster of T-cell receptors and a peripheral coat that contains $\alpha_L\beta_2$-integrins. In addition, T-cell receptors recruit their effectors, such as kinases, adaptors, and guanine nucleotide-exchange factors. Cytokines and cytolytic molecules are released into the immunological synapse between T lymphocytes and antigen-presenting cells that occurs within minutes after activation of T-cell receptors (Table 3.20).

3.10.3 T-Lymphocyte Maintenance

Interleukin-7 promotes T-cell survival, as it elevates expression of the anti-apoptotic molecule B-cell leukemia–lymphoma molecule (BCL2). Low-level activation of the TCR signaling participates in the maintenance of T-cell homeostasis for both protective immunity and limitation of autoimmunity.[109] IL7 (IL7R) and T-cell (TCR) receptor signaling are coordinated via cell division cycle CDC42 GTPase [222]. In the absence of CDC42, expression of growth factor-independent transcription repressor GFI1 augments, that of IL7Rα decays, and aberrant activity of ERK1 and ERK2 kinases enhances TCR-mediated T-cell proliferation. Regulation mediated by CDC42 of T-cell activity relies on CDC42 effector P21-activated kinase PAK1.[110]

Therefore, T-cell homeostasis is maintained through a synergystic regulation of IL7 cytokine responsiveness and survival under the control of GFI1 and IL7R on the one hand and strength of ERK-mediated TCR signaling regulated by the CDC42–PAK1 axis on the other [222].

[109] Interactions between MHC molecules and TCR initiate rearrangement of the actin cytoskeleton, formation of immunological synapses, and activation of signaling effectors (ZAP70, LAT, MAPK, etc.). Three CDC42 effectors, PAK1, WASP, and Par6, are involved in immunological synapse formation and/or T-cell activation.

[110] Once bound to CDC42, stimulated P21-activated kinase activates PI3K, JNK, and P38MAPK, inhibits pro-apoptotic BCL2-associated agonist of cell death (BAD), and provoke actin reorganization. Deficiency of CDC42 enhances TCR ligation-mediated T-cell proliferation due to elevated ERK activation, independently of Raf and MAP2K1, in the absence of CDC42 inhibition [222].

Table 3.20. Proteins in immunological synapse function (Source: [221]). The microtubule-organizing center (MTOC) reorients with minus- and plus-ends directed inward and outward, respectively. Formin-like-1 (FmnL1) and diaphanous-1 (Dia1) are recruited to the immunological synapse to participate in MTOC polarization. Formins stabilize microtubules at the plasma membrane and induce formation of actin bundles. Diaphanous-1 associates with IQ motif-containing GTPase-activating protein (IQGAP) that binds to microtubule tips and links them to cortical actin. Agent IQGAP also connects to CDC42 and Rac. Microtubule nanomotor dynein interacts with cytoskeletal adhesion and degranulation-promoting adaptor protein (ADAP). The ADAP–dynein complex associates with lymphocyte cytosolic protein (LCP; or adaptor SLP76) and is recruited to the immunological synapse. After MTOC polarization, cytokines and cytolytic agents are secreted into the immunological synapse within minutes of T-cell receptor stimulation. Perforin and granzyme are carried into lytic granules (secretory lysosomes) attached to microtubules. Between-cell communication is achieved via T-cell receptors (TCR). Unlike cytolytic factors, cytokines (interleukin-2 and -10 and interferon-γ) are newly produced after TCR stimulation and released through the immunological synapse with a time scale of hours. On the other hand, inflammatory tumor-necrosis factor is transported away from the immunological synapse (multidirectional path) and forms transmembrane trimers that are released from the plasma membrane by metallopeptidases. Chemokines, such as CCL3 and CCL5, are also secreted using the multidirectional path rather than synaptic route. Vesicular transport and membrane fusion are mediated by soluble ^{N}ethylmaleimide-sensitive factor accessory protein receptors vSNARE and tSNARE, SNARE-binding protein Unc13d, Adaptor protein AP3, and Rab GTPase. Efficient degranulation requires calcium-regulated proteins such as paxillin that is phosphorylated by ERK kinase to transit into the immunological synapse. Wiskott-Aldrich syndrome protein (WASP) and SNARE proteins SNAP23 and syntaxin-11 participate in cytokine release. Effectors of TCRs, such as kinases ζ-associated protein ZAP70, leukocyte-specific non-receptor Tyr kinase (LCK), and Fyn, adaptors Linker of activated T lymphocytes (LAT) and LCP, and guanine nucleotide-exchange factor Vav are required in MTOC polarization and molecular secretion, hence thymic development of T cells.

Process	Agents
MTOC polarization	ADAP, CDC42, Dia1, dynein, ERK, FmnL1, HDAC6, IQGAP, PYK2, Rac
Lytic granule release	AP3, Munc13-4, paxillin, PKCδ, Rab27a, SLP1/2, syntaxin-11, WIP
Cytokine secretion	Rab3d/19, SNAP23, syntaxin-4/6, WASP
All	LAT, LCP, PLCγ, Src, Vav, ZAP70

3.10.4 T-Lymphocyte Subpopulations

Among the several types of functionally distinct T-cell subsets produced in the thymus, 2 fundamentally different T-cell lineages exist (Table 3.21): (1) conventional $\alpha\beta$ and (2) unconventional $\gamma\delta$ T-cell lineages that are generated from common thymocyte progenitor cells [223]. Stronger TCR-mediated signals promote $\gamma\delta$ T-cell lineage development.[111] In addition to TCR signaling, T-cell specification toward

[111]Inhibitor of DNA-binding protein ID3 activated by strong TCR signals downstream from the extracellular signal-regulated kinase and early growth response axis polarizes toward the $\gamma\delta$ T-cell lineage [223]. Protein ID3 favors the development of $\gamma\delta$ T cells and limits that of $\alpha\beta$ T cells.

Table 3.21. Lineages $\alpha\beta$ and $\gamma\delta$ T cells (Source: [223]; ID3: Inhibitor of DNA-binding protein-3; Sox: SRY-related high-mobility group box transcription factor). Signals from T-cell receptors are strengthened by simultaneous binding of MHC molecules to a specific coreceptor. The coreceptor corresponds to CD4 and CD8 on CD4+ and CD8+ T cells, such as helper and cytotoxic T cells, respectively. Coreceptors CD4 and CD8 tether class-2 and -1 MHC molecules, respectively. Signals are delivered by TNFSF2- and TNFSF3-secreting, $\alpha\beta$, DP thymocytes and received via TNFRSF3 to developing $\gamma\delta$ T-cell progenitors to control the competence of $\gamma\delta$ T cells rather than their commitment.

	$\alpha\beta$ T cells (conventional)	$\gamma\delta$ T cells (non-conventional)
Site	Predominantly thymus and lymph nodes	Predominantly epithelia and mucosa-associated lymphoid tissues
Upregulation	CD4/8	
Downregulation	CD25 (IL2Rα)	CD24
Regulators		Sox13
Receptors	IL7Rα^{low}	IL7Rα^{hi}
TCR signal	Weak	Strong
EGR1 expression	Small	Great
EGR3 expression		Sustained
ID3	Repression	Promotion
Environment	Notch CXCR4–CXCL12	Notch (initial cell stages) TNFRSF3

$\alpha\beta$ and $\gamma\delta$ lineages relies on the thymus microenvironment, e.g., TCR and Notch ligands and lymphotoxin-mediated transconditioning. Extrinsic signals arise from numerous interactions of T-cell progenitors with stromal cells and other thymocytes.

The lineage divergence occurs at the transition from double-negative DN3a (preselection phase) to DN3b (post-selection phase) [223].[112] In adults, the population of DN2 thymocytes can be decomposed into IL7Rα^{high} and IL7Rα^{low} subsets that polarize toward the $\gamma\delta$ and $\alpha\beta$ T-cell lineages, respectively. Selection between $\alpha\beta$ and $\gamma\delta$ fate is associated with the upregulation of T-cell surface glycoprotein CD5 and TNFRSF7 and cell growth. In the $\alpha\beta$ T-cell lineage, expression of IL2Rα is downregulated, whereas that of TCR coreceptors CD4 and CD8 is upregulated to generate double-positive (DP) cells. Double-positive cells undergo TCRα gene rearrangement. The resulting $\alpha\beta$ TCR heterodimer then undergoes MHC-mediated selection to yield mature CD4 or CD8 single-positive (SP) T cells. On the other

[112]The double-negative (DN) thymocyte compartment is subdivided into precursor subsets, such as DN1a to DN1e, DN2, DN3a and DN3b (a.k.a. DN3e and DN3l, respectively). Transition from the DN1 to DN2 stage marks the initiation of gene rearrangement at the TCRβ, TCRγ, and TCRδ gene loci, a process that is completed at the DN3 stage. At the DN3 stage, preT-cell receptors or $\gamma\delta$ TCRs impede apoptosis as well as provoke thymocyte proliferation and further differentiation.

Table 3.22. Markers and transcription factors of T-lymphocyte populations (**Part 1**; Source: [225]; BCL6: B-cell lymphoma protein-6, a.k.a. zinc finger and BTB domain-containing protein ZBTB27 and zinc finger protein ZNF51; BLIMP: B-lymphocyte-induced maturation protein, a.k.a. β-interferon gene positive regulatory domain I-binding factor, positive regulatory domain I-binding factor PRDIBF1, and positive regulatory domain-containing protein-1 with zinc finger domain PRDM1; BMI1: polycomb group RING finger proto-ongogene product, a.k.a. protein CCR: CC-chemokine receptor; MBD: methyl-CpG-binding domain protein [demethylase]; Runx: Runt-related transcription factor; TCR: T-cell receptor; THPOK: T_H-inducing POZ–Krüppel-like factor, a.k.a. zinc finger and BTB domain-containing protein ZBTB7b [or ZBTB15] and zinc finger protein-67 homolog ZFP67).

Subset	Plasmalemmal markers and transcription factors
CD4+ αβ	αβTCR, CD3, IL7R, IL16R (CD4), CCR7, L-selectin^{high}, PTPRc (CD45); THPOK
CD8+ αβ	αβTCR, CD3, CD8, IL7R, CCR7, L-selectin^{high}, PTPRc, Runx3
CD8αα	αβ or γδTCR, CD3, PTPRc, CD8αα; KGF
γδ	γδTCR, CD3
Ifnγ- and IL17a-producing, TNFRSF7- and CCR6-expressing subpopulations	
Central memory	TCR, CD3, IL7R, IL15R, CCR7^{high}, CD44, L-selectin^{high}, BCL6/6b, MBD2, BMI1 (RNF51)
Effector memory	TCR, CD3, IL7R, IL15R, CCR7^{low}, CD44, L-selectin^{low}, BLIMP1 (PRDM1)

hand, most of the γδ T cells remain double-negative cells, but downregulate expression of CD24 during maturation.

γδ T cells give rise to at least 2 functionally distinct subsets: Ifnγ- and IL17-producing γδ T cells. In the thymus, TNFRSF7+ γδ thymocytes express TNFRSF3, TBx21, and Ifnγ, whereas TNFRSF7− γδ thymocytes express RORγ2 (NR1f3-2) and polarize to IL17-producing γδ T cells [223].

3.10.4.1 *αβ* T-Lymphocyte Subsets

Several αβ T lymphocyte subsets exist (Tables 3.22 to 3.27). Inflammatory CD4+ T lymphocytes recruit macrophages and neutrophils to the site of tissue damage. CD8+ cytotoxic T lymphocytes (CTL) kill cells infected by viruses, bacteria, or protozoans, allograft, and tumor cells. They have cytoplasmic granules that contain perforin and granzymes [224]. Regulatory T lymphocytes dampen excessive inflammation. Helper T cells enhance the production of antibodies by B lymphocytes.

Table 3.23. Markers and transcription factors of T-lymphocyte populations (**Part 2**; Source: [225]; AHR: aryl hydrocarbon receptor, or bHLHe76; BATF: basic leucine zipper transcription factor, ATF-like; BCL6: B-cell lymphoma protein-6, a.k.a. zinc finger and BTB domain-containing protein ZBTB27 and zinc finger protein ZNF51; BLIMP: B-lymphocyte-induced maturation protein, a.k.a. β-interferon gene positive regulatory domain I-binding factor, positive regulatory domain I-binding factor PRDIBF1, and positive regulatory domain-containing protein-1 with zinc finger domain PRDM1; BMI1: polycomb group RING finger proto-ongogene product, a.k.a. protein PCGF4 and RING finger protein RNF51. CCL: CC-chemokine ligand; CCR: CC-chemokine receptor; CRTH2: chemoattractant receptor homologous molecule expressed on T_{H2} cells; CXCR: CXC-chemokine receptor; GATA: GATA-binding protein; IκBζ: inhibitor of NF-κB-ζ; ICOS: inducible T-cell costimulator; Ifn: interferon; IL: interleukin; IRF: interferon-regulatory factor; KGF: keratinocyte growth factor [KGF1 and KGF2 are FGF7 and FGF10, respectively]; KLRb1: killer cell lectin-like receptor-B1, or CD161; MAF: musculoaponeurotic fibrosarcoma oncogene; PD: programmed cell death; ROR: retinoic acid receptor-related receptor; SAP: SLAM-associated protein; SLAM: signaling lymphocytic activation molecule; STAT: signal transducer and activator of transcription; TCR: T-cell receptor; T_{FH}: follicular helper T cell; T_H: helper T cell).

Subset	Plasmalemmal markers and transcription factors
T_{H1}	αβTCR, CD3, IL12R, IL16R, IfnγR, CXCR3; TBx21, STAT1/4
T_{H2}	αβTCR, CD3, IL4R, IL16R, IL33R, CCR4, IL17Rβ, CRTH2; DEC2, GATA3, IRF4, MAF, STAT6
T_{H9}	αβTCR, CD3, IL16R; PU.1, IRF4
T_{H17}	αβTCR, CD3, IL1R, IL16R, IL23R, CCR6, KLRb1; AHR (bHLHe76),, BATF, IκBζ, IRF4, RORα, RORγ2, STAT3
T_{H22}	IL16R, CCR4/6/10; AHR
T_{FH}	αβTCR, CD3, IL16R, IL21R, CXCR5, SLAM, ICOS, PD1, TNFSF4/5; SAP; BCL6 (ZBTB27), STAT3

Activated CD4+ helper T cells interact with dendritic cells. Natural killer cells and B and other T lymphocytes coordinate their activities to optimize the immune response. CD8+ T lymphocytes mature into cytotoxic T lymphocytes that secrete cytokines and can kill infected cells.

Naive CD4+ T cells activated by pathogens in a specific cytokine environment at initial engagement of their T-cell and costimulatory receptors differentiate into different subsets that mobilize other cell types to destroy invading pathogens. Naive CD4+ T cells are able to differentiate into various subsets of helper T cells,[113] as well as into FoxP3+ induced T_{Reg} cells (iT_{Reg}) in the presence of TGFβ.

CD4+ helper and CD8+ cytotoxic T cells constitute most of the mature αβ T lymphocytes. However, a small proportion of double-positive CD4+, CD8+ thymocytes is selected for differentiation into mature cells that home to peripheral secondary lymphoid organs. Coordinated activation and/or repression of genes of

[113]Differentiation into T_{H1} and T_{H2} cells depends on Ifnγ and IL4, respectively. Complete differentiation of T_{H17} cells requires 3 steps [226]: (1) induction by IL6 and TGFβ; (2) amplification by IL21; and (3) maintenance by IL23.

Table 3.24. Markers and transcription factors of T-lymphocyte populations (**Part 3**; Source: [225]; BLIMP: B-lymphocyte-induced maturation protein; BTLA: B- and T-lymphocyte attenuator; CBL: Casitas B-lineage lymphoma; CD3(8): cluster of differentiation-3(8), or T-cell coreceptor; CKI: cyclin-dependent kinase inhibitor; CTLA: cytotoxic T-lymphocyte antigen [CTLA4 is also called CD152]; EOMES: eomesodermin; Fox: forkhead box; GITR: glucocorticoid-induced TNF-receptor related protein, or TNFRSF18; GRAIL: gene related to anergy in lymphocytes; J: joining region; LAG: lymphocyte activation gene; MAIT: mucosal-associated invariant T cell; MR: MHC-related protein; NEDD: neuronal precursor cell-expressed developmentally downregulated; NKT: natural killer T cell; PD: programmed cell death; PLZF: promyelocytic leukemia zinc finger protein, a.k.a. zinc finger and BTB domain-containing protein ZBTB16 and zinc finger protein ZNF145; SAP: SLAM-associated protein; SLAM: signaling lymphocytic activation molecule; SLAMF: signaling lymphocytic activation molecule family member; SMAD: small mothers against decapentaplegic homolog; STAT: signal transducer and activator of transcription; TBx: T-box transcription factor; TCR: T-cell receptor; TIM: T-cell immunoglobulin and mucin domain protein; V: variable region).

Subset	Plasmalemmal markers and transcription factors
Cytotoxic	αβTCR, CD3, CD8, EOMES, TBx21, BLIMP1 mainly CD45Ro (PTPRc, $CD45_{lMW}$), some CD45Ra (PTPRc; $CD45_{hMW}$)
T_{R1}	αβTCR, CD3, IL16R
Natural T_{Reg} (nT_{Reg})	αβTCR, CD3, IL16R, CD25, CTLA4, TNFRSF18; FoxO1/O3/P3, STAT5
Inducible T_{Reg} (iT_{Reg})	αβTCR, CD3, IL16R, CD25, CTLA4, TNFRSF18; FoxO1/O3/P3, SMAD2–SMAD4, STAT5
NKT	CD1-restricted TCR; KLRb1c, SLAMF1/6, TβR; PLZF
MAIT	MR1-restricted; Vα7.2, Jα19
Anergic	αβTCR, CD3, BTLA; CKI1b
Exhausted	CD3, CD8, PD1, TIM3, LAG3; SAP; BLIMP1

the differentiation program in selected thymocytes lead to the specific lineages of mature T cells. The development of αβ T cells requires rearrangement of the loci that encode T-cell antigen receptor-α and -β and appropriate signals via interactions between TCRs on double-positive CD4+, CD8+ thymocytes and peptide in the context of major histocompatibility complex (MHC) molecules on thymic stromal cells. Double-positive CD4+, CD8+ thymocytes that successfully rearrange genes encoding T-cell antigen receptors in fact mature into single-positive CD4+ helper or CD8+ cytotoxic T cells whether TCRs are specific for MHC molecules of class-2 or -1, respectively. In addition, TCR signal duration during selection determines lineage type, as signals of short and long duration cause CD8+ and CD4+ differentiation, respectively. Prolonged engagement of MHC class-2-restricted TCRs followed by sustained cellular signaling is needed for the development of the helper lineage. MHC class-1 and -2 molecules lead to signals of short and long duration, as TCR signals in double-positive thymocytes block MHC class-1-dependent signaling, but not MHC class-2-dependent signaling.

Table 3.25. Role of T-lymphocyte populations (**Part 1**; Source: [225]; ↑: upregulation; APC: antigen-presenting cell; CTL: cytotoxic T lymphocyte; NOS2: inducible nitric oxide synthase). T_{H22} cell identity as an independent T_H subset needs to be confirmed.

Subset	Function
CD4+ $\alpha\beta$	Scouting inside lymph nodes for liganded class-2 MHC on APCs Differentiation into effector, regulatory, and memory T cells
CD8+ $\alpha\beta$	Scouting inside lymph nodes for liganded class-1 MHC on APCs Differentiation into CTLs and memory T cells
CD8$\alpha\alpha$	Intra-epithelial lymphocytes Intra- or extrathymic development
$\gamma\delta$	Pro- and anti-inflammatory effects according to $\gamma\delta$TCR and context Enriched at epithelial surfaces
T_{H1}	Target intracellular pathogens Macrophage activation; NOS2 ↑
T_{H2}	Humoral immune response Target extracellular parasites (e.g., helminths) Potentiation of allergic responses
T_{H9}	Target extracellular parasites Promote allergic inflammation
T_{H17}	Target extracellular bacteria and fungi Involved in autoimmune and inflammatory diseases
T_{H22}	Inflammatory skin diseases
T_{FH}	Germinal center responses; B-cell class switching

Table 3.26. Role of T-lymphocyte populations (**Part 2**; Source: [225]; ↑: upregulation; APC: antigen-presenting cell; CTL: cytotoxic T lymphocyte).

Subset	Function
Central memory	In secondary lymphoid organs, recall responses Differentiation into effector T cells
Effector memory	In peripheral tissues, production of cytokines
Cytotoxic	Killing of infected and tumor cells
T_{R1}	Immunosuppression
nT_{Reg}	Immunosuppression Generation in the thymus
iT_{Reg}	Immunosuppression Generation from naive T cells in the periphery
NKT	Pro- and anti-inflammatory effects
Anergic	Generation on TCR activation without costimulatory signals
Exhausted	Chronic antigen-mediated TCR stimulation

Table 3.27. Agents for development and maintenance of T-lymphocyte populations (Source: [225]; Itch: Itchy homolog [ubiquitin ligase]; TGF: transforming growth factor). Interleukin-25 regulates both T_{H2} and T_{H9} cell populations.

Subset	Released effectors Agents for development and maintenance
$\gamma\delta$	Released Ifnγ, IL17a/f, IL22
CD8$\alpha\alpha$	Released IL10, TGFβ
T_{H1}	Released Ifnγ, IL2, TNFSF2
	Developmental regulator IL12
T_{H2}	Released ILs4/5/10/13
	Developmental regulators ILs4/25/33
T_{H9}	Released ILs9/10
	Developmental regulators ILs4/25, TGFβ
T_{H17}	Released IL17a/f, IL21/22/26, CCL20
	Developmental regulators IL6/21, TGFβ
	Maintenance regulators IL1/23
T_{H22}	Released IL22
T_{FH}	Released IL21
Central memory	IL2, TNFSF5, low levels IL4, Ifnγ, IL17a
Cytotoxic	Perforin, granzyme, Ifnγ
T_{R1}	Released IL10
	Developmental regulators TGFβ, IL27
nT_{Reg}	Released IL10/35, TGFβ
iT_{Reg}	Released IL10, TGFβ
NKT	Released IL4/17a, Ifnγ
Anergic	GRAIL, CBLb, Itch, NEDD4

The transcription factor T_H-inducing POZ/Krüppel-like factor THPOK[114] is required after initial lineage specification for the differentiation of CD4+ helper T cells from double-positive CD4+, CD8+ thymocytes. Overexpression of THPOK (ZBTB7b) converts MHC class-1-restricted CD8+ T cells into MHC class-2-restricted CD4+ T cells, whereas THPOK repression redirects CD4+ T cells to the CD8+ lineage [227]. Regulator THPOK represses the CD8+ lineage commitmemt.

Regulator GATA3 acts upstream from THPOK to initiate CD4+ specification, whereas THPOK serves to stabilize the CD4+ fate [228, 229]. Factor GATA3 is necessary for THPOK expression in double-positive thymocytes.

The development of CD8+ cytotoxic T-cell requires Runx1 and Runx3 transcription factors. Both CD4 and THPOK are inhibited in CD8+ cytotoxic T cells. Conversely, THPOK represses the expression of Runx3 and other molecules of

[114]A.k.a. zinc finger and BTB domain-containing protein ZBTB7b, zinc finger protein ZNF857b, and Krüppel-related zinc finger protein KRox.

the CD8+ lineage [229]. Moreover, THPOK controls a positive feedback loop for further THPOK expression by binding to and antagonizing the activity of THPOK and CD4 silencer [227].

3.10.4.2 Unconventional $\gamma\delta$ Effector T Lymphocytes

$\gamma\delta$ T cells represent a small subset of T lymphocytes that possess a plasmalemmal T-cell receptor distinct from those of the majority of conventional T cells that are composed of 2 glycoproteic α and β TCR chains. This group of uncommon T cells lodges at its highest level in the gut mucosa, where it coexist with *intra-epithelial lymphocytes* (IEL). A $\gamma\delta$ T-cell subpopulation resides in the epidermis, the so-called *dendritic epidermal* $\gamma\delta$ *T cells* (DETC).

T lymphocytes expressing $\gamma\delta$ T-cell receptors possess several features that make them more similar to innate than adaptive immune effectors. $\gamma\delta$ T cells use only a tiny fraction of their potentially highly diverse TCR repertoire to recognize a restricted set of antigens. In addition to recognition of a limited set of monomorphic ligands, $\gamma\delta$ T cells share with innate effectors the capacity to rapidly exert their effector functions upon antigen encounter, even naive individuals [230]. They indeed strongly and early on produce pro- or anti-inflammatory cytokines, such as Ifnγ, IL4, or IL17, that can control the early recruitment of innate effectors as well as participate in the functional polarization of conventional $\alpha\beta$ T cells and clearance of eliciting pathogens.

$\gamma\delta$ Effector T cells develop their function under the influence of innate programming acquired during thymocyte maturation (that can be remodeled) and environmental signals. They interact with immune and non-immune cells. They bridge innate and adaptive immunity. They rapidly produce cytokines to regulate pathogen clearance and inflammation [206]. They promote tissue repair and wound healing. They recognize conserved non-peptide antigens. They have a tropism for epithelia and mucosae such as those of the respiratory tract.

$\gamma\delta$ T cells yield distinct cytokine responses in different lymphoid organs. In mice, experienced splenic $\gamma\delta$ T cells that are predominantly $CD122^{high}$ (interleukin-2 receptor β) manufacture Ifn-γ [231]. On the other hand, naive lymph node $\gamma\delta$ T cells that are mainly $CD122^{low}$ make interleukin-17. Therefore, $\gamma\delta$ T cells acquire pro-inflammatory functions without antigen encounter upon sensing of danger signals. In addition, thymic selection determines the effector fate of $\gamma\delta$ T cells rather than constrains their antigen specificities.

$\gamma\delta$ Effector T cells detect target molecules by T-cell (TCR), Toll-like (TLR), and natural killer (NKR) receptors that act separately, synergistically, or additively [206]. Their $\gamma\delta$ TCRs differ from one tissue to another for tissue-specific antigen-recognition repertoires.[115] $\gamma\delta$ T cells carry out diverse functions, but cells of subpopulations have more restricted effector properties, according to expressed TCRs, cell location, and activation status.

---

[115]Similarly to B- and $\alpha\beta$ T-cell receptors, $\gamma\delta$ TCRs undergo somatic rearrangement of V (variable), D (diversity), and J (joining) gene segments.

3.10.4.3 Mucosal-Associated Invariant T Lymphocytes

Subsets of non-conventional $\gamma\delta$ T cells restricted by the non-polymorphic MHC class-1-like molecules CD1d and MHC-related protein MR1 correspond to *invariant NKT cells*, or CD1d-restricted natural killer T cells, and *mucosa-associated invariant T cells* (MAIT), respectively. These lymphocytes are termed non-conventional, innate-like, or transitional T cells, owing to some properties shared with innate immunocytes.

Mucosal-associated invariant T cells constitute a subpopulation of T lymphocytes that preferentially reside in the lamina propria and Peyer's patches of the gut (1–4% of total human blood T-cell population). They share some features of invariant NKT cells, hence their other name, mucosal NKT cells (mNKT). They target highly conserved microbial antigens [232]. They are restricted to the monomorphic major histocompatibility complex class-1b molecule MR1 (MHC-related molecule-1).

Mucosa-associated invariant T cells express a semi-invariant TCR rearrangement composed of an invariant TCRα chain (Vα7.2–Jα33 in humans), hence their third name, MR1-restricted Vα19i T-cells. This invariant α chain associates with a limited number of TCRβ chains (Vβ2 and Vβ13 chains in humans). These T-cell receptors recognize MHC-related molecule MR1. MAIT cells are activated by cells infected with various bacteria and yeast, but not virus [233].

3.10.4.4 Memory T Lymphocytes

Development of an adaptive immune response specific for an invading pathogen involves activation and proliferation of T lymphocytes and subsequent regulated migration to infection sites. Effector T lymphocytes that are attracted to inflammation sites can enter any inflamed non-lymphoid tissue to prime an adequate immune response.

Immunological memory is done either by circulating antibodies or expanded populations of antigen-specific B and T lymphocytes with effector functions, such as recognition of previously encountered pathogens and reactivity toward allergens and other non-infectious antigens.

During an infection, the immune system not only produces a rapid pathogen-specific response to destroy an invading pathogen, but also generates among effector T cells recruited to the infection site long-lived, antigen-specific memory T cells that can persist at peripheral sites for many months.

Memory T lymphocytes in peripheral tissues are more resistant to apoptosis than circulating homologous cells. After meeting a known antigen, circulating antibodies bind the corresponding antigen for elimination by phagocytic cells, whereas certain types of immunocytes present this antigen and secrete cytokines for expansion of antigen-specific memory B and T lymphocytes.

After the resolution of an immune response, antigen-specific memory T cells reside in many body's organs, where they provide a first line of defense against secondary pathogen infection. During a secondary infection, memory T lymphocytes in peripheral tissues (i.e., outside of draining lymphoid nodes) are activated by pro-inflammatory cytokines (IL12, IL18, and type-1 interferons) produced by innate immunocytes. Resident memory CD8+ T cells can also be directly activated by dendritic cells together with CD4+ T cells recruited to the infection site. Antigen-specific memory T cells then interact with antigen-bearing dendritic cells to generate a local, secondary effector T-cell response to limit pathogen replication prior to arrival of secondary effector T cells from lymph nodes [235]. Because these tissular sentinels release cytokines and chemokines, peripheral memory T cells alert immunocytes. They quickly recruit circulating memory T cells (antigen-specific or not) before attracting effector T cells from draining lymph nodes to the inflammation site.

The number of antigen-specific memory T cells decreases at peripheral sites after pathogen removal. Antigen-specific memory T cells are relatively quiescent. Nevertheless, they undergo changes in phenotype, function, and location. Memory T-cell pool specific for a given antigen is maintained by interleukins IL7 and IL15 [235].

Subpopulations of Memory T Lymphocytes

The antigen-specific memory T-cell pool can be decomposed into peripheral and lymphoid tissue-resident populations that do not live in similar environments. In lung alveoli, memory T cells are exposed to near atmospheric oxygen concentrations and surfactant proteins, such as SPa1 and SPd.[116]

Memory T lymphocytes of adaptive immunity ensure antigen-specific protection and rapid recall response against repeated invasions of pathogens. Both viral and bacterial infections stimulate Ifnγ+, CD4+, CD8+ T lymphocytes that can differentiate into long-lived memory T cells. The long-lived CD8+ memory T-cell population is larger than that of CD4+ memory T lymphocytes. Both CD4+ and CD8+ memory T-cell subsets with distinct functional potentials arise from Ifnγ+ T-cell progenitors [234].

Memory T lymphocytes determined after infection can be divided into 2 groups based on expression of the lymphoid homing receptors L-selectin and CCR7 [235]: (1) central memory L-selectin+, CCR7+ T cells (T_{CM}) that preferentially localize in lymphoid tissues and (2) effector memory L-selectin−, CCR7− T cells (T_{EM}) that preferentially reside in peripheral tissues. Antigen-specific memory T lymphocytes gradually convert to central memory T cells that are less able to migrate into

[116]A.k.a. SftPa1 and SftPd surfactant proteins. Antigen-dependent and -independent mechanisms contribute to the maintenance of peripheral memory CD8+ T lymphocytes in lung airways. Circulating memory CD8+ T cells can migrate to airways upon stimulation by antigen located in lung-draining lymph nodes, whereas a small number of systemic memory CD8+ T cells migrate to airways even in the absence of specific antigen.

peripheral tissues. Effector and central memory T lymphocytes have distinct effector functions in response to antigen stimulation. Associated with dendritic cells, they both produce large amounts of interferon-γ and tumor-necrosis factor. However, T_{CM} cells synthesizes more interleukin-2 and much less lytic granzyme-B and perforin than T_{EM} cells. In addition, a small T_{EM} fraction expresses low-affinity IgG FcγR3a receptor.

Vaccines against chronic infections and tumors aim at inducing effective memory T-cell responses, especially memory CD8+ T cells. Target of rapamycin is a major regulator of memory CD8+ T-cell differentiation [236]. Rapamycin increases the number of memory precursors during the expansion phase of the T-cell response. During the contraction phase of the T-cell response that corresponds to a transition from effector T cells to memory T cells, rapamycin accelerates the memory T-cell differentiation program. Kinase TOR acts via the TOR complex-1 to regulate memory T-cell differentiation.

Distribution of Memory T Lymphocytes

Following the clearance of an invading pathogen, memory T lymphocytes arise in both lymphoid and peripheral exposed organs (mostly skin and digestive, respiratory, and genital tract) that are major portals of pathogen entry into the body. Although memory T lymphocytes are widely disseminated throughout the body, the distribution of pathogen-specific memory T cells depends on the type of initially infected tissue. Sites of restricted entry, such as the brain, lung airways, intestine lamina propria, and dermis, are characterized by a greater number of memory T lymphocytes, whereas numbers of memory T cells in sites of common entry, such as lung parenchyma, lymph nodes, spleen, liver, and bone marrow, are often similar [235].

Migration of Memory T Lymphocytes

Migration of circulating memory T lymphocytes into non-lymphoid peripheral tissues for effective surveillance is more restricted than that of common effector T cells. Memory T cell migration into peripheral tissues is controlled by combinations of adhesion molecules and chemokine receptors expressed by T cells and complementary adhesion molecules and chemokines expressed by cells in peripheral sites (Table 3.28). Effector T lymphocyte expression of adhesion molecules and chemokine receptors is determined during initial antigen contact by molecular cues governed by local environment [235]. Antigen-presenting cells such as dendritic cells from lymph nodes together with local non-hematopoietic peripheral cells and local draining lymph node-resident stromal cells are necessary and sufficient to program T cells with a tissue-tropic phenotype.

Table 3.28. Adhesion molecules and chemokine receptors in peripheral memory T-cell migration (Source: [235]; CCL: CC-chemokine ligand; CCR: CC-chemokine receptor; CLA: cutaneous lymphocyte-associated antigen; ICAM1: intercellular adhesion molecule-1; MADCAM1: mucosal vascular addressin cell adhesion molecule 1; PNAd: peripheral lymph node addressin; VCAM1: vascular cell adhesion molecule 1).

Molecule	Receptor	Tissue tropism
CD44	Hyaluronate and fibrinogen	General
CD43	E-selectin	General
Integrin-$\alpha_L\beta_2$	ICAM1	General
P- and E-selectin ligands	P- and E-selectin	General
Integrin-$\alpha_4\beta_1$	VCAM1 and fibronectin	General
Integrin-$\alpha_1\beta_1$	Collagen and laminin	Lung airways
L-selectin	PNAd	Lymph node
CCR7	CCL19 and CCL21	Lymph node
CLA-modified ligands	E-selectin	Skin
CCR4	CCL17	Skin
CCR10	CCL27 and CCL28	Skin
Integrin-$\alpha_4\beta_7$	MADCAM1	Gut
CCR9	CCL25	Small bowel

3.10.4.5 Regulatory T Lymphocytes

CD4+ regulatory T cells (T_{Reg}) are immunosuppressive lymphocytes that permit immunological tolerance. They maintain self tissue tolerance to avoid autoimmunity and allow fetomaternal tolerance. Pregnant women are indeed able to allow the growth during 9 months of fetuses that are antigenically foreign bodies.

Fetal and adult T cells arise from different populations of hematopoietic stem cells. The human fetal T-cell compartment begins to develop at approximately gestational week 10. The majority of fetal T cells at mid-gestation (~ 16–24 gestational wk) have a surface phenotype comparable to that of conventional naive T cells in neonates and adults [237]. In particular, many 18-to-22-wk-old fetal CD4+ T cells of mesenteric lymph nodes many have a phenotype (PTPRc+, CCR7+, TNFRSF6−, IL2Rα−) similar to that of naive adult CD4+ T cells. The fetal T-cell lineage is directed toward immune tolerance. Tolerance in the human fetus is partly mediated by an abundant population of fetal regulatory T cells. The developing human fetus actually possesses a greater percentage (~ 15%) of total peripheral CD4+ T cells than that in healthy infants and adults (~ 5%) [237]. Fetal naive CD4+ T cells are much more responsive than adult naive T cells to allogeneic cells [237]. Unlike adult T cells, fetal T cells have an enhanced proliferation rate after exposure to maternal alloantigens. This T-cell pool gives rise to T_{Reg} cells upon stimulation. This process depends on transforming growth factorβ. Hence, in humans, fetal exposure to foreign antigens, notably maternal alloantigens, can lead to the generation of immune tolerance.

T_{Reg} cells also contribute to the maintenance of a balance with commensal microbial flora. They participate in the regulation of immune responses to pathogens and allergens. Regulatory T cells actually hamper excessive immune reactions to limit chronic inflammation. In addition, they intervene in allergy, microbial infection, tumor immunity, and organ transplantation. They also control metabolic function in adipose tissue. Regulatory T cells represent an actively dividing cell population [238].

The control of particular effector functions of conventional T cells (T_{Conv}) requires distinct programs in T_{Reg} cells. These programs involve the same controlling factors in T_{Reg} lymphocytes and regulated T_{Conv} cells, such as IRF4 in T_{H2} lymphocytes and their T_{Reg} controllers [239]. A given transcriptional program promotes a particular type of effector T cell (T_{Eff}) differentiation, to efficiently restrain the corresponding type of the immune response. In the above-mentioned example, IRF4 expression endows T_{Reg} cells with a T_{H2} suppressing function.

T_{Reg} Cell Markers

Regulatory T cells originate primarily in the thymus. Yet, they can be formed from conventional T cells on exposure to low-dose antigen. T_{Reg} cells are characterized by a stable expression of the FoxP3 transcription factor. FoxP3+ T_{Reg} population density can be maintained by the differentiation of precursor cells, proliferation of terminally differentiated cells, and long-life programming. Factor FoxP3 regulates the bulk of transcriptional program indirectly through a set of target transcriptional regulators.

FoxP3+ cells can be generated in the thymus as an alternative lineage during the selection among CD4+ T-cell lineage. FoxP3+ thymocytes have a TCR repertoire that distinguishes them from T_{Conv} cells. Alternatively, mature CD4+ T cells from peripheral lymphoid organs can be converted to FoxP3+ cells under chronic suboptimal stimulation by agonist peptide or during lymphopenia-driven homeostatic expansion [240]. In addition, naive CD44+ T_{Conv} cells activated in vitro by IL2 and TGFβ express FoxP3 and acquire some features of T_{Reg} cells.

T_{Reg} signature with over- or underexpressed genes distinguishes T_{Reg} from T_{Conv} cells, at least in their resting states in lymphoid organs (Fig.1 from [241]).

Molecules involved in T_{Reg}-cell effector function, chemokine receptors, and transcription factors have expression patterns that define subpopulations. Surface expressions of α_E-integrin (CD103), killer cell lectin-like receptor KLRg1 (or CLec15a), or CXCR3 vary as well as cell functions [240]. These subsets of regulatory T cells (nT_{Reg}, iT_{Reg}, and T_{R1} cells) check the immune system and impede unrestricted expansion of effector T-cell populations.

Modes of Immunosuppressive Action

CD4+, CD25+ T_{Reg} cells suppress the function of CD4+ and CD8+ T_{Eff} cells via a T-cell receptor engagement and cell contact. T-cell receptor-dependent T_{Reg} activity controls T_{Eff}-cell function. Signaling from TCR leads to the formation of the immunological synapse within seconds of T-cell activation. Both T_{Reg}

Table 3.29. T_{Reg} cell mechanisms of immunosuppression (Source: [244]; ↑: increase; ↓: decrease; CTLA: cytotoxic T-lymphocyte antigen; ENTPD: ectonucleoside triphosphate diphosphohydrolase; LAG: lymphocyte-activation gene; LAP: latency-associated peptide; NT5E: ecto-5′-nucleotidase; TGF: transforming growth factor).

Mediator	Effect
	Contact-dependent suppression
CTLA4	Antigen-presenting cell costimulatory function ↓, Interaction with CTLA4 counter-receptor CD80 and CD86 on conventional T cells
ENTPD, NT5E	Hydrolysis of extracellular ATP
Granzyme-A	Lysis of conventional T cells
LAG3	Inhibitory signaling via MHC class-2 molecules
TNFRSF6	Apoptosis of conventional T cells
	Cytokine-mediated suppression
Galectin-1	Cell cycle arrest and apoptosis in conventional T cells
IL2Rα	Adsorption of IL2
IL10	Dendritic cell function ↓, Conversion of conventional T cells into T_{R1} cells
TGFβ, LAP	Induction of FoxP3 in conventional T cells

and T_{Eff} cells can form immunological synapses with antigen-presenting cells, where TCR signalosomes are assembled. However, the formation of immunological synapses with T_{Reg} cells is not associated with the recruitment of Src kinases, protein kinase-Cθ, and caspase recruitment domain family member CARD11 [242]. In addition, T_{Reg} cells receive inhibitory signals via TCRs. These receptors activate protein kinase-B to reduce T_{Reg} action. On the other hand, T_{Reg} activity is inhibited by the inflammatory cytokine tumor-necrosis factor-α.

Although T_{Reg} and T_{Eff} cells have many common signaling components, their T-cell receptors cause different outcomes. Protein kinase-Cθ is a pro-inflammatory kinase in both T_{Reg} and T_{Eff} cells. Whereas PKCθ is recruited to the immunological synapse for full T_{Eff} activation, T_{Reg} cells exclude it from the synapse to maintain the suppression against CD4+ and CD8+ T_{Eff} cells. However, costimulation enhances PKCθ recruitment and lowers immunosuppression, as PKCθ precludes T_{Reg}-mediated suppression [242].

On the other hand, PKCθ promotes T_{Reg}-cell differentiation via caspase recruitment domain (CARD)-containing NFκB activator B-cell lymphoma BCL10 that complexes with Mucosa-associated lymphoid tissue lymphoma translocation peptidase MALT1 and CARD-containing protein of the MAGUK family CARD11 [241]. T_{Reg}-cell differentiation is promoted by the TCR-activated NFκB pathway and the IL2–STAT5 axis, but hindered by the PKB–TOR pathway.

Regulatory T cells can have several basic modes of immunosuppressive action (Table 3.29) [243]: (1) release of inhibitory cytokines, such as interleukin-10 and -35 and transforming growth factor-β; (2) cytolysis via granzyme and perforin, e.g., CD4+ (CD4: coreceptor for MHC class-2 molecules), CD45Ra+ ($PTPRc_{hMW}$) regulatory T cells stimulated with CD3- (CD3: signaling component of the T-cell

Table 3.30. Development and phenotype of regulatory T cells (T_{Reg}; Source: [245]; MALT: mucosa-associated lymphoid tissue T_{R1}: type-1 regulatory T lymphocyte; iT_{Reg}: inducible T_{Reg} lymphocyte; nT_{Reg}: natural T_{Reg} lymphocyte). Forkhead box FoxP3+ T_{Reg} cells are potent mediators of self tolerance, thereby preventing autoimmunity. They promote cancer progression when they repress antitumor immunity. In humans, T_{Reg} cells have diverse functions and phenotypes.

Locus	Progenitor	Produced cells
Thymus	Uncommitted thymocyte	CD4+ thymocyte
		CD8+ thymocyte
	CD4+ thymocyte	CD4+ FoxP3+ nT_{Reg}
Periphery	CD4+ naive T cell	CD4+ FoxP3+ iT_{Reg}
	CD8+ naive T cell	CD8+ FoxP3+/− T_{Reg}
MALT	CD4+ naive T cell	CD4+ FoxP3+ iT_{Reg}
		TGFβ-secreting T_{Reg} (T_H3)
		IL10-secreting T_{Reg} (T_{R1})

receptor complex) and CD46-specific (CD46: ubiquitous inhibitory complement receptor) antibodies; (3) apoptosis mediated by cytokine deprivation that depends on IL2 receptor-α (or CD25); (4) cell inhibition by intracellular transfer of cAMP through gap junctions; (5) immunosuppression via A_2 adenosine receptor stimulated by pericellular generation of adenosine due to concordant action of ectoenzymes — ectonucleoside triphosphate diphosphohydrolase ENTPD1 (or CD39) and ecto-5′-nucleotidase (NT5E or CD73); (6) inhibition of dendritic cell maturation by lymphocyte-activation gene product LAG3 (or CD223) that binds to MHC class-2 molecules and are expressed by immature dendritic cells to suppress immunostimulation of dendritic cells ; and (7) production by dendritic cells of immunosuppressor indoleamine dioxygenase due to cytotoxic T-lymphocyte antigen CTLA4 that interacts with costimulators T-lymphocyte activation antigens CD80[117] and/or CD86[118] that are CTLA4 counter-receptors.

T_{Reg} Cell Subpopulation

Two most important subsets of T_{Reg} cells express forkhead box FoxP3 transcription factor. They develop in the thymus or can be induced in peripheral sites such as the mucosa-associated lymphoid tissue (MALT; Table 3.30). Functional activated human FoxP3+ T_{Reg} cells express a unique pattern of cell surface markers [245].[119]

[117] A.k.a. B7.1, BB1, CD28LG1, and LAB71.

[118] A.k.a. B7.2, B70, CD28LG2, FUN1, and LAB72.

[119] Markers include IL2Rα (CD25), CD45Ro (isoform generated by CD45 transcript lacking A, B, and C exons; i.e., low-molecular weight form [$CD45_{lMW}$]), TNFRSF18, or glucocorticoid-induced TNF-receptor-related protein (GITR), latent TGFβ, glycoprotein-A repetitions predominant (GARP; a.k.a. LRRC32), and IL1R1 and -2 (CD121a and CD121b) [245].

Table 3.31. Regulatory T-cell types (Source: [246]). Most of FoxP3+ T_{Reg} cells are CD4+ T cells that express interleukin-2 receptor α chain can suppress the activation, proliferation, and effector functions (e.g., cytokine production) of various populations of immunocytes, such as CD4+ and CD8+ T cells, natural killer (NK) and NKT cells, B cells and antigen-presenting cells.

Markers	Name	Possible origin
CD4+ regulatory T-cell populations		
CD4+CD25+FoxP3+	Natural regulatory T cell	Thymus
CD4+IL10+FoxP3−	Adaptive regulatory T cell, Induced regulatory T cell, T regulatory-1 cell (T_{R1})	Periphery
CD4+TGFβ+	T_H3 cells	Periphery
CD8+ regulatory T-cell populations		
CD8+CD25+		Thymus
CD8+CD28−		Periphery
CD8+CD62L+CD122+		?
CD8+IL10+		Periphery

A third important type of T_{Reg} cell secretes the immunosuppressive cytokine interleukin-10. It may develop from conventional CD4+ T cells subjected to IL10, or T_{H1} or T_{H2} cells [245]. Other T-cell subpopulations such as natural killer T, γδ T, and CD8+ T cells can also exert potent suppressor functions in certain circumstances.

CD4+, CD25+, FoxP3+ regulatory T cells develop in the thymus and exhibit a set of T-cell receptors that are specific for self antigens. Regulatory T cells can also be generated from effector T cells; these cells then usually do not express forkhead box protein FoxP3 [245].

Under certain conditions such as inflammatory environments with high levels of cytokines (e.g., IL6 and Ifnγ) that are normally involved in the polarization toward effector T cells, FoxP3 expression can be downregulated in FoxP3+ T_{Reg} cells. These cells then lose their suppressor function and manifest some of the functions of conventional effectors, such as T_{H1}, T_{H2}, T_{H17}, and T_{FH} cells [245]. In addition, deletion in T_{Reg} cells of certain transcription factors that are shared between T_{Reg} and effector cells such as the T_{H2}-specific factor interferon-regulatory factor IRF4 causes impaired suppression of T_{H2} responses by T_{Reg} cells [245]. Such transcriptional submodules rely also on IRF4 for T_{H17} cells and TBx21 for T_{H1} [241]. Therefore, transcription factors of T_{Reg} cells match those of their regulatees. This matching can help for cohoming.

In any case, regulatory T cells must preserve useful pathogen-specific immune responses as well as antitumor immunity. Several regulatory T-cell populations exist (Table 3.31). Forkhead box FoxP3+, CD4+ T cells can be divided into several functional populations based on their expression of CD45Ra (human, restricted form, i.e., synthesized from transcript subtype with exon A), CD45Ro ($CD45_{lMW}$: low-molecular-weight form, the transcript lacking exons A, B, and C), HLADR (major histocompatibility complex MHC class-2 plasmalemmal [hetero]dimeric receptor encoded by the human leukocyte antigen [HLA] complex) and FoxP3 [244]. Subpopulations of human FoxP3+ T_{Reg} cells change their levels of certain markers, such as CD45Ra+, FoxP3^{low} naive T_{Reg}, CD45Ra−, FoxP3^{high} activated T_{Reg}, and CD45Ra−, FoxP3^{low} cytokine-secreting T_{Reg} cells [245].

As for activated conventional T lymphocytes, T_{Reg} markers of activation or memory allow homing to locations other than the secondary lymphoid organs. In particular, $\alpha_E\beta_7$-integrin (CD103–Itgβ7) synthesized by 20 to 30% of FoxP3+ cells in secondary lymphoid organs and a higher percentage of T_{Reg} cells in tissues, such as the skin, intestinal lamina propria, and lung, binds E-cadherin [241]. The activated, memory, CD103+ T_{Reg}-cell subset can be further subdivided by markers of natural killer cells, such as KLRg1, cyclic ^{ADP}ribose hydrolase-1 (CD38), and $\alpha_2\beta_1$-integrin (CD49b–CD29).

Chemokine receptor CCR4 is not expressed on thymic T_{Reg} cells, but on a high percentage of extralymphoid FoxP3+ cells in the skin. A distinct population of FoxP3+ cells that express only a subset of the T_{Reg} signature reside in adipose tissue [241].

Natural T_{Reg} Cells

Naturally occurring, CD4+, CD25+, FoxP3+, regulatory T cells that maintain self-tolerance originate from thymocytes. These naturally occurring (natural), thymus-derived, T_{Reg} lymphocytes can be detected in the periphery about 3 days after birth, at least in mice [244].

In humans, immature FoxP3+ thymocytes have been identified in the thymus. The human thymus produces mature T cells from gestation week 13. From week 14, about 7% of mature CD4+ thymocytes express high levels of CD25 (IL2Rα) that produce CD4+, CD25^{high} T_{Reg} cells [244].

In mice, thymic cortical and medullary thymic epithelial cells and dendritic cells, contribute to T_{Reg}-cell differentiation and selection [244]. Thymic development of T_{Reg} cells indeed needs interactions between their T-cell receptors and self-peptide–MHC complexes presented by thymic stromal cells. Both IL2 and IL7 are involved in T_{Reg}-cell development.

In humans, *Hassall's corpuscles* in the thymic medulla may form a specialized microenvironment to support nascent T_{Reg} cells. Hassall's corpuscles release thymic stromal lymphopoietin (TSLP) that activates immature, migratory, α_X-integrin (CD11c)+ dendritic cells in the thymus and upregulates their production of costimulators [244]. Dendritic cells activated by TSLP may provoke FoxP3 expression in immature CD4+, CD8−, CD25− thymocytes, but not in mature naive CD4+ T cells derived from peripheral blood, hence supporting thymocyte differentiation into mature FFoxP3+ T_{Reg} cells [244].

Naive and Effector T_{Reg} Cells

In humans, the population of CD4+, FoxP3+, thymus-derived, naturally occurring T cells is composed of at least 3 phenotypically and functionally distinct subpopulations [247]: (1) CD45Ra+, $FoxP3^{low}$ resting T_{Reg} cells (rT_{Reg}); (2) CD45Ra−, $FoxP3^{high}$ activated T_{Reg} cells (aT_{Reg}); and (3) cytokine-secreting CD45Ra−, $FoxP3^{low}$ non-suppressive T cells (non-T_{Reg}).

Some FoxP3+ cells in peripheral blood are phenotypically naive, such as CD45Ra+ cells, whereas other FoxP3+ cells phenotypically resemble memory T cells such as CD45Ra− cells that may originate from peripheral memory FoxP3−, CD4+ T cells. Expression of CD45Ra without concomitant expression of CD45Ro is a marker for naive T cells. (The label CD45Ro is a marker of conventional memory T cells.) CD45Ra+, $FoxP3^{low}$ naive T_{Reg} cells that are characterized by plasmalemmal CD45Ra and low levels of intranuclear FoxP3 are in a quiescent stage, hence their name, resting T_{Reg} cells [244]. Naive T_{Reg} cells proliferate after TCR stimulation. They are highly resistant to apoptosis. On the other hand, CD45Ro+, $CD25^{high}$, FoxP3+ T_{Reg} cells tend to be hyporesponsive and undergo apoptosis, once activated.

Effector CD45Ro+ ($PTPRc_{lMW}$), $FoxP3^{high}$ T_{Reg} cells derive mainly from CD45Ra+ ($PTPRc_{hMW}$), $FoxP3^{low}$, resting T_{Reg} cells. In vitro, they are potent suppressor and hyporesponsive after activation. They express IL2Rα, TNFRSF6a, TNFRSF18, and cytotoxic T-lymphocyte-associated protein CTLA4 that are markers of recently activated T cells [244]. Effector T_{Reg}-cell subpopulation is heterogeneous in the expression of ICOS and HLADR markers. Effector ICOS+ and ICOS− (inducible T-cell costimulator or CD273) T_{Reg} cells produce the suppressive cytokines IL10 and TGFβ, respectively. Receptor HLADR enables to identify a terminally differentiated subpopulation of effector T_{Reg} cells [244].

Role of FoxP3 Transcription Factor

Regulatory T cells that promote immune tolerance and protect against autoimmunity develop in the thymus. Their differentiation and acquisition of suppressive function can rely on FoxP3 expression. However, FoxP3 that is a major regulator of T_{Reg} cells is insufficient and unnecessary to specify all aspects of this cell lineage. FoxP3+ T_{Reg} cells use the $\alpha\beta$-T-cell antigen receptor for antigen recognition with a TCR repertoire similar in size, but distinct in composition w.r.t. that of CD4+ conventional T cells.

The transcription factor FoxP3 primes the expression of T_{Reg}-cell markers, such as IL2Rα (CD25), cytotoxic T-lymphocyte-associated protein (antigen) CTLA4, and TNFRSF18 (Table 3.32) [244].

In humans, 2 main FoxP3 isoforms confer regulatory function when they are strongly expressed. Both FoxP3a and FoxP3b possess a proline-rich region that inhibits NFAT-mediated transcription and interacts with histone deacetylases, zinc finger sequence, Leu zipper motif for homodimerization, and DNA-binding forkhead domain. The 2 last domains interact with NFAT and NFκB factors [244].

Table 3.32. FoxP3 and T_{Reg}-cell markers (Source: [244]; CCR: CC-chemokine receptor; CTLA: cytotoxic T-lymphocyte antigen; GARP, glycoprotein-A repetition predominant protein; GITR, glucocorticoid-induced TNFR-related protein, or TNFRSF18; ICOS, inducible T-cell costimulator [CD278]; IL, interleukin; LAP: latency-associated peptide (region of TGFβ precursor); MS4a4b: membrane-spanning 4-domain, subfamily A, member 4B; PdCD: programmed cell death). Short aliases CD*i* are suitable for small-sized tables, but do not display their function. CD25 corresponds to low-affinity receptor IL2Rα; CD27 to tumor-necrosis factor receptor superfamily member TNFRSF7; CD31 to PECAM1; CD39 to ectonucleoside triphosphate diphosphohydrolase ENTPD1; CD45 to PTPRc phosphatase; CD62L to L-selectin; CD69 to C-type lectin domain family member CLec2c; CD73 to ecto 5′-nucleotidase; CD95 TNFRSF6a; CD103 to α_E-integrin; and CD304 to neuropilin-1.

	Activation	Homing	Effector	Cell fate
FoxP3	CD45	CD31/62L/103/304,	CD39/73	CD6/27/95
	CD25/69,	CCR4/6/9	CD2/80/86,	IL1R
	HLADR		IL10/17,	GARP, GITR
			Granzyme-B,	MS4a4b,
			Galectin-1/10	Galectin-3
			LAP, TNFSF11,	PdCD1, TNFRSF4
			ICOS, CTLA4	

The main isoform FoxP3b lacks the Leu–X–X–Leu–Leu motif encoded by exon 2 that serves to bind to retinoic acid receptor-related orphan receptor-γ as well as N-terminal residues that may mediate the interaction with nuclear factor of activated T cells [244].

The regulator of cytokine signaling, SUMo (small ubiquitin-related modifier) ligase PIAS1 (protein inhibitor of activated signal transducer and activator of transcription STAT1), represses FoxP3 expression by epigenetic inhibition, i.e., maintaining a repressive chromatin state of the FOXP3 promoter [248]. Ligase PIAS1 recruits methyltransferases, thereby promoting the methylation of the FOXP3 promoter. Enzyme PIAS1 also selectively blocks the binding of NFκB and STAT1 to gene promoters [249]. Recruitment of PIAS1 to chromatin requires its phosphorylation (Ser90) upon various immune regulatory stimuli, such as tumor-necrosis factor and activated T-cell receptor. Inhibitor of NFκB kinase-α, but not IKKβ, phosphorylates PIAS1 [249].

Role in Antiviral Immunity

Regulatory T cells promote early antiviral immune response at a site of infection [250]. Regulatory T cells accumulate, are activated, and proliferate both at the infection site and in draining lymph nodes.

Transcription factor FoxP3, transforming growth factor-β (particularly activated by integrins of dendritic cells), and IL2 support a T_{Reg} cell type.

The presence of regulatory T cells in the lymph node prevents influx and activation of effector T cells, natural killer cells, and dendritic cells that thus migrate into the site of viral infection.

Table 3.33. Helper T-cell differentiation and remodeling (Source: [251]; T_H: helper T lymphocyte). In addition to natural, thymic-derived, CD4+, CD25+ regulatory T cells, naive T cells in the periphery can become inducible, regulatory, immunosuppressive T lymphocytes (iT_{Reg}). CD4+ follicular helper T cells (T_{FH}) help B lymphocytes to produce antibody responses to T-cell-dependent antigens, hence aiding in the development of germinal centers within secondary lymphoid tissues to promote immunoglobulin class switch recombination and affinity maturation.

Cell source	Cell products
CD4+ naive T cell	$T_{H1/2/17/22}$, T_{FH}, iT_{Reg}
T_{H1}	T_{FH}
T_{H2}	$T_{H1/9}$, T_{FH}
T_{H17}	T_{H1}, T_{FH}
T_{FH}	$T_{H1/2/17}$
iT_{Reg}	$T_{H1/17}$, T_{FH}

3.10.4.6 Helper T Lymphocytes

CD4+ helper T cells coordinate adaptive immune defense against specific pathogens. They secrete given sets of cytokines that signal to other subsets of immunocytes, thereby helping B lymphocytes, CD8+ cytotoxic T cells, as well as other innate immunity cell types to combat given types of pathogens. T_H cells constitute many groups that do not correspond to terminally differentiated cells, or different cell lineages, but different functional groups, as they can remodel (Table 3.33) [251].

Type-1 (T_{H1}) and -2 (T_{H2}) helper T-cell subpopulations as well as interleukin-17-secreting helper T-cell (T_{H17}) subset are the most well known. In fact, antigen-inexperienced CD4+ T cells specialize not only into various clones of helper T cells, but also inducible regulatory T cells in response to infectious agents, commensal microorganisms, and self-antigens (Table 3.34).

T_{H1} cells produce interferon-γ to protect against intracellular pathogens. T_{H2} cells synthesize interleukins IL4, IL13, and IL25 to clear extracellular pathogens. T_{H17} cells manufacture IL17 (or IL17a), IL17Fs, IL21, and IL22 to target extracellular pathogens not effectively handled by other T_H cells.[120]

T_{H1} Cell

T_{H1}-specific transcription factor is TBx21 factor. T_{H1} polarization is primarily due to interleukin-12 and interferon-γ.

[120] Interleukin-17 family includes 6 known members (IL17a–IL17f). Both IL17a and IL17f induce the production of pro-inflammatory cytokines, chemokines, and metallopeptidases from various cell types. T_H cells express only IL17a, IL17f, and IL17a–IL17f heterodimer [226]. Interleukin-21 is produced by T_{H17} cells and ICOS+, CXCR5+, CCR7+ follicular helper T cells in B-cell areas of secondary lymphoid tissue. Interleukin-22 is synthesized by terminally differentiated T_{H17} cells.

Table 3.34. Differentiation of naive CD4+ T cells into helper T-cell subsets (T_H) and inducible regulatory T cells (T_{Reg}). Following connection of antigen-presenting cells such as dendritic cells and naive CD4+ T cells via the MHC class-2 molecule–antigen peptide–TCR complex, activated CD4+ T cells can polarize into T_{H1}, T_{H2}, and T_{H17}, among others, as well as T_{Reg}, according to the local cytokine environment (FoxP: forkhead box-P; IL: interleukin; ROR: retinoic acid receptor-related orphan receptor [NR1f1–NR1f3]; STAT: signal transducer and activator of transcription; TBx: T-box transcription factor; TGF: transforming growth factor). Master transcription factor BCL6 of follicular helper T cells (T_{FH}) is a transcriptional repressor, unlike TBx21 and GATA3. Interleukin-10 can be produced by multiple cell clones (T_{H1}, T_{H2}, T_{H17}, and iT_{Reg}) in a given context, demonstrating remodeling capacity of T-cell subsets [251]. Similarly, T_{H2} and T_{H17} cells can acquire IL9-synthesizing capacity. T_{H17} cells can extinguish production of their cytokines to become selective interferon-γ synthesizers. Expression of transcription activators and repressors can also change.

Trigger cytokine	Product	Signature cytokines and main cellular markers
IL12	T_{H1}	Ifnγ; STAT4, TBx21
IL4	T_{H2}	IL4/13; STAT6, GATA3
TGFβ, IL6	T_{H17}	IL17a/f; STAT3, RORγ2
	T_{FH}	IL21; STAT3, BCL6
TGFβ	iT_{Reg}	IL10/35, TGFβ; STAT5, FoxP3

Table 3.35. Cytokines in the initiation and amplification of T_{H2}-mediated responses (Source: [252]; ASMC: airway smooth myocyte; TSLP: thymic stromal lymphopoietin).

Cytokine	Inducers	Sources	Targets
IL2	Antigen, IL4/13, allergens, helminth products	Activated CD4+ T cell, epithelial cells, basophils, mastocyte,	T cells, dendritic cells, basophils, mastocytes
IL4	Antigen, TSLP, IgE, IL2/3/4/25	Naive CD4+ T cell, T_{H2}, NKT, basophil	T/B cells, macrophages, epithelial cells
IL25	Allergens, IL4, helminth products	T_{H2}, basophil, eosinophil, epithelial cells,	T cells, nuocytes
IL33	Allergens, helminth products	Epi/endothelial cells, ASMC, adipocyte	T_{H2}, macrophage, dendritic cells, nuocytes, basophil, mastocyte

T_{H2} Cell

Helper-2-specific transcription factor is GATA-binding GATA3 protein. T_{H2}-cell differentiation indeed relies on GATA3 and signal transducer and activator of transcription STAT5 that can be activated by IL2 (Table 3.35). T_{H2}-cell differentiation is promoted by interleukin-4 and -25 via STAT6 factor. The expression of GATA3 is upregulated by IL4, T-cell receptor, Notch, and Wnt signalings [252].

T_{H2} cells operate via their production of type-2 cytokines (IL4, IL5, IL9, IL13, and IL25) and homing to specific tissues. T_{H2} cells control immunity against extracellular parasites and allergic responses. Epithelial and dendritic cells and basophils sense allergens and helminth products and initiate T_{H2}-type immune responses [252]. During the initiation stage of T_{H2}-cell development, epithelial cells produce T_{H2}-promoting cytokines, such as thymic stromal lymphopoietin and interleukin-25 and -33, and basophils manufacture TSLP, IL4, and IL25. Dendritic cells and basophils can serve as T_{H2}-inducing antigen-presenting cells.

During sensitization to allergen, activated allergen-specific CD4+ helper-2 T cells produce T_{H2} cytokines that trigger a switch to ϵ-immunoglobulin heavy chain and IgE production by B cells via IL4 [253]. Immunoglobulin-E stimulates mastocytes and basophils by binding to high-affinity IgE FcϵR1 receptor. The allergen–IgE–FcϵR1 complex provokes degranulation of mastocytes and basophils, thereby releasing heparin, histamine, serotonin, leukotrienes, prostaglandins, peptidases, chemokines, and cytokines that prime the immediate phase of the allergic reaction with smooth myocyte constriction, edema that results from vascular permeability, and inflammatory cell recruitment. Immunoglobulin-E also binds to FcϵR1 of dendritic cells and monocytes, as well as low-affinity IgE receptor FcϵR2 of B cells. Allergen uptake by antigen-presenting cells and subsequent presentation of allergen-derived peptides to specific CD4+ T cells lead to the late phase of the allergic reaction.

T_{H2} cells migrate to lungs, where they recruit eosinophils via IL5 and mastocytes via IL9 [252]. Moreover, they act on epithelial cells via IL4, IL9, and IL13 and smooth myocytes via IL4 and IL13 to cause mucus production, goblet cell metaplasia, and airway hyperresponsiveness. Type-2 cytokines also target T cells via IL4, IL9, and IL25, macrophages via IL4 and IL13, as well as IL5- and/or IL13-producing, non-B, non-T cells via IL25 [252]. Furthermore, T_{H2} cells produce amphiregulin of the epidermal growth factor family and IL24, in addition to IL2, IL10, and IL21 [252].

T_{H9} Cell

Interleukin-9 targets B and T lymphocytes, mastocytes, as well as erythroid progenitors. CD4+, IL9-producing T cells constitute a subset distinct from other types of helper T cells.

T_{H9} cells depend on transforming growth factor-β, IL4, and IL25 [254]. Agents TGFβ and IL4 render T cells IL25-responsive, as they stimulate IL17Rb expression. Factor TGFβ reprograms T_{H2} cells so that they switch their synthesis from T_{H2} cytokines to IL9 and IL10. Interleukin-25 does not initiate T_{H9} polarization from naive CD4+ T cells. Interleukin-25 operates via both IL17Ra and IL17Rb that promote the production of IL4, IL5, and IL13. Interleukin-25 provokes IL9 and IL10 production in differentiated T_{H9} cells. In fact, IL25 promotes IL9 expression independently of IL4, but dependently on TGFβ.

T_{H17} Cell

T_{H17} cells are involved in host defense against specific pathogens, in tumor immunology, autoimmunity, and tissue inflammation. They interact with regulatory T cells to prevent tissue inflammation and mediate self-tolerance.

Transcription factors Activator protein AP1, STAT3, and retinoic acid-related orphan receptors RORα (or NR1f1) and RORγ2 (RORγT or NR1f3-2) promote the generation of T_{H17} cells. Production of IL17 relies on RORγ2. The STAT3 factor activated by IL6, IL21, and IL23 binds to Il17 and Il21 loci and regulates the expression of IL23 receptor, RORα, and RORγ2.[121]

Differentiation of naive T cells into T_{H17} and T_{Reg} cells depends on the local cytokine environment and antigen-presenting cells. Differentiation of T_{H17} cell is induced by transcription factor RORγ2 in collaboration with STAT3 in response to the combined effect of IL6, IL23, and TGFβ. The transcription factor forkhead box-P3 can hinder NR1f3-2 activity. The development and functioning of FOxP3+ T_{Reg} cells require the cooperation of FoxP3 and nuclear factor of activated T cells (NFAT) to regulate FoxP3 target genes.[122] T_{H17} cells also express CCR6 receptor.

Differentiation into T_{H17} cells can occur in the absence of TGFβ [255]. Interleukin-6 and -23 in combination with IL1β cause IL17 production in naive precursors. Therefore, T_{H17} cells constitute diverse subpopulations, such as TBx21+, RORγ2+, T_{H17} cells.

Basic leucine zipper ATF-like transcription factor (BATF), or B-cell-activating transcription factor (BATF), that contains only a basic region and leucine zipper domain (and not transcriptional activation domains of Activator protein-1)[123] is involved in the differentiation of T_{H17} cells [256]. The BATF factor binds conserved intergenic elements in the Il17a–Il17f locus and to the Il17, Il21 and Il22 promoters.[124]

T_{H17} cells synthesize several cytokines, such as IL2, IL17, granulocyte–macrophage colony-stimulating factor (CSF2), interferon-γ, and tumor-necrosis factor in the human tumor environment for tumor eradication [257].[125] On the other hand, tumor-infiltrating T_{Reg} cells suppress T_{H17}-cell expansion in tumors.

---

[121] The Il17 locus is composed of Il17A and Il17F genes. Encoded IL17a and IL17f can form homo- and heterodimers.

[122] Most (but not all) FoxP3+ T_{Reg} cells express IL2Rα (CD25). A subset of CD25+ (IL2Rα+) T cells comprises FOxP3− effector T cells without T_{Reg} function. Calcium ions are needed for nuclear translocation of NFAT and cytokine production. Stromal interaction molecules StIM1 and StIM2 detect intracellular calcium storage depletion and activate calcium release-activated calcium channels for calcium entry.

[123] Early response transcription factor Activator protein-1 is a dimer of Jun, Fos, Maf, Jun dimerization protein (JDP), and activating transcription factor family members (ATF1–ATF7).

[124] Factor BATF is also expressed in T_{H1} and T_{H2} cells, but it is not required for T_{H2} development.

[125] Interleukin-17, or IL17a, also called cytotoxic T-lymphocyte-associated protein CTLA8, belongs to a the IL17 family with IL17b to IL17f.

T_{H22} Cell

CD4+, IL22-producing helper T cells (T_{H22}) constitute an additional T_H cell subset. Production of interleukin-22 depends less on the transcription factor retinoic acid receptor-related orphan receptor-gamma2 (or NR1f3-2) than that of IL17 cytokine. Synthesis of IL22 relies on transcription factor aryl hydrocarbon receptor (AHR or class-E basic helix–loop–helix protein bHLHe76).

T_{H22} cells can be generated in vitro from naive CD4+ T cells subjected to IL6 and tumor-necrosis factor [258]. Plasmacytoid dendritic cells support more efficiently T_{H22} cell differentiation than conventional dendritic cells. Binding partners of bHLHe76 factor favor the differentiation into T_{H22} cells from naive T cells [259].

This subset of human T cells is characterized by the expression of skin-homing chemokine receptors, such as CCR4 and CCR10. Expression of CCR10 depends on the presence of active vitamin-D3 metabolites. However, T_{H22} cells do not synthesize interleukin-17 and interferon-γ [258].[126]

3.10.4.7 Cytotoxic T Lymphocytes

Cytolytic T lymphocytes (CTL) express T-cell receptor and its coreceptor cluster of differentiation CD8 for recognition of a complex formed by antigen and class-1 major histocompatibility complex molecule. After antigenic stimulation, CD8+ CTLs enter into a transient state characterized by reduced antigen–HLA binding despite normal plasmalemmal amounts of TCR and CD8. This decrease in activity several days after antigenic stimulation results from the loss of colocalization of T-cell receptor and CD8.

Galectin–glycan lattices on the cell surface formed from galectin-3 and β-galactoside regulate the formation of focal adhesions[127] and disrupt the association of T-cell receptor with CD8 in cytolytic T lymphocytes [260].[128] T-cell receptor becomes colocalized with galectin-3.

---

[126]Interleukin-22 is produced mainly by T_{H17} cells and some natural killer cell-like subsets (NK22) subset of natural killer cells that acts on epithelial cells and keratinocytes of mucosal surfaces.

[127]Migrating cells communicate with the extracellular matrix via focal adhesions that are formed by integrins, phosphorylated caveolin-1, Glc^{NAc} transferase-5 (GAT5), and focal adhesion kinase. Caveolin-1 serves as plasmalemmal adaptor that promotes integrin signaling via Src and FAK kinases. Galectin plasmalemmal nanodomain recruits glycoproteins depending on ^{N}glycan number and branching, a feature determined by Glc^{NAc} transferase-5. Galectin-3 bound to GAT5-processed ^{N}glycans and phosphorylated caveolin-1 cooperate to stabilize focal adhesion kinase [261].

[128]Galectin–glycan lattices restrict the lateral mobility of cell-surface glycoproteins, thereby inducing stable membrane structures. The formation of galectin–glycan lattices is regulated by ^{N}acetylglucosaminyltransferases that synthesize β-galactosides, which serve as terminal carbohydrates of ^{N}glycans.

3.10.4.8 Natural Killer T Lymphocytes

Natural killer T cells (NKT; ~ 0.2% of blood T cells) constitute a group of unconventional T lymphocytes that share properties of natural killer cells. They differ from conventional αβ T cells because their TCRs target a limited number of molecules. They actually express a limited T-cell receptor repertoire and recognize lipidic rather than proteic antigens presented by the non-classical major histocompatability complex molecule CD1d. Antigen-presenting molecule CD1d binds self and foreign lipids and glycolipids rather than MHC-associated complexes.

Natural killer T cells express the natural killer cell marker, killer cell lectin-like KLRb1c receptor.[129] Natural killer T cells share other features with NK cells, such as expression of FcγR3[130] and neural cell adhesion molecule NCAM1[131] and production of granzyme (granule enzyme).[132]

During infections, NKT cells proliferate rapidly and synthesize cytokines. Furthermore, they form long-lasting contacts with relevant antigen-presenting cells, essentially macrophages.

Activated NKT cells produce large quantities of interferon-γ, interleukin-4, and granulocyte–macrophage colony-stimulating factor (CSF2), as well as other cytokines and chemokines, such as IL2 and TNFα. These substances are rapidly released to promote or suppress different types of immune responses.

Natural killer T cells are classified into 3 groups. Group-1 NKT cells include classical, invariant (iNKT), and Vα24i NKT cells. Group-2 NKT cells comprises non-classical NKT cells and other types that do not belong to group 3. *NKT-like* set is composed of $KLRb1^+$ (CD161c+) T cells and CD3+ (T-cell coreceptor), NCAM1+ (CD56+) T cells.

NKT cells promote interactions between dendritic cells and CD8+ T cells. They operate upstream from the contact between dendritic cells and CD8+ T cells

---

[129] A.k.a. NK1.1 and CD161.

[130] Two genes — FCGR3A and FCGR3B — encodes low-affinity receptors FcγR3a (CD16a) and FcγR3b (CD16b).

[131] Neural cell adhesion molecule (or CD56) is a marker for natural killer cells and some T-lymphocyte types, in addition to neurons.

[132] Granzymes are serine peptidases released from granules (specialized secretory lysosomes) in natural killer cells and cytotoxic T cells to destroy virus-infected cells. In humans, 5 granzyme types exist (A–B, H, K, M) [262]. These 5 isoforms are synthesized as zymogens that are processed by cathepsin-C within granules or outside. Transfer to granules requires modifications by mannose 6-phosphate that then allows transport through mannose 6-phosphate receptor (of IGF2R). In addition, the granzyme-B–granulysin–perforin complex can enter a cell through the mannose 6-phosphate receptor or other types of receptors. Inside granules, granzymes link to serglycin, a hematopoetic chondroitin sulfate proteoglycan core protein or secretory granule proteoglycan core protein. At the site of cell contact, granule content is released in the intracellular medium, where granzymes are stored in vesicles. Perforin assists both transfer into and out these vesicles. Granzyme-A and -B can also be secreted and act in the extracellular milieu, where they can provoke target cell death. In healthy humans, granzymes can be detected in plasma (granzyme-A: 15–35 pg/ml; granzyme-B: $\leq$ 15 pg/ml).

(signal 0) and guide the priming of cytotoxic T lymphocytes [207]. NKT cells stimulate cognate dendritic cells to produce CCL17 chemokine to recruit effector CD8+ T cells that possess CCL17 receptor CCR4 [263].

Invariant Natural Killer T Lymphocytes

Type-1 NKT cells, or invariant NKT cells (iNKT), prime an innate immunity-like response. They are able to modulate the function of other cell types. In particular, iNKT cells provoke an effective antitumor immune response [264].

Invariant NKT cells express a single invariant T-cell receptor α chain ($V_{\alpha 24}$–$J_{\alpha 18}$) that combines with an β chain ($V_{\beta 11}$), hence the $V_{\alpha 24}$+, $V_{\beta 11}$+ phenotype. This T-cell receptor on $V_{\alpha 24}$+, $V_{\beta 11}$+ NKT cells reacts specifically and potently to CD1d tetramers[133] loaded with the glycosphingolipid antigen α-galactosyl ceramide on antigen-presenting cells such as CD1d+ dendritic cells. $V_{\alpha 24}$+, $V_{\beta 11}$+ NKT cells activate conventional T and NK cells as well as B lymphocytes. $V_{\alpha 24}$+ iNKT cells mediate antitumor activity via killing of tumor-associated macrophages [267].[134]

Invariant NKT cells also specifically synthesize transcription regulator zinc finger and BTB domain-containing protein ZBTB16 that is required for maturation (e.g., secretion of large amounts of interleukin-4 and interferon-γ after activation) [268].

Invariant NKT cells derive from CD4+, CD8+, double-positive thymocytes. Transcription factor myeloblastosis viral proto-oncogene product homolog MyB launches DP thymocytes into the iNKT lineage, as it simultaneously regulates the expression of CD1d loaded on DP thymocytes, the half-life of DP cells to allow $V_{\alpha 24}$–$J_{\alpha 18}$ rearrangement, and production of signaling lymphocytic activation molecule family members SLAMF1 and SLAMF6 and SH2 domain-containing protein SH2D1a, also called SLAM-associated protein (SAP) [269].

Mucosal-associated invariant T lymphocytes are similar to CD1d-restricted NKT cells, but are preferentially located in intestine. Their activation requires MHC class-1-related protein. Few NKT cells exist in the nasopharynx-associated lymphoid tissue and regional cervical lymph nodes. α-Galactosyl ceramide

---

[133]Cluster of differentiation-1 constitute a family of glycoproteins on the surface of various types of human antigen-presenting cells. These glycosylated class-1 MHC molecule-like molecules presents lipid antigens to T cells. The CD1 glycoproteins can be classified into 2 groups according to their lipid anchoring [265]. Group-1 CD1 molecules (CD1a–CD1c) are expressed on antigen-presenting cells. Group-2 CD1 molecules (CD1d) reside on a wider variety of cells. Group-1 and -2 CD1 present both self and microbial lipid antigens to T cells. Group-1 CD1-restricted T cells recognize mycobacterial glycolipid antigens; group-2 CD1-restricted T cells are regulatory T cells that act in innate and adaptive immunity [266]. CD1d-presented antigens activate natural killer T cells. Intracellular CD1e forms an additional category.

[134]CD68+ tumor-associated monocytes and macrophages represent the majority of CD1d+ cells, at least in some types of cancers. CD68+, tumor-associated monocytes and macrophages produce IL6 in primary tumors that stimulates tumor growth [267].

coadministrated with intranasal vaccine stimulates NK-cell migration in these lymphoid tissues and promotes protective mucosal immunity against influenza viruses [270].[135]

3.11 Innate Lymphocytes

Innate lymphoid cells (ILC) characterized by a prompt delivery of their effector functions constitute a set of developmentally related cells, such as natural killer (NK) and lymphoid tissue-inducer (LTi) cells as well as cells that produce interleukins IL5, IL13, IL17, and/or IL22, such as ILC22 cells[136] and nuocytes [272].[137] ILC17 (or LTi) and ILC22 (NKR+ LTi-like) cells produce interleukin-17 and/or -22; ILC2 cells, or nuocytes, synthesize IL5 and IL13 agents.

Different ILC cell types was grouped into 3 main subsets: natural killer, natural helper, and NR1f3-2+ (RORγ2+ or RORγT+) cells (Table 3.36). The IL15-dependent NK category includes: (1) conventional NK (cNK) CD16− and CD16+ cells, which have spontaneous cytotoxicity, and (2) Ifnγ+, CD56^{high}, CD16− cell subset, which includes a cytokine-polarized subset of CD56^{high} cells in humans [272]. ILC cells depend on cytokines that use the common γ chain of the receptors for IL2, IL4, IL7, IL9, IL15, and IL21. Different ILC subsets need distinct members of this cytokine set (Table 3.36).

These hematopoietic effectors operate in innate immune responses to microorganisms as well as in lymphoid tissue formation, in tissue remodeling after damage by injury or infection, and in stromal cell homeostasis. In particular, NK cells limit the propogation of viruses before the initiation of the adaptive immune response [272]. NK cells recognize pathogen ligands using specific receptors and eliminate infected or stressed cells using perforin- and granzyme-containing cytotoxic granules, TNFSF1, TNFSF6, TNFSF10, and interferon-γ.

---

[135]NKT cells colocalize and interact with dendritic cells that have taken up and store α-galactosyl ceramide. The flux of NKT cells to nasopharynx-associated lymphoid tissue and regional cervical lymph nodes is mediated by CXC-chemokine receptor CXCR6 of NKT cells. Accumulation of NKT cells in nasopharynx-associated lymphoid tissue and regional cervical lymph nodes stimulates IgA production by a mechanism that depends on interleukin-4. Molecule IgA is the most abundant antibody in the intestinal mucosa, where it provides immune protection against commensal bacteria and ingested pathogens. B lymphocytes produce IgA after undergoing class-switch recombination [271]. B lymphocytes undergo IgA class switching through both T-cell-dependent and T-cell-independent stimulations, which generate high- and low-affinity IgA antibodies, respectively. T-cell-dependent IgA class switching in the germinal centers of gut organized lymphoid structures requires T-cell help for B cells via TNFSF5 and transforming growth factor-β1. T-cell-independent IgA class switching in non-organized lymphoid tissue of the lamina propria involves microbial Toll-like receptor ligands and TNFSF5-related dendritic cell mediators (B-cell-activating factor and proliferation-inducing ligand).

[136]A.k.a. NK22 cells, natural cytotoxicity receptor-22 (NCR22) cells, and NK receptor NKR+ LTi-like cells.

[137]A.k.a. natural helper (NH) cells.

Table 3.36. Categories of innate lymphoid cells (ILC; Source: [272]). The ILC1 subset includes $CD56^{high}$, CD16− natural killer (NK) cells. The ILC2 subset comprises natural helper (NH) cells, or nuocytes. The ILC17 and ILC22 subsets encompass IL17-producing lymphoid tissue-inducer (LTi) cells and NKR+ LTi-like (NK22 or NCR22) cells, respectively.

Subset	Regulators	Function and targets Plasmalemmal and released agents
Thymic NK (ILC1)	IL7/15	Intracellular pathogens, inflammation Ifnγ
Conventional NK CD16− $CD16^{high}$	IL15	Intracellular pathogens, inflammation Ifnγ Ifnγ
Nuocyte (ILC2, NH)	common γ chain IL25/33	Extracellular parasites, allergy IL13
NR1f3-2+ LTi	IL7 TNFSF11	Formation of lymph nodes and isolated lymphoid follicles TNFSF2–TNFSF3 trimer, TNFSF1
NR1f3-2+ ILC17	IL7 IL23	Extracellular bacteria IL17
NR1f3-2+ ILC22	IL7 IL1β	Extracellular bacteria IL22

The functional differentiation of ILC subsets is controlled by certain transcription factors. Their formation needs the transcriptional repressor inhibitor of DNA binding ID2 (or bHLHb26) and signaling via the common IL2R γ chain [272]. Whereas the development of B and T cells and dendritic cells requires members of the E2A family ($bHLHb21^{E12}$, $bHLHb21^{E47}$, bHLHb19, and bHLHb20), E2A family proteins preclude the development of several ILC subsets This inhibition is overcome by inhibitors of DNA binding that heterodimerize with E2A proteins, thereby inactivating these transcription factors. Conversely, in human hematopoietic precursor cells, ID proteins promote the development of ILC cells, but impede the development of B and T cells and plasmacytoid dendritic cells.

Innate lymphoid cells constitute subpopulations of immunocytes that synthesize cytokines also expressed by helper T cells, but lack markers of mature lymphoid cells such as T-cell antigen receptors. Like some helper T-cell lineages, a proportion of innate lymphocytes express the transcription factor retinoic acid receptor-related orphan receptor RORγ2 (or NR1f3-2).

Innate lymphocytes may be generated by SCA1+, PTPRc+, IL7Rγ+, SCFR+ progenitors that produce the transcriptional regulator inhibitor of DNA-binding protein ID2 [273]. Topographical and functional compartmentation of NR1f3-2+ ILCs depends on CCR6 chemokine receptor that is targeted by CCL20 and β-defensins produced by epithelial cells [274].

Table 3.37. Precursors of innate lymphoid cells (ILCP) and transcription factors that specify hematopoietic lineages (Source: [272]; GATA: DNA sequence GATA-binding transcription factor; ID2: inhibitor of DNA-binding protein-2, or bHLHb26; NFIL3: interleukin-3-regulated nuclear factor; NR: nuclear receptor; Tox: thymocyte selection-associated high mobility group box). Transcription factors and cytokines enable the survival, proliferation, and differentiation of lineage-specific precursor cells from multipotent progenitors.

Precursor	Regulators	Generated cell type
ID2+ ILCP	NFIL3, Tox	Bone marrow NK precursor (bmNKP)
ID2+ ILCP	GATA3	Thymic NK precursor (thyNKP)
ID2+ ILCP		ILC2 precursors (ILC2P)
ID2+ ILCP	NR1f3-2, Tox	LTi precursors (LTiP)
ID2+ ILCP	NR1f3-2	ILC17 precursors (ILC17P)
ID2+ ILCP	NR1f3-2, bHLHe76	ILC22 precursors (ILC22P)

Lymphoid tissue inducer cells derive from hematopoietic progenitor cells in the fetal liver [273]. They seed the developing lymphoid tissue during embryogenesis. They initiate the signals that recruit lymphocytes to lymphoid tissue during development. In adult intestines, LTi-like cells express RORγ2 and lodge mainly in cryptopatches and isolated lymphoid follicles. LTi-like cells, as well as cryptopatches and lymph nodes can develop in the intestine in the absence of gut bacteria. On the other hand, isolated lymphoid follicle LTi-like cells are created only after gut colonization by bacteria [273]. Mesenteric LTi-like cells support the production of antibodies by B cells and prevent intestinal epithelium damage by bacteria. Like effector helper T_{H2} cells, mesenteric innate lymphocytes produce large amounts of interleukin-5, -6, and -13. However, some intestinal innate lymphocytes do not rely on RORγ2 for development, whereas other types of innate lymphocytes in the colon express RORγ2. The latter ILC subpopulation secretes cytokines of the T_{H2} set as well as interferon-γ and IL17 and IL22.

3.11.1 Lymphoid Tissue-Inducer Cells

Distinct subpopulations of innate lymphocytes include embryonic lymphoid tissue inducer cells (LTi) required for fetal lymphoid tissue organogenesis and a population of natural killer-like cells that operate in gut immune responses, in particular in intestinal defense before the gut is colonized by intestinal microbiota [273]. Lymphoid tissue inducer cells and stromal organizer cells are implicated in the development of lymphoid structures, e.g., lymph nodes and gut-associated lymphoid tissue that encompasses Peyer's patches, cryptopatches, and multiple isolated lymphoid follicles.

Lymphoid tissue inducer cells express cytokines TNFSF2 and TNFSF3[138] that form TNFSF2–$TNFSF3_2$ complexes.[139] They then activates local stromal cells via the TNFRSF3 receptor [274].

Developing lymph nodes collect CD4+, CD3−, TNFSF2+, TNFSF3+ cells that can differentiate into antigen-presenting cells and NK-like cells, but not into B and T cells [272]. LTi cells stimulate lymph node formation via LTi cell–stromal cell clustering. Subsequent binding of TNFRSF3 receptor on stroma-organizer cells to TNFSF2–$TNFSF3_2$ heterotrimer on LTi cells upregulates expression of adhesion molecules, such as VCAM1, ICAM1, and MAdCAM1. Moreover, this binding provokes the secretion of many chemokines (CXCL13, CCL19 and CCL21). Consequently, hematopoietic cells, such as B and T cells and dendritic cells, are recruited to form the lymph node.

In humans, LTi cells have been detected in fetal mesenteric lymph nodes. They synthesize the transcriptional regulators ID2 (or bHLHb26) and RORγ2 (or NR1f3-2) as well as several types of plasmalemmal antigens (e.g., CD117, or SCFR, and CD127, or IL7Rα chain). Fetal LTi cells may persist after birth. Postnatal LTi cells are involved in the formation of isolated lymphoid follicles in the gut in response to pathogen-associated patterns as well as in repair of damaged lymph nodes after acute viral infection, at least in adult mice [272]. In addition, LTi cells communicate not only with stromal cells, but also with other immunocytes.

3.11.2 ILC17 Cells

IL17-producing ILC cells reside mainly in the intestinal tract, especially the colon. They express and require RORγ (NR1f3) for development and function [272]. Production of IL17 by ILC17 cells is regulated by IL23 cytokine.[140] ILC17 cells are recruited to the inflammed intestine.

ILC17 cells have a CD4−, CD117− (SCFR), NCR1− phenotype. They can then be distinguished from CD4+, CD117+ LTi cells and NCR1+ ILC22 cells. ILC17 cells also express thymocyte differentiation antigen Thy1 (CD90), which is also produced by NK and ILC22 cells.

---

[138] A.k.a. lymphotoxins LTα and LTβ, respectively.

[139] Secreted protein TNFSF2 forms a homotrimer. Transmembrane protein TNFSF3 anchors to the cell surface the TNFSF2–TNFSF3 complex. The predominant form of surfacic TNFSF2–TNFSF3 heterotrimer is the TNFSF2–$TNFSF3_2$ heterotrimer.

[140] In humans, fetal CD117+, CD127+ LTi cells also express IL17, but LTi cells express mostly IL22 and very little IL17 in the postnatal tonsil.

3.11.3 NKR+ LTi-like Cells – ILC22 Cells

ILC22 cells, also called NKR+ LTi-like cells, NK22 (NK-like cell that produce large amounts of IL22), NCR22, and ILC22 cells, are mainly observed in the lamina propria of the intestine, Peyer's patches, mesenteric lymph nodes, and palatine tonsils.

In humans, they express CD56 (140-kDa isoform of neural cell adhesion molecule) and natural cytotoxicity receptor NCR2,[141] but have low NCR1[142] expression [272]. These cells differ from conventional NK cells, as they are non-cytotoxic, lack killer inhibitory receptors, and do not produce or only tiny amounts of Ifnγ. Human IL22-producing ILCs also synthesize CD117, or SCFR, and CD127, or IL7Rα chain, as well as IL5 and IL13; in addition to IL22. Human ILC22 cells isolated from tonsils secrete many cytokines, such as IL2, IL5, IL8, IL13, and TNF agents. They also secrete large amounts of B-cell-activation factor; they may thus contribute to the regulation of T-cell-independent antibody production.

Like LTi cells, ILC22 cells can induce expression of ICAM1 and VCAM1 on mesenchymal stem cells in vitro [272]. Soluble IL23 can regulate IL22 production by ILC22 cells in humans. Cytokines of the common γ-chain family (IL2, IL7, and IL15) can also activate the proliferation and cytokine production of human ILC22 cells.

In the adult small intestine, natural killer-like cells express RORγ2 (NR1f3-2) and secrete IL22. Cells of the family of pro-inflammatory RORγ2+ innate lymphoid cells differentiate from distinct fetal liver RORγ2+ precursors [274]. They develop in the absence of microbiota.

T_{H2}-like innate lymphoid cells have multipotent progenitor capacity. They generate mastocytes, basophils, and macrophages when stimulated by interleukin-25 [274].

3.11.4 Nuocytes

Well-known leukocyte effectors of innate and adaptive immunity include T and B cells, natural killer T cells, natural killer cells, dendritic cells, neutrophils, eosinophils, mastocytes, basophils, and macrophages. Some types of innate immunocytes, such as basophils, eosinophils, and mastocytes, release type-2 cytokines. Interleukin-25 (or IL17e) and IL33 activate the T_{H2} response. Type-2 immunity reacts to parasitic infections. In particular, IL13 cytokine mediates host response to helminth intruders and allergy. Type-2 cytokines cause eosinophilia, goblet cell hyperplasia, mucus secretion, and IgE production.

---

[141] A.k.a. natural killer cell P44 protein (NKp44).

[142] A.k.a. NKp46.

In addition to T_{H2} cells, another type of innate type-2 immunity effector, the non-T, non-B cells, also called nuocytes and natural helper (NH) cells, respond to IL25 cytokine. The latter causes their proliferation and production of T_{H2} cytokines such as IL13 agent. In fact, nuocyte responds to 2 potent initiators of type-2 immunity, interleukin-25 and -33, as it expresses IL33R (or IL1RL1) and IL17Rb, which are receptors for IL33 and IL25, respectively. It represents the predominant early source of IL13 during helminth infection, hence their name (ν is letter 13 in the Greek alphabet) [275].

Nuocytes constitute lymphoid structures associated with adipose tissue in the mesentery (peritoneal cavity that connects the small intestine [jejunum and ileum] to the posterior wall of the abdomen). Nuocytes can be detected in very small quantities in mesenteric lymph nodes.

Nuocytes (or ILC2 cells) differ from LTi and ILC22 cells, as they are RORγ2− (NR1f3-2−) cells. Nuocytes synthesize SCFR and IL7Rα chain, as well as Ly6a in mice. These innate cells also produce large amounts of T_{H2} cytokines, not only IL13, but also IL5.

Nuocytes expand rapidly in response to IL25 and IL33 cytokines. They operate as sentinels. In response to helminth invasion, they secrete high levels of IL5 and IL13 on the one hand and enhance T-cell activity on the other.

3.11.5 Natural Killer Cells

A small fraction (~2%) of the circulating lymphocytes are *natural killer* (NK) cells (cell turnover in blood ~ 2 wk). Natural killer cells are distinct from T and B cells. They can distinguish normal cells from abnormal cells, such as cancer, stressed (e.g., after genotoxicity), and infected cells. Natural killer cells recognize stressed cells endowed or not of antibodies that increase or decrease the expression of various ligands for NK-cell receptors, i.e., that downregulate MHC class-1 molecules or upregulate stress-induced self molecules. On the other hand, NK cells spare healthy cells that express self-MHC class-1 molecules as well as low amounts of stress-induced self molecules.

Upon engagement of its plasmalemmal receptors, i.e., once they are activated, NK cells trigger killing of virus-infected and tumor cells and synthesize and secrete cytokines and chemokines, type of which depends on the context. They interact with dendritic cells. They either boost or dampen activity of macrophages and T cells via interferon-γor interleukin-10, respectively [276].

As they have a constitutive cytolytic function, cells of this separate lymphocyte lineage were originally defined as effector lymphocytes of innate immunity that exert rapidly their function using a limited set of receptors. However, they can also have antigen-specific immunological memory [276]. In adaptive immunity, 2 types of lymphocytes — B and T cells — produce a large repertoire of antigen receptors (B- and T-cell receptors), once they encounter antigens in specialized lymphoid organs and undergo cell division and maturation.

Table 3.38. Examples of plasmalemmal receptors of NK cells (Source: [276]; CEACAM1: carcinoembryonic antigen-related cell adhesion molecule-1, a.k.a. biliary glycoprotein BGP1 and CD66a; DAP: DNAX-activation protein; DNAM: DNAX accessory molecule; KIR: killer-cell immunoglobulin-like receptor; KLR: killer-cell lectin-like receptor; LAIR1: leukocyte-associated immunoglobulin-like receptor-1, or CD305; NCR: natural cytotoxicity-triggering receptor; NKG2: NK-receptor group-2 member; SLAMF: signaling lymphocytic activation molecule family member; TIGIT: T-cell immunoreceptor with Ig and ITIM domains that binds CD155 [a.k.a. PVR (poliovirus receptor) and nectin-like protein NecL5] and CD112 [a.k.a. PVRL2 (poliovirus receptor-related-2), herpesvirus entry mediator-B (HVEb), and nectin-2]). These receptors can be categorized into stimulatory (here given with corresponding adaptor molecules), inhibitory, cytokine, chemotactic, and adhesion receptors. In addition to MHC class-1-specific receptors, other NK cell inhibitory receptors specific for non-MHC ligands also regulate NK-cell reactivity. Agent CD16 is the low-affinity IgG Fc receptor-3.

Receptor category	Members (adaptors)
Stimulatory	NCR1/3 (CD3ζ, FcγR), NCR2 (DAP12), KLRf1, CD16 (CD3ζ, FcγR), KIRs (DAP12), NKG2c (DAP12), NKG2d (DAP10), SLAMF3 (SAP), SLAMF4 (SAP, EAT2, ERT), SLAMF5 (SAP, EAT2), SLAMF6 (SAP), SLAMF7 (SAP, EAT2)
Inhibitory	KIRl, LIRb1, NKG2a (KLRc1), KLRg1, TIGIT, CEACAM1, LAIR1
Cytokine	IL1R, IL2R, IL12R, IL15R, IL18R, IL21R IfnαR
Chemotactic	CCR2/5/7, CXCR1/3/4/6, CX3CR1, $S1P_5$
Adhesion	CD2, DNAM1, $\beta_{1/2}$-integrins

Natural killer cell synthesizes a set of activatory and inhibitory receptors to ensure: (1) an efficient struggle against viral infection and tumor development as well as (2) self tolerance (Table 3.38). It adapts to its environment. It thus has attributes of both innate and adaptive immunity. NK cells sense the density of various types of plasmalemmal molecules at the surface of interacting cells. The integration of these signals dictates the quality and intensity of the NK-cell response.

Natural killer cells release cytotoxic enzymes and cytokines to recruit other effector immunocytes. These cytolytic effectors kill virus-infected and mutant cells by antibody-dependent cellular cytotoxicity via granule exocytosis and subsequent release of perforin and granzymes, in the absence of specific immunization. They secrete immunoregulatory growth factors and cytokines (interferon-γ, TNFα, TGFβ, gCSF [CSF3], gmCSF [CSF2], TNFSF2, and interleukins IL1 to IL3, IL5, IL10, and IL13) and chemokines (CC-chemokine ligands CCL2 to CCL5, CXC-chemokine ligand CXCL8 (or IL8), and XCL1) that, in particular, influence the differentiation of cells of adaptive immunity [276].

They colocalize with other hematopoietic cells such as dendritic cells in inflammation sites. They also interact with macrophages and neutrophils. Conversely,

NK cells require priming by various factors, such as IL15 transmitted by dendritic cells or macrophages, IL12, or IL18, to achieve their full effector potential [276]. Cytokine receptors IL2R, IL15R, and IL21R are involved in NK-cell development and maturation.

Natural killer cells express NK-cell markers and possibly a T-cell receptor, but not the B-cell receptor. The NK complex of the genome encodes for NK-cell receptors and type-2 C-type lectin-like molecules that are involved in cell recognition.

3.11.5.1 NK-Cell Stimulatory and Inhibitory Receptors

Detection of NK-cell targets relies on a set of plasmalemmal stimulatory and inhibitory receptors. Natural killer cells express inhibitory receptors (tolerance to self) for both MHC class-1 or non-MHC ligands.

The body's immunological tolerance to its own cells relies on cells of both innate and adaptive immunity. Natural killer cells express various inhibitory receptors that recognize self MHC molecules that prevent cytolysis. *Autoreactive T cells* do not experience the negative selection in the thymus that cause deletion of most developing T cells with specificity for self-antigens. NK cells use inhibitory receptors to gauge the absence of constitutively expressed self molecules on susceptible target cells.

NK-Cell Inhibitory Receptors

Killer inhibitory receptors comprise killer cell lectin-like receptors, such as KLRa3[143] that detects the MHC class-1 molecule H2 K^b, KLRc1,[144] as well as killer cell (KIR; CD158a, CD158b1–b2, CD158c–CD158d, CD158e1–e2, CD158f–CD158k, CD158z) and leukocyte (LIR or LILR; CD85a–CD85m) immunoglobulin-like receptors [277, 278].

Proteins of the KIR and LIR families are structurally and functionally similar. Killer cell immunoglobulin-like receptors are expressed not only by NK cells, but also a subpopulation of T cells.[145] Receptors of the LIR family have a more widespread cellular distribution. They are expressed on various hematopoietic cell types, such as NK, T, and B lymphocytes, as well as macrophages, mastocytes, and dendritic cells.

Members of the KIR and LIR families possess isoforms with different cytoplasmic signaling functions. Receptors of the KIR and LIR families with long

---

[143] A.k.a. lymphocyte antigen Ly49c.

[144] A.k.a. member A of group 2 of NK receptors (NKG2a).

[145] Genes that encode the KIR receptors that preclude the innate cytotoxic activity of NK lymphocytes constitute a genetic region on chromosome 19, the so-called *human leukocyte receptor complex* (LCR).

cytoplasmic tails, such as killer cell immunoglobulin-like receptor, 2 domains, long cytoplasmic tail (KIR2DL1–KIR2DL5) and those with 3 domains (KIR3DL1–KIR3DL3), inhibit cellular activation, as they recruit PTPn6 phosphatase. Receptors of the KIR and LIR families with short cytoplasmic tails, such as killer cell immunoglobulin-like receptor, 2 domains, short-cytoplasmic tail (KIR2DS1–KIR2DS5) and 3 domains (KIR3DS1), transmit activation signals via adaptor protein tyrosine kinase-binding protein TyroBP[146] Despite their similarities, KIR and LIR receptors have distinct functional characteristics. Only LILRb1 and LILRb2 receptors bind to various MHC class-1 ligands.

NK-Cell Stimulatory Receptors

NK cells have numerous stimulatory receptors that cooperate to activate T and B cells. Killer stimulatory receptors encompass killer cell lectin-like receptors KLRa8,[147] KLRb1,[148] KLRb1c,[149] KLRf1,[150] and KLRk1;[151] CD2, low-affinity IgG Fc receptor-3 (CD16), DNAX accessory molecule DNAM1 [or CD226];[152] natural cytotoxicity-triggering receptors, such as NCR1,[153] NRC2,[154] NCR3,[155] and NKG2d-DAP10 receptor complex.[156] Receptors used by NK cells in tumor surveillance include KLRf1, KLRk1, NCR1 to NCR3, DNAM1,

[146]A.k.a. DNAX-activation protein DAP12, killer-activating receptor-associated protein [KARAP], and polycystic lipomembranous osteodysplasia with sclerosing leukoencephalopathy protein [PLOSL]).

[147]A.k.a. lymphocyte antigen Ly49h.

[148]A.k.a. NK-cell receptor protein-1 [NKRP1]).

[149]A.k.a. natural killer cell-surface protein NK1.1 or NKRP1c and CD161.

[150]A.k.a. NKP80.

[151]A.k.a. natural killer group-2, member-D [NKG2d] and CD314. Receptor KLRk1 recognizes several self ligands that are often expressed on cancer cells and causes a protective immune response. This lectin-like type-2 transmembrane homodimer is expressed by almost all NK and activated CD8+ T cells, as well as subsets of $\gamma\delta$ T and NKT cells, and, in some conditions, CD4+ T cells. Ligands of the KLRk1 receptor are related to self-proteins that are similar to MHC class-1 molecules, such as MHC class-1 (I) polypeptide-related sequence A (MICa) and B (MICb) as well as retinoic acid early transcript RAET1 in humans. Normal cells do not express these ligands at substantial levels. Ligands of the KLRk1 receptor are upregulated in cancerous or stressed cells [279].

[152]Receptor DNAM1 is constitutively expressed by most NK cells, T lymphocytes, macrophages and dendritic cells [279]. Its ligands include nectin-2 (a.k.a. CD112 and poliovirus receptor-related PVRL2) and nectin-like protein NecL5 (a.k.a. CD155 and poliovirus receptor PVR).

[153]A.k.a. natural killer cell P46-related protein NKP46 and CD335.

[154]A.k.a. natural killer cell P44-related protein NKP44 and CD336.

[155]A.k.a. natural killer cell P30-related protein NKP30 and CD337.

[156]Natural cytotoxicity receptors comprise not only NCR1 to NCR3, but also KLRf1 (or NKP80). In humans, NRC1, NCR3, and KLRf1 are expressed by all NK cells, whereas NRC2 is only produced by activated NK cells [279].

and signaling lymphocyte activation molecule (SLAM)-related family member SLAMF4 [279].[157]

Integration of signals that emanate from stimulatory and inhibitory receptors during target cell recognition determines whether NK cells will eliminate their targets or not. Normal cells display stimulatory ligands, but their MHC class-1 molecules engage inhibitory receptors that counteract stimulatory signaling to avoid killing.

3.11.5.2 Licensed NK-Cell Receptors

Target cell recognition, a process that is called "licensing", "arming", or "education" leads to full functional maturation of NK cells and promotes the reactivity of mature NK cells [280]. Natural killer cells, the inhibitory receptors of which bind to major histocompatibility complex class-1 molecules are "licensed", or supposed to be more responsive to stimulation. On the other hand, "unlicensed" NK cells that lack receptors for MHC class-1 molecules are assumed to be hyporesponsive [281].

Stimulatory receptors often recognize self-ligands that are expressed selectively by transformed, infected, or damaged cells. Inhibitory receptors recognize MHC class-1 molecules on target cells. Cells that lack MHC class-1 molecules are much more sensitive to killing primed by NK cells than those with normal amounts of MHC class-1 molecules [280].

NK receptors NKG2d target stress-induced self ligands produced by stressed cells (cytomegalovirus UL16-binding protein [ULBP] and MHC class-1 (I) polypeptide-related sequence molecules [MIC]) [282]. Other alert molecules include infectious non-self ligands and Toll-like receptor ligands. A priori, binding of NKG2a (or KIRc1) and KIR by specific MHC class-1 molecules impedes killing by NK cells, whereas ligation of activating receptors with corresponding ligands triggers killing.

In addition, certain inhibitory receptors can bind to their ligands before target cell encounter, thereby influencing NK-cell responsiveness during subsequent target cell recognition [280]. In some circumstances, unlicensed NK cells are more efficient than licensed cells at eliminating abnormal cells. Therefore, NK cells devoid of inhibitory receptors for MHC molecules can exert a predominant role [280]. For example, unlicensed NK cells are the main contributors of the control of cytomegalovirus infection in mice [281].

[157]The SLAMF4 receptor, a.k.a. natural killer cell receptor NKR2b4 and CD244, is expressed by all NK cells, $\gamma\delta$T cells, a subset of CD8+ T cells, and all human CD14+ monocytes [279]. Its unique ligand is SLAMF2 receptor (or CD48). It operates as either a stimulatory or inhibitory receptor depending on the expressed isoform that results from alternative splicing.

3.11.5.3 Interactions between NK Cells and Other Cell Types

The balance between cell death and coexistence results from stimulatory and inhibitory signals delivered by plasmalemmal receptors. Natural killer cells are able to discriminate abnormal cells from cells of their normal environment. NK cells are tolerant to cells expressing MHC class-1 molecules. They circulate in blood vessels and lymphatics and patrol into tissues to check possible presence of transformed or infected cells that downregulate the expression of MHC class-1 molecules. NK cells then form synapses with target cells.

NK cells directly or indirectly interact with dendritic, NKT, B, and CD4+ T cells. In addition, NK cells can inhibit or promote the activation of autoreactive T cells during the initiation of autoimmunity [277]. NK cells are controlled by regulatory T cells and can operate as regulators, as they can kill macrophages, dendritic cells, and activated T and B lymphocytes and interact with endothelial cells.

Three NK-cell receptors promote adhesion to endothelial cells: (1) $\alpha_4\beta_1$-integrin that binds VCAM1; (2) L-selectin that links to addressins; and (3) CX3CR1 that associate with CX3CL1.

NK cells either cause vascular damage by killing endothelial cells or promote angiogenesis in other circumstances, such as during pregnancy [282].[158] NK cells promote vascularization of embryos.

3.11.5.4 NK-Cell Precursors and Subpopulations

NK cells lodge in different tissue types. They are phenotypically and functionally diverse. The NK-cell developmental niche with its set of stromal cells that secrete growth factors can determine the fate of NK precursors.

Precursors of NK cells can differentiate into NK-effector subsets according to the cytokine types. **NK1** cells stimulated by IL12 produce Ifnγ, whereas **NK2** cells stimulated by IL4 synthesize IL5 and IL13 [277]. NK1 cells kill target cells by a process based on TNFRSF6a (or CD95), perforin, and granzyme. NK2 cells use a mechanism that relies on TNFSF10 agent.

In humans, 2 populations of NK cells can be distinguished using CD56 marker, a 140-kDa isoform of neural cell adhesion molecule (NCAM). NK cells indeed comprise $CD56^{low}$ and $CD56^{high}$ subsets [282].

Most blood and spleen NK cells are $CD56^{low}$, CD16+ (low-affinity receptor for immunoglobulin-G, FcγR3) NK cells that are cytotoxic and produce CXCR1, perforin, and Ifnγ; they have enhanced killing activity. Excited with a proper stimulus, $CD56^{low}$, CD16+ NK cells synthesize abundantly cytokines.

---

[158] During pregnancy, endometrium is enriched with uterine natural killer cells that correspond to a specific subset of NK cells. They secrete vascular endothelial growth factor, placental growth factor, and NKG5 to remodel the arterial network and support the placental development.

Most NK cells in lymph nodes and tonsils are $CD56^{high}$, CD16− with CCR7 expression, but lack KIRs and have a low intracytoplasmic perforin content. They secrete large amounts of cytokines (Ifnγ, CSF2, and TNFSF1). Some human $CD56^{high}$, CD16− NK cells express IL7Rα chain (CD127). They are characterized by an enhanced cytokine production with a T_{H1} cytokine expression pattern.

3.11.5.5 Regulators of NK-Cell Activity

Various substances stimulate or inhibit NK-cell activity. Type-I interferons and interleukin-12, -15, and -18 are potent activators of NK cells. Interleukin-2 promotes NK-cell proliferation, cytotoxicity, and cytokine secretion. Transforming growth factor-βand IL10 are inhibitors. Memory T and NK cells that express IL23R and RORγ2 (NR1f3-2) constitutively synthesize IL17 rapidly and independently of IL6, in contrast with T_{H17} cells [283].

3.11.5.6 Immunological Memory

Immunological memory is characterized by the long-term persistence of memory cells that rapidly undergo clonal expansion and enhanced effector function in response to new challenges, especially viral infections. About 2 months after initial infection, memory NK cells can be detected in nearly all organs (spleen, lymph nodes, liver, lung, and kidney) [276].

3.12 B Lymphocytes

Bone marrow (B) lymphocytes, or B cells, are responsible for the humoral (antibody-mediated) immunity. B lymphocytes are component of the adaptive immunity via synthesized, relevant antibodies. B lymphocytes produce immunoregulatory antibodies against antigens. Each B lymphocyte possesses a unique plasmalemmal B-cell receptor (BCR), a membrane-bound immunoglobulin, that can bind to a cognate antigen.

B lymphocytes develop in the liver of fetuses and in the bone marrow of adults from hematopoietic stem cells. All B lymphocytes derive from a particular cell (clone). They recognize and bind the same component of a given antigen. The huge number of clones explains the wide range of detected antigens.

Naive B lymphocytes will meet antigens for the first time. In the absence of antigen, B lymphocytes are at rest, but they are not inactive. These antigen-presenting cells mature after activation by antigen. Antigens recruit antigen-specific cells from the pool of mature B lymphocytes. Once a B lymphocyte encounters its cognate antigen and receives an additional signal from a helper T lymphocyte, it differentiates into either a plasmocyte or memory B cell.

3.12.1 B-Lymphocyte Subpopulations

B-lymphocyte types include: (1) short-lived plasma B lymphocytes, or *plasmocytes* (plasma cells), that have been exposed to antigen and produce and secrete large amounts of antibodies; (2) long-lived *memory B lymphocytes* that derive from activated B cells and are specific to the antigen encountered during the primary immune response; (3) *B1 lymphocytes* that express IgM in greater quantities than IgG [285], hence having polyspecific, low-affinity receptors, and reside in lymph nodes and spleen in small amounts, but lodge predominantly in the peritoneal and pleural cavities; (4) *B2 lymphocytes* that synthesize IgG1 precursors more ($>$ 7 times) than IgM precursors [285]; (5) *marginal-zone* (MZ) *B lymphocytes* that are mature, naive, non-circulating B cells; and (6) naive *follicular B lymphocytes* that circulate until they encounter antigen or die.

Immature B cells produce plasmalemmal immunoglobulin-M. They leave the bone marrow and differentiate into B1 and conventional B2 cells with distinct locations and activation modes. B-cell clones that reach the mature long-lived pool include the B1 subset, $CD21^{high}$ compartment (mostly marginal zone), whereas others recirculate primarily among B-cell follicles. This segregation depends on B-cell receptors and B-lymphocyte coreceptor CD19 [284]. B-lymphocyte subpopulations B1a, B1b, B1c ($\alpha_M\beta_2$-integrin− and CD5+), and B2 have different roles during infections.

3.12.2 B-Lymphocyte Maturation Stages

B lymphocytes mature into naive B lymphocytes that express plasmalemmal IgM and IgD via 2 transitional stages. During the second maturation stage, some B lymphocytes home to and remain in the marginal zone of the spleen to form a pool of mature, naive, non-circulating, *marginal-zone B cells*.

The remaining part of the B-cell population (most B cells) mature into naive *follicular B cells* that circulate to the spleen follicles and lymph nodes via the bone marrow until they encounter cognate antigen or die. Follicular B cells that are activated by antigen, possibly in cooperation with T cells, form *germinal centers* of secondary lymphoid follicles and their antibody-encoding genes are modified. Some germinal center B cells express an immunoglobulin IgG-type BCR; others immunoglobulin IgM-type BCR. These 2 BCR types prime different signaling.

When follicular B cells receive costimulation from activated T cells, they can develop into *plasmablasts* or *long-life memory B cells*. Memory B cells that can quickly react to presentation of other antigen types express antibodies that are not secreted.

Germinal centers are thus sites where memory B cells and plasmocytes are generated for long-lasting, antibody-mediated (humoral) immunity. T-cell-dependent immunity relies on the selection of high-affinity B cells and depends on high-rate somatic mutation of immunoglobulin V regions to generate variants. B cells that possess high-affinity antigen receptors are selected for survival and proliferation (*affinity maturation*). B-cell receptors are devoted to either BCR signaling or antigen presentation. Higher-affinity BCRs more effectively capture antigens, which are subsequently presented to helper T cells, resulting in higher-affinity B cells that can then receive more T-cell-emitted survival or proliferative signals. Germinal center B cells with high-affinity B-cell receptors are selectively expanded, as B-cell receptors promote the selective survival or expansion of higher-affinity germinal center cells.

Spontaneous and induced signalings are associated with increased phosphatase activity [286]. Both SH2 domain-containing phosphatase PTPn6, and SH2 domain-containing inositol 5-phosphatase are hyperphosphorylated in germinal center cells and colocalize with liganded B-cell receptors.

3.12.3 Antibodies and B-Cell Receptors

Antibodies are glycoproteins. Their basic functional unit is an immunoglobulin monomer. Immunoglobulin comprises 4 polypeptidic chains, 2 identical, heavy and 2 identical, light chains connected by disulfide bonds.

Five types of Ig heavy chain exist (α, γ, δ, ϵ, and μ) that define the class of antibody (IgA, IgD, IgE, IgG, and IgM). Two types of Ig light chain exist (κ and λ). Secreted antibodies can be monomers (IgD, IgE, and IgG), dimers (IgA), and pentamers (IgM).

Immunoglobulin monomer is Y shaped. The Y arms contain the antigen-binding fragment (F_{AB}) region that is composed of a constant and variable (F_V) domain from each heavy (V_H) and light (V_L) chain of the antibody. The Y stem corresponds to the crystallizable fragment (F_C) that can bind to a specific class of Fc receptors as well as complement components.

Immunoglobulin-M is the first antibody to be produced during an immune response. It exists as [287]: (1) a *membrane-bound dimer* on the surface of B cells, i.e., the B-cell receptor, and (2) a *secreted pentamer* ($_S$IgM), mainly in blood (occasionally also as a hexamer that lacks a joining [J] chain).

Secreted IgM can be divided into natural (i.e., IgM in foreign antigen-free blood) and immune types. *Immune* $_S$IgM *type* is secreted during exposure to pathogens; it is antigen specific. *Natural* $_S$IgM *type* is produced by B1 and B2 cells. Immune IgM is mainly produced by B2 cells, but B1a and B1b cells can also secrete IgM in response to immune signals.

Natural $_S$IgM is characterized by its polyreactivity, as it protects against viral, bacterial, fungal, and parasitic infections, and enhances pathogen neutralization and agglutination [287]. It promotes pathogen clearance in synergy with complement component C1q, as it stimulates their engulfment by phagocytes.[159] It also increases the presentation of pathogen-derived antigens. Furthermore, it facilitates the removal of apoptotic cells [287].

Natural $_S$IgM is particularly secreted by peritoneal B1 lymphocytes [287]. It recognizes and binds self antigens such as phospholipids. It interacts with members of the complement system, mannose-binding lectin, and IgM-devoted Fc receptors (FcμR as well as Fcα–FcμR expressed by follicular dendritic cells, macrophages, and B cells that bind both IgA and IgM). Natural IgM promotes the generation of mature B cells in the spleen, but reduces the survival of peritoneal B1a cells [287].

B-cell receptors are members of the family of multichain immune-recognition receptors (MIRR) involved in adaptive immunity that also includes T-cell receptors and the high-affinity receptor for IgE (FcεR1) [288]. These receptors localize to membrane rafts. The engagement of antigens by BCR, TCR, and FcεR1 causes receptor clustering for signaling.[160]

3.12.3.1 Antibody Diversification

Clonal selection of specific B cells results from antigen binding to B-cell receptors owing to somatic hypermutation and class switching. In cooperation with helper T cells and cells of the innate immunity, antigens provoke proliferation and differentiation of naive B cells into memory B cells and antibody-secreting plasmocytes. Maturation with increased antibody affinity and class switching occur at the earliest step of BCR signaling [288].

Antibody diversification enables the humoral response. B lymphocytes diversify their antibody repertoire via genetic modifications during 2 distinct phases of B-cell maturation. In the *antigen-independent phase of B-cell maturation*, B-cell precursors in the bone marrow generate antigen recognition diversity by assembling exons that encode variable immunoglobulin heavy and light chains via gene recombination. Plasmalemmal IgM is then expressed by new B lymphocytes. After further stages of differentiation and expression of IgD, B lymphocytes exit from the bone marrow and migrate to secondary lymphoid organs, where they initiate the *antigen-dependent phase of B-cell development.*

In the presence of antigen, mature B lymphocytes in germinal centers of secondary lymphoid follicles diversify their antibody repertoire via somatic hypermutation and class switching [271].[161]

---

[159] Monomeric IgM does not bind complement component C1q.

[160] Antigen-engaged TCR clusters recruit leukocyte-specific cytosolic Tyr kinase (LCK), TCR coreceptor CD2, and adaptor Linker for activation of T cells (LAT). Antigen-committed BCR clusters attract spleen Tyr kinase (SYK).

[161] During an immune response, B lymphocytes in germinal centers undergo somatic hypermutation that introduces point mutations at high rates into the variable regions of immunoglobulin genes, thereby allowing selection of B cells with higher antibody affinity. Mature B lymphocytes also sustain class-switch recombination to rearrange DNA segments encoding the constant region

B cells can be activated independently from T cells by bacterial polysaccharides and microorganism-derived Toll-like receptor ligands, to generate short-lived, low-affinity, extrafollicular plasmocytes. However, interactions between T and B cells in secondary lymphoid tissues[162] are required for antibody responses specific for foreign antigens and generation of high-affinity memory B cells and long-lived plasmocytes. Such a process particularly involves *follicular helper T cells* (T_{FH}) that express chemokine receptor CXCR5,[163] like activated CD4+ T cells in the tonsils and memory CD4+ T cells in blood [289].[164] The lymphocytes that become T cells migrate from the bone marrow to the thymus, where they mature.

---

of the immunoglobulin genes, so that B lymphocytes change antibody isotypes from IgM to IgG, IgA, or IgE. Somatic hypermutation introduces point mutations into VHDJH and VLJL exons that encode immunoglobulin heavy (H) and light (L) chain variable regions from individual variable (V), diversity (D), and joining (J) gene segments. Class switching substitutes the IgH constant region $C\mu$ and $C\delta$ genes encoding primary IgM and IgD with $C\gamma$, $C\alpha$, or $C\epsilon$ genes (class-switch recombination) to generate secondary IgG, IgA, and IgE that have the same antigen specificity as IgM and IgD, but different functions. Circulating IgAs that include mostly members of the IgA1 class are predominantly present as monomers. They bind to various receptors expressed by granulocytes, monocytes, macrophages, dendritic cells, hepatocytes, hepatic Kupffer cells, and renal mesangial cells. Mucosal IgAs exist as dimers and oligomers and comprise both IgA1 and IgA2 types encoded by $C\alpha1$ and $C\alpha2$ genes.

[162]After binding antigen, B cells interact with activated T cells in the T-cell zones of secondary lymphoid tissues. B cells upregulate expression of chemokine receptor CCR7 and are attracted by a gradient of CCL21 chemokine to the outer T-cell zone of secondary lymphoid tissues, where B cells interact with T cells. B cells then differentiate using 2 pathways: (1) a follicular pathway that creates germinal centers and (2) an extrafollicular pathway. In the *extrafollicular pathway* of both T-cell-dependent and -independent responses, B-cell precursors migrate to the splenic bridging channel that connects the T-cell zone with the spleen red pulp or medullary cords in the lymph nodes. They then differentiate into short-lived plasmocytes that produce low-affinity antibody and form extrafollicular foci of such cells. In the *follicular pathway*, B-cell precursors of germinal centers give birth to high-affinity memory B cells and long-lived plasmocytes. Germinal centers of follicles, at least in tonsils, are composed of: (1) a dark zone with *centroblasts* (dividing germinal-center B cells that undergo mutation of their immunoglobulin genes); (2) a light zone with *centrocytes* (dark zone-exiting centroblasts), and follicular dendritic and T cells (with a large proportion of CD4+, CD57+ T cells); (3) an outer zone with CD4+, CD57− T cells; and (4) a mantle zone with IgD+ resting B cells and CD4+ T cells. The interaction with follicular helper T cells supports selection of high-affinity centrocytes, leads to the differentiation into long-lived plasmocytes and memory B cells, and stimulates centrocytes to recycle into centroblasts.

[163]Receptor CXCR5 is a receptor for chemokine CXCL13, also named B-cell-attracting chemokine BCA1, a chemokine secreted by follicular stromal cells, T_{FH} cells, and myeloid and plasmacytoid dendritic cells.

[164]T_{FH} cells differ from type-1 (T_{H1}) and -2 (T_{H2}) helper T cells. Various lineages of helper T cells have distinct differentiation pathways, such as CD4+, IL17+ cells implicated in autoimmunity.

3.12.4 B-Cell Activation and Signaling

Mature B lymphocytes are activated by antigen binding to B-cell receptors that trigger various signaling cascades.[165] B-cell receptors are needed for the survival of B lymphocytes, whatever the developmental stages. B-cell receptors may yield constitutive, low-level, tonic signals that bypass ITAM motif phosphorylation [288]. B lymphocytes recognize both soluble antigens and antigens attached on the surface of antigen-presenting cells [288].

Then, B-cell receptors cluster independently of activated signaling axes under the control of B-cell coreceptors and initiate signaling. Inactive BCR dimers can coexist with active BCR monomers [288]. Crosslinking between BCRs is dispensable, as a single antigen-committed BCR triggers signaling. Clusters of BCRs grow with time after antigen binding. The formation of BCR oligomers reduces their motility in the plasma membrane.

B-cell receptor responses are regulated by various coreceptors, such as IgG receptor FcγR2b, a potent inhibitor of BCR signaling, and CD19, a component of the B-cell coreceptor and a BCR activator. Receptor FcγR2b reverses the action of BCR-activated PI3K, as it recruits the lipid phosphatase SHIP to the coligated complex [288]. Molecule CD19 is phosphorylated in response to BCR–antigen binding. It then acts as an adaptor that recruits guanine nucleotide-exchange factor Vav, phosphatidylinositol 3-kinase, and protein Tyr kinase Lyn to enhance BCR signaling. In addition, CD19 can complex with the complement component receptor CR2[166] to enable complement-coupled antigen to connect to both the CD19–CR2 complex and B-cell receptor [288].

Upon antigen binding, ITAM motif of B-cell receptor is phosphorylated primarily by recruited Lyn kinase and then spleen Tyr kinase (SYK) [288]. The growth of BCR cluster beyond a certain size requires Lyn and SYK activity.

B lymphocytes engage antigen via membrane protrusions that provoke the formation of BCR clusters and an immune synapse [288]. Membrane protrusion in response to membrane-bound antigens relies on the cooperative action of phospholipase-Cγ2, Vav guanine nucleotide-exchange factor, and Rac2 and Rap GTPases.

[165] A B-cell receptor comprises a surface-bound immunoglobulin (IgD or IgM) and transmembrane, Igα–Igβ heterodimer for signal transduction, as their intracellular domains contain an immunoreceptor tyrosine-based activation motif (ITAM). A typical B lymphocyte can display $\mathcal{O}[10^4]$) immunoglobulins on its surface.

[166] A.k.a. C3d complement component receptor (C3dR) and CD21.

3.12.5 Main Regulators of B-Cell Proliferation in Germinal Centers

Transcription factors myeloblastosis viral oncogene homolog MyB and forkhead box FoxM1 are synergistic master regulators of B-cell proliferation in germinal centers, where antigen-stimulated B cells highly proliferate, undergo somatic hypermutation of immunoglobulin genes and are selected according to synthesized high-affinity antibodies [290]. Germinal center-resident B cells, or centroblasts, derive from naive B cells, from which they differ due to the activation of genetic programs that control cell proliferation, such as Aurora kinase-A and budding uninhibited by benzimidazoles homolog BuB1β (a.k.a. mitotic checkpoint kinase MAD3L and BUBR1), and pro-apoptotic programs and repress anti-apoptotic, cell cycle arrest, DNA repair, and those that regulate signal transduction primed by cytokines and chemokines.

3.12.6 B Cell–T Cell Interactions

CD4+ T cells participate in humoral immunity, as they help B cells, and enhance cellular immunity by producing cytokines. Conversely, effector and regulatory B cells not only intervene in humoral immunity against pathogens and autoantigens, but also regulate CD4+ T-cell responses.

Antibody production by B cells to protect against many pathogens is the B-cell's most important effector function. However, effector and regulatory B-cell subsets can enhance effector and memory CD4+ T-cell responses, independently of antibody production, as they regulate the development, proliferation, and maintenance of these CD4+ T cells. They indeed yield costimulation and produce inflammatory and polarizing cytokines that direct the proliferation and effector functions of responding T cells, in addition to antigen presentation [291]. B cells optimize proliferation of CD4+ T cells and can enforce or stabilize the differentiation of T cells into polarized effector cell subsets in cooperation with dendritic cells. B cells promote the development of CD4+ T-cell memory and intervene in the reactivation of memory CD4+ T cells.

Moreover, B cells may also attenuate immune responses, as they can improve the development, survival, and proliferation of regulatory T cells [291]. In addition, IL10-producing regulatory B cells can suppress effector T-cell responses.

3.12.7 Plasmocytes

Plasmocytes (size $\sim$13 μm), or plasma cells (plasma B cells), are effector B cells located in the bone marrow and lymphoid organs. Plasmocytes can survive for long periods in survival niches in secondary lymphoid organs, bone marrow, or inflamed tissue [292].

Plasmocytes yield humoral immunity by secreting antibodies against antigens that confer a continuous existence of potential specific response. They constitute a special pool of immunological memory cells. Short- and long-life plasmocytes derive from memory B lymphocytes.

Stimulation of B cells by antigen causes either cell death or proliferation and differentiation into *plasmablasts*.[167] Plasmablasts can develop from any type of activated B cells: naive marginal-zone, naive follicular, activated germinal-center, and memory B cells.

Plasmablasts express chemokine receptors CXCR4 to migrate to the bone marrow, spleen red pulp, and medullary cords of the lymph nodes, where is produced CXCL12 chemokine. On the other hand, plasmablasts repress synthesis of chemokine receptors CXCR5 and CCR7 that control germinal-center migration, hence avoiding a return to the sites of immune reactions.

Plasmocytes synthesize adhesion molecules, such as E- and P-selectin ligands, α_L-integrin (or CD11a), β_2-integrin (or CD18), and heparan sulfate proteoglycan receptor epican.[168]

Certain plasmablasts migrate to survival niches to maintain a concentration of antibodies following resolution of the immune response, thereby protecting against subsequent challenges. CD4+ helper T cells bind antigens presented by B cells and participate in clones of antibody-secreting plasmocytes.

In the spleen, 2 populations of plasmocytes exist: a long-life plasmocyte pool and a set of rapidly renewing short-life plasmocytes. Most long-life plasmocytes reside in the bone marrow. The development and survival of plasmocytes depend on several factors: interleukins, such as IL6 and IL21; tumor-necrosis factor; the transcription factor B-lymphocyte-induced maturation protein BLIMP1 encoded by the positive-regulatory domain-containing Prdm1 gene; stable isoform of the transcription factor X-box-binding protein XBP1; and suppression of transcription factor of B-cell development paired box protein PAX5 [292].

3.13 Monocytes and Macrophages

The myeloid cell lineage of innate immunity includes 2 distinct lineages of phagocytes: (1) monocytes and macrophages and (2) dendritic cells. Bone marrow-derived myeloid cells — blood-circulating monocytes and tissue-resident macrophages — are effectors and regulators of inflammation and immediate innate immunity. Dendritic cells initiate and regulate the pathogen-specific adaptive immune responses and major effectors of the development of immunological memory and tolerance.

---

[167]Plasmocyte precursors arise from terminal differentiation of B cells that is defined as the switch from expression of plasmalemmal antibody, which acts as an antigen receptor, to antibody secretion.

[168]A.k.a. CD44, extracellular matrix receptor-3, and lymphocyte homing–adhesion receptor GP90.

Blood monocytes, many macrophage subsets, and most dendritic cells originate from common myeloid (CMP), granulocyte–macrophage (GMP), and monocyte–macrophage–dendritic cell (MDP) progenitor and precursor cells.[169] In humans, like all lymphoid cell types, myeloid lineages — monocytes, macrophages and dendritic cells — originate from multilymphoid progenitors, a distinct population of $Thy1^{neg/low}$ (thymocyte differentiation antigen-1 or CD90), CD45+ cells of the CD34+, CD38− stem cell compartment [293].

Monocytes are circulating precursors for tissue macrophages and dendritic cells. However, many types of dendritic cells and macrophages, such as dendritic cells of lymphoid organs, plasmacytoid dendritic cells, cutaneous Langerhans cells, and brain microglial cells, originate from the common monocyte, macrophage, and dendritic cell precursor independently of monocytes [294]. Although inflammatory monocytes with the GPI-anchored myeloid differentiation marker lymphocyte antigen-6 complex, locus G (Ly6G in mice), i.e., Ly6G+, $Ly6C1^{high}$ monocytes, generate macrophages and dendritic cells, many types of dendritic cells and macrophages do not originate from monocytes in normal condition [301]. Furthermore, in some cases, they can develop independently of the bone marrow.

The generation of monocytes, macrophages, and to some extent, dendritic cells depends on cytokine and hematopoietic growth factor receptor CSF1R expressed exclusively on mononuclear phagocytes (monocytes and macrophages) and their precursors [296]. Known CSF1R ligands include CSF1[170] and interleukin-34.

Colony-stimulating factor-1 contributes to the control of survival, proliferation, and differentiation of mononuclear phagocytes. Liganded receptor protein Tyr kinase CSF1R stabilizes and provokes the nuclear translocation of β-catenin that activates genes involved in the cell cycle.[171] Activity of CSF1 relies on its associated adaptor, protein Tyr kinase-binding protein TyroBP that, once phosphorylated, acts as a docking site for spleen Tyr kinase (SYK) and 70-kDa TCR ζ-chain-associated protein kinase (ZAP70), and leads to phosphorylation of glycogen synthase kinase GSK3β [297]. Owing to TyroBP, calcium-dependent focal adhesion kinase-2 phosphorylates β-catenin that then accumulates in the nucleus.

3.13.1 Monocytes

Monocyte (Mo; size 15–25 μm) is the largest leukocyte. Its nucleus is most often U- or bean-shaped. These bone marrow-derived circulating leukocytes account for 3–9% of all leukocytes.

[169] In humans, hematopoietic stem cells are associated with the Lin−, CD34+, CD38− compartment [293]. In adults, hematopoietic stem cells and precursors recirculate between peripheral organs and the bone marrow.

[170] A.k.a. macrophage colony-stimulating factor.

[171] Liganded CSF1R dimerizes, autophosphorylates, and signals via phosphorylation of PTPn1 phosphatase and PLCγ, as well as Src, PI3K, PKB, and ERK kinases. The CSF1–CSF1R complex is subsequently internalized and degraded in lysosomes.

Monocytes originate from a myeloid progenitor shared with neutrophils in the bone marrow and the common monocyte, macrophage, and dendritic cell precursor (MDP). The latter generates monocytes as well as preclassical and plasmacytoid dendritic cells via a common dendritic cell precursor. Monocytes exit into the blood stream by diapedesis, where they circulate for several days.

Monocytes circulate in the blood or reside in the bone marrow and spleen. They do not proliferate in a steady state. These immunocytes are equipped with chemokine and adhesion receptors that are used during their migration from blood to tissues triggered by infection. They produce inflammatory cytokines and take up toxins and cells.

Circulating monocytes give rise to mature tissue-resident *macrophages*. Both monocyte and macrophage lineages are characterized by heterogeneity. Phagocytic monocytes defend the body against viruses and bacteria. Monocytes also differentiate into dendritic cells, mainly during inflammation. Migration to tissues and differentiation to inflammatory dendritic cells and macrophages are likely determined by the inflammatory milieu and pathogen-associated pattern-recognition receptors [296].

Monocytes are effectors of the innate immunity and inflammation. They kill pathogens using phagocytosis and production of reactive oxygen species, nitric oxide, myeloperoxidase, and inflammatory cytokines. They can trigger or suppress T-cell responses. CD11b+, Ly6G− naive monocytes can inhibit T-cell proliferation, at least in vitro, upon cell contact, partly via NO and possibly independently of FoxP3+ regulatory T cells [295]. They may contribute to tissue repair and angiogenesis [294].

In human monocytes, Wnt3a activates the canonical Wnt axis and decreases monocyte adhesion to endothelial cells (as well as to fibronectin, laminin, and collagen) and subsequent transendothelial migration [298]. Monocytes synthesize different Frizzled receptor types.

3.13.1.1 Monocyte Types

Circulating monocytes are variable in size, granularity, and nuclear morphology. Monocytes can be identified by expression of large amounts of monocyte differentiation antigen CD14, a lipopolysaccharide receptor.

In human blood, several monocyte populations exist [299]: (1) classical, $CD14^{high}$, CD16− (∼90%); (2) non-classical, CD14+, $CD16^{high}$; and (3) CD14+, CD16+ (FcγR3, a low-affinity receptor for IgG), and CD64+ (Fcγ receptor-1A, a high-affinity receptor for IgG) monocytes.

The low-affinity IgG Fc receptor FcγR3 on the surface of monocytes and macrophages, as well as natural killer cells and neutrophils, possesses 2 subtypes Fcγ receptor-3A and -3B (CD16a and CD16b) encoded by the FCGR3A and FCGR3B genes. In humans, CD14+ monocytes consist of CD16+ and CD16− cells specialize in phagocytosis, ROS production, and recognition of TLR ligands.

Classical CD14^{high}, CD16− monocytes express CCR2 chemokine receptor. Non-classical CD14+, CD16^{high} monocytes express higher amounts of MHC class-2 molecules, low-affinity IgG Fc receptor-2A encoded by the FCGR2A gene (or CD32), and CCR5 chemokine receptor [300]. CD14+ monocytes that respond to TLR stimuli are involved in inflammation and leukocyte recruitment. CD14+ monocytes can either activate or suppress T-cell responses.

Human CD14^{low}, CD16+ monocytes sense tissue damage and reply to viruses and nucleic acids via endosomal TLR7 and TLR8 receptors [294]. They produce tumor-necrosis factor, interleukin-1β, and CCL3 chemokine.

Another subpopulation of smaller scouting monocytes with longer half-life, the Ly6G−, Ly6C1^{low} monocytes,[172] lack CCR2 and L-selectin, but strongly produce chemokine receptor CX3CR1 and $\alpha_L\beta_2$-integrin [301].

3.13.1.2 Monocyte Development

Monocytes develop from a common myeloid progenitor cells (i.e., granulocyte–macrophage colony-forming unit) in the bone marrow in response to macrophage colony-stimulating factor (Vol. 2 – Chap. 3. Growth Factors and Chap. 2).[173] Monocytes are released from the bone marrow into flowing blood and migrate into tissues to mature into long-lived tissue-specific macrophages (Sect. 3.13.2).

During monocyte differentiation into macrophage, lipid and steroid metabolism changes. Synthesis of Apolipoprotein-C1, -C2, and -E, and lipoprotein lipase augments [302]. Production of related enzymes also rises, such as cholesterol 27-hydroxylase (CyP27a1), acyl coenzyme-A, cholesterol acyltransferase-1 and -2 that esterify cholesterol, 3-hydroxy 3-methylglutaryl coenzyme-A reductase, and lysosomal phospholipase-A2, as well as that of partners, such as fatty acid-binding proteins (FABP3–FABP5), scavenger receptors ScaRa, ScaRb2, ScaRd1, and LXRα (nuclear receptor NR1H3). Production of some transcription factors also rises, such as CCAAT/enhancer binding protein-α, sprouty homolog-2, interferon regulatory factor IFR2, and musculoaponeurotic fibrosarcoma homolog (MAF) [302].

Conversely, certain genes have a decreased activity in macrophages, such as 3 members of the nuclear receptor subfamily 4 (NR4a1–NR4a3) derived from immediate-early genes, Activator protein-1 components Fos, FosB, JunB, and JunD, E26 transcription factor ETS2, CCAAT/enhancer binding protein-δ, interferon regulatory factors IRF1 and IRF7, the basic leucine zipper (bZIP) transcription factor musculoaponeurotic fibrosarcoma proto-oncogene homolog-F (MAFf), FoxO1a and FoxO3a, Runx1 and Runx3 (AML1 and AML3), SMAD3 and SMAD6, and many

---

[172] In mice, these monocytes do not express murine marker Ly6G, unlike granulocytes, but can be distinguished by Ly6C marker.

[173] This precursor gives rise to many cell types: neutrophils, eosinophils, basophils, monocytes, macrophages, dendritic cells, and mastocytes.

members of the Krüppel-like factor (KLF2, KLF4–KLF6, and KLF10–KLF11) and early growth response families [302].

Colony-stimulating factor-1 promotes the differentiation of macrophages from circulating monocytes, but additional factors, such as fibronectin and its receptor, are required for optimal differentiation.

3.13.2 *Macrophages*

Macrophages are resident phagocytic cells in lymphoid and non-lymphoid tissues. Cells of the phagocytic lineage differentiate into many different types of macrophages. Specialized macrophage types include *alveolar macrophages* in lung alveoli, *microglial cells* in the central nervous system, *osteoclasts* in bones, *histiocytes* in connective tissues, *Kupffer cells* in liver, metallophilic macrophages in spleen (among other types), as well as dendritic cells (Table 3.39). Macrophages are equipped with numerous pathogen-recognition receptors that are used for phagocytosis.

In macrophages, endocytosis, i.e., receptor-mediated *pinocytosis* and non-specific internalization of substrates via micro- or macropinocytosis (fluid-phase pinocytosis) as well as *phagocytosis*, contributes to the immune response. Mannose receptor-mediated endocytosis yields uptake of numerous mannosylated glycoproteins and phagocytosed particles.

3.13.2.1 Macrophage Production and Action

Macrophages synthesize growth factors, among which are inflammatory cytokines. On the other hand, cytokines secreted by activated T cells regulate endocytosis in macrophages. Interleukin-4 and -13 promote tubular vesicle formation underneath the plasma membrane and at pericentriolar sites, but reduce vesicle sorting to lysosomes [303]. Whereas IL4 and IL13 enhance fluid-phase pinocytosis and mannose receptor-mediated uptake by activation of phosphatidylinositol 3-kinase, interferon-γ as well as interleukin-10 diminish these processes. Interferon-γ favors vesicle sorting to perinuclear lysosomes, whereas IL10 attenuates this event [303].

Macrophages operate in immunity, apoptotic-cell clearance, wound healing, and tissue development. Macrophages actually are involved in branching morphogenesis (e.g., in pancreas and mammary glands), neuronal patterning, angiogenesis, bone morphogenesis, as well as adipogenesis, and development and growth of myocytes [304]. Macrophages also promote malignancy, as they enhance tumor growth and invasive capacity as well as stimulate angiogenesis.

Macrophages not only operate as immune effectors, but also have homeostatic roles independent of other immunocytes. Macrophages indeed process dead erythrocytes and recycle iron and hemoglobin. Red pulp macrophages actually

Table 3.39. Tissue macrophages and dendritic cells (Source: [304]; NA: not applicable). During conduit formation, macrophages reside along collagen fibers of the stroma immediately adjacent to the growing terminal end bud, where they contribute to collagen fibrillogenesis, as well as within these multilaminate bulbous termini, where they phagocytose dead epithelial cells that occur during lumen formation. Macrophages that produce Wnt7b participate in the regulation of angiogenesis during wound healing. Macrophages phagocytose dead endothelial cells (due to angiopoietin-2 released from adjoining pericytes that antagonizes survival angiopoietin-1) for vascular regression and proper vessel patterning.

Tissue	Cell name	Function
Bone	Osteoclast	Bone remodeling, stem cell niche
	Bone marrow macrophage	Erythropoiesis (degradation of ejected nucleus)
Brain	Microglial cell	Neuronal survival
		Neuronal connectivity (hypothalamic–pituitary–gonadal axis)
		Tissue damage repair
Epidermis	Langerhans cell	Immune surveillance
Eye	NA	Vascular remodeling
Intestine	Crypt macrophage	Immune surveillance
Kidney	NA	Duct development
Liver	Kupffer cell	Debris clearance
		Tissue damage repair
Mammary gland	NA	Duct outgrowth and branching
Muscle	NA	Regeneration
Ovary	NA	Steroid hormone production
Pancreas	NA	Islet development
Testis	NA	Steroid hormone production
Uterus	Uterine dendritic cell	Angiogenesis
		Uterus wall decidualization
	Uterine macrophage	Cervical ripening

constitute a distinct splenic cell subset that is involved in removing senescent red blood cells.[174]

Macrophages are also involved in the clearance of cell debris after apoptosis. Macrophage receptors for homeostatic clearance include scavenger, phosphatidyl serine, thrombospondin, and complement receptors, as well as integrins [299].

On the other hand, removal of cellular debris after cell necrosis modifies macrophage function with production of cytokines and pro-inflammatory mediators and change in plasmalemmal protein expression. Macrophages detect debris of

[174]Transcription factors spleen focus-forming virus (SFFV) proviral integration proto-oncogene product SPI1, or PU1, and C/EBPα are involved in myelomonocytic development. The SPI1-related transcription factor SPIc controls selectively the development of red pulp macrophages [305]. Factor SPIc of the SPI subfamily of ETS transcription factor family (SPI1, SPIb, and SPIc) is highly expressed in red pulp macrophages, but not monocytes, dendritic cells or other tissue macrophages.

necrotic cells via Toll-like receptors, intracellular pattern-recognition receptors, and interleukin-1 receptor, as well as adaptor MyD88 (myeloid differentiation primary-response gene product).

Breathing and eating are associated with penetration of foreign substances and pathogens to the gut and lungs. The immune system must then be poised for attack against pathogens and damage to the body's tissues.

Alveolar macrophages in the lower respiratory tract abundantly express receptor CD200R for membrane glycoprotein CD200 of airway epithelial cells to restrain cytokine secretion and inflammation [306].

In the intestinal mucosa, the immune system must tolerate antigens of the intestinal commensal flora and food, whereas it must fight against pathogens. An intestinal subset of dendritic cells that expresses $\alpha_E\beta_7$-integrin[175] activates T lymphocytes that transform into T_{Reg} cells and produce less interferon-γ.

In macrophages, supplementation of arachidonic acid, eicosapentaenoic acid, and docosahexaenoic acid[176] and activation of Toll-like (TLR4) and purinergic ($P2X_7$) receptors involved in short- and long-term inflammation, respectively, lead to the release of 22-carbon fatty acids. These molecules can indeed be incorporated in membrane phospholipids under arachidonic and eicosapentaenoic acid supplementation via rapid elongation. These fatty acids inhibit the cyclooxygenase pathways (more strongly COx1 than COx2) and shunt arachidonic acid metabolism to the lipoxygenase (Lox1) axis, thereby augmenting leukotriene synthesis [308].

---

[175]Integrin-$\alpha_E\beta_7$ is expressed almost exclusively by mucosal T lymphocytes. E-cadherin on epithelial cells is a ligand of $\alpha_E\beta_7$-integrin. Production of $\alpha_E\beta_7$-integrin often rises during inflammation. Integrin-$\alpha_E\beta_7$ also mediates the specific connection to human intestinal microvascular endothelial cells, thus intestinal T-cell homing, independently of E-cadherin [307].

[176]Dietary fish oil that contains ω3-fatty acids, eicosapentaenoic acid, and docosahexaenoic acid has cardioprotective and anti-inflammatory effects. Eicosapentaenoic and docosahexaenoic acids are important ω3-polyunsaturated fatty acids with beneficial effects. A lower mortality rate results from intake of polyunsaturated fatty acids rather than saturated fatty acids, and more precisely ω3-polyunsaturated fatty acids (with a carbon–carbon double bond (C=C) in the −3 position [position 3 from the methyl (-CH_3) end, or ω end]) rather than ω6-polyunsaturated fatty acids (with a carbon–carbon double bond in the −6 position [position 6 from the ω end]). Arachidonic acid ($CH_3(CH_2)_{18}COOH$), a 20:4 fatty acid (the lipid number •:• referring to the numbers of carbon atoms [20 C] and of double bonds in the fatty acid) is the most common highly unsaturated fatty acid. α-Linolenic acid is an essential ω3-fatty acid ([18:3],[−3] fatty acid); linoleic acid ([18:2],[−6]) is the shortest ω6-fatty acid. ω3-Polyunsaturated fatty acid supplementation can lower the ratio of arachidonic acid to ω3-polyunsaturated fatty acids in cell membrane phospholipids. Membrane arachidonic acid can be released by phospholipase-A2 upon various stimuli. Free arachidonic acid can then be oxygenated by cyclooxygenases, lipoxygenases, and cytochrome-P450 enzymes to generate numerous lipid mediators, the eicosanoids, or icosanoids. Eicosanoids can initiate the innate immune response and operate in inflammation. Like arachidonic acid-derived eicosanoids, eicosapentaenoic and docosahexaenoic acid-derived eicosanoids can be produced.

Table 3.40. Mononuclear phagocytic system (Source: [304]; CSF1: colony-stimulating factor-1, or mCSF: macrophage colony-stimulating factor; CSF2: colony-stimulating factor-2, or gm-CSF: granulocyte–monocyte colony-stimulating factor; Ifn: interferon; IL: interleukin; TNF: tumor-necrosis factor). Colony stimulating factor-1 receptor (CSF1R, mCFSR, or Cluster of differentiation CD115) is expressed by all cells of the mononuclear phagocytic system.

Cell	Differentiation agent
Bone marrow	
Pluripotent stem cell	
gmCFU	
mCFU	
Monoblast	CSF1
Promonocyte	CSF1
Monocyte	CSF1
Macrophage	CSF1
Blood	
$Ly6C^{high}$ monocyte	
$Ly6C^{low}$ monocyte	
TIE2+ monocyte	
Tissues	
Dendritic cell	
Tissue-resident dendritic cell	IL4, CSF2, CSF1
Tissue-resident macrophage	CSF1
Inflammatory macrophage (M1)	CSF2, Ifnγ, TNF
Alternatively activated macrophage (M2)	IL4, IL13
TIE2+ macrophage	

3.13.2.2 Macrophages before and after Birth

Embryonic macrophages differentiate directly from mesenchymal progenitor cells. Hematopoietic progenitor cells first populate the primitive liver and a second wave of progenitor cells from the aorta–gonads–mesonephros region of the embryo follows. After birth, bones are formed and the main site of hematopoiesis resides in the bone marrow.

The mononuclear phagocytic system of macrophage classification includes: (1) bone marrow-derived precursor cells; (2) monocytes circulating in blood; and (3) mature macrophages in tissues; as well as (4) some dendritic cells that differentiate from monocytes and macrophages. Monocyte- and macrophage-derived dendritic cells often reside in precise regions, such as Langerhans cells in the skin. In addition, myeloid precursor cells can differentiate into endothelial cells.

Cells of the mononuclear phagocytic system arise in the bone marrow from pluripotent stem cells and multipotent progenitors that comprise granulocyte–macrophage colony-forming unit (gmCFU), macrophage CFU (mCFU), monoblast, and promonocyte [304] (Table 3.40).

In bones, osteoblasts develop from promonocytes owing to colony-stimulating factor CSF1. Osteoblasts and osteoblast-derived monocyte-osteoclast progenitors then differentiate into osteoclasts in response to TNFSF11 receptor. Bone marrow monocytes and macrophages also arise from promonocytes stimulated by CSF1 factor.

Monocytes that are released into the blood circulation include several subpopulations; for example, in rodents: $Ly6C^{high}$ (lymphocyte antigen-6 complex), $Ly6C^{low}$, and TIE2-expressing monocytes. $Ly6C^{high}$ monocytes differentiate into inflammation-associated and tissue-resident dendritic cells. $Ly6C^{low}$ monocytes differentiate into tissue-resident macrophages, inflammatory macrophages (or M1 macrophages) at sites of infection and injury, and alternatively activated macrophages (or M2 macrophages) in response to parasitic infection or allergic condition and during tissue repair. TIE2-expressing monocytes give rise to TIE2+ macrophages that are involved in tumoral angiogenesis.

3.13.2.3 Macrophage Types

Macrophages exhibit functional flexibility, as they change their activity in response to environmental signals and give rise to distinct cellular populations. Macrophages can be classified based on cell-specific biochemical markers as **M1** cells[177] and **M2** cells.[178] However, the M2 population includes macrophages that differ in expression and function.

Tumor-associated macrophages release growth and angiogenic factors. Tumor-associated macrophages release tumor-necrosis factor, interleukin-1, IL6, CXC-chemokine ligand-8, vascular endothelial growth factor, and colony-stimulating factor-1 (CSF1). Production of reactive nitrogen species after NOS2 stimulation can drive additional DNA mutations. Immunocyte contribution to tumorigenesis may depend on the balance between tumor-promoting cytokines (interleukin-6), and tumor-curtailing cytokines (interleukin-10 and transforming growth factor-β).

Macrophages can also be categorized according to their markers (Table 3.41), main function, and activation mode: (1) host defense (*classically activated macrophages*) upon interferon-γ and tumor-necrosis factor stimulation;[179] (2) wound healing (*non-classically targeted macrophages* or *alternatively activated*

---

[177] M1 cells produce interleukin-12, granulocyte–macrophage colony-stimulating factor, interferon-γ, and tumor-necrosis factor.

[178] M2 cells secrete interleukin-10, transforming growth factor-β, vascular endothelial growth factor, and inducible nitric oxide synthase (iNOS or NOS2) [309].

[179] Interferon-γ synthesized by helper-1 T cells or CD8+ T cells creates in synergy with tumor-necrosis factor released by antigen-presenting cells classically activated macrophages that secrete interleukins IL1, IL6, IL12, and IL23 [299]. Interferon-γ is transiently produced by natural killer cells. A sustained population of activated macrophages requires Ifnγ synthesis by helper-1 T cells. TLR ligands can also cause Ifnβ production to replace Ifnγ.

macrophages) upon interleukins IL4 and IL13 excitation;[180] and (3) immune regulation (regulatory macrophages) upon activation by immune complexes, prostaglandins, G-protein-coupled receptor ligands, glucocorticoids, apoptotic cells, or interleukin-10 [299].[181]

Classically activated macrophages produced during cell-mediated immune responses have microbicidal and tumoricidal activity. The non-classically targeted macrophage family comprises wound-healing macrophages. Regulatory macrophages produce high IL10 level to suppress immune responses. Macrophages adapt their activity to innate and adaptive immune responses. Endogenous stimuli generated by innate immunocytes after injury or infection exert a strong, transient effect on macrophages. On the other hand, prolonged signals produced by antigen-specific immunocytes induce long-term changes in macrophages.

3.14 Dendritic Cells

Dendritic cells can differentiate from tissue-resident progenitors, such as bone marrow-derived myeloid precursors, or blood monocytes. Therefore, monocytes differentiate not only to macrophages, but also dendritic cells.

Dendritic cells are major antigen-presenting cells. They collect and process exogenous antigens in peripheral tissues. After exposure to antigen, they enter lymphatic vessels and migrate to lymphoid organs to present antigen to and activate CD4+ and CD8+ T lymphocytes against these antigens.[182] Immature and mature dendritic cells possess a high phagocytic activity and high cytokine-producing capacity, respectively.

---

[180]Interleukin-4 and/or -13 are produced by helper-2 T cells that are primarily stimulated in disturbed mucosal layers [299]. Basophils and mastocytes are important early sources of IL4 production. Interleukin-4 stimulates arginase activity in macrophages to convert arginine into ornithine, a precursor of polyamines and collagen. These macrophages not only secrete components of the extracellular matrix, but also indirectly regulate the immune response because polyamines can influence cytokine synthesis. In addition, wound-healing macrophages synthesize resistin-like molecule RELMα.

[181]Regulatory macrophages dampen the immune response and inflammation during later stages of adaptive immune responses. Many types of regulatory macrophages express high levels of costimulatory molecules CD80 and CD86 and can present antigens to T lymphocytes. All regulatory macrophage subpopulations need 2 stimuli to prime anti-inflammatory activity. Interleukin-10 produced by regulatory T cells generates a population of regulatory macrophages that act as antigen-presenting cells, produce immunosuppressive IL10, reduce IL12 expression, and can expand helper-2 T-cell population. Glucocorticoids released by adrenal cells can impede macrophage-mediated host defense and inflammation, but not phagocytosis of apoptotic cells.

[182]In inflammatory sites, dendritic cells upregulate homing CCR7 receptors. They can then enter in draining lymph vessels that express the CCR7 ligands, CCL19 and CCL21 chemokines. In lymph nodes, antigen-loaded dendritic cells activate immature T cells that migrate to the inflammation site.

Table 3.41. Markers of main macrophage populations (Source: [299]). Tissue macrophages include many cell types that differ in location, surface markers and function. Markers of classically activated, non-classically targeted, and regulatory macrophages are controlled by interferon-γ, interleukin-4, and Toll-like receptors, respectively (CCL: CC-chemokine ligand; CXCL, CXC-chemokine ligand; IGF1: insulin-like growth factor-1; IL: interleukin; IL27R: IL27 receptor; NOS2: inducible nitric oxide synthase; RELM: resistin-like molecule; SphK1, sphingosine kinase-1; NK: natural killer cell; T_H: helper T cell).

Marker	Function
	Classically activated macrophage
CCL15	Recruitment of monocytes, lymphocytes, and eosinophils
CCL20	Dendritic cell and T-lymphocyte chemoattraction
CXCL9	T-cell migration
CXCL10/11	Recruitment of NK and T cells
IL12	T_{H1}-cell development
NOS2	Nitric oxide synthesis
	Non-classically targeted macrophage
CCL17	Recruitment of T lymphocytes and macrophages
CCL18	Recruitment of lymphocytes, immature dendritic cells, and monocytes
CCL22	Recruitment of T_{H2} cells
IGF1	Stimulation of fibroblast survival and proliferation
IL27Rα	Inhibition of pro-inflammatory cytokine production
RELMα	Deposition of extracellular matrix
	Regulatory macrophage
CCL1	Recruitment of eosinophils and T_{H2} cells
IL10	Anti-inflammatory cytokine
SphK1	Sphingosine-1 phosphorylation

Dendritic cells possess the receptor plexin-A1 for class-3 and -4 semaphorins [310]. Plexin-A1 localizes to the trailing edge of migrating dendritic cells. Semaphorins are involved in dendritic cell entry into lymphatics, their migration toward lymph nodes, extravasation across the lymphatic endothelium, and accumulation in lymph nodes. Semaphorin-3A enhances the entry of dendritic cells into lymphatics initiated by CCL21 chemokine [310]. Moreover, it promotes activation of the Rho GTPase effector, RoCK kinase, and hence myosin light chain phosphorylation and subsequent myosin-2-based stress fiber contraction.

Classical dendritic cells are highly migratory cells that can move from tissues to the T- and B-cell zones of lymphoid organs via afferent lymphatics and high-endothelial venules. They are generally short-lived and replaced by blood-borne precursors during infection as well as in normal condition, except epidermal dendritic cells, the so-called *Langerhans cells*, that are not replaced by blood-borne cells in the absence of infection [296].[183]

Dendritic cells produce the lysosomal cysteine peptidase, cathepsin-K, that regulates Toll-like receptor-9 and release pro-inflammatory cytokines to mediate immune responses [312]. In particular, they attenuate activation of helper-17 T cells.

FoxP3+, CD4+, regulatory T cells that prevent overproduction of reactive immunocytes to avoid autoimmune disease preclude T-lymphocyte activation and production of CD80[184] and CD86[185] on dendritic cells, which are costimulatory or coinhibitory molecules of CD28 on T lymphocytes. Both CD80 and CD86 coreceptors present foreign antigens to immunocytes. Cytotoxic T-lymphocyte antigen CTLA4 produced by regulatory T cells enables these cells to repress CD80 and CD86 expression on dendritic cells [313].

3.14.1 Myeloid and Plasmacytoid Dendritic Cells

Several dendritic cell subsets exist with distinct immunological activities, tissue distribution, and migratory properties. Different subsets of dendritic cells are defined according to cellular origin, location, function, and lifespan in peripheral tissues, particularly in the spleen (where they have a relatively short life) and lymph nodes.

Conventional dendritic cells originate from bone marrow-derived myeloid precursors for monocytes and dendritic cells [311]. *Inflammatory dendritic cells*, born from monocytes during inflammation, migrate to lymphoid organs. *Plasmacytoid dendritic cells*, which produce large amounts of type-1 interferon, may derive at least partially from myeloid progenitors. Antigen-presenting *Langerhans cells*, located in the skin, may be generated from skin-resident progenitors.

Myeloid and plasmacytoid dendritic cells are the main producers of interferon-α. The fetal liver kinase FLK2, or stem cell receptor protein Tyr kinase STK1[186] is required for the development of classical and plasmacytoid dendritic cells [296].

In lungs, among other sites, 2 distinct populations of dendritic cells exist [296]: (1) CD103+ (α_E-integrin) and (2) slight T-cell stimulatory CX3CR1+ dendritic cells.

In the bone marrow, macrophage–dendritic cell progenitors (MDP) generate common dendritic cell precursors (CDP) that give rise to preclassical and

[183]Langerhans cells develop from an embryonic precursor that colonizes the epidermis before birth, differentiates in situ, and then proliferates during the first week of life to constitute the pool of Langerhans cells [296]. Langerhans cells self-renew and can massively proliferate during inflammation.

[184]A.k.a. CTLA4 counter-receptor B7.1.

[185]A.k.a. CTLA4 counter-receptor B7.2.

[186]Not Ser/Thr kinase-1. A.k.a. Fms-related Tyr kinase FLT3 and CD135.

plasmacytoid dendritic cells. Preclassical dendritic cells circulate in blood and enter lymphoid tissue, where they give rise to CD8α+ (a coreceptor for MHC class-1 molecules) and CD8α− classical dendritic cells, as well as non-lymphoid tissues, where they may give rise to CD103+ *lamina propria dendritic cells* (lpDC).

In addition, in normal condition, Ly6C1− monocytes (Ly6C1 in Mus musculus: lymphocyte antigen-6 complex, locus C1) leave the bone marrow to enter blood. They can then form alveolar macrophages. On the other hand, Ly6C1+ monocytes can become CX3CR1+ lamina propria dendritic cells in non-lymphoid tissues [296].

During inflammation, Ly6C1+ monocytes engender monocyte-derived dendritic cells such as TNF- and NOS2-producing dendritic cells (tipDC),[187] inflammatory macrophages, and myeloid-derived suppressor cells [296].

During infection, Ly6G+,[188] Ly6C1^{high} monocytes differentiate into tip dendritic cells that produce tumor-necrosis factor-α, nitric oxide, and reactive oxygen species as well as mucosal macrophages that synthesize a different set of inflammatory mediators.

Relatively long-lived plasmacytoid dendritic cells that lodge in the bone marrow and all peripheral organs differ from classical dendritic cells. A proportion of these cells carry characteristic immunoglobulin rearrangements They are specialized in immune responses to viral infection with a massive production of type-1 interferons. They also can act as antigen-presenting cells and control T-cell responses [296].

Plasmacytoid dendritic cells are specialized for viral infections. They use Toll-like receptors TLR7 and TLR9 in endosomes to sense viral nucleic acids. Receptor TLR9 then triggers 2 different signaling pathways either from early endosomes or specialized lysosome-related organelles, hence NFκB-dependent production of pro-inflammatory cytokines or IRF7-dependent synthesis of antiviral type-1 interferons owing to Adaptor protein-3-dependent TLR9 transfer, respectively [314].[189]

3.14.2 Thymic Dendritic Cells

After migration into the thymus from blood, *thymic dendritic cells* inhibit self-reactive thymocytes and activate T-regulatory cells for central tolerance in vivo [315].

---

[187] In the tipDC alias, letter i stands for inducible NOS NOS2).

[188] Homo sapiens possesses genes lymphocyte antigen-6 complex, loci G5B to G5C and G6C to G6F.

[189] Both pathways depend on myeloid differentiation primary response gene product MyD88, an adaptor for TLRs. The AP3 complex interacts with cleaved TLR9 in proteolytically active VAMP3+, LAMP2−, PI(3,5)P_2+, NFκB endosomes and facilitates its transfer to the LAMP2+, lysosome-related organelle (LRO), also called IRF7 endosomes, where TLR9 can engage the TRAF3–IRF7 pathway to cause transcription of IFN genes [314].

Thymic dendritic cells include 2 conventional dendritic cell subtypes: $CD8^{low}$, Sirpα+ and $CD8^{high}$, Sirpα−, which have different origins. $CD8^{low}$, Sirpα+ thymic dendritic cells have a greater capacity to activate T-regulatory cells, at least in vitro.

3.15 Myeloid-Derived Suppressor Cells

Myeloid-derived suppressor cells constitute a special compartment of the immune system of myeloid origin that includes myeloid progenitor cells and immature myeloid cells (immature macrophages, granulocytes, and dendritic cells). In healthy subjects, immature myeloid cells generated in the bone marrow differentiate into mature granulocytes, macrophages, or dendritic cells. On the other hand, in pathological conditions, a partial immature myeloid cell pool differentiates into mature myeloid-derived suppressor cells.

Myeloid-derived suppressor cells suppress T-cell responses, once they build intercellular contact. In addition to suppression of adaptive immunity, myeloid suppressors regulate innate immunity, as they modulate the cytokine production of macrophages. In addition, these cells promote tumor angiogenesis and metastasis.

Suppressive myeloid cells mature from immature myeloid cell population (that arises from common myeloid progenitor cells) during inflammation, infection, trauma, and some autoimmune diseases, as well as cancer.[190] In normal conditions, immature myeloid cells do not give rise to suppressor cells. They reside in the bone marrow, but not in secondary lymphoid organs. In normal conditions, immature myeloid cells differentiate into dendritic cells, granulocytes, or macrophages. This differentiation is promoted by numerous factors (stem cell factor, colony-stimulating factors [CSF1–CSF3], vascular endothelial growth factor, interleukins IL3 and IL6, stem cell protein Tyr kinase receptor STK1 [or fetal liver kinase FLK2], Notch, and prostaglandins).

In pathological conditions, this usual differentiation of immature myeloid cells that accumulate in abnormal tissues is prevented. Instead, resident immature myeloid cells are transformed into activated myeloid-derived suppressor cells. In addition to usual expansion agents (CSF1–CSF3, SCF, VEGF, etc.) this abnormal transformation is caused by many factors, such as interferon-γ, Toll-like receptor ligands, interleukins IL4, IL10, IL12 and IL13, transforming growth factor-β, calcium-binding proteins S100A8 and S100A9, complement component-5A, CC- and CXC-chemokine ligands CCL2, CXCL5, and CXCL12, cyclooxygenase-2, and matrix metallopeptidase MMP9 [316]. Most of these factors trigger the JaK2–STAT3 axis.

[190] In tumors, myeloid-derived suppressor cells can be distinguished from tumor-associated macrophages by their granulocytic morphology, lower expression of F4/80, and elevated expression of both arginase-1 and NOS2.

Myeloid-derived suppressor cells are also activated by the JaK1–STAT1 and –STAT6 signaling, the JaK3–STAT6 pathway, as well as Toll-like receptor-mediated activation of nuclear factor-κB. Calcium-binding proteins S100A8 and S100A9 stimulated by STAT3 bind to the NADPH oxidase complex that generates reactive oxygen species. Signal transducers and activators of transcription STAT1, STAT3, and STAT6 upregulate production of arginase-1, NOS2, and TGFβ, as well as survival agents, such as MyC, BCLxL, and cyclin-D1.

Once activated, myeloid-derived suppressor cells are characterized by an increased production of nitric oxide and reactive oxygen and nitrogen species, as well as heightened activity of arginase-1[191] and inducible nitric oxide synthase (NOS2) [316]. Both ROS and NO are involved in the suppressive function of myeloid-derived suppressor cells. Furthermore, myeloid-derived suppressor cells are able to induce development of FoxP3+ regulatory T cells.

In humans, myeloid-derived suppressor cells have a CD33+ or CD11b+, CD14−, CD33+ phenotype.[192] Myeloid-derived suppressor cells lack expression of markers of mature myeloid and lymphoid cells and MHC class-2 HLADR molecule. They have also been detected within a CD15+ population in human peripheral blood.

Two subsets of myeloid-derived suppressor cells exist according to morphology and plasmalemmal molecule expression. In mice, the granulocytic subset has a CD11b+, Ly6G+, Ly6C^{low} phenotype, whereas the monocytic subset has a CD11b+, LY6G−, LY6C^{high} phenotype [316].[193] Several other surface molecules identify additional subsets of myeloid-derived suppressor cells, such as CD80, CD115 (CSF1R), and CD124 (IL4α chain).

---

[191] Arginase-1 that is encoded by the ARG1 gene converts ^{L}arginine to urea and ^{L}ornithine.

[192] CD11b is also designated as α_M-integrin. Molecule CD33 is a common myeloid marker.

[193] In mice, markers lymphocyte antigen-6 complex Ly6G and Ly6C that are encoded by different genes are antibody epitopes specific for the myeloid cell lineage differentiation antigen anti-granulocyte receptor-1 (GR1) that binds to Ly6G on neutrophils and to Ly6C on neutrophils, dendritic cells, and subpopulations of lymphocytes and monocytes. The GR1 antigen, or RB6-8c5, was subsequently identified as a member of the Ly6 gene set. Myeloid cell immunophenotyping in mouse spleen relies on markers, such as CD11b, CD11c, Gr1, Ly6C, and Ly6G to identify various splenic cell myeloid populations. Both Ly6G and Ly6C markers are better than GR1 for the identification of splenic neutrophils, eosinophils, and subsets of monocytes and macrophages [317].

Chapter 4
Lymph Drainage

The circulatory system has a specialized compartment that transports the *lymph*,[1] Lymph originates from the interstitial fluid that enters the lumen in *lymphatic vessels*, or lymph vessels (capillary filtration). Interstitial fluid is the part of plasma leaking into tissues through thin walls of blood capillaries. Plasma escaping blood capillaries contains oxygen, glucose, amino acids, and other nutrients.

Lymph is convected unidirectionally from tiny lymphatic capillaries that infuse almost every body's tissue toward large thoracic veins close to the heart, i.e., subclavian veins on left and right sides of the neck base via left and right thoracic ducts (length 38–45 cm; caliber ∼5mm) of the lymphatic system. Lymph return a fraction of carried particulate matter to the blood stream. Therefore, lymph enables to recycle plasma.

The lymph is carried along the lymph vascular network by intrinsic contractions of mural cells of mid-size and large lymphatic vessels and by extrinsic compression of lymphatic vessels by skeletal muscle activity, among other sources of increase in external pressure such as breathing. Valves prevent lymph from flowing backward. Lymph flow can cease during long periods of complete physical inactivity.

Lymphatic vessels convey lymph from nearly all the body's tissue, except the central nervous system, tiny vessel walls, and avascular tissues, such as the transparent structures of the eye (central cornea, lens, vitreous humor)[2] and articular cartilage,[3] in addition to epithelia and endothelia. Lymphatic vessels are not detected in cortical bone, but in connective tissue overlying the periosteum [318]. On the

---

[1]Latin lympha: eau.

[2]The cornea has some peripheral blood vessels. The lens remains avascular even after injury. Ocular cells involved in avascular repair synthesize thrombospondin-1, a regulator of early wound repair. The cornea is innervated, but not the lens.

[3]Cartilage is classified in 3 types — elastic, hyaline, and fibrocartilage, according to the relative amounts of constituents (chondroblasts and chondrocytes on the one hand and collagen-1 and -2, elastin fibers, and proteoglycans on the other.

M. Thiriet, *Tissue Functioning and Remodeling in the Circulatory and Ventilatory Systems*, Biomathematical and Biomechanical Modeling of the Circulatory and Ventilatory Systems 5, DOI 10.1007/978-1-4614-5966-8_4,

other hand, the lung, gut, genitourinary tract, and dermis of the skin contain a dense network of lymphatic vessels.

The lymphatic network not only maintains the fluid balance, but also carries fat-soluble vitamins, fatty acids, and other lipids from the digestive tract and delivers these materials to the body's cells. *Chyle*[4] is an emulsion[5] of lymph (continuous phase) and immiscible free fatty acids (dispersed phase). In villi of the small intestine during digestion, fatty foods processed by enzymes of digestive juices are taken up by enterocytes. In enterocytes, absorbed lipids are again processed and incorporated in chylomicrons. These lipoproteins enter *lacteals* (lymphatic capillaries) and form chyle. The relative low pressure of the lacteals enables entry of large molecules, whereas the higher pressure in veins allows only smaller digestion products (e.g., amino acids and sugars) to enter directly blood.

Lymph drainage is involved in immunity. The lymphatic network comprises the lymphoid tissue, particularly lymph nodes. The lymphoid system also includes all structures devoted to the production and circulation of lymphocytes (bone marrow, thymus, spleen, lymphoid follicles such as tonsils, and mucosa-associated lymphoid tissue, especially in the digestive and respiratory tracts). The lymphatic system transports leukocytes to and from the lymph nodes. Antigen-presenting cells such as dendritic cells are transferred by lymph drainage to lymph nodes where an immune response is launched.

Fluid carried out of the blood stream during normal blood circulation is filtered not only from blood capillaries, but also lymph nodes. In lymph nodes, filtered fluid permits to remove microorganisms, abnormal cells, and other materials. This fluid is then transported back into the blood stream via lymphatic vessels.

Therefore, the lymphatic system assists the immune system in removing waste, debris, dead blood cells, pathogens, and toxins. However, it also conveys cancer cells.

Lymph vessels collect and drain excess tissue fluid and transport lymph into veins, thereby returning it to blood. Lymphatic capillaries are slightly larger than blood capillaries. Their wall structure enables interstitial fluid to flow into them, but not out. Ends of mural endothelial cells overlap. When the interstitial pressure is greater than the lymphatic capillary pressure, the cells separate slightly and interstitial fluid enters the lymphatic capillary. When the lymphatic luminal pressure is greater than the interstitial pressure, endothelial cells strongly, thereby preventing lymph exit into the interstitial medium.

Unlike the blood circulation, the collecting lymphatic system is an open circuit with closed upstream ends, as the smallest, thin-walled, lymphatic capillaries located between cells are closed at one end.

---

[4]Greek χυλός: sap, juice, digestive juice.

[5]Latin emulgere: to milk (milk is an emulsion of lipids and water).

4.1 Lymphatics

Lymphatic vessels are associated with lymph nodes, oval-shaped filters enclosed by a fibrous capsule, generally organized into clusters. Lymphatic capillaries are thin walled, composed of endothelium with intercellular gaps.[6] The surrounding basement membrane is small and permeable or absent. The endothelium of terminal lymphatics lacks a continuous basement membrane and tight junctions.

Lymph sacs are the earliest elements of the lymphatic vasculature out of which lymphatic vessels grow. Lymph sacs are formed by budding and differentiation toward lymphatic lineage of endothelial cells under the control of Prospero homeobox gene product Prox1 [319].

Lymphatic capillaries are similar in size to venules, without a sheath of pericytes or smooth myocytes. Larger collecting lymphatic vessels have smooth myocytes and are similar to veins with thinner walls. The wall is surrounded by a basement membrane. Lymphatic vessels have numerous semilunar valves every few millimeters that prevent back and forth lymph motion.

Lymph from the upper right quarter of the body drains into the right lymphatic duct, and subsequently into the right subclavian vein, otherwise into the thoracic duct, and then into the left subclavian vein. Additional lymphaticovenous anastomoses occur in renal, hepatic, and adrenal veins, as well as lymph nodes.

Collecting lymphatic vessels convey interstitial fluid at low pressure down to veins in which luminal pressure is higher. Between valved segments, lymphatic vessels are made of *lymphangions* that contain smooth myocytes aimed at cyclically pumping lymph against a pressure gradient. Using a mathematical model of lymphangion, optimal lymphangion length downstream from a symmetrical junction is equal to 1.26 times upstream lymphangion length [320].

Lymphatics are able to generate myogenic constriction and dilation in response to intraluminal pressure changes, especially to prevent overdistension in edema. Substance-P increases both magnitude and rate of lymphatic vessel constriction, as well as the pressure range over which constriction occurs [321].

4.2 Lymphoid Tissues

Several types of lymphoid tissues exist [322]. Genetically preprogrammed anatomically distinct lymphoid organs include lymph nodes and spleen. Prepatterned environment-dependent mucosal-associated lymphoid tissues are represented by Peyer patches and gut-associated lymphoid tissue,[7] tonsils, adenoids,

---

[6]Lymphatic vessel endothelial hyaluronan receptor-1 (LyVE1) is a marker for lymphatic endothelial cells. Plasmalemmal podoplanin is also assigned to lymphatic endothelial cells.

[7]M cells of Peyer's patches of the intestine take up antigen that is then transferred to cells in the underlying gut-associated lymphoid tissue (macrophages and dendritic cells).

and nasal-associated and bronchial-associated lymphoid tissues. These 2 groups of secondary lymphoid organs trap and concentrate antigens at many loci throughout the body to initiate an adaptive immune response. The spleen detects blood-borne pathogens. T-cell maturation occurs in the thymus.

During embryonic development, circulating hematopoietic cells gather at predestined sites throughout the body and form B- and T-cell-specific regions of secondary lymphoid organs. Tertiary lymphoid structures appear with local, prolonged inflammation. Paracrine signaling regulates the organization and induction of immunocyte populations that lead to the formation of secondary and tertiary lymphoid structures.

4.2.1 Thymus

The thymus is located in the upper anterior region of the chest cavity behind the sternum and partly in the neck. In the thymus, lymphocyte precursors from the bone marrow become thymocytes to subsequently mature into T lymphocytes. The thymus becomes atrophic during puberty.

The thymus possesses 2 lobes. Each lobe is enclosed in a capsule. It is composed of numerous lobules that contain multiple small follicles. Each follicle is made of a medullary and a cortical region. Each follicle is surrounded by a vascular plexus. The cortex is mainly composed of lymphoid lineage cells supported by a network of epithelial reticular cells.

Early and late events in thymocyte development occur in the cortex and medulla, respectively. The medulla has a coarser reticulum and concentric corpuscles of Hassall filled with lymphoid lineage and granular cells and enclosed by epithelioid cells.

4.2.2 Spleen

The spleen and tonsils are large lymphoid organs that have functions similar to lymph nodes, although the spleen filters blood cells rather than lymph. The spleen is located in the left upper abdomen below the rib cage. It destroys old erythrocytes and serves as a center of the reticuloendothelial system.

Like the thymus, the spleen possesses only efferent lymphatic vessels. It is subdivided into 2 regions — the red and white pulp. The red pulp that operates as a blood filter contains blood sinuses (or sinusoids), splenic cords of reticular fibers, and a marginal zone along the white pulp. The white pulp that participates in fighting infections is composed of Malpighian corpuscles made of lymphoid follicles rich in B lymphocytes and peri-arteriolar lymphoid sheaths rich in T lymphocytes.

In the spleen, follicular dendritic and fibroblastic reticular cells represent B- and T-cell zone stromal cells, respectively. Fibroblastic reticular cells form continuous sheaths around a mesh of collagen fibers and act as a backbone for the T-cell zone.

Blood-convected pathogen antigens are trapped and processed in the marginal zone by dendritic cells, macrophages, and marginal-zone B lymphocytes. In the B-cell zone, antibody responses and germinal center reactions are initiated, whereas in the T-cell zone, mature dendritic cells activate naive T lymphocytes. The territory of B- and T-cell zones is determined by chemokines. Chemokine CXCL13 on the one hand and CCL19 and CCL21 on the other form the B- and T-cell zones, respectively. Chemokines CCL19 and CCL21 also contribute to macrophage positioning in the marginal zone.

4.2.3 Other Secondary Lymphoid Tissues

Secondary lymphoid organs during embryonic and early postnatal period develop owing to interaction of CD3−, CD4+, CD45+ (PTPRc+) *lymphoid-tissue inducer cells* with stromal *lymphoid-tissue organizer cells*.[8] Tumor-necrosis factor receptor superfamily member TNFRSF3[9] of stromal lymphoid organizer cells interact with the lymphokine and TNFRSF3 heterotrimeric ligand TNFSF2–$TNFSF3_2$[10] of lymphoid-tissue inducer cells to organize the architecture of secondary lymphoid organs and preserve their integrity.

The structure of secondary lymphoid organs maximizes the efficacy of immune responses to viral infection. Later in life, structural integrity of secondary lymphoid organs can be altered by infection, as antiviral cytotoxic T cells destroy stromal cells of infected T-cell zone. Reacquisition of immune responsiveness requires rebuilding of architecture of secondary lymphoid organs owing to crosstalk between lymphoid-tissue inducer cells and stromal lymphoid-tissue organizer cells. Secondary lymphoid organs are repaired when lymphoid-tissue inducer cells proliferate and accumulate in secondary lymphoid organs during peak infection and then interact with fibroblastic reticular cells of the T-cell zone [323].

In adults, lymphoid-tissue inducer cells mainly contribute to the restoration of the network of *fibroblastic reticular cells*. Fibroblastic reticular cells produce chemokines involved in generating the T-cell zone of secondary lymphoid organs. Therefore, restoration of the network of fibroblastic reticular cells allows chemokine secretion after a transient loss. However, other cells in secondary lymphoid organs can provide redundant cues for their restoration.

Lymphocytes B and T are sources of TNFSF2–$TNFSF3_2$ heterotrimers. Consequently, reactivation of crosstalk between lymphoid-tissue inducer cells and stromal lymphoid-tissue organizer cells is only mandatory when lymphoid organ integrity is strongly impaired.

---

[8]CD3 accessory receptor of the immunoglobulin superfamily is also called T-cell surface glycoprotein T3. It forms CD3–TCR complex.

[9]A.k.a. lymphotoxin-β receptor.

[10]The TNFSF2–$TNFSF3_2$ heterotrimer, or lymphotoxin-$\alpha\beta_2$, is made of 2 lymphotoxin-β subunit that anchor a lymphotoxin-α (or TNFβ) subunit to the cell surface.

Lymphoid neogenesis is defined by cellular assemblies during chronic inflammation in any region of the human body. These tertiary lymphoid organs have structural similarity to lymph nodes. Hence they are able to carry out angiogenesis and lymphangiogenesis.

4.2.4 Tertiary Lymphoid Tissue – Mucosa-Associated Lymphoid Tissues

4.2.4.1 Bronchus-Associated Lymphoid Tissue

Bronchus-associated lymphoid tissues (BALT) are ectopic organized structure with B- and T-lymphocyte regions supported by specialized vascular networks, high endothelial venules in the T-cell zone, and an overlying lymphoepithelium that contains cells similar to microfold cells of Peyer's patches [324].

Bronchus-associated lymphoid tissues reside, in general, along bifurcations of large bronchi and between an artery and bronchus. However, BALTs do not always possess an overlying epithelium. Furthermore, BALTs are not always associated with a bronchus and can form in the lung parenchyma.

Inducible bronchus-associated lymphoid tissue yields a niche for T-cell priming and B-cell learning to assist in the clearance of respiratory pathogens. Non-classical BALT that lacks contact with the epithelium is observed in most smokers (84%), but also a small fraction of non-smokers (14%) [324].

Bronchus-associated lymphoid tissue pertains to the integrated mucosal system that includes the gut-associated and nasal-associated lymphoid tissue (NALT), among others. The development of this tertiary lymphoid tissue commences postnatally. Formation of BALT results from various diseases characterized by chronic inflammation, infection, or autoimmunity, hence its preferred name of inducible BALT (iBALT).

Many of the molecular and cellular factors necessary for the genesis of secondary lymphoid organs intervene during the formation of tertiary lymphoid tissue. However, classical factors of secondary lymphoid organogenesis, such as TNFSF2 and chemokines CCL19, CCL21, and CXCL13 are dispensable for iBALT formation, even though they influence its organization and function [324].

Some cell populations may participate in the BALT genesis rather than lymphoid tissue-inducer cells, such as mature CD3+, CD4+ and TNFSF14+ T lymphocytes and TNFSF2–TNFSF3_2+ B lymphocytes. In addition, dendritic cells that, once activated, interact with stromal cells are required to form, organize, and sustain iBALT tissue. Integrin-α_X+ (CD11c+) conventional and complement component receptor CR2+ (CD21+) follicular dendritic cells reside in T- and B-cell regions of BALT and within B-cell follicles, respectively. Dendritic cells synthesize CCL19, CCL21, CXCL12, and CXCL13, thereby fostering entrance and survival of lymphocytes within iBALT [324]. Moreover, activated plasmacytoid dendritic cells that can produce high concentrations of TNFα and IL8 accumulate in lymphoid follicles of COPD patients.

On the other hand, regulatory T cells in the draining mediastinal lymph nodes can attenuate iBALT development. The majority of cells within BALT are IgM^{low}, IgD^{high} B lymphocytes. Nonetheless, IgG1+,[11] IgA+, and IgE+ plasmocytes can also lodge in BALT according to the nature of the antigen. Follicular dendritic cells in BALT present antigen to B lymphocytes and yield costimulatory signals that enhance B-cell activation and proliferation in the germinal centers [324].

Activated pulmonary epithelial cells synthesize many pattern recognition receptors (e.g., TLR1–TLR9) and secrete multiple cytokines, chemokines, and growth factors that trigger recruitment and local proliferation of immune cells. Therefore, at least 3 events can initiate BALT formation [324]: (1) activation of innate pattern-xrecognition receptors such as TLRs; (2) inflammatory cytokines, such as TNFα (TNFSF1), TNFSF2, and TNFSF3, and lymphoid chemokine production by hematopoietic or stromal cells; and (3) activation and maturation of antigen-presenting cells and development of high endothelial venules. Interleukin-6 with its cognate receptor IL6R also leads to BALT formation.

Bronchial epithelial cells and T-cell-zone fibroblastic reticular cells synthesize CCR7 ligands CCL19 and CCL21 [324]. Chemokine CCL21 localizes to endothelial cells and reticular cells on the periphery of follicles or in the interfollicular zones.

4.2.4.2 Nasal-Associated Lymphoid Tissue

In humans, inducible nasal-associated lymphoid tissue (iNALT) is a structure distinct from the pharyngeal lymphoid ring[12] with pharyngeal (adenoids when infected), palatine, lingual, and tubal (where Eustachian tube opens in the nasopharynx) tonsils. This tissue is disseminated in the nasal mucosa. It belongs to the set of mucosa-associated lymphoid tissue (MALT) with larynx- (LALT) and bronchus-associated lymphoid tissue in the respiratory tract.

This organized lymphoid tissue in the nasal mucosa is characterizd by aggregates of follicles in different parts of the nasal cavity that serve in local defense. It contributes not only to the development of the local mucosal immunity, but

[11] Four IgG categories exist in humans (IgG1–IgG4) with different amounts in blood (IgG1: $\sim$ 66%, IgG2: $\sim$ 23%, IgG3: $\sim$ 7%, IgG4: $\sim$ 4%). These IgG types can also be sorted according to their as complement activator potency (IgG4 $<$ IgG2 $<$ IgG1 $<$ IgG3) and affinity to IgG Fc receptors (Fcγ R) on phagocytic cells (IgG2 $\ll$ IgG4 $<$ IgG3 $\sim$ IgG1) [325]. Phagocytic effector cells express 2 types of phagocytic Fcγ receptors: high-affinity FcγR1 and low-affinity FcγR3. Hetero-oligomers arise from connection of ligand-binding α chains to the common γ chain that is required for the assembly of both FcγR1 and FcγR3 as well as triggering of phagocytosis. In mice, IgG2a, IgG2b, and IgG3 subtypes, but not IgG1, efficiently activate complement. The affinity of IgG to Fcγ receptors on phagocytic cells depends on the antibody type, in addition to its category (for a high-affinity antibody, IgG2a = IgG2b $>$ IgG3 $>$ IgG1, whereas for a low-affinity antibody, IgG2a $>$ IgG1 $>$ IgG2b $>$ IgG3) [325].

[12] A.k.a. Waldeyer's tonsillar ring and Waldeyer's-Pirogov ring.

also distal immunity via antigens introduced from nasal cavities. Nasal-associated lymphoid tissue induces mucosal immune responses, such as the generation of T_{H1} and T_{H2} helper T cells, and IgA-committed B lymphocytes.

Addressins, extracellular ligands of lymphocyte-homing receptors in venular endothelium,[13] are used to bind naive lymphocytes. High endothelial venules in NALT contain peripheral node addressin (PNAd), either associated with mucosal vascular addressin cell adhesion molecule MAdCAM1 or alone, whereas NALT follicular dendritic cells express both MAdCAM1 and VCAM1 [326]. Naive lymphocytes bind to NALT high endothelial venules predominantly via L-selectin and PNAd rather than using $\alpha_4\beta_7$-integrin–MAdCAM1 interactions. Moreover, MAdCAM1 and VCAM1 expressed by NALT follicular dendritic cells may recruit and sequester lymphocytes [326].

Lymphoid tissue-inducer cells (LTi) that express TNFSF2–TNFSF3 heterotrimer interact with local stromal organizer cells for NALT formation.[14] Consequently, the production of adhesion molecules, such as vascular cellular adhesion molecule VCAM1, intercellular adhesion molecule ICAM1, and mucosal addressin cell adhesion molecule MAdCAM1, as well as the secretion of lymphocyte-homing chemokines and cytokines rise. The TNFSF2–$TNFSF3_2$ complex engages its cognate receptor TNFRSF3 on stromal cells and triggers the production of CCL19 and CCL21 as well as CXCL12 and CXCL13 chemokines [324]. The resultant accumulation of B and T lymphocytes produces a second production wave of TNFSF2–$TNFSF3_2$ heterotrimer that primes a positive feedback loop to further enhance chemokine expression and cell recruitment. Chemokine CXCL13 primarily

[13]Addressin is a member of the immunoglobulin superfamily. It interacts preferentially with $\alpha_4\beta_1$-integrin (a.k.a. very late antigen VLA4 and CD49d) on myeloid cells, $\alpha_4\beta_7$-integrin (a.k.a. lymphocyte Peyer patch adhesion molecule LPAM) on leukocytes, and L-selectin to attract immunocytes into mucosal and inflamed tissues.

[14]CD3−, CD4+, PTPRc+ LTi cells derive from fetal liver progenitors. They yield the earliest cell type to colonize the lymphoid anlagen (cell aggregations in the embryo that serve as first traces of a given type of structure; from German die Anlage [pl. Anlagen]: disposition, arrangement (synonym: die Anordnung [formation], die Einrichtung); construction (synonym: die Konstruktion); installation (synonym: die Installation), or primordium (initial clustering of embryonic cells from which an organ develops; from Latin primus: first [chronologically] and ordior: to begin, to initiate, to start), of secondary lymphoid organs. LTi-cell differentiation depends on at least 2 transcription factors — inhibitor of DNA binding ID2, or bHLHb26, and RORγ2, or NR1f3-2 [327]. In the fetal liver, a subset of common lymphoid progenitors that expresses the integrin-$\alpha_4\beta_7$ generates LTi cells under the control of the transcriptional repressor bHLHb26 and the nuclear receptor NR1f3-2 that are sequentially upregulated during 2 consecutive stages of differentiation with opposite requirements for Notch signaling. Both bHLHb26 and Notch are required for the generation of $\alpha_4\beta_7$-integrin+, NR1f3-2− fetal progenitors, thereby engaging the LTi developmental axis, but Notch subsequently blocks the progression to the NR1f3-2+ stage and final maturation of LTi cells, thus needing to be later repressed to avoid diversion to the T-cell fate. Lymphoid tissue development is initiated during embryogenesis by the migration of lymphoid tissue-inducer cells from the fetal liver to form lymph nodes and Peyer's patches. The expression pattern changes during the transition from lymphoid tissue-inducer cells to lymphoid tissue-organizer (LTo) cells in the lymph node anlagen, but not in the Peyer's patch primordia.

produced by follicular dendritic cells is a B-cell chemoattractant that serves to regulate the compartmentation of CXCR5+ B lymphocytes within the follicle. It also recruits lymphoid tissue-inducer cells and follicular helper T lymphocytes. Stromal cells are the major producers of CCL21 in NALT.

4.2.5 Lymph Nodes

Lymph nodes (few mm–2 cm) filter the lymph and serve as phagocytosis sites of toxins and microorganisms. In addition, they are maturation loci for some lymphocytes and monocytes. Lymphatic vessels process the lymph in lymph nodes via dendritic cells.

4.2.5.1 Lymph Node Formation

Connective tissue protrudes into lymph sacs to form precursor structures of lymph nodes. Mesenchymal cells differentiate into specific lymph node organizer cells. Each set of lymph nodes (cervical, thoracic, mesenteric, etc.) develops according to a combination of various regulators.

Establishment of lymph nodes relies on interactions between TNFRSF3+ organizing cells[15] and TNFSF2–$TNFSF3_2$+, CD3−, CD4+, PTPRc+ inducer cells [319]. This cooperation primes 2 sequential NFκB classical and non-canonical pathways (Vol. 4 – Chap. 10. Other Major Signaling Mediators) that initiate the production of adhesion molecules and chemokines to recruit circulating cells. In addition, receptor-activator of NFκB ligand TNFSF11 and its receptors TNFRSF11a and TNFRSF11b are essential regulators of lymph node organogenesis, among other functions.

Lymph Node Structure

Lymph nodes are surrounded by a fibrous capsule and an underlying subcapsular sinus (Fig. 4.1). The inner region of lymph nodes can be decomposed into 3 parts: cortex, paracortex, and medulla. Elastin and reticular fibers form a supporting network within lymph nodes that delimit follicles of packed leukocytes in the cortex.

Primary *cortical follicles* are clusters of B lymphocytes and follicular dendritic cells. After antigen stimulation, B cells undergo intense proliferation and give rise to secondary follicles, the so-called *germinal centers*. The *paracortex* is composed of T lymphocytes and dendritic cells. The *medulla* consists of medullary cords separated by lymph-filled medullary sinuses. Medullary sinuses contain histiocytes (immobile macrophages) and reticular cells.

[15] I.e., lymphotoxin-β receptor-expressing cells.

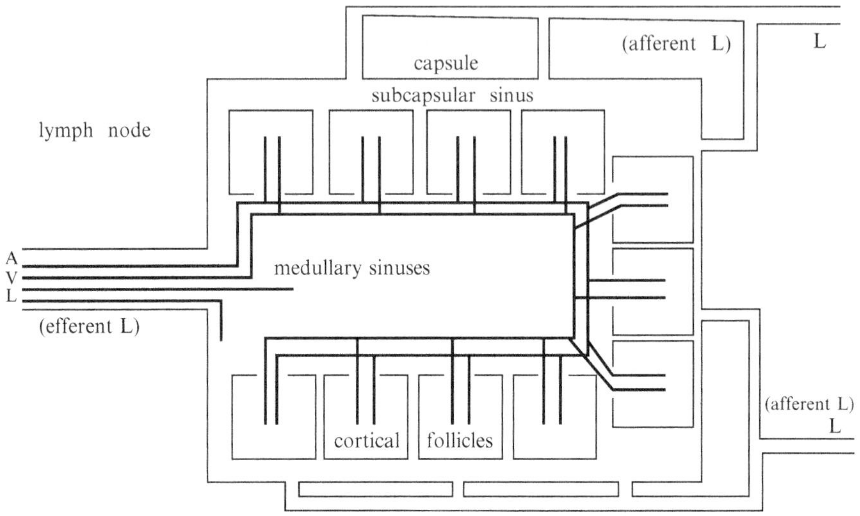

Fig. 4.1 Lymph node (A: artery; L: lymph vessel; V: veinule)

Lymph nodes contain lymphatic vessels and venules. Afferent lymphatics enter a lymph node at several sites to fill the subcapsular sinus. The subcapsular sinus drains into trabecular and cortical sinuses, then into the medullary sinuses, from which lymph flows into the efferent lymph vessel.

Lymphocytes B and T circulate inside the lymph node via venules. They can cross the endothelium and enter the lymph node. Afterward, B cells migrate to the cortex and medulla, whereas T cells move to the paracortex. Otherwise, they exit via efferent lymphatics.

Follicular dendritic cells and dendritic cells are antigen presenting. Dendritic cells present antigens in the context of major histocompatibility complex class-1 and -2 molecules to stimulate T cells. Follicular dendritic cells target B cells with antigen–antibody complexes.

The recruitment of lymphocytes and dendritic cells in the lymph node is regulated by lymphoid chemokines. Non-hematopoietic stromal cells form the scaffold of architecture of secondary lymphoid organs by producing chemokines. The lymphoid chemokine set is composed of 4 ligands — CCL19, CCL21, CXCL12, and CXCL13 — and 2 receptors — CCR7 and CXCR5. Chemokines CCL19 and CCL21 as well as CCR7 receptor are expressed by stromal cells of T-cell region. Chemokine CCL21 is produced by the endothelium of venules of the lymph node, as well as of lymphatics of non-lymphoid tissues. Chemokine CXCL13 is synthesized by follicular stromal cells.

Fibroblastic reticular cells not only constitutively produce chemokines, but also constitute support structures for migrating T lymphocytes and dendritic cells and distribute antigens throughout the parenchyma of secondary lymphoid organs.

In addition, fibroblastic reticular cells enhance the survival of naive T lymphocytes by releasing interleukin-7, presenting antigen to T lymphocytes, and supporting differentiation of regulatory dendritic cells.

4.3 Lymph

The lymph has a composition close to that of the plasma. The lymph-to-plasma protein concentration ratio is equal to about 0.75. The compositions in lipid (free cholesterol, cholesterol ester, triacylglycerol, and phospholipid) of plasma and lymph lipoproteins are similar or differ according to lipoprotein types.

The protein-rich lymph contains cells such as extravasated leukocytes and activated antigen-presenting cells. In the small intestine, lymphatic vessels of the intestinal villi absorb the dietary fat released by enterocytes that form chylomicrons.

Chapter 5
Cardiomyocytes

Cardiomyocytes (length 70–150 μm; width 20–35 μm) are striated, nucleated cells that are electrochemically excited to rhythmically contract and relax.

Cardiomyocytes are filled by aligned sarcomeres, i.e., a meshwork of contractile myofibrils enveloped in a network of Ca^{2+}-storing sarcoplasmic (endoplasmic) reticulum and mitochondria. The sarcolemma, or plasma membrane, invaginate perpendicularly to the cell axis at every Z disc. Transverse tubules reduces the distance between the sarcolemma and the cell axis (<1 μm). Transverse tubule surface area represents about 60% of the total cell surface area [328].

In the rat ventricular myocardium, about 1 capillary exists per cardiomyocyte couple for optimized irrigation and oxygenation [329]. However, contracting myocardium creates a pressure field within and outside cardiomyocytes responsible for elevated intramyocardial and intraventricular pressures that affects the lumen of intramural coronary vessels [330].

5.1 External Organization of Cardiomyocytes

Cardiomyocytes are surrounded by a trellis of collagen and elastin that supports cells and limits dilation (Fig. 5.1). The helical mesh of collagen prevents excessive cardiomyocyte stretching. Moreover, cardiomyocytes are related to each other by collagen–elastin cross-struts. The 2 sets of fibers develop recoil forces and store mechanical energy during myocardium contraction that can be restored during the following diastole [331].

The collagen network of the left ventricle wall includes 3 major components: (1) a woven collagen network that surrounds groups of cardiomyocytes; (2) array of myocyte-to-myocyte collagen struts (bore 120–150 nm) that extend from the basal lamina of a cardiomyocyte to those of contiguous cardiomyocytes; and (3) array of myocyte-to-capillary collagen struts (of similar caliber) from basal laminae of capillaries to those of contiguous cardiomyocytes [332]. Collagen connections

M. Thiriet, *Tissue Functioning and Remodeling in the Circulatory and Ventilatory Systems*, Biomathematical and Biomechanical Modeling of the Circulatory and Ventilatory Systems 5, DOI 10.1007/978-1-4614-5966-8_5,

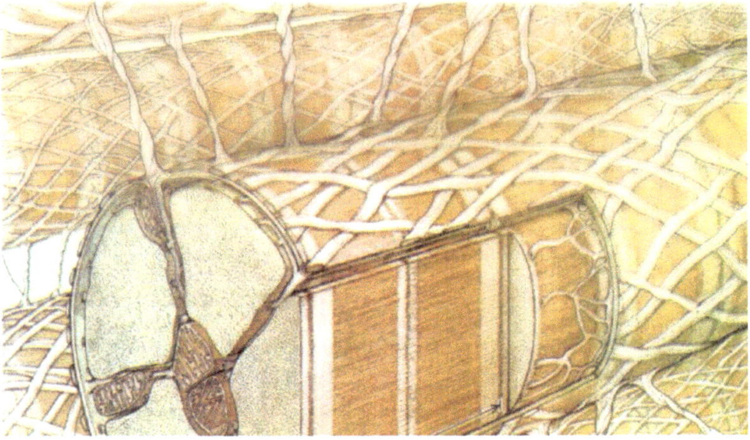

Fig. 5.1 Cardiomyocyte, its collagen envelope and struts; the sarcolemma and its T tubules; inside, the sarcomere, mitochondria and sarcoplasmic reticulum (from [331]).

between cardiomyocytes and between them and perfusing capillaries allow bulk motions during the cardiac cycle.

Four types of connective tissues can thus be distinguished: (1) an endomysium-like connective tissue made of beams of collagen fibers that connect cardiomyocytes; (2) connective tissue that links cardiomyocytes and capillaries; (3) perimysium-like connective tissue that forms a sheath of spiral collagen fibers around cardiomyocytes; and (4) epimysium-like connective tissue, i.e., layers of elastin and collagen fibers that limit the endocardic and epicardic surfaces.

5.2 Internal Organization of Cardiomyocytes

Cardiomyocytes contain structural and functional compartments with the nucleus, array of myofibrils, sarcoplasmic reticulum, and mitochondria. These densely packed organelles communicate via functional junctions (e.g., between the sarcoplasmic reticulum and T tubules for electromechanical coupling) and structural contacts (e.g., between the sarcoplasmic reticulum and mitochondria for the control of Ca^{2+} homeostasis).

5.2.1 Sarcolemma

Cardiomyocyte sarcolemma contains many ion channels, pumps, and exchangers that shape the action potential, the depolarization wave that is received and transmitted to trigger the contraction. Many receptors for extracellular ligands such as neutotransmitters released by the sympathetic and parasympathetic nerves reside in the sarcolemma. Liganded receptors initiate signaling cascades to control the activity of ion carriers. In addition, cardiomyocytes are connected by adhesion sites and gap junctions that propogate electrochemical excitation waves.

5.2.1.1 Intercalated Discs

Cardiomyocytes are joined by regions of interdigitating sarcolemmas, the so-called intercalated discs. They support intercellular communication necessary for proper cardiac function.

These electromechanical junctions contain clusters of gap junctions to allow electrochemical impulse, or action potential (APl), to spread rapidly and in a orderly way so the cell contraction is almost synchronized. Cardiomyocytes then act as a syncytium. Stimulation of an individual cell causes the contraction of the entire myocardium.

Intercalated discs contain not only gap junctions for chemical communications, but also mechanical junctions, composed of adherens junctions (fascia adherens) with N-cadherin, catenins, and vinculin, as well as desmosomes that contain desmin, desmoplakin, desmocollin, and desmoglein [333].

Adherens junctions and desmosomes dock terminal ends of the myofibrils, intermediate filament network, and microtubules, thereby stabilizing the cytoskeleton and allowing a mechanical coupling between cardiomyocytes. Adherens junctions especially anchor myofibrils and desmin cytoskeleton on cadherin membrane sites. They occupy between-cell interdigitating regions transversely with respect to the sarcomere. On the other hand, desmosomes and gap junctions are located in membrane parts in the direction of the sarcomere axis.

At intercalated discs, zonula occludens protein ZO1 regulates the organization of both gap and adherens junctions [334]. Zonula occludens protein ZO1 interacts with Cadherin-2 (or N-cadherin); it regulates localization of gap junctions via its association with the cadherin-2 complex.

At intercalated discs, $Na_V1.5$ and $K_V1.5$ channel subunits interact with junctional proteic complexes (*intercalated disc interactome*) [335]. Plakophilin-2, a desmosomal molecule, influences gap junction-mediated coupling as well as $Na_V1.5$ channel function. Cadherin-2 modulates gap junction and $K_V1.5$ functioning. Gap junction protein connexin-43 interacts with the sodium channel complex component ankyrin-G, which is required for proper intercellular adhesion and electrical coupling. Connexin-43, which is needed for normal functioning of sodium and potassium channels, regulates ankyrin-G transfer along microtubules into the intercalated disc and conversely.

5.2.1.2 Transverse Tubules

Ventricular excitation–contraction coupling relies on sarcolemmal, transversely oriented, deep, cylindrical invaginations — transverse (T) tubules — that are juxtaposed beside terminal cisternae of the sarcoplasmic reticulum.

Excitation–contraction coupling actually requires Ca^{2+} ions. Calcium mediator is mainly supplied by the sarcoplasmic reticulum that stores Ca^{2+} ions from diastolic uptake for systolic release.[1]

Sarcolemma repeatedly invaginates to form T tubules that plunge into the sarcoplasma (Fig. 5.1). Both sides of T tubules are close to the sarcoplasmic reticulum. In addition, T tubules limit both ends of the sarcomere. Density of T tubules influences Ca^{2+} influx. Transverse tubules allow to localize sarcolemmal voltage-gated calcium channels and sarcoplasmic reticulum ryanodine receptors[2] into close proximity. Cardiac contraction actually results from opening of sarcolemmal voltage-gated calcium channels that are mostly located in T tubules. These channels trigger Ca^{2+} influx with local accumulation that activates calcium release from its sarcoplasmic reticulum stores, especially the close junctional regions of this organelle. A decreased number of T tubules, particularly in atrial myocytes and Purkinje cells, is correlated to local slow rate of Ca^{2+} ingress [337]. The dominant Ca^{2+} ligand in the sarcomere is troponin-C attached to actin filaments.

Action potential sweeps quickly along the sarcolemma and T tubules to trigger Ca^{2+} release from the sarcoplasmic reticulum. Owing to its speed, action potential arrives almost simultaneously at all tubules of the T-tubule network to ensure that all sarcomeres have a coordinated contraction. Calcium ions then diffuse to bind to troponin.[3] This bond leads to the interaction between actin and myosin and sarcomere contraction. When the process ends, Ca^{2+} is pumped back into the sarcoplasmic reticulum (Sect. 6.6).

The formation and maintenance of T tubules require myotubular myopathy protein MTM1, or myotubularin-1 [338]. This phosphoinositide 3-phosphatase acts on both phosphatidylinositol 3-phosphate and (3,5)-bisphosphate. Myotubularin activity on PI(3)P and/or $PI(3,5)P_2$ may interfere with channel activaty.

On the other hand, atriomyocytes of many mammalian species do not express extensive T-tubule networks. Coupling of Ca_V channels and ryanodine receptors occurs at dyadic junctions at the cell periphery. Peripheral Ca^{2+} ions do not diffuse into the center of atriomyocytes. Three-dimensional non-junctional lattice of ryanodine receptors may sense cortical Ca^{2+} signals and transmit them to

[1] In the cytoplasm around myofibrils, free calcium concentration [Ca] $\sim 10^{-6}$ mol/kg and $[Ca^{2+}] \sim 10^{-9}$ mol/kg. These amounts are insufficient for contraction. Calcium influx $\sim 10^{-12}$ mol/cm^2 cannot explain contraction. Most Ca^{2+} comes from the sarcoplasmic reticulum (SR; SR $[Ca^{2+}] \sim 10^3$ sarcolemma $[Ca^{2+}]$) [336].

[2] The plant Ryania speciosa ensures its defense against insects by producing the alkaloid ryanodine. The latter binds specifically to Ca^{2+}-releasing channel of the sarcoplasmic reticulum Ca^{2+} store.

[3] When $[Ca^{2+}]$ is high, Ca^{2+} binds to troponin that then acts on tropomyosin, initiating conformational changes.

Table 5.1. Ion channels in caveolae and membrane rafts of the sarcolemma of nodal cells and cardiomyocytes (Source: [339]). Among hyperpolarization-activated and cyclic nucleotide-gated (HCN) channels, HCN4 is the main isoform in the sinoatrial node; HCN1 and HCN2 are also synthesized.

Channel	Membrane localization
$Ca_V1.2$	Caveolae
IP_3R	Caveolae
$K_V1.4$	Membrane raft
$K_V1.5$	Caveolae (ventriculomyocyte)
	Membrane raft (atriomyocyte)
$K_V2.1$	Membrane raft
$K_V4.2$	Membrane raft
$K_{IR}2.1$	Membrane raft
$K_{IR}3.1$	Membrane raft
$K_{IR}6.1$	Membrane raft
$K_{IR}6.2$	Membrane raft
K_{Ca}	Membrane raft
$Na_V1.5$	Caveolae (ventriculomyocyte)
HCN4	Caveolae (nodal cell of sinusal node)

the cell center via a Ca^{2+} release from the sarcoplamic reticulum. This lattice becomes active and allows a centipetal diffusion of Ca^{2+} ions when a greater atrial contraction is required.

5.2.1.3 Plasmalemmal Nanodomains

Membrane rafts and caveolae (50–100 nm plasmalemmal invaginations) constitute dynamic platforms that recruit ion channels and their partners to build channelosomes. Certain cardiac channel subunits that shape the cardiac action potential (Sect. 5.10) actually localize to membrane rafts and/or caveolae, such as those of $Ca_V1.2$, K_{ATP} $K_{IR}6.2$), $K_{IR}2$, $K_{IR}3$, $K_V1.4$, $K_V1.5$, $K_V2.1$, and $Na_V1.5$ (Table 5.1) [339]. Ion channel partners, receptors of nervous control, and protein kinases and phosphatases that coreside in channelosomes modulate localization and/or activity of ion channels.

Caveolin-3 associates with dystrophin and β-dystroglycan that are components of the dystrophin-associated protein complex. Dystrophin and α-actinin anchor caveolae to the actin cytoskeleton [339]. Sodium channels that localize in cardiomyocyte T-tubular caveolae interact specifically with syntrophin, another constituent of the DAPC complex [339].

The cAMP-dependent regulation of voltage-gated calcium channels $Ca_V1.2$ relies on interaction between A-kinase anchoring proteins and protein kinase-A. A-kinase anchoring proteins recruit protein kinase-A near ion channels in membrane rafts and caveolae. Channel $Ca_V1.2$ assembles with β2-adrenoceptor, Gs subunit of G protein, adenylate cyclase, AKAP, PKA, phosphatase PP2, and caveolin-3 [339]. Protein kinase-A as well as calmodulin-dependent protein kinase CamK2 phosphorylate $Na_V1.5$ and $Ca_V1.2$ channels, ryanodine-sensitive Ca^{2+}-release channel RyR2,[4] and phospholamban regulator of sarco(endo)plamic reticulum calcium ATPase.

Disc large homolog DLg4 is required for $K_V1.4$ localization in membrane rafts [339]. In the atrial myocardium, $K_V1.5$ channels lodge in intercalated discs that lack caveolin-3. Disc large homolog DLg1 that binds to cadherin promotes $K_V1.5$ localization in membrane rafts. In intercalated discs, connexin-43 resides in membrane rafts, where it links to N-cadherin.

Channel $K_{IR}2$ complexes with DLg1 and DLg4 adaptors [339]. In membrane rafts, DLg1 and DLg4 tether other scaffold proteins, such as calcium–calmodulin-dependent protein Ser kinase (CASK) of the membrane-associated guanylate kinase (MAGUK) family, vertebrate Lin7 homolog VeLi1, and Munc18-1-interacting protein MInt1[5] to form a complex that recruits $K_{IR}2$ to the plasma membrane [339].

Both DLg1 and DLg4 connect also to Ca^{2+}-activated calmodulin-dependent kinase CamK2 that regulates numerous ion channels according to intracellular Ca^{2+} concentration. Most channels regulated by CamK2 localize in membrane rafts and caveolae with CamK2, such as $Ca_V1.2$, $K_V1.5$, IP_3R, and Na^+ channels [339].

In addition, DLg1 and DLg4 link to A-kinase anchoring proteins. Channel $K_V7.1$ associates with both PKA and PP1phosphatase via AKAP9 [339]. The AKAP proteins interact also with β2-adrenoceptors. Therefore, channelosomes can be controlled by the sympathetic system.

Like β-adrenergic receptor and its G protein, adenylate cyclase, and protein kinase-A, α1 adrenergic receptor, caveolin-3, Gq subunit of G protein, and phospholipase Cβ subtypes (PLCβ1 and PLCβ3) are confined to caveolae [340]. A decrease in caveolin-3 density may contribute to the development of cardiac maladaptive hypertrophy.

5.2.2 *Sarcoplasmic Reticulum*

The sarcoplasmic reticulum is a smooth endoplasmic reticulum in myocytes. Whereas the endoplasmic reticulum synthesizes molecules, the sarcoplasmic reticulum stores, pumps, and expels calcium ions. The sarcoplasmic reticulum sequesters a large amount of calcium that can be released on the arrival of the action potential owing to special connections between the sarcoplasmic reticulum and the sarcolemma with T-tubules.

---

[4]Ryanodine receptor is a sarcoplasmic reticulum Ca^{2+} channel that is able to bind to ryanodine.

[5]A.k.a. adaptor X11α.

The sarcoplasmic reticulum located throughout the myoplasm can be divided into 3 main components: junctional (JSR), corbular (CSR), and network (NSR). The sarcoplasmic reticulum indeed widens at multiple sites to form junctional SR cisternae tightly coupled to the sarcolemma of T tubules. The junctional compartment is positioned within 12 to 15 nm to T tubules. Together with the T tubule, JSR defines a restricted subspace in which Ca^{2+} can reach high concentration. Junctional sarcoplasmic reticulum contains the majority of the ryanodine receptors.

Calcium release from sarcoplasmic reticulum storage occurs at discrete loci in the JSR–T tubule subspace — Ca^{2+} sparks, although other secondary Ca^{2+} transient currents are initiated by other ion carriers. The corbular region is located near the T tubules at a distance of several micrometers from the sarcolemma. Corbular sarcoplasmic reticulum releases Ca^{2+} without interaction with sarcolemmal $Ca_V1.2$ channels. Sarcoplasmic reticulum Ca^{2+} ATPases are predominantly located in network sarcoplasmic reticulum.

In cardiomyocytes, the cardiac Ca^{2+} release unit is a JSR proteic complex that comprises $Ca_V1.2$ juxtaposed to ryanodine-sensitive, Ca^{2+}-release channel RyR2, thereby achieving excitation–contraction coupling. This coupling is controlled by RyR2-associated proteins and RyR2 phosphorylation. Junctional sarcoplasmic reticulum unit includes ryanodin receptor RyR2, calsequestrin Casq2, junctin, junctophilin-1 and -2, and triadin. Triadin-1 is the predominant isoform in cardiomyocytes. The JSR complex is involved in Ca^{2+} release from the sarcoplasmic reticulum by Ca^{2+} release units to trigger myocyte contraction, as well as buffering of stored Ca^{2+} in the lumen of the sarcoplasmic reticulum. In addition, Ca^{2+}-induced Ca^{2+} release units exert a negative feedback on $Ca_V1.2$ channels. Conversely, muscle relaxation results from Ca^{2+} reuptake into the sarcoplasmic reticulum by sarco(endo)plasmic reticulum Ca^{2+} ATPases.

Calsequestrin-2 sequesters Ca^{2+} inside the sarcoplasmic reticulum lumen, avoiding Ca^{2+} leaks from its store during myocyte relaxation. Junctin docks Casq2 to RyR2. Junctophilin-1 and -2 promote colocalization of $Ca_V1.2$ with RyR2. In addition, Aspartyl–asparaginyl β-hydroxylase (AspH) participates in calcium storage in and release from the sarcoplasmic reticulum. Triadin anchors calsequestrin Casq2 to RyR2 ryanodine receptors. Triadin maintains the structural and functional integrity of the cardiac Ca^{2+} release units of the sarcoplasmic reticulum [341].

Ca^{2+}-binding calmodulin and Ca^{2+}–calmodulin-dependent CamK2 kinase regulate activity of ryanodine receptors and phospholamban. Calmodulin mediates Ca^{2+}-dependent channel inactivation.

Calcium-induced calcium release from the sarcoplasmic reticulum through ryanodine receptors initiated by Ca^{2+} influx through Ca_V channels is an amplification process relying on a positive feedback mechanism that governs cardiac inotropy. Each cardiomyocyte produces a Ca^{2+} transient. The amplitude of Ca^{2+} transients is determined by recruitment level of Ca^{2+} sparks (elementary Ca^{2+} sarcoplasmic reticulum release signals), i.e., by increasing the number of RyR channels clustered in spark-forming couplons. Sensitization and inhibition of RyR clusters shorten and lengthen the interval between consecutive Ca^{2+} sparks, respectively [342].

Cytosolic calcium concentration rises and falls very rapidly upon opening and closing of RyR channels to ensure adequate sarcoplasmic reticulum emptying and refilling through SERCA pumps, hence proper inotropy and lusitropy.

Calcium spark termination may be induced by junctional Ca^{2+} concentration within the sarcoplasmic reticulum lumen that enables RyR desensitization and inactivation, once a given threshold is reached [342]. Increased or decreased buffering capacity of the sarcoplasmic reticulum prolongs or abbreviates the duration of Ca^{2+} release, respectively. Sensitization of RyRs with caffeine or inhibition of RyRs with tetracaine heightens or diminishes the rate of triggering probability recovery, respectively [342]. Sensitivity of RyR channels does not influence Ca^{2+} spark amplitude recovery; the latter relies on local sarcoplasmic reticulum refilling. On the other hand, Ca^{2+} spark triggering depends on both refilling and RyR sensitivity.

Calcium release refractoriness, which reduce the probability of channel reopening during a given time interval after closure, permit to avoid arrhythmogenesis. Calcium release refractoriness can relies, at least partly, on the rate-limiting conformational change of a Ca^{2+} flux controller such as calsequestrin. Recovery of Ca^{2+} release through RyRs depends on the Ca_V channel trigger, between-beat changes in sarcoplasmic reticulum SR Ca^{2+} content, and β-adrenoceptor acitivity that controls cardiac frequency [342].

5.2.3 *Mitochondria*

Cardiomyocytes have a richer supply of mitochondria and greater dependence on adenosine triphosphate (ATP) than skeletal myocytes. Mitochondria cluster in the vicinity of the nucleus and enter into contact with it to support energy requirement for nuclear function. Mitochondria are also close to the sarcoplasmic reticulum and sarcomere to minimize molecular transfer.

5.2.3.1 Mitochondria Populations

In the cardiomyocyte, 2 functionally and biochemically distinct populations of mitochondria exist [343]: (1) subsarcolemmal mitochondria that are located beneath the sarcolemma and (2) interfibrillar mitochondria that are situated among myofibrils. Activities of succinate dehydrogenase and citrate synthase are higher in interfibrillar mitochondria with respect to subsarcolemmal mitochondria, whereas those of carnitine palmitoyltransferase and α-glycerophosphate dehydrogenase are nearly the same in both mitochondrium types. Interfibrillar mitochondria oxidize tested substrates with a rate about 1.5 times faster than that of subsarcolemmal mitochondria.

Table 5.2. Electron transport chain and ATP synthase (CytC: cytochrome-C; IMS: inner mitochondrial membrane; MM: mitochondrial matrix; O_2^-: superoxide anion; P_i: inorganic phosphate). The inner mitochondrial membrane is impermeable to most metabolites and ions, thereby creating a transmembrane potential and ionic gradient. Reactive oxygen species are formed at complex-I and -III of the respiratory chain. Superoxide anion produced from the transfer of an electron from the partially reduced ubiquinone to oxygen is then reduced to hydrogen peroxide (H_2O_2) by superoxide dismutase. Hydrogen peroxide is eliminated by glutathione peroxidase or reacts with ferrous ions and other transition metals to form hydroxyl radical.

Complex	Alias Reaction
I	NADH–coenzyme-Q oxidoreductase, NADH dehydrogenase $NADH + UQ + 5H^+_{MM} \rightarrow NAD^+ + UQH_2 + 4H^+_{IMS}$ $O_2 + e^- \longrightarrow O_2^-$
II	Succinate–coenzyme-Q oxidoreductase (made of heme, iron–sulfur clusters, and flavin adenine dinucleotide) $Succinate + UQ \rightarrow fumarate + UQ_2$
III	Coenzyme-Q–Cytochrome-C oxidoreductase, cytochrome-BC1 complex, cytochrome-C reductase (made of iron–sulfur cluster, cytochrome-C1, and 2 cytochrome-B) $UQH_2 + 2\,CytC_{ox} + 2H^+_{MM} \rightarrow UQ + 2\,CytC_{red} + 4H^+_{IMS}$ $O_2 + e^- \longrightarrow O_2^-$
IV	Cytochrome-C oxidase $4\,CytC_{red} + O_2 + 8H^+_{MM} \rightarrow 4\,CytC_{ox} + 2\,H_2O + 4H^+_{IMS}$
V	ATP synthase, F_1F_0 ATPase (F_O proton channel and F_1 synthase domain) $ADP + P_i + 4\,H^+_{IMS} \rightleftharpoons ATP + H_2O + 4\,H^+_{MM}$

5.2.3.2 Electron Transport Chain

The electron transport chain, composed of 4 main proteic complexes (Table 5.2) in the mitochondrion inner membrane, enables oxidative phosphorylation. Reduced nicotinamide adenine dinucleotide (NADH)[6] and succinate generated in the citric acid cycle, also called tricarboxylic acid and Krebs cycle, are oxidized to release protons. In redox reactions, electrons are transferred from electron donors such as NADH to electron acceptors such as oxygen. Redox reactions release energy that powers ATP synthase. Electrons carried through the electron transport chain are used to transport protons across the inner mitochondrial membrane. This process creates a potential energy in the form of a proton concentration (pH) gradient and electrical potential across the inner mitochondrial membrane. Protons move back

[6]Oxidized form of nicotinamide adenine dinucleotide (NAD^+) is an electron acceptor (oxidizing agent, oxidant, or oxidizer). Reduced NADH form is a reducing agent (reductants or reducers) that donates electron(s).

across the membrane and down their concentration gradient through ATP synthase that uses stored energy. Within the inner mitochondrial membrane, the lipid-soluble electron carrier coenzyme-Q10 carries both electrons and protons from its oxidized ubiquinone form (UQ; a.k.a. quinone Q) to its reduced ubiquinol form (UQH_2 or QH_2). In the coenzyme-Q10 redox cycle, oxidized ubiquinone UQ accepts 2 electrons and 2 protons, whereas reduced ubiquinol UQH_2 releases 2 electrons and 2 protons.

The electron transport chain is composed of 4 major complexes: NADH–ubiquinone reductase (complex-I), succinate–ubiquinone reductase (complex-II), ubiquinol–cytochrome-C reductase (complex-III), and cytochrome-C oxidase (complex-IV). Electron transfer between these complexes is ensured by interactions between electron carriers coenzyme-Q10 and cytochrome-C.

Cardiolipin can link complex-III and -IV. In addition, complex-I is stabilized by binding to complex-III in human mitochondria. Like complex-I, -III, and -IV (but not complex-II), ATP synthase (complex-V) is able to form supercomplexes by oligomerization [344]. A dimeric state of ATP synthase is specific for mitochondria. *Respirasomes* refers to respiratory chain supercomplexes such as the aggregate made by complex-I, -III dimer, and -IV (respirasome-I–III–IV). These supercomplexes reduce electron diffusion distance.

5.2.3.3 PGC1α and PGC1β

Peroxisome proliferator-activated receptor-γ (or nuclear receptor NR1c3) coactivators PGC1α and PGC1β are highly expressed in the heart. The expression of PGC1α rises when ATP production must increase, i.e., when the cardiac activity increases. Regulator PGC1α coactivates PPARα (nuclear receptor NR1c1), estrogen-related receptor ERRα (NR3b1), and other nuclear factors that control the transcription of genes involved in cardiac fatty acid oxidation and mitochondrial functional capacity [345].

Moreover, PGC1α stimulates the mitochondrial genesis in cardiomyocytes, although it is not essential for mitochondrial generation. In the absence of PGC1α, mitochondrial density remains normal, but expression of the genes of oxidative phosphorylation decays [346]. Homeostasis of ATP and phosphocreatine is disturbed. The Gs-coupled βAR–cAMP pathway, calmodulin-binding catalytic subunit of PP3, calcium–calmodulin-dependent protein kinase, AMP-activated protein kinase, P38MAPK, and nitric oxide activates PGC1α expression.

5.2.3.4 Mitochondrial Permeability Transition Pore

Mitochondrial permeability transition pore (mPTP) is a non-selective, large-conductance channel that remains normally closed. It is partly constituted of:

(1) a dimer of voltage-dependent anion channels (VDAC),[7] or porins, in the mitochondrial outer membrane (MOM); and (2) a dimer of adenine nucleotide translocases (ANT)[8] linked to (3) mitochondrial inorganic phosphate (Pi) carrier SLC25a3[9] in the mitochondrial inner membrane (MIM) [347]. Both VDAC and ANT serve as regulators of the mPTP in cooperation with the matrix protein, cyclophilin-D (CyPd).[10]

Adenine nucleotide translocase interacts with cyclophilin-D and SLC25a3 carrier. Binding of cyclophilin-D to mPTP increases mPTP sensitivity to Ca^{2+} ion. Elevation of matrix Ca^{2+} concentration triggers mPTP opening and cytosolic-open ANT conformation, thereby favoring Ca^{2+} sensitivity of the mitochondrial permeability transition pore.

Opening of mitochondrial permeability transition pore causes mitochondrial swelling, hence rupture of the mitochondrial outer membrane and release of cytochrome-C into the cytosol, triggering apoptosome formation and myocyte death. Multiple prosurvival signals prevents mPTP opening (Table 5.3). On the other hand, mPTP remains closed during ischemia and opens shortly after reperfusion. Moreover, ischemia–reperfusion events promote CyPd–PiC–ANT interaction in cardiomyocytes.

[7]Three VADC isoforms (VDAC1–VDAC3) are encoded by distinct genes. Subtype VDAC1 is the most abundant isoform in most cells. It interacts with multiple proteins involved in energy production, formation of cytoskeletal networks, and apoptosis. In the mitochondrial outer membrane, VDAC binds to hexokinase-2 that supports ADP transport into mitochondria and ADP production from phosphorylation of glucose in a nearby location [347].

[8]The ANT family comprises 4 isoforms encoded by different nuclear genes. Isoform ANT1 is predominant in the heart and muscle [347]. Subtype ANT3 is produced at low levels whatever the cell type. Both ANT2 and ANT4 are synthesized during cell proliferation and in the brain, liver, and testis, respectively. Whereas ANT1 and ANT3 are antisurvival factors, ANT2 and ANT4 are prosurvival agents. Adenine nucleotide translocase exchanges ADP^{3-} and ATP^{4-} between the mitochondrial matrix and intermembrane space. It is stimulated by ADP^{3-}, ATP^{4-}, and BCL2.

[9]Two SLC25a3 isoforms (SLC25a3a–SLC25a3b) are created by alternative splicing from the SLC25A3 transcript. Isoform SLC25a3a is synthesized in cardiac and skeletal muscles; SLC25a3b is ubiquitous. Carrier SLC25a3 acts as a Pi–H^+ symporter and/or Pi–OH^- antiporter [347]. In any case, it carries cytosolic inorganic phosphate to the mitochondrial matrix, where inorganic phosphate is used for ATP synthesis.

[10]Cyclophilin-D is a peptidyl prolyl cis–trans isomerase encoded by the PPIf gene. It is the single isoform that lodges in the mitochondrial matrix. It precludes elevation of mitochondrial Ca^{2+} concentration. It interacts with SLC25a3 and ANT, thus modifying mPTP sensitivity to Ca^{2+} and inorganic phosphate [347]. It also links to mitochondrial ATPase and reduces its activity.

Table 5.3. Prosurvival signaling pathways triggered by G-protein-coupled, protein Tyr kinase, and cytokine receptors that target mitochondria (Source: [347]; GSK: glycogen synthase kinase; HK: hexokinase; JaK: Janus kinase; mK_{ATP}: mitochondrial ATP-sensitive K^+ channel; PKC: protein kinase C; ROS: reactive oxygen species; STAT: signal transducer and activator of transcription; ⊖: inhibition). Certain mitochondrial protein kinases in cardiomyocyte ensure protection against ischemia–reperfusion damage. On the other hand, GSK3β phosphorylates porin, thereby releasing HK2 from mitochondria and provoking an oxidative stress, in addition to activation of mitochondrial permeability transition pore. On the other hand, PKCϵ phosphorylates porin, but HK2 remains attached to porin. Kinase PKA phosphorylates also porin, thereby supressing its opening.

Cytosol	Mitochondrion
JaK–STAT3	STAT3 (mPTP ⊖)
MAP2K1/2–ERK1/2–NOS3–PKG	PKCϵ–PKB–HK2 (mPTP ⊖)
PI3K–PKB–ERK1/2–NOS3–PKG	PKCϵ–PKB–GSK3β (GSK3β ⊖)
PKCϵ–PKB–ERK1/2–NOS3–PKG	PKCϵ–mK_{ATP}–ROS-
	-ROS–ERK1/2–GSK3β (GSK3β ⊖)
	-ROS–PKCϵ–GSK3β (GSK3β ⊖)
	-ROS–PKCϵ–AldH2

The mPTP regulation depends on hexokinase-2,[11] glycogen synthase kinase-3β,[12] signal transducer and activator of transcription STAT3,[13] and sirtuin-3 [347].[14]

Brief periods of ischemia and reperfusion, the so-called ischemic preconditioning, protect the myocardium against injury by a subsequent sustained ischemia. Binding of HK2 to mitochondria rises after ischemic preconditioning and insulin stimulation that activate PI3K–PKB axis. Deacetylation of cyclophilin-D (Lys166) by sirtuin-3 reduces its PPIase activity [347]. Cardioprotection is also ensured by STAT3 phosphorylation (Tyr705) [347]. Factor STAT3 can attenuate ischemia-induced ROS production in cardiomyocytes.

[11]Four hexokinase isoforms exist (HK1–HK4). Subtypes HK1 and HK2 reside in both the cytosol and mitochondria of cardiomocytes. They have different affinities for ATP and glucose.

[12]Kinase GSK3β is constitutively active. its activity rises by phosphorylation (Tyr216). On the other hand, phosphorylation at Ser9 by PKA, PKC, and ERK inactivates GSK3β. Phosphorylation of Thr43 and Ser389 and Thr390 by ERK and P38MAPK, respectively, facilitate Ser9 phosphorylation by other kinases. Cardioprotection by activated erythropoietin receptor or mitochondrial ATP-sensitive K^+ channel before ischemia is associated with mitochondrial GSK3β phosphorylation (Ser9; inactivation) [347].

[13]Once it is phosphorylated, STAT3 binds to promoters of target genes. It localizes to the mitochondrial matrix.

[14]Sirtuins couple lysine deacetylation to NAD^+ hydrolysis. Three subtypes (SIRT3–SIRT5) lodge in mitochondria. Once it is imported into mitochondria, the full-length sirtuin-3 is cleaved to generate an active short form. It activates acetylCoA synthetase-2 (Lys642), long-chain acylCoA dehydrogenase (LCAD; Lys42), and NDUF9 subunit of complex-I of mitochondrial electron transport chain [347].

In cardiomyocytes, GSK3β is predominantly located in the cytosol. Ischemia–reperfusion events cause GSK3β translocation from the cytosol to mitochondria. Cardioprotection by mitochondrial ATP-sensitive K^+ channel before ischemia relies on mitochondrial GSK3β phosphorylation (Ser9; inactivation) [347].

In ventriculomyocytes, connexin-43 phosphorylation at Ser or Tyr residues regulates gap junction permeability, hence electrochemical coupling of cardiomyocytes. Connexin-43 also localizes to the inner membrane of subsarcolemmal mitochondria. It increases mK_{ATP} channel activity, leading to elevated ROS production.

Glycogen synthase kinase-3β operates via mitochondrial inner membrane connexin-43[15] and mitochondrial inner membrane ATP-sensitive K^+ channel [348]. cardioprotection relies on opening of mK_{ATP} channel. Activated PKC kinase and inhibited PP2 phosphatase increase the open probability of mK_{ATP} channels via GSK3β and mitochondrial connexin-43 hemichannels, which can carry ATP. Phosphorylation of mitochondrial connexin-43 increases its activity. Ischemic preconditioning is supported by phosphorylation (inhibition) of glycogen synthase kinase-3β, thereby repressing mitochondrial permeability transition pore opening.

In cardiomyocytes, connexin-43, together with Gβ subunit of G protein, participates in $PI3K_{c1\gamma}$ activation, hence in the cytoprotective PI3K–PKB–GSK3β pathway [349].[16] In the myocardium, the PI3K–PKB–GSK3β axis enables ischemia tolerance supported by insulin, erythropoietin, bradykinin, and adenosine receptors. The PI3K–PKB pathway stimulates ROS production by opening of mitochondrial ATP-sensitive K^+ channel that requires a certain mitochondrial concentration of connexin-43. Endothelin-1 that acts via G-protein-coupled receptors and signals in cooperation with connexin-43, and insulin-like growth factor-1 that targets receptor protein Tyr kinase, without the assistance of connexin-43, cause phosphorylation of PKB and GSK3β. In other words, connexin-43 is involved in cell protection ensured by endothelin-1, but not that primed by IGF1 agent. Endothelin-1 can elicit ROS production in mitochondria using connexin-43, but not IGF1 factor. Connexin-

[15]Sarcolemmal and mitochondrial connexin-43 have different functions. The former is involved in the formation of gap junctions, the latter in the production of reactive oxygen species for redox signaling (Vol. 4 – Chap. 10. Other Major Signaling Mediators).

[16]In cardiomyocytes, both class-1A and -1B PI3Ks that are coupled with different types of receptors trigger PKB–GSK3β signaling. Receptor protein Tyr kinases and cytokine receptors operate via class-1A PI3Ks ($PI3K_{c1\alpha}$–$PI3K_{c1\beta}$ and $PI3K_{c1\delta}$), whereas G-protein-coupled receptors activate class-1B PI3K ($PI3K_{c1\gamma}$) using Gβγ dimer. Kinase GSK3β is phosphorylated in response to activation of G-protein-coupled receptors, such as δ-opioid receptor and endothelin-1 receptor, when connexin-43 is available. On the other hand, activated IGF1 receptor does not need the assistance of connexin-43 to cause GSK3β phosphorylation [349]. Moreover, PKCε, ERK, and P38MAPK are phosphorylated upon stimulation by endothelin-1, even in the absence of connexin-43. Therefore, connexin-43 does not influence all of the signaling types initiated by G-protein-coupled receptors [349].

43 is involved in the activation of the PKB–GSK3β axis by $PI3K_{c1\gamma}$ possibly via its interaction with the Gβ subunit of G protein, but not via mitochondrial ROS production.

5.2.3.5 Myoglobin

Myoglobin is an iron- and oxygen-binding protein of myocytes that functions in oxygen transport and storage. In addition, as a reductase, deoxygenated myoglobin generates nitric oxide from circulating nitrite. On the other hand, oxymyoglobin acts as a dioxygenase to convert NO into nitrates [350]. Low concentration of nitrite ions (NO_2^-) cause vasodilation and suppression of mitochondrial activity, thereby reducing generation of reactive oxygen species, especially after ischemia–reperfusion events. Owing to nitrite reductase, nitrite ions generate nitric oxide, a vasorelaxant that decreases cardiomyocyte contractility and cardiac frequency by repressing the electron transport chain in mitochondria.

5.2.3.6 Mechanotransduction

Mitochondrial depolarization and K^+ accumulation in the mitochondrial matrix by activated ATP- or Ca^{2+}-dependent K^+ channels and inhibited H^+–K^+ exchanger strongly increase contractile force in rat ventriculomyocytes [351]. Change in the mitochondrial ionic balance causes mitochondrial swelling that can compress the myofibrillar and nuclear compartments of cardiomyocytes that are sensitive to extracellular mechanical stimuli. A relatively small increase in mitochondrial volume can cause a relatively large reduction in sarcomeric lattice spacing in the absence of strong stretch of cardiomyocytes and concomitant increase in passive force [351].

When myosin and actin filaments are closer, both myofilament Ca^{2+} sensitivity and maximal Ca^{2+}-induced tension increase. In addition, mechanical stresses applied on the nucleus can influence both the DNA accessibility for gene transcription and the nucleocytoplasmic transport of molecules. Therefore, generated internal pressure that affects morphological and functional properties of organelles contributes to the regulation of cell functioning [351].

5.2.3.7 Cardioprotection

Cardioprotection relies on [352]: (1) activation of numerous cardioprotective kinases, such as extracellular signal-regulated protein kinase, phosphatidylinositol 3-kinase, and protein kinase B; (2) translocation to the mitochondria of kinases (e.g., PKB, PKCϵ,[17] glycogen synthase kinase GSK3, glyceraldehyde 3-phosphate dehydrogenase, and hexokinase, among others), phosphatases, and other signaling proteins;[18] (3) post-translational modifications of mitochondrial proteins, such as phosphorylation, S-nitrosylation,[19] acetylation, ^{O}linked β^{N}acetylglucosamine glycosylation, glutathiolation, sumoylation, and ubiquitination; and (4) inhibition of the mitochondrial permeability transition pore.

The mitochondrial permeability transition pore comprises outer membrane voltage-dependent anion channel, inner membrane, regulatory adenine nucleotide transporter, and matrix cyclophilin-D, in addition to mitochondrial phosphate carrier.

Under hypoxia, mitochondria generate toxic reactive oxygen species. The electron transport chain is the main ROS source. Nevertheless, cardiomyocytes subjected to hypoxia keep mitochondrial activity under control, as they are equipped with oxygen sensors prolyl hydroxylases and hypoxia-inducible factors [353].

During hypoxia, increased glycolysis causes a lactic acid accumulation, hence intracellular pH reduction. Consequently, Na^{+}–H^{+} antiporter is activated and raises intracellular Na^{+} concentration that cannot be corrected by Na^{+}–K^{+} ATPase due to the decline in ATP concentration. Increased intracellular Na^{+} concentration leads to elevated intracellular Ca^{2+} concentration because Na^{+}–Ca^{2+} exchanger is inhibited or reversed [353].

Reactive oxygen species are formed at complex-I and -III of the electron transport chain and mainly via xanthine oxidase on xanthine formed by adenosine degradation. Reactive oxygen species can inhibit ATP synthase and adenine nucleotide translocase. They can also cause peroxidation of unsaturated fatty acids in

---

[17]Protein kinase-Cϵ can complex with voltage-dependent anion channel (porin), adenine nucleotide transporter, and hexokinase-2 as well as another with mitogen-activated protein kinase. It phosphorylates mitochondrial aldehyde dehydrogenase-2 and α-ketoglutarate dehydrogenase to reduce generation of reactive oxygen species. It also phosphorylates cytochrome-C oxidase subunit-4, activates mitochondrial K_{ATP} channel, and inhibits the mitochondrial permeability transition pore. Protein kinase-B phosphorylates glycogen synthase kinase GSK3. Unphosphorylated (active) GSK3 phosphorylate cyclophilin-D to promote activation of the mitochondrial permeability transition pore. Inhibition of GSK3 also reduces ATP entry into the mitochondria via voltage-dependent anion channel, thereby reducing consumption of glycolytically generated ATP.

[18]Connexin-43 translocates to mitochondria owing to heat shock protein HSP90 and translocase of the mitochondrial outer membrane complex (TOM) to stimulate the mitochondrial K_{ATP} channel. Signal transducer and activator of transduction STAT3 can also enter mitochondria to regulate the activity of the electron transport chain.

[19]S-nitrosylation corresponds to addition of an nitric oxide group to protein thiols. ^{S}Nitrosothiols derive from several mitochondrial proteins, such as complex-I, α subunit of the F_1F_0 ATPase, α-ketoglutarate dehydrogenase, aconitase, creatine kinase, malate dehydrogenase, heat shock proteins HSP60 and HSP70, electron transfer flavoprotein-α and -β, electron-transferring flavoprotein dehydrogenase, subunit-5A of cytochrome-C oxidase, aldehyde dehydrogenase-2, succinate dehydrogenase subunit-A flavoprotein, very-long- and short-chain acylCoA dehydrogenase, carnitine palmitoyltransferase-2, enoylCoA hydratase, and isocitrate dehydrogenase. Nitric oxide targets complex-I, thereby reducing ROS generation. ^{S}Nitrosothiols inhibit complex-V, thus causing ATP degradation by reverse mode of the ATPase. They also block irreversible oxidation of proteins.

membrane phospholipids such as cardiolipin at the inner mitochondrial membrane that then prevents respiratory chain activity. The combined effects of ROS and augmented Ca^{2+} concentration in the mitochondrial matrix causes opening of the non-specific mitochondrial permeability transition pore, hence free passage of small cytosolic molecules (<1.5 kDa). Therefore, pore opening increases the mitochondrial matrix volume, disrupts mitochondrial outer membrane, and provokes cell death (Vol. 2 – Chap. 4. Cell Survival and Death).

Ischemic postconditioning that consists of alternating, brief periods of ischemia and reperfusion is effective when applied at the onset of reperfusion. It includes formation and release of autacoids and cytokines, maintenance of acidosis during early reperfusion, activation of protein kinases, and repression of opening of mitochondrial permeability transition pore.

During hypoxia, as HIFα is not targeted by the oxygen sensor prolyl hydroxylase that enables it interact with von Hippel-Lindau ubiquitin ligase for proteasome degradation, this subunit accumulates and forms HIFαβ heterodimer.[20] Following the recruitment of CBP and P300 coactivators, HIF causes target gene transcription. Hypoxia-inducible factor provokes [353]: (1) expression of glucose transporters GLUT1 and GLUT4; (2) that of glycolytic enzymes to balance decayed cell respiration; (3) that of lactate dehydrogenase-A that converts pyruvate into lactate, hence diverting pyruvate away from the mitochondria; (4) that of pyruvate dehydrogenase kinase that impedes conversion of pyruvate into acetyl coenzyme-A, which enter the Krebs cycle; (5) switch form COx4-1 to COx4-2 subunit in complex IVto optimize the efficiency of electron transport; and (6) mitochondrial autophagy. In addition, to prevent mitochondrial permeability transition pore opening, HIF1α elicits [353]: (1) adenosine production to stimulate the PI3K–PKB pathway; (2) NO generation by NOS2 to activate protein kinase-G for opening of mitochondrial K_{ATP} channels that promotes production of low levels of ROS messengers; as well as (3) activation of inner mitochondrial membrane uncoupling proteins UCP2 and UCP3 to limit mitochondrial generation of ROS concentrations. Produced ROS messengers oxidize (inactivate) PTen, thus activating survival PKB kinase.

5.2.4 *Nuclear GPCRs*

G-protein-coupled receptors, such as α1- and β-adrenergic, angiotensin-2 AT_1 and AT_2, apelin, bradykinin B_2, endothelin, metabotropic glutamate, lysophosphatidic acid, platelet-activating factor, and prostaglandin receptors target the nuclear membrane [354].

Furthermore, numerous GPCRs contain nuclear localization sequences that enable them to cross nuclear pore complexes, hence to move from the outer to the inner nuclear membrane, thereby delivering signals toward the cytoplasm or the nucleoplasm.

---

[20]Three HIF-α isoforms (HIFα1–HIFα3) are encoded by distinct genes.

Heterotrimeric G proteins transmit stimuli from plasmalemmal GPCRs to various intracellular effectors, such as enzymes and ion channels. Associated signaling mediators, such as heterotrimeric G proteins, adenylate cyclases, phospholipase-A2, -Cβ, and -D, regulators of G-protein signaling, β-arrestin-1, GPCR kinases, A-kinase-anchoring proteins, and protein kinase-A, among others, can be transferred to the nuclear membrane and nucleus. Moreover, enzymes involved in the synthesis and metabolism of phosphoinositides as well as in the processing of peptide ligands, such as angiotensin-converting enzyme and endothelin-converting enzyme-1 can localize to the nucleus.

Both plasmalemmal and nuclear GPCRs can activate effectors, such as G proteins, β-arrestins, and nuclear protein kinases that modulate the activity of transcription factors. Gβγ subunits as well as effectors PKB and MAPK interact with transcription factors, chromatin regulators such as histone deacetylases, and DNA elements [354].

In addition, Gβγ subunits regulate cellular signaling and organize the assembly and transfer of receptor-based complexes in intracellular compartments, such as the endoplasmic reticulum and Golgi body. Moreover, Gβ1γ2 dimer can localize in the nucleus, where it interacts directly with HDAC5 histone deacetylase [354]. The Gβ5–RGS7 complex localizes to both the cytosol and nucleus. Protein RGS6 can also reside in the nucleus. In the nucleus, Gβγ5 dimer complexes with the transcriptional repressor adipocyte enhancer-binding protein AEBP1 and attenuates its transcriptional repression. In addition, Gβγ colocalizes with activating protein AP1 in the nucleus and recruits HDACs to inhibit AP1 transcriptional activity. Prenylation increases Gβγ amount associated with the glucocorticoid receptor (NR3c1) in the nucleus.

Nuclear GPCRs launch classical pathways that produce and/or activate a second messengers or stimulate the MAPK module or the PI3K–PKB axis, like those initiated at the plasma membrane. Nonetheless, these receptors can directly regulate DNA synthesis, histone modification, and gene transcription [354].

5.2.4.1 Nuclear β-Adrenergic Receptors

In cardiomyocytes, plasmalemmal β1-adrenergic receptors launch the Gs–AC–cAMP pathway responsible for the chronotropic and inotropic effects of sympathetic stimulation. β-Adrenergic receptors can also interact with other heterotrimeric G-protein subunits such as Gi subunit. Other signaling pathways rely G-protein-coupled receptor kinases and β-arrestins. β-Arrestins are implicated in GPCR desensitization, hence suppression of the first wave of signals triggered at the plasma membrane by G proteins, and internalization, which is associated with an (eventually G-protein-independent) intracellular (endosomal) second wave of cues.

In cardiomyocytes, whereas plasmalemmal β1-adrenergic receptors play a predominant role in the regulation of cardiomyocyte contractility, plasmalemmal β2-adrenergic receptors play a more modest role in regulation of cardiac inotropy and lusitropy.

β-adrenoceptors can colocalize with NuP62 nucleoporin. In fact, Gs-coupled-β1- and Gi-coupled-β3-adrenoceptors, but not β2-adrenoceptors, can lodge in the nuclear membrane, at least in rodent adult ventriculocardiomyocytes [354]. Whereas β1AR activates adenylate cyclases, β3AR modulates transcriptional initiation. Stimulated βARs target rRNA and mRNAs, such as transcripts of NFκB and components of its signaling pathways. Extracellular noradrenaline is taken up and can move to the nucleus within hours.

5.3 Sarcomere

Myofibrils, formed by packed contractile proteins (actin, myosin, and associated proteins), fill a large fraction of the cytosol, whereas the rest is occupied by other organelles such as mitochondria.

The striated appearance of myofibers is created by a transverse pattern of alternating dark A bands,[21] bisected by the H zone[22] with a mid M line,[23] and light I bands,[24] divided by Z lines, or Z discs,[25] (Fig. 5.2).

The sarcomere (length $\sim 2\,\mu m$), the array of thick and thin filaments between the Z lines, is the anatomical unit of muscular contraction, and the hemisarcomere is the functional unit. Repeating sarcomere units are precisely aligned in the cardiomyocyte.

The sarcomere length is a predictor of muscle function. Non-invasive measurements have been developed to assess the sarcomere structure. Tissue scattering is related to refractive index distributions associated with the morphology and composition of the explored tissue. Sarcomere structural changes can be monitored by optical reflectance on muscle, a strong correlation existing between muscle optical scattering properties and sarcomere length used as a predictor of muscle function [355]. Visualization of individual sarcomeres and analysis of their dynamics can now be obtained owing to optical microendoscopy [356].[26]

---

[21] A stands for anisotropic.

[22] In German, hell: bright.

[23] In German, Mittelscheibe: disc in the middle (of the sarcomere).

[24] I stands for isotropic.

[25] In German, Zwischenscheibe: intermediary disc (in the middle of I bands).

[26] Optical microendoscope is composed of gradient refractive index microlenses (bore 350–1,000 μm). The microendoscopy associated with a laser-scanning microscope detects 2 intrinsic optical signals: (1) autofluorescence from nicotinamide adenine dinucleotide and flavoproteins that are concentrated in mitochondria along sarcomeric Z discs and (2) second-harmonic generation, coherent frequency-doubling of incident light that arises from myosin rods.

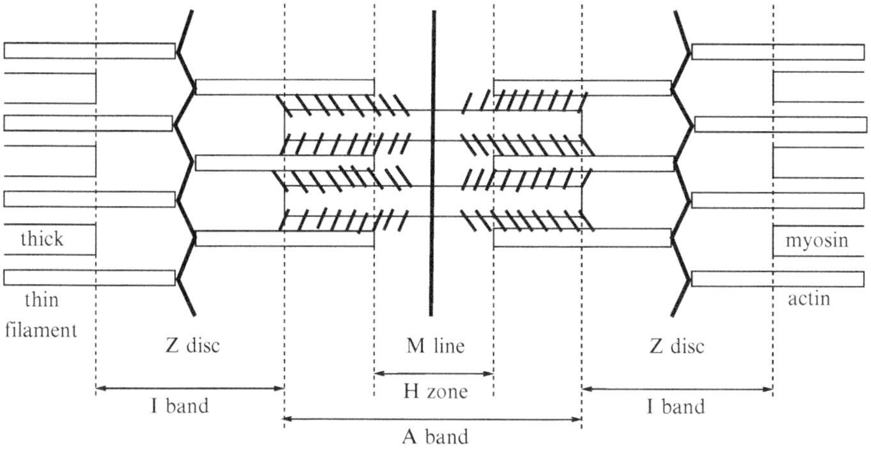

Fig. 5.2 The sarcomere, its microscopic structure and the myofilament arrangement. The M line and the Z discs are proteic scaffolds that orientate myosin and actin filaments, respectively. The H zone (bisected by M line) contains only thick myosin filaments, the I band (bisected by Z disc) only thin actin filaments, and the region of the A band close to the Z disc both filament types.

The sarcomere includes 4 filament complexes: (1) thin filament that is predominantly made of ^{F}actin,[27] (2) thick filament, mainly consisting of myosin[28] (3) titin,[29] and (4) nebulin. The lateral boundaries of the sarcomeric units, Z discs, relate contractile apparatus to cytoskeleton. Sarcomere shortening of striated myocyte results from actin and myosin filaments sliding over each other, as the lengths of these filaments remain constant.

Sarcomere Z discs that anchor myofilaments and stabilize their assembly, are densely packed cellular structures that include actin, titin, and nebulin, as well as smaller proteins, such as α-actinin and telethonin. Two giant *titins* are assembled into a complex by *telethonin*, a Z-disc ligand that has a symmetric structure [358].

---

[27]Striated myocytes express 2 actin isoforms, cardiac actin and skeletal actin. In adult myocardium, cardiac actin is the major isoform (80%) [357].

[28]There are three isoforms of non-muscle myosin, myosin-2A, myosin-2B, and myosin-2C. Myosin-2 is also observed in muscles. Myosin-2B is the single isoform found in adult cardiomyocytes.

[29]Titin is also named connectin.

The molecular chaperone, heat shock protein HSPb5,[30] abounds in the myocardium. It may act in the assembly and remodeling of sarcomeres. Under normal conditions, HSPb5 lodges in the cytosol; it is only slightly linked to the cytoskeleton. During stress, it moves to the Z disc of the sarcomere and can interact with actin, myosin, desmin, titin and vimentin, hence stabilizing sarcomeric and cytoskeletal structures. Once phosphorylated by P38MAPK and MAPKAPK2 and then translocated to mitochondria, it protects against acute metabolic stresses, such as ischemia and reperfusion, as it also maintains mitochondrial integrity, hence impeding H_2O_2-induced cytochrome-C release [360]. Chaperones HSPb5 and HSP47 might play complementary roles in cross-bridging actin filaments in the Z disc of the sarcomere.

5.3.1 *Myofibrils*

Each myofibril is made of arrays of parallel filaments. In the H zone, the thick and thin filaments never overlap (Figs. 5.3).

[30] A;k.a. heat shock protein-β5, 20-kDa heat shock-like protein, and αB-crystallin. Chaperones are aimed at protecting proteins from misfolding and aggregation. Most chaperones are stress inducible. They belong to the HSP family. In addition to proper folding of newly synthesized proteins, chaperones are also required during stress to assist in the maintenance of proteins structure and prevent aggregation. Oxidative stress and increase in intracellular calcium concentration can indeed lead to protein unfolding and aggregation. In mammals, crystallins are divided into α-, β-, and γ-crystallins. Members of the α- and β-families are subdivided into acidic (e.g., αA-crystallin, heat shock protein-β-4, or HSPb4) and basic (e.g., αB-crystallin or HSPb5) components. (Heat shock protein-β6, or α-crystallin-related HSPb6, is also called heat shock protein HSP20; heat shock protein-β1, or HSPb1, 27 or 28-kDa stress-responsive protein SRP27 (HSP25/27/28); heat shock protein-β2, or HSPb2, 27-kDa heat shock protein [HSP27]; heat shock protein-β3, or HSPb3, 17-kDa heat shock protein [HSP17]. Other α-crystallin-related heat shock proteins encompass cardiovascular HSPb7, or cvHSP, HSPb8, or small stress protein-like protein HSP22, HSPb9, and HSPb10, or outer dense fiber of sperm tails ODF1, ODF2, and ODF27.) αB-Crystallin interacts with numerous related small heat shock proteins (sHSP), such as HSPb1, HSPb6, and HSPb8, to form multimers. α-Crystallins are not only constituents of the ocular lens with both structural and functional homology with small heat shock proteins, but also reside in various types of tissues, especially with a high concentration in the myocardium, particularly at Z discs and intercalated discs. Whereas αA-crystallin is preferentially restricted to the lens, αB-crystallin is widespread. αB-Crystallin is produced in the brain, lung, kidney, skeletal muscle and heart. Nodal cells of the sinoatrial and atrioventricular nodes and His bundle contain a lower amount of αB-crystallin than cardiomyocytes, whereas nodal cells of the left and right bundle branches possess a greater quantity [359].

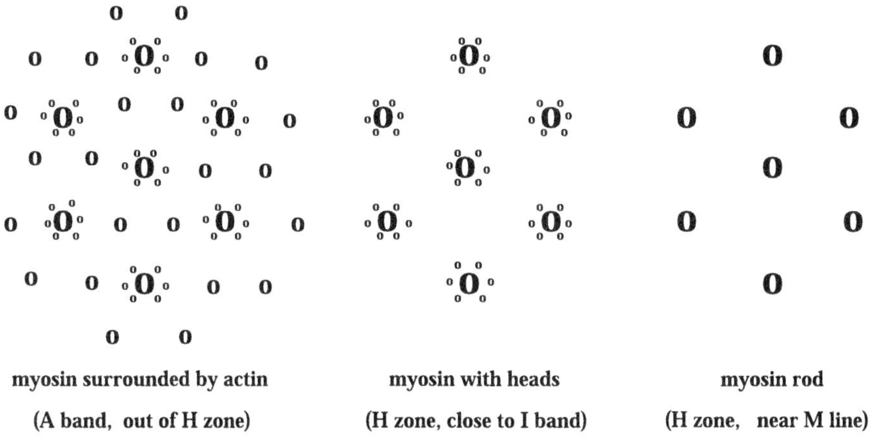

Fig. 5.3 Sarcomere cross-section and myofilament arrangement. The inner region of the H zone (in the M line and its vincinity) contains the rod part of the myosin filament (bare zone). Myosin heads are located in the outer regions of the H zone and A band. Myosin filaments are surrounded by actin filaments in the outer region of the A band. The I band (not represented) is composed only by actin filaments.

5.3.2 Myosin Thick Filaments

The thick filament (diameter ~ 15 nm) is composed of *myosin*[31] (Fig. 5.4). Myosin filaments produce the A bands.

The myosin structure is commonly split into 2 regions, light (LMM) and heavy (HMM) meromyosin. The former represents part of the tail and is responsible for the assembling of myosin (Fig. 5.4). The latter contains the 2 heads and the remaining part of the tail. Heavy meromyosin is further divided into globular fragment-1 (S1) and rod fragment-2 (S2). The latter contains binding sites for essential and regulatory light chains (MLC1 and MLC2). The former (S1) has ATP hydrolysis and actin-binding domains; it has helical arrangements around the filament axis.

The thick filament also contains M-line proteins myomesin and myosin-binding protein-C (cardiac isoform cMyBPc or MyBPc3), or C-protein, that crosslink myosins and the M-band part of titin. Myosin-binding protein-C maintains the width of myosin-containing thick filament, as it restricts them to 200 to 400 molecules each.

Titin is involved in the connections between the ends of myosin filaments and Z discs; it can span the entire sarcomere length. It controls both the length and elasticity of the sarcomere.

[31] The myosin molecule (length 1.55 μm, diameter 12–15 nm) can be divided into 2 heavy chains and 4 light chains [361].

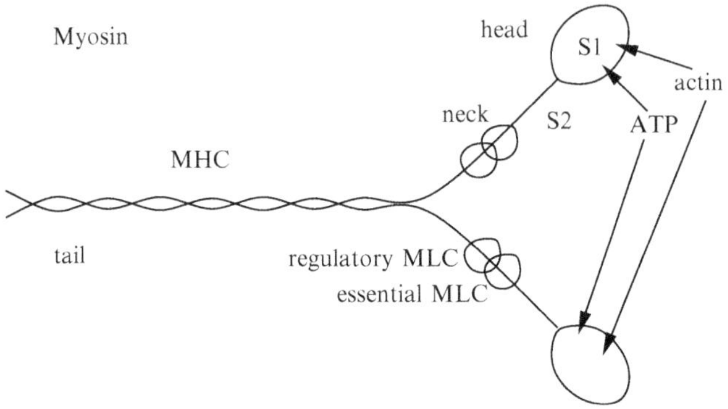

Fig. 5.4 Myosin filaments (MHC: myosin heavy chain, MLC: myosin light chain). The neck has binding sites for light chains, and the head for ATP and actin. For example, myosin-2 consists of 2 head segments: S1, composed of heavy chains, connected to a long dimeric rod by a flexible coiled-coil segment S2 (S2δ in β-myosin-2). The coiled-coil light meromyosin rod tails of myosin-2 are packed side by side, whereas both heads point away from the myosin filament surface (or light meromyosin dimer axis). S1 subfragments act as ATP-driven motors, the conformation and actin affinity of which change according to the status of ATPase sites. Actomyosin cross-bridges use chemical energy liberated by the hydrolysis of cardiac actomyosin Ca^{2+}-stimulated MgATPase.

Myomesin acts for lateral registration. This structural protein of the M line assists titin and myosin in the maintenance of their three-dimensional structure.

Cardiac myosin isoform-5.1 has higher ATPase and contractile activities than isoform-5.3 [362]. Myosin isoforms may have a heterogeneous regional distribution.

5.3.3 *Actin Thin Filaments*

The thin filament (diameter ∼5 nm) is composed of *actin*, with small amounts of 2 regulatory proteins, *troponin* (TN) and *tropomyosin* (TMy,).[32] Actin filaments extend in direction normal to the Z lines from each side. They create the I band where they do not overlap the thick filaments.

Heterodimeric troponin is composed of 3 subunits: TnC, TnI, and TnT. Troponin-C has a binding site for Ca^{2+} (calcium receptor/handling protein). Troponin-I[33] inhibits actin–myosin binding in the absence of Ca^{2+} ion. Troponin-I is regulated by

[32]τροποσ: belt. ^{G}actin polymerizes in the presence of Ca^{2+}, Mg^{2+}, and ATP to form ^{F}actin in a double-helix chain with a pitch length of 35–36 nm. One tropomyosin runs in the groove of ^{F}actin. One troponin exists for each tropomyosin.

[33]Troponin-I has a key role in ischemia-induced contractile dysfunction. Ischemia leads to intracellular acidosis. Troponin-I modifies the Ca^{2+} sensitivity of actin. The fetal cardiac isoform of troponin-I protects more from acidosis than the adult cardiac isoform, enhancing Ca^{2+} activation of sarcomere activity. The composition of troponin-I thus affects cardiac performance [363].

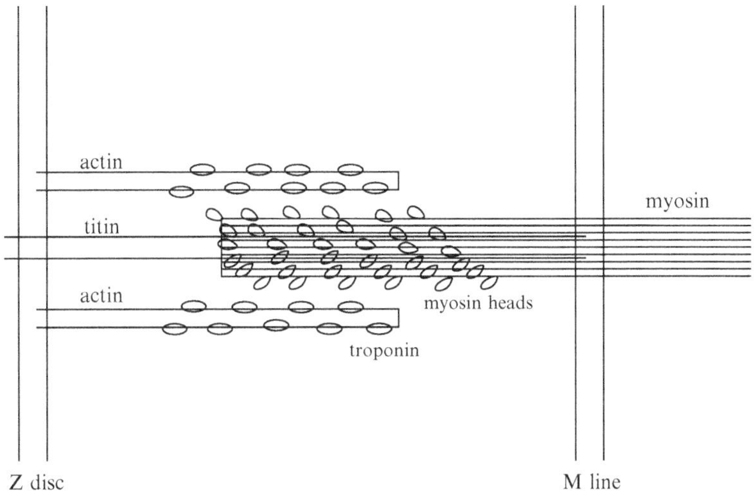

Fig. 5.5 Interacting myosin and actin in the sarcomere. The thick filaments are composed of myosin and its partners (myosin-binding proteins-C, -H, and -X). The thin filaments contain cardiac actin, α-tropomyosin, and TnT, TnI, and TnC troponins. Myocardium contraction is generated by cyclic interactions between actin and myosin. Conformational changes in the actin–myosin binding interface are related to ADP release and ATP binding. Nanoscale conformational changes are converted into macroscale myocardium displacements. The dissociation rate of actomyosin bonds was postulated to depend on strain in the Huxley model. The lifetime of the actomyosin bond indeed depends on the instantaneous load and loading history.

TnC that relieves the inhibition once TnC binds Ca^{2+}. Troponin-T is a tropomyosin-binding protein. The troponin-C–Ca^{2+} complex deforms TnC–TnI element and weakens TnI–TnT, exposing the myosin-binding site on actin.

Troponin-I and myosin-binding protein-C are substrates of protein kinase-A. As it phosphorylates troponin-I, PKA reduces the Ca^{2+} sensitivity, but not the cross-bridge cycling rate [364]. Protein kinase-C targets troponin-I, troponin-T, myosin light chain-2, and myosin-binding protein-C. Protein kinase-C may reduce the cross-bridge cycling rate.

Two actin polymers placed side by side form a double helix that behaves like a moving rack (Fig. 5.5). The screw-shaped myosin has a twisted tail and a head that emerges to form grooves in which the actin filaments slide like a wrench jaw.

5.3.3.1 Actin–Myosin Interactions

Contraction is induced by the 4-time nanomotor composed of interdigitated myosin and actin filaments (Fig. 5.6) [365]: (1) the myosin head detaches from actin and fixes ATP; (2) ATP is hydrolized and the myosin head binds to actin; (3) the myosin head releases the phosphate and undergoes a conformational change. Actin displacement is generated by a bascule motion of the myosin head that pulls on

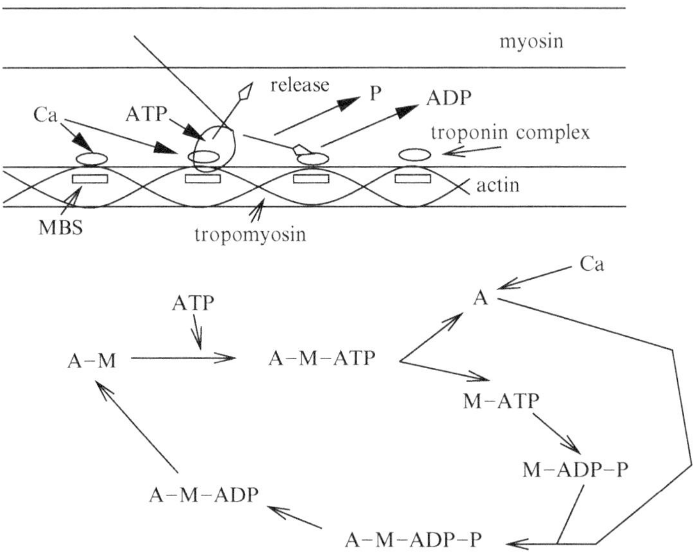

Fig. 5.6 Sliding-filament model of actin–myosin interactions: actin and myosin fixation, sliding, and detachment cycle (nanomotor cycle). In the absence of Ca^{2+}, tropomyosin locks the myosin-binding sites of actin. When Ca^{2+} binds to the troponin complex, troponin and tropomyosin free the mysosin binding site (MBS) of actin. The myosin head then binds to actin, forming a cross-bridge. Myosin detachment from actin is caused by ATP binding. ATP hydrolysis induces a rotation and a weak then strong rebinding of myosin to adjoining myosin-binding site of actin, associated with release of phosphate (P) and then ADP (phosphate release induces a rotation in myosin head). ADP unbinding leads to a new cycle if Ca^{2+} and ATP are available.

the actin filament. The myosin head rotates with an angle of about 45 degrees. This rotation leads to an actin displacement of about 10 nm. (4) The myosin releases ADP and remains anchored to actin. The interaction force is equal to about 1 pN [366]. Interdigitated actin and myosin filaments slide over each other to shorten the sarcomere during contraction (sliding-filament model).

As a sarcomere contracts, Z lines are closer, and the widths of the I bands and H zone decrease, whereas the width of the A band remains constant. Conversely, as a myofibril is stretched, the width of the I bands and H zones increases (constant A-band width). Sarcomere shortening generates myofibril and muscle shortening.

Each molecule of myosin contains a *myosin head* that is a binding site for actin and ATP (Fig. 5.7). Troponin and tropomyosin allow actin to interact with myosin heads in the presence of Ca^{2+} ions. Activation of the muscle fiber causes the myosin heads to bind to actin. Actin is drawn a short distance ($\sim$ 10 nm) past myosin. When linkages break, bonds are reformed farther along actin to repeat the process.[34] The sarcomere length is maximum near the endocardium [367].

[34]The filaments are pulled past each other in a ratchet-like action.

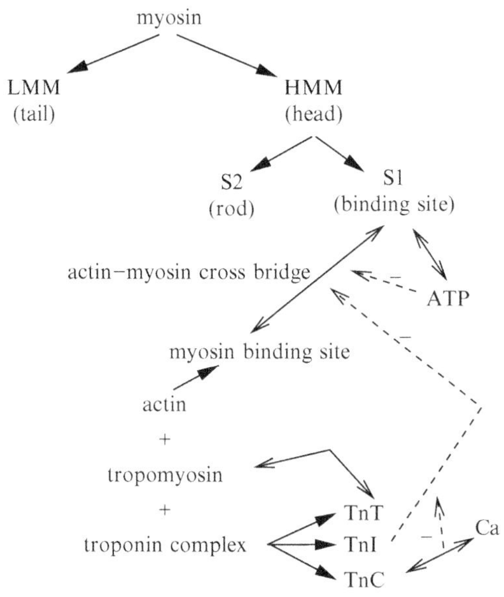

Fig. 5.7 Actin and myosin filaments. The thick filament is composed of many myosin units lined up in a bidirectional arrangement, each unit being made of 2 myosin chains wrapped around each other like a twisted golf club. The combination of calcium ions to troponin-C (TnC) that liberates the myosin-binding site on actin, followed by the inactivation of the tropomyosin–troponin complex, allows myosin head displacement and actin sliding. The heterotrimeric troponin complex is formed by regulatory troponin-C, inhibitory troponin-I (TnI), and tropomyosin-binding troponin-T (TnT). It is positioned along the actin filament at every seventh actin monomer. Troponin-C and calmodulin are the main targets of calcium ions. Both TnC and Cam contain 4 Ca^{2+}-binding domains. Calcium binding to the regulatory site of the cardiac troponin-C (cTnC) affects both its interaction with TnI and the interaction of TnI with TnT and tropomyosin. In the absence of Ca^{2+}, TnI inhibits the interaction of myosin-2 with actin. Ca^{2+} binding to cTnC relieves the TnI inhibition, ensuring contraction. Ca^{2+}-induced conformational changes of troponin-C are imparted to all actin monomers via TnI and TnT acting via 2 tropomyosin threads on both sides along the entire actin filament. cTnI phosphorylation by either PKA or PKC lowers the TnC affinity for calcium ions (positive lusitropy). Two main mechanisms change contractibility: (1) modifications in the amplitude and/or duration of transient Ca^{2+} fluxes and (2) modifications in the sensitivity of the contractile filaments to calcium by phosphorylation of the myosin light chain and/or troponin-I. Moreover, sarcomere compression due to stretching associated with rising diastolic filling brings the contractile filaments closer together, thus facilitating actin–myosin interaction and increasing the TnC affinity for calcium.

5.3.3.2 Myofibrils Scaffolding

Costamere

Myofibrils are held in position by scaffolds of *desmin* filaments, anchored by *costameres*, the cortical parts of the cytoskeleton, enriched in *vinculin*[35] along the sarcolemma (Fig. 5.8). The costameres maintain the spatial structure of sarcomeres and couple the cardiomyocytes to the extracellular matrix. The membrane skeleton is made from *spectrin* and *dystrophin*, adapting the sarcolemma to CMC functioning and contributing to the force transmission. The costameres, membrane skeleton, and cytoskeleton are linked to the extracellular matrix by integral membrane proteins such as integrins, dystrophin–glycoprotein complexes and β-dystroglycan–laminin bounds. Dystrophin and β-*dystroglycan* are membrane proteins that are distributed in the sarcolemma and T tubules.

Myosin-Binding Protein-C: Actin–Myosin Connector and Speed Reducer

The cardiac isoform of myosin-binding protein-C MyBPc3, a thick filament-associated sarcomeric protein member of the immunoglobulin superclass, is located in C zone of A bands.

Cardiac MyBPc binds to actin, myosin, and titin filaments, thereby participating in the organization of the sarcomere structure. Modulator MyBPc3 of sarcomere contractility stabilizes actin in its filamentous state, links actin and myosin filaments, and modulates actin–myosin interaction [368].

Protein MyBPc occupies a portion of the overlap zone — C zone. It reversibly binds via its C-terminus to the myosin filament. Its N-terminus reversibly links to the actin filament, thereby creating a secondary actomyosin connection. It contributes to regulating contraction velocity [369].

The MyBPc3 N-terminus slows actomyosin motion in parts of the thick filament within the C zone. The cardiac isoform contains a domain adjacent to its N-terminus endowed with several serine residues. Contraction slowing is tuned by phosphorylation of 4 serines (Ser273, Ser282, Ser302, and Ser307) [369]. Dephosphorylation of MyBPc3 can lead to its proteolytic cleavage.

Phosphorylations of accessory protein MyBPc3 at different sites by Ca^{2+}–calmodulin-activated kinases and protein kinases PKA and PKC maintain thick filament spacing [370]. β-Myosin-2 flexible coiled-coil segment S2δ has a charge distribution with 3 prominent rings of negative potential [371]. Charge interactions can regulate cardiomyocyte activity via myosin-binding protein-C bound to S2 [372].

[35] Vinculin is also observed in intercalated discs, T tubule membrane, and adherens junctions, between the myofibrils and sarcolemma.

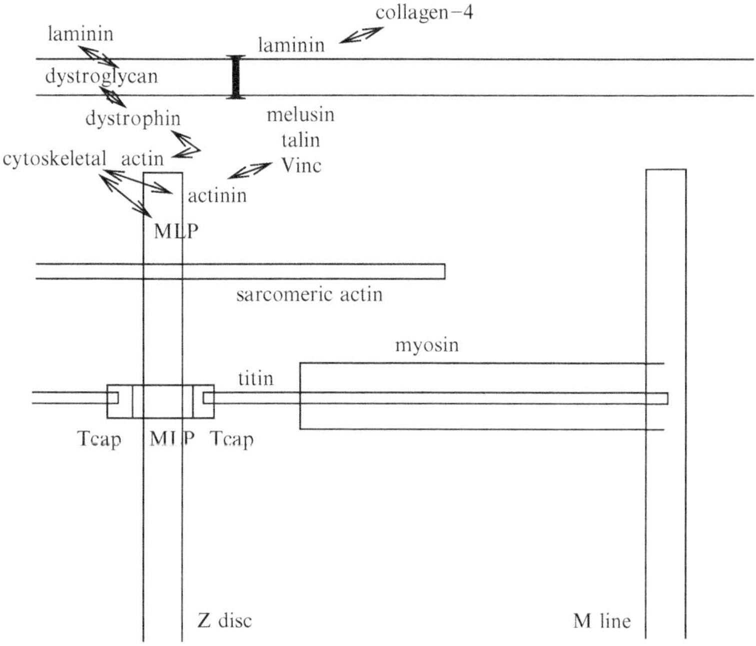

Fig. 5.8 Costamere is the junctional structure connecting the sarcomere to the sarcolemma and extracellular space. Costameres contain 3 main proteic complexes: (1) focal adhesions, (2) spectrin-based complex, and (3) dystrophin-associated complex. Costameric complexes, like the proteic complex of Z disc, contain α-actinin and muscle LIM protein (MLP). α-Actinin binds directly or indirectly vinculin (Vinc), talin, and melusin (a possible stretch sensor that links the costamere to integrins), among other proteins. α-Actinin is thus connected to the sarcoplasmic domain of muscle-specific β_{1D}-integrin (I). Integrin extracellular domain links to laminin. The costamere hence coordinates the deformation of the sarcomere, sarcolemma, and extracellular medium (telethonin). In addition, dystrophin binds actin as well as the dystrophin-associated protein complex of the sarcolemma and sarcoplasma (dystrobrevin, dystroglycans, sarcoglycans, sarcospan, and syntrophins). Integral membrane proteins interact with components of the extracellular matrix via α/β-dystroglycan–α2-laminin complexes. Therefore, dystrophin also connects the sarcomere to the sarcolemma and extracellular matrix. Focal adhesions are composed of sarcoplasmic proteins (i.e., paxillin, talin, tensin, vinculin, and zyxin) that connect with cytoskeletal actin filaments and transmembrane proteins (α- and β-dystroglycans, α-, β-, γ-, and δ-sarcoglycans, dystrobrevin, and syntrophin).

Adrenergic stimulation of the myocardium leads to quick phosphorylation of the cardiac isoform of myosin-binding protein-C by protein kinase-A, as well as to phosphorylation of troponin-I and phospholamban. These phosphorylations contribute to the positive inotropic effect of adrenergic substances. Phosphorylation of MyBPc3 decreases during ischemia–reperfusion injury and pathological hypertrophy (Vol. 6 – Chap. 7. Heart Pathologies), reducing the binding quality between actin and myosin.

5.3.4 Z Disc, a Stretch-Sensing Structure

The structure of sarcomeres is highly ordered. Nonetheless, sarcomeres permanently exchange molecules with surrounding cellular structures. The cytoskeletal support system surrounds and supports myofibrils.

The Z discs not only serve as crosslinkers of thin filaments and transmitters of force generated by the myofilaments, but also as stretch receptors and signal transducers [373]. Various signaling molecules interact with Z-disc proteins. Several interacting substances shuttle between the Z disc and other cellular compartments.

The Z discs are mainly composed of α-actinins assembled in successive layers (width ∼20 nm) that provide a skeleton for the insertion of actin filaments, titin, and nebulin (Fig. 5.9). Z discs of neighboring sarcomeres are aligned in parallel and connected via desmin. The sarcomere is coupled to the sarcolemma by costameres that comprise peripheral Z-disc molecules and subsarcolemmal proteins. Z disc and costameric proteins are involved in mechanotransduction.

5.3.5 Main Sarcomeric Constituents

5.3.5.1 α-Actinin

α-Actinin-2 is the main component of the Z disc. α-Actinin-2 forms homodimers aligned in an antiparallel fashion. It contains 3 domains: actin-binding sequence, central rod, and domain with 2 calcium-binding motifs. α-Actinin-2 interacts with multiple molecules, such as the Rho effector protein Ser/Thr kinase PKN, phospholipase-D2,[36] and G-protein-coupled receptor kinases.[37] α-Actinin interacts with Z-line filamentous actin-capping protein (CapZ). Dimer CapZ is composed of α and β-subunits.

5.3.5.2 Titin

Titin, the largest human protein, spans the sarcomere from the M line to the Z line, where it is anchored. Titin serves not only as a structural component, but also as a

[36]Phospholipase-D2, mainly located at the sarcolemma, regulates Ca^{2+} fluxes. It is inhibited by α-actinin and ARF1 [374].

[37]Phospholipase-D2 modulates GRK activity. In the presence of calmodulin and α-actinin, GRK5 phosphorylates soluble, but not plasmalemmal substrates. In the presence of phosphatidylinositol (4,5)-bisphosphate and α-actinin, GRK5 phosphorylates plasmalemmal, but not soluble substrates [375].

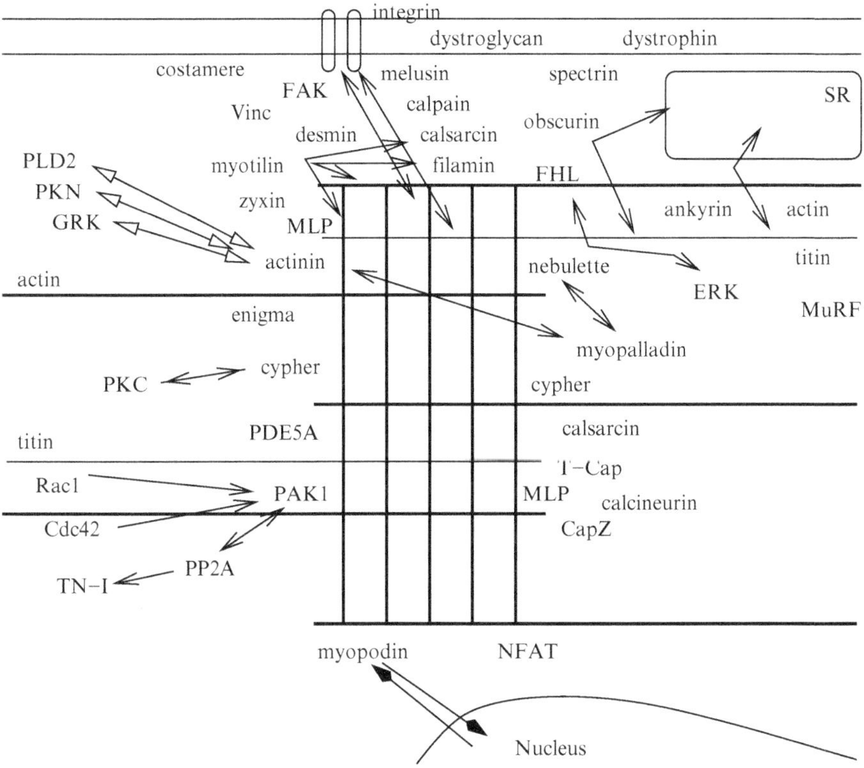

Fig. 5.9 Z-disc proteins and partners, and their interactions (Source: [373]). Z disc is a lattice of interdigitating proteins, such as nebulette, myopalladin, myotilin, filamin, CapZ, cypher, myozenin-1, etc. This proteic complex particularly contains α-actinin, titin-cap (T-cap or telethonin), muscle LIM protein (MLP, CLP, or CSRP3) that binds to actin and titin. Titin extends over half a sarcomere. Titin crosses the Z disc. Titin then associates the Z disc to the M line, both serving as sarcomere scaffolds (supporting and organizing actin and myosin filaments in parallel arrays by crosslinking antiparallel actin and titin). Desmin filaments surround the Z disc, connecting it to cortical (subsarcolemmal) costameres. The sarcomere is also linked to costameres via sarcomeric α-actinin and non-sarcomeric actin (i.e., γ-actin).

strain sensor using its C-terminal protein Ser/Thr kinase domain at M line[38] to adapt myocyte to changes in strain.

The titin kinase domain bears self-inhibition of its C-terminal regulatory tail that blocks the ATP-binding site as well as auto-inhibition of the catalytic domain. Strain

[38]Titin is attached at its C-terminus to M line by a complex formed by obscurin, obscurin-like protein-1, and myomesin. Obscurin kinase, titin kinase, and myosin light chain kinase belong to the family of cytoskeletal, autoregulated, and calcium–calmodulin-regulated protein kinases. M band that is deformed only during contraction is much more compliant than Z disc, and thus is a much more suitable location for a strain sensor.

provokes a conformational change that removes the C-terminal auto-inhibitory tail, thereby allowing access to the ATP-binding site and simultaneously relieving auto-inhibition by partial unfolding that stimulates the kinase domain for subsequent autophosphorylation [376].

Titin is responsible for myocyte elasticity by acting as a spring for sarcomere retraction of stretched cardiomyocytes. Titin near-Z-disc and near-M-line segments bind to actin and myosin and myosin-binding protein-C MyBPc3, respectively. In the relaxed state, titin I-band segment is highly folded [377]. When the sarcomere is stretched, titin extensible I-band segment elongates and develops a tension. During contraction below slack length, myosin filaments move into the near-Z-disc segment of titin that then generates a restoring force useful for diastolic filling. Titin I-band segment is composed of many subsegments of different rheological properties. Among these subsegments, N2a and spring element N2b in the cardiomyocyte sarcomere determine titin isoforms (N2b and N2ba). Titin subtype N2ba contains both N2a and N2b segments. Cardiomyocytes coexpress long (N2ba) and short (N2b) titin isoforms in half-sarcomeres. The N2ba-to-N2b ratio changes to adjust stiffness.[39] The A-band and near-Z-disc segments are inextensible.

Titin N2b subsegment interacts with members of the LIM family that bind creatine kinase and adenylate cyclase for appropriate compartmentation. Titin N2b subsegment is a substrate for protein kinase-A [378]. Ventricular stiffness may then be modulated by adrenergic signaling via titin phosphorylation.

Calcium binding to titin, especially to N2ba, also affects the titin rheology. Calcium also alters interactions between titin and actin. Ca^{2+}-binding protein S100A1 that exists in high concentration in the myocardium inhibits the interaction between titin and actin.

5.3.5.3 Telethonin

Titin crosses the Z disc up to its binding partner *telethonin*, or titin-cap (T-cap), at the periphery of the adjacent Z disc. Telethonin links titin to signaling and structural molecules, such as delayed rectifier K^+ channel in the T tubule and small ankyrin-1 that couples the T tubule to the sarcoplasmic reticulum. Titin also binds obscurin that associates with ankyrin.

Interactions between the Z-disc segment and various compounds may act in the positioning of the sarcoplasmic reticulum and T tubule close to the I band and ensure structure mobility according to sarcomere activity.

[39] The N2ba distribution depends on the cardiac wall region (atria express more N2ba than ventricles), apicobasal location, and transmural layer within the ventricular myocardium. Titin subsegment isoform N2ba is less extensible than N2b subsegment.

Telethonin also interacts with cysteine- and glycine-rich protein CSRP3,[40] a nuclear regulator of myogenic differentiation and possibly a stretch sensor that mediates natriuretic peptide release.

5.3.5.4 Obscurin

Obscurin is a large (~820 kDa) cytoskeletal calmodulin- and titin-interacting protein. It contains 68 immunoglobulin, 2 fibronectin-3, 1 calcium–calmodulin-binding (IQ motif), 1 SH3, 1 DH, 1 RhoGEF (Rho guanine nucleotide-exchange factor), 1 PH, and 2 Ser/Thr kinase MLCK-like domains [379,380]; hence, its other names RhoGEF30 and obscurin^{MLCK}. Multiple immunoglobulin and fibronectin domains enable interactions with myosin and other sarcomeric and cytoskeletal components. Its RhoGEF domain interact selectively with RhoA at the M band and activate this small GTPase [381]. Its Ig domain-binding sites mediate self- and hetero-assembly. They particularly connect to titin as well as slow isoform-1 of myosin-binding protein-C (MyBPc1) and myomesin [380].

Obscurin pertains to the family of giant sarcomeric signaling proteins that includes titin and nebulin that are needed for the assembly of myofibrils and the structural integrity of the sarcomere. Some titin isoforms also contain a protein kinase domain similar to that of myosin light chain kinase. Members of the MLCK family act in the regulation of actin–myosin organization, especially sarcomere assembly.

This protein of striated myocyte concentrates at M bands and Z discs of the sarcomere, where it can sense contractile activity. However, obscurin localizes preferentially to the M band rather than the Z disc [380].

Several obscurin isoforms (~ 70–870 kDa) are encoded by the single OBSCN gene. Alternative splicing generates: (1) conventional obscurin^{GEF}[41] that contains only the GEF domain (~800 kDa) and (2) kinase domain-containing obscurin^{MLCK} (~900 kDa).[42] Smaller obscurin isoforms that possess the RhoGEF domain accumulate at the Z disc.

In addition, in normal epithelia, obscurins localize to cytoplasmic puncta, plasma membrane, and nucleus (110- and 120-kDa nuclear isoforms) [382].

---

[40] A.k.a. Cys-rich protein CRP3, cardiac LIM protein (CLP), and muscle LIM protein (MLP).

[41] A.k.a. obscurin-A. This isoform contains a calmodulin-binding (IQ), several Ig, a Src homology SH3, a RhoGEF, a tandem pleckstrin homology (PH), 2 additional Ig domains, and a non-modular region sequence at the C-terminal end with several putative phosphorylation sites for extracellular signal-regulated kinases as well as a binding site for a small, integral membrane form of ankyrin that concentrates in the sarcoplasmic reticulum.

[42] A.k.a. obscurin-B. This larger isoform of obscurin contains Ser/Thr kinase domains, instead of the C-terminal Ig domains and non-modular sequence of obscurin-A.

Titin is connected to the sarcoplasmic reticulum by the giant muscular protein obscurin and *ankyrin*. Obscurin also binds to calmodulin in a calcium-independent fashion. It possesses also binding sites for different isoforms of ankyrin-1 and -2 from the sarcoplasmic reticulum as well as signaling mediators, in addition to sarcomeric proteins. In addition, obscurin and ankyrin regulate ryanodine channel distribution.

Striated muscle preferentially expressed gene product (SPEG)[43] is closely related to obscurin, with which it shares Ser/Thr kinase MLCK-like domains.

Obscurin helps to keep extrasarcomeric structures in lateral alignment with the sarcomere during contraction–relaxation cycles. Moreover, obscurin-related proteins such as obscurin-like protein-1 (ObsL1) serves also as cytoskeletal adaptor. Protein ObsL1 links the cytoskeleton to the cell membrane, particularly at sites of between-cell and cell–matrix adhesions [383]. Multiple isoforms arise from alternative splicing. Agent ObsL1 localizes to intercalated discs, perinuclear region, and M and Z lines. This cytoskeletal linker positions and stabilizes intercellular contacts and organelles within the cytoskeletal framework.

The titin kinase and zinc finger scaffold protein Neighbor of BRCA1 gene NBR1[44] form a signaling complex at the M band that can recruit the ubiquitin ligase Tripartite motif-containing protein TriM55[45] Enzyme TriM55 binds to serum response factor involved in the synthesis of many muscle proteins.

5.3.5.5 Nebulette and Filamin and Its Partners

Nebulin is an actin-binding protein. *Nebulette* is the cardiac-specific small isoform of nebulin. *Filamin-2*[46] is the filamin isoform specific to striated myocyte. Filamin-2 binds to several partners, such as calsarcin, myotilin, sarcoglycan, and β_1 integrin.

Myotilin binds to α-actinin-2, filamin-2, and actin. It also interacts with calsarcin-1 and -2. *Myopalladin* interacts with manifold proteins, such as nebulette, α-actinin, and cardiac ankyrin repeat protein. *Myopodin*[47] moves from Z discs to the nucleus.

---

[43]A.k.a. aortic preferentially expressed protein-1 (APEG1). It is preferentially expressed in differentiated vascular smooth myocytes. Its production falls during dedifferentiation of vascular smooth myocytes from a quiescent and contractile to a proliferative and synthetic phenotype, a feature of atherosclerosis. Subtypes SPEGα and SPEGβ can be detected in skeletal and cardiac myocytes.

[44]A.k.a. cell migration-inducing gene-19 protein (MIG19).

[45]A.k.a. muscle-specific RING finger MuRF2 and RING finger protein-29 (RNF29).

[46]Filamin-2 is also called filamin-C, ABP280, and γ-filamin.

[47]Myopodin is also called synaptopodin-2.

Table 5.4. The PDZ–LIM family of scaffold proteins or kinases. In the heart, members of this family intervene during embryogenesis as well as adaptive remodeling in response to mechanical stresses and strains.

Type	Alias(es)
	Class-1
PDLIM1	Elfin, CLIM1, CLP36
PDLIM2	Mystique, SLIM
PDLIM3	ALP
PDLIM4	RIL
LMO7	FBx20
	Class-2
LIMK1	
LIMK2	
	Class-3
PDLIM5	EnH (EnH1)
PDLIM6	Cypher, ZASP, LDB3, Oracle
PDLIM7	Enigma, LMP1
LMP4	

5.3.5.6 Members of the PDZ–LIM Family

The PDZ–LIM protein family[48] contains 10 known members that serve as scaffold proteins. These cytoplasmic proteins bind to the cytoskeleton via α-actinin. The PDZ–LIM family can be subdivided into 3 classes according to the number of LIM domains (Table 5.4). PDZ and LIM domain-containing protein PDLIM1 possesses a single C-terminal LIM domain. On the other hand, PDLIM5 (or Enigma homolog) and PDLIM7 (Enigma) contain 3 LIM domains. In the heart, PDZ–LIM family members act as adaptors that connect mechanical sensors embedded in specialized cytoskeletal structures such as the Z disc, to signaling mediators.

Categories of PDZ–LIM Proteins

Group-1 PDZ–LIM proteins that have a single PDZ domain and a single LIM domain include: (1) PDLIM1,[49] (2) PDLIM2,[50] (3) PDLIM3,[51] (4) PDLIM4,[52] and

[48] Aliases PDZ and LIM correpond to the first 3 identified PDZ proteins — PSD95 (a.k.a. DLg4 and SAP90), septate junction proteins Disc large homolog, and zonula occludens ZO1 — and the first 3 described LIM domain-containing proteins — Lin11, Isl1, and Mec3, respectively.

[49] A.k.a. Elfin, C-terminal LIM domain-containing protein CLIM1 and CLP36.

[50] A.k.a. Mystique and SLIM.

[51] A.k.a. actinin-associated LIM protein (ALP).

[52] A.k.a. reversion-induced LIM protein (RIL).

(5) LIM domain-only protein LMO7.[53] Actinin-associated LIM protein is expressed at lower levels in the heart than skeletal muscles.

Group-2 PDZ–LIM proteins are represented by 2 protein kinases: LIM kinase LIMK1 and LIMK2. They contain a PDZ domain, 2 LIM regions, and a Tyr kinase sequence.

Group-3 PDZ–LIM proteins – the Enigma subfamily – include 4 members that possess a single N-terminal PDZ domain and 3 tandem C-terminal LIM domains. This class encompasses: (1) Enigma,[54] (2) Enigma homolog (EnH or EnH1),[55] (3) Cypher,[56] and (4) LMP4 scaffold proteins.

Enigma Group

In humans, 4 spliced variants of the Enh transcript produce 4 isoforms (EnH1–EnH4). Isoform EnH1 is expressed in embryonic and neonatal hearts, whereas, in adults, EnH2 to EnH4 are predominant [384]. Isotype EnH1 possesses an N-terminal PDZ domain and 3 C-terminal LIM domains, whereas EnH2 to EnH4 do not have LIM domains, but retain the N-terminal PDZ domain.

Enigma is able to bind actin, β-tropomyosin, and protein kinases. Agent EnH1 is an adaptor for protein kinase-C. It also links to protein kinase-D1 to regulate the activity of Ca_V1 channels. In normal adult hearts, EnH4 does not recruit PKC and PKD enzymes. During sustained arterial hypertension, EnH1 can be re-expressed and scaffold PKC and PKD kinases [384]. Subtype EnH1 promotes hypertrophy, whereas other isotypes prevent hypertrophy.

Cypher interacts also with α-actinin-2. Its isoform Cypher-2 binds to various, but not all types of protein kinase-C.

LIM–PDZ domain-containing protein LMP4 is coexpressed with and directly binds to heart-specific T-box TBx5 transcription factor. Unbound to LMP4, Tbx5 shuttles between the nucleus and cytoplasm; in a complex with LMP4, it localizes to actin filaments [385]. The TBx5–LMP4 complex undergoes assembly–disassembly cycles in response to external stimuli. Protein LMP4 represses TBx5 activity.

5.3.5.7 Four-and-a-Half LIM-Only Proteins

The family of four-and-a-half LIM (abnormal cell lineage LIN11, insulin gene enhancer protein Islet-1 (Isl1), and mechanosensory abnormality protein MEC3)-

[53] A.k.a. F-box-only protein FBx20.

[54] A.k.a. PDZ and LIM domain-containing protein PDLIM7 and LIM domain-containing protein LMP1.

[55] A.k.a. PDLIM5.

[56] A.k.a. PDLIM6, Z-band alternatively spliced protein (ZASP), LIM domain-binding protein LDB3, and Oracle.

only proteins (FHL), also named skeletal muscle LIM domain-containing proteins (SLIM), includes FHL1, or SLIM1,[57] FHL2, or SLIM3,[58] and FHL3, or SLIM2.[59] Four-and-a-half LIM domain-containing protein is a LIM-only protein that is characterized by 4 LIM domains (double zinc finger sequences) and a single zinc finger domain (LIM half-motif) in the N-terminus. Members of the FHL family can homo- or heterodimerize.

Members of the FHL family are regulators of cell differentiation and remodeling of the cytoskeleton. The FHL proteins act in extracellular signal-regulated kinase signaling. In addition, at least some FHL proteins are transcriptional coactivators, but with distinct affinity for CREB and CREM transcription factors [386]. These members of the subclass-1A basic helix–loop–helix leucine zipper (bHLHzip) transcription factors regulate proliferation and differentiation of various cell types. They homodimerize and heterodimerize (CREB–CREM heterodimers) and then can connect to the cAMP-responsive element (CRE) in the regulatory region of diverse target genes.[60]

Protein FHL2 can associate with α_{3A}-, α_{3B}-, α_{7A}-, and several β-integrin subunits.

5.3.5.8 LIM-Only Protein Zyxin

Zyxin, a ubiquitous member of the LIM-only protein family, is involved in cytoskeletal organization. It is mainly located in focal adhesion complexes close to the Z disc. It binds to α-actinin and cardiac (muscle)-specific LIM protein (MLP or CLP),[61] a member of the C-reactive protein family that belongs to the cardiomyocyte cytoskeleton.

5.3.5.9 Muscle/Cardiac-Specific LIM Protein

The Z disc of cardiomyocytes contains a stress sensor made of muscle-specific LIM protein, titin-cap, and titin. Cardiac LIM protein binds to α-actinin, telethonin, nebulin-related actinin-binding protein NRAP, PP3 phosphatase, and β-spectrin. Cultured cardiomyocytes deficient for MLP do not express natriuretic peptides ANP and BNP in response to stretch.

---

[57] Diverse FHL1 isoforms with different numbers of LIM domains can be generated by alternative splicing.

[58] A.k.a. downregulated in rhabdomyosarcoma LIM protein (DRAL).

[59] Protein FHL5 corresponds to activator of cAMP-responsive element modulator (CREM) in testis (ACT). Modulator CREM is a transcription factor that operates in the differentiation of spermatids into mature spermatozoa. This zinc ion-binding molecule serves as transcription coactivator.

[60] Protein FHL1 has a higher affinity for CREM expressed at high levels in spermatid cells. Both FHL2 and FHL3 that are expressed mainly in the heart and skeletal muscle, respectively, interact specifically with CREB. Protein FHL2 also binds to the androgen receptor.

[61] A.k.a. cysteine-rich protein-3 (CRP3) or cysteine- and glycine-rich protein-3 (CSRP3).

5.3.5.10 Calsarcins

Calsarcins are calcium–calmodulin-dependent phosphatases, PP3-associated sarcomeric proteins that are involved in cardiac hypertrophy. Calsarcin-1 is exclusively expressed by the adult myocardium and slow-twitch skeletal muscle, calsarcin-2,[62] and calsarcin-3 by fast-twitch skeletal muscle fibers. Calsarcins interact with α-actinin, telethonin, filamin-2, cypher, and myotilin.

5.3.5.11 PP3 Phosphatase

Phosphatase PP3, or calcineurin, a calcium–calmodulin heterodimer, interacts with calsarcin-1 at the Z disc. It induces myocardial hypertrophy (Sect. 5.8). It dephosphorylates transcription factors of the NFAT family for translocation into the nucleus. Cytoplasmic PP3-dependent NFATs can be detected at the Z disc. After dephosphorylation by PP3, nuclear translocation of NFAT3 allows coactivation of muscle-specific adenylosuccinate synthetase AdSS1 gene [387].[63] Electrical pacing of cultured cardiomyocytes that serves as a model of cardiac hypertrophy activates muscle-specific Adss1 gene, but represses non-muscle Adss2 gene.

5.3.5.12 ARGBP2

Ubiquitous Abelson-related gene product (ARG)-binding protein-2 forms a complex with CBL ubiquitin ligase, a regulator of the actin cytoskeleton and apoptosis.[64] Agent ArgBP2 associates with and is a substrate of cytosolic Tyr kinases of the Abelson family Abl1 and Abl2 (or ARG).

Protein ArgBP2 abounds in cardiomyocytes, where it localizes to Z discs of sarcomeres. This adaptor assembles signaling complexes at Z discs. In particular, ARG-binding protein-2 binds to α-actinin at Z discs. It influences the contractile or elastic properties of cardiac sarcomeres.

5.3.5.13 Calpains

Calcium-dependent cysteine peptidases, calpains, target sarcomeric proteins and focal adhesions during cell migration. Calpain-3, the isoform specific to striated muscle, is located at the Z discs, costameres with desmin and ezrin, the M line, and the I band. Calpain-3 activates NFκB.

---

[62]Calsarcin-2 is also designated as myozenin or FATZ.

[63]Cardiac transcription factor GATA4 is also required for Adss1 gene transcription. Electrical pacing activates the pp3–NFAT3 and GATA4 pathways.

[64]Abelson-related gene product is a non-receptor Tyr kinase of the Abelson family (Abl2).

5.3.5.14 Desmin

Desmin, a muscle-specific intermediate filament protein, links the costameres to the Z disc. It is a component of the intercalated disc in cardiomyocytes. Desmin interacts with nebulin, spectrin, ankyrins, calpain-3, and the intermediate filament proteins synemin and syncoilin.

5.3.5.15 Integrins

Integrins couple the cytoskeleton, especially costameres, to the extracellular matrix. Integrins also link to other cell adhesion-associated signaling proteins and operate as mechanotransducers. Prolonged mechanical loading of the myocardium causes maladaptive hypertrophy as well as changes in expression of integrin and extracellular matrix proteins. Mechanical stress applied to integrins elevates cytoskeletal stiffness.

Sarcolemmal α_1- and α_2-integrins, expressed in the embryonic heart, can be re-expressed in an adult heart bearing unusual mechanical loading. Integrin-β_{1D} is specific to cardiomyocytes and skeletal myocytes. Integrins associate with focal adhesion kinase, integrin-linked kinase and its partner affixin, calreticulin, and melusin.

Cardiomyocytes synthesize at least 4 integrin types: $\alpha_1\beta_1$-, $\alpha_3\beta_1$-, $\alpha_5\beta_1$-, and $\alpha_7\beta_1$-integrins. Integrins-$\alpha_3\beta_1$ and -$\alpha_5\beta_1$ interact with extracellular fibronectin. This interaction is modulated by the contractile state of cardiomyocytes [388].

5.3.5.16 Melusin

Muscle-specific melusin, or integrin-β_1-binding protein Itgβ_1BP2 (encoded by the ITGB1BP2 gene), is expressed in skeletal and cardiac, but not smooth myocytes. It localizes to costameres in the vicinity of Z disc. It senses mechanical stresses. It promotes the phosphorylation (inactivation) of glycogen synthase kinase GSK3β, thereby preventing transcription factor activity [389].

5.3.5.17 Some Types of Sarcomeric Enzymes

Protein Kinase-C

Protein kinase-Cϵ, located at the Z disc, regulates cardiomyocyte contractility and hypertrophy. Activated by GPCRs, PKCϵ binds anchoring proteins, receptors for activated C-kinases (RACKs). PKCϵ induces normal growth (with preservation of the cardiomyocyte contractility and without fibrosis), rather than maladaptive hypertrophy.

P21-Activated Kinase-1

P21-activated kinase-1, a Ser/Thr protein kinase activated by Rac1 and CDC42 GTPases, is located at the Z disc, sarcolemma, intercalated discs, and nuclear membrane. Kinase PAK1 interacts with protein phosphatase-2, decreasing the phosphorylation of cardiac troponin-I, and enhancing calcium sensitivity.

Phosphodiesterase-5A

cGMP-specific phosphodiesterase-5A localizes to the Z disc. It dissociates from the Z disc during heart failure.

5.4 Cardiac Adenylate Cyclases and the cAMP Pathway

In cardiomyocytes as well as vascular endothelial and smooth muscle cells, cAMP second messenger is involved in signaling transduction by acetylcholine, adenosine, catecholamines, endothelin, glucagon, glucagon-like peptide-1, prostaglandins (PGe1 and PGe2), and urocortin, among others. Signaling compartmentalization and use of distinct effectors can explain distinct effects of cAMP signaling caused by different messenger types.

First messengers target G-protein-coupled receptors, which subsequently activate transmembrane adenylate cyclase. Afterward, cAMP operates via 4 types of effectors: (1) protein kinase-A; (2) RapGEF3 and RapGEF4 ($_{\text{cAMP}}$GEFs or EPACs); (3) cyclic nucleotide-gated ion channels (CNG); and (4) phosphodiesterases (PDE). The latter catabolize cAMP into AMP; however, PDE10a can be inhibited by cAMP that acts either as a competitive antagonist or as an allosteric regulator [390].

Protein kinase-A contains 2 regulatory and 2 catalytic subunits. Once catalytic subunits are liberated, PKA phosphorylate multiple effectors, such as enzymes (e.g., glycogen phosphorylase), transcription factors (e.g., CREB), nanomotors (troponin-I, myosin-binding protein-C), ion channels ($Ca_V1.2$, ryanodine receptor), and other signaling components.

On the other hand, soluble adenylate cyclase resides inside the nucleus and mitochondria, as well as at microtubules and centrioles. In mitochondria, it couples the electron transport chain with cellular respiration. It is activated by HCO_3^- and/or calcium. Once they are internalized, plasmalemmal receptors of extracellular messengers can signal via soluble adenylate cyclase. The latter is involved in apoptosis in coronary endothelial cells and cardiomyocytes [390].

Table 5.5. Cardiac calcium-dependent ion fluxes upon increased cytosolic Ca^{2+} concentration (Source: [391]; APD: action potential duration; $\oplus$: Ca^{2+} increases conductance of ion carrier; $\ominus$: inhibition). Stronger Ca^{2+} transients generate longer and briefer action potentials when ionic currents are inward and outward, respectively. Because cardiomyocytes contain several types of calcium-activated or -modulated ion carriers, the net effect depend on the balance between these carriers.

Inward flux (prolonged APD)	Outward current and reduced inward flux (shortened APD)
Na^+–Ca^{2+} exchanger ($\oplus$)	Ca_V channel ($\ominus$)
Non-selective Ca^{2+}-gated monovalent cation channel ($\oplus$)	Slow delayed rectifier K^+ channel ($\oplus$)
	Ca^{2+}-activated Cl^- channel ($\oplus$)
	Small-conductance Ca^{2+}-activated K^+ channel ($\oplus$)

5.5 Calcium Signals

Cardiomyocyte growth and contractility are spatially and temporally regulated by a set of ion channels, pumps, and exchangers responsible for Ca^{2+} shuttling between the cytosol, intracellular stores, and extracellular milieu.

5.5.1 Calcium-Dependent Ion Carriers

Calcium ion inactivates voltage-dependent calcium channel, but activates other ion carrier types. Calcium-activated ion channels in the heart include [391]: (1) calcium-dependent chloride channels, which can contribute to the initial repolarization; (2) calcium-activated non-selective monovalent cation channels, which carry inward current at negative potentials; and (3) small-conductance calcium-activated potassium channels (Table 5.5). Whereas the rapid delayed rectifier K^+ current ($i_{K,r}$), which is involved in repolarization phase 3, may be purely voltage dependent, the slow delayed rectifier K^+ current $i_{K,s}$ is more efficient at high cytosolic Ca^{2+} concentration [391].

The cardiac action potential triggers opening of $Ca_V1.2$ channels. Resulting Ca^{2+} entry subsequently causes Ca^{2+} release from the endoplasmic reticulum through ryanodine receptors. Calcium ions are removed from the cytosol, particularly by endoplasmic reticulum uptake via sarco(endo)plasmic reticulum Ca^{2+} ATPases (SERCA) and extrusion through the plasma membrane via Na^+–Ca^{2+} exchangers.

5.5.2 *Inositol Polyphosphate Messengers – Phosphoinositide Sources*

Therefore, whereas inositol (1,4,5)-trisphosphate is a major regulator of Ca^{2+}-based responses in non-excitable cells, excitable cells such as cardiomyocytes employ a coupled system of Ca^{2+} import (entry from the extracellular space and influx from the intracellular stores through ryanodine receptors) rather than Ca^{2+} release from the endoplasmic reticulum through IP_3R receptors.[65] In ventriculomyocytes, IP_3R2 localizes mainly in the perinuclear region and nuclear envelope [392]. In atriomyocytes, IP_3R2 content is 6 to 10 times larger than in ventriculomyocytes. These receptors are detected in regions of the sarcoplasmic reticulum adjacent to the sarcolemma, near ryanodine receptors.

Nonetheless, in human cardiomyocytes, a slight (1.5- to 2-fold) overall elevation of PLC activity results from activation of α1-adrenergic or endothelin ET_A receptors (20- to 50-fold increase in non-excitable cells in response to angiotensin-2 and acetylcholine) [392].

Inositol phospholipids, phosphatidylinositol, and its phosphorylated derivatives lodge in cardiomyocyte sarcolemma in concentrations similar to those in other cell types. Stimulated α1-adrenoceptors and ET_A receptors in cardiomyocytes involve primarily the generation of $I(1,4)P_2$ cleaved by PLC from PI(4)P agent [392]. Under normoxia, IP_3 production is limited. Whereas the IP_3 pool does not change significantly upon stimulation,[66] concentrations of $I(1,4)P_2$ and IP_1 isomers[67] increase markedly upon receptor activation [392]. Messenger IP_3 is formed from $PI(4,5)P_2$ by phospholipase-C. On the other hand, $I(1,4,5)P_3$ is rapidly processed by $I(1,4,5)P_3$ 3-kinases and membrane-linked $I(1,4,5)P_3$ 5-phosphatases that deactivate the IP_3 signal.

---

[65] Receptor IP_3R is a Ca^{2+} channel on the endoplasmic reticulum, as Ca^{2+} concentration is much greater than that of other cations. On the plasma membrane, it has a lower selectivity for divalent than for monovalent cations; it may thus act as Na^+ or K^+ channels [392]. Plasmalemmal TRP channels are much more selective for Ca^{2+} and enable Ca^{2+} entry into the cytosol. Receptor IP_3R can be phosphorylated by SRC family kinases, PKA, PKC, and Ca^{2+}–calmodulin-dependent protein kinase CamK2. Most cell types contain multiple IP_3R isoforms. Three IP_3R isoforms exist (IP_3R1–IP_3R3); IP_3R1 possesses 3 splice variants; IP_3R2 splice variants are not functional. Affinity for IP_3 of IP_3R2 is higher than that IP_3R1, which is greater than that of IP_3R3. In the heart, IP_3R1 predominates in nodal (conducting) myocytes; IP_3R2 in contracting myocytes. Yet, IP_3R in adult myocardium is 50-fold less abundant than ryanodine RyR2 receptor [392].

[66] Except in ischemia upon receptor activation (e.g., after noradrenaline release from sympathetic nerves) and transiently upon reperfusion. Production of $I(1,4,5)P_3$ can cause arrhythmias [392]. In atriomyocytes, synthesized $I(1,4,5)P_3$ can bind to subsarcolemmal IP_3R close to its generation site, causing local cytosolic Ca^{2+} level increases that disturb adjacent ion carriers, specifically $Ca_V1.2$ channel and Na^+–Ca^{2+} exchanger.

[67] Compounds with the same molecular formula but different structural formulas. Geometrical positioning of atoms varies. Inositol monophosphates include I(1)P, I(2)P, and I(4)P isomers.

Phospholipase-C has generally a low activity in the heart. Four widespread PLC isozymes exist (PLCβ–PLCϵ). Isoform PLCβ is activated by subunits of the Gq subclass or by Gβγ subunits of the Gi subclass members; PLCγ by receptor protein Tyr kinases; PLCδ by Ca^{2+}; PLCϵ by the G12 subclass members.

5.5.3 *Mitochondrial Calcium Uniporter*

Mitochondrial Ca^{2+} uniporter contributes to Ca^{2+} signal shaping. In adult cardiomyocytes, a fraction of mitochondria is apposed to terminal cisternae of the sarcoplasmic reticulum, where ryanodine receptors are located to generate Ca^{2+}-induced Ca^{2+} release. Hence, these mitochondria can transiently experience high Ca^{2+} concentration during systolic, high-speed Ca^{2+} transients that can overcome the relatively low affinity of mitochondrial Ca^{2+} uniporter for Ca^{2+} ion. In neonatal cardiomyocytes, the intramitochondrial Ca^{2+} level undergoes beat-to-beat oscillations; the average peak magnitude in mitochondrial Ca^{2+} level is quite small (∼ 50% smaller than that in the cytoplasm [393]. At least in neonatal cardiomyocytes, mitochondria significantly contribute to buffering the amplitude of systolic cytosolic Ca^{2+} import [393]. Calcium peaks (20–30 μmol) are formed on the outer mitochondrial membrane during systole; overexpression of mitochondrial Ca^{2+} uniporter significantly reduces them.

5.5.4 *Mechanotransduction*

Mechanical stretch during diastolic filling triggers mechanotransduction that regulates local production of reactive oxygen species. A small diastolic stretch (8%) of a ventriculomyocyte indeed rapidly causes a burst of Ca^{2+} sparks due to the release of free Ca^{2+} ions from intracellular stores. The microtubule network links stretching to changes in concentration of reactive oxygen species and intracellular free calcium ions. Cardiomyocyte stretching increases ROS production by sarcolemmal NADPH oxidase NOx2 using the microtubule network that then sensitizes nearby ryanodine receptors, thereby causing Ca^{2+} sparks [394].[68] NADPH oxidase NOx2 localizes to the junctional sarcoplasmic reticulum, thereby allowing a rapid RyR2 oxidation and tuning RyR2 sensitivity to Ca^{2+} ion. Stretch-dependent oxidation of RyR2 actually is a rapid and reversible control mechanism

[68] Stretch-activated Ca^{2+} sparks are not influenced by stretch-activated channels, Ca^{2+} import, Na^{+} influx, nitric oxide signaling, and acute stretch-induced increases in sarcoplasmic reticulum Ca^{2+} concentration. On the other hand, stretch-gated ryanodine receptors have an augmented Ca^{2+} sensitivity rapidly and reversibly in healthy cardiomyocytes when reactive oxygen and nitrogen species react with their cysteine residues.

of its Ca^{2+} sensitivity. However, in diseases, identical stretch creates transient cytosolic accumulation of free Ca^{2+} ions that can propagate through the cell and arrhythmogenic Ca^{2+} waves may result.

5.5.5 *Sodium–Calcium Exchangers and Sodium–Potassium Pumps*

In cooperation with plasma membrane Ca^{2+} ATPase (PMCA), sarco(endo)plasmic reticulum Ca^{2+} ATPase (SERCA), and Na^+–K^+–Ca^{2+} exchangers (SLC24a1–SLC24a6), Na^+–Ca^{2+} exchangers (NCX1–NCX3, or SLC8a1–SLC8a3) enable recovery of intracellular calcium concentration to basal levels after stimulation. Subtype NCX1 that exchanges 3 extracellular Na^+ for 1 intracellular Ca^{2+} is ubiquitous. Both NCX2 and NCX3 are restricted to the brain and skeletal muscle. However, NCX2 is also expressed in aortic and intestinal smooth muscles. On the other hand, during cardiomyocyte depolarisation, Na^+–Ca^{2+} exchanger can operate in the reverse direction to allow calcium import and sodium export from the cytosol.

Na^+–K^+ ATPase regulates cytosolic Na^+ concentration (main carrier for Na^+ extrusion in cardiomyocytes), hence Ca^{2+} level via Na^+–Ca^{2+} exchanger. This ATPase is composed of 2 subunits, a catalytic α subunit (NKAα1–NKAα4, or ATP1α1–ATP1α4) and a smaller glycoproteic β subunit (ATP1β1–ATP1β3). In the heart, different α subunit isoforms of Na^+–Ca^{2+} exchanger exist, NKAα1 is the dominant subtype, whereas NKAα2 and NKAα3 are present in smaller amounts in human hearts [395]. Both NKAα1 and NKAα2 are connected to NCX in cardiomyocytes. However, in mice, α2 subunit plays a greater role than α1 subunit in modulating Ca^{2+} transients and cardiomyocyte contractility [395].

5.5.6 *Normal and Impaired Calcium Signaling*

Calcium sparks can be triggered by 3 mechanisms: (1) by opening of Ca_V channels in Ca^{2+}-release couplons during an action potential (Ca^{2+}-induced Ca^{2+} release [CICR]); (2) by spontaneous openings of ryanodine receptors, especially when sarcoplasmic reticulum Ca^{2+} load is high; and (3) by Ca^{2+} diffusion from nearby Ca^{2+}-release couplons that have just liberated Ca^{2+} from the sarcoplasmic reticulum, (spark-induced sparks), a minor process during normal excitation–contraction coupling.

5.5.6.1 Calcium-Release Couplon

A typical cardiomyocyte possesses 10,000 to 20,000 Ca^{2+}-release couplons; each couplon consists of 5 to 20 $Ca_V1.2$ channels and 50 to 300 ryanodine receptors.

5.5.6.2 Modeling

The model of a Ca^{2+}-release unit characterized by a given spatial distribution on the cell cortex contains 6 to 7 compartments through which Ca^{2+} moves with given fluxes at interfaces: the bulk and junctional sarcoplasmic reticulum; cytosol, possibly a cortical subdomain (with sodium–calcium exchanger), and dyadic space; plasma membrane; and extracellular space.

The differential equations describing the Ca^{2+} concentrations in different compartments are [396]:

$$\begin{aligned} \mathsf{B}_{C_{1j}}(c_{C_{1j}})\,\partial_t c_{C_{1j}} &= \mathcal{D}_{C_{1j}} \nabla^2 c_{C_{1j}} + J_{C_{1j}}, \\ \mathsf{B}_{C_{2\ell}}(c_{C_{2\ell}})\,d_t c_{C_{2\ell}} &= J_{C_{2\ell}}, \end{aligned} \tag{5.1}$$

where type-1 compartment C_{1j} ($j = 1,2$) represents the cytosol (C_C) and sarcoplasmic reticulum (C_{SR}) and type-2 compartment $C_{2\ell}$ ($\ell = 1,2$) dyadic space (C_{DS}) and junctional sarcoplasmic reticulum (C_{jSR}), $c_{C_i}(\mathbf{x},t)$ compartmental Ca^{2+} concentrations [mol/L^3], B_{C_i} compartmental Ca^{2+} buffering functions (of the Ca^{2+} concentration [L]), J_{C_i} resulting Ca^{2+} flux for each compartment [mol/L^2/T], and $\mathcal{D}_{C_i}$ compartmental Ca^{2+} diffusivity [L^2/T].

Numerical simulations of a ventriculomyocyte couplon network are aimed at studying: (1) influence on Ca^{2+} signals of various parameters, such as sensitivity of ryanodine receptors of Ca^{2+}-release units (couplons), sarco(endo)plasmic reticulum Ca^{2+} load and uptake rate, Na^+–Ca^{2+} exchange magnitude, and Ca^{2+} buffer levels; and (2) stochastic and deterministic aspects of intracellular calcium dynamics.

5.5.6.3 Calcium Wave and Alternans

Transitions from Ca^{2+} spark to Ca^{2+} wave can occur when spark-induced sparks becomes significant, i.e., when sarcoplasmic reticulum Ca^{2+} load progressively increases. Because both Ca_V channels and ryanodine receptors open randomly, Ca^{2+} sparks can happen stochastically and sparsely at low Ca_V open probability and/or under strong Ca^{2+} buffering, but more regularly at higher Ca_V open probability and weak Ca^{2+} buffering [397].

Random Ca^{2+} sparks from Ca^{2+}-release couplons that interact collectively provoke entire-cell *Ca^{2+} alternans*. Cellular calcium transient alternans are beat-to-beat variations in the peak cytosolic Ca^{2+} concentration in cardiomyocytes during rapid electrical stimulation or under pathological conditions.

Beat-to-beat fluctuations in cardiac action potential duration (*APD alternans*, or *voltage alternans*) caused by beat-to-beat changes in amplitude of calcium transients (Ca^{2+} alternans) can generate arrhythmias [391].

When a cardiomyocyte is periodically stimulated, periodic changes in plasmalemmal voltage and cytosolic Ca^{2+} concentration enable a normal excitation–contraction coupling. However, for high-frequency and abnormal stimulations, the periodic behavior can degenerate into 2-period dynamics. The action potential duration alternates in a long–short—long–short trace; the peak cytosolic calcium concentration alternates in a large–small—large–small pattern [398].

At high pacing cycle periods, alternans have the same phase in every cell. However, at low pacing cycle durations (period 180–210 ms), alternans may be out of phase in neighboring cells as well as may appear in a cell region, but not another, and may be out of phase in 2 regions of the same cell (*subcellular alternans*) [391]. According to the sign and magnitude of calcium–APD coupling, calcium alternans can be spatially synchronized or desynchronized, in or out of phase with action potential duration alternans [398].

Spatially concordant Ca^{2+} alternans are Ca^{2+} alternans with the same phase throughout the cardiomyocyte. The situation in which Ca^{2+} and APD alternans are in phase, i.e., a large–small calcium pattern coupled to a long–short action potential duration pattern is the most common (electromechanically concordant alternans) [398]. *Spatially discordant Ca^{2+} alternans* result from uniform APD alternans in the cell due to a fast voltage change with slow calcium diffusion that enable Ca^{2+} alternans with opposite phase in different regions of the cell. Calcium and APD alternans can be out of phase, but spatially synchronized over the entire cell. Electromechanically discordant alternans refers to a small–large calcium pattern coupled to a long–short action potential duration pattern [398].

Irregular, asynchronous Ca^{2+} release generates Ca^{2+} alternans. Calcium alternans are determined by 3 features of Ca^{2+} sparks from Ca^{2+}-release couplons [396, 397]: (1) randomness of Ca^{2+} spark; (2) refractoriness of a couplon after a Ca^{2+} spark during a refractory period; and (3) recruitment of Ca^{2+} sparks from neighboring couplons (spark-induced sparks). The steep non-linear dependence of systolic Ca^{2+} release from the sarcoplasmic reticulum on diastolic Ca^{2+} load immediately preceding the release results from the collective behavior of Ca^{2+} sparks [397]. Calcium alternans may be explained by this steep non-linear relationship, when diastolic Ca^{2+} loading alternates concomitantly with Ca^{2+} release.

5.6 Endocrine Function

The cardiac pump has secondary functions. The heart actually behaves as an endocrine organ that secretes different types of hormones directly into the blood stream, thereby ensuring a remote control.

5.6.1 Natriuretic Peptides

A- and B-type natriuretic peptides reside not only in specific granules of myoendocrine cells, but also in the brain, adrenal medulla, and kidney.

5.6.1.1 Natriuretic Peptides Receptors and Effects

Atrial natriuretic peptide (ANP) regulates arterial blood pressure and cardiomyocyte growth. It stimulates angiogenesis and metabolism. Its hypovolemic effect is exerted on both the kidney and systemic microcirculation. Natriuretic peptides also contribute to the nervous regulation of the cardiac frequency.

In patients with maladaptive cardiac hypertrophy and heart failure, ANP and BNP levels markedly increase, without elevated cGMP formation and signaling. Desensitization of GC2a results from post-translational modifications, particularly dephosphorylation within the kinase homology (KH) domain [399].

Natriuretic Peptides Receptors

Transmembrane natriuretic peptide receptors (NP_1–NP_2) that are particulate guanylate cyclases act similarly to nitric oxide, which targets soluble guanylate cyclases, as they enhance acetylcholine release and vagal-triggered bradycardia [400].

Receptor GC2a consists of an extracellular ligand-binding domain, a short membrane-spanning region, and an intracellular part with a kinase homology (KH) and a coiled-coil dimerization domain as well as the C-terminal catalytic guanylate cyclase (GC) region. In the absence of ligand, the KH domain is highly phosphorylated and the catalytic activity of GC2a is repressed. Upon ANP binding, a conformational change occurs that activates the cyclase domain [399].

Another receptor NP_3 that lacks the guanylate cyclase domain operates via inhibitory subunit (Gi) of guanine nucleotide-binding protein and likely also as clearance receptors for circulating peptides.

cGMP-Dependent Signaling of Guanylate Cyclase-A

Via its particulate guanylate cyclase-2A (pGC2a) receptor and intracellular cGMP messenger, atrial natriuretic peptide supports endothelial permeability [401].

Under normal ANP concentration, the ANP–cGMP axis activates cGMP-dependent protein kinase PKG1 that prevents Ca^{2+} influx into cardiomyocytes triggered by angiotensin-2-bound AT_1 receptor. Repression of Ca^{2+} entry results from stimulation of regulator of G-protein signaling RGS2 and inhibition of transient receptor potential canonical TRPC3 and TRPC6 channels [399].

At least in guinea pig sinoatrial nodes, NP_1 and NP_2 are synthesized. Both BNP and CNP facilitate vagal neurotransmission and bradycardia possibly via a cGMP–PDE3 pathway that activates the cAMP–PKA pathway, hence supporting phosphorylation of presynaptic $Ca_V2.2$ channels.

Hormones BNP and CNP heighten vagal neurotransmission, as they increases the release of acetylcholine via a presynaptic cGMP-dependent pathway. Acetylcholine binds to postsynaptic muscarinic M_2 receptors on sinoatrial node pacemaker cells to decrease the cardiac frequency following effects on several pacemaking currents (calcium, hyperpolarization-activated, and ACh-sensitive potassium currents).

In addition, natriuretic peptides can directly augment the cardiac frequency independently of the autonomic nervous system via particulate guanylate cyclase NP_2, cGMP-primed inhibition of PDE3, hence elevated cAMP level. Messenger cAMP acts directly on the hyperpolarization-activated cyclic nucleotide-gated channel, shifting its activation to more positive potentials [400]. The tachycardia triggered by CNP can also result from a rise of the rate of diastolic depolarization.

cGMP-Independent Signaling of Guanylate Cyclase-A

Whereas the cGMP-dependent signaling pathway counteracts the Ca^{2+}-enhancing action of hypertrophic factors such as angiotensin-2, in chronic cardiac hypertrophy, when ANP level increases and GC2a receptor is desensitized, ANP binds to GC2a and stimulates a cGMP-independent signaling pathway. This signaling is characterized by an abnormally elevated cytosolic Ca^{2+} concentration that result from activation of TRPC3 and TRPC6 cation channels by GCa receptor [399]. The resulting cation influx through TRPC channels within a pre-existing stable GC2a–TRPC complex stimulates $Ca_V1.2$ channels.

5.6.1.2 Prohormones and Active Natriuretic Peptides

Cardiac natriuretic peptides constitute a family of peptides that are synthesized from different genes and then stored as distinct prohormones (Table 5.6), in particular: 126-amino acid atrial (A-type; proANP; ANP_{1--126}), 108-amino acid brain (B-type; proBNP; BNP_{1--108}), and 103-amino acid C-type (proCNP; CNP_{1--103}) natriuretic peptide prohormones.

The natriuretic peptide family consists of 5 known peptides (Table 5.7): *A-type* (atrial; ANP [$ANP_{99--126)}$], or cardiodilatin; plasma concentration ~ 10 fmol/ml [402]); *B-type* (brain; BNP; plasma concentration ~1 fmol/ml; 3.5 pg/ml); *C-type* (CNP); and *D-type* (DNP) natriuretic peptides; as well as *urodilatin*, or $ANP_{95--126)}$.

Both ANP and urodilatin are produced from the same type of precursor (prepro-hormone) itself cleaved into ANP_{1--126} prohormone, stored in membrane-bound secretory granules in the atria. Urodilatin targets NP_1 in renal tubules and vascular

Table 5.6. Prohormone and active members of the natriuretic peptide category. Prohormones are formed upon cleavage by peptidases of preprohormones. Preprohormone of each natriuretic peptide is encoded by a distinct gene such as proproANP, or prepronatriodilatin, by the NPPA gene. Natriuretic peptides have an antihypertensive function, as they lower plasma renin and aldosterone concentrations and stimulate natriuresis and diuresis, hence reducing the blood volume, in addition to a vasorelaxant activity. Cardiac natriuretic peptides operate as counter-regulators of sympatheticoadrenal and neurohormonal axes. In addition, they can have an anti-inflammatory and pro-apoptotic action in endothelial cells.

Type	Name
ANP_{1--126}	ANP prohormone (proANP)
	C-terminal active forms
$ANP_{99--126)}$	ANP, atriopeptin, cardiodilatin
$ANP_{95--126)}$	Urodilatin
Active fragments of inactive $proANP_{1--98}$ N-terminus	
ANP_{1--30}	Long-acting natriuretic peptide
ANP_{31--67}	Vessel dilator
ANP_{79--98}	Kaliuretic peptide
BNP_{1--108}	BNP prohormone (proBNP)
$BNP_{77--108}$	BNP
CNP_{1--103}	CNP prohormone (proCNP)
$CNP_{51--103}$	CNP_{53}
CNP_{82-103}	CNP_{22}
DNP_{1--38}	DNP

Table 5.7. Site of synthesis of cardiac natriuretic peptides.

Type	Gene	Number of amino acids	Producing cells	Secretion stimulus
ANP	NPPA	28	Atriomyocyte	Stretch Sympathetic stimulation (β-adrenoceptor) Hypernatremia Angiotensin-2, endothelin
BNP	NPPB	32	Ventriculomyocyte	Tension
CNP	NPPC	22	Endothelial cells Central nervous system	Shear
DNP	NPPD	38	Atriomyocyte	Stretch

smooth muscle clls in the kidney [403]. Then, cGMP-dependent protein kinase-G inhibits Na^+ reabsorption via ENaC channel. Relaxation of renal arterial smooth myocytes results from a decrease in intracellular Ca^{2+} concentration.

Prohormone ANP_{1--126} generates 4 peptide hormones that regulate blood pressure and maintain plasma volume: (1) *long-acting natriuretic peptide* (LANP; amino acids 1–30 of ANP prohormone [ANP_{1--30}]); (2) *vessel dilator* (amino acids 31–67 of ANP prohormone [ANP_{31--67}]); (3) *kaliuretic peptide* (amino acids 79–98 of ANP prohormone [ANP_{79--98}]); and (4) *atrial natriuretic peptide* (amino acids 99–126 of ANP prohormone [$ANP_{99--126)}$]). On the other hand, the Bnp and Cnp genes synthesize a single known peptide hormone.

These peptide hormones circulate in blood. Plasma concentrations of vessel dilator and LANP are 17- to 22-fold higher than that of ANP, 33- to 48-fold larger than that of BNP, 124- to 177-fold greater than that of CNP, and 170- to 177-fold higher than that of urodilatin that is synthesized only in kidney (amino acids 95–126 of ANP prohormone [$ANP_{95--126)}$]).

Clearance of natriuretic peptides from the blood circulation involves receptor binding, uptake especially by the clearance receptor NP_3, and metabolism in target cells, and degradation by zinc metallopeptidases — neutral peptidases — in vascular and renal tubular cells and excretion via urine.

Atrial and B-type natriuretic peptides are synthesized predominantly in atria and the left ventricle, respectively, in response to elevated wall tension. Plasma levels of natriuretic peptides correlate positively with cardiac filling pressures.

Atriocardiomyocytes are long, slender, and poor in T tubules. Atrial cardiomyocytes contain small granules, especially in the right atrium. These granules secrete atrial natriuretic peptide immediately when atriocardiomyocytes are excessively stretched. Ventricular cardiomyocytes synthesize B-type natriuretic peptide. Levels of BNP reflect long-term overload [404].

5.6.1.3 Natriuretic Peptide Receptors

Three types of receptors (NP_1–NP_3; Vol. 3 – Chap. 6. Receptors) have been described for the natriuretic peptides (Table 5.8). Factors ANP and BNP target guanylate cyclases (NP_1–NP_2) that produce cyclic guanosine monophosphate. Mediator cGMP then acts on cGMP-dependent protein kinases PKG, cGMP-gated ion channels, and cGMP-regulated phosphodiesterases.

5.6.1.4 Angiotensin-2 and Endothelin-1

Angiotensin-2 is not only a vasoconstrictor and mitogenic peptide, but also an inducer of atrial natriuretic peptide expression via extracellular signal-related kinase in ventriculomyocytes.[69]

---

[69]Hypertrophied heart releases natriuretic peptides ANP and BNP. Regulator ANP has both direct and indirect effects on the heart. It inhibits cardiomyocyte growth by lowering activation of ERK1 and ERK2 and upregulating phosphatase MKP1. It reduces hemodynamic load by educing diuresis

Table 5.8. Natriuretic peptide receptors and rank of affinity of the peptides to these receptors (Source: [402]). Most natriuretic peptides are secreted from cardiomyocytes in response to stretch of atrial or ventricular walls. The effects of cardiac natriuretic peptides include natriuresis, diuresis, and inhibition of the activities of several regulators, such as the renin–angiotensin–aldosterone system, endothelins, cytokines, and the sympathetic nervous system. Active peptides target natriuretic peptide receptors, NP_3 acting as a clearance receptor upon action of neutral endopeptidase. Hormone DNP activates NP_1 rather than NP_2 receptor. Messenger cGMP is the second messenger formed by activated NP_1 and NP_2 guanylate cyclases.

Receptor	Rank of affinity
NP_1	ANP $\geq$ BNP $\gg$ CNP
NP_2	CNP $>$ ANP $\geq$ BNP
NP_3	ANP $>$ BNP $>$ CNP

On the other hand, atrial natriuretic peptide opposes the reduction of angiotensin-converting enzyme-2 (ACE2) synthesis initiated by angiotensin-2 and endothelin-1 (Fig. 5.10) [405]. In cardiomyocytes, ACE2 responds to various vasoactive peptides and hormones. It preferentially forms heptapeptide angiotensin_{1-7} from angiotensin-2.[70] Conversely, in cardiomyocytes and cardiac fibroblasts, angiotensin-2 lowers ACE2 synthesis and activity, thereby preventing the conversion into angiotensin_{1-7}.

Endothelin-1 also significantly attenuates ACE2 expression in cardiomyocytes. Both angiotensin-2 and endothelin-1 that are released from cardiomyocytes in response to mechanical strain and/or stress, work via extracellular signal-regulated kinases ERK1 and ERK2.

5.6.1.5 A-Type Natriuretic Peptides

Once it is released, the ANP propeptide is cleaved by a heart-specific transmembrane serine peptidase corin into a 28-amino acid C-terminal active form — ANP — and a 98-amino acid, N-terminal, inactive form (proANP_{1--98}]). The latter is processed by peptidases to generate several active peptides: (1) *long-acting natriuretic peptide*, i.e., proANP_{1--30}; (2) *vessel dilator*, i.e.,

and natriuresis. Factor ANP binds to its receptors NPR-A and NPR-B to stimulate guanylate cyclase and generate cGMP.

[70]Enzyme ACE2 also hydrolyzes dynorphin-A and apelin. It is expressed in both endothelial and smooth muscle cells of coronary vessels as well as in cardiomyocytes and cardiac fibroblasts. Level of ACE2 rises in heart failure and ischemic and idiopathic dilated cardiomyopathy. Angiotensin_{1-7} operates via a specific receptor AT(1-7). Angiotensin_{1-7} hampers cardiomyocyte growth. Whereas ANP lowers effects of ERK1 and ERK2 via cGMP and MKP1, angiotensin_{1-7} decreases activity of ERK1 and ERK2 via cAMP.

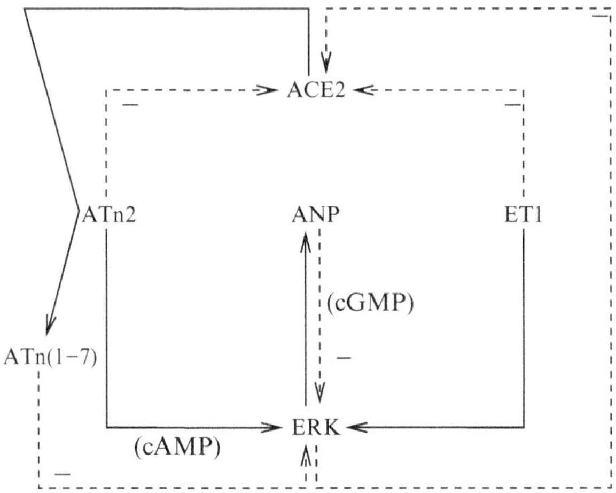

Fig. 5.10 Feedback loops between atrial natriuretic peptide (ANP), angiotensin-2 (ATn2), and endothelin-1 (ET1). Angiotensin-2 is converted to angiotensin$_{1-7}$ by angiotensin-converting enzyme-2 (ACE2). Angiotensin-2 precludes ACE2 activity, thereby avoiding its conversion. Whereas angiotensin-2 and endothelin-1 stimulate extracellular signal-related kinases ERK1 and ERK2, atrial natriuretic peptide and angiotensin$_{1-7}$ inhibit ERK1 and ERK2 via cGMP and cAMP, respectively.

proANP$_{31--67}$; and (3) *kaliuretic peptide*, i.e., proANP$_{79--98}$. Peptides ANP, vessel dilator, and long-acting natriuretic peptide bind to their specific receptors.

Cardiac hormone ANP is secreted from myoendocrine cells of atria in response to acute or chronic stretch of the atrial myocardium. It causes diuresis, natriuresis, and vasodilation. Atrial natriuretic peptides is indeed an endocrine regulator of blood volume, as it increases excretion of water and sodium and potassium ions by the renal distal convoluted tubule (Vol. 6 – Chap. 4. Cardiovascular Physiology).

In addition, circulating ANP also decreases blood pressure by impeding the secretion of *renin* by the kidney and *aldosterone* by the zona glomerulosa of the adrenal cortex. Furthermore, it prevents action of vasopressin as well as vasoconstriction mediated by angiotensin-2 [402].

It operates via cGMP messenger that remove salts and water in the collecting ducts. In addition, cGMP antagonizes the renin–angiotensin–aldosterone axis. Messenger cGMP is also a vasodilator that increases the permeability of blood vessels.

The antihypervolemic and antihypertensive factor ANP provokes vasodilation as well as a concomitant leakage of fluid from blood vessels to the interstitium by increasing the endothelial permeability to plasma proteins such as albumin [401]. Once the glycocalyx crossed, plasma proteins and solutes are transported across the endothelium either via the transcellular route, i.e., via caveola-mediated vesicular transport, or via the paracellular path, i.e., via interendothelial junctions (Sect. 9.6). Hormone ANP primarily stimulates vascular permeability in the skin and skeletal muscle [401]. It enhances caveolar transcytosis of albumin via its GCa receptor. Both GCa receptor and cGMP-dependent protein kinase-G1 partly colocalize with caveolin-1 in caveola nanodomains (caveolin-1-containing endothelial membrane rafts).

In addition, ANP lowers the sympathetic tone in the vasculature by direct suppression of sympathetic signaling from the central nervous system and release of catecholamines from nerve endings, thereby attenuating tachycardia and vasoconstriction to reduce postload.

In the kidney, ANP prohormone is cleaved between amino acids 95 and 96, unlike between amino acids 98 and 99 in the heart. *Urodilatin*, a 32-amino acid peptide, is secreted in response to increased mean arterial pressure and blood volume from cells of the distal tubule and collecting duct in the lumen of the nephron. It then interacts with receptors at distal segments of the nephron to regulate Na^+ and water reabsorption. Intrarenal regulator urodilatin may be more involved in the regulation of the body's fluid volume and water–electrolyte excretion and circulating ANP in blood pressure regulation.

Natriuretic peptides and urodilatin can operate in synergy with G-protein-coupled receptors (e.g., β2-adrenergic receptor in bronchial smooth myocytes) to lower intracellular calcium concentration, thereby reducing the smooth muscle tone.

5.6.1.6 B-Type Natriuretic Peptides

Mature, circulating, 32-amino acid BNP ($BNP_{77--108}$) is formed from a 134-residue preprohormone cleaved into a prohormone, proBNP ($BNP_{77--108}$), predominantly in cardiac ventricles. (Peptide BNP is synthesized in the atria and ventricle.) Its plasma concentration is a clinical marker for left ventricular overload.

Once proBNP is released into the blood circulation, it is cleaved into a C-terminal, active fragment — BNP — (plasma half-life $\sim$ 20 mn), and inactive, 76-amino acid, N-terminal fragment ($proBNP^{NT}$; plasma half-life $\sim$ 120 mn) [402]. Ventricular BNP production is transcriptionally regulated by cardiac wall stretch (volume overload).

5.6.1.7 C-Type Natriuretic Peptides

Member CNP is produced in vascular endothelia and human ventricle. This 22-amino acid peptide has a much wider distribution than that of ANP and BNP hormones. Two mature forms derived from a 126-amino acid preprohormone (preproCNP) processed into a 103-amino acid prohomone (proCNP; CNP_{1--103}) [402]. They include CNP_{53} ($CNP_{51--103}$), which predominates in tissues, and CNP_{22} ($CNP_{82--103}$) mainly in plasma and cerebrospinal fluid.

This hormone regulates the local vascular tone. It precludes angiotensin-2-stimulated release of endothelin [402]. Conversely, endothelin antagonizes CNP action.

In addition, CNP prevents proliferation of cardiomyocytes. It also impedes vasoconstriction caused by angiotensin-1, but not that of angiotensin-2, as it may be a local regulator of vascular angiotensin-converting enzyme [402].

5.6.1.8 D-Type Natriuretic Peptides

D-Type (or Dendroaspis) natriuretic peptide (DNP) is derived from the venom of Dendroaspis augusticeps (green Mamba snake). The 38-amino acid peptide DNP can be detected in human plasma (average concentration ~6 pg/ml [range 2–100 pg/ml]) [402]. It can be identified in the atrial myocardium. It provokes natriuresis and diuresis via the GCase–cGMP axis. It may lower sodium reabsorption by antagonizing action of angiotensin-2 in the proximal tubule. It is a potent vasoactive agent. It has a greater resistance to degradation by neutral endopeptidase than other natriuretic peptides.

5.6.2 Hemojuvelin–Hepcidin Regulation

Hemojuvelin is mainly expressed in skeletal and cardiac muscles and, to a lesser extent, the liver. It modulates the expression of the HAMP gene that encodes hepcidin, via the bone morphogenetic protein pathway [406, 407]. Hepcidin is a liver-derived iron-regulatory peptide acting on cells to limit iron flux toward the plasma. Hepcidin deficiency induces iron overload, whereas hepcidin excess induces anemia.

5.7 Energy Supply and Reserve

Energy provider ATP is required for both cell viability and myocardial pump activity. Most ATP is synthesized by substrate oxidation in mitochondria that have a high density in cardiomyocytes. Mitochondria-derived ATP is mainly used for cardiac contraction, whereas ATP produced by glycolysis intervenes in activity of kinases and ion pumps and channels (e.g., Ca^{2+}–calmodulin-dependent kinase-2, ATP-sensitive K^+ channels, Na^+–K^+ ATPase, and Ca^{2+} ATPase SERCA2).

At rest, fatty acid oxidation covers more than 70% of energy need. The remainder comes from oxidation of carbohydrates, principally glucose. Although oxidation of glucose is more efficient than that of fatty acid, fatty acid oxidation provides much more ATP than glucose. In healthy heart, selection among lipids and carbohydrates is governed by plasma substrate availability and hormonal regulation (insulin, catecholamines). When heart activity raises, additional sources of ATP synthesis, such as glycogenolysis, glycolysis, and phosphotransferase reactions catalyzed by creatine kinase and adenylate kinase, are involved (Fig. 5.11).

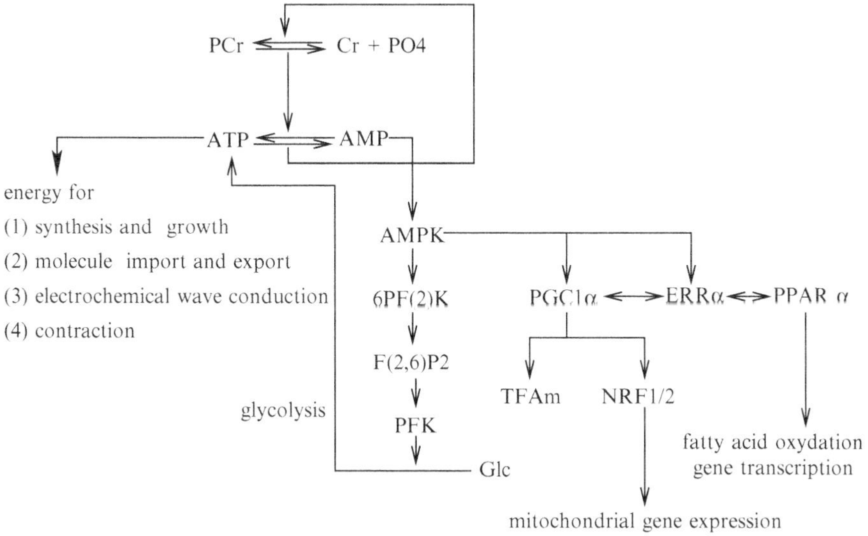

Fig. 5.11 Energy supply and reserve in cardiomyocytes. At rest, fatty acid oxidation covers most of energy need and the remainder comes from oxidation of carbohydrates, principally glucose. When heart activity rises, additional sources of ATP synthesis such as phosphotransferase reactions catalyzed by creatine kinase (CK) are involved. ATP production by phosphotransferase reactions is about 10 times faster than ATP synthesis in mitochondria. Phosphocreatine (PCr) actually is the primary energy reserve element in cardiomyocytes. Glycolysis is the set of reactions that converts glucose (Glc) into pyruvate. Phosphofructokinase (PFK) that controls glycolysis is stimulated by AMP and fructose (2,6)bisphosphate (F(2,6)P2). Attenuated phosphocreatine concentration and elevated AMP level activate AMP-activated protein kinase (AMPK) that increases glucose transport by upregulating transporter GluT4 expression and activates PFK by stimulating 6-phosphofructo 2-kinase (6PF(2)K) activity. Kinase AMPK also activates mitochondrial estrogen-related receptor ERRα and peroxisome proliferator-activated receptor-γ coactivator PGC1α that regulates fatty acid oxidation. ERRα, ERRβ, and ERRγ regulate cardiac energy metabolism. Factor PGC1α stimulates mitochondrial genesis, as it increases the production of nuclear respiratory factors NRF1 and NRF2 that regulate mitochondrial transcription factor-A TFAm and ERRα (NR3b1) that coordinates the expression of genes encoding mitochondrial proteins. Moreover, PGC1α regulates the expression of genes encoding proteins involved in fatty-acid oxidation via coactivation of PPARα and ERRα.

Phosphocreatine (Vol. 1 – Chap. 4. Cell Structure and Function) is the primary energy reserve element in cardiomyocytes. Production of ATP by creatine kinase is about 10 times faster than ATP synthesis in mitochondria, which is approximately 20 times quicker than ATP generation via glycolysis.

Glycolysis is the set of reactions that converts glucose (Glc) into pyruvate. Glycolysis in aerobic condition is the prelude of the tricarboxylic acid cycle and mitochondrial electron transport chain, where most of the free energy in glucose is harvested. Consumption of ATP is a preamble to ATP synthesis. Phosphofructokinase (PFK) is the most important controller in glycolysis. It is inhibited by high ATP level and stimulated by AMP and fructose (2,6)-bisphosphate ($F(2,6)P_2$). Attenuated phosphocreatine concentration and elevated AMP level activate AMP-activated protein kinase that increases glucose transport and activates phosphofructokinase for glycolysis. In the heart, AMPK stimulates 6-phosphofructo 2-kinase activity and glucose transporter GluT4 expression.

AMP-activated protein kinase also activates mitochondrial estrogen-related receptor ERRα (or nuclear receptor NR3b1) and peroxisome proliferator-activated receptor-γ (or NR1c3) coactivator PGC1α. ERRα, ERRβ (NR3b2), and ERRγ (NR3b3) regulate cardiac energy metabolism. ERRα interacts with PGC1α. Fatty acid oxidation is regulated by PPARs and their coactivator PGC1α. The complex formed by ERRα and PGC1α targets a set of genes that encode proteins involved in energy production that results from oxidative phosphorylation (e.g., fatty acid and glucose uptake, tricarboxylic acid cycle, and electron transport chain) and transfer (e.g., mitochondrial creatine kinase and adenine nucleotide transporter). Factor PGC1α stimulates mitochondrial genesis, as it increases the production of nuclear respiratory factor NRF1 and NRF2 that regulate mitochondrial transcription factor-A TFAm and ERRα (NR3B1) that coordinate the expression of genes encoding mitochondrial proteins. Moreover, PGC1α regulates the expression of genes encoding proteins involved in fatty-acid oxidation via coactivation of PPARα and ERRα.

5.8 Cardiomyocyte Adaptive Growth

Developmental growth deals with normal growth of the heart after birth. Cardiac hypertrophy is defined by an increase in size of cardiomyocytes. The heart is capable of acute and chronic *hypertrophy*. In response to long-term regular exercise and pregnancy, the heart undergoes physiological growth, or adaptive hypertrophy. Cardiomyocyte size increase in response to a hypertrophic stimulus is associated with augmented sarcomerogenesis and expression of natriuretic peptides. The cardiomyocyte length elevation is larger than the width rise. Both developmental growth and physiological hypertrophy are characterized by coordinated myocardial growth matched with heart chamber size, normal organization of sarcomeres, and enhanced cardiac function.

Myocardial hypertrophy can be either adaptive or maladaptive according to cardiomyocyte contractility, the former being associated with normal or even improved function, and the latter with impaired contractile function. Maladaptive hypertrophy is induced by long-term sustained cardiac stresses and diseases such as systemic arterial hyperpressure and aortic valve stenosis. Remodeling leads to a greater increase in cardiomyocyte width than length. Maladaptive hypertrophy is defined by a myocardial dysfunction.

Increase in the number of cardiomyocytes means *hyperplasia*. Hyperplasia is characterized by an increased myocardial dry mass, normal cardiomyocyte size, and the absence of an abnormal myocardial component.

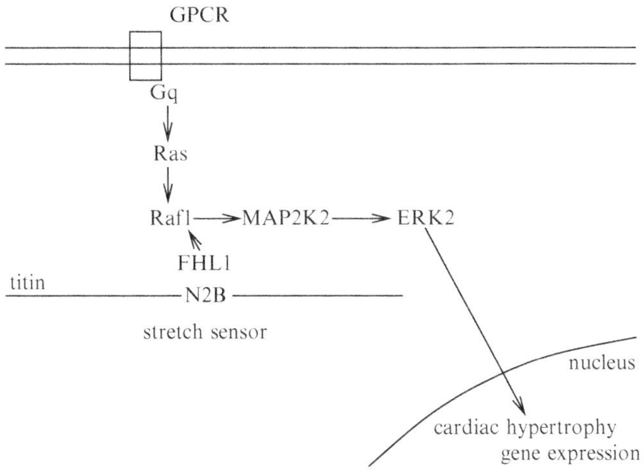

Fig. 5.12 Cardiomyocyte stretch sensor and signaling (Source: [408]). Protein FHL1 is associated with titin N2b subsegment in the sarcomere and serves as scaffold protein for the Raf1–MAP2K2–ERK2 module downstream from Gαq stimulation.

5.8.1 Mechanotransduction in CMC Hypertrophy

Mechanical stresses are sensed by cardiomyocyte sarcomeres and transduced to generate hypertrophy. Dysfunction of these pathways leads to maladaptive hypertrophy. Although cardiac adaptive hypertrophy allows an increase in myocardial contractility, initially compensatory, sustained hypertrophy indeed serves as an index for cardiac morbidity and mortality.

Cytoskeletal proteins of the LIM domain-containing protein family intervene in myocyte response to mechanical stress. Four-and-a-half LIM domain-containing proteins (FHL1–FHL3) have multiple binding partners. They are involved in striated myocyte development and function.

Protein FHL1 is the single FHL member that is upregulated in mouse heart subjected to hypertrophic excitation. In humans, FHL1 level is strongly increased in hypertrophic cardiomyopathy. Protein FHL1 belongs to the complex of cardiomyocyte sarcomere that senses mechanical stress via the cardiac-specific N2b stretch sensor domain of sarcomeric titin within the sarcomeric I band [408]. Titin N2b subsegment contributes to myofibrillar passive tension generated upon active stretch imposed by the sarcome during systole that is used for diastolic relaxation. Scaffold FHL1 connects to myocyte stretch sensor at titin N2b region to the Gq–cRaf–MAP2K2–ERK2 cascade (Fig. 5.12). It not only determines subcellular location of MAPK module, but also enhances intensity and duration of MAPK hypertrophic signaling.

Cardiomyocytes detect deformation possibly via plasmalemmal integrins and their partners, such as melusin (or Itgβ1BP2), and cardiac (muscle)-specific LIM protein in the sarcomeric Z disc. Melusin primes the phosphorylation (inactivation) of glycogen synthase kinase-3β. Cardiac LIM protein may trigger the PP3–NFAT pathway. Both molecules negatively control cardiomyocyte growth.

Ubiquitous focal adhesion kinase is activated by integrins and neurohumoral factors that lead to extracellular signal-regulated kinase activation and Fos expression. Focal adhesion kinases are implicated in compensatory hypertrophic remodeling [409].

The heart undergoes hypertrophy to accommodate loading modifications. Hypertrophy is initiated by mechanical causes and/or biological factors (hormones, cytokines, chemokines, and other growth factors). Secreted proteins, the so-called cardiokines, or cardiomyokines, regulate cardiac hypertrophy via endo-, para-, juxta-, or autocrine mechanisms.

The G-protein-coupled receptor APJ can mediate stretch-primed maladaptive hypertrophy (apelin-independent function) as well as opposite apelin-initiated cardioprotection. Binding of apelin to APJ receptor activates $G\alpha_i$ and actually protects against cardiac maladaptive hypertrophy induced by pressure overload. Apelin also has positive inotropic effect, in addition to be a vasodilator. On the other hand, loss of APJ mechanosensor confers resistance to chronic pressure overload [410]. Signaling from APJ activated by stretch and apelin involves extracellular signal-regulated kinases. Apelin, but not stretch, prevents cAMP synthesis via Gi, thereby antagonizing β-adrenergic receptor stimulation responsible for cardiac hypertrophy. In addition, it causes the accumulation of inositol phosphate-1 via the G16–PLC axis,[71] On the other hand, stretch favors the recruitment of β-arrestin-1 and -2 to APJ receptor (in the absence or presence of apelin [G-protein independent action]).

5.8.2 *Chemical Regulators of CMC Hypertrophy*

Animal models have shown that cardiac hypertrophy involves the MAPK, PP3–NFAT, and IGF1–PI3K–PKB–TOR pathways, as well as cyclin-dependent kinases CDK7 and CDK9, and class-2 histone deacetylases [411].

Krüppel-like factor KLF15, a member of the subset of the zinc finger family of transcriptional regulators expressed in cardiomyocytes, hinders cardiac hypertrophy [412]. Factor KLF15 acts, at least partly, by inhibiting the nuclear factor GATA-binding protein GATA4 and myocyte enhancer factor MEF2.

A- and B-type natriuretic peptides and nitric oxide produced in the heart activate guanylate cyclases that generate cGMP, thereby activating PKG1 kinase. The latter hampers cardiac hypertrophy.

[71] Stretch reduces apelin-induced accumulation of inositol phosphate-1 ($I(1)P_1$). Molecule I(1)P1 can be formed by PLC-mediated hydrolysis of phosphatidylinositol or by dephosphorylation of polyphosphate inositols such as $I(1,3)P_2$ by inositol polyphosphate 3-phosphatase.

Protein kinase-C, protein kinase-D, and calmodulin-dependent kinase phosphorylate histone deacetylases for translocation from the nucleus to the cytoplasm. In the nucleus, HDAC2 is recruited to genes involved in the regulation of cardiomyocyte growth. This enzyme prevents hypertrophy, as it inhibits myocyte enhancer factor, whereas HDAC1 favors cardiac hypertrophy.

5.8.2.1 Hypertrophy Regulators and Calcium Signaling

Cardiomyocyte hypertrophy is induced by angiotensin-2, aldosterone, endothelin, and noradrenaline. These hypertrophic substances can modify frequency, amplitude, duration, and/or cellular location of Ca^{2+} signaling.

Angiotensin-2 and endothelin generate Ca^{2+} influx from stores via IP_3 agent. Angiotensin-2 induces a moderate elevation in diastolic Ca^{2+} concentration lasting 1 to 2 mn. This minor effect leads to long-term consequences by prolonged occupancy of cognate receptors.

Catecholamines increase the frequency and amplitude of calcium transients. Aldosterone affects time- and space-dependent calcium activity. Certain hypertrophic agents generate nuclear translocation of NFAT, thereby modulating the frequency of spontaneous Ca^{2+} oscillations in neonatal rat cardiomyocytes in vitro. [413].[72]

5.8.2.2 Signaling Pathways

Signaling axes and mediators involved in cardiac hypertrophy include the PP3–NFAT, PI3K–PKB–GSK3, and MEF2–HDAC cascades, and NFκB activated by G-protein-coupled receptors of endothelin-1 and angiotensin-2, or cytokine receptors of TNFα and interleukin-1, with crosstalk at various levels.

This integrated signaling network regulates gene expression via many transcription factors, such as NFκB, MEF2, NFAT, and GATA4, as well as histone deacetylases. Long-term inhibition of cGMP phosphodiesterase PDE5a deactivates signaling pathways of cardiomyocyte hypertrophy.

[72]The small duration of spontaneous Ca^{2+} oscillations ($\sim$ 200 ms) does not allow substantial NFAT dephosphorylation and translocation into the nucleus. If the interval between 2 successive Ca^{2+} spikes is repeatedly shortened, dephosphorylated NFAT will tend to slowly accumulate. In addition, repeated Ca^{2+} spikes of higher amplitude can increase NFAT translocation. Each Ca^{2+} transient has a tiny effect, but integration over time produces a significant change. Near-nucleus Ca^{2+} increase also favors NFAT translocation. However, Ca^{2+} sparks are observed throughout the cytosol between Ca^{2+} transients without increase in perinuclear Ca^{2+} spark frequency or spatial spreading of Ca^{2+} sparks. Angiotensin-2, noradrenaline, and aldosterone (via both mineralcorticoid receptor and AT1R) increase Ca^{2+} oscillation frequency.

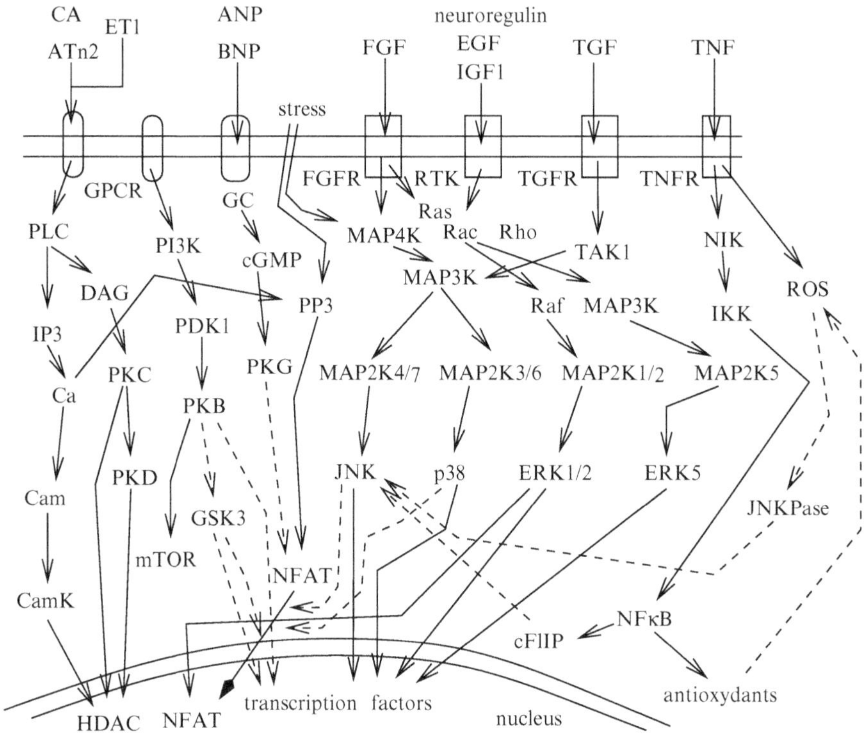

Fig. 5.13 Signaling pathways that initiate the cardiac hypertrophy (Source: [411]). Signaling starts at the sarcolemma by ligands (natriuretic peptides ANP and BNP, angiotensin-2 [ATn2], endothelin-1 [ET1], fibroblast growth factor [FGF], epidermal growth factor [EGF], insulin-like growth factor-1 [IGF1], transforming growth factor-β [TGF], and tumor-necrosis factor-α [TNF]) that bind to receptors (G-protein-coupled receptors [GPCR] and receptor Tyr kinases [RTK]), activating cytosolic effectors (guanylate cyclase-A [GC]; protein kinases [PKB, PKC, PKD, and PKG]; phospholipases [PLA2 and PLC]; mitogen-activated protein kinase pathways [MAP4K, MAP3K, Raf, MAP2K, ERK, JNK, and P38MAPK]; small GTPases [Ras, Rac, and Rho]; diacylglycerol [DAG] and inositol trisphosphate [IP_3]; phosphatidylinositol 3-kinase [PI3K]; phosphoinositide-dependent kinase [PDK]; glycogen synthase kinase-3β [GSK3]; calmodulin-dependent kinase [CamK]; target of rapamycin [TOR]; nuclear factor of activated T-cells [NFAT]; NFκB-inducing kinase [NIK]; inhibitor of NFκB kinase [IKK]; and cyclin-dependent kinase), then transcription factors (nuclear factor-κB [NFκB], and nuclear factor of activated T-cells [NFAT]), and nuclear components (histone deacetylases [HDAC]). NFκB impedes prolonged JNK signaling induced by TNFα via anti-oxidants and cellular flice-inhibitory protein (cFlIP), thus subsequent apoptosis. Transient MAPK stimulation by TNFα favors cell growth.

Catecholamines, angiotensin-2, and endothelin-1 bind to specific G-protein-coupled receptors (α1-adrenoceptor and angiotensin-2 and endothelin receptors) of the cardiomyocyte sarcolemma (Fig. 5.13) [411].

Calcium ions then stimulate the PP3–NFAT pathway and inactivate the CamK–HDAC axis. Activated proteins of the Gq/11 set stimulate MAPK modules. These biochemical cascades mediate cardiomyocyte growth and possible hypertrophy.[73]

5.8.2.3 Angiotensin-2

Angiotensin-2 causes cardiomyocyte hypertrophy via the nuclear factor of activated T cells, after activating phospholipase C that produces inositol trisphosphate and diacylglycerol (Fig. 5.13). The former generates calcium influx via IP_3R, emptying the intracellular calcium stores; the latter enhances the frequency of calcium oscillations triggered by membrane depolarization via the activation of store-operated diacylglycerol-sensitive TRPC channels, especially TRPC3 and TRPC6 channels [414]. The resulting Na^+ and Ca^{2+} entries in the cytosol change sarcolemmal potential and subsequently stimulate $Ca_V1.2$ channels.

^{cADP}Ribose is a Ca^{2+}-mobilizing second messenger produced from NAD^+ by ^{ADP}ribosyl cyclase. ^{ADP}ribosyl cyclase is activated by angiotensin-2 (that promotes cardiomyocyte hypertrophy) and subsequent, sequential stimulation of Src, phosphatidylinositol 3-kinase, protein kinase-B, phospholipase-Cγ1, and IP_3-mediated Ca^{2+} [415]. Angiotensin-2 generates a biphasic intracellular Ca^{2+} concentration increase with a rapid transient Ca^{2+} elevation via inositol trisphosphate receptor and sustained Ca^{2+} influx mediated by ^{ADP}ribose through activated $Ca_V1.2$ channels and ryanodine receptors.

5.8.2.4 GPCR–G-Protein Signaling

Cardiomyocyte stretch acts on the angiotensin-2 type-1 receptor (AT_1) that links to Janus kinase-2, hence translocating G-proteins into the cytosol, activating the extracellular signal-regulated kinase, and leading to hypertrophy.

α1-Adrenergic receptors coupled to $G\alpha_q$ play a role in cardiac hypertrophy.

Subunits of the $G\alpha_{q/11}$ subclass of guanine nucleotide-binding protein (G protein, i.e., Gαβγ trimer) is coupled to phospholipase-Cβ. Activated phospholipase-C produces diacylglycerol that binds to and activates protein kinase-C as well as inositol (1,4,5)-trisphosphate. The latter binds to intracellular IP_3 receptor, leading to the release of internal Ca^{2+} stores. Diacylglycerol also augments the concentration of cytosolic free calcium in cardiomyocytes. Calcium liberated from its intracellular stores regulates PP3 that controls NFAT transcription factor. Subunit $G\alpha_q$ also stimulates the mitogen-activated protein kinase modules and PKC–PKD pathway.

[73]Physiological cardiac hypertrophy over a limited time preserves the ventricular function. Pathological hypertrophy responds to sustained long-term pressure overload.

5.8.2.5 Receptor Protein Tyrosine Kinases

The receptor protein Tyr kinase human epidermal growth factor receptor HER2 links to HER3 or HER4 and binds neuregulin, thereby hampering cardiomyocyte growth (Fig. 5.14).

Exercise activates the IGF1–$PI3K_{c1\alpha}$–PKB pathway. Kinases PKB activated by IGF1R and $PI3K_{c1\alpha}$ promote growth and survival of cardiomyocytes. On the other hand, PKB stimulated by GPCR (ET1R) and $PI3K_{c1\gamma}$ induces maladaptive hypertrophy.

The receptor Tyr kinases of insulin-like growth factor-1[74] and fibroblast growth factor, receptor Ser/Thr kinases of transforming growth factor-β, and cardiotrophin-1, as well as G-protein-coupled receptors and stresses, stimulate the MAPK cascades.

Receptor HER2 is involved in cardiogenesis. It acts as a survival factor in adult myocardium. Inhibition of HER2 causes mitochondrial dysfunction in cardiomyocytes. Certain HER2 inhibitors (herceptin) favor cardiomyopathy. Other HER2 inhibitors (GW2974) activate AMP-activated protein kinase and protect the cardiomyocyte against TNFα-induced death [417]. Inhibition of HER2 blocks HER3 transactivation by HER2, increasing intracellular calcium level.

5.8.2.6 PI3K–PKB Axis

Activated PI3K leads to the sarcolemmal recruitment of protein kinase-B and phosphoinositide-dependent kinase-1 that phosphorylates (activates) protein kinase-B.

Isoform $PI3K_{c1\alpha}$ is strongly implicated in exercise-induced hypertrophy (Fig. 5.15) [418]. Moreover, elevated $PI3K_{c1\alpha}$ activity improves cardiac function and limits fibrosis; hence, it attenuates maladaptive cardiac hypertrophy in mouse model of dilated cardiomyopathy subjected to pressure overload caused by ascending aorta constriction.

Glycogen synthase kinase GSK3β, targeted by protein kinase-B, represses cardiac hypertrophy transcription effectors. However, protein kinase-B also activates the Ser/Thr protein kinase target of rapamycin, thereby favoring cardiac hypertrophy.

In addition, growth and differentiation factor GDF8 (or myostatin) precludes striated myocyte hypertrophy, as it prevents P38MAPK and PKB phosphorylation, but does not suppress PKB phosphorylation induced by IGF1 [419]. Furthermore, $PI3K_{c1\alpha}$ impedes signaling from G-protein-coupled receptor, $PI3K_{c1\gamma}$, and PKB in isolated cardiomyocytes (Fig. 5.15). $PI3K_{c1\alpha}$ hinders ERK1 and ERK2 activation, acting upstream from ERK1 and ERK2 kinases.

---

[74]Binding of IGF1 to its receptor activates $PI3K_{c1\alpha}$ that converts plasmalemmal phosphatidylinositol (4,5)-bisphosphate to phosphatidylinositol (3,4,5)-trisphosphate, hence activating signaling.

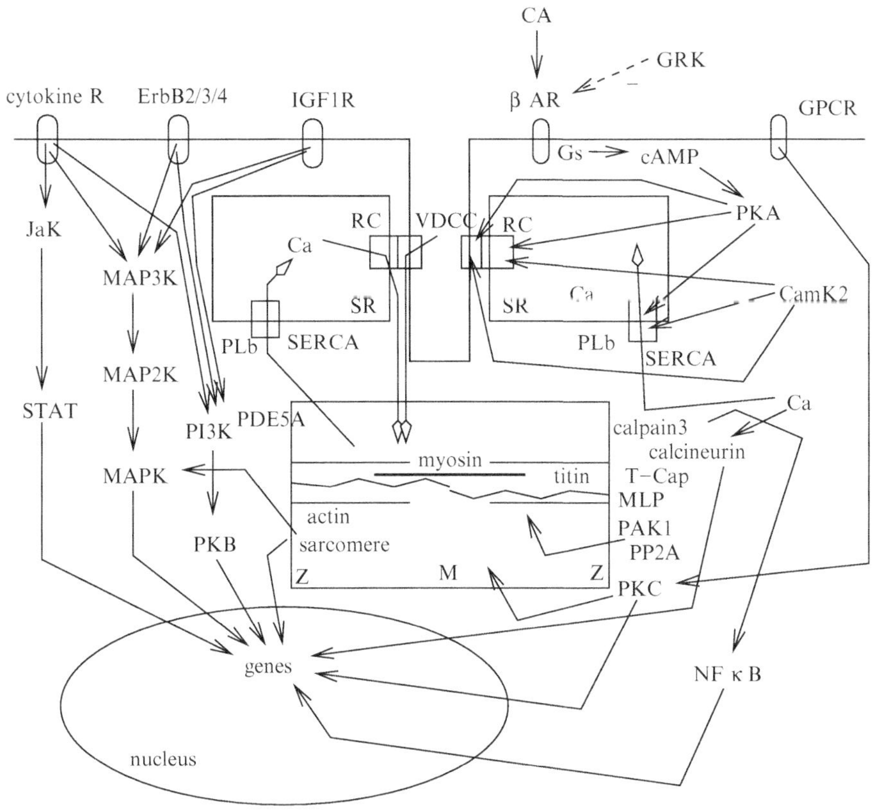

Fig. 5.14 Cytokine receptors, HERs, and insulin-like growth factor-1 receptor (IGF1R) activate the Janus kinase–signal transducer and activator of transcription (JaK–STAT) pathway, mitogen-activated protein kinase (MAPK) pathway, and phosphatidylinositol 3-kinase–protein kinase-B (PI3K–PKB) pathway, to regulate gene expression (Source: [416]). Calcium-activated PP3 (calcineurin) dephosphorylates NFAT transcription factors involved in cardiac remodeling. Mechanical stresses stimulate Z-disc sensors. The titin–telethonin–muscle-specific LIM protein (MLP) sensor binds PP3 for mechanotransduction. Excitation–contraction coupling requires calcium ions. Calcium fluxes govern myofilament contraction and relaxation, particularly via $Ca_V1.2$, or L-type voltage-dependent calcium channel (VDCC), ryanodine channel (RC), and sarcoplasmic reticulum calcium ATPase (SERCA) associated with its inhibitor phospholamban (PLb). The activity of these 3 calcium channels is regulated by protein kinase-A (PKA) and calcium–calmodulin-dependent kinase-2 (CamK2). PKA is stimulated by G-protein-coupled receptors, such as the β-adrenergic receptor (βAR). Protein kinase PKCϵ (PKC), located at the Z disc, regulates cardiomyocyte contractility and hypertrophy. p21-Activated kinase-1 (PAK1) interacts with protein phosphatase-2A (PP2A), targeting troponin-I and enhancing calcium sensitivity. Calpain-3 activates NFκB. Phosphodiesterase-5A (PDE5A) dissociates from the Z disc during heart failure. In heart failure, GRK2 desensitizes G-protein activation of ACase.

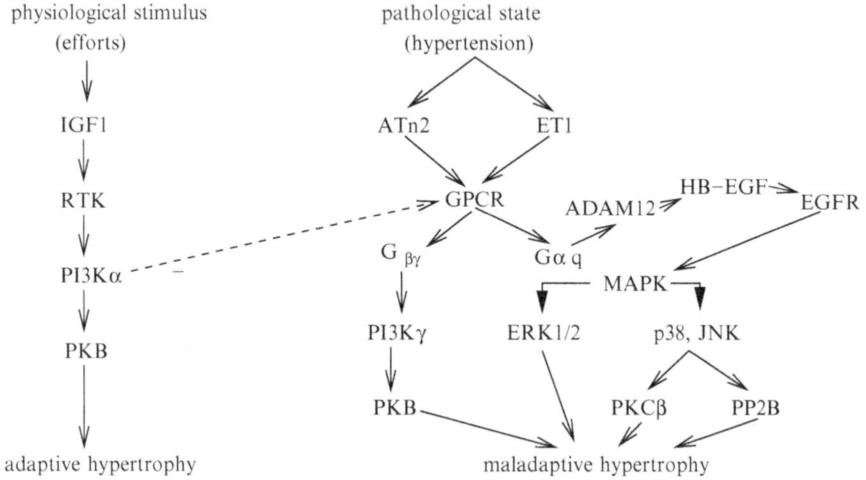

Fig. 5.15 Signaling cascades implicated in adaptive and maladaptive hypertrophy. Different pathways involve different isoforms of phosphoinositide 3-kinase with different effects in the cardiomyocyte (Source: [418]).

5.8.2.7 CamK–AMPK Axis

Calcium-activated calmodulin-dependent kinase kinase-β phosphorylates (activates) AMPK enzyme. AMP-activated protein kinase switches the cardiomyocyte from ATP consumption (fatty acid, cholesterol, and protein synthesis) to ATP production from fatty acid and glucose oxidation.[75]

5.8.2.8 MAPKs

Activated P38MAPK, JNK, and ERK phosphorylate multiple intracellular substances such as transcription factors.[76] Signaling launched by ERK1 and ERK2 causes hypertrophy in vivo. Signaling primed by ERK5 is also implicated in cardiac growth. On the other hand, P38MAPK and JNK provoke cardiomyopathy and hinder cardiomyocyte hypertrophy. Cardiomyopathy results from the overexpression of Ras GTPases.

[75] AMP-activated protein kinase inhibits acetylCoA carboxylase, 3-hydroxy 3-methyl glutarylCoA reductase, TOR, and eEF2 kinases. Enzyme AMPK activates mitochondrial ERRα and PGC1α. Fatty acid oxidation is indeed regulated by PPARs and their coactivator PGC1α. Estrogen-related receptor ERRα, ERRβ, and ERRγ regulate cardiac energy metabolism. ERRα interacts with PGC1α.

[76] MAP2K1 and MAP2K2 activate ERK1 and ERK2, MAP2K6 and MAP2K3 target P38MAPKs, and MAP2K4 and MAP2K7 stimulate JNK kinases.

5.8.2.9 AKAPs

Several AKAPs are detected in cardiomyocytes, such as AKAP1, AKAP5, AKAP7α, and AKAP8 to AKAP13, and related proteins that act as AKAPs, such as ezrin, BIG2, sphingosine kinase type-1-interacting protein (SKIP), and the intermediate filament protein that localizes to Z lines, desmuslin (or synemin) [420, 421].

A-kinase anchoring protein-13 is a Rho-specific guanine nucleotide-exchange factor (RhoGEF13). It activates RhoA involved in certain pathways of cardiomyocyte hypertrophy, transducing hypertrophic signals from α1-adrenergic receptors coupled to Gα12 [422]. Inactivation occurs via AKAP13 phosphorylation by anchored PKA and subsequent recruitment of 14-3-3 regulator.

5.8.2.10 Prostaglandins

Prostaglandin-E2 acts via its EP_4 receptor coupled to the Gs–AC–cAMP–PKA axis. It causes hypertrophy of neonatal ventricular myocytes via ERK1 and ERK2 kinases. The EP_4–PKA pathway stimulates Rap GTPase that activates ERK1 and ERK2 and then 90-kDa ribosomal S6 kinase (RSK) [423]. The EP_4–PKA–ERK1/2–RSK cascade stimulates Fos and synthesis of B-type natriuretic peptide. In addition, ERK1, ERK2, and RSK also contribute to PGE2 regulation of the transcription factor early growth response-1 (EGR1).

5.8.2.11 Epoxyeicosatrienoic Acids

Epoxyeicosatrienoic acids impede NFκB activation, thus having beneficial effects on cardiac hypertrophy. Soluble epoxide hydrolase (sEH) catalyzes the conversion of cardioprotective epoxyeicosatrienoic acids, mainly (14,15)EETs that act via STAT3 factor [424], into dihydroxyeicosatrienoic acids. Inhibitors of soluble epoxide hydrolase that indirectly block NFκB activation, preserve cardiomyocytes and can prevent development of cardiac hypertrophy as well as cardiac arrhythmias associated with cardiac hypertrophy in murine model with aortic banding [425].

5.8.2.12 Cardiotrophin

Cardiotrophin-1, a member of the IL6 cytokine family and one of the most potent cardiac myocyte survival factors, acts via GP130 cytokine receptor. Cardiotrophin-1 stimulates the JaK2–STAT3 pathway and induces cardiac myocyte hypertrophy. Cardiotrophin-1 phosphorylates STAT3 factor. Activated STAT3 increases angiotensinogen expression in cardiomyocytes (autocrine regulation) [426]. Cardiotrophin-1 binds to IL6R–GP130 dimer and triggers the signaling cascade that also involves MAPK enzymes.

5.8.2.13 Follistatins

The balance between members of the TGFβ superfamily (TGFβ, activins, BMPs, and GDFs; Vol. 2 – Chap. 3. Growth Factors) and their inhibitors (follistatins, inhibins, and nodal) regulates various physiological and pathological processes (e.g., cardiac fibrosis, myocardial ischemia–reperfusion injury, pressure-overload-induced cardiac hypertrophy, and heart failure). Proteins of the follistatin family may bind to and modify the function of members of the TGFβ superfamily (activin-A, BMP2, and GDF8) that can either prevent or promote growth by signaling via SMAD transcriptional regulators. The SMAD proteins (SMAD2–SMAD4) are implicated in the regulation of cardiac hypertrophy. In particular, Follistatin-like protein FStL3 binds to activin-A and intervenes the heart's response to ischemia and pressure overload. On the other hand, follistatin-like protein FStL1 can rapidly activate intracellular pathways in cardiovascular cells in the absence of TGFβ family proteins [427].

Follistatin-like protein FStL1, an extracellular glycoprotein, protects the heart from ischemia–reperfusion injury. The FSTL1 transcript is produced in cardiomyocytes as well as vascular endothelial and smooth muscle cells [427]. Its expression is upregulated in the myocardium, mainly in cardiomyocytes, in response to pressure overload. The secreted cardiokine FStL1 does not contribute to heart growth in normal conditions. However, it precludes cardiac growth in pressure overload-induced hypertrophy [427]. Protein FStL1 promotes a rapid AMPK phosphorylation (activation).[77] Activated AMPK phosphorylates NOS3 (Ser1179); NO impedes maladaptive cardiac hypertrophy and favors angiogenesis.

Activin-A and GDF8 impede CMC hypertrophy; their effect is antagonized by follistatin (FSt) and cardiac follistatin-like protein FStL3. The latter, an extracellular, stress-induced regulator of activin-A (which precludes SMAD2 activation), binds to activin-A with a much higher affinity than the former. Amount of FStL3 rises in cardiomyocytes of patients with heart failure.

5.9 Cardiomyocyte Orientation

Cardiac myofibers run in the heart wall as helices around the ventricles, strengthening them. Myofiber geometry enables an efficient wall reinforcement, as they render stiffness uniform in all directions parallel to their localization plane. Bundles of myofibers are usually represented into sheets across the wall thickness with a given orientation.

[77] Adiponectin is another peptide regulators of cardiac AMPK kinase. Protein FStL1 may operate via the receptor Disco-interacting protein-2 homolog-A (DIP2a).

Dog cardiomyocytes form myofibers with a given orientation that varies from the endocardium to the epicardium with an angle from +70 to −70 degrees with respect to the horizontal (equatorial) plane, the major axis of the truncated ellipsoid-like left ventricle being vertical [429].

The wall circumferential sheet thickness is given by the cross size of four to six cardiomyocytes. Subepicardial and subendocardial myofibers follow helical paths, whereas midmyocardial myofibers are roughly circumferential. The myofiber orientation can be quantified by helix[78] and transverse[79] angles.

Cardiac myofiber postmortem orientation has been measured in 5 healthy goat hearts using diffusion tensor imaging (DTI; Vol. 6 Chap. 6. Images, Signals, and Measurements) [430]. The helix angle is evaluated at several sectors[80] in the equatorial LV slice.[81] From the endocardium to the epicardium, the helix angle varies from nearly 90 to −60 degrees in the anterior sector, from nearly 90 to −30 degrees in the posterior sector, and from nearly 50 to −80 degrees in the septal sector. The transverse angle was determined in the midmyocardial free wall for all slices from apex to base and averaged over the circumference. The mean transverse angle varied from 12 ± 4 degrees near the apex to 9 ± 4 degrees near the heart base.

In fetal and neonatal human heart, the direction of muscular fibers has been provided for each studied slice parallel to the equatorial plane by elevation and azimuthal angles with respect to a given horizontal axis [431]. The technique is based on polarized light transmission and resulting image processing. In slices near the heart base, myocardial fibers cross the myocardium from the endocardium to the epicardium in a counterclockwise manner, whereas the reverse is observed in slices near the apex. Near the equatorial plane, the elevation angle continuously ranges from −50 degrees near the epicardium to +70 degrees near the endocardium. Near the heart base and near the apex, a discontinuity in elevation angle variation is observed at mid-thickness, the angle varying from +40 to +70 degrees near the endocardium for both base and apex, and from −40 to −50 or to −65 degrees near the epicardium in apex and base slices, respectively. Myocardial fibers are

[78]The helix angle is the angle between the myofiber direction and the plane perpendicular to the local major axis.

[79]The transverse angle is the angle between the local circumferential direction and the projection of the fiber direction on the plane normal to the local major-axis direction.

[80]The helix angle is determined in anterior, interpapillary muscle, posterior, and septal angular wall sectors of 20 degrees.

[81]The equatorial slice is defined as the slice positioned at one-third of the major-axis length from the heart base; the base corresponds to the slice nearest to the outflow tract, but outside it; the apex is the first slice showing the cardiac wall from the chest wall.

geodesic curves running on surfaces as demonstrated by quantitative polarized light microscopy in the whole left ventricle [432].[82] The Clairaut number indeed remains invariant on the geodesic of the revolution surface.

Sheets of geodesic fibers between 2 left ventricle slices have been reconstructed. An anatomical model of toroidal surfaces, fitted into each other inside the ventricular wall, over which run the myocardial fibers with a spiral pattern has also been proposed [433].

Spatial distribution of ventricular wall stresses is sensitive to the orientation of the muscular fibers [434]. The structure of the left ventricle is designed for maximum homogeneity of fiber strain during ejection [435]. Cardiomyocyte orientation may result from an optimization process (minimization of muscular tension for a given pressure).

Transmural difference in myofiber orientation equals to about 20 degrees per mm in human heart [436]. Therefore, transmural discretization of wall computational models must be small enough to compute strains and stresses.

Arrangement of cardiac myofibers takes the form of a special minimal surface, the *generalized helicoid*, a minimal surface maintained during the cardiac cycle, hence minimizing myofiber length and optimizing ventricular ejection volume [437]. The arrangement of myofibers in generalized helicoids characterizes their orientation in the entire cardiac wall.

5.10 Ion Carriers

The main involved Na^+, K^+, and Ca^{2+} ion channels span the sarcolemma and membranes of cell organelles, especially the sarcoplasmic reticulum (Table 5.9 and Fig. 5.16) [438]. Their permeability depends on the protein conformation. Sarcolemmal Na^+–K^+ ATPase slightly contributes to the resting membrane potential by admitting 2 K^+ into the cell and expelling 3 Na^+ from it to the extracellular fluid for every ATP consumed. To avoid local acidosis, H^+ is expelled from the cell by Na^+–H^+ exchanger.

[82] Each cardiac slice is discretized into 130×130 μm squares over which 2 square-averaged angles are measured: (1) an elevation angle γ_e between the fiber and slice plane and (2) an azimuth angle γ_a between the fiber projection in the slice and the east–west direction. The fiber direction $\mathbf{f}$ is then given by: $f_x = \cos\gamma_e \cos\gamma_a$, $f_y = \cos\gamma_e \sin\gamma_a$, and $f_z = \sin\gamma_e$. Angles close to degree 0 and greater than 75 degrees cannot be accurately measured. The measurement error is estimated to be equal to 1 degree.

Table 5.9. Cardiomycyte and its main ion carriers in sarcolemma and membrane of the sarcoplasmic reticulum as well as ion fluxes between the cytosol and either the extracellular space or the sarcoplasmic reticulum.

Carrier	Ion flux (with respect to cytosol)
	Sarcolemma
Ca_V channels	Ca^+ influx
Ca^{2+} ATPase	Ca^{2+} efflux
Na^+ Ca^{2+} exchanger	Ca^{2+} efflux (forward mode)/influx (reverse mode)
	Na^+ influx (forward mode)/efflux (reverse mode)
Na_V channels	Na^+ influx
Na^+–K^+ ATPase	Na^+ efflux
	K^+ influx
K^+ channels	K^+ efflux
	Sarcoplasmic reticulum
Ryanodine receptor	Ca^{2+} influx
Ca^{2+} ATPase	Ca^{2+} efflux

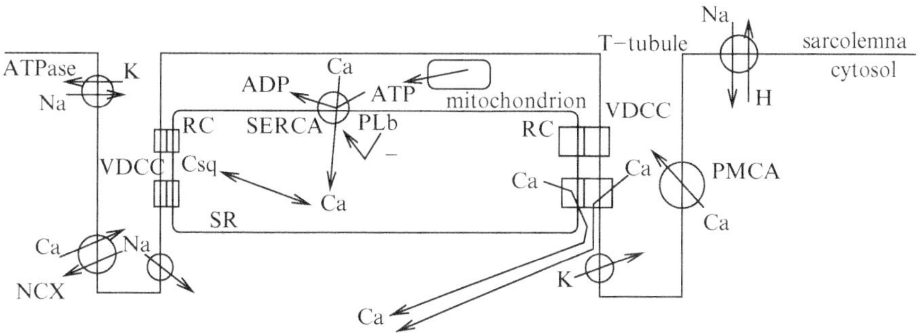

Fig. 5.16 The cardiomyocyte and its main ion carriers. Voltage-dependent L-type Ca^{2+} channels (VDCC, Ca_V1.2) of the sarcolemma trigger ryanodine calcium channels (RC) of the sarcoplasmic reticulum (SR), the main cellular storage source of Ca^{2+} ions. The cytosolic concentration of calcium determines the number of actomyosin cross-bridges per time unit, hence the maximal velocity of sarcomere shortening. At the end of systole, calcium dissociates from troponin-C and is transported into the sarcoplasmic reticulum by the sarco(endo)plasmatic reticulum Ca^{2+} ATPase pump (SERCA). Calcium efflux from the myocyte to the extracellular space also balances the calcium amount that has entered due to the action potential. During relaxation, Ca^{2+} is exchanged for Na^+ by the Na^+–Ca^{2+} exchanger (NCX) and sarcolemmal calcium ATPase pump (PMCA). The frequency-dependent control of Ca^{2+} flux is due to faster systolic calcium release from SR and faster diastolic calcium reuptake by SR (positive lusitropy). Voltage-gated Ca^{2+} channels are short-term regulators of intracellular Ca^{2+} level. An accumulation of cytosolic calcium increases the force of contraction.

5.10.1 Voltage-Gated Sodium Channels

In mammals, 10 α subunits of voltage-gated sodium channels are encoded by 10 genes (SCN1A–SCN5A and SCN7A–SCN11A). In some body's tissues, these α subunits associate with accessory β subunits, which are encoded by 4 genes (SCN1B–SCN4B). The number of Na_V channels can be extended by alternative mRNA splicing. Differences in functional properties of Na^+ channel isoforms result from unique conductances in specific cell types [439].

In mammalian hearts, $Na_V1.5$ isoform (encoded by the SCN5A gene) determines Na^+ flux in atrial and ventricular myocytes. It is characterized by rapid activation and inactivation kinetics. It opens and closes at specific stages of cardiac activity in response to voltage change. Entry into the inactivated state is very fast (< 1 ms). It controls the magnitude and duration of voltage-dependent Na^+ currents (i_{Na}) in cardiomyocytes.

In humans, splice variants of cardiac $Na_V1.5$ channel exist in both embryonic and adult hearts [440]. Although alternative splicing during $Na_V1.5$ synthesis is age- and gender-independent, it is species-dependent, as $Na_V1.5$d is not detected in mouse heart. Voltage-dependent Na^+ currents generated by $Na_V1.5$d is markedly reduced with respect to that through $Na_V1.5$ channel, without difference between intracellular transfer of these 2 isoforms.

Cardiac voltage-gated sodium channels $Na_V1.5$ cause action potential upstroke (fast depolarization). Once inactivated, $Na_V1.5$ channels may reopen, thereby generating a depolarizing inward current after the plateau phase of action potential. Glycosylated membrane protein $Na_V1.5$ interacts with β subunits (β1–β4).

In sinoatrial node pacemaker cells, $Na_V1.5$ activity couples pacemaker cells to cells surrounding them within the sinoatrial node, as any impairment of impulse initiation causes sinus bradycardia [441]. Furthermore, it is involved in impulse propagation through the sinoatrial node as well as between the sinoatrial node and atrial cells.

5.10.1.1 Na_V1.5 Regulation in Cardiomyocytes

In cardiomyocytes, protein kinase-A is mainly activated by β-adrenergic receptors that raise intracellular cAMP level. Upon phosphorylation by cAMP-dependent protein kinase-A, cardiac $Na_V1.5$ channel gating remains unchanged, but the bulk conductance rises, as the number of functional channels increases. This elevation in plasmalemmal $Na_V1.5$ density results from heightened exocytosis of these channels from multiple intracellular storage pools [442].

In addition to protein kinase-A and -C, $Na_V1.5$ channels are regulated by calmodulin and calmodulin-dependent protein kinase CamK2, particularly cardiac-specific CamK2δ isoform (^{C}CamK2δ).[83] Intracellular Ca^{2+} can modulate functioning of Na^+ channel. Calmodulin regulates Na^+ channel gating on binding. Kinase CamK2 exerts acute effect on $^{C}Na_V$ channel gating that is independent of calmodulin.

Calmodulin and CamK2 differ in their mode of inactivation of cardiac $^{C}Na_V$ channels [443]. Enzyme CamK2α shifts the voltage dependence of Na_V channel by approximately +5 mV, hastens recovery from inactivation, lowers entry into intermediate or slow inactivation, and heightens persistent (late) flux [443]. Calmodulin alone reproduces CamK2-induced changes, but does not modify recovery from inactivation and entry into slow or intermediate inactivation in normal guinea pig ventriculomyocytes.[84]

Regulatory β1 and β2 subunits link to ankyrin-2 and -3 (or ankyrin-B and -G). Interactions between β1 subunits and ankyrins are regulated by Tyr phosphorylation of β1 subunits. Phosphorylated β1 subunits localize to intercalated discs with connexin-43 and N-cadherin; non-phosphorylated proteins to other parts of the sarcolemma [445]. Epidermal growth factor receptors and cytosolic protein Tyr kinase Fyn as well as protein Tyr phosphatase (non-receptor) PTPn3 target $Na_V1.5$ channels.

In cardiomyocytes, ankyrin-2 and -3 anchor integral membrane proteins, such as ion channels, exchangers, and transporters to the actin and spectrin cytoskeletons. Ankyrins may not only act as adaptors for $Na_V1.5$ anchoring to the plasma membrane, but also mediators of adequate transfer to intercalated discs and other parts of the sarcolemma [445].

A pool of $Na_V1.5$ channels colocalizes with caveolin-3 and dystrophin in caveolae. This pool may be recruited upon β-adrenergic stimulation, in a PKA-independent, but Gs-dependent manner [445].[85]

5.10.2 Sodium–Calcium Exchangers

Sodium–calcium exchangers (NCX; Vol. 3 – Chap. 3. Main Classes of Ion Channels and Pumps) yield inward or outward Ca^{2+} motions and outward or inward Na^+ fluxes, respectively, with the ratio of 3 Na^+ for 1 Ca^{2+}, whether the intracellular concentration $[Na^+]_i$ or $[Ca^{2+}]_i$ is high. Sodium–calcium exchangers induce slow contractions. The unitary conductance for Ca^{2+} of Na^+–Ca^{2+} exchanger is smaller

[83] Enzyme CamK2α is mainly synthesized in neurons.

[84] On the other hand, using a different experimental setting in rabbit and mouse ventriculomyocytes, acute and chronic overexpression of ^{C}CamK2δ, hence Ca^{2+}- and CamK2-dependent phosphorylation of $Na_V1.5$, shift voltage dependence by −6 mV, enhance intermediate inactivation, slow recovery from inactivation, and raise the persistent, inward current [444].

[85] In Duchenne muscle dystrophy, caveolin-3 is overexpressed, but concentrations of dystrophin and its associated proteins decay in skeletal muscles as well as the heart.

than that of $Ca_V1.2$ channel. However, Na^+–Ca^{2+} exchangers can induce a Ca^{2+} influx able to trigger release of Ca^{2+} from the sarcoplasmic reticulum [446]. Furthermore, Na^+–Ca^{2+} exchangers have limited activity, because the intracellular Na^+ concentration decays during depolarization.

In cardiomyocytes, *phospholemman*[86] regulates Na^+–Ca^{2+} exchangers NCX1 and Na^+–K^+ ATPases, hence Na^+–Ca^{2+}-exchange (i_{NaCaX}) and Na^+–K^+-pump ($i_{NaK_{ATP}}$) currents. Phosphorylated phospholemman (Ser68) disinhibits Na^+–K^+ ATPase, but mainly inhibits NCX1 [447]. Association of phospholemman with Na^+–K^+ ATPase does not require NCX1 exchanger. Phospholemman is phosphorylated in response to β-adrenergic stimulation. Under stress conditions with high $[Na^+]_i$, β-adrenergic-induced phosphorylation of phospholemman causes a reduction in inotropy due to the relief of inhibition of Na^+–K^+ ATPase [448].

5.10.3 Myocardial Calcium Channels

5.10.3.1 Voltage-Gated Calcium Channels

Cardiomyocytes possess 2 classes of voltage-gated Ca^{2+} channels: $Ca_V1.2$ (L-type) and Ca_V3 (T-type). Although these Ca^{2+} channels have similar unitary Ca^{2+} conductances, $Ca_V1.2$ trigger more efficiently cardiomyocyte contraction. Calcium influx through $Ca_V1.2$ produces greater amplitude and faster contractions.

$Ca_V1.2a$

Channel $Ca_V1.2a$[87] localizes mainly to junctions between the sarcolemma and sarcoplasmic reticulum close to ryanodine receptors. Many of the $Ca_V1.2$ channels reside in T tubules.

Channel $Ca_V1.2$ is composed of several types of subunits: a pore-forming α1 and auxiliary β, γ, and α2δ subunits (α2δ1–α2δ4), in addition to channel-resident calmodulin [449]. Subunit α2δ1[88] is not necessary to direct the plasmalemmal $Ca_V1.2$ density, but boosts its activation and inactivation kinetics.

Channel $Ca_V1.2$ triggers Ca^{2+}-induced Ca^{2+} release from the junctional sarcoplasmic reticulum through $Ca_V1.2$-coupled ryanodin recptors ($Ca_V1.2$-RyR *couplon*). A single small cluster of $Ca_V1.2$ channels provokes a local Ca^{2+} *sparklet* that activates a small cluster of closely apposed ryanodine receptors on the sarcoplasmic reticulum membrane to create a Ca^{2+} *spark*. The synchronous occurrence of multiple Ca^{2+} sparks in the entire cardiomyocyte after action potential arrival causes a massive release of Ca^{2+} from the sarcoplasmic reticulum to trigger contraction.

[86] A.k.a. FXYD domain-containing ion transport regulator-1 (FXYD1).

[87] A.k.a. dihydropyridine receptors.

[88] Subunit α2δ1 is also synthesized at high levels in vascular smooth myocytes.

Major myocardial α2δ1 and β2 subunits regulate the excitation–contraction coupling in the cardiomyocyte [450]. Depolarization-induced Ca^{2+} entry through $Ca_V1.2$ channels contributes to the duration of the action potential plateau.

Channel $Ca_V1.2$ releases a relatively small and localized amount of calcium through the sarcolemma. The amount of Ca^{2+} influx during depolarization is limited by Ca^{2+}-dependent inactivation of the cytosolic side of the channels by calmodulin bound to the channels [451]. However, Ca^{2+} currents through $Ca_V1.2$ control the main calcium flux by the release from the sarcoplasmic reticulum through ryanodine channels that interact with closely apposed $Ca_V1.2$ channels. In turn, Ca^{2+} release from the sarcoplasmic reticulum contributes to the Ca^{2+}-dependent inactivation of the calcium current (double-negative feedback) [452]. These feedbacks require interaction with Ca^{2+}–calmodulin and a channel motif IQ of α1 subunit [449].

In ventriculomyocytes, small clusters of $Ca_V1.2$ channels are functionally and structurally coupled, hence enabling concerted openings of adjacent channels. Calcium spark may be higher in couplons with clustered $Ca_V1.2$ channels than in couplons with independent $Ca_V1.2$ channels. Sarcolemmal $Ca_V1.2$ channels form clusters of various sizes to coordinate their gating upon membrane depolarization. Binding of $Ca_V1.2$ channels via their C-termini enables allosteric activation of adjoining channels [453]. Coordinated opening of $Ca_V1.2$ clusters, together with coincident activation of independently gated channels, amplifyies Ca^{2+} influx for an optimized excitation–contraction coupling. Protein AKAP5 may determine the size and promote the stability of $Ca_V1.2$ clusters. In addition, like protein kinase-A, oligomerization of $Ca_V1.2$ channels shifts the voltage dependence of activation of $Ca_V1.2$ to more hyperpolarized potentials to enhance the signaling robustness.

A subgroup of $Ca_V1.2$ channels localizes to caveolae with caveolin-3 in ventriculomyocytes. They form a signaling complex with β2-adrenergic receptors, adenylate cyclase, protein kinase-A, and protein phosphatase-2 [454].

Ca_V3

Density of Ca_V3 channels is large in ventricular Purkinje cells (involved in action potential transmission rather than contraction) and atriomyocytes (atrial contraction does not play a major role in atrium emptying), but it is small in most ventriculomyocytes. Channel Ca_V3 inactivates very rapidly. Contractions initiated by Ca^{2+} entry in ventricular Purkinje cells via Ca_V3 are characterized by a longer delay to the onset of sarcomere shortening and slower rates of shortening and relaxation [455].

Another mechanism, voltage-activated Ca^{2+} release, activated directly by the membrane depolarization without Ca^{2+} entry, can cause Ca^{2+} transient currents [456]. This additional mechanism that is depressed in heart failure depends on internal cAMP concentration [457]. Phosphorylations can then occur via the ACase–PKA and Ca^{2+}–calmodulin-dependent kinase pathways to regulate the cardiac contraction strength.

5.10.3.2 Ryanodine Receptors

Three types of ryanodine receptors, i.e., ryanodine-sensitive Ca^{2+}-release channels, exist, 2 being cardiac isoforms [458]. They are arranged on the sarcoplasmic reticular membrane in arrays at junctions between the sarcoplasmic reticulum and sarcolemma on cell surface and in T tubules beneath $Ca_V1.2$ channels [459]. These coupled channels exhibit simultaneous gating due to FKBP1a [460].[89]

Clusters of ryanodine receptors (30–260 channels acccording to mammal species) in regions of close apposition between the sarcoplasmic reticulum and sarcolemma forms functional *couplons* [461]. Couplon distribution (mean transverse and longitudinal spacing ~ 0.8 and $\sim 1.8\,\mu m$, respectively) explains the difference in Ca^{2+} diffusion during Ca^{2+} sparks in axial and transverse directions. Ryanodine receptor clusters exist around T tubules and sarcomeres, in particular around myofilament edges (spacing of $0.78 \pm 0.07\,\mu m$ in humans). Each bundle of myofibrils is supplied by approximately 6 clusters.

Ryanodine channels RyR2 form junctional proteic complexes that include: (1) calmodulin, which exerts Ca^{2+}-dependent modulation of channel function; (2) peptidyl prolyl cis–trans isomerase FKBP1b, or calstabin-2, (3) protein kinase-A; (4) protein phosphatases PP1 and PP2; (5) calcium-binding modulator of calcium release sorcin; as well as (6) other proteins at the luminal surface of the sarcoplasmic reticulum, triadin, junctin, and calsequestrin, that participate in Ca^{2+} buffering and modulation of Ca^{2+} release. The cardiac ryanodine receptor complex RyR2 consists of 4 monomers and also phosphodiesterase-4D, among various other components.

In normal hearts, the opening of RyR2 channels is regulated by FKBP1b and PDE4d. Loss of PDE4d from RyR2 is associated with increased local cAMP levels. Therefore, activity of protein kinase-A rises, hence phosphorylating RyR2 monomers with loss of FKBP1b from the complex. Consequently, calcium leaks, thereby leading to heart failure and arrhythmia [462].

Overexpression of the transcription factor early growth response EGR1 impedes calsequestrin synthesis, but does not influence that of calreticulin, triadin, SERCA2a pump, and phospholamban [463]. Moreover, calcium release from its cellular store via calsequestrin-dependent ryanodine channels decreases, whereas calcium influx from calreticulin-dependent inositol trisphosphate channels remains unaffected. Transcription factor EGR1 binds to the CSQ2 gene promoter to control excitation–contraction coupling at the level of the sarcoplasmic reticulum.

5.10.3.3 IP_3 Receptors

Inositol (1,4,5)-trisphosphate receptors release Ca^{2+} from the sarcoplasmic reticulum. Type-1 and -2 IP_3Rs (IP_3R1 and IP_3R2) lodge predominantly in atria.

[89]FK506-binding protein FKBP1a, or FKBP12, was originally identified as the cytosolic receptor for the immunosuppressors FK506 (a.k.a. tacrolimus and fujimycin) and rapamycin.

Table 5.10. Calcium efflux in ventricular myocytes during relaxation. Relative participation of the involved carriers. The Ca^{2+} amount entering the mitochondria is insignificant with respect to excitation–contraction coupling, but cumulative changes in intra-mitochondrial $[Ca^{2+}]_i$ can stimulate dehydrogenases that produce nicotinamide adenine dinucleotide (NADH) and ATP to match increased energetic demands [464]. The relative activity (%) of the carriers depends on their concentrations as well as animal species, here given for rabbit and rat. Calcium efflux in ferret, dog, cat, guinea pig, and human ventricle is quantitatively similar to the rabbit. (Rat ventricles do not mimic the human heart with respect to Ca^{2+} efflux.)

Carrier	Rabbit	Rat
Sarcoplamic reticulum Ca^{2+} ATPase	~70	~92
Sarcolemmal Na^+–Ca^{2+} exchanger	~28	~7
Sarcolemmal Ca^{2+} ATPase and Mitochondrial Ca^{2+} uniporter	~1	~1

IP_3-dependent Ca^{2+} release contributes to atrial Ca^{2+} signaling. Receptor IP_3R interacts with membrane-associated adaptor ankyrin-2. It localizes to caveolae with hyaluronan[90] and cholesterol.

5.10.3.4 Sarco(endo)plasmic Reticulum Calcium ATPases

The cytoplasmic Ca^{2+} concentration is controlled by a pump in the membrane of the calcium-storing sarcoplasmic reticulum that temporarily sequesters Ca^{2+} ions. *Sarco(endo)plasmic reticulum* Ca^{2+} *ATPases* (SERCA) transport calcium ions from the sarcomere and cytosol into the sarcoplasmic reticulum.[91]

Plasma membrane Ca^{2+} *ATPases*, Na^+–Ca^{2+} exchangers[92] in their forward mode, and mitochondrial Ca^{2+} uniporters also participate in Ca^{2+} removal from the cytosol of cardiomyocytes (Table 5.10).

5.10.3.5 Regulators of Calcium Flux and Calcium Handling

Upon stimulation of G-protein-coupled receptors by their agonists, such as endothelin-1 or angiotensin-2, members of the heterotrimeric G-protein class provoke acute and chronic cardiac responses. β-Adrenergic receptors control the

[90] A.k.a. hyaluronic acid and hyaluronate. It is an anionic, non-sulfated glycosaminoglycan. It is ubiquitous. It is synthesized by integral membrane hyaluronan synthases (HAS1–HAS3) and degraded by hyaluronidases. It also resides in the extracellular matrix.

[91] Diverse SERCA types exist: SERCA1 in fast-twitch skeletal muscles; SERCA2a in the myocardium and in slow-twitch skeletal muscles; SERCA2b in smooth myocytes; and SERCA3 in endothelial cells.

[92] NCX carrier competes with PMCA for Ca^{2+} efflux, but it may induce less than one-third of Ca^{2+} efflux.

Table 5.11. Main ion channels determine the typical shape of action potential in the cardiomyocyte. Fast depolarization (phase 0) is induced by an inward sodium flux. Next, a partial early repolarization (notch; phase 1) is caused by outward potassium flux through rapidly activating and inactivating K^+ channels. The plateau phase (phase 2) results from a balance between inward depolarizing (mainly Ca^{2+} influx) and outward repolarizing (mainly K^+ efflux) ion fluxes. Potassium ion motions strongly contribute to phases 3 (rapid final repolarization) and 4 (resting potential). In addition to transient outward current ($i_{K,to}$) through rapidly inactivating $K_V4.2$ and $K_V4.3$ ($i_{K,to(r)}$: fast component) and $K_V1.4$ ($i_{K,to(s)}$: slow component), a rapidly activating, slowly inactivating delayed rectifier current ($i_{K,slow1}$) and another component ($i_{K,slow2}$) through $K_V2.1$ with slower inactivation kinetics than $K_V1.5$ explain repolarization in ventricular myocytes [467].

Phase	Event	Involved ion channels (currents)
0	Fast depolarization	$Na_V1.5$
1	Early repolarization	$K_V1.4/4.2/4.3$ ($i_{K,to}$)
2	Plateau	$K_V1.5/2.1$ ($i_{K,ur}$ and $i_{K,slow2}$); $Ca_V1.2$
3	Final repolarization	$K_V11.1/7.1$ ($i_{K,r}$ and $i_{K,s}$)
4	Rest	$K_{IR}2.1/3.1$ (i_{K1})

heart behavior via $G\alpha_s$-mediated signaling. Members of the $G\alpha_{q/11}$ subclass are involved in the development of pressure overload-induced maladaptive cardiac hypertrophy. This effect is counteracted by RGS4 GTPase-activating protein.

Activated $G\alpha_q$ proteins trigger phospholipase-Cβ activation, and subsequent processing of phosphatidylinositol (4,5)-bisphosphate into diacylglycerol, an activator of PKC, and inositol (1,4,5)-trisphosphate, a ligand of IP_3 receptors. In cardiomyocytes, the plasmalemmal PIP_2 modulates activity of ion carriers, such as $K_{IR}6.2$ channel ($i_{K(ATP)}$ current) and Na^+–Ca^{2+}–K^+ exchanger.

Modulation of $G\alpha_{q/11}$ subclass protein concentrations influences dynamics of Ca^{2+} transients (appearance, amplitude, and recovery), hence Ca^{2+} handling, as well as electrophysiological properties of cardiomyocytes [465]. In the absence of $G\alpha_q$ and in the presence of $G\alpha_{11}$, recovery is accelerated; in the absence of both $G\alpha_q$ and $G\alpha_{11}$, recovery is decelerated.

5.10.4 Myocardial Potassium Channels

Approximately 60 K^+-selective channel subunits exist in the myocardium that carry outward currents in the physiological range of potentials to repolarize excitable cardiac cells and set the resting potential that governs excitability [466] (Table 5.11). Depolarization-activated K^+ currents determine the amplitude and duration of action potentials in cardiomyocytes.

Table 5.12. Major voltage-gated potassium channels in the heart (Source: [466, 468]). Inward-rectifier (K_{IR}), voltage-dependent (K_V), and calcium-activated (K_{Ca}) K^+ channels also regulate the caliber of arteries and arterioles, especially in the cerebral territory. In the endocardium of human heart, a small $i_{K,to}$ component with slow recovery kinetics may correspond to $K_V1.4$ channel. Several inwardly rectifying currents in mammalian hearts include quasi-instantaneous rectifier i_{K1}, ATP-inhibited $i_{K_{ATP}}$, and muscarinic receptor-stimulated $i_{K_{ACh}}$ currents through $K_{IR}2.1$, $K_{IR}6.2$–SUR1, and G-protein-regulated $K_{IR}3.1$–$K_{IR}3.4$ (or GIRK1–GIRK4) complexes, respectively.

Type	Current	Subunits (Genes)
Ultrarapid delayed rectifier (phases 1, 2)	$i_{K,ur}$	$K_V1.5$ (KCNA5)
Rapid delayed rectifier (plateau end, phase 3)	$i_{K,r}$	$K_V11.1$ (hERG, ERG1, KCNH2) minK, MiRP1 (KCNE1, KCNE2)
Slow delayed rectifier (plateau end, phase 3)	$i_{K,s}$	$K_V7.1$ (K_VLQT1, KCNQ1) minK (KCNE1)
Rapid transient outward (phase 1)	$i_{K,to(r)}$	$K_V4.2/4.3$ (KCND2, KCND3) KCNIP2
Slow transient outward (phases 1, 2)	$i_{K,to(s)}$	$K_V1.4$ (KCNA4) MiRP1 (KCNE2)
Inward rectifier (phase 3, rest)	i_{K1}	K_{IR} $K_{IR}2.1$ (KCNJ2) $K_{IR}2.2$ (KCNJ12)

The long duration of cardiac action potentials (Sect. 6.5) that results from a coordinated action of numerous channels with diverse properties (gating mode, activation and inactivation rate, ionic conductance, etc.) is needed to prevent premature excitation. Three major potassium channels, with a varying concentration in the sarcolemma according to the transmural location, contribute to ventricular repolarization: *inward rectifier* (K_{IR}, i_{K1}) that closes during depolarization, *rapid* ($i_{K,r}$) and *slow* ($i_{K,s}$ current), *delayed rectifiers*, and *transient outward* K^+ ($i_{K,to}$ current; Table 5.12) *channels*.

The first group of K^+ channel subunits pertains to the 2TM and 1P class (K_{2TM1P}) of inwardly rectifying K^+-selective channels $K_{IR}2.i$, $K_{IR}3.i$, and $K_{IR}6.i$ that cause

i_{K1}, $i_{K_{ATP}}$, and $i_{K_{ACh}}$ currents, respectively (Vol. 3 – Chap. 3. Main Classes of Ion Channels and Pumps). They conduct substantial current near the resting potential, but carry little or no ions at depolarized potentials [466].

The second group of K^+ channel subunits contains 6 transmembrane segments (S1–S6; 6TM and 1P class [K_{6TM1P}]). The S5–P–S6 region resembles the 2TM–1P subunit. The additional membrane-spanning segments, especially the charged S4 segment, yield voltage-dependent gating (K_V channels). They are closed at resting potential, but activate at different rates upon depolarization to drive transient outward and several sustained delayed rectifying currents that control the duration of action potentials [466].

Various subunits of different K_V subfamilies are sources of functioning diversity in cardiac cells (nodal cells and cardiomyocytes). A special set of K_{6TM1P} subunits is responsible for the pacemaker current, as they activate upon hyperpolarization. They carry both Na^+ and K^+ ions.

Channel subunits of both K_{2TM1P} and K_{6TM1P} classes assemble into homo- or heterotetrameric channels. In addition, concatenation of 2 K_{2TM1P} subunits ($[K_{2TM1P}]_2$) form members of the K_{4TM2P} class. These dimers yield voltage-independent leak channels [466].

5.10.4.1 Voltage-Gated Potassium Channels (K_V)

Voltage-gated potassium channels are characterized by a functional diversity relevant to action potential repolarization in the myocardium. Multiple cardiac voltage-gated K^+ currents contribute to variations in shape of action potentials in different regions of the myocardium (Sect. 6.5) and control of cardiac excitability.

Voltage-gated K^+ channels are targeted by various endogenous neurotransmitters and neurohormones as well as exogenous drugs to modulate cardiac functioning.

Myocardial voltage-gated potassium channels are made of several subunit sets: K_V1 ($K_V1.1$–$K_V1.8$); K_V2 ($K_V2.1$ and $K_V2.2$); K_V3 ($K_V3.1$–$K_V3.4$); K_V4 ($K_V4.1$–$K_V4.3$); K_V5; K_V6; K_V7; and $K_V11.1$ [469]. Functional voltage-gated K^+ channels comprise 4 subunits. Members of this channel family combine various kinds of subunits with different time and voltage-dependent properties.

Retrograde microtubule-dependent transport modulates surface expression of several potassium channels in ventricular myocytes. The number of $K_V2.1$, $K_V3.1$, $K_V4.2$, $K_V11.1$, and $K_{IR}2.1$ increases when dynein is inhibited [470].

In adult mouse ventriculomyocytes, 4 kinetically and pharmacologically distinct voltage-gated K^+ currents exist [472]: (1) a rapidly activating and inactivating, transient outward K^+ current ($i_{K,to(r)}$) through K_V4 channel; (2) a rapidly activating, slowly inactivating, transient outward K^+ current ($i_{K,to(s)}$) through $K_V1.4$ channel; (3) a rapidly activating, very slowly inactivating K^+ current ($i_{K,slow}$) through $K_V1.5$ channel; and (4) a slowly activating, non-inactivating K^+ current ($i_{K,ss}$).

Table 5.13. Activation and inactivation rates of voltage-gated potassium channels (from [471]). There are more transient outward K^+ channels in epicardium than in endocardium, and in right than in left ventricle epicardium.

K^+ channel	Activation	Inactivation
Fast transient outward	Fast	Fast
Slow transient outward	Fast	Slow
Ultrarapid delayed rectifier	Fast	No
Rapid delayed rectifier	Moderate	Fast
Slow delayed rectifier	Very slow	No

The 2 last current types are detected in all ventriculomyocyte types, but not the 2 first types (either $i_{K,to(r)}$ in ventriculomyocytes isolated from the apex or $i_{K,to(s)}$) in ventriculomyocytes isolated from the septum) [472]. In ferret left ventricular epicardial and endocardial myocytes, transient outward K^+ currents results from activity of $K_V4.2$ and $K_V4.3$ (epicardial $i_{K,to(r)}$) and $K_V1.4$ (endocardial $i_{K,to(s)}$), respectively [473].

In mice, a truncated $K_V1.1$ protein attenuates the slow inactivating outward K^+ current ($i_{K,to(s)}$), thereby increasing action potential duration and QT interval on electrocardiogram traces and causing spontaneous ventricular arrhythmias [474]. A $K_V4.2$ pore mutant eliminates the rapidly inactivating transient outward current ($i_{K,to(r)}$), thus markedly prolonging action potential duration, without generating arrhythmias [474]. The higher the prolongation of action potential duration and QT interval, the lower the risk of spontaneous ventricular arrhythmias.

Transient Outward K^+ Currents

Transient outward K^+ channels set the rate of the early phase of repolarization. They are quickly inactivated (Table 5.13). Two subtypes can be identified, fast and slow, according to the rates of inactivation and recovery from inactivation. Rapidly and slowly inactivating transient outward K^+ currents are associated with $K_V4\alpha$ subunits and $K_V1.4$, respectively.

Atrial, rapidly inactivating, transient outward K^+ current ($i_{K,to(r)}$) through $K_V4.3$ contributes to phase-1 rapid repolarization of cardiac action potentials. Current $i_{K,to(r)}$ varies according to the intramural location (endocardium, midmyocardium, and epicardium) of ventriculomyocytes of the human heart.

Rapidly inactivating transient outward heterotetrameric K^+ channels are made of: (1) pore-forming α subunits, i.e., $K_V4.2$ and predominant $K_V4.3$; (2) accessory β subunits, i.e., K_V channel-interacting protein KChIP2; and (3) the high-affinity, low-capacity, calcium-binding, calcium sensor, frequenin homolog, or neuronal calcium sensor NCS1 [469].[93]

Transcription factor Irx5 from the Iroquois homeobox genes controls the action potential. It represses expression of the KCND2 gene that encodes $K_V4.2$ subunit of voltage-gated potassium channel via the recruitment of muscle-restricted, histone deacetylase-dependent transcriptional repressor, SET and MYND domain-containing protein SMYD1, or zinc finger protein BOp (for CD8b opposite; also called ZMYND18) [475].[94] Protein Irx5 (but neither Irx3 nor Irx4) increases $K_V4.2$ promoter activity in cardiac fibroblasts,[95] whereas it decreases $K_V4.2$ promoter activity in ventricular myocytes [476]. Transmural gradient of i_{Kto} current density across the left ventricular wall results from difference in transcription factors that control the KCND2 and KCNIP2 genes that encode $K_V4.2$ or KChIP2 according to mammalian species.[96]

Activity of human atrial $K_V4.3$ channel is modulated by [477]: (1) plasmalemmal epidermal growth factor receptor (EGFR) protein Tyr kinase (phosphorylation of Tyr136) and (2) cytosolic SRC family kinases (phosphorylation of Tyr108). Phosphatase PP2, an inhibitor of SRC family of protein Tyr kinase markedly reduces $K_V4.3$ current.[97]

Various molecules also regulate the cardiac transient outward potassium current [477]. α-Adrenergic receptor hinders $i_{\mathrm{K,to(r)}}$ in rabbit ventriculomyocytes. In human cardiomyocytes, protein kinase-C reduces $i_{\mathrm{K,to(r)}}$ current. Activated Ca^{2+}–calmodulin-dependent protein kinase CamK2 slows $K_V4.3$ inactivation and accelerates the recovery rate from inactivation. Nitric oxide inhibits human atrial $K_V4.3$ channel; adenylate cyclase activates PKA kinase and PP2 phosphatase.

---

[93]Frequenin binds to many protein types (either in a calcium-dependent or -independent fashion), such as protein phosphatase-2, phosphatidylinositol 4-kinase-3β (PI4K3β), phosphodiesterases, nitric oxide synthase, adpribosylation factor ARF1, IP_3 receptor, $K_V4.3$ channel, TRPC5 channel, G-protein-coupled receptor kinase GRK2, dopamine D_2 receptor, and interleukin-1 receptor accessory protein-like protein (IL1RAPL).

[94]Protein SMYD1 intervenes in cardiomyocyte differentiation and cardiac morphogenesis. The BOP alias is also used for Eye absent homolog EyA1 of cardiac and skeletal myocytes that can act as a transcriptional activator. Mutations in the EYA1 gene cause autosomal dominant type-1 branchio-oto-renal (BOR1) dysplasia syndrome, type-1 branchio-otic syndrome (BOS1), congenital cataracts, and ocular anterior segment anomalies. These syndromes constitute branchio-oto-renal spectrum disorders.

[95]Protein Irx4 inhibits Irx5-induced increase in channel promoter activity in fibroblasts.

[96]Transmural gradient is caused by expression gradient of $K_V4.2$ and/or $K_V4.3$ in small mammals and KChIP2 subunits in large mammals.

[97]Protein Tyr kinases modulate activity of many ion channels, such as $Ca_V1.2$, Na_V, and volume-sensitive Cl^- channels in cardiomyocytes, in addition to several types of K^+ channels in different cell types and TRPC channels. Cardiac volume-sensitive Cl^-, Na_V, and $K_V11.1$ are regulated by both EGFR and SRC family kinases. Cardiac $K_V7.1$, $K_{IR}2.1$, and $K_{IR}2.3$ are controlled by EGFR kinase.

Delayed K^+ Rectifiers

Delayed K^+ rectifiers activate later and inactivate more slowly during the repolarization than transient outward K^+ channels. Rapid delayed K^+ rectifiers have faster activation and deactivation kinetics and more negative threshold potential than slow delayed K^+ rectifiers. A third type exists, ultrarapid delayed K^+ rectifiers. These 3 delayed K^+ rectifiers underlie repolarization phase 3.

Subunits $K_V11.1$ and $K_V7.1$ are related to the rapid and slow delayed K^+ rectifiers, respectively (Table 5.12). $K_V11.1$ channel is responsible for the rapid delayed rectifier K^+ current during the third phase of repolarization. It exists not only in cardiomyocytes, but also in smooth myocytes and other cells. At negative membrane potentials, $K_V11.1$ channels are in a closed state. Membrane depolarization slowly opens the channel that is then rapidly inactivated. $K_V1.5$ channel underlies the ultrarapid component of outward delayed K^+ rectifiers in the atrium.

5.10.4.2 Inward Rectifier Potassium Channels (K_{IR})

Cardiac GIRK Channel

G-protein-dependent K^+ *inward rectifier* (GIRK or K_G), activated by $G\beta\gamma$ subunit,[98] has negative chronotropic effect and explains acetylcholine-induced bradycardia ($i_{K_{ACh}}$ current). These channels of the sinoatrial node indeed decelerate the pacemaker activity. Potassium flux depends on the electrochemical gradient for K^+ across the sarcolemma.

GIRK channels are heterotetramers of K^+ inward rectifier subunits of the $K_{IR}3$ group combined in a cell-specific manner. GIRK channel of the atriomyocyte is composed of $K_{IR}3.1$ and $K_{IR}3.4$. Each subunit has one $G\beta\gamma$ binding site. The subcellular localization of GIRK channels involves PDZ domain-containing anchoring proteins.

Cell membrane hyper- and depolarization lead to fast[99] and slow ("relaxation")[100] increase and decrease in channel activity, respectively [478].

During sarcolemma repolarization, the GTPase-accelerating protein action of regulators of G-protein signaling is inhibited by phosphatidylinositol (3,4,5)-trisphosphate that binds to RGS4 [479]. Voltage-dependent formation of Ca^{2+}–calmodulin leads to its binding to regulators of G-protein signaling. Thereby, Ca^{2+}–calmodulin relieves the inhibition by phosphatidylinositol trisphosphate and

[98]The availability of each channel subunit is dictated by $G\beta\gamma$ concentration, the available status being characterized by a higher affinity for $G\beta\gamma$ subunit.

[99]The fast channel flux is due to the blockade of outward K^+ motion through the channel by intracellular Mg^{2+} and polyamines. This feature is common to almost all K_{IR} channels.

[100]The slow channel flux is specific to GIRK channel. It is due to the voltage-dependent action of regulators of G-protein signaling that accelerate the intrinsic GTP hydrolysis of the $G\alpha$ subunit.

Table 5.14. Main features of CMC potassium carriers.

Inward rectifier	Small efflux at positive potential
	Open at negative potential
	Repolarization down to −80 mV
Delayed rectifier	Phase 3 repolarization
	Slow activation beyond −40 mV
	Slow inactivation
Transient outward	Activation by depolarization
	Phase 1 repolarization
Muscarinic (ACh)	Inward motion
	Hyperpolarization
	Background current in SAN
ATP	Opening by low [ATP],
	by increase in [ADP],
	and by adenosine, H^+

decreases free $G\beta\gamma$ level due to the restoration of the activity of regulators of G-protein signaling. Consequently, the number of active GIRK channels decays during cardiomyocyte contraction.

Cardiac ATP-Sensitive K^+ Channel

Cardiac sarcolemmal ATP-sensitive K^+ channel (that causes $i_{K_{ATP}}$ current) is composed of $K_{IR}6.2$ and SUR2a subunits, as well as glyceraldehyde 3-phosphate dehydrogenase.[101] It is regulated by intracellular ATP [480]. It couples glycolysis to membrane excitability. ATP-sensitive K^+ channel is responsible for ST-segment elevation on ECG during ischemia [481].

5.10.5 Calcium-Activated Chloride Channels

Calcium-activated Cl^- current[102] occurs in the same action potential period as transient outward K^+ current and contributes to ventricular repolarization. Main features of ion carriers are summarized in Tables 5.14 to 5.16.

[101] Enzyme GAPDH produces (1,3)-bisphosphoglycerate, an opener of ATP-sensitive K^+ channels. Hence, it regulates the channel activity.

[102] Calcium-activated, temperature-dependent Cl^- currents can be observed in atrial and ventricular myocytes, as well as cardiac Purkinje cells, at least in the rabbit heart.

Table 5.15. Main features of CMC sodium carriers.

Fast activating	Influx
	Voltage- and time-dependent opening
	Scarce in sinoatrial node
Hyperpolarization activated	Slow opening at $[-60, -40\,\text{mV}]$
	Depolarization of pacemaker
	Stimulation by catecholamines and cAMP
	Inhibition by ACh
Background	Inward flux
	Attenuation of resting membrane potential
NCX	Activated by increase in cytosolic $[Ca^{2+}]_i$
	3 Na^+ vs. 1 Ca^{2+}
	Early Ca^{2+} influx
	Na^+ influx in late plateau
Na^+–K^+ pump	3 Na^+ vs. 2 K^+
	Minor contribution to resting potential

Table 5.16. Main features of CMC calcium and chloride carriers.

Calcium carriers	
Voltage-gated	Ca_V3 activated at more negative potential than $Ca_V1.2$ channel
	Early plateau current ($Ca_V1.2$)
	Ca-Induced Ca Release ($Ca_V1.2$)
	SAN and AVN action potential ($Ca_V1.2$)
	Pacemaker depolarization (Ca_V3)
	Fast inactivation (Ca_V3)
Ca^{2+} ATPase	Efflux
Chloride carriers	
cAMP-dependent	Stimulation by cAMP, β1 agonists
	Phase 1 repolarization after β1 activation
Ca^{2+}-dependent	Stimulation by cytosolic Ca^{2+}
	Phase 1 repolarization
	Pacemaker potential
Stretch-activated	Opening by osmotic swelling

Chapter 6
Heart Wall

The heart lodges in the mediastinum. It is usually situated in the middle of the thorax with its largest part in the left side; its apex points toward in the inferior and left direction. However, sometimes it resides on the right thoracic region (dextrocardia situs inversus totalis).

The heart is made of 2 synchronized left and right pumps with a common wall, the septum between the left and right atria at its upper part and left and right ventricles at its lower part. These twin pumps in series are composed of 2 chambers — upstream atrium and downstream ventricle — to adapt to the pressure difference between the low-pressure, upstream, venous compartment and high-pressure, downstream, arterial bed.

Under sedentary conditions, atria are minor contributors to blood pumping. During exercise, atrial contraction rises to increase the amount of blood that fills the ventricles (*atrial kick*). This blood-pumping reserve falls during aging as well as during atrial fibrillation.

Arteries leave the cardiac pump, without relation with the blood gas content; veins run toward atria. The left ventricle propels blood through the *systemic circulation* that carries arterial oxygenated blood under high arterial pressure; the aorta is the arterial trunk that receive blood from the left ventricle during the ejection phase of its contraction. The right ventricle expels blood through the *pulmonary circulation* that transport arterial deoxygenated blood with a lower arterial pressure; the pulmonary arterial trunk corresponds to the exit of the right ventricle.

Cardiac muscle cells, or cardiomyocytes (CMC; Chap. 5), are joined both directly and indirectly via fibers of the extracellular matrix to form muscular bundles (myofibers) connected to neighboring coronary vessels and nerves (Vol. 6 – Chaps. 2. Anatomy of the Cardiovascular System and 4. Cardiovascular Physiology). These myofibers form muscular layers within the heart wall thickness that are characterized by their orientation (Sect. 5.9), which influences spreading of the electrochemical command (cardiomyocyte depolarization) wave and cardiac wall deformation associated with cardiomyocyte contraction.

M. Thiriet, *Tissue Functioning and Remodeling in the Circulatory and Ventilatory Systems*, Biomathematical and Biomechanical Modeling of the Circulatory and Ventilatory Systems 5, DOI 10.1007/978-1-4614-5966-8_6,

Diverse cell populations are structural determinants of heart function. Cardiomyocytes represent the main cell population ($\sim 75\%$) in the myocardium. Cardiac fibroblast is the main producer of the extracellular matrix with its collagen and elastin bundles and interconnected basement membranes.

Vascular cells are close to cardiomyocytes to efficiently supply nutrients. Parasympathetic and sympathetic nerve fibers extend within the myocardium. The right and left vagus nerves primarily innervate the sinoatrial and atrioventricular nodes responsible for the genesis and proper propagation of the electrochemical wave, respectively. Sympathetic nerves are located inside the atria and ventricles.

6.1 Cardiogenesis and Cardiac Remodeling

The heart is the first organ to form during embryogenesis. This first functional organ has a mesodermal origin. Growth and remodeling of the embryonic myocardium adapt dynamically to fit the changing needs of a developing body. Its contraction supports heart shaping via a feedback between the hemodynamic stress field and cardiac cells.

Soon after gastrulation, the heart starts to develop from cells of the mesoderm (Chap. 11). The main stages of cardiogenesis are briefly described in Table 6.1. In vertebrates, the heart is initially built from the fusion of the 2 endocardial tubes that arise in the splanchnic mesoderm. The resulting primitive heart tube at the ventral midline undergoes a series of movements and remodelings to form the chambered organ. Disturbances of cardiogenesis cause cardiac malformations.

Markers of initial cardiac differentiation include Heart and neural crest derivatives expressed HAND1 and HAND2 transcription factors, or class-A basic helix–loop–helix proteins bHLHa27 and bHLHa26, T-box transcription factor TBx5, as well as myosin regulatory light chain MLC2.[1]

Among the first features of anteroposterior patterning, atrial (aMHC1) and ventricular (vMHC1) myosin heavy chains are restricted to the anterior and posterior poles of the heart tube, respectively.

Two myosin heavy chain genes are expressed in the mouse heart: βMHC, or MyH7, in embryonic cardiomyocytes, and αMHC, or MyH6, postnatally. Efforts cause hypertrophy and a shift to fetal βMHC with negative repercussion on the cardiac function.

Organ morphogenesis involves the orchestration of cell differentiation, proliferation, and migration (Fig. 6.1). During early embryogenesis, the heart is a single, relatively straight tube that bends and twists. The cardiac tube is composed of an outer layer of myocardium and an inner lining of endocardial cells, separated by an extensive extracellular matrix, the so-called cardiac jelly. Local protuberances of the cardiac jelly and associated mesenchymal cells form cardiac cushions.

---

[1] A.k.a. regulatory myosin light chain MyL9.

Table 6.1. Main stages of cardiac development. The precardiac mesoderm forms a primitive tubular heart that starts beating at about 3 weeks of gestation in humans. The heart maintains its pumping function despite its continuous remodeling into a 4-chamber organ. The heart has an asymmetric structure with anterior–posterior, cranial–caudal, and left–right polarity.

Primary heart fields	Two mesoderm lateral regions that appear shortly after gastrulation with precursors of cardiac lineages (early myocardium)
Cardiac crescent	Merging of heart fields at the embryo midline
Linear heart tube	Endothelial tube surrounded by a single layer of myocardial cells
Looped heart	Tube elongation and looping with gain of progenitors from the secondary heart field
Atrioventricular canal development	Tube remodeling with a segment between atrial and ventricular regions
Endocardial cushion formation	Outflow tract and atrioventricular cushions, i.e., precursors of heart valves; epithelial-to-mesenchymal transition of endocardial cells
Heart wall trabeculation	Myocardium differentiation into 2 layers, an outer compact zone and an inner trabeculated zone; coordinated proliferation and differentiation of myocytes; crosstalk between trabecular myocytes and endocardial cells
Chamber and outflow tract septation	Formation of 4 heart chambers and aorta and pulmonary artery

Different myocardial progenitors form the various regions of the heart. The embryonic heart begins to pump blood before the development of valved chambers. The pumping action results from suction into the heart tube [482]. The blood stream in the developing heart favors the directed fluid motion in a stressed gel-like medium. The embryonic heart is transformed from a simple tube into a 4-chambered pump via multiple transluminal septation.

Two migratory cell populations invade the heart-forming region to regulate cardiogenesis. The cardiac neural crest originates from the neural tube. After migration into the cardiac outflow tract, it is involved in the septation of the aortic sac [483]. Epicardial cells that originate from the proepicardial serosa epithelium migrate onto the surface of the heart and create the epicardium and subepicardial connective tissue, the coronary vascular system (endothelial cells, smooth myocytes, and perivascular fibroblasts), intermyocardial and subendocardial connective tissue, as well as a small number of endocardial cells close to the sinus venosus [483].

The proepicardium gives rise to the myocardial stromal and vascular smooth muscle cells, but few coronary endothelial cells. Coronary vessels arise from angiogenic sprouts of the sinus venosus that returns blood to the embryonic heart [484]. Sprouting venous endothelial cells dedifferentiate as they invade the myocardium. These cells differentiate into arterial and capillary cells. Cells on the surface redifferentiate into venous cells.

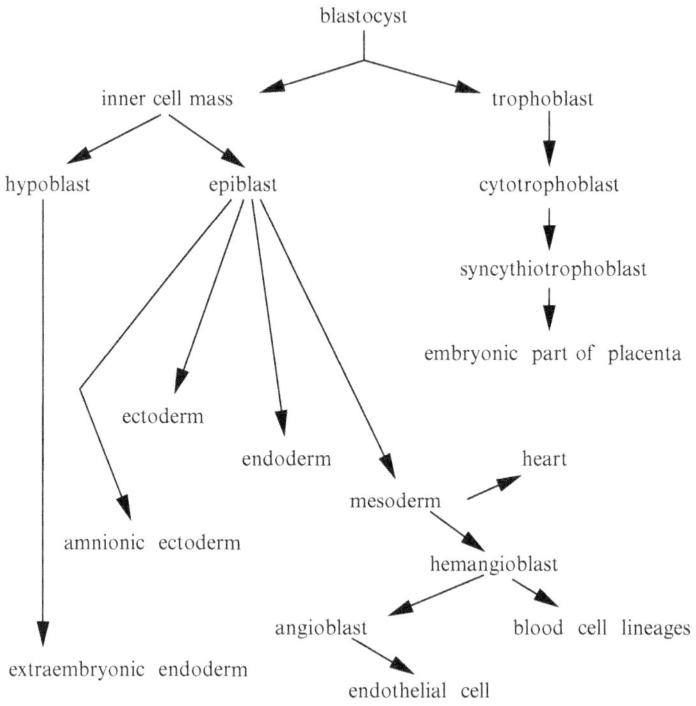

Fig. 6.1 The cardiovascular organs develop from the mesoderm. The blood vessels form independently from the heart. The heart is the first functional body's organ.

After heart looping, the cardiac jelly that overlies the future atrioventricular canal and outflow tract expands into cardiac cushions. The formation of the cardiac cushions is characterized by endothelial–mesenchymal transformation of a subset of endothelial cells. During heart valve formation, a subset of endothelial cells that corresponds to future valve sites delaminates, differentiates, and migrates into the cardiac jelly [485].

6.1.1 *First and Second Heart Fields*

Soon after gastrulation, heart formation starts in the precardiac anterior mesoderm (first and second heart fields) under the coordinated action of inductive and repressive signals from neighboring endoderm and ectoderm (Table 6.2). On the other hand, posterior mesoderm is a blood-forming region.

Mesodermal cells that differentiate into cardiac cells during cardiogenesis segregate into 2 distinct populations of cardiac progenitors: the primary heart and

Table 6.2. Examples of signaling molecules involved in cardiac specification (Source: [486]; FGF: fibroblast growth factor; GATA: DNA sequence GATA-binding transcrition factor; Myocd: myocardin; NKx2: NK2 transcription factor-related homeobox protein; SHh: sonic Hedgehog; SRF: serum response factor; TBx: T-box transcription factor; TH: tyrosine hydroxylase). Inhibitors of bone morphogenetic proteins (BMP) — noggin and chordin — secreted from the notochord limit the amount of developing cardiac tissue. They also retard the differentiation of a population of cardiac precursors that will form the outflow and inflow tracts of the mature heart. Cerberus-related protein is an extracellular, multivalent antagonist that binds to growth factors of the transforming growth factor (TGFβ) superfamily, Nodal and BMPs, as well as Wnt morphogens.

Ectoderm	Wnt1/3
Endoderm	BMP2, FGF8, SHh
(cardiogenic inducers)	Cerberus
	Mesoderm
Notochord	Noggin, chordin
Cardiac mesoderm	GATA4/5/6, MyoCd, NKx2.5 SRF, TBx20, TH, Wnt11

anterior heart fields that contribute to the left ventricle and right ventricle and outflow tract, respectively. Unlike the first heart field, second heart field cells remain undifferentiated until they migrate into the forming heart.

Both heart field cells are tagged by the homeobox NK2 transcription factor-related locus-5 protein NKx2-5 encoded by homeobox NKX2-5 gene,[2] whereas transcription factor Islet-1 (Isl1)[3] is only involved in the differentiation of second heart field cells. Second heart field cells persist in the postnatal heart. These progenitors can differentiate into several types of cardiac cells.[4] In human fetal hearts at 11 and 18 weeks of gestation, Isl1+ cells lodge in the right atrium and outflow tract that are both second heart field-derived structures as well as left atrial wall and appendage [487].

[2]The homeobox gene NKX2-5 (or NKX2.5) is also called cardiac-specific homeobox CSx. The homeobox gene set includes many members (NKX2-1–NKX2-6). Homeodomain protein NKx2-1 corresponds to thyroid transcription factor-1. It is also expressed in lungs. Factor NKx2-5 is also produced in the primitive pharyngeal endoderm that secretes the heart inducer.

[3]Homeobox protein Isl1 as well as Pdx1 of endocrine cells of the gastric mucosa that control the production of gastrointestinal hormones were discovered in pancreatic islets.

[4]Damaged heart components (valves, vessels, nodal conduction tissue, and myocardium) must be regenerated. The mobilization of endogenous progenitors and differentiation of exogenous stem cells participate in tissue repair. Adult stem cells are genetically equivalent to embryonic cells. Differential gene expression results from reversible epigenetic changes gradually imposed on the genome during development. However, adult cells have a reprogramming capacity toward an undifferentiated embryonic state, as differentiation is reversible.

6.1.2 *Heart Growth Patterns*

During cardiogenesis, heart chambers accommodate changes in load associated with the body's development to fulfil increasing needs. Certain cell subpopulations participate to a greater extent to cardiogenesis characterized by marked changes in size (heart chambers having specific dimensions) and tissue architecture. Hence, cardiogenesis does not follow a gradual expansion during which all involved cardiac cell types proliferate and contribute equally to heart remodeling and reshaping.

In the zebrafish heart, expansion of the surface area results from proliferation of a subset of cells from the embryonic wall, each of which creating a muscular patch of given size and shape. During embryonic maturation, morphogenesis continues from a primitive structure. The ventricular wall thickens, not by proliferation of the juvenile wall (original inner layer), but by a new outer layer of muscle (cortical muscle) generated by some dominant clones of proliferating cells that may arise from internal trabecular muscle. In fact, cortical cardiomyocytes initially emerge from internal myofibers that breach the juvenile ventricular wall and then expand over the surface [488]. Both inner and outer layers persist in the adult heart. Therefore, 3 ventriculomyocyte lineages localize to the adult organ: primordial, trabecular, and cortical myocardial lineage, created in this order from each other [488].

6.1.3 *Epithelial–Mesenchymal Transition*

Epicardium, coronary beds, and conduction paths interact during cardiogenesis. Cells of the *proepicardial organ* (PEO), a more or less diffuse cluster of extracardiac mesothelial cells, are precursors for epicardial epithelial, myocardial connective tissue, coronary smooth muscle, and endothelial cells. After proliferation, PEO cells travel over the myocardial surface and differentiate into an epithelium to form the primitive epicardium. A set of epithelial cells migrate into the subepicardium and undergo an epithelial-to-mesenchymal transition.[5]

Epithelial-to-mesenchymal transformation is implicated in manifold steps of cardiovascular morphogenesis, such as heart valve development, coronary artery formation, and inflow and outflow tract septation [489]. Epithelial–mesenchymal transition is the first step in morphogenesis that transforms the mesenchyme into functional structures, such as septa and valves, required for unidirectional pumping activity.

[5]During epithelial-to-mesenchymal transformation, epicardial and endocardial (endothelial) cells transform into mesenchymal cells that generate different tissues. Epithelial-to-mesenchymal transformation involves fibroblast growth factor, epidermal growth factor, vascular endothelial growth factor, transforming growth factor-β (Vol. 2 – Chap. 3. Growth Factors), and transcription factor WT1 encoded by the Wilms tumor-suppressor Wt1 gene.

A set of epithelial cells in the cardiac wall generate cardiac fibroblasts and participate in coronary vasculogenesis. Another set interacts with myocardioblasts for myocyte differentiation into cardiac Purkinje cells. Certain epicardium-derived cells migrate in the core of the endocardial cushions to produce atrioventricular valve leaflets. Epithelial-to-mesenchymal transformation occurs also in adult life, especially in cardiovascular diseases such as atherosclerosis, intimal hyperplasia, and aneurysms.

Ubiquitous periostin is involved in epithelial–mesenchymal transformation and subsequent mesenchymal maturation, with differentiated mesenchyme and condensed matrix in embryonic heart valves. Periostin is also highly expressed in the myocardium in patients with heart failure. Periostin inhibits cardiomyocyte spreading and adhesion of cardiac fibroblasts [490].

During normal heart development, epicardial-derived cells (EPDC) undergo an epithelial–mesenchymal transition in response to BMP, FGF, and TGFβ released from the myocardium (Vol. 2 – Chap. 3. Growth Factors). Secreted myocardial thymosin-β4 then induces EPDC migration into the myocardium, where they respond to angio- (VEGF and FGF2) or arteriogenic (PDGF and TGFβ) growth factors and differentiate into endothelial and smooth muscle cells, respectively, leading to a capillary plexus and stabilizing coronary vessels.

6.1.4 Vasculo- and Angiogenesis

Embryonic coronary vessel development involves the activation and proliferation of epicardial cells, followed by an epithelial–mesenchymal transition. Mural and mesenchymal cell markers such as PDGFRβ are upregulated [491]. Signaling by PDGF that causes the formation of stress fibers and loss of intercellular contacts of epicardial cells is needed for epicardial cell proliferation.

Endothelium-lined tubes construct a functional circulatory network. The formation of endothelial vacuoles followed by intra- and intercellular fusion drives vascular lumen genesis [492]. These endothelial pinocytic vacuoles quickly occur, fuse together, and enlarge to create the tube lumen. Endothelial cells form extensive contacts between themselves and merge their intracellular vacuoles without cytoplasmic mixing. The vacuolar compartments in adjacent endothelial cells associate and form a continuous lumen in developing blood vessels.

During embryogenesis, the generation of new cardiomyocytes from progenitor cells and existing cardiomyocytes is associated with the development of the coronary vasculature from progenitors that originate from the endothelium as well as proepicardial organ and neural crest [493]. Coupling of myocardial and coronary development is mediated by bone morphogenetic proteins and fibroblast (FGF) and vascular endothelial (VEGFa) growth factors.

The myocardial development relies [493]: (1) in cardiomyocytes, on growth factors and their receptors, such as VEGFa, BMPR1a (ALK3), FGFR1 and FGFR2; MAPK14 (P38MAPKα), TxnRd2 thioredoxin reductase, and transcription factors bHLHa26 (or HAND2), bHLHa27 (or HAND1), GATA4, ZFPM2,[6] and SRF; (2) in the proepicardial organ, on retinoid X receptor RXRα (or NR2b1) and Wilms tumor suppressor WT1; (3) in the endothelium, on ephrin-B2, endothelial receptor TIE2, MAPK7 kinase, PTen phosphatase, neurofibromin NF1, and caspase-8; and (4) in neural crest cells, on BMPR1a and Paired box Pax3 protein.

6.1.5 Neurogenesis

The autonomic nervous system involved in the body's homeostasis and stress defense includes not only the sympathetic and parasympathetic neurons and their centers in the central nervous system, but also adrenal medulla. The sympatheticoadrenal lineage corresponds to a common progenitor of adrenergic sympathetic neurons and medullary cells of the adrenal gland. Progenitors of sympathetic ganglia originate from neural crest cells, which constitute an embryonic progenitor cell population emigrating from the dorsal aspect of the neural tube. Neural crest cells generate sympathetic and medullary cell lineages during embryogenesis. The dorsal aorta acts as a morphogenetic center that coordinates migration and segregation of neural crest cells [494]. Bone morphogenetic proteins produced by the dorsal aorta stimulate the production of CXCL12 chemokine and neuregulin-1[7] (Sect. 6.1.8.3) in the para-aortic region. These molecules are chemoattractants. Later, BMP signaling instructs sympatheticoadrenal specifications.

Cardiac sympathetic noradrenergic neurons need nerve growth factor for survival and development. Neurturin (Nrtn), a member of the family of glial cell line-derived neurotrophic factors (GDNF), is required for cardiac parasympathetic cholinergic neurons [495]. This neurotrophic factor controls neuron survival and differentiation.

However, because a certain fraction of the population of cardiac cholinergic neurons persists in the absence of neurturin, additional neurotrophic factors intervene. Members of the GDNF family target their cognate coreceptors (GDNFRα1–GDNFRα4) coupled to transmembrane receptor protein Tyr kinase Rearranged during transfection (ReT; Vol. 3 – Chap. 8. Receptor Kinases).[8] The neurturin

---

[6]Alias ZFPM stands for zinc finger protein multitype. A.k.a. zinc finger protein ZnF89b and Friend of GATA protein-2 (FOG2).

[7]Neuregulin-1 is involved in the development of the nervous system and heart. It acts via receptors of the HER set. It has numerous isoforms produced by alternative splicing. Isoform NRg1 type-1 (heregulin), -2 (glial growth factor GGF2), and -3 (sensory and motor neuron-derived factor [SMDF]) are synthesized in excitatory and inhibitory neurons and astrocytes. Neuregulin-1 is a cardioactive growth factor released by endothelial cells that ensures cardiac development as well as structural maintenance and functional integrity. Effectors of the NRg1–HER axis include cardiac-specific myosin light chain kinase (cMLCK), protein phosphatase-1 (PP1), sarcoplasmic reticulum Ca^{2+} ATPase SERCA2, and focal adhesion kinase (FAK).

[8]Glial cell-derived neurotrophic factor and neurturin preferentially bind to GFRα1 and GFRα2, respectively.

receptor, i.e., the GFRα2–ReT complex, is expressed by cardiac parasympathetic neurons. Receptor GDNFRα2[9] is a glycosylphosphatidylinositol (GPI)-anchored cell surface receptor for both GDNF and Nrtn that primes activation of ReT receptor. Neurturin binds preferentially to GDNFRα2 coreceptor compared to GDNFRα1.[10]

6.1.6 Valvulogenesis

Cardiac cushions protrude from the cardiac jelly and underlying myocardium to form thin, tapered leaflets with a single endothelial cell layer and a central matrix with collagen, elastin, and glycosaminoglycans [485]. Endothelial–mesenchymal transformation allows the generation of the endocardial cushion, the primordia of valves, and septa of the heart. The final atrioventricular valves derive from endocardial cushions in atrioventricular canal. The development of ventriculoarterial valves from the outflow tract relies on a population of neural crest cells from the branchial arches that migrates to the distal outflow tract and contributes to aortopulmonary septation. Nevertheless, resident cells of leaflets of all the cardiac valves have an endothelial origin [485].

Developing cardiac valves, like other avascular tissues, express transcription factors, such as sex-determining region Y (SRY)-box Sox9, nuclear factor of activated T cells NFAT8, Runt-related transcription factor Runx2, and Msh homeobox gene product Msx2, as well as growth factors, such as BMP2 and TGFβ2.

Numerous signaling axes regulate endothelial proliferation and differentiation in developing and postnatal heart valves, such as VEGF–NFAT2, Notch–TGFβ, Wnt–β-catenin, BMP–TβR, HBEGF–HER, and NF1–Ras [485] (Table 6.3).

The Wnt–β-catenin signaling may control production of the growth factors such as bone morphogenetic proteins as well as structural elements, such as versican and hyaluronic acid, that are needed for epithelial–mesenchymal transition [485].

Vascular endothelial growth factor controls the development of endocardial cushion and cardiac valve. Subtype $VEGF_{165}$ signals via the VEGFR1–VEGFR2–neuropilin-1 complex on endothelial cells and their effectors, such as inositol triphosphate, diacylglycerol, extracellular signal-related kinases, and, in valve endothelial cells, transcription factor NFAT2 (or NFATc1) [497]. Factor VEGF acts via VEGFR2 and promotes NFAT2 nuclear localization to increase the proliferation of valvular endothelial cells. In addition to Ca^{2+} release from its store upon IP_3 stimulation, Ca^{2+} may enter endothelial cells in developing cardiac cushions through connexin-45-based hexameric gap junctions and activate protein

[9] A.k.a. GFRα2 and ReTL2.

[10] A.k.a. GFRα1 and ReTL1. Coreceptor GDNFRα1 binds to GDNF tightly to form a membrane-associated complex that then interacts with ReT. On the other hand, GDNFRα2 forms a high-affinity complex with GDNF or Nrtn only in the presence of ReT [496].

Table 6.3. Factors of cardiac valve development and remodeling (Source: [485]). Various signaling pathways and transcriptional regulators coordinately regulate valvulogenesis (BMP: bone morphogenetic protein; HA: hyaluronic acid; HAS: hyaluronan synthase; HBEGF: heparin-binding epidermal growth factor-like growth factor; HER: human epidermal growth factor receptor; NF1: neurofibromin-1; NFAT: nuclear factor of activated T cells; TGF: transforming growth factor; VEGF: vascular endothelial growth factor).

Factor	Effect
BMP2/4/6	Epithelial–mesenchymal transition
HA	Epithelial–mesenchymal transition, migration via HER2 and HER3
HBEGF	Inhibition of cell proliferation
NF1	Inhibition of Ras GTPase (of cell proliferation)
Notch	TGFβ signaling
TGFβ	Cell mobilization
VEGF	Endothelial cell proliferation via NFAT2
Wnt	Production of BMP4, versican, HAS2

phosphatase-3 [485]. Phosphatase PP3, in turn, dephosphorylates NFAT2 that then moves into the nucleus, where it interacts with transcriptional regulators such as Activator protein-1 to trigger gene transcription.

Activated Notch is converted from a transmembrane receptor to a transcriptional coactivator that participates in the control of epithelial–mesenchymal transition of endocardial cushions. Notch signaling increases the level of transforming growth factor-β2. The latter increases the activity of Snail transcription factor, thereby downregulating cadherin-5 and endothelial cell separation. On the other hand, SMAD6 attenuates TGFβ signaling in developing valves [485].

The major TGFβ family members implicated in cardiogenesis include TGFβ2, TGFβ3, BMP2, and BMP4 factors. Transforming growth factor-β3 is synthesized in transforming endothelial and invading mesenchymal cells, whereas bone morphogenetic protein-2 is expressed in the myocardium. Myocardial signals upregulate the expression of endothelial TGFβ3 that acts synergistically with BMP2 in the atrioventricular canal to prime the epithelial–mesenchymal transition [498]. The production of BMP2 and BMP4 is stronger in the myocardium underlying developing atrioventricular canal and outflow tract, respectively. Isotype BMP6 is expressed in endocardium, myocardium, and mesenchyme of the developing heart, especially in the atrioventricular canal [485].

Heparin-binding epidermal growth factor-like growth factor (HBEGF) is strongly expressed in the endocardium overlying the cushion-forming area. It is cleaved from the transmembrane precursor proHBEGF by ADAM17 metallopeptidase. It can bind human epidermal growth factor receptors HER1 and HER4. It limits mesenchymal cell proliferation once bound to HER1–HER2 heterodimer [485].

On the other hand, activation of HER2–HER3 heterodimer by hyaluronic acid increases the epithelial–mesenchymal transition and migration via Ras signaling [485]. Hyaluronic acid is produced by hyaluronan synthases (HAS1–HAS3). Enzyme HAS2, the main enzyme for hyaluronic acid production during embryogenesis, is required for the formation of the cardiac jelly. This glycosaminoglycan of the extracellular space regulates ligand availability. It interacts with numerous matrix constituents such as versican, a major proteoglycan of the cardiac jelly.

Small Ras GTPase is activated by receptor protein Tyr kinases to heighten mesenchymal proliferation. It can then signal via MAPK modules. Neurofibromin-1 is a Ras-specific GTPase-activating protein (RasGAP) that inactivates Ras GTPase. Neurofibromin-1 can diminish endothelial and/or mesenchymal cell proliferation. Small Ras GTPase can also interact with NFAT2 to control gene transcription [485]. Neurofibromin-1 modulates the timing of NFAT2 intranuclear localization to avoid hyperplasia of cardiac cushions.

Transcription factor Sox9, in conjunction with Sox5 and Sox6, targets the CHM1 gene in avascular tissue that encodes chondromodulin-1.[11] In the interstitial space, chondromodulin-1 inhibits endothelial cell proliferation and angiogenesis [499]. Chondromodulin-1 expression persists in normal cardiac valves throughout life. It is downregulated in degenerative cardiac valves. Loss in chondromodulin-1, indeed, leads to vascularization, calcification, lipid deposition, and inflammation of cardiac valves.

With its structural and functional heterogeneity, the heart achieves a coordinated contraction of its myofiber population in the 2 in-series pumps with its twinned chambers to eject blood in both circulations during each cycle. This composite material is optimally organized. The activation phase of the myofibers matches their mechanical heterogeneity for a suitable electromechanical coupling.

6.1.7 *Postnatal Growth*

During the fetal life, cardiomyocytes proliferate. Although from birth to adulthood heart mass rises, cardiomyocyte proliferation slows around birth [500]. Most of the postnatal increase in cardiomyocyte mass results from hypertrophy rather than cell division, whereas other cardiac cell types give rise to multiple new cells during postnatal growth of the heart. Nonetheless, long-lived cardiomyocytes can synthesize DNA after birth. In humans, postnatal cardiomyocytes can actually produce DNA without nuclear division (polyploid nucleus). In addition to nuclear diploidy and even sometimes polyploidy, a certain fraction (about 25%) of the cardiomyocyte population undergoes a final nuclear division without cell division. Bi-, tri-, and tetranucleated cells represent about 25.5, 0.4, and 0.1% of the cardiomyocyte population [501].

Cardiomyocytes are able to partly repopulate human myocardium, but the limited functional recovery after myocardial injury clearly demonstrates insufficient

[11] Chondromodulin-1 is also called leukocyte-derived chemotaxin LeCT1.

regeneration of cardiomyocytes. Cardiomyocytes renew with about 1% annual turnover at 25 years old that gradually decays down to 0.45% at 75 years old [502].

Myocardin (Myocd) is a potent transcriptional coactivator of serum response factor expressed exclusively in cardiac and smooth myocytes during postnatal development. It regulates maintenance and adaptation of the heart, especially cardiomyocyte structure and organization of intercalated discs and sarcomeres [503]. Myocardin links to serum response factor to synergistically activate gene transcription for heart adaptive response to growth factors and hemodynamic stress.

Myocardin-regulated genes encode cardiac cytoskeletal and myofibrillar structural proteins, such as cardiac α-actin-1 (ACTC gene), α-actinin-2 (ACTN2 gene), cardiac myosin heavy chain-7 (i.e., slow isoform βMHC; MYH7 gene), desmin (DES gene), dystrophin (DMD gene), and tropomyosin-1 (tropomyosin-α chain; TPM1 gene).

Growth factors activate extracellular signal-regulated kinases ERK1 and ERK2. The latter, in turn, phosphorylate myocardin isoform-B (Ser812, Ser859, Ser866, and Thr893) [504].

The synthesis of protein Ser/Thr kinase PIM1 in the myocardium decreases during postnatal development, but re-emerges after acute injury. Cardioprotection associated with PKB activation induces PIM1 expression (downstream from PKB) to impede cardiomyocyte apoptosis and enhance calcium dynamics [505].

6.1.8 Regulators

Cardiac specification requires a permissive environment. Anterior endoderm adjacent to the heart-forming region, the anterior mesoderm, provokes and maintains cardiogenic fate. This endoderm region secretes many messengers, such as fibroblast growth factors, activins and bone morphogenetic protein BMP2 of the TGFβ superfamily, and insulin-like growth factor-2 that promote cell survival and proliferation of cardiogenic cells and differentiating myocytes [483].

Cardiogenesis relies on few recycled functional modules that integrate into protein networks according to a given spatiotemporal control to drive the stages of organ development [506]. Complicated organ are thus built using simple building blocks that coordinate morphogenesis.

6.1.8.1 Transcription Factors

Homeobox gene products, such as members of the homeobox HOX and NK2 families, as well as transcription factors of the helix–loop–helix superfamily such as the myogenic genes (e.g., myogenin [Myog], or myogenic factor-4, myogenic differentiation factor-1 [MyoD or MyoD1], myogenic factor-5 [MyF5], and myogenic regulatory factor MRF4) participate in cell lineage commitment.

Cardiac and endothelial cell lineages arise from a common progenitor. Synthesis of transcription factor E-twenty six (ETS)-related protein ETSRP71 is transiently induced in the endocardium and endothelium of the embryo by NKx2-5 that binds to the NKX2-5 response element of the Etsrp71 gene promoter [507]. This factor targets TIE2 receptor to specify cell fate in the developing heart.

Transcription factors, such as NKx2-5, DNA sequence GATA-binding protein GATA4, and myocyte-specific enhancer factor-2C (MEF2c), are required for normal cardiogenesis that is regulated by signaling molecules (e.g., bone morphogenetic proteins, fibroblast growth factors, and Wnts). These signals induce or prevent the early stage of cardiomyocyte differentiation.

In vertebrates, 3 members of the GATA family of zinc finger transcription factors (GATA4–GATA6) operate in the heart in addition to products of the NK family of homeobox genes [483]. Serum response factor, a member of the MADS box family, mediates the transcription launched by various extracellular stimuli. Serum response factor is inhibited by a homeodomain-only protein HOP that is unable to bind to DNA, the synthesis of which is regulated by NKx2-5. Myocardin of the family of SAP domain-containing nuclear proteins activates promoters of cardiomyocyte genes via its association with serum response factor.

High-mobility group, nuclear protein HMGA2[12] regulates cardiomyocyte differentiation, as it associates with BMP-responsive transcription factors SMAD1 and SMAD4 to synergistically activate the promoter of the cardiac NKX2-5 gene [508].

Transcription factors nuclear factor of activated T-cells NFAT3 and NFAT4 are required for normal enzymatic activity of complex-2 and -4 of the respiratory chain and mitochondrial oxidative activity [509].

Transcription factor E2F4 activates proliferation of developing cardiomyocytes during cardiogenesis.[13] It operates as both a repressor and activator of cell proliferation. It accumulates in the nucleus at the end of the S phase of the cell division cycle, remains nuclear during mitosis, and disappears at the end of cytokinesis [510]. Proliferation of cardiomyocytes ceases during the first weeks

---

[12]The high-mobility group of non-histone chromatin proteins regulates the transcription of many genes by chromatin remodeling and proteic complex formation on promoter and enhancer regions. The HMG family of nuclear proteins includes 3 categories: HMGA, HMGB, and HMGN. Proteins of the HMGA category are ubiquitous and abound during embryogenesis. The HMGA category comprises 4 members: 3 splice variants (HMGA1a–HMGA1c) produced from the HMGA1 gene transcript and HMGA2 encoded by HMGA2 gene.

[13]The E2F family includes 9 members encoded by 8 genes (E2F1–E2F2, E2F3a–E2F3b, and E2F4–E2F8). Three members are activators (E2F1–E2F3a), 6 are suppressors (E2F3b–E2F8). In most quiescent cells, E2F4 is primarily nuclear. It localizes to the cytoplasm during entry into the S phase. Nuclear E2F4 is mainly bound to Retinoblastoma-like protein-2 (or P130) in G0 phase and Retinoblastoma protein and Retinoblastoma-like protein-1 (or P107) during the G1–S transition.

after birth. In cardiomyocytes, expression of E2F4 declines during embryogenesis. Cardiomyocyte hypertrophy then occurs. Reinduction of proliferation of postnatal cardiomyocytes relies on E2F4 production.

Transcriptional Control of Chamber Formation

Two Hairy enhancer of Split (HES)-related transcription repressors — HRT1 and HRT2 — are produced in atrial and ventricular progenitor cells [483]. They are controlled by the Notch signaling pathway.

Iroquois-related homeobox protein Irx4 labels ventricular progenitor cells in the cardiac crescent. Once chambers are formed, Irx4 is exclusively expressed in ventricles [483].

The transcription factors, heart and neural crest derivatives expressed proteins HAND1 and HAND2 (or bHLHa27 and bHLHa26, respectively) are also involved in ventricle formation. Both factors are coexpressed in the primary heart tube. Afterward, bHLHa26 and bHLHa27 accumulate in the right and left ventricle, respectively [483].

Myocyte enhancer factor-2C, which is phosphorylated by P38MAPKα and ERK4 (or ERK5) kinases, is needed for right ventricle construction.

On the other hand, the atrial myocardium is stamped by GATA4 and TBx5 factors.[14] Nuclear receptor NR2f1 is restricted to the inflow tract myocardium [483].

The allocation of cells to the atrial and ventricular lineages is also regulated by retinoic acid. Retinoic acid-synthesizing enzyme retinaldehyde dehydrogenase RAlDH2 is specifically synthesized in the posterior heart region, where retinoic acid participates in the formation of atria [483].

Specialized regions that demarcate the future working myocardium of atria and ventricles produce connexin-40 and -43, atrial natriuretic peptide, and X-linked small muscular protein[15] (SMPX) [483].

6.1.8.2 Tyrosine Hydroxylase

Tyrosine hydroxylase, ot tyrosine 3-monooxygenase, the rate-limiting enzyme in catecholamine synthesis, is strongly produced in the cardiac mesoderm. Tyrosine hydroxylase catalyzes the conversion of ^{L}tyrosine to ^{L}(3,4)-dihydroxyphenylalanine

[14] Shortly after early heart formation, GATA4 is downregulated in the anterior part of the heart field.

[15] A.k.a. stretch-responsive skeletal muscle protein and Chisel in mice.

(^{L}DOPA)[16] that generates dopamine, a catecholamine neurotransmitter and hormone, by aromatic amino acid decarboxylase.

Catecholamines present during cardiogenesis comprise ^{L}DOPA (predominantly) and dopamine. Molecule ^{L}DOPA enhances BMP2 production. It may be an important modulator of cardiac pacemaker activity, as it leads cardiac precursor cells to a sinoatrial fate [511].

Tyrosine hydroxylase regionalizes the production of atrial myosin heavy chain (aMHC1) and TBx5 factor to the posterior region of the heart tube during initial cardiomyocyte differentiation [511]. Retinoic acid synthesis is required for Tyr hydroxylase activity, hence expression of aMHC1 and TBx5 proteins.

6.1.8.3 Growth Factors

FGF

Fibroblast growth factor-8 (FGF8) that promotes cardiogenesis lodges in the anterior heart field and a subset of the primary heart field cells [512, 513]. Autocrine signaling launched by FGF8 is required in the anterior heart field to promote cell proliferation and survival.

Neuregulin-1

Signaling triggered by neuregulin-1, a member of the epidermal growth factor (EGF) family (Sect. 6.1.5), and its heterodimeric receptor protein Tyr kinase HER2–HER4 participates in cardiogenesis as well as proper structure and function of the adult heart.[17]

---

[16]Molecule ^{L}DOPA is a precursor to dopamine, noradrenaline, and adrenaline. Dopamine can be converted to noradrenaline by dopamine β-hydroxylase and adrenaline in the nervous system and adrenal glands.

[17]Neuregulin-1 carries out diverse functions in different cell types. It operates in the differentiation of neurons, thereby acting as a neuronal differentiation factor, accumulation of acetylcholine receptors in skeletal muscle, and proliferation of glial cells, hence serving as a glial growth factor. Neuregulin-1 is encoded by a single gene (NRG1). It has numerous isoforms produced by alternative splicing. Isoform-1 (heregulin), -2 (glial growth factor GGF2), and -3 (sensory and motor neuron-derived factor [SMDF]) are processed as transmembrane proteins that contain EGF-like domains. Heregulin and glial growth factor GGF2 that possess a heparin-binding domain (HBD) are cleaved by adamlysins into soluble forms that act as paracrine factors. Sensory and motor neuron-derived factor is also cleaved by adamlysins; its EGF motif-containing extracellular domain ($SMDF^{ECD}$) remains attached to the transmembrane domain through the cysteine-rich domain (CRD) to signal as a juxtacrine messengers. Neuregulin-1 acts via receptor protein Tyr kinases HER2 to HER4.

Cardiomyocytes in the adult myocardium produce HER2 and HER4 receptor protein Tyr kinases. Neuregulin-1 binds to HER4 with high affinity; HER4 dimerizes preferentially with HER2 receptor.[18] Receptor HER2 of the HER2–HER4 heterodimer acts as a non-autonomous amplifier. In the heart, neuregulin-1 primes [514]: (1) the Src–FAK, (2) SHCa–MAPK, and (3) PI3K–PKB pathway.

Neuregulin-1 contributes to cardiomyocyte differentiation from stem and progenitor cells, migration, and proliferation during cardiogenesis as well as during the postnatal myocardial development that involves the population of mononucleated cardiomyocytes and cardiac regeneration in damaged adult hearts [514].[19]

TGFβ

Transforming growth factor-βreceptor TβR3, a coreceptor for members of the TGFβ superfamily, resides in the heart fields [483]. It acts as a permissive factor in cardiogenesis. Coreceptor TGFBR3 is involved in Nodal signal transduction. Nodal signaling contributes to endoderm formation [483].

Member of the tumor-necrosis factor superfamily TNFSF12 regulates proliferation of several cell types, among which are cardiomyocytes. This type-2 transmembrane glycoprotein is processed into a soluble cytokine. In neonatal cardiomyocytes, TNFSF12 upregulates the expression of activators of cardiomyocyte proliferation, such as nuclear marker of cellular proliferation Ki67, mitotic marker phosphohistone-3 ($H3^P$), cyclin-D2, and Aurora-B kinase, as well as represses that of cyclin-dependent kinase inhibitor CKI1b [516]. Cytokine TNFSF12 targets TNFRSF12a receptor that signals via activated extracellular signal-regulated kinases and phosphatidylinositol 3-kinase as well as inhibition of glycogen synthase kinase-3β. In adult cardiomyocytes, induction of cell cycle re-entry requires TNFRSF12a re-expression.

6.1.8.4 Morphogens

The NKX2-5 promoter contains several SMAD binding sites. Bone morphogenetic protein BMP2 can elicit the expression of myocardial lineage markers. On the other hand, noggin that binds to and inhibits BMPs (as it precludes TGFβ family ligands

---

[18]Receptor HER4 also binds NRg2 to NRg4, HBEGF, and epiregulin. Receptor HER4 has multiple isoforms generated by alternative splicing [515]. Alternatively spliced variants can be categorized into juxtamembrane domain variant-A and -B ($HER4^{JMa}$ and $HER4^{JMb}$) and cytoplasmic tail variant-1 and -2. Variant $HER4^{Cyt1}$ is produced predominantly in the heart and breast. Variant $HER4^{Cyt2}$), synthesized mainly in the nervous system and kidney, lacks a binding site for ^{P85}PI3K subunit, thus being unable to stimulate PI3K, but remaining functional, as it binds neuregulin-1.

[19]Daughters of cardiomyocytes needs to have electromechanical connections to cooperate electrically and mechanically with the surrounding myocardium.

from binding to their receptors) prevents myocardial differentiation. In addition, several members of the BMP family with overlapping functions are expressed in the cardiac mesoderm, such as BMP5 and BMP7 [483].

Hedgehog

Both sonic and indian Hedgehog are expressed in the pharyngeal endoderm during cardiac crescent formation [483]. They operate as permissive rather than inductive factors via Smoothened receptor during cardiogenesis.

Wnt

Wnt morphogens operate as positive and negative regulators of the primary heart field. Subtype Wnt1 in the heart field blocks myocardial differentiation and induces blood formation. Canonical Wnt signals that operate via β-catenin, such as Wnt2a, Wnt2b, Wnt3, and Wnt8 regulate specification of cardiac cell types from the early mesoderm [517]. On the other hand, Wnt5a and Wnt11 trigger the non-canonical Wnt pathways during cardiogenesis. Subtype Wnt11 activates, via G protein, PKC, JNK, and CamK2 kinases.

An alternative branch of the Wnt pathway uses Jun N-terminal kinases for gastrulation movements. Diversin[20] interacts with 2 components of the canonical Wnt pathway, casein kinase-1ϵ, and axin. It suppresses Wnt signals mediated by the canonical β-catenin pathway and stimulates signaling via Jun N-terminal kinases [518]. Diversin controls fusion of heart precursors [519]. Diversin binds Disheveled to activate Rac and Rho GTPases and then Jun N-terminal kinase.

Morphogen Wnt11 controls early assembly of ventriculomyocytes, especially coordinated organization with intercellular contacts that rely on colocalized cell adhesion components, such as N-cadherin and β-catenin [517].[21]

6.1.8.5 Chromatin Remodelers

The large ATP-dependent chromatin-remodeling switching/sucrose non-fermenting (Swi/SNF)-like BAF (Brahma-related gene [BRG1]/Brahma [Brm]-associated factor) complex uses the energy of ATP hydrolysis to modify the chromatin structure made of DNA and histones, thereby activating or repressing gene expression.

---

[20] A.k.a. ankyrin repeat domain-containing protein-6 (AnkRD6).

[21] In mice, Wnt11 is synthesized in or in close proximity to the precardiac mesoderm, and later in the myocardium of the primitive heart tube, thereby overlapping the first and secondary heart fields. At later stages, Wnt11 as well as Wnt5a are expressed in the outflow tract.

The chromatin-remodeling Swi/SNF-like BAF complex that is involved in cardiogenesis is characterized by a combinatorial assembly of subunits that produces distinct functional complexes.

In particular, the cardiogenic BAF complex comprise Swi/SNF-related, matrix-associated, actin-dependent regulators of chromatin SMARCd1 to SMARCd3 subunits.[22] Cardiac-enriched BAF complex subunit SMARCd3 cooperates with GATA4 to provoke NKx2-5 production. Moreover, SMARCd3 and GATA4 cooperate with TBx5 cardiogenic transcription factor to activate genes that encode regulators of cardiogenesis, such as MEF2, NKx2-5, HAND, and Isl1 [522]. Factors SMARCd3, GATA4, and TBx5 promote differentiation into contracting cardiomyocytes and repress non-cardiac mesodermal genes.

Chromatin-remodeler and transcriptional activator SMARCa4[23] participates in the regulation of gene expression during cardiac growth and differentiation. The SMARCa4-containing BAF complex comprises 12 subunits. In embryos, SMARCa4 promotes myocyte proliferation, as it maintains expression of BMP10 and suppresses that of cyclin-dependent kinase inhibitor CKI1c [523]. It interacts with histone deacetylase and poly^{ADP}ribose polymerase to repress αMHC (MyH6) and activate βMHC (MyH7) isoform. The chromatin-modifying HDAC–PARP–BAF complex regulates cardiac hypertrophy. It assembles on MHC promoters, where it can interact with transcription factors such as thyroid hormone receptors TRα1 and TRβ1 (a.k.a. nuclear receptors NR1a1 and NR1a2), transcriptional enhancer TEA domain-containing family member TEAD1,[24] myocyte enhancer factor MEF2, serum response factor (SRF), GATA4, and NFAT to control MHC expression [523]. In adults, SMARCa4 is silenced in cardiomyocytes. It is reactivated by cardiac stresses and forms a complex with its embryonic partners HDAC and PARP1 to induce a pathological MyH6-to-MyH7 shift.

Developing myocardial cells respond to signals from the endocardial layer to form a network of trabeculae that characterize the ventricles of vertebrate hearts. Trabeculation requires the chromatin remodeler SMARCa4 to repress ADAM metallopeptidase with thrombospondin-1 motif ADAMTS1 produced in and secreted by the endocardium that overlies developing trabeculae to achieve an extracellular environment in the cardiac jelly that supports trabecular growth [524]. Later, ADAMTS1 degrades the cardiac jelly and prevents excessive trabeculation.

---

[22] A.k.a. 60-kDa Brm-associated factors (BAF60a–BAF60c) encoded by 3 genes (Baf60A–Baf60C). Both SMARCd3a (BAF60C1) and SMARCd3b (BAF60C2) isoforms are widely expressed [520]. These isoforms bind to several nuclear receptors and transcription factors. A variant of the BAF complex in embryonic stem cells (esBAF) interacts with octamer-binding transcription factor Oct4 and sex-determining region Y (SRY)-box transcription factor Sox2 [521]. Specialized chromatin-remodeling complexes exist in neural progenitors (npBAF) and neurons (nBAF), in addition to cardiac BAF complexes (cBAF) [521].

[23] A.k.a. ATP-dependent helicase brahma homolog Brg1.

[24] A.k.a. transcriptional enhancer factor TEF1.

6.1.8.6 Matrix Proteoglycans

Perlecan is a heparan sulfate proteoglycan in basement membranes that binds to integrins and dystroglycan, interacts with extracellular matrix components, and recruits growth factors and modulates their activity. Perlecan stabilizes the heart wall during cardiogenesis as well as in adults after injury. Basement membranes that lack perlecan deteriorate and lead to cell detachment in the ventricle and outflow tract and blood leakage into the pericardial cavity [525].

At the onset of cardiogenesis, perlecan serves as an adhesive substrate for cardiomyocytes with laminin and collagen-4 and maintains mechanically stressed basement membrane surrounding cardiomyocytes to ensure mechanical stability. At later stages, heart stability is achieved by formation and maturation of intercellular contacts such as intercalated discs composed of desmosomes and adherens junctions that are composed of γ-catenin[25] and cadherin, respectively.

6.1.8.7 MicroRNAs

Cardiogenesis depends on the spatiotemporal expression of microRNAs. MicroRNAs control the expression of messenger RNAs that are translated into regulators of cardiogenesis, propagation of electrochemical waves, and myocardium hypertrophy and contractility (Table 6.4).

MicroRNAs commonly bind to mRNAs and inhibit mRNA translation with or without mRNA cleavage. The most abundant microRNAs in the myocardium are miR1, miR133, miR126-3p, miR30c, miR26a, and the (lethal) Let7 class [526]. MicroRNAs MiR1, miR133, and miR208 are specific to striated myocytes. The microRNA-1 set comprises 2 transcripts miR1-1 and miR1-2 that are encoded by distinct genes. The miR133 set includes miR133a-1, miR133a-2, and miR133b. MicroRNA-208 is encoded by an intron in the MYH6 (αMHC) gene.

Muscular microRNAs, such as miR1 and miR133, are mainly controlled by myogenic transcription factors. MicroRNA-1 targets histone deacetylase HDAC4, a transcriptional repressor of muscle gene expression. It also controls cardiogenesis in mice, as it regulates cardiac transcription factor bHLHa26 [531]. MicroRNA-133 enhances myoblast proliferation by repressing serum response factor that acts in sarcomere organization.

Serum response factor is controlled by other regulators, such as GATA4, NKx2-5, and myocardin, as well as microRNAs, such as miR1-1, miR1-2, miR21, miR133, miR206, miR214 [531]. Conversely, serum response factor regulates microRNAs such as miR1-1, miR1-2, miR133, and miR129 in cardiomyocytes.[26]

---

[25] A.k.a. plakoglobin.

[26] Regulator of muscular development MEF2, in cooperation with MyoD, activates transcription of miR1-2 and miR133a-1 in skeletal myocytes.

Table 6.4. MicroRNAs and heart structure and activity in mice (Sources: [526–530]). MicroRNAs regulate cardiogenesis and growth, conduction of the electrochemical signal, and myocardium contraction. They include muscle-specific microRNAs (miR1, miR133, and miR208) and other microRNAs (e.g., miR18b, miR21, and miR195). Cardiac miR1 targets a gap junction protein and cardiac transcription factor Irx5 that regulates the nodal tissue. Overexpression of miR1 is associated with cardiac arrhythmia. Loss in miR1-2 leads to ventricular septum defects (septal holes), cardiomyocyte hyperplasia, and abnormal conduction in the ventricular nodal tissue. MicroRNA-133 prevents cardiac hypertrophy in response to mechanical and chemical stimuli. Heart-specific miR208 is transcribed from a sequence of the intron of the gene of the heavy chain of α-myosin. MicroRNA-208 is required for the expression of β-myosin heavy chain during stress or hypothyroidism. Hyperpolarization-activated and cyclic nucleotide-gated channels (HCN) are responsible for depolarizing pacemaker "funny" current i_f.

Type	Cardiac effect
MicroRNA-1	Growth, conduction (GJA1 [Cx43] and KCNJ2 [K_{IR}2.1]), automaticity (HCN2 and HCN4 [i_f])
MicroRNA-1-2	Morphogenesis, CMC number, conduction
MicroRNA-133	Growth, repolarization (KCNQ1 [K_V7.1] and KCNE1 [minK; $i_{K,s}$]) (KCNH2 [K_V11.1; $i_{K,r}$]), automaticity (HCN2 [i_f])
MicroRNA-208	Growth

MicroRNA-138 is required to establish distinct identity of cardiac structures and appropriate chamber-specific gene expression patterns during cardiogenesis [532]. MicroRNA-138 represses transcripts specific to atrioventricular canal in the developing ventricle.

Action of miR1 is dose sensitive [528]. MicroRNA-1-2 regulates not only cardiogenesis, but also the cardiomyocyte cycle. In mouse embryos, homozygous deletion of muscle-specific miR1-2 induces ventricular septal defects, possibly due to an increased level of bHLHa26 transcription factor or HAND2 regulator). Surviving mice exhibit cardiac arrhythmias caused by defects in potassium channels resulting from increased abundance of the transcription factor Irx5, a repressor of the KCND2 gene that encodes K_V4.2 channel. Moreover, they have an increased rate of cardiomyocyte mitosis.

Muscle-specific miR1 and miR133 hinder myocardium growth. MicroRNA-133 targets transcripts of RhoA and CDC42 GTPases that regulate cytoskeletal dynamics during hypertrophy, as well as nuclear factor Negative elongation factor-A (NElFa)[27] that is involved in cardiogenesis [529].

Cardiac contractility depends on the expression of 2 genes that encode myosin heavy chain. Isoforms α and βMHC are contractile proteins of cardiomyocytes that are regulated in an antithetical manner by various signals. The slower adenosine

[27] A.k.a. Wolf-Hirschhorn syndrome candidate WHSC2.

triphosphatase βMHC is the dominant isoform expressed in embryonic cardiomyocytes, whereas faster ATPase αMHC is upregulated in postnatal cardiomyocytes.[28]

Thyroid hormone T_3 stimulates αMHC and inhibits βMHC transcription after birth. Cardiac-specific miR208, encoded by an intron of the αMhc gene, controls the expression of MYH7 gene that is elevated by physical stress and hypothyroidism [530]. MicroRNA-208 targets thyroid hormone receptor-associated protein THRAP1[29] a regulator of the thyroid hormone receptor that hampers βMHC expression.

During cardiac hypertrophy, miR23a, miR23b, miR24, miR195, miR199a, and miR214 are upregulated, whereas miR150 and miR181b are downregulated [531]. In failing heart,[30] miR21, miR29b, miR129, miR210, miR211, miR212, and miR423 are overexpressed, whereas miR30, miR182, and miR526 are underexpressed. The miR1 level increases during cardiac infarctions specifically in ischemic regions [527]. Overexpression of miR1 increases the occurrence of postinfarct arrhythmias and promotes arrhythmias in healthy hearts. Overexpression of miR1 slows conduction and depolarizes the plasma membrane, as it represses transcript of the KCNJ2 gene that encodes the $K_{IR}2.1$ subunit of K^+ channel, which sets the resting membrane potential, and of GJA1, which encodes connexin-43 of gap junctions.

Many microRNAs that are mainly produced by cardiac non-myocytes participate in wall remodeling. In a mouse model of cardiac failure, miR21 is upregulated and decreases the percentage of apoptotic fibroblasts, thereby favoring cardiac fibrosis. MicroRNA-21 has an anti-apoptotic effect that results from activation of extracellular signal-regulated kinase following repression of an inhibitor of the Ras–ERK pathway, sprouty homolog Spry1 [533]. Expression of Spry1 is precluded in failing human myocardium. Subsequently, enhanced activity of the ERK pathway promotes fibroblast survival and growth factor secretion that can cause interstitial fibrosis and cardiomyocyte hypertrophy due to growth factors released from fibroblasts. In addition, miR21 is upregulated in cardiac fibroblasts after myocardial ischemia–reperfusion injury [534]. MicroRNA-21 represses PTen phosphatase that precludes PI3K–PKB signaling and MMP2 expression in cardiac fibroblasts.

[28] The βMHC level rises during cardiac diseases when αMHC is downregulated.

[29] A.k.a. mediator complex subunit MED13, mediator of RNA polymerase-2 transcription subunit-13, 250-kDa component of activator-recruited cofactor ARC250, 250-kDa vitamin-D3 receptor-interacting protein complex component DRIP250, and 240-kDa thyroid hormone receptor-associated protein complex TRAP240.

[30] Heart failure is a defect in proper blood filling and/or ejection. This progressive disorder is initiated by myocardial injury, most commonly caused by coronary artery disease, hypertension, and genetic factors.

6.1.8.8 RNA-Binding Proteins

Cytoplasmic and nuclear RNA-binding proteins (RBP) bind to either double- (dsRNA) or single-stranded (ssRNA) RNAs using RNA recognition motif (RRM),[31] type-1 and -2 K homology (KH) domains, zinc finger region, RGG (Arg–Gly–Gly) box, DEAD/DEAH box (associated with RNA helicase activity), Pumilio/FBF (PUF) domain, double-stranded RNA-binding motif (DSRBD), Piwi/Argonaute/Zwille (PAZ) sequence (associated with binding to short single-stranded RNAs such as microRNAs),[32] Sm region (associated with small nuclear snRNA binding), cold-shock domain (CSD; involved in transcriptional and post-transcriptional regulation), among others [535, 536]. RNA-binding proteins can link to RNAs via RNA bases, ribose, and phosphate groups.

Many RBPs contain 1 or more copies of a given RNA-binding domain and auxiliary domains, whereas others have 2 or more different domains. Interdomain linkers facilitate the recognition of substrates [535]. RNA-binding modules can be combined with catalytic domains to regulate the activity of the enzyme. Many RBPs have modular structures, various modules defining the structural unit; they are composed of multiple repeats of a small number of basic domains arranged in different manners that create both RNA-binding and functional diversity [535]. Combination of a given RNA-binding domain with distinct auxiliary functional motifs adds further diversity. In addition, RBPs undergo post-translational modifications, such as phosphorylation, methylation, and sumoylation, which control RNA binding as well as their function and localization [536]. Moreover, RNA-binding proteins can dimerize [535].

Ribonucleoprotein complexes (RNP) result from association of RNAs with RNA-binding proteins that have different RNA-sequence specificities and affinities. In vertebrates, cells encode hundreds to thousands of RBPs, each with unique RNA-binding activity and between-protein interaction characteristics. A relatively small number of RNA-binding scaffolds can be involved in RNA processing.

RNA-binding proteins influence the structure and interactions of RNAs. RNA-binding proteins can be classified according to target RNA types, such as messenger RNA (mRNA), ribosomal RNA (rRNA), or transfer RNA (tRNA).

Heterogeneous ribonucleoproteins (hnRNP), or pre-mRNPs, and messenger ribonucleoproteins (mRNP) are connected to pre-mRNAs (originally called hnRNAs) and mRNAs, respectively. Several hnRNPs shuttle between the nucleus and cytoplasm, whereas others localize only in the nucleus. They operate in transcription, stability, function, packaging (i.e., preparing pre-mRNAs for post-transcriptional processes), transport, localization, processing, and turnover, hence controlling many post-transcriptional steps in gene expression. In particular, they regulate pre-mRNA

[31] A.k.a. RNA-binding domain (RBD) and RNP domain.

[32] For example, induced pluripotent stem cell factor Lin28 is a repressor of microRNA processing and a post-transcriptional regulator of a mRNA set.

splicing, editing,[33] and polyadenylation.[34] Heterogeneous ribonucleoproteins bind to different sets of nascent transcripts (pre-mRNAs) at distinct times during RNA processing.[35]

Therefore, as RBPs are involved in the transcriptional and post-transcriptional control of RNAs, they regulate gene expression patterns during the body's development and remodeling from embryogenesis to adulthood. RNA-binding proteins also participate in the maintenance of the stem cell state and cell-fate specification.

The signal transduction and activation of RNA (STAR) set of RBPs can be subdivided into 3 categories: SAM68, SF1, and Quaking-related (QR) subsets. Members of the STAR set function in pre-mRNA splicing, export, stability, and translation. Quaking proteins recognizes and binds to a specific sequence in the 3′UTR of target mRNAs. They control muscle development via the Hedgehog pathway [538].

Heterogeneous ribonucleoprotein Embryonic lethal, abnormal vision protein-like molecule ELAV1, or human antigen-R (HuR), selects mRNAs to be exported to the cytosol for translation, hence contributing to the regulation of myoblast differentiation [539].[36] The RNA-binding protein Hermes (portmanteau for heart, RNA recognition motif (RRM)-expressed sequence) connects to mature RNAs and represses cardiogenesis [540].

The assembly of sarcomeres requires proper amounts of multiple types of proteins at a given time in a coordinated manner. Contractile proteins assemble into actin-containing thin filaments and myosin-containing thick filaments. Non-contractile proteins anchor these filaments to Z discs at sarcomere ends.

---

[33]RNA editing is a type of RNA modification that results from the deamination of adenosine to inosine catalyzed by the ADAR proteins. Target transcripts localize mainly in the nervous system, where they generate ion channels and G-protein-coupled receptors, such as glutamate and serotonin receptors [536].

[34]Polyadenylation of mRNAs using polynucleotide adenylate transferase (a.k.a. adenosine triphosphate (ATP):ribonucleic acid adenylate transferase, RNA adenylase, polyadenylate synthase, poly(A) polymerase, and poly(A) hydrolase) that has 2 substrates, ATP and RNA, and 2 products, diphosphate and RNA with an extra adenosine nucleotide at its 3′end. Polyadenylation of mRNAs strongly influences their nuclear transport, translation efficiency, and stability. Cleavage and polyadenylation-specific factor (CPSF) is a ribonucleoproteic complex (with CPSF1–CPSF4, CPSF2L–CPSF4L, CPSF6–CPSF7, and nudix, or CPSF25 possible subunits) required in polyadenylation [536]. In cooperation with nuclear polyadenylate-binding protein (PABPn1), it activates polyadenylate polymerase.

[35]RNA-binding protein hnRNPa1 is involved in pre-mRNA splicing and nuclear export; hnRNPa2b1 in splicing and mRNA transfer; hnRNPc in pre-mRNA packaging, splicing, stability, and nuclear retention; hnRNPd in mRNA stability; hnRNPf in splicing; hnRNPh in splicing and polyadenylation; hnRNPk in transcription, pre-mRNA splicing, translation, and regulation, and mRNA stability; hnRNPl in mRNA export and stability; and hnRNPu in nuclear retention [537].

[36]Cleavage of ELAV1 generates 2 fragments: ELAV1 cleavage products ELAV1CP1 and ELAV1CP2. The former tethers to transportin-2, thereby allowing non-cleaved ELAV1 to lodge in the cytoplasm and promoting myogenesis.

The Z disc is a nodal point for signaling, in addition to its structural role in the maintenance of sarcomeres. The thin filament is also composed of tropomyosin and troponin.

Alternative splicing is involved in cardiac adaptation. Isoform switch of sarcomeric titin participate in adjustment to ventricular filling. RNA-binding motif-containing protein RBM20 regulates titin splicing [541]. Abnormal mRNA splicing that results from mutations in the Rbm20 gene produces pathological titin isoform and heart failure.

The RNA-binding protein, RBM24,[37] contributes to the regulation of cardiac gene expression and sarcomeric assembly, hence cardiac contractility [542]. It controls the expression of several genes that encode constituents and regulators of the Z disc. Tissue-specific protein RBM24 is upregulated during the differentiation of human embryonic stem cells into cardiomyocytes. Agent RBM24 also intervenes in skeletal muscle differentiation via the myogenic regulatory factor, myogenin. Whereas the relatively small set of myogenic regulatory factors, such as MyoD and myogenin, governs skeletal muscle differentiation, the myocarde development is controlled by numerous transcription factors. In addition, RBM24 is also involved in vasculo- and angiogenesis.

6.1.9 Cardiac Progenitor Cells and Precursors

Cardiomyocytes come from the primary and secondary heart fields in the embryo [543]. Multipotent progenitor cells generate the 4 major cell types of the heart: cardiomyocytes as well as nodal, smooth muscle, and endothelial cells.

Cardiac progenitors express the transcription factors NKx2-5 and insulin gene enhancer LIM homeobox protein Islet-1 (Isl1). Factor NKx2.5 is activated in multipotent cardiac progenitor cells of early embryos. NKx2.5+ cells and their progeny populate the precardiac mesoderm located dorsal to the cardiac region and developing heart tube. Factor Isl1 is expressed transiently in multipotent cardiac progenitor cells before their differentiation. Very-low-frequency electromagnetic field raises metabolic activity in cardiac stem cells in vitro [544].

Distinct cardiac lineages, such as first and Isl1-expressing second lineage, exist according to timing of entry into the heart and differentiation. Second heart field progenitors generate cardiomyocytes that reside in the outflow tract, ventricles, and atria. Proepicardium and epicardium are sources of coronary progenitors and cardiac fibroblasts.[38]

Both embryonic and postnatal cardiovascular progenitor cells characterized by the expression of the transcription factor Islet-1 can differentiate into cardiac smooth

[37] RMB24 has 2 known variants (Rbm24a and Rbm24b).

[38] Epicardial cells migrate from the proepicardium, an outgrowth of the septum transversum, and spread over the heart surface.

muscle or endothelial cells. The adult heart contains a pool of resident SCFR+ pluripotent cardiovascular progenitor cells for its regeneration [556].[39]

Tripotent progenitor cells from the secondary heart fields express transcription factors Islet-1 and NKx2-5 as well as vascular endothelial growth factor receptor VEGFR2 [545]. Certain cardiac tripotent progenitor cells express the marker brachyury at a specific stage of embryogenesis [546]. SCFR+, NKx2-5+, differentiated embryonic stem cells that serve as cardiac progenitors create cardiac and smooth myocytes, but not endothelial cells [547]. Factor Isl1 is required for differentiation of multipotent cardiac progenitor cells into cardiac and smooth myocytes, but not endothelial cells [548].

Human pluripotent stem cells, such as embryonic and induced pluripotent stem cells generate cardiomyocytes. Cardiomyocyte specification requires Wnt signaling [549]. However, cardiomyocyte differentiation is sensitive to the timing and dose of supplied Wnt.

Vascular endothelial growth factor is one of the most important angiogenic growth factors. It has pleiotropic activities, as it influences non-endothelial cell types derived from all 3 embryonic lineages, such as pneumocytes, renal podocytes, cardiomyocytes, various neuron types, hematopoietic and cardiac stem cells, among others. The Wnt pathway triggers production of various developmental messengers, such as Nodal, BMP2 and BMP4, Noggin, Wnt3a, an Wnt8a, as well as transcription factors involved in cardiomyocyte differentiation, such as NKx2-5, TBx5, and MEF2c factors, as well as insulin gene enhancer protein Isl LIM domain-containing protein IsL1, a marker for cardiac progenitors of the secondary heart field, and Mix paired-like MixL1 homeodomain-containing protein.

Cardiac implantation of multipotent mesenchymal stem cells that express VEGF stimulates revascularization, thereby reducing infarct size. These stem cells secrete numerous paracrine factors, such as VEGF, FGF2, IGF1, HGF, and CXCL12 chemokine. They can differentiate into cardiomyocytes and endothelial cells. Their liberated paracrine factors support survival of implanted mesenchymal stem cells as well as mobilization and migration of stem cells. VEGF+ mesenchymal stem cells are characterized by PKB activation that enables mesenchymal stem cell engraftment in the ischemic heart and cardiac stem cell migration [550]. Moreover, VEGF+ mesenchymal stem cells recruit resident, adult, cardiac stem cells to ischemic regions. Factor VEGF fosters cell migration via VEGFR1 and VEGFR3, but not VEGFR2, although cardiac stem cells synthesize the 3 receptor types. In addition, VEGF recruits CXCR4+ cardiac stem cells via CXCL12 that targets its CXCR4 receptor [550].

[39] Among markers that identify cardiac stem cells (VEGFR2, also called KDR and Flk1, and transcription factors Islet-1 and NKx2-5), stem cell factor receptor (SCFR or KIT) is also expressed during various stages of cell lineage commitment in germ, mast, stellate, epithelial, endothelial, and smooth muscle cells [557].

6.1.9.1 Resident Coronary Vascular Progenitors

VEGFR2+, SCFR+, resident coronary progenitor cells reside in vascular niches inside the human myocardium [551]. *Vascular progenitor cells* are located in the proepicardium, from whence they migrate into the myocardium. They are connected by gap junctions to vascular endothelial and smooth muscle cells as well as fibroblasts. These self-renewing cells differentiate predominantly into endothelial and smooth muscle cells and partly into cardiomyocytes.

On the other hand, *myocyte progenitor cells* distributed in the cardiac crescent and pharyngeal mesoderm differentiate into cardiomyocytes to progressively constitute, together with coronary vessels, the 4-chambered heart [551].

6.1.9.2 Notch and Wnt Morphogens

Niche-resident cardiac progenitors and supporting cells express Notch-1 receptors and Notch ligand Jagged-1, respectively. The intracellular domain of Notch-1 ($Notch^{ICD}$) translocates to the nucleus and activates the NKX2-5 gene [552]. $Notch^{ICD}$ connects to RBPJκ that binds to NKX2-5 promoter, hence initiating transcription and cardiomyocyte differentiation, whereas Jagged-1 activation of the Notch1 pathway downregulates transcription factors of vascular cells.

During early stage of development, when the transcription factor NKx2-5 is not expressed, the Notch pathway mediates a coordinated regulation between bone morphogenetic proteins, such as BMP2, BMP6, and BMP7, that are required for cardiogenesis, and Wnt pathways.[40] This action can generate cardiomyocytes from VEGFR2+ hemangioblasts rather than cells of the hematopoietic, vascular endothelial, or smooth muscle lineages [553].

Wnt signaling regulates NKx2-5+, Isl1+ multipotent cardiac progenitor cells. The Wnt–β-catenin signaling promotes expansion of multipotent cardiac progenitor cells. Notch-1 impedes accumulation of phosphorylated β-catenins in these progenitors [548]. Notch-1 and β-catenin regulate positively and negatively, respectively, the expression of cardiac transcription factors Isl1, myocardin, and SET and MYND domain-containing protein SMYD1.

During later stages of cardiogenesis, Notch-4 inhibits the maturation of the cardiac lineage. Active Notch-1 receptor in embryonic stem cells reduces their cardiac potential at late differentiation stage. Notch-4 receptor, the expression of which is restricted to the heart endothelium, promotes cardiac development from cardiac mesoderm and is able to respecify VEGFR2+ hemangioblasts to a cardiac cell fate by activating BMP signaling and inhibiting the Wnt–β-catenin pathway.

Certain cardiomyocyte progenitors derive from proepicardial cells that express T-box TBx18 transcription factor. These progenitors migrate onto the outer cardiac

[40]Protein Wnt2 induces cardiac differentiation from mesoderm, whereas Wnt4 acts via the β-catenin-independent pathway that inhibits the β-catenin-dependent pathway for cardiac specification of mesoderm.

surface to form the epicardium, and then contribute to cardiomyocyte generation in the ventricular septum and atrial and ventricular walls [554]. However, adult TBx18 expressing epicardial cells do not yield cardiomyocytes. TBx18+ cardiac progenitors also create to cardiac fibroblasts and coronary smooth myocytes. First cells from the proepicardium–epicardium to enter the heart give rise to cardiomyocytes, whereas subsequent TBx18+ epicardial lineages generate vascular support cells and fibroblasts.

Growth factors of cardiac myogenesis include members of the bone morphogenetic protein and Wnt families. Both canonical and non-canonical Wnt pathways can promote cardiomyocyte formation. Transcription factor Sox17 acts on transcription factors of cardiac specification in primitive mesoderm, mesoderm posterior homologs MesP1 and MesP2 [560]. Transcription factor MesP1 is specifically synthesized in almost all cardiovascular precursors and considered as the earliest marker of the cardiovascular lineage. Subtype MesP1 targets Dkk1 that impedes Wnt signaling during cardiovasculogenesis [561].

Wnt signaling can act as an activator of cardiogenesis in a cell-autonomous manner (i.e., within the mesoderm) after initial cardiac commitment during discrete periods of development, although it can also inhibit early cardiac engagement in a non-cell-autonomous fashion, likely by acting on adjacent endoderm. Canonical Wnt–β-catenin pathway is required for development and differentiation of cardiac progenitors of the second heart field [562]. Wnt signaling upregulates cyclin-D2.

In addition, the Wnt–β-catenin signaling exhibits developmental stage-specific effects not only on cardiomyogenesis, but also on hematopoiesis and vasculogenesis [563]. In the early phase during embryoid body formation, it enhances differentiation of embryonic stem cells into cardiomyocytes and suppresses differentiation into hematopoietic and vascular cell lineages. In the late phase after embryoid body formation, it prevents bone morphogenetic protein signaling, thus inhibiting differentiation into cardiomyocytes and enhancing differentiation into hematopoietic and vascular cell lineages.

Coronary progenitor cells of the proepicardium at the septum transversum near the venous pole of the heart are responsible for coronary arteriogenesis (but not formation of the venous system and microvasculature) [564]. β-Catenins do not influence the formation of the proepicardium, migration of proepicardial cells to the heart, and formation of the primitive epicardium, but participate in the expansion of the subepicardial space and differentiation of epicardium-derived mesenchymal cells into coronary smooth myocytes.

6.1.9.3 Transcription Factor WT1

Another cardiomyocyte precursor of the epicardium expresses transcription factor Wilms tumor protein WT1 [555].[41] WT1+ proepicardial cells arise from progenitors that express GATA4, Isl1, and NKx2-5. WT1-derived cardiomyocytes are

[41] Factor WT1 is expressed in proepicardium and epicardium, but not in myocardium.

located in the myocardium of all cardiac chambers and interventricular septum (7–10% and 18% of ventricular and atrial cardiomyocytes, respectively). They are coupled to non-WT1-derived cardiomyocytes.

6.1.9.4 ZFPM2 (FOG2) and Thymosin-β4

Regulators of coronary vasculogenesis are also involved in repair of ischemic heart damage resulting from vascular insufficiency. Coronary vasculogenesis from epicardium involves ZFPM2, or Friend of GATA2 (FOG2) and VEGF [558].[42]

Thymosin-β4, a ^{G}actin monomer-binding protein, promotes vascularization from the embryonic epicardium. Thymosin-β4 stimulates TIE2+ progenitor cells derived from quiescent adult epicardial cells and triggers their migration and differentiation into fibroblasts and smooth myocytes when stimulated by PDGF and TGFβ, and endothelial cells, in response to a combination of VEGF and FGF2 [559]. Thymosin-β4 product of peptidase cleavage, the pro-angiogenic ^{N}acetyl seryl-aspartyl-lysyl-proline (AcSDKP), stimulates differentiation from adult epicardial cells into endothelial cells.

6.1.9.5 Cysteine-Rich LIM-Only Proteins

Smooth muscle α-actin exists in cardiomyocyte progenitors. During cardiomyocyte differentiation, α-actin gene activity is downregulated as soon as rhythmic contractility starts. LIM-only proteins of the CRP family are also called cysteine and glycine-rich proteins. They support the organization of proteic complexes both in the cytoplasm and in the nucleus.

Cysteine-rich LIM-only proteins CRP1 and CRP2 are expressed during the development of the cardiovascular apparatus. They bridge GATA transcription factor to serum response factor. The SRF–CRP–GATA complex strongly activates smooth muscle genes. Together with SRF and GATA proteins, CRP1 and CRP2 facilitate smooth muscle differentiation. They are able to convert pluripotent fibroblasts into smooth myocytes. On the other hand, muscle (cardiac) LIM protein CRP3 impedes this conversion.

The level of CRP2 protein is correlated with smooth muscle gene activity [565]. Agent CRP2 is a transcription coadaptor that recruits serum response factor and remodels silent chromatin in adult cardiomyocytes toward smooth muscle gene activity.

---

[42] FOG2(−/−) hearts are characterized by a thin ventricular myocardium, common atrioventricular canal, tetralogy of Fallot, and absence of coronary vasculature. Transgenic re-expression of FOG2 in cardiomyocytes leads to coronary vessel development.

6.2 Wall Structure

The heart wall is composed of 3 layers: the internal thin *endocardium*, the thick muscular *myocardium*, and the external thin *epicardium*. Cardiac muscle is an involuntary striated muscle. The heart is enclosed in a double-wall sac, the pericardium.

The double-layered *pericardium* is constituted of the outer parietal pericardium and the inner visceral pericardium, or epicardium, which is thus attached to the myocardium. The pericardial cavity separates the 2 pericardium layers. It contains a lubrificating fluid ($V \sim 25$–35 ml). The parietal pericardium consists of an outer layer of thick, fibrous connective tissue and an inner serous layer with a *mesothelium* and connective tissue. The anchoring parietal pericardium is attached to the diaphragm and fuses with the outer wall of blood vessels entering and leaving the heart. The visceral pericardium has an external layer of flat mesothelial cells lying on a support tissue that contains elastic fibers. The coronary arteries reside in the epicardium and send branches into the myocardium.

Heart cellular population includes cardiomyocytes (at least two-thirds of all cardiac cells), nodal cells, endothelial and vascular smooth muscle cells, and fibroblasts. These various cell types are embedded in the collagen-rich extracellular matrix.

Collagen forms diverse connected structures, such as cables (collagen fibers), struts (larger crosslinked collagen bundles), nets (porous sheets of collagens), and gels (filamentous collagens associated with glycosaminoglycans). These structures bind cardiomyocytes via integrins (Vol. 1 – Chap. 7. Plasma Membrane) that interact with the cytoskeleton. In fact many components of the extracellular matrix bind to plasmalemmal integrins (collagen-1 to $\alpha_3\beta_1$-, fibronectin to $\alpha_3\beta_1$- and $\alpha_5\beta_1$-, and laminin to $\alpha_1\beta_1$-, $\alpha_3\beta_1$-, and $\alpha_7\beta_1$-integrins).

Integrins are constituents of adhesomes and signalosomes, particularly in mechanotransduction and electromechanical coupling, as integrins colocalize with gap junction proteins. Moreover, connections between integrins and the cytoskeleton regulate the functioning of mechanosensitive ion channels. The cytoskeleton can regulate the activity of channels implicated in cardiomyocyte depolarization and subsequent repolarization [566].

Collagenous connections between adjacent cardiomyocytes register sarcomere landmarks to ensure synchronous contraction and relaxation. Furthermore, the collagen network maintains gap junctions between contiguous cardiomyocytes as well as between cardiomyocytes and nodal cells to favor the electrical connectivity.

The collagen turnover is regulated by: (1) stimulators, such as angiotensin-2,[43] endothelin-1, catecholamines, aldosterone, fibroblast growth factor FGF2, insulin-like growth factor, among others, as well as (2) inhibitors, such as prostaglandins, nitric oxide, natriuretic peptides, etc. [567].

---

[43] Angiotensin-2 favors the production of collagen-1 and -3.

The myocardium contains mainly cardiomyocytes. In both ventricles, papillary muscles and chordae tendinae tether the valve. The extracellular matrix couples the different cardiac cell types and conveys chemical, mechanical, and physical signals that regulate cell functioning.

The endocardium is composed of 3 layers. The outer layer is made of collagen fibers, which merge with collagen surrounding adjacent cardiomyocytes. This layer contains the *Purkinje fibers* (cardiac Purkinje cells) of the nodal tissue. The thickest middle endocardial layer contains more regularly arranged collagen fibers and elastic fibers arranged in parallel. The inner layer is constituted of endothelial cells (EC), which are continuous with the vessel endothelium. The endocardium is thicker in the atria than in the ventricles.

The heart has a *fibrous skeleton* with its central fibrous body, which prevents early propagation of action potential. The central fibrous body provides extensions: (1) the valve rings, to which are inserted the cardiac valves and (2) the membranous interventricular septum,[44] that can bulge into the right ventricle to form an aneurysm.

6.2.1 Cardiofibroblasts

Activated fibroblast, a mature cell, synthesizes constituents of the extracellular matrix, such as collagen, elastic, and reticular fibers, as well as glycosaminoglycans and glycoproteins. Fibrocyte, a less active cell type, is involved in matrix maintenance.

Cardiac fibroblasts reside in the cardiac wall, where they align in the direction of myofibers. In the adult human heart, fibroblasts represent about 60% of total cell population, whereas myocytes occupy about 75% of the myocardium volume [568]. They not only synthesize extracellular matrix constituents, but also respond to mechanical and electrical stimuli, and secrete hormones and growth factors involved in cardiac homeostasis.

6.2.1.1 Heterocellular Connections

Heterocellular gap junctions may couple cardiomyocytes and nodal myocytes (especially sinoatrial nodal myocytes) to fibroblasts [568]. Connexin-43 and -45 are produced by both cardiac myocytes and fibroblasts.

In addition, thin, large conductance membrane nanotubes (caliber 50–200 nm), which are ephemeral connections, enable long-distance intercellular interactions between cardiac myocytes and fibroblasts [569]. Membrane nanotubes contain actin

[44]The pars membranacea septi correspond to the upper part of the interventricular septum.

filaments and microtubules that can be used for selective molecular transfer. They indeed allow mitochondria exchange and calcium signal propagation between the 2 cell types.

6.2.1.2 Myofibroblast–Myocyte Coupling

Cardiofibroblasts contribute to the regulation of the myocardial function as well as myocardial remodeling in maladaptive hypertrophy of cardiomyocytes during hypertension and after myocardial infarction, among other heart diseases. Myocardial remodeling results from fibroblast proliferation and increased collagen synthesis.

In response to injury, fibroblasts proliferate and differentiate into *myofibroblasts*. The latter secrete collagen and synthesize smooth muscle α-actin, stretch-sensitive ion channels, and connexins.

Myofibroblasts can interrupt normal between-myocyte coupling; cardiac fibrosis promotes arrhythmias. Moreover, myofibroblasts can form gap junctions with myocytes, at least in cocultures. The lower negative resting potential of fibroblast plasma membrane can then contribute to depolarization of myocytes. Myofibroblast–myocyte gap junction current comprise an early, transient outward ($i_{K,to}$)-like component and a late, sustained component. Coupling of myocytes to myofibroblasts promotes early afterdepolarizations due to the former, especially when myocyte repolarization reserve iss reduced by oxidative or ionic stress [570].

6.2.1.3 Effects of Natriuretic Peptides

Numerous growth factors and vasoactive peptides, such as transforming growth factor-β, angiotensin-2, and endothelin-1, participate in maladaptive cardiomyocyte hypertrophy, fibroblast proliferation, and augmented interstitial deposition. Conversely, antagonists of myocyte hypertrophy and fibrosis exist, such as type-A, -B, and -C natriuretic peptides.

Natriuretic peptides, such as ANP and BNP, are released from cardiomyocytes. They then act as auto-, juxta-, and paracrine factors to reduce cardiomyocyte maladaptive hypertrophy and exert antifibrotic and antiproliferative effects on cardiofibroblasts.

Brain natriuretic peptide limits the production of collagen-1 and fibronectin, myofibroblast conversion estimated by α-smooth muscle actin, fibroblast proliferation assessed by PDGFa and IGF1, and inflammation evaluated by COx2 and IL6 levels. C-type natriuretic peptide released from endothelial cells and cardiofibroblasts also has antifibrotic effects.

Cardiofibroblasts synthesize guanylate cyclase receptors NP_1 and NP_2 (NPRa and NPRb, or GC2a and GC2b) as well as the natriuretic peptide clearance receptor NP_3 (NPRc; Vol. 3 – Chap. 6. Receptors).

All 3 natriuretic peptides trigger the production of cGMP, an inhibitor of cardiomyocyte hypertrophy. Receptor NP_3 is involved in the regulation of fibroblast proliferation.[45] On the other hand, ANP impedes the activity of ERK and P38MAPK in smooth myocytes in response to angiotensin-2 [571].

Myocardial and coronary development is coupled via bone morphogenetic proteins as well as fibroblast (FGF) and vascular endothelial (VEGFa) growth factors [493]. The transcription factor GATA4 regulates cardiac gene expression, thereby modulating cardiomyocyte differentiation and adaptation in adult hearts. In cardiomyocytes, ERK1, ERK2, and P38MAPK kinases phosphorylate (activate) GATA4 transcription factor that promotes myocyte survival and hypertrophy. Factor GATA4 regulates the transcription of numerous peptides synthesized in cardiomyocytes, such as ANP, BNP, α- (MyH6) and β-myosin (MyH7) heavy chains, and endothelin-1 that causes cardiomyocyte hypertrophy [572]. Cardiac lineage-promoting GATA4 factor produced by cardiomyocytes stimulates angiogenic factor VEGFa that targets endothelial cells [573].

Both embryonic and adult cardiofibroblasts produce GATA4 factor. In cocultures of embryonic cardiac fibroblasts and myocytes, the former provoke proliferation of the latter [574]. Endothelin-1 serves as an autocrine stimulator of fibroblast proliferation. In cardiofibroblasts, ANP produced in neighboring cardiomyocytes prevents the synthesis of endothelin-1, as it suppresses GATA4-dependent transcriptional activity [571].

6.2.1.4 Effects of Angiotensin-2

Angiotensin-2 is involved in matrix constituent accumulation, especially during cardiac remodeling after long-term exposure of pressure overload. Angiotensin-2 promotes the synthesis of periostin, a matrix protein and regulator of cardiac fibrosis [575].[46] Angiotensin-2 provokes activation of the RasGRP1–Ras–P38MAPK–CREB pathway. cAMP response element-binding protein responds to different stimuli and mediates the fibrotic response. In addition, ERK1 and ERK2 participate in angiotensin-2-induced periostin expression via the TGFβ1–SMAD2/3 axis.

[45] Second messenger cGMP inhibits TGFβ1-induced phosphorylation of SMAD3, hence precluding myofibroblast transformation and proliferation as well as synthesis of extracellular matrix proteins.

[46] Periostin is secreted primarily by osteoblasts and fibroblasts. It is produced in the bone and, to a lesser extent, in the lung, kidney, and heart valves under normal conditions. However, periostin expression rises in heart failure.

6.2.2 Telocytes

Telocyte, originally called interstitial Cajal-like cell,[47] localizes to the myocardial interstitium, myocardial sleeves of pulmonary veins, as well as subendothelial layer of endocardium [576–578]. This cell differs from other interstitial cell types, such as fibroblasts, fibrocytes, fibroblast-like cells, and mesenchymal cells.

Telocyte is endowed with 2 or 5 long, slender, moniliform (many dilations throughout the length) prolongations, the so-called telopodes (length 0.1–0.4 mm; thickness 0.1–0.5 μm) [578]. These cytoplasmic processes establish close contacts between each other as well as with capillaries, nerve endings, cardiomyocytes, and other connective tissue cells. Telopodes of endocardial telocytes can penetrate the myocardium and extend among the cardiomyocytes and surround them. Telocyte synthesizes stem cell factor receptor [577].

Telocytes may be involved in: (1) intercellular signaling; (2) cardiac repair and remodeling; and (3) stem cell nursing, as it contacts cardiac stem cells in subepicardial cardiogenic niches [579]. A heterogeneous population of round stem and progenitor cells (size 6–10 μm) lodges in mammalian adult heart, more precisely in epicardial stem cell niches. Telocytes make a supportive interstitial network for stem cells and progenitors in the stem cell niche.

6.2.3 Epicardial Adipose Tissue and Adipokines

The epicardial adipose tissue, a source of free fatty acid, generates various substances such as *adipokines* that include *adiponectin*, *leptin*, *resistin*, *apelin*, and *visfatin*. Adipokines exert cardiovascular effects either directly or by central stimulation of the sympathetic nervous system. Epicardial adipose tissue may serve as a cardiac risk marker and be implicated in the development of cardiac pathology [580].

---

[47] Interstitial cells of Cajal are considered to be pacemakers of the motility of the gastrointestinal tract, where they have been discovered. Peculiar interstitial cells have also been detected in the urinary tract, blood vessels, pancreas, male and female reproductive tracts, mammary glands, placenta, and heart.

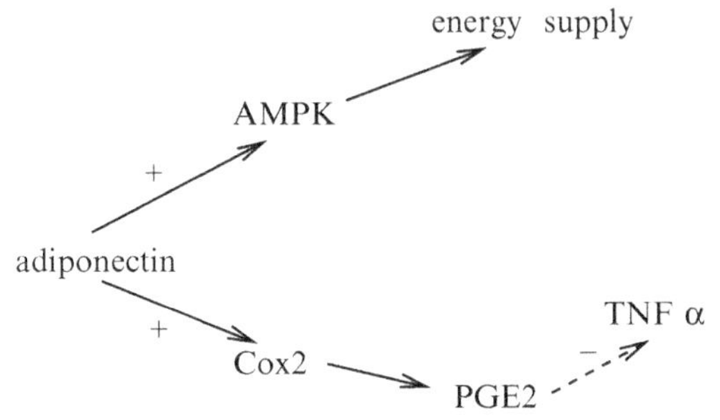

Fig. 6.2 Adiponectin activities. Adiponectin stimulates cyclooxygenase-2 (Cox2) and the synthesis of prostaglandin-E2 (PGE2), and thereby inhibits tumor-necrosis factor TNFα. It also acts via adenosine monophosphate-activated protein kinase (AMPK).

6.2.3.1 Adiponectin

Adiponectin[48] (Apn; Vol. 2 – Chaps. 1. Remote Control Cells and 3. Growth Factors and Vol. 3 – Chap. 6. Receptors) produced by adipocytes has cardioprotective effects, via its anti-inflammatory and anti-atherogenic properties (Fig. 6.2). Adiponectin binds to adiponectin receptors AdipoR1 and AdipoR2, which also link adaptors. Adiponectin is an insulin-sensitizing hormone, as it increases sensitivity to insulin in muscle and liver.

The adiponectin receptors — AdipoR1 and AdipoR2 — are expressed in the heart. Stimulated AdipoRs increases AMP-activated protein kinase, mediates peroxisome proliferator-activated receptor-α ligand activities, glucose uptake, and fatty acid oxidation [581]. Low levels of circulating adiponectin, as well as AdipoR1 and -R2, are measured in obesity, which lead to insulin resistance.

In cultured neonatal rat ventriculomyocytes, insulin decreases production of adiponectin receptor AdipoR1, but not that of AdipoR2 receptor via the PI3K–PKB axis and FoxO1 phosphorylation and translocation from the nucleus to the cytosol [582]. Factor FoxO1 binds to the promoter of the ADIPOR1 gene.

Adiponectin acts either via AMPK or via cyclooxygenase COx2, involved in prostanoid synthesis (AMPK-independent pathway; Fig. 6.2). Kinase AMPK, which is also activated by an increased ratio of AMP to ATP, supports fatty acid catabolism and glucose uptake and inhibits glucose synthesis. Enzyme AMPK also protects cardiomyocytes from ischemic injury, particularly hypoxia–reoxygenation-induced apoptosis. Suppression of cardiac AMPK activity indeed leads to increased injury when myocardium perfusion is reduced. Phosphorylation (activation) of AMPK after cardiac ischemia–reperfusion injury is attenuated in hearts of adiponectin-deficient mice [583]. Adiponectin prevents TNFα production by upregulating

[48] Adiponectin is also called ACRP30, AdipoQ, or gelatin-binding protein-28. It is involved in the regulation of lipid and glucose metabolism, as well as in differentiation of adipocytes. Its plasma concentration ranges from 3 to 30 μg/ml and accounts for 0.01% of total plasma protein. Adiponectin expression and subsequent release from adipocytes are stimulated by activated peroxisome proliferator-activated receptor PPARγ.

COx2, and subsequently prostaglandin PGE2. It thus reduces inflammation. Intravenous Apn injection, either shortly before or after the induction of ischemia in mice reduces ischemic injury [583].

6.2.3.2 Leptin

Leptin[49] (Vol. 2 – Chaps. 1. Remote Control Cells and 3. Growth Factors and Vol. 3 – Chap. 6. Receptors) is synthesized not only by adipocytes, but also by cardiomyocytes. Leptin production by cultured neonatal rat ventriculomyocytes increases upon stimulation by endothelin-1 and angiotensin-2 [584].

Leptin binds to its receptor (LR, a.k.a. OBR and LEPR) that is expressed as splice variants classified as short (LRa, LRc–LRd, and LRf), secreted (LRe), and long (LRb). Leptin receptors abound on the cardiomyocyte surface. The intracellular domain of LRb primes the JaK2–STAT3 axis. Leptin activates the RhoA–RoCK and PI3K–PKB pathways, as well as ERK1, ERK2, P38MAPK, and PKC [584].

Leptin modulates myocardial metabolism. It operates diversely according to the cell type on fatty acid oxidation and glucose uptake [584].[50] Leptin could also limit lipotoxicity that occurs when fatty acid uptake exceeds oxidative capacity. Leptin has negative inotropic effect via nitric oxide. Leptin is a predictor of poor outcome in patients with coronary heart disease or heart failure [584]. Leptin induces cardiac hypertrophy via the RhoA–P38MAPK pathway.

6.2.3.3 Resistin, Apelin, Visfatin, and Other Adipokines

Apelin binds to G-protein-coupled APJ receptor and exerts a positive inotropic effect in both normal and failing myocardium via protein kinase-C and Na^+–H^+ exchangers [584]. Apelin exerts a cardioprotection against ischemia–reperfusion injury.

Resistin is produced almost exclusively in white adipose tissue. It circulates as a trimer (monomeric form of resistin) or hexamer (dimeric form). Resistin is associated with insulin resistance. The dimerized form of resistin is more effective in antagonizing insulin-stimulated glucose uptake in adult murine cardiomyocytes [584].

---

[49]Circulating leptin concentration is related to body mass and adiposity. Leptin plasma level ranges from 5 to 15 ng/ml in non-obese individuals to reach more than 100 ng/ml in obese subjects. Leptin is convected by blood mainly bound to plasma proteins, whereas it flows primarily in free form in obesity.

[50]Fatty acid oxidation accounts for about 70% of ATP generated under normoxic conditions, whereas carbohydrates become the major substrate in anaerobic conditions.

Visfatin associates with the insulin receptor to induce hypoglycemia. Visfatin level is elevated in patients with coronary artery disease [584]. Other adipokines comprise vaspin, omentin, and chemerin, but they do not have identified function in healthy or diseased heart.

6.2.4 Heart Valves

The cardiac valves are sheets of connective tissue that begin at the *annulus fibrosus*. The valve cusps are passive soft tissues attached to the wall at the insertion line.

The ventriculoarterial — aortic and pulmonary — valves have smooth ventricular and wavy arterial faces. The free edge is indented. The middle part of the free edge is thicker with the *Arantius nodule*, characterized by a high concentration of collagen fibers.

These geometric features influence both valve motion during the cardiac cycle,[51] as well as flow in the aortic sinus when it is open, and thus the transvalvular pressure. Effective boundary conditions or wall laws for laminar pattern can be introduced. Roughness elements are supposed to be periodic with a length scale greater than the flowing cell size [585].

The cardiac valves, covered by an endothelium and containing smooth myocytes, are reinforced by many internal bundles of collagen and elastic fibers (Table 6.5). Sparsely distributed valvular interstitial cells have extensive contact with proteoglycans of the extracellular matrix, as well as with elastin and collagen fibers.

Collagen-1 forms a dense content beneath the endothelium. Collagen fibers of the outer layer of the cusp are mainly oriented in the azimuthal direction. Collagen fibers are more randomly oriented in the cusp central part. Elastin fibers are more abundant near the fibrous ring. In the fibrous ring, collagen fibers form massive wavy twisted bundles and elastin fibers have different directions. Also in the commissures, collagen fibers form massive twisted bundles.

Chordae tendinae mingle with valvular connective tissue. The myocardium penetrates into the valve leaflets. The valve components and their estimated elastic moduli are given in Vol. 7 – Chap. 5. Rheology. Whatever the transvalvular pressure at which the porcine aortic valves are fixed (< 12 kPa), the fibers mainly run along the circumferential direction [586]. Moreover, strong collagen fiber bundles that travel through the valve cusps and attach to the arterial walls behave like suspension cables for force transmission.

The cusp is a multilayer structure. The 3 layers are the *fibrosa*, the *spongiosa* (which is absent in the coaptation region), and the *ventricularis* (Fig. 6.3). The fibrosa, located toward the arterial wall, contains a large number of circumferential[52]

[51] In addition to the valve cross curvature of the open leaflet, the Arantian nodule induces axial curvature due to stiffer medial thickening.

[52] Collagen bundles run from commissure to commissure, spreading out near the cusp belly and combining again toward the opposite commissure.

Table 6.5. Structural elements of heart valves, with the 3 layers: (1) the ventricularis (rich in elastin); (2) the central spongiosa; and (3) the fibrosa (with packed collagen). Forces applied on the valve leaflets (pressure, shear, tension, and bending) are borne by structural rearrangement (fiber reorientation, folding, uncrimping, and compaction). Collagen and elastin fibers cooperate. Alignment of packed collagen bundles transfers forces between the cusps and the wall. Corrugations of the cusp face in front of the wall (fibrosa) help size and shape changes during the cardiac cycle. Microscopic collagen foldings (crimps) enable lengthening from low stresses. The anisotropic behavior is responsible for required differences in radial and circumferential extensibility. Cusp coaptation prevents valve prolapse. The trileaflet semilunar ventriculoarterial valves are inserted and supported by aortic and pulmonary roots. The atrioventricular valves are inserted on an annulus. Chordae tendineae and papillary muscles maintain the atrioventricular valve closed during blood ejection from the ventricles into the corresponding arteries (Source: [587]).

Component	Function
Cells	Interstitial cells synthesize and remodel ECM (fibroblasts, smooth myocytes, myofibroblasts) Valvular endothelial cells regulate the behavior of flowing and mural cells
Matrix	Elastin fibers extend in closed valve and shrink in open valve Collagen fibers maintain coaptation in closed valve Proteoglycans associate with matrix fibers for stress accommodation
Blood vessels	Valve cusps are avascular Nutrient/waste transport from heart cavity and myocardial vessels
Nerves	Regulation of coordinated activity of heart components

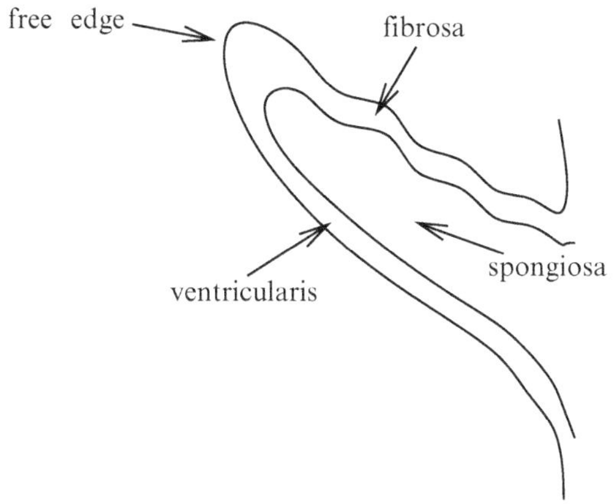

Fig. 6.3 Ventriculoarterial valve layers, according to the internal collagen network of the valve leaflets: fibrosa, spongiosa and ventricularis from the wavy arterial face to the smooth ventricular face. Collagen fibers provide strength, elastin fibers elasticity and proteoglycans swelling. Collagen architecture is mediated by mechanical loading.

wide collagen bundles. These circumferential bundles are arranged in a corrugated manner for radial expansion. A matrix with elastin surrounds the collagen bundles. The ventricularis contains collagen and elastin fibers, which are less organized than in the fibrosa. However, elastin sheets provide recoil, which retains the folded shape of the fibrosa. The loosely organized spongiosa, between the fibrosa and ventricularis, contains glycosaminoglycans, collagen, and elastin mostly in the radial direction. The spongiosa could act as load buffer for the fibrosa and the ventricularis.

The heart structure provides the 3 properties of contractility, automatism, and conduction due to 2 types of cardiac muscular cells: myocardial and nodal myocytes.

6.2.4.1 Mathematical Histology

Structure–function features of the heart have been mathematically investigated. Two fiber networks have been particularly studied: (1) the network of collagen fibers of the aortic valve cusp, and (2) the myofibers of the left ventricle wall, using a simple model of mechanically loaded fibers. The structure of the aortic leaflet has been derived from its function, which is assumed to consist of supporting a uniform pressure load undergone by a single family of fibers under tension [588]. The equation of equilibrium for the fiber structure is solved to determine its architecture. The computed fiber architecture resembles the real one. Assuming a constant myofiber cross-sectional area, symmetry with respect to the ventricle axis, small wall thickness with respect to the other dimensions, and a stress tensor resulting from hydrostatic pressure and myofiber stress, the bundles of myofibers have been shown to be located on approximate geodesics on a nested set of toroidal surfaces centered on a degenerate torus in the equatorial plane of the cylindrical part of the left ventricle [589].

6.2.5 Conduction Paths of the Depolarization Wave

The sinoatrial node act as the primary cardiac pacemaker. Together with nodal cells in the atrioventricular node (secondary cardiac pacemaker) and those of the ventricular nodal circuit, they constitute the cardiac conduction network of electrochemical waves (or action potentials).

6.2.5.1 Nodal Cells

Nodal cells are small ($80–100 \times 10–20\,\mu$m), specialized cardiomyocytes with few myofibrils that create or quickly spread depolarization waves in the myocardium. The conducting cells are wider than actual cardiomyocytes. They are not branched.

Table 6.6. Auto-excitation frequency (Hz) in nodal tissue (AVN: atrioventricular node; SAN: sinoatrial node). SAN frequency is decreased by the vagal influence (cranial nerve X).

SAN	$2.0–2.3 \xrightarrow{X} 1.2–1.3$
AVN	0.5–1
His bundle	0.3–0.6
Purkinje fibers	0.2–0.5

In the absence of neural and hormonal control, nodal cells naturally create action potentials, i.e., electrochemical signals at a given frequency that varies along the nodal tissue. When an upstream element becomes dysfunctional, a group of cells in a downstream part of the nodal tissue takes the relay, but at a lower frequency.

Tight junctions (Vol. 1 – Chap. 7. Plasma Membrane) facilitate action-potential propagation. Gap junctions also transmit electrochemical signals in the excitable myocardium, in which response speed and tissue activity synchronization are crucial for ion transfer.

Nodal myocytes are sparse with dispersed gap junctions that have high-conductance connexin-40 and low-conductance connexin-45.[53] Nodal Purkinje cells have abundant gap junctions with connexin-40 and -45 [333].

6.2.5.2 Architecture of the Nodal Tissue

Nodal cell automaticity relies on spontaneous, rhythmic, local depolarization associated with ionic fluxes. The *sinoatrial node* (SAN) that has the highest emission frequency of action potentials (Table 6.6) constitutes the "natural pacemaker".

The nodal tissue permits electrochemical conduction through the heart from the sinoatrial node, both atria, to the atrioventricular node. Electrochemical signals then propagate across the atrioventricular septum and ventricles using the His bundle and its 2 branches — left and right bundle branches — that ramify into Purkinje fibers organized in many fascicles. The latter transmits the depolarization wave to the inner layers of cardiomyocytes. Afterward, the electrochemical wave spreads through the surface and the width of the ventricular myocardium in a given time.

An automatism hierarchy exists between different regions of nodal tissue, according to the emission frequency of the action potential (Table 6.6). The propagation speed of the electrochemical signal also varies with the heart region (Table 6.7).

[53]Many types of connexins aggregate to form gap junctions between adjacent cells. They form homo- or heteromeric hemichannels and homotypic, heterotypic, or heteromeric channels with different conductance, permeability, and gating properties. Connexin type expression varies according to the heart site. Cells that express human connexin Cx31.9 exhibit much faster transport than cells expressing other connexin types [590].

Table 6.7. Estimated conduction speeds (m/s) in nodal tissue (AVN: atrioventricular node; SAN: sinoatrial node).

SAN	0.05
Atrium	0.3–1
AVN	0.02–0.10
His bundle	1–2
Purkinje network	3–5
Ventricle (axial)	0.6–1
Ventricle (transverse)	0.2–0.5

6.2.5.3 Sinoatrial Node

The electrochemical signal starts with automatic self-excitation of cells of the sinoatrial node (SAN; length ∼8 mm, width ∼2 mm). This node is located at the top of the right atrium, more precisely at the junction of the superior vena cava and right atria, between the orifice of the superior vena cava and the auricle. In the majority of humans, the sinoatrial node is irrigated by the right coronary artery.

The sinoatrial node is composed of different cell types, stellate and elongated spindle cells. Stellate myocytes, the primary pacemaker cells, exhibit a faster functioning rate and longer action potential duration.

In adult mouse hearts, the density of connexin-30 is low in the sinoatrial node, where it contributes to the regulation of the cardiac frequency, in addition to hyperpolarization-activated cyclic nucleotide-gated potassium channel HCN4 and other ion carriers [591].

The sinoatrial node is endowed by conduction barriers to be functionally insulated from the surrounding atrial myocardium as well as conduction paths to transmit action potentials. Anatomical barriers, i.e., connective tissue layers, adipose tissue, and coronary arteries, as well as functional barriers, which results from gap junction connexin paucity, protect from hyperpolarizing influence of the surrounding atrial cells [592]. Specialized paths — sinoatrial conduction pathways — represent preferential lines of conduction of action potentials. They transmit electrochemical impulses from the sinoatrial node to the atrial wall. They consist of elongated and branching fibers of transitional cells.

In dogs, the sinoatrial node, which is a more realistic animal model for the human SAN than that of small mammals, is isolated from the surrounding atrium by 3 branches of the coronary arterial bed and fibrosis [592].

The sinoatrial node is a heterogeneous structure with many pacemaker compartments and conduction pathways, electrophysiological properties of which vary. Major pacemaker compartments comprise head, center, and tail of the sinoatrial node [592]. A pacemaker hierarchy within the SAN exists with the fastest pacemakers at the head and the slowest at the tail. Electrochemical coupling enables the synchronization of pacemaker clusters inside the sinoatrial node. Excitation originates in the quickest pacemaker compartment and slowly propagates through the SAN tissue. It then runs via different conduction paths and excites atria (beginning of the P wave on the electrocardiogram).

Both canine and human sinoatrial nodes have at least 4 preferential conduction paths, such as the superior and inferior conduction paths, that deliver depolarization waves simultaneously to different regions of the atrial myocardium. According to the preferential conduction paths, action potential spreads from the right atria to the left atria either via the Bachmann's bundle and the coronary sinus myofibers or via anterosuperior and posteroinferior intra-atrial septum connections [592].

In addition to hyperpolarization-activated inward "funny" current (i_f) through hyperpolarization-activated cyclic nucleotide-gated channels, various other inward currents can contribute to cardiac automaticity, such as i_{NaCaX} current through Na^+–Ca^{2+} exchanger (*calcium clock* mechanism). Blockers of i_f slow, but do not stop automaticity, whereas ryanodine that depletes intracellular Ca^{2+} stores can cause a cessation of automaticity. The Ca^{2+} influx regulates Na^+–Ca^{2+} exchanger. Spontaneous, cyclical, local Ca^{2+} release from the sarcoplasmic reticulum can indeed participate in the generation of both basal and adrenergically stimulated depolarization rate, in addition to i_f current [593]. In fact, Ca^{2+} dependence of SAN automaticity, i.e., the calcium clock, is directly and indirectly linked to the plasmalemmal ion flux mechanism of SAN automaticity. The sinoatrial node produces a Ca^{2+}-stimulated adenylate cyclase (AC1 or AC8) rather than the typical cardiac isoform (AC5 or AC6).

6.2.5.4 Atrial Paths

Since sinoatrial fibers fuse with the surrounding atrial cardiomyocytes, the action potential spreads through the atria at a rate at least of ~ 0.3 m/s and produces atrial contraction.

However, several nodal bundles conduct action potentials with a greater speed of about 1 m/s. Three preferred conduction paths have been observed in the right atrium wall: the anterior Thorel, the mid Bachman, and the posterior Wenckebach bundles.

6.2.5.5 Atrioventricular Node

The action potential reaches the *atrioventricular node* (AVN; ~ 5 mm long) that localizes to the right atrium near the lower part of the interatrial septum. The atrioventricular node is adjacent to the coronary sinus.

The atrioventricular node is subdivided into 3 regions: atrionodal, nodal, and nodal-His, according to histological and electrophysiological criteria. Ovoid cells are observed in nodal and nodal-His regions, rod-shaped cells in the atrionodal region. The conduction speed is slower in the nodal region.

The atrioventricular node imposes a delay at most of about 100 ms in impulse transmission to the ventricles, when atria have completed their contraction (Table 6.8). Hence, it travels at a speed of 1.5 to 4 m/s.

Table 6.8. Approximate depolarization instants in the heart wall (ms). The myocardium depolarization takes $\sim$ 50–100 ms after reaching the junction between the Purkinje fibers and the subendocardial layer of the myocardium.

Atria	$5 \rightarrow 85$
AVN	50 (impulse arrival)
	125 (activation departure)
His bundle	130
His branches	145
Purkinje fibers	150
Endocardial myofibers	$\rightarrow \sim 190$
Epicardial myofibers	$\rightarrow \sim 250$

Table 6.9. Left ventricle activation timing (ms; Source: [336]; LV, RV: left, right venticle).

Mid-left part of ventricular septum	0–10
Inner part of LV wall (septum, apex)	10–20
Ventricular septum, RV and LV inner apex	20–30
Heart apex, external RV, LV lateral walls	30–45
RV and LV bases	45–80

Atrioventricular delays can be assessed by the sum of intervals between atrial and nodal cell firing and between nodal and ventricular cell firing. The conductance time across the atrioventricular node is equal to about 10 ms (conduction velocity $\sim$160 mm/s) [594]. The conduction barrier is located between the atrium and the atrioventricular node.

6.2.5.6 His Bundle

The action potential runs in the *His bundle*, or atrioventricular bundle, beneath the endocardium in the interventricular septum. The His bundle divides after a short distance into 2 branches – the right and left bundle branches – that descend along respective sides of the interventricular septum.

6.2.5.7 Purkinje Fibers

His bundle gives birth to the *Purkinje fibers* that remain subendocardial. Although they localize just beneath the endocardium to end on cardiomyocytes, they can penetrate into the myocardium. Cardiac Purkinje cells allow a very rapid and almost simultaneous impulse distribution to activate the bulk of the ventricular myocardium.

The action potential propagates through the ventricular myocardium at 0.3 to 0.5 m/s. Ventricular depolarization duration is equal to about 80 ms (Table 6.9). Excitation begins in the subendocardial layer in the interventricular septum.

Like atrial pacemaker myocytes, cardiac Purkinje cells are devoid of transverse tubules, but have a much larger width. They possess a specialized triplicate layer of inositol (1,4,5)-trisphosphate-sensitive Ca^{2+} channels under the sarcolemma juxtaposed to a layer (~ 5 μm) of sarcoplasmic reticulum RyR3 channels, whereas RyR2 channels lodge in the inner cell region.

In canine Purkinje cells, 3 types of Ca^{2+} transients are observed [595]: (1) non-propagating Ca^{2+} transients that originate directly under the sarcolemma and lead to (2) small Ca^{2+} wavelets in a subsarcolemmal region (thickness 6 μm) that cause (3) large Ca^{2+} waves that travel to sarcomeres. Purkinje cells contain a Ca^{2+} release structure that translates Ca^{2+} transients under the sarcolemma into rapid Ca^{2+} release in the subsarcolemmal region, itself initiating a large-amplitude long-lasting Ca^{2+} liberation. Three layers of Ca^{2+} channels actually exist: (1) channels associated with type-1 IP_3 (IP_3R1) and type-3 ryanodine receptors (RyR3) directly under the sarcolemma; (2) RyR3 receptors are dominant Ca^{2+} release channel in a subsarcolemmal layer (between 2 and 4 μm below the sarcolemma); and (3) RyR2 receptors are observed throughout the cell, but not in the subsarcolemmal layer (between 2 and 4 μm below the sarcolemma). The threshold of RyR3 receptors that cause propagating wavelets transients is lower than that of RyR2 receptors responsible for large Ca^{2+} waves.

6.2.5.8 Innervation of the Nodal Tissue

Both the sinoatrial and the atrioventricular nodes are supplied with sympathetic and parasympathetic postganglionic fibers from the intrinsic cardiac ganglia. Parasympathetic nerve terminals are more numerous in the atrioventricular node. When stimulated, the sympathetic and parasympathetic fibers release *noradrenaline* (NAd) and *acetylcholine* (ACh), respectively.

On the other hand, the main bundle and bundle branches lack parasympathetic nerve endings in some mammalian species such as dogs, but not monkeys [596]. Purkinje fibers receive a sympathetic input. Terminal axons run in the space between the Purkinje fibers and the fibrocytic envelope, either naked or accompanied by Schwann cell processes. The width of the synaptic cleft between an axonal varicosity and Purkinje cell is about 60 nm.

6.3 Cardiac Regulation of Energy Homeostasis

The liver, adipose tissue, and skeletal muscle are major regulatory organs and tissues of body's energy expenditure and metabolism, especially of carbohydrates and lipids. The heart that has the highest rate of oxygen consumption per gram of tissue, participates to the control of energy homeostasis locally as well as remotely possibly via cardiac-derived endocrine factors either acting directly or via the central nervous system.

The heart adapts to changes in diet and stress by regulating metabolism and gene expression to maximize energy efficiency. The heart regulates energy homeostasis, i.e., the balance between energy consumption and expenditure, via heart-specific Med13, a regulatory subunit of the Mediator complex, which controls transcription by nuclear hormone receptors, especially NR1a1 and NR1a2,[54] in addition to other nuclear receptors, such as NR2b1 to NR2b3 and NR1c3, as well as well as their coactivators and corepressors, such as NCoR1 and SREBP proteins [597].

Cardiac expression of Med13 is inhibited by a heart-specific microRNA, miR208a [597].[55] In mice, overexpression of Med13 or inhibition of miR208a confers resistance to high-fat diet-induced obesity and improves insulin sensitivity and glucose tolerance. Subunit Med13 enhances the body's metabolism and causes an increased energy expenditure, as, although food intake remains normal, oxygen consumption and carbon dioxide production rise. Conversely, in the absence of cardiac Med13 activity, lipids accumulate without changes in food intake. Changes in energy consumption can affect thermogenesis.

6.4 Main Ion Currents

Efficient cyclic heartbeat is achieved by fast and brief contraction followed by a rapid and short-duration relaxation triggered by quick Ca^{2+} ions flux in and out the cytosol of cardiomyocytes, that enables blood ejection from ventricules and venous blood admission in cardiac chambers.

Activated ion channels generates ion fluxes (Table 6.10) in and out nodal cells and cardiomyocytes. Ionic currents and resulting action potentials vary among mammal species, and within an animal species, cell types.

Mathematical models of ion fluxes across the plasma and organelle membranes of nodal cells or cardiomyocytes as well as ion motions and sequestration within the cytosol and certain organelles are aimed at restoring the conduction velocity and action potential duration of the explored region of the heart wall.

In the heart, MinK-related peptide MiRP1 is an auxiliary subunit of $K_V11.1$ responsible for the rapid delayed rectifier K^+ current ($i_{K,r}$). It modulates ionic flux amplitude via channel conductance and/or gating kinetics. It also influences $K_V11.1$ stability. Upon phosphorylation (Ser98), it accelerates $K_V11.1$ degradation [598]. Its density is higher in ventricles than in atria in spontaneously hypertensive rat and guinea pig hearts. Mutations in the KCNE2 gene that encodes MiRP1 can cause long QT syndrome (LQT6).

Adaptive and maladaptive cardiac hypertrophy is associated with a change in density of potassium channels. In maladaptive hypertrophy in humans, angiotensin-2 or endothelin-1 reduces abundance of $K_V4.3$ that is responsible for the transient

[54]Thyroid hormones regulate the metabolic rate, energy expenditure, and cardiac contractility.

[55]MicroRNA-208A is encoded by an intron of the cardiac-specific α-myosin heavy-chain gene.

Table 6.10. Main inward anion currents and corresponding channels in the heart (i_f: non-selective cation current; Source: [468]). Sodium current (i_{Na}) due to cardiac voltage-gated sodium channels controls action potential upstroke and cell excitability.

Current	Channel	Genes
i_{Na}	$Na_V1.5$	SCNA5
		SCN1B, SCN2B
$i_{Ca,L}$	$Ca_V1.2$	CACNA1C, CACNL1A1
		CACNB1–CACNB2, CACNA2D1
$i_{Ca,T}$	$Ca_V3.1$	CACNA1H
i_f		HCN1, HCN2, HCN4
		KCNE2

outward current, via reactive oxygen species produced by NADPH oxidase and stress-activated kinases such as MAP3K5 and its main targets, P38MAPK and JNK kinases [599]. Therefore, the MAP3K5–P38MAPK/JNK pathway participates in cardiac ion channel remodeling.

Sustained elevation in cytosolic Ca^{2+} concentration stimulates PP3 that dephosphorylates NFAT3 and promotes its translocation to the nucleus, thereby reducing several repolarizing K^+ flux types ($i_{K,to}$ and rapidly activating and slowly inactivating K^+ currents $i_{K,slow1}$ and $i_{K,slow2}$) [599]. Therefore, another pathway, the Ca^{2+}–CamK2–PP3–NFAT3 axis, is involved in cardiac ion channel remodeling.

In mouse ventriculomyocytes with constitutive activation of $PI3K_{c1\alpha}$ as well as in exercise-trained mice, K^+ current amplitudes and cell size increase, but K^+ current density does not significantly differ from that in wild-type and untrained mice, because K_V channel expression rises in proportion to myocyte size in adaptive cardiac hypertrophy [600]. Therefore, the PI3K–PKB pathway contributes to cardiac ion channel remodeling.[56]

6.4.1 *Pacemaker Cells*

Major currents in sinoatrial nodal as well as atrial cells correspond to fluxes of sodium, potassium, and calcium ions (Tables 6.11 to 6.13; Fig. 6.4) [601–603]. Sinoatrial nodal cells are heterogeneous, hence exhibiting different types of action potentials. A mathematical model that relies on a non-linear system of 10 first-order ordinary differential equations has been proposed to represent the functioning of a cardiac pacemaker cell [604].

[56]Transcription factors of the FOXO family that are effectors of the $PI3K_{c1\alpha}$–PKB axis regulate the promoter activity of several K^+ channel genes, such as KCNJ8 ($K_{IR}6.1$), KCNJ11 ($K_{IR}6.2$), and AbcC8 (SUR1), as well as KCND2 ($K_V4.2$), KCNB1 ($K_V2.1$), KCNK3 (TASK1), and KCNIP2 (KChIP2) [600]. Furthermore, glycogen synthase kinase GSK3β, another mediator of $PI3K_{c1\alpha}$, also controls the activity of several transcription factors, such as NFAT, GATA4, myocardin, MyC, Jun, and β-catenin, many of which may bind to promoters of several K^+ channel subunit genes.

Table 6.11. Ion currents that underlie the pacemaker function. Action potential of sinoatrial cells can be decomposed into 3 phases: slow and rapid depolarization and repolarization (i_f: hyperpolarization-activated inward Na^+ and K^+ current; i_{sus}: sustained inward Na^+ (essentially) current; $i_{Ca,L}$, $i_{Ca,T}$: inward L- and T-type Ca^{2+} currents; $i_{Ca,b}$: inward Ca^{2+} background current; i_{Kto}: transient outward K^+ current; $i_{Kto(r/s)}$: rapid/slow K^+ delayed rectifier current; i_{Ksus}: sustained K^+ current; $i_{Na,b}$: inward Na^+ background current; i_{NaCa}: Na^+–Ca^{2+} exchanger current; i_{NaK}: Na^+–K^+ pump current). The bulk background current comprises background Ca^{2+}, -K^+, and -Na^+ currents as well as i_{NaK} current. Background Ca^{2+} current is needed to keep the diastolic level of the intracellular free calcium concentration. Background conductance involves also spontaneous opening of ACh- and ATP-sensitive K^+ channels ($i_{K_{ACh}}$ and $i_{K_{ATP}}$ as well as Na^+–Ca^{2+} exchangers (i_{NaCa}). Depolarization activates Ca_V1 channels ($i_{Ca,L}$). This event is followed by a gradual activation of the delayed rectifier K^+ current. The major component of the delayed rectifying K^+ current (i_K) in pig and rabbit sinoatrial cells are i_{Ks} and i_{Kr}, respectively. Sustained K^+ current (i_{Ksus}) occurs through $K_V1.5$ in human atriomyocytes that underlie the ultrarapid delayed rectifier K^++ current (i_{Kur}). Transient outward K^+ current (i_{Kto}) is related to $K_V4.2$ and $K_V4.3$ sununits.

Stage	Ion currents
Slow depolarization	i_f, i_{sus}, $i_{Ca,T}$; $i_{Na,b}$, $i_{Ca,b}$, $i_{NaK_{ATP}}$, i_{NaCaX}
Rapid depolarization	i_{Na}, $i_{Ca,L}$
Repolarization	$i_{K,to}$, $i_{K,r/s}$, $i_{K,sus}$, $i_{Ca,L}$

Hyperpolarization-activated current, an inward depolarizing current, is due to flux of Na^+ and other ions. Hyperpolarization-activated cyclic nucleotide-gated cation channel HCN4 is involved in the Cl^--sensitive, Ca^{2+}-dependent, inward "funny" current (i_f) of cardiac pacemaker activity.[57]

In mouse embryos, HCN4 serves as a powerful cardiac pacemaker when it is stimulated by cAMP [606]. Moreover, HCN4 channel activity initiated by cAMP is required for embryo viability.[58]

Acute stress stimulates β-adrenergic receptors that activate HCN4 via cyclic adenosine monophosphate and provokes increased heart rate, as sinoatrial nodal cells generate faster action potentials. However, in some circumstances, HCN4 can be dispensable for fight-or-flight response. Increase in intracellular Ca^{2+} concentration by a release from sarcoplasmic reticulum stores can contribute to an i_f-independent mechanism that accelerates cardiac frequency during β-adrenoceptor stimulation. Ca^{2+}–calmodulin-dependent protein kinase-2 and phosphatase modulate the function of sarcoplasmic reticulum in nodal cells and cardiomyocytes. Kinase CamK2 is a mediator of β-adrenoceptors that contributes to speeding up

[57]During development, HCN4 location becomes restricted to the dorsal wall of the right atria, then to the junction of the right atrial appendage and the superior vena cava, where the sinoatrial node appears [605]. HCN4 channels are highly expressed in the adult sinoatrial node.

[58]In adult mice, HCN4 channel does not contribute to cardiac pacemaking, but prevents sinus pauses during and after adrenergic stimulation stress. In humans, the role of HCN4 channels is different, as the cardiac frequency is higher in mice than in humans.

Table 6.12. Ion channels, genes, and currents of the sinoatrial node (Source: [441]).

Current	Channel subunits	Genes
	Voltage-gated Na^+ channels	
i_{Na}	$Na_V(\alpha)1.1/2/5$	SCN1A/2A/5A
	$Na_V\beta1/2/3$	SCN1B/2B/3B
	Voltage-gated Ca^{2+} channels	
$i_{Ca,L}$	$Ca_V(\alpha)1.2/1.3$	CACNA1C/D
	$Ca_V\beta1/2/3$	CACNB1/2/3
	$Ca_V\alpha2\delta1/3$	CACNA2D1, CACNA2D3
	$Ca_V\gamma5$	CACNG5
$i_{Ca,T}$	$Ca_V(\alpha)3.1$	CACNA1G
	Voltage-gated K^+ channels	
$i_{K,to}$	$K_V(\alpha)1.4$	KCNA4
	$K_V(\alpha)4.2/4.3$	KCND2/3
	KChIP2	KCNIP2
	MiRP1	KCNE2
$i_{K,ur}$	$K_V(\alpha)1.5$	KCNA5
	$K_V\beta1/2$	KCNAB1/2
$i_{K,r}$	$K_V(\alpha)11.1$	KCNH2
	MiRP1	KCNE2
$i_{K,s}$	$K_V(\alpha)4.3$	KCNQ1
	Mink	KCNE1
$i_{K,b}$	$K_{2P}2.1$ (TREK1)	KCNK2
	$K_{2P}3.1$ (TASK1)	KCNK3
$i_{K_{ACh}}$	$K_{IR}3.1/4$	KCNJ3/5
$i_{K_{ATP}}$	$K_{IR}6.2$	KCNJ11
	Hyperpolarization-activated, cyclic nucleotide-gated channels	
i_f	HCN1/2/4	Hcn1/2/4
	MiRP1	KCNE2
	Na^+–K^+ pump	
$i_{NaK_{ATP}}$	$\alpha1/2/3$	ATP1A1/2/3
	$\beta1$	ATP1B1
	Na^+–Ca^{2+} exchangers	
i_{NaCaX}	NCX1	SLC8A1

the cardiac frequency, hence yielding an additional mechanism with respect to i_f current [607]. A threshold of βAR activity is required to engage CamK2 kinase. In ventriculomyocytes, CamK2 acts on Ca^{2+} flux from and into sarcoplasmic reticulum. It is able to increase Ca^{2+} filling into and release from the sarcoplasmic reticulum, as it phosphorylates SERCA inhibitor phospholamban (positive lusitropic effect) and RyR2 (positive inotropic effect), respectively. Kinase CamK2 thus allows maintenance of mechanical performance of ventriculomyocytes during sustained (24 h) catecholamine stimulation. In sinoatrial cells, CamK2 is necessary for

Table 6.13. Regulation of pacemaker currents (Source: [601]).

Effector	Target	Effect
Ca^{2+}	Na^+–Ca^{2+} exchanger	Activation
	Delayed rectifier K^+ channel	Activation
H^+	Inward rectifier K^+ channel	Inhibition
	Ca^{2+} channel	Inhibition
	Na^+–HCO_3^- cocarrier	
β-Adrenergic agonist	Gs–ACase–cAMP–PKA	
	Ca_V1.2 channel	Activation
	Delayed rectifier K^+ channel	Activation
α1-Adrenergic agonist	Gq–PLC–DAG–PKC	
	Delayed rectifier K^+ channel	Activation
Muscarinic agonist	Gi–ACase	Inhibition
	Ach-activated K^+ channel	Activation
Adenosine	ACase	
	Adenosine-sensitive K^+ channel	Activation
ATP	ATP-sensitive K^+ channel	Inhibition
	Ca^{2+} pump	Activation
	Na^+–K^+ pump	Activation

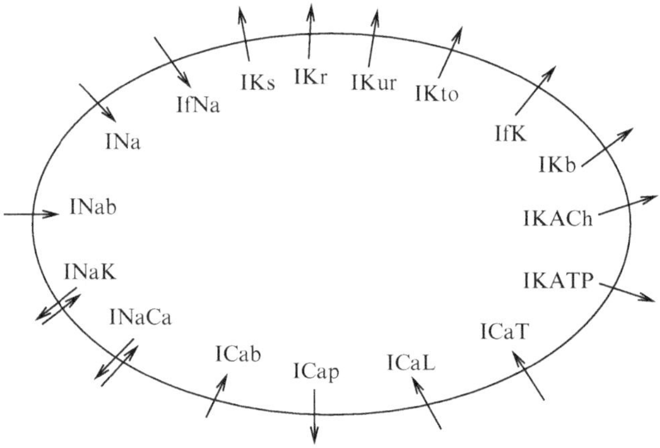

Fig. 6.4 Main ionic currents across the sarcolemma of sinoatrial node and atrial cells, associated with the action potential (Sources: [602, 603]; ICaL [$i_{Ca,L}$]: L-type Ca^{2+} current; ICaT [$i_{Ca,T}$]: T-type Ca^{2+} current; ICap [$i_{Ca,p}$]: sarcolemmal Ca^{2+} pump current; ICab [$i_{Ca,b}$]: background Ca^{2+} current; IfK [i_{f_K}]: hyperpolarization-activated K^+ "funny" current; IKur, IKr, IKs [$i_{K,ur}$, $i_{K,r}$, $i_{K,s}$]: ultrarapid, rapid, and slow K^+ delayed rectifier current; IKb [$i_{K,b}$]: background K^+ current; INa [i_{Na}]: inward Na^+ current; INab [$i_{Na,b}$]: background Na^+ current; INaCa [i_{NaCaX}]: Na^+–Ca^{2+} exchanger current; INaK [$i_{NaK_{ATP}}$]: Na^+–K^+ pump current; IfNa [$i_{f_{Na}}$]: hyperpolarization-activated Na^+ "funny" current).

maximal increase in βAR-mediated cardiac frequency. Greater rates in Ca^{2+} uptake and discharge from the sarcoplasmic reticulum can help sinoatrial nodal cells trigger shorter depolarization waves.

The other important ion channel in cardiac pacemaking is voltage-dependent Ca^{2+} channel $Ca_V1.3$ that carries $i_{Ca,L}$ current activated during the early phase of the diastolic depolarization. In addition to i_f and $i_{Ca,L}$, other ionic currents contribute to the regulation of the heart frequency, such as $i_{Ca,T}$ that is almost exclusively due to $Ca_V3.1$ channels, sustained inward current, and electrogenic Na^+–Ca^{2+} exchange that are caused by rhythmic cortical Ca^{2+} oscillations due to spontaneous opening of ryanodine receptors.

Sodium fluxes lead to several currents, a small fast inward Na^+, a Na^+–Ca^{2+} exchange , a Na^+–K^+ pump, and a Na^+ background current. Among the voltage-dependent K^+ currents, the outward K^+ delayed rectifier current plays a major role in pacemaker activity. Only a small fraction of nodal cells of the sinoatrial node yields a transient outward K^+ current. Furthermore, inward rectifier K^+ current, although observed in the atrium and part of the atrioventricular node, is not found in the sinoatrial node, whereas it is the dominant background current in ventriculomyocytes. Additional K^+ currents are provided by ligand-gated channels (ATP-, adenosine-, and acetylcholine-sensitive channels).[59] Transient and long-lasting Ca^{2+} currents have been recorded in the sinoatrial node through Ca_V3 and Ca_V1 channels. A stretch-activated Cl^- channel can provide an inward background current. The decay in outward K^+ current and inward background current leads to initial cell depolarization to the threshold. The hyperpolarization-activated current modulates pacemaker depolarization. Ca^{2+} current is a main factor during late depolarization. ATP-sensitive K^+ channels are activated by ATP depletion. Na^+–K^+ pumps and Na^+–Ca^{2+} exchangers influence pacemaker activity. Intracellular ion concentrations and regulators affect the functioning of pacemaker ion carriers (Table 6.13).

6.4.1.1 Parasympathetic Control

In sinoatrial cells, cholinergic signals activate Gi-coupled M_2 cholinergic receptors that decrease cAMP level, thus suppressing activation of protein kinase-A and slowing pacemaker rate. In addition, these receptors activate acetylcholine-activated potassium channels ($K_{IR}3$).

On the other hand, adrenergic cues activate β-adrenergic receptors in pacemaker cells, thereby increasing cAMP concentration, activating PKA, and raising the pacemaker frequency. Protein kinase-A regulates numerous proteins in nodal cells, especially ion channels such as Ca_V1 channel and phospholamban.

---

[59]The activity of both adenosine- and acetylcholine-sensitive K^+ channels involves GTP-binding proteins. Both ATP and Mg^{2+} are cofactors of acetylcholine-sensitive K^+ channels.

In addition, cAMP stimulates the "funny" pacemaker hyperpolarization-activated current (i_f). This current results from fluxes of Na^+ and K^+ ions through hyperpolarization-activated, cyclic nucleotide-gated tetrameric channels (HCN). Each monomer has 6 transmembrane domains with voltage sensors and binding sites for cyclic nucleotides such as cAMP. The HCN channel is predominantly produced in the sinoatrial node. However, several HCN isoforms are also synthesized in cardiac Purkinje cells and atrio- and ventriculomyocytes. The HCN channel is slowly activated by hyperpolarization at a range of voltages that arise from constitutive pacemaker depolarization. This activation triggers the next electrochemical wave generation and propagation.

A-kinase-anchoring proteins control PKA activity. In humans, AKAP10 variants are associated with increased cardiac frequency and decreased heart rate variability, hence decaying sensitivity to external stimuli. Protein AKAP10 modulates the sensitivity of cardiomyocytes to vagus nerve stimuli [608].

6.4.1.2 Spontaneous, Rhythmic, Subplasmalemmal Ca^{2+} Oscillations

In the absence of sarcolemma depolarization, sinoatrial nodal pacemaker cells are able to generate localized, rhythmic, submembrane Ca^{2+} oscillations due to extraction via sarcoplasmic reticulum Ca^{2+} pumps and Ca^{2+} release via ryanodine receptors. Calcium cycling by the sarcoplasmic reticulum, the so-called sinoatrial Ca^{2+} clock, generates these rhythmic, spontaneous, local Ca^{2+} releases that depend on the cAMP–PKA axis.

Spontaneous membrane depolarizations interrupt these spontaneous Ca^{2+} oscillations, as action potential activates Ca_V1 channels to trigger Ca^{2+} release from the sarcoplasmic reticulum. During the later phase of the depolarization, ryanodine receptors are inactivated and Ca^{2+} is pumped back into its store to replenish it. Once the inactivation of ryanodine receptor wanes, the spontaneous release of Ca^{2+} via ryanodine receptors restarts.

The local increase in submembrane Ca^{2+} concentration generates an inward current via Na^+–Ca^{2+} exchangers that elevates the depolarization slope, ignites plasmalemmal ion channels, and triggers the occurrence of action potential.

Activated β-adrenoceptors increase the submembrane Ca^{2+} oscillation amplitude and reduce the period between sarcoplasmic reticulum Ca^{2+} release triggered by prior action potential and onset of the local Ca^{2+} release during the subsequent depolarization, hence causing earlier influx through Na^+–Ca^{2+} exchangers and arrival of the next action potential [609].

Regulation of cardiac automaticity in sinoatrial cells is attributed to the modulation of sarcolemmal ion currents by G-protein-coupled receptors of the β-adrenergic and cholinergic receptor classes (Table 6.14). Modulation of GPCRs of sinoatrial cell automaticity involves coupling via transmembrane voltage and Ca^{2+} flux of sarcolemmal and intracellular proteins, mainly those that are responsible for [610]: (1) the diastolic Na^+–Ca^{2+} exchanger current; (2) magnitude of spontaneous, rhythmic, local Ca^{2+} releases from the sarcoplasmic reticulum; and

Table 6.14. Model of the autonomic modulation of a cardiac pacemaker cell (Source: [610]; AChR: cholinergic receptor; AR: adrenergic receptor; $i_{Ca,T}$: T-type Ca^{2+} current ($Ca_V3.1$); $i_{K,r}$: rapid inwardly rectifying K^+ current ($K_V11.1$); i_{NaCaX}: Na^+–Ca^{2+} exchanger current; i_{sus}: sustained (steady), essentially Na^+ current). Varying intracellular cAMP concentration couples intracellular and plasmalemmal clocks via phosphorylation by cAMP-dependent protein kinase A (PKA) of phospholamban and $Ca_V1.3$ channels that generate L-type Ca^{2+} current ($i_{Ca,L}$) and trigger activity of ryanodine receptor on the sarcoplasmic reticulum. Acetylcholine slows the cardiac clock via ACh-activated K^+ current ($i_{K_{ACh}}$) via $G\beta\gamma$ dimeric subunit of G protein. Both amplitude and phase of spontaneous, local Ca^{2+} release can be assigned mainly by functioning of ryanodine receptors and sarco(endo)plasmic reticulum Ca^{2+} ATPase (SERCA) subjected to the inhibition of phospholamban on the one hand and $Ca_V1.3$ channels and Na^+–Ca^{2+} exchanger on the other that all regulate subplasmalemmal Ca^{2+} fluxes.

Components	Elements
Intracellular Ca^{2+} clock	SERCA–phospholamban (reuptake), Ryanodine receptor (release) Subsarcolemmal Ca^{2+} current that influences the transmembrane potential
Sarcolemmal clock	$G\beta\gamma$-stimulated $i_{K_{ACh}}$ cAMP-activated i_f Refueling PKA-activated $i_{Ca,L}$ Clearance i_{NaCaX} βAR-stimulated $i_{K,r}$ $i_{Ca,T}$ i_{sus}
Nervous control	Gs-coupled βAR Gi-coupled AChR Activation or inhibition of the cAMP–PKA pathway

(3) spontaneous action potential firing rate that is modulated by β-adrenergic and cholinergic receptors (firing rate increase and decrease, respectively). In addition to intracellular Ca^{2+} transient triggered by the action potential, spontaneous, local Ca^{2+} releases from the sarcoplasmic reticulum beneath the sarcolemma creating multiple, relatively synchronous, locally propagating Ca^{2+} wavelets during the late part of diastolic depolarization. The sarcoplasmic reticulum behaves as an intracellular Ca^{2+} oscillator. Spontaneous Ca^{2+} release from the sarcoplasmic reticulum depends on the Ca^{2+} pumping capability of the sarcoplasmic reticulum, hence phospholamban that is regulated by the cAMP–PKA axis. Phospholamban phosphorylation by protein kinase-A relieves its inhibition on SERCA pumps. Protein kinase-A also phosphorylates ryanodine receptors. Local Ca^{2+} releases activate the electrogenic operation of the sarcolemmal Na^+–Ca^{2+} exchanger (NCX) and generate an inward current that yields a steep rising depolarization phase [610]. β-Adrenergic and cholinergic receptors shorten or prolong the period of local Ca^{2+} releases, as assessed by the duration between peak of Ca^{2+} transient that results from action potential and maximum of spontaneous, local Ca^{2+} releases.

In rabbits, the threshold of cholinergic agonist (carbachol) concentration to reduce the beating rate is approximately 10 nmol [611]. Half-maximal inhibition that is achieved at 100 nmol lowers phospholamban phosphorylation by PKA by approximately 50%. In addition, this dose causes a time-dependent reduction in the number and magnitude of rhythmic, spontaneous, local Ca^{2+} releases, but an augmentation of the period of local Ca^{2+} releases. A dose of 1,000 nmol stops spontaneous beating. On the other hand, i_f blockade does not affect the reduction of cardiac frequency caused by cholinergic signals at any cholinergic agonist concentration.

Initiation of acetylcholine-activated current $i_{K_{ACh}}$ happens from cholinergic agonist concentration greater than 30 nmol. Numerical simulations point out that Ca^{2+} cycling is modulated by cholinergic input via activation of i_{NaCaX} current primed by local Ca^{2+} releases [611]. The cholinergic control operates via Gi activation. The extent of Gi coupling to Ca^{2+} cycling results from PKA signaling as well as activation acetylcholine-activated K^+ current when the stimulation reaches a given threshold (> 30 nmol).

6.4.2 Ventriculomyocytes

Mathematical models of the ventricular action potential are aimed at reproducing the different action potential shapes corresponding to the 3 transmyocardial regions (endo-, epi-, and midmyocardia) of ventricles and their different rate dependencies [603,612]. Basal models of ventricular myocyte functioning comprise 3 major components: (1) membrane ion currents (fast sodium, L-type calcium, transient outward potassium, rapid and slow delayed rectifier potassium, inward rectifier potassium, sodium–calcium exchanger, sodium–potassium pump, calcium and potassium plateau, and background currents; Fig. 6.5) [612]; (2) intracellular calcium compartments (subspace between the T tubule and junctional sarcoplasmic reticulum,[60] junctional and network sarcoplasmic reticulum, sarcomere, and bulk myoplasm); and (3) calcium buffers (mainly the sarcoplasmic Ca^{2+} buffer calmodulin, sarcomeric Ca^{2+} buffer troponin, and sarcoplasmic reticulum Ca^{2+} buffer calsequestrin).

Very small inward currents of sodium and calcium ions maintain a long plateau of depolarization. The action potential plateau is also maintained by the activity of the sodium–calcium exchanger, driven by transmembrane voltage and Na^+ and Ca^{2+} concentration gradients. During the plateau phase of the action potential, the Na^+–Ca^{2+} exchanger initially works in reverse mode, importing Ca^{2+} and exporting Na^+, afterward it switches to forward mode, extruding Ca^{2+}.

[60] Sarcoplasmic reticulum calcium channel respond to increase in calcium ion concentration in the subspace caused by the local sarcolemmal influx.

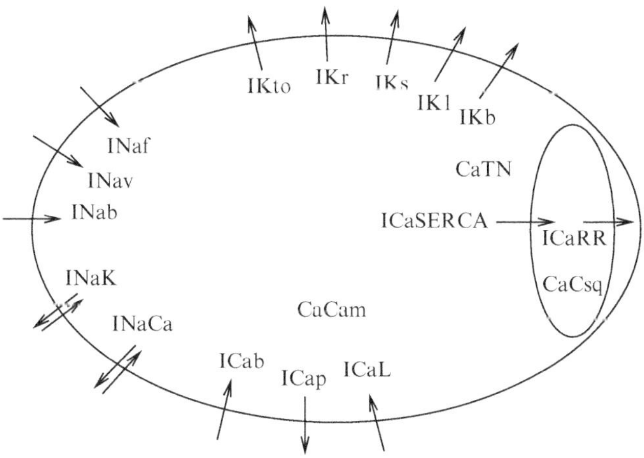

Fig. 6.5 Main ionic currents associated with the action potential and calcium buffers in the ventricular cardiomyocyte (Sources: [612–617]; IcaL: L-type Ca^{2+} current; ICap: sarcolemmal Ca^{2+} pump current; ICab: background Ca^{2+} current; ICaRR: Ca^{2+} ryanodine release current; ICaSERCA: sarcoplasmic reticulum Ca^{2+} pump (uptake) current; INaCa: Na^+–Ca^{2+} exchanger current; INaK: Na^+–K^+ pump current; INaf: fast Na^+ current; INab: background Na^+ current; IKto: transient outward K^+ current; IKr, IKs: rapid and slow K^+ delayed rectifier current; IK1: inward rectifier K^+ current). The 3 main transient buffers are the complexes of Ca^{2+} with calsequestrin (Ca–Csq) in the sarcomplasmic reticulum, troponin (Ca–TN) in the sarcomere, and calmodulin (Ca–Cam) in the sarcoplasma. The ventricular cell model can also include accumulation and depletion of calcium ions in tiny spaces between the sarcolemma and sarcoplasmic reticulum.

The excitation–contraction process is triggered by calcium and involves especially the complexes (couplons or dyads) formed by sarcolemmal Ca_V1 channels and ryanodine receptors of the junctional region of the sarcoplasmic reticulum. Reverse-mode Na^+–Ca^{2+} exchangers and Ca_V3 channels also trigger the Ca^{2+} release, but with less efficiency [618].

6.4.2.1 Electrophysiological Compartmentation and Gap Junctions

Gap junctions are involved in synchronization of propagation of electrochemical impulses and cardiomyocyte repolarization. Heterogeneous expression of the principal ventricular connexin-43 creates a dispersion of action-potential duration across the ventricular wall. An electrophysiological regionalization of heart exists according to the anteroposterior axis, at least in canine hearts [619].

Transmural dispersion of action potential duration is significantly greater in anterior than posterior left ventricle [619]. Conduction velocity distribution is also markedly wider across the anterior than posterior left ventricle. Expression of Cx43 varies more strongly in the anterior than posterior left ventricle. Local underexpression of Cx43 is correlated to transmural electrophysiological heterogeneity on the anterior left ventricle.

6.5 Action Potential

The ionic composition, in particular sodium, potassium, and calcium contents, of cardiomyocyte sarcoplasma is controlled by ion channels, pumps, and exchangers that maintain steep ion concentrations and electrical gradients across the thin sarcolemma ($\sim$ 6 nm) that surrounds myofibrils (Table 1.4). The sarcolemma has a high and low conductance for K^+ and Na^+, respectively. The resting membrane potential is about equal to -88 mV (inside negative).

Strong, acute changes in intracellular calcium concentration that are related to release from and reuptake into Ca^{2+} stores determine action potential features such as channel conductance possibly via Ca^{2+}-binding proteins such as calmodulin, therefore adapting ion channel functioning to the body's need. Calcium ions actually modulate various ion channels in ventriculomyocytes. For example, Ca^{2+} regulates the magnitude of ion flux (i_{Na}) through voltage-gated sodium channels without affecting kinetic properties such as gating [620]. Voltage-gated Na^+ channels control action potential upstroke. Cardiac action potentials result from time course of depolarizing inward and repolarizing outward ionic currents through a set of ion carriers (channels, exchangers, and pumps; Table 6.15).

Action potential shape changes between neonatal and adult cardiomyocytes. Neonatal cardiomyocytes indeed have action potentials with brief plateau and longer duration than that of adults due to [621]: (1) decreased density and modified inactivation of transient outward K^+ currents; (2) increased delayed rectifier K^+ currents; (3) Ca^{2+} influx through T-type (Ca_V3) and L-type (Ca_V1) Ca^{2+} channels; (4) increased Ca^{2+} influx through Na^+–Ca^{2+} exchangers; and (5) Ca^{2+} transients from Ca^{2+} entry rather than release from the sarcoplasmic reticulum.

The heterogeneity of action potential shapes according to nodal cell type is illustrated in Fig. 6.6. Purkinje and ventricular action potentials exhibit a plateau (phase 2). Atrial action potentials have a blunt triangular shape (absence of phases 1 and 2). The upstroke is slow because the fluxes elicited by Ca^{2+} channels are smaller and these channels are activated more slowly than Na^+ channels (triangle-shaped action potential). Activated non-selective cation channels induce a slow increase in transmembrane voltage after the end of the action potential, up to the triggering threshold for action potential.

The features of the action potential (initial depolarization rate, duration, etc.) depend on the activities of involved ion channels of the nodal cells in the different compartments of the conduction network and cardiomyocytes in walls of atria and ventricles, and within the ventricle, according to the myocardium layer (subendocardial, midmyocardial, and subepicardial; Table 6.16). The duration of the action potential depends on the nodal tissue territory with the following sequence of increasing order: (1) distal Purkinje fibers ($\sim$400 ms, with upstroke of 500–600 V/s), (2) proximal Purkinje fibers and His bundle, (3) midmyocardial layer, (4) subendocardial layer, (5) subepicardial layer, and (6) atria, SAN, and AVN ($\sim$ 150 ms, with upstroke of 2–10 V/s) [468].

Table 6.15. Ionic fluxes that direct the cardiac action potential with depolarizing inward and repolarizing outward currents (Source: [466]; $i_{K,to}$: transient outward K^+ current; $i_{K,r}$: rapid delayed rectifier K^+ current; $i_{K,s}$: slow delayed rectifier K^+ current; $i_{K,ur}$: ultrarapid delayed rectifier K^+ current; i_{K1}: inward rectifier K^+ current; $i_{K_{ATP}}$: ATP-sensitive channel; $i_{K_{ACh}}$: acetylcholine-activated channel; phase 0: rapid depolarization; phase 1: initial, slight, fast repolarization; phase 2: slow repolarization (plateau); phase 3: final, rapid repolarization; phase 4: resting potential [gradual depolarization of pacemaker cells]).

Ion	Channels	Motion	Current	Action potential phase
Na^+		Inward	i_{Na}	0 (1–3)
Ca^{2+}	Ca_V1	Inward	$i_{Ca,T}$	1–3
	Ca_V3	Inward	$i_{Ca,T}$	1
Na^+ and	NCX	Outward (Ca^{2+} influx)	i_{NaCaX}	0–1
Ca^{2+}		Outward (Na^+ influx)		2, 4
K^+	$K_V4.2/3$	Outward	$i_{K,to}$	1–2
	$K_V1.5$	Outward	$i_{K,ur}$	0–3
	$K_V11.1$	Outward	$i_{K,r}$	1–4
	$K_V7.1$–minK	Outward	$i_{K,s}$	2–4
	$K_{IR}2$	Outward	i_{K1}	3–4
	$K_{IR}3.1/4$	Outward	$i_{K_{ATP}}$	3–4
	(or GIRK1/4)		$i_{K_{ACh}}$	
	$K_{IR}6.2$–SUR1	Outward	$i_{K_{ATP}}$	3–4
Na^+, K^+, Ca^{2+}	HCN2/4	Inward	i_f	4
Cl^-	CFTR, TWIK	Outward	i_{Cl}	0–3

Non-pacemaker cells have a resting membrane potential (electrical diastole, phase 4), whereas pacemaker cells generate regular, spontaneous action potentials (Fig. 6.7).

6.5.1 Sinoatrial Node Action Potential

Cardiac pacemaker cells exhibit automaticity associated with membrane depolarization and repolarization. Atrial pacemaker potential is characterized by 2 distinct phases: an initial gradual depolarization (`D1`) followed by a steep slope (`D2`). The sinoatrial node yields spontaneous action potentials. In comparison with standard action potentials exhibited by ventriculomyocytes, only 3 phases can be defined. Spontaneous depolarization (equivalent to phase 4 of standard action

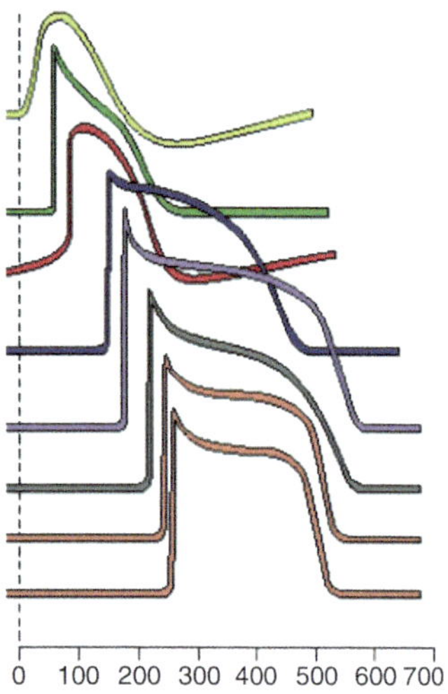

Fig. 6.6 Varying shapes (time in ms) of the action potential during its propagation in the nodal tissue. From top to bottom: sinoatrial node, atrial myocardium, atrioventricular node, His bundle and its branches, Purkinje fibers, and ventricular myocardium (Source: [622]).

Table 6.16. Nodal tissue and ionic currents (Source: [468]). According to the types of involved ion channels, the initial depolarization ramp (upstroke; V/s) is more or less steep, action potential duration (APD) more or less long (ms), and firing frequency more or less high (beats/mn). Action potential duration is longer in the crista terminalis than in pectinate muscles in the right atrium, and longer in the right auricle than the left atrium. The short action potential in atria is partially explained by $i_{K,ur}$ current. Currents $i_{Ca,L}$ and $i_{Ca,T}$ are responsible for phase 0 in the sinoatrial node (SAN) and $i_{Ca,L}$ in the atrioventricular node (AVN). Non-selective cation "funny" current i_f in the atrioventricular node is responsible for a secondary pacemaker activity (A: atrium; V: ventricle; H: His bundle; P: Purkinje fibers; endo: subendocardial; epi: subepicardial; mid: midmyocardial layers of the myocardium).

Current	SAN	A	AVN	H	P	V
$i_{K,ur}$	0	+	0	0	0	0
$i_{K,r}$	+	+	+	+	+	+
$i_{K,s}$	+	+	<SAN	+	+	+
$i_{K,to}$		+	+	+	+	Mid, epi > endo
i_{K1}	∼0	Low	Low	<V	<V	+
$i_{K_{ACh}}$	+	+	+			+
$i_{K_{ATP}}$		+		+	+	+
i_f	+	+	+	+	+	+
i_{Na}	0	+	+	+	+	+
$i_{Ca,L}$	+	+	0	<V	<V	+
$i_{Ca,T}$	+	+	+	+	+	+
Frequency	70–80	50	30–60	20–40	15–30	15–30
Upstroke	2–5	150–250	10–20	400-600	500-600	
APD	150	150	150	400	400	epi < endo < mid

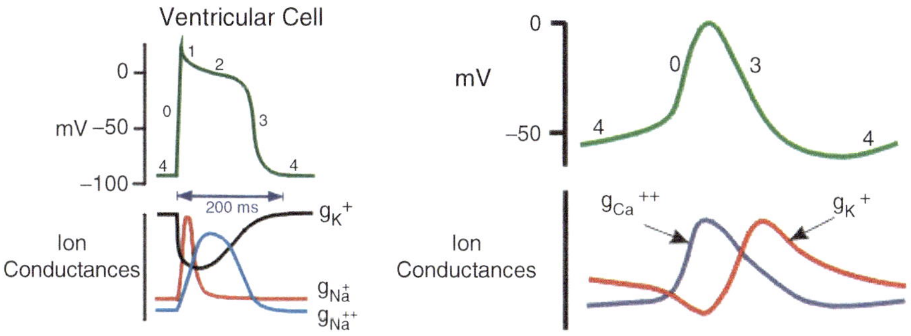

Fig. 6.7 (**Left**) action potential of the ventriculomyocyte (classical model) with its five phases 0–4, phase 4 being the resting membrane potential. Phase 0, the rapid depolarization phase, is due to opening (influx) of fast Na^+ channels (with quick rise in membrane conductance g_{Na^+}, current INa) and closure of K^+ channels. Phase 1 is due to closure of the fast Na^+ channels and transient outward fluxes of K^+ and Cl^-. Phase 2 (plateau) is caused by a balance between Ca^{2+} influx using long-lasting channels and K^+ outflux. Phase 3, repolarization down to the resting membrane potential, is induced by a decrease in Ca^{2+} influx and K^+ outflux. During phases 0, 1, 2, and part of phase 3, the cell is refractory to the initiation of a new action potential. (**Right**) action potential spontaneously generated (automaticity) in pacemaker cells of the sinoatrial node. A continuous K^+ outflux causes slow depolarization. Phase 0, fast depolarization due to Ca^{2+} influx. Phase 3, repolarization, is due to inactivation of voltage-gated Ca^{2+} channels, decaying Na^+ flux, and rising K^+ flux (from [761] (web site) with author permission).

potentials; Fig. 6.7) is due to a K^+ outflux associated with Na^+ influx and a small Ca^{2+} influx. Once the depolarization reaches a threshold of about −40 mV, a new action potential is triggered. Quick depolarization (corresponding to phase 0) is mainly caused by an augmented Ca^{2+} influx. Repolarization occurs (corresponding to phase 3) when Ca^{2+} influx decreases and K^+ outflux increases.

Involved mechanisms include the modulating hyperpolarization-activated inward Na^+ current,[61] inward T- and L-type Ca^{2+} currents, time-dependent decay of K^+ conductance, Na^+–Ca^{2+} exchange, and low background K^+ conductance. Voltage-activated Ca^{2+} release (Ca^{2+} sparks) significantly contributes with the activity of multiple sarcolemmal ion channels to heart electrical excitations that originate from the sinoatrial node as well as latent atrial pacemakers. This voltage-gated intracellular Ca^{2+} release is triggered by the activation of T-type Ca^{2+} channels (Ca_V3) [623]. Atrial pacemaker cells do not contain T tubules. Therefore, subsarcolemmal cisternae of the sarcoplasmic reticulum are located along the cell periphery with a subspace (size ∼25 nm) rich in Ca_V3 channels.

[61] Pacemaker cells lacking hyperpolarization-activated inward Na^+ current have a lower frequency.

Table 6.17. Phases of action potential and main ionic motions.

Phase	Ion motion	Path
0	Na^+ influx	Na^+ channels (quick and short opening)
1	K^+ efflux	Transient outward K^+ channels
2	Ca^{2+} influx	Ca_V1 channel, RyR
3	K^+ efflux	Delayed rectifier K^+ channels
4	K^+ efflux	Inwardly rectifying K^+ channels

6.5.2 Ventricular Action Potential

The action potential is initiated by depolarization of the sarcolemma, disturbing the ionic gradient across the sarcolemma. Separate voltage-sensitive channels[62] for Na^+, K^+, and Ca^{2+} exist in the sarcolemma that transiently open and increase the passage of these ions.

Sodium channels open first and then rapidly inactivate (Phase 0; Table 6.17). The depolarization, indeed, inactivates these Na^+ channels. The quick cellward Na^+ motion increases transmembrane potential up to about +30 mV. The conductance of the inward K^+ rectifiers strongly decreases from the onset of depolarization.

Phase 1 occurs after action potential peak. It corresponds to the first stage of rapid repolarization associated with a transient outward motion of K^+. Sarcolemmal repolarization, especially the phase-1 rate that influences the number of activated Ca_V1 channels and ion flux rate through them, modulates the recruitment of ryanodine channels to Ca_V1 channels, hence Ca^{2+} flux to the sarcomere. A second transient flux for repolarization is induced by Ca^{2+}-activated Cl^- channels.

Phase 2 is associated with a slow inward Ca^{2+} current (i_{Ca}) for several hundred milliseconds (plateau). Channel Ca_V1 is indeed secondarily activated with 2 opening modes: mode 1 with short opening bursts and mode 2 with long-duration opening. Calcium channels inactivate more slowly than Na^+ channels. Mode 2 Ca^{2+} channels are involved during phase 2 of action potential with Ca^{2+} flux from extracellular space to sarcoplasm. They elicit additional Ca^{2+} release from the sarcoplasmic reticulum for contraction with slow inactivation.

The plateau of ventricular action potentials is prolonged because K^+ currents occur slowly and corresponding channels have a reduced conductance at positive transmembrane potentials. K^+ effluxes from rapid (K_V11) and slow (K_V7) delayed K^+ rectifiers balance the Ca^{2+} import.

The kinetics and conductance of Ca_V1 channels and delayed K^+ rectifiers determine plateau duration. A long action potential ensures adequate time for entry

[62]Channel protein conformations are influenced by the membrane potential, so that voltage-gated ion channels are only permeable over a narrow range of membrane potentials. On the other hand, ligand-gated ion channels require specific chemical activators.

of extracellular Ca^{2+} into the cardiomyocyte for optimal excitation–contraction coupling. Delayed repolarization also yields cardiomyocyte refractory to premature excitation.

Other ionic fluxes arise from Na^+–Ca^{2+} exchanger, with Na^+ and Ca^{2+} transport reversal during repolarization, and from Na^+–K^+ pumps. The transition between phases 2 and 3 is characterized by an increase in repolarization rate due to the beginning in activity of inward K^+ rectifiers, expelling K^+ ions.

During phase 3, delayed rectifier K^+ channels induce rapid repolarization by K^+ current that counters depolarization induced by Na^+ and Ca^{2+} inflow. These membrane channels remains closed in the presence of ATP, but as soon as ATP levels fall, they open for K^+ efflux that leads cardiomyocytes to a hyperpolarized inactive state.

Throughout phase 4, the resting membrane potential is regulated by a background K^+ current. A Na^+ efflux can balance the K^+ influx generated by the inward K^+ rectifiers to ensure a resting cell potential. This process can be modulated by voltage-sensitive Cl^- channels activated by catecholamines via cAMP messenger. The action potential is shortened by Cl^- influx and lengthened by adrenergic effect on Ca^{2+} channels.

6.5.3 *Electrophysiological Myocyte Types*

Transmural differences in electrochemical properties of the ventricular myocytes are attributed to changes in both quantity and quality of sarcolemmal (K^+ transient outward, rapid and slow delayed K^+ rectifiers [624], inward K^+ rectifier, Na^+–Ca^{2+} exchanger with its reverse and forward modes, and late Na^+ channel) as well as sarcoplasmic reticulum carriers such as sarcoplasmic reticulum Ca^{2+} ATPase.[63] Such variations lead to electromechanical coupling differences, which can also be due to variations in intracellular Ca^{2+} buffering, as well as in contractile protein isoform expression.[64]

These transmural differences may be required for coordinated contraction and relaxation of the ventriculomyocyte population across the ventricular wall stimulated at different instants by the propagating electrochemical wave with a delay on the order of few tens of milliseconds. The ventricular myocardium contains at least 3 electrophysiologically distinct myocyte types: epicardial, midmyocardial, and endocardial myocytes.

---

[63]The expression of sarcoplasmic reticulum Ca^{2+} ATPase is greater in epicardial than midmyocardial and endocardial myocytes [625]. Consequently, epicardial myocytes exhibit a quicker decay of intracellular Ca^{2+} content than endocardial myocytes. Transmural expression of Na^+–Ca^{2+} exchanger does not significantly vary [625]. Regional differences in ryanodine channel expression and/or its binding protein might exist.

[64]Myosin isoforms do not have similar ATPase and contractile activities [626].

6.5.3.1 Epicardial Myocytes

The epicardial action potential exhibits a prominent early notch (spike-and-dome morphology). The action potential notch is due to a transient outward K^+ current [627] that decays in amplitude from the epicardial to endocardial surfaces (the notch is smaller in midmyocardial myocytes and absent in endocardial myocytes).

The transient outward K^+ current is also larger in right than left ventricular epicardial myocytes. Higher transient outward K^+ currents lead to faster rate of repolarization, hence causes larger L-type Ca^{2+} current through Ca_V1 channel and greater increase in $[Ca^{2+}]_i$ than slower currents [628].

Epicardial and, to a lesser extent, midmyocardial myocytes indeed exhibit faster sarcomere shortening kinetics than endocardial myocytes. The number of inward K^+ rectifiers and characteristics of delayed K^+ rectifiers differ between endocardial and epicardial cells. Because the amount of sarcoplasmic reticulum Ca^{2+} ATPase is greater in epicardial myocytes, Ca^{2+} reuptake is quicker in ventricular epicardium than endocardium.

There is no transmural difference in rapid-activating delayed rectifier K^+ currents and inward rectifier K^+ currents, but apicobasal differences can exist.

6.5.3.2 Midmyocardial Myocytes

Midmyocardial, or M myocytes, have been observed in the human heart [629]. They localize between the deep subendocardium and midmyocardium in the anterior wall, between the subepicardium and midmyocardium in the lateral wall, throughout the wall of the right ventricle outflow tracts, as well as in papillary muscles, trabeculae, and interventricular septum [630].

Midmyocardial myocytes that can extend to the deep subendocardium exhibit longer action potential duration due to smaller slow components of the delayed K^+ rectifier density and/or unitary conductance [624] as well as a larger density in Na^+–Ca^{2+} exchangers [631] and higher late Na^+ current [632]. In the midmyocardium, higher density of sodium channels, larger channel conductance, and greater probability of reopening yield stronger late Na^+ currents.

Prolonged action potential augments myocyte contraction by increased intracellular Ca^{2+} concentration. Midmyocardial myocytes have a greater rate of depolarization (higher conduction velocity than in other cardiomyocyte types, but lower than in nodal cells). Midmyocardial myocytes are more sensitive to frequency changes.

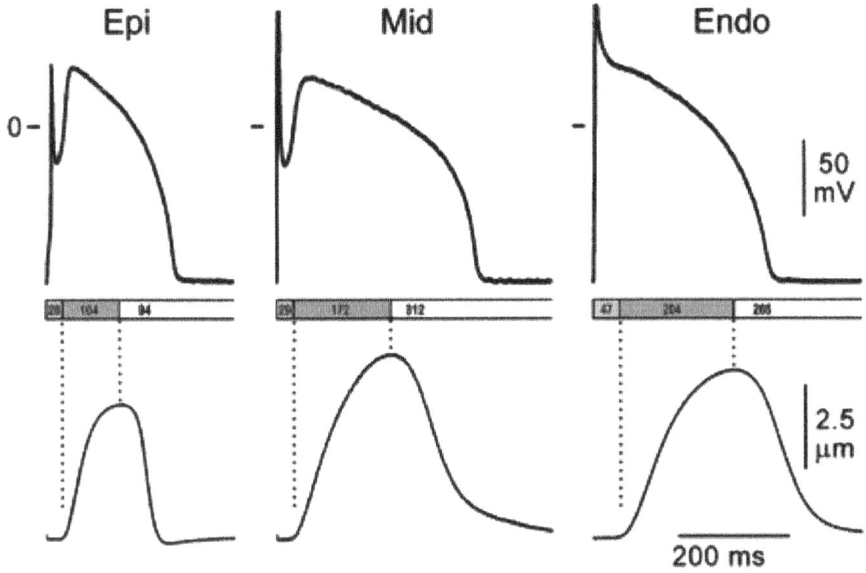

Fig. 6.8 (**Top**) Action potentials in epicardial (Epi), midmyocardial (Mid), and endocardial (Endo) myocytes (from [633] with permission). (**Bottom**) Corresponding cell shortening.

6.5.4 Repolarization

Following depolarization and ventricular contraction, repolarization initiates cardiac relaxation. Ventricular repolarization takes a longer time in the heart apex than base, and near endocardium than near epicardium. Endocardial myocytes have smaller rates of repolarization than epicardial myocytes, because of a lower density of voltage-gated potassium channels (K_V4; transient outward K^+ flux) in endocardial and septal myocardia than in epicardial myocardium.[65] A space gradient in voltage-gated K^+ channels is required for coordinated cardiac repolarization.

6.5.5 Electrophysiology and Contraction Kinetics

Differences in electrophysiological characteristics among cardiomyocytes are associated with variations in shortening kinetics [633] (Fig. 6.8). Features of epi-, endo-, and midmyocardial myocytes are summarized in Table 6.18.

[65]Loss in repolarization heterogeneity can lead to polymorphic ventricular tachycardia.

Table 6.18. Main features of action potentials and cell shortening in unloaded epi-, endo-, and midmyocardial myocytes (Sources: [625, 633]).

	Epicardial myocyte	Midmyocardial myocyte	Endocardial myocyte
	Action potential		
Spike-and-dome configuration	+	+	No
Duration	Shortest	Longest	
Peak time			Longest
	Ca^{2+} carriers		
Ca^{2+} influx onset		Similar time	
Ca^{2+} level decay rate	++	+	+
SR Ca^{2+} content	Largest		Smallest
$Ca_V1.2$		Similar density	
SERCA	++	+	+
NCX		Similar density	
	Cell shortening		
Contraction onset	Shortest		Greatest
Peak time	Shortest		Longest
Relaxation	Fastest		

6.6 Excitation–Contraction Coupling

Cyclic blood ejection from cardiac ventricules arises from the synergy between heart electrical and mechanical functions. Myocardium contraction and relaxation are influenced by many governing parameters, such as intracellular ion concentrations and membrane potential, in addition to pre- and afterload and direction and velocity of movement [634].

The structure that enables rapid contraction and relaxation of cardiomyocytes is the connection between transverse tubules and terminal cisternae of the sarcoplasmic reticulum. Sodium and calcium channels, Na^+–Ca^{2+} exchangers, and Na^+–K^+ pumps aggregate on the T-tubular membranes and colocalize with ryanodine-sensitive Ca^{2+} channels clustered on the sarcoplasmic reticulum membrane. In particular, many tens of $Ca_V1.2$ channels and RyRs form couplons that couple sarcolemmal action potential to release of Ca^{2+} messenger (Ca^{2+} spike) by the sarcoplasmic reticulum.

Calcium ions in clefts between transverse tubules and terminal cisternae, i.e., in dyad clefts,[66] bind to regulatory calmodulin sites on the cytosolic domains of $Ca_V1.2$ channels and rapidly, but partially, inactivate the channels terminating the

[66]The so-called dyad is constituted by a T-tubule and a single terminal cisterna.

Ca^{2+} spike, but leaving a tonic Ca^{2+} flux until myocytes repolarize to -40 mV. Rapid partial repolarization to 0 mV by the transient outward K^+ current ($i_{K,to}$) through K_V4 channels provokes a maximal Ca^{2+} spike (~ 0.5 nA) [328]. The tonic component of Ca^{2+} current via T tubules loads the sarcoplasmic reticulum with Ca^{2+} ions.

Moreover, binding of Ca^{2+} to Na^+–Ca^{2+} exchangers rapidly activates these exchangers, thereby adding Ca^{2+} entry. Voltage-gated $Ca_V1.2$ channels provoke Ca^{2+} sparklets, RyRs cause Ca^{2+} sparks and waves. In fact, the transient Ca^{2+} entry through sarcoplasmic reticulum ryanodine receptors yields the main Ca^{2+} input. Adrenergic command modulate the magnitude and the dynamics of Ca^{2+} fluxes via Ca^{2+}–calmodulin-dependent kinase CamK2 that targets all Ca^{2+} carriers.

Released Ca^{2+} ions ($\sim 15 \times 10^3$ ions/half-sarcomere) diffuse over a short distance ($\sim 1\,\mu$m) to adjacent myofibrils and then bind to troponin-C ($\sim 25 \times 10^3$ mol/half-sarcomere) of the troponin–tropomyosin complex on actin filaments [328]. Subsequently, tropomyosin moves away from myosin-binding sites, thus enabling actin–myosin cross-bridges ($\sim 90 \times 10^3$/half-sarcomere) and converting the chemical energy of ATP into force development (~ 2 pN/bridge for a maximal force of ~ 100 mN/mm^2), hence allowing filament sliding and sarcomere shortening. Synchronization of sarcomeres of normal left ventricle enables a fast pressure soar and subsequent blood ejection.

Calcium release by the sarco(endo)plasmic reticulum during a heartbeat rises the free cytosolic Ca^{2+} concentration from ~ 0.1 to $\sim 1\,\mu$mol, above which the contractile response saturates (at $\sim 2\,\mu$mol) [328]. Troponin-C has a binding site for Ca^{2+} ion in this range of Ca^{2+} concentration. The Ca^{2+}–troponin-C molecule enables sarcomere contraction by actin–myosin cross-bridges. At saturating $[Ca^{2+}]_i$, force increases with stretch over the range of sarcomere length at which cardiomyocyte operates (1.5–2.3 μm). At submaximal $[Ca^{2+}]_i$, the sensitivity of the contractile response depends on sarcomere length (length-dependent activation [LDA]). Several mechanisms may contribute to LDA process. The first (minor) mechanism supposes that sarcomere operates at constant volume, i.e., length changes are associated with width (spacing between actin and myosin lattice) changes, hence modulation of cross-bridge probability. Other mechanisms relies on cooperativity between actin, troponin, and tropomyosin [328]: (1) Ca^{2+}–troponin-C enhances binding of Ca^{2+} to neighboring troponin-C; (2) interaction between troponin–tropomyosin units influences length-dependent effect of Ca^{2+}–troponin-C activation; (3) cross-bridging improves binding of Ca^{2+} to adjacent troponin-C; and (4) force exerted by cross-bridges on actin deforms this thin filament, hence actin–troponin–tropomyosin complex, with enhanced binding of Ca^{2+} ion to troponin-C.

Released Ca^{2+} ions also rapidly reaches sarco(endo)plasmic Ca^{2+} ATPase SERCA2 of the sarcoplasmic reticulum membrane that envelops the sarcomere. Calcium ions are then sequestered back into the sarcoplasmic reticulum for efficient lusitropy (myocardium relaxation). In addition, a moderate amount ($\sim 30\%$) of cytosolic Ca^{2+} is simultaneously extruded through NCX carriers, which enables Na^+ entry.

The combined compliance of titin inside cardiomyocytes and of collagen network surrounding cardiomyocytes enables sarcomere stretch during diastole. During relaxation, sarcomere length and $[Ca^{2+}]_i$ remain constant.

6.6.1 Intercellular and Cell-Matrix Adhesions

Mechanosensitive adhesions between cardiomyocytes enable their functioning as an electromechanical syncytium. In the adult heart, healthy ventriculomyocytes are characterized by cell–matrix adhesions at their sides (transversal edges) and intercellular junctions at axial ends (between-cell interfaces) [635].

Focal adhesions stabilize cardiomyocytes between them and to the matrix during assembly of sarcomeres and intercalated discs. During maturation of intercalated discs of cardiomyocytes subjected to cyclic loadings, focal adhesions disassemble [635].

6.6.2 Troponins and Tropomyosin

Cardiomyocyte contraction and relaxation are effected by cyclic association and dissociation of actin and myosin. This cycle is regulated by Ca^{2+}, the troponin complex, and tropomyosin. Troponin-C binds Ca^{2+}, troponin-I links to actin and blocks interaction with myosin when $[Ca^{2+}]_i$ is low, and troponin-T connects to tropomyosin. When $[Ca^{2+}]_i$ rises (mainly after Ca^{2+} release from the sarcoplasmic reticulum through voltage-dependent Ca^{2+} channels is insufficient to cause contraction), Ca^{2+} binds to troponin-C. The Ca^{2+}–TnC complex moves TnI away from myosin-binding sites on actin, alters the interaction between TnT and tropomyosin, and frees the binding sites between myosin and actin to initiate a new cross-bridge cycle at many points along the filament, which promotes contraction. Supply of ATP yields energy for contraction. Hydrolysis of ATP stops as soon as Ca^{2+} is removed from cytosol.

6.6.3 Calcium Ion and Its Partners

Heart inotropy depends on intracellular (cytosolic) Ca^{2+} concentration and sensitivity of contractile proteins to Ca^{2+} ions. A maximal stress of about 1 kPa suffices to generate the systolic pressure in the normal left ventricle for blood ejection into aorta [328]. In fact, 4 major factors dictate the active force: $[Ca^{2+}]_i$, phosphorylation status of troponin, sarcomere length, and velocity of sarcomere shortening.

Table 6.19. Main calcium-handling proteins in cardiomyocytes. The intracellular calcium concentration is regulated by ion carriers and intracellular Ca^{2+}-binding proteins.

Ca_V1	Influx across sarcolemma	Contraction (inotropy)
RyR	Release from SR	Contraction
Troponin-C	Ca^{2+} binding	Contraction
Calmodulin	Ca^{2+} binding	Influx limitation
NCX	Influx across sarcolemma (high $[Na^+]_i$)	Contraction
	Efflux across sarcolemma (high $[Ca^{2+}]_i$)	Relaxation
SERCA	SR sequestration	Relaxation (lusitropy)
Calsequestrin	Storage	Relaxation

Calcium ions are a major factor with ATP in cardiac excitation–contraction coupling. Calcium influx from the extracellular medium and mainly cellular stores to the sarcomere, together with ATP from cytolosic sources and mitochondria, as well as outflux from the sarcomere to the sarcoplasmic reticulum, mitochondria, and extracellular fluids are strongly regulated. Moreover, effects influence causes. The membrane potential modulates Ca^{2+} transport that, in turn, modifies the membrane potential.

The magnitude and duration of myocyte contraction is regulated by $[Ca^{2+}]_i$ that depends on protein carriers on cell and sarcoplasmic reticulum membranes (Fig. 5.16, Tables 5.9 and 6.19). Four main Ca^{2+} compartments exist in the cardiac micropump: (1) the extracellular space; (2) sarcoplasmic reticulum for capture and release; (3) troponin-C transiently; and (4) calmodulin. Oxygen is the combustible molecule. Cytoplasmic $[Ca^{2+}]_i$ (hence CMC contraction–relaxation) is strongly controlled.

During the systole, 2 mechanisms – voltage-sensitive release and Ca^{2+}-induced Ca^{2+} release – induce Ca^{2+} flux from the sarcoplasmic reticulum via ryanodine channels (SR Ca^{2+} release channels) [636]. The voltage-sensitive release mechanism is activated by a potential of about -60 mV and triggers Ca^{2+} release from the sarcoplasmic reticulum into the cytoplasm. Calcium-induced Ca^{2+} release deals with Ca^{2+} flux through Ca_V1 channels that elicits Ca^{2+} release from ryanodine channels. Influx of Ca^{2+} from the sarcoplasmic reticulum sufficiently increases the cytosolic $[Ca^{2+}]_i$ to cause contraction.[67] Na^+–Ca^{2+} porter which exchanges 3 Na^+ for 1 Ca^{2+} operates during phase 2. It stabilizes intracellular $[Ca^{2+}]_i$ by a repeated entry of Ca^{2+}.

Relaxation requires Ca^{2+} removal from the cytosol. During the relaxation, cytosolic $[Ca^{2+}]_i$ decreases due to Ca^{2+} uptake: (1) into the sarcoplasmic reticulum via sarco(endo)plasmic reticulum Ca^{2+} ATPases and Na^+–Ca^{2+} exchangers;

[67] About 10% of the calcium needed for cardiac contraction enters from the extracellular space. This trigger calcium induces release of the remainder from the sarcoplasmic reticulum.

Table 6.20. Myofilament Ca^{2+} sensitivity (Source: [464]). CMC stretching enhances the actin–myosin interaction partly by transverse compression of the filament lattice.

Increased	Decreased
Stretching	Acidosis (ischemia)
Caffeine	Elevated phosphate concentration (ischemia)
Certain inotropic drugs	Elevated Mg^{2+} concentration (ischemia)
	β-Adrenergic activation

(2) into mitochondria by mitochondrial Ca^{2+} uniporters; and (3) by extrusion for a small amount via sarcolemmal Ca^{2+} ATPases and Na^{+}–Ca^{2+} exchangers. Sodium–calcium exchanger that functions in its forward mode as well as sarcolemmal and sarcoplasmic reticulum Ca^{2+} ATPases restore cytosolic Ca^{2+} concentration to resting levels and have a lusitropic effect.

Most Ca^{2+} returns to the sarcoplasmic reticulum, where Ca^{2+} is stored by bonds with *calsequestrin*.[68] Agent ATP is not only associated with pumping of Ca^{2+} back into SR and removing Ca^{2+} surplus, but is also required for relaxation of the actin–myosin cross-bridges. The stronger the contraction, the greater the Ca^{2+} reuptake.

Phospholamban (PLb) associated with SERCA inhibits this pump. Its phosphorylation[69] accelerates Ca^{2+} uptake by the SERCA pump; it is enhanced by β-adrenergic stimulation that thus has a positive lusitropic effect. Phosphorylation by protein kinase-A of Ca_V1 channels increases the number of functional channels and opening duration, as well as decreases the channel activation threshold and deactivation rate. Phosphorylation by PKA of delayed rectifier K^{+} channels increases the activity and shortens action potential duration. Protein kinase-A is stimulated by activated Gs protein.

Catecholamines increase inotropy in particular by phosphorylating sarcolemmal Ca^{2+} channels and increasing Ca^{2+} influx into the cytosol during the action potential plateau (phase 2).

Cellular acidosis decreases the sensitivity of contractile elements to Ca^{2+} (Table 6.20). Inorganic phosphate at a concentration achieved during ischemia reduces maximum developed force and sensitivity to Ca^{2+} ions.

Whereas phospholamban abounds in ventricles, *sarcolipin* (Sln) abounds in atria, where it regulates SERCA2a pump. Sarcolipin reduces the affinity of SERCA for Ca^{2+} and maximal velocity of Ca^{2+} uptake rate [637]. In addition, sarcolipin decreases Ca^{2+} transient amplitude, hence atrial contractility, and mediates β-adrenergic responses in atria.

[68]Two calsequestrin isoforms exist, one in the skeletal muscle, the second in the myocardium [458].

[69]Phosphorylation of phospholamban by cAMP-dependent (PKA) or calmodulin-dependent (CamK2) protein kinases relieves the inhibition of SERCA by phospholamban.

6.6.4 *Zinc Ion and Its Partners*

Zinc serves as a structural constituent of proteins, among which thousands of enzymes involved in cell signaling and transcription factors. On the other hand, reactive free Zn^{2+} ion localizes in the nucleus (30%) and in the cytosol and organelles (50%, the remainder being bound to proteins). Free Zn^{2+} ion amount depends on the redox state and the level of reactive oxygen species in cells [638]. Zinc is cytotoxic. Excess Zn^{2+} cation in cells is buffered by binding to cysteine, histidine, or glutamate residues of metalloproteins.

In cardiomyocytes at rest, the cytosolic Zn^{2+} concentration is low (< 1 nmol, i.e., 100-fold less than $[Ca^{2+}]_i$). Oxidants cause approximately a 30-fold increase in $[Zn^{2+}]_i$, but only 2-fold in $[Ca^{2+}]_i$ in cardiomyocytes.

Zinc sparks and *transients* result mostly from an increased cytosolic Ca^{2+} concentration [638]. In neurons, Ca^{2+}-dependent Zn^{2+} release originates from mitochondria.

In mastocytes, *Zinc wave* that corresponds to Zn^{2+} release from the endoplasmic reticulum and/or the surrounding area after cross-linking of the high-affinity immunoglobulin-E receptor also depends on Ca^{2+} influx, without contribution of extracellular Zn^{2+} ion [638].

Calsequestrin is a Zn^{2+}-binding protein of the endoplasmic reticulum that can bind up to 200 mol Zn^{2+}/mol as well as 50 mol Ca^{2+}/mol [638].

Zinc homeostasis in the cell results from a balance between uptake, excretion, and intracellular storage [638]. Zinc enters the cardiomyocyte through SLC39 transporter and different channels such as $Ca_V1.2$ channel. Zinc efflux is carried out by SLC30 transporter (or ZnT). Furthermore, Na^+–Zn^{2+} exchanger enables Zn^{2+} extrusion. A Zn^{2+}–H^+ antiporter may contribute to Zn^{2+} accumulation in cellular organelles.

Zinc provokes autophosphorylation of Ca^{2+}–calmodulin-dependent protein kinase camK2 and inhibition of protein Tyr phosphatases and voltage-gated Ca^{2+} channels [638].

Soluble cytosolic Zn^{2+}-binding proteins comprise anti-oxidant metallothionein. Zinc ion can be released after metallothionein nitrosylation [638]. In cardiomyocytes, mitochondria, which occupy 30 to 40% of the cell volume, constitute the major source of intracellular reactive oxygen species. Reactive oxygen species cause changes in metal-binding properties of metallothioneins and other redox-active proteins. Reactive oxygen species stimulate inducible nitric oxide synthase NOS2 that, in turn, supports the release of Zn^{2+} from Zn^{2+}–metallothioneins.

In cardiomyocytes, nitric oxide mobilizes intracellular Zn^{2+} via the cGMP–PKG pathway and mitochondrial ATP-sensitive K^+ channel [638]. Mitochondrial NO production rises with increasing Ca^{2+} concentration that can result from Ca^{2+} uptake by Ca^{2+} uniporter.

In addition to metallothioneins, carnosine (or β-alanine–histidine) complexes with various heavy metals such as Zn^{2+} cation. It has anti-oxidative activity. However, it abounds much more in skeletal muscles than in the myocardium [638].

Thiol is responsible for about 30% of the Zn^{2+}-buffering capacity and contributes to redox buffer capacity. In cardiomyocytes, $[Zn^{2+}]_i$ can be greatly increased by thiol-reactive oxidants [638].

In cardiomyocytes during the excitation–contraction cycle in physiological conditions, Zn^{2+} sparks are controlled by changes in intracellular free Ca^{2+} [638]. They result mainly from ionic exchange on intracellular binding sites; this process is highly sensitive to the oxidative state and pH inside the cell. In pathophysiological conditions, free Zn^{2+} cation can contribute to oxidant-induced alterations of excitation–contraction coupling. An optimal intracellular Zn^{2+}/Ca^{2+} ratio in cardiomyocytes, particularly in mitochondria, must be preserved to avoid oxidative stress.

6.6.5 Cyclic Adenosine Monophosphate

Cyclic adenosine monophosphate, a mediator of the sympathetic system, operates via cyclic nucleotide pacemaker channel as well as protein kinase-A and RapGEF3 and RapGEF4[70] to regulate the excitation–contraction coupling.

Under β-adrenergic stimulation, protein kinase-A modulates cardiac contractility via intracellular Ca^{2+} fluxes, as it targets sarcolemmal $Ca_V1.2a$ channel and sarcoplasmic reticulum ryanodine receptor. In addition, PKA phosphorylates troponin-I, thereby reducing myofilament Ca^{2+} sensitivity and accelerating relaxation. Furthermore, it phosphorylates myosin-binding protein-C, in addition to titin.

Activators RapGEF3 and RapGEF4 are guanine nucleotide-exchange factors for Rap1 and Rap2 GTPases. Protein RapGEF3 is highly expressed in the heart. Unlike PKA, it decreases the amount of Ca^{2+} released by the sarcoplasmic reticulum, hence reducing Ca^{2+} transient amplitude, enhances myofilament sensitivity to Ca^{2+}, and increases cell shortening [639]. It provokes phosphorylation of cardiac troponin-I and myosin-binding protein-C by protein kinase-C and Ca^{2+}–calmodulin-dependent kinase CamK2.

[70] A.k.a. exchange proteins directly activated by cAMP EPAC1 and EPAC2, respectively.

6.6.6 S100 Protein

Protein S100a1[71] is expressed in the myocardium, where it interacts with contractile filaments and proteins of the sarcoplasmic reticulum (titin, SERCA, phospholamban, and ryanodine channel). Calcium-binding protein S100a1 increases Ca^{2+} release from the sarcoplasmic reticulum by interacting with ryanodine channels. It also contributes to the regulation of Ca^{2+}-induced Ca^{2+}-release.

Cardiomyopathies are associated with altered S100 protein levels. Protein S100a1 is upregulated in right ventricular hypertrophy that is an adaptation to a higher load [642] and downregulated in end-stage heart failure [643]. Administration of the S100A1 gene restores the contractile function of the failing myocardium [644].[72]

Heart failure is characterized by a decrease in calcium removal into the sarcoplasmic reticulum associated with SERCA2 activity and dysregulated calcium release from the sarcoplasmic reticulum via ryanodine channels (Fig. 5.14). Many types of heart failure are associated with downregulation of SERCA2 expression, impaired SERCA2–phospholamban ratio, and decreased phosphorylated phospholamban. Heart failure can therefore be stopped and even reversed by correcting sarcoplasmic reticulum calcium release and storage (SR Ca^{2+} cycling). Inhibition of phospholamban enhances Ca^{2+} uptake in the sarcoplasmic reticulum and reduces heart failure in several animal models of dilated cardiomyopathy and postmyocardial infarction chronic heart failure [416].

6.6.7 Nervous Influences

Adrenergic signaling in the myocardium contributes to the positive control of the heart rate (chronotropy+), contraction strength (inotropy+), and relaxation rate (lusitropy+) by changes in levels of intracellular calcium ions or modifications in sensitivity of regulatory proteins to calcium ions. Inotropic effects are due to increased Ca^{2+} current and greater availability of stored Ca^{2+} ions. Sympathetic control of the heart via β-adrenergic stimulation regulates Ca^{2+} handling by

[71]Multifunctional proteins of the S100 category are diversified according to the structure, ion-binding property (Ca^{2+}, Zn^{2+}, or Cu^{2+}), spatial distribution in the intra- or extracellular space, and ability to homo- and heterodimerize [640]. Several S100 proteins (S100a4, S100a8–S100a9, S100a12–S100a13, and S100b) are secreted and have cytokine-like functions, as they interact with the receptor for advanced glycation end (RAGE) products of the immunoglobulin superfamily. Protein S100a4 has angiogenic effects [641].

[72]Adenovirus-mediated S100A1 gene delivery normalizes S100a1 protein expression in a postinfarction rat heart failure model and reverses contractile dysfunction of failing myocardium. Moreover, S100A1 gene transfer decreases elevated $[Na^+]_i$ to levels detected in non-failing cardiomyocytes and restores energy supply in failing cardiomyocytes.

Ca_V1 channels, ryanodine receptors, Ca^{2+}-activated Ca^{2+} release channels, and sarcoplasmic reticulum Ca^{2+} SERCA2 ATPase. The latter controls Ca^{2+} reuptake into the sarcoplasmic reticulum, after dissociation from phospholamban.[73]

Adrenaline and noradrenaline released by the sympathetic system increase calcium currents conducted by heteromeric $Ca_V1.2$ channels of cardiomyocytes.[74] β-Adrenergic signaling causes a 3- to 4-fold rise in calcium current due to the activation of adenylate cyclase and inhibition of phosphodiesterases. Both activated β1- and β2-adrenergic receptors increase the channel activity via phosphorylation by protein kinase-A in association with A-kinase-anchoring protein AKAP15. However, the calcium current predominantly results from activation of β1-adrenergic receptors.[75]

The sympathetic system stimulates the heart through β-adrenergic receptors. About 75% of cardiac β-adrenergic receptors are β1-receptors. They are distributed throughout the sarcolemma and coupled to Gs-protein. Less abundant β2-adrenergic receptors reside predominantly in caveolae. Several other ion carriers exist in caveolae of cardiomyocytes: voltage-dependent Na^+ channel, $K_V1.5$ channel, Na^+–Ca^{2+} exchanger, pacemaker HCN4 channel, and Ca_V1 channels.[76] β2-Adrenergic receptors signal via the set of cardiac Ca_V1 channels of caveolae in ventriculomyocyte [454]. Channel Ca_V1 colocalizes with caveolin-3, a major structural protein of caveolae in striated myocytes.[77]

β-Adrenergic receptor stimulates Gs protein, which activates adenylate cyclase, producing cAMP, which in turn activates protein kinase-A (Fig. 6.9). The latter phosphorylates SERCA inhibitor phospholamban (lusitropic effect), Ca_V1 channel, ryanodine receptor, troponin-I (speed up dissociation of Ca^{2+} from myofilaments), and myosin-binding protein-C. Increased L-type Ca^{2+} currents through $Ca_V1.2$

[73]Dephosphorylated phospholamban binds to SERCA2 and suppresses SERCA2 activity. PKA-mediated phosphorylation of phospholamban unbinds it from SERCA2 to relieve SERCA2 inhibition.

[74]$Ca_V1.2$ channels are responsible for the major calcium current in cardiomyocytes which triggers release of stored calcium from the sarcoplasmic reticulum for contraction.

[75]β2-Adrenergic receptor-mediated phosphorylation depends on intracellular calcium concentration that must be increased. β2-Adrenergic receptors do not significantly participate in the calcium flux at basal calcium levels [645]. β2-Adrenergic receptors associated with adenylate cyclases, AKAP15, and calcium channels in the cardiomyocyte sarcolemma stimulate a small set of $Ca_V1.2$ calcium channels. Protein kinase-C and calcium–calmodulin-dependent protein kinase-2 can also phosphorylate voltage-gated calcium channels to increase the calcium influx.

[76]These cardiac L-type voltage-gated Ca^{2+} channels outside of junctional complexes do not strongly intervene in excitation–contraction coupling but regulate cellular functions.

[77]Caveolae (Vol. 1 – Chap. 9. Intracellular Transport) are tiny invaginations of the plasma membrane with cholesterol, sphingolipids, and caveolin. Many proteins involved in cellular Ca^{2+} activity are located in caveolae. They include IP_3 receptor, Na^+–Ca^{2+} exchanger, Ca^{2+} ATPase, members of the transient receptor potential family. Caveolin-3, $Ca_V1.2$ channel, β2-adrenergic receptor (but not β1-adrenoceptor), $G\alpha_s$ and $G\alpha_i$ proteins, adenylate cyclase, protein kinase-A, and protein phosphatase-2 form signaling complexes in caveolae of cardiomyocytes.

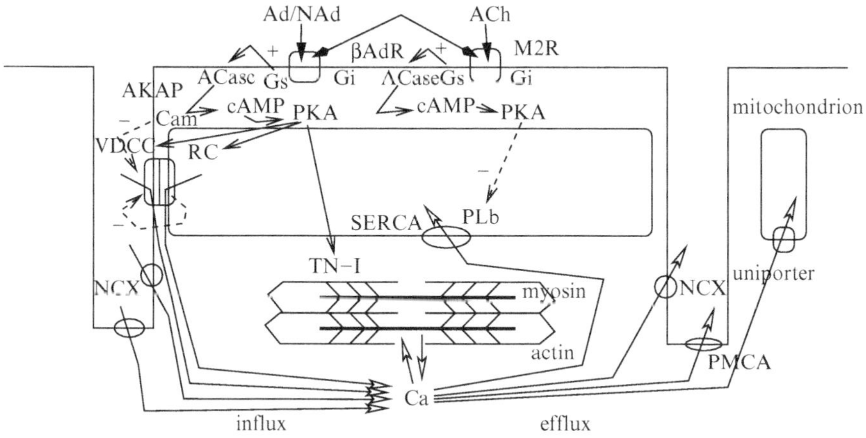

Fig. 6.9 β-Adrenergic and M_2 muscarinic interacting receptors and CMC functioning (Source: [464]).

channels by activated β-adrenergic receptors require phosphorylation by protein kinase-A anchored to the channel via an AKAP15 protein. Proteins AKAP15 and PKA form an inhibitory complex [646].

A-kinase-anchoring protein splice variant AKAP18α complexes with Ca_V1 and ryanodine channels. Another proteic complex that consists of SERCA2, phospholamban, AKAP18δ, and PKA coordinates PKA phosphorylation of phospholamban and β-adrenergic regulation of Ca^{2+} reuptake into the sarcoplasmic reticulum [647].

Muscarinic M_2 receptor activation can either decrease or increase cAMP concentrations, whether these receptors interfere with β1- or β2-adrenergic receptors, respectively [648].

Both compartmentation and dynamics in cyclic adenosine monophosphate and protein kinase-A signaling in cardiomyocytes influence cardiac inotropy. Activation cAMP by protein kinase-A corresponds to a rate-limiting node of PKA pathway downstream from the β-adrenergic receptor.[78] Prostaglandin E1 stimulates higher PKA activity in the cytosol than at the sarcolemma, likely due to differences in cAMP diffusion [649]. Restricted diffusion, cAMP degradation by phosphodiesterases, and cAMP accumulation near protein kinase-A contribute to both spatial and temporal signaling variations.

[78]Synthesis of cAMP can greatly exceed requirements for PKA activation.

Table 6.21. CMC excitability and refractory periods. Cardiomyocytes and nodal cells are inexcitable during absolute refractory period (ARP). Even strong stimuli do not produce any excitation. The cell is refractory to the initiation of new action potentials. The absence of excitability appears because Na^+ channels remains inactivated until sufficient repolarization occurs in the initial part or middle of phase 3. At the end of ERP, a stimulus generates an action potential that can propagate. The recovery of Na^+ channels allows cell excitablity recovery. The recovery of excitability is delayed in the atrioventricular node because Ca^{2+} channels rather than Na^+ channels elicit the upstroke.

ARP	No excitability
RRP	No conduction
ERP	Conditional propagation
SPN	Maximal excitability

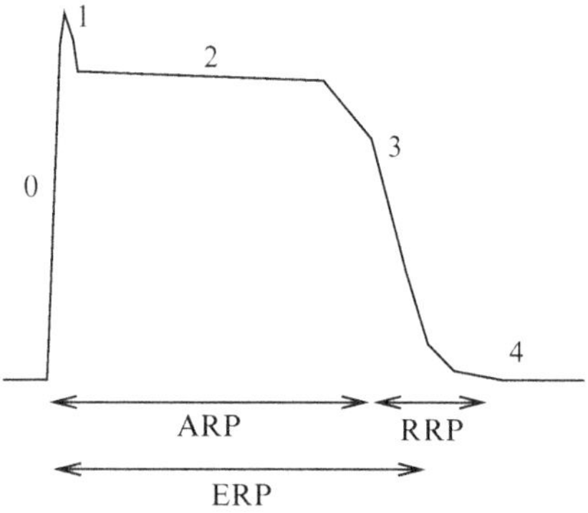

Fig. 6.10 Refractory periods and recovery of excitability. No stimulus can generate another action potential during absolute refractory period (ARP). During the relative refractory period (RRP), an excitation stronger than the one for the electrical diastole is required to evoke an action potential, and a higher threshold potential is set up. Effective refractory period (ERP) extends from the upstroke beginning to the middle of RRP. ERP corresponds to the time interval of the action potential without conduction.

6.6.8 Cell Excitability Adjustment

The cardiomyocyte adapts its sensitivity to repeated stimulations. Moreover, the *electrochemical delay*, i.e., the duration between the membrane depolarization and contraction (cross-bridge formation) is small enough (< 50 ms) to limit asynchrony of the myocardial activity.

The *refractory period* is a protective mechanism to maintain efficient successive blood fillings and ejections. The CMC refractory period is longer than contraction and relaxation periods. Several periods, which overlap each other, starting from the beginning of phase 0, have been defined (Table 6.21, Fig. 6.10). From phase 0 to initial and mid-phase 3, i.e., during the *absolute refractory period* (ARP), the CMC unexcitability is complete. During the *effective refractory period* (ERP), a strong stimulus causes an evoked excitation, but the amplitude response is too low for propagation. During the *relative refractory period* (RRP), the potential propagates when the stimulus is supraliminal. The *supernormal period* (SPN) is a short phase at the end of repolarization, during which the potential initiated by an infraliminal stimulus propagates. The excitability is maximal and decreases to its rest value at the end of the *total recovery period*. Refractory periods vary with excitation rate.

6.6.9 Electromechanical Modeling

Advances in signal and image acquisition and processing enable measurements of electrical and mechanical (cardiac wall motion, cardiac chamber flow patterns, and coronary perfusion) quantities and improve clinical outcome. Nevertheless, anatomical and functional data must be combined in a consistent framework. In addition, personalized treatment remains difficult to optimize.

Mathematical models and associated multimodel, multiscale, and multiphysics computational coupling platforms are aimed at integrating produced measurements, visualizing evolution of quantities of interest during the cardiac cycle, as well as deriving indicators that are not directly observable to assist treatment decisions. Continuum-based simulation codes rely on high-performance computing.

Three-dimensional ventricular electromechanical models primarily rely on 2 submodels that treat the electrical and mechanical activity, respectively. Ventricular geometry is obtained from high-resolution magnetic resonance imaging and computerized tomography. The cardiac geometry is extracted using automatic and semi-automatic segmentation techniques that can be based on atlases computed from population data or statistical shape models that can be applied to heart images. Manual correction tools are available when the segmentation is not accurate enough.

Once the ventricular geometry is reconstructed, an appropriate spatial discretization of the cardiac geometry yields the computational mesh. Mesh adaptation and adaptivity allow to match physical requirements and to follow proper propagation of physical quantities. As the governing equations of electrical and mechanical problems differ, mesh element type and size fit each problem requirements.

The electrical model simulates the propagation of a depolarization wave in the cardiac wall. The electrical model describes the ionic current using a simplified representation with a low number of parameters (Vol. 7 – Chap. 8. Numerical Simulations). High-resolution tetrahedral meshes can be employed for detailed electrophysiological analysis. Further information on the nodal network can be added. An organ-level problem is described by reaction–transport equations. A reduced electrophysiological model such as the one proposed by Mitchell and Schaeffer can be used. Incorporating heterogeneous electrical conductivities of the thorax and its content, the propagation of depolarization wave can be computed on the body surface to yield an ECG trace. A sensitivity analysis of the effect of mesh size, initial stimulation points, and conduction velocity, among other parametyers, can be carried out. Moreover, the inverse problem can be treated to estimate adequate activation times. Electromechanical coupling can incorporate changes in electrical conductivity and membrane capacitance caused by heart deformations.

The mechanical model simulates the myocardium contraction mainly using a phenomenological approach and considering 2 main chemical mediators, ATP and calcium ion. Diffusion tensor magnetic resonance imaging can be used to estimate the myofiber orientation in the heart wall that determine the local electrical conductivities and mechanical properties of the myocardium. The mechanical model simulates ventricular contraction and associated deformation. The finite element method is commonly used to solve the electrical and mechanical problems. The solid mechanics problem relies on a large-element mesh with higher order of interpolation.

As the heart is aimed at propelling blood, a blood convection model is added with blood perfusion of the myocardium for nutrient uptake. Lumped parameter models (0D models) such as 2- to 4-element windkessel models can be coupled to detailled flow models.

Validation of the electrical model relies on comparision of simulation results to electrical depolarization maps obtained from epicardial and/or endocardial electrode sets as well as from ECGs. The mechanical model is validated by wall motion or local strain measurements using magnetic resonance or ultrasound techniques. Flow simulations can be compared to pressure and volume measurements.

6.7 Nervous Inputs

The control of the function of the cardiovascular system involves both fast-acting responses and long-lasting adaptations acting on gene expression. The cardiovascular system is quickly regulated by a set of mechanisms, which mainly involve the central nervous system (fast neural command) and the release of hormones from endocrine organs (delayed humoral command, triggered particularly by catecholamines, vasopressin, angiotensin-2, and natriuretic peptides).

The heart is innervated by both components of the autonomic nervous system, parasympathetic and sympathetic nerves, which send signals to the cardiac cells via plasmalemmal receptors, thus triggering signaling pathways (Tables 6.22 and 6.23).

Signaling pathways involve sequential protein phosphorylation cascades that ultimately activate transcriptional factors driving gene expression. Protein phosphatases terminate nervous signaling. Within the cells, particularly in cardiac and vascular smooth myocytes, calcium ions trigger contraction, metabolism, and growth. Cell activity, in particular cardiomyocyte contraction, ends when calcium ions are removed from regulatory proteins, such as calmodulin and troponin.

Coronary vessels also receive both sympathetic and parasympathetic innervation. The cardiovascular centers of the brainstem get signals from various sensors

Table 6.22. Parasympathetic and sympathetic (α) neurotransmitters, receptors, and effectors. Protein Gs activates adenylate cyclase (ACase), which triggers the PKA pathway, leading to positive inotropy (faster, stronger, shorter contraction) and lusitropy (faster relaxation). Protein Gs stimulates Rho GTPases and mitogen-activated protein kinases (MAPK). Trimer Gi–Gβγ inhibits ACase and stimulates PLC, NOS, and PI3K pathways, thus targeting PKC (cell conditioning), PKG (antiproliferative and negative inotropy), and PKB (anti-apoptotic), respectively. Trimer Gq–Gβγ also activates the PLC, NOS, and PI3K pathways. PKC stimulates and PKA inhibits the activity of adenylate cyclases (positive and negative feedback loops). Reduced arginine can form superoxide and reactive nitric oxide species (RNOS). Superoxide rapidly converts NO to RNOS (Source: [650]).

	Parasympathetic	Sympathetic (α)	
Neurotransmitter	Acetylcholine	Noradrenaline	
GPCR	M_2	α1	α2
G Protein	Gi/Gβγ	Gi/Gβγ Gq/Gβγ	Gs Gi/Gβγ
First effectors	ACase (−) PLC NO/GCase PI3K	ACase (−) PLC NO/GCase PI3K	ACase (−) PLC NO/GCase PI3K
Second messengers	 IP_3/Ca/DAG RNOS/cGMP PDK1	 IP_3/Ca/DAG RNOS/cGMP PDK1	cAMP IP_3/Ca/DAG RNOS/cGMP PDK1
Protein kinases	PKB/PKC/PKG	PKB/PKC/PKG	PKA/PKB/PKC/PKG

distributed throughout the body. Efferent nerves connect to nodal tissue and myocardium, controlling primarily the heart frequency and contractility. Efferent nerves to the smooth muscles in coronary vessel walls control arterial tone, arteriolar resistance, and blood volume in the large veins.

Normally, parasympathetic innervation is the dominant neural influence on the heart. However, the adrenergic system, with its various specific G-protein-coupled receptors and their corresponding cell signaling cascades, yields the major regulatory system, whereas the parasympathetic system plays a minor role in responses to environmental stimuli.

Galanin (Gal), a neuropeptide that is cleaved from preprogalanin, inhibits neurotransmitters. Galanin increases mean arterial blood pressure and heart rate, inhibiting cholinergic activity in the heart in cooperation with neuropeptide Y (NPY). Activated Gal1 and NPY-Y2 receptors of parasympathetic terminals of vagal neurons in the heart reduce acetylcholine release, especially during prolonged attenuation of parasympathetic slowing following activation of the cardiac sympathetic nerve [651].

Table 6.23. Sympathetic (β) neurotransmitters, receptors and effectors (see caption of Table 6.22; Source: [650]).

	Sympathetic (β)		
Neurotransmitter	Noradrenaline		
GPCR	β1	β2	β3
G Protein	Gs	Gs Gi/Gβγ Gq/Gβγ	Gs Gi/Gβγ
First effectors	ACase(+/−)	ACase(+/−) PLC NO/GCase PI3K	ACase(+/−) PLC NO/GCase PI3K
Second messengers	cAMP	cAMP IP_3/Ca/DAG RNOS/cGMP PDK1	cAMP IP_3/Ca/DAG RNOS/cGMP PDK1
Protein kinases	PKA	PKA,PKB, PKC,PKG	PKA,PKB PKC,PKG
Effectors	Ras, MAPK		

Adrenergic G-Protein-Coupled Receptors

The myocardial systolic and diastolic function are mainly governed by β1- and β2-adrenergic receptor activities. However, α-adrenergic receptors leads to positive and negative inotropy. The family of adrenergic receptors consists of 9 subtypes (α1a–α1b, α1d, α2a–α2c, β1 (∼70%),[79] β2 (∼10%), and β3 (∼1%; Table 6.24).

Dimerization can occur between adrenergic receptors, such as between α2- and β1- or β2-adrenergic receptors. Scaffold A-kinase-anchoring proteins interact with adrenergic receptors, especially β2-adrenoceptors.

Stimulated adrenergic receptors activate pathways leading to cytosolic protein Ser/Thr kinases (PKA, PKB, PKC, and PKG).[80] Adrenergic G-protein-coupled receptors can also activate Ras GTPase and MAPK cascades. Phosphorylation by G-protein-coupled receptor kinases, with arrestin and adaptor AP2, desensitizes nervous signaling.[81]

---

[79] β1-Adrenergic receptors are downregulated in hypertrophic and ischemic cardiomyopathy.

[80] Inactive PKA binds to cAMP, then dissociates into cAMP-bound dimer and active PKAs. PKC family members are activated via phospholipase-C. PKCα, PKCβ1/2, and PKCγ require a phospholipid, Ca^{2+}, and DAG for activation; PKCδ, PKCϵ, PKCη, and PKCθ needs DAG but not Ca^{2+}; PKCζ and PKCλ do not have Ca^{2+}- and DAG-binding sites.

[81] Kinases GRK2 and GRK5 are expressed in cardiomyocytes.

Table 6.24. Adrenergic receptors; corresponding signaling pathways; and functions in nervous, cardiac, and vascular tissues (Source: [650]). Effectors include adenylate cyclase (ACase), cyclic adenosine monophosphate (cAMP), protein kinase-A (PKA), guanylate cyclase (GCase), cyclic guanosine monophosphate (cGMP), protein kinase-G (PKG), nitric oxide synthase (NOS), phospholipase-C (PLC), diacylglycerol (DAG), inositol trisphosphate (IP_3), calcium ions, protein kinase-C (PKC), phosphoinositide 3-kinase (PI3K), phosphoinositide-dependent kinase-1 (PDK1), protein kinase-B (PKB), Rho GTPase, and mitogen-activated protein kinases (MAPK).

Receptor	Effectors	Functions
α1a	Gq–PLC–IP_3–Ca^{2+} Rho–RoCK MAPK (ERK, JNK, P38MAPK)	Inotropy +, myocyte growth, vasoconstriction
α1b	Gq–PLC–IP_3–Ca^{2+} ERK, P38MAPK	Inotropy +, myocyte growth, vasoconstriction
α1d	Gq–PLC–IP_3–Ca^{2+} MAPK (weakly)	Vasoconstriction
α2a	Gi (presynaptically)	Sympathetic output reduction, hypotension
α2b	Gi (postsynaptically)	α2a counteraction, peripheral vasoconstriction, hypertension
α2c	Gi (presynaptically, at low frequency nerve activity) (postsynaptically)	Sympathetic output reduction, cAMP antagonist Vasoconstriction (cold), hypotension
β1	Gs–ACase–cAMP–PKA	Inotropy +, lusitropy +, metabolism, growth
β2	Gs–ACase–cAMP–PKA Gi/Gβγ–PLC–DAG–PKC Gβγ–PI3K–PDK1–PKB	Inotropy +, lusitropy +, metabolism, growth, CMC survival, relaxation of bronchial and vascular SMC
β3	Gi/Gβγ–PLC–DAG–PKC Gi/Gβγ–NOS–NO–GCase–cGMP–PKG	Inotropy −, CMC survival

Activated α1-adrenoceptors interact with $G\alpha_q$, which signals via phospholipase-C leading to: (1) diacylglycerol, then activating protein kinase-C; and (2) inositol trisphosphate and calcium influx mainly from the sarcoplasmic reticulum. α2-Adrenoceptors coupled to $G\alpha_i$ inhibit plasmalemmal adenylate cyclase,[82] activating K^+ channels and inhibiting sarcolemmal Ca_V1 channels.

[82]Among the 9 known isoforms (ACase1–ACase9), adenylate cyclase-5 is specifically expressed in cardiomyocytes; adenylate cyclase-6 is expressed in other cardiac cells. Other isoforms are expressed in the heart to a lesser extent. The fine-tuning in ACase regulation controls the adrenergic

$\beta 1$-Adrenoceptors are coupled to $G\alpha_s$, inducing positive inotropy and lusitropy via adenylate cyclase and protein kinase-A. Both $\beta 2$- and $\beta 3$-adrenoceptors activate $G\alpha_s$, $G\alpha_i$, or $G\alpha_q$ according to the humoral context. $G\alpha_s$ and $G\alpha_q$ protein sets work with adenylate cyclase and PLCβ pathways, respectively. Members of the $G\alpha_i$ set can act via $G\beta\gamma$. Dimer $G\beta\gamma$ activates phosphoinositol 3-kinase.

$\beta 2$-Adrenoceptors are preferentially coupled to $G\alpha_s$, but they also bind $G\alpha_i$- or $G\alpha_q$-coupled receptors. When stimulated by noradrenaline, $\beta 2$-adrenoceptors interact much faster with $G\alpha_s$ than with $G\alpha_i$ or $G\alpha_q$. Besides, PKA-dependent phosphorylation of $\beta 2$-adrenoceptors switches the G-coupling from $G\alpha_s$ to $G\alpha_i$, thus inhibiting ACase and activating the MAPK cascade.

$\beta 3$-Adrenoceptors lack phosphorylation sites. They thus are not desensitized by GRKs, PKA, or PKC kinase. Adrenoceptors-$\alpha 1$-, -$\alpha 2$, and -$\beta 1$ to $\beta 3$ exist postsynaptically in cardiomyocytes, smooth myocytes, and endothelial cells. β-Adrenoceptors are major regulators of the cardiac activity.

Positive inotropic and lusitropic effects of $\beta 1$- and $\beta 2$-adrenoceptors are mediated by PKA phosphorylation of phospholamban, troponin-I, Ca_V1 channel, and ryanodine receptor RyR2. Cardiac phosphodiesterases degrade cAMP, thus limiting the magnitude of β-adrenoceptor stimulation. At least 4 phosphodiesterase types (PDE1–PDE4), mostly in the cell cortex, exist in cardiomyocytes.[83]

Positive chronotropy can be achieved by shifting the activation curve of the ion channels to more positive voltages. Such a process is used by cAMP and cGMP to increase HCN channel activity. Cyclic nucleotides bind to HCN channels. The HCN channels are activated during membrane hyperpolarization after the termination of an action potential and yield Na^+ influx, which slowly depolarizes the sarcolemma. Several types of voltage-gated HCN cation channels exist (HCN1-HCN4). Isotype HCN4 is the predominant type in the sinoatrial node. Sympathetic stimulation of the sinoatrial node raises cAMP concentrations, hence accelerating depolarization and the cardiac frequency. Stimulation of muscarinic receptors slows the heart rate by an inverse action.

---

signal transmission rate. Elevated intracellular calcium levels during sustained activity inhibit adenylate cyclases ACase5 and ACase6.

[83]Activity of PDE1 depends on the calcium–calmodulin complex, the quantity of which increases concomitantly with elevation in cytoplasmic calcium level. Subtype PDE2 is stimulated by cAMP messenger. Subtype PDE3 is inhibited by cAMP agent. Subtype PDE4 is insensitive to cAMP mediator.

Chapter 7
Vessel Wall

Blood and lymphatic vessels form networks for the transport of fluids, gases, macromolecules, particularly nutrients and signaling compounds, and circulating cells between the body's tissues. The lymphatic network is involved in the immune surveillance of the body. Lymphocytes and antigen-presenting dendritic cells travel through lymphatics between peripheral tissues and lymphoid organs, where immunocytes mature.

Blood and lymphatic networks are connected by left and right lymphatic ducts that drains lymph into blood circulation. *Left lymphatic duct*,[1] or the *thoracic duct*, is the largest lymphatic vessel (length 38–45 cm; diameter ~5 mm). It originates in the abdomen from the confluence of the right and left lumbar and intestinal trunks to terminate close to the junction of the left subclavian and left internal jugular veins. Lymph from the right arm, right region of the thoracic cavity, right side of the neck and head, and possibly lower left lobe of the left lung, is collected by the *right lymphatic duct* (length ~12.5 mm), which commonly ends in the right subclavian vein, near its junction with the right internal jugular vein.

Both vascular structures are lined with endothelial cells that acquire tissue-specific specialization (Chap. 9). Endothelium ensures formation, maintenance, growth, and repair of vessel walls (Chap. 10).

7.1 Physical Environment

Vascular cells adapt to mechanical stimuli and remodel, thereby changing geometry, structure, and properties. Major stresses applied to blood vessels include: (1) axial stress induced by longitudinal tension; (2) circumferential stress ($c_\theta = p_i R_i / h; R_i$: internal hydraulic radius; h:wall thickness [Laplacian equation for

[1] A.k.a. alimentary duct, chyliferous duct, and Van Hoorne's canal.

M. Thiriet, *Tissue Functioning and Remodeling in the Circulatory and Ventilatory Systems*, Biomathematical and Biomechanical Modeling of the Circulatory and Ventilatory Systems 5, DOI 10.1007/978-1-4614-5966-8_7, © Springer Science+Business Media New York 2013

thin, homogeneous wall]) that results from perpendicularly applied blood pressure p_i; (3) radial compressive stress developed by blood pressure that has a much smaller magnitude with respect to axial and circumferential stresses; and (4) wall shear stress exerted by blood flow at the wetted surface of the vessel wall, i.e., friction on the luminal surface of the endothelium.

Stress distribution within large vessel wall is heterogeneous, as the wall is composed of 3 layers (intima, media, and adventitia) of composite materials. Moreover, stress field bears large temporal and spatial variations, as blood flow is three-dimensional, developing (or unestablished, as flow develops at the entrance of short curved vessels between 2 successive branchings),[2] and unsteady.[3]

Elastic properties of vascular wall are usually represented by rheological parameters defined at specific pressures, as the pressure–cross-sectional area relationship is non-linear: (1) either vascular distensibility $((\partial A_i/A_i)/\partial p_i)$ or wall compliance and (2) elastic moduli (Vol. 7 – Chap. 5. Rheology). These parameters deal with the bulk wall behavior and do not represent elastic properties of wall material.

7.2 Vasculature Development

Mesodermal cells in the early embryo differentiate into endothelial precursors — angioblasts — and form blood islands. Fusion of blood islands leads to the primary capillary plexi. When blood circulation is established, primary plexi remodel into a hierarchical network of arteries, arterioles, capillaries, venules, and veins.

Vascular smooth myocytes (SMC) are associated with large and mid-sized vessels, whereas capillaries are covered by pericytes. First lymphatic endothelial cells sprout from embryonic veins, then they migrate to form lymphatic sacs. Lymphangiogenesis also involves sprouting, branching, and remodeling. Closed-end lymphatic capillaries drain into collecting vessels and ducts.

The branchial arch artery apparatus is a transient structure that appears in 6- to 7-week-old human embryos. Five branchial arches exist, each of which contains an arch artery (1–4 and 6). Blood exits from the heart via branchial arch arteries to circulate throughout the embryo. The branchial arch artery apparatus is initially symmetrical with equivalent amounts of blood flow through all of these arteries.

---

[2]Because of the presence of the vessel wall, a boundary layer develops in the entry segment of any vessel, at least in the inner wall after branching. In other words, a velocity gradient from wall to flow core occurs and develops along the wetted surface of the conduit wall, whereas the blood stream in the central region of the duct lumen appears to have an inviscid-flow type, i.e., the velocity remains uniform. The thickness of the boundary layer, in which momentum diffuses toward the vessel axis, grows in the streamwise direction until it reaches a critical length, the so-called entry, or entrance, length (L_e). From the entry length, flow becomes established, or fully developed. Downstream from L_e, the pressure gradient that must overcome the resistance to flow decreases linearly and the cross-sectional velocity distribution is invariant.

[3]The action of the cardiac pump imposes a time-dependent flow evolution that is quasi-periodic.

This arterial apparatus afterward matures and sustains an asymmetrical remodeling. The left fourth and sixth arch arteries persist and generate the aortic arch and pulmonary trunk, whereas the right fourth and sixth branchial arch arteries regress, as blood predominantly flows through the left arch arteries.

This event is concomitant with increased expression of growth factors, such as platelet-derived (PDGF) and vascular endothelial (VEGF) growth factors. It follows rotation and realignment of the ventricular outflow tract that forms part of the left and right ventricles. Outflow tract rotation is required for reshaping of the right sixth arch artery into a long, narrow vessel with reduced blood flow.

The transcription factor pituitary paired-like homeodomain-containing protein Pitx2 (encoded by the homeobox PITX2 gene) stimulated by Nodal, a member of the TGFβ growth factor superfamily involved in cell differentiation, mesoderm formation, and subsequent organization of left–right axial structures, regulates the asymmetric growth. It abounds in cells of the secondary heart field that includes the myocardium of the left wall of the outflow tract. Pitx2+ cells participate to the control of rotation of the outflow tract. Factor Pitx2 is not synthesized in cell types that belongs to or surrounds sixth branchial arch arteries [652]. Although branchial arch artery remodeling requires Pitx2, it is not directly governed by Pitx2+ cells. The anterior heart also requires FGF8 that is overexpressed in the absence of Pitx2 protein. Factor FGF8 may be involved in arch artery maturation. Besides, the Nodal–Pitx2 pathway and abnormal rotation of the outflow tract do not strongly affect branchial arch artery laterality.

Vascular growth factors, such as PDGFa protomer and VEGF, contribute to branchial arch artery apparatus remodeling, as their inhibition specifically yields loss of both left and right sixth arch arteries [652]. Expression of PDGFa is maintained on the left side but decays on the right. The concentration of phosphorylated VEGFR2 in endothelial cells also decreases in the right sixth branchial arch artery, whereas in the case of Pitx2 loss, PDGFa and VEGFR2^{P} levels remain symmetrical. In addition, ligated left sixth branchial arch artery in 11.5-day-old mouse embryos (before the onset of remodeling; i.e., approximately 35-day-old human embryos) leads to their regression and development of right arteries [652]. In 4 among 6 embryos with ligated left sixth branchial arch artery, PDGFR and VEGFR2^{P} signalings are reduced on the left side and maintained on the right side. Therefore, hemodynamical stress shapes the asymmetrical arterial system by operating via growth factors.

7.3 Wall Structure

As blood pulses in an artery, its wall alternatively stretches and recoils. In 1733, S. Hales described blood systolic storage and diastolic restitution. In 1899, O. Frank modeled arterial capacitance by a *windkessel* element, i.e., a damping air chamber incorporated in a rigid pipe network. Blood vessels are pressure and volume buffers.

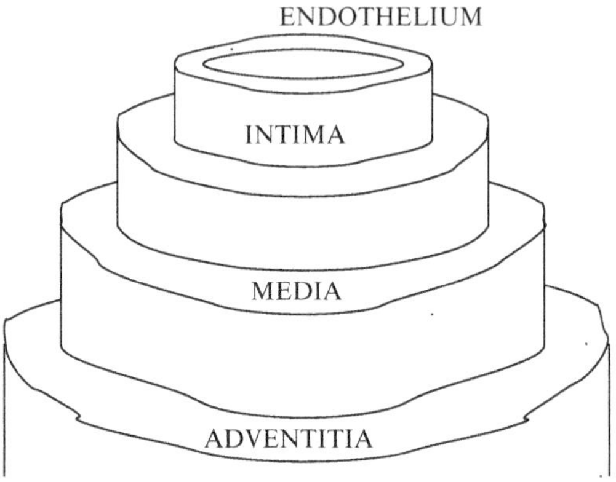

Fig. 7.1 Wall tunicae. The media (intermediate layer) is bounded by the elasticae interna and externa from the intima (inner layer) and adventitia (outer layer).

Veins that are the major blood storage compartments, accommodate volume changes by shrinking to possibly reach a collapse state.

Wall expansion and relaxation depend on the rheology of the vessel wall, and thus on its composition and structure. In particular, distension and subsequent recoil are achieved by the architecture and interactions between coupled elastin and collagen fibers. In addition to passive caliber changes, the vessel wall undergoes active variations due to vasomotor tone. Vascular smooth muscle activity fits the blood flow to the tissue requirements.

Structural components exist in every kind of blood vessel, except capillaries, although the element amount and structure vary between vessel types. Endothelial cells with a hemodynamic stress-dependent shape ($50–100 \times 10 \times 0.5–2\,\mu m$), line the blood–wall interface of the whole vasculature (Chap. 9). Smooth myocytes are responsible for vessel tone and hence lumen size (Chap. 8). They thus influence the vascular impedance.

Specific cells of connective tissue (Vol. 1 – Chap. 1. Cells and Tissues) include fibroblasts (FB; Vol. 1 – Chap. 2. Cells of the Blood Circulation)) that produce ground matrix and fibers in growing tissues and fibrocytes (FC). The ground matrix fills the space between cells and fibers. It is composed of interstitial fluid that contains electrolytes and plasma molecules, proteoglycans, and glycoproteins. *Elastin* forms low-stiffness elastic fibers, taut in the unstretched configuration that can stretch up to 60% with elastic behavior. *Collagen* makes high-stiffness fibers that are tortuous in the unstretched configuration and are responsible for structural integrity. Thereby, elastin provides vessel distensibility and collagen tensile strength.[4]

The wall structure is characterized by a tunica set that comprises intima, media, and external adventitia in large arteries and veins (Fig. 7.1). Tunica number and structure vary according to vessel type and size.

[4] Wall distensibility increases when tension rises. Elastin is involved at low stresses and collagen is progressively recruited when tension increases.

7.3.1 *Large Vessels*

The wall of large blood vessels has a circumferentially layered structure. The internal thin *intima* is composed of the inner *endothelium* and a subendothelial tunica of connective tissue. The endothelium is a monolayer of endothelial cells supported at its abluminal surface by a basal lamina. The endothelium modulates wall structure and functions.

The *internal elastic lamina* (IEL) made of numerous azimuthal bundles of elastic fibers delimits the intima from the media. The fenestrated internal elastic lamina is characterized by a quasi-uniform distribution of pores. Water and solutes are transported through these fenestrae. Cells communicate and migrate via these pores. The number of fenestrations is greater at branching points.

The middle muscular tunica — *media* — is formed by layers of circumferential smooth myocytes and connective tissue with azimuthally arranged elastic fibers. Elastin sheets, or lamellae, are fenestrated to facilitate material diffusion. Action potentials trigger SMC contraction after a delay ranging from 80 to 100 ms that lasts from 10 to 15 s [653]. According to the vessel type, media contains varying amounts of collagen and elastic fibers, elastic lamellae, and proteoglycans secreted by smooth myocytes. Media of arteries is larger than that of veins of similar size.

The *external elastic lamina* (EEL), a thick elastic circumferential band, is located between the media and adventitia.

The external *adventitia* consists mainly of connective tissue with predominant fibroblasts, some smooth myocytes, many macrophages, elastic fibers and predominant longitudinally arranged collagen fibers. In large vessels, the adventitia contains nerves, vessels — *vasa vasorum*, and possibly lymphatic vessels. Media of arteries is much thinner than that of veins of similar size.

7.3.1.1 Elastic Arteries

The media is the main site of histological specializations of artery walls (Table 7.1). Vessels proximal to the heart are elastic arteries.

Wall distension during pressure wave propagation for efficient windkessel effect is facilitated by thin concentric fenestrated lamellae[5] of elastin in a thick media, between which are smooth myocytes and collagen fibers (Table 7.2). The lamella number depends on vessel size, between-lamella space being nearly constant. Inside the elastic lamella, elastic fibers have a helical orientation.[6]

---

[5]Lamella: diminutive of lamina, or lamna: thin plate, membrane, layer.

[6]The orientation of elastin fibers in a given rabbit artery slice is parallel to the local artery axis or nearly perpendicular to it, whether the elastin fibers belong to the fenestrated internal elastic lamina or medial elastic bundles (P. Farand, personnal communication). Therefore, elastin fibers of the internal elastic lamina and media sustain longitudinal and circumferential loadings, respectively. Bridges exist between axially and circumferentially oriented elastin fibers.

Table 7.1. Artery wall structure. Downstream from the heart, the arterial tree successively comprises elastic, transition, and muscular arteries, and arterioles. The different artery types act as conducting, distributing, and resistive arteries with increasing distance from the heart. During the systole, elastic arteries store a part of the blood volume ejected by the left ventricle for diastolic restitution owing to wall recoil. Peripheral muscular arteries and arterioles locally control the lumen size and the vessel resistance.

Layer	Elastic artery	Muscular artery	Arteriole artery	Cerebral artery
Caliber	Great	Mid	Small	Mid
Intima			Absence of subendo- thelium	
IEL	Thin elastic lamellae	Important, scalloped	Thin Thin	Well developed
Media	Thick Numerous elastic lamellae	Thin Numerous layers of SMCs	1–5 layers of SMCs	Almost no EnFs
EEL	Badly defined	Clearly defined	Badly defined	Absent
Adventitia	Thin	Thick		

Table 7.2. Wall structure (EC: endothelial cell; FB: fibroblast; SMC: smooth muscle cell; Col: collagen; Eln: elastin; FN: fibronectin; Lam: laminin; PoG: proteoglycan; Tsp: thrombospondin; vWF: von Willebrand factor).

Wall layer	Components
Endothelium	EC
Basement membrane	Col4, Lam
Subendothelium	Col3/4/5, Lam, FN, PoG, Tsp, vWF, FB, ±SMC
IEL	Eln
Media	Col, Eln, PoG, FB, SMC
EEL	Eln
Adventitia	FB, Eln, Col Vasa vasorum, nerve endings, lymphatics

The external elastic lamina is not very well defined and the adventitia is thinner. The orientation and extension of the collagen fibers of the adventitia of human aortas that undergo uniaxial tension are correlated to the applied stress [654]. After straightening and suitable orientation, collagen fibers can bear increasing loads.

Table 7.3. Wall composition in elastin and collagen (dog, % dry defatted tissue) in different segments of proximal arteries and elastic modulus range (*kPa*; Sources: [655, 656]).

	Collagen	Elastin
Ascending aorta	15–20	40–50
Thoracic aorta	15–30	20–40
Abdominal aorta	25–35	15–20
Femoral artery	35–45	15–25
Elastic modulus	10^2–10^6	10^2–10^3

7.3.1.2 Muscular Arteries

Muscular arteries have thinner intima. The media is characterized by numerous concentric layers of smooth myocytes with elastic fibers and few collagen fibers. The external elastic lamina can be clearly observed. The wall content in elastin and collagen varies according to the artery type, elastic, transitional, or muscular (Table 7.3). The fibers have a helical orientation with a variable pitch angle from the inner to the outer vessel wall. Cerebral arteries are thin walled without external elastic lamina.

Adventitia that surrounds most blood vessels allows cell migration into and out of the artery wall. Adventitial cells participate in vascular growth and remodeling. Adventitia is a residence zone for progenitor cells that can adopt both vascular and non-vascular fates. Endothelial progenitor cells form clusters between the media and adventitia in arteries.

The morphogen Sonic hedgehog (Vol. 3 – Chap. 10. Morphogen Receptors) stimulates production of angiogenic factors (e.g., VEGFa and angiopoietin-1) by fibroblasts, and promotes angiogenesis. It also specifies arterial or venous identity of endothelial cells and recruits mural cells. Sonic Hedgehog restricted to the adventitia of artery wall favors maintenance of resident SCA1+ vascular progenitor cells [657].[7] Some progenitors self-renew, whereas others differentiate.

These progenitors express transcription factors for SMC differentiation, including serum response factor and myocardin family members[8] (i.e., myocardin and myocardin-related transcription factors MRTFa and MRTFb), as well as transcriptional corepressors of the SRF-dependent transcription (Krüppel-like factor KLF4, homeodomain-containing protein Hox7,[9] and forkhead box transcription factor FoxO4) for progenitor maintenance.

[7] Adventitial sonic Hedgehog has low levels in adult mouse arteries for postnatal growth and remodeling of the artery wall.

[8] Myocardin is a smooth and cardiac myocyte-specific transcriptional coactivator of serum response factor. Members of the myocardin family of coactivators activate genes involved in cell proliferation and migration and myogenesis.

[9] A.k.a. Msh homeobox gene product Msx1.

Table 7.4. Comparison between a vein and its associated artery walls.

	Artery	Vein
Shape	Circular	Elliptical
Wall	Thick	Thin
Intima		Very thin
Media	Prominent	Weak
IEL		Quasi-absent

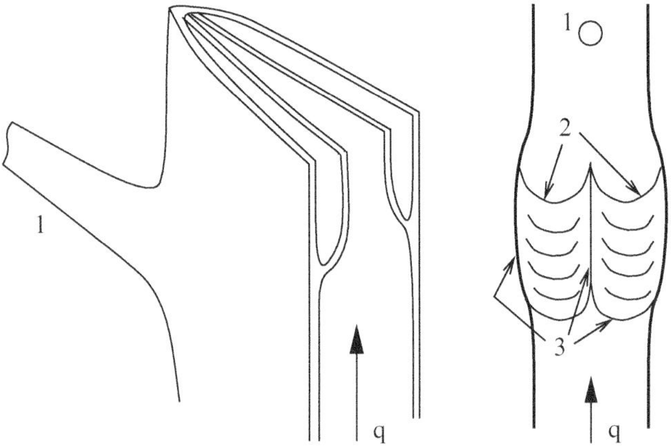

Fig. 7.2 Bicuspid venous valves. (**Left**) Cross and axial cut of a vein. (**Right**) Open and spread vein after axial incision of a wall edge (1: orifice of communication vein at the vein edge; 2: cusp free border in front of the vein face; 3: valve insertion lines). The valvular sinuses are behind the cusps. The luminal face is smooth; the parietal face is rough.

7.3.1.3 Veins

Vein walls are thinner than artery walls, whereas the bore is larger (Table 7.4). The intima is very thin. Internal and external elastic laminae are either absent or very thin. The media is thinner than the adventitia.

Medium-sized veins are characterized by the presence of *valves* to prevent transient blood return to upstream segments during muscular compression (Figs. 7.2 and 7.3).

The largest veins of the abdomen and thorax have very thick adventitia that contains bundles of longitudinal SMCs and vasa vasorum. Valves are absent. The main features of the wall are given for different kinds of veins, sorted by their bore, in Table 7.5.

Different gene expression occurs between arterial and venous walls that are exposed to different mechanical stress magnitudes. γ-Catenin (or plakoglobin),

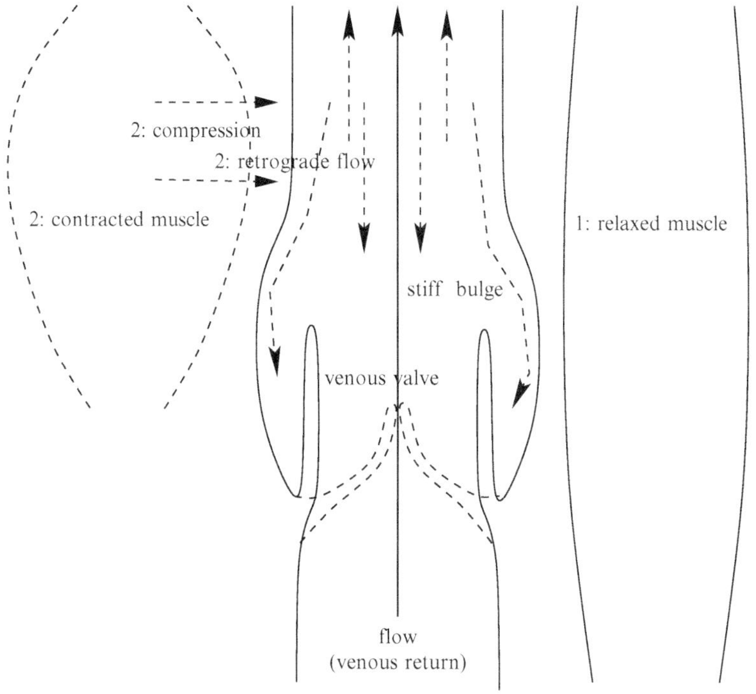

Fig. 7.3 Venous valves are fully open when neighboring muscles are relaxed. Adjoining muscle contraction compresses many more avalvular segments of veins than valvar parts that are slighly dilated and stiffer. Consequently, venous return is favored, but is associated with retrograde flow. The latter is impeded by valve closure due to blood stream between the vein wall and the valve.

galactoside-binding, soluble lectin LGalS7 (or galectin-7),[10] sciellin (Scel), and small proline-rich protein SPrR3,[11] are expressed in arteries but not veins [658]. γ-Catenin, sciellin, SPrR3, and galectin-7, as well as periplakin (PPl)[12] and envoplakin (EvPl),[13] are arterial intima-enriched components [659].[14]

[10]Galectins are β-galactoside-binding proteins that modulate between-cell and cell–matrix interactions. It is specifically expressed in keratinocytes.

[11]A.k.a. pancornulin and cornifin-β.

[12]A.k.a. 195-kDa cornified envelope precursor protein.

[13]A.k.a. 210-kDa cornified envelope precursor protein.

[14]In saphenous vein coronary artery bypass grafts, sciellin and periplakin are upregulated after arterialization [659]. Synthesis of sciellin, SPrR3, and periplakin rises upon prolonged exposure of vascular smooth myocytes in vitro to cyclic strain (24–72 h).

Table 7.5. Vein wall structure.

Vein type	Caliber	Wall
Microvenule	15–100 μm	Endothelium, thin layer with CnFs, ±FBs, SMCs if caliber > 50 μm
Minivenule	100–300 μm	Continuous layer of SMCs, ±postcapillary sphincter
Venule	300–500 μm	Media, thin adventitia (EnFs, CnFs, FBs)
Drainage vein	0.5–2 mm	Intima with subendothelium, ±EnF network, adventitia
Collecting vein	2–9 mm	Intima sometimes bounded by EnF network, media poorly developed, except in inferior limbs, thick adventita, valves in zones of greater caliber
Central vein	20–30 mm	Adventitia of vena cava with CMCs in the heart vicinity

7.3.1.4 Venous Valves

Perforating veins usually contain venous valves that prevent reflux of blood from the deep veins into the superficial veins (*ostial valve*).

Although venous drainage in the human head and neck commonly does not counter gravity, venous valves are also found throughout veins of these body parts [660].

Vieussens valve is the ostial valve of the great cardiac vein located near the coronary sinus that is provided at its ostium by the *Thebesian valve* [661].

Venous valves most often are bicuspid and located in stiffer expanded segments (Figs. 7.2 and 7.4). Venous valves are made of connective tissue with elastic fibers and few smooth myocytes, covered by the endothelium.

Veins usually have elliptical cross-sections. Valve leaflets are inserted on each face of the vein wall,[15] whereas vein edges receive perforating and tributary veins [665].

[15]Valvular free borders of the superficial veins of inferior limbs are parallel to the skin surface.

Fig. 7.4 Limb valved vein network, with a superficial vein and a deep vein anastomosed by a communicating vein equipped with ostial valves for one-way circulation from superficial to deep vein.

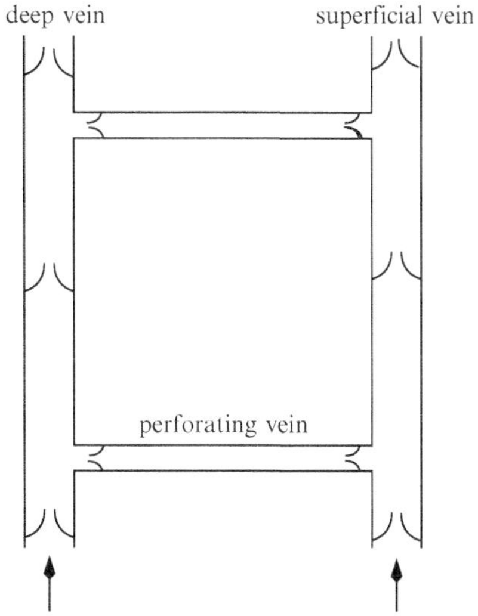

Endothelial cell density is significantly higher on the surface of valve leaflets (about 10 cells per 1000 μm^2) than on other venous surfaces (about 7 cells/1000 μm^2) [666]. Endothelial cells on the medial surface and on the lateral surface of the valve leaflet are oriented parallel to the vein axis and perpendicular to it, respectively.

7.3.2 *Small Vessels*

Small blood vessels of the microcirculation comprise perfusing arterioles, capillaries (caliber 5–10 μm), and draining venules (caliber 8–100 μm).

7.3.2.1 Arterioles

Arterioles are composed of an endothelium surrounded by one or a few concentric layers of smooth myocytes that regulate blood flow (Table 7.6). Arterioles thus have a thin wall. Arterioles receive both sympathetic and parasympathetic innervation.

Table 7.6. Wall structure difference in blood macro- and microcirculation (ColF: collagen fiber; EC: endothelial cell; ElnF: elastin fiber; FB: fibroblast; vSMC: vascular smooth muscle cell).

Macrocirculation	Microcirculation
Intima Endothelium (glycocalyx) Subendothelium (vSMC, ElnF)	Endothelium (monolayer of ECs, blood–wall interface)
IEL	
Media (vSMC, ElnF)	1–2 layers of vSMC in arterioles, pericytes in capillaries
EEL	Elastic externa (±)
Adventitia (FB, ElnF, ColF) Nervi vascularis (end on vSMC) Vasa vasorum (cap plexi)	Absent in arterioles

Capillaries

Capillaries[16] are small exchange vessels composed of endothelium surrounded by a basement membrane.

A capillary bed can encompass 2 types of vessels: (1) true capillaries which branch mainly from metarterioles[17] and yield mass exchange between cells and blood and (2) a short vascular shunt that directly connects the arteriole and venule at opposite ends of the capillary bed.

Three structural types exist. *Continuous capillaries* in muscles, skin, lung, and central nervous system are defined by a continuous basement membrane and tight intercellular clefts and have the lowest permeability. *Fenestrated capillaries* in exocrine glands, renal glomeruli, and intestinal mucosa are characterized by perforations in endothelium and thus by relatively high permeability. *Discontinuous capillaries* in the liver, spleen, and bone marrow, are defined by large intercellular and basement membrane gaps and, consequently, very high permeability.

Precapillary sphincters are rings of smooth myocytes at the origin of true capillaries that regulate blood flow into tissues. Mass flux across capillaries depends mainly on 6 variables: (1) capillary and interstitial hydrostatic pressure; (2) capillary and interstitial oncotic pressure; and (3) filtration and reflection coefficients.

[16] In the second half of the seventeenth century, M. Malpighi identified the capillaries that close the blood circulation loop between arteries and veins, demonstrated earlier by W. Harvey. About at the same time, C. Aselius discovered the lymphatic vessels.

[17] Terminal short segment of artriole with smooth myocytes forming a precapillary sphincter that controls the blood entry into that capillary bed.

7.3.2.2 Venules

Many venules unite to form a vein. Venules are composed of an endothelium surrounded by a basement membrane for the postcapillary venules and smooth muscle for the larger venules (Table 7.5). Larger venules possess a middle layer of muscular and elastic tissue and an outer tunica of fibrous connective tissue. Venules have thinner walls than arterioles.

High endothelial venules (HEV) are specialized postcapillary venules characterized by simple cuboidal cells. They enable circulating lymphocytes to enter a lymph node.

7.4 Blood–Brain Barrier

Vascular endothelial cells in the central nervous system form a barrier that restricts molecule fluxes between the blood and brain. This blood–brain barrier (BBB) ensures proper neuronal functioning and protects the central nervous system (CNS). The blood–brain barrier results from interactions between endothelial cells and neural cells. Its main constituents include endothelial cells, astrocytes, and pericytes, in addition to the extracellular matrix.

The average distance from capillary to neuron ranges from 8 to 20 μm. The microvasculature of the central nervous system is aimed, in addition to its usual goals, at maintaining the homeostasis of the internal medium. A constant microenvironment is indeed required for neuronal activity.

The blood–brain barrier was first described by P. Ehrlich in 1903. It corresponds to a specialized endothelium of cerebral capillaries that avoids any fluctuation in substance concentration in the extracellular space, especially ions, whereas blood supplies the brain with the required nutrients (Fig. 7.5). Unlike peripheral capillaries that allow substance filtration across and between endothelial cells, the blood–brain barrier strictly limits transport into the brain through cerebral microvascular endothelial cells sealed by tight junctions[18] (Vol. 1 – Chap. 7. Plasma Membrane) and plasmalemmal carriers. The endothelial cells of the blood–brain barrier are characterized by transport restriction due to tight junctions and polarized carriers, as well as unique metabolic properties.

Walls of cerebral capillary is thinner than that of the muscle capillary [668]. The wall thickness could modulate the restrictive permeability of the blood–brain barrier with a shorter transendothelium transport time. A greater mitochondrion number might exist in BBB endothelial cells than in other endothelial cells, in relation to energy-dependent transcapillary transport [669].

---

[18]The intercellular cleft width is equal to about 20 nm. External layers of adjacent cell surfaces fuse. Cerebral capillary endothelial cells contain tight junctions that strongly tether adjacent endothelial cells. Tight junctions between cerebral endothelial cells lead to high endothelium electrical resistance.

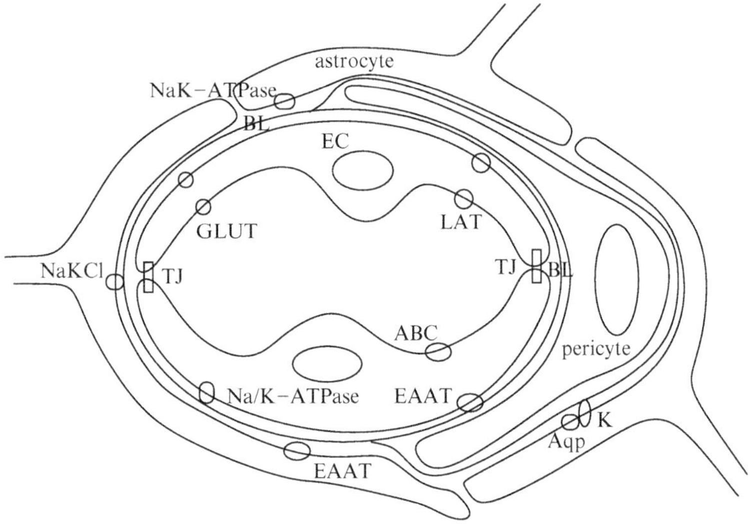

Fig. 7.5 Schematic drawing of the blood–brain barrier, with endothelial cells (EC) of the cerebral capillary, surrounded by basal lamina (BL), perivascular extensions of astrocytes and pericytes (Source: [667]). Endothelial cells are joined by tight junctions (TJ). Endothelium contains carriers for glucose (glucose transporter [GLUT]), amino acids (L system for large neutral amino acids [LAT]), glutamate (excitatory amino acid transporters [EAAT]), ions (Na/K-ATPase), and ABC transporters. Astrocyte extensions facing the capillary wall have patches with aquaporins (Aqp4) and K^+ channels (K). They contain also Na^+–K^+ ATPase, Na^+–K^+–Cl^- cotransporters, and EAATs.

The restrictive blood–brain barrier with its enzymes and transporters develops during embryogenesis. Endothelial cells characterized by a reduced vesicular transport form nascent vessels in the central nervous system to which pericytes are recruited, before astrocytes are generated [678]. Pericytes regulate the function and proper structure of the blood–brain barrier, in particular the formation of tight junctions and vesicle transfer in endothelial cells of the central nervous system [678]. They inhibit the expression of molecules that increase vascular permeability and immunocyte infiltration. Pericytes reduces the permeability of the blood–brain barrier to water and low- and high-molecular-mass molecules [671]. In addition, pericytes may mediate the attachment of astrocyte end-feet to endothelial cells [671]. The PDGFb–PDGFRβ complex is required in pericyte recruitment during angiogenesis.

7.4.1 BBB Restricted Permeability

The brain endothelium has a much lower capacity for endo- and transcytosis than that of peripheral endothelia. Furthermore, intra- and extracellular enzymes yield a metabolic barrier. Ectoenzymes, such as peptidases and nucleotidases, are able to metabolize peptides and ATP, respectively. Intracellular enzymes such as monoamine oxidase and cytochrome-P450 can inactivate many neuroactive and toxic compounds. Renin, angiotensin-converting enzyme, and aminopeptidase-A acting on angiotensin belong to degrading enzymes of the blood–brain barrier. Certain substances that are not carried by the endothelium, are metabolized in endothelial cells such that they cannot enter into the brain. The cerebral endothelium has enzymes that mainly target neuroactive substances, found in low concentration or absent in other capillaries.

7.4.1.1 Intracellular Permeability

The luminal and abluminal membranes of brain capillaries differ chemically and functionally. Asymmetrical distribution of carriers is observed between luminal and abluminal membrane regions, due to amount difference or absence (transport polarity).

Small gaseous molecules such as O_2 and CO_2 can diffuse freely through the lipid membranes, as well as small lipophilic agents. Specific carriers on the luminal and abluminal membranes regulate transcellular transport. The endothelium contains carriers for glucose, amino acids (such as bidirectional L systems for large neutral amino acid transporter LAT1, or SLC7a5), purine bases, nucleosides, choline, and other substances. Efflux carriers depend on available energy.

Whereas alkaline phosphatase activity is located in both the luminal and antiluminal cytoplasmic membranes of brain capillary endothelial cells, the K^+-dependent phosphatase activity associated with Na^+–K^+ ATPase resides only in the abluminal membrane [672]. The main glucose carrier isoform GluT1 at BBB that is bidirectional, has a 1:4 ratio of luminal–abluminal distribution [673]. Na^+-dependent glutamate[19] carriers, such as excitatory amino acid transporters EAAT1 to EAAT3 that allow glutamate efflux against concentration gradient, are mostly abluminal.

7.4.1.2 Extracellular Permeability – Tight Junctions

Tight junctions (or zonula occludens), in combination with ubiquitous adherens junctions, form junctional complexes that restrict the paracellular permeability of the brain capillary wall. Tight junctions separate cell membranes into apical and basolateral regions. Tight junctions are indeed characterized by completely occluded areas between external membrane sheets of adjoining cells.

---

[19]Glutamate is a cerebral excitatory transmitter. Astrocytes recycle glutamate. Glutamate influx occurs particularly via excitatory amino acid transporter-1 and -2 is followed by conversion into glutamine to be taken by adjoining neurons.

Tight junctions are composed of several transmembrane proteins, especially occludin and claudins. Occludin extracellular domains span intercellular cleft. Their intracellular domains connect the cell membrane to the cytoskeleton by association with zonula occludens proteins (ZO1–ZO3) of the MAGUK family. Claudin extracellular loops form the primary seals of tight junctions. Occludins and claudins assemble into heteropolymers to form intramembranous strands along endothelial cell contacts that allow regulation of selective diffusion of ions and hydrophilic molecules. Extracellular loops of these heteropolymers interact with junctional adhesion molecules. Claudin-1, -3, -5, and -12 contribute to the high transport resistance of the brain tight junctions. Junctional adhesion molecules (JAM1–JAM3) maintain tight junctions.

Adaptors ZO1 to ZO3, Ca^{2+}-dependent protein Ser/Thr kinase, membrane-associated guanylate kinase inverted proteins MAGI1 to MAGI3, multi-PDZ domain-containing protein-1, partitioning defective proteins Par3 and Par6 participate in the cytoplasmic plaques associated with tight junctions.

The cytoplasmic plaque is associated with another proteic complex of regulatory and signaling molecules, small GTPases, regulator of G-protein signaling RGS5, ZO1-associated nucleic acid-binding protein, junction-associated coiled-coil proteins, etc.

Brain endothelial cells produce specific proteins,[20] such as solute carrier family 35 member SLC35f2, adpribosylation factor-like ARL5b, transforming growth factor-β1-stimulated clone 22 (TSC22) domain family member TSC22D1,[21] integral membrane protein ITM2a,[22] and epithelial membrane protein EMP1.[23] Epithelial membrane protein-1 colocalizes with occludin. This BBB tight junction-associated protein is particularly expressed during cerebral ischemia [674].[24]

7.4.2 Neurovascular Units

Continuous capillaries with transport systems for nutrients and wastes are characterized by: (1) between-endothelial cell tight junctions that occlude the intercellular

---

[20] Blood–brain barrier control proteins, such as glucose transporter, P-glycoprotein, and transferrin receptor, disappear in cultured endothelial cells that undergo dedifferentiation.

[21] Member 1 of the TSC22 domain-containing family (TSC22D), also called transforming growth factor-β1-induced protein TGFβ1I4, is a tumor suppressor and transcriptional repressor of C-type natriuretic peptide promoter. This transcription factor belongs to the family of early response gene products. It homo- and heterodimerizes with TSC22D4 protein.

[22] Integral transmembrane protein ITM2a is involved in T-cell activation and thymocyte development.

[23] Epithelial membrane proteins EMP1 to EMP3 belong to the peripheral myelin protein PMP22 gene family. In epithelia, PMP22 is coexpressed with occludin and ZO1 in or near tight junctions. Protein EMP2 regulates surface expression of $\alpha_V\beta_3$-integrins, caveolins, and glycosylphosphatidyl inositol-anchored proteins, hence modulating cell–matrix and between-cell adhesions.

[24] In cerebral ischemia model, BBB disruption occurs 1 to 24 h after reperfusion due to changes in ion and nutrient transport, inflammation, and vascular remodeling.

clefts more effectively than in other endothelia;[25] (2) absence of interstice, except in the hypothalamic–pituitary axis;[26] and (3) astrocyte extensions[27] that surround the capillary endothelium. Astrocytes provide the cellular connection to the neurons.

Therefore, transit across the blood–brain barrier involves transport successively through the endothelium wetted membrane, cytoplasm, antiluminal membrane, basement membrane, and astrocyte extension.

The endothelium forms functional neurovascular units with astrocytes and neurons characterized by strongly regulated blood–brain barrier functioning. The distance between any neuron and brain capillary usually is lower than 20 μm.[28] The brain endothelium supplies the brain with essential nutrients and mediates efflux of waste products. Specific ion carriers regulate ionic motions to produce appropriate brain interstitial fluid, avoiding damaging fluctuations in ionic composition, and then disturbances in axonal signaling.

Biochemical bridges exist between the endothelium and astrocytes. Basal laminae of cerebral capillaries contain *agrin*. This heparin sulfate proteoglycan binds to *α-dystroglycan*, a member of the dystrophin–dystroglycan complex that links to aquaporin-4 of astrocyte extension via *α-syntrophin*, another member of the dystrophin–dystroglycan complex. α-Syntrophin also binds to the $K_{IR}4.1$ channel of astrocyte extension.

Astrocytes, pericytes, perivascular macrophages, and neurons regulate BBB features, such as tight junction properties, endothelium polarity, and enzymatic activity. Astrocyte-derived factors, such as transforming growth factor-β, glial-derived neurotrophic factor, basic fibroblast growth factor, and angiopoetin-1 can act on endothelial receptors [667]. Conversely the endothelium acts on the growth and differentiation of perivascular cells, particularly via endothelium-derived leukemia inhibitory factor.

Cell interactions occur not only for development, but also for functioning modulation. Astrocytes, after sensing neuronal cues, can send signals to the capillaries to supply nutrients, using inositol (1,4,5)-trisphosphate, glutamate, and ATP through gap junctions or via receptor stimulation. Moreover, water and ions are redistributed between the components of the functional neurovascular units.

Active neurons release neurotransmitters and K^+ and take up Na^+. They generate water during glucose metabolism. Neurotransmitters and ions are mostly recycled. The perivascular space acts as a K^+ space buffer for recycling, as the astroglial

---

[25]The brain tight junction restricts even the movement of small ions such as Na^+ and Cl^- ions.

[26]The hypothalamus, above the pituitary gland, or hypophysis, is located in the middle of the base of the brain, encapsulating the ventral portion of the third ventricle. The hypothalamus controls many body functions. The anterior pituitary (adenohypophysis) secretes protein hormones, under control by hypothalamic hormones. The posterior pituitary (neurohypophysis) is an extension of the hypothalamus. They release several hormones controlling endocrine gland function.

[27]Astrocyte extensions facing the capillary wall have peculiar features, such as high density of patches with aquaporin-4 and $K_{IR}4.1$ channels, that are involved in ion and volume regulation.

[28]The distance between a neuron and cerebrospinal fluid is on order 1 millimeter or even centimeter.

plasma membrane has a high density of K^+ carriers on its perivascular extension. Astrocyte extensions particularly possess inwardly rectifying $K_{IR}4.1$ channels, Na^+–K^+ ATPase, and Na^+–K^+–Cl^- cotransporters (SLC12a1–SLC12a2). The cerebral endothelium has a low K^+ and water permeability. Excess water is removed from the perivascular spaces, using lymph and cerebrospinal fluid.

Activated protein-C is a serine peptidase with anticoagulant,[29] anti-inflammatory, and cytoprotective activity[30] that crosses the blood–brain barrier via endothelial protein-C receptor to fulfill its neuroprotective effect [676].[31] Its interaction on the endothelium is needed, as the protein-C precursor is not synthesized or at very low amount in the central nervous system. It is conveyed by blood down to the brain. Its transfer from blood requires not only its binding to endothelial protein-C receptor, but also activation of peptidase-activated PAR_1 receptor.

7.4.3 Functional Hyperemia

Blood flow is locally increased in active cerebral regions. Neural activity consumes energy (*functional hyperemia*). Neural activity cues are transmitted to the vasculature by astrocytes. During neuronal activity, glutamate released by the neuron activates glutamate receptors of the astrocyte plasma membrane and triggers a calcium wave. Calcium ions reach the vascular extension of the astrocyte, where they stimulate phospholipase-A2 for arachidonic acid production. Arachidonic acid is then metabolized by cyclooxygenase COx1[32] into a prostaglandin vasodilator PGe2 (Fig. 7.6) [675]. Enzyme COx1 that is highly expressed in astrocyte vascular endfeet is the main mediator of astrocyte-induced vasodilation. Moreover, Ca^{2+} stimulates production of inositol trisphosphate that crosses gap junctions and triggers calcium motions and prostaglandin synthesis in neighboring astrocyte extensions that ensheath the vessel wall.

[29] Anticoagulant activity of activated protein-C is done by irreversible inactivation of coagulation factors FVa and FVIIIa.

[30] Anti-apoptotic and anti-inflammatory activity of activated protein-C requires activation of peptidase-activated receptor PAR_1.

[31] Activated protein-C inhibits neuron death, protects endothelial barriers from injury, and blocks tissue-plasminogen activator-mediated breakdown of the blood–brain barrier after stroke. Endothelial protein-C receptor binds both protein-C and activated protein-C selectively and saturably. However, endothelial protein-C receptor has a significantly lower affinity for protein-C than activated protein-C.

[32] Enzyme COx1 is constitutively produceed by astrocytes, whereas COx2 is expressed in response to inflammatory stimuli.

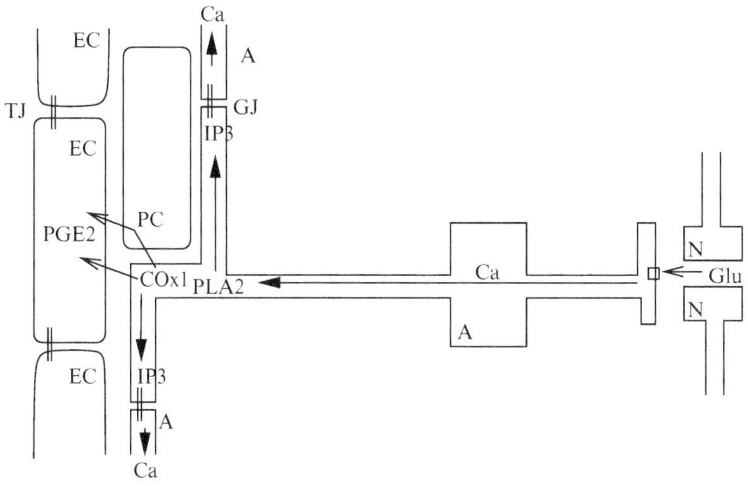

Fig. 7.6 Interactions between neurons (N), astrocytes (A), pericytes (PC), and endothelial cells (EC) for functional hyperemia (TJ: tight junction, GJ: gap junction; Source: [675]). With their extensions to both the vasculature and the nerves, astrocytes are anatomical links between the circulatory and nervous systems. Induced vasodilation involves cyclooxygenase COx1, and inositol trisphosphate (IP_3).

7.4.4 Drug Delivery

The blood–brain barrier prevents the delivery of most therapeutic molecules to their targets within the brain parenchyma, such as neurons and glial cells. A solution for large molecules consists of triggering transcytosis via cell receptors of the blood–brain barrier (low-density lipoprotein receptor, transferrin receptor, insulin-like growth factor receptor) that bind ligands to facilitate their transport to the central nervous system by adding the receptor-binding domain of the targeted ligand (so-called molecular Trojan horses) [677]. A non-replicating lentivirus vector system is used to deliver a substance to neurons and astrocytes by fusing it to the binding domain of ApoB for low-density lipoprotein receptor and adding a secretory sequence to allow its release. Such a method provides a source (from either a central or peripheral organ, such as liver, spleen, or muscle, in the cells of which genes are introduced) for prolonged production and release in the blood flow of a BBB-crossing drug, but fusion proteins must be immunocompatible.

7.4.5 BBB Formation

During embryogenesis, the blood–brain barrier forms when endothelial cells invade the central nervous system from the surrounding vascular plexus and PDGFRβ+,[33] CSPG4+[34] pericytes are recruited to the nascent vessels, before astrocyte generation [678].[35] Pericyte–endothelial cell interactions control the developing blood–brain barrier. However, even in the absence of pericytes, embryonic CNS endothelial cells express many BBB-specific molecules via interactions with neural progenitors. Neural progenitors secrete Wnt morphogens for angiogenesis as well as production of BBB-specific transporters.

Pericyte coverage determines vascular permeability. Pericytes regulate the formation of tight junctions and vesicle transfer in CNS endothelial cells [678]. Pericytes prevent the synthesis of molecules that increase vascular permeability and leukocyte infiltration.[36] Pericytes may thus provide signals to limit immune surveillance in the central nervous system. Afterward, astrocytes likely regulate the maintenance of the blood–brain barrier as well as repair after injury. The functional integrity of the blood–brain barrier is thus controlled by pericytes during embryo- and fetogenesis and by astrocytes in adulthood [678].

7.5 Other Cerebral Epithelial Barriers – Choroid Plexi

In addition to the blood–brain barrier endothelium, 2 other cell layers with tight junctions provide a barrier: (1) arachnoid epithelium that forms the middle layer of the meninges and (2) choroid plexus epithelium.

Choroid plexi form an active barrier between blood and *cerebrospinal fluid* (CSF).[37] They secrete the cerebrospinal fluid and ensure its chemical stability. They localize to the surface of the lateral, third, and fourth ventricles. They are made of 2 cell layers that are bound by tight junctions: a capillary endothelium and a

---

[33]PDGFR: platelet-derived growth factor receptor.

[34]CSPG: chondroitin sulfate proteoglycan-4, or neuron–glial antigen NG2.

[35]Neural cells are produced from progenitors in a defined sequence, with neurons generated before glial cells. Oligodendroglia progenitor cells and PDGFRα+, CSPg4+ glial cells migrate throughout the cortex.

[36]In the absence of PDGFRβ in the vasculature, the transcription of several leukocyte adhesion molecules, such as intercellular adhesion molecule ICAM1, activated leukocyte cell adhesion molecule (ALCAM), and lectin, galactoside-binding, soluble LGalS3, is upregulated [678].

[37]Cerebrospinal fluid is located within the head ventricles, spinal canal, and subarachnoid spaces. Its volume ranges from 140 to 150 ml, with 30 to 40 ml in the ventricles. Hence, CSF is mainly in the subarachnoid space, basal cisternae, and around the spinal cord. The production rate is equal to about 21 ml/h and the turnover rate of total CSF is smaller than 7 h. Cerebrospinal fluid cushions the brain, regulates brain extracellular fluid, distributes neuroactive substances, and collects brain wastes.

choroid epithelium that are separated by the brain interstitial fluid. Epithelial cells produce 90% of CSF ($\sim$ 0.4 ml/mn/g) [679]. Cerebrospinal fluid is composed of water (99%), vitamins, ions, glucose, amino acids, etc. The choroid plexi also have a CSF purification role.

7.6 Perivascular Nerves

Vascular tone undergoes a dual regulation by perivascular nerves and endothelial cells. Endothelial cells can provoke both vasodilation and -constriction. Endothelial cells indeed produce, store, and release vasoactive substances, either vasodilators, such as nitric oxide and acetylcholine, or vasoconstrictors such as endothelin (Sect. 9.10). Perivascular nerves and endothelial cells interact.

Perivascular nerves are not limited to adrenergic or cholinergic axons. Four types of perivascular nerves exist: sympathetic, parasympathetic, sensory–motor, and projection neurons from intrinsic ganglia of the heart and other organs [680].

Cotransmission usually occurs, as: (1) sympathetic nerves use the transmitters noradrenaline, adenosine triphosphate, and neuropeptide-Y (NPY); (2) parasympathetic nerves signal via acetylcholine and vasoactive intestinal peptide (VIP); and (3) messages of sensory–motor nerves[38] are coded by calcitonin gene-related peptide (CGRP), substance-P, and adenosine triphosphate. (4) Neurons from organ ganglia release many types of neurotransmitters, such as peptides (peptidergic nerves [transmission]),[39] purines (purinergic nerves [transmission]), nitric oxide (nitrergic nerves [transmission]), and classical transmitters.

Artemin (Artn) that is expressed in vascular smooth muscle cells, serves as a guidance factor for the growth of sympathetic nerve fibers along blood vessels [681]. Artemin is a member of the glial cell line-derived neurotrophic factor (GDNF) family of ligands.[40] These ligands signal via a receptor complex composed of Ret receptor protein Tyr kinase and a glycosyl phosphatidyl inositol-anchored coreceptor of the receptor-α set for members of the GDNF family .

7.7 Pericytes

Pericytes surround the capillaries. They can also encircle precapillary arterioles and postcapillary venules. These periendothelial cells are rich in stress actin bundles.

---

[38] Certain afferent nerves have an efferent function and axonic reflex control of vascular tone.

[39] Peptidergic nerves contain substance-P, CGRP, and VIP transmitters.

[40] The glial cell line-derived neurotrophic factor family ligands include glial cell-derived neurotrophic factor (GDNF), neurturin (Ntn), persephin (Pspn), and artemin (a.k.a. enovin [Evn] and neublastin [Nbn]).

Differences in pericyte morphology and distribution among vascular bed types can be related to tissue-specific functions (heterogeneous functionality). The pericyte coating (wrapping degree and pericyte density) depend on the perfused organ. The greater the pericyte density, the stronger the microvascular barrier.

Among the types of pericytes, precapillary arteriolar, capillary, and postcapillary venular have different features. Pre- and postcapillary pericytes have α-actinin (SMC-pericyte transitional cells), whereas capillary pericytes do not.

Pericytes contain smooth muscle (like SMCs) and non-smooth muscle (like endothelial cells) isoforms of actin and myosin. The former has higher levels in pre- and postcapillary pericytes, whereas the concentration of the latter is greater in capillary pericytes.

A basement membrane separates the endothelial cells and pericytes. However, tight and gap junctions can develop between endothelial cells and pericytes. A basement membrane can also be found along the outer surface of the pericytes.

Pericytes, by their contractile function, can regulate capillary bore, and then the tissue perfusion, as well as transport from the blood via pericytic processes at interendothelial clefts. Furthermore, pericytes mainly secrete vasoactive autoregulating substances and release structural constituents of the basement membrane and interstitial matrix.

Pericytes produce: (1) matrix structural proteins (collagen, fibronectin, laminin); (2) agents involved in clotting and vasculogenesis, such as tenascin, thrombospondin,[41] and plasminogen activator inhibitor; (3) cell adhesion molecules; (4) vasomotor factors such as endothelin-1; and (5) prostaglandins (TXa2, PGi2, PGe2, PGf2α).

Pericytes have plasmalemmal receptors for growth factors (e.g., epidermal growth factor receptor [EGFR] and platelet-derived growth factor receptor PDGFRβ) and various other ligands (e.g., endothelin-1 and histamine receptors) [682].

Adenosine receptor A2R stimulates adenylate cyclase activity. β-Adrenergic and muscarinic receptors have also been observed. Signaling leads to the activation of cAMP, transient increase in IP_3, and Ca^{2+} influx. In opposition to smooth myocytes, the membrane potential of the pericyte does not significantly vary after stimulation by ET1, vasopressin, and acetylcholine.

Pericyte coverage is required during angiogenesis. Contacts between endothelial cells and pericytes may control vascular growth, via platelet-derived growth factor and transforming growth factor-β, and endothelium functioning. In cocultures, pericytes modulate endothelial cell proliferation and the converse is also true.

Pericytes are recruited by endothelial cells that express PDGFb to stabilize the vessel. Pericytes protect the endothelial cell, as they reduce their apoptosis rate.

[41] Thrombospondins interact with blood coagulation and anticoagulant factors. They are involved in cell adhesion, platelet aggregation, cell proliferation, angiogenesis, tissue repair, and tumor metastasis.

Cerebral blood flow is strongly regulated to provide 20% of body energy consumption, even though the brain accounts for only 5% of total weight, and to match the changing metabolic needs of specific brain regions. Cells that control the blood flow include endothelial and smooth muscle cells, pericytes, neurons, and astrocytes. Contraction and relaxation of the vascular smooth myocytes decreases and increases the artery bore, respectively. Increase in intracellular calcium concentration in astrocytes regulates the arteriole caliber. In the absence of surrounding smooth myocytes, blood flow through capillaries is, indeed, regulated by pericytes that contain contractile machinery [683].[42] Groups of pericytes along capillaries and at capillary junctions of the central nervous system can constrict and relax according to local neuronal activity. Pericyte constriction can hence redirect the blood flow at the capillary level.

7.8 Nervous Inputs

Activated α1-adrenergic receptors interact with Gαq that signals via phospholipase-C leading to: (1) diacylglycerol that activates protein kinase-C and (2) inositol trisphosphate that primes calcium influx mainly from the endoplasmic reticulum (Fig. 7.7). In vascular smooth myocytes, G-protein-coupled receptor kinase GRK2 that phosphorylates and desensitizes ligand-bound G-protein-coupled receptors impedes vasoconstriction caused by α1d-adrenoceptors [684]. β2-Adrenergic receptors induce relaxation of vascular smooth myocytes (Table 7.7).

Smooth muscle myosin-2 consists of 2 heavy chains and 2 types of light chains. One part of the heavy chains folds into the globular head that forms the motor domain, the other constitutes the myosin tail that connects to others to produce the myosin filament. Contraction initiation and termination are switched by myosin phosphorylation by smooth muscle myosin light chain kinase, and dephosphorylation by myosin light chain phosphatase. The myosin light chain can also be phosphorylated by PKC and CamK2 kinases.

Phosphorylation magnitude determines shortening velocity and tension development. Isoforms of MLCK in smooth and non-myocytes differ from MLCK in striated myocytes. Myosin light chain kinase can be phosphorylated by PKA, PKC, CamK2, and PAK kinases.

Myosin light chain phosphatase is also phosphorylated (inhibited) by RoCK2 kinase, ensuring a sustained vasoconstriction. Integrin-linked protein Ser/Thr kinase phosphorylates myosin light chain phosphatase as well as myosin light chain.

Regulator caldesmon complexes with actin, tropomyosin, and a Ca^{2+}-binding protein. In the absence of calcium, caldesmon restricts the interaction of the myosin

[42] About 65% of noradrenergic innervation of brain blood vessels ends near capillaries rather than near arterioles. Pericyte constriction is associated with a rise in intracellular calcium concentration.

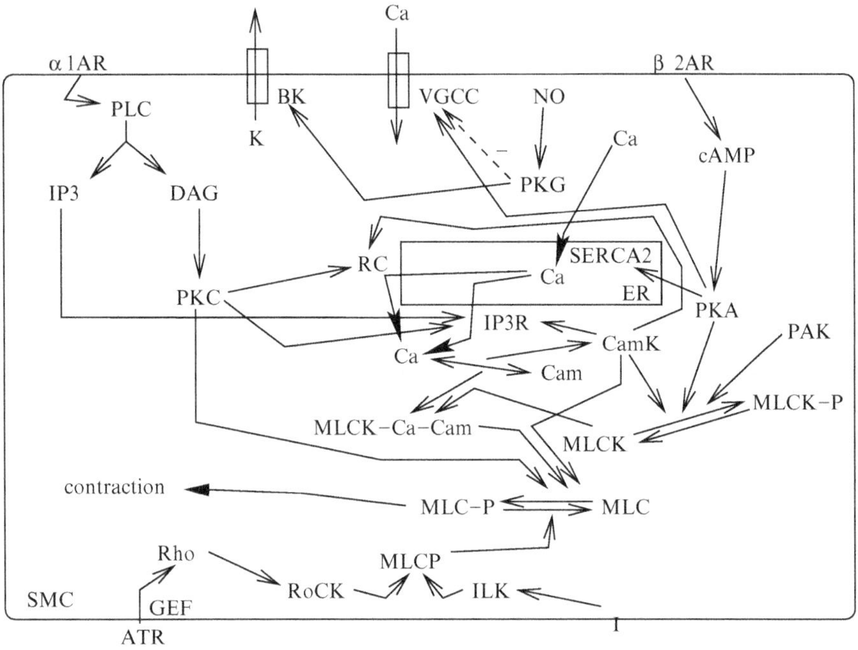

Fig. 7.7 Contraction of the vascular smooth myocyte and its regulation (Source: [650]; BK: large-conductance Ca^{2+}-activated K^+ channel; Cam: calmodulin; DAG: diacylglycerol; GEF: GDP-to-GTP-exchange factor; IP_3(R): inositol trisphosphate (receptor); I: integrin; PKA, PKC: protein kinase-A, -C; PLC: phospholipase-C; VGCC: L-type voltage-gated Ca^{2+} channel [Ca_V1.2]). Adrenergic receptors-α1 (α1AR) and -β2 (β2AR) cause contraction via the PLC–PKC pathway and relaxation via the cAMP–PKA cascade, respectively. The myosin light chain kinase (MLCK) can be phosphorylated by PKA, PKC, CamK2, and PAK kinases. Myosin light chain phosphatase (MLCP) can be phosphorylated by RoCK and ILK kinases. Phosphorylation of MLCK, which requires Ca^{2+}–Cam, leads to activation, whereas MLCP phosphorylation leads to inactivation. Calcium ATPase SERCA2 is targeted by PKA, ryanodine-sensitive channel (RC) with its regulatory protein FKBP and IP_3R are activated by PKA, PKC, or CamK2 kinase. Nitric oxide-induced PKG inhibits Ca_V1.2 channel and activates BK channel.

Table 7.7. Contraction and relaxation features in the cardiomyocyte and smooth myocyte. The cardiomyocyte is characterized by fast, reversible calcium binding to cardiac troponin-C. Smooth myocyte activity is regulated by reversible phosphorylation of myosin and/or actin components. Vasomotor tone depends on the activity ratio between MLCK and MLCP. β-Adrenergic receptor signaling increases intracellular calcium concentration in the cardiomyocyte, whereas its activity in vascular smooth myocyte is independent of calcium.

	CMC	SMC
Contraction/relaxation	Fast	Slow
ATP consumption	High	Low
β-Adrenergic signaling	I+	Relaxation

Table 7.8. Approximated extracellular ATP levels in various tissues (Source: [685]).

Tissue	Bulk amount
Guinea pig heart endothelium	20–100 pmol/10^6 cells
Rat caudal artery endothelium	27 pmol/10^6 cells
Bovine aortic endothelium	2–20 nmol
Vascular endothelium	7–105 nmol
Airway epithelium	0.1–20 nmol
Human renal epithelium	350–2,000 nmol

heads with actin. Calcium binding to Ca^{2+}-binding protein relieves the inhibition. Calponin that binds to actin and calmodulin may enhance PKC–ERK1/2 signaling.

Smooth muscle-specific $Ca_V1.2b$ channel is phosphorylated by PKA, PKC, and CamK. Most of the Ca_V channels are packed with Na^+–Ca^{2+} exchanger NCX1, Na^+–K^+ pump, plasmalemmal Ca^{2+} ATPase and large-conductance Ca^{2+}-activated K^+ channel ($BK_{V,Ca}$).

β-Adrenoceptor stimulation activates PKA that, in cooperation with AKAP, phosphorylates Ca_V channels. The cAMP–PKA axis can be counteracted by nitric oxide and its effector protein kinase-G. The latter inhibits Ca_V channels, and activates $BK_{V,Ca}$ and MLCP proteins.

7.9 ATP Signaling in Blood Vessel Walls

The nucleotide adenosine 5′-triphosphate is a cell's energy source[43] and auto- and paracrine signaling regulator that targets its cognate plasmalemmal receptors on nearly all cell types (Vol. 3 – Chaps. 1. Signal Transduction, 2. Membrane Ion Carriers, and 7. G-Protein-Coupled Receptors; Table 7.8).

Extracellular ATP is involved in physiological (e.g., blood vessel tone regulation [functional reactive hyperemia, hypoxia-induced vasodilation, and α-adrenoceptor-mediated vasoconstriction]) and pathophysiological (e.g., atherosclerosis) processes. Various carriers are involved in the controlled ATP release from vascular cells (vascular endothelial and smooth muscle cells and perivascular sympathetic neurons) as well as circulating red blood capsules, such as energy-consuming transmembrane F_0–F_1 ATP synthase complex, ATP-binding cassette (ABC) transporters, connexin hemichannels, and pannexin channels, in addition to vesicular exocytosis [686].

[43]Mitochondrial ATP synthase produces ATP using electron carrier NADH generated in the citric acid cycle and electron transfer through the electron transport chain, inorganic phosphate, ADP, and oxygen. The cell uses ATP for active transport of molecules across the plasma membrane, contraction of actin–myosin filaments, and protein phosphorylation by kinases.

The matching of oxygen supply with demand requires change in blood perfusion upon sensing tissue oxygen need. The red blood capsule can sense oxygen level and release ATP when oxygen content falls and its hemoglobin becomes desaturated [687]. Intraluminal ATP is a vasodilator. In the blood vessel lumen, ATP concentration increases during hypoxia [688]. Red blood capsules also secrete ATP when exposed to increased temperature, reduced pH, hypercapnia, elevated stresses and strains. Hypoxia supports the activation of K_{ATP} channels that causes hyperpolarization and vasodilation [690]. In arterioles of rat cremaster muscle that possess both α1d- and α2d-adrenergic receptors, the latter is functionnaly coupled to K_{ATP} channel; α2d-adrenoceptors inhibits and hypoxia activates K_{ATP} channels.

In addition, augmented hemodynamic stresses on endothelial cells prime ATP release into the vessel lumen [689]. Therefore, luminal ATP sources encompass circulating red blood capsules and endothelial cells.

At perivascular sympathetic nerve endings, ATP is released as a cotransmitter with noradrenaline. Messenger ATP then links to $P2X_1$ receptors on adjacent smooth myocytes and launches vasoconstriction.

7.9.1 Extracellular ATP Processing

Extracellular ATP concentration is regulated by enzymes that degrade ATP to ADP, AMP, and adenosine. The latter can binds to its cognate receptors, thereby regulating the vascular tone. However, ATP can elicit signaling on vascular cells before its catabolism, although ATP is rapidly and sequentially hydrolyzed to adenosine.

The degradation of ATP to AMP is catalyzed either by ectonucleotidases (ectonucleotide triphosphate diphosphohydrolases ENTPD1–ENTPD6) or ectoapyrase (ectonucleoside triphosphate diphosphohydrolase ENTPD1);[44] AMP is processed to adenosine by ecto-5′-nucleotidase (Nt5E). The plasmalemmal, Ca^{2+}- and Mg^{2+}-dependent ectoenzyme ATPase on endothelial cells hydrolyzes extracellular ATP. Membrane-bound ectonucleotidases process extracellular tri- (ATP and UTP) and diphosphonucleosides (ADP and UDP). Cells also release soluble nucleotidases, or exonucleotidases, into the extracellular medium. Therefore, especially ectonucleotidases regulate the concentration of ATP both in the blood vessel lumen and extracellular space surrounding vascular smooth myocytes, hence amounts of ATP and its catabolites available for activation of receptors.

[44] A.k.a. ectoATPase-1 and ectoATP diphosphohydrolase-1 (EATPD1).

Table 7.9. Nucleoside and nucleotide receptors (Sources: [691,692]).

Ligand	Receptors
	Nucleoside
Adenosine	A_1, A_{2A} A_{2B}, A_3
	Purine nucleotides
ATP	$P2X_1$, $P2X_7$,
	$P2Y_1$, $P2Y_2$, $P2Y_4$, $P2Y_{11}$
ADP	$P2Y_1$, $P2Y_{12}$, $P2Y_{13}$
	Pyrimidine nucleotides
UTP	$P2Y_2$, $P2Y_4$
UDP	$P2Y_6$, $P2Y_{14}$
^{UDP}glucose	$P2Y_{14}$

7.9.2 Nucleotide Receptors

Purinergic (nucleoside and nucleotide) receptors include G-protein-coupled P1 receptors that bind the nucleoside adenosine (A_1, A_{2A}, A_{2B}, and A_3) and P2 receptors, either ionotropic P2X (ligand-gated ion channels $P2X_1$–$P2X_7$) or metabotropic P2Y receptors (G-protein-coupled receptors $P2Y_1$–$P2Y_2$, $P2Y_4$, $P2Y_6$, and $P2Y_{11}$–$P2Y_{14}$; Table 7.9).

Activation of P2X receptor channel by ATP triggers the influx of extracellular cations, such as Na^+ and Ca^{2+} ions. The subsequent cell depolarization can excite voltage-gated channels. Activation of G-protein-coupled P2Y receptors can provoke changes in concentrations of intracellular cAMP or Ca^{2+} messengers via the Gs–ACase, Gq–PLC, and Gi–ion channel pathways.

Entry of Ca^{2+} ions through $P2X_1$ receptors in vascular smooth myocytes causes vasoconstriction. On the other hand, release of Ca^{2+} ions from the endoplasmic reticulum triggered by $P2Y_1$ and $P2Y_2$ receptors in vascular endothelial cells results from the production of inositol triphosphate that stimulates IP_3 receptors; it elicits activation of endothelial nitric oxide synthase NOS3, thereby synthesizing the gasotransmitter nitric oxide that induces vasodilation. Similarly, activation of P2X receptors on the endothelial cell surface leads to vasodilation; excitation of P2Y receptors on the smooth myocyte surface causes vasoconstriction.

7.9.3 ATP Release

Release of ATP and other nucleotides by non-excitable cells occurs in 3 circumstances: (1) upon mechanical stresses sensed by endothelial and smooth muscle cells among others; (2) following chemical stimuli; and (3) in resting cells. Mediator ATP is liberated out of vascular cells (vascular endothelial and smooth muscle cells and perivascular sympathetic neurons) as well as circulating red blood capsules using various mechanisms: vesicular exocytosis, ATP-binding cassette (ABC) transporters, and connexin- and pannexin-based channels.

7.9.3.1 Vesicular Exocytosis

Exocytosis is the release of soluble intracellular molecules into the extracellular space. For example, perivascular sympathetic nerve terminals liberate from synaptic vesicles the neurotransmitter noradrenaline as well as ATP that targets α-adrenergic receptors of postjunctional membranes of innervated vascular smooth myocytes and primes vasoconstriction. Endothelial cells may also release ATP from stores via exocytosis during changes in blood flow.

Neurons secrete many neurotransmitters, such as catecholamines (dopamine and noradrenaline), amines (acetylcholine, serotonin, and histamine), and amino acids (glutamate, glycine, and GABA). Vesicular transporters are generally non-selective: vesicular monoamine transporters VMAT1 and VMAT2 (SLC18a1–SLC18a2) carry serotonin, noradrenaline, and dopamine; vesicular inhibitory amino acid transporter (VIAAT, or SLC32a1) conveys GABA and glycine. However, vesicular acetylcholine transporter (VAChT, or SLC18a3) transports only acetylcholine; glutamate transporters VGluT1 to VGluT3 (SLC17a6–SLC17a8) are selective for glutamate.

Vesicular nucleotide transporter (VNuT, or SLC17a9) is a secretory vesicle protein responsible for the vesicular storage of ATP as well as ADP and GTP, using ATP and proton via vacuolar ATPase to create an electrochemical potential and a pH gradient across the vesicular membrane, and subsequent exocytosis [693]. It also carries divalent cation, such as Ca^{2+} and Mg^{2+}, when ATP is present [694].

Extracellular ATP and ATP-containing specialized vesicles can be detected in most non-excitatory tissues. Vascular endothelial cells may store ATP bound to peptides in vesicles [695]. Messenger ATP can be secreted from endothelial cells as well as airway epithelial cells, among others, which are subjected to mechanical stresses [685]. Aortic endothelial and smooth muscle cells stimulated by thrombin release ATP messenger. Endothelial cells release nucleotides in response to bradykinin, acetylcholine, and serotonin [685].

Basal ATP release by resting cells enables autocrine activation of nucleotide P2Y receptor, hence basal activity of some proteins such as the calcium- and cAMP-sensitive chloride (CFTR) channels in airway epithelial cells, vascular smooth muscle and endothelial cells, red blood capsules, and platelets.

7.9.3.2 ABC Transporters

ATP-binding cassette transporters use the energy of ATP hydrolysis to support the motion across the plasma membrane of macromolecules such as cholesterol (Vol. 3 – Chap. 4. Membrane Compound Carriers). Cystic fibrosis transmembrane conductance regulator (CFTR, ABC35, or ABCc7), multidrug resistance gene product (MDR, P-glycoprotein-1, ABC20, or ABCb1), and sulfonylurea receptor (SUR, or ABCc8), which functions as a modulator of K_{ATP} channels and insulin release, can not only utilize ATP as an energy source for active transport, but also export the purine nucleotide out of the cell, enabling auto- and paracrine purinergic signaling [686].

However, these ATP-binding cassette transporters expressed in endothelial cells are not involved in ATP release [695]. Nonetheless, CFTR contributes to ATP release from red blood capsules upon activation of PKA that fosters CFTR activity [686].

7.9.3.3 Plasmalemmal F_1–F_0 ATPase

ATP synthase is composed of 2 fractions: F_0 membrane-anchoring subunit, which forms the pore for proton transport, and F_1 catalytic subunit. This enzyme resides not only on the inner mitochondrial membrane, but also on the plasma membrane of numerous cell types such as vascular endothelial cells [686].[45] Synthesis of ATP at the endothelial cell surface requires the existence of a proton gradient across the plasma membrane.

7.9.3.4 Connexin Hemichannels

Connexins form hexameric connexons in the endoplasmic reticulum and Golgi body, which are then transferred to and inserted into the plasma membrane to build gap junctions between apposed cells or hemichannels between the cytosol and the extracellular space. Glial cells in the central nervous system, granulocytes, monocytes, and vascular endothelial and smooth muscle cells exhibit connexin hemichannels [686]. At rest, connexin hemichannels remain impermeable; under strong membrane depolarizations and mechanical stress, these hemichannels open.

Four connexin isoforms abound in vascular smooth muscle and endothelial cells (Cx37, Cx40, Cx43, and Cx45). They construct gap junctions between adjacent vascular smooth myocytes on the one hand and endothelial cells on the other, as well as between these 2 cell types at the myoendothelial junctions (Sect. 9.1.2.3). Connexin hemichannels at the non-junctional plasma membrane of endothelial cells may release ATP agent.

Macula densa cells of the juxtaglomerular apparatus coat the wall of the distal tubule at the glomerular vascular pole. They detect changes in distal tubular flow rate and sodium chloride concentration. They release ATP through gap junctions into the interstitium of the juxtaglomerular apparatus to control the glomerular

[45] Angiostatin, an anti-angiogenic product of plasminogen cleavage in the fibrinolytic system, has a high binding affinity for α and β subunits of F_1 fraction of ATP synthase and prevents its activity, hence secretion of ATP messenger.

filtration rate, hence supporting the *tubuloglomerular feedback*. Both secreted ATP and its catabolite adenosine on afferent arteriole smooth myocytes cause vasoconstriction upon binding to nucleoside (A_1) and nucleotide (e.g., $P2X_1$) receptors and generating Ca^{2+} signals [696]. An intercellular Ca^{2+} wave through gap junctions causes a coordinated contraction of both intra- and extraglomerular mesangial cells, specialized smooth myocytes. Therefore, the tubuloglomerular feedback relies on Ca^{2+} waves primed by extracellular ATP that propagate from macula densa cells toward proximal segments of the afferent arteriole and corresponding glomerulus as well as all intraglomerular elements, including the most distant podocytes [696]. The strongest vasoconstriction occurs in the terminal, intraglomerular segment of the afferent arteriole and propagates to proximal segments, thereby reducing the glomerular filtration rate.

In addition, connexin-40 hemichannels, extracellular ATP, and P2 receptors enable the motion of calcium waves in glomerular endothelial cells from a single glomerular endothelial cell subjected to mechanical stimuli [697]. Hence, autoregulation that maintains renal blood flow and glomerular filtration rate involves a propagating calcium wave in the juxtaglomerular apparatus via the release of ATP and nucleotide signaling.

7.9.3.5 Pannexin Channels

Three pannexin isoforms exist (Panx1–Panx3). Whereas Panx1 and Panx3 may assemble into hexamers, Panx2 may form heptamers or octamers. Once assembled in the endoplasmic reticulum and Golgi body and transported to the plasma membrane, they do not form gap junctions, but pannexin channels open to the extracellular space. The latter is permeable to ATP messenger. They are observed on glial cells and neurons in the central nervous system, T cells, airway epithelial cells, red blood capsules, and vascular smooth muscle and endothelial cells [686]. Pannexin-1, but neither Panx2 nor Panx3, lodges in mouse thoracodorsal resistance arteries [698]. Pannexins mediate paracrine communication via release of ATP or UTP nucleotides.

Pannexin-1 channel in resistance arteries participates in the response triggered by α1d-adrenergic receptor bound to noradrenaline liberated from nerve terminals innervating the blood vessel wall that coordinates vascular smooth myocyte contraction via release of ATP, a potent vasoconstrictor [698]. Pannexin-1 and α1d-adrenoreceptor form a proteic complex.

7.10 Vascular Oxygen Sensing

Vascular oxygen sensors in smooth myocytes respond to changes in oxygen concentration by stimulating redox mediators that, in turn, regulate suitable effectors. Vascular smooth myocytes use aerobic glycolysis that generates 2 mol of ATP per molecule of glucose, whereas mitochondrial glucose oxidation produce 36 mol of ATP per molecule of glucose. Aerobic glycolysis has an advantage, as compartmentation between the cytoplasm and mitochondria in ATP production is associated with ion channel functioning and SMC excitability. There is also a compartmentation of glucose metabolism in vascular smooth myocytes. Most exogenous glucose is consumed in glycolysis and produced lactate, avoiding mitochondrial oxidation, whereas in constricted cells intrinsic glucose released by glycogenolysis is oxidized in mitochondria to fit the ATP demand during contraction. Acute hypoxia increases mitochondrial and cytoplasmic NAD(P)H, but has opposite effects on mitochondria and cytoplasmic redox systems [699]. Mitochondrial and cytoplasmic generation of reactive oxygen species is augmented and lowered, respectively.

During hypoxia, oxygen-supplying pulmonary and placental arteries constrict, whereas systemic arteries for oxygen-consuming organ perfusion dilate to achieving optimal oxygen supply. Difference in mitochondrial features between vascular smooth myocytes of pulmonary and systemic arteries could explain this opposed behavior. Mitochondria of renal artery smooth myocytes are more hyperpolarized than those of pulmonary artery smooth myocytes. Compartmentation between the cytoplasm and mitochondria and difference in magnitude of metabolic enzyme activity in both compartments between pulmonary and systemic arteries could explain such a discrepancy.

7.11 Wall Adaptability

Proliferation of smooth myocytes can be modulated by growth factors and other mediators, released from endothelial cells. Endothelial platelet-derived growth factor-A that acts synergistically with endothelin-1, transforming growth factor-β, and nitric oxide are secreted by endothelial cells, the secretion level depending on flow conditions.

Angiotensin-2, formed by angiotensin-converting enzyme in a flow-dependent manner, is a growth factor for vascular smooth muscle cells. Both serotonin and thromboxane-A2, released by endothelial cells, or aggregating platelets at sites of vascular injury, have a mitogenic effect from a given concentration via their respective specific receptors [700]. These substances can synergistically interact; SMC proliferation occurs at subthreshold concentrations.

Endothelium is able to adapt to ischemia. Nitric oxide contributes to cardiomyocyte "hibernation" by reducing oxygen consumption and preserving calcium sensitivity and contractile function without an energy cost during ischemia [701]. Hypoxemia induces the production of PGe2 and LTc4 by cardiomyocytes. Delayed increase in NO production by NOS3 is a mediator of endothelial preconditioning, with brief periods of ischemia–reperfusion that induce delayed protection of coronary endothelial cells against reperfusion injury (ischemic preconditioning) [702].

Hypertension induces wall remodeling. The endothelial cells display structural changes associated with increased media thickness due to hypertrophy of smooth myocytes [703]. The amount of elastin does not change significantly, whereas the collagen concentration strongly increases. Concentration of TGFβ produced by aortas of hypertensive rats rises up to 3-fold with respect to those of normotensive rats [704].

Chapter 8
Smooth Myocytes

Smooth muscle cells, or smooth myocytes (SMC), similarly to any myocyte, contract and relax. Yet, unlike striated myocytes, they can withstand prolonged stretch. Smooth myocytes can contract *phasically* with rapid contraction and relaxation such as in the digestive tract, where they create *peristaltism*. Smooth myocytes can also contract *tonically* with slow and sustained contraction, such as in the cardiovascular and respiratory apparatus.

Transient and sustained contraction of smooth myocytes relies on 2 different pathways. (1) *Transient, calmodulin-dependent contraction* results from activation of G-protein-coupled receptors by their specific ligands, such as substance-P and cholecystokinin, that trigger production of inositol (1,4,5)-trisphosphate and calcium influx in the SMC cytosol. (2) *Prolonged contraction* is also caused by chemical messengers, such as bombesin and ceramide, that prime extracellular calcium influx and a protein kinase-C-dependent, calmodulin-independent axis.

Smooth myocytes pertain to involuntary (i.e., without conscious control), non-striated myocytes. They control the dimension and mechanical function of hollow organs and anatomical conduits. In particular, smooth myocytes lodge in walls of both the vasculature (arteries, veins, and large lymphatic vessels) and large and mid-sized segments of the ventilatory tract that experience large deformations during the cardiac and respiratory cycles. In both structures, they reside beneath the cell layer that lines the conduit lumen, i.e., the vascular endothelium and respiratory epithelium, respectively.

Smooth myocytes are innervated by the autonomic nervous system.[1] According to the unitary or collective innervation of SMC sheets or bundles, smooth myocytes constitute 2 groups. (1) In single-unit smooth muscles, the autonomic nervous system innervates a single cell within the SMC set. The command electrochemical wave, or action potential, propagates through gap junctions to apposed cells, such

[1]Involuntary and visceral nervous system, a part of the peripheral nervous system. Autonomous means independent of cerebrospinal nervous centers.

M. Thiriet, *Tissue Functioning and Remodeling in the Circulatory and Ventilatory Systems*, Biomathematical and Biomechanical Modeling of the Circulatory and Ventilatory Systems 5, DOI 10.1007/978-1-4614-5966-8_8, © Springer Science+Business Media New York 2013

that the entire cell set contracts in bulk, as a syncytium. Single-unit smooth muscles line blood vessels, except large elastic arteries. (2) In multi-unit smooth muscles, such as in the trachealis muscle that connects the ends of U-shaped cartilaginous rings of the trachea and media of large arteries, the autonomic nervous system innervates many cells to finely tune nervous control and cause gradual responses.

Smooth myocytes can contribute to the formation of the extracellular matrix, as they secrete collagens, mostly type-1 and -3, elastin, glycoproteins, and proteoglycans. They possess on their surface elastin and collagen receptors that interact with these extracellular proteins.

The plasma membrane, or sarcolemma, possesses specialized nanodomains devoted to cell signaling with ion carriers and receptors for hormones, growth factors, neurotransmitters (e.g., acetylcholine and cathecholamines), and regulators of the SMC tone (e.g., nitric oxide and endothelin). However, smooth myocytes in various regions of the vascular bed, airways, and kidneys differ in their production of ion carriers and receptors, as well as other proteins that determine their function.

8.1 Markers of Smooth Myocytes

Smooth myocytes of walls of the vasculature and ventilatory tract possess various contractile proteins. Contractility of smooth myocytes relies on smooth muscle α-actin, myosin heavy chain, calponin,[2] caldesmon$_{hMW}$,[3]. β-tropomyosin, and myosin light chain kinase (MLCK).

Hypercontractility markers include myosin light chain kinase, smooth muscle-specific myosin heavy chain (smMHC) isoforms, and 22-kDa actin-binding, smooth muscle protein (SM22), or transgelin (TaGln). In addition to cytoskeleton constitutents, calmodulin is synthesized in smooth myocytes to participate in the control of cytoskeleton functioning.

Myosin-2 and actin generate cell contraction via stress fibers. In contracting smooth myocytes, β-actin network is conserved, whereas α-actin and myosin-2 filaments remodel. Myosin-2 is activated by Ca^{2+}–calmodulin myosin light chain kinase and inhibited by myosin light chain phosphatase.

Contraction of smooth myocytes is regulated not only by phosphorylation of the myosin light chain by Ca^{2+}–calmodulin-dependent myosin light chain kinase, but also by the caldesmon–calmodulin complex, protein kinase-C, and calponin.

[2]Calponin is a calcium-binding inhibitor of the ATPase activity of myosin in smooth muscle. Phosphorylation of calponin by a calcium–calmodulin-dependent protein kinase relieves calponin-induced inhibition.

[3]High-molecular-weight caldesmon$_{hMW}$ is predominantly synthesized in smooth myocytes. The low-molecular-weight isoform, caldesmon$_{lMW}$ is widespread in non-myocytes. Caldesmon alternatively binds (flip-flop binding) either to calmodulin or ^{F}actin according to the Ca^{2+} concentration.

8.1.1 Actin

Two classes and 3 main groups of actin isoforms exist: class-1, non-muscle actin, which encompasses β and γ groups, and class-2, muscle actin, which corresponds to α group. In myocytes, α actins are constituents of the contractile apparatus.

Smooth myocytes possess class-2 α2- and α3- (or γ2)-actins, in addition to class-1 actins. Isoforms β and γ coexist in most cell types as components of microfilaments of the cytoskeleton that support signal transduction, cell adaptation to mechanical loads, intracellular transport, and cell motility, among other functions (Vol. 1 – Chap. 6. Cell Cytoskeleton).

8.1.2 Tropomyosin

Tropomyosin polymers (TpM1–TpM4) are actin-binding proteins that regulate the binding of myosin, as they block access to myosin-binding sites on actin. Tropomyosins reside along each of the 2 grooves of actin filaments to yield structural stability and modulate filament function.

Many isoforms are produced by alternative promoters of 4 genes (TPM1–TPM4) as well as RNA processing. Many of these isoforms have a tissue- and filament-specific distribution.

8.1.3 Myosin

Myosin are adenosine triphosphatases that constitute a superfamily (Vol. 1 – Chap. 6. Cell Cytoskeleton). Non-conventional myosins, such as myosin-5 and -6 that travel toward the barbed (or +) and pointed (or −) ends of actin filaments, respectively, are devoted to cargo transfer within the cell. Conventional myosin, i.e., myosin-2, contains 2 heavy and 2 pairs of light chains. Myosin heavy chains contain N-terminal heads and C-terminal tails that hold the 2 heavy chains together in a coiled-coil manner. Two myosin regulatory light chains per heavy chain N-terminus bind the region (neck) between the head and tail.

Four known isoforms of smooth muscle myosin heavy chain (smMHC) are formed by 2 independent, alternative splices of the transcript of a single gene (smMHC1a–smMHC1b and smMHC2a–smMHC2b).[4] Isoforms of the mMHC*i*b type contain an additional 7 amino acids that double ^{Mg}ATPase activity as well as the displacement rate on actin filaments [705].

[4]Alternative splicing in the region that encodes the heavy chain tail produces smMHC1 and smMHC2 isoforms. Alternative splicing in the region that encodes the 25–50-kDa junction generates the smMHCa and smMHCb isoforms. The proportion of smMHCa and smMHCb differs in tonic and phasic smooth muscles.

The ratio between actin and myosin is greater in smooth myocytes (15/1) than striated myocytes (7/1). This ratio is very large in non-myocytes. α-Smooth muscle actin (αsmA), localized in microfilament bundles, is typical of smooth myocytes.

8.1.3.1 Myosin Light Chain Kinase and Phosphatase

The myosin regulatory system is more important in smooth muscles than striated muscles. The myosin light chain (MLC) phosphorylation state that determines actin polymerization associated with smooth muscle contraction is tightly controlled by the relative activities of antagonist enzymes myosin light chain kinase (MLCK) and phosphatase (MLCP; Fig. 8.1).

Myosin light chain kinase activated by calcium–calmodulin increases sensitivity of myosin to calcium and regulates actin–myosin interactions. Phosphorylation of MLC leads to cross-bridge formation between myosin heads and actin filaments, and hence to smooth muscle contraction. Dephosphorylation of myosin light chains by myosin light chain phosphatase causes smooth muscle relaxation.

8.1.3.2 Small RhoA GTPase and RoCK Kinase

The RhoA–RoCK pathway inhibits myosin phosphatase. Stimulation of GPCRs generates smooth myocyte contraction via Ca^{2+}–calmodulin-dependent activation of MLCK and subsequent myosin phosphorylation. Kinase RoCK also regulates MLC phosphorylation. Myosin phosphatase- and RhoA-interacting protein (MPRIP)[5] is detected with protein kinase-G and myosin phosphatase and colocalize with myofibrils in smooth myocytes.

8.1.4 Calmodulin

Calmodulin (i.e., calcium-modulated protein) is an ubiquitous, multifunctional regulator and calcium sensor. It controls numerous cell functions, such as metabolism (calcium, cyclic nucleotide, and glycogen), secretion, smooth muscle contraction, intracellular transport, and gene expression.

[5]A.k.a. Rho-interacting protein RhoIP3.

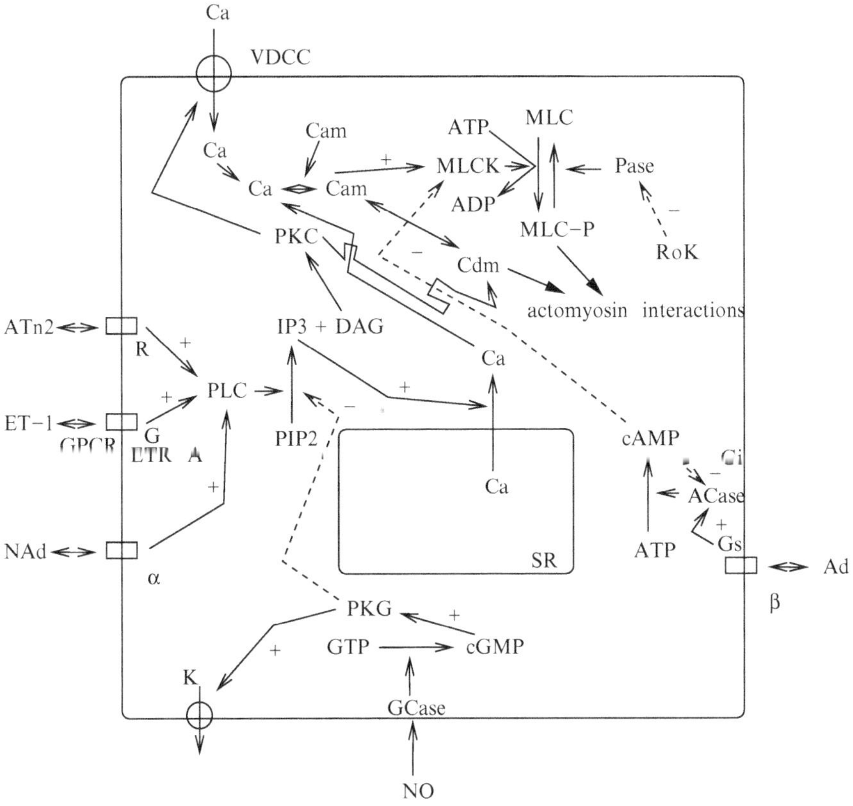

Fig. 8.1 The smooth myocyte function and its regulation (Source: [761]). Two pathways are involved: (1) the phosphatidylinositol and (2) Gs-protein axes. A ligand activates a GPCR, which stimulates phospholipase-C (PLC), producing inositol trisphosphate (IP_3) and diacylglycerol (DAG) from phosphatidylinositol bisphosphate (PIP_2). IP_3 releases calcium from the sarcoplasmic reticulum (SR) and DAG activates protein kinase-C (PKC). PKC acts on the contractile proteins and ion channels. Calcium and calmodulin (Cam) stimulate a myosin light chain kinase (MLCK), leading to actin–myosin interactions. Noradrenaline (NAd), acting via α1-adrenoceptors, angiotensin-2 (ATn2) via its cognate receptors, and endothelin-1 (ET1) via ETRa receptors (ET_A; in fugure ETR-A) activate phospholipase-C (PLC). Gs protein activated by a β2-adrenoceptor agonist stimulates adenylate cyclase (ACase), which catalyzes the formation of cAMP messenger. The latter inhibits MLCK and causes vasodilation.

Calmodulin isoforms (Cam1–Cam3) are encoded by 3 genes in humans. In addition to these isoforms, the calmodulin family contains calmodulin-like proteins (CalmL1 and CalmL3–CalmL6)[6]

[6]Calcium-binding calmodulin-like protein-3 is also called calmodulin-like protein (CLP), CalmL5 calmodulin-like skin protein (CLSP), and CalmL6 calglandulin-like protein (CalgP or CagLPL).

Table 8.1. Regulatory proteins of contraction. Caldesmon can be sequestered away from actin by calcium–calmodulin complexes; it is phosphorylated by PKC.

CMC	SMC
Troponin	Caldesmon

8.1.5 Caldesmon

Binding partners of caldesmon include actin, Ca^{2+}–calmodulin, myosin, tropomyosin, and phospholipids. Caldesmon also binds caltropin [706] and tubulin [707]. Alternatively spliced transcript variants generate diverse isoforms.

Caldesmon is implicated in the regulation of actomyosin interactions in smooth myocytes as well as non-myocytes (Table 8.1). Caldesmon precludes the ATPase activity of myosin in smooth myocytes, in addition to the regulation of non-myocyte contraction.

Caldesmon reversibly binds to ^{F}actin and actin–tropomyosin filaments. In particular, the interaction of high-molecular-weight caldesmon$_{hMW}$ with actin–tropomyosin filaments hinders myosin ATPase activity. Calmodulin binds to caldesmon$_{hMW}$ at activating calcium concentrations, thereby hampering the interaction of caldesmon$_{hMW}$ with actin–tropomyosin to activate the myosin ATPase activity and cause contraction.

Mitogen-activated protein kinase that is regulated by protein kinase-C phosphorylates caldesmon, especially under smooth myocyte stretching [708].

8.1.6 Calponin

Calponin is a calcium- and calmodulin-binding protein that prevents the ^{Mg}ATPase activity of myosin in smooth myocytes [709]. Calponin also connect to tropomyosin, independently of Ca^{2+}, but links to calmodulin in the presence of Ca^{2+} ion (Table 8.2). Calponin is phosphorylated by protein kinase-C and Ca^{2+}–calmodulin-dependent protein kinase CamK2 to relieve calponin inhibition of myosin.

8.2 Glycocalyx

Heparan sulfate proteoglycans of the glycocalyx (thickness 5–11 μ [711]), such as syndecan-1 and -4, can cooperate with integrins to launch cell spreading and assembly of focal adhesions and stress fibers.

In humans, calmodulin-like protein is a multifunctional calcium sensor that can regulate the Ca^{2+}-induced Ca^{2+} release.

Table 8.2. Calponin-binding proteins (Source: [710]). Addition of calponin to actin–tropomyosin complex provokes formation of actin bundles.

Partner	Function
Actin	Modulation of actin–myosin-2 ATPase activity
α-Actinin	Stabilization of the α-actinin–actin complex
Caldesmon	Myosin-2 binding regulation
Caltropin	Weakening of caldesmon–actin interaction
Calmodulin	Inhibition of actin binding
Myosin-2	Stimulation of myosin-2 ATPase activity
Tropomyosin	Actin stability and myosin-2 regulation
Tubulin	Microtubule binding
HSP90	Prevention of actin bundle formation Decrease in calponin-induced polymerization of ^{G}actin
CamK2 ERK1/2 PKC	Phosphorylated Scaffolding for ERK signaling Phosphorylated

Interstitial flow upregulates matrix metalloproteinase expression in rat vascular smooth myocytes, fibroblasts, and myofibroblasts via the FAK–ERK1/2-Jun pathway [712]. At the surface glycocalyx of rat vascular smooth myocytes, heparan sulfate proteoglycans sense interstitial flow and transduce this mechanical signal. Subsequent phosphorylated (activated) of mechanosensitive focal adhesion kinase enable ERK1/2 activation, interstitial collagenase (MMP13) expression, and smooth myocyte motility.

8.3 Intrinsic Regulators of Smooth Myocyte Tone

8.3.1 Caldesmon and Calmodulin

Caldesmon may act on ^{F}actin by either phosphorylation by protein kinase-C and Ca^{2+} or formation of a caldesmon–calcium–calmodulin complex that is unable to bind to actin. Moreover, caldesmon enhances the binding of ATP-bound heavy meromyosin (HMM), as well as (but to a lesser extent) the subfragment S1–ATP (like HMM–ATP in skeletal myocytes), to actin [713]. This effect is hindered by Ca^{2+}–calmodulin. Caldesmon also interacts with myosin in a process regulated by casein kinase-2 [714].

8.3.2 Arachidonic Acid

Arachidonic acid stimulates ATPase activity of unphosphorylated smooth muscle myosin, heavy meromyosin, and subfragment S1, thereby causing contraction of smooth myocytes [715].

8.3.3 G-Protein-Coupled Receptors

In smooth myocytes, signaling from G-protein-coupled receptors controls contraction and growth as well as tissue remodeling and inflammation. Gq-coupled receptors operate via phospholipase-Cβ that hydrolyzes phosphatidylinositol (4,5)-bisphosphate into intracellular messengers diacylglycerol and inositol (1,4,5)-trisphosphate.

Cytosolic Ca^{2+} concentration increases upon the release from sarcoplasmic reticulum stores through IP_3 receptors and influx from plasmalemmal IP_3 channels (Vol. 3 – Chap. 3. Main Classes of Ion Channels and Pumps). Elevated intracellular calcium level causes Ca^{2+} binding to calmodulin. Calcium–calmodulin then activates myosin light chain kinase. The latter phosphorylates myosin light chains to enable actin to activate myosin ATPase activity for contraction of actomyosin filaments.

Diacylglycerol that remains bound to the plasma membrane initiates the translocation of protein kinase-C from the cytoplasm to the membrane and its subsequent activation. Protein kinase-C phosphorylates calponin to relieve its inhibition of myosin ATPase. It also phosphorylates components of MAPK modules (Vol. 4 – Chap. 6. Mitogen-Activated Protein Kinase Modules) that activate transcription factors involved in cell growth. In addition, PKC stimulates phospholipase-A2 that releases arachadonic acid.

Gq-coupled receptors also interact with guanine nucleotide-exchange factors of Rho GTPases (RhoGEFs) to regulate actin network reorganization. Monomeric Rho GTPase activates RoCK kinase that phosphorylates the myosin-binding subunit of myosin light chain phosphatase to inhibit its activity, thereby heightening phosphorylation of myosin light chain.

8.3.4 Guanylate Cyclases, Nitric Oxide, and Natriuretic Peptides

Nitric oxide and atrial natriuretic peptide relax smooth myocytes via soluble and particulate guanylate cyclases, cGMP messenger and cGMP-dependent protein kinase isozymes PKG1α and PKG1β.[7] Both isoforms have several common substrates, such as inositol (1,4,5)-trisphosphate receptor (IP_3R)-associated cGMP kinase substrate (IRAG),[8] myosin-binding $PP1_{r12a}$ regulatory subunit, and regulator RGS2 of G-protein signaling.

Enzyme PKG1β phosphorylates $PP1_{r12a}$, hence causing calcium desensitization of myosin regulatory light chain. It also targets RGS2, thereby modulating receptor-mediated signaling via Gq/11 class of G-protein subunits. It interacts specifically with IRAG protein that is highly expressed in the membrane of the endoplasmic reticulum of smooth myocytes. It allows stable interaction of IP_3R1 receptor with PKG1β enzyme [716].

8.3.5 *Inositol Trisphosphate Receptors and Calcium Release*

Entry of extracellular calcium can occur through plasma membrane ion carriers, such as voltage-dependent $Ca_V1.2$ channels, non-selective cation channels, and Na^+–Ca^{2+} exchangers. Intracellular calcium signals are also produced by Ca^{2+}2+ release from intracellular stores (sarco[endo]plasmic reticulum and mitochondria).

In smooth myocytes, calcium signals include Ca^{2+} *flashes*,[9] *oscillations*,[10] *puffs*,[11] *ripples*,[12] *sparklets*, *sparks*,[13] *waves*,[14] and global intracellular Ca^{2+} concentration elevation (Table 8.3).[15] Activated IP_3R generates these cues. However, non-selective cation, store-operated Ca^{2+}, ryanodine-sensitive Ca^{2+} release, transient receptor potential, and voltage-dependent Ca^{2+} channels also contribute to and modulate these calcium signals.

---

[7] A.k.a. cGK1α and cGK1β, respectively.

[8] A.k.a. murine retrovirus integration site homolog MRVI1.

[9] Rapid calcium flashes observed during rhythmic phasic contractions of smooth myocytes rely on voltage-dependent Ca^{2+} channels, IP_3Rs, and ryanodine-sensitive Ca^{2+} channels.

[10] Repetitive, non-propagating, global elevations of intracellular Ca^{2+} concentration generated by periodic release from sarco(endo)plasmic reticulum of Ca^{2+} in smooth myocytes due to cyclical positive and negative feedback of intracellular Ca^{2+} concentration on IP_3R channel activity with contributions from ryanodine-sensitive Ca^{2+} and canonical transient receptor potential channels.

[11] Calcium puffs arise from the synchronous opening of about 30 IP_3R channels (cluster size ~400 nm) with a minor contribution from ryanodine-sensitive Ca^{2+} channels.

[12] Spontaneous, low-amplitude, propagating, IP_3R-mediated Ca^{2+} signals following PLC activation and Ca^{2+} release.

[13] Localized, intracellular Ca^{2+} transients that result from the concerted opening of ryanodine-sensitive Ca^{2+} channels. IP_3R receptors can contribute to Ca^{2+} sparks.

[14] Propagating elevations in intracellular Ca^{2+} concentration due to release from sarco(endo)plasmic reticulum upon activation of IP_3R and RyR receptors.

[15] These events result from IP_3R activation that stimulates plasmalemmal TRPC1, TRPM4, and voltage-dependent Ca^{2+} channels.

Table 8.3. Calcium signals regulated directly or indirectly by IP_3Rs in smooth myocytes (Source: [717]; Ca_V: voltage-dependent Ca^{2+} channel; NEx: Na^+–Ca^{2+} exchanger; NSCC: non-selective cation channel; RyR: ryanodine-sensitive Ca^{2+} release channel; SOC: store-operated Ca^{2+} channel; TRPC: transient receptor potential canonical channel; TRPM: melastatin-related transient receptor potential channel).

Calcium signals	Frequency (Hz)	Rise time (s)	Decay half-time (s)	Spatial spread (μm)	Involved Ca^{2+} channels
Flash	0.2–0.5	0.2–0.8	0.5–1.5		Ca_V, RyR
Global spread					Ca_V, NSCC, RyR, SOC TRPC1/3, TRPM4
Oscillation	0.07–0.4	2–6	3–18		RyR, TRPC6
Puff	0.02–0.04	0.06–0.16	0.11–0.25	1.89–2.51	RyR
Ripple	0.05–0.19	1–2	4–6		
Spark	0.07–0.59	0.01–0.07	0.03–0.15	0.52–4.7	RyR
Wave	0.04–1.41	0.5–2	2–10	8.8–50	Ca_V, NSCC, NEx, RyR, SOC

Plasmalemmal Gq/11-coupled receptors stimulate phospholipase-C that hydrolyzes plasmalemmal phosphatidylinositol (4,5)-bisphosphate and generates diacylglycerol and inositol (1,4,5)-trisphosphate.

Inositol trisphosphate receptors constitute a set of tetrameric intracellular Ca^{2+} release channels on the endoplasmic reticulum membrane that generate Ca^{2+} signals (Vol. 3 – Chap. 3. Main Classes of Ion Channels and Pumps). Three IP_3R isoforms (IP_3R1–IP_3R3; Table 8.4) encoded by distinct genes are responsible for signaling differences with given frequencies, amplitudes, and spatiotemporal properties. These 3 isoforms are synthesized in smooth myocytes. Subtype IP_3R1 is the predominant isoform in vascular smooth myocytes; IP_3R2 and IP_3R3 densities are higher in proliferating vascular smooth myocytes [717]. In the tracheal smooth muscle, IP_3R3 is the principal isoform, IP_3R1 and IP_3R2 being also expressed [717].

The cytosolic N-terminus possesses an IP_3-binding domain, a suppressor motif that inhibits IP_3 binding, and a regulatory sequence with binding sites for ATP and Ca^{2+} ion as well as phosphorylation sites and a coupling region for interacting with transient receptor potential canonical (TRPC) channels [717]. The transmembrane region and cytosolic C-terminus enable IP_3R tetramerization.

Table 8.4. Main features of IP_3R isoforms (Source: [717]). In vascular smooth myocytes, IP_3R receptors localize to the central sarco(endo)plasmic reticulum around the nucleus as well as its peripheral region beneath the plasma membrane. They thus can regulate Ca^{2+}-dependent gene expression as well as local signaling to membrane proteins. IP_3R channel is activated at a cytosolic Ca^{2+} level <300 nmol; an elevation in cytosolic Ca^{2+} level together with calmodulin inhibit IP_3R channel. Agent ATP is not required for IP_3R activation, but enhances sensitivity to IP_3 and Ca^{2+} ligands. Binding of IP_3 to IP_3R in aortic and tracheal smooth myocytes is weak at pH<7 and maximal at pH ranging from 8 to 9. In portal vein smooth myocytes, a pH increase from 6.7 to 7.0–7.3 reduces the Ca^{2+} concentration required for IP_3R activation. Reactive oxygen species stimulate IP_3R-mediated Ca^{2+} release in aorta smooth myocytes, whereas they impede SERCA pump in coronary artery smooth myocytes, but do not affect IP_3R activity in mesenteric artery smooth myocytes.

	IP_3R1	IP_3R2	IP_3R3
IP_3 affinity	Intermediate (~270 nmol)	Highest (~100 nmol)	Lowest (~400 nmol)
Activation by Ca^{2+} Inhibition by Ca^{2+}	~170 nmol ~370 nmol	~150 nmol ~160 nmol	~60 nmol ~170 nmol
Sensitivity to ATP	High (~130 μmol)	None	Low (~2 mmol)
Influence of pH (IP_3 binding, Ca^{2+} sensitivity)	6.8		7.5
Alternative splicing	Yes	None described	None described

Density of IP_3Rs results from the balance between gene transcription and protein degradation. Transcription factor MyB binds to the Ip_3r1 promoter and stimulates transcription, whereas retinoic acid precludes it via Activator protein-2 [717]. Hydrogen peroxide promotes proteasomal degradation of IP_3R1 and IP_3R3, whereas JaK2 kinase phosphorylates IP_3R1 and prevents its proteasomal degradation. Vasopressin impedes IP_3R1 gene transcription and favors protein degradation [717].

Several ligands (IP_3, cytosolic and sarco(endo)plasmic reticulum luminal Ca^{2+}, and ATP), kinases (inhibitory PKA and PKG and stimulatory PKC and Tyr kinases), and other types of regulators (FKBP12 via inhibitory PP3 and stimulatory TOR, scaffold enhancer RACK1) and modulators (cellular pH[16] and reactive oxygen species) control IP_3R activity [717].

In addition, IP_3R communicates with adjoining ryanodine-sensitive Ca^{2+} channels of the sarco(endo)plasmic reticulum as well as mitochondria to influence Ca^{2+} release from its main store as well as reactive oxygen species generation. It can couple to plasmalemmal transient receptor potential channels (e.g., TRPC1 and

[16]Cellular pH influences IP_3 binding to IP_3R and sensitivity of IP_3R to Ca^{2+} ion.

TRPC3),[17] store-operated Ca^{2+} channels,[18] and large-conductance Ca^{2+}-activated K^+ channels (BK),[19] which lodge in caveolae where the sarco(endo)plasmic reticulum and plasma membrane are in close proximity [717].

In airway and vascular smooth myocytes, proximity (distance ~30 nm) between mitochondria and sarco(endo)plasmic reticulum enables local signaling between these 2 organelles. In particular, IP_3R regulates mitochondrial Ca^{2+} concentration. Mitochondrial Ca^{2+} uniporter, the major mitochondrial Ca^{2+} uptake axis, is sensitive to micromolar Ca^{2+} levels. IP_3R-mediated Ca^{2+} waves increase mitochondrial Ca^{2+} concentration, stimulate ROS generation and NFκB, and foster $Ca_V1.2$ gene expression in cerebral artery smooth myocytes [717].

Conversely, mitochondria regulate IP_3R in smooth myocytes, as they: (1) buffer local and global elevation in cytosolic Ca^{2+} level; (2) control ATP synthesis;[20] (3) participate in ROS signaling;[21] and modulate ion channels that regulate membrane potential and cytosolic Ca^{2+} level. In vascular smooth myocytes, mitochondria regulate the activity of several plasmalemmal ion channels, such as voltage-dependent Ca^{2+}, Ca^{2+}-activated K^+, Ca^{2+}-activated Cl^- (ClCa), and store-operated Ca^{2+} channels, thereby influencing membrane potential and Ca^{2+} signals [717].

Inositol trisphosphate receptors controls not only cell contraction, but also gene expression and cell migration and proliferation. In cerebral artery smooth myocytes, IP_3R-mediated Ca^{2+} release activates NFκB directly and indirectly via mitochondrial ROS production; NFκB stimulates $Ca_V1.2$ channel α1c subunit expression [717]. Cell proliferation correlates with an elevation in IP_3R1 production in aortic and carotid artery smooth myocytes and upregulation of all IP_3R isoforms in mesenteric artery smooth myocytes [717].

---

[17] An elevation in intracellular IP_3 concentration supports coupling of IP_3R1 to transient receptor potential canonical channel TRPC3 in arterial smooth myocytes, hence causing a depolarization that stimulates Ca_V channel. Subsequent global intracellular Ca^{2+} concentration elevation provokes vasoconstriction. IP_3R-mediated Ca^{2+} release also activates Ca^{2+}-sensitive PKC isoforms that stimulate PKC-dependent TRPC1 in portal vein smooth myocytes [717]. In cerebral artery smooth myocytes, Ca^{2+}-sensitive TRPM4 is regulated by released Ca^{2+}, but not by direct interaction with IP_3R in the absence of C-terminal calmodulin- and IP_3R-binding (CIRB) domain.

[18] I.e., Orai protein, in addition to TRPC1 and TRPC5 to TRPC7 channels in smooth myocytes of coronary and mesenteric arteries, pulmonary arteries, inferior vena cava, and portal vein.

[19] Activated IP_3R1 increases the Ca^{2+} sensitivity of nearby BK channels, especially in cerebral artery smooth myocytes, thereby facilitating channel activation at lower Ca^{2+} concentrations. Resulting hyperpolarization inhibits Ca_V channel, thereby amplifying stimulation by released Ca^{2+} to oppose IP_3-induced smooth myocyte contraction and favoring vasodilation.

[20] Agent ATP enhances sensitivity of IP_3R to IP_3 and Ca^{2+}. Moreover, it is required for SERCA activity.

[21] Mitochondrial ROS may modulate IP_3R expression and activity.

8.4 Airway Smooth Myocytes

8.4.1 Mechanical Loading-Induced Cytoskeleton Remodeling

Airway smooth myocyte (aSMC) is able to modify its shape, stiffness, and contractility in response to mechanical loading after adaptation of actomyosin interaction and integrins at adhesion junctions [718].

In the tracheal muscle, strains elevate phosphorylation level of both focal adhesion kinase and paxillin. Once phosphorylated by focal adhesion kinase, adhesion complex component paxillin couples to neuronal Wiskott-Aldrich syndrome protein (nWASP) via CRK adaptor to launch nWASP activation by activated CDC42 GTPase that binds to recruited nWASP (Vol. 1 – Chaps. 6. Cell Cytoskeleton and 7. Plasma Membrane). The latter stimulates the ARP2–ARP3 complex that polymerizes actin filaments.

Integrin-linked kinase and its binding partners coordinate cytoskeleton remodeling, especially formation of structural links between integrins and the actin cytoskeleton. Integrin-linked kinase tethers to β_1-integrins and can then transduce signals from integrin receptors to regulate actin polymerization. This scaffold forms a heterotrimer with LIMS1 adaptor (or PINCH) and α-parvin that binds to actin filaments. Enzyme ILK also connects to paxillin. In addition, LIMS1 adaptor may couple to nWASP via NCK adaptor.

8.4.2 Contractile and Proliferative Phenotype

Airway smooth myocyte can evolve from a contractile to a proliferative phenotype. When aSMCs proliferate, the content in smooth muscle α-actin, high-molecular-weight caldesmon$_{\text{hMW}}$ isoform, calponin, desmin, smooth muscle myosin heavy chain, myosin light chain kinase, and β-tropomyosin decays. On the other hand, the content of non-muscle myosin heavy chain, low-molecular-weight caldesmon$_{\text{lMW}}$, vimentin, protein kinase-Cα and -Cβ, and CD44 heightens (1–6 fold) when cells divide [719]. The content in cytoskeletal proteins is greater in the trachealis muscle than that in the media of the pulmonary artery [719]. Airway smooth myocytes shorten faster than arterial ones.

8.4.3 Calcium Influx

Plasma membrane depolarization activates Ca_V channels. Subsequent Ca^{2+} entry causes rapidly (in a few milliseconds) local Ca^{2+} release through ryanodine receptors from Ca^{2+} store in myocytes.[22] All 3 types of ryanodine receptors (RyR1–RyR3) are expressed in smooth myocytes.

In airway smooth myocytes, membrane depolarization does not prime instantaneous Ca^{2+} release. However, voltage depolarization during sufficiently long duration (>1 s) can induce local Ca^{2+} release through ryanodine receptors, independently of Ca_V channels and associated extracellular Ca^{2+} import into the cytosol [720].

Resulting Ca^{2+} sparks prime cell contraction. The long latency for depolarization-induced Ca^{2+} release in airway smooth myocytes is explained by involvement of GPCR that can serve as voltage sensors and their initiated signaling. This type of mechanism is illustrated by activated M_3 muscarinic receptors in the absence of exogenous agonists that trigger the Gq–PLC–IP_3 axis. This signaling pathway provokes a Ca^{2+} release through IP_3Rs from the sarcoplasmic reticulum that is then enhanced by that through RyR2 (local Ca^{2+}-induced Ca^{2+} release), but neither RyR1 nor RyR3 [720].

8.4.4 Airway Smooth Myocyte Products

Airway smooth myocytes synthesize growth factors, cytokines, and chemokines, as well as lipid mediators, in addition to receptors, ion carriers, enzymes, and constituents of the extracellular matrix (Table 8.5).

Airway remodeling relies on changes in phenotype of airway smooth myocytes and fibroblasts, especially in asthma. Mitogens are released by epithelial cells, fibroblasts, smooth myocytes, mastocytes, and eosinophils. In addition to mechanical stress, several growth factors can be involved in the change in aSMC phenotype, such as epidermal (EGF), fibroblast (FGF2), and platelet-derived (PDGF) growth factors.

8.4.4.1 G-Protein-Coupled Receptors

In humans, airway smooth myocytes are endowed with multiple types (>350 [721]) of G-protein-coupled receptors that participate in the intra-, auto-, juxta-, and paracrine regulation. Furthermore, alternative splicing diversifies the recepterome and augments the number of GPCR types encountered on their surface, hence heterogeneity in GPCR signaling. More than 190 GPCRs have, on average,

---

[22] In skeletal myocytes, membrane depolarization activates Ca_V channels for Ca^{2+} influx from the extracellular medium. Newly imported Ca^{2+} opens RyR1 ryanodine receptors (voltage-mediated Ca^{2+}-induced Ca^{2+} release). In cardiomyocytes, Ca_V channels activated by depolarization carry a small amount of extracellular Ca^{2+} that, nonetheless, causes RyR2 opening and subsequent large Ca^{2+} from the sarcoplasmic reticulum.

Table 8.5. Examples of molecules synthesized by airway smooth myocytes, such as growth factors, including cytokines and chemokines, as well as receptors and enzymes (Main source: [739]; A_i: adenosine receptor; AR: adrenergic receptor; B_2 bradykinin receptor; CCL: chemokine CC-motif ligand; COx: cyclooxygenase; CSF: colony-stimulating factor; CysLT: cysteinyl leukotriene receptor; ECM: extracellular matrix; ETR: endothelin receptor; H_1: histamine receptor; ICAM: intercellular adhesion molecule; Ifnγ: interferon-γ; IfnγR: Ifnγ receptor; IL*i*: interleukin type-*i*; IL*i*R: IL*i* receptor; $K_{Ca}1.1$: voltage-sensitive, calcium-activated large conductance potassium channel (BK); LIF: leukemia inhibitory factor [a.k.a. cholinergic differentiation factor (CDF) and differentiation-stimulating factor (DIA)]; LPAR: lysophosphatidic acid receptor; mAChR: muscarinic acetylcholine receptor; MHC: major histocompatibility complex; NKR: neurokinin receptor; NOS: nitric oxide synthase; PAR: peptidase-activated receptor; PDGF: platelet-derived growth factor; PGe2: prostaglandin-E2; PGi2: prostacyclin; sPLA2: secreted phospholipase A2; PR: prostanoid receptor; RAR: retinoic acid receptor; RXR: retinoid X receptor; S1PR: sphingosine 1-phosphate receptor; SCF: stem cell factor; VCAM: vascular cell adhesion molecule; VIPR: vasoactive intestinal peptide receptor; 5HTR: serotonin receptors).

Type	Molecules
Growth factors	IGF2, PDGFa, PDGFbb, SCF, VEGFa
Cytokines	CSF2, IL2/5/6/8/11/12, Ifnγ, LIF
Chemokines	CCL2/5/7/8/11
Receptors	$A_{1/2/3}$, AR, B_2, CysLT, ETR, H_1, LPAR, mAChR, NKR, PAR, PR, S1PR, VIPR, 5HTR EGFR, IGFR1/2, PDGFRα/β, IL2R, IL12R, IL13R, IfnγR, TNFRSF5, CCR3, Class-2 MHC, CD44
Adhesion molecules	ICAM1, VCAM1
Lipid mediators	PGe2, PGi2
Enzymes	COx1/2, MMP1/2, NOS, sPLA2
Ion carriers	IP_3R, RyR, SERCA, $Ca_V1/3$ $K_{IR}2.1$, $K_{Ca}1.1$
ECM components	Tenascin-C (during healing)
Nuclear receptors	RARα/β/γ, RXRα/β

5 different receptor isoforms that result from various types of alternative splicing of pre-mRNAs, such as alternative splice donors and acceptors, novel introns, intron retentions, exon(s) skips, and novel exons [721].

In airways, GPCRs expressed on smooth myocytes contribute to the regulation of contraction and relaxation state. Histamine, leukotriene, prostaglandin, and muscarinic Gq-coupled receptors support contraction, whereas Gs-coupled β2-adrenergic receptors foster relaxation.

In addition, in airway smooth myocytes, G-protein-coupled receptors modulate secretion of growth factors and other signaling mediators such as eicosanoids that intervene in airway inflammation (Table 8.6).

Table 8.6. G-protein-coupled receptors and their effects in airway smooth myocytes (Source: [722]; A_i: adenosine receptor; AR: adrenergic receptor; B_i: bradykinin receptor; $CysLT_i$: cysteinyl leukotriene receptor; EP_i: prostanoid receptor; ET_x: endothelin receptor; H_i: histamine receptor; IP: prostanoid (prostacyclin) receptor; LPA_i: lysophosphatidic acid receptor; M_i: muscarinic acetylcholine receptor; NK_i: neurokinin receptor; PAR_i: peptidase-activated receptor; RasGRF: Ras guanosine nucleotide-releasing factor [RasGEF]; RasGRP: Ras guanosine nucleotide-releasing protein [RasGEF]; $S1P_i$: sphingosine 1-phosphate receptor; TP: prostanoid (thromboxane A2) receptor; VIPR: vasoactive intestinal peptide receptor; $5HT_i$: metabotropic serotonin (or 5-hydroxytryptamine) receptors).

GPCR Type	Main G protein partner	Function
$5HT_{2C}$	Gq Gi	PLC–IP_3–Ca–RasGRF1–Ras–ERK1/2 axes Contraction, growth potentiation
α1AR β2AR	Gq Gs	 Relaxation, growth inhibition, regulation of cytokine synthesis
A_1 A_{2B} A_3	Gi Gs Gq	Contraction Relaxation
B_2	Gq	Contraction
$CysLT_1$	Gq	Contraction
EP_2 IP TP	Gs Gs Gq	Relaxation, growth inhibition, regulation of cytokine synthesis Growth inhibition Contraction, growth potentiation
ET_A–ET_B	Gq	Contraction, growth potentiation
H_1	Gq	Contraction, growth potentiation
LPA_1–LPA_4 $S1P_1$–$S1P_3$	Gq, Gi, G12/13 Gq, Gi, G12/13	Growth stimulation, Regulation of cytokine synthesis
M_2 M_3	Gi Gq	 Contraction, growth potentiation
NK_1–NK_2	Gq	Contraction, growth potentiation
PAR_1–PAR_3	Gq, Gi, G12/13	Growth stimulation and potentiation
P2Y	Gq	
VIPR	Gs	Relaxation, growth inhibition

Rapid desensitization results from receptor phosphorylation by G-protein-coupled receptor kinases (GRK). Arrestin binding mediated by GRKs initiates receptor endocytosis for resensitization after recycling to the plasma membrane.

In addition, activated intracellular kinases, such as protein kinases PKA and PKC also phosphorylate GPCRs and lead to GPCR–G-protein decoupling.

RasGRF Activators

The RasGRF proteins are coupled to G-protein-coupled receptors. Three main families of guanosine nucleotide-exchange factors (GEF) are associated with Ras GTPases and link plasmalemmal receptors to Ras protein activation: Son of sevenless (SOS1–SOS2), Ras guanosine nucleotide releasing factors (RasGRF1–RasGRF2), and Ras guanosine nucleotide releasing proteins (RasGRP1–RasGRP4; Vol. 4 – Chap. 9. Guanosine Triphosphatases and Their Regulators). Activators of the SOS and RasGRP sets link (via GRB2 adaptor and phospholipase-Cγ-dependent generation of diacylglycerol, respectively) membrane or cytoplasmic protein Tyr kinases to Ras GTPases. Calcium influx promotes RasGRF translocation to the plasma membrane, where it can be phosphorylated (activated).

Gq-Coupled Receptors and Bronchoconstriction

Other Gq-coupled receptors that can cause contraction include A_3 adenosine, B_2 bradykinin, ET_A endothelin, H_1 histamine, NK1 and NK_2 neurokinin, and P2 purinergic receptors [722]. Both thrombin and lysophosphatidic acid that operate via Gq-coupled receptors PARs and LPARs provoke cell proliferation. Endothelin, histamine, leukotriene-D4, sphingosine 1-phosphate, and thromboxane potentiate mitogenic effects of receptor protein Tyr kinases via their cognate Gq-coupled receptors.

High-Affinity Receptors for Leukotriene-B4

For example, high-affinity receptors for leukotriene-B4 (LTb4R), i.e., BLT_1 and BLT_2 (Vol. 3 – Chap. 7. G-Protein-Coupled Receptors), are expressed on neutrophils, eosinophils, and T lymphocytes. They promote leukocyte migration to the lung and participate in their activation and degranulation. Moreover, they are produced by human airway smooth myocytes. They are coupled to $G\alpha_i$ or $G\alpha_q$ subunit. Three forms of the transcribed LTb4R and 2 additional variations in the 5′ UTR exist [721].

Gs-Coupled Receptors and Bronchodilation

Gs-coupled receptors elicit relaxation of airway smooth myocytes. They comprise β2 adrenergic, A_{2B} adenosine, prostanoid EP_2 and IP (for prostaglandin-E2 and prostacyclin, respectively), and vasoactive intestinal peptide (VIP) receptors [722].

Activated Gs stimulates membrane-bound adenylate cyclase AC5 and AC6 isoforms. Activated adenylate cyclases create second messenger cAMP from ATP to activate protein kinase-A. This enzyme can phosphorylate some Gq-coupled receptors, phospholipase-C, as well as IP_3Rs to prevent phosphoinositide generation and calcium influx. It also phosphorylates myosin light chain kinase, thereby reducing its affinity to calcium–calmodulin and activity. It phosphorylates calcium-activated potassium channel to cause K^+ efflux and hyperpolarization. It phosphorylates the transcription factor cAMP-responsive element-binding protein (CREB) to control the synthesis of numerous immunomodulatory proteins.

VIP Receptor

Vasoactive intestinal peptide (VIP), a relaxant neuropeptide, impedes translocation of cRaf to the plasma membrane induced by ceramide as well as that of PKCα [723]. The Ras-mediated signaling links plasmalemmal receptors to mitogen-activated protein kinase modules that phosphorylate (activate) several target proteins such as 27-kDa heat shock protein (HSP27). In smooth myocytes, HSP27 colocalizes with actin filaments and proteins of the contractile cytoskeleton, such as myosin, tropomyosin, and caldesmon to modulate the interaction between these proteins. Phosphorylation of HSP27 promotes actin polymerization and stress fiber formation. In airway smooth myocytes, phosphorylated HSP27 stabilizes the actin cytoskeleton [724].

Gi-Coupled Receptors

A third group of GPCRs preferentially couples to members of the Gα_i family (Gi1–Gi3). Activated Gi hampers the activity of adenylate cyclases. On the other hand, Gi and/or G$\beta\gamma$ subunits stimulate Rho guanine nucleotide-exchange factors, hence Rho GTPases for actin polymerization. Dimer G$\beta\gamma$ inhibits AC1, but can enhance the activity of certain adenylate cyclase isoforms, such as AC2, AC4, and AC7, as well as sensitization of adenylate cyclases to stimuli [722]. In addition, G$\beta\gamma$ can activate phospholipase-Cβ.

G12/13-Coupled Receptors

A forth set of G-protein-coupled receptors associate with members of the G12/13 subclass. Their effectors encompass members of a family of guanine nucleotide-exchange factors for small Rho GTPases. G12/13-coupled receptors may include LPA, S1P, peptidase-activated (PAR), and prostanoid TP (for thromboxane-A2) receptors [722].

Table 8.7. α-Adrenoceptors in the lung and their function (Source: [725]). The tone of airway smooth myocytes is mainly regulated by members of the Gq/11 and Gs subclasses of G-protein subunits that cause bronchoconstriction and bronchodilation, respectively.

Site	Predominant subtype	Effect
Airway smooth myocyte	α1	Contraction
Vascular smooth myocyte	α1	Contraction
Autonomic nerves	α2	Neurotransmitter release decrease
Secretory glands	α1	Secretion increase
Mastocyte		Mediator release increase

Adrenergic Receptors

In human lungs, the amount of β-adrenoceptors is greater than that of α-adrenoceptors [725].

α-Adrenoceptors

Airway smooth myocytes synthesize both α1- and α2-adrenoceptors [725]. α-Adrenoceptors cause bronchoconstriction (Table 8.7). α1a-Adrenoceptors primarily act on calcium channels, thereby fostering entry of extracellular Ca^{2+} ions. α1b-Adrenoceptors activates phospholipase-C that processes phosphatidylinositol (4,5)-bisphosphate in the plasma membrane to form inositol (1,4,5)-trisphosphate and diacylglycerol, which triggers Ca^{2+} release from intracellular stores and activates Ca^{2+}- and phospholipid-dependent protein kinase-C. The latter may contribute to smooth muscle contraction by increasing the Ca^{2+} sensitivity of contractile proteins. α2-Adrenoceptors inhibit adenylate cyclase via guanine nucleotide inhibitory protein (Gi).

Both noradrenaline and adrenaline increase very weakly the bronchial tone after pretreatment with β-adrenoceptor antagonists in bronchi isolated from healthy human lungs; the response is not enhanced in the presence of histamine [725].

β-Adrenoceptors

Both β1- and β2-adrenoceptors lodge on airway smooth myocytes (~20 and 80%, respectively) in human lungs [725]; hence, in human bronchial smooth myocytes, β-adrenoceptors belong mainly to the β2-subtype. β-Adrenoceptors can also be detected in respiratory epithelial as well as mucous and serous gland cells of the trachea; however, most of the β-adrenoceptor pool of the human lung localizes in alveolar cells.

Gs-coupled β2-adrenergic receptors cause a bronchorelaxation and inhibit proliferation of human airway smooth myocytes. β2-Adrenoceptors also mediate vasodilation and influence mucus production and transport and activity of cholinergic nerves [725].

Relaxation of airway smooth muscles results from Gs stimulation and subsequent elevation of intracellular cAMP level and activation of protein kinase-A. The latter phosphorylates myosin light chain kinase, thereby reducing MLCK affinity for calcium–calmodulin, myosin phosphorylation, and actin–myosin coupling. Smooth myocyte relaxation can also result from the decrease in concentration of free calcium ions. In fact, β2-adrenoceptors operate via cAMP and RapGEF3 and RapGEF4,[23] but not protein kinase-A [726].

Muscarinic Acetylcholine Receptors

Airway smooth myocytes are innervated by postganglionic parasympathetic nerves that release acetylcholine. Muscarinic acetylcholine M_2 and M_3 receptors (mAChR) are coexpressed in many types of smooth myocytes, such as those in walls of blood vessels and airways.

The density of M_2 receptors is greater than that of M_3 receptors (relative density $\geq$4:1) [727]. In smooth myocytes, M_2 receptor inhibits AC5 and AC6 adenylate cyclases via $G\alpha_{i3}$, whereas Gq-coupled M_3 activates AC5 and AC6 via $G\beta\gamma$ [728].

In airway smooth myocytes, activated Gi-coupled M_2 receptor inhibits β-adrenergic-mediated relaxation; activated Gq-coupled M_3 receptor initiates their contraction. Gq-coupled receptors are the main mediators of contraction of airway smooth myocytes. Neurotransmitter acetylcholine acts on M_3 receptors to control the resting tone.

8.4.4.2 Receptor Protein Tyrosine Kinases

Epidermal Growth Factor Family Members

Bronchial epithelial cells produce members of the epidermal growth factor family, such as EGF, transforming TGFα and heparin-binding EGF-like (HBEGF) growth factors, and amphiregulin, that can stimulate cell migration, proliferation, differentiation, and survival, as well as repair of bronchial epithelia.

Both epidermal (EGF) and heparin-binding EGF-like (HBEGF) growth factor promote repair of human bronchial epithelial cells [729].

In normal subjects, the concentration of epidermal growth factor receptor (HER1 or EGFR) is lower than that in asthmatic subjects [729]. The more severe the asthma, the higher the EGFR level. Agent EGF contributes to hypertrophy and hyperplasia of airway smooth myocytes during airway remodeling in asthma [730].

[23] A.k.a. EPAC1 and -2 as well as cAMPGEF1 and -2, respectively.

Insulin-like Growth Factor and Partners

Insulin-like growth factor IGF2, IGF-binding protein IGFBP2 (but neither IGF1 nor IGFBP1), and both IGF receptors IGFR1 and IGFR2 are produced in airway smooth myocytes [732]. In addition, IGF2 and IGFBP2 are secreted.

Platelet-Derived Growth Factors

Platelet-derived growth factors (PDGF) are mitogens and recruitment factors for smooth myocytes. Protomer PDGFb signals via the receptor protein Tyr kinase PDGFRβ.[24] Protomers PDGFa and PDGFb are synthesized by various cell types, including vascular smooth myocytes. Isotype PDGFa is required during the body's development in the recruitment of smooth myocytes to alveolar sacs and their compartmentation into alveoli [733]. Smooth myocytes express both PDGFR types, but usually higher levels of PDGFRβ.

Vascular Endothelial Growth Factors

In humans, airway smooth myocyte synthesizes vascular endothelial growth factor VEGFa isoforms, such as $VEGFa_{121}$, $VEGFa_{165}$, $VEGFa_{189}$, and $VEGFa_{206}$ splice variants [731]. The constitutive secretion of VEGFa increases on bradykinin exposure, activation of B_2 bradykinin receptor and protein kinase-C, and generation of prostanoids. Activated cyclooxygenase-2 produces predominantly prostaglandins PGe2 and prostacyclin (PGi2).[25]

8.4.4.3 Epican

Plasmalemmal CD44 glycoprotein, or epican, a heparan sulfate proteoglycan, is also synthesized by airway smooth myocytes. This molecule interacts with hyaluronic acid, collagens, and matrix metallopeptidases; its glycosylated form with fibrinogen and fibrin; its sialyfucosylated glycoforms with selectins. Matrix metallopeptidase MMP2 is an autocrine factor required for airway smooth myocyte proliferation.

---

[24]During angiogenesis, sprouting tip endothelial cells secrete PDGFb that promotes recruitment of PDGFRβ+ vascular smooth myocytes to the endothelium.

[25]Densities of VEGFR1, a modulator of VEGF activity, and VEGFR2, the active pro-angiogenic receptor, increase in asthma [734]. Receptor VEGFR1 abounds on vascular smooth myocytes. In addition, VEGF maintains alveolar cell survival, hence avoiding alveolar wall destruction (emphysema) [735].

8.4.5 Hyperresponsiveness and Hypersensitivity

Asthma is characterized by hyperplasia and hypertrophy of airway smooth myocytes that strongly (hyperresponsiveness) contract in response to low-magnitude stimuli (hypersensitivity), in addition to tissue inflammation with edema and thickening of the subepithelial basement membrane. Both wall edema and bronchoconstriction narrow lumens of ventilatory ducts, thereby reducing air flow as demonstrated by lung function tests. Smooth myocytes colocalize with mastocytes. In addition, released interleukin-8 is a chemoattractant for neutrophils and eosinophils. Mastocytes do not affect the rate of proliferation, but prevent CCL11-mediated CCR3-primed migration of airway smooth myocytes [736].

Basement membrane thickening is due to the deposition of interstitial collagen-1, -3, and -5 as well as fibronectin. Collagens are synthesized by myofibroblasts, the number and activity of which rise in asthma due to fibroblast (FGF2), insulin-like (IGF1), dimeric platelet-derived (PDGFbb), and transforming (TGFβ) growth factor, as well as endothelin-1 [737, 738].

Airway smooth myocytes contribute to the pathogenesis of asthma because of [739]: (1) proliferation beyond the ordinary rate provoked by mitogens and (2) recruitment of progenitors to airway smooth myocyte bundles, thereby causing hyperplasia in addition to hypertrophy; (3) interactions of airway SMCs with inflammatory cells, such as T lymphocytes and eosinophils; and (4) their secretion of pro-inflammatory cytokines, in particular chemokines (Table 8.5).[26]

Sphingosine 1-phosphate operates as an inflammatory mediator in asthma. It provokes airway smooth muscle contraction and growth as well as production of pro-inflammatory cytokines such as interleukin-6 to trigger bronchoconstriction and airway inflammation and remodeling [740].

8.4.6 Deep Inspiration as Bronchodilation Inducer

Deep inspiration stretches airway smooth myocytes, thereby softening, or fluidizing the cytoskeleton, especially stress fibers. Rapid fluidization followed by a slow resolidification enables cells to reorganize their stress fibers and adhesion plaques in response to mechanical stresses [741]. Resulting heightened mobility of cytoskeleton constituents and partners generates a significant bronchodilation.

In asthma, this phenomenon disappears due to remodeling of the cytoskeleton and surrounding matrix [741]. However, in an acute, mild asthma attack, bronchospasm can be counteracted by a slow, complete inspiration. Deep inspirations

[26]Bradykinin, an asthmatic mediator, stimulates cyclooxygenase COx2 that produces prostaglandin-E2. The latter causes interleukin-8 release [731].

that are too fast or delayed remain ineffective. In severe asthmatic crises, slow and complete inspirations do not relieve bronchospasm. The failure of deep inspiration-generated bronchodilation may cause airway hyperresponsiveness.

8.4.6.1 Contractile Cytoskeleton Adaptivity

Mural smooth myocytes have a large functioning length range, in opposition to striated myocytes that works at most at 20% of their resting length. The malleability of the myofilament lattice of smooth myocytes results from remodeling of contractile filaments of the cytoskeleton in response to stress and strain exerted on these cells [742].

Smooth muscle length is determined not only by the balance between active force that it generates and passive reaction developed by the external load, but also, due to the breathing time scale, by an outside-equilibrium behavior [743]. In normal conditions, tidal fluctuations of muscle length suffice to avoid bronchoconstriction.

The malleable cytoskeleton behavior is similar to that of soft, colloidal materials [743].[27] The elastic modulus ranges from the order of 1 Pa to 1 kPa. The stiffness over a wide range of frequencies does not exhibit any characteristic molecular relaxation time or resonant frequency. Frictional stress within the network is typically smaller than elastic stress. Both stress types are scale-free, as material rheological responses are not associated with any particular relaxation time and rise with frequency. On application of shear of sufficient magnitude, these materials become fluid-like bodies. When shear is stopped, they return to their original solid-like state. Colloidal dispersions can actually exhibit both fluid-like and solid-like behavior, with various types of transition between these two states. Colloidal particles with an attractive interparticle energy display a fluid-to-solid transformation due to gelation, at a critical volume fraction that depends on the magnitude of attractive interparticle energy on contact. Gelation occurs when interconnected colloid-rich regions solidify. Other features of soft, glassy materials include the rheological pattern that results from the competition between the spontaneous restructuration (slow scale-free recovery of rheological properties, the so-called aging) and destruction of the internal structure by shear ("shear rejuvenation"). Intermittent motions result from nanoscale rearrangements. Sub- and superdiffusive displacements rely on interactions between material constituents. Material transfer creates to low- and high-density regions of given types of constituents. Slowing of dynamics and more prominent intermittency of structural rearrangements indicate approach of a transition and kinetic arrest of colloidal particles, when crowded particles are trapped by their nearest neighbors (gelation).

---

[27] A colloidal solid, liquid, or gaseous body consists of a dispersed phase in a continuous dispersion medium. Dispersed-phase particles have a typical size that ranges from 5 to 200 nm. Blood illustrates a solid dispersed phase in a continuous liquid medium. Inhaled aerosols are mostly liquid dispersed phase in a continuous gaseous medium.

Smooth myocytes adapt to a wide range of cell lengths and maintain optimal contractility. Assembly and disassembly of cytoskeletal filaments in smooth myocytes enable adaptation to large changes in cell geometry. The ability of actomyosin filaments to form and dissolve within the myofilament lattice at adequate sites and suitable instants yields the basic mechanism of cell adaptation [742]. Formation of myosin filaments is promoted by various filament-stabilizing proteins such as caldesmon. Many proteins may be assigned, removed, or reassigned in the context of reversible reorganization of the cytoskeleton. Myosin filament length is determined by the size of the actin filament network [742].

In airway smooth myocytes, the amount of contractile filaments can change rapidly to accommodate the context. Actin filament content actually increases during contractile activation. Myosin filament content also varies. In relaxed smooth myocytes, the density of myosin thick filaments is low, although of significant number. Myosin filaments indeed undergo rapid cycles of assembly and disassembly. Upon contractile stimulation, the content of myosin filaments in airway smooth myocytes rises strongly [742]. In solution, assembly of non-phosphorylated myosin monomers into filaments depends on monomer concentration. Once a critical concentration is reached, self-assembly occurs. The critical concentration increases when ATP is available and decreases when the myosin regulatory light chain is phosphorylated. In addition, C-terminal isoforms SM1 and SM2 of smooth muscle myosin heavy chain influence stability of filament assembly. Isoform SM1 confers a greater stability [742]. Yet, myosin filament lability is required for adaptive cell remodeling. Synthesis of SM1 and SM2 isoforms differs among different types of smooth myocytes as well as during development. Isoform SM2 may enhance SMC adaptation to large changes in length. In the presence of ATP^{Mg}, phosphorylation of the myosin regulatory light chain as well as other facilitating factors in smooth myocytes may regulate thick filament assembly.

Upon large-amplitude length oscillation applied to relaxed trachealis muscle strips, myosin filaments can partially disassemble [742]. These filaments reassemble when smooth muscle is repeatedly activated isometrically, following phosphorylation of regulatory myosin light chains. Moreover, myosin filament formation increases in the presence of actin filaments. Caldesmon that is detected interspersed in the actin filament network where myosin filaments are located can also provoke reassembly of long myosin filaments.

8.4.6.2 Cell Mechanoresponsiveness

Every cell exerts traction on its matrix and experiences more or less large time-dependent stretch exerted by its environment. The cytoskeleton can stiffen, hence increasing traction (*reinforcement response*), or can soften (*fluidization response*) [744]. Therefore, cells respond to imposed stretch using 2 procedures. (1) Reinforcement refers to rapid actin polymerization and increased focal adhesion assembly that cause heightened cytoskeletal stiffness and traction forces. Actin polymerization and action of nanomotors can indeed generate a stress field and traction forces. However, in cells that bear cyclically significant stretch (e.g., cells of heart and lung), cell reinforcement that causes cell stiffening thus progressively impedes normal organ deformation during the functioning cycle. Hence, the reinforcement response corresponds to an inadequate strategy. These cell types then select the opposing compensatory mechanism. Fluidization means prompt reduction in cytoskeleton stiffness and elevation in molecular mobility.

When they are subjected to repetitive homogeneous stretch, human airway smooth myocytes rapidly soften, fluidize, and decrease traction force, with subsequent slow recoveries [744]. After completion of a transient stretch–unstretch maneuver of 4-s duration, the traction field quickly decays and slowly recovers. In addition, in substrates of different stiffness, the greater the matrix stiffness, the larger the static traction. On the other hand, cell response to dynamic traction is similar, whatever the matrix stiffness [744]. Therefore, dynamic responses to stretch are governed by a bulk mechanism with a time scale of the order of 1 s that does not consider changes in substrate stiffness. In addition, uni- and biaxial, isotropic stretches cause identical fluidization responses. Whereas homogeneous stretch induces fluidization, heterogeneous stretch induces reinforcement. Although fluidization response does not depend on stretch homogeneity. subsequent recoveries markedly differ [744]. Cell response is influenced by the duration of transient stretch. Traction recovery after a homogeneous biaxial stretch of 4-s duration never exceeds the prestretch value, whereas traction recovery following a homogeneous stretch of 30-s duration overtakes it. However, this reinforcement response disappears when homogeneous biaxial transient stretches are applied repetitively. On the other hand, repetitive, heterogeneous, transient stretches generate upon load removal a rapid fluidization followed by reinforcement [744]. Other smooth myocyte types as well as human umbilical vein endothelial cells exhibit the same behavior.

In summary, in repetitive homogeneous cell stretch, whether isotropic or not, the fluidization response prevails. The reinforcement response to repetitive load transients in human airway smooth myocytes occurs in the case of heterogeneous cell deformations. Transient stretch causes [744]: (1) a quick detachment of the myosin nanomotor from actin; (2) a transient decline in ^{F}actin concentration; and (3) a change in interactions among cytoskeletal proteins.

8.5 Vascular Smooth Myocytes

Vascular smooth myocytes are implicated in the maintenance of vessel wall, in the regulation of blood pressure and flow, and response to wall damage. Mature smooth myocytes are not terminally differentiated because they can change their phenotype.

Vascular smooth myocytes are responsible for vasomotor tone using actin–myosin filaments. The contractile apparatus of differentiated smooth myocytes encompasses smooth muscle isoforms of α-actin, myosin heavy chain, calponin-1,[28] transgelin-1, and smoothelin.

The vasomotor tone depends on sympathetic innervation and neurohumoral vasoactive command as well as various types of mechanotransduction processes according to the region within the arterial compartment (large arteries to minimize cardiac postload, small arteries and arterioles to maintain a constant nutritive blood flow, and small arterioles to fit the local metabolism needs).

Arterial smooth myocytes regulate blood pressure and flow according to cardiac postload and tissue metabolism needs. In the microvasculature, smooth muscles construct *precapillary sphincters* that regulate blood flow in capillary beds.

Vascular smooth myocytes include phasic and tonic varieties. The *tonic type* does not generate action potentials, whereas the *phasic type* in the portal system and lymphatics creates contraction waves to propel luminal fluid toward the heart.

Vascular smooth myocytes also differ according to vascular compartment (whether they pertain to an artery or a vein, to a large artery or resistance arteriole), embryological origin, and organ-dependent microenvironment.

Due to blood pressure-dependent radial hydraulic conductance, the arterial circulating plasma molecules are outwardly convected. Transferred macromolecules within the vessel wall act as extrinsic stimuli of mural cells. Vascular smooth myocytes can reprogram their expression pattern in response to acute and chronic stimuli. In addition, mechanical signals can directly (blood pressure-derived intramural tension) or indirectly (shear rate-induced endothelial command) influence the vSMC phenotype. In particular, arterial smooth myocytes respond to hypertension and resulting increase in wall tension by hypertrophy and secreting extracellular matrix constituents via RhoA, ERK1, and ERK2 [745].

Differentiated vascular smooth myocytes are also able to modify their phenotype in response to vessel wall inflammation or injury. In healthy vasculature, smooth myocytes remain quiescent and highly contractile; in disorders of the arterial wall, smooth myocytes adopt a proliferative phenotype and loose their non-contractile phenotype. Vascular diseases (aneurysm, atherosclerosis, hypertension, and post-stenting restenosis; Vol. 6 – Chap. 8. Vascular Diseases) are generally characterized by smooth myocyte phenotype switching. In response to vessel damage, smooth myocytes dedifferentiate and cause intimal hyperplasia. They proliferate, migrate

[28]Calponins that abound in smooth myocytes are encoded by 3 genes. The most abundant form is calponin-1, also called α-calponin, basic calponin, and H1-calponin (292 amino acids, i.e., a long form [calponin-1_S]), which is encoded by the CNN1 gene. β-Calponin that is also encoded by the CNN1 gene represents a short isoform (252 amino acids; calponin-1_S). Two other genes encode calponin-2 (neutral calponin or H2-calponin), and calponin-3 (acidic calponin), which are expressed at lower levels in smooth and non-muscle cells. Other members of the calponin family that are more widely synthesized include transgelin-1, or SM22α, and transgelin-2, or SM22β, encoded by the TAGLN and TAGLN2 genes.

from the media to intima, secrete remodeling enzymes such as metallopeptidases, and increase matrix protein synthesis, but decrease the production of contractile proteins (smooth muscle α-actin, smooth muscle myosin heavy chain, transgelin).[29]

8.5.1 Vascular Smooth Myocyte Subtypes

Several subpopulations of smooth myocytes (length 20–60 μm, center width ~4 μm) can be described in the arteries, such as spindle-shaped smooth myocytes (sSMC) and rhomboid SMCs (rSMC). The SMC morphology can be modulated by growth factors and environmental stimuli [746]. A large proportion of rSMCs is observed after stent implantation (Vol. 6 – Chap. 10. Treatments of Cardiovascular Diseases), with high proliferative and migratory activities.

8.5.2 Mesangial and Juxtaglomerular Cells

In kidneys, mesangial cells are smooth muscle-like cells that participate in the formation of glomeruli of nephron entries. The glomerulus filters blood supplied by an afferent arteriole that drains into an efferent arteriole to yield urine into the Bowman's capsule, the basic filtration unit of nephron. The blood–urine filtration barrier is constituted by endothelial cells of the glomerulus, negatively charged glomerular basement membrane (thickness 250–350 nm), and podocytes of the Bowman's capsule that control the filtration of proteins. Glomerular mesangial cells lodge in the interstitium between endothelial cells. They control the filtration surface area and the local blood flow, as they can contract in response mainly to stretch. They also secrete extracellular matrix constituents, prostaglandins, and cytokines. They scavenge molecules entrapped in the glomerular basement membrane.

Specialized mural, granular smooth myocytes, the so-called *juxtaglomerular cells*, reside in the wall of the afferent arteriole of each nephron. The *juxtaglomerular apparatus* localizes between the vascular pole of the renal corpuscle (initial blood-filtering component of the nephron with the glomerulus and Bowman's capsule) and the distal convoluted tubule of the nephron. Juxtaglomerular cells synthesize, store, and secrete renin. Renin activates the angiotensin cascade to regulate blood volume and pressure. Juxtaglomerular cells are endowed with β1-adrenergic receptors. Once stimulated by cathecholamines, these receptors provoke renin secretion.

[29]Promoter of genes of differentiated SMC markers contain common cis elements, among them multiple CC(A/T-rich)6GG (CArG) motifs and a TGFβ control sequence.

Table 8.8. Vascular smooth myocyte phenotypes (Sources: [747, 748]). These cells can reprogram their expression pattern according to environmental conditions. These changes can trigger chromatin remodeling especially using acetylation and methylation of histones that modifies DNA accessibility to transcription factors, such as Krüppel-like factor KLF4, P300, myocardin, and serum responsive factor (SRF). In fact, numerous chromatin regulators act at SMC-specific marker sites to silence or enable access to the transcriptional machinery. Various stimuli influence the SMC phenotype, such as transforming growth factor-β, platelet-derived growth factor, retinoic acid, and oxidized phospholipids.

Phenotype	Features
Contractile quiescent	Synthesis of contractile proteins Low proliferative rate
Synthetic (matrix) and proliferative	Production and secretion of matrix constituents and metallopeptidases Interplay between KLF4, KLF5, and miR146a
Endocytic and phagocytic	Uptake of modified LDLs in atheroma Pinocytosis of small components of the fluid phase Phagocytosis of molecular complexes, cell particles, or apoptotic cells
Organizer of advential response	Acquisition of ability to migrate and proliferate Production of inflammatory signals
Osteoblastic	Calcification in atheroma

8.5.3 Vascular Smooth Myocyte Phenotypes

Vascular smooth myocytes have various subtypes that prevent easy identification in histological sections. Moreover, their fate oscillates from quiescent, differentiated, contractile cells to proliferative cells, according to types of local environmental signals (Table 8.8). The vSMC phenotype is controlled by interactions between general and SMC-specific factors. Phenotype change is a rapid adaptation to fluctuating environmental cues.

Vascular smooth myocyte reprogramming relies on chromatin remodeling that results from acetylation and methylation of histones.[30]

[30]Nucleic DNA is packaged into chromatin, the core structure of which is nucleosome. The chromatin state mainly depends on histone modifications (*epigenetic reprogramming*). Modifications of histones in nucleosomes can determine gene expression. In nucleosomes, DNA is wrapped by an octamer that contains 2 molecules of core histones H2a, H2b, H3, and H4, which is stabilized by the linker histone H1 (Vol. 1 – Chap. 4. Cell Structure and Function). N-termini of core histones contain flexible protease-sensitive tails that are sites of post-translational modifications, such as acetylation, adpribosylation, methylation, phosphorylation, and ubiquitination. These modifications allow or forbid access to machineries involved in DNA replication, chromatin assembly, and gene transcription. Histone acetylation can activate transcription; histone

Expression of vSMC-specific genes is regulated by interactions of multiple transcription factors that are either ubiquitous (e.g., serum response factor) or specific of smooth myocytes (e.g., myocardin). The activity of these transcription factors and cofactors is regulated by kinases (CamK, ERK, JNK, P38MAPK, RoCK,[31] and PKB) [745].

Factor KLF4 represses SMC marker genes. Phosphorylation of KLF4 and its subsequent acetylation by P300 enable SMC differentiation. Proliferative stimuli cause KLF4 dephosphorylation and deacetylation by HDAC2, association with ELk1, ejection of SRF, and repression of SMC markers.

Myocardin acts as a transcriptional SRF coactivator. Histone acetyltransferase P300 is involved not only in skeletal and cardiac myogenesis, but also in SMC differentiation. Gene activation by myocardin-SRF is enhanced upon P300 binding. Enzyme p300 can also enhance myocardin activity independently of SRF factor. Myocardin-related transcription factors MRTFa and MRTFb are needed in SMC differentiation.

In fact, numerous chromatin regulators act at SMC-specific marker sites to silence or enable access to the transcriptional machinery. Various stimuli influence the SMC phenotype, such as transforming growth factor-β, platelet-derived growth factor, retinoic acid, and oxidized phospholipids.

Retinoic acid that fosters SMC differentiation targets RARα and RXRα receptors; protein arginine (R) methyltransferase PRMT2 bind to these receptors and interacts nuclear receptor coactivator NCoA1,[32] another histone acetyltransferase. On the other hand, PDGFbb and oxidized phospholipids are dedifferentiating stimuli that causes histone deacetylation and demethylation in SMC marker CArG-box chromatin.

In the differentiated state, they express high concentrations of smooth muscle-specific isoforms of contractile proteins. In the proliferative state, they secrete large amounts of extracellular matrix components and synthesize low levels of characteristic isoforms of contractile proteins. The biomarker smooth muscle α-actin does not necessarily identify a smooth myocyte because it is also produced by myofibroblasts.

Vascular smooth myocyte reprogramming results from rapid adaptation associated with protein turnover. The ubiquitin–proteasome system participates in rapid turnover of short- and medium-lived proteins via degradation of mediators involved in cell signaling,[33] especially during cell differentiation and proliferation,

---

mono-, di-, and trimethylation either repress or stimulate transcription according to positions of modified histone tail lysine residues. Epigenetic regulators include histone acetyltransferases, deacetylases, and methyltransferases as well as DNA methyltransferases, among others.

[31] The SRF–myocardin complex is activated by the Rho–RoCK pathway.

[32] A.k.a. steroid receptor coactivator SRC1.

[33] For example, both NOS2 and NOS3 isoforms are degraded by the ubiquitin–proteasome system. Proteasome inhibition favors NOS3 dephosphorylation upon translocation from the cytosol to the plasma membrane of ubiquitinated PP2 phosphatase. Nitric oxide hinders vSMC proliferation, as it decreases polyubiquitination and degradation of cyclin-dependent kinase inhibitor CKI1a

that become unnecessary. Inhibition of cell cycle-controlling ubiquitin ligases or proteasome represses vSMC proliferation and prevents change to the synthetic phenotype [749].

When the endocytic phenotype is achieved, cationic proteins are mainly endocytosed by an absorptive mechanism, involving interactions with the negative charges of the glycocalyx; anionic proteins such as LDLs interact mainly with LDLR receptors and modified LDLs with scavenger receptors.

In atherosclerosis characterized by an osteoblastic phenotype, foci of calcification are likely linked to cell death and membrane–annexin-rich particle release, which supports calcium–phosphate concentration; extracellular free DNA triggers precipitation.

8.5.3.1 Mechanical Signals in Phenotype Change

Regulation of cell functions, such as proliferation and differentiation, relies not only on chemical compounds such as growth factors, but also on physical and mechanical signals, in particular geometrical and rheological features of the extracellular matrix and eventual anisotropy that influence cell spreading and curvature.[34]

Deformations of the plasma membrane activates mechanosensitive immersed proteins, such as mechanically gated ion channels, receptors, and cell adhesion molecules. In particular, integrins that cluster to build focal adhesions tether the intracellular actin cytoskeleton to the extracellular matrix. Focal adhesions are signalling hubs that contain kinases, such as FAK, MAPKs, and Src, and small GTPases such as Ras. The actin cytoskeleton also connects to the nucleus and its nucleoskeleton with intermediate filaments and lamins, via nestins (Vol. 1 – Chap. 4. Cell Structure and Function). Nuclear strains and stresses exerted on nuclear lamins and other nucleoskeletal constituents and changes in interactions of nuclear structural proteins with chromatin may modulate the chromatin organization and interplays with DNA-binding proteins, thus controlling gene expression patterns. Furthermore, mechanical stresses are converted into chemical signals that can be transmitted down to to the nucleus to modify gene expression.

In particular, effectors of the Hippo pathway, or kinase (STK3 and STK4) cassette, Yes-associated protein (YAP) and transcriptional coactivator with PDZ-binding motif (TAZ; or WWTR1) relay mechanical cues to the nucleus [750].

---

and impedes its phosphorylation (that can target substrates for ubiquitination) due to enhanced phosphatase activity. Nitric oxide inhibits proteasome via thiol nitrosylation as well as UbCh10 ubiquitin conjugase [749].

[34] Myogenic stem and precursor cells cultured on substrates mimicking muscle rheological properties have a greater capacity to self-renew than those cultured on stiffer or softer matrices [750]. In addition, contractility of cultured vascular smooth myocyte depends on substrate rigidity.

The regulation of YAP and TAZ mediators requires RhoA GTPase and actin–myosin contractility. Transmitter YAP precludes expression of the contractile phenotype of smooth myocytes, as it impedes activity of the myocardin–SRF complex.

8.5.3.2 Differentiation of Vascular Smooth Myocytes

Chromatin remodeling enables SMC differentiation, in particular demethylation of dimethylated Lys9 of histone-3 ($H3K_9me^2$) and Lys20 of histone-4 ($H4K_{20}me^2$) that allows decompaction of SMC marker gene chromatin in SMC precursors [748]. Differentiated vascular smooth myocytes possess several specific histone tail modifications (acetylation [$H3K_9ac$, $H3K_{14}ac$, and H4ac) and methylation ($H3K_4me^2$, $H3K_{79}me^2$) at SMC marker gene CC(A/T-rich)6GG (CArG) boxes. These histone modifications enable access of the myocardin–SRF complex and coactivator P300 histone acetyltransferase to SMC-specific CArGs, thereby activating gene transcription. Dedifferentiation of vascular smooth myocytes (phenotype change) results from removal of these histone modifications from chromatin at SMC marker gene CArG boxes, except $H3K_4me^2$ that may help maintain SMC marker genes as euchromatin and permit redifferentiation under adequate conditions [748].

During embryo- and fetogenesis, vascular smooth myocytes have various sources (e.g., neural crest, somites, and proepicardium, in particular mesoangioblasts of the dorsal aorta in mammals, in addition to endothelial cells upon endothelial–mesenchymal differentiation), according to the vascular compartment, in addition to a common population of progenitor cells, such as TIE1+, $PECAM1^{low}$, Cdh5−, PTPRc− precursors [751]. These precursors also express smA and homeodomain-containing factor Msx2.

Notch and Wnt control vSMC differentiation and modulate their phenotype according to exerted stimuli via transcription factors and cofactors that determine gene expression program.

Notch signaling in TIE1+ precursor cells is involved in developing arteries, whatever the vascular compartment, but not in mature vascular smooth myocytes, i.e., once the artery is stabilized. Notch upregulates transcription of the mesenchymal markers smA and PDGFRβ; it supports endothelial–mesenchymal transdifferentiation.

8.5.3.3 Notch Signaling in vSMC Specification

Smooth myocytes of major arteries arise from several cell lineages. Smooth myocytes of the developing arch artery, ductus arteriosus, and proximal regions of the major aortic arch branches derive from the cardiac neural crest; those of the root of the aorta and pulmonary artery from the second heart field; those of the the descending aorta from somites [752].

Smooth myocytes of organ arteries have also diverse origins. The neural crest gives rise to smooth myocytes and pericytes of most brain vessels; the mesothelium and proepicardium generate mesenteric and coronary vessels, respectively [752].

Notch stimulates various aspects of vSMC specification, differentiation, and maturation during development. Notch positively regulates differentiation of cardiac neural crest precursor cells into smooth myocytes. Notch is also required in the second heart field for proper patterning of the outflow tract [752]. Notch controls the differentiation of epicardium-derived cells into coronary smooth myocytes.

In classical Notch signaling, the extracellular domain of Notch receptor on the surface of signal-receiving cells interacts with the extracellular domain of Notch ligand on apposed signal-sending cells (Vol. 3 – Chap. 10. Morphogen Receptors). Four Notch receptors (Notch-1–Notch-4) on signal-receiving smooth myocytes tether to 5 transmembrane Delta-like and Jagged ligands (DLL1, DLL3–DLL4, and Jag1–Jag2) on apposed signal-sending endothelial and smooth muscle cells. Notch-1 and Notch-4 reside primarily in endothelial cells; Notch-1 pulmonary artery smooth myocytes during development; Notch-2 in mesenchymal cells and aorta and pulmonary artery smooth myocytes; Notch-3 in smooth myocytes and pericytes of the microvasculature of the central nervous system [752]. Ligands DLL1, DLL4, and Jag2 lodge on endothelial cells; Jag1 on both endothelial and smooth muscle cells. In addition, Notch receptors can interact with other ligand types.[35]

Notch receptor is transported to the cell surface as a heterodimer. Notch receptor precursor is actually cleaved by furin during transport to the plasma membrane into a large ectodomain ($Notch^{ECD}$) and a membrane-tethered intracellular domain ($Notch^{MTIC}$) that remain connected by a juxtamembrane heterodimerization domain.

Upon ligand binding, Notch is successively cleaved by adamlysins such as ADAM17 at the extracellular juxtamembrane region and by the γ-secretase complex that liberates the Notch intracellular domain ($Notch^{ICD}$) from the plasma membrane. On the one hand, $Notch^{ECD}$ is endocytosed into the signal-sending cell. On the other, $Notch^{ICD}$ translocates to the nucleus, where it forms an active transcriptional complex with RBPJκ transcription factor and coactivators such as Mastermind-like coactivator (MamL). In the absence of Notch, RBPJκ represses transcription, as it recruits histone deacetylases and other corepressors. Nuclear $Notch^{ICD}$ displaces the histone deacetylase–corepressor complex from RBPJκ, enabling transcriptional activation of Notch target genes such as class-B basic helix–loop–helix proteins Hairy enhancer of Split (HES1–HES7 or bHLHb37–bHLHb43) and HES-related transcriptional regulators (HRT1–HRT3). In addition, other Notch target includes PDGFRβ, the microRNA cluster miR143–miR145, and components of the Notch pathway (e.g., Notch-3 and Jag1).

[35] In vitro, microfibril-associated glycoproteins MAGP1 and MAGP2, which associate with fibrillins in elastic fibers, can bind to the extracellular domain of Notch-1 and activate Notch [752].

Intimal hyperplasia is characterized by a temporospatial change in Notch-1 to Notch-3 activity [752]. In smooth myocytes, Notch-1, but not Notch-3, mediates proliferation, migration, and survival, hence neointima formation.

Pulmonary arterial hypertension results from a remodeling of small pulmonary arteries with proliferation of smooth muscle and endothelial cells that causes wall thickening and lumen narrowing. The Notch-3-HES5 pathway is involved [752].

Platelet-derived growth factor-BB and transforming growth factor-β mediate vSMC phenotype switching. The former is involved in the repression of vSMC-specific gene expression; the latter promotes the contractile phenotype. Notch cooperates with PDGF and TGFβ to regulate vSMC migration and differentiation [745]. On the other hand, Wnt inhibitory factor-1 hampers PDGFbb-induced vSMC proliferation.

8.5.3.4 Wnt Signaling in vSMC Specification

Secreted, cysteine-rich Wnt glycoproteins that are encoded by 19 genes regulate SMC proliferation, differentiation, migration, and survival. They interact with 10 Frizzled receptors as well as coreceptors, such as low-density lipoprotein-related proteins LRP5 and LRP6 for the canonical, β-catenin dependent Wnt pathway and receptor protein Tyr kinase-like orphan receptor ROR2 for the non-canonical, β-catenin independent Wnt axis.

Wnt proteins regulate the synthesis of several matrix constituents, such as fibronectin and versican, as well as matrix metallopeptidases (MMP2, MMP3, MMP7, MMP9, MMP13, MMP14, and MMP26).

In the conventional pathway, β-catenin translocates to the nucleus, where it assists in the TCF–LEF-mediated transcription of target genes. In arterial and venous smooth myocytes, the βCtn–TCF axis upregulates cyclin-D1 and downregulates CKI1a cyclin-dependent kinase inhibitor, thus increasing the proliferative rate. Wnt1 and Wnt3a can trigger cyclin-D1 production in arterial smooth myocytes in vitro. Under growth factor stimulation, Wnt4 supports arterial SMC proliferation in vitro; it also contributes to pathological intimal thickening in vivo [753].

In unconventional pathways, Wnts initiate the PLC–Ca^{2+} axis, thereby activating via calmodulin CamK2 and PP3 that stimulate NFκB and NFAT transcription factors on the one hand and PKC that stimulates NFκB and CREB transcription factors on the other. Furthermore, Wnt can bind to ROR2 in the absence of Frizzled to release Ca^{2+} from intracellular stores and activate JNK kinase. In addition, Wnt can also activate Rho and Rac GTPases, hence RoCK on the one hand, and PKA, thus CREB on the other. Factor NFAT1 promotes cyclin-D1 gene expression, hence arterial SMC proliferation [753]. In addition, CamK2 favors arterial SMC proliferation. Moreover, JNK is activated in proliferating smooth myocytes.

Table 8.9. Control of phenotype of vascular smooth myocytes (Source: [745]). The synthetic, proliferative phenotype can result from increases in angiotensin-2, intracellular Ca^{2+} level, or stretch. On the other hand, the serum response factor (SRF)–myocardin complex maintains the contractile state. Cell contractility is associated with Ca^{2+} ion and RhoA signaling. Signaling initiated by Wnt and MAPK determines cell shape and promotes cell differentiation. Elongation of smooth myocytes supports the contractile phenotype.

Controllers	Mediators
PDGF	KLF4, ELk1
TGFβ	KLF4, ELk1
Notch	HERP1
Wnt	MAPK
Stretch	RhoA
Calcium channels	Calcium
Angiotensin-2	RhoA

The canonical Wnt pathway is controlled by Dickkopf inhibitors that preclude Wnt–LRP interaction. Both canonical and non-canonical Wnt pathways are inhibited by secreted Frizzled-related proteins and Wnt inhibitory factor-1 that prevent Wnt-Fz linkage. Nevertheless, Dickkopf-1 promotes SMC proliferation in rat mesenteric microvasculature [753].

Differentiation of venous and arterial smooth myocytes may also be regulated by Wnt proteins. Wnt3a promotes the expression of the early myofibroblast marker SM22α; Wnt7b may intervene during differentiation of pulmonary smooth myocytes [753].

8.5.3.5 Proliferation of Vascular Smooth Myocytes

MicroRNAs contribute to the regulation of vascular smooth myocyte fate. The microRNA cluster miR143–miR145 as well as miR21 and miR1 support the contractile phenotype. On the other hand, miR221, miR146a, miR26a, and miR24 promote cell proliferation after vascular injury [745]. The microRNA cluster miR143–miR145 may be activated by Jagged-1–Notch and myocardin–SRF axes. In young hypertensive patients, the expression of angiotensin AT_1 receptor is negatively correlated with miR155 agent.

Krüpple-like factors KLF4 and KLF5, ETS-like transcription factor ELk1, HES-related repressor protein HERP1, FoxO4, and RelA subunit of NFκB hinder vSMC-specific gene expression. These mediators propagate effects of PDGFbb, TGFβ, and Notch, at least partly by disrupting SRF–myocardin interaction, thereby supporting the synthetic and proliferative phenotype of vascular smooth myocytes (Table 8.9) [745]. MicroRNA-143 and -145 are targets of SRF and myocardin, which maintain the contractile phenotype of vascular smooth myocytes and target KLF4 and ELk1 transcription factors.

MicroRNA-146a targets the 3′-untranslated region of the Krüppel-like factor Klf4 transcript, thereby promoting vSMC proliferation in vitro and intimal hyperplasia in vivo [754].[36] Conversely, Krüppel-like factor KLF4 binds to and impedes activity of the miR146a promoter (mutual inhibition with feedback loop), in competition with its KLF5 antagonist that promotes vSMC proliferation via miR146a regulator.[37]

Chronic elevation of angiotensin-2 concentration is observed in atherosclerosis and restenosis. Angiotensin-2 causes SMC migration and proliferation, oxidative stress, endothelial cell death, and infiltration of inflammatory cells into the arterial wall, hence contributing to intimal hyperplasia. Angiotensin-2 binds to AT_1 receptor and provokes synthesis of proinflammatory cytokines such as pleiotropic interleukin-18, chemokines, metallopeptidase MMP9, NADPH oxidase NOx1, and superoxide [755]. Angiotensin-2 activates redox-sensitive transcription factors NFκB and Activator protein AP1. In fact, the gene that encodes promitogenic and promigratory interleukin-18 is targeted by both NFκB and AP1 factors. Interleukin-18 launches the production of other pro-inflammatory cytokines, chemokines, MMP9, and adhesion molecules. Therefore, interleukin-18 stimulates similar promitogenic and promigratory pathways, thereby amplifying angiotensin-2-induced, redox-dependent inflammation. Indeed, AT_1 receptor connects to NOx1[38] in smooth myocytes; angiotensin-2 enhances this association. Generation of second messengers reactive oxygen species by NOx1 leads to activation of oxidative stress-responsive NFκB and AP1 factors. Angiotensin-2, a potent inducer of oxidative stress at least in rat carotid artery smooth myocytes, primes IL18 production via NOx1–ROS-mediated IKKβ–RelA and JNK–Jun axes.

---

[36]Transcription factors KLF4 and KLF5 constitute a pair of negative and positive regulators of cell proliferation. MicroRNA-146a is upregulated upon balloon-injury as well as in oxidized low-density lipoprotein-stimulated human macrophages.

[37]In vascular smooth myocytes, KLF4 decreases miR146a production; KLF5 exerts an opposing effect on the miR146a promoter. The KLF4–miR146a mutual inhibition includes miR146a-mediated post-transcriptional prevention of KLF4 synthesis and KLF4-induced transcriptional blockage of miR146a expression. Hence, KLF4 expression is regulated indirectly by KLF4 and KLF5 through miR-146a-mediated post-transcriptional prohibition. On the other hand, KLF4 production is directly regulated by interaction of KLF4 and KLF5 with the same element on the KLF4 promoter.

[38]Among NADPH oxidase subtypes (NOx1–NOx5 and DuOx1-DuOx2), human vascular smooth myocytes manufacture NOx1 and NOx4 isoforms. Inducible NOx1 localizes to the plasma membrane. Constitutively active NOx4 resides in several subcellular compartments (nucleus, endoplasmic reticulum, and mitochondria). Whereas NOx1 promotes SMC proliferation, Nox4 maintains SMC phenotype. Angiotensin-2 induces NOx1 expression in smooth myocytes, but represses that of NOx4.

8.5.3.6 Renin Phenotype

Renin is involved in the regulation and maintenance of blood pressure and fluid–electrolyte balance. During embryogenesis, before the development and maturation of the kidney vasculature, the undifferentiated metanephric blastema contains renin-expressing cells [756]. During fetogenesis, renin+ cells are distributed throughout the intrarenal arterial tree, inside glomeruli and in the interstitium. As maturation continues, renin production is restricted to myoepithelioid, granulated juxtaglomerular cells in walls of afferent arterioles at entrances to renal glomeruli. Non-juxtaglomerular cells of afferent arterioles and within glomeruli that descend from renin+ progenitor cells differentiate into arteriolar smooth muscle, mesangial, and interstitial cells, as well as a subset of tubular cells. Hence, in adults, renin is synthesized and released only by juxtaglomerular cells of the kidney.

Renin production and secretion rely on the cAMP pathway. Renin transcription is launched when cAMP responsive element (CRE)-binding (CREB) protein links to CREB-binding protein (CBP) and P300 coactivators that are histone acetyltransferases, which can remodel chromatin and facilitate access of repressor or enhancer transcription factors to gene promoters. The resulting CREB–CBP–P300 complex connects to CRE in the renin gene promoter.

Dedifferentiation and reacquisition of the renin phenotype in arteriolar smooth myocytes that originate from renin-expressing cells during in utero life, which is required when the fluid–electrolyte balance is disturbed, i.e., during sustained hypotension, dehydration, hemorrhage, and sodium depletion, enable increase of circulating renin level. In fact, increase in the number of renin-expressing cells results from: (1) augmentation of the number of juxtaglomerular apparatus producing renin; (2) elevation of the number of renin+ cells per juxtaglomerular apparatus; and, mainly, (3) retransformation of vascular smooth myocytes of afferent arterioles upstream from glomeruli into renin+ cells. Both CBP and P300 enable the maintenance of identity of cells from the renin lineage, i.e., to retain memory of renin phenotype in renin+ cell descendants, and regain the expression of the renin gene [756].

8.5.3.7 Macrophage-like Phenotype

Vascular smooth myocytes can acquire a scavenger receptor ScaRd1+, macrophage-like phenotype[39] with phagocytic activity and endocytosis of modified LDLs in atherosclerosis [747].[40] These cells can indeed take up low-density liproteins

---

[39]Macrophages produce various scavenger receptors, such as ScaRa1, ScaRa2, ScaRb1, ScaRb3, ScaRd1, and ScaRe1.

[40]Endocytic and phagocytic activities of vascular smooth myocytes are associated with synthesis of ScaRd1, a scavenger receptor of the lysosomal-associated membrane protein family, S100a9,

that are convected from plasma to arterial wall and modified in the intima. Apolipoprotein-B100 binds to sulphated proteoglycans, fibronectin, and collagens, that are all secreted by intimal vascular smooth myocytes.

Vascular smooth myocytes also possess ATP-binding cassette transporter ABCa1 that controls cholesterol efflux, transferring excess cholesterol to newly formed HDL particles. The ABCa1 production that differs in arterial smooth myocytes and macrophages depends on liver X receptor (nuclear receptor NR1h).

Lipid uptake comprises pinocytosis of small components in the fluid phase and phagocytosis of nanometric molecular complexes, cell particles, or apoptotic cells [745]. Cationic proteins are mainly internalized by absorption via interactions with the negative charges of the glycocalyx; anionic proteins, such as LDLs, interact mainly with LDLR receptors and modified LDLs with scavenger receptors.

Vascular smooth myocytes take up more modified LDLs than native LDLs due to LDLR downregulation in response to LDLs. Scavenger receptors and lectin-like oxidized low-density lipoprotein receptor-1 on vascular smooth myocytes VSMCs are mainly involved in the endocytosis of acetylated or oxidized LDLs; low-density lipoprotein receptor-related protein LRP1 in the internalization of aggregated LDLs. In addition, minimally modified LDLs may activate Toll-like receptor TLR4 [745].

8.5.4 Vascular Smooth Myocyte Uptake

Sphingolipids, such as C2-, C8-, and C16-ceramides, and sphingosine, and neutral sphingomyelinase (nSMase) participate in the regulation of the intracellular concentration of free magnesium ions from the extracellular Mg^{2+} pool in vascular smooth myocytes. These sphingolipids rapidly increase in a sphingolipid and Ca^{2+} concentration-dependent manner the intracellular Mg^{2+} level [757]. Activity of phospholipase-C, of certain PKC isozymes, Ca^{2+}–calmodulin, and Ca^{2+} ATPase may be involved in Mg^{2+} uptake.

8.5.5 Vascular Smooth Myocyte Activity

8.5.5.1 Vasomotor Tone

Vascular smooth myocyte carries out slow and sustained contractions, whereas cardiomyocyte contractions are rapid and of short duration ($\mathcal{O}[100\,\text{ms}]$). Myogenic activity has been demonstrated in arteries, arterioles, venules, veins, and lymphatics. Unlike in cardiomyocytes, actin and myosin in smooth myocytes are not arranged

S100a10, calprotectin, CD14, a receptor of bacterial lipopolysaccharide and coreceptor of TLR4, galectin-3, among others.

Table 8.10. Calcium transport in myocytes. Phosphorylation by cAMP-dependent protein kinase acts for channel opening. PKA also phosphorylates phospholamban to increase SERCA activity.

Transport path	Activity	Influence factor
	Sarcolemma	
Voltage-dependent channel	Ca^{2+} influx	PKC
Ca_V1	Slow inactivation	PKA
Ca_V3	Quick inactivation	
Strech-activated channel	Ca^{2+} influx	
Receptor-operated channel	Ca^{2+} influx	α_1-Adrenergic receptor
ATPase pump	Ca^{2+} efflux	Cam
Na^+–Ca^{2+} exchanger	Ca^{2+} efflux	
	Endo(sarco)plasmic reticulum	
Ryanodine-sensitive	Ca^{2+} influx	
Ca^{2+}-sensitive	Ca^{2+} influx	
IP_3-sensitive channel	Ca^{2+} influx	
SERCA	Ca^{2+} resequestration	PLb
	Mitochondrion	
Ca^{2+} uniporter	Ca^{2+} uptake	

into distinct bands, but organized for maintainance of tonic contractions. Vascular smooth muscle contraction can be initiated by mechanical, electrical, and chemical[41] stimuli, in particular SMC stretching causes contraction (myogenic response).

Manifold isoforms of actin [758] and myosin light and heavy chains [759] exist. The difference in isoforms of contractile proteins between small and large arteries may induce different characteristics. SMC contraction speed is very low, but can achieve a much greater degree of shortening.

Activity of smooth myocytes is regulated by cytosolic calcium concentration ($[Ca^{2+}]_i$) due to: (1) membrane depolarization and Ca^{2+} entry from the extracellular medium via plasmalemmal, voltage-gated Ca_V1 and Ca_V3 channels (Fig. 8.1); (2) Ca^{2+} release through ryanodine receptors from endoplasmic reticulum; and (3) receptor-dependent Ca^{2+} channels (Table 8.10).

[41] Noradrenaline, angiotensin-2, vasopressin, endothelin-1, and thromboxane-A2 cause SMC contraction.

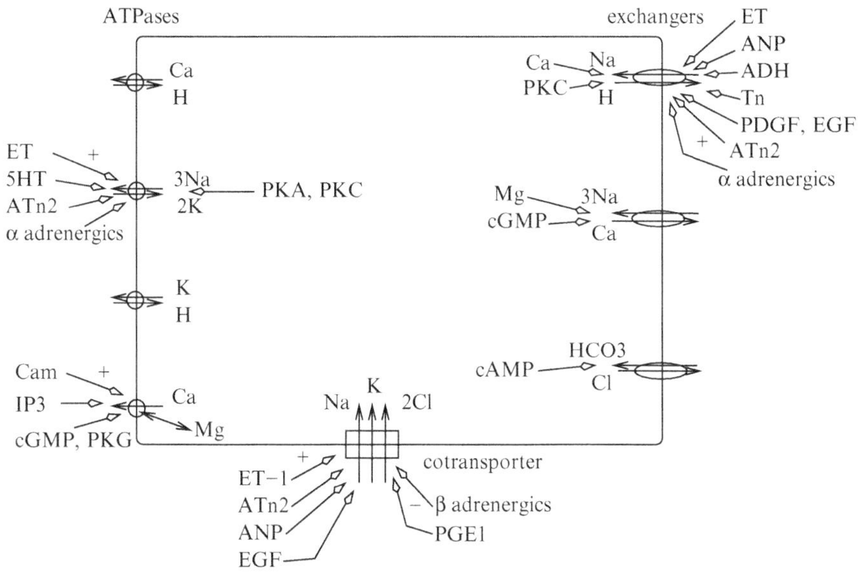

Fig. 8.2 Interacting cocarriers of vascular smooth myocytes: primary active ion pumps and secondary active carriers (adapted from [762]).

Calcium is resequestered in its intracellular stores through sarco(endo)plamic reticulum calcium ATPase (SERCA). Calcium motion from the intra- to the extracellular space use pumps and exchangers. In addition, passive calcium leak from the intracellular stores can be observed. Although *translocon* (a.k.a. translocator or translocation channel) can act as a calcium leak factor in the ribosomal endoplasmic reticulum, it is normally closed [760].

Smooth myocytes often have a graded response to membrane depolarization. The activity of ion pumps and exchangers on the SMC membrane is modulated by several factors, such as vasoactive substances and growth factors (Fig. 8.2). Moreover, gap junctions between smooth myocytes are required.

Vasodilator and -constrictor influences are exerted upon a basal vascular tone. A vasomotor tone is indeed spontaneously developed in most arterioles [763]. In isolated arterioles and arteries, the basal tone is developed for vessel physiological pressure [764]. In addition to the average pressure, pulsatility may be involved in veins [765] and arteries [766].

Electrical coupling in the vessel wall associated with the vasomotor activity can be provided by connexins between endothelial cells on the one hand, smooth myocytes on the other (homocellular coupling), and endothelial and smooth myocytes (heterocellular coupling). Calcium ions can cross gap junctions [767]. Such processes can contribute to vasomotor response spreading. Many ion channels (Tables 8.11 and 8.12) and transporters (Table 8.13) do not seems to be modulated by cell tension.

Table 8.11. Ion channels and vasomotor tone with influence factors for involved channels (Sources: [768–771]).

Channel	Effect	Influence factors
	K^+ channels	
Inward rectifier (K_{IR})	No significant role	
ATP-sensitive (K_{ATP})	No significant role	
Voltage-dependent (K_V)	Repolarization	PKC
Ca^{2+}-activated (K_{Ca})	Repolarization	20 HETE (⊖)
	Cl^- channels	
Selective (Cl^- efflux)	No significant role	
Ca^{2+}-activated	No effect	
Volume-gated	No effect	
	Stretch-activated channels (SAC) Non-selective cation channels (Na^+, K^+, Ca^{2+})	

Protein Tyr kinases are implicated in vascular smooth muscle contraction, both upon Ca^{2+} influx and sensitization. Calcium sensitization corresponds to increased myosin light chain phosphorylation, and hence cell contraction. Kinase RoCK is activated by binding to $RhoA^{GTP}$. Ca^{2+} Sensitization and the Rho–RoCK pathway contribute to contraction of smooth myocytes in pulmonary artery in response to acute or chronic hypoxia as well as agonist-mediated Ca^{2+} sensitization. Kinases of the SRC family are activated by myristoylation and/or palmitoylation, translocation to the plasma membrane, and subsequent autophosphorylation. Members of the SRC family, such as Fyn, Src, and Yes, are activated by PGf2α in rat small distal pulmonary artery [772]. They can regulate Rho activation via RhoGEF phosphorylation. Kinases Fyn and Src have opposing effects on the formation of intercellular junctions and focal contacts [773]. Kinase Fyn contributes to Ca^{2+} sensitization and stress fiber formation in smooth muscle contraction in coronary artery mediated by sphingosylphosphorylcholine–RoCK signaling.

Protein kinase PKG1α that mediates SMC relaxation binds directly to the myosin-binding subunit (MBS) of MLC phosphatase. Disruption of the PKG1–MBS complex prevents NO–cGMP action from myosin phosphatase activation [774]. Phosphorylation of 20-kDa myosin regulatory light chain (MLC20; a.k.a. smooth muscle isoform of myosin regulatory light chain MyLC2B and MRLC2) triggers actin–myosin cross-bridges and actin polymerization in arterial smooth myocytes [775].

8.5.5.2 Production and Liberation of Peptidases

Vascular smooth myocytes produce *tissue inhibitors of metallopeptidases* and several types of *serpins*, such as serpin-E1, or plasminogen activator inhibitor-

Table 8.12. Main features of certain SMC ion carriers.

	Potassium carriers
Inward rectifier	Mainly in arterioles; opening at resting potential; outward current
Delayed rectifier (outward rectifier)	Repolarization; slow activation beyond -30 mV
ATP-gated	Opening by low [ATP], increased [ADP], [GDP], adenosine A_1 receptors, and H^+; low basal activity; stimulation by the cAMP–PKA axis
Ca^{2+}-gated	Stimulation by cytosolic Ca^{2+} and by depolarization; high conductance; repolarization; NO effector
	Calcium carriers
Voltage-gated	Influx; high (Ca_V1) or low (Ca_V3) conductance; long (Ca_V1) or transient (Ca_V3) opening; abundant in small arteries (Ca_V1)
Receptor-operated	Non-selective cation channel; effector of NAd, ATn, vasopressin, 5HT, histamine; linked to α1-adrenoceptor and ATPR
	Chloride carriers
Ca^{2+}-gated	Efflux Modulation of basal potential
Non-selective cation channel	
	Permeability to Ca^{2+}, Na^+, K^+ Stimulation by DAG Depolarization
Stretch-sensitive cation channel	
	Permeability to Ca^{2+}, Na^+ Activation of voltage-gated Ca^{2+} channel Depolarization

1, and serpin-E2, or protease nexin-1. The synthesis and secretion of TIMPs and PAI by vascular smooth myocytes are mainly controlled by the TGFβ–SMAD pathway [745].

In arterial aneurysms and dissections, vascular smooth myocytes can express and release peptidases such as *matrix metallopeptidases* [745]. Vascular smooth

Table 8.13. Membrane carriers with regulation factors for mechanosensitive carriers (PKA: cAMP-dependent protein kinase; PKG: cGMP-dependent protein kinase; Sources: [762, 771]).

Type	Regulation
Active transporters	
Na^+–K^+ ATPase	PKC, PKA, $[Na^+]_i$, $[K^+]_e$
Ca^{2+} ATPase	Cam, PKG
K^+–H^+ ATPase	
Cotransporters	
Na^+–Cl^-–HCO_3^-	
Na^+–K^+-Cl^-	
Cl^-–HCO_3^-	
Exchangers	
Na^+–H^+	
Na^+–Ca^{2+}	PKC, PKG

myocytes constitutively synthesize and secrete MMP2, but MMP9 production is inducible under the control of NFκB factor. These cells also fabricate MMP3[42] and MMP7 peptidases.[43] Vascular smooth myocytes secrete MMPs as inactive precursors. Plasmin activates proMMPs.

Vascular smooth myocytes produce *tissue peptidase inhibitors*. They also constitutively express and secrete several *serine peptidases* such as tissue-type plasminogen activator (tPA); tPA synthesis can be enhanced by numerous stimuli.

Like any cell type, vascular smooth myocytes manufacture intracellular *cysteine peptidases* associated with phagolysosomes. On the plasma membrane, cathepsin-G colocalizes with $\alpha_V\beta_3$-integrin. It participates in vSMC migration. Unlike other cathepsin types, *cathepsin-S* can be secreted and act in the extracellular milieu at neutral pH. Inducible cathepsin-S synthesis is triggered by IL1β and other cytokines.

8.5.5.3 Clearance of Blood-Conveyed Proteases

Blood plasminogen is a zymogen secreted by the liver. It circulates at a micromolar concentration and can be convected through the vascular wall. Tissue plasminogen activator that, together with plasminogen, binds to Lys residues of the annexin-A2–S100a10 heterotetramer on the vSMC surface and forms an activator for plasmin [745]. Pericellular plasmin degrades fibronectin and causes vSMC

[42] A.k.a. stromelysin-1.

[43] A.k.a. matrilysin-1.

Table 8.14. Examples of products of vascular smooth myocytes (ABP: actin-binding protein; ASP: actin-severing protein).

Contractile proteins	Actin, ABP, ASP, myosin
Adhesion molecules	Cadherin, integrin
Matrix components	Elastin, collagen-1/3 Proteoglycans (PGe2, PGi2) Glycoproteins (emilin)
Growth factors Chemokine Cytokines	PDGF, CSF1–CSF3 CCL2 IL1/6, TNF
Enzymes	NOS, COx, PLCβ1/3, PLCγ1, PKC, MAPKs

detachment and apoptosis. Proteoglycan-bound pericellular serpin-E2 forms a complex with plasmin and tPA that can be endocytosed by LRP1 scavenger[44] present on the vSMC plasma membrane.

Similarly, the complex hyaluronan-binding protein HABP2 made of clotting factor-VII and activator serine protease (a.k.a. FSAP), also synthesized by the liver, circulating (~10 μg/ml) as a zymogen, is activated by autocatalysis after binding to anionic (sulphated) glycosaminoglycans or nucleic acids. It cleaves and activates coagulation factor-VII, PDGFbb homodimer, and prourokinase [745]. The proteic complex HABP2 is inhibited by circulating serpins as well as protease nexin-1 (PN1) in the vSMC pericellular space. The HABP2–PN1 complex is then cleared by vascular smooth myocytes via LRP1-dependent endocytosis.

8.5.6 Vascular Smooth Myocyte Production

The vascular smooth myocyte undergoes direct and indirect stimuli, essentially from its extracellular matrix, cells of the adjacent layer of the vessel wall, and blood flow. It then produces several types of substances (Table 8.14).

Myocardin-related transcription factors, coactivators of the transcription factor serum response factor, are required in vascular smooth myocyte as well as cardiomyocyte differentiation. Serum response factor regulates genes involved in cell proliferation, migration, cytoskeletal dynamics, and myogenesis [776].

[44] A.k.a. α2-macroglobulin receptor.

MicroRNAs can regulate the differentiation of multipotent stem cells into a specific lineage as well as reprogramming of cells. MicroRNAs miR143 and miR145, targets of serum response factor, myocardin, and NK2 transcription factor-related homeobox protein NKx2-5, are cotranscribed in murine multipotent cardiac progenitors that give rise to vascular smooth myocytes [777].[45] MicroRNA-145 is necessary for myocardin-induced reprogramming of fibroblasts into smooth myocytes and sufficient to provoke differentiation of multipotent neural-crest stem cells into vascular smooth myocytes. MicroRNAs miR143 and miR145 cooperatively works on a set of transcription factors, such as Krüppel-like factor KLF4, myocardin, and member of ETS family ELk1, to promote differentiation and repress proliferation of smooth myocytes.

In coronary artery smooth myocytes, adenosine A_1 receptor coupled to G-protein subunits of the $G\alpha_{i/o}$ subclass activates the PLCβ3–PKCα pathway that causes phosphorylation of extracellular signal-regulated protein kinases ERK1 and ERK2 [778].

The glycoprotein elastin microfibril interfacer-1 (emilin-1) is secreted in the extracellular matrix. Emilin-1 coexpressed with transforming growth factor-β in vascular smooth myocytes is a negative regulator of transforming growth factor-β signaling. Emilin-1, once bound to proTGFβ1 in the extracellular space prevents the production of active TGFβ1 [779]. Moreover, emilin-1 regulates blood vessel caliber and peripheral resistances, hence blood pressure. In the absence of emilin-1, the blood vessel bore is reduced; peripheral resistances and blood pressure thereby increase.

Coupling factor-6 (CF6) is a component of mitochondrial ATP synthase. It is released from endothelial cells by tumor-necrosis factor-α and hemodynamic stresses via activation of nuclear factor-κB. Adenosine triphosphate synthase bound to the plasma membrane of vascular endothelial cells acts as a CF6 receptor. Coupling factor-6 increases the concentration of intracellular protons by activating ATPase. In smooth myocytes of resistance arterioles, CF6 causes vasoconstriction by inhibiting phospholipase-A2 and lowering endothelial generation of prostacyclin. However, direct vasoconstriction occurs in an auto- or paracrine fashion, independently of decrease in endothelium-induced vasodilation. Coupling factor-6 also induces a monophasic increase in intracellular calcium concentration in smooth myocytes and enhances angiotensin-2-induced Ca^{2+} spikes via activated Src kinase [780].

Protein Tyr kinase Src regulates numerous mediators of vasoconstriction and intervenes in intracellular acidosis. Activated Src enhances $Ca_V1.2$ channels and Ca^{2+} release from intracellular stores in response to angiotensin-2, as it phosphorylates Ca^{2+} channel and phospholipase-Cγ1.

Anticoagulant and anti-inflammatory thrombomodulin is not detected in quiescent vascular smooth myocytes, but is expressed by these cells in diseased vessels. In cultured vascular smooth myocytes, thrombin, cAMP, and COx2-derived

---

[45]Multipotent cardiac progenitors can differentiate into cardiomyocytes, vascular smooth myocytes, and endothelial cells.

prostaglandins, such as PGe2 and PGi2, upregulate thrombomodulin expression. Platelet-derived growth factor-BB, a potent mitogen and chemoattractant of smooth myocytes, induces thrombomodulin expression via the Src–PI3K–PKB–TOR signaling pathway and transcription factor E26-transformation-specific sequence ETS1 that activates the thrombomodulin gene promoter [781].

Prorenin receptor (PRR) is expressed in smooth muscle cells of human mammary arteries among others. Prorenin, an inactive proenzyme, can thus function independently of angiotensin-1 production from angiotensinogen (secreted out of the liver) by renin aspartyl protease (angiotensinogenase). Prorenin binds PRR 3-times more efficiently than renin. Prorenin causes SMC migration. It indeed increases via PRR the intracellular levels of RhoA^{GTP} and Rac1^{GTP} [782]. In addition, tethering of prorenin to PRR triggers activation of extracellular signal-regulated kinases ERK1 and ERK2 and phosphorylation of heat shock protein HSP27,[46] thereby promoting cell proliferation and contractility. Prorenin also triggers the formation of larger focal adhesions as well as cleavage of focal adhesion kinase FAK1, a regulator of cell migration, into 2 fragments by calpain. Processing of FAK1 enzyme is a marker of focal adhesion disassembly required for cell rear retraction during motion.

Arterial smooth myocytes express surfactant protein SPd involved in innate immunity, especially in defense against lung pathogens, hence directly protecting the vessel wall [783]. Surfactant protein-D attenuates the release of interleukin-8 stimulated by lipopolysaccharide and tumor-necrosis factor-α, thereby repressing inflammation. Interleukin-8 indeed supports the recruitment of monocytes and neutrophils. Conversely, both lipopolysaccharide and TNFα promote SPd expression.

8.5.7 *Vascular Smooth Myocyte Migration*

Migration of vascular smooth myocyte occurs during angiogenesis and wound healing, as well as atherogenesis and after stent implantation.[47] Numerous pro- and antimigratory molecules influence SMC movement (Table 8.15). Migratory stimuli activate signaling cascades that trigger remodeling of cytoskeleton (actin filaments, microtubules, and intermediate filaments) and focal adhesions as well as activation of cellular nanomotors such as those associated with actomyosin filaments (Vol. 2 – Chap. 6. Cell Motility).[48]

[46]Protein HSP27 is involved in SMC contractility, cytoskeleton reorganization, and cell migration.

[47]During atherogenesis, vascular smooth myocytes migrate to intima from the media. In addition, CD34+ hematopoietic progenitor cells can migrate from blood circulation to give rise to smooth myocyte progenitors.

[48]During cell movements, the nucleus moves toward the cell rear. The PI3Kγ–PKB–GSK3β axis acts via adenomatous polyposis coli protein that interacts with microtubules [784].

Table 8.15. Stimulators and inhibitors of SMC migration (Source: [784]).

Stimulators	Inhibitors
Hormones, biogenic amines, growth factors, and cytokines	
Angiotensin-2	Adrenaline
Noradrenaline	Dopamine
Histamine	Sphingosine-1 phosphate
Serotonin	Transforming growth factor-β1
Fibroblast growth factor-2	Tumor-necrosis factor-α
Heparin-binding epidermal growth factor	Hypoxia-inducible factor-1
Insulin-like growth factor-1	Interleukins IL1β, IL10
Platelet-derived growth factor	
Sphingosine-1 phosphate	
Transforming growth factor-β1	
Tumor-necrosis factor-α	
Vascular endothelial growth factor	
Interleukins IL1β, IL6	
Extracellular matrix components	
Collagen-1/4/8	Heparin
Fibronectin	Tissue inhibitors of metallopeptidases
Laminin	
Hyaluronan	
Osteopontin	
Thrombospondin	
Vitronectin	
Matrix metallopeptidases	
Miscellaneous	
ATP, UTP	cAMP, Epoxyeicosatrienoic acids
Thrombin	Peroxisome proliferator-activated receptor-γ
Urokinase plasminogen activator	

Extracellular molecules that control SMC migration include biogenic amines as well as growth factors such as platelet-derived growth factor[49] and cytokines, in

[49]Homodimer PDGFaa made of 2 PDGFa isoforms corresponds to a short-term antimigratory signal, whereas PDGFbb homodimer yields a short-term promigratory cue. In addition, PDGF contributes to long-term activation of signaling pathways primed by other growth factors, as it indirectly provokes synthesis of epidermal (EGF) and fibroblast (FGF2) growth factors in smooth myocytes.

addition to extracellular matrix components [784].[50] Multiple migration regulators can intervene simultaneously, such as sphingosine-1 phosphate and platelet-derived growth factor.[51]

Intracellular molecules that contribute to SMC migration comprise kinases, especially calcium-dependent and Rho-activated protein kinases, phosphatidylinositol 3-kinase,[52] its antagonist PTen,[53] SRC family kinases, such as Fyn, Lyn, Src, and Yes,[54] P21-activated protein kinases,[55] LIM kinases,[56] mitogen-activated protein kinases,[57] and integrin-linked kinase,[58] as well as their main targets, such

---

[50]Collagen-1 and -4 as well as laminin participate in SMC displacements. Osteopontin is a matrix glycoprotein that enhances SMC migration via $\alpha_V\beta_3$-integrins. Conversely, heparin and tissue inhibitors of metallopeptidases inhibit cell motility.

[51]Activated $S1P_1$ and $S1P_5$ receptors foster and impede cell migration, respectively. In airway smooth myocytes, activated PDGFR provokes release of sphingosine kinase. Synthesized S1P then activates $S1P_1$–PDGFRβ complexes. In addition, a transactivation of S1P receptors by PDGF can occur [784].

[52]Activated PDGFRβ targets phosphatidylinositol 3-kinase and phospholipase-Cγ, thereby causing PIP_2 hydrolysis, Ca^{2+} influx, and activation of mitogen-activated protein kinases, such as ERKs and P38MAPKs. Kinase PI3K also participates in microtubule remodeling via PKB, GSK3β, and adenomatous polyposis coli protein [784].

[53]Phosphatase and tensin homolog deleted on chromosome 10 dephosphorylates PIP_3 into PIP_2. During cell migration, it contributes to disassembly of focal adhesions at the trailing cell edge via ILK inactivation and FAK1 dephosphorylation.

[54]Kinase Src is stimulated by PDGF during SMC migration. It can then target ERK, FAK, BCAR, and paxillin.

[55]Cyclin-dependent kinase inhibitor CKI1a (or P21)-activated protein kinase phosphorylates LIM kinase that, in turn, phosphorylates cofilin, cortactin, caldesmon, and myosin light chain for SMC contraction. Activated PAK by the small GTPase Rac1 stabilizes microtubules, as it precludes activity of microtubule-destabilizing stathmin [784]. Kinase PAK also phosphorylates desmin and vimentin to enhance intermediate filament turnover [784].

[56]LIM kinase activated by small GTPase RhoA and its effector RoCK as well as small GTPase Rac and p21-activated protein kinase PAK1 phosphorylates and inhibits cofilin to enhance actin polymerization.

[57]In response to growth factors, ERK activation leads to 2 sequential distinct effects: an early and later stimulation that elicit cell migration and proliferation, respectively. In addition, a signaling cascade with MAP2K3, P38MAPK, MK2, and HSP27 that can be activated by PDGF, angiotensin-2, S1P, and thrombin contributes to cell migration [784]. Overexpressed JNKs can also participate in cell migration induced by PDGFbb, but prevent angiotensin-2-induced migration. The JNK kinases can be activated by PKC, FAK2, and RoCK kinases. They can phosphorylate microtubule-associated proteins and paxillin.

[58]Integrin-linked kinase binds several integrin–actin linkers, such as affixin and paxillin, and adaptor.

Table 8.16. Signaling pathways involved in migration of vascular smooth myocytes (Source: [784]).

Molecular cascade	Effect
Ca^{2+}–calmodulin–MLCK–myosin-2	Actomyosin fiber contraction
CDC42–WASP–ARP2/3	Actin polymerization
MAP3K3/6–P38MAPK–MAPKAPK–HSP27	Actin polymerization
Rac–WaVe–ARP2/3	Actin polymerization
Rac–PAK1–MLCK–myosin-2	Actomyosin fiber contraction
Ras–Raf–ERK–MLCK–myosin-2	Actomyosin fiber contraction
RhoA–RoCK–MLCP–myosin-2	Actomyosin fiber contraction
RhoA–RoCK–LIMK–cofilin	Actin polymerization
RhoA–Dia1	Actin polymerization

Table 8.17. Integrin partners (Source: [784]).

Partners and targets	Effect
BCAR1–CRK–RapGEF1/SOS	Activation of small GTPases
FAK1–Src–BCAR1	Signaling
Paxillin–PAK–PIX–GIT1	Formation of focal adhesion complex
Vinculin–talin–tensin–α-actinin	Anchorage of actomyosin fibers

as myosin-2, actin capping and severing proteins,[59] formins,[60] profilin,[61] cofilin,[62] and actin-related protein ARP2–ARP3 complex[63] (Table 8.16) [784].

Plasmalemmal proteins such as integrins and their actin-binding partners that serve as adaptors between the cytoskeleton and focal adhesion plaques, such as α-actinin, paxillin, talin, and vinculin, regulate formation of nascent focal adhesions (Table 8.17).[64]

[59]Gelsolin acts as both a severing and capping protein during actin filament turnover. Actin-filament polymerization and depolymerization by gelsolin is promoted by Ca^{2+} and PIP_2, respectively.

[60]Formins Dia1 and Dia2 associate with the fast-growing barbed end of actin filament and contribute to actin polymerization. They operate together with profilin for actin filament extension. Formins Dia1 and Dia2 are activated by RhoA and CDC42, respectively.

[61]Profilin tethers to actin monomers to add ^{G}actin at barbed (plus) filament ends.

[62]Cofilin promotes ^{F}actin depolymerization at pointed (minus) filament ends.

[63]The ARP2–ARP3 complex serves as a nucleation site for actin filament branching.

[64]α-Actinin attaches actin filaments to dense bodies in smooth myocytes. Paxillin is an adaptor that is phosphorylated to regulate cell spreading and migration. Talin links integrins to the actin cytoskeleton either directly or indirectly via α-actinin and vinculin. Talin is a substrate for Ca^{2+}-activated peptidase calpain-2. Vinculin anchors ^{F}actin to the plasma membrane.

Cell migration actually requires degradation and reformation of focal adhesions. Major components of focal adhesions include focal adhesion kinase, paxillin,[65] tensin, vinculin, BCAR1 docker, and SRC family kinases. Cell migration is actually associated with phosphorylation of focal adhesion components as well as proteolysis by metallopeptidases and calpain. Integrin ligation and aggregation stimulates local accumulation and binding of structural proteins, such as α-actinin and vinculin. Integrin clustering recruits not only structural proteins, but also signaling components, such as small GTPases, p21-activated kinases, extracellular signal-regulated kinase, and focal adhesion kinase. Integrin-mediated cell adhesion indeed initiates signaling cascades characterized by phosphorylation of focal adhesion proteins, such as focal adhesion kinase and inter-protein interaction mediator BCAR1 that is phosphorylated by Src family kinases recruited by autophosphorylated focal adhesion kinase [785].[66]

Monomeric GTPases (CDC42, Rac, Ras, and Rho) regulate actin-binding proteins (ARP2–ARP3, Dia, VASP, and WASP) to promote actin nucleation and filament extension at barbed ends, particularly at the cell cortex.[67] Small RhoA GTPase activates RoCK1 and RoCK2 in vascular smooth myocytes. In addition, activated RhoA stimulates its effector RoCK that can phosphorylate myosin-2, in addition to its Ca^{2+}-dependent activation by myosin light chain kinase. Kinase RoCK also phosphorylates (inactivates) the myosin-binding subunit of myosin light chain phosphatase. Moreover, RoCK promotes actin polymerization, as it activates LIM kinase. Activated Rac1 and Rac2 also favor SMC migration, as they both stimulate PAK kinase.

8.5.8 *Vasomotor Tone*

The vascular response, especially in arterioles, to changes in intravascular pressure (p_i) is vasoconstriction or vasodilation, whether p_i heightens or decays [786]. Blood flow regulates vessel caliber: (1) quickly by vasomotor tone changes; (2) later by stimulated gene expression and structural reorganization when loading is maintained, and (3) chronically by wall remodeling (adaptative changes). Studies have been aimed at identifying the sequence of cellular events, from sensing to reaction, responsible for vasomotor activity and SMC responsiveness, independently of

---

[65] Paxillin joins nascent focal adhesion complex formed by G-protein-receptor kinase-interacting protein GIT1, p21-activated kinase, and RhoGEF6 and/or RhoGEF7.

[66] Docking protein BCAR1 (CAS or P130CAS), together with CRK adaptor, also connects guanine nucleotide-exchange factors RapGEF1 and SOS once integrin is activated.

[67] Members of the Wiskott-Aldrich syndrome protein family are involved in signal transduction from receptors to the actin cytoskeleton. They interact with small GTPase CDC42 and cytoskeletal organizing ARP2–ARP3 complexes. Diaphanous is a Rho effector involved in actin polymerization. Vasodilator-stimulated phosphoprotein binds both ^{G}actin and ^{F}actin. It regulates integrin–matrix interactions. It is regulated by PKA and PKG kinases.

neural and humoral factors. In particular, mechanical signals can be transformed into electrical cues by modulations of ion channel activity in endothelial cells and smooth myocytes.

Membrane potential is an important determinant of vascular tone. The activity of many ion channels (e.g., chloride and potassium channels) that is involved in setting the membrane potential is targeted by vasoactive substances.

8.5.8.1 Vasomotion

The cGMP-dependent Ca^{2+}-activated Cl^- flux in vascular smooth myocytes through bestrophins, especially bestrophin-3, is involved in the synchronization of SMC activity, i.e., rhythmic contractions in the vascular wall, or vasomotion [787]. Vasomotion may participate in blood pressure control and optimization of transfer between tissue and capillary blood. Vasomotion indeed enhances dialysis of the extravascular fluid.

Bestrophin-3 localizes to both smooth muscle and endothelial cells. Bestrophin-3 resides in the sarcolemma and can be internalized [787]. Bestrophin-1 and -2 also lodge in the sarcolemma. Bestrophins may be located at the endoplasmic reticulum where they yield countercurrent necessary for Ca^{2+} release. Bestrophin-3 supports synthesis of bestrophin-1 and -2 [787].[68] Calcium-activated chloride channel TMEM16, or anoctamin-1, may also contribute to vasomotion.

In small mesenteric arteries of rats, intermittent release of Ca^{2+} ions from intracellular stores activates a Ca^{2+}- and cGMP-dependent Cl^- channel in smooth myocytes; the ensuing Cl^- flux entrains the Ca^{2+} oscillations. Bestrophin is implicated in rhythmic, but not in tonic contractions, such as those induced by noradrenaline and vasopressin [787].

8.5.8.2 Myogenic Response

The myogenic response is caused by a change in arterial pressure, in particular in cerebral, coronary, and renal arteries. Blood flow *autoregulation* indeed results from the ability of small resistance arteries and arterioles[69] (bore $\mathcal{O}[100\,\mu m]$) embedded within organs to reduce or increase their calibers in response to changes in intravascular pressure. This mechanism thus maintains a constant blood flow to organs over a given range of pressures. It was described in 1902 by Bayliss.

The myogenic response is independent of the action of the vascular endothelium. This mechanotransduction process relies on the intrinsic property of smooth

---

[68] Similar crossinteractions at the synthesis level exist for caveolin-1 and-2 as well as connexin-37 and -40.

[69] Large and mid-sized irrigation arteries carry blood to organs. Small arteries and arterioles convey blood to the capillary bed; they distribute blood within organs.

myocytes to contract in response to stretch. It actually couples the contraction–relaxation state to smooth myocyte deformation with length-dependent changes in contractility associated with possible modifications in ion channel activity, cell membrane polarization, and/or cell metabolism [788]. It is characterized by depolarization of the smooth myocyte plasma membrane and rise in intracellular calcium concentration.

Stresses can actually act on: (1) plasmalemmal ion channels, exchangers, and transporters and (2) membrane-bound enzymes that modulate activity of contractile proteins. Stretch-activated channel gating depends on stresses transmitted by the cytoskeleton.

Plasma Membrane Depolarization – Mechanosensitive Ion Carriers

Stretched smooth myocytes respond to pressure-dependent membrane depolarization associated with spike generation [789, 790], the resting potential being mainly determined by K^+ ions [769]. The membrane depolarization provokes a pressure-induced myogenic contraction of arterial smooth muscle and reduces lumen caliber.

Stretch-activated calcium channels[70] that modulate intracellular $[Ca^{2+}]_i$,[71] can recruit Ca_V channels to raise $[Ca^{2+}]_i$. Ligand-gated channels on SMC membrane may be excited by vasomotor factors. Voltage-gated (K_V) and Ca^{2+}-activated (K_{Ca}) K^+ channels, especially large-conductance ones (BK), yield membrane repolarization that counteracts effect of SMC stretch. Consequently, stretching effect on SMC polarization is maintained in presence of inhibitors of K_{Ca} channel, due to activation by Ca^{2+} ions.

Plasma Membrane Depolarization – Triggering Chemicals

On the other hand, Agent 20-hydroxyeicosatetraenoic acid, a metabolite of arachidonic acid, can inhibit large conductance Ca^{2+}-activated K^+ channel [794]. Cerebral arterial myocytes synthesize cytochrome-P450-4A that catalyzes the formation of 20HETE from arachidonic acid. The 20HETE production in cerebral arterial myocytes increases with elevated intravascular pressure. The potent vasoconstrictor 20HETE activates protein kinase-C. The resulting inhibition of BK channels depolarizes arterial smooth muscle cell membrane, thereby activating Ca_V1 channel, increasing intracellular Ca^{2+} level, and priming vasoconstriction [795].

---

[70]Smooth myocytes can directly contract in response to stretch after activation of its stretch-sensitive ion channels [791–793], contraction being maintained by a sustained deformation.

[71]Plasmalemmal stretch-activated calcium channels on endothelial cells modulate $[Ca^{2+}]_i$, from which depend syntheses of vasomotor substances.

G-Protein-Coupled Receptors

Immersed in strain and stress fields, mechanosensors at the surface of vessel wall cells detect these signals and relay them via chemical reaction cascades that control the phosphorylation state of myosin light chain, hence actin–myosin interactions. In addition to membrane depolarization primed by mechanically gated ion channels, certain types of integrins (e.g., $\alpha_V\beta_3$- and $\alpha_5\beta_1$-integrins) can operate as mechanotransducers that may cause phosphorylation of Ca_V and BK ion channels. Furthermore, among G-protein-coupled receptors that are commonly activated by chemical compounds (peptides, lipids, nucleotides, and amines; Vol. 3–Chap. 7. G-Protein-Coupled Receptors), mechanosensitive GPCRs trigger mainly $G\alpha_{q/11}$ and $G\alpha_{12/13}$ intracellular signaling pathways [796].

Gq/11-Coupled Receptors and the PLC–IP_3/DAG–Ca^{2+}/PKC Pathway

Receptors coupled to the Gq/11 subclass activate phospholipase-Cβ that generates IP_3, thereby causing Ca^{2+} influx from endoplasmic reticulum store through IP_3 receptors. Calcium–calmodulin causes phosphorylation of myosin light chain kinase. The other generated messenger, diacylglycerol, stimulates protein kinase-C that phosphorylates $PP1_{r14a}$ inhibitory subunit,[72] and inhibits myosin light chain phosphatase. In addition, diacylglycerol directly activates TRPC3 and TRPC6 channels, thereby causing Ca^{2+} entry, in addition to a coordinated interaction of IP_3R and TRPC3 channels [796]. Moreover, PKC phosphorylates and, in synergy with intracellular ATP and calmodulin, activates TRPM4 channels. Both TRPP1 and TRPP2 may also be involved in pressure sensing.

G12/13-Coupled Receptors and the RhoA–RoCK Axis

Receptors coupled to the G12/13 subclass activate Rho guanine nucleotide-exchange factor (RhoGEF) that stimulates RhoA GTPase. The latter targets RoCK kinase that phosphorylates MLCP, thus impeding MLC dephosphorylation.

Prostanoid Receptors

In response to smooth myocyte stretching resulting to increased wall tension due to elevated luminal pressure, phospholipase-A2 creates arachidonic acid that is metabolized into thromboxane-A2 and 20-hydroxyeicosatetraenoic acid by cyclooxygenase and cytochrome-P450, respectively. Agent 20HETE not only inhibits

[72] A.k.a. 17-kDa substrate phosphatase inhibitor of PP1 (CPI17).

BK channels, thereby limiting plasma membrane hyperpolarization, but also, as well as thromboxane-A2, act as autocrine regulators by binding to prostanoid TP (TxA2–PGh2) receptor [796].

S1P Receptors

Sphingosine kinase-1 translocates to plasma membrane in response to stretch and phosphorylates sphingosine to generate sphingosine 1-phosphate. Once secreted, this messenger activates $S1P_2$ and $S1P_3$ receptors (S1PR) on the surface of vascular myocytes to reinforce vSMC contraction via the RhoA–RoCK pathway [796].

P2Y Receptor

Vascular myocytes can be a source of uracil nucleotides. Nucleotides are liberated by stretched cells. Autocrine activation of P2Y receptor by extracellular UTP and UDP nucleotides supports vasoconstriction. In addition, P2Y-dependent activation of TRPC3 channel further increases elevation of intracellular Ca^{2+} level [796]. Moreover, ionotropic $P2X_1$ receptor may participate in autoregulation in kidney afferent arterioles.

Mechanosensitive Angiotensin-2 Receptor

Mechanosensitive angiotensin-2 receptor AT_1 may be directly activated by stretch (independently of its agonist). It enhances actin–myosin interaction and activates TRPC6 channels [796].

Second Messengers

Deformations activate second messengers other than Ca^{2+} ions. Phospholipid metabolism is linked to cytoskeleton activity.[73] Effects of phospholipase-C activation and release of IP_3 is immediate via Ca^{2+} influx from intracellular stores [798]. The phosphatidylinositol metabolism stimulated by mechanical stress also causes a translocation of protein kinase-C from the cytosol to the plasma membrane [799]. Diacylglycerol indeed activates PKC, especially in the microcirculation [800]. Effects related to PKC activation are delayed with respect to that caused by Ca^{2+} messenger. Protein kinase-C enhances myogenic reactivity by modulating Ca^{2+} availability [801]. It may cooperate with Ca^{2+} for full contraction [802].

[73] Association of the actin-binding profilin and phosphoinositids inhibits phospholipase-C [797].

Mechanotransduction may also be mediated by cAMP messenger. Repeated cyclic strain on cultured smooth myocytes of coronary arteries reduces adenylate cyclase activity and cAMP [803]. Decrease in activity of adenylate cyclase, hence cAMP level, can result from applied cyclic stretching on SMC in vivo [804].

Owing to Ca^{2+} influx, myosin light chain kinase activates myosin light chains to initiate contraction. Blood vessels rely on auto- and paracrine mechanisms to immediately react to hemodynamical changes.

8.5.8.3 Endothelium-Dependent Regulation of the Vasomotor Tone

Endothelial regulators of the vasomotor tone are vasoactive substances released from the endothelium in response not only to humoral, but also mechanical stresses, to act on smooth myocytes (Chap. 9). In addition, mechanical stresses can provoke charge changes in the subendothelial layer that can modify cell membrane polarization. Cultured endothelial cells are sensitive to environmental charges as well as frequency and modulation rate of charge changes.

Blood friction along the wall luminal surface is sensed by endothelial cells and transduced into chemical reactions. Wall shear stress can excite plasmalemmal G-protein-coupled receptors via membrane tension [805]. Adaptation to flow forces, in paricular, results from cytoskeleton reorganization that enables a redistribution of intracellular forces [806]. Actin-based stress fibers align with flow direction. Mechanical forces are transmitted by the plasma membrane and cytoskeleton to the entire cell, and then to adjacent cells, basement membrane, and the entire extracellular matrix. Actin filaments anchored to the nucleus membrane via transmembrane proteins as well as intercellular and cell–matrix adhesion proteins,[74] are the main transmission elements.

Interactions between integrins and matrix constituents determine tensional integrity (tensegrity) of the cells. Tensegrity models have been proposed for active cytoskeleton reaction [808]. Actin filaments are also required in signal transduction that involves adenylate cyclases and stretch-activated ion channels.

Hemodynamic forces imposed on endothelial wetted surface trigger the synthesis of either vasodilators, such as nitric oxide and prostacyclin, or vasocontrictors, such as endothelin-1[75] and PGf2α (local flow-induced regulation of the blood circulation). Stress increase causes SMC relaxation.

Molecules released by endothelial cells act locally on underlying smooth myocytes to modulate the vasomotor tone [809]. Several biochemical systems are involved: (1) activation of plasmalemmal ion channels[76] and receptors; (2) reaction

---

[74] Actin filaments bind to integrins by α-actinin, talin, and vinculin [807].

[75] Endothelin, a strong vasoconstrictor, also causes myocardium contraction, ANP secretion, and SMC division.

[76] For example, the release of vasodilators is associated with activation of plasmalemmal K^+ channels of endothelial cells.

Table 8.18. Vasomotor tone-regulating substances (ANP: atrial natriuretic peptide; DHET: dihydroxyeicosatrienoic acid; EDHF: endothelial-derived hyperpolarizing factor; EET: epoxyeicosatrienoic acid; ET: endothelin; H_2S: hydrogen sulfide; HEET: hydroxyepoxyeicosatrienoic; NO: nitric oxide; NPY: neuropeptide-Y; PGf2: prostaglandin-F2; PGi2: prostacyclin; S1P: sphingosine 1-phosphate; THET: trihydroxyeicosatrienoic acid; TxA2: thromboxane-A2; UP_4A: uridine adenosine tetraphosphate; 5HT: serotonin). These chemical messengers function in blood vessels as directly or indirectly acting vasoconstrictors or vasodilators. Numerous vasoactive substances are synthesized by endothelial cells.

Vasoconstrictor	Vasodilator
ET1	NO
TxA2	H_2S
PGf2α	PGi2
UP_4A	EDHF (EET, DHET, HEET, THET)
ATP	Acetylcholine
5HT	ANP
NPY	Adrenomedullin
20HETE	Urocortin
Apelin	Ghrelin
Motilin	Nociceptin
Neuromedin-U	
S1P	
Urotensin-2	

cascades; and (3) transcription regulation [810]. Several endothelial genes are regulated by mechanical stress [811]. Transcriptional response to wall shear stress target only certain promoters.

8.5.8.4 Vasoactive Substances

Several substances, once bound to specific surface receptor, locally cause either SMC contraction or relaxation (Table 8.18). Among these substances, some are released by neighboring endothelial cells, thereby acting as indirect vasoconstrictors or vasodilators. In human arteries, in addition to the action of the sympathetic nervous system, these locally produced mediators regulate the vasomotor tone. Nitric oxide and endothelin-1 are the major endothelial-derived opposing vasodilator and vasoconstrictor, respectively.

G-Protein-Coupled Receptors and Vasomotor Tone

Among vasomotor regulators that bind to G-protein-coupled receptors (Vol. 3 – Chap. 7. G-Protein-Coupled Receptors), some GPCRs mediate vasoconstriction (Tables 8.19 to 8.21), whereas others prime directly or indirectly vasodilation (Table 8.22 and 8.23). The expression of receptors varies among distinct vascular

Table 8.19. Direct vasoconstrictors that activate GPCRs. (**Part 1**) Gi/o-coupled receptors. G-protein-coupled receptors on vascular smooth myocytes can generate a vasoconstriction upon ligand binding (Source: [812]). Among these, some receptors are linked to several signaling cascades via different G-protein subunits — $G\alpha_{i/o}$, $G\alpha_{q/11}$, and/or $G\alpha_{12/13}$ — that inhibit adenylate cyclase and activate phospholipase-C and monomeric GTPase Rho, respectively. Activation of members of the subclass of $G\alpha_{i/o}$ subunits inhibits adenylate cyclase, but stimulates via $G\beta\gamma$ phospholipase-C, hence causing Ca^{2+} influx.

GPCR Type	Agonist
α2a/2b/2c	Adrenaline
$5HT_{1B}$	Serotonin
A_1	Adenosine
$P2Y_{12}$	ADP
APJ	Apelin
ETB	Endothelin-1
NK_2	Neurokinin-A
Y_1	Neuropeptide-Y
Sst_2	Somatostatin
EP_3	Prostaglandin-E2
BLT_1	Leukotriene-B4
$S1P_2/_3$	Sphingosine 1-phosphate
PAR_1	Thrombin

beds as well as between arterial and venous and central and peripheral blood vessels [812]. Vasoactive substances encompass nucleotides and -sides, acetylcholine, biogenic amines and other peptides, eicosanoids, and lipid messengers.

Eicosanoids

Eicosanoids (or icosanoids) include prostaglandins, prostacyclins, thromboxanes, leukotrienes, and lipoxins (Tables 8.24 to 8.26). Leukotrienes regulate smooth muscle tone, hence wall permeability, as well as cell adherence and chemotaxis. Endothelial cells cannot synthesize leukotrienes, but cooperate with neutrophils to metabolize leukotrienes.

Prostacyclin is synthesized by endothelial cells in response to inflammatory mediators. Cultured endothelial cells metabolize arachidonic acid not only to prostaglandins, but also to hydroxyeicosatetraenoic acids (HETE) and epoxyeicosatrienoic acids (EET; Fig. 8.3). In the presence of EETs and DHETs, isolated porcine coronary arteries relax.

Leukotrienes are local messengers generated primarily in leukocytes. Activated cytosolic phospholipase-A2 liberates arachidonic acid from membrane phospholipids. 5-Lipoxygenase interacts with 5LOx-activating protein (FLAP), or arachidonate 5LOx-activating protein (ALOx5AP), that facilitates the recruitment of transfer of arachidonic acid to 5LOx [815]). (1) At the nuclear envelope, incorporation

Table 8.20. Direct vasoconstrictors that activate GPCRs. (**Part 2**) Gq/11-coupled receptors. Activation of members of the group of $G\alpha q/11$-protein subunits stimulates phospholipase-Cβ, hence provoking activation of myosin light chain kinase via inositol (1,4,5)-trisphosphate, Ca^{2+} influx, and Ca^{2+}–calmodulin, as well as inhibition of myosin light chain phosphatase via diacylglycerol, protein kinase-C, protein phosphatase-1, a phosphorylation-dependent inhibitor protein of smooth muscle myosin phosphatase.

GPCR Type	Agonist
M_1, M_2	Acetylcholine
α1a/1b/1c	Adrenaline, noradrenaline
$P2Y_1$	ADP
$P2Y_2$	ATP, UTP
$P2Y_6$	UDP
$5HT_{2A}$	Serotonin
H_1	Histamine
AT_1	Angiotensin-2
$ET_{A/B}$	Endothelin-1
NK_2	Neurokinin-A
NMU_1	Neuromedin-U
UT	Urotensin-2
V_{1A}	Vasopressin
EP_1	Prostaglandin-E2
TP	Thromboxane-A2
$CysLT_1$	Leukotriene-D4
$S1P_{2/3}$	Sphingosine 1-phosphate
PAR_1	Thrombin

Table 8.21. Direct vasoconstrictors that activate GPCRs. (**Part 3**) G12/13-coupled receptors. Activation of members of the group of $G\alpha 12/13$-protein subunits activates small GTPase Rho, hence Rho kinase that inhibits myosin light chain phosphatase.

GPCR Type	Agonist
AT_1	Angiotensin-2
ET_A	Endothelin-1
TP	Thromboxane-A2
$S1P_{2/3}$	Sphingosine 1-phosphate
PAR_1	Thrombin

of molecular oxygen into arachidonic acid happens in a 2-step reaction. The first reaction yields 5-hydroperoxyeicosatetraenoic acid (5HPETE). Product 5HPETE can form 2 types of 5LOx metabolites: either unstable LTa4 (half-life<3 s at physiological pH) or stable 5-hydroxyeicosatetraenoic acid (5HETE) that has a slight biological activity. Derivative 5HETE is the precursor of potent 5-oxo-ETE eicosanoid involved in eosinophil chemotaxis. On the other hand, soluble LTa4 hydrolase produces LTb4, a powerful pro-inflammatory agent. LTb4 Mediator is secreted through the LTb4 transporter or can act in the nucleus as a transcription modulator. Messenger LTb4 targets BLT_1 and BLT_2 receptors. (2) At the plasma

Table 8.22. Direct vasodilators that activate GPCRs on vascular smooth myocytes (Source: [812]). G-protein-coupled receptors on vascular smooth myocytes are predominantly coupled to Gs. Ligand binding to these GPCRs activates adenylate cyclase, thereby increasing the intracellular cAMP concentration that stimulates protein kinase-A to reduce the sensitivity of stress fibers to calcium ions and favor vasodilation. In addition, $G\beta\gamma$ can cause membrane hyperpolarization by K^+ efflux.

GPCR Type	Agonist
$\beta 1/2/3$	Adrenaline
$5HT_7$	Serotonin
D_1	Dopamine
H_2	Histamine
$A_{2A}/_{2B}$	Adenosine
Ghrelin receptor	Ghrelin
AM	Adrenomedullin
$CGPR_1$	Adrenomedullin, amylin, CGRP
NOP	Nociceptin
CRF_2	Urocortins
V_2	Vasopressin
VPAC	Vasoactive intestinal peptide
DP	Prostaglandin-D2
EP_4	Prostaglandin-E2
IP_1	Prostacyclin

membrane, 5-lipoxygenase and 5LOx-activating protein can associate with integral membrane LTc4 synthase to generate LTc4. The latter is exported by multidrug resistance-associated protein to the extracellular medium, where it is cleaved into LTd4 and LTe4 by extracellular peptidases. In addition to 5-lipoxygenase, (12,15)LOx lodges in some tissue-resident macrophages, but not in circulating monocytes.

Apelins

Apelins are potent vasoconstrictors in endothelium-denuded arteries and veins that can mediate vasodilation via vasodilators released from endothelial cells that operate on nearby smooth myocytes [812]. $Apelin_{36}$ localizes to endothelial cells of small arteries.

Motilin

Vasoconstrictor motilin is able to provoke endothelium-dependent relaxation of coronary arteries [812]. Motilin receptors on smooth myocytes can be activated by circulating motilin or that released from nerve terminals of arteries and veins.

Table 8.23. Indirect vasodilators that activate GPCRs on vascular endothelial and smooth myocytes (Source: [812]). A large number of GPCRs localize to endothelial cells. Their activation creates an increase in the concentration of free cytosolic Ca^{2+} that can generate and release prostacyclin, nitric oxide, or endothelium-derived hyperpolarizing factors. These agents diffuse onto the underlying smooth myocyte to activate their cognate GPCRs, soluble guanylate cyclase, and protein kinase-G, or produce membrane hyperpolarization (CGRP: calcitonin gene-related peptide).

GPCR Type	Agonist
$M_1/_3$	Acetylcholine
α2a, β2/3	Adrenaline
H_1	Histamine
$P2Y_1$	ADP
$P2Y_2$	ATP, UTP
AMR	Adrenomedullin
$CGPR_1$	Adrenomedullin, CGRP
$B_1/_2$	Bradykinin
ET_B	Endothelin-1
Motilin receptor	Motilin
NK_1	Substance-P
$V_1/_2$	Vasopressin
VPAC	Vasoactive intestinal peptide
FP	Prostaglandin-F2α
PAFR	Platelet-activating factor
$CysLT_1$	Leukotriene-D4
$CysLT_2$	Leukotriene-C4
$S1P_1/_3$	Sphingosine 1-phosphate
PAR_1	Thrombin
PAR_2	Serine peptidases

Urotensin-2

Urotensin-2 has antagonist effects in distinct vascular networks, as it can behave as a potent vasoconstrictor or a vasodilator. Besides, it can have no vasoactivity. Its vasodilation effect requires a functional endothelium. In humans, resistance coronary arteries are much more sensitive to urotensin-2 than epicardial distribution arteries [812]. Urotensin-2 as well as vasodilator ghrelin reside in endothelial cells and their receptors have been identified on vascular smooth myocytes [812]. Urotensin-2 is also observed in regions of infiltrating macrophages within the atheromatous lesion.

Sphingosine 1-Phosphate

Sphingosine 1-phosphate operates as a vasoconstrictor via $S1P_2$ and/or $S1P_3$ receptors, whereas endothelium-dependent vasodilation is mediated via $S1P_1$ and possibly $S1P_3$ receptors [812]. Vasoconstricting effect of S1P may be more important in small arteries.

Table 8.24. Eicosanoids associated with blood circulation (taken from [813], with permission). (**Part 1**) Prostaglandins and prostacyclins (Lφ: lymphocyte; TC: thrombocyte [platelet]; WBC: white blood cell [leukocyte]; Bdk: bradykinin; His: histamine; IL: interleukin). Prostaglandin-H2 (PGh2) is a precursor of 5 types of prostanoids using specific prostanoid synthases (PGdS, PGeS, PGfS, and PGiS) that connect to their specific binding sites in PGh2 molecule. The number (2) refers to the number of carbon–carbon double bonds. Synthesis of prostanoids prostaglandins, which are mediators of inflammatory and anaphylactic reactions, thromboxanes, which are vasoconstrictors, and prostacyclins, which operate in the resolution of inflammation, involves a sequence of 3 stages: (1) release of arachidonic acid from phospholipids by secretory and/or cytoplasmic phospholipase A2 (sPLA2 and cPLA2); (2) oxygenation of arachidonic acid by cyclooxygenases to form PGh2; and (3) conversion of PGh2 to prostaglandin-D2 (PGd2), prostaglandin-E2 (PGe2), prostaglandin-F2α (PGf2α), prostacyclin (PGi2), and thromboxane-A2 (TxA2) via 7 specific synthases, including thromboxane synthase (TxAS). Two PGdS subtypes comprise hematopoietic (hPGdS) and lipocalin (lPGdS) synthases. Three isoforms of PGE synthase (PGeS) encompass microsomal mPGeS1 and mPGeS2 and cytoplasmic cPGeS enzymes.

Type	Synthesis site	Activity
PGd2	Mastocyte	Vasodilation Inhibition of TC and WBC aggregation Reduction in T-cell proliferation Decrease in Lφ migration Secretion of IL1a/2 ↓
PGe2	Heart, kidney, spleen	Vasodilation TC aggregation Reduction in T-cell proliferation Decrease in Lφ migration Secretion of IL1a/2 ↓ Enhancement of Bdk and His effects
PGf2α	Heart, kidney, spleen	Vasoconstriction, bronchoconstriction
PGh2	Widespread precursor	Vasoconstriction TC aggregation
PGi2	EC	Vasodilation Inhibition of TC and WBC aggregation Reduction in T-cell proliferation Decrease in Lφ migration Reduced secretion of IL1a and IL2

Reactive Oxygen Species

The pulmonary circulation is very sensitive to changes in the blood partial pressure of oxygen. Pulmonary vessels contract in response to low oxygen content in blood, whereas systemic vessels dilate [816]. Reversible vasoconstriction of pulmonary artery smooth myocytes is a compensatory response that redirects blood to better ventilated lung regions. Pulmonary vessel constriction also results from an

Table 8.25. Eicosanoids associated with blood circulation (taken from [813], with permission). (**Part 2**) Thromboxanes and leukotrienes (Bφ: basophil; Eφ: eosinophil; Lφ: lymphocyte; Mo: monocyte; TC: thrombocyte [platelet]; WBC: white blood cell [leukocyte]; Ifn: interferon; IL: interleukin). The lipoxygenase pathway is active in leukocytes (basophils, eosinophils, neutrophils, mastocytes, and monocytes). The enzyme 5-lipoxygenase converts arachidonic acid into 5-hydroperoxyeicosatetraenoic acid (5HPETE) using arachidonate 5-lipoxygenase-activating protein (ALOx5AP). Metabolite 5HPETE spontaneously reduces to 5-hydroxyeicosatetraenoic acid (5HETE). The latter is processed by 5-lipoxygenase into leukotriene-A4 (LTa4; 4 double bonds). In cells equipped with LTa4 hydrolase, such as neutrophils and monocytes, Unstable LTa4 is converted to leukotriene-B4 (LTb4). In cells endowed with LTc4 synthase, such as mastocytes and eosinophils, LTa4 forms the first of the cysteinyl-leukotrienes, leukotriene-C4 (LTc4). Cysteinyl leukotrienes (that contain amino acids) include LTc4, LTd4, LTe4, and LTf4 peptidoleukotrienes.

Type	Synthesis site	Activity
TxA2	TC	Vasoconstriction, bronchoconstriction, TC aggregation, Lφ proliferation
TxB2	TC	Vasoconstriction
LTb4	Mo, Bφ,Eφ, Nφ, mastocyte, epithelial cells	Increase in vascular permeability WBC chemotaxis and aggregation Secretion of Infγ, IL1/2
LTc4	Mo, Bφ,Eφ, mastocyte, alveolar macrophage, epithelial cells	Vasodilation, increase in vascular permeability, secretion of Infγ, bronchoconstriction, mucus secretion
LTd4	Mo, Eφ, mastocyte, alveolar macrophage, epithelial cells	Vasodilation, increase in vascular permeability, secretion of Infγ bronchoconstriction, mucus secretion
LTe4	Mastocyte, Bφ	Vasodilation

imbalance between vasodilators and vasoconstrictors, especially those released by the endothelium (nitric oxide, endothelin, prostanoids, and reactive oxygen species).

Reactive oxygen species that are generated by many sources (mitochondrial electron transport, xanthine oxidase, cytochrome-P450, cyclooxygenases, lipoxygenases, and NADPH oxidases, as well as uncoupled NO synthase), which interact via feedback and feedforward, participate in the control of the pulmonary vascular tone. In normal conditions, the redox state is maintained in an overall reduced state via a balance between oxidants (reactive oxygen and nitrogen species) and anti-oxidants.

Reactive oxygen species produced in the pulmonary vasculature encompass superoxide ($O_2^{\bullet -}$), hydrogen peroxide (H_2O_2), hydroxyl radical ($OH^{\bullet}$), and

Table 8.26. Eicosanoids associated with blood circulation. (**Part 3**) Lipoxins (CXCL: CXC (C-X-C motif; α) chemokine ligand; TGF: transforming growth factor; TNF: tumor-necrosis factor). The lipoxygenase pathway is active in leukocytes (basophils, eosinophils, neutrophils, mastocytes, and monocytes). Synthesis of lipoxins from arachidonic acid occurs via cellular interactions, using 3 pathways. The classical pathway involves 5-lipoxygenase (5LOx) in leukocytes followed by 12LOx in platelets. In airway epithelial cells, 15LOx action is followed by that of 5LOx in leukocytes (second lipoxin synthesis pathway). In epithelial and endothelial cells as well as monocytes, aspirin acts on COx2 to produce 15-epilipoxins.

Type	Synthesis site	Activity
LxA4	Platelet, endothelial cells, mucosal epithelial cells, and other leukocytes via interactions with granulocytes	Reduction of granulocyte infiltration to inflammation sites, monocyte recruitment, phagocytosis by macrophage stimulation of TGFβ, repression of CXCL8 production, inhibition of TNFα release
LxB4	Platelet, endothelial cells, mucosal epithelial cells, and other leukocytes via interactions with granulocytes	Reduction of granulocyte infiltration to inflammation sites, monocyte recruitment, phagocytosis by macrophage stimulation of TGFβ, repression of CXCL8 production, inhibition of TNFα release

hydroperoxyl radical ($HO_2^{\bullet}$), among others. Reactive nitrogen species comprise nitric oxide (NO), nitrogen dioxide ($NO_2^{\bullet}$), dinitrogen trioxide (N_2O_3), and peroxynitrite ($ONOO^-$).

Anti-oxidants include enzymes, such as superoxide dismutase, catalase, glutathione peroxidase, peroxiredoxins, heme oxygenase, glutaredoxin, and thioredoxin, and non-catalytic agents, such as glutathione, β-carotene, retinol, and vitamin-C and -E (Vol. 4 – Chap. 10. Other Major Signaling Mediators).[77]

The first NOX group comprises NOx1 to NOx4; the second group DuOx1 and DuOx2; the third group NOx5 with its 4 splice variants (NOx5α–NOx5δ) and a truncated isoform (NOx5ε). Most NOx enzymes are multisubunit enzymes with cytochrome-B245, organizer, and activator components; in the absence of stimulation, they are unassembled and inactive.

In airway epithelia, reactive oxygen species are mainly produced by dual oxidases (DuOx1 and DuOx2) [816]. Their synthesis is regulated by cytokines.[78]

[77] Superoxide is dismuted to hydrogen peroxide by superoxide dismutase. This derived product can be converted to water by catalase or glutathione peroxidase. Superoxide can also react with nitric oxide to generate peroxynitrite that may be converted into nitrogen dioxide, which reacts with tyrosine residues to generate ^{NO_2}Tyr that limits NO availability.

[78] Interleukin-4 and -13 produced by T_{H2} cells increase moderately DUOX1 transcription; interferon-γ produced by T_{H1} cells elevates markedly DUOX2 transcription.

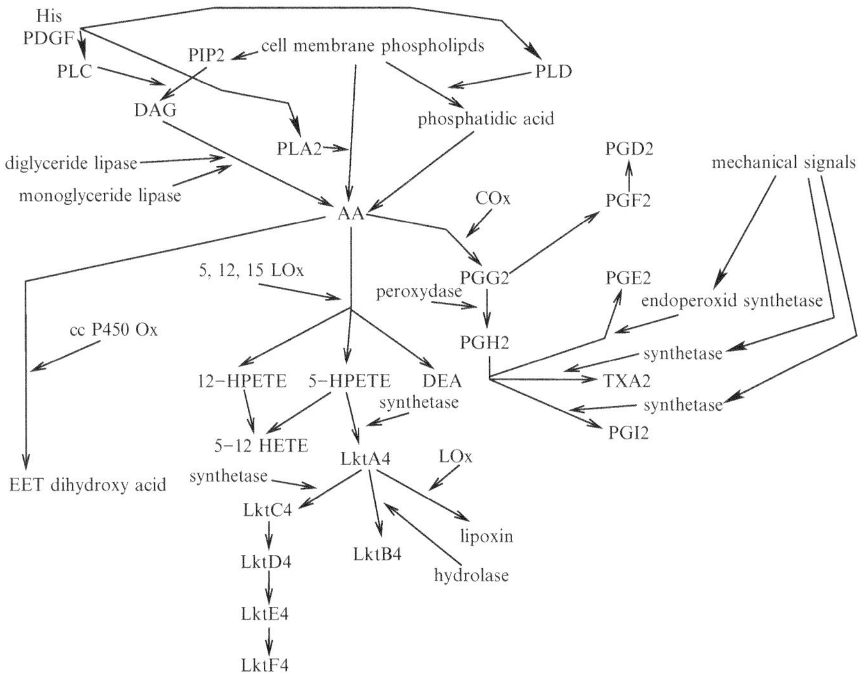

Fig. 8.3 Arachidonic acid (AA) metabolism and stimuli (molecules, hemodynamic stresses, pressure, and WSS, and their modulation rates; Source: [814]). Arachidonic acid is a constituent of membrane phospholipids that is released from these phospholipids by phospholipase-A2. Many of the lipids involved as second messengers in cell signaling pathways arise from the arachidonic acid pathway. The 3 main AA-release pathways involve membrane-bound phospholipase-C (PLC), phospholipase-D (PLD), and phospholipase-A2 (PLA2). The 3 main AA-metabolism pathways include: (1) cyclooxygenase (COx) that leads to eicosanoids from prostaglandins-G2 to -H2, then, prostacyclin (PGi2) and thromboxane-A2 (TxA2), etc.; (2) lipoxygenase (LOx) that forms hydroperoxyeicosatetraenoic acids (HPETE) and dihydroxyeicosatetraenoic acid (DEA), converted into hydroxyeicosatetraenoic acids (HETE), leukotrienes (Lkt) and lipoxins; and (3) cytochrome P450 epoxygenase (CyP450-Ox) that manufacture epoxyeicosatrienoic acid (EET).

Other NOx isozymes in respiratory epithelia contribute to lung development, host defense, production of epithelial mucins, cell death, response to mechanical stress, MMP expression, and regulation of MMP12 activity (Table 8.27) [816]. In type-1 pneumocytes, NOx2 concentration is higher than in type-2 alveolar cells that also produce DuOx1 enzyme. Pulmonary fibroblasts manufacture NOx1 and NOx4 (mainly) isotypes. Vascular endothelial cells express NOx1, NOx2, NOx4, and NOx5 isozymes. In the pulmonary vasculature, NOx subtypes are activated by numerous stimuli (angiotensin-2, endothelin, serotonin, thrombin, thromboxane-A2, tumor-necrosis factor-α, transforming growth factor-β, and mechanical forces). Endothelial NOx activity rises in response to pulmonary ischemia and hyperoxia [816]. In lungs, NOx-derived ROS are involved in various types of pulmonary diseases (Table 8.27).

Table 8.27. Isozymes of NADPH oxidase (NOx) and reactive oxygen species in the pulmonary vasculature: their functions and lung diseases (Source: [816]; DuOx: dual oxidase; EC: endothelial cell; MFb: myofibroblast; REpC: respiratory epithelial cell; SMC: smooth myocyte; ARDS: acute respiratory distress syndrome; CGD: chronic granulomatous disease; COLD: chronic obstructive lung disease; PAH; pulmonary arterial hypertension). Isozyme NOx4 is highly expressed in endothelial and pulmonary artery smooth muscle cells.

Subtype	Location	Function	Lung disease
NOx2/4	Arterial EC	Control of vascular tone, remodeling, angiogenesis, inflammatory cytokine production	ARDS PAH
NOx2	Macrophages, neutrophils		CGD
NOx3	EC		Emphysema
NOx4	Arterial SMC	Cell differentiation, hypoxia-induced proliferation	PAH
	MFb	Cell differentiation,	Fibrosis
DuOx1/2	REpC	Host defense, mucin production, cell migration and differentiation	COLD, asthma
	Lung carcinoma cell		Cancer

Hypoxia increases ROS signaling in pulmonary arterial smooth muscle cells [816]. Hypoxia upregulates NOx4 expression via hypoxia-inducible factor-1α. Hypoxia-induced ROS may inhibit $K_V1.5$ and $K_V2.1$ channels (mainly), thereby causing membrane depolarization and subsequently activating $Ca_V1.2$ channels. In addition, hypoxia causes activation of protein kinase-Cϵ. Hydrogen peroxide provokes constriction of pulmonary arteries via PKCα. In fact, several PKC isoforms are involved in hypoxic pulmonary vasoconstriction. In particular, PKCζ contributes to NOx activation. On the other hand, superoxide triggers the RoCK pathway, hence constriction of pulmonary arteries [816]. In addition, ROS stimulate cyclooxygenases.

8.5.8.5 Vasodilators

The 3 main endothelial processes that generate vascular smooth muscle relaxation—nitric oxide, prostacyclin (PGi2), and endothelial-derived hyperpolarizing factor (EDHF) — involve fatty acid cycling. Endothelial nitric oxide synthase is reversibly palmitoylated. Palmitoylation allows NOS3 cycling between the plasma membrane, where NOS3 is inactive, and the cytosol, where it is active.

β-Adrenoceptors

Vascular smooth myocytes synthesize multiple adenylate cyclase subtypes; AC3, AC5, and AC6 have the highest expression. Adenylate cyclase isozymes are effectors of β-adrenoceptors signaling that modulates activity of certain ion channel to initiate a vasodilation. In vascular smooth myocytes, AC6 is the predominant isozyme involved in the Gs–ACase–cAMP–PKA axis triggered by β-adrenoceptors (~60% of the response) [817]. Activated cAMP-dependent protein kinase PKA phosphorylates many types of K^+ channels, in particular ATP-sensitive K^+ channel. The latter causes the hyperpolarization primed by stimulated β-adrenoceptors. Isoform AC3, which is involved in signaling primed by PGe2 in vascular smooth myocytes, has a minor role in β-adrenoceptor signaling; AC5 subtype does not play a significant role.

Nitric Oxide

Nitric oxide is a vasodilator that inhibits vasoconstrictor influences. Nitric oxide is produced by vascular endothelial and smooth myocytes, among others.

Nitric oxide synthase catalyzes the conversion of ^{L}arginine to NO and ^{L}citrulline. Two major NOS types exist: constitutive (cNOS [NOS1 and NOS3]) and inducible (iNOS or NOS2).

Calcium–calmodulin-dependent NOS localizes to endothelial cells (constitutive endothelial eNOS or NOS3) and smooth myocytes. Nitric oxide synthase NOS3 is a plasmalemmal protein that yields a basal release. The continuous NO production is enhanced by multiple stimuli. Inducible NOS2 is expressed in endothelial cells in the presence of cytokines (cytokine-inducible NOS2).

Pro-inflammatory cytokines increase the activity of cyclohydrolase^{GTP}, the rate-limiting enzyme for tetrahydrobiopterin production, a cofactor for nitric oxide synthase. In arteriolar smooth myocytes, nitric oxide operates via cGMP and cGMP-dependent protein kinase-G.

The NO–cGMP–PKG axis influences Ca^{2+} signaling. Calcium-induced Ca^{2+} release can indeed be attenuated because [818]: (1) NO inhibits adenosine diphosphoribose cyclase and (2) activated $K_{Ca}1$ (BK) channel counteracts KCl-induced depolarization.

In uterine vascular smooth myocytes, large-conductance, Ca^{2+}-activated K^+ channel ($K_{Ca}1.1$) is activated by the upregulated NO–cGMP pathway to adapt the uteroplacental blood flow during the last trimester of pregnancy [819]. In addition, changes in α-subunit isoform of $K_{Ca}1.1$ channels and α:β1-subunit stoichiometry modify the functional features of uterine vascular smooth muscle.

Nitric oxide can be produced during the reduction of nitrite by deoxygenated erythrocytes to cause vasodilation. However, nitrite-induced vasodilation caused by red blood cells during hypoxemia can rely on release of adenosine triphosphate to produce vasodilation [820]. Once liberated, ATP can target endothelial nucleotide receptors that stimulate endothelial nitric oxide synthase (NOS3).

Prostacyclin

Prostacyclin (PGi2) is another endothelium-derived vasodilator. Arachidonic acid is a polyunsaturated fatty acid in membrane phospholipids, especially phosphatidylethanolamine, -choline, and -inositides. Synthesis of PGi2 depends on a balance between liberation of arachidonic acid by phospholipase-A2 from membrane phospholipids and incorporation into membrane phospholipids of arachidonate metabolites by acyl coenzyme-A synthase.

Endothelium-Derived Hyperpolarizing Factor

Endothelium-derived hyperpolarizing factor can be a prominent vasodilator [821]. Endothelial-derived hyperpolarization has been attributed to K^+ ions, arachidonic acid metabolites, and ionic communication via gap junctions. This process operates via calcium-sensitive, small-conductance (SK or $K_{Ca}2.3$) and intermediate-conductance (IK or $K_{Ca}3.1$) potassium channels in the vascular endothelium.

Metabolites of arachidonic acid produces by cytochrome-P450 epoxygenase such as epoxyeicosatrienoic acids can be released from endothelial cells to provoke hyperpolarization of neighboring vascular smooth myocytes. Released epoxyeicosatrienoic acids activate Ca^{2+}-dependent K^+ channels and Ca^{2+}–K^+ ATPase (Fig. 8.4).

In vascular endothelial and smooth muscle cells, enzymes of the cytochrome-P450 set metabolize arachidonic acid to vasoactive substances. Epoxygenases of the arachidonic acid metabolism in the vasculature belong to cytochrome-P450-2B, -2C, and -2J. ω-Hydroxylases are members of the cytochrome-P450-4A and -4F classes. These enzymes produce epoxyeicosatrienoic acids (EET) and 20-hydroxyeicosatetraenoic acid (20HETE) [822]. Soluble epoxide hydrolase catalyzes the conversion of epoxyeicosatrienoic acids into dihydroxyeicosatrienoic acids (DHET). Cytochrome-P450 thus regulates blood pressure.

Members of the cytochrome-P450 superfamily are inhibited, i.e., EDHF production is damped by nitric oxide. When NO synthesis is impaired, alleviation of EDHF inhibition by NO, at least partly, maintains endothelial vasodilator function [823].

Epoxyeicosatrienoic and dihydroxyeicosatrienoic acids are quickly incorporated into phospholipids in cardiomyocytes, endothelial cells, and platelets due to acyl coenzyme-A synthase (at least in endothelial cells).

Signaling via $G\alpha_s$ subunit and Src kinase operates via epoxyeicosatrienoic acids. Epoxyeicosatrienoic acids act as auto-, juxta-, and paracrine effectors that modulate ion transport and gene expression to regulate the vasomotor tone, hinder inflammation, and favor fibrinolysis. Acute increase in transmural pressure in vessels that express cytochrome P450-4A can stimulate stretch-activated Ca^{2+} channels, and subsequently enhance 20HETE formation.

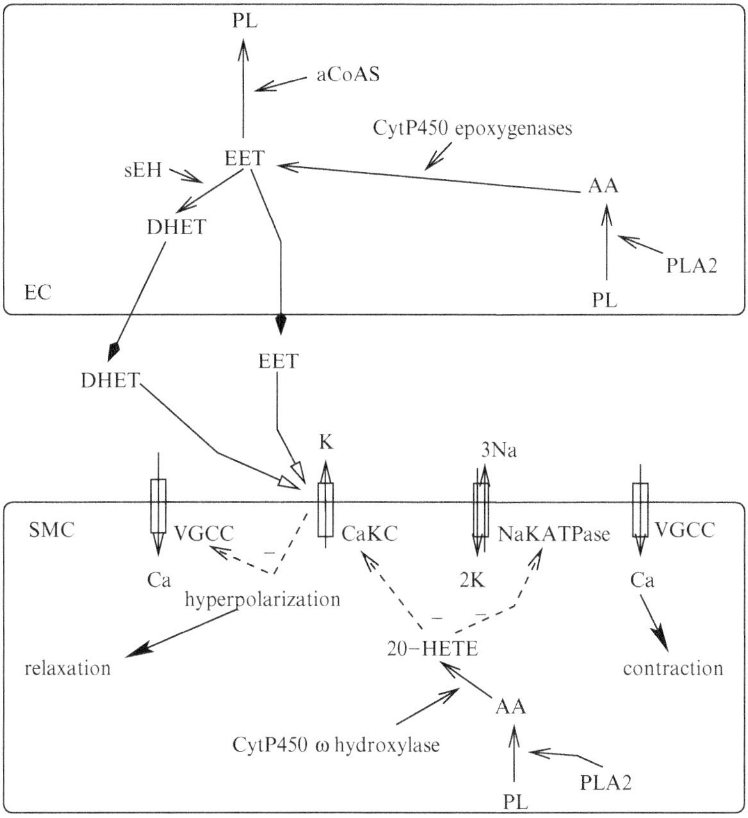

Fig. 8.4 Cytochrome-P450 (CytP450) synthesizes vasoactive metabolites of arachidonic acid (AA; Source: [822]). Activated by endothelin-1 or angiotensin-2 receptor, phospholipase-A2 releases arachidonic acid from plasmalemmal phospholipids (PL). Eicosanoids, such as epoxyeicosatrienoic acids (EET) produced by cytochrome-P450 epoxygenase and 20-hydroxyeicosatetraenoic acid (20HETE) by ω-hydroxylase, regulate blood pressure. Soluble epoxide hydrolase (sEH) catalyzes the conversion of EETs into dihydroxyeicosatrienoic acids (DHET). Conversely, EETs are incorporated into plasmalemmal phospholipids by acyl coenzyme-A synthase (aCoAS). Potential of the intracellular edge of the plasma membrane is negative with respect to that of the extracellular face. Agent 20HETE produced by smooth myocytes (SMC) is a vasoconstrictor that hampers the activity of Ca^{2+}-activated K^+ channels (CaKC) and Na^+–K^+ ATPase, thus preventing K^+ and Na^+ efflux that causes cell depolarization to activate Ca_V1 channels and elicit SMC contraction. Agents EETs and DHET released by endothelial cells (EC) activate Ca^{2+}-activated K^+ channels of SMC sarcolemma, hence provoking a K^+ efflux and cell hyperpolarization that inhibits Ca^{2+} influx via voltage-gated Ca^{2+} channels (VGCC or Ca_V), leading to SMC relaxation.

In endothelial cells, cytochrome-P450 epoxygenase causes, independently of nitric oxide and prostacyclin, vasodilation of several vascular beds, particularly in the heart and kidney, by the production and release of epoxyeicosatrienoic and dihydroxyeicosatrienoic acids (Fig. 8.4).

In smooth myocytes, after calcium influx, cytochrome-P450 ω-hydroxylase generates 20-hydroxyeicosatetraenoic acid from arachidonic acid. Signaling mediated by 20HETE is activated by endothelin-1 and angiotensin-2. Vasoconstriction caused by 20HETE, an antagonist to Ca^{2+}-activated K^+ channel, is attenuated by epoxyeicosatrienoic acids, agonists of Ca^{2+}-activated K^+ channel.

Products of cytochrome-P450 epoxygenase and ω-hydroxylase as well as reactive oxygen species derived from NADPH are intracellular signal transducers for proliferation of vascular cells (via extracellular-regulated protein kinase-1 and -2) as well as angiogenesis.

In the kidney, epoxyeicosatrienoic acids decrease sodium excretion and renin release. The cerebral blood flow is regulated by the antagonistic effects of EET and 20HETE, EETs being produced by adjoining astrocytes. 20HETE can reduce the vascular tone and dilate pulmonary arteries [824]. Agent 20HETE lowers the pulmonary vasoconstriction due to hypoxia. Effects of EETs (vasodilation or constriction) on pulmonary vessels vary according to mechanical and chemical environment.

In vascular endothelial cells subjected to acetylcholine, bradykinin, and shear stress, activated phospholipases produce arachidonic acid that is processed by cyclooxygenases, cytochrome-P450s, and lipoxygenases. In vascular endothelium of some arteries, a substantial component of vasodilation depends on lipoxygenase-induced arachidonic acid metabolites. Arachidonate 15-lipoxygenase (ALOx15) synthesizes vasoactive metabolites such as 15-hydroxy (11,12)-epoxyeicosatrienoic acid that is hydrolyzed by soluble epoxide hydrolase to (11,12,15)-trihydroxyeicosatrienoic acid. Hydroxyepoxyeicosatrienoic and trihydroxyeicosatrienoic acids are endothelium-derived hyperpolarizing factors that activate calcium-activated, small-conductance $K_{Ca}2$ channels [825]. In vascular endothelium of other arteries, arachidonate 12-lipoxygenase (ALOx12) produces 12-hydroxyeicosatetraenoic acid that relaxes smooth myocytes via calcium-activated, large-conductance $K_{Ca}1$ channel [825].

Adrenomedullin

Adrenomedullin activates nitric oxide synthase. It then causes vasodilation and possible hypotension. Adrenomedullin is a peptide similar to calcitonin gene-related peptide (CGRP). Hence, it belongs to the calcitonin–CGRP–amylin peptide family [826].

Adrenomedullin is manufactured from a precursor, the preproadrenomedullin. The adrenomedullin gene is highly expressed in endothelial cells. It also functions in many cell types, such as vascular smooth myocytes, cardiomyocytes, and blood cells. Adrenomedullin circulates in blood (concentration: 2–4 pmol).

Adrenomedullin increases the cellular cAMP level via Gs activation. It modifies the intracellular amount of calcium and cGMP messengers.

Adrenomedullin has many effects. It regulates cellular growth and differentiation. It impedes the secretion of ACTH, aldosterone, and cortisol. It is involved in the regulation of fluid and electrolyte balance (diuresis and natriuresis). Adrenomedullin also enhances cardiac contractility (inotropic effect).

In isolated porcine ciliary arteries, adrenomedullin induces arterial relaxation via $CGRP_1$ receptors. This vasodilation can be mediated either by the release of EDHF and NO and activation of calcium- and ATP-dependent K^+ channels or be independent of NO and EDHF [827].

ADP, ATP, Acetylcholine, and Substance-P

Endothelial cells synthesize the vasodilator *acetylcholine* [828]. Fluid flow releases of *adenosine triphosphate*, acetylcholine, and substance-P from cultured endothelial cells [829].

Adenosine diphosphate, a P2Y ligand, produces vasodilation. Endothelial $P2X_4$ channels are strongly involved in flow-sensitive mechanisms that regulate local blood pressure and vascular remodeling [830].

Substance-P is a vasodilator that activates nitric oxide synthase and releases nitric oxide from the endothelium. It also acts as a neurotransmitter and modulator. It belongs with neurokinin-A[79] and -B to the tachykinin neuropeptide family, members of which bind to 3 known G-protein-coupled receptors (NK_1–NK_3) that activate phospholipase-C.

Ghrelin

Ghrelin is the endogenous peptide ligand for the ghrelin receptor (a.k.a. growth hormone secretagogue receptor GHSR1a). In humans, it is a potent vasodilator. It exerts a direct effect on vascular smooth myocytes, with a potency comparable to that of natriuretic peptides [812].

Ghrelin is secreted by parietal oxyntic cells of the stomach epithelium and circulates at relatively high concentrations. Ghrelin is also detected in endothelial cells throughout the human vasculature, where it can act as an endothelium-derived vasoactive peptide.

Relaxin

Relaxin, a heterodimeric peptide hormone produced from prorelaxin, is a renal vasodilator. It favors angiogenesis. Relaxin relaxes the uterine musculature. It inhibits collagen synthesis and enhances matrix metallopeptidase activity. Relaxin

---

[79]Neurokinin-A is also called substance-K. Neurokinin-A mediates contraction of the colon.

Table 8.28. Vasoactive peptide receptors and their main targeted G proteins (Source: [831]).

Type	Main transducer
	Galanin receptors
GalR1/R3	Gi/o
GalR2	Gi/o, Gq/11
	Neuromedin-U receptors
$NMU_1/_2$	Gq/11
	Relaxin peptide family receptors
$RXFP_1/_2$	Gs, Gi/o
$RXFP_3/_4$	Gi/o
	Tachykinin receptors
NK_1	Gq/11 (substance-P>neurokinin-A>neurokinin-B)
NK_2	Gq/11 (neurokinin-A>neurokinin-B>substance-P)
NK_3	Gq/11 (neurokinin-B>neurokinin-A>substance-P)

interacts with G-protein-coupled relaxin–insulin-like family peptide receptors $RXFP_1$ (LGR7) and $RXFP_2$ (LGR8) used by insulin-like peptide-3 and -7 (Table 8.28). Relaxin receptors exist in the heart, smooth muscle, connective tissue, and central and autonomous nervous systems.

Nociceptin

Nociceptin is a potent peptidic vasodilator that targets nociceptin receptor, i.e., κ-opioid receptor KOR3 (a.k.a. Nop1, Op_4, and opiate receptor-like ORL1 or OpRL1). It has a direct effect on smooth myocytes. It may also act via the endothelium, i.e., the NO–cGMP pathway. In cultured endothelial cells, liganded KOR3 activates the mitogen-activated protein kinase cascade [832].

In the human trigeminal ganglion, nociceptin colocalizes with calcitonin gene-related peptide, substance-P, nitric oxide synthase, or pituitary adenylate cyclase-activating peptide [833]. Nociceptin is not synthesized in human cerebral blood vessels. On the other hand, opioid KOR3 receptor is synthesized in vascular endothelial cells (of rat brain microvessels as well as human aortic artery and umbilical vein) [832].

Urocortin-1

Urocortin-1 and corticotrophin-releasing factor bind to receptors CRF_1 and CRF_2 with high affinity, whereas urocortin-2 and -3 are selective for CRF_2 receptors. Receptor CRF_2 have been identified in vascular smooth myocytes in humans. All urocortins cause a potent vasodilation, independently of the endothelium [812].

8.5.8.6 Vasoconstrictors

In addition to relaxing factors, endothelial cells can produce vasoconstrictors, such as endothelin-1, superoxide anions [834], endoperoxides, and thromboxane-A2.

Some G-protein-coupled receptors on the surface of smooth myocytes respond to vasoconstrictors, such as nucleotides or nucleosides and eicosanoids, that are released locally from the endothelium or perivascular nerves as well as circulating vasoconstricting hormones such as adrenaline. Other locally secreted (generated and/or stored) vasoconstrictors include angiotensin-2 produced from its precursor angiotensin-1, a cleavage product from angiotensinogen, in endothelial cells; monoamine serotonin synthesized in enterochromaffin cells of the gut, then stored in and released from platelets; and histamine released from mastocytes. The sympathetic nervous system activates smooth muscle α-adrenoceptors mainly by releasing biogenic amine noradrenaline from perivascular nerves to constrict blood vessels, especially small arteries that are responsible for systemic vasculature resistance.

Nucleotides

Extracellular nucleotides, such as adenosine triphosphate (ATP), uridine di- (UDP) and triphosphate (UTP), regulate pulmonary vasomotor tone via P2X and P2Y receptors. The nucleotide *uridine adenosine tetraphosphate* (UP_4A) vasoconstricts likely via $P2X_1$ receptors (probably also $P2Y_2$ and $P2Y_4$). Stimulation by adenosine triphosphate, uridine triphosphate, acetylcholine, endothelin, and mechanical stress releases UP_4A from endothelium [835]. Agent UP_4A, UDP, and UTP are equipotent in the endothelium-denuded artery, but UP_4A is much more potent than UDP and UTP in endothelium-intact vessels [836]. Vasoconstriction primed by UP_4A involves the entry of extracellular Ca^{2+} and release of Ca^{2+} from intracellular stores.

Uridine triphosphate provokes actin polymerization via RoCK kinase that suppresses activity of delayed rectifier K^+ channel, thereby facilitating depolarization and constriction of vascular smooth myocytes of cerebral arteries [837].

Adenosine triphosphate, quickly released by the endothelium when the flow increases, is a vasoconstrictor that binds to the 2 P2X and P2Y purinergic receptor types. However, ATP bound to endothelial $P2X_1$ receptors induces a constriction followed by a vasodilation [838].

Bradykinin

Wall shear stress enhances bradykinin secretion [839]. Like nitric oxide and endothelin, certain vasoactive substances, such as ATP, angiotensin-2, and substance-P, can cause proliferation of endothelial and/or smooth myocytes.

Neuropeptide-Y

Neuropeptide-Y (NpY) is a neurotransmitter. It augments the vasoconstrictor effects of noradrenergic neurons. It forms the lipostat with leptin and corticotropin-releasing hormone to reduce feeding and increases energy expenditure. Neuropeptide-Y operates via G-protein-coupled receptors.[80]

Neuromedin-U

Neuromedin-U[81] (NMU) is a neuropeptide acting on smooth myocytes independently of endothelium. However, neuromedin-U is synthesized in vascular endothelial cells and, once released, can operate as a paracrine messenger. It is also widely distributed in the gut and central nervous system.

Neuromedin-U activities include contraction of smooth muscle, increase of blood pressure, control of local blood flow, control of ion transport in the gut, and regulation of adrenocortical function.

Neuromedin-U binds to G-protein-coupled receptors NMUR1 (NMU_1 or FM3) in peripheral tissues (predominantly in the gastrointestinal tract) and NMUR2 (NMU_2 or FM4) in specific regions of the brain (Vol. 3 – Chap. 7. G-Protein-Coupled Receptors). Neuromedin-U and its receptor also localize to endothelial cells. Neuromedin-U receptor also resides on vascular smooth myocytes.

Neuromedin-U preferentially attenuates mesenteric blood flow, thus acting as an important regulator of intestinal blood flow [812]. Yet, neuromedin-U is a vasoconstrictor for human large arteries and veins via a direct effect on vascular smooth myocytes. Both neuromedin-U and -S[82] can act as potent vasoconstrictors in human arterial and venous networks [840].

---

[80]Receptor subtypes Y_1 and Y_5 stimulate feeding, whereas Y_2 and Y_4 favor satiety.

[81]In humans, this peptide contains 25 amino acids, hence its name neuromedin-U25. It abounds particularly in the gastrointestinal tract and some regions of the central nervous system. Neuromedin-U has anorexigenic effects.

[82]Neuromedin-S is another ligand for NMU_1 and NMU_2 receptors. Neuromedin-S that is a structurally and functionally peptide related to neuromedin-U is encoded by the Nms gene (distinct from the Nmu gene). Neuromedin-S is detected in the left ventricle [840] together with NMU_1 receptor in both vascular endothelial and smooth myocytes of intramyocardial vessels.

Chapter 9
Vascular Endothelium

Vascular endothelium constitutes the interface between the flowing blood and the deformable solid wall. The endothelium is a thin layer of connected and anchorage-dependent cells that are subjected to chemical, physical, and mechanical stimuli. They are directly exposed to molecules that circulate in the blood stream.

Vascular endothelium has several functions, as it is involved in: (1) blood–wall exchange control; (2) vasomotor tone modulation;[1] (3) coagulation regulation; (4) vessel wall growth and remodeling; and (5) inflammation and immune defense owing to leukocyte adhesion and transmigration [841]. Last, but not least, vascular endothelium is required for angiogenesis (Sect. 10.2). During angiogenesis, the arteriovenous differentiation drives vessel maturation.

Although endothelial cells of blood and lymph vessels share many features (strong apicobasal polarity and expression of certain endothelial markers), they are specialized according to vessel function and convected fluid loading. The endothelium of terminal lymphatics lacks a continuous basement membrane and intercellular spaces are not tightly sealed by junctional complexes.

Fluorescent plant virus can be used to image small-bore vessels of the macrocirculation in deep tissues using fluorescence microscopy [842], as virus is incorporated in vascular endothelial cells. This technique identifies arterial and venous compartments, as fluorescent virus uptake in the veins occurs at a much higher rate than in the arteries.

The semi-permeable endothelial barrier acts in molecule exchange between blood and vessel wall or interstitial space of perfused tissues (Sect. 9.6), thereby regulating tissue fluid homeostasis. Endothelial permeability of transported molecules depends on molecular size as well as intercellular junction nature and pattern. This size-selective transport governs fluid balance of tissues.

---

[1] Among vascular endothelium's active functions, the secretion and modification of vasoactive substances induce contraction and relaxation of adjoining vascular smooth myocytes (Chap. 8, Sect. 8.5.8).

M. Thiriet, *Tissue Functioning and Remodeling in the Circulatory and Ventilatory Systems*, Biomathematical and Biomechanical Modeling of the Circulatory and Ventilatory Systems 5, DOI 10.1007/978-1-4614-5966-8_9, © Springer Science+Business Media New York 2013

Vascular endothelium determines the vasomotor tone as well as growth and proliferation of vascular smooth myocytes via the release of several compounds (Sect. 9.10). Vasoactive substances include vasoconstrictors such as endothelin-1 and vasodilators such as nitric oxide. In particular, NO gas is produced by nitric oxide synthases, especially endothelial NOS3 isoform. Messenger NO diffuses from vascular endothelial cells to smooth myocytes, where it activates NO-sensitive soluble guanylate cyclase to initiate signaling.

Vascular endothelium regulates blood coagulation as well as thrombolysis (Sect. 9.8). Furthermore, it controls adhesion and extravasation of flowing leukocytes (Sect. 9.7), thus inflammation and immune defense. In normal conditions, the vascular endothelium has anti-inflammatory and -thrombotic activities. It responds by synthesizing multiple molecule types. In particular, endothelial adenosine triphosphate diphosphohydrolase hydrolyzes ADP and ATP into AMP molecule.

Endothelial cells ensure formation of the blood vessel network into a hierarchical set of arteries, arterioles, capillaries, venules, and veins that enables the transport of fluid, nutrients, circulating cells, hormones, and gasses to organs.

Endothelial cells detect hemodynamic stresses via mechanosensors, such as adhesion molecules (mainly integrins), ion channels, and plasmalemmal receptors (GPCRs and RTKs). Signaling pathways (MAPK, PKB, PKC, and ROS) augment the activity of transcription factors (Activator proteins AP1 and AP2, cAMP response element, early growth response protein EGR1, and NFκB) with a magnitude that depends on the cell type, i.e., vascular region.

Time-dependent hemodynamic stresses applied on and within the vessel wall (wall shear stress, axial and circumferential tensions within the wall) are implicated in: (1) the secretion of vasoactive substances (nitric oxide, endothelin, and prostacyclin, among others) and regulation of the vascular tone that determines the vessel bore according to the magnitude of sensed mechanical stresses; (2) short-term wall adaptation and long-term remodeling, as they influence cell signaling that directs cell growth (growth inhibitor heparin and growth factors such as platelet-derived growth factor), differentiation, migration, and apoptosis; (3) expression of proteins entailed in coagulation and fibrinolysis (tissue plasminogen activator, plasminogen-activator inhibitor, tissue factor, etc.), in cellular adhesion (vascular cell adhesion molecule VCAM1 and intercellular adhesion protein ICAM1), in diapedesis (CCL2 chemokine); and (4) vasculature diseases because they affect the cell functioning and transport processes.

Wall shear stress, in particular, upregulates the connexin expression and entail calcium influx that activate protein kinases and nitric oxide synthase for fast nitric oxide release.

9.1 Endothelial Cell

Endothelial cells are flat with a central, elongated, projective nucleus that yields a wavy wetted surface at the microscopic level. The endothelium surface, either from fresh arterial walls mounted on an appropriate holder and kept in physiological buffer or from cultures, can be assessed by a scanning force microscope associated with a phase-contrast microscope [843].

In addition to endothelial cells that cover blood and lymph vessels, mature endothelial cells (circulating endothelial cells [CEC]) as well as endothelial progenitor cells (circulating endothelial progenitor cells [CEPC]; Sect. 9.3) circulate in the blood flow at a very low concentration. Resident endothelial cells may detach during the normal turnover after apoptosis, with clearance by the reticuloendothelial system, in the absence of endothelial damage. Circulating endothelial progenitor cells are able to form patches at sites of endothelial discontinuity to ensure the integrity of the vessel wall [844].[2]

Flattened endothelial cells (thickness 1–2 μm, width 10–15 μm, length in the streamwise direction 60–100 μm) have rest and stretched (flow-adapted) configurations. The height variation due to projective nucleus has been measured along the endothelium with a maximum of 750 nm.

Observed plasmalemmal granules and ring-like structures of various sizes have been assumed to be associated to the cytoplasmic layer of the plasma membrane rather than the outer one. Moreover, fibers can be seen, likely associated with actin filaments, that are bound to the cytoplasmic face of the plasma membrane.

The cell membrane is covered by a thin glycocalyx. The between-cell space width ranges from 10 to 20 nm with tight and gap junctions.

Endothelial cells contain Weibel-Palade bodies, long rod-shaped storage organelles (length 1–5 μm, caliber $\sim$ 200 nm). These storage and secretory granules are filled with von Willebrand factor.[3] The von Willebrand factor recruits platelets to the site of injury.[4] The architecture of the Weibel-Palade body and tubular folding of von Willebrand factor requires a low pH [846]. The tubules must not be disassembled prior to exocytosis. Once released, von Willebrand factor unfolds rapidly and efficiently at neutral pH to trap circulating platelets, as it forms platelet-catching filaments (length $\sim$ 100 μm). Weibel-Palade bodies also contain P-selectins.

---

[2]Circulating endothelial progenitor cells migrate from the bone marrow into the blood circulation. They can differentiate into mature endothelial cells. They operate not only in re-endothelialization and vascular healing, but also in vasculogenesis and angiogenesis. $PTPRc^{low}$ (low CD45 expression level), $VEGFR2^{+}$ endothelial progenitors can be mobilized by growth factor such as erythropoietin [845].

[3]Multimeric von Willebrand factor is a glycoprotein (mass > 20 MDa; length > 4 μm). Precursor of von Willebrand factor is synthesized by endothelial cells (mass $\sim$ 350 kDa). The von Willebrand factor contains binding sites for factor-VIII, platelet membrane glycoprotein-1b, collagen-1, -3, and -6, and $\alpha_{2B}\beta_3$ integrin.

[4]Large vWF polymers stored in Weibel-Palade bodies are released by Ca^{2+}–calmodulin to bind coagulation factor-VIII (hence avoiding its cleavage), platelet glycoprotein GP1b, and $\alpha_{2B}\beta_3$-integrin that associate to collagen, heparin, and vitronectin.

Endothelial cells monitor cell internal state as well as environmental actions. Endothelial cells especially sense hemodynamic and hormonal stimuli and respond by secreting various mediators. Endothelial cells experience blood pressure, axial and circumferential tension from connecting endothelial cells, and blood friction on the wetted surface, i.e., the wall shear stress[5] (WSS). The applied forces are unsteady with noticeable spatial and temporal magnitude gradients as well as possible direction changes. The stress distributions in the membrane and cytosol affect endothelial functions.

9.1.1 Glycocalyx

The glycocalyx forms a thin layer[6] between the circulating blood and endothelium. It is a hydrated mesh of negatively charged glycosaminoglycans,[7] proteoglycans, glycoproteins, and glycolipids secreted by endothelial cells.

The glycocalyx is the first barrier to molecular transport from the flowing blood to the vessel wall, providing hydraulic resistance to mass transport through this sieve [847]. The transport conductance in the glycocalyx depends on molecular size.[8] Both the hydrostatic and osmotic pressures act on transport across the glycocalyx. However, albumin (molecular mass 67 kDa) and fibrinogen (molecular mass 340 kDa) cross the glycocalyx at about the same rate. Charge restriction imposed by the glycocalyx also determine accessibility of proteins. Nonetheless, the glycocalyx generates nanodomains associated with a heterogeneous distribution of negative charges that modulate transendothelial transport.

At the basal state, the rheological properties of the glycocalyx might induce a lift that prevents cell adhesion [848]. In inflammatory sites, heparan sulfate on the surface of endothelial cells is a potential ligand for P-selectins (expressed by endothelial cells) and L-selectins (expressed by leukocytes), which are involved in initial attachment and rolling of leukocytes on the endothelium [849]. It also binds chemokines for stable adhesion of leukocytes on the endothelium.

The glycocalyx, which is a polyelectrolyte coating, has been modeled as a semi-infinite, doubly periodic array of parallel charged cylinders [850]. Only the luminal layer part of the glycocalyx model markedly influences transport.

---

[5]The wall shear stress is the tangential force produced by blood moving along the endothelial surface. At the endothelial surface, most momentum and vorticity transports occur between fast-moving fluid particles and those slowed down by wall friction. The boundary layer develops. The wall shear stress depends on the local velocity gradient.

[6]The glycocalyx thickness is very difficult to determine accurately ($h \sim 100$ nm in the microcirculation and $h \sim 250$ nm in the arteries). However, using a recent observation technique, the glycocalyx thickness was shown to range from 5 to 11 μ [711]).

[7]Among glycocalyx glycosaminoglycans, syndecans and glypicans bind to the cell membrane.

[8]The sizes of albumin and LDL are nearly 3.5 and 11 nm, respectively.

^{N}acetylglucosamine is one of the main components of glycocalyx oligosaccharides. It interacts with endothelial Glc^{NAc}-recognizing lectins of the luminal surface, among which some may participate in flow sensing such as glycosylated endothelial Na^+ channel (ENaC) [851]. Connection of hyaluronan to lectin-bearing substance such as ENaC channel.

9.1.2 Endothelial Cell Adhesions

Endothelial intercellular junctions are necessary for formation and integrity of the interface between blood and vessel wall. The main junctions between endothelial cells are adherens and tight junctions that are composed of transmembrane adhesion molecules linked to cytoskeletal-binding proteins and intracellular signaling partners.

Intercellular junctions not only yield attachment sites, but also transfer signals for morphogenesis and stabilization of the vessel wall architecture. Vascular homeostasis relies also on cellular adhesion with the extracellular matrix. Integrins link endothelial cells with constituents of the extracellular matrix, such as fibronectin or vitronectin. Their cytosolic domains are associated with actin cytoskeleton via talin and vinculin.

9.1.2.1 Junctions between Endothelial Cells

Intercellular junctions include tight, adherens, and gap junctions. Adherens junctions are mainly required for correct vasculo- and angiogenesis and remodeling, whereas tight junctions essentially control the endothelial barrier. Endothelial cells do not have desmosomes. In addition, $\alpha_2\beta_1$- and $\alpha_5\beta_1$-integrins have also been identified in endothelial clefts.

In endothelial cells, actomyosin filament contraction can generate forces (up to 120 nN) that pull perpendicularly to the face of intercellular contact, the so-called *tugging forces*. Mechanical loading at cell–matrix and –cell adhesions causes focal adhesion growth. In particular, the size of adherens junctions enlarges when tugging forces rises and conversely [852]. Actomyosin-dependent regulation of adherens junction size is supported by small Rac1 GTPase.

Adherens Junctions

Adherens junctions are mainly composed of cadherins (Vol. 1 – Chap. 7. Plasma Membrane). Cadherin cytoplasmic tail contains 2 binding sites. The first binding site associates with β- and γ-catenins in a mutually exclusive manner. The second links to catenin-δ1, an inhibitor of Rho GTPases. Cadherin then indirectly connects to α-catenin via β- or γ-catenin. Catenin-α links the cadherin–catenin complex to the actin cytoskeleton. Vascular endothelial (or Cdh5) and possibly neuronal (or Cdh2) cadherins thus bind to intracellular partners that contribute to signaling and dynamics of the actin cytoskeleton.[9] Catenin-γ[10] can also tether cadherin-5 to intermediate filaments via desmoplakin to form the *complexus adhaerentes*, a special type of cell junctions. The desmoplakin–vimentin complex corresponds to an additional agent of mechanical stability.

Catenins-β are able to bind to junctional proteins, such as IQ motif-containing GTPase-activating protein IQGAP1, platelet–endothelial cell adhesion molecule PECAM1, casein kinase-2, as well as signaling and transcription factors, such as Wnt, adenomatous polyposis coli Ub ligase, and T-cell factor [854]. Scaffold IQGAP1 not only links to β-catenin and E-cadherin, but also to small GTPases CDC42 and Rac, as well as actin, calmodulin, and microtubule-associated cytoplasmic linker integral protein CLIP170.[11] Adhesion molecule PECAM1 that is concentrated in endothelial clefts interacts with Tyr-phosphorylated β-catenin and phosphatase PTPn11. Molecule PECAM1 may participate in modulating adherens junction assembly and restoring endothelial barrier integrity after injury. In addition, PECAM1 binds $\alpha_V\beta_3$-integrins and regulates the function of $\alpha_4\beta_1$- and β_2-integrins to possibly mediate transendothelial migration of leukocytes.

α-Catenin that links to β-catenin and actin-polymerizing proteins, such as α-actinin, vinculin, vasodilator-stimulated phosphoprotein, and formin, as well as actin microfilaments, promotes actin bundling, thereby stabilizing adherens junction and cleft. Furthermore, actin polymerization is needed for adherens junction assembly. Non-muscle myosin heavy chain-2A is an another regulator of adherens junction formation [854]. Cadherin-5 can regulate intercellular permeability by modulating GTP binding to and GTP hydrolysis of small GTPases CDC42, Rac, and RhoA [854]. Cadherins can interact with actin-related proteic ARP2–ARP3 complex that associates with Wiskott-Aldrich syndrome protein (WASP), cortactin, and vinculin. Wiskott-Aldrich syndrome protein is an effector of CDC42 GTPase. The CDC42–ARP2/3–WASP pathway increases actin polymerization. Cadherin-5 can also activate GTPase Rac via Rac-specific GEF T-cell lymphoma invasion and metastasis Tiam1. By activating small GTPases, cadherins can control actin polymerization at intercellular junctions and modulate paracellular permeability.

---

[9] Vascular endothelial (VE)-cadherin, or cadherin-5, is the single member of the cadherin family restricted to the endothelium. Neuronal (N) cadherin, or cadherin-2, is also expressed in endothelial cells. Cadherin-2 mainly has a dispersed distribution on plasma membrane. In the presence of cadherin-5, cadherin-2 is excluded from cell junctions, except during early stages of cellular confluency [853]. In addition, cadherin-2 may also be involved in junctions between endothelial cells and pericytes.

[10] A.k.a. plakoglobin.

[11] Calmodulin and E-cadherin compete for binding to IQGAP agent. Calmodulin thus impedes E-cadherin-mediated homophilic adhesion. Activated CDC42 and Rac1 preclude interaction of IQGAP1 with β-catenin. Activated CDC42 and Rac strengthen endothelial barrier by freeing IQGAP from β-catenin.

Catenin-δ1 regulates cadherin-5 expression and insertion into the plasma membrane. The Ctnnδ1–Cdh5 complex precludes binding of ubiquitin ligase Hakai, hence preventing cadherin-5 degradation. Catenin-δ1 also has many partners, such as microtubule nanomotor kinesin and regulatory kinases and phosphatases (e.g., SRC kinase family member Fyn, FRK family kinase Fer, and plasmalemmal PTPRm and cytosolic PTPn6 protein Tyr phosphatases) [854]. Unlike β-catenin and plakoglobin, catenin-δ1 does not associate with the actin cytoskeleton, but with microtubules. However, catenin-δ1 regulates the contractile apparatus of endothelial cells and modulates endothelium permeability. The Ctnnδ1–Cdh5 complex actually impedes the activity of RhoA GTPase that mediates myosin light chain phosphorylation and actin stress fiber formation.

Several receptor and cytoplasmic protein Tyr phosphatases localize to adherens junctions and can dephosphorylate components of the cadherin–catenin complex and control Rho activity.

Vascular endothelial protein Tyr phosphatase PTPRb restricted to the endothelium associates specifically with and dephosphorylates VE-cadherin. Moreover, the PTPRb–Cdh5 complex can contribute to strengthening of adherens junction barrier by a phosphatase-independent mechanism [854].

Phosphatase PTPRj[12] abounds in endothelial cells at least in arteries and capillaries of several organs [853]. When it associates with the Cdh5–βCtn complex, it dephosphorylates VEGFR2 vascular endothelial growth factor receptor.

Receptor-like protein Tyr phosphatase PTPRm is strongly present in endothelial junctions of arteries and continuous capillaries. Phosphatase PTPRm dephosphorylates catenin-δ1, thus modulating interaction of catenin-δ1 with cadherin-5 and controlling catenin-δ1 regulation of RhoA activity.

Protein Tyr phosphatase PTPn11 is an additional component of adherens junctions that links to VE-cadherin. Dissociation of PTPn11 can expose junctional proteins to phosphorylation by kinases and cause adherens junction disassembly. In addition, PTPn11 prevents small GTPase RhoA activity to stabilize adherens junction.

Small GTPases CDC42, Rac, and RhoA contribute to the regulation of adherens junctions. Bradykinin, histamine, platelet-activating factor, and thrombin that heighten endothelial permeability provoke disassembly of adherens junctions via RhoA GTPase. On the other hand, Rac GTPase promotes endothelial barrier by stabilizing adherens junctions. Moreover, activated CDC42 is involved in reformation of adherens junctions in endothelial cells during recovery from abnormal permeability induced by permeability-increasing agents. Small GTPases CDC42 and Rac modulate interactions between α-catenin and the cadherin–catenin complex to favor formation of adherens junctions. Activated CDC42 and Rac indeed interact with β-catenin-sequestering protein IQGAP to free β-catenin that can then binds with its partners cadherin and α-catenin [854].

[12] A.k.a. PTPη, RPTPη, CD148, and Density-enhanced phosphatase DEP1.

Permeability-increasing factors, such as thrombin and VEGF, induce phosphorylation of cadherin, β-catenin, and catenin-δ1 to disrupt adherens junctions [854]. Phosphorylation by Src or PKC of adherens junction constituents can modify affinity of catenins for cadherin-5, as well as Cdh5–actin interactions. Conversely, phosphatases stabilize adherens junctions. PTPn11 Phosphatase associates with and protect the cadherin–catenin complex.

Tight Junctions

Tight junctions contain occludin, junctional adhesion molecules, and claudins that interact directly or indirectly with cytoplasmic partners, such as cingulin and zonula occludens adaptors. Zonula occludens proteins (ZO1–ZO3) are members of the family of membrane-associated guanylate kinases (MAGUK). Zonula occludens proteins and cingulin contribute to interaction of tight junction with the actin cytoskeleton.

Occludins form homotypic bonds. Their cytoplasmic C-terminus associates with zonula occludens protein ZO1, hence with the actin cytoskeleton to stabilize tight junction. Arterial and blood–brain endothelial barriers that are the least permeable of the vasculature contain a much greater number of occludins than do other compartments.

Among 24 known members of the claudin family, only claudin-5 is specifically expressed by endothelial cells. Yet, several claudins are synthesized in endothelial cells (claudin-1, -3, -5, and -12). Claudin-5 that is particularly produced by cerebral endothelial cells is a major regulator of the blood–brain barrier function. Claudin extracellular regions form homo- and heterotypic bonds. Its cytoplasmic part also binds to zonula occludens protein ZO1, and therefore indirectly to ZO1 partners.

Junctional adhesion molecules, endothelial cell-selective adhesion transmembrane glycoproteins, and coxsackievirus and adenovirus receptors associate with tight junctions, but do not build strands per se [853]. However, they modulate leukocyte diapedesis across the endothelium.

Junctional adhesion molecules comprise 3 main types: JAM1 (or JAMa), JAM2 (or JAMb), and JAM3 (a.k.a. JAMc and veJAM). Isotype JAM1 resides in epithelial and endothelial cells; JAM2 in high endothelial venular cells, i.e., endothelial cells of postcapillary venules of lymphoid tissues that form the leakiest endothelium of lymphatic vessels; JAM3 exclusively in endothelial cells. Junctional adhesion molecules are able to bind partitioning-defective protein Par6, small GTPase CDC42, and PKCζ, thereby recruiting these signaling mediators to tight junctions.

Zonula occludens proteins interact directly or indirectly via bridging proteins with claudins, occludins, and junctional adhesion molecules. They intervene in spatial organization of tight junction constituents, particularly occludins. Zonula occludens proteins link tight junction proteins to the actin cytoskeleton and recruit signaling molecules. In addition, ZO1 also binds to adherens junction protein α-catenin, gap-junction component connexin-43, and actin-polymerizing proteins vasodilator-stimulated phosphoprotein and spectrin, hence linking tight junctions to the actin cytoskeleton. Therefore, in endothelial cells, adherens and tight junctions that are intermingled interact via common partners such as ZO1 for their formation, maintenance, and remodeling. In addition, nectins and their intracellular partners such as afadin contribute to the organization of adherens and tight junctions.

Endothelial barrier requires endothelial markers cadherin-5 and claudin-5 of adherens and tight junctions, respectively. Homotypic Cdh5-based adhesions control claudin-5 expression, as they prevent the nuclear accumulation of transcriptional repressors FoxO1 and β-catenin that inhibit claudin-5 promoter [855].[13] This crosstalk relies on Z01 and JAMs. The crosstalk is mutual as tight junction molecules such as junctional adhesion molecules can regulate cadherins in endothelial cells.

The RhoA–RoCK pathway that causes actin stress fiber formation can then induce a loss in junctional occludin, disrupt tight junctions, and increase endothelial permeability. Histamine, lysophosphatidic acid, thrombin, and vascular endothelial growth factor disassemble tight junctions via the RhoA–RoCK pathway that phosphorylates occludin and ZO1, possibly via kinase Src, protein kinase-C, and casein kinase-2. Conventional PKC isoforms (α, β, and δ) also phosphorylate junctional adhesion molecules.

Gap Junctions

Each gap junction is made up of 2 connexons that correspond to contribution of each of the 2 partner cells. Connexon is constituted of 6 connexins. The intercellular pore formed by connexons (caliber ~ 2 nm) has an open or closed configuration. Gap junction gating is regulated by connexin phosphorylation. Both Ser–Thr and Tyr phosphorylation of connexins induce channel closure. Phosphorylation also regulates the rate of gap junction assembly and turnover.

Endothelial cells express connexins Cx37, Cx40, and Cx43. As connexins form gap junctions, they allow rapid exchange of low-molecular-mass messengers, such as calcium ions and inositol trisphosphate, between contiguous cells. Junction protein ZO1 binds connexin-43 and facilitates communication between tight, adherens, and gap junctions.

Connexins contribute to nitric oxide production and endothelial barrier integrity. Moreover, endothelial Cx43 can also associate with Cx40 of neutrophils during their extravasation and vascular smooth myocytes.

---

[13]Cadherin-5 transfers cues either directly by committed signaling effectors, such as phosphatidylinositol 3-kinase or growth factor receptors, or indirectly by tethering and retaining transcription factors such as β-catenin at the plasma membrane, thereby limiting their translocation to the nucleus. Cadherin-5 adhesion provokes a sustained activation of the PI3K–PKB pathway that phosphorylates FoxO1 for cytosolic sequestration, thereby avoiding its interaction with DNA in cooperation with β-catenin and T-cell factor TCF4.

9.1.2.2 Cell–Matrix Junctions

Focal adhesions are sites of adhesions of endothelial cells to the extracellular matrix mediated by integrins. Endothelial cells express mainly on their abluminal surface numerous heterodimeric integrins ($\alpha_1\beta_1$, $\alpha_2\beta_1$, $\alpha_3\beta_1$, $\alpha_5\beta_1$, $\alpha_6\beta_1$, $\alpha_V\beta_3$, $\alpha_1\beta_5$, and $\alpha_V\beta_5$) [854]. These transmembrane glycoproteins interact with matrix proteins, such as fibronectin, fibrinogen, vitronectin, and collagen. Association between integrins and extracellular matrix constituents restricts the passage of macromolecules across the endothelial barrier.

Integrin cytoplasmic domain connects to actin-binding proteins, such as α-actinin, filamin, paxillin, talin, tensin, vinculin, and zyxin. α-Actinin that binds zyxin targets vasoactive stimulatory phosphoprotein and profilin. Filamin links to small GTPases CDC42, Rac, RhoA, and Ral1, as well as RhoA-associated kinase, Trio with its 3 enzymatic domains,[14] and caveolin-1. Paxillin interacts with P21-activated kinase (PAK) and PAK-interacting exchange factors RhoGEF6 and RhoGEF7, Abelson Tyr kinase (Abl), and Ras GTPase-activating protein RasA1. Tensin binds multiple phosphotyrosine proteins, such as BCAR1 (or CAS) and Src kinase. Vinculin interacts with the actin-related proteic ARP2–ARP3 complex and phosphatidylinositol 4-phosphate 5-kinase. These proteins coordinate signals between focal adhesions and the actin cytoskeleton.

These proteic interactions define the adhesome, a signaling platform integrator of many signaling axes. Adhesome regulates actin polymerization and focal adhesion function. For example, ligation of $\alpha_V\beta_3$-integrin with matrix proteins causes: (1) tyrosine phosphorylation of focal adhesion kinase, paxillin, cortactin, and ezrin; (2) Ca^{2+} influx; and (3) activation of phospholipase-A2 and Rac GTPase.

Focal adhesion constituents are recruited into focal adhesions upon tyrosine phosphorylation by focal adhesion kinase and Src kinases. Integrin clustering provokes FAK autophosphorylation (activation). Kinase Src further phosphorylates activated FAK enzyme. Activated FAK phosphorylates various substrates, such as paxillin, tensin, PI3K, and BCAR1, that aggregate to form focal adhesions.

Paxillin recruits other focal adhesion constituents, as it associates with vinculin, CRK, C-terminal Src kinase, and Src, as well as ArfGAPs APAP1 and APAP2 [854]. Activated APAP1 regulates CDC42 and Rac, as it can complex with RhoGEF6 and RhoGEF7 that serves as CDC42- and Rac1-GEF to maintain or restore endothelial barrier.

Focal adhesion kinase not only associates with paxillin and talin, but also with SRC family kinases Src and Fyn, guanine nucleotide-exchange factor Ras-GRF1, growth factor receptor-bound protein GRB2, adpribosylation factor GAP-containing, SH3, ankyrin repeats, and PH domain protein ASAP1, BCAR1 adaptor, RhoGAP26, and actin-polymerizing proteins ezrin and WASP [854]. Focal adhesion kinase is phosphorylated (activated) in response to integrin activation.

---

[14]Molecule Trio contains 2 GEF sites for RhoA and Rac and a Ser–Thr kinase region, hence its name.

Various mediators, such as thrombin, histamine, hydrogen peroxide, and VEGF, are able to cause FAK phosphorylation and promote focal adhesions to maintain endothelial barrier integrity. Furthermore, focal adhesion kinase restores endothelial barrier function after intervention of inflammatory mediators that causes cell junction disassembly.

On the other hand, increased FAK activity allows hyperosmolarity-induced strengthening of adherens junctions. Focal adhesion kinase hinders activity of GTPase RhoA, as it associates with RhoA inhibitors RhoGAP26 and P190RhoGAP. Furthermore, focal adhesion kinase activates APAP1 protein. In addition, FAK is required for normal vascular development.

Kinases of the SRC family can be stimulated by activated integrins. Activated Src kinase interacts with focal adhesion kinase for full kinase activity. Among SRC family members, Src kinase contributes to increased endothelial permeability in response to superoxide anion (O_2^-), thrombin, and VEGF [854].[15] Kinase Src not only elevates endothelial junction permeability, but also caveola-mediated transcytosis.

Proline-rich Tyr kinase PYK2, a Ca^{2+}-dependent cytosolic kinase, also binds integrins. It is highly expressed in pulmonary endothelial cells. It is rapidly phosphorylated upon angiotensin and mechanical stimuli. It could regulate endothelial barrier by activating RhoA GTPase. Activated $RhoA^{GTP}$ participates in the formation of focal adhesions via its effectors RoCK and Diaphanous as well as PIP5K recruitment.

Transmembrane domains of integrins also interact with tetraspanins, GPI-anchored urokinase-type plasminogen activator receptor, and caveolin-1. Caveolin-1 is involved in formation of focal adhesions, development of basement membrane, paracellular endothelial permeability, and integrin signaling [854].

9.1.2.3 Myoendothelial Junctions

Myoendothelial junctions (MEJ) correspond to endothelial projections that protrude and cross holes of the endothelial basement membrane and internal elastic lamina to reach adjacent smooth myocyte membranes.

The principal regulators of vascular tone are neural, endothelial, and mechanical stimuli that initiate vasodilation or vasoconstriction. Three primary vasodilatory signals include messengers produced by endothelial cells, nitric oxide and prostaglandins, and endothelium-dependent vasodilatory hyperpolarization. The latter refers to the transfer of an endothelium-derived electrochemical current through

[15]Inhibition of Src activity by phosphatase PP1 prevents TNFα augmentation of thrombin-induced endothelial permeability by impeding Ca^{2+} entry and internalization of cadherin–catenin complexes. Moreover, Src phosphorylates MLCK and cortactin, thereby provoking endothelial cell contraction and increasing endothelium permeability.

Table 9.1. Mechanisms of endothelium-derived hyperpolarization (Source: [856]). Endothelium-derived hyperpolarization (EDH) relies, at least partly, on cytosolic calcium influx upon liganded receptors and/or hemodynamic stresses and activation of potassium channels, such as small, intermediate, and, in some cases, large-conductance calcium-activated potassium channels (EC: endothelial cell; MEES: myoendothelial microdomain extracellular space; SMC: smooth myocyte.

Diffusible factors	
Potassium ion (transient, localized cue)	Efflux from EC to MEES through K_{Ca} and activation of SMC Na^+–K^+ ATPase and K_{IR} channel
Epoxyeicosatrienoic acids (EETs)	Opening of EC and SMC $K_{Ca}1.1$ (BK), Ca^{2+} entry through TRPV4
Hydrogen peroxide (H_2O_2)	Vasoconstriction, activation of SMC K_{Ca}, K_{ATP}, and Na^+–K^+ ATPase; no effect on EDH in human radial and subcutaneous arteries
C-type natriuretic peptide (CNP)	Vasorelaxation, activation of NP_2 and NP_3 receptors; role questionable
Contact-mediated mechanisms – Gap junctions	
Electrochemical coupling	Eventual modulation by K^+, EETs, H_2O_2, CNP

direct coupling between endothelial and smooth muscle cells via myoendothelial gap junctions (contact mechanism) and activity of ion carriers located in myoendothelial microdomains (diffusible endothelial factor release; Table 9.1).

Myoendothelial projections (MEP) contain myoendothelial gap junctions composed of connexins, thereby enabling direct signaling between vascular endothelial cells and smooth myocytes. Endothelial hyperpolarization is directly transmitted to adjacent smooth myocyte via the myoendothelial gap junction. Endothelium-dependent hyperpolarization of adjacent smooth myocytes closes their $Ca_V1.2$ channel.

Hydrogen peroxide (H_2O_2) can influence gap junctional coupling in addition to modulating the sensitivity of the contractile apparatus to calcium and activating smooth muscle Na^+–K^+ ATPase and BK_{Ca} and K_{ATP} channels [856]. However, H_2O_2 does not play a significant role in endothelium-dependent hyperpolarization [857].

Endothelium-dependent hyperpolarization factors (EDHF) comprise actions of K^+ ion, nitric oxide, prostaglandins, cytochrome-P450 products epoxyeicosatrienoic acids, and myoendothelial electrical coupling, but is neither prostacyclin nor nitric oxide.

Two types of Ca^{2+}-activated K^+ channels are involved in endothelium-dependent hyperpolarization (Table 9.2): (1) small conductance Ca^{2+}-activated K^+ channels (SK or $K_{Ca}2.3$) that are widely distributed over the endothelial plasma membrane and (2) intermediate conductance Ca^{2+}-activated K^+ channels (IK or

Table 9.2. Ion carriers of the myoendothelial microdomain between an endothelial cell (EC) and adjoining smooth myocyte (SMC; source: [858]). Signaling in the myoendothelial microdomain through gap junctions and calcium-activated potassium channel enables endothelium-dependent vasodilation. Three types of nanodomains on myoendothelial microdomains (myoendothelial projections) can be defined: (1) myoendothelial gap junction that facilitates ion transfer; (2) K_{Ca} channel residence; and (3) transient receptor potential (TRP) channel (non-selective cation carrier) site (TRPC3: type-3 TRP canonical; TRPV4: type-4 TRP vanilloid). The 2 latter nanodomains can merge.

Gap junction (myoendothelial feedback)	
Molecular transfer	Ions, IP_3
Endothelial projection membrane	
$K_{Ca}3.1$ (IK)	K^+ export (from EC cytosol)
TRPC3	Ca^{2+} import (into EC cytosol)
TRPV4	Ca^{2+} import
Endothelial projection endoplasmic reticulum membrane	
IP_3R	Ca^{2+} release
Smooth myocyte membrane	
Na^+–K^+ ATPase	K^+ entry (into SMC cytosol)
	Na^+ efflux (from SMC cytosol)
K_{IR}	K^+ influx

Table 9.3. Features of myoendothelial microdomains in human mesenteric arteries (Source: [857]; MEGJ: myoendothelial gap junction).

Number of SMC layers	~ 7
Adventitial thickness	10.8 ± 1.1 µm
MEGJ density	$1.9 \pm 0.7 \times 10^3$ µm^2
$K_{Ca}3.1$	$4.1 \pm 0.6 \times 10^3$ µm^2
Connexin-37	$2.2 \pm 0.5 \times 10^3$ µm^2

$K_{Ca}3.1$) that mainly localize to the myoendothelial projections. The contribution of SK_{Ca} and IK_{Ca} channels varies between species as well as in different vascular beds of the same species [857].

The opening of endothelial Ca^{2+}-sensitive $K_{Ca}3.1$ channels on myoendothelial projections elevates the extracellular K^+ concentration in the myoendothelial space that activates inwardly rectifying K^+ channels, which may lodge exclusively on the endothelial surface (in rat mesenteric artery), and Na^+–K^+ pumps on smooth myocytes adjacent to myoendothelial projections, at least in small resistive arteries, thereby hyperpolarizing these cells [859].

In fact, endothelium-dependent vasodilation, at least in human mesenteric arteries, is primarily mediated by [857]: (1) nitric oxide; (2) NO- and PGi2-independent endothelium-derived hyperpolarizing current through IK_{Ca} channel; and (3) NO- and PGi2-independent material transfer through gap junction connexin-37 (Table 9.3).

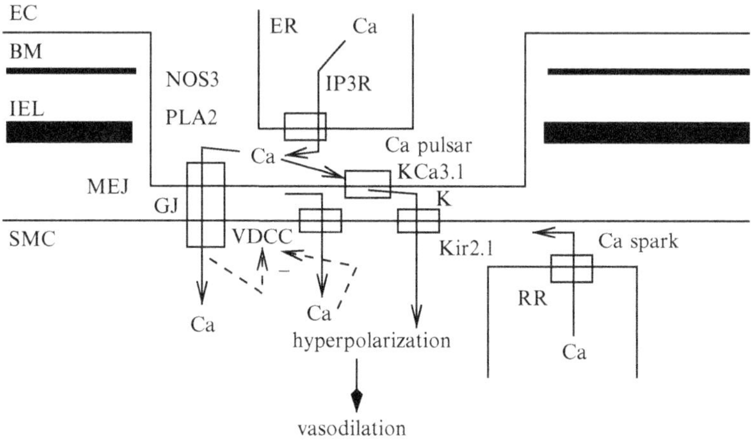

Fig. 9.1 Intercellular functional unit of the myoendothelial junction (MEJ; Source: [860]). Myoendothelial junction is an endothelial domain that crosses the basement menbrane (BM) and internal elastic lamina (IEL) to reach adjacent smooth myocyte membrane. The intercellular functional unit is composed of: (1) connexins that form gap junctions (GJ); (2) inositol trisphosphate receptors (IP_3R) and calcium-sensitive $K_{Ca}3.1$ channels (IK) in endothelial cell (EC); and (3) inward rectifier potassium channels ($K_{IR}2.1$) and voltage-dependent calcium channels (VDCC; $Ca_V1.2$) in smooth myocytes (SMC). Calcium pulsar is an endothelial Ca^{2+} signal that has a restricted localization to MEJ. Its activity is regulated by biological and mechanical agents. Other endothelial calcium-dependent enzymes (e.g., endothelial nitric oxide synthase [NOS3] and phospholipase-A2 [PLA2]) can be activated by calcium pulsars or waves. In smooth myocytes, calcium sparks from clusters of ryanodine receptors of the endoplasmic reticulum closely juxtaposed to the plasma membrane activate Ca^{2+}-sensitive, large-conductance $K_{Ca}1.1$ channels, thereby causing a transient hyperpolarization that promotes vasodilation.

An increase in cytosolic Ca^{2+} concentration in endothelial cells serves as a vasodilatory signal, whereas, in smooth myocytes, it triggers vasoconstriction by targeting the actin–myosin stress fibers.

In smooth myocytes, elementary calcium release (*calcium sparks*) from clusters of ryanodine receptors of the endoplasmic reticulum closely juxtaposed to the plasma membrane activate calcium-sensitive large-conductance $K_{Ca}1.1$ channels, thereby causing a transient hyperpolarization that reduces vasoconstriction.

Calcium ion can also be locally released through inositol trisphosphate receptors of endothelial endoplasmic reticulum to create the so-called endothelial *calcium pulsars* in myoendothelial junctions to transmit vasoregulatory signals [860]. Calcium pulsars that encode signals between vascular endothelial cells and smooth myocytes differ from Ca^{2+} sparks generated by ryanodine receptors. One target of calcium pulsars is $K_{Ca}3.1$ channels in endothelial projections to relax adjoining smooth myocytes.

An intercellular functional unit can thus be defined that is composed of: (1) connexins that form gap junctions; (2) endothelial IP_3Rs and $K_{Ca}3.1$ channels; and (3) inward rectifier $K_{IR}2.1$ channels and voltage-dependent $Ca_V1.2$ channels in smooth myocytes (Fig. 9.1).

Two types of Ca^{2+} signals can cause the opening of IK_{Ca} channels in myoendothelial projections [858]. Opening of TRPV4 channels on endothelial membrane causes localized Ca^{2+} influxes — Ca^{2+} sparklets — that activate IK_{Ca} and SK_{Ca} channels. This mechanism ensures response to acetylcholine, at least in mouse small mesenteric arteries. On the other hand, Ca^{2+} pulsars caused by the opening of clusters of IP_3 receptors on parts of the endoplasmic reticulum within the myoendothelial projections may be generated mainly in these microdomains. Calcium pulsars can be triggered by IP_3 entering myoendothelial projections from either the endothelial or smooth myocyte cytosol. The former contributes to endothelium-dependent hyperpolarization; the latter to the *myoendothelial feedback*, by which smooth myocyte contraction is autolimited via K^+ efflux from the myoendothelial projection, i.e., by which stimulation of vascular smooth myocytes activates endothelial cells to repress vasoconstriction.

In addition, calcium wavelets, another IP_3-dependent endothelial Ca^{2+} signal, play an important role in myoendothelial feedback [858].

Several types of transient receptor potential channels may be involved in calcium entry and myoendothelial signaling: in rat cerebral artery, type-1 TRP ankyrin (TRPA1) and type-3 and -4 TRP vanilloid (TRPV3 and TRPV4) channels; in rat carotid artery, endothelial TRPV4 channel can trigger NO-dependent relaxation as well as, in rat gracilis muscle arterioles and in mouse mesenteric artery, both NO- and EDH-mediated dilation.

Myoendothelial projections are capable of generating localized Ca^{2+} pulsars via non-selective cation channel TRPC3, which lodges on the endothelial plasma membrane, in close proximity to both IP_3R on the sarcoplasmic reticulum and IK_{Ca} on the plasma membrane as well as gap junction, but not on that of vascular smooth myocytes, at least in rat mesenteric artery [861].[16] This channel mainly localizes to myoendothelial projections. Protein TRPC3 is distributed throughout the endothelium, but with approximately 5-fold higher density at myoendothelial contact sites. The K_{Ca}-mediated endothelial-based vasodilation relies on TRPC3 channels [861].

9.1.3 *Vascular Permeability*

Angiogenic activity of VEGFa is mainly exerted by upregulating expression of testicular receptor TR3, or NR4a1 nuclear receptor, in vascular endothelial cells. Transcription factor NR4a1 contributes to the regulation of vascular permeability in 3 contexts [862]: (1) basal vascular permeability that suffices to bring nutrients to

[16] A spatial and functional association between TRPC3, IP_3R, and K_{Ca} channels exists in various cell types (vascular endothelial and smooth muscle cells, pancreatic β cells, gastric smooth muscle cells, etc.).

cells; (2) acute vascular hyperpermeability, especially in postcapillary venules, in response to short-term exposure to vascular permeabilizing agents such as VEGFa; and (3) chronic vascular hyperpermeability in pathological angiogenesis.

Production of NR4a1 is regulated not only by VEGFa, but also by small vascular permeabilizing agents, such as histamine, platelet-activating factor, and serotonin. Unlike VEGFa that connects to VEGFR2 receptor protein Tyr kinase, histamine, serotonin, and PAF tether to their cognate G-protein-coupled receptors. Like VEGF, NR4a1 acts, at least partly, by increasing indirectly the NOS3 synthesis and decreasing indirectly that of several endothelial cell junction constituents such as cadherin-5.

9.2 Endothelium Types

9.2.1 High Endothelial Venules

High endothelial venules are specialized postcapillary venules of lymphoid tissues, such as lymph nodes and intestine-associated Peyer's patches [863]. These venules serve as entry of blood-convected lymphocytes into lymphoid organs. Lymphocytes indeed migrate across high endothelial venules for immune surveillance.

High endothelial venules are lined by quasi-cuboidal (plump) endothelial cells rather than flat, thin (except in the nucleus region) ones. These endothelial cells allow lymphocyte extravasation into tissues, using concerted action of integrins, selectins, and chemokines. Lymphocyte capture is initiated by L-selectin and $\alpha_4\beta_7$-integrin. Binding of CCL21 chemokine to its CCR7 receptor activates $\alpha_L\beta_2$-integrin that mediates lymphocyte arrest and $\alpha_4\beta_1$-integrins. Intercellular adhesion molecule-1 and CCL21 chemokine are upregulated during fever [864].

Member ENPP2 of the ectonucleotide pyrophosphatase/phosphodiesterase family, or autotaxin (Atx), is a secreted enzyme with lysophospholipase-D activity. It converts lysophosphatidylcholine into lysophophatidic acid. It is released by endothelial cells of high endothelial venules. It facilitates the entry of lymphocytes into secondary lymphoid organs. After chemokine activation, naive lymphocytes that search for antigens exit blood stream to lymph node by binding to autotaxin via activated $\alpha_4\beta_1$-integrin on lymphocytes. Extracellular autotaxin produces lysophophatidic acid that enhances lymphocyte motility by binding to cognate G-protein-coupled receptors. Lysophophatidic acid stimulates actin polymerization in primary lymphocytes to trigger their motility [865].

In the spleen, lymphocytes exit blood stream through terminal arterioles that open into the marginal sinus of the spleen. Autotaxin is strongly expressed in central arterioles and spleen marginal zones.

Autotaxin also synthesizes sphingosine 1-phosphate that control lymphocyte egress from various lymphoid tissues.

9.2.2 *Lymphatic Endothelium*

The lymphatic system has many functions. It conveys immunocytes. It drains fluids from the interstitial spaces. It transports absorbed dietary lipids to metabolism sites. Lymphatic vessels develop from specialized venous endothelial cells. During embryogenesis, subpopulations of venous endothelial cells form lymphatic sacs in the region of the primitive subclavian, inferior vena cava, and iliac veins. These sacs then divide to create lymphatic networks. Mediators of embryonic lymphatic development include the transcription factor Prospero-related protein-1 (Prox1) and VEGFc. Separation of lymphatic and blood microvasculature in the intestinal mucosa continues beyond fetal life.

Fasting-induced adipose factor (FIAF),[17] produced by enterocytes of the small intestine, is required for separation between postnatal intestinal lymphatic and blood vessels [866]. Signaling by FIAF implicates Prox1 effector in the postnatal intestinal lymphatic endothelium. However, lymphaticovenous partitioning also uses Prox1-independent pathways.

9.2.3 *Endothelial Fenestrae*

Fenestrae (caliber 60–70 nm) exist in the capillary endothelium, where large molecule exchanges occur between flowing blood and perfused tissues. Fenestrae hence increase the endothelium permeability for water, electrolytes, and small macromolecules, especially in the nephron glomerulus, gastrointestinal tract, liver sinusoids, ocular choriocapillaris, and endocrine glands.

Fenestrae form an array characterized by regular spacing, the so-called sieve plate. The fenestra density in sieve plates can reach about 30 fenestrae per μm^2. The fenestra pore is made of 5- to 6-nm openings delineated by a diaphragm with radial fibrils from a central node. Fenestrae are composed of the diaphragm protein PV1, which is required for fenestra formation, as well as actin-filament remodeling [867].

9.3 Endothelial Progenitor Cells

The uninterrupted endothelial lining is maintained and regenerated by both proliferation of endothelial cells and migration of (blood) circulating cells and undifferentiated cells from the subendothelial space. In particular, some blood mononuclear CD34+ cells can acquire endothelial-like characteristics and can home to angiogenesis sites [868].

---

[17]Fasting-induced adipose factor is also called angiopoietin-like protein 4. It inhibits lipoprotein lipase involved in the storage of triglycerides in adipocytes. It promotes endothelial cell survival. It reduces VEGF-induced microvascular permeability.

Circulating endothelial progenitor cells can be incorporated into ischemic tissues as well as at the border of infarcted regions and facilitate neovascularization, as they secrete paracrine factors [868]. They release growth factors and chemokines that stimulate endothelial regeneration by resident endothelial cells. They also contribute to the endothelial lining of microvessels during wound healing.

Circulating endothelial progenitor cells may be involved in re-endothelialization after mechanical vascular injury that prevents intimal hyperplasia. Mobilization of endothelial progenitor cells EPCs may contribute to endothelial regeneration promoted by estrogen, exercise, heme oxygenase-1, and statins [868]. However, homing of circulating endothelial progenitor cells is a minor factor with respect to endothelium growth from the edges of the injured region.

Mobilization, circulation, homing, and local differentiation of bone marrow-derived leukocytes intervene in evolution of most arterial diseases characterized by inflammation.

In adults, the name "endothelial progenitor cell" applied to different cell types. *Early endothelial progenitor cells*, or endothelial-like cells, have a myelomonocytic origin. They have paracrine effects in neovascularization in vivo. *Late endothelial progenitor cells*, blood-outgrowth endothelial cells, or endothelial colony forming cells, are highly proliferative. They also participate in neovascularization. Few circulating late endothelial progenitors are CD31+, CD34+, CD146+, PTPRc−, prominin-1−[18] cells similar to mature circulating and resident endothelial cells [868].

Endothelial progenitor cells or angioblasts can be isolated from blood [869]. Circulating endothelial progenitor cells (CEPC) differ from mature circulating endothelial cells (CEC) by their markers (Table 9.4). Endothelial progenitors and hematopoietic stem cells share numerous surface markers, but the former also express VEGFR2 receptor.

Bone marrow is a reservoir of stem cells that can regenerate the bone marrow as well as other tissues. Bone marrow is constituted by different types of stem and progenitor cells, such as multipotent adult progenitor cells that can, at least in vitro, generate many cell types as well as mesenchymal stem cells, and hemangioblasts. However, the contribution of bone marrow-derived progenitor cells to repair of vascular damages is rather limited.

9.3.1 Hemangioblast

Hemangioblast is a common precursor for hematopoietic and endothelial cells under the influence of growth factors. Hematopoietic stem cells give birth to lymphoid

[18]Pentaspan transmembrane glycoprotein prominin-1 is also called macular dystrophy, retinal protein McDR2, Stargardt disease protein StgD4, hematopoietic stem cell antigen CD133, and AC133.

Table 9.4. Identification markers of endothelial progenitor (EPC) and circulating endothelial cells (CEC; Source: [844]; MCAM: melanoma cell adhesion molecule [a.k.a. cell-surface glycoprotein MUC18 and CD146]; PTPRc: protein Tyr phosphatase, receptor type C [a.k.a. CD45]; SCA1: stem cell antigen-1 [a.k.a. lymphocyte antigen-6A]; SCFR: stem cell factor receptor; vWF: von Willenbrand factor). CD14 is a pattern-recognition receptor. CD133 is the prototypic member of pentaspan transmembrane glycoproteins. $CD45^{dim}$ is a marker of endothelial progenitor cells. During maturation, the CD45 hematopoietic marker disappears to be replaced by endothelial labels.

Marker	EPC	CEC
Stem cell markers		
CD34	2+	±
CD133	2+	−
SCFR	+	−
Ataxin-1	+	−
Endothelial markers		
VEGFR2	2+	+
PECAM1	+	2+
vWF	+	2+
MCAM	+/−	2+
NOS3	+	+
Leukocyte markers		
PTPRc	dim/-	−
CD14	+/−	−

and myeloid progenitor cells;[19] vascular stem cells are precursors of endothelial progenitors and, secondarily, endothelial cells and pericytes.

Hemangioblasts are mobilized by several factors mainly via activated endothelial nitric oxide synthase and matrix metallopeptidase MMP9 produced in bone-marrow niches [870]. Nitric oxide availability rises on signaling by growth hormone and insulin-like growth factor-1. On the other hand, NO inhibitors (e.g., asymmetric dimethylarginine [ADMA]) prevent mobilization and differentiation of endothelial progenitor cells, their incorporation into endothelial tube-like structures, and formation of colony-forming units from cultured peripheral blood mononuclear cells [870]. Estrogens mobilize endothelial progenitors via estrogen receptor ERα and ERβ (a.k.a. nuclear receptors NR3a1 and NR3a2), MMP9, and NOS3 [870]. Smoking increases oxidative stress and reduces NO availability.

Hemogenic endothelial cells that line blood vessels in the embryo give rise to progenitors of blood cells in fetal liver and adult bone marrow. Hematopoietic progenitor cells can actually form clusters attached to the endothelium of the ventral wall of the abdominal aorta during embryogenesis.

[19]Lymphoid progenitor cells differentiate into B, T, and NK lymphocytes; myeloid progenitor cells into granulocytes, monocytes, erythrocytes, thrombocytes, and dendritic cells.

Table 9.5. Factors of endothelial progenitor cell (EPC) mobilization (Source: [870]).

Factor	Effect
Age	$\downarrow$ EPC mobilization, survival, activity
Estrogen	$\uparrow$ EPC concentration
Exercise	$\uparrow$ EPC concentration
Growth factors	$\uparrow$ EPC mobilization
Hypercholesterolemia	$\downarrow$ EPC proliferation, migration, survival
Hypertension	$\uparrow$ EPC proliferation
	$\downarrow$ EPC survival
Smoking	$\downarrow$ EPC density

9.3.2 Endothelial Progenitor Cell – Circulating Angiogenic Cell

Endothelial progenitor cells, or circulating angiogenic cells,[20] can reside in the bone marrow and blood as well as adventitia and endothelium [868].

Endothelial progenitor cells are activated by stimuli for tissue regeneration, such as vascular endothelial growth factor, placental growth factor, granulocyte–monocyte colony-stimulating factor (CSF2), granulocyte colony-stimulating factor (CSF3), erythropoietin,[21] angiopoietin-1, and CXCL12 chemokine, are recruited from the bone marrow into blood flow to be convected toward angiogenesis sites (Table 9.5) [870, 871].[22] Circulating endothelial progenitor cells yield protection by their innate ability to replace dysfunctional or damaged endothelium.

However, during angiogenesis, slight recruitment of bone marrow-derived cells, in particular VEGFR2+ precursors, does not contribute to vascular endothelium [872]. Bone marrow-derived cells that express platelet–endothelial cell adhesion molecule PECAM1, VEGFR1, and VEGFR2 are always stromal or perivascular cells. Perivascular hematopoietic cell populations that can produce endothelial markers are not bone marrow-derived endothelial cells.

---

[20]The term "endothelial progenitor cell" initially designated immature precursor cells capable of differentiating into mature endothelial cells in vivo. However, endothelial progenitor cells can comprise circulating angiogenic cells without an endothelial fate. Blood-derived pro-angiogenic cells with paracrine activity boosting local endothelial cells can be an adequate, although long term that avoids confusion with immature progenitor and precursor cells that differentiate into a given mature cell type [868].

[21]Receptor EpoR that stimulates proliferation of early erythroid precursors and differentiation of late precursors of the erythroid lineage localizes to endothelial cells.

[22]After endothelial injury, platelet adhesion to the vascular wall causes a cytokine-mediated release of CXCL12 chemokine, or stromal cell-derived factor SDF1, that, in turn, recruits and provokes proliferation of CD34+, CXCR4+, VEGFR1+ endothelial progenitor cells for re-endothelization [870]. Chemokine CXCL12 that targets receptor CXCR4 is a major actor of tissue engraftment of endothelial progenitor cells.

9.3.2.1 Early and Late Endothelial Progenitors

Two populations of endothelial progenitor cells exist — early and late endothelial progenitors — with distinct growth patterns and secretion modalities of angiogenic factors. Spindle-shaped early endothelial progenitors are also called early-outgrowth endothelial progenitors, early-outgrowth culture-expanded endothelial progenitor cells, endothelial cell-like cells, colony-forming unit (CFU) of endothelial cells, circulating angiogenic cells, attaching cells, and culture-modified mononuclear cells [870].

In fact, this population of endothelial progenitor cells can be subdivided into 2 classes. The first reported progenitor cell type — the colony-forming unit-Hill cells — originates from cultures of non-adherent peripheral blood mononuclear cells that are not able to form vascular structures in vivo [873]. Circulating angiogenic cells — early-outgrowth cells — are descendants of the monocyte–macrophage subset and operate in initiation of angiogenesis during wound healing and tissue remodeling [873]. Early-outgrowth endothelial progenitors express VEGFR2, PECAM1, cadherin-5, CD34 (generally at a low level), and von Willebrand factor, as well as monocytic marker CD14 (bacterial lipopolysaccharide receptor component) and panleukocytic marker PTPRc [870].[23]

Late endothelial progenitor cells are also named late-outgrowth endothelial progenitors, endothelial colony-forming cells (ECFC), and blood-derived outgrowth endothelial cells. They start to proliferate only after 2 to 3 weeks in culture [870], but possess a high proliferative capacity [873]. These cells can spontaneously form blood vessels. They express all typical properties of endothelial cells. They express endothelial markers, like VEGFR2, melanoma cell adhesion molecule (MCAM or CD146), and cadherin-5, in addition to CD34, but not hematopoietic markers, such as PTPRc and monocyte differentiation antigen CD14 [870].

Circulating endothelial progenitor cells that include cells that generate both early- and late-outgrowth endothelial progenitors may be represented by CD34+, VEGFR2+ cells [870].

Hemangioblasts do not possess receptor protein Tyr phosphatase PTPRc (or CD45), a common leukocyte antigen. Circulating endothelial progenitors are defined by markers, such as VEGFR2, hematopoietic progenitor cell glycoprotein and intercellular adhesion factor CD34, PTPRc, and pentaspan transmembrane glycoprotein prominin-1.[24] Monocytes can also provide a source of endothelial progenitors that do not proliferate, but release angiogenic growth factors.

---

[23]Molecule CD14 serves as a component of the bacterial lipopolysaccharide receptor with TLR4 and lymphocyte antigen Ly96. Endothelial cells express both CD14 and TLR4, but to a lower extent than monocytes and macrophages. Lipopolysaccharides can then bind to endothelial cells and upregulate ahesion molecules and coagulation factors.

[24]Prominin-1, a.k.a. CD133 and AC133, is a primitive hematopoietic stem cell marker that is not produced by mature endothelial cells. Both endothelial progenitors and mature endothelial cells express similar endothelial-specific markers, such as VEGFR2, TIE1, TIE2, and VE-cadherin (a.k.a. Cdh5 and CD144). Subpopulations of monocytes, macrophages, lymphocytes, and

9.4 Endothelial Cell Migration

Endothelial cells migrate during angiogenesis (Chap. 10; Vol. 2 – Chap. 6. Cell Motility). Chemokines of the CXC set enhance the migratory capacity of endothelial cells and facilitate homing of endothelial progenitor cells into ischemic tissues. Chemokine ligand CXCL12[25] promotes tubulogenesis of microvascular endothelial cells via enhanced expression of growth factors VEGF and FGF2. Hypoxia-inducible factor HIF1 primes CXCL12 synthesis in ischemic tissue for cell recruitment and homing, especially of vascular endothelial cells and circulating progenitor cells that express chemokine CXCR4 receptor,[26] thereby boosting tissue regeneration.

Chemokine CXCL12 activates the PKB–NOS3 axis as well as mitogen-activated protein kinases, such as ERK1 and ERK2, JNK, and P38MAPK, in different cell types. Activated nitric oxide synthase NOS3 then produces nitric oxide that subsequently nitrosylates (inactivates) mitogen-activated protein kinase phosphatase MKP7. Phosphatase MKP7 then cannot inhibit Jun N-terminal kinase JNK3 that binds to its adaptor β-arrestin-2 [874].[27] In bovine aortic endothelial cells, CXCL12 activates an NOS3-independent pathway that targets ERK1 and ERK2 to initiate cell migration.

9.5 Molecular Expression in the Vascular Endothelium

The vascular endothelium synthesizes numerous molecules to regulate its behavior and to respond to environmental cues, as well as to control its surrounding (Table 9.6, Fig. 9.2).

Vascular endothelial cells, like vascular smooth myocytes, possess enzymes of the cytochrome-P450 superfamily, such as epoxygenases and ω-hydroxylases that metabolize arachidonic acid released from phospholipids of the plasma membrane by activated phospholipase-A2 into vasoactive substances.

20-Hydroxyeicosatetraenoic acid (20HETE) is produced by cytochrome-P450 in vascular smooth myocytes as well as uniquely in endothelial cells of pulmonary arteries. Compound 20HETE enhances reactive oxygen species production by

---

primitive hematopoietic progenitors share markers such as VEGFR1, VEGFR2, TIE2, PECAM1, type-1 membrane glycoproteic endoglin (or CD105), and cadherin-5 with endothelial cells.

[25] A.k.a. stromal cell-derived factor SDF1. It is produced in 2 forms by alternative splicing of the same Cxcl12 gene transcript: SDF1α (or CXCL12a) and SDF1β (or CXCL12b).

[26] A.k.a. Fusin or leukocyte-expressed 7-transmembrane-domain receptor (LESTR). With its coreceptor CXCR7, it is required for vascular development.

[27] Activity of JNK1 is also suppressed after nitrosylation (Cys116) by nitric oxide.

Table 9.6. Examples of endothelial cell production. Many manufactured molecules are stored and released upon calcium stimulation. All substances stored in granules (e.g., vWF, tPA, TFPI, protein-S, and ET1 and ET3) are released by regulated exocytosis. Certain substances such as tissue plasminogen activator (tPA) are constitutively released, but acute exocytosis from small tPA-containing vesicles and caveolae is primed by compounds, such as histamine, endothelin (ET), and cytokines. Endothelial cells control hemostasis, as they produce both urokinase-like (uPA) and tissue-type plasminogen activators for fibrinolysis as well as plasmalemmal thrombomodulin (TM that links to thrombin and cleaves carboxypeptidase-B2, or thrombin-activable fibrinolysis inhibitor [TAFI], into its active form, in addition to be a cofactor that enhances the thrombin-catalyzed activation of anticoagulant protein-C), and antithrombin-3 (a proteolytic inhibitor), in addition to secretion of a specific thrombin receptor and peptidase nexin (an inhibitor that binds to thrombin and plasminogen activators). Endothelial cells also secrete matrix constituents, such as fibronectin (FN), various types of collagens, laminin (Ln), tropoelastin, von Willebrand protein (vWF), and thrombospondin (Tsp).

Structure	Proteoglycans (hyaluronic acid, DS, HS, KS) Glycoproteins (Ln, FN, Tsp, vWF) Collagens-4/8,
Survival	HETE
Growth	Growth factors (PGDF, gmCSF, gCSF, mCSF) Growth mediators (NO, 5HT, PGI2, TXA2, Ang2, ROS, TNFα) Hormones (CNP)
Motility	Chemokines (Lkt, HETE) Semaphorins, plexins
Guidance	Delta and Notch, EPH and ephrins, netrin and Unc, Slit2 and Robo4
Vasomotor	ET, NO, H_2S, PGH2, PGI2, TXA2, EDRF, HETE, EET, KLF2, adrenomedullin, ATn2, 5HT, ACh, ATP, ACE, ROS, CNP
Adhesion	Integrin, selectin, cadherin, ICAM1, ELAM1, 13HODE, TxnIP
Inflammation	His, Bdk, Ang2, NO, ROS, TNFα
Coagulation	PAI1, PGI2, TM, tPA, uPA, TFPI, NO, Protein S, TXA2, vWF, FV, FIII, PAI, heparin sulfates, ectonucleotidases
Hormone	Adiponectin, CNP, prolactin, growth hormone, placental lactogen

pulmonary arterial endothelial cells using NADPH oxidase and promotes angiogenesis. Agent 20HETE activates NADPH oxidase. It protects from apoptosis, as it activates phosphatidylinositol 3-kinase [876].

In human dermal microvascular endothelial cells, 20HETE provokes a rapid and sustained increase in superoxide synthesis by NADPH oxidase. It also raises the

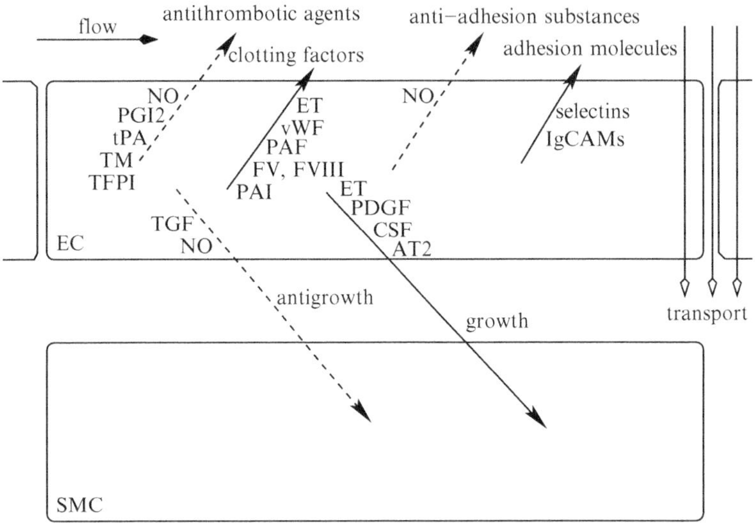

Fig. 9.2 Effects of the endothelial cell on its environment (Source: [875]). (1) Endothelial cells regulate endothelial permeability for plasma substances and adhesion of blood cells (promigration substances: selectins, IgCAMs; antimigration molecules: nitric oxide). They sequester leukocyte-interactive proteins, such as P-selectin and chemokines. They repress the synthesis of certain adhesion molecules, such as E-selectin, vascular cell adhesion molecule VCAM1 and, intercellular adhesion molecule ICAM1. (2) Endothelial cells release clotting factors and anticoagulation agents, such as nitric oxide (NO), prostacyclin (PGI2), tissue factor pathway inhibitor (TFPI), tissue plasminogen activator (tPA), and thrombomodulin (TM). Agent TFPI prevents the initiation of coagulation, as it inhibits factor VIIa–tissue factor complex. Heparan sulfate proteoglycans bind anti-thrombin-3 to inactivate thrombin. Thrombomodulin binds thrombin and diverts its activation activity from fibrinogen to protein-C that, in coordination with protein-S, inactivates several clotting components. Both NO and PGI2 synergistically impede platelet adhesion and aggregation. (3) Endothelial cells produce growth regulators, either progrowth, such as angiotensin-2 (ATn2), platelet-derived growth factor (PDGF), colony-stimulating factor (CSF), and endothelin (ET), or antigrowth molecules, such as NO and transforming growth factor-β(TGF), especially for smooth myocytes. (4) Endothelial cells synthesize vasoactive substances.

production of vascular endothelial growth factor that, in turn, upregulates hypoxia-inducible factor HIF1α via the ERK1/2 pathway [877]. Moreover, it heightens the expression of erythropoietin receptor and angiopoietin-2 via HIF1α.

Products of cytochrome-P450 epoxygenase and ω-hydroxylase as well as reactive oxygen species derived from NADPH are intracellular signal transducers for proliferation of vascular cells (via extracellular-regulated protein kinases ERK1 and ERK2) as well as angiogenesis.

In vascular endothelial cells subjected to acetylcholine, bradykinin, or shear stress, activated phospholipases produce arachidonic acid that is processed by cyclooxygenases, cytochrome-P450s, and lipoxygenases. In endothelium of some arteries, a substantial component of vasodilation depends on lipoxygenase-induced arachidonic acid metabolites. Arachidonate 15-lipoxygenase (ALOx15) synthesizes vasoactive metabolites such as 15-hydroxy (11,12)-epoxyeicosatrienoic acid that is hydrolyzed by soluble epoxide hydrolase to (11,12,15)-trihydroxyeicosatrienoic acid. Hydroxyepoxy- and trihydroxyeicosatrienoic acids are endothelium-derived hyperpolarizing factors that activate calcium-activated, small-conductance $K_{Ca}2$ channels [825]. In vascular endothelium of other arteries, arachidonate 12-lipoxygenase (ALOx12) produces 12-hydroxyeicosatetraenoic acid that relaxes smooth myocytes via calcium-activated, large-conductance $K_{Ca}1$ channel [825].

Insulin activates the PI3K pathway not only to stimulate glucose uptake, but also to promote synthesis of nitric oxide in the endothelium. Insulin hence favors endothelium-dependent relaxation of vascular smooth myocytes (Sect. 9.10).

9.5.1 *Caveolae*

Caveolae are specialized, invaginated membrane rafts, i.e., dynamic assemblies of sphingolipids and cholesterol, that contribute to vesicular transport and signaling. They indeed concentrate or segregate receptors and signaling intermediates. These nanodomains constitute platforms on which kinases and phosphatases can operate. Caveola-mediated endocytosis differs from other types of endocytosis by chemical sensitivities, cargos, adaptors, and signaling proteins (Vol. 1 – Chap. 9. Intracellular Transport).

The organization and function of caveolae depend on coat caveolins and adaptor cavins (cavin-1–cavin-4) that promote membrane remodeling and caveolin-derived structure transfer. Caveolae can form transendothelial channels and vesiculovacuolar organelles and cavicles.

Caveolin-1 and -2 lodge in most cell types of the cardiovascular apparatus, whereas Cav3 is expressed primarily in myocytes (cardiac, skeletal, and smooth myocytes). Caveolin-1 or -3 is needed for the formation of caveolae, but not Cav2. Caveolin-1 resides in the plasma membrane and Golgi body. Exocytosis from the Golgi body to the plasma membrane is regulated by amyloid-β protein and insulin [878]. In addition, Cav1 endocytosis is controlled by Na^+–K^+ ATPase. Caveolin-2 supports caveola assembly via its hetero-oligomerization with Cav1.

9.5.1.1 Regulators of Caveolae

Cavins (cavin-1–cavin-4) regulate caveolin density as well as caveola morphology [878].[28] Cavins possess leucine zipper-like domains for between-protein

[28]Cavin-1 is also known as polymerase transcript release factor (PTRF); cavin-2 serum deprivation protein response (SDPR); cavin-3 as protein kinase-Cδ-binding protein (PKCδBP) and serum deprivation response-related gene product (SDPR) that binds to C-kinase (SRBC); and cavin-4 as muscle-restricted coiled-coil protein (MURC).

interactions, PEST domains for protein turnover, and phosphorylation motifs. They bind phosphatidylserine. They are phosphorylated upon insulin stimulation [878].

Cavin-1 colocalizes with Cav1 in membrane rafts, where they interact with the cell cytoskeleton, especially cortical microtubules and actin filaments [878]. Cavin-1 sequesters caveolins into caveolae. Cavin-2 binds to and recruits cavin-1 to the plasma membrane. The cavin-1–cavin-2 complex stabilizes Cav1-containing structures. Cavin-2 is a PKC substrate that is involved in PKC compartmentation in caveolae. Cavin-3 supports budding and formation of cavicles. Cavin-4 is predominantly in myocytes.

Dynamin-2 binds to Cav1 during Cav1-mediated endocytosis. Insulin receptor as well as Src kinase phosphorylate Cav1 (Tyr14), in response to growth-factor stimulation and cellular stress. Caveolin-1^P interacts with C-terminal Src kinase to preclude Src action [878].

9.5.1.2 Caveolae in Endothelium Functions

Caveolin-1 regulates microvascular permeability, Ca^{2+} influx, vascular remodeling, and angiogenesis. Many G-protein-coupled receptors, receptor and cytosolic Tyr kinases (e.g., EGFR), small GTPases, and components of the MAPK module (e.g., Raf, ERK1, and ERK2) interact with Cav1 and reside in caveolae [878]. In particular, caveolin-1 sequesters ERK1, and ERK2, thereby impeding the activity of this pathway.

In addition, caveolin-1 contributes to integrin signaling, particularly β_1-integrin localization to caveolae upon IGF stimulation. Caveolin-1 also tethers cyclo-oxygenase-2 to the endoplasmic reticulum, hence promoting its degradation.

In vascular endothelial cells, caveolae participate in the regulation of the vascular tone, as caveolin-1 in caveolae inhibits endothelial nitric oxide synthase (NOS3) activity, but not caveolin-1 in non-caveolar membrane rafts [878]. Caveolin-1 and Ca^{2+} ions antagonistically regulate NOS3 in the microcirculation. However, caveolin-1 participates in Ca^{2+} import into endothelial cells.[29]

[29] Many processes rely on Ca^{2+} ions, in particular inflammation and angiogenesis. Ca^{2+} signaling can comprise 3 phases [878]: (1) rapid, initial Ca^{2+} influx by release from endoplasmic reticulum store that can result from stimulated G-protein-coupled receptors and receptor protein Tyr kinases via phospholipase-C activation and inositol trisphosphate; (2) sustained entry of extracellular Ca^{2+} ions through Ca^{2+} release-activated Ca^{2+} channels and receptor-operated Ca^{2+} channels; and (3) tonic Ca^{2+} entry that permits long-term Ca^{2+} signal propagation and replenishes intracellular Ca^{2+} stores.

9.5.2 Nuclear Receptors

Drug catabolism ensures the body's protection against toxics. The nuclear receptor NR1i2[30] controls drug clearance via transcription of genes involved in drug transport (multidrug resistance transporter MDR1) and metabolism (conjugation [glutathione transferase] and oxidation [cytochrome-P450 CyP2b, CyP2c, and CyP3a, and glutathione peroxidase]). In human vascular endothelial and smooth muscle cells, activated NR1i2 stimulates expression of the Mdr1, CYP3A, CYP2B, and CYP2C genes as well as increases cellular level of glutathione and activity of glutathione peroxidase to protect the vasculature, in particular against oxidative stress [879].

9.5.3 Examples of Endothelial Receptors

9.5.3.1 Endothelial Protein-C Coreceptor

Endothelial cells express endothelial protein-C receptor (EPCR) to regulate the protein-C anticoagulant and anti-inflammatory pathway via the thrombin–thrombo modulin complex. Cytoprotective activated protein-C can also upregulate anti-apoptotic and anti-inflammatory gene expression.

Vascular smooth myocytes also express EPCR [880]. In smooth myocytes, activated protein-C induces phosphorylation of extracellular signal-regulated kinases ERK1 and ERK2 via peptidase-activated PAR_1 receptor. Effect of activated protein-C is significantly enhanced in the presence of thrombin.

However, thrombin does not engage EPCR coreceptor; it cleaves completely PAR_1 that is subsequently internalized and degraded [881]. Thrombin binds to PAR_1 and activates preferentially subunits of the Gq and G12/13 subclasses of G protein heterotrimer to initiate Ca^{2+} mobilization and PKC activation on the one hand and to disrupt the endothelial barrier using monomeric RhoA GTase, on the other.

On the other hand, activated protein-C connects to EPCR and causes a limited cleavage of PAR_1 receptor [881]. Moreover, activated protein-C and thrombin stimulate Rac1 and RhoA GTPases using G protein and β-arrestin, respectively [882]. Therefore, activated protein-C operates as a *biased agonist* that activate β-arrestin signaling.[31] Activated protein-C releases β-arrestin-2 from PAR_1, which then

[30] A.k.a. pregnane X receptor (PXR) and steroid and xenobiotic receptor.

[31] Biased agonists activate β-arrestin signaling. Ubiquitous adaptors β-arrestin-1 and -2 desensitize G-protein-coupled receptors, thereby blocking G-protein signaling as well as supporting receptor endocytosis, thus mediating G-protein-independent signaling. On the other hand, ubiquitinated β-arrestin-2 that acts as a ubiquitin ligase adaptor, stabilizes GPCR signalosomes.

Table 9.7. Activation by thrombin and APC of peptidase-activated receptor PAR_1 and opposite effects (Source: [882]; Arr: arrestin; Dvl: Disheveled; EPCR: endothelial protein-C receptor).

Messenger	Effect	Pathway
Thrombin	Endothelial barrier disruption	PAR_1–$G\alpha_q$–Ca^{2+}–PKC PAR_1–$G\alpha_{12/13}$–RhoA
Activated protein-C	Endothelial barrier maintenance	PAR_1–EPCR–Cav–βArr– –Dvl2–Rac1

interacts with Disheveled-2, a scaffold and mediator of the Wnt–Frizzled signaling, which polymerizes and protect the endothelial barrier.[32]

9.5.3.2 Peptidase-Activated Receptors

Peptidase-activated receptor PAR_1 on endothelial cells contributes to cell responses that trigger or prevent blood coagulation and ensures cell protection [882]. Both thrombin and activated protein-C stimulate PAR_1, but cause opposite effects.

Serine peptidase thrombin binds to PAR_1 and cleaves its extracellular domain to form a tethered ligand to activate PAR_1-mediated inflammation and increase endothelial barrier permeability.

The anticoagulant peptidase activated protein-C stimulates a subpopulation of PAR_1 that colocalizes with their coreceptor, endothelial protein-C receptor, PAR_1, in membrane nanodomains enriched in caveolin, to promote endothelial barrier protection (Table 9.7). Receptor PAR_1 localizes to caveolae connected to β-arrestin in unstimulated cells. The PAR_1–EPCR couple supports cytoprotection. Activated protein-C recruits and activates Disheveled-2 [882].

Hypoxia primes an angiogenic phenotype in endothelial cells. Hypoxic cancer cells upregulates in endothelial cells protease-activated receptor PAR_2 and pro-angiogenic heparin-binding EGF-like growth factor (HBEGF) and increases phosphorylation of ERK1 and ERK2 [883]. Tissue factor that triggers PAR signaling is induced by hypoxia in several types of cancer cells; however, tissue factor remains undetectable in hypoxic endothelial cells, although several stimuli (e.g., shear stress and growth factors) transiently cause induce tissue factor production TF in endothelial cells.

[32]Disheveled–β-arrestin intervenes in Wnt5a–Rac1 signaling. β-Arrestin is required for internalization of Frizzled-4 by Wnt5a ligand.

9.5.3.3 I-Peptide Receptor

Endothelial cells express carbohydrate I-peptide receptor (IPR) that is responsible for lung colonization of cancer cells [884]. Receptor IPR corresponds to alternatively spliced variants of Arg/Ser-rich splicing factors (SFRS1, SFRS2, SFRS5, and SFRS7). Like many carbohydrate-binding proteins of the C-type lectin family, SFRS protein that is not a C-type lectin requires calcium to bind to carbohydrates, especially fucosylated oligosaccharides.

9.5.3.4 FGFRs

Fibroblast growth factor FGF2 provokes endothelial cell migration and angiogenesis via 2 types of receptors: high-affinity receptor protein Tyr kinases such as FGFR1 and heparan sulfate proteoglycans such as transmembrane syndecan-4, a FGFR1 coreceptor. The latter determines the kinetics and magnitude of FGF2-induced MAPK signaling (ERK1 and ERK2) by promoting the macropinocytosis of the FGFR1–syndecan-4–FGF2 complex using RhoG and Rab5 GTPases [885]. Small RhoG GTPase promotes membrane ruffling and macropinocytosis. Monomeric Rab5 GTPase is involved in early signaling endosomes. Signaling from FGFR1 initiates MAPK activation; syndecan-4-dependent FGFR1 macropinocytosis modulates the kinetics of MAPK activation.

9.5.3.5 VEGFRs

Angiogenic vascular endothelial growth factor VEGFa is also a potent vascular permeabilizing factor, whereas endothelial growth factors, such as FGF2 and PDGF, do not affect vascular permeability.

The VEGFR receptors participate in signaling pathways that control and coordinate transcriptional, post-transcriptional, and post-translational mechanisms involved in the control of endothelial cell behavior during angiogenic sprouting, branching with endothelial leading *tip* and trailing *stalk cells*, and tubulogenesis. Stalk cells support the extension of sprouting vessels, generate the trunk of new vessels, build a vascular lumen in growing vessels, and maintain connection with the parental vessel.

During angiogenesis, signaling launched by pro-angiogenic ligands (i.e., autocrine VEGFa and VEGFc regulators) of VEGFR2 and VEGFR3 select tip cells for sprouting [886]. Angiogenic sprouting is guided by gradients of pro-angiogenic growth factors and various guidance cues, such as semaphorins and ephrins. The navigators Uncoordinated-5 homolog Unc5b (receptor of secreted netrins), Roundabout homolog Robo4, plexin-D1, neuropilins, ephrin-B2, and EPHb4 receptor are major conductors of angiogenesis.

On the other hand, the decoy receptor VEGFR1 limits tip cell formation; Delta-like ligand DLL4 of Notch receptor prevents tip cell fate in endothelial cells adjacent to tip cells [886]. In addition, Roundabout homolog Robo4, and Wnt signaling in stalk cells repress tip cell behavior to maintain the hierarchical organization of sprouting endothelial cells.

Vascular tubulogenesis is initiated by the acquisition of the apicobasal polarity of endothelial cells that is regulated by cell–matrix interactions and signaling via partitioning defective protein PAR3 and VEGFR receptor [886].

9.5.3.6 TGFβ Receptors

Endothelial cells produce receptor protein Ser/Thr kinases TβR2, TβR1 (or ALK5), and ALK1 activin receptor-like kinase. Unlike inhibition in quiescent endothelium of cadherin-5 on the VEGF–ERK1/2 pathway, cadherin-5 at adherens junctions that is stimulated by TGFβ recruit TβR2 and enhance the assembly of TβR2–TGFβR1 heteromers [887].

Activated TGFβR1 phosphorylates receptor-associated SMAD proteins (rSMAD). Receptors ALK1 and ALK5 phosphorylate SMAD1, SMAD5, and SMAD8 and SMAD2 and SMAD3, respectively. Cadherin-5 is needed for potent, sustained phosphorylation of SMAD1/5 and SMAD2/3 by TβR1 receptor. Phosphorylated cytoplasmic rSMAD dissociates from the receptor and complexes with SMAD4 mediator. The rSMAD–SMAD4 complex accumulates in the nucleus and interacts with specific DNA-binding proteins to regulate transcription. Antiproliferative and antimigratory signaling by TGFβ thus contributes to Cdh5-dependent stabilization and remodeling of the vascular endothelium. Activation by SMAD2 and SMAD3 is indeed stronger than SMAD1 and SMAD/5 stimulation. Consequently, the TGFβ–ALK5–SMAD2/3 pathway is more efficient than TGFβ–ALK1–SMAD1/5/8 signaling that promotes cell growth and motility.

Proliferating endothelial cells synthesize a large amount of endoglin, a modulator of ALK1 and ALK5, which is bound by TGFβ. Lack in endoglin or ALK1 causes defects in cardiovascular development. In particuler, endothelial cells of the endocardium do not undergo mesenchymal transition required for their migration into the atrioventricular cushion.

Vascular endothelial cells can transform into multipotent stem-like cells using ALK2 receptor [888]. Constitutively active ALK2 (upon mutations of the Alk2 gene) in endothelial cells or exposure to its ligands TGFβ2 or BMP4 causes endothelial-to-mesenchymal transition and acquisition of a stem cell-like phenotype. These mesenchymal stem-like cells can differentiate into osteoblasts, chondrocytes, and adipocytes.[33]

[33]Mesenchymal tissues comprise bone, cartilage, muscle, and tendon.

9.5.3.7 BMPRs

Bone morphogenetic proteins link to BMPR1 and BMPR2, thereby causing phosphorylation of BMPR-specific SMADs (SMAD1, SMAD5, and SMAD8). Blood flow provokes BMP4 synthesis in endothelial cells.

Endothelial cells exposed to time-dependent shear stress have higher turnover (and proliferation) rate than those bearing steady shear stress. The SMAD proteins can modulate the cell cycle, which is controlled by cyclin-dependent protein kinases coupled to their regulatory subunits cyclins and inhibitors CKI1a and CKI1b. In endothelial cells, SMAD1 and SMAD5 can be activated from BMPR in the absence of BMP ligands to promote cell proliferation, unlike BMP4-dependent ICAM1 expression by nuclear factor-κB in the context of inflammation to support monocyte adhesion to the endothelium [889]. Time-dependent shear stress actually causes a sustained activation of SMAD1 and SMAD5 and subsequently of TOR and S6K (or ^{P70}RSK) kinases, thereby upregulating cyclin-A and downregulating CKI1a and CKI1b [889]. Shear-mediated activation of TOR and S6K is modulated by the interplay between AMPK inhibitor and PKB activator.

9.5.3.8 Neurotrophin Tyr Kinase Receptors

Neurotrophins, such as nerve growth factor, brain-derived neurotrophic factor (BDNF), and neurotrophin-3 and their high-affinity neurotrophin Tyr kinase receptors (NTRK2–NTRK3)[34] and low-affinity nerve growth factor receptor (NGFR, or TNFRSF16)[35] localize to pulmonary vasculature and airways in humans. Both ligands and receptor are produced in the same cell type to create auto-, juxta-, or paracrine effects. In addition, endothelial cells of the systemic circulation produce NTRK receptors. Moreover, BDNF circulates in blood (15–30 ng/ml).

Neurotrophins rapidly trigger (6–8 mn) nitric oxide production in pulmonary endothelial cells [890]. Neurotrophins BDNF and NT3 act predominantly via high-affinity NTRK2 and NTRK3 receptors, respectively, with partial involvement of low-affinity TNFRSF16 receptor. Upon NTRK autophosphorylation, both BDNF and NT3 increase phosphorylation of PKB that, in turn, phosphorylates NOS3 kinase. As, in smooth myocytes, neurotrophins increase intracellular Ca^{2+}

[34]Nerve growth factor preferentially binds to NTRK1, brain-derived neurotrophic factor to NTRK2, and neurotrophin-3 to NTRK3. In lungs, neurotrophins and their receptors reside in nerves, immunocytes, epithelial and smooth muscle cells, fibroblasts, and pulmonary vascular cells. In human airway smooth myocytes, BDNF and neurotrophin-3 can rapidly increase and decrease intracellular Ca^{2+} level, respectively. Both NGF and BDNF are involved in cardiovascular development and ischemia [890].

[35]Receptor TNFRSF16 is expressed in both endothelial and smooth muscle cells in mouse pulmonary arteries.

concentration that promotes vasoconstriction, their net effect on vascular tone is determined by the relative contribution of neurotrophins on endothelial cells with NO generation that elicits vasodilation.

9.5.3.9 Angiotensin-2 Receptors

Angiotensin-2 is a vasoconstrictor and progrowth factor that operates via AT_1 receptor. On the other hand, AT_2 receptor transmits signals via bradykinin and nitric oxide to cause vasodilation.

Angiotensin-2 receptors AT_2 enables increased expression of prolylcarboxypeptidase, a serine carboxypeptidase and a plasma prekallikrein activator, using PTPn6 phosphatase, activation of the prekallikrein-kallikrein axis,[36] and release of bradykinin by mouse coronary artery endothelial cells [891].

9.5.3.10 Roundabout Receptors and Slit Ligands

The Slit–Robo signaling complex is required for the development of the heart and blood vessels. Endothelial-specific Robo4 maintains vascular integrity [892]. Activation of Robo4 by Slit2 inhibits VEGF-induced migration along angiogenic stimulus gradients and vascular leak.

Slit proteins steer the migration of many cell types once bound to Robo receptors. Signaling incorporates Rho GTPase-activating proteins (RhoGAPs). Protein RhoGAP39 promotes hydrolysis of Rac^{GTP} and, less efficiently, of $CDC42^{GTP}$ [893]. Guided cell migration also intervenes in the development of the respiratory tract. In addition, Rac and possibly CDC42 act downstream of Robo in midline repulsion of brain axons.

9.5.3.11 Cytokine Receptors

Endothelial cells express IL20 receptors. Interleukin-20 of the interleukin-10 family causes proliferation of endothelial cells, thereby promoting angiogenesis. Interleukin-20 favors phosphorylation of ERK1, ERK2, P38MAPK, and JNK kinases [894].

---

[36] When the complex of high-molecular-weight kininogen (HMWK) and plasma prekallikrein binds to the endothelial membrane, prolylcarboxypeptidase (PrCP) converts plasma prekallikrein to kallikrein, which then cleaves HMWK and generates bradykinin that targets kinin B_2 and inducible B_1 receptors.

Table 9.8. Role of endothelial ion channels. Ion channel expression depends on cell culture features and growth conditions, as well as endothelial cell type. Endothelial cells express numerous ion channels that intervene in many processes between endothelial cells and between endothelial cells and smooth myocytes. Two major functions of ion channels are the control of calcium influx and the modulation of membrane potential. The membrane potential provides the electrochemical driving force for Ca^{2+} influx. In endothelial cells, Ca^{2+}-dependent K^+ currents are modulated by $\alpha_V\beta_3$-integrin attachment to fibronectin. Potassium outflux through K_{IR} channels enhances cell migration.

Function	Examples of produced mediators
Synthesis of vasoactive substances	NO, ET, PGi2, TxA2
Production of hemostatic factors	vWF, PAF, TPFI
Secretion of fibrinolytic factors	tPA
Generation of growth factors	EGF, FGF2/4, HGF, IGF1/2, PDGF, VEGF
Molecular transport	Molecular machinery
Mechanotransduction	NOS activation
Cell proliferation	NO
Cell extravasation	Cell adhesion molecules
Endothelial cell–matrix adhesion	Cell adhesion molecules
Endothelial cell migration	Chemokines
Angiogenesis	Angiomotin, angiopoietin, angiokines, growth factors, Notch, DCC, ephrin, EPH, semaphorins, Slit-2/3, SlitR, VEGFR

9.5.4 Ion Carriers of Endothelial Cell

Ion channels abound in the plasma membrane of non-excitable endothelial cells [895]. They control ion influx necessary to produce vasoactive molecules, pro- and anticoagulants, growth factors, etc., that are released for storage or immediate action (Table 9.8). They are involved in the regulation of cell transport. Angiogenesis entails mediation of ion channels.

Endothelial cells undergo slow and more or less small changes in membrane potential associated with the activity of plasmalemmal ion channnels and pumps. The membrane potential of vascular endothelial cells is negative compared with the blood and tissue. The membrane potential is in general more negative in macrovascular than microvascular endothelial cells [895].[37]

Ion fluxes between endothelial cells, between endothelial and smooth muscle cells across the basement membrane and myoendothelial junctions, and between endothelial cell and migrating lymphocyte or monocyte, also occur through gap junctions. The large variability in gap junctions depends on the connexin isoforms

[37] The resting potential is mainly determined by K^+ channels, the expression of which varies greatly among endothelial cell types. Vasoactive stimuli, such as acetylcholine, bradykinin, and histamine, can cause membrane hyperpolarization.

Table 9.9. Calcium channels and pumps of endothelial cells (Source: [895]). (**Part 1**) Non-selective cation channels (NSC). Endothelial cells respond to most stimuli with a calcium influx. The depolarizing action of Ca^{2+} influx is compensated by activation of Ca^{2+}-dependent K^+ channels. In addition, non-selective cation channels, permeable to Ca^{2+} or not, can generate a negative feedback on calcium influx by membrane depolarization.

Ion channels	Features
Purinoceptor	ATP and shear stress-induced
($P2X_4$)	Ca^{2+} entry
Superoxide-gated NSC	Na^+, K^+, and Ca^{2+} permeability
Receptor-activated cation channels	Activation by PLC, IP_3, and Ca^{2+}
(Ca^{2+}-permeable; RACC)	(positive feedback)
(TRP3/4/5/6 ?)	NOS3 stimulation
Amiloride-sensitive NSC	Blood–brain barrier
Cyclic nucleotide-activated NSC	Negative feedback on Ca^{2+} influx
CGN2	Activation by the NOS3–NO–cGMP axis
Hyperpolarization-activated	Blood–brain barrier
cyclic nucleotide-gated channels	NO- and cGMP-dependent cell processes
(HCN; Ca^{2+} permeable)	Modulation by NO, cGMP, and cAMP

(Cx37, Cx40, and Cx43) that are expressed in endothelial cells.[38] The electrical coupling via high-conductance gap junctions between endothelial cells and smooth myocytes is particularly important in arterioles.

9.5.4.1 Calcium Channels and Calcium Signaling

Calcium influx channels are important for long-lasting responses that are required for the release of various compounds (e.g., nitric oxide, platelet-activating factor, tissue factor pathway inhibitor, tissue plasminogen activator, and von Willebrand factor; Sect. 9.8; Tables 9.9 to 9.11).

Two types of Ca^{2+} signals include: (1) *calcium oscillations* and (2) *biphasic increase* in intracellular calcium level with a fast peak followed by a long-lasting plateau. Fast transient Ca^{2+} peak is caused by IP_3-induced Ca^{2+} release from intracellular stores, whereas the plateau phase results from Ca^{2+} influx from the extracellular space activated by store depletion.

Calcium oscillations are mainly due to periodic discharges of intracellular Ca^{2+} stores. They are triggered by low concentrations of ATP, acetylcholine, bradykinin, histamine, and substance-P. Biphasic increase in intracellular calcium concentration

[38]Connexin-43 is more abundant in macrovascular than in microvascular endothelia. Aortic and pulmonary arterial endothelia express all three types of connexin, whereas coronary artery endothelium expresses only Cx40 and Cx37. Gap junctions are made of same or different isoforms. (only Cx40–Cx37 and Cx43–Cx37 complexes).

Table 9.10. Calcium channels and pumps of endothelial cells (Source: [895]). (**Part 2**) Transient receptor potential channels (TRP). Channels TRPC1 and TRPC6 are highly expressed by endothelial cells, whereas TRPC3, TRPC4, and TRPC7 are weakly produced. Channels TRPC1, TRPC4, and TRPC5 are store-operated channels, whereas receptor-operated channels TRPC3 and TRPC6 are activated independently of store depletion. Multiple heterologous combinations of TRPCs (e.g., TRPC1–TRPC4, TRPC1–TRPC5, and TRPC3–TRPC6) generate tetrameric channels. Both TRPV1 and TRPV4 are involved in endogenous cannabinoid-dependent vasorelaxation. Channel TRPC4 participates in ATP-primed vasoconstriction and acetylcholine-induced vasorelaxation.

Ion channels	Features
(TRPC1–TRPC7; TRPV1/2/4; TRPM1–TRPM8 except TRPM5, and TRPP1/2 are expressed in endothelial cells)	
TRP	Variable expression among EC type Ca^{2+} influx (except TRPM4)
TRPP1–TRPP2 complex	Mechanosensing
TRPC1/6, TRPV2, TRPM4	Stretch (pressure) activation
TRPV4	Shear activation (stretch-insensitive)
TRPC1–TRPC7	Regulation by GPCRs and RTKs
TRPC1/4/6, TRPV1	Control of vascular permeability
TRPC4, TRPV1/4, TRPP1/2, TRPM4	Regulation of vascular tone
TRPC4	Hypoxia-induced vascular remodeling
TRPC3/4, TRPM2/7	Oxidative stress response
TRPC1/4/6, TRPV1, TRPM6/7	Angiogenesis
TRPM6/7	Mg^{2+} transport (EC proliferation)
TRPV1–4, TRPM8	Temperature sensitivity

is primed by higher concentrations of these chemical stimuli. Calcium oscillations are frequently accompanied by activation of $K_{Ca}1$ channels (BK) that provoke oscillatory changes in membrane potential. They require non-selective cation channels, such as TRP channels,[39] Na^+–Ca^{2+} exchangers, and SERCA pumps.

Vasoactive agents generate Ca^{2+} waves that travel through the cell at a speed of 5 to 60 μm/s according to stimulus concentration and cellular region [895]. These waves spread through gap junctions.

Members of the transient receptor potential (TRP) family,[40] may be involved. Short canonical TRP channels (STRPC) include: (1) receptor-activated cation channels (RACC) that are commonly activated by phospholipase-C;

[39]Channel TRPC4 is markedly required in ATP- and acetylcholine-induced Ca^{2+} influx in aortic endothelial cells and absolutely needed in thrombin-induced Ca^{2+} influx in pulmonary artery endothelial cells [896].

[40]The family of canonical TRP channels (TRPC) comprises different subsets, such as STRPCs (short), LTRPCs (long), and OTRPCs (osmotic), among others (Vol. 3 – Chap. 2. Membrane Ion Carriers). The OTRPC channels respond to changes in osmolarity. The STRPC channels contain several calmodulin-binding regions. They are activated by stimulated PLC-coupled receptors.

Table 9.11. Calcium channels and pumps of endothelial cells (Source: [895]). (**Part 3**) Store-operated (SOC) and other channels. Activated VEGFR2 triggers activation of Ca^{2+}-release-activated Ca^{2+}-channel (CRAC) channels.

Ion carriers	Features
Store-operated channels (SOC)	
Ligand-gated Ca^{2+}-selective channels CRAC	Activation by IP_3, ATP, and bradykinin Smaller Ca^{2+} selectivity and higher conductance than usual CRAC
Ryanodine receptors	Ca^{2+} influx
Na^+–Ca^{2+} exchanger	Ca^{2+} entry Modulator (Ca^{2+} efflux)
Plasma membrane Ca^{2+} pump	Modulator (Ca^{2+} efflux)
Sarco(endo)plasmic reticulum Ca^{2+} pump (SERCA3)	Ca^{2+} efflux Modulator of plateau phase of Ca^{2+} influx
Mitochondrial Ca^{2+} uniporters	Modulator (Ca^{2+} efflux)
Mechanosensitive Ca^{2+}-permeable channels Shear stress-activated NSC	Na^+, K^+, and Ca^{2+} transport Inhibition by PKG Negative feedback by NOS3 activation Permeability: $Ca^{2+} > Na^+$

(2) non-selective cation channels (NSC);[41] and (3) Ca^{2+}-selective store-operated channels.[42] Channels TRPC1 and TRPC6 are highly expressed in endothelial cells, but TRPC3, TRPC4, and TRPC7 are weakly produced. Channels TRPC1,

[41]Calcium influx via Ca^{2+}-permeable non-selective cation channels activated by vasoactive agonists, depends on inositol trisphosphate. Constitutively open, non-selective Ca^{2+}-permeable outwardly rectifying cation channels exist in both luminal and abluminal membranes of endocardial endothelial cells. Superoxide-gated non-selective cation channels exist. Cyclic nucleotide-gated NSCs bind to either cAMP or cGMP messenger.

[42]Store-operated channels (SOC), or capacitative Ca^{2+} entry (CCE) channels, modulate Ca^{2+} entry (in the cytosol). Capacitative and non-capacitative Ca^{2+} entry (NCCE) involve different types of Ca^{2+}-permeable channels. Channel TRPC1 contributes to CCE process. Platelet activation relies on an increase in intracellular Ca^{2+} concentration following Ca^{2+} release from intracellular Ca^{2+} stores and subsequent store-operated Ca^{2+} entry as well as Ca^{2+} entry, independent of Ca^{2+} store depletion, through TRPC6, especially upon thrombin stimulation. Filled stores prevent Ca^{2+} influx through store-operated Ca^{2+} channels, whereas empty stores promote influx of extracellular Ca^{2+}. In vascular smooth myocytes, CCE channel contributes to vasoconstrictor and mitogenic effects of vasoactive hormones. In endothelial cells, highly Ca^{2+}-selective agonist-activated channels are gated by ATP, bradykinin, and IP_3 agents. Calcium-release-activated Ca^{2+} channel (CRAC) pertains to the SOC set. Growth factor VEGFe activates CRAC channels.

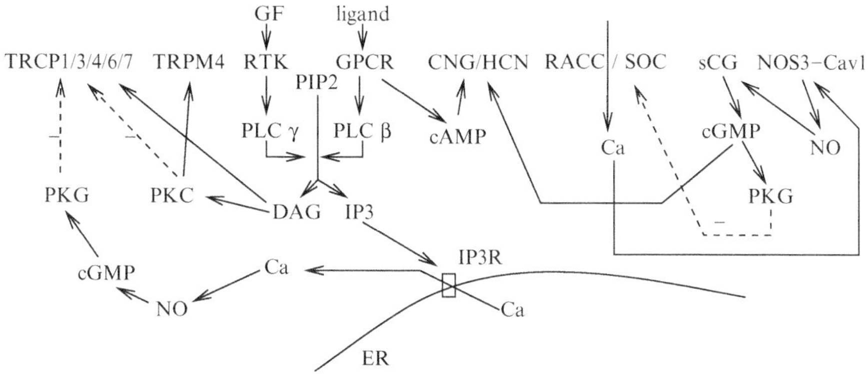

Fig. 9.3 Negative feedbacks resulting from: (1) membrane depolarization on Ca^{2+} entry caused by cyclic nucleotide-gated (CNG) and/or hyperpolarization-activated CNG K^+ (HCN) channels and (2) receptor-induced activation of endothelial TRP channels (Sources: [895, 896]). Calcium entry via receptor-activated cation channels (RACC) and/or store-operated Ca^{2+} (SOC) channels stimulates endothelial nitric oxide synthase (NOS3). Under resting conditions, NOS3 is inactivated due to its binding to caveolin-1. Nitric oxide activates soluble guanylate cyclase (sGC) that produces cGMP messenger. Liganded G-protein-coupled receptors (GPCR) generate cAMP via Gs and adenylate cyclase. Messengers cGMP and cAMP activate CNG and/or HCN channels that cause membrane depolarization, which exerts a negative feedback on RACC and SOC channels. In addition, PKG inhibits RACC and SOC channels. Endothelial TRP channels can be activated by different stimuli to regulate vascular permeability, modulate vascular tone, and cell proliferation and apoptosis. Ligands bind to GPCRs (ATP, bradykinin, acetylcholine, etc.) or receptor protein Tyr kinases (RTK; growth factors) to activate PLCβ or PLCγ that produce DAG and IP_3 agents. Messenger IP_3 binds to IP_3R on the membrane of the endoplasmic reticulum to release Ca^{2+} from its store. Messenger DAG activates TRPC3, TRPC6, and TRPC7, as well as PKC. PKC inhibits TRPC3, TRPC4, and TRPC6, but stimulates TRPC1 and TRPM4. The NO–cGMP pathway primed by augmented cytosolic calcium level stimulates PKG that inactivates TRPC3, TRPC6, and TRPC7, yielding a negative feedback.

TRPC4, and TRPC5 are store-operated channels, whereas receptor-operated channels TRPC3 and TRPC6 are activated independently of store depletion. Multiple heterologous combinations of TRPCs generate tetrameric channels (e.g., TRPC1–TRPC4, TRPC1–TRPC5, and TRPC3–TRPC6 complexes). In bovine aortic endothelial cells, glucose enhances synthesis of canonical transient receptor potential TRPC1 channel, but not that of TRPC3, TRPC4, and TRPC6 [897].

Endothelial Ca^{2+}-permeable receptor-activated cation channels, activated by an increase in $[Ca^{2+}]_i$, provide a positive feedback on their own activation, whereas negative feedbacks are provided by mechanosensitive Ca^{2+}-permeable and/or cyclic nucleotide-gated channels via NO and non-selective cation channels (Fig. 9.3). Ligand-gated Ca^{2+} channels, such as cyclic nucleotide-gated channels (CNGC),[43]

[43]Cyclic nucleotide-gated channels are non-selective cation channels that open on binding to cyclic nucleotides, cyclic adenosine monophosphate (cAMP) and cyclic guanosine monophosphate (cGMP). Although CNGC activity exhibits a small voltage dependence, CNGCs belong to

P2X receptors,[44] and probably Na^+–Ca^{2+} exchangers, provide alternative routes for Ca^{2+} entry in endothelial cell. Voltage-gated Ca^{2+} channel is a potential candidate for a sustained Ca^{2+} influx. Stretch-sensitive Ca^{2+} channel is inhibited by PKG kinase. Calcium pumps PMCA and SERCA are responsible for efflux and intracellular sequestration.

9.5.4.2 Calcium-Gated Potassium Channels

Certain autacoids that are released in response to chemical or mechanical stimuli prime the endothelium-derived hyperpolarizing mechanism that contributes to endothelial control of vascular caliber, particularly in small resistive arteries. This process not only raises the vascular lumen, but also spread the electrochemical signal to coordinate cell behavior along the vessel length (Sect. 9.10).

Potassium influx can hyperpolarize the membrane potential in endothelial cells or smooth myocytes that are coupled via myoendothelial gap junctions. Electrochemical signals that originate from a cell type spread in the blood vessel wall in all directions via gap junctions. Calcium-activated, large-conductance K^+ (BK) channels hyperpolarize vascular smooth myocytes, thus decreasing vascular tone. Calcium-activated small- (SK) and intermediate-conductance (IK) K^+ channels control NO synthesis in vascular endothelial cells. Furthermore, they cause depolymerization of cortical actin cytoskeleton. On the other hand, activated epithelial sodium channels (ENaC) depolarize the membrane potential and provoke polymerization of ^{G}actin monomers in the cell cortex.

Endothelium-derived hyperpolarization mediates vasodilation in the skeletal microcirculation generated by acetylcholine. Acetylcholine can then test the respective contribution of endothelial Ca^{2+}-dependent channels $K_{Ca}3.1$ (IK) and $K_{Ca}2.3$ (SK) in smooth myocyte relaxation by endothelium-derived hyperpolarization in arterioles [898]. Channel IK is mainly involved in endothelium-derived hyperpolarization, as blockade of SK channel does not have significant effect, although it participates in vasodilation in the absence of IK channel. Moreover, $K_{Ca}3.1$ can not only initiate a response, but also contribute to signal propagation, although it is not mandatory.

---

the superfamily of voltage-gated ion channels. CNGC ligand sensitivity and selectivity, ion permeation, and gating are determined by the subunit composition of the channel complex. CNGC activity is modulated by Ca^{2+}–calmodulin and phosphorylation. Channel CNGC1 is found in arteries. Nitric oxide-induced cGMP can activate CNGC2 that exerts a negative feedback on Ca^{2+} entry.

[44]Nucleotide (purinergic) ATP-gated receptor–channels (P2X family) are involved in mechanical stress-induced Ca^{2+} influx.

9.5.4.3 Voltage-Gated Ion Channels

Endothelial cells produce voltage-gated ion channels [895]. Voltage-gated Na^+ channels are found in endothelia of arteries, veins, and microvessels. Voltage-gated Ca^{2+} channels localize to capillary endothelium. They belong to Ca_V1 and Ca_V3 channel family, as well as voltage-dependent, tiny-conductance, long-opening, SB (sensitive to calcium channel agonist BAY K8644) Ca^{2+} channels.[45] Channel $Ca_V2.3$ is activated by a long-lasting depolarization. Hyperpolarization-activated, non-selective cation (K^+ and Na^+) channels are modulated by nitric oxide. Voltage-dependent, transient outward A-type K^+ ($K_V4.2$) and delayed rectifier K^+ channels have been observed.

9.5.4.4 Mechanosensitive Ion Channels

Mechanogated ion channels (MSIC) are required in mechanotransduction, as endothelial cells sense both shear and stretch. Mechanosensitive ion channels can be characterized by their closed-state conformation and gating transition. Ion channel status can be dictated by a coupling with integrins, cortical cytoskeleton, and G proteins. Opening of shear- or stretch-activated ion channels leads to Ca^{2+} influx and stimulation of Ca^{2+}-dependent signaling pathways. Stretch-activated non-selective cation channels of the endocardium and microvascular endothelia increase the intracellular Ca^{2+} concentration that activates Ca^{2+}-dependent BK channels, hence hyperpolarizing the endothelial membrane [895].

Mechanosensitive inwardly rectifying K^+ (K_{IR}), Ca^{2+}-activated K^+[46] (K_{Ca}), and probably voltage-dependent K^+ (K_V) channels participate in the setting of the membrane potential, with volume-regulated anion (VRAC), Ca^{2+}-activated Cl^- (CaCl), cystic fibrosis transmembrane conductance regulator channels (CFTR), and Ca^{2+}-impermeable non-selective cation (NSC) channels (Tables 9.12 and 9.13).

Potassium channels participate in endothelium-mediated regulation of the vasomotor tone in adjoining smooth myocytes (Fig. 9.4). ATP-sensitive K^+ channels may be involved in mechanotransduction.

Volume-regulated anion channel, mainly permeable to Cl^-, is activated by shear stress. The complex made of annexin-2 and a regulatory light chain, S100 calcium-binding protein S100a10,[47] an element of the cortical cytoskeleton involved in the formation of caveolae, is required for endothelial VRAC activation.

---

[45] The SB channel generates the low-threshold, sustained current above a membrane potential of -40 mV.

[46] Different types of Ca^{2+}-activated K^+ channels include $BK_{V,Ca}$, responsible for K^+ efflux, and intermediate-conductance (IK_{Ca}), which are inwardly rectifying, and small-conductance (SK_{Ca}) channels. The $BK_{V,Ca}$ channels are targeted by the endothelium-derived nitric oxide.

[47] A.k.a. annexin-2 ligand, calpactin-1, light polypeptide (aliases CLP11, Cal1L, and Anx2Lg). It modulates aggregation and membrane fusion of annexin-2. It lacks a calcium-binding site. Annexin-2 binds to a S100a10 dimer, interacting directly with both monomers.

Table 9.12. Potassium channels of endothelial cells (Source: [895]). K^+ channels induce hyperpolarization, in particular Ca^{2+}-sensitive K^+ channels after Ca^{2+} entry in endothelial cells.

Ion carriers	Features
Inwardly rectifying K^+ channels ($K_{IR}2.1$)	Expression in the macrocirculation; Location on wetted side of endocardium; K^+ sensor (potential threshold based on extracellular K^+ level) Downregulation by reduction in intracellular ATP and hypoxia Inhibition by phosphatase PP2a, angiotensin-2, vasopressin, vasoactive intestinal polypeptide, endothelin-1, and histamine Hemodynamic stress gating
Ca^{2+}-dependent K^+ channels	EC type-dependent expression; Ca^{2+}- and voltage-gating; Fluid flow sensitivity (SK and BK)
BK	Activation by ryanodine receptor
IK	Activation by IP_3-induced Ca^{2+} release (ATP, acetylcholine, bradykinin)
SK	Voltage-independent, calmodulin-dependent
ATP-sensitive K^+ channels	Activation by low ATP cellular level; Shear stress-mediated vasodilation (?)
Stretch-activated K^+-permeable channels	Carriers for K^+ and Na^+

Voltage-activated Na^+ channels are stimulated by PKC kinase. Epithelial Na^+ channels may be connected to water transport via aquaporin-4.

Calcium-dependent Cl^- channels, outward rectifiers, require ATP agent. Endothelial cell expresses other Cl^- channels, including high-conductance Cl^- channels, which are quasi-silent in intact cells, and cAMP-stimulated CFTRs.

9.5.5 *Hormones and Vasoactive Substances*

Functions of the endothelium are affected by hemodynamics. The intracellular calcium concentration in cultured endothelial cells strongly depends on flow pattern.[48] Although oscillatory flow does not change $[Ca^{2+}]_i$, steady and pulsatile

[48]Cultured bovine aortic endothelial cells have been subjected to steady shear stress (0.02–7 N/m^2), non-zero mean sinusoidal shear stress, either with a strong mean component (4 +/− 2 N/m^2) or

Table 9.13. Chloride and sodium channels of endothelial cells (Sources: [895,896]; CFTR: cystic fibrosis transmembrane conductance regulator channel). Chloride channels prevent hyperpolarization. Volume-regulated anion channels (VRAC) are associated with RhoA activation that restricts the endothelial barrier function. They favors endothelial cell growth and angiogenesis.

Ion carriers	Features
	Chloride channels
Ca^{2+}-activated Cl^- channels (CaCC)	EC type-dependent effect on membrane potential
High-conductance Cl^- channels	Activation by β-adrenergic stimuli
Volume-regulated anion channels (VRAC)	Activation by shear and stretch; participation of annexin, caveolin-1, and RhoA; mechanosensor; role in endothelium growth and angiogenesis; decrease intracellular pH
CFTR	Reduction in VRAC and CaCC activity
Shear stress-activated Cl^- channels	Sustained membrane depolarization
	Sodium channels
Amiloride-sensitive	Microvascular endothelium
Na_V1	Shear stress sensitivity; ERK1/2 activation
Na^+–K^+ ATPase	Resting potential
Na^+–Ca^{2+} exchanger	Ca^{2+} flux

flows increase $[Ca^{2+}]_i$, with response dynamics that depend on the flow pattern [899]. The release of the substances active on vasomotor tone depends on shear stress applied on the endothelium (Sect. 9.10). Shear stress might also regulate the secretion of molecules involved in coagulation and fibrinolysis or inhibiting thrombosis (Sect. 9.8). High wall shear stress hinders leukocyte adhesion on the endothelium (Sect. 9.7). The endothelium adapts the vessel wall to the local flow pattern (Sect. 7.11).

A set of signaling messengers and mediators that involves 2 secreted molecules, PDGFbb and TGFβ1, and 4 other proteins, lamin-A, lysyl oxidase, ERK1, and ERK2, contributes to the response of cultured vascular endothelial cells according to the magnitude of applied mechanical stress [900]. Factor PDGFbb that acts on both

with a high unsteady component (2 +/− 4 N/m^2), and oscillatory shear stress (+/− 2 N/m^2), using few flow frequencies (0.4, 1, and 2 Hz).

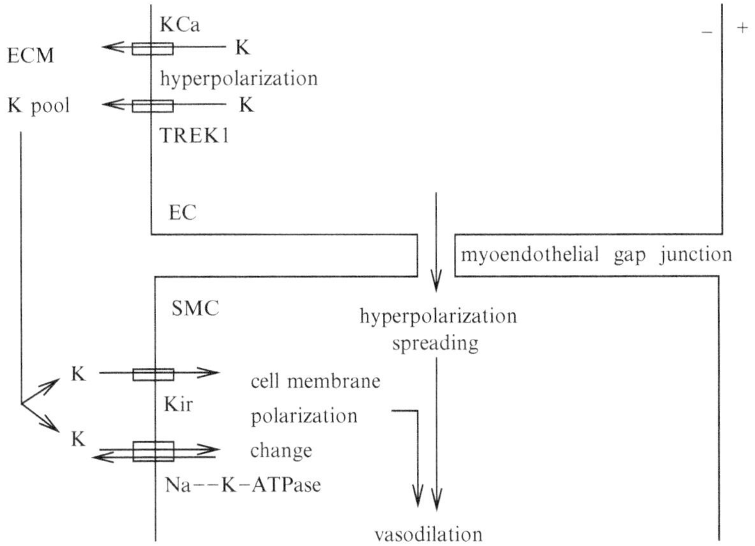

Fig. 9.4 Potassium channels in vascular endothelial cells and smooth myocytes cooperate to regulate the vasomotor tone either via the extracellular matrix (ECM) or myoendothelial junctions.

vascular endothelial cells and smooth myocytes participates in the paracrine control of smooth myocytes by endothelial cells; TGFβ that acts on endothelial cells, but not on smooth myocytes, is implicated in the feedback from smooth myocytes to endothelial cells.

9.5.5.1 Acetylcholine

Acetylcholine is synthesized and stored by choline acetyltransferase in endothelial cells of small brain vessels [901] and rat coronary arteries [904]. Acetylcholine is an NOS3 activator.

In coronary arteries, acetylcholine effect varies according to its concentration and the context (apparently normal or observed atheroma). In normal coronary arteries, vasodilation mediated by NO released from endothelial cells subjected to low acetylcholine concentrations via endothelial muscarinic M_3 receptors,[49] whereas vasoconstriction mediated by a direct action on smooth myocytes via their

[49] Muscarinic M_2 receptors localize mainly in the acrdiac atria, whereas M_3 receptors reside in vascular smooth muscle and endothelial cells. Muscarinic M_3 receptor is coupled to Gq subunit and phospholipase-C. The latter splits plasmalemmal phosphatidylinositol (4,5)-bisphosphate (PIP_2) into inositol trisphosphate and diacylglycerol. The former targets IP_3 receptors that release Ca^{2+} from the endoplamic reticulum. In smooth myocytes, Ca^{2+} primes contration. In endothelial cells, Ca^{2+} activates nitric oxide synthase.

muscarinic M_3 receptors appears at higher concentrations. At relatively high doses, its direct constrictor effect on smooth myocytes dominates over its vasodilator effect on endothelial cells.

In fact, acetylcholine initiates 2 competing responses in mouse arteries: (1) endothelium-dependent vasodilation mediated predominantly by nitric oxide and (2) endothelium-dependent vasoconstriction mediated by thromboxane-A2 [902].

In atheromatous coronary arteries, vasoconstriction and -dilation occur at low concentration and vasoconstriction happens at high concentrations. In 53% of patients with normal coronary arteriograms, chest pain, and risk factors for coronary artery disease, infusion of acetylcholine causes both constriction and dilation of proximal and distal segments of epicardial coronary arteries [903]. Vasoconstriction and -dilation coexist in different coronary arteries as well as in different segments of a given artery (acetylcholine concentration range 10^{-7}–10^{-6} mol/l). In any case, change in epicardial coronary artery caliber in response to acetylcholine does not correlate with change in intramyocardial coronary blood flow.

9.5.5.2 Serotonin and Substance-P

Serotonin and substance-P are observed in endothelial cells [905].[50] Atheromatous arteries that constrict in response to acetylcholine may dilate upon exposure of substance-P, an endothelium-dependent vasodilator, although dysfunction of the endothelium action is not restricted to muscarinic receptors, but also to other receptors such as tachykinin (substance-P) receptors [903]. In addition, the endothelial response to chemical stimuli is not related to the degree of coronary atherosclerosis.

9.5.5.3 Adrenomedullin

Adrenomedullin is widespread, but at higher concentrations in endothelial cells and vascular smooth myocytes. Adrenomedullin circulates in the plasma at picomolar levels. Elevated plasma concentration of adrenomedullin occur during pregnancy. Strongly vascularized tissues, such as the heart, lungs, and kidneys, release adrenomedullin during inflammation, hypoxia, sepsis, and cardiovascular diseases (myocardial infarction, heart failure, and atherosclerosis), particularly during hypertension.

Adrenomedullin synthesis is regulated via estrogen-responsive elements Adrenomedullin type and concentration vary according to the genetic ground, especially gender, because the adrenomedullin gene is regulated by estrogen.

[50]Levels of serotonin and substance-P are similar in endothelial cells of femoral and mesenteric arteries of rats. Substance-P is also localized in endothelium of rat coronary arteries [904].

The genetic ground then hinders or favors development of cardiac and renal damages, but in the latter case does not necessarily change the blood pressure [906].

9.5.5.4 Steroid Hormones

Estrogens favor nitric oxide production in the endothelium to protect against inflammation, as nitric oxide can induce cysteine S-nitrosylation[51] of anti-inflammatory proteins. 17β-Estradiol ($E_{2\beta}$), the major estrogen form in the body, causes protein S-nitrosylation in endothelial cells via estrogen receptor ERα (nuclear receptor NR3a1) and endothelial nitric oxide synthase (NOS3) [907].

In endothelial cells, both basal NO production and 17β-estradiol activation of the PI3K–PKB–NOS3 pathway (estrogen-mediated vascular protection) require Src kinase [908]. In the plasma membrane, Src kinase is associated with a truncated, relocalized estrogen receptor-α and NOS3 enzyme. Endogenous 17β-estradiol activates its cognate ERα that stimulates Src kinase. In turn, Src kinase phosphorylates ERα for subsequent plasmalemmal recruitment of Src kinase and ERα and assembling with NOS3, followed by NOS3 activation.

Some steroid hormones modify endothelial cell behavior. Endothelial cells are insensitive to progesterone and testosterone, but they are softened by estrogens that activate plasmalemmal sodium–proton antiporter. Aldosterone elevates endothelial cell volume and causes growth and stiffness by activating epithelial sodium channels ENaCs, whereas glucocorticoid hormones do not have such effects. In the presence of aldosterone, when plasma sodium level is in the high physiological range, small physiological changes in extracellular sodium concentration stiffens endothelial cells and reduces the release of nitric oxide. On the other hand, acute increase in extracellular potassium softens vascular endothelium and augments nitric oxide release [909]. Change in cell rheology mediated by potassium and sodium ions mainly involves cortical actin cytoskeleton that switches from gelation to solation according to ambient sodium and potassium concentrations. Potassium causes cortical actin depolymerization, i.e., cortical filamentous actin is transformed into monomeric globular actin that colocalizes with endothelial NOS (NOS3) and heightens its activity.

[51] S-nitrosylation is a post-translational protein modification induced by nitric oxide that refers to the attachment of a nitrosyl group to the thiol (SH) group of cysteine residues. Intracellular protein S-nitrosylation regulates various endothelial functions, such as ion channel activity, reduction in oxidative stress, suppression of pro-inflammatory pathways, and inhibition of apoptosis. S-nitrosylation of NADPH oxidase reduces the generation of free radicals. In addition, S-nitrosylation activates the free radical scavenger thioredoxin. S-nitrosylation prevents pro-inflammatory NFκB activation. S-nitrosylation that results from action of 17β-estradiol also precludes angiotensin-2-induced upregulation of intercellular cell adhesion molecule ICAM1.

9.5.5.5 Gasotransmitters

Endogenous gaseous diffusible messengers, or gasotransmitters, include carbon monoxide (CO), nitric oxide (NO), and hydrogen sulfide (H_2S), in addition to several reactive oxygen species, that act as intra- and intercellular regulators.

In addition, feedbacks exist among different gasotransmitters that enable the regulation of the action of a given gas by others gaseous mediators [910].

Carbon Monoxide

Carbon monoxide (CO) is a stable, lipid-soluble gas, produced by heme oxygenase (HO) with iron and biliverdin, using NADPH from heme degradation [911] (Vol. 4–Chap. 10. Other Major Signaling Mediators). Although CO can bind to heme-containing proteins for transient storage and later release, the predominant source is de novo synthesis. It then diffuses according to the local partial pressure gradient. Carbon monoxide binds to the ferrous iron of heme and, then, influences the activity of all heme proteins.

Carbon monoxide not only causes vasodilation, but also supports action of vasodilators and attenuates that of vasoconstrictors such as platelet-activating factor. Endothelial nitric oxide and prostacyclin (PGi2) cooperate with the HO–CO axis to regulate the blood circulation. On the other hand, H_2S can inhibit heme oxygenase-2 in endothelial cells [911].

Nitric Oxide

Nitric oxide (NO) a highly diffusible gas that runs very rapidly from its site of synthesis. However, NO is a labile vasodilator with a limited half-life of NO ($\mathcal{O}$[1 s]); its effect is thus localized close to the site of release. This free radical is highly reactive with other chemical species, notably oxygen, superoxide, and iron-containing hemes that act as NO scavengers. Nitric oxide is able to serve as an auto-, para-, and endocrine signal (Vol. 4 – Chap. 10. Other Major Signaling Mediators).

Nitric oxide can complex with all transition metals, the so-called metal nitrosyls. Nitric oxide interacts with metalloproteins, especially those that contain hemes, and with thiols. Nitric oxide can act in the form of free radical (NO or $NO^{\star}$) or as chemical species, such as nitrosyl (NO^+) or nitroxyl (NO^-) ions or ^{S}nitrosothiols (SNO). Hemes either transform NO into nitrate or promote its transformation into ^{S}nitrosothiols.

Additional sources of nitric oxide than nitric oxide synthases arise from the cycling of nitrate, nitrite, and nitric oxide. Several pathways modulate the NO_3^-–NO_2^-–NO cycling, such as oxidation or reduction by hemoglobin, myoglobin, neuroglobin, xanthine oxidoreductase (XOR), nitric oxide synthase, carbonic anhydrase, cytochrome-C oxidase, cyclooxygenase, microsomal cytochrome-P450, mitochondrial aldehyde oxidase, and cytochrome-C, as well as bacteria, in different tissues under different conditions [912, 913]. Hemoglobin operates at normal pH and oxygen saturation in the vasculature.

Hydrogen Sulfide

Hydrogen sulfide (H_2S) is a smooth myocyte relaxant. Whereas cystathionine γ-lyase is the major H_2S-producing enzyme in smooth myocytes, 3-mercaptopyruvate sulfurtransferase (3MST) and cysteine aminotransferase (CAT) synthesize H_2S from cysteine and α-ketoglutarate in vascular endothelial cells [914].

9.5.6 Adhesion Molecules

Platelet–endothelial cell adhesion molecule PECAM1,[52] an inhibitory receptor, contains an intracellular C-terminus with 2 immunotyrosine-based inhibitory motifs (ITIM) that launch inhibitory signaling upon transhomophilic engagement in intercellular interaction (CD31+ cell–CD31+ cell adhesion). Its extracellular motifs are also required for signaling [915]. Molecule PECAM1 participates in the regulation of angiogenesis and prevents leukocyte–endothelial cell adhesion and activation of platelets and T lymphocytes. Once PECAM1 is eliminated after cleavage and secretion, inflammation can occur. A truncated extracellular PECAM1 fragment on T lymphocyte surface cannot preclude inflammation, but a PECAM1-derived peptide is able to stimulate this fragment, thereby restoring the PECAM1 inhibitory pathway, thereby repressing inflammation and suppressing atherosclerosis and aneurysm evolution [916]. Molecule PECAM1 primes phosphorylation of PTPn11 phosphatase in activated leukocytes.

In given environments, cells may express different forms of PECAM1 that exert different functions. For example, it loses its intercellular junction role when endothelial cells are exposed to inflammatory cytokines [915]. In addition, neutrophils can lose PECAM1 surface expression after extravasation.

Oxidized low-density lipoprotein (oxLDL) is an autoantigen that reduces the existence of extracellular PECAM1 and increases the presence of its soluble form in cultures of CD4+ T lymphocytes [915]. In atherosclerotic plaques and tissues surrounding aneurysmal aortas, macrophages have an attenuated activity with reduced production of intracellular peptidases and release of cytokines and chemokines.

[52]Initially identified on endothelial cells and platelets, PECAM1 also lodges on blood leukocytes (granulocytes, monocytes, and a lymphocyte subset.

In humans, both CD4+ and CD8+ T lymphocytes lose PECAM1 with aging, whereas thrombosis risk augments. Conversely, thrombosis reduces PECAM1 expression, thereby increasing T-cell proliferation. In humans, numbers of peripheral CD4+, PECAM1+ and CD8+, PECAM1+ T lymphocytes decrease markedly in patients with abdominal aortic aneurysms, whereas populations of CD4+, PECAM1− and CD8+, PECAM1− T lymphocytes increase [915].

9.5.7 *Miscellaneous*

Endothelial cells can also store ^{Arg}vasopressin, angiotensin-2, histamine, atrial natriuretic peptide, calcitonin gene-related peptide, neuropeptide-Y, and vasoactive intestinal peptide [680].

9.5.8 *Vessel Wall Homeostasis*

9.5.8.1 Apoptosis Regulators

Apoptosis regulator through modulating IAP expression (ARIA) synthesized by vascular endothelial cells induces endothelial apoptosis and hampers angiogenesis via proteasomal degradation of inhibitor of apoptosis IAP1 and IAP2 [917].

X-box binding protein XBP1 that is involved in endoplasmic reticulum stress response is highly expressed at areas of atheroma [918]. Sustained XBP1 activation causes endothelial cell apoptosis and favors atherosclerotic lesion (Vol. 6 – Chap. 8. Vascular Diseases).

Humanin (Hn) is a 24-amino acid peptide synthesized in vascular endothelial cells (of both arteries and veins) that protects against oxidative stress and apoptosis [919].[53] Exogenous administration of humanin (0.1 μmol) to endothelial cell cultures reduces oxidized LDL-induced formation of reactive oxygen species and apoptosis [919].[54] The binding of exogenous humanin to plasmalemmal receptors involved in cell survival pathways and intracellular humanin that binds to BCL2-associated X protein (BAX) can impede BAX pro-apoptotic effects.

---

[53] During early stages of atherosclerosis, the arterial intima is damaged after LDL penetration and accumulation into the subendothelial space. Endothelial cells and smooth myocytes are the initial source of reactive oxygen species that oxidize LDL particles. Oxidized LDLs amplify ROS generation.

[54] Humanin limits the production of ceramide, a lipid second messenger involved in apoptosis. Ceramide activates BCL2-associated X protein. Oxidized LDLs increase the concentration of cellular ceramide.

9.5.8.2 Angiomotin

In endothelial cells, angiomotin is a double-pass transmembrane protein with an extracellular angiostatin-binding region and intracellular N- and C-termini.

Angiomotin constitutes a family with angiomotin-like-1 and -2 (AMotL1–AMotL2). Members of the angiomotin family possess an angiomotin coiled-coil homology (ACCH) domain that has a high affinity for the monophosphorylated PI(3)P and PI(4)P phosphatidylinositols. Similar to the Bin/amphiphysin/Rvs (BAR) domain, the ACCH domain binds preferentially to curved membranes and induces vesicle tubulation. The angiostatin-binding domain is not present in AmotL1 and AmotL2. The angiomotin C-terminus residues form a PDZ-binding motif.

Angiomotin localizes to tight junctions and endosomes. It is involved in the maintenance of tight junctions and migration of endothelial cells, as well as insulin sensitivity in skeletal myocytes [920].

Angiomotin is involved in the maintenance of tight junctions. Angiomotin binds to multiPDZ domain-containing protein MuPP1 and its paralog, protein (PALS1) associated with tight junctions (PATJ). These partners couple AMot to tight junction transmembrane proteins, pleckstrin homology (PH) domain-containing RhoA guanine nucleotide-exchange factor PlekHg5, and a large complex involved in the maintenance of apicobasal polarity. Other members of the complex such as protein associated with Lin7 (PALS1) and Lin7 proteins.

Angiomotin colocalizes with zonula occludens protein ZO1 and to a lesser extent with cadherin-5[55] at adherens junctions [920].

Angiomotin interacts with GTPase-activating RhoGAP17 protein. In the absence of angiomotin, endothelial cells extend lamellipodia simultaneously in several directions, rather than forming a single major lamellipodium at the leading edge [920].

Angiomotin (AMot) was initially identified as a receptor of angiostatin, an anti-angiogenic protein. Angiomotin is required to ensure action of FGF2, VEGFa, and lysophosphatidic acid [920].

Two splice variants are synthesized, short ^{P80}AMot[56] and ^{P130}AMot,[57] both produced in vascular endothelial cells [920]. These variants colocalize with tight junction protein ZO1 in endothelial cells.

Variant ^{P80}Amot mediates migration stimulation by FGF2 and its inhibition by angiostatin; ^{P130}Amot does not intervene. Higher expression of ^{P80}Amot corresponds to a migratory phenotype prevalent during angiogenesis; larger production of ^{P130}Amot is observed in mature blood vessels [920]. Variant ^{P80}Amot causes translocation of ^{P130}Amot from tight junction, thereby repressing its stabilizing effect.

---

[55] In epithelial cells, cadherin-1, or E-cadherin.

[56] A.k.a. $AMot_S$ that was identified originally.

[57] A.k.a. $AMot_L$.

Only ^{P130}Amot localize to adherens junction. It binds to scaffold membrane-associated guanylate kinase-related protein with inverted domain organization MAGI1b. Only ^{P130}Amot binds to actin [920].

Isoforms ^{P80}AMot and ^{P130}AMot homo- and heterodimerize [920]. In addition, ^{P80}AMot colocalizes with the early endosome marker EEA1 (or ZFYVE2) as well as with the RhoGEF PlekHg5, with ARF6 and Rab11 in perinuclear recycling endosomes, and with Rab13 GTPase.

9.5.8.3 EGFL7

Multiple EGF-like domain-containing protein EGFL7, or veStatin, is exclusively secreted by endothelial cells to modulate smooth myocyte migration. It may also be involved in the spatial organization of endothelial cells in the angiogenic sprout. It impedes deposition and organization of elastic fibers [921].[58] This inhibitor of vascular elastogenesis interacts with lysyl oxidase.[59]

9.5.8.4 Endosialin

Angiogenic endosialin[60] expressed in endothelium binds to fibronectin[61] and collagen-1 and -4 to enhance cell–matrix adhesion and cell migration [922]. Endosialin in cells of tumor stroma and endothelium favors tumor angiogenesis and invasion. Endosialin enhances MMP9 activity. In addition, endosialin regulates pericyte–endothelial cell interactions, as its expression is also associated with pericytes.

---

[58]Elastic fibers are stable components of the extracellular matrix that strongly participate in rheology of blood vessel walls. Elastic are fibers composed of a core of elastin assembled onto a microfibrillar scaffold made of several proteins, such as fibrillins, fibulins, emilins, and MAGPs. Fibroblasts, smooth myocytes, and endothelial cells, among other cells, secrete tropoelastin, which is deposited on the pre-assembled microfibrillar scaffold. Adjacent tropoelastins spontaneously aggregate then crosslink to form elastin polymers owing to the copper-dependent family of lysyl oxidases.

[59]The family of lysyl oxidases comprises 5 known members (Lox, LoxL1–LoxL4). They crosslink several ECM components, such as elastin and collagens.

[60]A.k.a. tumor endothelial marker TEM1, CD164 sialomucin-like protein-1 (CD164L1), and CD248.

[61]Fibronectin interacts with many plasmalemmal receptors and components of the extracellular matrix. Integrin-$\alpha_4\beta_1$ and -$\alpha_5\beta_1$ also mediate fibronectin-dependent cell adhesion.

9.5.8.5 Transglutaminase

Many proteins of the extracellular matrix are substrates of transglutaminase-2 (TG2) produced by endothelial cells. Transglutaminase-2 affects both matrix deposition and turnover. It promotes collagen-1 synthesis and resistance of the extracellular matrix to degradation by matrix metallopeptidase MMP1, but it downregulates the expression of collagen-3 and -4 [923].

9.5.8.6 Heparan Sulfate Proteoglycans

Different structural types of heparan sulfate are secreted on the cell surface, such as syndecan and glypican, as well as in the basement membrane (thickness of 20–100 nm), such as perlecan, agrin and collagen-18 [924].[62] Several heparan sulfate proteoglycans have heparan sulfate chains only under specific circumstances, the part-time heparan sulfates, such as epican (CD44), β-glycan, and testican.

Heparan sulfate proteoglycans interact with manifold proteins, such as growth factors, cytokines, chemokines, peptidases, lipases, and cell adhesion molecules. The vascular endothelium can modulate inflammation by changing the heparan sulfate structure, a given type of heparan sulfate being able to bind or not to a given ligand. Heparan sulfate, in association with interacting proteins, inhibits SMC proliferation [925].

9.5.8.7 Neuregulins

Neuregulins (Nrg1–Nrg4) are cell–cell signaling proteins. They are ligands of receptor Tyr kinases of the HER family. Isotype Nrg1[63] operates in the nervous system, heart, and breast. Neuregulin-1 possesses different isoforms from alternative splicing of the transcript of the NRG1 gene, such as type-1 neuregulin-1,[64] type-2 neuregulin-1, or glial growth factors (GGF1–GGF3),[65] and type-3 neuregulin-1, or sensory and motor neuron-derived factor (SMDF), in addition to type-4 to type-6 neuregulin-1.

[62]These proteoglycans link to other basement membrane components, such as laminin, collagen-4, and entactin. Five types of heparan sulfate proteoglycans always carry heparan sulfate chains, the so-called full-time heparan sulfates. The full-time heparan sulfates on cell surfaces consist of 4 integral-membrane syndecans and 6 glypicans, which are attached to the plasma membrane. The 3 matrix heparan sulfate proteoglycans are perlecan, agrin, and collagen-18.

[63]Neuregulin-1 isoforms are produced in various cell types, including cardiac endothelial cells.

[64]A.k.a. heregulin (Hrg), NEU differentiation factor (NDF), and acetylcholine receptor-inducing activity protein (ARIA).

[65]Glial growth factors promote Schwann cell migration and proliferation.

Neuroprotection is based on the *neurovascular unit* that includes neurons and associated glial cells, vascular endothelium and support cells, and extracellular matrix. Endothelial cells of the human brain microcirculation synthesize neuregulin-1.

Upon ligand binding, Nrg1 receptors HER2 and HER3 are phosphorylated and trigger the PI3K–PKB, Ras–ERK, and JaK-STAT pathways to ensure cytoprotection, particularly during oxidative stress [926]. Because HER2 and HER3 lack ligand-binding and efficient kinase activity, respectively, they form an active HER2–HER3 heterodimer. Other Nrg1 mediators comprise phospholipase-Cγ and protein kinase-C. Neuregulin-1 enhances survival of endothelial cells and can contribute to neuron survival during stroke. Furthermore, neuregulin-1 promotes angiogenesis.

9.5.8.8 Histone Deacetylases – Epigenetic Control by Hemodynamic Stress

Certain class-1 (i.e., HDAC1–HDAC3) and -2a (i.e., HDAC5 and HDAC7) histone deacetylases are sustainably upregulated and accumulate in the nucleus in cultured endothelial cells subjected to a low-steady-component pulsatile flow (modulation rate 8) [927]. On the other hand, a high-steady-component pulsatile flow (modulation rate 1/3) causes phosphorylation-dependent nuclear export of class-2a HDACs.

Histone deacetylases regulate the activity of NFE2-related factor NRF2 and Krüppel-like factor KLF2 that target many mechanical stress-responsive genes and cell cycle. Transcription factor NRF2 binds to the anti-oxidant response element (ARE) in promoters of many anti-oxidant genes.

Steady flow causes phosphorylation-dependent nuclear export of HDAC5, hence its dissociation from myocyte enhancer factor MEF2, as well as expression of anti-inflammatory Krüppel-like factor KLF2 and endothelial nitric oxide synthase (NOS3) via sirtuin-1 [927]. Nitric oxide-mediated nuclear shuttling of HDAC4 and HDAC5 relies on the association of protein phosphatase PP2 with the phosphorylated calcium–calmodulin-dependent kinase CamK4^{P}–HDAC complex. Phosphorylation of HDAC enzymes by CamK kinases causes HDAC export from the nucleus and cytosolic sequestration by 14-3-3 proteins.

A low-steady-component pulsatile flow uses the PI3K–PKB pathway. It enables the association of class-1 histone deacetylases with NRF2 and that of HDAC3 and class-2a histone deacetylases with MEF2 [927]. Subsequently, NRF2 and MEF2 are deacetylated, thus being unable to bind to ARE element. Consequently, expression of anti-oxidant gene NAD(P)H quinone oxidoreductase-1 (NQO1) and KLF2 is reduced. Moreover, this type of unsteady flow increases the production of cyclin-A, but decreases that of CKI1a cyclin-dependent kinase inhibitor in endothelial cells, thereby inducing their proliferation. In addition, inhibition of KLF2 expression enables elevated VCAM1 production.

Because a high-steady-component pulsatile flow causes export of class-2a HDACs from the nucleus, it precludes HDAC5/7–MEF2 association, thereby promoting KLF2 expression. Factor KLF2 prevents VCAM1 synthesis, hence leukocyte adhesion. Therefore, VCAM1 expression depends on the modulation rate of the periodic flow [927].

9.5.9 Growth Factors and Regulators of Growth Factor Signaling

Endothelial cells synthesize a huge number of factors (Tables 9.14) to 9.18). They release numerous paracrine growth factors, also called *angiocrine factors*. In fact, angiocrine factors comprise secreted molecules as well as plasmalemmal and membrane-anchored compounds that are cleaved and generate liberated variants.

Some angiocrine factors can act on stem and progenitor cells, as endothelial cells achieve an instructive vascular niche. Endothelial cells are indeed involved in self-renewal of hematopoietic stem cells. Specific angiocrine factors produced by endothelial cells in response to different stimuli either support self-renewal and proliferation using the TOR–PKB pathway or, due to MAPK coactivation, lineage-specific differentiation of hematopoietic stem and progenitor cells [929]. In particular, Notch ligands by endothelial cells promote HSC proliferation.

Angiocrine factors secreted from endothelium also regulate tumor cell proliferation and motility. Angiocrine factor Slit-2, which is inhibited by endothelial EPHa2 receptor, represses tumor growth.

Endothelial cells from different compartments of the vasculature, such as arteries, veins, microcirculation segments, and lymphatic vessels, differ in their gene expression patterns (different phenotypes). In addition, tumor endothelial cells also differ by expression of specific markers.

9.5.9.1 Connective Tissue Growth Factor

Connective tissue growth factor production in endothelial cells can be induced by bioactive lipids, such as lysophosphatidic acid and sphingosine 1-phosphate, and platelets [930].

9.5.9.2 VEGF and Cadherin-5

Endothelial cells produce vascular endothelial (VE)-cadherin, or Cdh5, that binds to cortical partners, such as β-, δ1-, and γ-catenin. Endothelial-specific cadherin-5 clusters build homotypic adherens junctions that enable endothelium stabilization, as they prevent cell growth, proliferation, and apoptosis.

Table 9.14. Endothelial cells are able to synthesize multiple growth factors (Source: [928]). (**Part 1**) Growth factors (CTGF: connective tissue growth factor; PlGF: placental growth factor; RBP: retinol binding protein; S1PR: sphingosine 1-phosphate receptor). Protein production is influenced by cell localization, type, maturation level (immature or fully-differentiated cell), activity status (resting or activated cell), context (healthy or disease) in a given animal species. Weibel-Palade bodies are secretory granules that store hormones, growth factors, and cytokines, in addition to enzymes, receptors, and adhesion molecules, and release them in response to adequate stimuli. Sprouty proteins (Spry1–Spry4) produced by endothelial cells antagonize fibroblast (FGF), epidermal (EGF), hepatocyte (HGF), and vascular endothelial (VEGF) growth factors, as it is a negative feedback regulator of multiple receptor Tyr kinases.

Type	Receptors
BMP1–BMP7, BMP9 BMP14 (GDF5) GDF15	BMPR1/2
CTGF	CTGFR
CSF1/2/3	CSF1R/2R/3R; erythropoietin EpoR
HBEGF, neuregulin-1, TGFα	HER1–HER4
ephrin-A1–A5/B1–B2	EPHa1–EPHa8, EPHb1–EPHb4, EPHb6
FGF1/2/4/5/7/8/12/16/18/22	FGFR1/2/3/5
HGF	HGFR
IGF1/2	Insulin receptor; IGFR1/2
NGF Neurotrophin-3/4	TNFRSF16 NTRK
PDGFa–PDGFd	PDGFR
SCF	SCFR
TGFβ1–TGFβ3 follistatin, FSL1/2, inhibins	ALK1, TβR1 (ALK5), TβR2, Cryptic
VEGFa–VEGFd $VEGFa_{121/165/189}$ PlGF1/3/4	VEGFR1–VEGFR3 Neuropilin-1/2 (VEGF coreceptors)
S1P	S1PR1

In sparse cells, VEGF quickly induces VEGFR2 clathrin-dependent endocytosis for a long duration.[66] Internalized VEGFR2 then associates with PLCγ and PKC to

[66]Receptor endocytosis controls the signaling magnitude, duration, and nature. Endothelial cells in contact do not respond to proliferation signals of vascular endothelial growth factor.

Table 9.15. Endothelial cells are able to synthesize multiple growth factors (Source: [928]). (**Part 2**) Peptides and other types of hormones and growth factors (C1qTNF: C1q and tumor-necrosis factor (TNF)-related protein; GHR: growth hormone receptor; NPR: natriuretic peptide receptor; RAMP: receptor activity-modifying protein). Angiogenic factor with G patch and FHA domains (AGGF1), or vasculogenesis gene on 5q protein (VG5q), is a potent angiogenic factor that may act as an autocrine growth factor promoting slightly endothelial cell proliferation. Mutations of the AGGF1 gene cause Klippel-Trenaunay syndrome characterized by an enhanced angiogenesis. Angiopoietin-1 and -2 and their receptor protein Tyr kinases TIE1 and TIE2 are involved in angiogenesis, Ang1 as a paracrine agonist of TIE2 and Ang2 as an autocrine TIE2 antagonist. In the absence of VEGF, Ang2 antagonizes Ang1 and causes vessel regression; in the presence of VEGF, Ang2 facilitates vascular sprouting. Whereas Ang1 is produced by perivascular cells and stabilize quiescent endothelium, Ang2 is secreted by endothelial cells during angiogenesis. Artemin, an auto- and paracrine factor, is a member of the GDNF family with neurturin, persephin, and glial cell-derived neurotrophic factor. It binds to a complex formed by GPI-anchored coreceptor GFRα3 and Ret receptor protein Tyr kinase Granins are also called chromogranins or secretogranins.

Type	Receptors
AGGF1	AGGF1 receptor (integrin-$\alpha_5\beta_1$)
Angiopoietin-2	TIE1–TIE2
Artemin	Adrenomedullin receptor (RAMP2/3)
Bradykinin	Bradykinin BR, bombesin BBR
Cartducin (C1QTNF3)	Cartducin receptor
CNP	(BNP/CNP) NPR (GuCy2)
CGRP, gastrin	Cholecystokinin CCKR
	Cortistatin receptor
	CRHR
Endothelin-1/2/3	ETR
^{Met}enkaphalin	Op_1, Op_4
Ghrelin	Ghrelin receptor
	GHR
Granin (ScG2 or CGc)	Granin receptor
Kisspeptin	GPR54

be phosphorylated and to activate the MAPK pathway. The Cdh5–βCtn complex is involved in the development of the cardiovascular apparatus. It links to VEGFR2 to reduce its activity due to coclustered PTPRj phosphatase [931].[67]

Confluent endothelial cells hence respond poorly to VEGF factor. Phosphatase PTPRj colocalizes with cadherin-5 and β-catenin at adherens junctions between endothelial cells. It dephosphorylates VEGFR2 and impedes its normal endocytosis

[67] Plasmalemmal protein Tyr phosphatase PTPRj is also called density-enhanced PTP (DEP1), HPTPη, and CD148. It is expressed on blood leukocytes, B and T cells, granulocytes, macrophages, certain dendritic cells, and mature thymocytes. In non-lymphoid tissues, it is found on many epithelial cell types with glandular and/or endocrine differentiation as well as on fibrocytes, melanocytes, and Schwann cells [932].

Table 9.16. Endothelial cells are able to synthesize multiple growth factors (Source: [928]). (**Part 3**) Peptides and other types of hormones and growth factors (Cont.; ApoER: apolipoprotein-E receptor; LRP: low-density lipoprotein receptor-related protein; PTHRP: parathyroid hormone-related protein; RXFP: relaxin family peptide receptor; VLDLR: very-low-density lipoprotein receptor). Midkine receptor is the complex formed by receptor protein Tyr phosphatase PTPRb, low density lipoprotein receptor-related protein LRP1, anaplastic leukemia kinase (ALK), and syndecan. Three pleiotrophin receptors exist (N-syndecan, receptor protein Tyr phosphatase PTPRb, and ALK). Neurturin, a member of the GDNF family, like artemin, binds to a complex formed by GFRα2 and Ret receptor protein Tyr kinase.

Type	Receptors
Midkine	PTPRb–LRP1–ALK–syndecan
	Pleiotrophin receptor
	Oxytocin receptor
Neurokinin-1 (substance-P)	NK_1
Neuromedin-U	Neurokinin-A/B receptors
Neuropeptide-Y	Neuropeptide-Y receptors
	Neurotensin receptors
Neurturin	GFRa2
Prokineticin-1/2	Prokineticin receptor (PKR)
Proopiomelanocortin	
PTHRP	
Reelin	Reelin receptors (VLDLR, ApoER2 [LRP8])
Relaxin	RXFP1
Salusin-α/β (torsin-2A)	
Somatostatin	SstR
Urocortin-1–urocortin-3	Urocortin (CRF) receptor
Urotensin-2	Urotensin-2 receptor
Vasohibin	
Vasopressin	Vasopressin receptor

that leads to activation of extracellular signal-regulated kinases ERK1 and ERK2 via PLCγ and PKC enzymes. The Cdh5–VEGFR2 complex limits cell proliferation by preventing activation of ERK1 and ERK2 kinases.

9.5.10 MicroRNAs

Specific microRNAs are involved in various developmental processes, especially mechanisms that support the formation of blood and its circulation, such as cardio- and angiogenesis and hematopoietic lineage differentiation. Dicer (Vol. 1 – Chap. 5. Protein Synthesis) that is constitutively expressed in endothelial cells is required for appropriate vasculo- and angiogenesis in embryos (Chap. 10) [933]. Drosha alters endothelial cell function and angiogenesis to a lesser extent than Dicer, althhough Drosha silencing in endothelial cells impairs capillary sprouting and vascular tube formation, two activities associated with endothelial cells.

Table 9.17. Endothelial cells are able to synthesize multiple growth factors (Source: [928]). (**Part 4**) Guidance molecules and morphogens (DLL: Delta-like ligand; EDIL: EGF-like repeats and discoidin-1-like domain-containing protein; EDIL3 is an integrin ligand and angiogenesis mediator; sFRP: secreted Frizzled-related protein).

Type	Receptors
Guidance molecules	
Netrin-4	Ntn4R
Slit-2/3	Robo1/4
Semaphorin-3a/3b/3e–3g,	Neuropilin-1/2 (semaphorin coreceptors)
semaphorin-4A/7A	Plexin-A1/12/B1/B3/D1
Morphogens and remodelers	
	Sonic Hedgehog receptors
Notch-1–Notch-4	DLL1, Jag1/2
	EDIL3
Wnt1–Wnt3, Wnt5a–Wnt5b, Wnt6,	Frizzled-1–Frizzled-7
Wnt7a/8a, Wnt10a–Wnt10b,	sFRP1–sFRP2
Wnt13a–Wnt13c, Wnt13 (Wnt2b),	
Wnt14 (Wnt9a), Wnt15	
Miscellaneous	
Gliomedin	Glucagon-like peptide-1 receptor
	Spondin-1 receptor
Neuritin	Neuritin receptor

Endothelial cells synthesize, at least in certain circumstances, more or less specific microRNAs.[68] Among them, let7f, miR27b, and miR130a are pro-angiogenic, whereas miR221 and miR222 preclude endothelial cell migration and proliferation, and thus angiogenesis [933].

MicroRNA-210 is produced during hypoxia in endothelial cells to enhance vascular tube formation as well as VEGF-caused migration of normoxic endothelial cells. A group of hypoxia-regulated microRNAs is elicited by hypoxia-inducible factor.

MicroRNA-130a that is produced at low levels in quiescent endothelial cells is upregulated during angiogenesis. It impedes the functioning of anti-angiogenic homeodomain-containing Gax[69] and HoxA5 proteins.

[68]Endothelial microRNAs encompass miR16, miR17, miR21, miR22, miR23a, miR27b, miR29, miR31, miR100, miR103, miR106, miR107, miR126, miR130a, miR181a, miR-210, miR221, miR222, miR320, let7b and let7f.

[69]Growth arrest homeobox gene product GAx prevents endothelial cell proliferation and migration, as well as vascular tube formation.

Table 9.18. Endothelial cells are able to synthesize multiple growth factors (Source: [928]). (**Part 5**) Cytokines, chemokines, and adipokines (CT1R: cardiotrophin-1 receptor; CNTFR: ciliary neurotrophic factor receptor; LIF: leukemia inhibitory factor). Cardiotrophin-1 (CT1), a member of the IL6 family, is a cytokine capable of inducing cardiomyocyte hypertrophy. Ciliary neurotrophic factor (CNTF) is a cytokine that assists in the differentiation and survival of various cell types in the peripheral and central nervous systems. A secreted complex made of cardiotrophin-like cytokine (CLC) and cytokine-like factor-1 (CLF), like CNTF, acts via CNTF receptor.

Type	Receptors
Cytokines	
CT1	CT1R (GP130–LIFR)
CNTF, CLC–CLF	CNTFR (LIFR–CNTFR–GP130
Ifnα/β/γ	IfnaR1, IfngR1
IL1–IL3, IL5–IL8,	IL1RA, IL1R2, IL2Rα–IL4Rα,
IL10–IL20, IL23–IL25,	IL6Rα–IL8Rα, IL10Rα–IL11Rα,
IL27–IL29, IL32–IL33,	IL13Rα, IL15Rα, IL17Ra, IL18R,
IL36	IL20R, IL22R1, IL27Rα, IL33R
LIF	LIFR
	Oncostatin-M receptor (OSMR)
Prolactin	Prolactin receptor
TNFSF1–TNFSF7, TNFSF10–TNFSF11,	TNFRSF1a–TNFRSF1b,
TNFSF13b (20), TNFSF15	TNFRSF3–TNFRSF5,
	TNFRSF6b/11b/12a/14/16/21/25 (12)
Chemokines	
CCL1–CCL5, CCL7–CCL8, CCL11,	CCR1–CCR9
CC15–CCL24, CCL26–CCL28	
CXCL1–CXCL7, CXCL9–CXCL13,	CXCR3–CXCR4, CXCR6–CXCR7
CXCL16–CXCL17	
CX_3CL1	CX_3CR1
XCL1	
Adipokines	
	Leptin receptor
	RBP4 receptor
	Resistin receptor
Visfatin	Visfatin receptor

On the other hand, anti-angiogenic miR221 and miR222 repress stem cell factor receptor and endothelial nitric oxide synthase (NOS3). Moreover, miR221 and miR222 control the synthesis of other microRNAs for more or less redundant activity.

MicroRNA cluster miR17-92 (miR17, miR18a, miR19a–miR19b, miR20a, and miR92a) is significantly upregulated in tumors to promote angiogenesis in a paracrine manner [933]. These miRs target anti-angiogenic proteins. MicroRNA-18 preferentially suppresses the expression of connective tissue growth factor. MicroRNA-19 impedes thrombospondin Tsp1 activity. The miR17-92 cluster is also involved in the regulation of B-cell differentiation.

MicroRNA-378 improves not only tumor cell growth, but also tumor angiogenesis, as it inhibits SuFu and Fus1 tumor suppressors.

Dicer depletion reduces the expression of microRNAs that control the production of high-mobility group (HMG)-box protein HMGB1 transcriptional suppressor, which targets NOx organizer NOxO2 of the NADPH oxidase complex. MicroRNAs, such as miR21, miR126, and miR155, contribute to vascular inflammation [933]. Overexpression of MiR21 causes neointima formation. MicroRNA-126 inhibits TNFα-stimulated VCAM1 that favors leukocyte adherence to the vascular endothelium. MicroRNA-155 represses the expression of angiotensin-2 AT1R receptor. MicroRNA-155 expression is induced in macrophages by TNFα and Ifnβ for proliferation of granulocytes and monocytes during inflammation.[70]

In normal adult swine, expression of endothelial microRNA-10a is lower in the inner aortic arch and aorto-renal branches than elsewhere; subsequently those of HoxA1, a miR10a target, and inflammatory biomarkers CCL2 chemokine, interleukin-6 and -8, VCAM-1, and E-selectin rises in the same regions [934]. Transcripts that encode 2 regulators of IκBα degradation, MAP3K7 and β-transducin repeat-containing ubiquitin ligase, contain a miR10a-binding site in their 3′UTR region. MicroRNA-10a thus prevents activation of NFκB and thus expression of a pro-inflammatory endothelial phenotype in the arterial bed.

9.5.11 Angiogenesis Guidance Molecules

New blood vessels arise from existing ones by emergence of tips with non-proliferating tip cells that have a guidance role and are followed by sprouting with proliferating stalk cells. Signaling involved in vessel (and axon)

[70]MicroRNA-155 is also required for functioning of B and T lymphocytes, as well as dendritic cells. Furthermore, miR424 is activated by transcription factor PU.1 to impede translation of transcription factor NFI-A that hinders M-CSF receptor actvity during monocyte differentiation [933]. MicroRNA-17-5p-20a-106a also controls monocytopoiesis via AML1 and CSF1 receptor. MicroRNA-146 is elicited in macrophages by pro-inflammatory cytokines via NFκB. Myeloid-specific miR223 regulates progenitor proliferation and granulocyte differentiation and activation during inflammation.

Table 9.19. Angiogenesis guidance pathways.

Pathway	Ligand and receptor types
EPH–ephrin	Ephrin ligands (transmembrane proteins) Erythropoietin-producing hepatocyte receptor kinase
Netrin–DCC/Unc5	Secreted netrin-1/3/4/5 Deleted in colorectal carcinoma receptor Unc5 receptor homologs
Netrin–NtnGR	GPI-anchored netrin-G1/G2 Netrin-G single-pass transmembrane receptor
Notch–DLL Notch–Jagged	Notch receptor (single-pass transmembrane heterodimer) Delta-like ligands (membrane-bound and soluble proteins) Jagged (membrane-bound and soluble proteins)
Slit–Roundabout	single-pass transmembrane Robo1–Robo4 molecules Secreted Slit ligands

guidance includes Delta-like ligand (DLL)–Notch,[71] EPH–ephrin,[72] netrin–Unc5,[73] and Slit–Roundabout pathways,[74] in addition to VEGF, Sprouty, and Hedgehog (Sect. 10.2; Table 9.19).

Robo activated by Slit2 is expressed in stalk cells, but not in tip cells [892]. Receptor Robo1 is expressed by endothelial cells [936]. According to [892], Robo4 suppresses the activity of vascular endothelial growth factor required in angiogenesis (as it induces endothelial cell proliferation and migration, as well as blood vessel wall permeability), hence stabilizing the endothelium. However, Robo4 can have a guidance role in developing vessels [935].

---

[71] Notch receptor is a single-pass transmembrane heterodimeric receptor. It is composed of a large extracellular subunit linked to a single-pass transmembrane subunit that has a short extracellular and a small intracellular region. Membrane-bound proteins Delta-like ligands (DLL1 and DLL4) and Jagged (Jagged-1–Jagged-2) are ligands for Notch receptors (endothelial Notch-1 and Notch-4 and smooth muscle Notch-1 and Notch-3). Soluble DLL and Jagged proteins can also bind to Notch receptors with similar or distinct binding affinities w.r.t. those of membrane-bound forms.

[72] Ephrin ligands (ephrins; a portmanteau for EPH receptor-interacting protein) and erythropoietin-producing hepatocyte receptor kinase (EPH) are both transmembrane proteins. Consequently, EPH–epherin signaling can only occur via direct between-cell interaction.

[73] Three secreted netrins (netrin-1 and netrin-3–netrin-4) and 2 GPI-anchored netrins (netrin-G1–netrin-G2) exist. Receptors for the secreted netrins include Deleted in colorectal carcinoma protein (DCC) and Unc5 homologs (Unc5a–Unc5d). The former is an attractant, whereas Unc5 homologs and, in some cases, DCC are repellants. Netrin-G bind netrin-G single-pass transmembrane protein (between-cell interaction). Netrins contribute to cell migration and regulate between-cell and cell–matrix adhesion.

[74] Members of the Roundabout family (Robo1–Robo4) are single-pass transmembrane cell adhesion molecules that serve as receptors for secreted Slit ligands (Slit1–Slit3).

9.5.12 Blood Group Antigens

α(1,2)-Fucosyltransferase Fut1 adds the terminal fucose to form blood group antigen glycan Lewis-Y (CD174) and its precursor glycan H2 (CD173) that are predominantly expressed on erythrocytes and subsets of vascular endothelial cells. Bone marrow and microvascular endothelial cells (but not macrovascular endothelial cells, such as aortic endothelial cells) stimulated by tumor-necrosis factor-α increase their production of blood group glycans-H2 and Lewis-Y. Fucosylated blood group glycans-H2 and Lewis-Y promote tumor angiogenesis [937]. Factors of malignant cells such as TNFα elicit endothelial cell migration to form new blood vessels. Glycans-H2 and Lewis-Y are concentrated on pseudopodial extensions of endothelial cells.

9.6 Transendothelial Mass Transfer

Vascular endothelium is a selective permeable barrier that regulates the transfer of molecules and cells from blood to underlying tissues. Transport across the endothelium depends on relative intramural pressure and concentration gradients. Molecule transport across endothelium is influenced by molecule characteristics: molecular size, charge, shape, and carbohydrate content. In fact, transport properties of the endothelial barrier are sensitive to both its chemical and mechanical environment.

Transport across the endothelium use intra- and paracellular routes. Intracellular route corresponds to vesicular transcytosis. Paracellular route is enabled by breaks in tight (size ~ 20 nm) and adherens junctions as well as interendothelial leaky cleft uncovered by the glycocalyx (size 80–1330 nm at cell division sites and 15–1000 nm in dying or dead cell zones [938]). Intercellular cleft can be strongly sealed by tight junctions. Mechanical cohesion of the endothelial monolayer is supported by adherens junctions. Adherens junction connects the actin cytoskeleton to the plasma membrane.

Endothelial permeability varies according to the state of the cytoskeleton, i.e., to the width of intercellular spaces that narrows or enlarges with endothelial cell relaxation or contraction.

Flowing blood shears the wetted (luminal) surface of endothelial cells endowed with a glycocalyx that yields the first barrier to transport of solutes and water between blood and tissue, hence modulating the permeability to solutes. Mechanical forces influence transendothelial molecular flux via both intra- and paracellular routes.

9.6.1 Flow-Dependent Transendothelial Transport

Reversible endothelial cell remodeling primed by hemodynamic stresses is associated with an initial increase followed by a gradual decline (~40% after 24 h) in electrical impedance of cultured endothelial monolayers. The endothelium permeability significantly increases after relatively short exposure (1 h) to steady flow, but markedly decreases (~1/3) after long exposure (7–9 d) [938]. When endothelial cell are subjected to sinusoidal flow with an amplitude ratio smaller than 1, the hydraulic conductivity increases similarly to its response to steady flow. When the modulation rate exceeds 1, i.e., when flow reversal occurs, the hydraulic conductivity does not heighten [938].

Mechanotransduction at the luminal (wetted) endothelial surface provokes nitric oxide synthesis by nitric oxide synthase. Heparan sulfate, the dominant glycosaminoglycan of the endothelial glycocalyx, participates in mechanosensing that mediates NO production in response to flow [939]. Oscillatory flow shear stress (modulation rate 3/2) provokes a greater production of NO_2^- and NO_3^- than steady flow. Hyaluronic acid, another glycosaminoglycan of the endothelial glycocalyx, enables mechanotransduction, whereas chondroitin sulfate has no effect on shear-induced NO production and increase in hydraulic conductivity.

In addition, mechanosignaling in response to flow changes in perfused lung microvessels is initiated in caveolae. Flow-preconditioned cells express a 5-fold increase in caveolin and other caveolar proteins at the luminal surface with respect to control (no-flow) condition [944]. Phosphorylation of luminal surface proteins (caveolin-1 and nitric oxide synthase NOS3, which are preferentially localized to caveolae) as well as activation of ERK1 and ERK2 rise upon flow step application in comparison to no-flow condition, more in flow-preconditioned cells than in the absence of flow preconditioning.

Under flow reversing conditions, large NO production limits endothelium permeability via cAMP formation and cAMP-dependent protein kinase (PKA) that controls cytoskeletal tension. On the other hand, the hydraulic conductivity rises under small increases in NO concentration via cGMP production and cGMP-dependent phophodiesterases and protein kinase (PKG) that targets junctional proteins such as tight junction occludin.

Mechanotransduction can also exert on the lateral edge of endothelial cells along the intercellular cleft due to convection generated by the pressure difference across the endothelium.

Mechanical stress heightens hydraulic conductivity by raising phosphorylated occludin level and lowering occludin content, hence by tight-junction disassembly [940]. However, exposure of endothelial cells to sustained equibiaxial cyclic strains (1 Hz, 24 h) strengthen tight junction, as it elevates amounts of both occludin and zonula occludens protein ZO1, a linker protein that connects the tight junction to the actin cytoskeleton, as well as reduces occludin phosphorylation, increases

ZO1 phosphorylation, and supports occludin–ZO1 association [941]. In addition, maintenance of adherens junction (zonula adherens or belt desmosome) under mechanical stress requires tethering of cadherins to γ-catenin [942]. β-Catenin as well as PECAM1 cannot compensate for the loss of γ-catenin to ensure endothelial barrier integrity under flow. The glycocalyx is implicated in endothelial junction remodeling launched by mechanical stress [943]. Endothelial cells respond rapidly to flow by increasing the amount of cadherin-5 in adherens junctions due to heparan sulfate proteoglycans. In any case, transport through endothelial monolayers subjected to flow depends on endothelial cell type. Nevertheless, the hydraulic conductivity also relies on mechanical stress field in excised arteries.

Most explorations are carried out ex vivo, i.e., in endothelial cells unaccustomed to flow. Experiments on cultured endothelial cells and excised blood vessels do not accurately mimic in vivo conditions.

9.6.2 Transport Mechanisms

Two different transport mechanisms can be defined (Fig. 9.5): (1) an intercellular transport through between-cell junctions for small molecules and (2) an intracellular transport for macromolecules (Vol. 1 – Chap. 9. Intracellular Transport).

Vascular wall permeability varies according to regions. Wall filtration measurements show that arterial and venous beds are much less efficient than microvascular compartments. Microvascular endothelia regulate tissue fluid balance and supply nutrients, as these semipermeable barriers control paracellular and transcellular transports between blood and interstitial space. Vascular endothelia carry solutes, the molecular radius of which range from 0.1 nm (sodium ion) to 11.5 nm (immunoglobulin-M) [854]. Endothelial permeability decreases by four orders of magnitude when molecular radius increases from 0.1 to 3.6 nm. Low-molecular-weight molecules (water, hexoses, such as glucose, mannitol, and fructose, amino acids, and urea) go up endothelial clefts, i.e., the paracellular permeation path of continuous endothelia. When molecular radius rises from 3.6 to 6 nm, endothelial permeability remains nearly constant, as the transport shifts to transcytosis.

9.6.2.1 Intercellular Transport – Normal and Leaky Regions

The junctions between endothelial cells act also as a selective barrier to the egress of water and hydrophilic solutes from blood circulation. Endothelium can be divided into 2 parts according to the state of endothelial cells and their junctions: (1) normal endothelial cells, the most numerous, with normal junctions; and (2) leaky endothelial cells, either mitotic or apoptotic endothelial cells of random location, with leaky junctions. The dynamic regulation of endothelial integrity relies

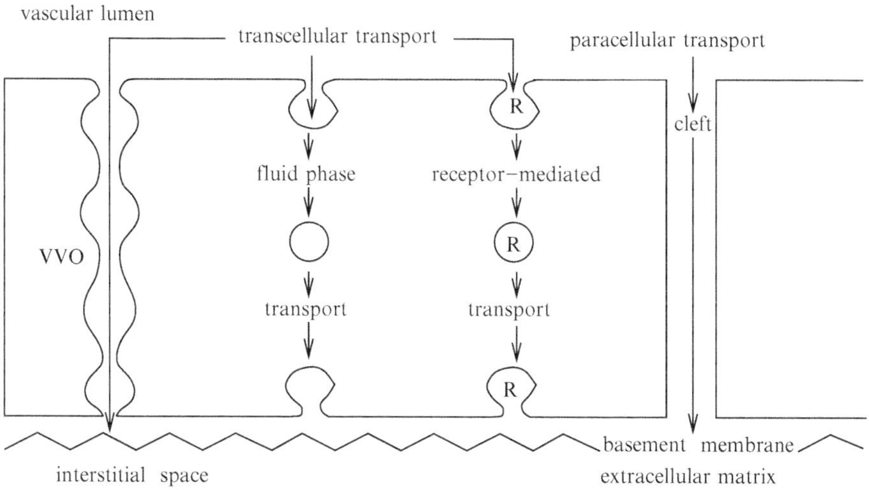

Fig. 9.5 Transport routes in continuous endothelium (e.g., pulmonary, coronary, skeletal muscle, and splanchnic endothelia) comprise para- and transcellular paths (Source: [854]). The paracellular route that uses interendothelial clefts that contain proteins connected to extracellular matrix constituents depends on signaling pathways. Transcytosis is carried out by vesicles that shuttle between the vascular lumen and subendothelial space. Plasma protein size-selective permeability of the endothelial barrier generates a transendothelial protein gradient, the so-called osmotic gradient. Water-insoluble substances are conveyed from blood to interstitium often by specific carrier proteins. Plasma proteins such as albumin serve as chaperones for hydrophobic substances, fatty acids, and hormones. Transvascular flux of solutes and fluid includes transport of small hydrophilic molecules via interendothelial spaces, the permeability of which is regulated, as well as vesicular transit of small hydrophobic solutes through endothelial cells, especially in response to intrinsic and extrinsic stimuli. Small molecules (equivalent radius < 3 nm) such as glucose are transported in interendothelial space, whereas larger plasma proteins such as albumin are conveyed by vesicles filled with receptor-bound (absorptive or receptor-mediated transport) or free solute (fluid-phase transport). Vesiculovacuolar organelles (VVO) formed by fusion of vesicles yield macromolecule transport of growth factors and other regulators. Aquaporins form channels across the lipid bilayer of luminal and abluminal endothelial membranes that are highly selective for water. Inflammatory and angiogenic mediators, such as thrombin, bradykinin, histamine, vascular endothelial growth factor, bind to their receptors and disrupt interendothelial junctions and integrin–matrix protein complexes to form intercellular gaps and open the interendothelial barrier.

on cell migration coordination and re-establishment of cellular junctions. These processes are regulated by the actin cytoskeleton, and thus, in particular, by Rho GTPases and their effectors.

Macromolecular permeability between endothelial cells is regulated by tight junctions, the most apical component of the intercellular cleft, and the thinnest intercellular space, making intimate contact between adjacent endothelial cells. Most water and hydrophilic solutes that are small enough (lower than the albumin dimension) can cross the orifices of tight junctions in the endothelium clefts.

The spaces between several adjoining endothelial cells such as *tricellular corners*, where the borders of three endothelial cells meet, are more permeable regions because endothelial tight junctions are discontinuous at the corners [945]. Discontinuities are observed for occludin, ZO1, cadherin, and β-catenin at the tricellular corners [946]. In addition, 70% of neutrophil transendothelial migration occurs at tricellular corners.

Leaky endothelial cells favor big molecule transport. Endothelium permeability is, indeed, enhanced when endothelial cells have high turnover rates such as in atherosclerosis [947]. Moreover, hypoxemia induces EC apoptosis, and thereby increases macromolecular transport across the vessel wall via leaky junctions. The normal junctions are modeled as circular section spaces around endothelial cells with a pore set in its central part, and leaky junctions as rings around leaky endothelial cells [948].

9.6.2.2 Transcellular Transport

The highest transfer flux between the vascular lumen and extracellular medium occurs by vesicles (size 80 nm). Vesicles can fuse to one another and possibly with cytosol organelles to form tubules. *Vesiculovacuolar organelles* (VVO; caliber 80–140 nm; length 1–2 μm) that are formed by fusion of vesicles or vacuoles (∼ 80–360 vesicles) in continuous endothelia of venules build channels through the cytoplasm. These channels link the luminal edge to abluminal or lateral endothelial cell surface. Vesiculovacuolar organelles convey macromolecules, such as ferritin, vascular endothelial growth factor, serotonin, and histamine, that can bind VVOs by means of receptors [949]. Moreover, vascular endothelial growth factor favors transcellular route via VVOs. Stomatal diaphragms between adjacent, fused vesicles within VVOs can act as barriers between ends of VVOS. Membrane-bound tubules can create transendothelial channels for transfer of large plasma molecules [950]. *Vesiculotubular structures* specialize according to molecule types.

A significant amount of water (up to 40%) crosses the endothelial barrier through an aquaporin-based transcellular route.

Circulating proteins can interact with the endothelial cell surface to enhance or hinder transcapillary exchange. Endothelial plasmalemmal receptors or transporters exist for plasma molecules, such as albumin, a multifunctional carrier (for fatty acids, sterols, amino acids, hormones, metal ions, etc.), insulin, orosomucoid, an acute-phase glycoproteic α-globulin, ceruloplasmin, a copper-carrying ferroxidase, transferrin, lipoproteins, and glucose, among others.

Many amino acid transporters exist in endothelial cells. Cationic amino acids, such as arginine (Arg), histidine (His), lysine (Lys), and ornithine (Orn),[75] are

[75]Ornithine is an amino acid that results from the action of arginase on L-arginine to create urea. It is not encoded by DNA, and thus not involved in protein synthesis. Ornithine is carried through cationic amino acid transporter CAT1 and y+LAT2 transporter, but not markedly through L-type amino acid transporters LAT1 and LAT2.

Table 9.20. Transmembrane transport of cationic amino acids. Amino acid transporters are made of a light and a heavy chain.

SLC alias	Transport system alias
L transport system	
SLC7a5	LAT1 (light chain)
SLC7a8	LAT2
SLC3a2	4F2HC (heavy chain)
y^+ transport system	
SLC7a1	CAT1 (light chain)
SLC7a2	CAT2
SLC7a3	CAT3
SLC7a4	CAT4
SLC3a2	4F2HC (heavy chain)
y^+L transport system	
SLC7a7	y^+LAT1 (light chain)
SLC7a6	y^+LAT2
SLC3a2	4F2HC (heavy chain)
b^{0+} and B^{0+} transport system	
SLC7a9	BAT1 (light chain)
SLC3a1	rBAT (heavy chain)
x_c^- transport system	
SLC7a11	xCT (light chain)
SLC3a2	4F2HC (heavy chain)
asc transport system	
SLC7a10	Asc1 (light chain)
SLC3a2	4F2HC (heavy chain)

transported through the cellular membrane by 4 transport systems (y^+, y^+L, b^{0+}, and B^{0+}; Table 9.20). Both y+ and y+-like transport systems are inhibited by arginine and lysine.

The Na^+-independent L transport system carries large neutral amino acids with branched or aromatic side chains, such as leucine, isoleucine, valine, phenylalanine, tyrosine, and tryptophan. L-type (large neutral) amino acid transporters include LAT1, or SLC7a5 carrier, LAT2, or SLC7a8 carrier.

The y^+ transport system is insensitive to neutral amino acids (NAA) and does not depend on pH. It mediates the bidirectional CAA transport. The y^+ transfer system is based on cationic amino acid (CAA) transporter (CAT) that carries ^{L}arginine, a substrate for nitric oxide [951]. The y^+ transport is mediated by ubiquitous plasmalemmal CATs (CAT1, or SLC7a1, CAT2a–CAT2b, CAT3, and CAT4).

The y^+L transport system is sensitive to neutral amino acids. It has a reduced affinity in the absence of Na^+ cation. It recognizes cationic amino acids in the absence of Na^+, but it requires sodium to interact with neutral amino acids. The exchangeable y^+-like transport is mediated by y^+L amino acid transporters (y^+LAT), such as y^+LAT1 (SLC7a7) and y^+LAT2 (SLC7a6).

The $B^{0,+}$ transport system, which is sensitive to cystine, is another route that use $B^{0,+}AT1$ carrier. The substrate specificity of $B^{0,+}$ and $b^{0,+}$ transport systems is similar, but the former also accepts alanine and serine. The $B^{0,+}$ transport is mediated by SLC3a1 carrier (cystine, dibasic and neutral amino acid transporters), or $B^{0,+}$-type amino acid transport[76] Activator of cystine, dibasic and neutral amino acid transport possesses variants (SLC3a1 variant-B2–SLC3a1 variant-G2).

The $b^{0,+}$ transport system recognizes cationic and neutral amino acids with a similar affinity. It carries amino acids independently of sodium. The $b^{0,+}$ transport system depends on both Na^+ and Cl^- ions. It conveys cationic and neutral amino acids with the highest affinity for hydrophobic amino acids. The $b^{0,+}$ transport is mediated by SLC7a9 carrier (glycoprotein-associated amino acid transporter light chain, $b^{0,+}$ transport system), a.k.a. cationic amino acid transporter of y^+ transport system and $B^{0,+}$-type amino acid transporter BAT1, $B^{0,+}AT$, and $b^{0,+}$-type amino acid transporter ($b^{0,+}AT$).[77]

Transcripts (mRNAs) of amino acid transporters CAT1, LAT1, LAT2, y^+LAT1, y^+LAT2, and $b^{0,+}AT$ are expressed together with the heterodimeric glycoproteins SLC3a2 carrier (activator of dibasic and neutral amino acid transport)[78] and rBAT transporter.

In fact, the amino acid transporter systems L, y^+L, x_c^-, asc, and $b^{0,+}$ represented by transporter light chains require interaction via a disulfide bridge with the type-2 membrane glycoproteins — heavy chains — for transmembrane transfer. System L, y^+L, x_c^-, and asc require SLC3a2, whereas system $b^{0,+}$ requires rBAT subunit.

The x_c^- transport system relies on SLC7A11 carrier, also called cystine/glutamate transporter xCT and anionic amino acid transporter light chain.[79] The asc transport system is based on neutral amino acid transporter SLC7a10 that corresponds to the light chain of the dimeric carrier.

In endothelial cells, arginine uptake does not depend on extracellular Na^+ and Cl^-, but is inhibited by neutral amino acid glutamine and leucine. In addition, molecules can be modified during endothelial transit.

[76] A.k.a. neutral and basic amino acid transport nbAT and rBAT as well as cystinuria protein-1 (Csnu1).

[77] A.k.a. cystinuria protein-3 (Csnu3).

[78] A.k.a. 4F2 cell-surface antigen heavy chain (4F2HC).

[79] A.k.a. calcium channel blocker resistance protein CCBR1.

Intracellular transport can begin with receptor binding.[80] Once bound to their receptors, the molecules assemble in a patch that then bulges to form a vesicle. From the starting membrane granulation to its fusion with target membrane, the vesicle carries the molecules. Such types of shuttles also transport manufactured molecules from the endoplasmic reticulum to the Golgi body, from which are launched vesicles for immediate extracellular release, vesicles for signal-dependent secretion, and digestive lysosomes.

The intracellular receptor-mediated transport is aimed at transfering molecules either to feed endothelial cell (*endocytosis*) or across the cytoplasm to supply wall tissues with nutrients and growth factors from the blood (*transcytosis*). Endocytosis allows not only nutrient uptake, but also membrane maintenance. After internalization, main routes lead to degradation, recycling,[81] transient sequestration in endosomes, or transcytosis. Pinching off of vesicles requires dynamin. Vesicular traffic is controlled.

Clathrin-mediated endocytosis, which involves actin, captures receptors, ligands and extracellular fluid. Caveolae-mediated endocytosis[82] is involved in receptor-mediated endocytosis and the transport of blood macromolecules. Caveosomes transport albumin across the endothelium. Vesicles used for intracellular transport are also involved in signal transduction. Caveolin of the caveolar membrane binds to certain signaling molecules and can be involved in angiogenesis and apoptosis. Actin and microtubules are implicated in raft-mediated endocytosis.

9.6.2.3 Caveolae in Transcytosis

Transcellular transport, or transcytosis, corresponds to endocytosis followed by exocytosis at apical and basal plasma membranes of endothelial cells, respectively. Caveolae correspond to a large fraction of the plasma membrane of endothelial cells (10,000–30,000 per cell; size 50–100 nm) [854]. Caveolae trigger transport of plasma proteins, fatty acids, hormones, and other signaling molecules.

Transcytosis involves a sequential series of events: budding, fission, translocation, docking, and fusion of caveolae to the abluminal membrane. Transcytosis is initiated by interaction of plasma proteins with specific docking molecules in plasmalemmal caveolae that are released upon scission. Caveola-mediated transcellular route uses multiple caveola-associated regulators, such as dynamin and intersectin.

---

[80]Glucose is a hydrophilic compound that requires specific carriers for its transport into the cytosol. Glucose GLUT transporters are expressed in endothelial cells [952]. These carriers are intrinsic transmembrane proteins with different tissue-specific isoforms. Cells have a continuous supply of glucose to be used either as a precursor of bigger molecules or an energy source by generating ATP through glycolysis.

[81]Expression of plasmalemmal receptors is regulated by recycling.

[82]Caveolae, invaginations of the plasma membrane that trap extracellular substances, abound in endothelial cells. Caveolae are main locations of PDGF receptors at the platelet surface.

Constituents of endocytic vesicles include caveolin in addition to dynamin and intersectin. The caveolin family includes 3 gene products (Cav1–Cav3). Caveolin-1 is the major component of caveolae. Caveolin-1 has 2 isoforms. Caveolin-1 and -2 are coexpressed in many cell types, especially in endothelial cells, smooth myocytes, fibroblasts, and adipocytes, whereas caveolin-3 is a muscle-specific type. Endothelial cells from brain microvessels as well as type-2 alveolar epithelial cells express caveolin-1, but have relatively few caveolae. Otherwise, caveola number is correlated with caveolin-1 concentration.

Caveola genesis that requires caveolin-1 occurs by plasmalemmal lipid remodeling into specialized membrane nanodomains rich in cholesterol and sphingolipids. Caveolin-1 binds with high affinity to cholesterol, sphingolipids, sphingomyelin, and glycosphingolipids [854]. Plasmalemmal cholesterol regulates expression of caveolin-1 and caveola assembly. Caveolin-1 forms oligomers of 14 to 16 monomers in caveolae.

Caveolin-1 self-assembly begins in the endoplasmic reticulum. Post-translational modifications of monomeric caveolin-1 (acylation, palmitoylation, and phosphorylation) and conjugation of glycosphingolipids facilitates caveolin-1 oligomerization. Cholesterol is needed for caveolin-1 oligomerization and recruitment to the plasma membrane. Oligomerized caveolin-1 stabilized by membrane cholesterol generates caveolae owing to cofactors such as small Ras GTPase, whereas large GTPase dynamin recruited by intersectin releases caveolae from the plasma membrane.

Caveolin-1 interacts with diverse signaling molecules, such as EGFR and PDGFR receptors, Gq and other GTP-binding proteins, Src and PKCα kinases, IP_3R and TRPC1 Ca^{2+} channels, and nitric oxide synthase NOS3. Caveolin-1 inhibits NOS3 enzyme. Caveolin-1 regulates activity of Rac GTPase and matrix metallopeptidase.

Caveolin-1 favors association of β-catenin with cadherin, thereby promoting the assembly of adherens junction complex and contributing to the endothelial paracellular barrier.

In endothelial cells, cortical actin barrier that is integrated with the plasma membrane must be crossed by vesicles that carry various types of molecules. However, actin participates in caveola-mediated transport, as localized actin polymerization occurs at sites of caveola fission. Dynamin interacts with actin-polymerizing proteins of cortical actin filaments, such as the ARP2–ARP3 complex, cortactin, filamin, profilin, and WASP, as well as adaptor GRB2. Therefore, dynamin is able to control actin dynamics required for transport of released caveolae. Actin-related nanomotor myosin serves to carry vesicles in the cell periphery in both anterograde and retrograde direction with respect to plasma membrane. Microtubules also drive caveola motion within the endothelial cell owing to their nanomotors kinesins and dyneins.

Dynamin causes caveolar fission. Homotetrameric dynamin undergoes GTP-dependent self-assembly that increases its intrinsic GTPase activity. Dynamin oligomers generate a constricting force around vesicle collars that leads to membrane fission and caveola internalization.

Intersectins regulate membrane fission and fusion of endothelial caveolae. Intersectins are dynamin partners and regulators. Intersectin isoforms comprise nerve-specific intersectin-1 and ubiquitous intersectin-2. The latter has 2 splice variants, short (intersectin-2_S) and long (intersectin-2_L) forms.[83] Intersectin-2 is located at necks of caveolae, where it can sequester dynamin dimers or tetramers [854].

Scaffold intersectin binds to Ca^{2+}, phospholipids, inositol phosphates, WASP, RasGEF Son of sevenless, CDC42 and Rac1 GTPase-activating protein RhoGAP31, soluble ^{N}ethylmaleimide-sensitive factor attachment protein receptor (SNARE), synaptosome-associated protein SNAP23 and SNAP25, and synaptojamin. Unlike dynamin, intersectin remains associated with vesicles following fission.

Caveolae require vSNAREs and tSNAREs on vesicle and target membranes, respectively. Vesicle docking and fusion involve syntaxin, Rab5, vesicle-associated membrane protein (VAMP),[84] and ^{N}ethylmaleimide-sensitive factor (NSF). Molecule NSF binds SNARE complex via soluble α- or βSNAP. Hexameric ATPase NSF and αSNAP are SNARE regulators that catalyze SNARE complex disassembly following membrane fusion and liberate SNARE monomers for recycling [854]. Small GTPases of the RAB family regulate cellular transport. Endothelial cells express Rab1 to Rab9, Rab11, Rab13 to Rab15, Rab18, Rab22, and Rab30 [854].

Caveola-associated proteins can regulate transcytosis. *Endothelial macromolecular transcytotic complex* that comprises caveolin-1, dynamin, Rab5 GTPase, ^{N}ethylmaleimide-sensitive factor, and syntaxin is involved in caveola fission, targeting, docking, and fusion with the plasma membrane.

Kinase Src allows formation of the caveolin-1–dynamin complex after phosphorylation of both dynamin and caveolin-1. Balance between tyrosine phosphorylation and dephosphorylation of caveolin-1 and dynamin controls endocytosis. Protein kinases PKCα, PKCβ, and PKCδ lodge in caveolae. Activation of these PKC isoforms as well as PKCϵ and PKCγ prevents caveola internalization [854]. Phosphatidylinositol 3-kinase is also involved in endocytosis.

Caveolae contain Ca^{2+}-signaling elements, such as IP_3R, store-operated Ca^{2+} channels, transient receptor potential channels TRPC1 and TRPC3, and Ca^{2+} ATPase [854]. Increase in intracellular Ca^{2+} concentration induces exocytosis of Weibel-Palade bodies.

---

[83]Intersectin-2 long form possesses a GEF domain that activates CDC42 GTPase, but neither Rac nor RhoA GTPse.

[84]A.k.a. synaptobrevin.

9.6.3 Endothelial Permeability and Vascular Compartments

Structural and functional differences among endothelia that coat successive compartments of the vasculature are associated with changes in transport rate. Endothelium structure is a first agent that can explain differences in transport along the vasculature. Postcapillary venules have intercellular gaps of 3.0 nm, whereas clefts of arterioles and capillaries are impermeable to molecules with size greater than 2.0 nm [854].

Endothelial cell expression and extracellular matrix composition vary according to the vascular territory. Expression of Lin1, Isl1, and Mec3 (LIM) kinase (LIMK), myosin light chain kinase (MLCK), guanine nucleotide-exchange factor Vav for Rho GTPase, and myosin is higher in endothelial cells of microvessels than in larger vessels [854]. Cytoskeletal reorganization and intercellular and cell–matrix attachment remodeling regulate endothelial transport. Moreover, caveola density is the highest in capillaries.

Endothelial cells are coupled to the extracellular matrix that influences transport. Degradation of fibronectin, collagens, or proteoglycans indeed increases ECM permeability. The extracellular matrix contains collagen-4, laminin, fibronectin, entactin, chondroitin sulfate, heparan sulfates (perlecan and syndecan), and matrix-associated proteins, such as thrombospondin and adhesion antagonist secreted protein acidic and rich in cysteine (SPARC). Matrix proteins laminin and collagen-4α1 and -4α2 are associated with microvessel endothelia, whereas a greater contribution of fibronectin and collagen-5α1 and -5α2 is observed in large vessel endothelia [854]. Specific interactions between endothelial cell adhesion molecules and extracellular matrix constituents can modify endothelial permeability from a vascular compartment to the other.

Endothelial cell responsiveness also influences transendothelial transport. Endothelial cells of microvessels exhibit higher basal cyclic nucleotide levels, distinct profile of thrombin-induced intracellular Ca^{2+} transients, and stronger cAMP-increasing agent reactivity and oxidant production with respect to that in large vessels [854].

9.6.4 Endothelial Permeability Regulation

The overall mass transfer is not significantly affected by cell shape, although the transport varies over the cell wetted surface, in a model of mass transfer coupled to a steady flow in a domain of height small enough to assume a homogeneous fluid (plasma flow in a diffusion boundary layer) [953]. However, change in cell configuration can associate with variations in intercellular cleft size.

9.6.4.1 Actin–Myosin Cytoskeleton and Partners

The endothelium permeability is enhanced by transient rise in $[Ca^{2+}]_i$ induced by various mediators that regulate arrangement of the cytoskeletal mesh and quality of cell junctions. The primary determinant of endothelial permeability is the state of the actin cytoskeleton. Myosin sliding along actin filaments generates contraction. Cytoskeleton contraction induces small gaps between endothelial cells. It then raises extravasation of macromolecules [954]. Two myosin light chains (MLC), one structural and one regulatory, are bound to the myosin heavy chain (MHC).

Actin polymerization is controlled by various processes, such as nucleation, elongation, severing, and crosslinking of actin filaments. These mechanisms involve a huge set of regulators ($\sim$160). Concomitant formation of actin stress fibers, actomyosin cross-bridging, and destabilization of microtubules increase endothelial permeability. Microtubule destablization induces activation of P38MAPK and microtubule-associated RhoGEF1 or RasGRF1 that targets RhoA GTPase [854]. Endothelium resistance to configuration change depends on adhesions of endothelial cells to both adjoining cells and matrix.

Actomyosin-mediated contraction that primes formation of intercellular gaps requires mono- or diphosphorylation of regulatory myosin light chain (MLC) by myosin light chain kinase. Ca^{2+}–calmodulin-dependent MLCKs phosphorylate regulatory MLCs that interact with actin filaments. On the other hand, myosin-1-associated protein phosphatase balances by dephosphorylation of MLCs the phosphorylation effect of MLCKs. Phosphatase must then be inhibited. Agent cAMP reduces MLC phosphorylation to preserve endothelial barrier integrity. In addition, PKA-anchoring proteins (AKAPs) and vasoactive stimulatory phosphoprotein (VASP) are required for cAMP-induced Rac1 activation that stabilizes endothelial barrier.[85]

Myosin light chain kinase can be phosphorylated by protein kinase-A and P21-activated kinase that reduce its catalytic activity, as well as extracellular signal-regulated kinases ERK1 and ERK2 and Src kinase (Table 9.21). Conversely, protein phosphatases dephosphorylate MLC^P. Cytosolic Ser/Thr phosphatase PP3, but neither PP1 nor PP2, is a Ca^{2+}-dependent phosphatase (a.k.a. calcineurin). Inhibition of phosphatase activity increases endothelial permeability. Phosphorylation of PP1 by RoCK kinase specifically targeted by GTPase RhoA hinders PP1 activity.

Actomyosin-mediated contraction results from an increase in cytosolic calcium concentration from endoplasmic reticulum stores followed by a sustained activity of transient receptor potential channels in endothelial cell plasma membrane (Sect. 9.6.4.2).

Calcium sensitization corresponds to increased myosin light chain phosphorylation, and hence cell contraction, in response to an external stimulus independently of changes in intracellular Ca^{2+} concentration. This process happens mainly by

[85]VASP may act as a component of a scaffold complex that controls GEFs or GAPs dedicated to Rac1 GTPase.

Table 9.21. Regulatory enzymes of the contractile state of endothelial cells (+: activation or stabilization of cell junctions; −: inactivation or inhibition of actin filament formation; AMPK: AMP-activated protein kinase; ERK: extracellular signal-regulated protein kinase; ERM: ezrin–radixin–moesin; LIMK: Lin1, Isl1, and Mec3 kinase; MLCK: myosin light chain kinase; MLCP: myosin light chain phosphatase; PAK: P21-activated kinase; PKA: protein kinase-A; PKG: protein kinase-G; PP: protein phosphatase; RoCK: Rho-associated, coiled coil-containing protein kinase; Src: sarcoma-associated Schmidt-Ruppin A2 viral oncogene homolog kinase; VASP: vasodilator-stimulated phosphoprotein). Endothelial permeability is regulated by myosin light chain phosphorylation and dephosphorylation that direct endothelial cell contraction and relaxation.

Enzyme	Substrates
AMPK	VASP (−)
ERK1/2	MLCK (+)
Fyn	Vav (RhoGEF)
LIMK	Cofilin (−)
MLCK	MLC
MLCP	MLC^P
PAK	MLC, MLCK (−)
PKA	MLCK (−), VASP (+)
PKG	VASP (−)
PP1	Src (−)
PP2	Src (−)
RoCK	ERM (+), LIMK (+), MLC, MLCP (−), PP1 (−)
Src	MLCK (+), FAK, paxillin,

phosphorylation of the myosin-binding protein phosphatase-1 inhibitory subunit $PP1_{r12a}$ and subsequent inhibition of myosin light chain phosphatase. At basal concentrations of intracellular Ca^{2+}, phosphorylation level of myosin light chain is modulated by $PP1_{r12a}$ subunit. Calcium sensitization of endothelial cell contraction is initiated by the activation of the small GTPase Rho and RoCK kinase.

Kinases of the SRC family do not change the strength of E-cadherin-mediated adhesion, but integrin-dependent enhancement of cadherin-mediated adhesion strength. Establishment of cadherin-based cell adhesion needs myosin-2 and MLCK and RoCK kinases. The regulation of intercellular adhesion strength by stimulated integrins requires activated SRC family kinases that target the Rho–RoCK pathway [955]. Kinase Src binds to newly formed actomyosin stress fibers linked to focal adhesions. Phosphorylation by Src kinase modulates endothelial cell contraction [956]. Similarly, in thrombocytes, Src is recruited to actin bundles. Enzyme Src activates PLCγ2 and causes MLC phosphorylation for clot retraction downstream from $\alpha_{2B}\beta_3$ integrin [957].[86]

[86]Activation of $\alpha_{2B}\beta_3$-integrin by inside–out signals promotes binding to fibrinogen and platelet aggregation. In turn, clustering of $\alpha_{2B}\beta_3$-integrin, once engaged by fibrinogen, mediates outside–in signals that elicit phosphorylation of β_3-integrin tail and activation of the SRC cascade that involves SYK, lymphocyte cytosolic protein LCP2, and PLCγ2. However, binding of SYK to

Small Rho GTPase and its effector RoCK kinase are implicated in the formation of stress fibers from bundles of actin microfilaments and myosin-2. Kinase RoCK is activated by binding to $RhoA^{GTP}$. Small RhoA GTPase also regulates actin polymerization by its effectors diaphanous protein (Dia) and RoCK kinase [854]. Effectors Dia and RoCK activate LIM kinase that phosphorylates (inactivates) actin-severing cofilin. In addition, RoCK also activates crosslinkers ezrin–radixin–moesin between cortical actin and plasma membrane. On the other hand, P21-activated kinase that is an effector of both small GTPases CDC42 and Rac, not only inhibits MLCK, but also phosphorylates MLC independently of MLCK kinase.

Actin crosslinkers of the family of actin-binding proteins, such as filamin and spectrin, stabilize cortical actin and control organization and distribution of actin filaments at intercellular junctions [854]. They then regulate endothelial barrier. In endothelial cells, ligation of spectrin with band-4.1 is required for TRPC4-induced Ca^{2+} entry. Filamin also links to E-selectin upon leukocyte binding to the luminal surface of endothelial cells.

Other actin-binding proteins, such as cortactin and WASP, bind to the heptameric ARP2–ARP3 complex. They not only foster actin polymerization, but also participate in the assembly of adherens junctions and focal adhesions, as ARP2–ARP3 interacts with E-cadherin and vinculin. Phosphorylation of WASP by focal adhesion kinase prevents its nuclear translocation and thus prolongs WASP-induced actin polymerization caused by CDC42 to restore barrier function.

Vasodilator-stimulated phosphoprotein (VASP) cooperates with Rac1, as it facilitates Rac1 activation downstream from PKA and PKG, for the maintenance of the endothelial barrier. Furthermore, it modulates the activity of capping (e.g., gelsolin) and actin-binding proteins (e.g., profilin).

In endothelial cells, VASP is phosphorylated by cAMP- and cGMP-dependent protein kinases PKA (Ser157) and PKG (Ser239), as well as by AMP-activated protein kinase (AMPK; Thr278) [958]. Phosphorylation of VASP (Ser157) causes VASP localization to the cell cortex, but has a minor impact on ^{F}actin assembly. Upon phosphorylation by PKA, VASP lodges in adhesion complexes with occludin, zonula occludens protein ZO1, and junctional adhesion molecule JAM1 to stabilize cell junctions. Phosphorylation of VASP (Ser239 or Thr278) impedes actin filament formation.

Synthesis of VASP is regulated by several transcription factors. During hypoxia, HIF1 binds to the VASP promoter [958]. Inflammatory cytokines, such as tumor-necrosis factor-α and interleukin-1 and -6 provoke NFκB binding to the VASP promoter and downregulate Vasp gene expression.

$\alpha_{2B}\beta_3$-integrin is impeded by tyrosine phosphorylation of the β_3-integrin tail. On the other hand, Src kinase activates PLCγ2 and provokes MLC phosphorylation.

Gelsolin regulates filamentous actin turnover, thus contributing in endothelial barrier. Ezrin, radixin, and moesin crosslink actin upon activation and can serve as signaling molecules.[87]

Heat shock protein HSP27 hampers actin polymerization. This inhibition is relieved by HSP27 phosphorylation that can be induced by P38MAPK as well as several edemagenic agents, such as TNFα, lipopolysaccharides, thrombin, and hydrogen peroxide (H_2O_2) [854].

9.6.4.2 Calcium Signaling in Endothelial Permeability Regulation

Part of solute exchange between blood and tissue is mediated by changes in the paracellular permeability of the endothelial layer that increases with rising level in endothelial calcium and cytoskeleton contraction owing to myosin light chain phosphorylation by Ca^{2+}–calmodulin activation of endothelial-specific myosin light chain kinase. Thrombin or histamine transiently increase $[Ca^{2+}]_i$ and vascular permeability, whereas growth factors induce a sustained change.

Store-operated channel activation (e.g., TRPC1 and TRPC4) regulates endothelial barrier function. Channels TRPC1, TRPC4, TRPC6, and TRPV1 participate in the control of vascular permeability [896]. TRPC1 overexpression augments thrombin- and VEGF-induced increase in endothelial permeability. Tumor-necrosis factor-α stimulates TRPC1 expression via NFκB. Small GTPase RhoA, activated by thrombin, can form complexes with IP_3R and TRPC1 that translocate to the plasma membrane [959]. Protein kinase-Cα phosphorylates TRPC1 for Ca^{2+} influx to increase endothelial permeability induced by thrombin [960].

Coupling between IP_3R and TRPC1 may be mediated via cytoskeletal linkage. The Gq–PLC–PKCα axis activates RhoA GTPase. The latter can then facilitate IP_3R–TRPC1 interaction by promoting actin polymerization. Calcium entry through SOC channels requires conjugation of actin crosslinker spectrin with protein-4.1. In addition, TRPC1 activation needs phosphorylation by PKCα.

Channel TRPC6 mediates VEGF-induced increase in endothelial permeability. Channel TRPV1 can control vascular tone and permeability via Ca^{2+} influx that enhances the phosphorylation of vasoactive stimulatory phosphoprotein.[88]

Actin-binding protein VASP stabilizes actin filaments and protects actin filaments from actin-severing protein gelsolin that causes actin filament disassembly [961]. Agent VASP is a substrate of cGMP-dependent protein kinase

[87]The ERM proteins contain a C-terminal actin-binding domain and N-terminal FERM protein- and lipid-binding domain (FERM stands for Four point-1, ezrin, radixin, and moesin). Ezrin binds focal adhesion kinase. Radixin can interact with Gα_{13} protein [854].

[88]Vasoactive stimulatory phosphoprotein (VASP) localizes to regions of high actin filament turnover. Agent VASP attracts profilin–actin complex and is related to high concentration of free profilin [962].

PKG1,[89] and PKA; its phosphorylation leads to vasodilation. Phosphorylated VASP level can serve as an indicator of the NO–sGC–PKG1 pathway that results in relaxation of smooth myocytes [964]. Once VASP is phosphorylated by PKA, $VASP^P$ localizes in endothelial junctional complexes with ZO1, occludin, and junctional adhesion molecule-1 [963]. Protein kinase-G phosphorylates P21-activated kinase, disrupts the NCK–PAK complex, thereby stimulating cell polarization and preventing angiogenesis, stimulates PAK–VASP binding, changes PAK localization, and causes remodeling of focal adhesions [965].

9.6.4.3 Microtubules

Microtubule network is another constituent of the endothelial cytoskeleton that operates in the barrier between flowing blood and interstitium beneath endothelial cell monolayer.

Microtubule heterodimers are formed by self-assembly of α- and β-tubulins. They generate a lattice of rigid hollow rods that span the cytoplasm from the nuclear region to the cell cortex. The faster growing end (plus-end) that is attached to the cell cortex is composed of β-tubulin that hydrolyzes GTP to GDP, whereas the minus-end that is fixed to the microtubule-organizing center consists of α-tubulin and binds to GTP without hydrolyzing GTP. Microtubules undergo dynamic instability, as they permanently shift between lengthening and shortening. Microtubule depolymerization increases endothelial permeability.

Microtubules sequester RhoGEFs and LIM kinase and thus control the state of tubulin and actin polymerization [854]. Activators RhoGEFs stimulate RhoA GTPase that disrupts microtubules. Microtubule-coupled nanomotor kinesin can associate with catenin-δ1. Microtubules hence promote intracellular transport of constituents of adherens junctions. Kinesin-1 is also implicated in component transfer to focal adhesions, Therefore, microtubules control turnover of endothelial cell adhesion sites to neighboring cells and extracellular matrix and, consequently, endothelial permeability.

9.6.4.4 Intermediate Filaments

Intermediate filaments (caliber 10–12 nm, i.e., between that of actin filaments [∼7 nm] and that of microtubules [∼,25 nm]) participate in cell mechanical integrity. They are constituted of vimentin and keratin. Vimentin can be phosphorylated by protein kinase-C. Moreover, it can link to cadherin-5 [854]. Therefore, intermediate filaments cooperate with the 2 other types of cytoskeleton elements to modulate endothelial permeability.

---

[89] cGMP-dependent protein kinase-1 is targeted by atrial natriuretic peptide and nitric oxide.

9.6.4.5 Extracellular Constituents

Endothelia are covered by the glycocalyx (thickness 20–3,000 nm depending on method and vessel type), a meshwork of interacting proteins, glycoproteins, proteoglycans, and glycolipids, at their luminal surface and rest on a basement membrane (thickness 40–60 nm) at their abluminal edge. Both structures influence transendothelial transport. In addition, endothelial cells sense and respond to signals from blood and neighboring cells that regulate transendothelial transport.

Endothelial permeability increases during inflammation, healing, and angiogenesis. Many agents, such as platelet-activating factor, thrombin, and vascular endothelial growth factor, raise endothelial permeability by increasing intercellular cleft size, i.e., damaging intercellular junctions. Histamine and serotonin provoke gaps (width 1–8 nm) between venular endothelial cells [854].

Matrix metallopeptidases and plasminogen activators such as urokinase-type plasminogen activator degrade the extracellular matrix for cell migration and release of sequestered growth factors during healing and angiogenesis.

Focal adhesion (Sect. 9.1.2.2), a specialized array of integrins, anchors the endothelial cell to the extracellular matrix. Because focal adhesions are connected to actin filaments, they transmit tension generated by actomyosin filaments to the extracellular matrix and conversely. Focal adhesions also serve as signaling platforms, the activity of which is mechanosensitive. Permeability-increasing mediators reorganize punctate focal adhesions into thick foci to strengthen cell–matrix adhesion.

9.6.4.6 Albumin

Plasma albumin maintains the transendothelial oncotic pressure gradient and regulates the transport of fatty acids, steroids, thyroxine, and amino acids. Albumin binding to endothelial surfaces initiates its transcytosis via plasmalemmal vesicles. Plasmalemmal albumin receptors in vascular endothelia include: (1) secreted protein, acidic, cysteine-rich protein (SPARC), or osteonectin; (2) 60-kDa albumin-binding protein located in caveolae (linked to calveolin-1) GP60, or albondin; and (3) smaller albumin-binding scavenger receptors GP30 and GP18 that target conformationally modified albumins [966].[90]

Albumin can be transported in caveolae in free form in the fluid phase or mainly bound to cell-surface albumin-binding protein that colocalizes with caveolin-1 and

[90] Albumin receptors GP30 and GP18, unlike GP60, are expressed by cultured rat fibroblasts, smooth myocytes, and endothelial cells, whatever the type of endothelial lining, in various organs (heart, lung, brain, kidney, adrenal, adipose tissue, skeletal muscle, diaphragm, duodenum, pancreas, and liver).

regulates transendothelial transport of albumin.[91] Albumin transcytosis happens within minutes. Albumin binding to its endothelial receptor is a specific, saturable process, but endothelial uptake of albumin is submaximal at normal plasma albumin concentration. Albumin-binding proteins GP18 and GP30 serve as scavenger receptors for denatured albumin that is carried toward lysosomes for degradation [854].

Albumin facilitates the transport of numerous hydrophobic molecules, such as enzymes and hormones, across the endothelium. Albumin carries not only myeloperoxidase and thyroid hormone, but also elementary building materials such as amino acid tryptophan. Albumin acts as a cargo chaperone for fatty acids that serve as energy sources, especially for myocytes, as well as other lipids such as growth factor sphingosine 1-phosphate [854]. Sphingosine 1-phosphate protects the endothelial barrier.

Albumin also contributes to the maintenance of endothelial barrier by interacting with the glycocalyx, as loss of adsorbed albumin on the glycocalyx can heighten tracer transport across the endothelium [854].

9.6.4.7 Endothelial Barrier Protectors

Certain proteins and phospholipids released from platelets enhance the integrity of vascular endothelium. Sphingosine 1-phosphate and angiopoietin-1 belong to the set of endothelial barrier-stabilizing factors that also restore endothelial barrier after intervention of agents that raise endothelial permeability.

ACase–cAMP–PKA Pathway

The cAMP–PKA pathway that regulates cytoskeletal and adhesive protein activity in endothelial cells assists the endothelial barrier [967]. At the abluminal region, focal adhesions between endothelial cells and the basement membrane contribute to about 20% of permeability resistance [968].

Activated protein kinase-A prevents endothelial contraction, as it impedes activation of RhoA GTPase and myosin light chain kinase. Moreover, PKA phosphorylates VASP, JAMs, and ZO1 to promote the formation of a restrictive paracellular barrier [854].

Messenger cAMP excites Rap1 GTPase that abounds at sites of intercellular junctions [854]. Concentration of cAMP in endothelial cells is regulated by adenylate cyclase and phosphodiesterases. Adenylate cyclase AC6 isoform that

---

[91] In endothelial cells, activation of plasmalemmal GP60 by albumin induces phosphorylation of GP60, caveolin-1, dynamin, and SRC family kinases Src and Fyn. Activated GP60 forms a complex with caveolin-1, Src, Fyn, and dynamin. Receptor GP60 stimulates Src kinases via activated $G\beta\gamma$ subunit of Gi protein. Activated Src then phosphorylates dynamin and caveolin-1.

Table 9.22. Regulation of myosin light chain phosphatase (MLCP), hence of the endothelial barrier maintenance, by cAMP signal (Source: [969]).

MLCP substrates	Involved mediator	cAMP signaling effector
$PP1_{r12a}$–PP1	RoCK	PKA (not the cAMP–RapGEF axis)
$PP1_{r14a}$–PP1		PKA (not the cAMP–RapGEF axis)

is synthesized in endothelial cells and colocalizes with SOCs and at sites of intercellular junctions possesses both high- and low-affinity Ca^{2+}-binding sites. Permeability-increasing factors that raise cytosolic Ca^{2+} concentration can transiently preclude AC6 activity. Upregulation of AC8 isoform in endothelial cells reverses Ca^{2+}-induced inhibition of cAMP production, hence impeding intercellular gap formation.

Activation of cAMP signaling suppresses thrombin-induced increase in endothelial permeability, as it inactivates endothelial contractile apparartus, mainly myosin light chain phosphatase [969]. Two cAMP effectors, PKA and RapGEF3, cooperate to promote intercellular adhesion; however, PKA, but not RapGEF3, impedes thrombin-induced phosphorylation of myosin light chain and $PP1_{r12a}$ regulatory subunit, as it precludes the RhoA–RoCK pathway. In addition, the MLCP catalytic subunit PP1 can be activated via dephosphorylation and dissociation of $PP1_{r14a}$ inhibitor,[92] thereby also increasing the free pool of active PP1 phosphatase; this mechanism that is also used by the cAMP–PKA axis does not intervene in the cAMP–RapGEF3 pathway (Table 9.22) [969].

Sphingosine 1-Phosphate

Sphingosine kinase phosphorylates sphingosine and generates S1P, whereas sphingosine phosphatase or sphingosine lyase catalyzes phosphate-dependent transformation of S1P to phosphoethanolamine. Sphingosine kinase, phosphatase, and lyase thus regulate S1P plasma level. Circulating S1P can target endothelial cells and binds to $S1P_1$ to $S1P_3$ receptors. These liganded receptors prime endothelial programs for cell migration and proliferation.

Platelets that lack sphingosine 1-lyase constitute a primary S1P storage. Once released from platelets, S1P binds to serum albumin. Sphingosine 1-phosphate also connects to apolipoprotein-M of HDL particles [970]. ApoM+ HDL causes $S1P_1$ receptor internalization, MAPK and PKB activation, endothelial cell migration, and formation of endothelial adherens junctions.

Protection of the endothelial barrier ensured by S1P at physiological level (250–500 nmol) results from activation of Gi-coupled $S1P_1$ and $S1P_3$ receptors that

[92] A.k.a. 17-kDa PKC-primed inhibitor of PP1 (CPI17).

stimulate Rac GTPase.[93] Small Rac GTPase favors assembly of cortical actin and adherens junction. Phospholipid S1P also induces assembly of focal adhesion complexes.

Sphingosine 1-phosphate mobilizes Ca^{2+} from intracellular stores and activated non-selective cation channels via Gi protein. Calcium influx from endoplasmic reticulum induces Rac translocation to cell junctions and its activation.

Angiopoietin-1

Angiopoietin-1 activates TIE2 receptor Tyr kinase in vascular endothelial cells during angiogenesis as well as vascular maturation and remodeling. Angiopoietin-1 is constitutively expressed and its receptor TIE2 is constitutively phosphorylated. It triggers endothelial cell sprouting and migration, as well as tube formation. Moreover, it prevents apoptosis. It also reduces leukocyte adhesion to the endothelium.

Angiopoietin-1 ensures protection of the endothelial barrier, avoiding endothelial leakage. Angiopoietin-1 thus opposes VEGF-induced increase in endothelial permeability, probably by inhibiting the RhoA pathway. Furthermore, angiopoietin-1 prevents heightening of endothelial permeability caused by bradykinin, histamine, platelet-activating factor, and thrombin. Angiopoietin-1 inhibits formation of endothelial gaps, strengthens endothelial adhesion, and restore the endothelial glycocalyx [971].

Nitric Oxide

Endothelial nitric oxide synthase (NOS3; Vol. 4 – Chap. 10. Other Major Signaling Mediators) regulates microvascular permeability. Nitric oxide augments cGMP concentration and consequently induces cytoskeleton relaxation and reduces endothelial permeability. Increase in $[Ca^{2+}]_i$ activates endothelial nitric oxide synthase. Furthermore, Ca^{2+}–calmodulin removes the inhibition by caveolin-1 on NOS3 synthase.

Plasma membrane-anchored, NOS3-containing caveolae prevent endothelial hyperpermeability induced by platelet-activating factor and vascular endothelial growth factor [972].[94] In addition, caveolar NOS3 tethered to the plasma membrane uncouples NOS3 phosphorylation (Ser1177) from NO production. Conversely, NOS3 endocytosis allows delivery of NO to subcellular compartments to modify junctional and/or cytoskeletal proteins and increase endothelial permeability to macromolecules.

---

[93]However, at high concentrations, S1P activates RhoA GTPase to form actin stress fibers possibly via Gq and G12/13 subclass members.

[94]Once myristoylated and palmitoylated, NOS3 targets caveolae in the plasma membrane and binds to caveolin-1. Activators of NOS3 acts by S-nitrosylating, phosphorylating (e.g., Ser1177), and dephosphorylating (e.g., Thr495) specific residues, as well as recruiting NOS3 to specific subcellular loci. Among them, platelet-activating factor and VEGF stimulate PKB to phosphorylate NOS3.

Klotho

Klotho is a circulating protein involved in the maintenance of the integrity of the endothelial barrier. Absence of Klotho actually causes hyperpermeability with increased endothelial cell apoptosis and downregulation of cadherin-5 that result from an increase in VEGF-mediated cytosolic Ca^{2+} concentration and activation of Ca^{2+}-dependent peptidases [973]. Klotho binds to VEGFR2 and TRPC1 Ca^{2+} channel; it fosters their co-internalization.

9.6.4.8 Permeability-Increasing Factors

Inflammatory mediators, such as thrombin, histamine, and platelet-activating factor, activate various signaling axes in endothelial cells to increase vascular permeability.

Agents that increase endothelial paracellular permeability, such as histamine and thrombin, cause distinct cell response features (magnitude and time of recovery) [854]. Thrombin is a procoagulant serine peptidase[95] that also regulates cell proliferation and smooth muscle contraction.

Thrombin

Thrombin binds to Gq- and G12/13-coupled peptidase-activating receptor PAR_1 and causes a transient increase in endothelial permeability. The predominant Gβ endothelial isoform Gβ1 is sequestered by receptor for activated C kinase RACK1. Thrombin dissociates Gβ1 from RACK1. Subsequently, Gβγ, Fyn kinase, and focal adhesion kinase form a complex in endothelial cells stimulated by thrombin. Subunit Gβγ activates Fyn kinase that phosphorylates FAK that can then be recruited to adherens junctions, thereby enabling recovery of endothelial barrier [974].

Upon stimulation of peptidase-activated receptor PAR_1,[96] thrombin provokes: (1) Ca^{2+} influx via, at least partly, receptor-operated channels (TRPC); (2) phosphorylation of endothelial Ca^{2+}–calmodulin-dependent myosin light chain kinase to increase its activity and myosin light chain phosphatase by RoCK kinase to attenuate its functioning; and (3) endothelial contraction. Resulting centripetal force develops within 5 mn after stimulus onset. Reversal of the response occurs within 2 h after removal of or in the continuous presence of thrombin [854].

[95] Thrombin is generated by proteolytic cleavage of inactive prothrombin by factors V and X with cofactors calcium ions and membrane phospholipids. Short-lived thrombin is eliminated by activated protein-C.

[96] Among peptidase-activated receptor isoforms in endothelial cells (PAR_1–PAR_4), only 2 (PAR_1 and PAR_2) are activated by thrombin. Magnitude of the permeability response mainly depends on PAR_1 number and affinity, as small and larger populations of distinct types exist with different dissociation constants.

Subunit $G\alpha_q$ of G protein coupled to PAR_1 activates phospholipase-C that catalyzes production of inositol trisphosphate and diacylglycerol from phosphatidylinositol (4,5)-bisphosphate. Inositol trisphosphate, in turn, activates IP_3R to rapidly release Ca^{2+} from its store. Depletion of endoplasmic reticulum elicits Ca^{2+} entry from the extracellular medium via store-operated channels. Subunits of the G12/13 subclass downstream of active PAR_1 can act on RhoA GTPase via PKCα-dependent phosphorylation of RhoGEF1 agent.

Elevated intracellular Ca^{2+} concentration also activates kinases, such as PKCα and Src, whereas it inhibits adenylate cyclase-6. Thrombin also: (1) transiently elevates RhoA activity via RhoGEF1 phosphorylation by PKCα for contraction of actomyosin filaments and adherens junction disassembly; (2) reduces Rac function; and (3) causes a delayed activation of CDC42 for endothelial barrier recovery [854].

In addition, thrombin stimulation of endothelial cells provokes cortactin phosphorylation by Src kinase. Cortactin^{P} associates with myosin light chain kinase. However, thrombin induces a sustained increase in endothelial contraction, although it generates a transient increase in myosin light chain phosphorylation. It indeed targets low-molecular-mass form of caldesmon (caldesmon_{lMW}) that is widely expressed in non-myocytes [854]. Thrombin provokes caldesmon translocation from the cytosol to the plasma membrane and cytoskeleton and induces caldesmon phosphorylation[97] to form stress fibers and stabilize contraction of actin–myosin filaments that causes a sustained increase in endothelial permeability.

Contraction of endothelial cells can lead to FAK activation. Focal adhesion kinase enhances focal adhesion formation and can suppress RhoA activity by activating p190RhoGAP to favor intercellular junction assembly.

Bradykinin

Bradykinin is a potent inflammatory and vasoactive substance generated by kallikreins at sites of tissue injury. It also disrupts endothelial barrier. Bradykinin is rapidly degraded (half-life ~30 s) by angiotensin-converting enzyme and dipeptidyl carboxypeptidase kininase-2 located on the endothelial cell surface. Bradykinin targets 3 receptor subtypes (B_1–B_3) that have distinct affinity for bradykinin.

Endothelial cells possess both inducible B_1 and constitutive B_2 receptors. Stimulated B_1 receptor causes a prolonged response. On the other hand, B_2 activation provokes a transient activity, as B_2 is rapidly internalized. In addition, receptor expression depends on endothelial cell types. Activated B_1 or B_2 signal via Gq and Gi to augment cytosolic calcium content and producte prostaglandins and nitric oxide. Bradykinin-induced endothelial permeability increase is independent of MLCK or the Rho pathway [854].

---

[97] Caldesmon can be phosphorylated by CamK2, ERK, P38MAPKs, and PKC.

Histamine

After its release from mastocytes, histamine excites endothelial cells, fibroblasts, and smooth myocytes by binding to 3 receptor subtypes (H_1–H_3). Both receptors H_1 and H_2 exist in endothelial cells, but histamine acts mainly via H_1 receptor. Histamine generates intercellular gaps (size 100–400 nm; half-life $\sim$2 mn). Histamine elicits actin stress fiber formation in a RhoA- and Rac-dependent manner [854].

Reactive Oxygen Species

Reactive oxygen species generated during inflammation and infection, i.e., at high concentrations, increase endothelial permeability. Endothelial barrier dysfunction can result from an increase in reactive oxygen species concentrations during an oxidative stress.

Oxygen-derived free radicals with one or more unpaired electrons, such as superoxide anion radical ($O_2^{\bullet -}$) and hydroxyl radical ($^{\bullet}OH$) constitute a ROS subset. Some reactive oxygen species are not radicals such as hydrogen peroxide (H_2O_2). Superoxide formed from molecular oxygen by mitochondrial electron-transport chains, NAD(P)H oxidases, xanthine oxidases, cyclooxygenases, lipoxygenases, and uncoupled nitric oxide synthases, is a precursor for several reactive oxygen and nitrogen species.[98] Reactive nitrogen species, such as peroxynitrite and nitric oxide, are other types of signaling molecules within blood vessels.

Endothelial NADPH oxidase complexes are major producers of O_2^- agent. Oxidants cause endothelial contraction by activating RhoA GTPase and MLCK kinase, increasing intracellular Ca^{2+} concentration, and stimulating phospholipases PLA2, PLC, and PLD, as well as PKC, MAPK, and Src kinases [854].

Hydrogen peroxide, a stable and permeant reactive oxygen species, even at low concentrations, modulates cytoskeletal reorganization with actin stress fiber formation, vasorelaxation, and vascular remodeling. It also increases endothelial permeability that can result from an augmented generation of stress fibers and a disruption of cortical actin. Among mediators of H_2O_2-primed permeability elevation, actin-binding myristoylated alanine-rich C-kinase substrate (MARCKS) expressed in endothelial cells mediates H_2O_2-induced changes in actin cytoskeleton architecture and endothelial permeability in bovine aortic endothelial cells [975].[99]

---

[98]Superoxide can be converted by superoxide dismutases to hydrogen peroxide. It can react with nitric oxide to form peroxynitrite ($ONOO^-$). It can also intereact with arachidonic acid to generate isoprostanes.

[99]Protein MARCKS links binds to actin, calcium–calmodulin, and membrane phospholipids (PIP_2). Once it is phosphorylated by PKC, MARCKS dissociates from the membrane and is unable to crosslink ^{F}actin. Agents that increase endothelial permeability, such as thrombin and diacylglycerol, also launch MARCKS phosphorylation in endothelial cells.

Agent H_2O_2 causes the phosphorylation and translocation of MARCKS from the cell membrane to the cytosol using a signaling cascade that involves successively Rac1, Abl1, PLCγ1, and PKCδ.

Vascular Endothelial Growth Factor

Vascular endothelial growth factor is also called vascular permeability factor as it increases both paracellular permeability via actin–myosin filament contraction and transcellular permeability via heightened vesicle fusion leading to vesiculovacuolar organelles. It is produced by perivascular cells and targets its endothelial receptor VEGFR2 to activate phospholipase-Cγ that synthesizes DAG and IP_3 second messengers. Moreover, VEGF enhances IP_3R–TRPC1 coupling and stimulates protein Gq/11 for RhoA activation [854].

Tumor-Necrosis Factor

Tumor-necrosis factor-α is an inflammatory cytokine released from activated monocytes and macrophages. It upregulates endothelial adhesion molecules ICAM1 and E-selectin to promote neutrophil adhesion. Subsequent ROS generation increases endothelial permeability. Moreover, TNFα: (1) induces expression of TRPC1; (2) excites PKCα that activates GTPase RhoA for actin stress fiber formation; (3) causes microtubule remodeling and adherens junction disassembly via P38MAPK activation; and (4) stimulates zinc-dependent MMP gelatinase that degrades the extracellular matrix [854].

Lipopolysaccharides

Lipopolysaccharides of bacterial outer membrane cause: (1) endothelial cell contraction by RhoA-dependent increase in MLC phosphorylation; (2) HSP27 phosphorylation; and (3) secondary release of TNFα and interleukin-1 and -8.

9.7 Extravasation

Flowing cells interact with one another and with vascular endothelia as well as subendothelial matrix during hemostasis, inflammation, healing, and thrombosis via intercellular and cell–matrix attachments. Leukocytes can efficiently infiltrate biological tissues to ensure the body's defense using innate and adaptive immunity. On the other hand, mesenchymal cells, such as fibroblasts and smooth myocytes, that are also able to migrate adhere strongly to the extracellular matrix. This adhesiveness determines cell shape. Mesenchymal cells align along fibers of the extracellular matrix.

Table 9.23. Extravasation-involved adhesion molecules (Source: [986] BM: basement membrane; ICAM: intercellular adhesion molecule;). E-, L-, and P-selectins stand for endothelium, lymphocyte, and platelet selectin, respectively. Sialomucins are secreted or membrane-associated mucins that either act as adhesion receptors or anti-adhesive agent. Sialomucin core protein-24, also called endolyn, multiglycosylated core protein MGC24, and CD164, is a cell adhesion molecule. Selectin-P ligand (SelPLg; a.k.a. P-selectin glycoprotein ligand PSGL1 or CD162) is a high-affinity receptor on myeloid cells and stimulated T lymphocytes for P-selectin on activated platelets or endothelial cells. It can tether to E- and L-selectins with lower affinity.

Extravasation step	Adhesion molecules
Capture tethering and rolling	Sialomucin, L-selectin, selectin-P ligand, Integrin-α_4 (leukocyte), E- and P-selectins, ICAM (endothelium)
Activation	Chemoattractants and GPCRs
Tight binding arrest	Integrin-β_2 and -α_4 (leukocyte), L-selectin, ICAM (endothelium)
Endothelium crossing	Cadherin-5
BM crossing Migration in matrix	Integrin-β_1 and -β_2, ICAM

Leukocytes are the largest flowing cells that must strongly deform to travel into small blood vessels and traverse biological media. Leukocytes change their shape permanently during migration. Once attached to the endothelium, they cross tiny gaps between endothelial cells or even cross endothelial cells. Leukocyte distribution in small vessel lumen depends on the interaction with the surrounding erythrocytes susceptible to aggregation [976, 977].

9.7.1 Flowing Cell Adhesion to Endothelium

Flowing cell adhesion to the endothelium starts with the penetration of the glycocalyx, mainly at the endothelium of postcapillary venules, where endothelial cells have weak tight junctions [978]. Circulating blood cells have adhesion receptors that enable the cells subjected to flow forces to adhere to the vessel wall (Table 9.23).

Between-cell interactions are initially reversible. Once tethered to the wall, a cell forms an irreversible adhesion. Following adhesion to the blood vessel wall, leukocytes change from an approximately spherical to a flattened shape due to a reorganization of the actin cytoskeleton and polarization. Cell polarization with a protrusive leading edge and a contractile rear is associated with a redistribution of intracellular signaling proteins and plasmalemmal receptors and adhesion molecules.

Table 9.24. Leukocyte integrins involved at the beginning of transendothelial migration (CR: complement receptor; ICAM: intercellular adhesion molecule; LFA: leukocyte function-associated antigen; Mac1: macrophage-1 antigen; VCAM: vascular cell adhesion molecule; VLA: very-late-activation antigen).

Integrin type	Other alias	Endothelial ligands
$\alpha_4\beta_1$	VLA4, CD29–CD49d	ICAM1, VCAM1
$\alpha_L\beta_2$	LFA1, CD11a–CD18	ICAM1
$\alpha_M\beta_2$	Mac1, CR3, CD11b–CD18	ICAM1

Prior to leukocyte transendothelial migration, heterotypic intercellular junctions are mediated by various types of adhesion molecules (Table 9.24; Vol. 1 – Chap. 7. Plasma Membrane). Activation of endothelial cells by inflammatory cytokines, such as tumor necrosis factor-α and interleukin-1, increases the density of selectins and integrin ligands, such as intercellular (ICAM1) and vascular (VCAM1) cell adhesion molecule, on the luminal, apical surface. Clusters of adhesion molecules form on aggregates of pre-existing plasmalemmal nanodomains that contain members of the tetraspanin family, such as Tspan29 and Tspan24 as well as eventually Tspan28. These clusters construct rings around bound leukocytes.

Clusters of integrin ligands recruit regulatory and structural elements such as filamentous actin, thereby connecting to the cortical actin cytoskeleton. These clusters can associate with protrusions of the apical endothelial membrane to constitute docking sites that can even encapsulate adherent leukocytes.

Endothelial cell adhesion molecules are connected to actin-binding adaptors such as α-actinin, cortactin, and filamin. Hence, a *diapedesis synapse* results that forms a strong intercellular adhesion and force-transduction platform to secure anchorage of migrating leukocytes to the cortical cytoskeleton of endothelial cells.

9.7.2 Cell Movement over the Wetted Endothelial Surface

Flowing cells then undergo an extravasation with sequential steps, the kinetics of which depends on shear. Leukocyte extravasation steps include [979]: (1) capture and tethering on the inflamed endothelium; (2) rolling; (3) activation of leukocytes by endothelium-bound chemokines; (4) arrest and firm adherence of activated leukocytes to the endothelium; and finally (5) locomotion (crawling) over the endothelial wetted surface; before they undergo diapedesis, or transendothelial migration (TEM), that needs degradation of the subendothelial basement membrane to continue motion through interstitium.

When cells are crawling on the wetted surface, they develop ventral membrane protrusions on the vessel wall to probe for permissive sites on the endothelial surface. In addition to podosomes and pseudopodia, chemokines and shear stress elicit the formation of filopodia composed of parallel bundles of actin filaments with actin-associated proteins [980].

9.7.3 Integrin-Based Cell Adhesion

The adhesive potential of a cell is defined by its repertoire of integrins that serve as adhesion receptors (Vol. 1 – Chap. 7. Plasma Membrane). These transmembrane proteins mediate adhesion of both resting and migrating cells. Coupling between actin and integrins serves as a hook for migrating cells.

Cell displacement is achieved partly through construction and reconstruction of adhesion foci. Moreover, migrating cells in the extracellular matrix remain bound to matrix constituents in preferential routes (haptokinesis and -taxis). Whereas mesenchymal cells are confined to matrix tracks, leukocytes prioritize guidance signals. They move along a chemoattractant gradient.

9.7.4 Intracellular Signaling in Endothelial Cells

Cytokines activate endothelial cells that then express adhesion molecules at their wetted surface such as vascular cell adhesion molecule VCAM1 to regulate leukocyte transmigration to sites of inflammation. In normal conditions, adhesion molecules are not produced. In particular, endothelial cells express microRNA-126 that inhibits VCAM1 expression [981].

The endothelium can either favor or inhibit flowing cell adhesion on its wetted surface. Released nitric oxide can inhibit adhesion of cells involved in inflammation or coagulation. In addition, endothelial cells continually produce 13-hydroxy-octadecadienoic acid (13HODE) that confers to normal endothelium resistance to platelet or monocyte adherence.

Conversely, adhesion molecules attract leukocytes for transmigration. Numerous adhesion molecules localize at junctions between adjacent endothelial cells to support leukocyte diapedesis, such as platelet endothelial cell adhesion molecule PECAM1, ICAM2, junctional adhesion molecules (JAM1–JAM3), T-cell surface glycoprotein-E2 (or CD99), and endothelial cell-selective adhesion molecule (ESAM) [980]. Some of these molecules are involved in the extravasation of specific leukocyte types or in response to specific stimuli.

E-selectin and intercellular adhesion molecule ICAM1 lodge also in focal adhesions, i.e., filamentous actin-linked integrin clusters that mediate cell–matrix adhesion [980]. Proteins that reside at both docking sites and focal adhesions include α-actinin, filamin, paxillin, and vinculin, as well as focal adhesion kinase.

Actin crosslinkers control connections between the plasma membrane and intracellular signaling machinery. Filamin binds small guanosine triphosphatases Rho and Ral, their activators such as RhoGEF Trio, and caveolin-1 to promote localized signal transduction [980].

In response to leukocyte binding, clusters of E-selectin and integrin ligands on the endothelial surface augment the intracytosolic concentration of Ca^{2+} in endothelial cells. This second messenger triggers cell contraction as well as transport of intracellular vesicles, especially those that carry adhesion molecules used for endothelial transmigration.

In addition, clusters of adhesion molecules activate Src and mitogen-activated protein kinases as well as monomeric Rac1, RhoA, and RhoG GTPases. Both Rac1 and RhoA promote contraction of actomyosin-based stress fibers. Both Rac1 and RhoG intervene in the formation of apical protrusions that generate to docking structures [980].

Small Rac1 GTPase favors the activation of endothelial NADPH oxidase NOx2 that generates reactive oxygen species. These substances, in turn, regulate the activity of protein kinases, such as protein kinase-C, focal adhesion kinase FAK2, as well as PTPn1 phosphatase, to control the phosphorylation state of junctional proteins [980].

Small GTPases participate in feedback loops. Binding of ICAM1 to leukocyte integrins launches recruitment of actin- and ICAM1-binding proteins cortactin and filamin that enhance ICAM1 clustering [980]. In addition, RhoA GTPase is activated on ICAM1 clustering and, in turn, promotes clustering of ICAM1, VCAM1, and E-selectin [980].

In endothelial cells, clusters of ICAM1 molecules linked to the cortical actin network via actin-binding adaptors (α-actinin, cortactin, ezrin, radixin, and moesin, as well as filamin, paxillin, and vinculin) reorganize the cytoskeleton and intercellular junctions.[100]

Filamentous actin-binding lymphocyte-specific protein LSP1[101] expressed by leukocytes and endothelial cells is a substrate of MAPK-activated protein kinase MAPKAPK2 subtype. Phosphorylation of LSP1 is impeded by P38MAPK, an upstream kinase for MAPKAPK2 [982].[102]

9.7.5 Intracellular Signaling in Leukocytes

Integrins are activated by chemokine receptors during adhesion of platelets and leukocytes onto activated endothelial cells. During the first stage of extravasation, blood cells interact owing to integrins.

[100] α-Actinin is a Ca^{2+}-regulated antiparallel crosslinker of actin filaments that participates in the maturation of small, nascent adhesions [980]. Filamin crosslinks actin filaments in orthogonal actin networks. α-Actinin and filamin cooperate to elevate actin mesh stiffness. They may also enhance the stability of docking structures.

[101] A.k.a. 47-kDa actin-binding protein and 52-kDa phosphoprotein (PP52).

[102] In neutrophils, $LSP1^{P}$ localizes to lamellipodium and stabilizes ^{F}actin polarization during neutrophil chemotaxis.

The rolling step along endothelial cells is a prerequisite for adhesion. Cells then bind firmly to endothelial cells via chemical interactions. Kindlin-3 that is exclusively expressed in hematopoietic cells binds to β-integrins (β_1–β_3) and regulates their activation [983].

Leukocyte adhesion, its directional motility, and protrusion formation are stimulated by chemokines. These processes rely on integrins as well as monomeric GTPases Rho and Ras. These GTPases regulate cell shape, adhesion, and motility, as they reorganize the cytoskeleton in coordination with integrins.

Activation of leukocytes by chemokines stimulates guanine nucleotide-exchange factors via local production of phosphoinositides, phosphatidic acid, or cAMP and phosphorylation by kinases such as Src protein. Involved GEFs include RacGEF T lymphoma invasion and metastasis-inducing protein TIAM1, RhoGEFs Vav1 to Vav3, and RasGEF dedicator of cytokinesis DOCK2 that activate Rho GTPases, as well as RapGEF1 and RapGEF3 that activate Rap GTPases [980]. The Vav RhoGEFs are involved in leukocyte adhesion, DOCK2 in actin polymerization and lateral mobility over activated endothelium, TIAM1 in crawling.

Small RhoA GTPase further contributes to ICAM1 and VCAM1 clustering, as it controls cortical actin polymerization, whereas RhoG supports cellular protrusion and formation of endothelial cell docking structures [980]. Small Rho GTPase initiates signaling to activate kinases and nitric oxide synthase NOS3 subtype. Small GTPase Rac1 primes the production of reactive oxygen species to reduce junctions between endothelial cells, thereby facilitating paracellular migration. Small GTPase Rac1 controls polarized actin polymerization in the leading edge by stimulating the actin-related protein ARP2–ARP3 complex. Small GTPases CDC42 and Rap1a control lymphocyte polarity via the PKCζ–PAR3 and –PAR6 polarity complexes.

9.7.6 Extravasation Steps

Rolling slows down circulating leukocytes and allows binding of chemokines (CC-chemokine ligand-5 and CXC-chemokine ligand-8) on inflamed endothelium to their G-protein-coupled chemokine receptors on leukocytes. Binding kinetics of adhesion molecules that allow leukocyte adhesion and rolling on endothelium depend on magnitude, rate, and history of mechanical stress applied by flowing blood. As mechanical stress can induce conformational change, thereby exposing binding domains of adhesion molecules, they can stabilize proteic interactions. Activated chemokine receptors trigger pathways of leukocyte integrin activation for firm adhesion of leukocytes. Interactions between β_2-integrins and intercellular adhesion molecule ICAM1 and between $\alpha_4\beta_1$-integrins and vascular cell adhesion molecule VCAM1 are also required for leukocyte adhesion.

Table 9.25. Main types of heparan sulfate proteoglycans (HSPG). The major plasmalemmal HSPGs are the glycosylphosphatidylinositol-anchored glypicans and transmembrane syndecans. Perlecan, agrin, and collagen-18 are the principal species in the extracellular matrix, especially basement membranes (Source: [924]).

	Location	Function
Syndecan-1	Monocytes, macrophages, neutrophils, eosinophils, lymphocytes, platelets, Endothelial cells	L-selectin ligand Chemokine receptor Signaling
Syndecan-2	Macrophages, endothelial cells	
Glypican	Endothelial cells, macrophages	
Perlecan	Basement membrane, interstitial matrix	Growth factor reservoir
Collagen-18	Basement membrane, interstitial matrix	L-selectin ligand Growth factor reservoir
Agrin	Basement membrane, interstitial matrix	Growth factor reservoir

Rolling of flowing blood cells is affected by the hemodynamic forces applied at the wetted surface of the vessel wall. The rolling mainly involves selectins. Hemodynamic forces can favor an extended conformation of certain domains of the selectin structure and thereby strengthen selectin–ligand bonds [984].[103]

During inflammation, resident macrophages produce cytokines, such as tumor-necrosis factor, which quickly induce appearance of preformed P- and E-selectins on the wetted surface of endothelial cells. Interactions between P-selectins and leukocyte ligands, such as P-selectin glycoprotein ligand PSGL1, initiates rolling on the endothelium. Rolling is stabilized when formation of new downstream receptor-ligand bonds balance dissociation of upstream bonds. Rolling is further stabilized by L-selectin binding to heparan sulfate proteoglycans on the endothelial wetted surface.[104]

Heparan sulfate proteoglycans participate in almost every stage of leukocyte diapedesis (Table 9.25). Like L-selectin, leukocyte $\alpha_M\beta_2$-integrin binds to heparan sulfate proteoglycans. They also present heparan sulfate-bound chemokines, such as CXCL8 produced by tissue-resident macrophages, to chemokine receptors on leukocytes. Chemokine binding activates leukocyte integrins; a stable interaction is then formed between leukocytes and endothelial cells.

[103]Lectin and epidermal growth factor-like domains of P-selectin show both curved and extended conformations. The extended configuration increases the affinity of P-selectin for its ligand P-selectin glycoprotein-1. It also enhances rolling adhesion on L-selectin under shear flow.

[104]The 3 selectin types interact with mucins produced by the endothelium. However, heparan sulfate proteoglycans are the dominant L-selectin ligand on the inflamed endothelium during acute inflammation.

Adherent cell motion between the point of tight adhesion and the junction between endothelial cells requires synergistic actions of $\alpha_L\beta_2$- and $\alpha_M\beta_2$-integrin with ICAM1 and ICAM2, respectively, and possibly of β_1 integrin with VCAM. Neutrophils express α_M-integrin and slighly $\alpha_2\beta_1$-integrin. Lymphocytes express α_L- and $\alpha_1\beta_1$-integrin [979].

Leukocyte crawling can cause transient weakening of endothelial cell junctions and formation of intracellular pores for para- and transcellular migration. Intercellular adhesion molecule ICAM1 clustering stimulates protein Tyr kinases to alter the function of vascular endothelial cell (VE)-cadherin (or Cdh5) [980]. Moreover, Cdh5-stabilizing Tyr phosphatase PTPRb (or VEPTP) dissociates from VE-cadherin. Subsequently, γ-catenin is phosphorylated to further reduce endothelial integrity.

Amine oxidase copper-containing AOC3,[105] expressed at the cell surface of endothelial cells, regulates leukocyte migration.[106]

Platelet endothelial cell adhesion molecule PECAM1 is recruited to sites of transmigration from the lateral border recycling compartment (LBRC) using microtubules and kinesin. Other adhesion molecules, such as T-cell surface glycoprotein-E2 as well as junctional adhesion molecules JAM1 and JAM3, are also enrolled [980].

The cell squeezes to penetrate the cell junction and cross the basement membrane using platelet–endothelial cell adhesion molecule-1 (PECAM1) and endothelial junctional adhesion molecules (JAM).

The transmigration takes 15 to 45 mn. Once leukocytes have ended their transendothelial migration, they interact with chemokines presented by heparan sulfate proteoglycans (perlecan, agrin, and collagen-18) in the basement membrane. Leukocytes secrete various peptidases, such as matrix metallopeptidases and heparanase. The degradation of the basement membrane results from a collaboration among endothelial cells, leukocytes, and platelets, as these 3 types of cells produce heparanase. Moreover, heparanase releases growth factors bound to basement membrane proteoglycans. These growth factors contribute to angiogenesis and tissue remodeling.

Once stably arrested by their integrin ligands on the TNFα-activated vascular endothelium, leukocytes most often undergo a cytoskeletal remodeling to move to the cleft between endothelial cells, resisting detachment from the vessel wall by the blood flow, before crossing the endothelium.

Activated ligand-bound $G\alpha_i$-coupled receptors are required in adhesion and rolling of leukocytes on the wetted surface of the endothelium in response to chemokine gradients.[107] However, VCAM1 expression level and α_4-integrin–VCAM1-dependent firm adhesion of lymphocytes on endothelial cells do not

---

[105] A.k.a. vascular adhesion protein-1 (VAP1).

[106] In AOC3-deficient mice, rolling velocity of polynuclear leukocytes and lymphocytes on the endothelium in inflammation sites is greater than in wild-type mice [985].

[107] G-protein subunits $G\alpha_{i1}$, $G\alpha_{i2}$, and $G\alpha_{i3}$ are expressed in granulocytes, lymphocytes, and airway smooth muscle, epithelial, and vascular endothelial cells.

depend on $G\alpha_i$ signaling. Subunit $G\alpha_{i2}$ in the vascular endothelium favors the diapedesis of firmly adherent lymphocytes [987]. Chemokine-stimulated G-protein-coupled receptors and their effectors, especially CDC42, RhoA, Rac, and Rap GTPases, promote actin reorganization.

In lymphocytes, Rac GTPase activation by integrins and chemokines depends on dedicator of cytokinesis (DOCK). T lymphocytes use 2 different chemokine-triggered actin processes [988]: (1) DOCK-dependent mechanism to move laterally along the endothelium and (2) DOCK-independent procedure to cross the endothelium.

9.7.7 Extravasation Modes

Leukocytes use both para- and transcellular extravasations. *Paracellular diapedesis* requires a transient loss of cadherin-5 that determines contact between apposed endothelial cells. Cadherin-5 is regulated by adaptor catenins that mediate interactions between cadherin-5 and the cortical actin cytoskeleton. Loss of cadherin-5 function results from a combination of stress fiber contraction by activated RhoA and phosphorylation of cadherin-5 and catenins. Cadherin-5 to which PTPRb phosphatase dissociates, is phosphorylated by Src and FAK2 kinases.

Transcellular diapedesis is associated with an endothelial transmigratory structure composed of docking microvilli that surrounds and guides migrating leukocytes (Sect. 11.5.1). This villus-like structure is constituted of intercellular adhesion molecules, particularly leukocyte integrins and endothelial IgCAMs [989]. Both B and T lymphocytes preferentially use the transcellular route.

Vesiculovacuolar organelles contribute to the lateral border recycling compartment for the transport of adhesion molecules such as PECAM1 to sites of transmigration [980]. Transcellular migration may require endothelial caveolin-1 for the formation of transcellular pores used by migrating cells.

Intermediate filament networks of both lymphocytes and endothelial cells contribute to lymphocyte migration. Intermediate filaments are implicated in the distribution of adhesion molecules, such as ICAM1 and VCAM1 on endothelial cells and $\alpha_V\beta_1$-integrin on lymphocytes. Intermediate filaments of both endothelial cells and lymphocytes form a dynamic anchoring structure at the between-cell contact loci. The initiation of this anchoring structure requires vimentin [990]. Vimentin in both the endothelial cell and lymphocyte stabilizes endothelial cell–lymphocyte interactions and reorganizes the intermediate filament network.

Leukocyte extravasation requires not only cell adhesion molecules and chemoattractants, but also plasmalemmal enzymatic reactions on both leukocytes and endothelial cells. *Ectoenzymes* (nucleotidases, cyclases, ^{ADP}ribosyltransferases, peptidases, and oxidases) that have catalytic domains outside the plasma membrane, regulate the cell recruitment [991].

Ectonucleotidases, such as ectonucleoside triphosphate diphosphohydrolase ENTPD1 and 5′-nucleotidase, and adenosine deaminase regulate ATP and adenosine concentrations. Adenosine inhibits adhesion molecules and decreases vascular permeability. Ecto^{ADP}ribosyl cyclases, such as ^{ADP}ribosyl cyclase-1[108] and -2[109] as well as ectopeptidases, such as membrane metalloendopeptidase,[110] alanyl membrane aminopeptidase (AnPep),[111] and dipeptidyl peptidase-4 (DPP4)[112] mediate chemotaxis. Amine oxidase, copper-containing enzyme AOC3,[113] an ecto-oxidase synthesized by endothelial cells, regulates leukocyte rolling, firm adhesion, and transmigration.

9.7.8 Migration across Pericyte Sheath and Basement Membrane

Venules are the primary site of leukocyte transmigration. In capillaries and postcapillary and collecting venules, pericytes that present gaps between adjacent cells and are embedded in the venular basement membrane surround endothelial cells and connect them via protrusions [980]. Leukocyte migration through the pericyte sheath can occur by both paracellular and transcellular routes. Neutrophils migrate across the pericyte sheath through gaps between apposed cells.

The venular basement membrane that is generated by both endothelial cells and pericytes comprises permissive sites. The composition of the basement membrane is determined by the degree of collagen-4 crosslinks and types of laminin isoforms. Regions that contain laminin-411 and -511 promote and prevent cell migration, respectively [980].[114] Yet, laminin-511 has a greater ability to promote human lymphocyte migration than other laminin isoforms such as laminin-411, at least in vitro. The influence of the composition of the vascular basement membrane on the regulation of leukocyte migration may depend on leukocyte species as well as types of inflammatory stimuli.

Cell migration through the basement membrane involves leukocyte-derived peptidases as well as endothelial cell-associated enzymes such as matrix metallopeptidases.

---

[108] A.k.a. cyclic^{ADP}ribose (cADPr) hydrolase-1 and CD38.

[109] A.k.a. cyclic^{ADP}ribose (cADPr) hydrolase-2, bone marrow stromal antigen BST1, and CD157.

[110] A.k.a. neutral endopeptidase, neprilysin, and CD10.

[111] A.k.a. aminopeptidase-M [APm] or -N [APn], microsomal aminopeptidase, myeloid plasma membrane glycoprotein GP150, and CD13.

[112] A.k.a. adenosine deaminase-complexing protein AdCP2 and T-cell activation antigen CD26.

[113] A.k.a. vascular adhesion protein-1 (VAP1).

[114] Specific laminin isoforms can serve as matrix to maintain embryonic stem cells, such as laminin-511, but neither laminin-111, -332, nor -411 [992].

9.7.9 *Migration through Subendothelial Layer and Interstitium*

Once the perivascular basement membrane is crossed, interstitial migration requires extracellular, soluble guidance molecules. Although leukocytes require integrins when they extravasate from the blood stream, their locomotion in the interstitium does not depend on these adhesion receptors.

In lymph node and splenic white pulp, lymphocytes migrate along the scaffold of fibroblastic reticular cell network [980]. In the dermis, lymphocytes move preferentially along thick collagen bundles. Guidance structures can then be used for haptotaxis along adhesive structures and/or chemotaxis according chemokine gradients that exist in these structures.

Among guidance cues, chemokines and lipid mediators support positioning of leukocytes. Activators of Rac GTPase, such as PIP_3- and ^{F}actin-binding RacGEF 70-kDa switching B-cell complex-associated protein SwAP70, Vav, and DOCKs are involved in migration through tissue interstitium [980]. Polymerizing actin causes the leading edge to protrude. The prototypic cellular protrusion is the lamellipodium (Vol. 2 – Chap. 6. Cell Motility). Filopodia also contain a growing mesh of actin filaments that help pull the leading edge forward [993].

In addition, modulators of the ARP2–ARP3 complex at the leading edge, such as cortactin homolog hematopoietic lineage cell-specific Lyn substrate protein HCLS1 (or HS1) and actin- and microtubule-binding coronins are required for efficient cell protrusion and migration in interstitium [980].[115]

Cell deformability supports squeezing locomotion in a porous medium, especially through narrow pores. Contractile and protrusive forces generated by the actin cytoskeleton and transmitted at the leading edge are transformed into traction forces. The leading edge pulls the cell forward.

Combined actin polymerization and myosin-2-induced contractility pull matrix-ligated integrins backward and the cell body forward. Actin filaments grow at the leading plasma membrane and existing filaments are pushed backward into the cell body, where the actin network is disassembled (treadmilling) [993]. Rearward actin flux generates friction between the cell and its environment that can be sensed by transmembrane force sensors. In addition, behind the leading edge, myosin-2 pulls the cortex backward (polymerization-driven retrograde movement). Transmembrane force receptors that couple the retracting actin cortex to the matrix create traction force to move the cell body.

[115]Hematopoietic lineage cell-specific protein HS1, a substrate of intracellular protein Tyr kinases such as SRC family kinases in hematopoietic cells, colocalizes with and connects to actin-related protein ARP2–ARP3 complex that causes actin assembly. It increases the rate of the actin assembly mediated by the ARP2–ARP3 complex [994]. Actin-associated coronins and coronin-like proteins participate in the remodeling of the cortical actin cytoskeleton, especially during phagocytosis and macropinocytosis [995].

Nanomotor myosin-2 is controlled by RhoA GTPase via MLCK and RoCK kinases. On the other hand, the filamentous septin mesh modulates cortical rigidity, hence lymphocyte motility [996].[116]

Leukocytes in 3D environments do not necessarily form leading edge protrusions that align with matrix fibers. Rather than emitting flat protrusions such as lamellipodium, they can build cylindrical or tongue-like extensions (lobopodia)[117] between matrix fibers. In addition, they use integrins and other adhesion molecules, such as syndecans, CD44, and discoidin domain receptors, to transmit traction forces, but not necessarily as anchors onto the matrix [993]. Locomotion can result from actin polymerization-driven cell deformation (ameboid-like migration). Lobopodia pull the cell forward. Slipping cells protrude and migrate with the same speed and shape as gripping cells.

However, in the absence of integrins, neutrophils that migrate through the interstitium decelerate by about 30% [993]. T lymphocytes that move in lymph nodes slow down by 10% when β_1- and β_2 integrins are lacking. Therefore, leukocytes can use alternative modes of movement, switching during their displacement from adhesion-dependent to deformation-based mode of motion, and conversely.

9.7.10 Regulation of Inflammation

Inflammation is an adaptive response to tissue damage or dysfunction. Acute inflammation triggered by infection or tissue injury caused by chemical or physical agents is achieved by the coordinated delivery of blood components (plasma proteins, platelets, and leukocytes) to the damaged site.[118]

In the case of infection, particularly due to bacteria, inflammation is triggered by pathogen recognition owing to receptors of the innate immune system (Toll-like and nucleotide-binding oligomerization-domain protein-like receptors; Vol. 3 – Chap. 11. Receptors of the Immune System), tissue-resident macrophages, and mastocytes. It leads to the release of numerous inflammatory mediators: chemokines, cytokines, vasoactive amines, eicosanoids, and products of 4 proteolytic cascades, i.e., the kallikrein–kinin, coagulation, fibrinolytic, and complement cascades. These mediators generate an exudate mainly across postcapillary venules.

---

[116]Septins constitute an additional family of cytoskeletal proteins distinct from microfilaments, microtubules, and intermediate filaments. Septin GTPases assemble in T-lymphocyte cortex and repress protrusions, hence restricting cell migration through narrow pores.

[117]λoβoζ: lobe; πoδoζ: foot.

[118]A cell is in a basal state when oxygen, nutrients, and growth factors are available in the absence of abnormal change in physicochemical parameters (temperature, osmolarity, etc.) of its environment. In noxious conditions, stressed tissues become dysfunctional. Mild dysfunction can be cured by tissue-resident cells, mainly macrophages and mastocytes that drive para-inflammation, i.e., a process between basal and inflammatory states, whereas extensive (sustained or excessive) dysfunctions require the recruitment of additional types of leukocytes.

When abnormal conditions are sustained, inflammation can become maladaptive. Chronic inflammation can result from: (1) defective elimination of pathogens with formation of granulomas (local accumulation of macrophages and lymphocytes) and tertiary lymphoid tissues; (2) undegradable foreign bodies (e.g., silica and asbestos particles); and (3) autoimmunity [997]. Chronic allergic inflammation occurs at sites of persistent or repetitive exposure to allergens.

The inflammatory pathway comprises [997]: (1) exogenous non-microbial (allergens, irritants, foreign bodies, and toxics) and microbial (pathogen-associated molecular patterns recognized by host pattern-recognition receptors and virulence factors) as well as endogenous (coagulation factor-XII and other inflammatory mediators, oxidized lipoproteins, crystals of monosodium urate, and calcium pyrophosphate dihydrate, products of matrix proteins, etc.) inducers; (2) sensors (TLRs, IgE, NALPs, etc.); (3) mediators (vasoactive amines [histamine and serotonin], vasoactive peptides [substance-P, kinins, fibrinopeptide-A and -B, and fibrin degradation products], fragments of complement components [anaphylatoxins C3a, C4a, and C5a], lipid mediators [eicosanoids and platelet-activating factors], cytokines [TNFα, IL1, IL6, etc.], chemokines, and proteolytic enzymes [cathepsins and matrix metallopeptidases]); and (4) effectors (endothelial and smooth muscle cells, leukocytes, neuroendocrine cells, etc.).

Activated endothelium allows a selective extravasation of leukocytes, as it impedes the transmigration of erythrocytes. Leukocyte diapedesis (Sect. 9.7) relies on adhesion molecules of leukocytes that connect to those of endothelial cells, and then to proteins of the extracellular matrix with its newly deposited plasma proteins.

In the case of infection, migrating neutrophils are activated by contact with pathogens or secreted cytokines to kill invading microbes by releasing granular content (reactive oxygen and nitrogen species, peptidase-3, cathepsin-G, and elastase) [997]. Resolution and repair phases follow the elimination of infectious agents that is mainly done by tissue-resident and recruited macrophages. Pro-inflammatory prostaglandins give rise to anti-inflammatory lipoxins for the transition from inflammation to resolution. Lipoxins impede the recruitment of neutrophils and promote that of monocytes to remove dead cells and initiate tissue remodeling in cooperation with resolvins, protectins, and growth factors produced by macrophages.

Endothelial cells of the microvascular bed at a site of inflammation regulate each stage of inflammation. Two modes of endothelial cell activation exist that [998]: (1) increase blood flow, hence leukocyte delivery; (2) cause leakage of plasma proteins into the tissue that creates a matrix for leukocyte migration; and (3) promote recruitment, binding, and activation of leukocytes for extravasation.

Endothelial cell activation mode 1 associated with acute inflammation arises rapidly in response to microbes or injured tissue and transiently (duration 10–20 mn). It is mediated by Gq-coupled receptors such as histamine H_1 receptors. It recruits and activates neutrophils to kill pathogens and remove cellular debris. Successful, acute inflammation restores normal tissue architecture or forms a connective tissue scar.

Endothelial cell activation mode 2 during persistent inflammation is mediated by cytokines, such as tumor-necrosis factor and interleukin-1. It is slow, as it depends on protein synthesis, and sustained (duration hours to days). It recruits monocytes and T lymphocytes. Moreover, intervention of innate immunocytes stimulates lymphocytes of adaptive immunity. Activated T lymphocytes then enhance the activity of macrophages and recruit eosinophils.

Prolonged antigenic stimulation by resistant microbes or tissue-derived auto-antigens leads to chronic inflammation. During chronic inflammation, endothelial cells respond to angiogenic factors to generate new blood vessels. Endothelial cells subjected to lymphotoxin-β acquire features of high endothelial venules and secondary lymphoid organs, characterized by the recruitment of naive T cells and central memory B and T cells (sometimes called tertiary lymphoid organs).

9.7.10.1 ED1L3

Whereas selectins, integrins, and immunoglobulin-like cell adhesion molecules promote inflammatory cell recruitment, EGF-like repeat- and discoidin-1-like domain-containing protein ED1L3[119] prevents leukocyte adhesion [999]. Inhibitor ED1L3 is synthesized and secreted by endothelial cells. This anti-adhesive factor is a ligand of $\alpha_L\beta_2$-integrin.

9.7.10.2 ELAV1

In human umbilical vein endothelial cells, the nucleocytoplasmic shuttling protein Embryonic lethal, abnormal vision homolog ELAV1[120] contributes to the regulation of inflammation by endothelial cells under mechanical and chemical stresses. It regulates stress-sensitive genes that encode Krüppel-like factor KLF2, endothelial nitric oxide synthase NOS3, and bone morphogenic protein BMP4 [1000].

Synthesis of ELAV1 is controlled by hemodynamic stress. Agent ELAV1 promotes inflammation supported by endothelial cells, as it upregulates production of ICAM1 and VCAM1 and causes NFκB phosphorylation and adhesion of monocytes.

9.8 Hemostasis

Hemostasis stops hemorrhage. Several processes participate in bleeding arrest when a blood vessel is wounded: (1) vasoconstriction to reduce blood input in the damaged region; (2) primary hemostasis induced by platelets bound to collagen that forms the hemostatic plug within seconds after an injury; (3) blood coagulation, or

[119] A.k.a. integrin-binding Developmentally regulated endothelial cell locus-1 (DEL1).

[120] A.k.a. Hu antigen-R.

Table 9.26. Antithrombogenic substances and clotting agents secreted by normal and damaged endothelium, respectively. Blood circulation is preserved by a balance between hemorrhage and thrombosis (HODE: hydroxyoctadecadienoic acid).

Antithrombotic agents	Clotting factors
Protein-C	Tissue factor
Protein-S	von Willebrand factor
Thrombomodulin	Multimerin
Heparan sulfate	Platelet-activating factor
Antithrombin (AT or AT3)	Adhesion molecules
Tissue plasminogen activator (tPA)	Fibronectin
Urokinase (uPA)	Collagen
Nitric oxide	Endothelin
13HODE	Clotting factors V and VIII
Prostacyclin (PGi2)	Factors IX- and -X receptors
Prostaglandin-E2	Plasminogen activator inhibitor-1 and -2
Tissue factor pathway inhibitor	

secondary hemostasis, i.e., a cascade of reactions that involves the set of coagulation factors down to fibrin production to generate a clot after several minutes; (4) repair, as the clot attracts and stimulates the growth of fibroblasts and smooth myocytes; and (5) fibrinolysis leading to the clot dissolution.

Hemostasis relies on 4 interacting components to staunch blood, repair blood vessel wall, and maintain adequate flow to downstream tissues: (1) vessel wall, especially endothelium; (2) flowing platelets; (3) plasma coagulation factors and their inhibitors; (4) fibrinolytic proteins; as well as (5) repair materials.

In normal vessels, endothelium prevents clotting (Table 9.26). Endothelial cells have anticoagulant properties because they: (1) form a barrier between blood with its platelets and clotting factors and vessel wall as well as (2) have thromboregulation and anticoagulation function, i.e., inactivate thrombin and promote fibrinolysis in normal conditions.

Endothelial plasmalemmal substances include thrombomodulin (TM), protein-C (PC), tissue factor pathway inhibitor (TFPI), plasminogen activator inhibitor-1 (PAI1 or serpin-E1; Table 9.27), and heparan sulfate proteoglycans.

On the surface of the endothelium, *thrombomodulin* is an integral membrane cofactor in the thrombin-induced activation of protein-C in the anticoagulant pathway. In fact, it interacts with multiple proteins to inhibit blood clotting and inflammation [1001]. This proteoglycan, once bound to thrombin, promotes thrombin-mediated activation of protein-C, which has anticoagulant, anti-apoptotic, and anti-inflammatory activities. With cofactor protein-S synthesized by endothelial cells, thrombomodulin activates protein-C that inhibits factors V and VIII as well as protein-C and -S.

Table 9.27. Plasminogen activator inhibitors (PAI) are serpins (serine peptidase inhibitor). Plasminogen activator inhibitor-1 is mainly produced by vascular endothelial cells, hence its other name endothelial plasminogen activator inhibitor (ePAI). It is also liberated from other cell types such as adipocytes. It is the principal inhibitor of the serine peptidases tissue plasminogen activator (tPA) and urokinase (uPA), the activators of plasminogen and hence of fibrinolysis. Peptidase nexin-1 (PN1), synthesized by most cell types, such as fibroblasts, monocytes, platelets, and vascular cells, is also called plasminogen activator inhibitor type-1 member 2. It strongly inhibits plasminogen activators (tPA and uPA), plasmin, and thrombin. It has a high affinity for glycosaminoglycans such as heparan sulfates. Upon heparin binding, it is a better inhibitor of blood coagulation factor XIa than C1 inhibitor, or serpin-G1. It interacts with low-density lipoprotein receptor-related proteins (LRP). During embryogenesis and postnatal development, it is expressed prominently in remodeling regions, in which cell fate specification is influenced by morphogens such as sonic Hedgehog. Plasminogen activator inhibitor-2 (PAI2) is secreted in significant amounts by the placenta (hence its name placental plasminogen activator inhibitor [pPAI]). However, it is synthesized in many cell types, especially monocytes and activated macrophages. Two forms of PAI2 exist: 60-kDa, extracellular, glycosylated and 43-kDa, intracellular form. Plasminogen activator inhibitor-3 (PAI3) is the acrosomal serine peptidase inhibitor, or protein-C inhibitor (ProCI). It has an affinity for heparin, oxidized phosphatidylethanolamine, and oxidized and unoxidized phosphatidylserine. Antithrombin-3 (AT3) is another heparin-binding serpin produced by the liver that inactivates several enzymes of the coagulation cascade, such as activated factor II and VII of the tissue factor (extrinsic) pathway as well as activated factor IX, X, XI, and XII of the contact activation (intrinsic) pathway. It also inactivates kallikrein and plasmin as well as trypsin and C1s subunit of the classical complement pathway. Its activity is increased by heparin.

Subtype	Aliases
PAI1	Serpin-E1, ePAI
PAI2	Serpin-B2, pPAI
PAI3	Serpin-A5, ProCI
Nexin	Serpin-E2, PN1
AT3	Serpin-C1

The thrombin–thrombomodulin complex activates carboxypeptidase-B2 (CPb2).[121] The latter is a plasminogen-bound zymogen. This enzyme inactivates vasoactive peptides such as complement C5a. When it is activated by proteolysis (Arg92) by the thrombin–thrombomodulin complex, CPb2 removes fibrin C-terminus that binds and activates plasminogen. Furthermore, thrombomodulin inhibits high-mobility group box 1 DNA-binding protein HMGB1[122] that has cytokine-like activity [1002]. Thrombomodulin thus hampers overstimulation of the

[121] A.k.a. thrombin-activable fibrinolysis inhibitor [TAFI], plasma carboxypeptidase-B [pCPb], and carboxypeptidase-U [CPu]).

[122] A.k.a. amphoterin.

innate immunity. Binding of thrombomodulin to HMGB1 prevents signaling from HMGB1 receptor, the receptor for advanced glycation end products (RAGE).[123] Thrombomodulin secretion depends on hemodynamic stress.

Protein-C[124] is a vitamin-K-dependent circulating zymogen. It is activated once bound to thrombin. Its activation is improved by thrombomodulin and *endothelial protein-C receptors* (EPCR). It then acts as a serine peptidase and inactivates activated clotting factors-Va and -VIIIa.

Thrombin, thrombomodulin, and endothelial protein-C receptor together activate protein-C. Both protein-C and activated protein-C interact with EPCR with a similar affinity. Both thrombin and activated protein-C cleave peptidase-activated receptor-1 (PAR_1), but PAR_1 cleavage by thrombin can be more efficient than that of activated protein-C.[125] Cleavage of PAR_1 by thrombin yields thrombotic and inflammatory responses. Both endothelial protein-C receptor and peptidase-activated receptor-1, colocalized in membrane rafts of endothelial cell plasma membrane, are implicated in protein-C signaling [1003]. Thrombomodulin that localizes with endothelial protein-C receptor and peptidase-activated receptor-1 in the same membrane raft, can recruit thrombin, then generates a thrombin–thrombomodulin complex to activate EPCR-bound protein-C. Anti-inflammatory and cytoprotective properties of activated protein-C are due to EPCR-dependent PAR_1 cleavage by activated protein-C in endothelial cells. Anticoagulant function of activated protein-C results from degradation of activated clotting factor-Va and -VIIIa, hence preventing thrombin generation by activated protein-C that is stimulated by protein-S.

Protein-S of the anticoagulation pathway is a vitamin-K-dependent plasma glycoprotein synthesized in endothelial cells. Protein-S circulates in 2 forms: free and bound to C4b complement protein. This free cofactor of protein-C participates in the inactivation of clotting factor-Va and -VIIIa.

Tissue factor pathway inhibitor, (TFPI), also termed lipoprotein-associated coagulation inhibitor (LACI) and extrinsic pathway inhibitor (EPI), can reversibly inhibit clotting factor-Xa and thrombin (or factor-IIa), in addition to tissue factor. The majority of tissue factor pathway inhibitor is bound to the endothelium, as plasma and platelet pools account for less than half the vascular content.[126] Tissue factor pathway inhibitor reduces activation of factor-X, without abolishing it, in

---

[123]Receptor RAGE is a pro-inflammatory transmembrane molecule of the immunoglobulin superfamily for non-enzymatically altered proteins and other ligands, in addition to HMGB1 protein.

[124]A.k.a. autoprothrombin-2A and blood-coagulation factor-XIV.

[125]Endogenous activated protein-C generated by thrombin on the endothelial cell surface has greater effects than exogenous activated protein-C.

[126]Platelets contain TFPI, the amount of which is estimated at 5–10% of the plasma level [1004]. In endothelial cells, TPFI-containing vesicles can be linked to plasmalemmal caveolae. Agent TFPI is released when intracellular calcium level rises.

the absence of antithrombin-3 (AT3).[127] It inhibits both clotting factor-Xa and the complex made of tissue factor and clotting factor-VIIa by rapid formation of TFPI–FXa complex, a potent inhibitor of the TF–VIIa complex [1006].

Antithrombin-3 is a major plasma inhibitor of FXa. Antithrombin-3 also inhibits thrombin and FIXa. Endothelial cells inhibit platelet aggregation, releasing inhibitors such as prostacyclins (PGI2) and nitric oxide.[128]

serpin-E1[129] inhibits urokinase, thrombin, and plasmin. It is mainly produced by endothelial cells. It is also secreted by other cell types, particularly adipocytes. It resembles antithrombin-3, but at and near the cell surface. It inhibits serine peptidases tissue- (tPA) and urokinase-type (uPA) plasminogen activators that activate plasminogen, hence impeding fibrinolysis (Sect. 9.8.2).

Endothelial cells activate fibrinolysis by binding plasminogen activator inhibitor PAI1. Serpin-E1 and -F2[130] are plasma enzymes that bind to fibrin and inhibit fibrinolysis. Serpin-F2 also deactivates plasmin.

Blood can be exposed to clotting stimuli without initiating coagulation. Clotting is initiated on loci larger than a threshold size in the absence of fluid flow [1007]. The magnitude of the threshold size can be described by the *Damköhler number*, which measures competition of reaction and diffusion.[131] Reaction produces activators at the site, and diffusion removes activators from the site.

Cytokines such as tumor-necrosis factor-α (TNFSF1) influence the expression of genes in endothelial cells that produce anti- and prothrombotic agents, such as plasminogen activator inhibitor PAI1, tissue- (tPA) and urokinase-type (uPA) plasminogen activators, thrombomodulin, nitric oxide synthase NOS3, and vascular cell adhesion molecule VCAM1.

Endothelial cells that coat large blood vessel lumen are permanently exposed to blood flow-induced pulsatile shear and tensile stress on their apical and basolateral surface, respectively. Interplay between cytokines and hemodynamic stress field can modify the sensitivity of endothelial cells to inflammatory cues.

The response of endothelial cells to TNFSF1 is modulated by shear rather than tensile stress [1009]. Mechanotransduction and TNFSF1 additively promote PAI1 production, but reduce cytokine effect on VCAM1, as well as impede that on tPA and uPA synthesis. These effects rely on mitogen-activated protein kinase modules as well as NFκB and AP1 transcription factors.

---

[127] Factor-Xa is an activator of the TF–FVII complex that is blocked by AT3 in the presence of heparin [1005].

[128] Prostacyclin release from endothelial cells is enhanced when they are subjected to pulsatile flow compared with steady flow.

[129] Clade-E, member-1 serine peptidase inhibitor is also named peptidase-nexin and type-1 plasminogen activator inhibitor (PAI1). Plasminogen activator inhibitor-2 (PAI2) is secreted by the placenta.

[130] A.k.a. α2-antiplasmin (α2AP), α2-plasmin inhibitor (α2PI), and plasmin inhibitor (PlI).

[131] The Damköhler number, a measure of the rate of reaction of a molecule relative to diffusion, is also called the Thiele modulus. The Biot number is a measure of the rate of external transfer of a substance with respect to the diffusion rate.

9.8.1 Blood Coagulation

Blood hemostasis, i.e., cessation of blood loss from an injured vessel, results from changes in blood flow pattern, vessel wall, and thrombogenic potential. Blood coagulation corresponds to the latter component of blood hemostasis. Blood coagulation begins immediately after an injury has damaged the endothelium. Plasmatic coagulation factors, or clotting factors, are involved in a complex reaction cascade to form a fibrin mesh that strengthens the platelet aggregate fixed onto the evolving, responsive vessel wall matrix.

Upon vascular injury, intracellular materials from damaged cells are released and thus exposed to blood components. Extracellular RNA can promote blood coagulation [1010]. It increases the activation of peptidases of the *contact pathway* of blood coagulation, such as factor-XII and -XI. *von Willebrand Factor* (vWF) is able to stop bleeding under high flow [1011]. The reversible unfolding of this glycoprotein extends its length up to 100 μm, thereby yielding a sufficient density of binding sites for collagen. Moreover, this conformational change occurs in a very short time. The strong increased probability of bound von Willebrand factor favors the formation of a mesh for platelet adhesion.

When the blood vessel wall, especially the endothelium, is damaged, blood must clot to halt bleeding. The coagulation cascade takes place at the site of vessel wall gap where platelets aggregate (Fig. 9.6). Blood clotting involves circulating platelets that are able to clump. Complete aggregation requires the release of adenosine diphosphate (ADP) and other mediators from dense granules and stimulation of $P2Y_{12}$ receptors. Among Rab GTPases that regulate vesicular transport during both endo- and exocytosis (Vol. 1 – Chap. 9. Intracellular Transport), Rab27 regulates the secretion of dense granules by platelets [1012].

The hemostatic process depends on stable adhesion and aggregation of platelets with constituents of the subendothelial matrix at the vessel injury site. Several clotting factors are required (Table 9.28). Endothelial cells synthesize coagulation factors, von Willebrand factor, and tissue factor that activates FVII.

The coagulation cascade is primed by blood vessel injury to produce a solid fibrin clot to cover the damaged blood vessel and stop hemorrhage. A blood clot is made of a branched scaffold of fibrin fibers stabilized by factor-XIIIa, platelet aggregates, and trapped erythrocytes.

Plasma fibrin precursor — fibrinogen — is converted by cleavage into fibrin. Fibrin monomers polymerize to form the fibrin clot, a gel generated by a network of fibers attached by crosslinks. Fibrin clot is associated with aggregation of platelets to form a plug that blocks the orifice across the blood vessel wall, thereby quickly restricting blood leakage before stopping bleeding.

The fibrin clot is a viscoelastic polymer. A blood clot needs to have suitable stiffness to obstruct the mural breach and remodeling capacity to remain degradable by enzymes to avoid thrombi and emboli. Changes in calcium level and pH can affect fibrinogen springiness. Fibrin fibers can be strained 180% (2.8-fold extension) without sustaining permanent lengthening and up to 525% (average 330%) before rupturing [1013]. Fibrin fibers are much stiffer for stretching than for flexion. Elastic

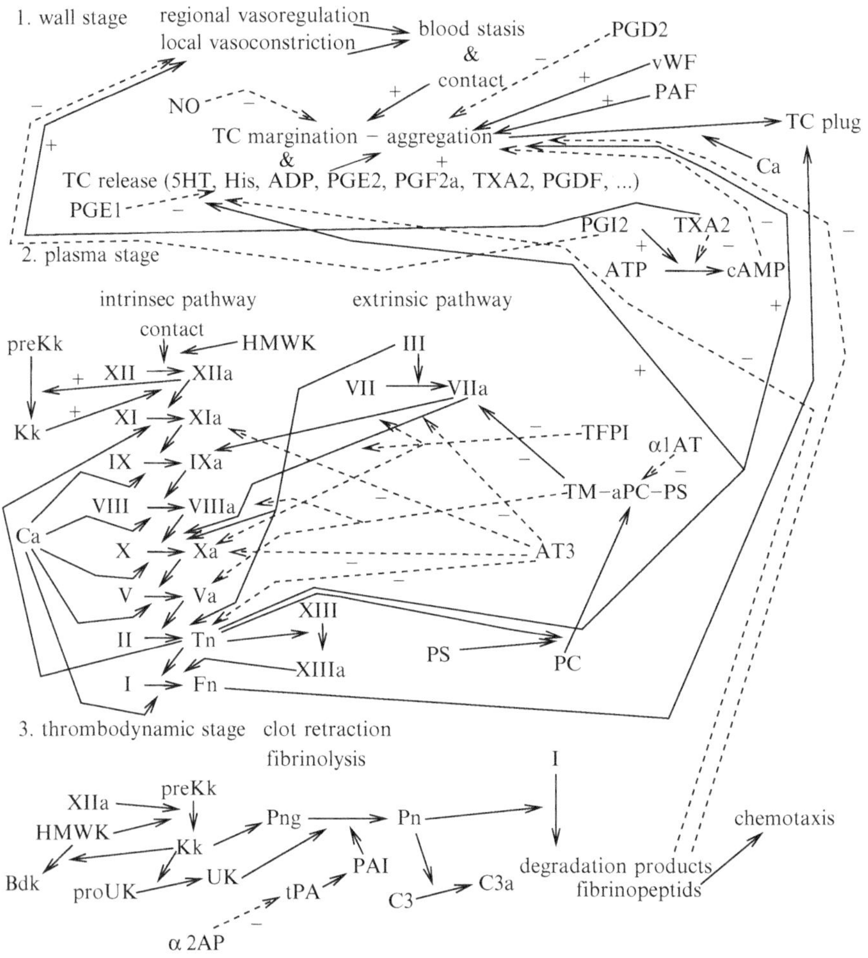

Fig. 9.6 Coagulation and fibrinolytic events. The contact-phase proteins include factors XII and XI, prekallikrein, and high-molecular-weight kininogen (HMWK).

moduli for individual fibrin fibers in plasma clots assessed using optical tweezers are 1.7 (±1.3) and 14.5 (±3.5) MPa for unligated and ligated fibers, respectively [1014]. At large strains, clot stiffness increases. Strain-stiffening behavior prevents large deformations that could threaten gel integrity. Theoretical networks of filamentous proteins arranged in an open crosslinked mesh invariably stiffen at low strains without requiring a specific architecture or different components with distinct intrinsic stiffness [1015]. Some domains of fibrins unfold with stress. A blood clot comprises platelets that control fibrin polymerization. Moreover, thrombocytes with their actomyosin cytoskeleton generate large contractile stresses on fibrin clots.

Table 9.28. Coagulation factors (Source: [1025]). Hepatic-derived, vitamin K-dependent procoagulants, i.e., coagulation factor-II (prothrombin), -VII, -IX, and -X, are substrates for the thrombin formation by activated factor-X (Xa) in the common pathway that participates in platelet agglutination.

	Coagulation factors	Targets
Number	Name (alias)	
I	Fibrinogen (Fng)	Fn
II	Prothrombin	I, V, VII, XIII, PC, platelets
III	Tissue factor (TF)	VII
IV	Ca^{2+}	V, VIII, IX, X
V	Proaccelerin	X, II
VI	Accelerin (FVa)	
VII	Proconvertin	IX, X
VIII	Antihemophilic factor-A	IX, X
IX	Antihemophilic factor-B	X
	Plasma thromboplastic component	
X	Stuart-Prower factor	II
	Prothrombin converting enzyme	
XI	Plasma thromboplastin antecedent (PTA)	XII, preKK
XII	Hageman or contact factor	XIa, preKK, Fn
XIII	Fibrin-stabilizing factor	Fn
XIV	Protein-C (PC)	V, VIII

Other clotting factors	Targets
von Willebrand factor (vWF)	VIII, platelet
Prekallikrein (preKk)	XIIa, preKK, HMWK
High-molecular-weight kininogen (HMWK)	XII, XI, preKK
Antithrombin-3 (AT or AT3; serpin-C1)	IIa, Xa
Heparin cofactor-2 (serpin-D1)	IIa
Protein-S (PS or PSα)	PC
Plasminogen (Png)	Pn, Fn
α2-Antiplasmin (serpin-F2)	Pn
Prourokinase (proUK)	Png
Tissue plasminogen activator (tPA)	Png
Plasminogen activator inhibitor	
PAI1 (serpin-E1)	Png, tPA
PAI2	Urokinase

9.8.1.1 Primary Hemostasis

Primary hemostasis refers to the plug formed by platelets that aggregate at injury site to seal damaged blood vessels and stop bleeding. Blood vessel injury at a more or less high blood pressure site according to the vasculature locus exposes collagen and tissue factor and triggers blood coagulation and thrombocyte activation, accumulation, and aggregation.

Initially arriving platelets stick to collagen, form a particle monolayer, and release platelet activators. Once activated, thrombocytes generate stable bonds to adhere stably to each other and promote a growing hemostatic clot. Fibrin mesh created by blood coagulation constructs a support to stabilize the platelet cluster. The platelet mass grows as additional platelets attach to previously aggregated thrombocytes. More or less rapidly moving, discoid platelets must slow down to reach the growing thrombus and become activated, spread platelets. Platelets accumulate downstream from a constricted vascular segment and developing clot that causes a stenosis upon local changes in blood flow [1016]. Platelets join the developing plug and emit extensions of their plasma membrane to attach to the growing clot. Then, they become activated.

During platelet adhesion and aggregation of primary hemostasis, platelets transmit contractile forces developed by the actomyosin cytoskeleton to the extracellular matrix that provoke a 30 to 40% reduction in plug volume to maintain the integrity of the primary hemostatic plug, independently of thrombin and fibrin generation [1017].

Platelets release thrombin and soluble agonists for further platelet activation (ADP, thromboxane-A2, etc.). Simultaneously, stimulated platelets trigger activation of $\alpha_{2B}\beta_3$-integrin and platelet aggregation. Platelet integrins are activated due to contact with a wounded vessel wall to ensure firm adhesion and aggregation. Platelet integrins interact with collagen and von Willebrand factor. Integrin-$\alpha_{2B}\beta_3$ interacts with collagen-bound von Willebrand factor, fibrinogen, and fibronectin, whereas $\alpha_2\beta_1$-integrin binds to collagen fibers. Various plasma clotting factors then respond to form fibrin strands that strengthen the platelet plug.

The kindlin family consists of 3 known members (Kind1–Kind3) that are located at integrin-associated adhesion sites. Whereas kindlin-1 and -2 are widely expressed, kindlin-3 is exclusively produced in hematopoietic cells. It abounds particularly in megakaryocytes and platelets. Similarly to cytoskeletal talin that is also involved and associates with the tail of β-integrin whatever the signaling pathway, kindlin-3 activates platelet β_1- and β_3-integrins by binding to their cytoplasmic domain (but at a more distal site than that targeted by talin) [1018].

Primary hemostasis is initiated when platelets adhere to collagen fibers, using $\alpha_2\beta_1$-integrin, a platelet collagen receptor.[132] This adhesion is stabilized by von Willenbrand factor that bridges platelet GP1b–GP9–GP10 complexes and collagen fibrils. Platelets are then activated and secrete granule contents into the plasma.

The GP1b–GP5–GP9 receptor complex consists of 4 components: GP1bα, GP1β, GP5, and GP9 glycoproteins. It allows interaction of resting platelets with activated leukocytes via $\alpha_M\beta_2$-integrin and activated endothelial cells via P-selectin.

[132] A.k.a. glycoproteic complex GP1a–GP2a. Tethering and rolling of platelet on the vascular wall depend on bonds between platelets and wall elements. Bond formation is firstly reversible with fast association and dissociation rates. Connections between von Willenbrand factor and glycoproteins initiate activation of $\alpha_{2B}\beta_3$-integrins (GP2b–GP3a) with slow rates of bond formation and destruction, and mediate irreversible adhesion [1019].

Glycoprotein GP1bα can associate with thrombin, high-molecular-weight kininogen, coagulation factor-XI and -XII, and thrombospondin-1. Ligand binding activates the GP1b–GP5–GP9 receptor complex for calcium mobilization, cytoskeleton rearrangement, granule release, and activation of $\alpha_2\beta_3$-integrin, the main integrin implicated in platelet aggregation that interacts with von Willebrand factor. Collagen receptor GP6 needs GP1bα for platelet adhesion. von Willebrand factor carries coagulation factor-VIII at sites of vascular lesions and protects it against proteolysis. Platelet adhesion receptor GP1bα that binds to von Willebrand factor can act independently of von Willebrand factor for platelet recruitment under high hemodynamic stress in sites of exposed subendothelium after vascular injury and adhesion to the extracellular matrix and growing thrombus [1020].

von Willebrand factor possesses binding sites for collagen and platelet glycoproteic subunit GPIbα for platelet crosslinking and platelet plug formation. Binding of von Willebrand factor to platelets must withstand strong hydrodynamic forces. The vWF–GPIbα bond is specialize in force resistance. Two states of the vWF–GPIbα bond indeed exist [1021]: (1) a state at low force and (2) another state that begins to be involved at 10 pN with aboout 20-fold longer lifetime and greater force resistance.

Fibrinogen (Fng) links adjacent platelets by binding with $\alpha_{2B}\beta_3$-integrins. Primary hemostasis thus involves a set of adhesion receptors (platelet plasmalemmal glycoprotein GP1b [CD42b],[133] GP9 [CD42a], $\alpha_{2B}\beta_3$-integrins, etc.) and proteins (von Willebrand factor, collagen, fibronectin, fibrinogen, laminin, homotrimeric thrombospondin TSP2, etc.). Once released from activated platelets, ADP is a potent inducer of platelet aggregation (self-perpetuating process). Platelet-activating factor and adrenaline act together to relieve the effect of inhibitors of platelet aggregation.

9.8.1.2 Secondary Hemostasis

Secondary hemostasis has 2 pathways — intrinsic and extrinsic — that join in a common pathway leading to fibrin (Fn or activated factor-I [FIa]) formation. This fibrillar protein polymerizes to form a mesh around platelets that stabilizes the hemostatic plug over the breach.

Intrinsic Pathway

The intrinsic pathway is characterized by the formation of the primary complex on collagen by high-molecular-weight kininogen (HMWK), prekallikrein (preKK),

[133]Glycoprotein-1b, or cluster of differentiation CD42b, complexes with glycoprotein-9 to function as a receptor for von Willebrand factor. It is a component of the GP1b–GP5–GP9 complex on platelets.

and coagulation factor-XII (Hageman factor).[134] Coagulation factor-XII cleaves into 2 fragments, one is fixed on the wound edge for molecule adhesion, the other becomes a peptidase that activates coagulation factor-XI. In turn, coagulation factor-XIa becomes a peptidase that initiates an activation cascade (Fig. 9.6) until fibrinogen is transformed into fibrin to form a clot. More precisely, coagulation factor-XII is activated (coagulation factor-XIIa). Activated coagulation factor-XIIa subsequently converts coagulation factor-XI; coagulation factor-XIa activates FIX (coagulation factor-IXa). Reacting platelet surface phospholipids accelerate intrinsic coagulation.

In addition to coagulation factor-XII, thrombin, the final enzyme of the coagulation cascade, also activates factor-XI (feedback activation). Coagulation factor-V serves as a cofactor in the activation of coagulation factor-XI by thrombin [1022]. Coagulation factor-XI specifically interacts with both coagulation factor-V and activated coagulation factor-V via multiple binding sites, but not other cofactors of coagulation, such as activated clotting factor-VIII, high-molecular-weight kininogen, or protein-S. Depletion of plasma coagulation factor-V and activated protein-C lower the activation of the intrinsic pathway by thrombin in plasma.

Extrinsic Pathway

In the extrinsic pathway, coagulation factor-VII is activated in the presence of calcium by *tissue factor* (TF), also called platelet tissue factor, coagulation factor-III, thrombokinase, tissue thromboplastin, or CD142, on plasma membrane surface or released tissue factor from cell membranes during injury.[135] This glycoproteic activator resides in the subendothelial layer and in adventitia of blood vessel walls as well as in platelets and leukocytes. Endothelial cells do not produce tissue factor when they are not exposed to cytokines such as tumor-necrosis factor-α.

Unlike other peptidases that act in the blood coagulation cascade and circulate as non-functional precursors, tissue factor is a functional initiator when expressed on cell surfaces. This transmembrane protein possesses an extracellular region that binds factor-VII and cytoplasmic domain. Binding of coagulation factor-VIIa to tissue factor can prime signaling for angiogenesis and apoptosis. The activity of the FVIIa–TF complex is precluded by tissue factor pathway inhibitor via a quaternary complex TF–FVIIa–FXa–TFPI [1023].

[134]In vitro, activation of the intrinsic pathway can be initiated by coagulation factor-XII on negatively charged surfaces.

[135]In fact, thromboplastin is the combination of both phospholipids and tissue factor that both trigger the extrinsic pathway. Hence, thromboplastin is an incorrect synonym for tissue factor.

In addition to plasmalemmal receptor for the serine peptidase FVIIa, a soluble form of tissue factor (TF^S) originates from an alternatively spliced transcript (TF_{AS}) that lacks a transmembrane domain [1024]. Compound TF^{AS} can circulate in blood. It has procoagulant activity when exposed to phospholipids.

Tissue factor initiates the blood coagulation cascade, as it operates as high-affinity receptor for coagulation factor-VII. Activated coagulation factor-VII (FVIIa) activates coagulation factor-IX and -X, ultimately causing thrombin formation.

Common Pathway

The common pathway begins with activation of coagulation factor-X by coagulation factor-IXa and/or -VIIa. The process requires coagulation factor-VIIIa. Thrombin (Tn) is then produced.[136] Activated coagulation factor-X activates thrombin, requiring coagulation factor-Va, via a positive feedback loop by thrombin itself like coagulation factor-VIII. Thrombin not only converts fibrinogen to fibrin, but also activates coagulation factor-VIII and -V as well as their inhibitor protein-C, in addition to platelets. Fibrin strands appear where the platelets adhere, change in shape, and aggregate.

Prothrombin is a 72-kDa single-chain glycoprotein. Its N-terminus is removed after proteolytic activation by factor-Xa. Its C-terminus contains Ser peptidase domain of thrombin. Carboxylation of its N-terminus in the presence of vitamin-K is followed by additional binding of calcium and phospholipid and proteolysis by FXa. Cleavage of prothrombin by coagulation factor-Xa creates prothrombin fragments of N-terminus and prethrombin-2 [1026]. Prethrombin-2 cleavage produces meizothrombin. Processing can generate α-thrombin or procoagulant prethrombin-2 that accumulates in the plasma. The latter binds to platelet-membrane glycoprotein GP4 and boosts platelet aggregation. Another substance, platelet-agglutinating protein PAP37 has similar characteristics to prethrombin-2 [1026].

Coagulation is a controlled process. Cofactors and inhibitors are required for the coagulation cascade. Calcium, phospholipids, and membrane constituents are cofactors for activation of coagulation factor-VII, -IX, -X, and -II. Such interactions are possible if vitamin-K is operating during coagulation factor synthesis.

Thrombin leads to a positive feedback of blood coagulation by activating platelets and coagulation factor-XI and inhibits coagulation by stimulating protein-C via thrombomodulin binding. Activated protein-C binds to protein-S and degrades coagulation factor-Va and -VIIIa.

Serine peptidase inhibitors (serpins) inhibit substrate peptidases by forming a stable complex with the enzyme. Several serpin–peptidase complexes are removed from the circulation by a clearance receptor in the liver. This clearance receptor is

---

[136]Thrombin binds to G-protein-coupled peptidase-activated receptors (PAR). Thrombin activation of PAR_1 requires ADP to stimulate Gi.

the LDLR-related protein (LRP) of the low-density lipoprotein receptor (LDLR) family. It binds numerous other ligands, such as matrix proteins as well as certain apolipoprotein E- and lipoprotein lipase-enriched lipoproteins, in addition to peptidase-inhibitor complexes [1027].

Antithrombin[137] is a serine peptidase inhibitor (serpin-C1) that degrades thrombin and coagulation factor-Xa, -XIIa, and -XIa. Its action is enhanced by heparan sulfate proteoglycans and heparin. This glycoprotein is produced by the liver. Antithrombin-α is the dominant plasmatic form of antithrombin; β-antithrombin is a minor type.

Tissue factor pathway inhibitor inhibits coagulation factor-VIIa-mediated activation of coagulation factor-IX and -X. Synthesis of TFPI occurs in the vascular endothelium (which also expresses the tissue factor).

Serotonin inhibits platelet release of ADP messenger. Degradation products of prostaglandin-E1 and fibrin preclude platelet material release. Furthermore, fibrin degradation products also impede platelet aggregation.

Serpin-D1[138] inhibits thrombin (FIIa). This minor antithrombin serves as a cofactor for heparin and dermatan sulfate proteoglycans. It indeed rapidly inhibits thrombin in the presence of dermatan sulfate proteoglycans or heparin.

The relative abundance of various blood cell types within the fibrin clot depends on the local hemodynamic stress field. The lower the shear, the higher the number of erythrocytes in the fibrin network. Platelets aggregate at sites of blood vessel wall injury and then serve as a surface for coagulation. They set the final structure of fibrin. Afterward, thrombocytes influence fibrinolysis via their proteins and phospholipids that modulate plasmin activity. Leukocytes that form mixed aggregates with platelets affect clot structure [1008]. Platelets actually interact with and capture neutrophils via P-selectins and integrins.

Platelets also secrete neutrophil and endothelial activators that generate production of inflammatory cytokines and release of neutrophil granule content with serine peptidases (elastase, cathepsin-G, and peptidase-3) and matrix metallopeptidases (MMP8 is stored in specific granules and MMP2 and MMP9 in small-storage gelatinase granules). Platelets secrete matrix metallopeptidases, such as MMP1, MMP2, and MMP9.

Upon activation, secreted leukocyte peptidases enhance von Willebrand factor-dependent platelet adhesion. Neutrophils can also bind to von Willebrand factor.

Red blood cells entrapped in clot influence hemostasis by secreting procoagulant substances and contributing to the phospholipid surface for assembly of

---

[137] A.k.a. antithrombin-3 (AT3). Antithrombin-1 refers to the absorption of thrombin onto a fibrin gel. Antithrombin-2 is a cofactor in plasma that, with heparin, interferes with the interaction of thrombin and fibrinogen. Antithrombin-4 is an antithrombin that is activated during and shortly after blood coagulation.

[138] A.k.a. heparin cofactor-2 and peptidase inhibitor leuserpin-2.

procoagulant complexes. Erythrocytes indeed promote platelet aggregation and degranulation by releasing ATP and ADP under hypoxia and acidosis, as well as in response to mechanical deformation.

Platelets contain cellular factor XIII and promote fibrin crosslinking by plasma coagulation factor-XIII, thereby contributing to formation of a resistant fibrin structure to avoid premature fibrinolysis.

9.8.2 Fibrinolysis

Fibrinolysis occurs after coagulation during healing. It involves plasmin, which cleaves the fibrin clot. Plasmin is produced from plasminogen, which is synthesized in the liver, by tissue plasminogen activator[139] (tPA). Tissue plasminogen activator that activates fibrinolysis and thrombolysis is stored in small vesicles in endothelial cells. It is released by thrombin, histamine, endothelin, and cytokines. α2-Antiplasmin and plasminogen activator inhibitor inhibit plasminogen activators (tPA and uPA), thereby preventing the conversion of plasminogen to plasmin. Inhibitor of fibrinolysis PAI1 is released by a calcium-dependent process primed by thrombin, histamine, endothelin, and cytokines, such as TNFα. Endothelial cells activate fibrinolysis by binding PAI. Activated protein-C deactivates PAI1. Moreover, the inhibitory effect of PAI1 is reduced when fibrin, plasmin, and tPA form ternary complexes.

During healing, activated clotting factor-XIIIa binds to fibrin and protects it from plasmin. Similarly, myosin released by platelets interacts with fibrin fibers and masks their plasmin-cleavage sites. Conversely, the free form of myosin binds both tPA and plasminogen and accelerates plasminogen activation. Like fibrin-bound plasmin, myosin-bound plasmin is protected against inhibitors of fibrinolysis. Elastase degrades clotting factor-XIII and inactivates clotting factor-VII, -VIII, -IX, and -XII [1008].

Trapped leukocytes in the clot modulate fibrinolysis. Leukocyte peptidases not only digest fibrin, but also indirectly control fibrinolysis, as they degrade zymogens and inhibitors of coagulation and fibrinolytic peptidases. Neutrophil elastase promotes fibrinolysis directly by degrading fibrin and indirectly by cleaving plasminogen into miniplasminogen that is more quickly activated to miniplasmin by plasminogen activators without cofactors [1008]. Miniplasmin is much less sensitive to inhibition by α2-antiplasmin than plasmin.

[139] When stroke occurs, glutamate is released by ischemic neurons. Glutamate induces an excess in receptor stimulation and a subsequent calcium ion influx. The latter leads to neuron death. tPA is produced by endothelial cells and neurons. tPA released by damaged neurons increases glutamate effects. Therefore, tPA administration after stroke must be carefully handled.

9.8.3 Embolus Extravasation

Small clots that form in blood vessels can be removed by hemodynamic forces and fibrinolysis. A third mechanism relies on endothelial cells that emit membrane projections that envelop emboli and carry them into the perivascular tissue [1028]. Embolus extravasation occurs within 2 to 7 days especially in the cerebral vasculature to avoid prolonged microvascular occlusion and relatively quickly recanalize microvessels. This process creates a breach in the blood–brain barrier. The rate of embolus translocation decays when activity of matrix metallopeptidases MMP2 and MMP9 is hindered.

9.8.4 Thrombosis

The vascular endothelium locally maintains blood flow, as it prevents blood coagulation (thrombosis) and promotes fibrinolysis of any possible thrombus that can incidentally appear. Endothelial cells lose their non-thrombogenic property upon exposure to inflammatory stimuli (e.g., TNFα, IL1β, and thrombin).

Transcription factor Krüppel-like factor KLF2[140] regulates inflammatory activation of endothelial cells and favors antithrombotic gene expression. Factor KLF2: (1) provokes production of thrombomodulin, an inhibitor of blood coagulation, and nitric oxide, an inhibitor of platelet aggregation, by NOS3; and (2) favors fibrinolysis, as it lowers synthesis of plasminogen activator inhibitor-1 and promotes tissue plasminogen activator; and (3) precludes cytokine-mediated induction of tissue factor [1029].

Thrombi are formed in blood vessels and cardiac cavities from clotting factors using the same procedure as coagulation. The contact between blood and an unusual material triggers the clotting reaction cascade quickly (in few seconds), activating platelets that adhere to the endothelial gap. Mural thrombi are dominantly composed of platelets. Platelets bind fibrinogen that forms fibrin strands, which are cofactors in platelet aggregation. Tissue factor and plasma fibronectin are also involved in thrombus growth.

When the blood flow is high enough, the microthrombi do not strongly adhere to the wall and can be destroyed by the fibrinolytic system. When the blood flow is low, the competition between activation and inhibition of coagulation favors thrombus formation. A thrombus can develop and block the vessel lumen and then induce ischemia of the irrigated tissues. Thrombus portions can detach; they are then carried in the bloodstream as emboli (embolization). Emboli can block a vessel segment downstream from the thrombus site. Local or remote blockades cause infarction of the tissues normally irrigated by the corresponding blood vessels.

---

[140] A.k.a. lung Krüppel-like factor (LKLF).

Platelet behavior with regard to the vasculature wall is also based on a balance between stimulation and inhibition of the different stages of the coagulation reaction chain. In particular, endoperoxides, precursors of prostaglandins, are used either by platelet enzymes to produce thromboxane-A2 (TxA2), which supports aggregation, or by endothelial enzymes to synthesize PGi2, which inhibits platelet aggregation.

Platelets express 2 receptors for subendothelial collagen, such as collagen-1 and -3: integrin-$\alpha_2\beta_1$ and GP6 glycoprotein. Integrin-$\alpha_2\beta_1$ is overexpressed in pathological clot formation [1030]. As its underexpression only causes a mildly prolonged bleeding time, it is a good target for antithrombotic therapy.

Type 1A phosphoinositide 3-kinase[141] $PI3K_{c1\beta}$ isoform regulates the formation and stability of $\alpha_{2B}\beta_3$-integrin adhesion bonds, involved in shear activation of platelets [1031]. Inhibitors of $PI3K_{c1\beta}$ have been developed to preclude thrombus formation.

Serpins, inhibitors of peptidases, include serpin-A1, or α1-antitrypsin, that reduces the activity of elastase, serpin-C1, or antithrombin, that prevents blood coagulation, serpin-E1, or plasminogen activator inhibitor-1, serpin-B2, or plasminogen activator inhibitor-2, etc. Peptidase binding cleaves an exposed serpin amino acid sequence (reactive-center loop [RCL]) that is incorporated into the central β-sheet. Some serpins can spontaneously undergo a similar conformational rearrangement without RCL cleavage to form an inactive protein. Serpins are able to swap domains as well as to misfold. In addition, domain swap can happen between different monomers, as an RCL portion of one monomer can be incorporated into the β-sheet of another [1032]. Serpins can actually polymerize via β-sheet linkages. Accumulation of misfolded serpin multimers create conformational diseases or serpinopathies that cause thrombosis (and also emphysema, liver cirrhosis, and neurodegenerative lesions such as Alzheimer's, Huntington's, and Parkinson's diseases, as well as prion-induced encephalopathy).

In atherothrombotic diseases, inhibition of peptidase-activated receptor PAR_1 of thrombin prevents thrombin-mediated platelet activation, but does not interfere with thrombin-dependent fibrin generation for hemostasis, i.e., does not raise the bleeding risk [1033].[142]

---

[141] Class-1 PI3Ks that produce second messengers phosphoinositide $PI(3,4,5)P_3$ and $PI(3,4)P_2$ regulate platelet responses, such as the activation of integrin-$\alpha_{2B}\beta_3$, a major platelet integrin that mediates platelet–vessel wall and platelet–platelet adhesions via multiple ligands, such as von Willebrand factor, fibrinogen, fibronectin, and TNFSF5 protein. Platelets contain all class-1 PI3K isoforms.

[142] Platelet activation that is needed for hemostasis is mediated by multiple agents. Platelet integrins bind to exposed extracellular matrix proteins, such as collagen and von Willebrand factor to ensure thrombocyte adhesion. The following release of adenosine diphosphate, thromboxane-A2, serotonin, and catecholamines (adrenaline and noradrenaline [1036]), as well as local production of thrombin prime autocrine and paracrine platelet activation. These agents connect to their cognate GPCRs, such as Gq-coupled $P2Y_1$ and Gi-coupled $P2Y_{12}$ for ADP, 2 TP receptor subtypes that are generated by alternative splicing of the primary transcript (Gq-coupled TPα and Gi-coupled TPβ) for TxA2, Gq-coupled $5HT_{2A}$ for serotonin, Gi/o-coupled α2a adrenoceptors for catecholamines, and Gi/z-, Gq-, and G12/13-coupled peptidase-activated receptor PAR_1 for thrombin. Resulting

9.8.5 Mathematical Modeling and Numerical Simulations

Mathematical multiscale[143] models of either clotting on a breach of the vessel wall or thrombosis after a rupture of an atherosclerotic plaque has been developed in the presence of a flow of an incompressible viscous fluid [1037, 1038]. In the explored fluid domain, the model set incorporates the platelets and coagulation factors, that are involved in the chemical reaction chain, each step being triggered by the activation of the corresponding zymogen into enzyme, and coagulation inhibitors. Compounds and platelet transport by convection and diffusion, are assumed to take place in a near-wall thin plasma layer. Competition occurs between the activation of the coagulation stages and removal by the flowing fluid of the clotting factors and cells away from the reaction site.[144] Numerical simulations use the immersed boundary method (Vol. 7 – Chap. 8. Numerical Simulations). Adhesion of platelets to the injured wall and cohesion between activated platelets are modeled by distributed elastic links.[145] Strain-dependent breaking of cohesive bonds between platelets and adhesive links between the platelet and the vessel wall are treated by a closure approximation of the evolution equations. The probability of platelet aggregation increases quickly after activation, remains nearly maximum for a significant time interval, and then declines [1039]. The fibrinogen concentration and density in surface receptors strongly affect the time constant of platelet aggregation. The thrombin production depends on available binding sites. Thrombus growth, with possible vessel occlusion, and embolus shedding from the thrombus can be predicted by the stress field exerted by the moving fluid on the thrombus.

Interactions between mechanical factors introduced by the flowing blood and the biochemical agents with their multiple positive and negative feedbacks and regulators have also been mathematically modeled, not only during clot formation, but also during degradation [1040, 1041]. Clot reaction kinetics are affected by shear

---

platelet shape changes are associated with: (1) expression of pro-inflammatory molecules, such as P-selectin, soluble CD40 ligand (also expressed by platelets and endothelial and smooth muscle cells); (2) stimulation of platelet procoagulant activity; (3) potentiation of aggregation by other prothrombotic factors; and (4) activation of $\alpha_{2B}\beta_3$-integrins. Activation of endothelial cells by TNFSF5 from activated platelets promotes the production of reactive oxygen species, chemokines, and cytokines as well as expression of adhesion molecules (selectin, ICAM1, and VCAM1). In addition, platelet integrins bind to fibrinogen and von Willebrand factor, thereby causing platelet aggregation and thrombus formation. Serotonin released from activated platelets: (1) accelerates aggregation of platelets; (2) promotes proliferation, migration, and contraction of vascular smooth myocytes; and (3) heightens and reduces procoagulant and fibrinolytic activities, respectively, of endothelial cells via $5HT_{2A}$ receptors, as it provokes expression of tissue factor and plasminogen activator inhibitor-1 [1034]. Thromboxane-A2 causes platelet aggregation in synergy with other secreted agonists [1035].

[143] The model requires several spatial scales: the nanoscale of the coagulation factors, the microscale of the platelets, and the macroscale of the blood vessel.

[144] The flux of activated coagulation factors and platelets is flow dependent.

[145] Platelets aggregate assuming that bridges form isotropically, linear springs modeling the clusters of binding sites.

stresses. Shear stress-dependent platelet transport and activation, the extrinsic pathway of coagulation leading to a viscoelastic clot and fibrinolysis[146] are represented by a set of coupled advection–diffusion–reaction equations, assuming a uniform distribution of platelets, reactants, and regulators in a shear-thinning viscoelastic blood for a given wall flux of stimulating agents.

9.9 Effects of High-Density Lipoproteins on Endothelium

High-density lipoproteins decrease the lipid peroxidation of LDL, due to the presence of paraoxonase-1 (PON1) [1042]. Furthermore, HDLs influence several functions of the vascular endothelium [1043].

High-density lipoprotein acts on the vascular tone via vasoactive substances released by endothelium. It activates constitutive NOS3 and inducible COx2, releasing NO and PGi2 agents. It triggers MAPK signaling via scavenger receptor ScaRb1 on endothelial cells. In vivo, the plasma concentration of HDLc is used as a predictor of NO-dependent vasodilation and correlates with the plasma concentration of the stable PGi2 metabolite 6-keto PGf1α. Moreover, HDL may inhibit ET1 secretion.

High-density lipoprotein intervenes in inflammation. It downregulates TNFα, and consequently decreases the synthesis of cell adhesion molecules, reducing leukocyte recruitment and extravasation. It reduces the production of platelet-activating factor. Furthermore, PAF is also degraded by HDL-bound enzymes, platelet-activating factor acetylhydrolase (PAFAH), LCAT, and paraoxonase. High-density lipoprotein then prevents PAF-induced adhesion of leukocytes to the endothelium. However, HDL may lose its anti-inflammatory properties during inflammation, part of apoA1, apoA2, paraoxonase, and PAFAH being removed.

High-density lipoprotein is involved in coagulation. It enhances the synthesis of antithrombotic NO and PGI2 agents. It limits PAF level. It may also repress the production of von Willebrand factor. It binds to tissue factor pathway inhibitor, hampering the extrinsic coagulation pathway. It enhances the anticoagulant activity of protein-C and -S.

High-density lipoprotein stimulates the proliferation of endothelial cells, in association with phospholipase-C and intracellular calcium. It protects cultured endothelial cells from TNFα-induced apoptosis, inhibiting caspase-3. It suppresses the mitochondrial pathway of apoptosis, activating PKB kinase.

---

[146]Fibrinolysis occurs when fibrin concentration decays below a given threshold or the local shear stress reaches a critical value that depends on concentrations of platelets and fibrin.

9.10 Chemical and Mechanical Regulation of the Vasomotor Tone

Any cell acts on its environment and vice versa. Mechanical forces, such as forces applied by adjoining cells or the surrounding matrix, as well as forces developed by a given cell on its matrix or contacting cells, are transmitted. The soft extracellular matrix contains a three-dimensional fibrous mesh that sends and receives mechanical stimuli.

Living cells not only sense applied forces and respond to received signals, but also evaluate mechanical properties of their matrix, in particular the stiffness of the extracellular matrix.[147] Moreover, the cell can recognize a local membrane curvature and recruit small GTPases to the curved parts of the plasma membrane [1044]. Fiber curvature can induce cell-membrane curvature. Besides, certain K^+ channels are opened by a convex curvature of the membrane.

Furthermore, endothelial cells have a wetted surface that faces flow forces. Blood vessel walls are permanently subjected to tension and shear. In endothelial cells, hemodynamic stress applied on the wetted apical surface produces fast remodeling of mechanical stress-responsive focal adhesions that orient in the streamwise direction at basal cell sites where resulting stress of transmitted forces concentrates [1045]. Moreover, increase in focal adhesion size and decrease in number result from coalescence of small adhesion sites. Focal adhesion remodeling is accompanied by redistribution of stress fibers.

Mechanotransduction starts with mechanosensing and leads to a relatively quick mechanoresponse. Molecular sensing of forces is transduced into biochemical signals that drive cell responses to properly shape cells and tissues and trigger suitable cell functions. The force can act not only by its magnitude and direction but also by the loading rate.

Mechanical forces regulate cell function so that the vessel wall restores basal stress conditions (Fig. 9.7). The cell changes its shape and adapts to the mechanical loading. The cytoskeleton of mural cells transmits and modulates stresses within the cell via cellular junctions and cell–matrix adhesion sites. Moreover, applied forces initiate mechanotransduction cascades leading to transcription factor activation and subsequent gene expression.

Mechanosensing occurs locally at the plasma membrane. It can be done by conformational changes in plasmalemmal proteins and can involve cytoskeletal filaments that propagate loadings down to the nucleus.[148] Force sensing can be processed by: (1) partial protein unfolding with gain or loss of binding sites or of enzyme function; (2) opening of mechanosensitive ion channels due to membrane tension or to stress applied by force-bearing filaments connected to channels; and (3) strained receptor–ligand interactions [1044].

---

[147] Different cell types respond differently to matrix rigidity.

[148] Cytoskeletal early responses take seconds to minutes.

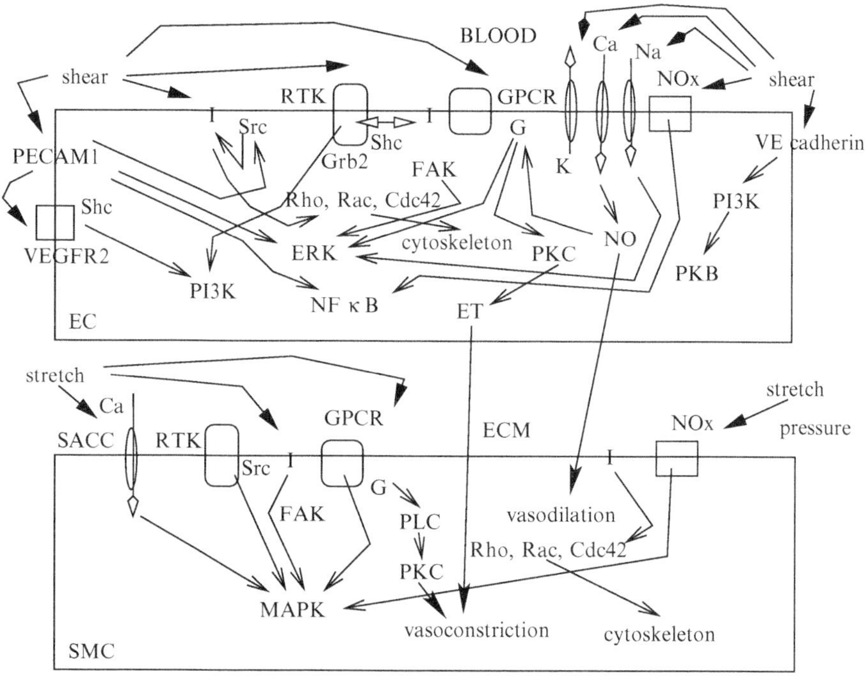

Fig. 9.7 Effects of blood pressure characterized by its large magnitude on the vessel wall cells and of three-dimensional shear of much smaller magnitude than the applied pressure, but with large spatial gradients, on the wetted surface of endothelial cells. Both forces undergo large-amplitude oscillations during the cardiac cycle. Directional changes can also occur (flow separation and flow reversal during the diastole). I: integrin.

The endothelium provides a link between blood flow and vessel responses, in particular the vessel caliber. Multiple kinds of mechanical stresses and various types of mechanical environments are associated with flow patterns and unsteadiness. The vessel wall is sheared by the moving blood particles on the one hand and stretched and compressed by the pressure applied by blood. Mechanical stresses applied to the endothelial wetted surface are indeed normal (mostly pressure) and tangential (shear). Wall friction fluctuates in magnitude and in direction at a given location and from point to point. Pressure and tensile reactions also undergo quasi-periodic fluctuations. The stresses applied by the blood continuum[149] are

[149]When the flow scale is much greater than the circulating cell size, the blood can be considered as a continuum with given apparent physical properties, which depend on the microstructure of the concentrated suspension. The microstructure depends on the local stress field.

Table 9.29. Circumferential tension T_θ per unit length estimated by the Laplace law at a first approximation, assuming a membrane-like behavior ($T_\theta \sim hc_\theta \sim p_i R$).

Vessel	p_i (kPa)	R (mm)	T_θ (Pa)
Aorta Large arteries	13.3	10	133
Small arteries	11.3	2	22.6
Arterioles	8	0.25	2
Capillaries	4	0.01	0.04
Venules	2.7	0.25	0.7
Veins	2	3	6
Vena cava	1.3	15	20

transmitted throughout the wall continuum, which reacts. Intramural stresses are composed of axial, radial, and circumferential components[150] (Table 9.29).

Vascular endothelial cells synthesize vasodilators, such as nitric oxide, prostacyclin, hydrogen peroxide, carbon monoxide, and epoxyeicosatrienoic acids, among others, as well as vasoconstrictors, such as endothelin-1, thromboxane-A2, and prostaglandin-H2. During inflammation, tryptophan is catabolized into *kynurenine* by indoleamine (2,3)-dioxygenase in some types of activated macrophages and other immunoregulatory cells.[151] Kynurenine generates an endothelium-dependent vascular relaxation [1047].

Three enzymes – *manganese superoxide dismutase*, *cyclooxygenase*-2, and *nitric oxide synthase* – have anti-oxidant, antithrombotic, and anti-adhesive activities. They react differently according to the type of mechanical stimulus [1048].

Mechanical removal of the endothelium alters the flow-dependent response, but not reaction of mural cells to acetylcholine and noradrenaline [1049]. Mechanical stress exerted on the wall by blood generates a basal tone in smooth myocytes in the absence of neurogenic and hormonal influences. Local metabolism and appropriate ion ([Ca^{2+}], [Na^+]) concentrations are required. The hemodynamic stresses act on the smooth myocytes via direct stress transmission or compound release by endothelial cells.

[150]For a cat thoracic aorta at a pressure of 16 kPa, with an axial stretch ratio of ~ 1.7 and an opening angle of ~ 70 degrees in the unstressed state, intramural longitudinal and circumferential tensile Cauchy stress and radial compressive stress at outer and inner walls have been found to be 95 and 71, 105 and 61, and −16 and 0, respectively [1046].

[151]Kynurenine is catabolized by kynureninase to anthranilic acid, by kynurenine–oxoglutarate transaminase to kynurenic acid, and by kynurenine 3-hydroxylase to 3-hydroxykynurenine.

The endothelial membrane is the first wall component to bear stresses from circulating blood.[152] The bulk cell content can be supposed to be a fluid-like material that contains solid particles and a cytoskeleton, wrapped into a solid-like membrane.[153] Membrane tension has been estimated to be 3 orders of magnitude higher than wall shear stress imposed by the flowing blood [1050]. Membrane tension is affected by transmembrane static and osmotic pressures. It also at least partially controls osmotic pressures. Membrane tension influences not only the cytosol[154] and nucleus configurations by stress transmission, the cytoskeleton being moored to anchoring membrane proteins, but also the mass transport across the cell membrane, via pores, channels, or vesicles, and, thus, cell metabolism. Interactions thus occur between cell internal structures and messengers using electrochemical and biochemical processes.

Mechanical stresses alter structure and function at both cellular and molecular levels. Stresses can directly act on stretch-activated channels (SAC), which are sensitive to tension imposed on cell membrane [1051].[155] Mechanosensitive ion channels can indeed be opened by membrane stretch. They convert external mechanical forces into electrochemical signals in the cell.

Mechanical stresses act not only on molecular transport from flowing blood to the endothelial surface, but also on transendothelial transfer and through the wall directly by interface deformation. Mechanical stresses cause conformational changes of membrane proteins (ion channels, adhesion molecules, etc.), of proteins associated with membrane proteins, and/or of proteins associated with the cytoskeleton.

Because arteries convey pulsatile flows, arterial walls are subjected to tension, with steady and time-dependent components. High steady intraluminal pressures (~ 20 kPa) activate focal adhesion kinases. Phosphorylated FAK binds to GRB2 adaptor and stimulate a pathway that involves extracellular signal-regulated kinase-1 and -2, Src kinase, and integrins [1053]. However, FAK is not implicated in cyclic stretches, although ERK1 and ERK2 are activated.

---

[152]The membrane can be subdivided into 3 parts according to its neighborhood. A luminal part, which is subjected to the wall shear stress, has a protruding segment above the average cell height due to the nucleus. The nucleus in the stress field bears deformation; it is elongated, like the whole cell, in the streamwise direction. The abluminal part, which adheres to BM, and sides, which have a given inclination and are connected to adjoining endothelial membranes, are subjected to wall tension imposed by the blood pressure and transmitted by the extracellular matrix and adjacent endothelial cells.

[153]The plasma membrane of endothelial cells can be assumed to be a solid-like material because it resists applied stresses by deformation and not by flow.

[154]The configuration of the marginal fiber network may be imposed by the direction of principal stresses undergone by the membrane to which it adheres.

[155]Other stretch-sensitive ion channels are implicated in mechanotransduction and osmoregulation. They include stretch-inactivated K^+ channels, which can be detected in neurons [1052].

Steady pressurization[156] of isolated rat cerebral and mesenteric arteries decreases Fos expression in the cerebral arteries subjected to 18.6 kPa compared with 10.6 kPa (control), but not in mesenteric arteries [1054].

Vascular endothelial cell are subjected to intercellular junctional and intracellular tensions. Under static conditions, cells in confluent monolayers are subjected to uniform intracellular tensions from cell junctions to cell edges [1055]. When a mechanical stress is applied with a mean forward component, focal adhesion sites change and stress fibers and intracellular tensions orientate mainly in the streamwise direction. Intracellular tension may follow the path of the stress fibers with adherens junctions at upstream and downstream ends [1055]. When the mechanical stress oscillates with a very low mean, intracellular tensions do not have a preferential orientation [1055].

Stretch-mediated signaling by *elastin–laminin receptors* reduces Fos expression and subsequent cellular proliferation. Mechanosensitive signaling via elastin–laminin receptors hence depends on the artery type and/or perfusion territory (e.g., cerebral vs. mesenteric circulation).

Mechanical stresses either promote or repress gene expression in endothelial cells. A small fraction of targeted genes respond to mechanical stresses [1056]. In general, mechanical stimuli promote expression of transcription activators or repressors Fos and Jun. Mechanical stresses also activate another transcription promoter, the *mechanical stress response element* [1057]. The latter is found in promoters of genes that encode platelet-derived growth factor-A and -B, tissue plasminogen activator, transforming growth factor-β1, and intercellular adhesion molecule ICAM1.

9.10.1 Transducers of Mechanical Stresses

Several types of mediators are targeted by mechanotransduction (Fig. 9.8). They locatize: (1) either on the endothelium wetted surface; (2) on the lateral walls (cell–cell junctions); (3) on the abluminal membrane (cell–matrix junctions), or (4) within the cytoplasm. Three-dimensional conformational changes of membrane proteins induced by applied stresses can initiate the cell response. Mechanical transduction uses a set of messengers and leads to gene transcription.

Various mechanosensors detect stresses and strains applied to the endothelial cells (Fig. 9.8). Transduction surface elements include: (1) mechanosensitive ion carriers, especially stress-gated ion channels; (2) plasmalemmal receptors, such as receptor Tyr kinases and G-protein-coupled receptors; (3) adhesion molecules of intercellular and matrix–cell junctions that are associated with cytoskeletal filaments;

[156]Isolated arteries are subjected to 8.6 kPa during 2 h, then to a step pressure increase of either 10.6 or 18.6 kPa during 30 mn.

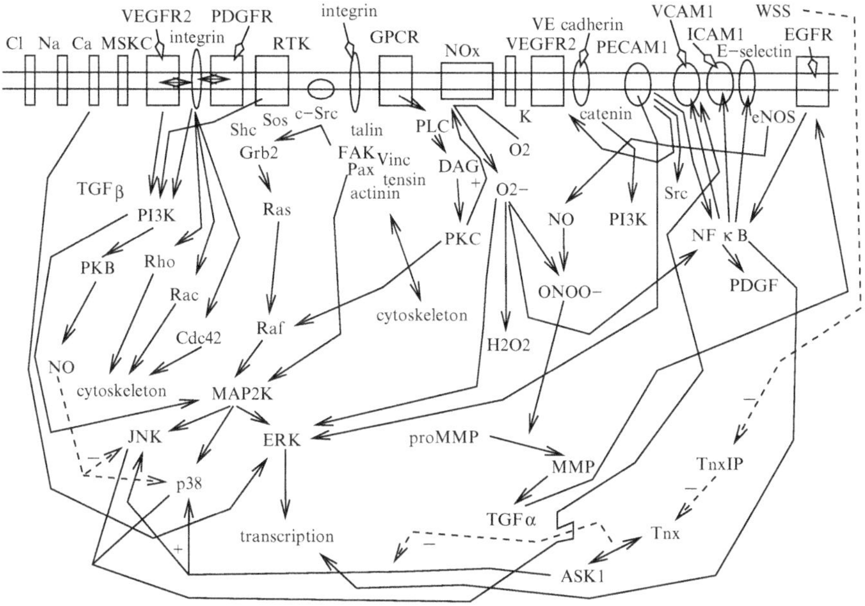

Fig. 9.8 Biochemical pathways stimulated by applied pressure and shear (Sources: [1058–1060]). Initiation is done via: (1) ion channels; (2) growth factor receptors and others Tyr kinase receptors (RTK); (3) G-protein-coupled receptors (GPCR); (4) membrane NADH/NADPH oxidase (NOx); and (5) adhesion molecules, mainly integrins, but also cadherin-5 and IgCAMs. Effectors include small GTPases, PI3K, PKB, PKC, NO, FAKs, TGF, and O_2^- (O2−). Main targets are the transcription factors for cell and tissue remodeling (with possible degradation via MMPs after a long-duration pressure rise), such as NFκB, the cytoskeleton, adhesion molecules, and ion channels.

(4) membrane rafts (caveolae); and (5) stress-sensitive enzymes connected to the plasma membrane. Stimulated mechanosensors initiate different signaling pathways to trigger responses.

Mechanotransduction generally occurs at numerous sites at the plasma membrane and eventually inside the cell because mechanical stresses are transmitted to multiple cellular structures, especially via the cytoskeleton on load-bearing molecules. All agents involved in mechanotransduction then act synergistically. Plasmalemmal, cytoskeletal, and nuclear scaffold proteins serve as mechanotransmitters.

9.10.1.1 Plasmalemmal Receptors and Carriers

The density in pressure-activated cation channels increases in endothelium subjected to high pressure, with complete reversal by antihypertensive therapy [1061]. Mechanosensitive potassium channels modulate plasmalemmal

NAD(P)H oxidase [1062]. Stretch-activated Ca^{2+} channels are located near integrins. Subunit $G\alpha_q$ of G proteins participates in flow-induced Ras activation [1063]. Heparan sulfate, the dominant glycosaminoglycan of the glycocalyx, participates in mechanosensing for NO synthesis in response to wall shear stress [939].

Plasmalemmal nanodomains can also be implicated in mechanotransduction. Caveolae recruit endothelial nitric oxide synthase, cyclooxygenase-2, prostacyclin synthase, and Src kinase, among other enzymes. Caveolin-1 mediates shear stress effects by activating extracellular-regulated kinase-1 and -2, but not Jun N-terminal kinase that, nonetheless, can also be activated by wall shear stress [1064].

Integrin–cytoskeleton couples act as mechanosensors. Moreover, forces applied to endothelial cells induce assembly and extension of adhesive structures. Cell adhesion molecules are targeted for transmembrane signaling, using various biochemical cascades, such as FAK–Src and Rho pathways.

Wall shear stress signals at the endothelial wetted surface via receptor Tyr kinases [1065, 1066], caveolin-1, extracellular signal-regulated kinases [1067], and guanine nucleotide-binding (G) proteins [805].

Wall shear stress activates mechanosensitive K^+ channels involved in TGFβ1 production [1068], Na^+ channels implicated in activation of ERK1 and ERK2 [1069], and Cl^- channels [1070]. Voltage-gated, inward rectifier K^+ channels ($K_{IR}2.1$) of endothelial cells respond to shear stress [1071]. Activity of $K_{IR}2.1$ can also be modulated by protein kinases PKA and PKC, FAK2, and phosphatases.[157] Shear-induced ionic currents are stopped by channel dephosphorylation by phosphatase, which thus favors a shear-unresponsive channel state.

Hemodynamic stresses applied by the flowing blood promote endothelium-dependent vasodilation not only via nitric oxide- and prostacyclin-dependent signaling, but also NO- and PGI2-independent mechanisms.

Hyperpolarization precedes release of vasoactive substances. *Endothelium-derived hyperpolarizing factor* (EDHF) is the component of endothelium-dependent relaxation of smooth myocytes that resists full blockade of nitric oxide synthases and cyclooxygenases. This mechanism can rely on opening of $K_{Ca}1.1$ channels on vascular smooth muscle and endothelial cells by epoxyeicosatrienoic acids generated by cytochrome-P450 epoxygenase that leads to Ca^{2+} entry and endothelial hyperpolarization by activating mechanosensitive transient receptor potential TRPV4 channel. The contribution of epoxyeicosatrienoic acids to flow-induced vascular response is significant only in arteries that express sufficient amount of functional TRPV4 channel. Activation of TRPV4 channels by flow requires not only active cytochrome-P450 epoxygenase, but also translocation of TRPV4 channel to the plasma membrane [1072].

---

[157]Protein kinase-A controls the links between $K_{IR}2.1$ channel and scaffold A kinase anchoring protein (AKAP) and SH2 domain-containing SH2D1a adaptor. Kinase FAK2 also associates with SH2D1a molecule. Scaffold proteins allow formation of proteic complexes with ion channels, receptors, and enzymes that quickly tranduce signals.

In endothelial cells, epoxyeicosatrienoic acids activate not only $K_{Ca}1.1$ channels, but also protein kinase-A. They amplify endothelial cell hyperpolarization by enhancing Ca^{2+} influx through TRPC3 and TRPC6, as they also promote their translocation to the plasma membrane.[158] Epoxyeicosatrienoic acids have not only intracrine, but also paracrine effects, as they can diffuse to vascular smooth myocytes.

In endothelial cells, activity of constitutive membrane-bound hemoproteic members of the cytochrome-P450 superfamily CyP1a1 and CyP1b1 depends on the pattern of hemodynamic stress [1073].[159] Ligand-activated basic helix–loop–helix transcription factor aryl hydrocarbon receptor (AHR or bHLHe76) increases CyP1a1 and CyP1b1 transcription. Conversely, CyP1a1 modulates AHR-regulated genes.[160] Nuclear localization of aryl hydrocarbon receptor that primes CyP1a1 transcription is observed in the descending thoracic aorta, whereas AHR nuclear localization and CyP1a1 expression is lower in the aortic arch. Steady regime and pulsatile flow reversal reduce expression of both CyP1a1 and CyP1b1.

During endothelial cell stimulation by hemodynamic stresses, sustained Ca^{2+} influx within endothelial cell cytosol via store-operated channels activated by Ca^{2+}-independent phospholipase-A2 isoform (iPLA2) contributes to release of vasodilators and stress-induced endothelium-dependent relaxation of adjoining smooth muscle cells [1074]. Sustained store-operated Ca^{2+} entry can be triggered in endothelial cells by circulating ATP owing to iPLA2 activation to govern amplitude and duration of ATP-induced endothelium-dependent vasodilation.

Acetylcholine-induced endothelium-dependent relaxation of smooth myocytes also depends on iPLA2 activation. Both iPLA2 metabolites, arachidonic acid and lysophosphatidylcholine, stimulate Ca^{2+} entry in endothelial cells. However, only lysophosphatidylcholine produced by iPLA2 upon Ca^{2+} store depletion elicits store-operated Ca^{2+} influx and subsequent NO production to trigger vasodilation after NO release.

9.10.1.2 Adhesion Molecules

Wall shear stress on the wetted cell surface and pressure-induced axial and circumferential stretch on the laterobasal plasma membrane are also transduced by

[158] Ion channels of the TRPC and TRPV families respond to distinct stimuli as well as different types of epoxyeicosatrienoic acids. Ion channel TRPV4 is primed by mechanotransduction. It is also activated by (5,6)- and (8,9)-EET, but not (11,12)-EET molecule. Channel TRPC6 is not influenced by shear, hypo-osmotic stress, and (5,6)-EET, but is stimulated by (11,12)-EET agent.

[159] Enzymes CyP1a1 and CyP1b1 modulate cellular levels of lipid signaling molecules, such as arachidonic acid and retinoic acid metabolites.

[160] Activated aryl hydrocarbon receptor also targets the thrombospondin-1 gene in endothelial cells. Enzyme CyP1a1 favors thrombospondin-1 expression.

adhesion molecules of intercellular and cell–matrix junctions (Fig. 9.8). Adhesion molecules that link endothelial cells either to apposed endothelial cells or basement membrane transmit sensed mechanical forces and adapt.

Vascular endothelial cell cadherins, platelet–endothelial cell adhesion molecule PECAM1, and $\alpha V\beta_3$ integrins are particularly involved [1059]. The stronger the changes in hemodynamic stress amplitude and direction, the more sustained the activation.

The pathways mediated by adhesion molecules use vascular endothelial growth factor receptor kinase (VEGFR2) that activates phosphatidylinositol 3-kinase, Src kinase, mitogen-activated protein kinases (ERK, JNK, and P38MAPK), and protein kinase-B, as well as transcription factors such as NFκB.

Cell adhesion molecules PECAM1 and cadherin-5 on the lateral membrane of endothelial cells respond to hemodynamic stress. Substance PECAM1 is phosphorylated when endothelial cells are exposed to flow. Phosphorylation of PECAM1 leads to ERK activation [1075]. Cadherin-5 at adherens junctions complexes with β-catenins, PI3K, and VEGFR2 and transmits signals to PKB and B-cell lymphoma protein BCL2 [1076]. The interaction of α-catenin with Cdh5-bound β-catenin is regulated by phosphorylation of stress-modulated β-catenin phosphatase PTPn11 associated with cadherin-5 complexes [1077].

Integrin-$\alpha_5\beta_1$ enhances stress-dependent cell migration [1078]. Integrin-$\alpha_V\beta_3$ is involved in strain-dependent phosphorylation of VEGFR2 and its binding to SHC adaptor [1066], and strain-dependent activation of IκB kinases[161] and NFκB [1080]. Activation of NFκB is mediated by time and magnitude changes in hemodynamic stress. Small Ras GTPase regulates intracellular transport of inhibitor IκB kinase-α [1081]. Last but not least, the cytoskeleton transduces force signaling to the different cell compartments [1082].

9.10.1.3 Cell Junctions

Cell communications exist between neighboring vascular cells, i.e., between: (1) apposed endothelial cells; (2) adjoining smooth myocytes; and (3) nearby endothelial and smooth muscle cells through fenestrae of the basement membrane and internal elastic lamina.

Mechanical stresses imposed on endothelial cells can increase the hydraulic conductivity of the endothelium. In particular, they act on the paracellular transport by altering the state of occludins at tight junctions (Sect. 9.6.1). After several minutes of flow exposure, occludin phosphorylation increases [940]. Stress fiber bundles in endothelial cells subjected to blood flow develop and align in the direction of the blood stream. The cytoskeleton fibers associated with tight junctions between endothelial cells are reinforced and elongated [1083].

---

[161] Inhibitors of NFκB (IκBα, -β, and -ε) sequester in the cytoplasm, hence precluding the binding of $P65_{NF\kappa B}$ (RelA) to DNA and transcription of NFκB-dependent genes [1079].

In addition, cadherin-bound γ-catenin and Ca^{2+} stabilize adherens junction between cultured neighboring endothelial cells subjected to flow [942]. Depletion in $[Ca^{2+}]_e$: (1) removes both cadherin and γ-catenin away from cell junctions; (2) inhibits γ-catenin expression; and (3) dissociates the endothelial layer subjected to flow. β-Catenin does not compensate for γ-catenin loss. In addition, PECAM1-mediated adhesion cannot ensure endothelium integrity under flow.

9.10.2 Flow Chambers

To understand the flow-dependent functions of the endothelium and evaluate gene and protein expressions, EC cultures are exposed to mechanical stresses in flow chambers. In vitro effects of flow over cultured EC layer and cyclic stretch of the culture support have been investigated to study the responses of endothelial cells in well-defined mechanical conditions.

Mechanical stresses applied on EC wetted or on abluminal surface affect: (1) the cell shape and orientation [1084],[162] as well as cell ultrastructure [1086];[163] (2) cell rheology [1087, 1088], endothelial cells becoming stiffer; (3) cell proliferation; (4) cell metabolism and transport; and (5) cell adhesion to its support as well as matrix content.[164] Time estimations of EC responses to imposed stresses are given in Table 9.30.

Endothelial cells respond, in particular, to space and time changes in wall shear stress imposed at the luminal surface. Fluctuating low-amplitude WSS are supposed to have a stronger effect on mechanosensitive plasmalemmal proteins than varying large-magnitude stretch applied to intramural plasmalemmal constituents directly or transmitted by matrix elements.

Magnitude and orientation of wall shear stress depend locally on both vessel axial and transverse configurations and flow pattern. Wall shear stress is equal to zero at separation and reattachment points. Change in direction of the WSS axial component occurs when the pulsatile flow generates a backflow in the whole vessel lumen or flow reversal in a layer of given (axis curvature-dependent) thickness near the wall. A WSS circumferential component appears when the vessel presents either a curvature of its axis and/or its walls caused by a transition upstream from branching (adaptation to caliber change between trunk and 2 major branches), taper, neighbor organ print, vessel collapse due to cyclic variation in external pressure such as in the thorax (breathing) or muscular compression (walk), etc.

---

[162]The orientation of the cultured endothelial cell under flow conditions is a shear minimization process [1085].

[163]EC elongation and orientation, as well as the remodeled cytoskeleton arrangment, reflect the local flow direction.

[164]The synthesis of junctional proteins increases. Hemodynamic stresses influence the production of growth factors, thereby wall remodeling, with a given elastin and collagen content.

Table 9.30. Mechanotransduction in cultured endothelial cells. Order of magnitude of response time (Sources: [1086,1089]; IP_3: inositol (1,4,5)-trisphosphate; LDL: low-density lipoprotein; NO: nitric oxide; PDGF: platelet-derived growth factor; PGf2α: prostaglandin-F2α; PGi2: prostacyclin; SAC: stretch-activated channel; tPA: tissue plasminogen activator).

Target	Time
K^+ channel	$\mathcal{O}$(ms)
SAC	
$[Ca^{2+}]_i$	$\mathcal{O}$(s)
NO	
IP_3	$\mathcal{O}$(10 s)
Adenylate cyclase	$\mathcal{O}$(mn)
PGi2, PGf2α	
PDGF	Hours
Endithelin	
tPA	
Pinocytosis	
Cell proliferation	
Cytoskeleton alignment	
Fibronectin	Day
LDL metabolism	
Cell alignment	
Collagen	Days
Cell growth	

Cross variations in wall curvature that are observed in collapsed tubes of uniform cross-section, due to wall transverse bending, induce transverse gradient of the WSS axial component [1090, 1091].

Many investigations have been carried out to explore either cell rheology or behavior of mechanical stress-subjected cells. Flow chambers coated with a homogeneous confluent monolayer of EC cultures are used for EC exposure to laminar-flow stresses [1092–1095]. A special flow chamber has been designed to expose endothelial cells to a given field of hemodynamic stresses that generates 2 forces applied at the cell inertia center: (1) a shear force that stretches the cell in the streamwise direction, e.g., the force arising from the shear stress at the cell inertia center; and (2) a shear torque that twists the cell perpendicularly to the cell plane, induced by the WSS transverse gradient (WSSTG) [1096]. Four regions of the wetted cell culture were defined by these investigators: (1) region I near the contact point where both WSS and WSSTG are small; (2) region II of low velocity flow where WSS is low but WSSTG is significant; (3) region III with a change in wall curvature sign where both WSS and WSSTG are great; and (4) region IV, where WSS reaches its highest values but WSSTG is small. These authors discussed the role of torsion in the EC detachment both from the wall of the flow chamber and cell layer, i.e., from adjoining endothelial cells, in region II. Wall shear stress varies along the endothelial surface between the peripheral region and the central cell bulging zone associated with the nucleus. Atomic force microscopy has been used to study the shear distribution at a subcellular scale; steady flow was simulated over the endothelial surface [1097].

Endothelial cells exposed to mechanical stress using a shaker react according to their location [1098]. Orbital shear stress increases endothelial cell proliferation and adhesion molecule expression, whereas it reduces PKB phosphorylation and E-selectin downregulation in central endothelial cells exposed to lower shear stresses, with respect to peripheral endothelial cells. When endothelial cells are subjected to step, impulse, ramp, inverse ramp, and pulsatile flows,[165] the time derivative of wall shear stress, but not the shear stress itself, is directly responsible for endothelial reactions [1099]. Both phosphorylation of extracellular signal-regulated kinases ERK1 and ERK2 and release of endothelin-1 are attenuated when the duration of the pulse cycle diminishes [1100].

In static conditions, the features of extracellular matrix affects cell adhesion and spreading, differentiation, migration, and apoptosis.[166] When endothelial cells are subjected to flow, the directionality of hemodynamic stresses governs their morphology, as well as their functioning and fate (survival or apoptosis).[167] In flow chambers, cultured endothelial cells on strips of fibronectin on silicone membrane react differently according to the flow direction with respect to strip axis. Pulsatile flow parallel to strip axis (pressure modulation rate of 1/3, after 12 h under static conditions) causes cell elongation, increase in stress fibers and in focal adhesion kinase phosphorylation, and apoptosis reduction [1101]. Pulsatile flow perpendicular to strip axis does not induce such changes. Constitutive GTPase RhoV augments the actin stress fiber formation and FAK phosphorylation,[168] and attenuates apoptosis under both static conditions and flow, whatever its direction (parallel or perpendicular to strip axis).

[165] Step increments in shear stress from 0 to 1.6 N/m^2 are followed by a sustained steady shear. Ramp shear is defined by a gradual increase in shear stress from 0 to 1.6 N/m^2 over 2 mn, the loading being sustained afterward. Impulse flow means a loading over 3 s of 1.6 N/m^2. Pulsatile flows correspond to repeated impulse flows.

[166] In human umbilical vein endothelial cells cultured on strips of fibronectin on silicone membrane, the cell behavior depends on the strip width [1101]. On broader strips (width 30 and 60 μm), endothelial cells create actin stress fibers with anchoring spots of phosphorylated focal adhesion kinases and do not present significant apoptosis. On narrower strips (width 15 μm), endothelial cells develop few stress fibers without spots of phosphorylated focal adhesion kinases, whereas the apoptosis rate is significant.

[167] The smaller the ability to form stress fibers and attachments with the extracellular matrix, the higher the apoptosis rate. Apoptosis leads to the disassembly of focal adhesion complexes and the assembly of actin into a peripheral ring.

[168] Rho GTPases cause FAK phosphorylation that augments the onset of stress fiber formation.

Support and perfusion media used in flow chambers can bring substances that can interfere with the cell response to investigated stimuli. Consequently, experimental testing and result interpretation must be carefully handled.

9.10.3 Nitric Oxide

Nitric oxide (Vol. 4 – Chap. 10. Other Major Signaling Mediators) is a vasodilator that inhibits vasoconstrictor influence, such as that of angiotensin-2 as well as sympathetic vasoconstriction. Acetylcholine and muscarinic agonists stimulate NO release from endothelial cells [1102]. Nitric oxide can also be released from the endothelium by activated α2-adrenoceptors and aggregating platelets that particularly release serotonin and leukotrienes [1103]. On the other hand, superoxide (O_2^-) and oxyhemoglobin (HbO_2) sequester nitric oxide.

9.10.3.1 Nitric Oxide and Vascular Cells

Nitric oxide impedes the adhesion of platelets and leukocytes to the vascular endothelium. It also precludes the proliferation of vascular endothelial and smooth muscle cells.

Nitric oxide binds to and activates guanylate cyclase that produce cyclic guanosine monophosphate from guanosine triphosphate. Messenger cGMP activates PKG kinase that subsequently inhibits calcium influx into smooth myocytes and decreases Ca^{2+}–calmodulin stimulation of myosin light chain kinase, thereby decaying smooth muscle tone in the vascular wall.

In endothelial cells, nitric oxide can prevent Ca^{2+} entry and lower intracellular Ca^{2+} concentration, as nitric oxide exerts a negative feedback. Nitric oxide can augment calcium uptake into intracellular stores by sarco(endo)plamic reticulum calcium ATPase to avoid detrimental effects on cell functioning [1104]. In aged cells, reduced phosphorylation of its active site (Ser1177) uncouples NOS3 enzyme.

Tissular hypoxia yields vasodilation in various vessel types. Endothelial cells act as local sensors for the regulation of O_2 supply, increasing $[Ca^{2+}]_i$ and releasing nitric oxide and prostacyclin in response to hypoxia [1105].

9.10.3.2 Nitric Oxide Production

Nitric oxide is produced by endothelial and smooth muscle cells as well as cardiomyocytes, among other cell types. The continuous production of nitric oxide is fostered by multiple stimuli, such as adenosine di- and triphosphate, acetylcholine, bradykinin, histamine, insulin, and substance P.

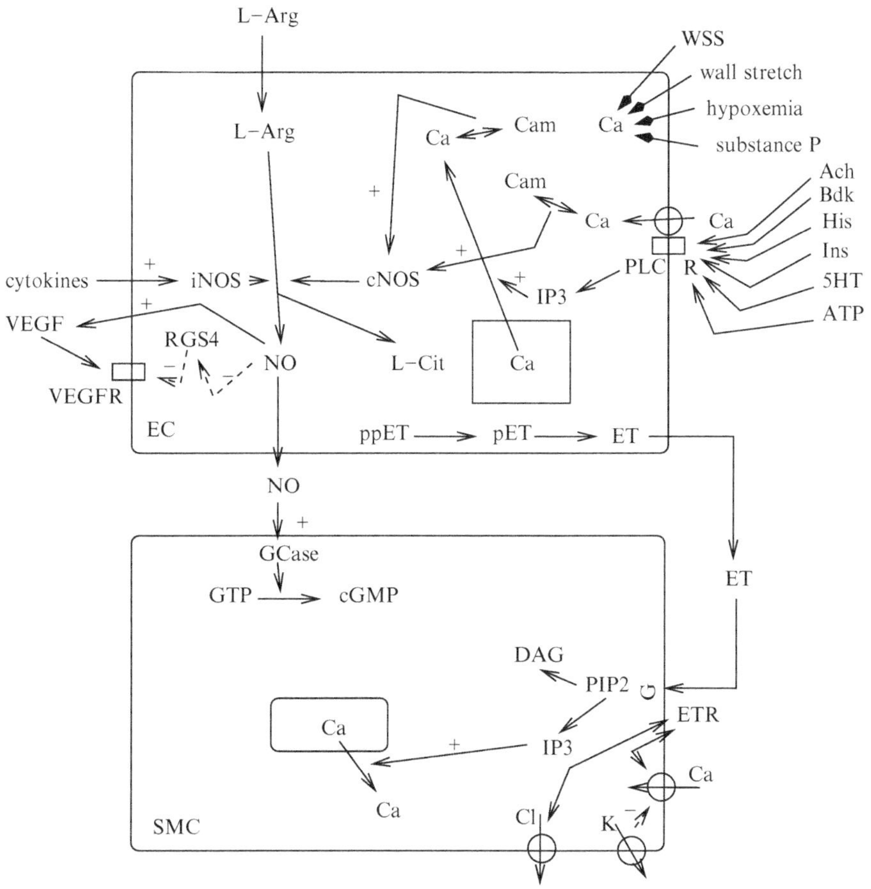

Fig. 9.9 Nitric oxide and endothelin-1 synthesis, function, and regulation (Sources: [761, 1106]). Nitric oxide, discovered in endothelial cells, is produced by NO synthase from ^{L}arginine. It binds to its guanylate cyclase receptors and increases the intracellular concentration of cyclic guanosine monophosphate (cGMP). Messenger cGMP stimulates the cGMP-dependent protein kinase-G (PKG) that precludes intracellular calcium release by impeding the activity of phospholipase-C and inositol trisphosphate.

Nitric Oxide Synthase

Nitric oxide synthase catalyzes the conversion of ^{L}arginine to NO and ^{L}citrulline (Fig 9.9). Two NOS categories exist: constitutive (cNOS) and inducible (iNOS). Ca^{2+}–calmodulin-dependent NOS resides in endothelial cells (endothelial constitutive NOS [NOS3 or eNOS]) and smooth myocytes. Isoform NOS3 yields a basal release. Enzyme NOS2 is expressed in endothelial cells in the presence of cytokines (cytokine-inducible NOS).

Endothelial nitric oxide synthase is a plasmalemmal protein. It can be confined to caveolae in endothelial cells, as it binds to caveolin, particularly caveolin-1 [1107]. Caveolin-1 inactivates NOS. Caveolae thus act as temporal and spatial regulators of NO release in endothelial cells.[169]

Regulation of the Cellular Concentration of Nitric Oxide

Cellular concentration of nitric oxide relies on cytochrome-C oxidase, the terminal enzyme (complex-IV) of the mitochondrial respiratory (electron transport) chain [1108]. Cytochrome-C oxidase catalyzes both the oxidation and reduction of nitric oxide. In its oxidized state, cytochrome-C oxidase inactivates nitric oxide. Conversely, nitric oxide inhibits cytochrome-C oxidase.

During hypoxia, cytochrome-C oxidase shifts from oxidized to reduced form, thereby lowering NO inactivation and fostering hypoxic vasodilation independently of the release of ^{S}nitrosothiols or of the reduction of nitrite anion (NO_2^-) to NO by hemoglobin. Other sources of nitric oxide may indeed exist [1108]. S-nitrosylated hemoglobin (Hb^{SNO}) releases ^{S}nitrosothiols during deoxygenation. Hemoglobin may operate as a nitrite reductase that reduces NO_2^- to NO as hemoglobin deoxygenates.

Inactivation of nitric oxide may also result from interaction with superoxide ions, hemoglobin, myoglobin, accelerated auto-oxidation favored by partition within cell membranes, and free radicals derived from eicosanoid lipoxygenase, cyclooxygenases, peroxidases, or catalase [1108].

However, the concentration of nitric oxide is mainly regulated by cytochrome-C oxidase in an O_2-dependent manner. The concentration of nitric oxide depends on the production rate by the main source, i.e., nitric oxide synthase as well as the redox state, turnover, and oxygen concentration at which cytochrome-C oxidase is working. The amount of nitric oxide increases at low oxygen concentration, when the activity of cytochrome-C oxidase decays.

The intracellular redox state that relies in particular on thioredoxin and the glutathione–glutaredoxin system[170] influences endothelial nitric oxide synthase in endothelial cells. Activity of glutathione reductase and cytosolic (TrxR1) and mitochondrial (TrxR2) thioredoxin reductase influences the intracellular ratio between reduced and oxidized glutathione, hence NOS3 function. Inhibition of glutathione and thioredoxin reductase reduces VEGF-induced NO production [1109]. In the absence of TrxR2, production of hydrogen peroxide in endothelial cells rises

[169] Muscle-specific caveolin-3 inhibits NOS3 in cardiomyocytes and NOS1 in skeletal myocytes.

[170] Glutaredoxins are small redox enzymes that use anti-oxidant glutathione as a cofactor. Glutaredoxins are reduced by the oxidation of glutathione. Oxidized glutathione is then regenerated by glutathione reductase that is constitutively active and inducible upon oxidative stress. In its reduced state, the thiol group is able to donate an electron to unstable molecules such as reactive oxygen species. Glutathione then becomes reactive and reacts with another reactive glutathione to form glutathione disulfide (oxidized state). Glutathione reductase reforms reduced glutathione.

markedly, but lack of glutathione reductase and TrxR1 heightens H_2O_2 production only slightly. In the absence of glutathione reductase, tetrahydrobiopterin supplementation prevents increase in H_2O_2 generation.[171]

Asymmetric Dimethylarginine

Asymmetric dimethylarginine (ADMA) is an endogenous inhibitor of nitric oxide synthase. It is metabolized by dimethylarginine dimethylaminohydrolase DDAH1 and DDAH2 isoforms. Sterol response element-binding proteins — SREBP1c repressor and SREBP2 activator — impede and promote transcription of the Ddah1 gene, respectively [1110].

9.10.3.3 Flow and Nitric Oxide Synthesis

Two signaling pathways can be involved in flow-mediated changes in NO production. The initial, rapid, G-protein- and Ca^{2+}-dependent phase of NO production depends on the rate of change of shear and not shear magnitude. The subsequent lower rate, G-protein- and Ca^{2+}-independent NO production that depends on shear magnitude, is involved in sustained NO release.

Wall shear stress, among other components of hemodynamic stress, regulates NO release. Ca^{2+} response to WSS depends on $[Ca^{2+}]_e$ [1111]. Stepwise increase in wall shear stress transiently raises $[Ca^{2+}]_i$ [1112]. Mechanically modulated $[Ca^{2+}]_i$ is associated with NO metabolism via a receptor-triggered signaling cascade.

Time gradient of wall shear stress yields a transient high-concentration burst of NO release. Time gradient of WSS causes a NO-mediated sustained activation of extracellular signal-regulated kinases ERK1 and ERK2 via G proteins [1099]. Low and high NO concentrations inhibit and activate ERK1 and ERK2 kinases, respectively. These gradients also generate a persistent production of *reactive oxygen species* that act independently and synergetically with NO to mediate activation of ERK1 and ERK2 kinases.

Sirtuins

Pulsatile flow over cultured endothelial cells increases mitochondrial genesis as well as the activity of sirtuin-1 (silent mating type information regulator; Vol. 2–Chap. 2. Cell Growth and Proliferation) that functions as an NAD^+-dependent

---

[171]Tetrahydrobiopterin is a cofactor of the 3 aromatic amino acid hydroxylases used in the degradation of phenylalanine and synthesis of neurotransmitters serotonin, melatonin, dopamine, adrenaline, and noradrenaline, as well as nitric oxide. Tetrahydrobiopterin is a cofactor for tryptophan, phenylalanine, and tyrosine hydroxylase, glyceryl ether monooxygenase, and nitric oxide synthase.

class-3 histone deacetylase and its interaction with AMP-activated protein kinase and endothelial nitric oxide synthase. Phosphorylation of NOS3 by AMPK (Ser633 and Ser1177) primes NOS3 deacetylation induced by Sirt1 to enhance NO production [1113].

9.10.4 Endothelin

Endothelin is a potent vasoconstrictor secreted by endothelial cells. Three endothelin isoforms (ET1–ET3) have been identified [1114]. Endothelin-1 is the single subtype produced in endothelial cells stimulated by ischemia or shear stress. Endothelin-1 production is also stimulated by thrombin, transforming growth factor-β, interleukin-1, angiotensin-2, and arginine vasopressin.

Endothelin-1 binds to 2 endothelin receptor types ETRA (ET_{A1}–ET_{A2}) and ETRB (ET_{B1}–ET_{B2}; Vol. 3 – Chap. 7. G-Protein-Coupled Receptors). Receptors ET_A and ET_B lodge mainly on vascular smooth myocytes to generate vasoconstriction and endothelial cells to release NO and prostacyclin, respectively (both types can be expressed by both cell types). Therefore, the net effect produced by ET1 is determined on the receptor localization and amount for each cell type, i.e., the resulting balance between ETRA and ETRB receptors. Endothelin acts locally. Once endothelin-1 is bound to ETRA of smooth myocytes, it provokes vasoconstriction. The predominant ETR type in endothelial cells, ET_B, is dose- and PKC-dependently upregulated by wall shear stress, and causes NO release. Under physiological conditions, ET1 is produced in small amounts, mainly in endothelial cells, acting as an auto- and paracrine mediator.

Mature ET1 is formed from precursor. Preproendothelin-1 is processed to prohormone big endothelin-1, which is subsequently cleaved by endothelin-converting enzyme to produce endothelin. Endothelin-1 can be produced by other enzymes such as chymases, endopeptidases, and other metallopeptidases [1115].

Locally secreted endothelin-1 binds to its G-protein receptors on smooth myocytes. This binding leads to the formation of diacylglycerol and inositol trisphosphate. The latter stimulates Ca^{2+} influx from the sarcoplasmic reticulum and causes vasoconstriction. Endothelin half-life ranges from 15 to 20 mn. There is no storage. Once released into blood circulation, ET is catabolized as soon as it reaches the lungs.

Like NO, endothelin has several biological actions. Vascular endothelium regulates vascular structure. Endothelin yields proliferation of endothelial cells. ET1 stimulates SMC proliferation. Endothelin thereby regulates extracellular matrix synthesis by stimulated vascular smooth myocytes. Smooth myocytes synthesize collagen-1 and -3 in particular. Although angiotensin-2 enhances collagen-3 synthesis (ET1 does not have significant effect), ET1 increases collagen-1 production more efficiently than angiotensin-2 in cultured smooth muscle cells [1116].

In the heart, endothelin-1 is produced by cardiomyocytes, fibroblasts, and endothelial cells. Cardiomyocytes predominantly express ET_A receptors. Binding of ET1 to both ETRA and ETRB on cardiomyocytes activates Gq subunit that increases intracellular calcium [1117]. In human myocardium in vitro, endothelin exerts a positive inotropic effect via sensitization of cardiac myofilaments to calcium and activation of calcium–sodium exchanger [1118]. The positive inotropic effect is associated with a prolongation of the duration of the action potential [1119], but endothelins induce coronary vasoconstriction and delayed negative inotropic effects, which balance and cancel the transient positive inotropic and chronotropic effects. Endothelin is also a growth factor for cardiomyocytes, inducing muscle-specific gene transcripts through the possible involvement of protein kinase-C activation or intracellular Ca^{2+} mobilization [1120].

The endothelin release is shear dependent. The release of ET1 from cultures of endothelial cells varies with the duration and level of shear [1121]. Sustained exposure to low levels of shear (0.18 Pa) or brief exposure (< 1 h) to 1 Pa stimulate ET1 release via activation of protein kinase-C. Exposure to 0.6 to 2.5 Pa for less than 6 h inhibits ET1 release, implicating either NO and/or cGMP agent. Transient and long-term steady application of wall shear stress at physiological levels (0.1–3 Pa) up- and downregulates preproendothelin-1 (ppET1) mRNA, respectively [1122]. Inhibition of endothelial NO synthase prevents ppET1 mRNA downregulation by shear stress. Increasing degrees of long-term shear stress upregulate endothelin receptor-B mRNA via NO and PKC kinase.

Under pathophysiological conditions, ET1 production increases in numerous cell types, such as endothelial cells, vascular smooth myocytes, cardiomyocytes, and macrophages (Fig. 9.10). Endothelin-1 production is stimulated in various cell types either under the influence of risk factors (hypercholesterolemia, type-2 diabetes) for cardiovascular diseases or during development of cardiovascular diseases, such as atherosclerosis and hypertension.

Endothelin-1 is considered as a developmental factor for vascular dysfunction that leads to cardiovascular diseases because it causes strong vasoconstriction, favors inflammation, has proliferative effects, and stimulates free radical production and platelet activation [1115]. Endothelial dysfunction is defined not only by impaired endothelial function associated with increased production and activity of vasooconstrictor and pro-inflammatory ET1, but also by reduced availability and subsequent altered activity of endothelium-derived relaxing factors, such as nitric oxide, prostacyclin, etc.

High ET1 levels impair endothelial NO production via PKC isoform that leads to inhibition of NOS3 expression [1123]. Endothelin-1 stimulates translocation of PKCα and PKCδ and inhibit that of PKCλ. Endothelin-1 increases cytosolic level of PKCλ, which is the isoform involved in ET1-mediated NO regulation.

9.10.5 Calcium Signaling in Vasomotor Tone Regulation

Calcium signaling in endothelial cells regulates the function of underlying vascular smooth myocytes. In endothelial cells of resistive arteries such as small

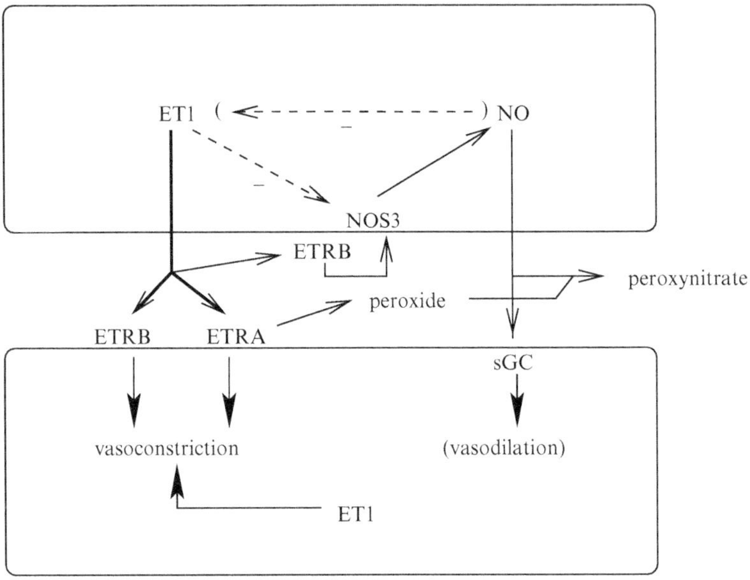

Fig. 9.10 Coupling between endothelial and smooth muscle cells in endothelial dysfunction (Source: [1115]). In healthy arteries, vasorelaxation is favored because of small ET1 production and NO activity. In dysfunctional arteries, ET1 synthesis increases in endothelial and smooth muscle cells as well as macrophages. In parallel, ETRB (ET_B) expression in smooth myocytes soars, hence generating vasoconstriction. Especially in endothelial cells, ET1 can decrease endothelial nitric oxide synthase (NOS3) expression, thus lowering NO synthesis. Endothelial cells thus slightly modulate the effects of ET1 in smooth myocytes, moderately reducing vasoconstriction. Receptor ETRB and predominantly ETRA in smooth myocytes mediate formation of superoxide (O_2^-). Superoxide forms peroxynitrite ($ONOO^-$), thereby reducing NO availability. ET1 can also reduce the expression and activity of endothelial NO synthase. ET1 activates macrophages and favors their accumulation, thereby causing release of pro-inflammatory and chemotactic substances, such as tumor-necrosis factor-α and interleukins IL1, IL6, and IL8. These processes collectively induce vasoconstriction, inflammation, and oxidative stress.

mesenteric arteries (caliber $\sim 100\,\mu$m), localized Ca^{2+} *sparklets* (elementary local calcium entry) generated by flux through a single transient receptor potential vanilloid TRPV4 cation channels are amplified by a small number of nearby TRPV4 channels [1124]. Local signaling initiated by a sigle TRPV4 channel leads to a cooperative gating of clustered TRPV4 channels (4-channel cluster) can yield a robust and rapid communication between apposed cells. The resulting signal indeed activates intermediate- (IK; $K_{Ca}3.1$) and small-conductance (SK; $K_{Ca}2.3$), Ca^{2+}-sensitive, K^+ channels that are boosted by the high Ca^{2+} sensitivity conferred by calmodulin. These K_{Ca} channels are involved in endothelial-derived hyperpolarization.

Therefore, a small number of active TRPV4 channels (3–8 per cell) mediate local Ca^{2+} cues that, upon cooperative activation of TRPV4 channels in clusters to produce stronger Ca^{2+} signals, activate IK (primarily)[172] and SK channels. These K_{Ca} channels trigger an electrochemical current that hyperpolarizes endothelial cells and spreads from endothelial cells to smooth myocytes, especially at myoendothelial junctions with possible gap junctions between endothelial and smooth muscle cells through the basement membrane and fenestrae of internal elastic lamina. Transmitted hyperpolarization causes maximal vessel dilation.

Another signaling pathway can be involved. In endothelial cells, activated IP_3R channels of the sarco(endo)plasmic reticulum generate local, high-intensity, cytosolic Ca^{2+} signals; calcium *pulsars* represent the release of calcium from intracellular stores through IP_3R channels to cause a local stationary signal in endothelial cells [1124]. They occur primarily in endothelial projections to the smooth myocytes. Like the first Ca^{2+} signaling process, local increase of cytosolic Ca^{2+} concentration due to calcium pulsars activates nearby Ca^{2+}-activated K^+ channels ($K_{Ca}2.3$ and $K_{Ca}3.1$) to hyperpolarize endothelial and, subsequently, smooth muscle cells.

Calcium pulsar- and sparklet-mediated hyperpolarization may work together with extracellular calcium in the nearby, restricted, extracellular milieu that can gate inward rectifier K^+ channel, thereby causing a K^+ efflux to augment smooth myocyte hyperpolarization that improve smooth myocyte relaxation.

9.10.6 Other Mechanotransduction Effects

Endothelial cells that can form a cellular layer in vitro can be subjected to steady and pulsatile flows in flow chambers. Hemodynamic forces develop a stress field onto and within the vascular wall, e.g., shear on the wetted surface and intramural stretch. The stress field acts permanently on vascular cells to modulate gene expression and release of newly and/or stored substances. In particular, vascular endothelia react to applied mechanical forces by gene expression.

9.10.6.1 Krüppel-like Factor-2

In addition to hemodynamic stresses, gene expression can be generated by many other agents, such as tumor-necrosis factor-α (TNFSF1) that may involve reactive oxygen species, interleukin-1β, transforming growth factor-β, vascular endothelial growth factor, among others. These reactions that are not specific to flow, belong to general adaptive responses.

Nonetheless, the endothelial transcription factor Krüppel-like factor-2 is uniquely stimulated by flow [1125]. Krüppel-like factor-2 regulates several genes

[172] IK channels localize to endothelial projections, where about 40% of TRPV4 reside and myoendothelial gap junctions are concentrated.

involved in the vascular tone in response to local flow pattern. Isotype KLF2 acts on the synthesis of nitric oxide as well as endothelin-1 and adrenomedullin. In steady and pulsatile flows, high hemodynamic stresses increase KLF2 expression, whereas cyclic stretch has no effect [1126].

9.10.6.2 P53 and MicroRNAs

In endothelial cells exposed to a prolonged (24 h), steady, laminar flow triggers an antiproliferative and anti-inflammatory response. Laminar flow-induced endothelial cell growth arrest involves CDK inhibitors and P53 transcription factor. In addition, mechanical stress mediates P53-dependent apoptosis of smooth myocytes via the P38MAPK pathway [1127].

In human umbilical vein endothelial cells subjected to steady laminar flow, miR19a attenuates the expression of a reporter gene controlled by a conserved 3′-untranslated region of the cyclin-D1 gene, hence synthesis of cyclin-D1 that causes a cell cycle arrest at the G1–S transition [1128].

Endothelial cell growth arrest also results from pulsatile flow. In response to 24-h pulsatile flow, 8 (e.g., miR23b and miR27b) and 13 microRNAs (e.g., miR16, miR17, and miR221) are up- and downregulated in cultured endothelial cells, respectively, w.r.t. static condition [1129]. MicroRNA-23b causes E2F1 reduction and hypophosphorylation of retinoblastoma protein, thereby eliciting G0–G1 arrest.

9.10.6.3 Small Monomeric GTPases

Small GTPase CDC42[173] is activated by hemodynamic stress via interactions between stimulated integrins[174] and the extracellular matrix via Par6 and PKCζ [1130] (Fig. 9.8). Furthermore, in flow-stimulated endothelial cells, Rac1 GTPase is also activated from integrin–matrix binding.

9.10.6.4 Thioredoxin

Endothelial cells subjected to blood stresses not only secrete regulators of the vasomotor tone, but also act on other local biological phenomena, such as inflammation (Sect. 11.5). *Thioredoxin-interacting protein* (TxnIP)[175] and thioredoxin

[173]Small GTPase CDC42 is implicated in cell polarity, inducing reorientation of the microtubule organizing center and Golgi body toward the motion direction.

[174]Hemodynamic stress activates particularly $\alpha V\beta_3$-integrins and increases integrin binding to the extracellular matrix that inhibits Rho GTPase. This transient inhibition is required for cytoskeletal alignment in the flow direction.

[175]Thioredoxin-interacting protein is also known as vitamin-D-upregulating protein-1.

(Txn) are involved in mechanotransduction associated with TNF signaling in endothelial cells and inflammation. Mechanical stress-responsive TxnIP prevents thioredoxin activity. Thioredoxin binds to and inhibits MAP3K5. When TxnIP binds to Txn, it hinders Txn binding to MAP3K5 (Fig. 9.8). Cytokine TNFα can then promote MAP3K5 phosphorylation. Activated MAP3K5 stimulates MAP2K, hence P38MAPK and JNK, that increase the expression of VCAM1 and elicit leukocyte adhesion. Short (1 h) and long (1 d) exposure to steady flow on rabbit and mice aortas and EC culture inhibits the TNF–MAP3K5–JNK/P38MAPK pathway and TNF-mediated VCAM1 [1060]. Wall shear stress decreases the TxnIP concentration in endothelial cells and favors the binding of Txn to MAP3K5, subsequently hampering P38MAPK and JNK inflammatory effects.

9.10.6.5 Oxidants

Oxidants release Txn from MAP3K5 enzyme. This effect of wall shear stress on the vascular endothelium is beneficial, owing to nitric oxide production and enhanced expression of anti-oxidant enzymes. The apoptosis-suppressive effects of wall shear stress are indeed mediated by *superoxide dismutase* (SOD) and nitric oxide synthase [1131].[176]

Hemodynamic stress and nitric oxide protects the vessels against oxidative reactions, including anti-oxidant defenses such as glutathione and activation of mitogen-activated protein kinase and inhibition of cytochrome-C release from mitochondria. Hemodynamic stress upregulates the expression of glutathione peroxidase (GPx1) mRNA in a time- and force-dependent manner in bovine aortic endothelial cells, and increased GPx activity [1132].[177] Wall shear stress can then protect the vessel wall against oxidative stresses. Furthermore, NO inhibits MAP3K5 activation of MAP2K [1133]. Nitric oxide then modulates redox cell signaling [1134].

9.10.6.6 Platelet-Derived Growth Factor

Smooth myocytes migrate during atherosclerosis and neointimal hyperplasia. Hemodynamic forces stimulate endothelial cells to secrete SMC chemoattractants such as platelet-derived growth factor. Smooth myocyte chemotaxis is associated with activation of SMC extracellular signal-regulated protein kinases ERK1 and -2

[176]Enzyme SOD converts superoxide anion to H_2O_2 that can prime vascular wall inflammation. Both SOD and NO that are upregulated by WSS hinder the caspase cascade in response to apoptosis-inducing stimuli.

[177]Glutathione peroxidase (GPx1) reduces H_2O_2 to H_2O.

[1135]. Hemodynamic stresses stimulate secretion of PDGFbb and IL1α by endothelial cells and subsequent SMC migration, whereas lower levels of shear stress and cyclic strain have no effect.

9.10.6.7 T2-Tryptophanyl-tRNA Synthase

^{T2}Tryptophanyl-tRNA synthase (^{T2}TrpRS), a natural fragment of TrpRS, inhibits flow-induced activation of PKB, ERK1 and ERK2, and NOS [1136]. It hinders endothelial cell elongation and alignment of stress fibers and microtubules in the flow direction. In addition, ^{T2}TrpRS is an antagonist of VEGF-induced angiogenesis. It binds at intercellular junctions of endothelial cells to cadherin-5 [1137], but complete TrpRS remains inactive.

9.10.7 Other Aspects of the Regulation of the Vasomotor Tone

9.10.7.1 Nitroglycerin

Nitroglycerin is a potent vasodilator at low dose (1–10 nmol) that activates nitric oxide synthase, although it can be converted into NO and/or other vasoactive compounds [1138].[178] Very small doses of nitroglycerin (comparable to basal level of free NO, i.e., 5–20 nmol) actually cause rapid vasodilation after administration (1–3 mn after administration), hence suggesting NO production by NOS1 and NOS3 rather than nitroglycerin conversion into vasoactive substances.[179] At higher dose, nitroglycerin exerts its effect via metabolites.

9.10.7.2 Secretoneurin

Secretoneurin belongs to the secretogranin-2 family. It enhances adhesion and transendothelial migration of monocytes. In porcine coronary arteries, after acute exposure, secretoneurin relaxes via activation of endothelial nitric oxide synthase (NOS3) and cyclooxygenase, nitric oxide being the dominant agent [1139]. During prolonged administration (24 h), secretoneurin enhanced the relaxation caused by bradykinin.

---

[178]Xanthine oxidase and mitochondrial aldehyde dehydrogenase mediate nitroglycerin reduction to nitrite and NO.

[179]Nitroglycerin induces phosphorylation of NOS1 at Ser852 and NOS3 at Ser1177 in rats and mice.

9.10.7.3 Reactive Oxygen Species

Reactive oxygen species (ROS) are second messengers that contribute to normal cell functioning. However, excessive ROS production or dysregulated metabolism (oxidative stress) elevates vascular smooth myocyte contractility and growth and causes inflammation as well as systemic and pulmonary hypertension.

Reactive oxygen species that oxidize their substrates are involved in endothelial cell signaling, especially that assigned to vascular remodeling and endothelium-regulated vasorelaxation.

Stress-activated NADPH oxidase produces reactive oxygen species in the vessel wall.[180] Reactive oxygen species interact with NO to produce peroxynitrite that activates matrix metallopeptidases, leading to vessel wall remodeling [1140].

Hydrogen peroxide activates AMP-activated protein kinase in cultured bovine aortic endothelial cells [1141]. Activated by elevated AMP/ATP ratio, AMPK switches ATP-generating catabolism on and anabolism off. Endothelial AMPK is activated by numerous stimuli, such as hypoxia, peroxynitrite ($ONOO^-$), adiponectin, vascular endothelial growth factor, and vasoactive mediators, such as sphingosine 1-phosphate, bradykinin, and thrombin. Enzyme AMPK phosphorylates (activates) NOS3 (Ser1177) [1141]. Hydrogen peroxide (H_2O_2) activates in a time- and dose-dependent manner Ca^{2+}–calmodulin-dependent protein kinase kinase-β that leads to AMPK phosphorylation. In addition, liver kinase-B LKB1, is phosphorylated on H_2O_2 administration, but LKB1 plays a minor role. On the other hand, Ca^{2+}–calmodulin-dependent endothelial nitric oxide synthase NOS3 prevents AMPK activation (negative feedback) [1141].

Hydrogen peroxide synthesized by nitric oxide synthase NOS1 is responsible for acetylcholine-induced endothelium-dependent relaxation of smooth myocytes of wall media [1142]. Isoform NOS1 is constitutively expressed in the endothelium of mouse aorta. Hydrogen peroxide produced by NOS1 can act as a coparticipant in endothelium-dependent vasodilation.

Reactive oxygen species cause vasoconstriction and -dilation according to vasculature compartment, vessel size, nature of preconstriction, and ROS type and amount. Superoxide anion and its product hydrogen peroxide actually have distinct actions on the vasculature. Superoxide precludes endothelium-dependent relaxation by NO scavenging in both pulmonary and systemic arteries. Hydrogen peroxide favors metabolic dilation in the coronary circulation via K^+ channels and subsequent hyperpolarization (predominantly via K_V and possibly also BK channels) [1143]. Both superoxide and hydrogen peroxide provoke vasoconstriction in rat pulmonary arteries, mainly acting via RoCK kinase that phosphorylates $PP1_{r12a}$, which inhibits myosin phosphatase, thus enhancing MLC20 phosphorylation.

[180] Subunit NOx organizer NOxO2 of NADPH oxidase upregulates production of reactive oxygen species and matrix metallopeptidases.

To match perfusion to ventilation, the pulmonary circulation responds differently to numerous stimuli, such as ROS and hypoxia. In response to hypoxia, pulmonary and systemic arteries generally constrict and dilate, respectively [1143]. In chronic obstructive pulmonary diseases, ROS can cause pulmonary hypertension.

In pulmonary arteries, superoxide anion causes RoCK-dependent Ca^{2+} sensitization; in systemic (mesenteric) and pulmonary arteries, superoxide anion provokes opening of voltage-gated K^+ channel, but only primes dilation in mesenteric arteries [1143]. Superoxide also relaxes preconstricted coronary and renal arteries. Vasoconstriction depends mainly on non-selective cation channels in pulmonary arteries, principally on voltage-gated Ca^{2+} channels in mesenteric arteries.

Reactive oxygen species are involved in the regulation of cerebral vascular tone [1144]. According to the context, ROS can either generate or prevent calcium sparks, i.e., can contract or relax vascular smooth myocytes. Moreover, ROS can interact with other substances, such as nitric oxide and arachidonic acid, thereby changing vascular tone.

Cerebral vasodilation results from action of products of univalent reduction of oxygen — superoxide anion radical, hydroxyl radical, and hydrogen peroxide — as well as peroxynitrite on cerebral arterioles [1145]. Application of hydrogen peroxide and peroxynitrite in anesthetized cats equipped with cranial windows induces dose-dependent dilation of cerebral arterioles via ATP-sensitive potassium channels. On the other hand, superoxide causes a dose-dependent dilation of cerebral arterioles via calcium-activated potassium channels. Although peroxynitrite can dilate cerebral arteries by activating potassium channels, it also may impair responses mediated by the same or other types of K^+ channels. Peroxynitrite may be responsible for impaired vasodilation during hypoxia–reoxygenation in response to the endothelium-dependent messenger acetylcholine [1146].

In cerebral arteries, superoxide causes vasodilation via potassium channels and vasoconstriction at low and high concentrations, respectively [1144]. Superoxide may act directly on smooth myocytes of cerebral arteries and indirectly via endothelium. It interacts rapidly with nitric oxide, hence governing nitric oxide half-life and attenuating NO-mediated vasodilation. Furthermore, peroxynitrite produced by the reaction of nitric oxide with superoxide influences vascular tone.

Exogenous hydrogen peroxide relaxes small cerebral arteries and arterioles via potassium channels [1144]. Very high H_2O_2 concentrations can provoke vasoconstriction followed by vasodilation. This relaxing factor may act as an endothelium-derived hyperpolarizing factor or not according to the arterial compartment.

Arachidonic acid and bradykinin produce dilation of cerebral arterioles via endothelium and reactive oxygen species, more precisely hydrogen peroxide [1144]. Bradykinin binds to B_2 receptor and activates cyclooxygenase-1 that processes arachidonic acid and produces superoxide. On the other hand, vasodilation of cerebral vessels caused by acetylcholine and ATP is not influenced by reactive oxygen species.

9.10.7.4 Reactive Nitrogen Species

Peroxynitrite anion ($ONOO^-$) is a short-lived (~10 ms; hence low-steady-state concentration) oxidant produced by the reaction of short-range diffusible nitric oxide (NO) with superoxide radical ($O_2^{\bullet-}$) formed by plasmalemmal NAD(P)H oxidases and the mitochondrial respiratory chain at diffusion-controlled rates ($\sim 1 \times 10^{10}$/(mol·s; estimated production rate of 50–100 μmol/mn [1147]).[181] Its toxic effects can drop upon fast reduction to nitrite (NO_2) or isomerization (same molecular content, but different structure) to nitrate (NO_3^-) by metalloporphyrins (e.g., manganese and iron metalloporphyrins) [1147]. Certain enzymes detoxify peroxynitrite. Various scavengers and neutralizers of peroxynitrite and peroxynitrite-derived radicals include peroxiredoxin-5 that reduces peroxynitrite to nitrite, thiol-based anti-oxidants, and selenocompounds (selenoprotein-P and glutathione peroxidase).

Peroxynitrite reacts in a pH-dependent manner mainly with carbon dioxide (CO_2) in aqueous phase (rapidly) as well as numerous types of molecules, among which many possess thiols and transition metal centers such as hemin.[182] Reaction of peroxynitrite and/or its derived radicals (e.g., carbonate [$CO_3^{\bullet-}$] and nitrogen dioxide [$NO_2^{\bullet}$] radicals) with targets causes 1- and 2-electron oxidations and nitration [1147].[183] Diffusion of peroxynitrite (over length 5–20 μm) from its formation site across membranes can provoke oxidative damage (oxidation and nitration of proteins [cytosolic and membrane protein Tyr nitration], membrane lipids, and DNA).[184]

Peroxynitrous acid (ONOOH) in equilibrium with peroxynitrite ($ONOO^-$) can undergo isomerization to generate nitrate (NO_3^-) or homolytic fission[185] to generate 1-electron oxidants hydroxyl ($^{\bullet}OH$) and ($^{\bullet}NO_2$ radicals [1147]. This reaction is slow in aqueous compartments, but peroxynitrous acid crosses lipid bilayers and gives rise in hydrophobic phases to $^{\bullet}OH$) and ($NO_2^{\bullet}$ radicals to initiate lipid peroxidation

[181] Mitochondrial electron-transport chain represents a set of redox carriers in the inner mitochondrial membrane that enable the flow of electrons from substrates to oxygen. The potential energy in the electron gradient is used to drive ATP synthesis, when protons move back across the membrane via ATP synthase.

[182] Transition metals include copper, iron, and manganese.

[183] Both $CO_3^{\bullet-}$ and $NO_2^{\bullet}$ are 1-electron oxidants that can oxidize amino acids, such as cysteine and tyrosine to form cysteinyl and tyrosyl radicals.

[184] Tyrosine nitration, dimerization, and hydroxylation by peroxynitrite form 3-nitrotyrosine, (3,3′)-dityrosine, and (3,4′)-dihydrophenylalanine, respectively. The reaction of peroxynitrite-derived radicals with lipids leads to peroxidation and formation of nitrito-, nitro-, nitrosoperoxo-, and nitrated lipid oxidation adducts.

[185] Rupture of a covalent bond in a molecule, in which the 2 resulting products keep one of the bond electrons.

Table 9.31. Effects of peroxynitrite (Source: [1147]). Tetrahydrobiopterin is an essential cofactor of nitric oxide synthase that carries electrons for redox reactions. Depletion of NOS substrates and cofactors can prime a feedback (additional peroxynitrite generation).

Target, effect	Examples of targets
Antioxidant enzyme (inhibition)	Superoxide dismutase, glutathione reductase, glutaredoxin
Anti-oxidant (depletion)	Glutathione, cysteine
Cytosolic enzyme (inhibition)	creatine kinase, Tyr phosphatases, Tyr hydroxylase
Protein aggregation	Microtubule-associated protein Tau
Enzyme activation	Fibrinogen, matrix metalloproteinases, cytochrome-C, glutathione-S-transferase, PKCϵ
Enzyme cofactor impairment	Tetrahydrobiopterin (BH4), NADH
Membrane channel (inhibition)	Calcium pumps, Na^+–K^+ ATPase, sodium channels, Ca_V1
Signaling messengers	FGF, IL8/10, NGF,
Receptors	β-adrenoceptors, A_1 (adenosine), HER, insulin receptor
Calcium	
Signaling mediators	Prostacylin synthase, cyclooxygenases, AMPK, MAPK, Src, PKC, PI3K
Transcriptional regulators	PPARγ (NR1c3), STAT3 (VEGF axis), HDAC2 (nitration [inactivation])
Adhesion molecules	ICAM1, selectins, integrins
Surfactant proteins	
Lipids, arachidonic acid	Low-density lipoproteins
DNA injury	Genotoxicity, cell death
Mitochondrial dysfunction	Impaired mitochondrial respiration

and lipid and protein nitration. Moreover, ONOOH in membranes can react with thiols (2-electron oxidation that forms NO_2^- and sulphenic acid derivatives [RSOH] often converted to disulphides [$R^{(1)}SSR^{(2)}$]) and metal centers.[186]

Transient $ONOOCO_2^-$ adduct that results from the reaction between $ONOO^-$ and CO_2 in aqueous phase also creates either NO_3^- (isomerization; ~ 2/3 reactions) or $^\bullet OH$) and $NO_2^\bullet$ radicals (homolysis; ~ 1/3 reactions) [1147].

Peroxynitrite can influence cell signaling (especially apoptosis) and mitochondrial function and cause cell death (Table 9.31). Although peroxynitrite exert various cytotoxic effects mainly due to oxidation and nitration reactions, it can lead to the

[186] Resulting oxometals can trigger secondary free-radical reactions.

production of secondary NO donors that can inhibit platelet aggregation and support vasodilation of cerebral arterioles [1145]. However, peroxynitrite alters not only myogenic response of cerebral arteries, but also vasodilation induced by calcitonin gene-related peptide [1144]. Products of peroxynitrite reactions with molecules can be detected in vascular diseases, ischemia–reperfusion injury, circulatory shock, and inflammation, among others [1147].

9.10.7.5 Isoprostanes

Isoprostanes are unconventional eicosanoids formed from non-enzymatic peroxidation of membrane fatty acids such as arachidonic acid by free radicals and other types of reactive oxygen species, without direct action of cyclooxygenases.[187] These inflammatory mediators serve as markers of oxidative stress, in particular in smokers, especially prostaglandin-like compound 8-iso-PGf2α.[188] Once isoprostanes are released into the blood circulation, they are quickly catabolized and eliminated. Free radical scavengers, such as vitamin-A, -C, and -E, β-carotene, selenium, superoxide dismutase, and catalase, reduce isoprostane levels [1148].

In endothelia of microvessels, 8-iso-PGf2α upregulates endothelin-1 expression as well as synthesis of TxA2 and IP_3, hence increasing cytosolic calcium concentration [1148].

Some isoprostanes constrict cerebral microvessels via plasmalemmal receptors or indirectly via thromboxane liberation from endothelial cells [1144]. 8-iso-PGf2α elicits vasoconstriction in the renal vasculature [1148]. In human small pulmonary arteries and veins, among isoprostanes of the E-ring set (8-iso-PGe1 and 8-iso-PGe2) and of the F-ring set (8-iso-PGf1α and -β, 8-iso-PGf2α and -β, and 8-iso-PGf3α), 8-iso-PGe2 are the most potent in pulmonary arteries and have a nearly equivalent efficiency than that to 8-iso-PGf2α in pulmonary veins. Isoprostanes 8-iso-PGe1, 8-iso-PGf1α, and 8-iso-PGf2β) are weakly effective; 8-iso-PGf1β and 8-iso-PGf3α are ineffective [1148]. The contraction of vascular smooth myocytes results from activation of the Rho–RoCK pathway. In addition,

---

[187]Four classes of F2-isoprostanes can arise from arachidonic acid. In addition, other polyunsaturated fatty acids targeted by reactive oxygen species produce isoprostanes. Peroxidation of eicosapentaenoic acid may generate 6 classes of F3-isoprostanes, that of α- and γ-linolenic acids to 2 classes of E1- and F1-isoprostanes, and that of docosahexaenoic acid to 8 classes of D4-isoprostanes and 8 classes of E4-isoprostanes [1148]. Cyclooxygenase also contribute to isoprostane production in vascular smooth myocytes, endothelial cells, platelets, monocytes, macrophages, and mesangial cells via COx-derived reactive oxygen species [1148].

[188]The level of 8-iso-PGF2α rises in healthy individuals exposed to cigarette smoke, allergen, ozone, and hyperoxia, as well as in patients with asthma, chronic obstructive pulmonary disease, interstitial lung disease, cystic fibrosis, acute lung injury such as acute respiratory distress syndrome and pulmonary hypertension and in patients with ischemia–reperfusion injury, unstable angina, heart failure, coronary heart disease, stroke, and atherosclerosis [1148]. Concentrations of free and esterified isoprostanes in plasma of smokers remain significantly elevated for weeks after smoking cessation.

isoprostanes can operate via endothelial cells by releasing endothelin. Isoprostanes also constrict lymphatic vessels [1148].

9.10.8 Hypoxic Vasoconstriction of Pulmonary Arteries

Hypoxic pulmonary vasoconstriction is a localized, acute, adaptive mechanism that optimizes ventilation–perfusion ratio, as it bypasses blood flow toward highly ventilated regions of lungs, away from poorly ventilated zones. Hypoxic pulmonary vasoconstriction is typically biphasic, with a large, transient phase and a sustained, gradual second stage that are associated with an acute relatively large then stabilized small Ca^{2+} influx [1149]. Hypoxic, transient Ca^{2+} response results from an influx through voltage-gated Ca^{2+} channels and then release from Ca^{2+} stores. In the second phase, the effect of low Ca^{2+} influx is enhanced by a RoCK-dependent Ca^{2+} sensitization. Calcium sensitization of contractile fibers results from inhibition of myosin light chain phosphatase and subsequent increased phosphorylation of 20-kDa myosin light chain (MLC20), independently of changes in intracellular Ca^{2+} concentration. Inhibition of MLCP results from phosphorylation of the myosin-binding regulatory (inhibitory) subunit of protein phosphatase-1 by RoCK kinase.

Among cytosolic protein Tyr kinases of the SRC family, Fyn, Src, and Yes kinases are highly expressed in pulmonary arteries, where they are involved in contraction of vascular smooth myocytes. Hypoxia enhances phosphorylation (activation) of Fyn, Src, and Yes kinases, causes translocation of RoCK from the nucleus to the cytoplasm, and subsequent phosphorylation of $PP1_{r12a}$ and MLC20 [1149].

Hypoxic pulmonary vasoconstriction involves the generation in mitochondria of reactive oxygen species, which stimulate SRC family kinases. Phosphorylation of $PP1_{r12a}$ by RoCK concomitant with a small Ca^{2+} influx greatly enhances MLC20 phosphorylation. On the other hand, PP2 inhibitor of SRC family kinases blocks the hypoxia-primed Ca^{2+} response in intraparenchymal pulmonary arteries.

In addition, hypoxia activates Src and Fyn kinases in cardiomyocytes. Among other protein targets, SRC family kinases phosphorylate (activate) focal adhesion kinases FAK1 and paxillin [1149]. Besides, prostanoids cause vascular smooth muscle contraction partly via RoCK-mediated $PP1_{r12a}$ phosphorylation. Kinases of the SRC family contribute to PGF2α-mediated Ca^{2+} sensitization [1149].

9.10.9 Hypoxic Vasodilation of Cerebral Arterioles

Functional hyperemia relies on activation neuronal nitric oxide synthase NOS1, in particular upon glutamate binding to NMDA receptors and increase in intracellular calcium content.

Adenosine triphosphate is an intra- and extracellular regulator. Red blood capsules respond to changes in their environment, in particular, by releasing ATP, especially upon small changes in pH (slight acidosis [pH 7.0–7.1] [1150]) and O_2 and CO_2 concentration (brief periods of hypoxia and hypercapnia [1151]), hence enabling matching of O_2 supply by the microvasculature to demand, as well as in osmotic pressure[189] and hemodynamic stress.

Red blood capsules experience variations in hemodynamic stress along the vascular network. Increased shear inside narrow capillaries and small arterioles causes RBC deformation. The latter triggers ATP release. The amount of ATP released depends on the shear magnitude and duration [1153]. The activity of cystic fibrosis transmembrane conductance regulator, or ATP-binding cassette transporter ABCc7, is required for deformation-induced ATP release from RBCs [1154]. The transfer ATPase CFTR does not convey ATP, but regulates other carriers, among which those that enable ATP egress. The AC–cAMP pathway is involved in ATP release [1156].

The paracrine regulator ATP causes vasodilation by releasing nitric oxide from endothelial cells via P2Y receptors. In addition to reduction of pulmonary vascular resistances,[190] liberation from red blood capsules of ATP causes vasodilation of cerebral arterioles during hypoxia.

The gaseous mediator carbon monoxide (CO) is a vasodilator in some compartments such as liver sinusoids, but in the cerebral circulation, CO elicits vasoconstriction. On the other hand, hydrogen sulfide, another gasotransmitter, is a vasodilator in both the peripheral and cerebral circulations. It causes S-sulfhydration (activation) of ATP-sensitive potassium channels, thereby hyperpolarizing the vascular endothelial and smooth muscle cells. Heme oxygenase HO2, an O_2 sensor, uses O_2 to generate carbon monoxide. The latter, but not NO, inhibits astrocytic cystathionineβ-synthase, which thus acts as a CO sensor, in addition to be a generator of hydrogen sulfide. During hypoxia, HO2 activity is prevented, CBS inhibition is relieved, thereby produced H_2S elicits vasodilation of cerebral arterioles [1157].

In fact, hypoxia via non-vascular cerebral components stimulates the production of various vasodilators, such as potassium and hydrogen ions, prostaglandins, and adenosine [1158]. *Reactive hyperemia*, i.e., the local increase in blood flow following an intensification of tissue metabolism, is related to metabolic vasodilation triggered by active cells that surround arterioles and release vasodilators. Vasodilatory tissue metabolites and ions include, in addition to hypoxia: (1) adenosine generated from AMP; (2) carbon dioxide formed during increased oxidative metabolism; (3) lactic acid produced by anaerobic metabolism; (4) H^+ ion associated with CO_2 augmentation, (5) liberated K^+ ion that hyperpolarizes and relaxes vascular smooth myocytes. In addition, hypoxia also promotes the release of excitatory amino acids by neurons. These amino acids augment the export of vasodilators. Direct vascular

---

[189] Extracellular ATP supports regulatory RBC volume decrease via Ca^{2+} and ATP-dependent K^+ efflux, thereby reducing hemolysis [1152].

[190] In primary pulmonary hypertension, RBCs have a reduced deformability and ATP is not released [1155].

effects of hypoxia involves the endothelium. In cerebral arteries, the net contribution of endothelial factors to hypoxic vasodilatation is moderate. In addition, hypoxia activates ATP-sensitive K^+ channels of vascular smooth myocytes of cerebral arteries, thereby causing membrane hyperpolarization and reducing calcium influx. Hypoxia also lower the calcium sensitivity of contractile proteins.

The cerebral circulation is similarly sensitive to hypoxia than the coronary circulation, but more sensitive to hypercapnia [1159]. Hypercapnia also activates K_{ATP} channel and supports NO synthesis [1160]. Two CO_2 regulatory mechanisms enable pH or CO_2 homeostasis to maintain activity of numerous enzymes and ion channels operating in neural activities [1161]: the central respiratory chemoreflex aimed at maintaining CO_2 level and repressing associated respiratory acidosis (hyperventilation) and cerebrovascular CO_2 reactivity that controls cerebral blood flow. The onset of the latter is faster than that of the former. At rest, cerebral vasodilation attenuates activation of the central chemoreflex and subsequently ventilatory response onset. The sensitivity of cerebral vasoreactivity to hypercapnia is restrained by augmented sympathetic activity [1162].

9.10.10 Vasomotor Tone during Pregnancy

Pregnancy is characterized by the development of the uteroplacental vasculature. Pregnancy potentiates endothelium-derived hyperpolarizing factor that enables vasodilation [1163]. A cumulative exposure of acetylcholine causes a rapid dilatation in non-pregnant and late pregnant rats. This vasodilation results from reduced cytosolic Ca^{2+} concentration and membrane hyperpolarization in smooth myocytes. On the other hand, endothelial cytosolic Ca^{2+} concentration increases. Small- (SK) and intermediate-conductance (IK) Ca^{2+}-activated K^+ channels are involved in the vasodilation of the maternal uterine resistive vasculature [1163].

9.10.11 Modeling of Mechanotransduction in Arteries

Vascular endothelial and smooth muscle cells at the interface between flowing blood and underlying layer of the vessel wall such as the media in arteries interact to locally regulate the vasomotor tone. This process pertains to the set of regulations carried out by endothelial cells (Fig. 9.11). The endothelial interface serves thus as a local controller of tissue homeostasis for both blood and constituent tissues of the vessel wall.

Vascular endothelial and smooth muscle cells can be supposed to be mainly shear and stretch sensor, respectively. Different types of mechanotransduction responses exist in the arterial bed. The *autoregulation*, also termed myogenic or Bayliss effect, is observed in small resistive arteries, i.e., relatively far downstream the heart exit. The autoregulation aims at ensuring tissue perfusion. In other words, the flow rate

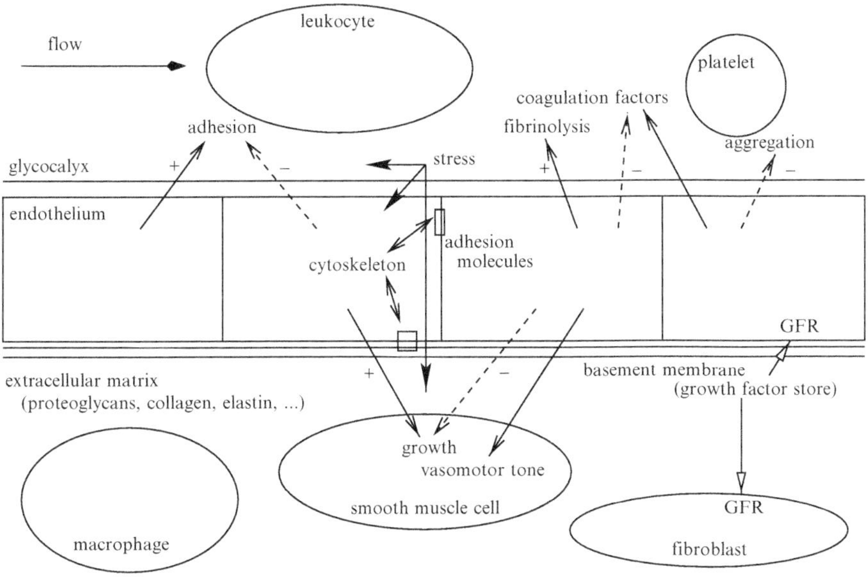

Fig. 9.11 Vascular cells are responsible for local regulations: (1) adaptive responses to hemodynamic stresses that direct the vessel caliber; (2) control of proliferation of vascular endothelial and smooth muscle cells; (3) prevention of blood coagulation and platelet aggregation in normal conditions; (4) prevention of leukocyte diapedesis in the abscence of inflammatory response. Vascular cells bear time-dependent 3D shearing and stretching from flowing blood. Both stress types vary strongly in magnitude during the cardiac cycle. Whereas stretching results from high-amplitude pressure applied perpendicularly to the wall, the wall shear stress exerted tangentially on the wall–blood interface has a low magnitude and its direction, besides its amplitude, can change during the cardiac cycle, in the case of flow reversal during diastole. Even in static conditions, the extracellular matrix influences cell adhesion and spreading, as well as differentiation, migration, and apoptosis.

(q) remains constant owing to adaptive response of the local artery bore, hence local resistance ($\mathtt{R}$): $q = p/\mathtt{R} = \text{cst}$. Therefore, when the blood pressure rises, the resulting vasoconstriction to maintain the flow rate further elevates blood pressure, hence the cardiac load.

On the other hand, large arteries can be supposed to modulate this reflex vasoconstriction in resistive arteries, as they can reset stress and strain field experienced in their walls, thereby reducing the cardiac postload. This additional mechanotransduction effect aims at maintaining the cardiac postload in the physiological range. In other words, when the blood pressure ($p = q\mathtt{R}$) soars, large arteries will dilate to lower local resistance. Vascular smooth myocytes that sense pressure-induced intramural stretch immediately react by fluidizing their stress fibers and relaxing.

Table 9.32. Plasmalemmal transducers of stresses applied by flowing blood to cells of the vessel wall (endothelial and smooth muscle cells; AKAP: A-kinase anchoring protein; ERK: extracellular signal-regulated protein kinase; ET: endothelin; FAK: focal adhesion kinase; IKK: inhibitor of NFκB kinase; KLF: Krüppel-like factor; MAPK: mitogen-activated protein kinase; NADPH oxidase: reduced form of nicotinamide adenine dinucleotide phosphate oxidases; NFκB: nuclear factor κ light-chain enhancer of activated B cells; NO: nitric oxide; PECAM: platelet endothelial cell adhesion molecule; PI3K: phosphatidylinositol 3-kinase; PK: protein kinase; PTPn: protein Tyr phosphatase non-receptor; RTK: receptor Tyr kinase; SHC: Src-homologous and collagen-like substrate; VEGFR: vascular endothelial growth factor receptor).

Transducer	Pathway
G proteins	PKC–ET, ERK
RTK	PI3K, MAPK
Ca^{2+} channels	NO
K^+ channels	PKA/C–AKAP–SAP–FAK2
NADP oxydase	ERK
Cadherin-5	PI3K–PKB (VEGFR2, β-catenin, PTPn11)
PECAM1	ERK; NKκB; PI3K (VEGFR2)
Integrin	Rho–cytoskeleton; VEGFR2 (SHC); IKK–NKκB
Caveolin	NOS; receptors
Occludin	Cell junction
KLF2	NO, ET, adrenomedullin

Table 9.33. Time scales of cell responses to hemodynamic stresses.

Event	Time scale
Stress-activated ion channel	$\mathcal{O}$(ms)
$[Ca^{2+}]_i$	$\mathcal{O}$(s)
NO	$\mathcal{O}$(ms)
IP_3	$\mathcal{O}$(10 s)
Adenylate cyclase	$\mathcal{O}$(mn)
PGI2, PGF2α	$\mathcal{O}$(mn)
PDGF, ET, tPA	hours
Cell proliferation	hours
Cytoskeleton alignment	hours
LDL metabolism	day
Cell alignment	day
Cell growth	days

Secondarily, but only after a short delay, vascular endothelial cells can release nitric oxide, among other vasoactive substances, that targets underlying smooth myocytes to maintain a relaxation state.

Artery wall adaptation to hemodynamical stresses not only results from local mechanotransduction by vascular endothelial and smooth muscle cells mainly via ion channels, receptors, and adhesion molecules (Tables 9.32 and 9.33), but also regulation by hormones and growth factors (Fig. 9.12).

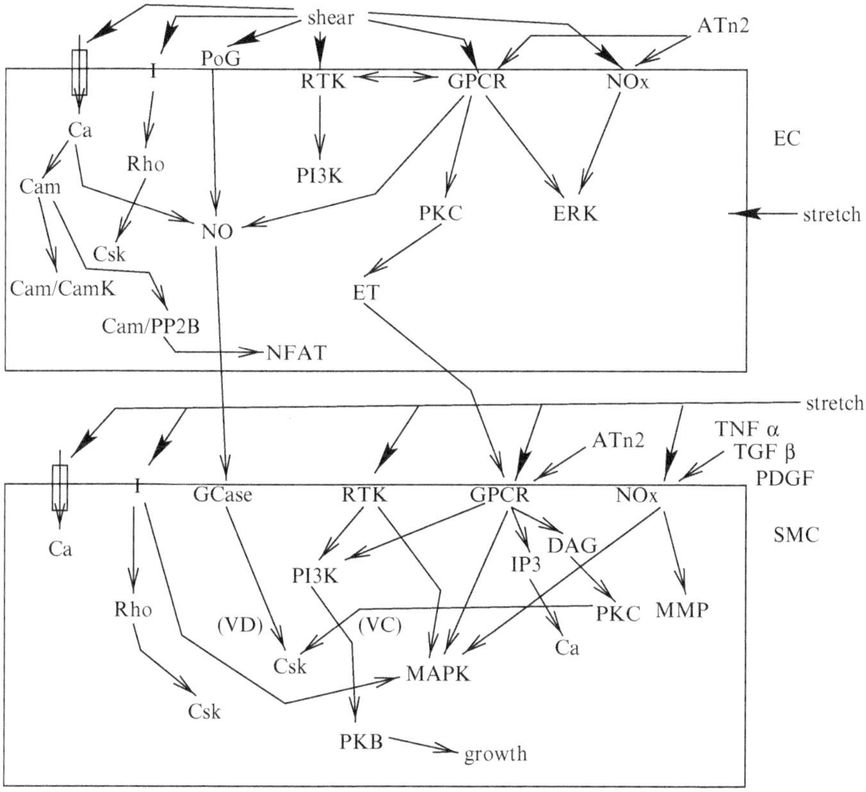

Fig. 9.12 Artery wall adaptation to hemodynamic stresses. The artery bore is locally controlled by mechanotransduction experienced by both endothelial and smooth muscle cells as well as hormones and growth factors. Angiotensin-2 acts on endothelial cells via NO to cause vasodilation that reduces vasoconstriction generated by direct angiotensin-2 effect on smooth myocytes. In other words, angiotensin-2 receptors have opposite effects according to cell type. Their number on endothelial cells is smaller than that on smooth myocytes (I: integrin; PoG: proteoglycan; NOx: reduced form of nicotine adenine dinucleotide phosphate (NAD(P)H) oxidase; NFAT: nuclear factor of activated T cells).

A fundamental couple at the microscopic scale is composed of an endothelial and a smooth muscle cell (Fig. 9.13). At the nanoscopic scale, numerous agents contribute to the local control of the vasomotor tone (Table 9.34). Yet, let us consider a fundamental couple at the nanoscopic scale: nitric oxide and endothelin that cause vasodilation and -constriction, respectively. The major regulators of the vasomotor tone are thus assumed to carry out the entire chemical control. Even using this assumption, reality remains complex. Endothelin-1 indeed targets 2 types of cognate receptors — ET_A and ET_B — that have opposite effects (Table 9.35). The simplified model of mechanotransduction takes into account these 2 types of endothelin receptors to be representative enough (Fig. 9.14).

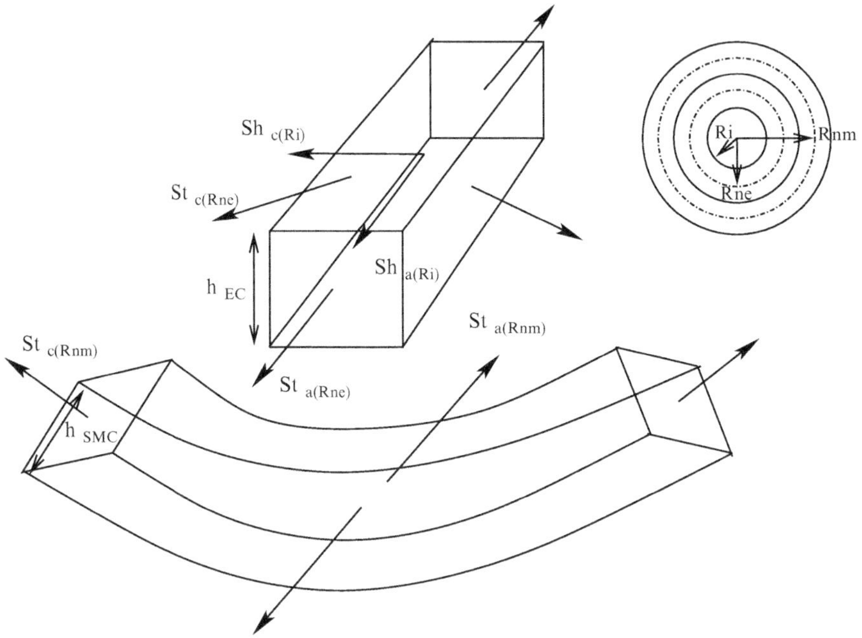

Fig. 9.13 Couple for mechanotransduction at the microscopic scale: endothelial and smooth muscle cells that govern the vasomotor tone.

Table 9.34. Regulators of the vasomotor tone.

Vasoconstrictors	Vasodilators
ET1	NO
TXA2	H_2S
PGF2α	PGI2
UP4A	EDHF/EET, DHET
ATP	Acetylcholine
5HT	ANP
NPY	Adrenomedullin
20HETE	Urocortin
Apelin (SMC)	Apelin (EC, via NO)
Motilin	Ghrelin
Neuromedin-U	Nociceptin
S1P	
Urotensin-2	

Two time-dependent stress types are considered: (1) small-amplitude wall shear stress (WSS) on wetted surface of the endothelial cell and (2) large-magnitude pressure-induced wall axial and azimuthal stretch (T) exerted on the basolateral surface of the endothelial cell and the plasma membrane of the smooth myocyte. A set of hypotheses is needed. The vasomotor tone is only regulated by 2 types

Table 9.35. Endothelin-1 receptors – ET_A and ET_B – have opposite effects in the regulation of the vasomotor tone as well as other processes.

ET_A	ET_B
Vasoconstriction	Vasodilation
(NO inhibition)	(NOS coupling)
ROS production	ROS decrease
(vSMC)	(EC)
Extravasation	Extravasation inhibition

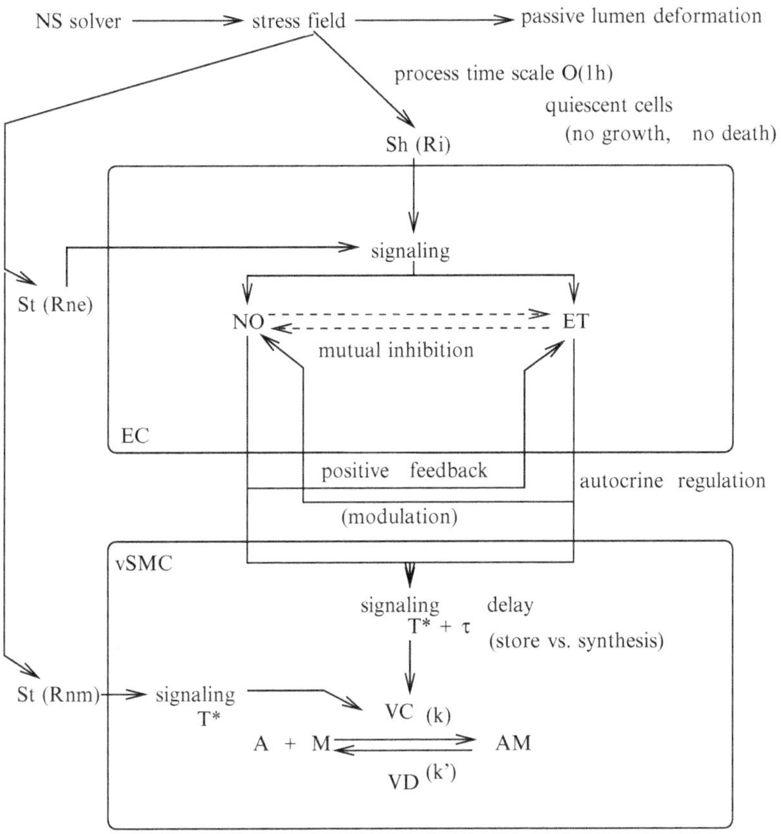

Fig. 9.14 Simple model of the complex mechanotransduction for regulation of the vasomotor tone. The microscopic couple includes synergistic endothelial and smooth muscle cells. The nanoscopic couple encompasses antagonists nitric oxide (NO), a vasodilator, and endothelin-1 (ET), a vasoconstrictor.

of processes: (1) endothelium-independent regulation via pressure-induced stretch action on mechanosensitive sensors **S** in the plasma membrane of the smooth myocyte and (2) endothelium-mediated regulation via release of NO or ET that targets their cognate receptors **R** with a given delay τ_1. On endothelial cell, wall shear stress effect is much greater than that of stretch (WSS $\gg$ $T_{\ell,c}$) Only much higher (lower) stress triggers signaling, i.e., WSS $>$ $WSS_{max} + \Delta$ (WSS $<$ $WSS_{min} - \Delta$) and $T_{\ell,c} > T_{\ell,c;max} + \Delta$ ($T_{\ell,c} < T_{\ell,c;min} - \Delta$). Blood flow is assumed to be periodic and strictly reproducible and, as only abnormal time-averaged stress (**C**) is considered, time integration over many cardiac cycles is unneeded. The density of responsive shear and stretch sensors on SMC surface is proportional to stress magnitude ($[S_{SMCa}] \propto \widehat{T}$ and $[S_{ECa}] \propto \widehat{WSS}$). The concentration of released vasoregulator (NO or ET) is proportional to that of mechanosensitive sensors on EC surface ($[NO]([ET]) \propto [S_{ECa}]$). The concentration of vasoregulator receptors on SMC surface is proportional to that of their agonist ($[R_{SMCa}] \propto [NO]([ET])$). All the involved mediators are produced at steady state and are highly diffusible. The permanent basal (constitutive) synthesis is large enough. All the mediators cycle between a pool of inactive forms and working state. Sensors and receptors are supposed to be immediately activated upon conformational change and ligand binding. Activation stage is the rate-limiting reaction. Inside the endothelial cell, NO and ET exert mutual inhibition. Both NO and ET prime paracrine (PR) and autocrine (AR) regulation, the latter with additional delay (τ_2). Autocrine regulation corresponds to a negative feedback on vasoregulator release. This feedback only prevents further release, hence deactivating receptors. Concentrations of signaling mediators are small, but effectors are close. Within the smooth myocyte, mechanosensitive sensors (S) and receptors (R) directly target MLCK (MLCP) for actomyosin filament contraction (relaxation). Delays τ_1 and τ_2 do not depend on stress types and regulators of the vasomotor tone (NO or ET). Rapid, transient activation of stretch sensor is followed by a quick recovery of inactive state upon continuous stress. The concentration of activated myosin-2 is proportional to that of activated MLCK. The length of the smooth myocyte is proportional to concentration of activated myosin-2. The vessel wetted perimeter is proportional to the SMC length.

The basic set of reactions is then as follows:

$$S_{Ta} + MLCK \underset{k_1^{\prime(T)}}{\overset{k_1^{(T)}}{\rightleftarrows}} S_{Ta} - -MLCK \underset{k_2^{\prime(T)}}{\overset{k_2^{(T)}}{\rightleftarrows}} S_{Ta} + MLCK_a$$

$$NO + R \underset{k_1^{\prime(WSS)}}{\overset{k_1^{(WSS)}}{\rightleftarrows}} NO - -R \underset{k_2^{\prime(WSS)}}{\overset{k_2^{(WSS)}}{\rightleftarrows}} NO + R_a$$

$$R_a + MLCK \underset{k_3^{\prime(WSS)}}{\overset{k_3^{(WSS)}}{\rightleftarrows}} R_a - -MLCK \underset{k_4^{\prime(WSS)}}{\overset{k_4^{(WSS)}}{\rightleftarrows}} R_a + MLCK_a$$

The corresponding set of equations is:

$$\partial_t [MLCK_a] = k_1^{(T)} [S_{Ta}][MLCK], t \in [0, \tau_1]$$

$$\partial_t [R_a] = k_1^{(WSS)} [NO][R], t \in [\tau_1, \tau_2]$$

$$\partial_t [MLCK_a] = k_3^{(T)} [R_a][MLCK], t \in [\tau_1, \tau_2]$$

This set of equations is typically used by chemists. Similarly to rheology, data mining can be inefficient and values of kinetic coefficients can remain unknown. Therefore, more abstract transport–reaction equations are preferred. The following hypotheses are used. The endoplasmic reticulum is close to the plasma membrane for fast calcium release triggered by IP_3. The local activation of MLCK primes actomyosin filament assembling. The GCase–cGMP–PKG axis inhibits IP_3 formation in smooth myocytes at the plasma membrane for efficient signaling (i.e., low activation threshold, long lifetime, and fast transmission) and local activation of MLCP. In vascular smooth myocytes, MLCK is transported from plasma membrane to a set of azimuthal planes to constitute stress fibers and ensure sustained contraction or conversely MLCP transport for actomyosin disassembling. Regulators of the vasomotor tone released from endothelial cells are transported to smooth myocytes instantaneously through myoendothelial junction and with a delay through subendothelial space.

Chapter 10
Vasculature Growth

Vasculogenesis, an embryological process, defines the formation of capillary plexi from endothelial precursor cells. A primitive vascular network is built via the assembly of angioblasts. Initial patterning of embryonic vascular network is independent of hemodynamic forces. The onset of blood circulation contributes to vasculature remodeling.

Angiogenesis corresponds to the maturation of the primary vascular network during embryo- and fetogenesis as well as the expansion of existing vasculature, i.e., generation of new branches from this network in utero as well as vasculature extension after birth during tissue development and repair. Angiogenesis enables delivery of oxygen and nutrients to ischemic regions. An intermittent, low-magnitude ischemia during exercise in normal conditions as well as during ischemic preconditioning activates angiogenesis for muscle adaptation.

Angiogenesis relies on the integration of hemodynamic and chemical signals. The vascular growth requires the coordinated proliferation and migration of endothelial cells. It involves vessel dilation or contraction, sprouting and branching, as well as intussusceptive involution and pruning (non-functional vessels). Intussusception, another type of remodeling of pre-existing vessels, i.e., insertion of tissular pillars into blood vessel lumens, enables vessel splitting.

Transcriptional, post-transcriptional, and post-translational mechanisms participate in the control of endothelial cell behavior during sprouting, branching with leading endothelial tip cells and trailing stalk cells, and tubulogenesis [886]. Specialized endothelial tip cells at the leading edge of vascular sprouts form filopodia in response to guidance signals.

Lymphangiogenesis is the development of new lymph vessels. *Arteriogenesis* deals with formation of mature arterioles and arteries with smooth myocytes, particularly for collateral development to bypass an obstructed artery.

Neovascularization involves 3 major processes stimulated by several chemical, physical, and mechanical (wall shear stress and intramural [circumferential and longitudinal] tension) factors: (1) postnatal vasculogenesis, i.e., formation of new

M. Thiriet, *Tissue Functioning and Remodeling in the Circulatory and Ventilatory Systems*, Biomathematical and Biomechanical Modeling of the Circulatory and Ventilatory Systems 5, DOI 10.1007/978-1-4614-5966-8_10,

blood vessels from bone marrow-derived endothelial progenitor cells; (2) angiogenesis; and (3) arteriogenesis, i.e., growth and remodeling of arterioles into arteries (collateralization).

10.1 Vasculogenesis

Mesodermal cells in the early embryo differentiate into endothelial progenitor cells — *angioblasts* — and aggregate. A functional vasculature develops from angioblasts.

During vasculogenesis, angioblasts determine "blood pockets" that lengthen to form irregular capillaries. Vascular plexi formed by aggregating angioblasts remodel and mature into organized vascular networks of large and small, ramified or merged, vessels. Parts of vascular plexi contain blood islands. Outer cells of the blood islands give rise to endothelia, inner cells to hematopoietic progenitors [1167]. Fusion of angioblast pockets forms *primary capillary plexi.* Ducts connect each other in a non-hierarchical, inhomogeneous network of primitive vessels.

Immediately after vasculogenesis, angioblasts and endothelial cells undergo specification to either arterial or venous fate. Angioblasts that aggregate to form blood islands, which fuse and remodel in response to hemodynamic stresses and genetic factors,[1] create a primitive interlaced network of arterial and venous plexi.

Angioblasts also acquire arterial or venous fates and coalesce to generate the first embryonic blood vessels: the dorsal aorta and cardinal vein. The dorsal aorta and cardinal vein are indeed directly formed by the coordinated sorting and segregation of arterial and venous angioblasts and their respective assembly.

After vasculogenesis and once associated with the heart pump, the primitive network that conveys blood remodels with branching. Angiogenesis from the dorsal aorta, cardinal vein, and vascular plexi builds a hierarchical network of arteries, arterioles, capillaries, venules, and veins. Subsequent recruitment pericytes and vascular smooth myocytes stabilizes nascent vessels and promotes vessel maturation. In addition, sprouting of lymphatic endothelial cells from venous endothelia gives rise to the closed-ended lymphatic network. Furthermore, a subset of hemogenic arterial endothelial cells generates hematopoietic stem cells.

10.2 Angiogenesis

Angiogenesis is the development of new branching vessels from existing vasculature. Angiogenesis thus involves migration and proliferation of endothelial cells

---

[1] E.g., Hairy enhancer of Split [HES]-related transcriptional regulators HRT1 and HRT2 promotes arterial differentiation; nuclear receptor NR2f2 favors venous differentiation [886].

from pre-existing vessels. Angiogenesis includes sequential events: (1) increased capillary permeability and endothelial cell and pericyte activation and hypertrophy; (2) destabilization with degradation of the vascular basement membrane and remodeling of the extracellular matrix; (3) endothelial cell proliferation and migration in the target extracellular matrix; (4) capillary lumenogenesis (Sect. 10.2.5); and (5) maturation with recruitment of pericytes, subsequent inhibition of endothelial proliferation, basement membrane reconstitution, and junctional complex formation, that stabilizes new vessels. Bone marrow-derived pericytes are also recruited for angiogenesis, particularly after ischemia [1168]. Angioblasts are able to differentiate into both blood and endothelial cells.

Both *chemotaxis* and *haptotaxis*[2] contribute to tissue development, defense, and repair. Various fibronectin-binding integrin types collaborate to yield cell adhesion and migration on fibronectin [1164]. In addition, disruption of epithelial layers instantaneously generates lateral electric fields [1165], directed toward the wound center, with sustained outward electric currents. The lesion shunts the transepithelial potential difference. These electric fields trigger cell migration during wound healing. *Electrotaxis* is controlled by $PI3K_{c1\gamma}$ and PTen phosphatase [1166].

Vascular sprouts develop and build a structure. Angiogenesis is tightly coupled to tissue development to supply the growing tissue with oxygen and nutrients and remove its metabolic waste. The three-dimensional, fractal-like network of branched vessels results from local growth gradients over small distances. Localized production of growth factors promotes tissue expansion and determination of the position of branching nodes. The process is adaptive, because vessel development, maturation, and regression coexist. Local control of cell responses to stimuli, such as the regulation of the extracellular matrix, avoids disorganization.

Angiogenic vessels differ from mature vessels. The wall structure is not well organized. The interactions between endothelial cells and pericytes are impaired. The wall is leaky. Angiogenic endothelial cells have altered surface markers and adhesion molecules.

Vessel maturation leads to a fully formed and functional network. After suppression of endothelial proliferation and sprouting, mural cells are incorporated into vessel walls and structural elements are constructed (valves, fenestrations or, conversely, tight junctions, etc.).

10.2.1 Stem, Progenitor, and Precursor Cells in Angiogenesis and Neovascularization

Fetal cells (trophoblasts, erythroblasts, leukocytes, hematopoietic progenitors, and mesenchymal stem cells) migrate in the maternal circulation during pregnancy.

---

[2]Haptotaxis corresponds to the adhesion gradient associated with the concentration of the constituents of the support medium, i.e., gradient in extracellular matrix density.

Fetal endothelial progenitors can then participate in maternal angiogenesis during pregnancy [1169]. After birth, angiogenesis participates in organ growth. During adulthood, angiogenesis in a healthy subject mainly occurs in the cycling ovary and the placenta during pregnancy. Otherwise, angiogenesis appears in trauma sites, wound healing, and developing tumors.[3]

Among subsets of mesenchymal stem cells, circulating endothelial progenitors contribute to angiogenesis, even during steady-state conditions in adult humans. Circulating angiogenic cells participate in tissue repair.

The sequential recruitment of mesenchymal stem cells begins by their mobilization from the bone marrow through rupture of the CXCL12–CXCR4 complex [1170]. Their migration to injury sites results from chemotaxis, not only via the CXCL12–CXCR4 complex, but also via CCL7 chemokine that targets CCR1, CCR2, and CCR3 receptors [1170].[4] Chemotaxis of hematopoietic progenitor cells and circulating angiogenic cells is also triggered by the CCL19–CCR7 complex, but the cell migration is weaker than that stimulated by CCL7 chemokine. In addition, activated macrophages produce interleukin-6 that regulates the migration of endothelial progenitor cells via interleukin-6 receptor-α chain. Chemokine CX3CL1 can also be involved, as its CX3CR1 receptor resides on endothelial progenitor cells.

Bone marrow-derived endothelial progenitors are recruited to tumoral growing vessels using transcriptional Class-B basic helix–loop–helix protein bHLHb24 factor. Attracted endothelial progenitors help in the progression of dormant micrometastases to lethal metastases [1171].

Recruitment of myelomonocytic cells from the bone marrow to tissues can serve as a source of pro-angiogenic cytokines after ischemia. The coexistence of myeloid lineage progenitors capable of endothelial differentiation and pro-angiogenic myeloid accessory cells then leads to 2 complementary mechanisms of angiogenesis. Vascular endothelial cells can differentiate from common myeloid progenitors and more mature granulocyte–macrophage progenitors [1172]. Bone marrow-derived progenitors of endothelial cells express PECAM1, von Willebrand factor, and TIE2, but not PTPRc, and pericyte marker desmin and smooth muscle actin.

[3]Tumors are hypoxic at some stage because of high oxygen consumption and inadequate blood supply. In response to hypoxia, tumor cells secrete angiogenic factors. Angiogenesis promotes tumor progression and metastasis.

[4]In addition to mesenchymal stem cells, basophils, eosinophils, neutrophils, T lymphocytes, NK cells, and monocytes use CCL7 chemoattractant.

10.2.2 Involved Factors in Angiogenesis

Numerous mechanisms are involved at various length scales: chemical signaling and genetic response, cell interactions, and environmental stresses.[5] Computational simulations have been proposed to provide insights into structure–function relationships at all involved scales [1173]. Vasculogenesis is modeled as traction-driven remodeling of an initially uniform tissue in the absence of blood flow, and angiogenesis as a flow-driven remodeling of a porous structure.

Development of vascular trees includes adaptation to mural stress field. Angiogenic molecules are generated in response to hypoxia and other stimuli. *Angiopoietins*, VEGF, and integrins regulate the vessel caliber. The expression of angiopoietin TIE2 receptor characterizes 3 cell types that have angiogenic activity: (1) endothelial cells; (2) TIE2+ monocytes and their hematopoietic progenitors, and (3) pericyte precursors of mesenchymal origin [1174].

10.2.3 Endothelial Sprouting: Tip and Stalk Cells

Certain endothelial cells that will form the distal end of the sprout — *tip cells*— strongly express vascular endothelial growth factor VEGFR2 receptor. Once endothelial cells are selected for sprouting, sprouting is controlled by the balance between angiogenic signals (e.g., VEGF and ephrin-B2; Tables 10.1 and 10.2) and antagonists (tight contacts with pericytes recruited by platelet-derived growth factor-B, certain extracellular matrix components, and VEGF inhibitors).

Endothelial sprouting for angiogenesis requires the coordinated behavior of involved endothelial cells that is regulated by the Notch and VEGFR signalings. Selection of endothelial cells for tip cell position depends on the ratio between VEGFR1 and VEGFR2 receptors [1175]. Receptor VEGFR1 has a high affinity for VEGFa, but a weak kinase activity. It modulates VEGFa signaling via VEGFR2 receptor. Whereas signaling launched by VEGFR2 and VEGFR3 supports tip cells, the decoy receptor VEGFR1 limits tip cell formation [886]. In addition, alternative splicing of VEGFR1 transcript generates a secreted, inactive isoform (soluble $VEGFR1^S$), which serves as a sink for free VEGFa factor.

Alternative splicing of the VEGFA transcript creates many variants with distinct functions. Migration of endothelial cells and, hence, vascular branching are promoted by heparan sulphate-binding $VEGFa_{165}$ variant. Free $VEGFa_{121}$ variant influences endothelial cell proliferation, but not migration [886]. Moreover, VEGFR2 output elicited by matrix-bound VEGFa differs from that primed by soluble VEGFa messenger.

---

[5] Vascular network development is regulated by local interactions between vascular cells and hemodynamic conditions.

Table 10.1. Endothelial sprouting (**Part 1**); Source: [886]; DLL: Delta-like ligand; EC: endothelial cell; ECM: extracellular matrix; MMP: matrix metallopeptidase; Nrp: neuropilin; Robo: Roundabout homolog; S1P: sphingosine 1-phosphate; Sema: semaphorin (Sema, Ig, transmembrane, and short cytoplasmic domain); TIE: Tyr kinase with Ig and EGF homology domains (angiopoietin receptor); Unc5b: Uncoordinated-5B homolog; VEGF: vascular endothelial growth factor; VEGFR: VEGF receptor). Endothelial cells produce platelet-derived growth factor PDGFb, transforming growth factor TGFβ1, angiopoietin-2 (Ang2), S1P and S1P receptor $S1P_1$, TIE2 receptor, and VEGFR1 to VEGFR3 receptors. Mural cells synthesize angiopoietin-1 (Ang1), S1P, activin receptor-like kinase ALK5, which heteromerizes with TβR2 receptor, and PDGFb. Signaling via VEGFR2, VEGFR3, or VEGFR2–VEGFR3 heterodimers is pro-angiogenic. Cleaved VEGFc and VEGFd interact with VEGFR2; VEGFa connects to VEGFR1 and soluble $VEGFR1^S$ (secreted extracellular domain) that then limit VEGFa availability. Vascular endothelial growth factor VEGFa and Ang2 support mural cell detachment and vessel destabilization. Receptor TIE2 tethers to matrix-associated Ang1 at adhesion sites between endothelial cells and matrix, thereby assisting migration. On the other hand, at EC–EC adhesions, the PTPRb–TIE2–Ang1–TIE2–PTPRb complex between apposed cells. In addition, Ang2 antagonizes Ang1 activity on TIE2 to foster angiogenesis. Ephrin-B2 links to VEGFR2 or VEGFR3 and promotes their internalization, thereby enhancing angiogenesis. Tip cell selection is associated with inhibition of tip cell formation laterally and migration of tip cells followed by stalk cells during sprout elongation with repression of tip cell fate in stalk cells. Tip and stalk cells can exchange their respective positions during sprouting elongation. The Robo4–Unc5b complex impedes VEGFR signaling. Endothelial cell migration continues in a given direction until anastomosis.

Event	Factors
Tip cell selection	VEGFa/c, VEGFR2/3 Ang2
Inhibition of tip cell fate in adjacent endothelial cells	DLL4–Notch, Robo4, Wnt, VEGFR1
Tip cell migration Sprout elongation	Ang1–TIE2 (EC–ECM adhesion sites) VEGFa/c/d–VEGFR2, VEGFc/d–VEGFR3/VEGFR2–VEGFR3 Cadherin-5 extraction from cell junctions MMP (matrix degradation)
Tip cell guidance	Ephrin-B2–EPHb4, Ephrin-B2–EPHb4–VEGFR2/3, VEGFa/c–Nrp1/2–VEGFR2/3, Sema3e–plexin-D1, Slit2–Robo4, Robo4–Unc5b, netrin–Unc5b

Tip cells contains a higher Delta-like (Notch) ligand DLL4 concentration with respect to *stalk cells* (Table 10.3). Messenger Notch, in turn, determines VEGFR concentration. As a feedback, VEGFR controls DLL4 expression. Endothelial tip cells are activated and guided by an extracellular VEGFa gradient. Activated Notch signaling in stalk cells impedes VEGFR, thereby repressing tip cells and maintaining the hierarchical organization of sprouting tip and stalk cells. The VEGF–VEGFR–DLL4–Notch–VEGFR feedback loop assigns position of endothelial cells to tip or stalk cells. However, tip cells can shift to stalk cells and conversely

Table 10.2. Endothelial sprouting (**Part 2**); Source: [886]; ALK: activin receptor-like kinase; Ang: angiopoietin; aPKC: atypical protein kinase-C; EPH: erythropoietin-producing hepatocyte receptor kinase Par: partitioning defective protein; PAK: P21-activated kinase; PDGF: platelet-derived growth factor; PDGFR: PDGF receptor; RasIP: RAS-interacting protein; S1P: sphingosine 1-phosphate; TβR: TGFβ receptor; TGF: transforming growth factor; TIE: Tyr kinase with Ig and EGF homology domains (angiopoietin receptor); VEGF: vascular endothelial growth factor; VEGFR: VEGF receptor; vSMC: vascular smooth muscle cell). Prior to tubulogenesis, or lumenogenesis, the apicobasal polarity is established in endothelial cells arranged in a string (precursor vessel). Protein PAR3 is a major determinant of cell polarity that influences lumenogenesis. Cell adhesion proteins (e.g., zonula occludens ZO1, claudin-5, and cadherin-5) move from the apical (wetted) surface to the basolateral segment of the plasma membrane; at the basal surface, integrins connect to matrix constituents. Integrin, PAR3, and RasIP1 promote the lateral redistribution of these junctional components. Lumenogenesis is initiated, at least partly, by relocalization of sialomucin CD34 and podocalyxin (Podxl) to the apical surface mediated by cadherin-5 and β_1-integrins. Kinase PKC phosphorylates moesin that links to apical Podxl–CD34 complexes and promotes the deposition of filamentous actin. Podocalyxin may cause an electrostatic repulsion of apical surfaces between endothelial cells. Lumenal expansion may result from vacuole exocytosis and fusion at the apical surface as well as signaling pathways, such as VEGFa–VEGFR2 axis that recruits of myosin-2 to the apical surface and those that activate RoCK to foster actomyosin filament contraction. Certain macrophage populations may act as cellular chaperones for vascular anastomosis.

Event	Factors
Tubulogenesis (lumenogenesis)	Par3/6, integrins, RasIP1, RhoGAP29, CDC42, Rac1, Src, aPKC, PAK2/4, Raf
Anastomosis	Macrophage
Stabilization and maturation	PDGFb, TGFβ1 (PDGFRβ+ pericyte and SMC recruitment), S1P–S1P$_1$, Ang1–TIE2, ephrin-B2–EPH (mural–endothelial cell attachment), TGFβ1–TβR2–ALK5 (vSMC differentiation), deposition of the basement membrane, strengthening of intercellular junctions)

according to the feature of VEGFR–DLL4–Notch signaling, which is constantly re-evaluated.

Endothelial sprouts are converted into functional vessels. Sprout extension involves the local proliferation and migration of the endothelial cells behind the tip that forms the sprout stalk. Tip cells do not proliferate.

Navigation cues are sensed by tip cells. The navigators Uncoordinated-5 homolog Unc5b, Roundabout homolog Robo4, plexin-D1, neuropilins, ephrin-B2, and EPHb4 receptor are major conductors of angiogenesis. The growing endothelial sprout is guided by attractive (e.g., netrins) and repulsive cues (e.g., semaphorin-3). MicroRNAs ensure the post-transcriptional control of angiogenesis.

Table 10.3. Mechanisms of endothelial leading, tip and trailing, stalk cell selection and their main features and roles (Source: [886]). Endothelial cells produce multiple Notch receptors (Notch-1, Notch-3, and Notch-4) and transmembrane Notch ligands (DLL1, DLL4, Jag1, and Jag2). Activated VEGFR2 induces DLL4 expression in tip cells that activates Notch on adjacent stalk cells. Ligand binding causes cleavage of Notch receptor by ADAM10, ADAM17, and presenilins, thereby liberating an intracellular fragment ($Notch^{ICD}$). In stalk cells, Notch precludes VEGFR2 activity, downregulates the expression of VEGFR3, and upregulates that of VEGFR1 and soluble VEGFR1, a sink for VEGFa that then represses VEGFR2 activity, thereby preventing tip cell fate. In addition, Notch induces expression of DLL4 and Notch-regulated ankyrin repeat-containing protein (NRARP). The latter supports Wnt signaling in stalk cells, thus maintaining adhesion sites between endothelial cells; it also fosters inhibitory feedback of Notch. In tip cells, Notch is blocked by Jagged-1 produced by stalk cells; it impedes DLL4–Notch connections on tip cells when Notch is glycosylated. Binding of VEGFc to its VEGFR3 receptor, which is highly expressed in tip cells, enables lymphangiogenesis.

Tip cell	Stalk cell
VEGF signaling	DLL4–Notch signaling
VEGFa/c	Low VEGFR signaling
VEGFR2/3	
Low Notch signaling	
PDGFb, Unc5b	
Highly motile	Weakly motile
Guide sprouting	Lumenogenesis
	Maintenance of between-cell junctions
	Connection to parent vessel

Establishment of blood flow requires the formation of a vascular lumen. Cellular structures are converted into tubes by vacuole formation and intracellular and subsequent intercellular fusion of large vacuoles. Vascular tubulogenesis is initiated by the acquisition of the apicobasal polarity of endothelial cells that is regulated by cell–matrix interactions and signaling via partitioning defective protein PAR3 and VEGFR receptor. Tubulogenesis is controlled by EGF-like domain-containing protein EGFL7 expressed by endothelial cells.

Interactions between tip cells regulate the fusion of adjoining sprouts to form a continous lumen. Tip cells abandon exploration upon encountering tips of other sprouts or existing capillaries, i.e., when they can form new vascular connections. Junctional contacts are then built at merging regions. Newly formed vessels are stabilized by recruitment of pericytes and deposition of matrix proteins into a basement membrane.

10.2.4 Other Modes of Vessel Formation and Remodeling

Changes in the local balance between pro- and anti-angiogenic factors can lead to the elimination of new connections (*pruning*). The splitting of vessels through the insertion of tissue pillars (*intussusception*) expends the vascular network.

Intussusception ("growth within itself"), a non-sprouting angiogenesis, represents a particular way of expanding and modifying a vessel network, as it creates and remodels blood vessels via the formation of transluminal pillars [1176].[6] Regions of locally changed stress sensed by endothelial cells can trigger growth of septa, ridges, pillars, and folds. Intussusception predominantly occurs in regions with accelerated blood flow. Intussusceptive vascular growth and remodeling are regulated by vascular growth factors, such as VEGF, PDGFb, and angiopoietins, and TIE receptors.

Maturation is related to the transition from a growing vascular bed to a functional network, characterized by stabilized vessels with mural cells and a basement membrane, valved veins, capillaries, and lymphatics with either fenestrations or tight junctions.

Perfusion reduces hypoxia-induced angiogenic factors and promotes vessel maturation. Bone marrow-derived circulating cells can be retained in the perivascular space due to CXCL12 chemokine in response to VEGFa and then enhance endothelial proliferation.

---

[6]Intussusception corresponds to pillar formation for capillary genesis. It is observed in the rapidly expanding postnatal lung capillary bed. The shape of these pillars is maintained by endothelial cells and pericytes. The pillars can grow in size and number. They can also merge to form folds and new vascular branches. There are 3 modes of intussusception [1176]: intussusceptive microvascular growth, arborization, and branching remodeling. Intussusceptive microvascular growth is characterized by insertion of transluminal pillars, which induces a rapid expansion of the capillary plexus. Vessels are generated from the capillary plexus by intussusceptive arborization by perpendicular pillar formation in rows, which delineate future vessels. Pillar reshaping and fusions determine the new vascular entity. Pillars and folds are then formed in the capillary sheet, which separate the new vessels from the old capillary plexus. Intussusceptive branching remodeling adapts the branching angle and the bores of the branches of new supplying and draining vessels by insertion of transluminal pillars at branching points (abundant small holes are observed at branching sites). The combined cross-sectional area of the branches is greater than that of the stem: $R_t^p = R_{b1}^p + R_{b2}^p$ (subscripts t and b for trunk and branch) with $2 < p < 4$, p depending both on the trunk bore and the vessel type (supplying or draining). The area ratio is indeed significantly different in arteries and veins [1176]. The distribution of blood flow at each bifurcation depends on the asymmetry ratio between the two branches. Strong asymmetry can lead to vascular pruning (non-perfusion and removal of branches at bifurcations). The time needed to complete pillar formation ranges from 40 to 120 mn.

10.2.5 Lumenogenesis

The *apical membrane initiation site* (AMIS) corresponds to one of the earliest intermediates in lumenogenesis [1177]. This cell-surface compartment as well as subjacent vesicles are characterized by the transient accumulation of numerous apical polarity and transfer proteins. The delivery of apical cargos to specialized zones of intercellular contacts, AMISs, i.e., the forming lumen, requires exocytosis.[7]

The apical membrane initiation site generates the *preapical patch*, a closed lumen formed by the newly established apical membranes of adjacent cells that is bound at its margins by junctional complexes. Opening of the lumen is a result of ion and water transport. Further organization of membranes, junctional complexes, cytoskeleton, and organelles creates the mature cyst, i.e., the tight junction-delineated lumen.

Lumenogenesis thus depends on Rab11a, Rab8a, and Rab8b GTPases and guanine nucleotide-exchange factor Rab8-interacting protein (RabIn8) [1178].[8] Lumen formation, at least in nephron epithelial cells, relies on a Rab11a–RabIn8–Rab8a axis that recruits Sec15a and cortical polarity GTPase CDC42[9] and promotes apical exocytosis by enrolling the partitioning-defective protein (Par) complex with members Par3 and atypical protein kinase-C and multimeric exocyst-tethering complex subunit Sec10, and exocyst subunit Sec8 to small vesicles and multiple rudimentary lumens and close to sites of intercellular contacts that constitute early apical membrane initiation loci.

Rab11a+ recycling endosomes are involved in the transfer of apical proteins.[10] Exocytosis that depends on Rab11a involves Rab8GEF RabIn8; the latter recruits

---

[7]The exocyst complex is composed of 8 proteins: Sec3, Sec5, Sec6, Sec8, Sec10, and Sec15, as well as Exo70 and Exo84.

[8]Protein RabIn8 is homologous to Rab3a-interacting protein (RabIn3), hence its name RabIn3-like protein. Protein RabIn8 stimulates nucleotide exchange on Rab8, but not Rab3a and Rab5, hence being a Rab8-specific activator [1179].

[9]Monomeric GTPase CDC42 is recruited to PIP_2-rich sites of forming lumens by annexin-2, where it recruits atypical protein kinase-C. Annexin-2 resides transiently in Rab11a+ vesicles during lumen formation [1178].

[10]In nephrons, apical proteins, such as podocalyxin and Crumbs polarity complexes are delivered to Rab11a+ recycling endosomes. Podocalyxin is a sialoglycoprotein of the CD34 family of transmembrane sialomucins. It is a constituent of the glycocalyx of podocytes that acts in podocyte morphogenesis. The apical Crumbs homolog complex keeps the atypical protein kinase-C apical during complex cellular shape changes. The apical-polarity determinant Crumbs complex is essential for lumen formation [1177]. The Crumbs complex associates with the Par complex directly or via the Crb complex protein Membrane protein, palmitoylated MPP5 of the P55-like MAGUK (membrane-associated guanylate kinase) subfamily (a.k.a. protein associated with Lin-7 PALS1). In addition, Crumbs homologs promote the dissociation of Par3 from Par6 and aPKC, hence allowing Par3 to localize to apicolateral borders during the transition from AMISs to preapical patches [1177].

and activates Rab8 GTPase.[11] Activation of Rab8 is antagonized by Rab8GAP TBC1D30 (Vol. 4 – Chap. 9. Guanosine Triphosphatases and Their Regulators).[12] Small Rab11a GTPase and possibly Rab8 recruit the exocyst subunit Sec15a that promotes the binding of the exocytic carrier to the Sec10 exocyst subunit on the emerging AMIS membrane. Exocytosis also permits the initial recruitment of the Par3–aPKC complex to AMIS sites. Active Rab8 stimulates CDC42 interaction with the exocytic carriers, probably via CDC42GEF DnmBP or Tuba.[13] Small GTPase CDC42 recruits atypical protein kinase-C that further promotes AMIS formation and exocytosis, as it recruits exocyst protein Sec10 and Sec8 as well as Par3 that colocalizes with Sec8 protein.

10.3 Arteriogenesis

Once the lumen of a main artery is strongly narrowed, the lumen of small arteries increases to form collaterals that can maintain the blood perfusion. Arteriogenesis is defined as the remodeling of a part of a pre-existing arterial network outside an ischemic region to form functional irrigation arteries.

Arteriogenesis is initiated by CCL2 chemokine. Chemoattractant CCL2 stimulates the formation of a collateral circulation on arterial occlusion [1181]. Arteriogenesis thus involves inflammation with monocyte recruitment.

Various substances are also required at different stages of arteriogenesis, among these, TGFβ, PDGF, FGF2, CSF2, and TNFα. Attracted monocytes produce fibronectin and proteoglycans as well as peptidases to remodel the extracellular matrix. These inflammatory cells then produce growth factors to stimulate proliferation of endothelial and smooth muscle cells.

[11]Rab8 GTPase modulates polarized membrane transport via reorganization of actin and microtubules. Rab8 GTPase regulates membrane transport associated with TNFα-mediated stress responses and cell morphogenesis [1179].

[12]TBC (Tre2 proto-oncogene [ubiquitin-specific peptidase USP6], BUB2 [budding uninhibited by benzimidazoles], and CDC16) motif-containing proteins pertains to GTPase activators of Rab and Rab-like GTPases.

[13]The CDC42-specific GEF dynamin-binding protein (DnmBP; a.k.a. Tuba according to the tradition of naming large synaptic proteins after musical instruments) concentrates at the apical region of cell junctions in epithelia via its interaction with zonula occludens protein ZO1. Scaffold Tuba links dynamin, Rho GTPase, and the actin cytoskeleton. It connects dynamin to numerous actin regulators [1180].

10.4 Vasculature Compartments

Hemangioblasts give birth to angioblasts that generate vascular endothelial cells. During vasculogenesis, endothelial precursor cells in response to local signals undergo proliferation, migration, differentiation, specification, and coalescence to form the lining of nascent vessels. During angiogenesis, the vascular network is remodeled into arteries, veins, and capillaries.

10.4.1 Lumen Size and Vessel Architecture

Whether blood flows with high or low flow rates and/or rapidly or slowly, vessels widen or narrow, respectively. Blood vessels that carry a high-speed, large-volume flow enlarge, whereas those with low-speed, small-volume flow regress. Local features of hemodynamics thus direct formation of main perfusion vessels, as endothelial cells respond to hemodynamical forces. Therefore, mechanical forces generated by blood flow modify gene expression in the developing embryo. Blood flow contributes to shaping a functional vascular architecture although vessel identity and developmental patterning are genetically predetermined.

The vessel network is organized according to neighboring tissue growth and structure. Vessel positioning depends on traction and compression forces exerted by the growing tissues.

10.4.2 Cost Functions

Optimal design of vessel branching is based on cost functions that are the sum of the rate at which work is done on blood and the rate at which energy is used, which are supposed to be proportional to the vessel volume for each vessel segment [1182]. Other cost functions have been proposed based on the minimal total surface area of blood vessels, minimal total volume, or the minimal total wall shear force on the vessel wall, or minimal power of the blood flow.

10.4.3 Arteriovenous Differentiation

The primitive vascular network progressively matures with arteries, capillaries, and veins [1183]. The specification of vascular cells in the different compartments (arteries, capillaries, and veins) of the vascular circuit is not only determined by applied mechanical forces and chemical factors (activators, inhibitors, and hypoxia),

Table 10.4. Molecules involved in arterial and venous differentiation of endothelial cells. Notch targets 2 families of transcriptional repressors: hairy and enhancer of split (HES) and HES-related (HRT) transcriptional regulators. Sonic Hedgehog and VEGF induce arterial cell fate. Forkhead box-C transcription factors FoxC1 and FoxC2 control arterial specification via Delta-like ligand DLL4, a Notch agonist. Nuclear receptor NR2f2 (a.k.a. chicken ovalbumin upstream promoter transcription factor COUPTF2) suppresses the Notch pathway and inhibits neuropilin-1 expression. Ephrin-B2 and ephrin receptor EPHb4 are markers for arteries and veins, respectively. Phosphatidylinositol 3-kinase inhibits the phospholipase-Cγ1–PKC–MAP2K–ERK pathway, an effector of VEGF signaling for arterial fate. Protein kinase-B induces venous fate.

Arterial fate	Venous fate
VEGFa	VEGFR3
Neuropilin-1	Neuropilin-2
Notch	NR2f2
Ephrin-B2	EPHb4
Extracellular signal-regulated kinase	Phosphatidylinositol 3-kinase

but it is also genetically programmed. Specific markers assigned to cells in each compartment can be detected before the onset of circulation [1167].

Growth and specialization of arteries and veins continues throughout development. Blood vessels adapt to experienced hemodynamic stresses. Arteriovenous differentiation is controlled by hemodynamic factors. Blood vessels remodel according to loading history. Concentric layers of vascular smooth myocytes and elastic fibers ensure resistance to arteries. The venous low-pressure system is provided with valves that open unidirectionally to prevent backflow.

Vasculature development depends on a combination of intrinsic prepatterning and extrinsic responses to environmental parameters. The direction of moving blood determines the differentiation in artery or vein, whether the lateral branch receives blood or provides it. Arterial endothelial cells characterized by a spindle-like shape in the streamwise direction have a low proliferation rate.

Some genes control the vascular specification into arteries and veins via signaling molecules (Table 10.4). Arterial endothelial cells express certain transcription factors, signaling molecules, gap-junction proteins, matrix molecules, and adhesion proteins that are absent from veins.

The arteriovenous differentiation is done via different processes. Notch guides arterial fate, as it targets Hairy and enhancer of Split (HES) and HES-related (HRT) transcriptional regulators. In arteries, Notch signaling is indirectly stimulated by VEGFa via VEGFR2 or neuropilin-1 and promoted by forkhead box transcription factors FoxC1 and FoxC2 [1184]. One the other hand, Notch pathway in the venous endothelium is suppressed by NR2f2 nuclear receptor.[14]

Arterial expression profile is defined by marker ephrin-B2, whereas venous pattern is identified by marker EPHb4 receptor. Neuropilin-1 is found in arteries, whereas Nrp2 is restricted to veins and lymphatic vessels.

[14] A.k.a. chicken ovalbumin upstream promoter transcription factor COUPTF2).

10.5 Extracellular Matrix in Vessel Formation

Angiogenesis depends on many growth factors and enzymes, hence different cell types and surrounding medium. The behavior of endothelial cells, especially their migration and proliferation as well as formation of tubular structures is influenced by the extracellular matrix (Table 10.5). Tip cells produce peptidases for cell migration, such as membrane-type matrix metallopeptidase mt1MMP (or MMP14). Nitric oxide, prostaglandin-E2, and CCL2 chemokine increase the cell-surface clustering and activity of mt1MMP [1185].

Interactions between endothelial and surrounding cells, such as pericytes and vascular smooth myocytes, regulate not only vessel stabilization and remodeling, but also vascular formation. Intercellular communications are based on multiple molecules, such as transforming growth factor-β, angiopoietins, platelet-derived growth factor, sphingosine 1-phosphate, and Notch ligands, among others [1186].

Cell shape is coupled to proliferation. Cell division depends more on the degree of possible extension than the level of matrix binding. The actin cytoskeleton state and activity of myosin contribute to cell and tissue growth. Laminin, fibronectin, and collagen-1, -3, -4, and -5 promote cell spreading.

Rho GTPases are involved in tension-dependent growth control, because they regulate cytoskeletal contractility. Matrix-associated cytoskeletal mechanics can explain tension-driven tissue modeling. Podosomes of endothelial cells, sites of MMP concentration, are involved in endothelial cell migration and angiogenesis.

10.5.1 Growth Factors

Many growth factors such as VEGF bind to matrix constituents.

Table 10.5. Sprout endothelial cells and inflammatory and angiogenic agents (Source: [1185]; $VEGFR1^S$: soluble VEGFR1). A nascent vascular sprout contains 3 types of endothelial cells. Tip cells produce VEGFR2, VEGFR3, DLL4, and angiomotin, among others, and navigate into the surrounding tissue, but proliferate poorly. Stalk and phalanx cells form the lumen and promote nascent vessel stabilization.

Type	Stimulators
Tip	VEGF, S1P, Bdk, TNFα
Stalk	EGFL7, FGF, VEGF, $VEGFR1^S$
Phalanx	Ang2, BMP9, FGF2, VEGF, Tsp

10.5.1.1 VEGF

Vascular endothelial growth factors connected to matrix constituents can be cleaved. A C-terminus of variable length according to the involved peptidase is then released. Released VEGF from matrix stores promotes angiogenesis.

Various VEGF isoforms have different affinities for given matrix proteins. However, soluble VEGF subtypes ($VEGF^S$) lack the matrix-binding region. They are either secreted as short alternative spliced forms such as $VEGF_{120}$ or cleaved by plasmin or matrix metallopeptidases, such as MMP3, MMP7, MMP9 (in the presence of heparin), and MMP19 isozymes.

Matrix-bound and soluble VEGFs can both trigger VEGFR phosphorylation, but differ in signaling. Matrix anchorage leads to clustering of VEGFR2, promotes receptor endocytosis, and raises downstream phosphorylation kinetics. Soluble VEGF has a tendency to build low-density, poorly branched vascular networks (vascular hyperplasia) [1185]. On the other hand, matrix-bound VEGF favors highly branched vessels.

10.5.1.2 FGF

Fibroblast growth factors interact with matrix proteins, particularly heparan sulfate proteoglycans. The extracellular matrix contributes to the regulation of FGF signaling. In pancreatic β cells, FGFR1 concentration is reduced by the binding of α_6-integrin to laminin [1185].

10.5.1.3 PDGF

Platelet-derived growth factor can link to several collagen types, laminin-1, nidogen, and perlecan. Binding to matrix constituents may serve for storage in cell's surrounding for later release and signaling.

10.5.1.4 TGF

Transforming growth factor-β can anchor to the extracellular matrix. This anchorage can contribute to its activation. Thrombospodin binding stimulates TGFβ, but thrombospodin-1 inhibits VEGF factor.

Table 10.6. Examples of matrikines in inflammation and angiogenesis (Source: [1185]; I: integrin; SPARC: secreted protein acidic and rich in cysteine; Tsp: thrombospondin). Matrikines are fragments of matrix constituents generated by proteolytic cleavage with functions distinct from those of parental proteins.

Matrikine	Source	Receptor	Function
Anastellin	Fibronectin		Anti-angiogenic
Angiostatin	Plasminogen	$\alpha_V\beta_3$, $\alpha_9\beta_1$-I	Anti-angiogenic
Arresten	Collagen-4	$\alpha_1\beta_1$-I	Anti-angiogenic
Canstatin	Collagen-4		Anti-angiogenic
Endorepellin	Perlecan		Anti-angiogenic
Endostatin	Collagen-18	$\alpha_V\beta_3$-I	Anti-angiogenic
Metastatin	Collagen-4		Anti-angiogenic
Neostatin	Collagen-18		Anti-angiogenic
Tumstatin	Collagen-4	$\alpha_V\beta_3$, $\alpha_5\beta_1$-I	Anti-angiogenic
Vastatin	Collagen-8		Anti-angiogenic
Restin	Collagen-15		Anti-angiogenic
Peptides from	Elastin		Pro-angiogenic
Peptides from	SPARC		Pro/anti-angiogenic
Peptides from	Tsp		Pro/anti-angiogenic
Peptides from	Collagens-1/4, elastin, fibronectin, laminins, entactin, Tsp, hyaluronan	Elastin-binding protein, L-selectin, integrins, CXCR1/2	Chemotactic

10.5.2 *Protein Fragments and Peptidases of the Extracellular Matrix*

Proteic constituents and peptidases of the extracellular matrix contribute to angiogenesis. Released active fragments of cleaved matrix constituents, the so-called *matrikines*, are involved in the control of angiogenesis and healing (Table 10.6). Most of the matrikines compete with intact matrix components for interaction with integrins.

Many matrikines have an anti-angiogenic effect. They derive from components of: (1) the basement membrane, such as collagen-4,[15] perlecan,[16] and fibronectin,[17]

[15] A.k.a. arresten, canstatin, metastatin, and tumstatin.

[16] A.k.a. endorepellin.

[17] A.k.a. anastellin.

Table 10.7. Peptides released from cleavage of matrix constituents and chemotaxis (Source: [1185]).

Peptidic fragments	Effect
Collagen	Neutrophil recruitment
Elastin	Monocyte recruitment
Laminin	Cell migration

Table 10.8. Peptides released from cleavage of matrix constituents and production of peptidases and cytokines (Source: [1185]; MMP: matrix metallopeptidase; PAI: plasminogen activator inhibitor; uPA: urokinase-type plasminogen activator [urokinase]).

Peptidic fragments	Effect
Collagen	Production of peptidases and cytokines
Elastin	Production of peptidases and cytokines
Fibronectin	Production of MMP9/12, cytokines
Hyaluronan	Production of MMP12, PAI1, cytokines
Laminin-111	Production of MMP9, uPA
Laminin-511	Production of MMP9/14, cytokines

as well as (2) the interstitial matrix that surrounds the vasculature, such as collagen-18,[18] collagen-8,[19] and collagen-15.[20]

Both thrombospondin-1 and -2 as well as their proteolytically derived fragments hinder angiogenesis during wound healing. Thrombospondin fragments are more effective than full-length thrombospondins.

In addition, peptidic fragments that are cleavage products of collagen-1 and -4, elastin, fibronectin, laminins, nidogens, thrombospondins, and hyaluronan operate as chemotactic substances for inflammatory cells (Table 10.7).

These fragments can also induce specific gene expression programs in leukocytes, in particular those of peptidases and cytokines (Table 10.8). In particular, fragments of fibronectin and laminin increase the production of various peptidases such as mt1MMP in monocytes and macrophages [1185].

Secreted protein acidic and rich in cysteine (SPARC)[21] is a member of the group of non-structural components of the extracellular matrix. This secreted Ca^{2+}-binding glycoprotein modulates cell adhesion and proliferation, as it intervenes in interactions between cells and their environment [1187].[22] It is highly expressed in

[18] A.k.a. endostatin and neostatin.

[19] A.k.a. vastatin.

[20] A.k.a. restin.

[21] A.k.a. basement-membrane protein BM40 and osteonectin.

[22] The Sparc gene is generally detected in cells with high rates of matrix production and proliferation. It can be identified in osteoblasts, odontoblasts, and chondrocytes, as well as megacaryocytes, steroid-producing cells of adrenal glands and gonads, and cells of kidney glomeruli, bronchi, skin, and large vessels [1188].

various cell types in remodeling tissues. It can interfere with the binding of angiogenic stimulators, such as vascular endothelial (VEGF), platelet-derived (PDGF), and fibroblast (FGF2) growth factor, to their receptors in endothelial cells. Schwann cells influence neuroblastoma growth by secreting inhibitors of angiogenesis, the most potent of which is SPARC protein. Full-length SPARC ($SPARC_{FL}$) and SPARC-derived peptides ($SPARC_f$) that correspond to the follistatin domain of the protein as well as its N- and C-termini block angiogenesis [1189].[23]

Involved peptidases include matrix metallopeptidases mt1MMP (MMP14) to mt3MMP (MMP16), MMP2, MMP3, MMP7, MMP9, and MMP13, and members of the adamlysine family, such as ADAM10, ADAM15, ADAM17, and ADAMTS1 [1185].

10.5.2.1 Endostatin and Endorepellin – Endogenous Angiogenesis Inhibitors

Endostatin, a fragment of collagen-18, and endorepellin, an inhibitor of angiogenesis derived from the basement membrane proteoglycan perlecan, have dual roles as structural constituents and functional regulators of tissue growth. They modulate the activity of growth factors. They can inhibit the growth of blood vessels and stabilize the basement membrane [1190].

Endostatin increases the production of nitric oxide in endothelial and smooth muscle cells [1191]. Hypertension is a major drawback of VEGF inhibitor administration, which necessitate to limit the dose. On the other hand, NO, a potent vasodilator, lowers blood pressure. Endostatin reduces blood pressure via the NOS3–NO axis and/or improved NO availability.

10.5.3 Cell Adhesion Molecules

Angiogenesis is regulated by endothelial cell adhesion molecules, such as vascular endothelial cadherin, integrins, and platelet–endothelial cell adhesion molecule. A mechanosensory complex composed of PECAM, cadherin-5, and VEGFR2 causes the response initiated by activated integrins [1192].

α-(2,6)-Sialic acid is necessary for the cell-surface residency and maintenance as well as homophilic interactions of platelet–endothelial cell-adhesion

[23] Peptides derived from SPARC domain-4 binds to endothelial cells and impedes endothelial cell proliferation, like SPARC. This inhibition also results from the action of a peptide derived from the follistatin-like domain-2. On the other hand, peptides from SPARC domain-1 and -3 have no effect on VEGF-stimulated DNA synthesis in endothelial cells [1187]. The inhibition of proliferation of human microvascular endothelial cells is caused by, at least partly, VEGF–SPARC binding.

molecules [1192]. This cell adhesion molecule is involved in angiogenesis once PECAMs are linked, and also in mechanical stress sensing and its anti-apoptotic function.

10.5.3.1 Integrins

Integrins are implicated as mediators of vascular formation and homeostasis, due to their role in cell adhesion, migration, positioning, proliferation, and differentiation [1193]. Endothelial cells express different vascular integrins involved in vasculo- and angiogenesis such as $\alpha_V\beta_3$-integrins.

Plasmalemmal integrins that interact with the extracellular matrix promote cell attachment to the extracellular matrix and transduce signals to the nucleus. Integrin-mediated adhesion to the extracellular matrix activates the ERK pathway. Activated extracellular signal-regulated kinases stimulate cyclin-D1. Cell division is controlled by cyclins, cyclin-dependent kinases, and CDK-inhibitors [1194] (Vol. 2–Chap. 2. Cell Growth and Proliferation).

Integrins are also required in VEGF signaling via Ras GTPase or phosphatidylinositol 3-kinase. Integrin-binding protein *lactadherin*[24] is involved in VEGF-dependent neovascularization [1195]. Lactadherin interacts with $\alpha_V\beta_3$- and $\alpha_V\beta_5$-integrins and alters VEGF-dependent PKB phosphorylation and neovascularization.

Activated $\alpha_V\beta_3$-integrin supports angiogenesis and growth of brain metastasis via upregulation of vascular endothelial growth factor under normoxic conditions [1196]. Synthesis of VEGF is post-transcriptionally controlled by phosphorylation (inhibition) of translational repressor 4E-binding 4eBP1 protein.

10.5.3.2 Tenascins

The tenascin (Ten) family of matrix, cell adhesion glycoproteins (Table 10.9) contributes to vasculogenesis and various normal and pathological processes of mature life, such as wound healing and vascular diseases [1197].

In association with matrix proteins and plasmalemmal receptors such as integrins, tenascins have opposite cellular functions, according to the mode of presentation, cell type, and differentiation state of the target tissues. Tenascins are regulated by growth factors, vasoactive peptides, matrix proteins, and mechanical factors [1197].

Tenascin-C interacts with integrins, collagens, proteoglycans, and fibronectin. It behaves either as an adhesive or anti-adhesive protein. It binds to annexin-2, a plasmalemmal receptor of endothelial cells [1198]. The TENC gene expression is mechanosensitive.

---

[24] Lactadherin is also known as milk fat globule-EGF factor-8 (MFGE8).

Table 10.9. Family of tenascins (Ten). These matrix, cell adhesion glycoproteins act via integrins, cell adhesion molecules of the immunoglobulin superfamily (IgCAMs), protein Tyr phosphatase receptor PTPRz1, annexin-2, fibronectin, and matrix proteoglycan lectican. The tenascin family includes TenC, TenR, TenW, TenX, and TenY. Tenascins can have anti-adhesive effects. Tenascin-C can promote integrin-dependent protein clustering at focal adhesions, as it interacts with RTKs and activates ERK. Like TenC, TenX (or flexilin) is expressed in the heart as well as vascular smooth myocytes. TenY is coexpressed with TenC in the lungs.

Tenascin	Alias
TenC	Cytotactin
TenR	Restrictin, janusin
TenW	
TenX	Flexilin, hexabrachion
TenY	

Table 10.10. Family of fibulins (Fbln [Fibl]; ARMD: age-related macular degeneration; AxPC1: ataxia, posterior column 1, with retinitis pigmentosa; DANCE: developmental arteries and neural crest EGF-like protein; DHRD: Doyne honeycomb retinal dystrophy (macular degeneration) protein; EFEMP: EGF domain-containing fibulin-like extracellular matrix protein). Fibulins regulate organ shape with growth factors and stromal cells. They are involved in cell proliferation, migration, differentiation, and survival.

Type	Other aliases
Fbln1	Fibl1
Fbln2	Fibl2
Fbln3	Fibl3, EFEMP1, DHRD
Fbln4	Fibl4, EFEMP2
Fbln5	Fibl5, DANCE
Fbln6	Fibl6, hemicentin-1 (Hmcn1), ARMD1, AXPC1
Fbln7	Fibl7

10.5.4 Fibulins

Fibulins (Fbln1–Fbln7; Table 10.10) constitute a family of secreted calcium-binding glycoproteins that are encoded by 7 genes. They interact with several matrix proteins, such as fibrillin, fibronectin, laminins, proteoglycans, and tropoelastin.

Fibulin-1 is associated with elastic fibers. It lodges particularly in basement membranes. It is also present in blood, where it binds to fibrinogen. It is incorporated in fibrin clots. Fibulin-1 suppresses fibronectin-mediated adhesion and motility [1199].

Fibulin-2 is involved in microfibril and elastic fiber organization [1200]. It can participate in clusters made of nidogen-1 and -2, fibronectin, and perlecan, a major heparan sulfate proteoglycan in the basement membrane that possesses immunoglobulin-like modules (IG2-IG15) with many binding sites for cellular and extracellular ligands [1201].

Fibulin-3 stimulates production of vascular endothelial growth factor, hence angiogenesis [1202]. Moreover, fibulin-3 hinders tumor cell apoptosis.

Fibulin-4 is an elastic fiber-associated protein that is required for formation of elastic fibers, especially in arteries, lungs, and skin. It regulates tropoelastin expression [1203].

Fibulin-5 that interacts with integrins is expressed in developing arteries. Its expression is reinduced in injured vessels and atherosclerosis in intimal smooth muscle and endothelial cells [1204].

10.5.5 Thrombospondins

Vessel formation requires a tuned balance between various interacting stimulatory and inhibitory signals. Thrombospondin inhibits blood vessel formation. Thrombospondin-1 binds to integrins, matrix metallopeptidase-2, transforming growth factor TGFβ1, proteoglycans, and fibronectin.

10.5.6 Matrix Metallopeptidases

Remodeling of the extracellular matrix by various molecules, especially extracellular and membrane-bound matrix metallopeptidases, is required during angiogenesis. The type of cleavage of matrix components can lead to opposite effects. Proteolysis of collagen-4 and -18, as well as plasminogen, releases angiogenesis inhibitors, whereas cleavage of collagen-4 by MMP9 stimulates migration of endothelial cells and angiogenesis.

10.5.7 Transglutaminase and Carboxypeptidases

Angiogenesis requires localized destabilization of the extracellular matrix. Afterward, transglutaminase-2 is downregulated in endothelial cells during capillary morphogenesis to secondarily stabilize the underlying basement membrane [923]. Glutamate carboxypeptidase-2[25] expressed in endothelial cells, contributes to the remodeling of the extracellular matrix for angiogenesis. Glutamate carboxypeptidase-2 acts on β_1-integrin and leads to activation of P21-activated kinase PAK1 [1205]. It also interacts with actin-binding filamin-A, leading to phosphorylation of

[25]Transmembrane glutamate carboxypeptidase-2 is also named prostate-specific membrane antigen.

P21-activated kinase-1.[26] Glutamate carboxypeptidase-2 thus participates in an autoregulatory loop.

10.5.8 Collagen Prolyl 4-Hydroxylase

Transcription factor P53 activates the gene that encodes collagen prolyl 4-hydroxylase. The latter is required for extracellular release of collagen-derived peptides, such as endostatin and tumstatin that inhibit angiogenesis [1206].

10.6 Mediators of Vascular Formation

The formation of new blood vessels, the patterning of the ramified network,[27] as well as the remodeling of existing blood vessels require the migration and proliferation of various cell types controlled by manifold substances, such as guidance and growth factors, and their inhibitors (Table 10.11). Microvascular growth and remodeling, which is influenced by the mechanical factors (local stress field), are variable. These phenomena are indeed based on sprouting, non-sprouting (intussusceptive), and mixed angiogenesis.

10.6.1 Navigation Signals

Angiogenesis is conducted by an initiation and patterning program. Guidance molecules regulate patterning of blood vasculature. The navigation of migrating angiogenic components requires guidance signaling via ligand–receptor complexes, such as: (1) ephrins with EPH receptors; (2) Slit ligands with Roundabout (Robo) receptors; (3) semaphorins with plexin and neuropilin receptors; and (4) netrins with plasmalemmal receptors Deleted in colorectal carcinoma (DCC), DCC-related neogenin (Ngn), and uncoordinated-5 (Unc5) receptors [1207, 1208].[28]

---

[26]Inhibition of P21-activated kinase-1 enhances the association of glutamate carboxypeptidase-2 with filamin-A.

[27]The vascular network is locally composed of arteriolar branched and venular junctional networks interconnected by the capillary bed. These arteriolar and venular networks are afferent to and efferent from irrigated tissues, or efferent from and afferent to the heart pump, respectively. As the heart can be defined as the center of circulating blood circuitry, the latter meaning is preferable.

[28]The ephrin-B2–EPHb4 signaling is involved in arteriovenous differentiation. Four Slit receptors exist, the Roundabouts (Robo). Receptor Robo4 and plexin-D1 are expressed by endothelium. Semaphorin-3 inhibits endothelial cell motility via plexin-D1. Neuropilins, in competition with Sema3a, associate with VEGF receptors and allow binding of heavy VEGF isoforms to endothelial

Table 10.11. Angiogenesis receptors and their ligands. Crosstalks between signaling effectors allow to coordinate their spatial and temporal activities.

Receptor	Ligand
VEGFR	VEGF (endothelial cell)
Neuropilin	Semaphorin
PDGFR	PDGF (smooth myocyte)
HER	EGF
FGFR	FGF
TGFR	TGF
TNFR	TNF
EPH	Ephrin
TIE	Angiopoietin
S1PR	S1P
Roundabout	Slit
Netrin	Uncoordinated-5
Galectin-1	Anginex

Nerve and vessel navigations to targets are subjected to a cooperative procedure, axonic and angiogenic signals guiding both vessels and axons. Ephrins, semaphorins, netrins, and slits[29] control vessel morphogenesis.

10.6.1.1 Semaphorins

Semaphorin-3 Subtypes

During blood vessel genesis, endothelial cells produce semaphorin-3 that controls integrin function for vascular reshaping [1209]. Semaphorin-3A expressed by endothelial cells of developing vessels inhibits endothelial cell migration at nascent adhesive sites of spreading endothelial cells, by interacting with integrins, such as endostatin and thrombospondin.

Neuropilin-1 is a membrane receptor for multiple ligands[30] such as semaphorin-3 that regulates developmental process guidance [1210, 1211].

cell receptors. Netrin-1 acts on both the endothelial cells and smooth myocytes of the vessel wall. It can have repulsive and attractive effects, according to the context. Neogenin is expressed in vascular smooth myocytes, whereas Unc5 is expressed in endothelial cells.

[29] Slits are proteic chemorepellants in axon guidance and neuronal migration via the Roundabout receptor. Robo1 is expressed by vascular endothelial cells. Slit2 attracts endothelial cells.

[30] Neuropilin-1 and -2 that bind semaphorin-3A (or collapsin-1) are involved in axon guidance. They are also receptors for heparin-binding VEGFa isoforms. Both Sema3a–Nrp1 and VEGF–Nrp1 signaling are required for heart development.

Like Sema3a, Sema3f influences vessel formation via neuropilin-1. Dimer Sema3e that targets plexin-D1 is a chemorepellant of endothelial cells [1212]. Agent Sema3e controls endothelial cell positioning and patterning of the developing vasculature. One the other hand, catabolized Sema3e can act as a chemoattractant.

Semaphorin-4 Isoforms

Semaphorin-4A, an activator of T-cell-mediated immunity, is also expressed in endothelial cells. Semaphorin-4A hinders endothelial cell migration and division stimulated by vascular endothelial growth factor, thus preventing angiogenesis in vivo [1213]. Semaphorin-4A uses different receptors according to the cell type. It targets plexin-D1 in endothelial cells with a lower affinity for plexin-D1 than Sema3e.

As an activator of T-cell-mediated immunity, Sema4a connects to T-cell immunoglobulin and mucin domain-containing protein TIMD2 in T lymphocytes. Semaphorin-4A also binds to Plexin-B1. It uses PI3K effectors in the VEGF–VEGFR2 pathway. It suppresses PKB phosphorylation and production of Rac^{GTP} that promotes endothelial cell motility.

Semaphorin-4D induces endothelial cell proliferation and tubule formation via plexin-B1 [1214]. Plexin-B1 activates HGFR receptor protein Tyr kinase.

10.6.1.2 Netrins

Netrins (sanskrit netr: scout) are secreted matrix-binding proteins that yield migration signals for axon guidance and motion of other cells in the developing central nervous system. Axon-guidance functions are mediated by 2 families of receptors: (1) Uncoordinated-5 (Unc5) receptor family and (2) Deleted in colorectal cancer (DCC) family that includes neogenin subclass and belongs to the immunoglobulin superfamily. Netrins provide attractive or repulsive signals by interacting with DCC and Unc5 receptors or Unc5–DCC heterodimers, respectively.

Calcium influx via plasmalemmal Ca^{2+} channels and from cytoplasmic stores is required. High and low cAMP-to-cGMP ratios favor attraction and repulsion, respectively. Netrins also regulate cell adhesion, motility, proliferation, differentiation, and survival in other biological tissues.[31]

Netrins are guidance cues for vascular endothelial and smooth muscle cells during angiogenesis. Netrin receptor Unc5b in endothelial tip cells and developing capillaries mediates repulsion. Integrin-$\alpha_6\beta_4$ and -$\alpha_3\beta_1$ operate as netrin-1 receptors in endothelial cells [1215]. Activated integrins recruit and phosphorylate adaptor SHC that, in association with the GRB–SOS complex, activates Ras, then

[31] Netrin-1 is particularly expressed in the developing lungs, pancreas, and mammary gland.

Table 10.12. Effectors of netrin-1 signaling (Source: [1215]).

GTPases	CDC42, Rac, Rho
Kinases	FAK, MAPK, PAK, PI3K, PKA, PKG, PLCγ, Src
Transcription factors	ELk1, NFAT, P53

Raf and MAPK module. Adaptor SHC also activates phosphatidylinositol 3-kinase and protein kinase-B.

Netrins activate small GTPases Rac, Rho, and CDC42 (Table 10.12). Netrin-1 bound to DCC recruits extracellular signal-regulated kinases ERK1 and ERK2 to a DCC receptor complex. Receptor DCC also interacts with focal adhesion kinases and Src kinases.

Netrin-1 acts a survival cue. Once bound to DCC and Unc5h, it activates caspase-3 that cleaves the death domains of these receptors, the latter activating apoptotic caspase-9. Moreover, Unc5b targeted by P53 mediates P53-dependent apoptosis until it binds netrin-1.

10.6.1.3 Ephrins

Ephrin-A1 mediates TNFα-induced angiogenesis in vivo [1216]. Ephrin-B2 and its receptor EPHb are arterial markers. Receptor EPHb activates Src kinase via a cleavage of ephrin-B2, releasing a cytoplasmic C-terminal fragment ($_{C_1}$ephrin-B2) that is further processed by the presenilin-1–γ-secretase complex to produce intracellular $_{C_2}$ephrin-B2 peptide.[32] Peptide $_{C_2}$ephrin-B2 binds Src kinase, thereby impeding Src association with CSK kinase, a Src inhibitor, and allowing Src autophosphorylation [1217].[33] Activated Src phosphorylates ephrin-B2, inhibiting its processing by γ-secretase and triggering the recruitment of GRB4 to ephrin-B2. Adaptor GRB4 controls actin dynamics and cell migration, thereby sprouting of endothelial cells.

Endothelial cells express guidance molecules for angiogenesis, such as EPHb4 and its ligand ephrin-B2 [1218]. EPHb4 is a negative regulator of blood vessel branching, leading to circumferential vessel growth rather than angiogenic sprouting and vessel interconnection. EPHb4 and ephrin-B2 restrict migration of endothelial cells. Moreover, EPHb4 reduces vascular permeability via angiopoietin-1–TIE2 activation at the endothelium–pericyte interface. EPHb4 reverse signaling via ephrin-B2 represents the predominant signaling pathway, being independent of EPHb4 RTK activity and EPHb4 forward signaling.

[32] Presenilin-1 is a widespread transmembrane protein. It is cleaved into N-terminal ($_N$PS1) and C-terminal ($_C$PS1) fragments that associate to form a functional heterodimer. It promotes γ-secretase processing of substrates.

[33] C-terminal Src kinase (CSK) binds Src kinase and prevents its activation by precluding its autophosphorylation (Tyr418).

10.6.1.4 Slit Ligands and Roundabout Receptors

Slit ligands (Slit1–Slit3) are involved in neuronal and vascular development. They bind to Roundabout receptors (Robo1–Robo4) that belong to the immunoglobulin superfamily of transmembrane signaling molecules. The Slit–Robo signaling induces repulsion. Slits are involved in heart morphogenesis, angiogenesis, and tumor metastasis.

Vascular endothelium-specific Roundabout-4 has an extracellular domain different from that of all other Robo family members. Receptor Robo4 activated by Slit2 maintains vasculature integrity [892]. Stabilizing factor Robo4 is expressed in developing blood vessels where it can neutralize VEGF, but it is not required for developmental angiogenesis. During sprouting, Robo4 is only expressed by more mature, stabilized endothelial cells that form sprout stems. The Robo4–Slit2 signaling counteracts hyperpermeability induced by VEGF that uses Src and Yes kinases. In addition, activated Robo4 impedes endothelial cell migration primed by fibroblast growth factor-2 [1219]. Angiogenic endothelial cells (as well as many other cell types) also express Robo1 receptor. Because Slit2 binding to Robo1 promotes angiogenesis, Slit2 positively and negatively regulates angiogenesis by binding to Robo1 and Robo4, respectively.

10.6.2 Transcriptional Regulators

10.6.2.1 Hairy and Enhancer of Split-Related Transcription Factors

Two genes — Hrt1 and Hrt2 — that encode Hairy enhancer of Split (HES)-related transcriptional regulators (HRT) are major contributors of vessel formation [1220]. The combined loss of transcription factors HRT1 and HRT2 does not influence initial vasculogenesis, but it does affect subsequent development of major vessels.

10.6.2.2 Homeobox Transcription Factors

Homeobox (Hox) gene-encoded transcriptional factors are temporally and spatially restricted regulators of tissue patterning during embryogenesis owing to their DNA-binding homeodomain. The Hox regulators are involved in cell differentiation, proliferation, and migration. In particular they participate in endothelial cell fate, especially the transcriptional control of genes responsible for angiogenesis and vascular remodeling. The Hox transcription factors regulate genes involved in cell– and cell–matrix interactions, such as the expression of $\alpha_V\beta_3$-integrin, matrix metallopeptidase MMP14, or urokinase plasminogen activator receptor [1221]. These proteins contribute to matrix remodeling during angiogenesis. The HOX genes are characterized by their clustered genomic arrangement. Each HOX gene can have up to 3 paralogs within independent clusters.

Class-3 Hox factors (i.e., associated with 3 clusters, HoxA3, HoxB3, and HoxD3) have a pro-angiogenic role. Thay can provoke endothelial cell migration. Class-4 Hox factors (HoxA4–HoxD4) are involved in hematopoesis. Factor HoxB5 upregulates VEGFR2 receptor [1221]. In addition, in vitro, HoxB5 enhances endothelial cell sprouting and modulates the expression of adhesion molecules [1221]. Nevertheless, HoxB5 may mainly regulate intussusceptive vascular growth. In vivo, HoxB5 can upregulate angiopoietin-1 and -2 and VEGF [1221]. Its pro-angiogenic effect is abolished by soluble TIE2 ($TIE2^S$), an angiopoietin antagonist. Factor HoxB5 also promotes the production of matrix metallopeptidase-1 and -2, but downregulates that of cell adhesion molecules such as β3-integrin.

10.6.2.3 DLx and NKx Factors

Brain development relies on concomitant vasculature development. In mice, telencephalic angiogenesis is governed by a time and space (ventrodorsal) gradient determined by compartment-specific homeobox transcription factors Distal-less homeobox protein DLx1 and DLx2, NK2 transcription factor-related homeobox protein NKx2-1, and paired box protein Pax6 [1222]. These transcription factors also regulate the development of telencephalic neuroepithelial domains and neurons. Factor NKx2-1 is stimulated by sonic Hedgehog that influences both neurogenesis and angiogenesis. Distal-less homeobox DLx1 and DLx2 hamper the expression of Delta-like ligand-1, a Notch ligand. Paired box protein Pax6 hinders cornea vascularization.

10.6.2.4 ETV6 Transcriptional Repressor

Transcriptional repressor ETS-related translocation variant ETV6[34] is needed for sprouting of human endothelial cells. Factor ETV6 binds to the corepressor C-terminal-binding protein [1223]. The ETV6–CTBP complex temporally restricts VEGF-mediated pulse of Notch ligand DLL4. Control of DLL4 expression, which depends on the ETV6–CTBP complex, by VEGF is not iterated when VEGF signals continuously. Whereas VEGFR activation stimulates angiogenesis, Notch signaling primed by DLL4 inhibits the process. In endothelial cells, VEGF induces expression of DLL4 that then triggers the Notch pathway in adjacent cells to attenuate VEGFR activity and prevent spontaneous sprouting in the absence of sufficient VEGF signaling. The ETV6–CTBP complex further controls branching by regulating expression of other factors that limit angiogenesis, such as Sprouty family members and VE-cadherin.

---

[34] A.k.a. translocation ETS leukemia protein (Tel or Tel1) and malignant proliferation, eosinophil (MPE).

10.6.2.5 Nuclear Factor-κB

Nuclear factor-κB in endothelial cells has both positive and negative effects on angiogenesis [1224]. According to the context, NFκB impedes or promotes apoptosis. Interactions of $\alpha_V\beta_3$-integrin with matrix components that occur during migration of endothelial cells activate NFκB factor.

Nuclear factor-κB upregulates VEGF expression as well as vascular endothelial growth inhibitor. Moreover, NFκB can bind to angiostatic thrombospondin-1 and -2. In addition, NFκB participates in activation of E-selectin in endothelial cells by CXCL4 chemokine.

Once NFκB is activated by reactive oxygen species and tumor-necrosis factor-α, it induces expression of plasminogen activator inhibitor PAI1 and can hinder tissue-type plasminogen activator.[35] On the other hand, NFκB promotes expression of several matrix metallopeptidases (MMP2, MMP3, and MMP9). Consequently, NFκB can either favor or impede the degradation phase of angiogenesis. Last but not least, the activity of many angiostatic compounds depends on NFκB activation.

10.6.2.6 PPARγ and PGC1α Coactivator

Vascular endothelial growth factor operates via both PKCα and PKCβ to activate cAMP-responsive element-binding protein and subsequently produce the pro-angiogenic enzyme cyclooxygenase-2 in endothelial cells. On the other hand, nuclear receptor NR1c3, or peroxisome proliferator-activated receptor PPARγ, precludes angiogenesis. Factor PPARγ suppresses membrane translocation of PKCα [1225].

Regular exercise improves blood circulation in the limbs, as it promotes the generation of new blood vessels in muscle. Numerous signaling pathways are activated during exercise that involve PP3 phosphatase, calmodulin-dependent kinases, AMP-activated protein kinase, stress-responsive P38MAPK, and reactive oxygen species. All of these pathways impinge on PGC1α [1226].

Transcriptional PPARγ coactivator PGC1α elicits angiogenesis in muscles [1226]. β-Adrenergic receptors increase expression of the Pgc1α gene and promote expression of vascular endothelial growth factor, thereby triggering angiogenesis. Factor PGC1α cooperates with the nuclear receptor estrogen-related receptor ERRα to regulate expression of the VEGF gene [1226].

In myocytes and adipocytes, the Pgc1α gene has an alternative promoter upstream from the proximal promoter that generates a PGC1α isoform with a few extra N-terminal amino acids. Changes in N-terminus of these proteins may confer specificity on the angiogenic program [1226]. Expression of mRNA from

[35] Plasmin hydrolyzes many extracellular proteins. Urokinase and tissue-type plasminogen activators have high affinity for inactive plasminogen to produce plasmin. Conversely, plasminogen activator inhibitors PAI1 and PAI2 prevent the activation of plasminogen into plasmin.

this promoter heightens 100-fold after exercise, whereas the transcription from the proximal promoter remains constant [1227].

10.6.2.7 Krüppel-like Factor

Krüppel-like factor KLF2 is produced in mature endothelial cells subjected to both shear stress and statins as well as circulating pro-angiogenic cells, formerly called endothelial progenitor cells, at a comparable level [1228]. It provokes endothelial functional differentiation. In mature endothelial cells, KLF2 prevents angiogenesis.

During senescence, KLF2 level decays [1228]. In aged mice, numbers of circulating SCA1+, SCFR+, Lin− progenitors and SCA1+, VEGFR2+ endothelial-committed progenitor cells lower. Moreover, KLF2 expression drops in spleen- and bone marrow-derived circulating pro-angiogenic cells.

10.6.2.8 Hypoxia-Inducible Factor

Hypoxia-inducible factor HIF1 intervenes in the compensatory angiogenesis to insufficient O_2 supply. It initiates gene expression for numerous angiogenic regulators, vessel remodeling, as well as recruitment of bone marrow-derived, circulating angiogenic cells.

This ubiquitous, heterodimeric transcription factor mediates adaptive responses to hypoxia and ischemia. Isotype HIF1 is made of O_2-regulated HIF1α and constitutively expressed HIF1β subunits. Subtype HIF2α is an HIF1α paralog. Under hypoxia, HIF1α translocates to the nucleus, dimerizes with HIF1β, and binds to hypoxia response elements (HRE).

Transcriptional activity of HIF1 functions on genes that encode vascular endothelial (VEGF), placental (PlGF), and platelet-derived (PDGF) growth factors, stem cell factor, angiopoietin-2, and CXCL12 chemokine. These angiogenic factors bind to their cognate receptors (VEGFR1 and VEGFR2 for VEGF; VEGFR1 for PlGF; PDGFRα and PDGFRβ for PDGFB; SCFR for SCF; TIE2 for Ang2, and CXCR4 for CXCL12) in the plasma membrane of vascular endothelial cells, pericytes, and smooth myocytes.

In addition, HIF1 regulates the expression of hundreds of genes that encode, in particular, plasmalemmal receptors of angiogenic cytokines and homing signals for the recruitment of pro-angiogenic cells, such as endothelial progenitor cells, hematopoietic stem and progenitor cells, and mesenchymal stem cells, in addition to bone marrow-derived myeloid cells [1229].

HIF in Pulmonary Arterioles

Whereas systemic arterioles dilate in response to local tissue hypoxia to increase O_2 delivery, pulmonary arterioles constrict in response to hypoxia to ensure proper local ventilation–perfusion ratios. Increased resistance of pulmonary arterioles results from elevated pulmonary arterial smooth myocyte tone as well as hypertrophy and proliferation. Agent HIF1 decreases the expression of voltage-dependent $K_V1.5$ and $K_V2.1$ channels and increases that of transient receptor potential TRPC1 and TRPC6 channels and Na^+–H^+ exchangers in pulmonary arterial smooth myocytes [1229]. Moreover, HIF1 triggers the production of endothelin-1 as well as angiotensin convertase and angiotensin receptor AT_1 in human pulmonary artery fibroblasts.

10.6.2.9 Histone Deacetylases

Phosphorylated class-2A histone deacetylases recruit 14-3-3 proteins for sequestration in the cytoplasm to overcome their repressor activity, because they cannot interact with their cognate transcription factors and corepressors. Dephosphorylation of class-2A HDACs allows nuclear import and gene activity repression.

Many kinases and phosphatase could phosphorylate and dephosphorylate HDAC7 according to location and signaling time for adequate and efficient cell reponses to stimuli, respectively. Phosphatase PP2 constitutively dephosphorylates HDAC7 histone deacetylase to control its function as a regulator of T-lymphocyte apoptosis and endothelial cell functioning during angiogenesis [1230].[36]

10.6.2.10 Histones

Histone-2A family member H2ax involved in DNA-damage repair promotes survival and proliferation of endothelial cells during hypoxia and hypoxia-induced angiogenesis [1231]. Hypoxia leads to H2ax phosphorylation ($H2ax^P$ or γ-H2ax) in endothelial cells, similarly to H2ax C-terminus phosphorylation that results from DNA damage by ataxia teleangiectasia mutated kinase (ATMK), ATM- and Rad3-related kinase (ATRK), and DNA-dependent protein kinase. Hypoxia-triggered neovascularization requires functional H2ax in endothelial cells.

10.6.3 MicroRNAs

Post-transcriptional control of angiogenesis is ensured by microRNAs. MicroRNAs can be divided into 2 sets: pro- and anti-angiogenic microRNAs. Pro-angiogenic microRNAs include Let7, miR17-2, miR27b, miR126, miR130a, miR210, miR296,

---

[36]Enzyme HDAC7 is transiently and strongly expressed in CD4+, CD8+ thymocytes, as it represses a pro-apoptotic gene that encodes nuclear receptor NR4a1 (a.k.a. testicular receptor TR3, nerve growth factor NGF1b, and Nur77). Enzyme HDAC7 in the vascular endothelium downregulates MMP10 matrix metallopeptidase.

Table 10.13. MicroRNAs in angiogenesis (Source: [886, 1235]; SpRED: Sprouty-related EVH1 domain-containing protein).

Type	Targets
MicroRNA-92a	Integrin-$\alpha 5$ (repression of angiogenesis) Sirtuin-1 (activation of Notch, repression of VEGF [?])
MicroRNA-126	$PI3K_{r2}$ (derepression of the PI3K axis) SpRED1 (derepression of the Raf1 axis) IGFBP2 repression (inactivation of endothelial IGF1R) Endothelial MerTK stimulation (cancer call MerTK inhibition)
MicroRNA-132	RasA1 (derepression of Ras GTPase)

and miR378, whereas anti-angiogenic microRNAs encompass miR15b, miR16, miR92a, miR214, miR221, miR222, and miR328 [1232].

Dicer that produces final forms of microRNAs participates in postnatal angiogenesis induced by various stimuli. Growth factors can regulate microRNA expression. Factor VEGF regulates the expression of several microRNAs, such as components of the MyC cluster miR17-92 [1233].

Angiogenic sprouting of aortic arch vessels relies on the mechanosensitive transcription factor Krüppel-like factor KLF2a that provokes synthesis of endothelial-specific miR126 to activate VEGF signaling [1234]. MicroRNA-126 represses Sprouty-related EVH1 domain-containing protein SpRED1 and $PI3K_{r2}$ subunit, which prevent MAPK and PI3K signaling, respectively [886]. Therefore, miR126 derepresses PI3K and cRaf axes to support VEGF-induced angiogenesis (Table 10.13).

MicroRNA-132 suppresses RasA1, a RasGAP, in the endothelium, thereby increasing Ras activity and promoting angiogenesis [1236]. Its expression is upregulated in a human embryonic stem cell model of vasculogenesis and in endothelia of human tumors and hemangiomas, but is undetectable in normal endothelium. On the other hand, RasA1 is produced in normal endothelium, but not tumor endothelium.

MicroRNAs encoded by the miR23–miR27–miR24 gene clusters[37] are involved in cell cycle control, proliferation, and differentiation of various cell types. They abound in endothelial cells. Their pro-angiogenic effect results from the repression of the anti-angiogenic mediators Sprouty-2 and semaphorin-6A [1237].

Microvescicles represent a mode of communication between cancer and endothelial cells to launch endothelial cell migration. MicroRNAs are indeed packaged into and carried by microvescicles that can be taken up by endothelial cells. In cancer cells, some microRNAs such as miR126 can impede recruitment of endothelial cells [1235], whereas others such as miR9 are tumor promoters. In endothelial

[37] Two miR23–miR27–miR24 clusters exist: an intergenic miR23a–miR27a–miR24-2 and intronic miR23b–miR27b–miR24-1 cluster.

cells, upon delivery by microvesicles, miR9 decreases the abundance of SOCS5, an antagonist of the JAK-STAT pathway [1238]. Activated Janus kinase-1 and -2 phosphorylate signal transducer and activator of transcription STAT1 and STAT3.

10.6.4 Morphogens

Morphogens govern positions of various cell types in a developing tissue. These signaling substances build concentration gradients to generate and maintain the transcription of target genes at given concentration thresholds. Hence, genes may determine the anatomical structure of an organism [1239].[38]

10.6.4.1 Notch

Notch components mediate the arterial gene program,[39] the venous specification via the nuclear receptor NR2f2,[40] and the lymphatic commitment via Prospero homeobox gene product Prox1 [1167].[41] Most lymphatics differentiate from veins.

The formation of endothelial tip cells at the leading edge of vascular sprouts is regulated by DLL4–Notch-1 signaling that is specific to the vasculature. Vascular-specific Delta-like ligand-4 regulates proliferation and differentiation of endothelial cells and vascular development. The Notch ligand DLL4 is expressed during vascular remodeling. Notch signaling initiated by DLL4 and restricted to the vascular system is involved in arteriovenous differentiation.

---

[38]Mathematical reaction–diffusion discretized and continuous models of pattern formation represent either cells as points or as local physicochemical characteristics in a continuum. In the cell form of the model, mechanical aspects are related to cell positions, masses, velocities, and elastic properties, as well as intercellular forces, and chemical aspects to the chemical composition and diffusivity of each separate cell. In the continuous form of the model, mechanical aspects are described by stress, velocity, density, and elasticity of the matter, and chemical aspects by concentration and diffusivity in each point. Chemical reactions include production, conversion, and degradation of morphogens.

[39]Delta-like ligand DLL4 is a Notch agonist via Notch-1 and Notch-4 receptors devoted to the specific development of the arterial endothelium.

[40]Nuclear receptor NR2f2, or chicken ovalbumin upstream promoter transcription factor COUPTF2, inhibits the activity of many transcription factors, such as intracellular receptors retinoic acid (RAR or NR1b), retinoid X (RXR or NR2b), and thyroid (TR or NR1a) and steroid hormone (SF1 or NR5a1) receptors. Transcription factor NR2f2 is able to bind to DNA either as a homo- or heterodimer. Two NR2f types are encoded by the NR2F1 and NR2F2 genes. Factor NR2f2 is highly expressed in the venous endothelium, but not the arterial one. Transcription factor NR2f2 is also involved in the formation of atrioventricular septum and coronary vessels.

[41]The homeobox gene product Prox1 is expressed in certain endothelial cells that yield buds from veins to give rise to lymphatics.

The VEGF and Notch pathways are coupled. Signaling triggered by VEGFa provokes DLL4 production prominently in tip cells of endothelial sprouts. Notch ligand DLL4 induced by VEGF in tip cells activates Notch in adjacent endothelial cells, where it suppresses the expression of VEGFR receptors, thereby restraining excessive endothelial sprouting. In fact, expression of DLL4 in endothelial cells is controlled by VEGF that promotes Notch signaling to suppress the formation of additional tip cells [1240, 1241].

Notch signaling primed by DLL4 hence acts as a negative regulator of VEGF-induced angiogenesis. Blockade of interaction between DLL4 and Notch receptor prevents Notch signaling and inhibits angiogenesis in tumors resistant to VEGF inhibitors [1242, 1243].[42] The DLL4–Notch pathway intervenes during active vascularization rather than during vessel maintenance [1243]. Ligand DLL4 is upregulated in tumor rather than in normal vessels [1242].

Sprouting blood vessels are organized with leader tip and follower stalk cells. Competition between endothelial cells for the tip position is regulated by glycosylation of Notch receptors and by the antagonism of Notch ligands Jag1 and DLL4 [1244]. In mice, whereas sprouting of endothelial tip cells is inhibited by the binding of Notch receptor to its ligand DLL4, the Notch ligand Jagged-1 is a potent angiogenic activator in cells that possess the Fringe family of glycosyltransferases. Notch glycosylation enhances DLL4-Notch signaling, whereas Jagged-1 has a weak signaling capacity.

Cerebral cavernous malformation protein CCM1 stabilizes endothelial junctions and keeps endothelium quiescent, thus acting as an anti-angiogenic substance [1245].[43] It primes the inhibitory DLL4–Notch signaling to prevent endothelial cell proliferation, migration, apoptosis, and lumenogenesis. It then promotes PKB phosphorylation, but represses that of ERK kinases.

In arteries, in which Notch is more active, DLL4 was strongly expressed and VEGFR3 concentration is low. On the other hand, VEGFR2 lodges in both arteries and veins. In retinal tip cells, DLL4 expression is only weakly modulated by VEGFR2 signaling. Therefore, VEGFR2 is not essential for DLL4 expression in tip cells [1246]. Moreover, activated Notch is a potent inhibitor of VEGFR3 synthesis, but not of VEGFR2 expression. Receptor VEGFR2 is indeed not or only weakly regulated by Notch [1246]. Activity of VEGFR3 is pro-angiogenic in endothelial

[42]Tumor growth rate is not correlated with tumor vascular density. Tumor perfusion could be improved by vasculature pruning. Repression of DLL4 results in excessive angiogenesis. Normally, endothelial cell stimulation by multiple growth factors leads to cell proliferation with controlled degradation of the extracellular matrix and migration using various chemoattractants and adhesion molecules. Endothelial cells then form a central lumen and mature into a functional vessel. Neutralized DLL4 leads to defective cell fate and impairs formation of tubular structures by endothelial cells. Most tumor vessels are not functional; they are either not perfused or convey a reduced blood flow.

[43]Protein CCM1 is produced in astrocytes and endothelial cells. It complexes with CCM2 and CCM3, as well as lipids and components of the cytoskeleton, cell junctions, and signal transduction pathways.

cells, in which Notch signaling is weak or absent. In low Notch activity, VEGFR3 upregulation allows ligand-independent, excessive, deregulated angiogenesis even in the absence of VEGF–VEGFR2 signaling.

Notch and VEGFR2 are antagonists in the regulation of VEGFR3 level. Notch regulates VEGFR3 activity independently of VEGFR2 receptor. Conversely, VEGFR3 kinase activity enables Notch-regulated sprouting. On the other hand, VEGFR2 strongly improves VEGFR3 density (but not DLL4) at the angiogenic front. Therefore, VEGFR3 may have an active, ligand-dependent, pro-angiogenic and a Notch-mediated, passive signaling modes [1246].

The non-canonical Notch ligand Delta-like homolog DLk1,[44] a transmembrane glycoprotein expressed in the endothelium that lacks the DSL domain, impedes angiogenic sprouting, endothelial cell migration, and tubulogenesis via Notch receptors [1247].

Transmembrane syndecans are heparan sulfate proteoglycans that function as coreceptors for growth factors. They are implicated in angiogenesis. Syndecans are implicated in communication between endothelial and smooth muscle cells aimed at ensuring proper structure and function of blood vessels and their maintenance. These mural cell types communicate via, at least partly, Notch-3 and syndecan-2 plasmalemmal receptors. The latter is produced in smooth myocytes upon contact with endothelial cells and Notch activation in smooth myocytes [1248]. Both Notch-2 and -3 contribute to the upregulation of syndecan-2. In particular, syndecan-2 interacts with Notch-3 and regulates expression of several Notch-3 target genes (only in coculture conditions). Conversely, expression of syndecan-2 in smooth myocytes needs Notch-2 and Notch-3 proteins.

10.6.4.2 Wnt

The Wnt–βCtn pathway favors DLL4 expression in the vasculature. Conversely, the DLL4–Notch complex activates Wnt in stalk cells. In addition, Notch promotes vascular stability via Notch-regulated ankyrin repeat-containing protein (NRARP) [886]. Agent NRARP limits Notch activity, but stimulates Wnt signaling to stabilize stalk cells.

The planar cell polarity mode of Wnt signaling participates in endothelial cell proliferation and angiogenesis. On the other hand, Disheveled-associated activator of morphogenesis DAAM1 that operates as a formin to cause actin polymerization and especially microtubule stabilization precludes endothelial cell proliferation and migration [1249].

Retinal myeloid cells (macrophages) produce Wnt ligands to suppress retinal (postnatal) angiogenesis via the non-canonical Wnt–VEGFR1 pathway [1250].

[44] A.k.a. pre-adipocyte factor-1 (PreF1). It precludes differentiation into adipocytes, neuroendocrine cells, osteoblasts, myocytes, and hematopoietic cells.

Table 10.14. Examples of growth factors and cytokines involved in angiogenesis and arteriogenesis (Source: [1252–1254]). Thymidine phosphorylase (TymP; a.k.a. gliostatin and platelet-derived endothelial cell growth factor [PDECGF or ECGF1]) that belongs to the family of glycosyltransferases stimulates endothelial cell growth and promotes angiogenesis.

Molecule	Target
Angiogenin	EC
Angiopoietin-1	EC
CCL2	EC, SMC
EGF	EC
FGF	EC, SMC
Interleukin-1	EC, SMC
Interleukin-20	EC
PDGF	EC, SMC
TGF	EC (TGF$\alpha\oplus$, TGF$\beta\ominus$), SMC, ECM
TNF	EC $\ominus$
TymP	EC, SMC
VEGF	EC

10.6.4.3 Hedgehog

Hedgehog signaling participates in blood vessel development. It regulates distinct vascular patterning events using VEGF-dependent and -independent mechanisms [1251]. Expression of VEGF in the mouse embryo depends on Hh signaling. Hedgehog activity in both endothelial tube formation and Notch-dependent arterial identity solely depends on VEGF regulation. Hedgehog effects are limited by Hh inhibitor Patched-1 that depends neither on VEGF nor Notch.

10.6.5 Growth Factors

Many angiogenic factors exert chemotactic and mitogenic activities on vascular endothelial and smooth muscle cells as well as on fibroblasts. Moreover, they control the metabolic functions of these cells. Numerous growth factors act on endothelial and/or smooth muscle cells (Table 10.14).

Certain growth factors, such as vascular endothelial growth factor and angiopoietin-1, can act both on angiogenesis and vascular permeability. Transforming growth factor-β stimulates extracellular matrix production to stabilize the structure of new vessels.

10.6.5.1 Vascular Endothelial Growth Factors

Vascular endothelial growth factors regulate the initial stages of blood vessel formation. Later, they are also implicated in association with other activators, such as angiopoietins, fibroblast (FGF), insulin-like (IGF), and transforming (TGFβ) growth factors, as well as TNFα, apelin, ephrin, and Notch. Most of these factors act on other cell types and complement or coordinate VEGF signaling rather than operating as independent regulators of angiogenesis.

In the absence of TGFβ1, VEGF favors endothelial cell survival via VEGFR2 and P38MAPK [1255]. In the presence of TGFβ1, VEGF promotes cell apoptosis. During angiogenesis, apoptosis is required for pruning the forming vascular network. Factor TGFβ1 promotes the expression of FGF2 by endothelial cell, which upregulates the expression of VEGF and acts via ERK1 and ERK2 kinases.

Vascular endothelial growth factor is a potent mitogen for endothelial cells derived from arteries, veins, and lymphatics [1256]. Moreover, VEGF stimulates cardiomyocytes. Factor VEGF exerts its angiogenic effects via its cognate VEGFR receptors (Fig. 10.1). Endothelial cells activated by VEGFa, assemble, and form tubes.

Vascular endothelial growth factors also act on the development of the nervous system[45] and protect nerve cells under stress conditions, as a neurotrophic and angiogenic factor.

Vascular endothelial growth factor-A also acts on pericytes in coordination with other growth factors such as the platelet-derived growth factor-BB homodimer secreted by endothelial cells [1259].

VEGF Receptors

Receptor Tyr kinase signaling, especially VEGFR and angiopoietin receptors TIE1 and TIE2, is regulated by Tyr phosphatases, such as receptor protein Tyr phosphatase PTPRb. Phosphatase PTPRb dephosphorylates TIE2 and cadherin-5, but not VEGFR2 receptor. It is expressed in both arterial (more strongly) and venous endothelium in embryos [1260]. It is highly produced in the developing outflow tract of the heart and developing heart valves. Although PTPRb does not intervene in early stages of vasculogenesis, it mediates suitable remodeling of vascular plexi into large vessels and branched vascular networks by pruning, coalescence, sprouting, and intussusception. Phosphatase PTPRb is also synthesized in adults in the vasculature and heart valves.

Binding of VEGF to its receptors induces receptor dimerization and autophosphorylation. This process then activates several downstream kinases: protein

[45] In vitro, VEGF is implicated in the survival, proliferation, and migration of Schwann cells, astrocytes, and microglia [1258]. In vivo, VEGF is required for migration of certain motoneurons.

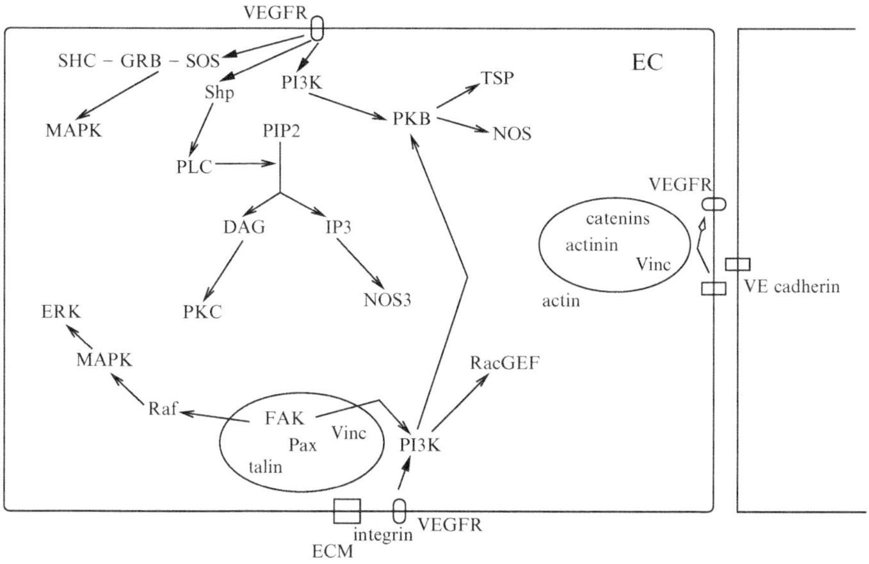

Fig. 10.1 Vascular endothelial growth factor (VEGF) stimulates angiogenesis, particularly proliferation of endothelial cells (EC; Source: [1257]). Factor VEGF binds to its receptors (VEGFR1–VEGFR3). It then activates cell type-dependent signaling cascade via adaptors SHC and GRB, cytosolic protein Tyr phosphatases PTPn6 and PTPn11, phospholipase PLCγ, phosphatidylinositol 3-kinase (PI3K), etc. Vascular endothelial (VE)-cadherins (cadherin-5) are involved in adherens junctions between neighboring endothelial cells. They interact with catenins, and subsequently with α-actinin and vinculin (Vinc), and the actin cytoskeleton. Cadherin-5 acts with VEGFR2 to control the PI3K–PKB pathway. Endothelial cells are linked to the extracellular matrix (ECM) via integrins such as $\alpha_V\beta_3$-integrin and focal adhesion molecules, such as focal adhesion kinase (FAK), talin, paxillin (Pax), and vinculin (Vinc).

kinase-C and -D (Vol. 4 – Chap. 5. Cytosolic Protein Ser/Thr Kinases),[46] phosphatidylinositol 3-kinase (Vol. 4 – Chap. 2. Signaling Lipids), and mitogen-activated protein kinase (Vol. 4 – Chap. 6. Mitogen-Activated Protein Kinase Modules).

Vascular endothelial growth factor-A binds to 2 receptor Tyr kinases on the plasma membrane of endothelial cells: VEGFR1[47] and VEGFR2.[48] However, VEGFa angiogenic effects are mainly mediated by VEGFR2 receptor. VEGFa Factor is required for chemotaxis and differentiation of endothelial precursors, as well as endothelial cell proliferation, migration, and assembly into vascular structures

[46]The PKD family includes PKD1 to PKD3. Kinases PKD1 and PKD3 are also called PKCμ and PKCν, respectively. PKD1 is a class-2a HDAC export kinase.

[47]Receptor VEGFR1 is implicated in chemotaxis and vascular permeability.

[48]Receptor VEGFR2 is involved via protein kinase-B and mitogen-activated protein kinase in endothelial cell proliferation and survival.

(sprouting and tube formation) and remodeling. Different VEGFa isoforms regulate blood vessel growth and patterning.

Factor VEGFb binds to VEGFR1.[49] Receptor VEGFR1 mediates the recruitment of implicated bone marrow-derived cells. It is also involved in activation of matrix metallopeptidases [1261] and paracrine release of growth factors from endothelial cells [1262].

VEGF Signaling

Signaling triggered by VEGF leads to activation of numerous genes, such as those encoding regulator of calcineurin (PP3) RCan2[50] and NR4a1 nuclear receptor.

Histone deacetylase HDAC7 controls endothelial function by repressing MEF2-dependent and -independent genes. Activated PKD phosphorylates HDAC7, hence creating docking sites for 14-3-3 proteins. Enzyme PKD thus promotes HDAC7 nuclear export to activate VEGF-responsive genes in endothelial cells [1263].

In cultured endothelial cells, VEGFa stimulates the synthesis of various molecules involved in different angiogenesis stages, such as urokinase-type (uPA) and tissue-type (tPA) plasminogen activators, uPA receptor (uPAR), as well as plasminogen activator inhibitor-1 and $\alpha_V\beta_3$- and $\alpha_V\beta_5$-integrins. Factor VEGFa also activates collagenase for a subsequent environmental degradation to prepare the ground for angiogenesis. Because VEGF promotes the expression of urokinase receptor in endothelial cells, it favors uPA–uPAR interaction that mediates cellular invasion and tissue remodeling.

In addition, VEGF provokes VCAM1 and ICAM1 production by endothelial cells. Factor VEGF increases vascular permeability with possible plasma protein leakage and formation of a fibrin gel used as a substrate by endothelial cells. Agent VEGF stimulates hexose transport in endothelial cells for increased energy demands during proliferation.

In endothelial cells, G proteins, particularly $G\alpha_q$ and $G\alpha_{11}$, interact with and mediate intracellular signaling stimulated by VEGFR2 [1264].

Activation of VEGFR2 by VEGF stimulates the ERK and P38MAPK pathways. Vascular endothelial growth factor-A induces endothelial cell migration after actin filament assembly and stress fiber formation [1265], via a P38MAPK pathway, activated by MAP2K3 and MAP2K6 enzymes.

MAPK-activated protein kinase MAPKAPK2, a downstream kinase of the P38MAPK cascade stimulated by VEGFa, phosphorylates heat shock protein HSP27, an actin-binding protein that, in its non-phosphorylated form, precludes actin polymerization; phosphorylated HSP27 released from capped actin filaments stimulates actin polymerization [1266]. Kinase MAPKAPK2 also

[49]Growth factors VEGFb and PlGF are selective ligands of VEGFR1 (Vol. 3 – Chap. 8. Receptor Kinases).

[50]A.k.a. calcipressin-2 [(Cps2).

phosphorylates (activates) LIMK1 kinase,[51] that, in turn, inactivates cofilin, an actin-severing factor required in cell migration [1267].[52]

Actin-binding proteins filamin-A and -B are synthesized in endothelial cells.[53] Filamin-B intervenes in endothelial cell migration [1270]. The signaling cascade includes filamin-B, Rac1, its downstream effectors P21/CDC42/Rac-activated kinases PAK4 to PAK6 and guanine nucleotide-exchange factor Vav2.

Upon activation of VEGFR2, Src kinase phosphorylates Vav2 to stimulate Rac GTPase. Activated Rac activates its effector P21-activated kinase that phosphorylates cadherin-5 associated with VEGFR2 receptor. Phosphorylated cadherin-5^P recruits and binds β-arrestin-2 involved in endocytosis via clathrin-coated vesicles. β-Arrestin-2 attracts Src near cadherin-5 for phosphorylation of cadherin–catenin complexes. Factor VEGF hence promotes cadherin-5 endocytosis and disassembling of intercellular adherens junctions to disrupt the endothelium [1271].

Mediators of VEGF Synthesis

Nitric oxide favors VEGF synthesis. Expression of VEGF is stimulated by growth factors and mitogens as well as hypoxia. It is also regulated by many transcriptional regulators.

Hypoxia stimulates several transcription factors, such as hypoxia-inducible factor (mainly), cAMP-response element-binding protein, early growth response EGR1, metal response element (MRE)-binding transcription MTF1,[54] NFκB, and activating enhancer-binding protein AP1.[55] Under hypoxic conditions, HIF1α is activated by the PI3K–PKB and ERK pathways. Factor HIF1α–HIF1β stimulates glycolysis and angiogenesis [1272].

Transcription factors NFκB and Activator protein AP1 are often activated by the same stimuli to target common genes. Hypoxia-inducible factor targets NFκB that stimulates AP1 subunit JunB that is required for basal and hypoxia-induced activation of VEGF transcription (Fig. 10.2) [1273].[56] The latter targets core-binding factor-β (non-DNA-binding CBFβ) that forms with Runt-related transcription factor Runx (CBFα) the core-binding heterodimeric transcription

---

[51] Kinase LIMK1 is also phosphorylated by RoCK and PAK, downstream from Rho GTPases.

[52] Cofilin phosphatases slingshot and chronophin dephosphorylate (activate) cofilin. Overexpression of LIMK1 suppresses cell motility [1268, 1269].

[53] Filamin-A, the most abundant, is widespread, whereas filamin-B, also broadly expressed, is produced mainly in endothelial cells. Filamin-C is primarily expressed in skeletal myocutes.

[54] A.k.a. metal-regulatory transcription factor-1 (MTF1).

[55] The AP1 family is composed of dimeric protein complexes formed by products of Jun, Fos, and ATF gene families. Hypoxia upregulates AP1 expression that activates target genes for endothelin-1 and PDGFb in endothelial cells.

[56] Hypoxia-responsive transcription factor JunB is activated by numerous stimuli, such as NFκB and inflammatory cytokines. Other AP1 subunits Jun and Fos act as partners of VEGF response to hypoxia.

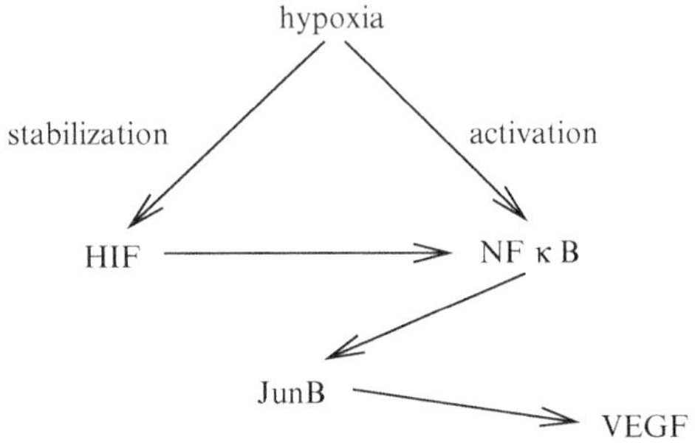

Fig. 10.2 Hypoxia-responsive transcription factor JunB is activated by NFκB triggered by HIF for hypoxia-induced VEGF transcription activation (Source: [1273]).

factor [1274]. Factor CBFβ also intervenes in synthesis of enzymes such as collagenase-3, or matrix metallopeptidase MMP13, that degrades the extracellular matrix, thus liberating space for angiogenesis (CBFβ–JunB–MMP13 cascade).

Interferon-γ released by T lymphocytes and natural killer cells activates monocytes and macrophages and suppresses the production of VEGFa by macrophages, as Ifnγ stimulates transcription of the VegfA gene, but inhibits translation [1275]. Heterotetrameric RNA-binding Ifnγ-activated inhibitor of translation (GAIT)[57] is a complex that contains ribosomal protein L13a, glutamylprolyl-tRNA synthetase, synaptotagmin-binding, cytoplasmic, RNA-interacting protein (SynCRIP or heterogeneous nuclear ribonucleoprotein hnRNPq), and glyceraldehyde 3-phosphate dehydrogenase. It binds to mRNA-bound eIF4g and precludes recruitment of the 43S pre-initiation complex. The GAIT complex also prevents translation of acute phase inflammatory ceruloplasmin (a copper-containing glycoprotein) secreted into plasma by hepatocytes and in inflammatory sites by stimulated macrophages. Inhibitor GAIT thus targets pro-inflammatory transcripts and initiate translational silencing.[58]

Placental Growth Factor

Placental growth factor, a VEGF family member, is dispensable for normal morphogenesis of the vasculature, but inhibits tumoral angiogenesis and lymphangiogenesis, as well as the recruitment of pro-angiogenic macrophages [1277].

---

[57] A cis-acting RNA element in the 3′UTR of the ceruloplasmin mRNA.

[58] Interferon-γ activates a kinase cascade. Death-associated protein kinase DAPK1 activates DAPK3 that phosphorylates L13a [1276]. Both Dapk1 and Dapk3 mRNAs contain a 3′UTR GAIT element. They are thus translationally repressed by the GAIT complex. This negative feedback prevents complete suppression of translation, thereby allowing a low-level translation of target transcripts such as VegfA mRNA.

10.6.5.2 Transforming Growth Factor-β

Nascent channels become covered not only by endothelial cells, but also by pericytes and vascular smooth muscles. Pericytes and smooth myocytes stabilize small and large blood vessels, respectively. Pericytes differentiate from perivascular progenitor cells in the bone marrow. The differentiation of progenitors into pericytes and vascular smooth myocytes is promoted by transforming growth factor-β1. However, the latter can lead to local destabilization and regression.

10.6.5.3 Platelet-Derived Growth Factors

Vascular endothelial and platelet-derived growth factors contribute to angiogenesis and vessel stabilization by activating vascular endothelial and smooth muscle cells, respectively. At an optimal angiogenic dose, PDGFbb homodimer or VEGF alone promotes angiogenesis. During PDGF–VEGF costimulation, VEGF and PDGF have antagonistic effects on angiogenesis. Vascular endothelial growth factor receptor-2 actually binds to and antagonizes platelet-derived growth factor receptor-β in mural cells to prevent pericyte recruitment to nascent vessels and limit neovascularization [1278].

Homodimer PDGFbb and angiopoietin-1 recruit mural cells around endothelial tubes. Protomer PDGFb on the endothelial cell surface ensures suitable coverage of blood vessels by pericytes in the microcirculation and smooth myocytes in the macrocirculation. Factor PDGFb is needed for recruitment of pericytes and maturation of the microvasculature. The PDGFa–PDGFRα pair recruit angiogenic stromal fibroblasts [1279].

Homodimer PDGFcc is secreted and then processed to bind its receptor.[59] It causes monocyte migration and upregulates expression of matrix metallopeptidases MMP2 and MMP9. It operates on multiple cell types, such as vascular endothelial and other mural cells, macrophages, as well as choroidal fibroblasts and retinal pigment epithelial cells [1280].[60] In addition, PDGFcc possesses many molecular angiogenic targets, such as PDGFbb factor and PDGFRα receptors, and provokes phosphorylation (inactivation) of glycogen synthase kinase GSK3β by protein kinase-B. The latter, once activated, impedes PDGFcc-induced angiogenesis [1280].

[59] Platelet-derived growth factors form dimers that bind and activate their receptor tyrosine kinases PDGFRα and PDGFRβ that also build dimers. Both PDGFc and PDGFd are produced as latent growth factors that require proteolytic removal of their N-terminal CUB (complement subcomponents C1r and C1s, fibropellin-1, or uEGF, from Strongylocentrotus purpuratus (Purple sea urchin), and bone morphogenetic protein BMP1) domain to become active.

[60] Platelet-derived growth factors PDGFaa, PDGFab, PDGFbb, and PDGFcc activate PDGFRαα homodimers, PDGFbb and PDGFdd stimulate PDGFRββ homodimers, and PDGFab, PDGFbb, PDGFcc, and PDGFdd activate PDGFRαβ heterodimers.

Platelet-derived growth factor receptor-β in pericytes and smooth myocytes is required for guided migration and incorporation into the vessel wall, then cell proliferation and maturation of the vasculature. Receptor PDGFRβ can cooperate with G-protein-coupled receptors S1PRs, as sphingosine 1-phosphate is secreted by endothelial cells to get the appropriate number of mural cells and form adhesion sites between them and endothelial cells. Perivascular progenitors expressing PDGFRβ are recruited from the bone marrow to perivascular sites, particularly in tumors [1281].[61]

10.6.5.4 Fibroblast Growth Factors

The initiation of angiogenesis particularly requires fibroblast growth factor-2 and paracrine vascular endothelial growth factor. Growth factor activity that favors either cell survival or apoptosis depends on activated signaling pathway. Both VEGF and FGF2, independently of VEGF, are survival factors for endothelial cells, whereas TGFβ1 that acts via SMADs and MAPKs induces transient endothelial cell apoptosis.[62] Growth factors FGF2, TGFβ1, and VEGF are often coexpressed in tissues in which angiogeness occurs.

10.6.5.5 Prokineticins

Prokineticins Prok1 and Prok2 are mitogens for endothelial cells which bind to cognate G-protein-coupled receptors PKR1 and PKR2. Prokineticin-2 is upregulated in the bone marrow by CSF3 colony-stimulating factor. They promote both angiogenesis and myeloid cell mobilization from the bone marrow [1282].

10.6.5.6 Midkine

Hypoxia increases midkine synthesis in human neutrophils, monocytes, M2 macrophages, and endothelial cells, as well as cancer cells [1283]. Midkine is a member of the family of heparin-binding growth factors with pleiotrophin. Its production results from binding of hypoxia inducible factor-1α to a hypoxia response element in the midkine gene promoter. Midkine causes angiogenesis during ischemia.

---

[61] During tumoral angiogenesis, pericytes have morphological abnormalities.

[62] The transient apoptotic effect on endothelial cells of TGFβ1 is followed by refractoriness of these cells to TGFβ1-induced apoptosis.

10.6.6 Hormones and Cytokines

Small hormones, such as prolactin, growth hormone, and placental lactogen, secreted by endothelia, promote angiogenesis.[63] Their cytokine receptors that belong to class-1 family activate the JaK–STAT pathway. However, proteolytic fragments of these small hormones that result from cleavage by extracellular metallopeptidases of the bone morphogenetic protein BMP1 subgroup have anti-angiogenic effects [1284].[64]

Interleukin-20[65] stimulates both macro- and microvascular endothelial cells via activation of IL20R1–IL20R2 or IL22R1–IL20R2 receptors and various signaling effectors, such as JAK2–STAT5, ERK1, ERK2, PKB, Rac and Rho GTPases, NO, and Ca^{2+} [1285]. In particular, IL20 activates NOS3 that can then move away from the plasma membrane. Synthesis of NO occurs mostly near the cortical and perinuclear region. Both activated PKB that phosphorylates NOS3 and rapid, strong Ca^{2+} influx favor NO synthesis. Interleukin-20 promotes vessel maturation and remodeling, but not endothelial cell proliferation and motility. Delivery of IL20 to ischemic muscles in a rat hind-limb model improves arteriogenesis.

10.6.6.1 Angiopoietins

Two main isoforms of angiopoietins — Ang1 and Ang2 — bind endothelial cell-specific Tyr kinase with domains TIE2 (Fig. 10.3).[66] Isoforms angiopoietin-3 and -4 are human and mouse orthologs that exert a negative or positive effect in angiogenesis according to the context.

---

[63]Prolactin operates in reproduction, growth, metabolism, immunity, osmoregulation, and behavior. Growth hormone not only stimulates growth, but also modulates metabolism, reproduction, and immunity, among other activities. Placental lactogen acts on mammary development, corpus luteum maintenance, and progesterone production.

[64]The bone morphogenetic protein BMP1 family comprises structurally similar metallopeptidases: BMP1, mammalian Tolloid, and mammalian Tolloid-like Tll1 and Tll2. BMP1-like peptidases participate in the formation of the extracellular matrix by processing precursors: procollagens-1, -3, -5, -7, and -11 into functional ECM components; prolysyl oxidase zymogen into enzyme that catalyzes crosslinks in collagens and elastins; prolaminin-5; precursors of proteoglycans biglycan and osteoglycin; etc. BMP1-like peptidases activate certain TGFβ complexes, including BMP2, BMP4, GDF8, GDF11, and TGFβ1.

[65]Interleukin-20 belongs to the IL10 family with IL10, IL19, IL22, IL24, and IL26. Many cell types produce IL20, such as monocytes and epithelial and endothelial cells.

[66]The endothelial receptor protein Tyr kinase with Ig and EGF homology TIE2 is also termed Tyr-protein endothelial kinase TEK. Receptor TIE1 may build TIE1–TIE2 heterodimers to contribute to TIE2 signal transduction. Besides, Ang1–TIE2 signaling operates in the maintenance of hematopoietic stem cells.

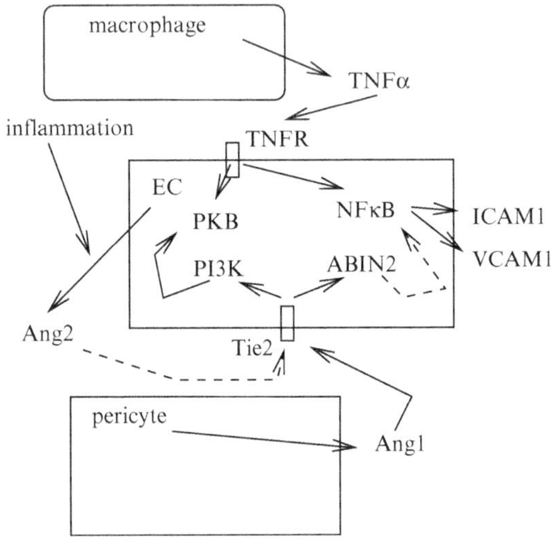

Fig. 10.3 Angiopoietins are involved in angiogenesis and inflammation. They have antagonist actions (Source: [1286, 1287]).

Angiopoietin-1

Angiopoietin-1 is expressed by vascular smooth myocytes and pericytes, as well as fibroblasts and several types of non-vascular cells, whereas angiopoietin-2 is primarily produced by endothelial cells.

Angiopoietin-1 regulates blood vessel maturation [1288]. During the maturation stage of blood vessel formation, endothelial cells that form the channel recruit supporting mural cells (pericytes and smooth myocytes) by releasing PDGFbb dimer. Mural cells subsequently adhere to endothelial cells to build stable blood vessels, as they stimulate TIE2 receptor. Paracrine angiopoietin-1 produced by mural cells stabilizes newly formed blood vessels and decreases vascular permeability. The Ang1–TIE2 pathway indeed favors the association of pericytes with endothelial cells to reduce vascular leakage. Angiopoietin-1 tightens blood vessels via adhesions between cells as well as between cells and the extracellular matrix.

Angiopoietin-1 induces formation of TIE2 clusters in quiescent endothelial cells at intercellular junctions to transduce survival signals [1289]. Receptor TIE2 activation primes several signaling pathways. The dominant pathway corresponds to the PI3K–PKB survival axis.

During angiogenesis, blood vessels change their caliber to adapt to the nutritional needs of tissues. The Ang1–TIE2 complex regulates the widening of blood vessels.

Angiopoietin-1 has anti-inflammatory and vascular permeability-inhibiting effects via sphingosine kinase-1 that catalyzes the formation of sphingosine 1-phosphate [1289]. Angiopoietin-1 collaborates with VEGF during blood and lymph vessel development.

Ligand-bound TIE2 leads to its phosphorylation and recruitment of adaptor proteins to activate different signaling pathways. Ang1–TIE2 complexes associate with protein Tyr^P phosphatase receptor PTPRb to signal via protein kinase-B to stimulate nitric oxide synthase NOS3 (Fig. 10.4) [1289].

The Ang1–TIE2 complex activates not only phosphoinositide 3-kinase, but also survivin, or baculoviral inhibitor of apoptosis repeat-containing protein BIRC5, whereas it inhibits caspase-9 and BCL2 antagonist of cell death (BAD), thus preventing apoptosis. Moreover, sequestration of Src kinase by RhoA effector Diaphanous stabilizes interendothelial junctions, as it prevents internalization mediated by Src kinase of vascular endothelial cadherin.

Angiopoietin-2

Synthesis of angiopoietin-2 is stimulated by hypoxia and VEGF, in particular, and inhibited by Krüppel-like factor-2 [1289]. Autocrine angiopoietin-2 disrupts mature vessels and initiates vessel regression. Angiopoietin-2 upregulated by hypoxia or VEGF impedes the interaction between Ang1 and TIE2 receptor.

In endothelial cells, Ang2 is stored in Weibel-Palade bodies with von Willebrand factor. Angiopoietin-2 storage level varies according to the vascular bed, thereby influencing the degree of responsiveness of a given vascular bed to inflammatory or angiogenic cytokines. Ang2 stores are much greater in cerebral microvessels than in vasculature of skeletal muscles and heart. Secretagogue thrombin or vasopressin can quickly liberate content of Weibel-Palade body to act on vessel permeability, coagulation, and inflammation.

Angiopoietin-2 can activate TIE2 on some cells (agonist), whereas it blocks TIE2 activation on others (antagonist). Ang2 acts as a TIE2 agonist in lymphatic vessels, but as an antagonist in blood vessels [1290]. Therefore, angiopoietin-2 promotes angiogenesis according to the tissue and context. In addition, Ang2 produced and stored by endothelial cells controls endothelial responsiveness to multiple factors, such as vascular endothelial growth factor, histamine, bradykinin, and inflammatory factors (e.g., tumor-necrosis factor-α).

TIE2 Receptor

Whereas VEGF and VEGFR2 intervene in the early stage of angiogenesis, angiopoietin-1 and -2, TIE2 receptor, and PTPRb phosphatase operate in the later stage of vessel remodeling. In particular, the Ang–TIE2 couple regulates tumor vessel size [1291]. Phosphatase PTPRb reduces signaling from the Ang–TIE2 pathway.

VEGFR2− embryos fail to develop any blood vessel structures. Nevertheless, murine embryonic stem (ES) cells, once implanted subcutaneously in mice, generate teratomas, differentiate into epithelial, muscle, as well as endothelial cells within teratomas that form occasional large-caliber blood vessels, even when they lack VEGFR2 receptor [1291]. Increased TIE2 phosphorylation in PTPRb− teratomas demonstrates that PTPRb represses TIE2 activity in tumor angiogenesis [1291].

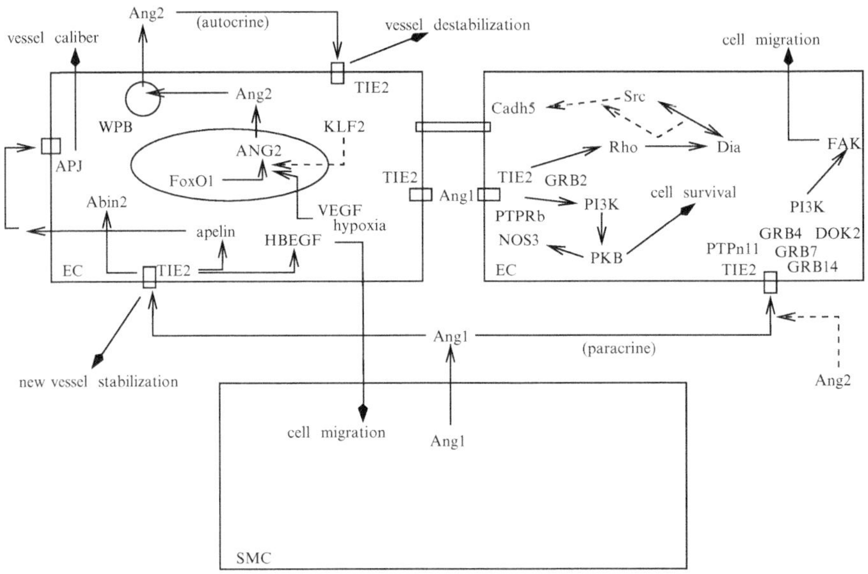

Fig. 10.4 Ang–TIE signaling pathways in vascular cells (EC; Source: [1289]). Isotype Ang1 produced by mural cells such as smooth myocytes (SMC) stimulates receptor TIE2. In endothelial cells, Ang2 synthesis is stimulated by forkhead box transcription factor FOxO1, VEGF, and hypoxia, but is impeded by Krüppel-like factor KLF2. Angiopoietin-2 is stored in Weibel-Palade bodies (WPB). Isoform Ang2 inhibits Ang1–TIE2 signaling to destabilize the resting endothelium, but activates TIE2 in already hastened endothelium. Receptor TIE2 activation indeed primes several signaling pathways. Activated TIE2 can recruit adaptors, such as growth factor receptor-bound proteins GRB2, GRB4, GRB7, and GRB14, Dok-related docking protein Dok2, as well as SH2 domain-containing Tyr phosphatase PTPn11 (SHP2) and mediator A20-binding inhibitor of NFκB Abin2. The dominant pathway corresponds to the PI3K–PKB axis that is responsible for survival signals (inhibition of caspase-9 and BAD and activation of survivin. Secreted angiopoietin-1 can bind to receptors TIE2 of adjacent cells, thus making trans-complexes into space between adjacent endothelial cells. This complex associates with vascular endothelial phosphotyrosine phosphatase (vePTP or PTPRb). Receptor TIE2 stimulates nitric oxide synthase NOS3, protein kinase-B (PKB), RhoA GTPase and its effector Diaphanous that maintains interendothelial stability by sequestering Src kinase, thus precluding Src-mediated VE-cadherin endocytosis. Phosphoinositide 3-kinase also phosphorylates focal adhesion kinase (FAK) for cell migration. Activated TIE2 upregulates endothelial heparin-binding EGF-like growth factor (HBEGF) and hepatocyte growth factor (HGF) that favor SMC migration and recruitment to ECs. Last but not least, PKB phosphorylates (inactivates) FOxO1 that targets the ANG2 gene (negative feedback loop).

TIE2 Receptor and Apelin

In endothelial cells, TIE2 activation generates apelin secretion [1292].[67] Apelin is a pleiotropic peptide that targets its cognate G-protein-coupled receptor

[67] Apelin localizes preferentially in blood vessels. An immature peptide, preproapelin, is secreted and cleaved by proteases to form apelin$_{13}$, apelin$_{17}$, and apelin$_{36}$, among others. These isoforms

APJ[68] on endothelial cells to regulate the lumen size of blood vessels. In vascular endothelial cells, the synthesis of apelin and APJ is regulated during angiogenesis. Under hypoxia, hypoxia inducible factor-1α binds to the hypoxia-responsive element of the apelin gene and increases apelin expression in vascular cells [1293]. Expression of the apelin gene is high in tip cells to modulate proliferation of APJ+ stalk cells.

The major signaling pathways of apelin are the Gq–PKC–ERK and Gi–PI3K–PKB axes in APJ+ endothelial cells. Both the PI3K–PKB–TOR pathway and MAPK module lead to the phosphorylation of S6K kinase [1293]. The PI3K–PKB pathway contributes to endothelial cell migration, whereas the ERK–S6K and PI3K–S6K axes control endothelial cell proliferation.

During embryogenesis, APJ is expressed on endothelial cells of new blood vessels sprouted from the dorsal aorta, but not on endothelial cells of the pre-existing dorsal aorta. Apelin enhances endothelial cell proliferation primed by vascular endothelial growth factor and promotes between-cell aggregation. In addition, apelin may stabilize intercellular junctions between of endothelial cells.

Angiopoietin-like Molecules

Seven angiopoietin-like molecules exist (AngL1–AngL7). These pleiotropic molecules are expressed by many cell types. Isoforms AngL2 to AngL4 and AngL6 are secreted and can then be detected in plasma. Angiopoietin-like molecules regulate lipid, glucid, and energy metabolism. Angiopoietin-like molecules bind neither to TIE1 nor TIE2 receptor. In endothelial cells, AngL1 to AngL4 and AngL6 contribute to the control of angiogenesis via survival and migration signals [1289].

10.6.6.2 Adrenomedullins

Adrenomedullin (Vol. 2 – Chap. 3. Growth Factors), as are VEGF, endothelin-1 and -2, and angiogenin, is stimulated by hypoxia. Angiogenic adrenomedullin tethers to its receptors calcitonin receptor-like receptor (CalcRL) connected to receptor activity-modifying proteins RAMP2 and RAMP3 (Vol. 3 – Chap. 7. G-Protein-Coupled Receptors) and activates protein kinase-B, mitogen-activated protein kinase, and focal adhesion kinases in endothelial cells, especially during acute ischemia [1294]. Moreover, adrenomedullin upregulates VEGF expression.

Intermedin, or adrenomedullin-2 (Adm2), is an angiogen. In cultured endothelial cells, this growth factor provokes cell migration and tube formation. Adrenomedullin-2 increases the phosphorylation of extracellular signal-regulated

---

have different receptor-binding affinities and activities. Apelin has a positive inotropic effect and can cause vasoconstriction and vasodilation (via NO) according to circumstances.

[68]Receptor APJ is widespread in embryos and adults, especially in the lung and heart, as well as in vascular and endocardial endothelial and smooth muscle cells.

kinase, protein kinase B, and nitric oxide synthase-3 [1295]. It heightens synthesis of vascular endothelial growth factor and VEGFR2 receptor.

10.6.6.3 Sphingosine 1-Phosphate

Sphingosine 1-phosphate released by platelets during blood clotting is a selective chemoattractant for endothelial cells. It favors angiogenesis during healing [1296]. It activates $S1P_1$ receptor and 3 subtypes of G-protein-coupled receptors (Vol. 3 – Chap. 7. G-Protein-Coupled Receptors) on endothelial cells to regulate angiogenesis.

Sphingosine 1-phosphate is also secreted by endothelial cells. Receptor PDGFRβ can cooperate with sphingosine 1-phosphate receptors ($S1P_1$–$S1P_5$). Receptors $S1P_1$ to $S1P_3$ serve to recruit mural cells (pericytes and vascular smooth myocytes). Receptor $S1P_1$ favors localization of N-cadherins in contact zones between endothelial and mural cells [1184].

10.6.6.4 Vasohibins

Whereas angiogenic factors, such as VEGF and FGF2, support vascular endothelial cell proliferation and organization into tubular networks, angiogenesis inhibitors restrain vessel formation. Angiogenesis inhibitors include angiostatin, endostatin, prolactin, serpin-F1, thrombospondins, and chemokines CXCL4 and CXCL10, as well as vasohibins. The latter functions in a negative feedback loop to repress VEGF- and FGF2-primed angiogenesis.

Vasohibin actually ensures an autocrine negative feedback, as it impedes their migration and proliferation [1297]. On the other hand, it does not influence the migration of smooth myocytes and fibroblasts.

Vasohibin is also produced by hematopoietic stem cells, but not hematopoietic progenitors or mature hematopoietic cells from adult bone marrow [1298]. Vasohibin production is stimulated by VEGF and FGF2 in a time- and concentration-dependent manner. The secreted protein antagonizes the angiogenic effects of VEGF. Vasohibin synthesis is attenuated by hypoxia, TNFα, IL1β, and Ifnγ [1299].

Vasohibin-1, but not most angiogenesis inhibitors, is synthesized by endothelial cells. An alternately spliced isoform lacks C-terminal half of the molecule ($VasH1^{\Delta CT}$) that suppresses FGF2-induced angiogenesis and mediates heparin binding [1299,1301]. Moreover, various vasohibin forms with distinct molecular weights result from proteolytic cleavage in the extracellular milieu. At least 2 cleaving sites exist in the N-terminus. The N-terminal truncated form ($VasH1^{\Delta NT}$) not only retains its inhibition on angiogenesis, but also has strong affinity to heparin [1301].

Another VEGF-inducible, endothelium-derived inhibitor of angiogenesis, vasohibin-2, or vasohibin-like protein [1300], is produced preferentially in

Table 10.15. Vasohibins (Source: [1302, 1303]). Vasohibin-1 is mainly synthesized in endothelial cells at the termination zone. Its production is indeed low in proliferating endothelial cells at the sprouting front, but is high in non-proliferating endothelial cells in the termination zone of angiogenesis. It halts angiogenesis in the sprouting zone. It also has an antilymphangiogenic activity. On the other hand, vasohibin-2 is mainly produced in agranulocytes mobilized from the bone marrow and localized to the sprouting front. It promotes angiogenesis.

Type	Effect
Vasohibin-1	Anti-angio- and -lymphangiogenesis
Vasohibin-2	Pro-angiogenesis

mononuclear leukocytes mobilized from bone marrow that infiltrate the sprouting front (Table 10.15). The full-length form of vasohibin-2 ($VasH2_{FL}$) and vasohibin have comparable inhibitory activity in in vitro angiogenesis. Different alternatively spliced variants include an isoform with a complete C-terminus and other subtypes with shorter C-termini.

Both vasohibin-1 and -2 are produced in endothelial cells during mid-gestation. Afterward, VasH1 and VasH2 persist in arterial endothelial cells from late gestation to neonate.

10.6.7 G-Protein-Coupled Receptors

The vasculature possesses particular features, especially architecture, in each perfused organ. The vasculature of the central nervous system is characterized by a blood–brain barrier, extensive pericyte coverage, and reciprocal interactions with neurons and glial cells. In addition, it can serve as neural stem cell niche. Brain angiogenesis occurs concomitantly with barrier creation. Many agents regulate angiogenesis in the central nervous system, such as the Wnt–β-catenin and VEGF–Nrp1 pathways, α_V- and β_8-integrins, and inhibitors of DNA binding ID1 (or bHLHb24) and ID3 (or bHLHb25) [1304].

G-protein-coupled receptor GPR124 of the adhesion GPCR family (Vol. 3 – Chap. 7. G-Protein-Coupled Receptors) is highly expressed in endothelium of the central nervous system as well as pericytes, during embryogenesis (and, to a lesser extent, in embryonic heart, liver, kidney, lung, and esophagus, as well as mesenchyme) and in adults with an exclusively vascular expression. It controls the development of the cerebral vasculature [1304]. In vivo, pro-angiogenic, endothelial GPR124 receptor regulates endothelial sprouting and CDC42-dependent cell migration as well as development of the blood–brain barrier GluT1 marker.

10.6.8 Regulators of Small GTPases

Angiogenesis is controlled by mechanical and chemical cues, i.e., physical interactions between cells and extracellular matrix as well as soluble agonist and antagonist regulators. A signaling pathway is sensitive to both extracellular matrix elasticity and VEGF that targets VEGFR2 receptor.

Protein RhoGAP35 that inhibits small GTPase Rho is indeed regulated by growth factors, matrix elasticity, and cytoskeletal distortion. It controls angiogenesis, as it modulates the balance between 2 mutually antagonistic transcription factors GTF2i and GATA2 that regulate expression of VEGFR2 [1305]. Mediator RhoGAP35 binds to GTF2i for cytosolic sequestration.

10.6.8.1 Regulators of G-Protein Signaling

Regulators of G-protein signaling shorten the duration of active $G\alpha_i$ and $G\alpha_q$ and act on cell migration, proliferation, and mitogen-activated protein kinase activities [1306]. Subtype RGS4 inhibits MAPKs and VEGFR2 expression [1307]. Other stimulators of G-protein-coupled receptors (thrombin, angiotensin-2, endothelin-1, prokineticin-1 and -2, etc.) are also implicated.

Tumoral angiogenesis leads to aberrant vasculature. Tumor vasculature is in fact characterized by dilated and fragile vessels, intensive vessel sprouting, and loss of hierarchical architecture. Protein RGS5 causes abnormal vessel morphology in tumors, as RGS5 loss leads to pericyte maturation and vascular structure normalization [1344]. Altered tumor vasculature prevents migration of immunocytes into tumor parenchyma. Vascular normalization that results from RGS5 loss promotes immune destruction.

10.6.9 Phosphoinositide 3-Kinase and Phospholipase-C

Angiogenesis is characterized by stabilization and destabilization stages, selecting the vessel able to develop for suitable perfusion. Angiogenesis can be regulated at the level of plasmalemmal lipids. Enzyme $PI3K_{c1\alpha}$ generates the lipid second messengers phosphatidylinositol (3,4)-bisphosphate and phosphatidylinositol (3,4,5)-trisphosphate that localize to the inner leaflet of the plasma membrane. Kinase $PI3K_{c1\alpha}$ regulates angiogenesis in the developing embryo and fetus as well as pathological situations.

Phosphatidylinositol (4,5)-bisphosphate coordinates the phases of angiogenesis. The angiogenic program is driven by phosphoinositide 3-kinase and antagonist phospholipase-$C\gamma$. The former promotes tube formation, the latter tube regression [1309]. Both substances compete for their common substrate, phosphatidylinositol (4,5)-bisphosphate. Enzyme $PLC\gamma$ controls tube formation by reducing the level of $PI(4,5)P_2$ that is targeted by protein kinase-C to produce diacylglycerol and inositol trisphosphate, and inhibits protein Tyr kinases.

Only activity of class-1A catalytic subunit $PI3K_{c1\alpha}$ that is preferentially induced by receptor protein Tyr kinase ligands such as VEGFa in endothelial cells is essential for vascular development [1310].[69]

About 50 PI3K effectors that contain pleckstrin homology (PH) domains specifically bind to $PI(3,4,5)P_3$. These effectors are regulated by $PI(3,4,5)P_3$ that causes translocation to the plasma membrane and/or a conformational change. Many $PI(3,4,5)P_3$-binding proteins are regulators of small guanosine triphosphatases of the RHO and ARF superfamilies.

Dual GTPase-activating protein ARAP3 is a $PI(3,4,5)P_3$- and Rap-activated stimulator for RhoA and ARF6 GTPases that is required for sprouting angiogenesis that depends on cell migration [1311]. Recruitment of ARAP3 to the plasma membrane as well as its ArfGAP activity depends on $PI(3,4,5)P_3$ produced by $PI3K_{c1\alpha}$. Both ARF6 and RhoA GTPases regulate cell motility.

10.6.10 Protein Kinase-B

Protein kinase-B regulates vascular permeability and angiogenesis by fine-tuning of signaling pathways in endothelial cells, including pro- and anti-angiogenic signalings [1312]. Kinase PKB has various roles at different stages of angiogenesis.

Coronary angiogenesis is enhanced during the acute phase of adaptive cardiac growth, implicating myocardial VEGF and angiopoietin-2 [1313].[70] On the other hand, neovascularization is reduced during maladaptive remodeling. Indeed, sustained PKB signaling causes microvascular malformations [1314]. Chronic PKB activation precludes PI3K activity (PI3K-dependent effectors are required for full cardioprotection) [1315]. Short-term PKB activity mediates hypoxia-induced expression of VEGF, whereas its prolonged activation decreases VEGF level. Protein kinase-B1 is required for migration and integrin-mediated adhesion of endothelial cells. Protein kinase-B is also necessary in pericyte recruitment for vessel maturation.

Protein kinase-B targets endothelial nitric oxide synthase. The NOS3 pathway is implicated in vascular maturation. Protein kinase-B also controls the expression of thrombospondins Tsp1 and Tsp2, inhibitors of angiogenesis, by endothelial cells.

---

[69] In endothelial cells, $PI3K_{c1\beta}$ operates downstream from liganded G-protein-coupled receptors such as CXCR4 α-chemokine receptor. In these cells, $PI3K_{c1\delta}$ is expressed at low concentration. Catalytic subunits of class-1A PI3K isoforms $PI3K_{c1\alpha}$ and $PI3K_{c1\beta}$ are ubiquitous, whereas $PI3K_{c1\delta}$ is only enriched in leukocytes.

[70] This transient signaling is associated with normal microvascular development, acute PKB activation of myristoylated PKB (PKB_M) being cardioprotective.

Reduced concentrations of thrombospondin Tsp1 and Tsp2 shift the PKB activity toward increased angiogenesis.[71]

The zinc-binding matrix metallopeptidase, a disintegrin and metallopeptidase with thrombospondin-1 motif ADAMTS1, modulates angiogenesis, cleaving matrix-bound thrombospondins TSp1 and Tsp2, and thereby releasing anti-angiogenic polypeptidic fragments [1317]. Enzyme ADAMTS1 mainly targets Tsp1 of the endothelial basement membrane.[72]

Protein kinase-B phosphorylates actin-binding protein girdin to promote VEGF-dependent migration of endothelial cells and tube formation [1318].[73] Both PKB1 and PKB2 isoforms phosphorylate girdin. However PKB2 concentration is very low in human umbilical vein endothelial cells. Girdin is required in cell–cell and cell–matrix adhesion, as well as in phosphorylation of other PKB substrates, such as NOS3 and GSK3α and -β, and in activation of small GTPases. Girdin is highly expressed in endothelial cells of capillaries in both normal and tumoral tissues. The PKB–girdin axis mediates VEGF-dependent cell migration, tube formation, and microvessel remodeling, as well as microvessel sprouting in postnatal angiogenesis. Girdin is localized on the actin stress fibers, especially at junctions between the actin filaments, but weakly found on cortical actin filaments. In large blood vessels, girdin is produced by smooth myocytes, but not endothelial cells. In hemangiomas, girdin is synthesized in both immature endothelial cells and mural cells. Phosphorylated girdin is located at the leading edge of migrating cells.

10.6.11 Cyclooxygenases and Prostaglandins

Cyclooxygenase-2 favors expression of angiogenic factors such as VEGF and FGF2. It can inhibit endothelial cell apoptosis, stimulating intracellular anti-apoptotic pathways.

Cyclooxygenase-2 has a pro-angiogenic activity owing to its metabolite prostaglandin-E2. A pathway crosstalk exists between prostaglandin-E2 and transforming growth factor-β in tumor angiogenesis [1319]. Prostaglandin-E2 activates the ALK5–SMAD3 pathway to induce neovascularization. This activated pathway releases active TGFβ from its latent form using clustered mt1MMP metallopeptidases.

[71]Thrombospondins are involved not only in endothelial cell proliferation and apoptosis, but also in collagen assembly and maintenance of vascular integrity. Thrombospondin-2 controls the level of matrix metallopeptidase MMP2 [1316].

[72]Substrates of ADAMTS1 also include aggrecan, versican, and nidogen. It is associated with catalytic modifiers such as fibulin-1. Moreover, ADAMTS1 has a non-catalytic function, as it sequesters VEGF factor.

[73]Girdin is also known as Akt (PKB) phosphorylation enhancer (APE), Gα-interacting vesicle-associated protein (GIV), and Hook-related protein-1 (HkRP1). Kinase PKB phosphorylates girdin to regulate its subcellular location and fibroblast migration.

10.6.12 Tissue Factor

Tissue factor, an initiator of coagulation,[74] mediates angiogenesis in interaction with coagulation factor VIIa. Interactions between tissue factor and activated clotting factor VIIa promote VEGF synthesis and secretion. Full-length tissue factor (TF_{FL}) complexed with factor VIIa influences angiogenesis via peptidase-activated receptor PAR_2 [1320]. This signaling axis promotes the synthesis not only of VEGF, but also matrix metallopeptidase MMP7 as well as CXCL1 chemokine and CXCR1 and CXCR2 receptors.

An alternatively spliced variant of tissue factor (TF_{AS}) can cause angiogenesis in vitro at low concentrations (1 nmol), independently of PAR_2 or clotting factor VIIa, whereas a soluble TF_{FL} form (TF^{S}_{FL}) has smaller promigratory effect [1320]. Proangiogenic action of asTF relies on integrin ligation. Migration of endothelial cell depends on interaction between TF_{AS} and $\alpha_V\beta_3$-integrin as well as P38MAPK and PI3K enzymes. Formation of capillaries depends on interaction between TF_{AS} and $\alpha_6\beta_1$-integrin as well as ERK1, ERK2, and PI3K.

10.6.13 Tissue Kallikrein and Kininogen

Tissue kallikrein that produces kinin from kininogen promotes angiogenesis and arteriogenesis. The signaling cascade downstream from tissue kallikrein that involves protein kinase-B and glycogen synthase kinase GSK3β in human coronary artery endothelium [1321] or endothelial nitric oxide synthase in human umbilical vein endothelium [1322] upregulates vascular endothelial growth factor and VEGFR2 receptor. In addition, GSK-3β protects cardiomyocyte against hypoxia-induced apoptosis. It is associated with the Wnt pathway in endothelium. Moreover, GSK-3β regulates energy metabolism of many tissues.

High-molecular-weight kininogen is a plasma protein that serves as a cofactor in the intrinsic coagulation cascade (Sect. 9.8) and inhibits cysteine peptidases. Proteolytic cleavage of high-molecular-weight kininogen by kallikrein in endothelial cells produces 2 active molecules: (1) bradykinin that mediates nitric oxide release and (2) cleaved 2-chain high-molecular-weight kininogen. The latter inhibits angiogenesis, whereas the former is an angiogenesis stimulator.

Ferritin that intervenes in iron storage binds to cleaved high-molecular-weight kininogen with high affinity to antagonize its anti-angiogenic effect [1323]. Therefore, plasma ferritin regulates vascular remodeling and angiogenesis.

In addition, cleaved high-molecular weight kininogen can stimulate inflammatory cytokine and chemokine secretion from human monocytes [1324]. It can upregulate tissue factor in human monocytes via its receptor $\alpha_M\beta_2$-integrin. It

[74]After blood vessel injury, tissue factor with factor VIIa activates factor Xa, thrombin, and fibrin to build a hemostatic plug.

triggers a paracrine pathway in monocytes that depends on both TNFα and IL1β, as well as ERK, JNK, and P38MAPK signaling pathways and NFκB transcription factor.

10.6.14 Gasotransmitters Hydrogen Sulfide and Nitric Oxide

Hydrogen sulfide belongs with nitric oxide and carbon monoxide to the gasotransmitter family. It is synthesized from ^{L}cysteine mainly by cystathionine γ-lyase (CSE), as well as cystathionine β-synthase (CBS) and 3-mercaptopyruvate sulfurtransferase in both endothelial and smooth muscle cells. It stimulates angiogenesis. It stimulates the proliferation of vascular endothelial cells, but impedes that of vascular smooth myocytes (cell type-specific effect).

Hydrogen sulfide and nitric oxide cause S-sulfhydration and S-nitrosylation, respectively. S-sulfhydration converts cysteine^{SH} group to hydropersulfide, thereby increasing protein activity [1325]. On the other hand, S-nitrosylation of cysteine can decrease protein activity.

However, during ischemia, NO modulates angiogenesis via S-nitrosylation of various proteins. In particular, VEGF stimulates NO liberation from endothelial cells, as it increases NOS3 expression and phosphorylation. Conversely, NO via the sGC–cGMP axis reinforce this signaling, as it heigthens VEGF and FGF2 expression. In addition, NO supports endothelial cell migration, as it stimulates the production of $\alpha_V\beta_3$-integrin, and raises FGF2-mediated degradation of extracellular matrix by MMP13 and MMP14 peptidases.

In vascular smooth myocytes, H_2S-mediated relaxation results from activation of ATP-sensitive potassium channel (K_{ATP}) [1326] (Table 10.16). Hydrogen sulfide signals via mitogen-activated protein kinase modules to activate ERK1 and ERK2, JNK1 and JNK2, and P38MAPK.[75]

In vitro, hydrogen sulfide (H_2S) enhances angiogenic potential of endothelial cells, as it improves cell proliferation, migration, and tube formation [1327]. Hydrogen sulfide operates via K_{ATP} channels and increases phosphorylation of PKB, ERK, and P38MAPK kinases in endothelial cells. Stimulation of endothelial cells by vascular endothelial growth factor heightens H_2S release. Therefore, endothelial H_2S production is involved in the pro-angiogenic VEGF effect. Hydrogen sulfide promotes the proliferation of endothelial cells, but primes apoptosis in smooth myocytes, as does nitric oxide. It also causes the phosphorylation of the regulator HSP27 of cell migration by mitogen-activated protein kinase-activated protein kinase (MAPKAPK), a P38MAPK substrate.

Hydrogen sulfide and nitric oxide are mutually required in vascular endothelial cells to trigger angiogenesis (hence proliferation and migration of endothelial cells)

---

[75] Hydrogen sulfide activates ERK1 and ERK2 in monocytes, smooth myocytes, and gastric epithelial cells [1327]. It stimulates P38MAPK in smooth muscle and β cells, but precludes P38MAPK phosphorylation in neutrophils and microglial cells.

Table 10.16. Individual and combined effects of gasotransmitters in vascular smooth myocytes (Source: [1325]; $\oplus\longrightarrow$: stimulation; $\ominus\longrightarrow$: inhibition; cGMP: cyclic guanosine monophosphate; BK_{Ca}, IK_{Ca}, SK_{Ca}: large- [$K_{Ca}1.1$], intermediate- [$K_{Ca}3.1$], and small-conductance [$K_{Ca}2$] calcium-sensitive K^+ channel; K_{ATP}: ATP-dependent K^+ channel [$K_{IR}6.2$]; sGC: soluble guanylate cyclase). Gasotransmitters (NO, CO, and H_2S) may antagonize or potentiate their production, molecular targets, and mutual interaction. Nitric oxide can play a more important vasorelaxant role in large arteries such as aorta; hydrogen sulfide in peripheral resistive arteries such as mesenteric and coronary arteries, acting as an endothelium-derived hyperpolarizing factor. Both NO and CO have a synergistic effect on BK_{Ca} channels in vSMCs, as they increase its opening probability.

Gasotransmitter	Effect
CO	sGC–cGMP–PKG PKG$\ominus\longrightarrow Ca^{2+}$ influx from endoplasmic reticulum PKG$\ominus\longrightarrow Ca_V1.2$ PKG$\oplus\longrightarrow BK_{Ca}$ $\oplus\longrightarrow BK_{Ca}$ (BK_{Ca} α subunit) Hyperpolarization
H_2S	$\ominus\longrightarrow$ PDE5 $\oplus\longrightarrow IK_{Ca}$, SK_{Ca} $\oplus\longrightarrow K_{ATP}$ Hyperpolarization
NO	sGC–cGMP–PKG $\oplus\longrightarrow BK_{Ca}$ (BK_{Ca} β subunit) Hyperpolarization

as well as vasodilatation [1328].[76] Nitric oxide binds to soluble guanylate cyclase that generates cGMP second messenger. The latter activates protein kinase-G; it is degraded by PDE5a phosphodiesterase. Protein kinase-G1 is needed for endothelial cell proliferation and migration. Hydrogen sulfide precludes PDE5a activity. Exposure of endothelial cells to H_2S increases intracellular cGMP level in a NO-dependent manner. Hence, these 2 gasotransmitters cooperate to elevate and maintain intracellular cGMP concentration. Hydrogen sulfide supports the phosphorylation of the PKG substrate vasodilator-stimulated phosphoprotein (VASP). In addition, H_2S increases NOS3 phosphorylation at Ser1177 (activation) and decreases that at Thr495 (inhibition). Moreover, H_2S activates the PI3K–PKB axis.

[76]Inhibition of endothelial NO synthase (NOS3) or PKG1 abolishes H_2S-triggered angiogenesis and attenuates H_2S-primed vasorelaxation. Conversely, inactivation of H_2S synthase cystathionine γ-lyase supresses NO-stimulated cGMP accumulation and angiogenesis and reduces acetylcholine-stimulated vasorelaxation [1328].

10.6.15 Oxidative Stress and Inflammation

Inflammation, oxidative stress, and angiogenesis are involved in numerous physiological and pathophysiological processes from healing and remodeling to tumorigenesis. Angiogenesis is not only associated with hypoxia, but also with inflammation-induced oxidation.

Angiogenesis and inflammation are related. Inflammatory cells release pro-angiogenic growth factors such as VEGF. In turn, newly formed blood vessels enhance inflammatory cell recruitment. Leukocytes, in particular myeloid cells, are guided by and contribute to oxidative stress and generation of oxidative products, such as hydroxy-ω-oxoalkenoic acids and their esters [1329]. Associated with oxidized phospholipids, these molecules are sensed by scavenger receptors ScaRb3 and contribute to platelet hyper-reactivity. Hydrolysis followed by reaction of the resulting unesterified hydroxy-ω-oxoalkenoic acids with proteins, or conversely reaction with proteins followed by hydrolysis, creates a family of carboxyalkylpyrrole protein adducts (CAP).[77] End products of lipid oxidation ω-2-carboxyethylpyrrole (CEP) and other related pyrroles are generated during inflammation and healing.

Carboxyalkylpyrroles are detected by Toll-like receptor TLR2, but neither by TLR4 nor by scavenger receptors on endothelial cells. They prime angiogenesis, independently of vascular endothelial growth factor [1329]. Toll-like receptor-2 and its adaptor MyD88 are required for CEP-induced stimulation of Rac1 GTPase and endothelial migration.

Reactive oxygen species are implicated in stress responses. Reactive oxygen species in stressed cells can be produced by NADPH oxidases, uncoupled nitric oxide synthases, and xanthine oxidases. NADPH oxidase NOx4 facilitates cardiac adaptation to chronic stress [1330]. Unlike other NOx proteins, NOx4 activity is regulated mainly by its expression level, which rises in cardiomyocytes subjected to mechanical and chemical stresses, such as pressure overload or hypoxia. Enzyme NOx4 enhances stress-induced activation of hypoxia-inducible factor-1 in cardiomyocytes as well as release of vascular endothelial growth factor for paracrine angiogenic signaling.

---

[77] These adducts localize to oxidized low-density lipoprotein and accumulate in atherosclerotic plaques as well as retina, where they promote choroidal neovascularization and age-related macular degeneration. These adducts, carboxyethylpyrrole (CEP) in particular, are transiently observed during wound healing. At injury sites, they reach a maximum concentration 3 days after damage and return to original levels when the healing is completed. High levels of CEP correlate with intense vascularization of damaged tissue.

10.6.16 Mechanical Signals

In addition to sprouting, intussusception and possibly vasculogenesis from nearby differentiating progenitor cells, vascularization during wound healing mainly results from translocation of intact vessels by mechanical forces mediated by myofibroblasts, i.e., non-angiogenic expansion of pre-existing vessels [1331]. Myofibroblasts and activated fibroblasts participate in the rapid revascularization of the wound by contraction of the granulation tissue that pulls nearby intact blood vessels. Resulting vascular loops can then expand (vessel enlargement and elongation). This process that relies on a mechanical stress field is initially independent of endothelial sprouting and proliferation. Secondarily, the neovascular network can be remodeled by splitting, sprouting, and regression of vessels.

In addition, tissue deformation modulates angiogenic signals. Mechanical signals regulate angiogenesis via both endothelial and smooth muscle cells. Cyclic tensile strain [1332]: (1) upregulates the secretion of angiopoietin-2 and PDGFbb dimer and (2) enhances endothelial migration and sprout formation. Applied strain permits a directed migration of smooth myocyte toward migrating endothelial cells. A mechanical stimulus such as a cyclic tensile strain can trigger a cascade of auto- and paracrine signaling between endothelial and smooth muscle cells.[78]

Hemodynamic forces cooperate with chemical messengers and coregulate angiogenesis. According to the nature of its local gradient (negative or positive), VEGF causes either vessel dilation or sprouting [1333]. Endoluminal flow attenuates VEGF-induced endothelial sprouting via NO, whereas interstitial flow encourages endothelial sprouting, irrespective of flow direction or VEGF gradient [1333]. Transverse flow can act on mechanosensors at intercellular adhesion sites. Intraluminal flow operates via wall shear stress and induced intramural tension, i.e., on both endothelial and smooth muscle cells. Endothelial tip cells preferentially protrude according to the direction of interstitial flow, toward vessels with higher internal pressure, in the direction of an increasing VEGF gradient.

Regions of high deformations are associated with a long-range VEGFa gradient generated by interstitial cells, a local overexpression of VEGFR2 receptor, and local differences in endothelial cell proliferation [1334]. In addition to the upregulation of VEGFR2 expression, cytoskeletal tension regulates the cell proliferation, via activity of CDK inhibitors such as CKI1b and activators such as cyclin-D.

In cultured tissues with engineered vascular networks, effects of matrix deformation on neovascular network formation and remodeling can depend on loading application time, as early and delayed mechanical loadings with respect to culture

---

[78]Factor VEGF during the early stage of angiogenesis activates cell migration and sprouting. Angiopoietin-1 mediates the interactions between endothelial and smooth muscle cells. Angiopoietin-2 disrupts these interactions to promote independent migration. Both Ang1 and Ang2 are produced by vascular cells, bind to receptor TIE2, and act synergistically with VEGF to regulate angiogenesis. Homodimer PDGFbb released by endothelial cells at a late stage of angiogenesis recruits smooth muscle cells to stabilize the nascent sprouts.

onset influence differently neovascular growth. Early mechanical loading impedes angiogenesis, whereas delayed loading enhances tissue formation and stimulates vascular remodeling [1335].

10.7 Angiogenesis-Targeted Therapies

Angio- and lymphangiogenesis can be either promoted or reduced according to the disease type (arterial lesions or cancers). Blood and lymph vessel formations are regulated by interacting promoters and inhibitors (Table 10.17). Imbalance between these factors can lead to disease.

10.7.1 Angiogenesis Promotion

Therapeutic angiogenesis by administration of growth factors (VEGF, PDGF, FGF, and angiopoeitin) or gene therapy is aimed at treating artery diseases, especially

Table 10.17. Stimulators and inhibitors of angiogenesis (Source: [1336]).

Stimulators	Inhibitors
Angiopoietin-1	Angiopoietin-2
Angiogenin	Thrombospondins (Tsp1–Tsp2)
Adrenomedullin	Cortisone
Vascular endothelial growth factor	Angiostatin
Platelet-derived growth factor	PTen
Fibroblast growth factor	
Epidermal growth factor	
Erythropoietin	
Insulin-like growth factor IGF1	
Hepatocyte growth factor	
Transforming growth factor TGFα, TGFβ	
Tumor-necrosis factor TNFα	TNFα
Hypoxia-inducible factor HIF1, HIF2	Interferons Ifnα, Ifnβ
Interleukins IL1, IL4, IL8	IL10
Matrix metallopeptidases	Tissue inhibitors of matrix Metallopeptidases TIMP1, TIMP2
Tissue factor	Endostatin
Urokinase plasminogen activator	
Platelet-activating factor	
Nitric oxide	
Protein kinase-B	Protein kinase-B
Cyclooxygenase-2	
Prostaglandins PGE1, PGE2, PGI2	

coronary obstructions, by endogenous bypasses. Endothelial precursor cells can be seen in adult blood. However, they are less efficient than umbilical cord blood-derived cells. The transplantation of cord blood-derived endothelial progenitor cells enhances neovascularization [1337]. Transplantation of L-selectin+, VEGFR2+ cells increases capillary density in ischemic tissues, via VEGFa release [1338].

During ischemia, endothelial PKA–NOS3 pathway promotes neovascularization. The PKA–NOS3 pathway indeed regulates endothelial cell survival, proliferation, and migration, tubulogenesis, relaxation and proliferation of vascular smooth myocytes, and recruitment and differentiation of endothelial progenitor cells, or circulating pro-angiogenic cells, to promote vasculogenesis.

Hypoxia and concurrent acidic pH foster the production of angiogenic factors, such as VEGF and placental growth factors, chemokines (e.g., CCL2 and CXCL12), and endothelial cell adhesion molecules (e.g., PECAM1). In addition, hypoxia provokes recruitment and homing to ischemia sites of endothelial progenitor cells [1339].

Protein kinase-A and -B are major regulators of NOS3 activation. The VEGF–VEGFR2 complex stimulates PKA and PKB, hence NOS3 enzyme. Produced nitric oxide primes the sGC–cGMP–PKG axis and subsequently the Ras–Raf–ERK1/2 cascades [1339]. Moreover, the PKA–NOS3 pathway interacts with the PI3K–PKB–NOS3 (NOS3 activation), AMPK–NOS3 (NOS3 activity enhancement), and PLC–PKC–NOS3 (NOS3 inactivation) axes.

In fact, numerous stimuli that influence NOS3 activation, hence NO synthesis, by activated PKA, are activators of neovascularization. They comprise vasoactive substances, such as bradykinin, prostaglandin-E2, and prostacyclin, in addition to growth factors such as VEGF [1339].

Moreover, nitric oxide mediates angiogenesis and arteriogenesis primed by neuropeptide-Y during ischemia via activation of NOS3 and upregulation of VEGF and PDGF expression [1339].

Nitrite (NO_2^-) and nitrate (NO_3^-) anions are not only metabolites resulting from NO oxidation, but also represent an alternative NO source to synthesis that maintains NO availability. Dietary or endogenous nitrate are converted into nitrite in the digestive tract; nitrate is reduced to nitrite by oral commensal bacteria (Vol. 4 – Chap. 10. Other Major Signaling Mediators). Nitrite can be reduced to NO by deoxyhemoglobin, deoxymyoglobin, xanthine oxidoreductase, ascorbate, polyphenols, and protons, under PKA stimulation [1339]. In addition, adenosine-dependent NOS3 can be activation by the ACase–PKA axis.

Both NADPH oxidase and uncoupled NOS3 are plasma membrane-bound ROS synthases.[79] Reactive oxygen species at low levels can regulate vascular growth and remodeling. Hypoxia-stimulated NADPH oxidase promotes VEGF production via JunB, activation of PKB, ERK1, and ERK2, and subsequent angiogenesis [1339].

[79]NADPH oxidase also localizes to the endoplasmic reticulum. Mitochondria are another major ROS sources.

10.7.2 Angiogenesis Inhibition

Most body tissues are irrigated by blood vessels, except cartilage, tendons, ligaments, retina, and cardiac valves. The *avascularity* of these tissues is needed for an appropriate functioning.

Excessive angiogenesis occurs in various diseases, such as cancer, macular degeneration, diabetic retinopathy, rheumatoid arthritis, atherosclerosis, diabetic nephropathy, pathologic obesity, asthma, cystic fibrosis, inflammatory bowel disease, psoriasis, endometriosis, vasculitis, etc. Aberrant angiogenesis in the cartilage contributes to arthritis. Angiogenesis could be involved in degenerative tendon and ligament diseases. Angiogenesis in the retina is associated with vision loss. The cornea expresses soluble VEGFR1 ($VEGFR1^S$), which binds to and sequesters corneal VEGFa, inhibiting its activity even during hypoxia [1340].[80] In atherosclerosis, rheumatic valvular heart disease or infective endocarditis, cardiac valves express angiogenic factors leading to neovascularization. Thickening of the adult cardiac valves, which leads to congestive heart failure, such as in aortic stenosis, occurs by dysregulation of angiogenesis inhibitors, such as endostatin and chondromodulin-1 [499]. Anti-angiogenesis therapy is used against cancer.

10.7.2.1 Peptidic Inhibitors of Endothelial Cell Proliferation and Migration

A large number of peptides inhibit proliferation and migration of endothelial cells. They include substances derived from members of families of collagen-4, thrombospondins,[81] thrombospondin motif-containing proteins, such as ADAM9 and ADAM12, ADAMTS1 to ADAMTS20, semaphorin-5A and -5B, and Wnt1-inducible signaling pathway protein WISP1 to WISP3, CXC chemokines, somatotropins, serpins, kringle-containing proteins, such as thrombin, coagulation factor-XII, tissue thromboplastin activator, and kremen-1 and -2, and peptides derived from pro-angiogenic proteins [1341].

Whereas vascular endothelial growth factor receptor-2 in endothelial cells activates phospholipase-Cγ1 during angiogenesis, casitas-B-lineage lymphoma ubiquitin ligase (CBL) forms a complex with PLCγ1 and VEGFR2 and precludes angiogenesis [1342].

---

[80] Soluble $VEGFR1^S$ also binds VEGFb and placenta growth factor. Agent $VEGFR1^S$ forms heterodimers with VEGFR2 receptor. Avascularity may be maintained by multiple redundant mechanisms associated with many anti-angiogenic molecules that exist in the cornea, but neutralization of $VEGFR1^S$ alone abolishes corneal avascularity. $VEGFR1^S$ also regulates the availability of VEGFa in cyclic vascularization and embryonic sprouting.

[81] Thrombospondins constitute 2 subsets: subset A with Tsp1 and Tsp2 homotrimers and subset B with Tsp3 to Tsp5 homopentamers.

Limitation in angiogenesis is provided by protein fragments, such as *angiostatin* (fragment of plasminogen), released by tumor-infiltrating macrophages, and endostatin, which specifically inhibits EC proliferation. The angiogenesis inhibitor angiostatin, which acts via its receptor angiomotin, is specific for endothelial cells.

Angioinhibins and other factors prevent angiogenesis, either by inhibiting EC proliferation (CXCL4, Ifnα and Ifnγ, and thrombospondin-1) [1343], hampering EC migration (serpin-F1 and IL4), or EC proliferation induced by other mediators (FGF2 and ILs). Uncontrolled growth is prevented by growth factors, intercellular and cell–matrix contacts, and mechanical forces.

10.7.2.2 Tumoral Angiogenesis

Inhibition of angiogenesis is useful to treat cancers. Angiogenesis is a regulated process necessary for tumor growth beyond 1 to 2 mm^3 (tissue oxygen diffusion limit is 100–200 μm, i.e., from 3 to 5 cell layers around a blood vessel), locoregional invasion, and metastasis.

Tumor endothelial cells divide up to about 40 times more frequently than normal cells. They overexpress $\alpha_V\beta_3$-integrin, E-selectin, endoglin, endosialin, and VEGF receptors.

Tumor vasculature is less organized and more fragile than that of normal tissues. Altered tumoral blood vessels are mainly due to abnormal pericytes and endothelial cells. Vascular wall abnormalities are caused by regulator of G-protein signaling RGS5 [1344].[82]

Inhibition of RGS5 induces a more mature tumor vasculature and an elevation in the number of CD4+, CD8+ T lymphocytes in the tumor parenchyme that leads to a strong antitumor immune response. Loss in RGS5 function thereby increases the effectiveness of T-cell immunotherapy in cancer.

The turnover of tumor endothelial cells is much greater than in normal endothelium. Activated endothelial cells express specific markers, like certain adhesion molecules and growth factors. Drugs can target activated endothelial cells. Plasmalemmal coreceptor CD276 is specifically overexpressed in endothelial cells of tumors. Costimulatory molecule CD276 can be targeted by anti-angiogenic therapies [1345].

Matrix metallopeptidase MMP9 produced by cancer cells and tumor-infiltrating neutrophils favor tumor angiogenesis. Angiogenesis induced by neutrophil zymogen proMMP9 released from neutrophil granules required activation of tissue inhibitor of metallopeptidases-free zymogen [1346]. The proMMP9–TIMP1 complex cannot cause angiogenesis.

MicroRNA-378 promotes tumor growth and angiogenesis. It reduces expression of 2 tumor suppressors, Suppressor of Fused and Fus1 [1347]. Cancer treatment can thus target oncogenic microRNAs.

[82] Regulator of G-protein signaling RGS5 is a marker for pericyte progenitor cells.

Numerous anti-angiogenic compounds exist, such as inhibitors of: (1) matrix remodeling, such as MMP inhibitors[83] (TIMP1–TIMP4) [1348]; (2) adhesion molecules, in particular antagonist of $\alpha_V\beta_3$ [1349]; (3) activated endothelial cells; (4) angiogenic mediators or receptors; and (5) signaling in endothelial cells. Therapies aimed at struggling against tumor angiogenesis must target a set of growth factors and mediators to be successful. Depletion in VEGF, in IL8, the expression of which is regulated by nuclear factor-κB, and in hypoxia-inducible factor-1 reduces tumor angiogenesis [1350].

Angiogenesis inhibitors can be used as single agents or in combination with cytotoxic drugs. Treatments that target multiple angiogenic mechanisms can increase the effectiveness of anti-angiogenic therapy. Direct angiogenesis inhibitors hinder the ability of endothelial cells to form new vessels. Indirect angiogenesis inhibitors block either the production of angiogenic factors or signaling pathways. Mixed angiogenesis inhibitors include kinase inhibitors, epidermal growth factor receptor inhibitors, inhibitors of cyclooxygenase, inhibitors of Ang2, etc.

Certain microtubule-disrupting agents used in cancer therapy, such as combretastatin A4 phosphate[84] (CA4P), have additional antitumor effects, inducing tumor vasculature regression, acting on the Cdh5–Ctnnβ–PKB pathway, thus hampering between-cell junctions, cell migration, and anchorage [1351]. CA4P not only blocks the formation of a capillary network, but also induces the regression of unstable tumor neovessels.

Angiogenic factors released by tumor cells promote activation, proliferation, and migration of endothelial cells to the tumor tissue for neovessel formation. Tumor cells produce angiogenic factors, such as VEGFa, VEGFc, FGF2, PlGF, angiopoietins, and interleukin-8 (CXCL8).

Tumors recruit not only neighboring vascular endothelial cells, but also VEGFR2+, bone marrow-derived, circulating endothelial precursors as well as VEGFR1+, pro-angiogenic, hematopoietic cells.

Tumor cells secrete stromal cell-recruitment factors, such as PDGFa, PDGFc, or transforming growth factor-β. Tumor-associated endothelial cells synthesize PDGFb, which promotes recruitment of pericytes after activation of PDGFRβ. Angiogenic factors such as CXCL12 chemokine, which can recruit bone marrow-derived angiogenic cells, are also released by stromal cells (fibroblasts and inflammatory and immune cells).

Tumor-associated fibroblasts produce hepatocyte growth factor and heregulin. Because PDGFa is involved in the recruitment of angiogenic stromal cells that produce angiogenic factors, PDGF and VEGF inhibitors can be combined for tumor treatment.

---

[83] Activation of matrix metallopeptidase-9 leads to the release of soluble SCF, which promotes the proliferation and motility of circulating endothelial precursors.

[84] Molecule CA4P binds to tubulin. It selectively targets endothelial cells, but not smooth myocytes.

Galectin-1 synthesized in endothelial cells of various tumors for vessel guidance and growth is a receptor for anginex, an angiogenesis inhibitor [1352]. Galectins localize to the nucleus, cytoplasm, or extracellular matrix according to the cell type.

Promyelocytic leukemia (PML) tumor suppressor[85] inhibits angiogenesis in ischemic and neoplastic tissues.[86] Agent PML prevents the synthesis of hypoxia-inducible factor-1α, interacting with target of rapamycin (TOR), thereby hampering TOR association with cytoplasmic RHEB GTPase [1353].

Vascular endothelial growth factor inhibitors destroy most of the tumor vasculature, but the remaining structural components of the local vasculature, the basement membrane and pericytes,[87] allow the rapid growth of new vessels once VEGF inhibitor administration is stopped [1354].

Growth factor signaling (e.g., EGF, IGF1, FGF2, PDGF, and VEGF) increases reactive oxygen species, such as hydrogen peroxide (H_2O_2), which oxidizes (inactivates) protein Tyr phosphatases, hence enhancing phosphorylation of ERK1/2 in endothelial cells. Superoxide dismutase SOD1 is an abundant copper/zinc enzyme in the cytoplasm (cytosol, nucleus, and intermembrane space of mitochondria)[88] that converts superoxide into hydrogen peroxide and molecular oxygen. Enzyme SOD1 thereby favors growth factor-mediated ERK1/2 phosphorylation. It can thus be targeted by anti-angiogenic and antitumor agent to repress signaling pathways in endothelial and tumor cells that promote growth [1355].

Gene silencing through RNA interference can serve as a treatment (specific action). In addition, therapies can be based on a non-specific process mediated by activation of cell surface receptors. In the case of choroidal neovascularization in age-related macular degeneration or cancers, specific small interfering RNAs that target VEGFa or VEGFR1 may be used. Furthermore, non-specific siRNAs can act without entering into cells via Toll-like receptor TLR3 [1356]. Any double-stranded siRNA of at least 21 nucleotides that binds to TLR3 impede angiogenesis in the mouse model of choroidal neovascularization. Efficient siRNAs triggers the TRIF–NFκB cascade to activate genes encoding interleukin-12 and interferon-γ that inhibit angiogenesis.

On the other hand, reducing the blood supply to cancer cells causes hypoxia and may favor metastasization. Repression of tumor cell oncogenes (Ras, PI3K, PKB, and HER2) diminishes production by tumor cells of several pro-angiogenic factors, thereby stabilizing vessel architecture and improving perfusion, oxygenation, and drug delivery [1357]. In tumors characterized by a BRAF oncogenic mutation

---

[85]A.k.a. RING finger protein RNF71 and tripartite motif-containing protein TRIM19.

[86]Promyelocytic leukemia, which regulates tumor suppressor transcription factors, is lost in various types of human cancers.

[87]Many pericytes survive after VEGF inhibitor administration. However, the pericyte phenotype reversibly changes during treatment, with a downregulation of the expression of α-smooth muscle actin. This change reverses when the treatment is stopped.

[88]Manganese-containing enzyme SOD2 (^{Mn}SOD) resides in the mitochondrial matrix. Secreted, copper-containing SOD3 (^{Cu}SOD) lodges in the extracellular matrix.

(valine-to-glutamate transition [V600E]), suppression of bRaf signaling switches off the ERK pathway and inhibits the synthesis of pro-angiogens needed for the growth, maintenance, and spread of solid tumors [1358].

10.8 Lymphangiogenesis

The lymphatic vascular system participates in the maintenance of tissue fluid balance, traveling of immunocytes, and resorption of dietary lipids in the gut (i.e., penetration in the blood circulation and transfer after absorption and crossing of the intestinal barrier). In body's tissues, Lymphatic vessels return interstitial fluid and proteins to the blood stream.

Lymph vessels form an unidirectional vasculature of close-ended terminal lymphatics that drain through a network of collecting lymphatics, lymph nodes, lymphatic trunks and ducts into the venous circulation. The mature lymphatic system consists of the lymphatic vasculature and lymphoid organs (lymph nodes, Peyer's patches, tonsils, spleen and thymus). The lymphatic vasculature resides in most of the body domains, except central nervous system, cartilage, epidermis, cornea, and retina.

Close-ended lymphatic capillaries do not have basement membranes and are not covered by smooth myocytes, but are tethered by anchoring filaments to collagen fibers of the extracellular matrix. As the surrounding interstitial pressure changes, these anchoring filaments extend or colapse, causing lymphatic capillaries dilation or constriction to propel lymph, respectively.

Lymphatic capillaries merge in precollecting lymphatics, themselves in larger secondary collecting and collecting lymphatics. Large collecting lymphatics are covered by smooth myocytes that contract to assist lymph flow. Collecting lymphatics contain bileaflet valves. Lymph is drained through the thoracic duct and right lymphatic duct into the blood circulation via left and right subclavian veins, respectively.

Lymphatic vessel walls are coated by lymphatic endothelial cells that are interconnected by tight and adherens junctions. Lymphatic endothelial cells differentiate under a proper transcriptional control from veins. Lymphatic vessels form an independent vascular tree with only few connections to the venous circulation.

Therefore, lymphatic vessels form when the blood vasculature is already functional. According to the current theory proposed by F. Sabin in 1902, the first lymphatic endothelial cells sprout from embryonic venous endothelial cells and migrate to form the primary lymphatic sacs [1359]. Circulating endothelial progenitor cells may contribute to lymphangiogenesis.

The primary lymph sacs and the primary lymphatic plexus appear in 6- to 7-week-old human embryos. Peripheral lymphatic vessels then form by sprouting from lymph sacs. Large collecting lymphatic vessels mature by deposition of a basement membrane, accumulation of mural cells, and formation of valves. Therefore, lymphangiogenic tissues arise by sprouting, branching, and remodeling.

Lymphatic vessel endothelial hyaluronan receptor LyVE1 in endothelial cells that coat anterior cardinal veins is the first marker of differentiation of blood endothelial cells into lymphatic endothelial cells [1360]. The population of LyVE1+ endothelial cells constitutes the first committed lymphatic endothelial cells that migrate from the cardinal vein to form the primary lymph sacs.

10.8.1 Transcription Factors in Lymphangiogenesis

Transcription factor prospero-related homeobox gene product Prox1 is required during initial lymphangiogenesis, i.e., endothelial precursor budding from the cardinal vein. Factor Prox1 regulates lymphatic endothelial cell differentiation [1361]. It indeed causes expression of specific genes of lymphatic endothelial cells and downregulates those of endothelial cells of blood vessels. It not only targets homeobox-containing genes, but also interacts with nuclear receptors such as NR5a2.[89]

Developmental transcription factor sex determining region-Y (SRY)-related high mobility group (HMG) box Sox18 intervenes in lymphatic development [1362]. Factor Sox18 is expressed in a subset of cardinal vein cells that later coexpress Prox1, differentiate into lymphatic endothelial cells, and migrate to form lymphatic vessels. Agent Sox18 activates Prox1 transcription by binding to its proximal promoter. During later stages of embryonic lymphangiogenesis, Sox18 expression decreases [1363]. In fact, Sox18 is not involved in Prox1 production in mature lymphatic endothelial cells. Homeobox gene product HoxD8 may then maintain Prox1 synthesis.

Forkhead box factor FoxC2 is involved in the differentiation of terminal and collecting lymphatics, as well as in the formation of valves. Transcription factor FoxC2 is expressed in lymphatic primordia, jugular lymph sacs, lymphatic collectors, and capillaries [1364]. Mutations in the FOXC2 gene cause the hereditary lymphedema-distichiasis syndrome. Factor FoxC2 is implicated in the differentiation between lymphatic capillaries and collecting, valvular lymphatic vessels. It is required not only for the formation of lymphatic valves, but also the establishment of pericyte-free lymphatic capillaries in cooperation with VEGFR3 [1365]. In FoxC2-deficient fetuses, lymphatic capillaries are abnormally covered with smooth myocytes and collecting lymphatic vessels lack valves. In addition, PDGFb and endoglin are upregulated and deposition of basement membrane collagen-4 rises.

Nuclear receptor NR2f2 is expressed in the venous and lymphatic endothelial cells [1363]. It directly activates Prox1 expression in venous lymphatic endothelial progenitors.

Both Sox18 and NR2f2 synergistically induce the Prox1 expression. Factor Prox1 downregulates venous blood vessel endothelial cell markers and upregulates

[89] A.k.a. liver receptor homolog-1.

specifiers of the lymphatic endothelial cell phenotype. Factor Prox1 collaborates with ETS2[90] to stimulate VEGFR3 production. On the other hand, NR2f2 and Prox1 cooperate for FGFR3 synthesis; NR2f2 binds to Prox1 factor [1363].

10.8.2 Growth Factors and Guidance Molecules

Sprouting, growth, survival, migration, and proliferation of lymphatic endothelial cells need vascular endothelial growth factor-C and -D that activate VEGFR3 [1361][91] and neuropilin-2.[92] Neuropilin-2 expressed in lymphatic vessels can interact with VEGFR3 and bind to lymphangiogenic VEGFc and VEGFd. Isotype VEGFd is less important than VEGFc isoform. Fibroblast growth factor-2, platelet-derived growth factor-B, and hepatocyte growth factor also stimulate lymphatic vessel growth [1367].

In adult mice, myeloid cells are sources of VEGFc and VEGFd factors. Monocytes M2 that express SYK kinase exhibit a lymphangiogenic activity, as they stimulate endothelial cell sprouting via VEGFc and VEGFd [1368]. Factors VEGFc and VEGFd have essential and modulatory functions, respectively, during lymphangiogenesis.

Angiopoietin receptors TIE1 and TIE2 are produced by lymphatic endothelial cells [1369]. Receptor TIE2 is expressed at lower levels in lymphatic vessels than in blood vessels, but Ang2 is synthesized in a greater amount in endothelial cells of lymph vessels than in blood vessels.

Angiopoietin-1 and VEGF collaborate during blood vessel development. Angiopoietin-1 activates TIE2 receptor. On the other hand, angiopoietin-2 either activates or blocks TIE2 according to the cell type [1370]. Angiopoietin-2 acts as an agonist in lymphatic vessels and antagonist in blood vessels. Isoforms Ang1 and -2 have a redundant role in lymphatics. Furthermore, angiopoietins bind to integrins in lymphatic endothelial cells. Angiopoietin-2 is required in patterning of the lymphatic network and recruitment of smooth myocytes to the collecting lymphatics.

Ephrin-B2 that signals via EPHb4 is involved in the remodeling of the lymphatic vasculature and valve formation [1371]. In the absence of the binding PDZ domain of ephrin-B2, pericytes and vascular smooth myocytes do not stably associate with blood vessels, albeit blood vessel defects do not occur. On the other hand,

---

[90]Nineteen ETS transcription factors are expressed in blood vessel endothelial cells.

[91]Receptor VEGFR3 exists in all vascular endothelia during embryo- and fetogenesis, whereas in adults, it is restricted to the lymphatic endothelium. However, VEGFR3 is observed in the microvasculature of tumors and wound healing. In adult humans, VEGFR3 is also synthesized in some fenestrated and discontinuous endothelia of blood vessels. Receptor VEGFR3 promotes the migration and proliferation of lymphatic endothelial cells.

[92]Neuropilin-2 is a semaphorin receptor of the nervous system and lymphatic capillaries. Combined inhibition of VEGFR2 and VEGFR3 prevents angiogenesis and tumor growth [1366].

lymphatic defects encompass chylothorax, hyperplasia of collecting lymphatic vessels, agenesis of valves, and abnormal recruitment of smooth myocytes into lymphatic capillaries [1371]. Moreover, specific ablation of ephrin-B2 on vascular smooth myocytes provokes their migration to cover lymphatics [1372]. Tissue-specific mutant mice display perinatal lethality with vascular defects in lung, kidney glomeruli, digestive tract, and skin. Receptor EPHb4 is expressed in collecting lymphatic vessels and lymphatic capillaries, while ephrin-B2 is only expressed in collecting lymphatic vessels in adults.

10.8.3 Other Mediators

Cytosolic protein Tyr kinase spleen tyrosine kinase (SYK) and adaptor lymphocyte cytosolic protein LCP2 control the separation of lymphatic and blood vessels [1184]. Both limit the number of lymphaticovenous anastomoses.

Podoplanin, a transmembrane glycoprotein highly expressed in glomerular podocytes and lymphatic endothelial cells, may intervene in lymphatic patterning.

Integrin-$\alpha_9\beta_1$, a receptor for vascular endothelial cell adhesion molecule-1, tenascin-C, and osteopontin, is highly expressed in lymphatic endothelial cells. Once bound to fibronectin, β_1-integrin interacts with VEGFR3 and can induce its phosphorylation to modulate cell migration, proliferation, and survival. Integrin-$\alpha_9\beta_1$ binds VEGFc and VEGFd subtypes.

10.9 Modeling of Angiogenesis

Angiogenesis is a interconnected set of events occurring in sequence and in parallel, on multiple levels, triggered by a major stimulus, hypoxia. In response to hypoxia, the transcription factor hypoxia-inducible factor HIF1 activates transcrition of numerous genes, especially vascular endothelial growth factor (VEGF). Therefore, the 2 prominent chemicals of angiogenesis models are O_2 trigger and VEGF messenger. The latter stimulate chemotaxis and proliferation in endothelial cells during sprouting, the main investigated process. Angiogenesis is modeled at different length scales, from molecule (eventually cell signaling cascades) to cells (proliferation and migration) and tissue (molecular transfer and microvascular network).

Models are mainly based on partial differential equations with spatially distributed variables and appropriate boundary conditions. The predominant strategy relies on transport–reaction differential equations based on the continuum concept. However, hybrid approaches are explored, in addition to discrete models and ordinary differential equations [1373].

Blood flow can be computed in a 3D geometrical model of a microvascular network; convection–diffusion–reaction partial differential equations governing oxygen transport are then solved to yield a 3D map of oxygen level in the computational domain.

Developing three-dimensional arterial tree based on sprouting angiogenesis upon chemotaxis mediated by growth factors secreted by ischemic cells to match the metabolic demand that can reorganize was modeled using morphometrical optimality principles [1374]. The signaling is reduced to a single messenger (VEGF). The iterative arterial tree generation that results from the balance between growth and degeneration is carried out on a gradually growing simulation domain. The geometry of the vasculature relies on node coordinates and connectivity. Each vessel segment assumed to be a rigid, straight, cylindrical tube (of constant radius R and given length L) is represented by a single directed (streamwise pointing) edge connecting 2 nodes. Vessel bending is defined by intermediary nodes (with radius invariance). Branches are defined as paths between 2 bifurcation nodes. Morphological constraints also prescribe the area ratio at bifurcations:

$$R_t^p = \sum R_b^p \tag{10.1}$$

($p \in [2,3]$) as well as bifurcation angles:

$$\cos\phi_i = \frac{R_t^4 + R_i^4 - R_j^4}{2R_t^2 R_i^2}. \tag{10.2}$$

The model relies on the mutual interplay of oxygen supply and released VEGF amount with a given oxygen consumption rate function of the local oxygen concentration, an arbitrary oxygen-dependent VEGF secretion rate function, and given diffusivities ($\mathcal{D}_{O_2} = 2.41\,\mathrm{m^2/s}$; $\mathcal{D}_{VEGF} = 1.04\,\mathrm{m^2/s}$). Reaction–transport equations allow to approximate the perfusion and to compute VEGF concentration. Capillary elongation or bifurcation is governed by a log-normal distribution of branch aspect ratios. Sprouting occurs at downstream end nodes (symmetrical bifurcations) and intermediary nodes (asymmetrical bifurcations), where the VEGF concentration is maximal and exceeds a threshold; the direction is dictated by the local VEGF gradient. Tree remodeling is ensured by pruning and rescaling.

Tumor growth characterized by its volume ($V_1(t)$) coupled to a developing tumor vasculature described by its carrying capacity ($K_1(t)$; tumor volume supported by blood input at time t) was modeled as well as effects of anti-angiogenic therapy with various drug administration modes [1375]. Hence, tumor vasculature growth depends on the balance between tumor-derived stimulatory and inhibitory factors (vessel loss function $\psi(V)$) and exogenous drugs, in addition to natural loss (N). Exogenous inhibitors of angiogenesis include cytostatic drugs that impede the formation of new blood vessels and cytotoxic drugs that destroy existing blood vessels (blood drug concentration $c(t)$; drug effect $\kappa c(t)$). Changes in blood vessel density have delayed effects and this delay can affect therapy efficiency. The specific growth rates of the tumor (V_2/V_1) and vasculature (K_2/K_1) depend on the ratio between the tumor volume and the carrying capacity of the vasculature (V/K), which is proportional to the inverse of the tumor vessel density:

$$\begin{aligned} V_2 &= V_1 f(V/K), \\ K_2 &= K_1 \Big(\Phi(V/K) - \psi(V) - \mathsf{N} - \kappa c(t) \Big). \end{aligned} \tag{10.3}$$

Chapter 11
Tissue Development, Repair, and Remodeling

Tissue growth as well as repair of tissue injury is controlled by many regulators such as growth factors that both reorganize the cytoskeleton and cell adhesions and trigger the action of transcription factors, thereby enabling proper cell migration and lodging as well as adaptation to environmental conditions. Cell wall extensibility and junctional communications are involved in tissue growth. In adults, cell differentiation happens during normal cell turnover as well as during tissue repair. Tissue remodeling caused by a sustained, time-dependent mechanical stress field and chemical cues, possibly associated with angiogenesis, rely on the coordination between regulated cell proliferation, differentiation, migration, and apoptosis to determine the correct tissue configuration [1376].

Repair, or healing, involves a series of rapid infiltrations and accumulations of various cell populations that deposit a new matrix and mature (Table 11.1). Once they have completed their tasks, some involved cell types are eliminated prior to the progression to the next phase of healing via apoptosis (Vol. 2 – Chap. 4. Cell Survival and Death). Apoptosis allows elimination of entire cell populations without tissue damage or inflammation.

During the proliferation phase of tissue repair, collagen-3 prevails. During the maturation, collagen-3 is gradually degraded and collagen-1 produced. Moreover, collagen fibers that are originally disorganized rearrange, crosslink, and align along tension lines [1377]. Blood vessel density decays as local activity attenuates. Some blood vessels are removed by apoptosis.

When cells undergo apoptosis, tiny regions of the plasma membrane pinch off and form vesicles, the so-called *apoptotic bodies*, that are shed and carry messages from dying cells to healthy ones to promote repair. Apoptotic bodies from dying endothelial cells that are taken up by healthy endothelial cells heightens the expression of the Cxcl12 gene [1378]. Chemokine CXCL12 counteracts apoptosis and recruits progenitor cells for repair. In apoptotic bodies, microRNA-126 prevents the synthesis of an inhibitor of CXCR4, the CXCL12 receptor. This GPCR then triggers a feedback loop that augments the production of CXCL12 to attract progenitor cells for new tissue generation.

M. Thiriet, *Tissue Functioning and Remodeling in the Circulatory and Ventilatory Systems*, Biomathematical and Biomechanical Modeling of the Circulatory and Ventilatory Systems 5, DOI 10.1007/978-1-4614-5966-8_11,

Table 11.1. Repair stages and timing (Source: [1377]).

Period	Events
Day 1	Hemostasis Vasoconstriction Cell influx (neutrophils)
Week 1	Vasodilation; inflammation Cell influx (macrophages, lymphocytes, fibroblasts) Growth factor signaling; cell proliferation Collagen deposition
Month 1	Remodeling Collagen crosslinking

11.1 Organogenesis

Organogenesis during embryonic development relies on differentiation and replication of stem and progenitor cells down to terminal differentiation and proliferation of differentiated cells. Organogenesis is achieved in sequential stages from the formation of an anlage, invagination of the primary bud, reiteration into various branches, to organization into specific proximal and distal structures.

11.1.1 Ecto-, Endo-, and Mesoderm

Organogenesis is the process by which the ecto-, endo-, and mesoderm develop into organs (between week 3 and 8 of in utero life; Table 11.2). Regulators control the formation of spatial patterns as well as timing of patterning events.

Differential adhesion and tension as well as directed cell migration determines progenitor sorting. *Differential adhesion hypothesis* of organization of cell layers during the development of embryonic tissues states that in 2 cell populations that aggregate the more cohesive one (with the stronger homotypic adhesion) distributes in the middle and the less cohesive one on the outside [1379, 1380]. The spreading of one embryonic tissue over another and the formation of intertissue borders are supposed to behave like immiscible liquids. Adhesive forces between aggregated cells play the same role in cell sorting as intermolecular attractive forces in liquid surface tension. Distinct expression of cell adhesion molecules, especially cadherins, regulates histotypic sorting out of cells of embryonic tissues when cells are intermixed.

Differential surface contraction theory of morphogenesis states that cell cortical tension developed by the cytoskeleton rather than adhesion between cells generates cell sorting [1381]. In addition, signaling between mesenchymal and epithelial cells

Table 11.2. Organogenesis (Source: Wikipedia). Cells that migrate inward along the archenteron form the inner layer of the gastrula, which develops into the endoderm. Mesoderm gives rise to: (1) chordamesoderm (axial mesoderm) along the central axis under the neural tube; (2) paraxial mesoderm on both sides of the neural tube that generates somites (origin of vertebral column, dermis, and skeletal muscles) and branchial arches (facial muscle and cartilage, etc.); (3) intermediate mesoderm between the paraxial mesoderm that forms the urogenital system; and (4) the lateral plate (or hypomere) at the periphery of the embryo that splits into 2 layers, the somatic (body wall) and splanchnic layer (circulatory system and gut wall). The ectoderm that emerges first and forms from the outermost of the germ layers, has 3 parts: external ectoderm (surface ectoderm), neural crest, and neural tube (neuroectoderm).

Germ layer	Organ products
Endoderm	Respiratory tract
	Endocrine organs
	Gastrointestinal tract
Mesoderm	Circulatory apparatus
	Mesenchyme (blood and lymph vessels, blood cells, bone, cartilage, and connective tissues)
	Mesothelia (pericardium, pleura, and peritoneum
	Muscles
	Digestive tract
	Reproductive apparatus
	Urinary tract
Ectoderm	Nervous system
	Skin
	Nose and mouth epithelia
	Eye

influences organ morphogenesis. Epithelial wrapping controls mesenchymal shape change and sorting of enclosed mesenchymal cells [1382].

Measurements of cell adhesiveness and cortical tension in the 3 germ layers of zebrafish embryos using atomic force microscopy show that the strength of homotypic adhesion decays from mesoderm, then endoderm, to ectoderm [1383]. The magnitude of measured cortical tension diminishes from ectoderm, then mesoderm, to endoderm.

Differential actomyosin-dependent cell-cortex tension regulated by Nodal–TGFβ signaling directs progenitor sorting. Ecto- and endoderm cells occupy the cell aggregate center and periphery, respectively. However, in vivo configuration is defined by outer ectoderm and innermost endoderm, with mesoderm in the middle. Strong interactions of germ-layer cells with enveloping layer cells can invert in vitro sorting to that attained in vivo. Furthermore, groups of cells do not behave like organized sheets of cells.

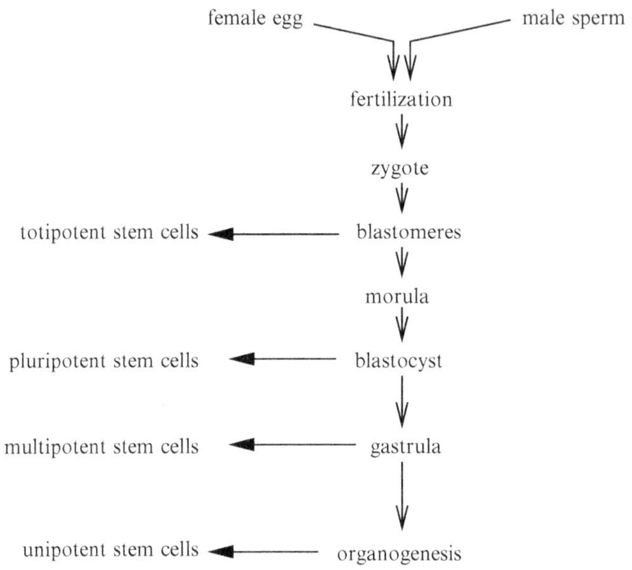

Fig. 11.1 Different types of stem cells. Fusion between female and male gametes (fertilization) leads to the zygote and embryogenesis. After fertilization, a set of fast mitoses occurs (cleavage) with formation of blastomeres building the blastula. In mammals, the blastula is called the blastocyst. The morula corresponds to an organized cell set with external and internal cells. After the mitosis rate has slowed down, the blastomeres move (gastrulation), forming the gastrula with 3 germ layers: ecto-, endo-, and mesoderm.

11.1.2 Precursor Cells

During the early stages of embryogenesis, the fertilized egg starts to divide into blastomeres. Embryonic stem cells (ESC), derived from the totipotent cells[1] of preimplantation embryos are pluripotent cells (Fig. 11.1).

Pluripotent stem cells either self-renew continuously or form the various types of mature, functional cells. Pluripotent embryonic stem cells derive at various stages of embryo development according to their type, embryonic germ cells occurring at an earlier phase than others.[2]

---

[1] A single totipotent cell leads to the development of several hundred cell types.

[2] Human embryonic stem cells can be removed either from the blastocyst or morula, or from later-stage embryos, using various isolation procedures. The optimal method consists in deriving pluripotent embryonic stem cells from the inner cell core of blastocysts, before implantation in uterus (preimplantation blastocyst), i.e., 5 to 6 days after in vitro fertilization. The inner cell mass is then large enough (outer trophoblasts differentiate into the placenta). A single blastomere can be extracted from the 8-cell (3-day) embryo for preimplantation genetic diagnosis. Cultured blastomeres prior to genetic analysis can divide, so the genetic check-up can be associated with the production of human embryonic stem cells without reducing embryo developmental potential.

The *morula* (morus: mulberry; 12–32 blastomeres within the zona pellucida) corresponds to an early stage of embryonic development produced by the division of the zygote. After reaching the 16-cell stage, cells of the morula differentiate.

The central, fluid-filled blastocyst cavity (a.k.a. blastocele and cleavage or segmentation cavity) appears and the morula becomes a hollow spheric blastocyst (blastulation). The inner and outer blastomeres form the inner cell mass and trophoblast, respectively. The *blastula* (blastosphere) that follows the morula consists of a spherical layer of about 128 cells surrounding the blastocele. At the blastula stage, the cell cluster contains nearly identical cells.

During the next stage, the *gastrula* is characterized by structured tissues. Gastrulation is the first stage in vertebrate development when different progenitor types sort out and assemble into 3 germ layers: ecto-, endo-, and mesoderm. The germ layers in tissue development involve 3 processes: folds, splits, and condensation.

After gastrulation, pluripotent embryonic stem cells give birth to multipotent stem cells that are irreversibly programmed for a given tissue. They form: (1) mesoderm, from which are derived connective tissue, muscles, and circulatory apparatus; (2) ectoderm (nervous system, skin, etc.); and (3) endoderm (respiratory apparatus, digestive tract, and endocrine glands).

Tissue and organ development includes several stages under the control of the genetic code: (1) acquisition of cell identity and transient activation of specific regulatory genes; (2) stabilization of the regulation and exclusion of alternative regulatory states by feedback; (3) reception of signals required for development of embryonic cell lineages with respect to neighboring lineages and lineage-specific activation of organogenic genes.

11.1.3 Organogenesis, Vascularization, and Innervation

Organ development requires the differentiation and coordinated development of nerves and blood vessels that contain many cell types. The peripheral parasympathetic nervous system innervates many organs during embryogenesis.

Human embryonic stem cells are self-renewing pluripotent cells, that provide a source for cell replacement therapies in certain pathologies (e.g., infarction sites). These cells are able to differentiate into sufficient quantities of a specific, transplantable, functional cell type. They behave normally within a blastocyst but can induce tumors when they are ectopically transplanted. Cultured human embryonic stem cells have the following main time-invariant characteristics: (1) in vitro proliferation on appropriate substrata; (2) maintenance of a normal karyotype throughout long culture durations; (3) expression of standard molecular markers of stem cells (stable expression of surface markers and transcription factors associated with an undifferentiated state); and (4) differentiation into cell lines of all 3 germ layers with an epigenetic status (autosomal imprinting and X-chromosome inactivation) that differently respond according to environmental stimuli [1384]. Signaling primed by FGF is strongly involved in the maintenance of a proliferative, undifferentiated state in cultured human embryonic stem cells.

In mouse embryos, the submandibular salivary glands appear at mid-gestation as small epithelial buds on the floor of the oral cavity. Buds and stalks of the developing submandibular gland contain multipotent stem cells. Nerves that originate from parasympathetic ganglia innervate epithelial buds to support their development via neurotransmitter acetylcholine and acetylcholine muscarinic M_1 receptor. At least in mouse embryonic salivary glands, parasympathetic innervation preferentially signals to keratin-5+, epithelial progenitor cell population via muscarinic receptor [1385]. Furthermore, cholinergic stimulation triggers the release of heparin-binding epidermal growth factor that binds to epidermal growth factor receptor. These progenitor cells are then maintained in the adult salivary gland. Therefore, innervation not only controls organ function, but also contributes to organ repair and regeneration, as it maintains a progenitor cell population.

11.1.4 Regulation of Organogenesis

Organogenesis regulation involves various DNA-binding proteins, such as transcription factors, nuclear receptors, and their coactivators and inhibitors, as well as signaling messengers, such as growth factors and hormones, in addition to mechanical constraints. Triggered pathways modulate cell functioning and secretion of constituents of the extracellular matrix and its organization. Various growth factors and hormones bind to their specific receptors and provoke protein synthesis, and post-translational modifications of proteins, as well as exocytosis of substances. Any morphogen influences the fate of its target cells in a dose-dependent manner to direct regional specialization within a developing tissue or organ.

Many other regulators influence embryo- and fetogenesis, as they modulate gene transcription for patterning factors, such as vitamin-A and its metabolites [1386]. Retinoic acid contributes to the control of cell division and differentiation in tissues of ecto-, endo-, and mesodermal origin via cognate transcription factor, the nuclear retinoic acid (RAR or NR1b) and retinoid-X (RXR or NR2b) receptors.

Retinoic acid, the active form of vitamin-A, has many functions during embryogenesis. In adults, it regulates fertility, maintains normal vision, and prevents tumor growth and degenerative diseases. This small lipophilic molecule binds to nuclear receptors that are heterodimers of retinoic acid receptors and retinoid X receptors (RAR–RXR complexes; Vol. 3 – Chap. 6. Receptors) to regulate the transcriptional activity of target genes. Retinoic acid is synthesized from inactive precursors (retinol or β-carotene) from food after digestive processing that leads to plasmatic transport and after placental transfer in embryos.[3] Retinoic acid controls the activity of fibroblast growth factor and sonic Hedgehog [1387].

Among transcription factors expressed predominantly in the heart, members of the GATA set — GATA4, GATA5, and GATA6 — require vitamin-A [1386]. Retinoid-regulated cardiogenesis also involves transcription factor of the Msh homeobox class Msx1 that also needs vitamin-A input. Cardiac abnormalities result from lack or excess of retinoids.

Specification of any embryonic cell lineage is controlled by the sum of activities of transcription factors — activators and repressors — that result from interactions between involved genes that encode these transcription factors following a given signaling. Many genes indeed participate in specification and differentiation of embryonic cell lineages and progressive organization of tissues and organs. Development genes coordinate the sequence of interactions between genes that encode transcription and signaling factors.

Long and small non-coding RNAs (ncRNAs) regulate organogenesis [1388]. Small ncRNAs are associated with heterochromatin formation via the RNA interference pathway. Long ncRNAs are linked to gene clusters. Non-coding RNAs regulate transcription, as they interact with transcription factors, RNA polymerases, or DNA. Short interspersed elements of the genome transcribe non-coding RNAs that repress specific genes by binding to and inhibiting RNA polymerase-2.

The cell's capacity to differentiate with specialized attributes is, at least partially, achieved using the genetic code. In the large majority of cases, cells acquire their fate owing to their lineage or signaling from adjoining cell lineages. In addition to the deterministic program, the differentiation pathway can be randomly fullfilled to cope with adverse changes in the environment [1389].

Cells interact with adjoining cells and the extracellular matrix owing to cell adhesion molecules such as integrins that serve as receptors of matrix constituents. Cell adhesion molecules link intracellular cytoskeletal proteins of adjacent cells as well as cytoskeleton to matrix constituents. Matrix constituents influence intracellular events via their receptors that lodge at the cell surface, thereby modulating cell differentiation, migration, and polarization. In addition, cells release matrix-degrading enzymes and their inhibitors to remodel their nearby environment.

In kidney genesis, epithelial–mesenchymal interactions require connections of $\alpha_8\beta_1$-integrins to extracellular matrix protein nephronectin [1390]. Nephronectin

[3]Retinol is taken up by retinol-binding protein (RBP), transferred into cells by the receptor protein stimulated by retinoic acid STRA6, and transformed into retinaldehyde by cytosolic alcohol dehydrogenases (ADH) and microsomal retinol dehydrogenases (RDH). Retinol to retinaldehyde conversion is catalyzed mainly by retinol dehydrogenase RDH10 (canonical synthesis). The final step is the oxidation of retinaldehyde into retinoic acid. Retinoic acid is synthesized by retinaldehyde dehydrogenases (RALDH1–RALDH3) and a group of cytochrome-P450s (CyP26a1–CyP26c1) with tissue specificity. In addition, β-carotene cleavage by β-carotene dioxygenase generates retinaldehyde (alternative synthesis). Cellular retinol-binding proteins (CRBP) and retinoic acid-binding proteins (CRABP) participate in retinoic acid acitivity.

and $\alpha_8\beta_1$-integrin are synthesized in the ureteric bud epithelium and metanephric mesenchyme, respectively.

11.2 Cell Differentiation

During cellular differentiation, an immature cell becomes a mature specialized cell. Cell differentiation occurs not only during embryogenesis, but also in adults. In particular, adult stem cells generate differentiated daughter cells for tissue repair and during cell turnover. Cell differentiation modifies cells size, shape, polarity, metabolic activity, and responsiveness to signals due to modifications in gene expression. Dedifferentiation allows differentiated cell to revert to an earlier developmental stage.

Three basic cell classes include germ, somatic, and stem cells. *Germ line cells* give rise to gametes (eggs and sperm). *Totipotent cells* (zygote and early embryonic cells) can differentiate into all cell lineages. *Pluripotent stem cells* (e.g., hematopoietic and mesenchymal stem cells) are able to differentiate into many cell types. *Multipotent stem cells* and *unipotent progenitors* give rise to functional cells.

Each differentiated cell type expresses a gene subset of the genome due to a peculiar pattern of gene expression. Growth factors control the switch from one gene expression pattern to another. Distinct cell types differentially retrieve the genetic information encoded in the genome. Transcription factors and microRNAs form the largest families of gene regulatory factors [1391]. Sets of combinatorially expressed transcription factors and microRNAs delineate cell types. Transcription factors and microRNAs share many similar features although they have their specific properties that determine specialized regulatory niches (Table 11.3). Some microRNAs are expressed in a cell- or tissue-specific manner and contribute to cell identity. MicroRNAs can control the expression of transcriptional regulators. They also regulate alternative splicing during tissue development.

11.3 Branching Morphogenesis

The vasculature is characterized by its branched architecture. In addition, to their own perfusion vasculature, organs can contain branched structures in a limited body region, such as the tracheobronchial tree, collecting parts of renal nephons, and glandular ducts.

Branched conduit networks connect body's environment to organs and/or conversely. Branching augments the cummulated functional surface area (e.g., trumpet shape of the cummulated surface area of the tracheobronchial tree from large bronchi to alveolar ducts).

The respiratory tract links atmospheric air to gas exchange membrane of pulmonary alveoli. This alveolocapillary membrane is crossed by blood gas. Blood

Table 11.3. Features of transcription factors and microRNAs (Source: [1391]).

	Transcription factors	MicroRNAs
Pleiotropy	+	+
Cooperation	Cooperative binding to cognate DNA sequences Cooperative recruitment of transcription cofactors	Cooperative activity
Recognition of site accessibility and binding	Nucleosome coating	RNA-binding proteins; folding of target mRNA sequences; other microRNAs
Regulation	Activation or repression; slower activity ($\pm$ translocation)	Repression mostly; fast, reversible repression; quick reactivation
Expression	MicroRNA dependency; feedback; post-translational regulation (phosphorylation)	Post-translational regulated maturation; RNA editing (adenosine-to-inosine conversion); microrna cofactor modifications

circulation then carries these gases inside the cardiovascular system that irrigates and drains body tissues. The 2 serial ventilatory and circulatory networks transport oxygen to cells and remove carbon dioxide from them.

11.3.1 Airway Morphogenesis

In mammals, air enters the organism through the nasal and oral cavities and crosses the pharynx, larynx, and trachea. The trachea branches into 2 stem bronchi that give rise to densely branched left and right bronchial trees inside lung parenchymas. The bronchial tree of the human lung contains more than 105 conducting and 107 respiratory airways [1392].

During mammalian embryogenesis, lungs develop from 2 primary buds that arise in the ventral foregut. These buds undergo extensive branching under the reciprocal control of the epithelial endoderm and surrounding mesoderm. Terminal buds may contain a population of multipotent epithelial progenitors [1393]. As airways extend, descendants of these cells give rise to progenitors of the major cell types of airways, such as ciliated and secretory Clara cells. Differentiation of epithelial cells begins proximally and proceeds distally. Once airway morphogenesis is complete, the lung continues to increase in size and new epithelial cells are created. Cells with Clara cell markers can give birth to ciliated cells. In the proximal airways, columnar epithelial cells give rise to basal cells. Basal cells can self-renew and contribute to other cell lineages. On the other hand, differentiated ciliated cells do not divide [1393].

Patterning information specifies branching time and loci during development as well as the length, shape and direction of outgrowth of each branch. Branching does not occur randomly to fill available space. This process is precisely controlled. Nevertheless, coding is simplified by repeated use of a branching mechanism, using an algorithm such as that in Mandelbrot's fractal model [1392].

11.3.2 Airway Morphogenesis Mediators

Several morphological processes with feedbacks control branching initiation and extension that rely on activation, competition, and interaction for tip and stalk assignment in growing branches, migration, proliferation, and survival of involved cell types [1394]. Moreover, cells implicated in lumenogenesis undergo an apicobasal polarization with an apical side building the lumen (Sect. 10.2.5).

Before birth, peripheral saccules are created and pulmonary capillaries grow in close apposition to the alveolar epithelium for gas exchange. Pulmonary immaturity and lack of pulmonary surfactant cause respiratory distress syndrome. Surfactant function relies especially on surfactant protein-B and -C, and ABCa3 transporter.

Budding and growth of the respiratory tree involve numerous growth factors, signaling proteins, and matrix components, as well as transcription factors (Sect. 11.3.4.2).

11.3.2.1 Growth Factors

In mice, lung formation begins at 9.5 days postcoitum (E9.5), when the foregut endoderm invades the splanchnic mesenchyme and undergoes dichotomous branching. Branching morphogenesis depends on mesenchymal–epithelial cell signaling using fibroblast (FGF10), hepatocyte (HGF), and transforming (TGFβ) growth factors, bone morphogenetic protein BMP4, sonic Hedgehog, and Wnt morphogen (Sect. 11.3.4.2). Factors EGF, FGF, and PDGF are involved in alveologenesis.

Fibroblast Growth Factors

Fibroblast growth factors are important regulators of the respiratory tract genesis. Fibroblast growth factor-7 and -10 diffuse from the mesenchyme and activate alternatively splice variant IIIb (of extracellular loop III) of protein Tyr kinase receptor FGFR2IIIb in the nearby epithelium and induce budding that is composed of an inner endodermal epithelium and an outer mesenchyme. The expression pattern of specific FGFs at distinct steps of bud emergence controls regional endodermal cell proliferation.

Isotype FGF10 is produced by the mesenchyme at the earliest stages of lung-bud formation with 2 primordial buds from the trachea, from which the bronchial tree arises using a process of elongation, budding, and branching. Agent FGF10 stimulates proliferation of endodermal cells in the apical region of developing buds [1395].

Epidermal Growth Factor

Epidermal growth factor, transforming growth factor-α, and amphiregulin are produced by the mesenchyme and act via the common EGFR receptor [1396].

Hepatocyte Growth Factor

Hepatocyte growth factor is produced by the mesenchyme and its receptor is confined to the epithelium [1396]. Agent HGF supports action of other growth factors in branching genesis.

Insulin-like Growth Factor

Concentrations of insulin-like growth factors IGF1 and IGF2 as well as that of their receptors do not change during gestation; IGF-binding proteins (IGFBP1–IGFBP6) are developmentally regulated [1396]. Receptor of IGF1 is necessary to avoid lung hypoplasia, but does not intervene in branching formation. Receptor of IGF2 is required for normal alveolus development.

Platelet-Derived Growth Factor

Production of platelet-derived growth factor is restricted to the epithelium, whereas its cognate receptors localize to the mesenchyme [1396]. Homodimer PDGFaa contributes to branching regulation. Protomer PDGFa is required for alveolus septation.

Transforming Growth Factor

Transforming growth factor-β1 to -3 as well as receptors TβR1 and TβR2 are involved in embryonic lungs. Agents TGFβ1 and TGFβ2 are produced by the mesenchyme and epithelium, respectively [1396]. They impede the generation of pulmonary branching via TβR2. On the other hand, TGFβ3 stimulates lung development.

Bone Morphogenetic Proteins

Bone morphogenetic proteins intervene in lung development, in particular in the proximal–distal patterning. Isotype BMP4 is highly produced by epithelial cells of bud tips and, to a lesser extent, in the surrounding mesenchyme [1396]. It precludes epithelial cell proliferation.

11.3.2.2 Guidance Cues

Members of the netrin and semaphorin families participate in budding and branching during lung morphogenesis. Netrin-1 and -4 are produced by epithelial cells of bud stalks. They are deposited in the surrounding basement membrane, where they preclude ectopic budding and inappropriate lateral branching [1395]. Both netrin-1 and -4 are expressed at their highest levels at non-branching proximal endoderm and stalk region of growing buds. On the other hand, they are excluded from growing bud tips, where most of the branching occurs. The spatial restriction of netrins permits efficient branching genesis. In addition, netrin-4 and, to a lesser extent, netrin-1 inhibit the formation of secondary buds induced by FGF7 or FGF10 [1395]. Semaphorin-3A is expressed by the mesenchyme that surrounds the ends of buds. It organizes the size and shape of the emerging buds.

11.3.2.3 Wnt Morphogens

Wnt morphogens signal in progenitor cells of the respiratory epithelium both during development and after injury. Activation of β-catenin leads to its translocation to the nucleus and formation of a transcriptional activation complex with either T-cell factor (TCF), sex-determining region Y (SRY)-box gene product Sox9, or pituitary (or Paired-like) homeobox transcription factors Pitx1 and Pitx2 to cause the transcription of target genes, such as neuroblastoma-derived MYC (NMYC), Bmp4, and Fgfr2b in epithelial cells, and MYC, PITX2, and Fgfr2c in mesenchymal cells [1397]. Once bound to the mesenchymal fibroblast growth factor FGF10, FGFR2b receptor triggers the β-catenin pathway. In the mesenchyme, mesothelial and epithelial fibroblast growth factor FGF9 connects to FGFR2c to also initiate the β-catenin pathway. β-Catenin in the mesenchyme controls the differentiation of angioblasts into endothelial cells as well as proliferation of parabronchial smooth muscle progenitors [1397].

β-Catenin reduces the expression of epithelial differentiation markers during the early pseudoglandular stage of lung development [1397]. During early lung development, β-catenin signaling triggers the proliferation of epithelial differentiation marker-negative (surfactant protein SPc− and secretoglobin ScGb1a1−)[4] cells that

---

[4]A.k.a. 10-kDa Clara cell phospholipid-binding secretory protein CC10.

may be precursors of SPc+, ScGb1a1+ cells. Very early SPc− progenitor cells localize to a slightly more proximal position than SPc+ cells at the distal tip of the developing lung.

In adult lungs that undergo slow turnover and possess regenerative capacity via activation of endogenous stem cell populations, SPc+, ScGb1a1+ cells at the bronchoalveolar junction represent lung epithelial progenitor cells that can give rise to both alveolar and bronchiolar cell lineages, especially bronchiolar Clara cells. β-Catenin controls the balance between progenitor expansion and epithelial differentiation. During lung epithelium regeneration, the canonical Wnt pathway is activated in niches at the bronchoalveolar duct junctions that contain bronchoalveolar stem cells [1398].

Transcription factor GATA6 synthesized in the respiratory epithelium regulates the appearance timing and amount of bronchoalveolar stem cells. Its absence causes the precocious occurrence of bronchoalveolar stem cells and lack in epithelial differentiation. It activates the synthesis of the epithelial Wnt receptor Frizzled-2 [1398]. Non-canonical Frizzled-2 receptor precludes early epithelial canonical Wnt–β-catenin signaling. Frizzled-2 also regulates effectors of non-canonical Wnt signaling, such as G-proteins and calcium–calmodulin-dependent protein kinase CamK2. Moreover, GATA6 is needed for proper lung epithelial regeneration and differentiation of bronchoalveolar stem cells.

11.3.2.4 P38MAPK

P38α mitogen-activated protein kinase activates transcription factors, such as CAAT/enhancer-binding protein C/EBPs and Forkhead box protein FoxA2 (a.k.a. hepatocyte nuclear factor HNF3β) that are necessary for the correct differentiation of the stem cells into alveolar type-2 and Clara cells [1399]. It controls self-renewal of the lung stem and progenitor cells, as it coordinates proliferation and differentiation signals.[5] In particular, it impedes the action of mitogenic factors, especially epidermal growth factor receptor, hence restraining the proliferation of stem and progenitor cells.

11.3.2.5 Forkhead Box Transcription Factors

Forkhead box protein FoxA2 operates in lung maturation and differentiation of goblet cells; FoxJ1 in proper development of ciliated cells; and FoxF1 in formation of pulmonary capillaries.

[5]Kinase P38MAPKα also regulates the differentiation and proliferation of myocytes, cardiomyocytes, neurons, adipocytes, and leukocytes. In addition, it prevents the occurence of lung adenocarcinomas caused by kRas GTPase.

Forkhead box factor FoxM1 does not alter lung growth, in particular epithelial proliferation and airway morphogenesis, but prevents lung maturation and causes postnatal respiratory failure, as it is required for adaptation to air breathing [1400]. It is necessary for alveolus epithelial type-1 cell differentiation and synthesis of mucin-type type-1 integral membrane glycoprotein podoplanin (Pdpn),[6] aquaporin-5, and surfactant protein-A to -D. Moreover, it is a transcriptional activator of the genes that encode surfactant protein-A and -B. In adddition, FoxP2 contributes to postnatal lung alveolarization.

11.3.2.6 Other Transcription Factors

Many other transcription factors are produced in cells of the primitive lung, such as NKx2-1, GATA5 and GATA6, members of the C/EBP family, HoxA5 and HoxB3 to HoxB5, and nMyC, as well as intracellular (nuclear) receptors such as the glucocorticoid receptor (GR or NR3c1) [1396].

11.3.3 Mechanical Basis of Branching

Mechanical forces contribute to tissue morphogenesis, as they can shape epithelial sheets. Cadherins engaged in homophilic complexes between apposed cells interact with cortical actin–myosin filaments. Intercellular adhesion sites transmit forces generated in- and outside involved epi- or endothelial cells. Moreover, cadherin-based adhesion loci create forces that tend to lengthen these cell contacts, whereas cortical tensions exerted by actin–myosin filaments tend to shorten them [1394].

Monolayers of epi(endo)thelial cells that organize to locally minimize its potential energy and reach a stable equilibrium have been modeled as sheets. Mechanical modeling will then have to be applied to more complex structures, such as capillaries covered by a single cell layer and conduits coated by pseudo- and stratified epithelia, as well as epithelia associated with a mucous layer.

11.3.4 Molecular Basis of Branching

11.3.4.1 Signaling in the Vasculature

In the vasculature, vascular endothelial growth factor, especially VEGFa via VEGFR2, causes angiogenic sprouting. Tip cells emit numerous filopodia and migrate in a given direction under the guidance of VEGFa (Chap. 10). Whereas tip

[6] A.k.a. aggrus, glycoprotein GP36, GP38, and GP40, and T1α.

cell migration depends on a gradient of VEGFa, the followers — stalk cells — proliferate under the control of VEGFa concentration [1394]. Selection of tip from stalk cells results from Delta-like ligand DLL4 and its receptor Notch-1. Ligand DLL4 is preferentially produced in tip cells that reside close to a high VEGF concentration. Endothelial cells at tips of new tubes compete for leader positions. They can suppress the Notch-restricted, tip-cell phenotype of their immediate neighbors when they synthesize DLL4 and present DLL4 to adjacent migrating cells [1394]. Many other primary messengers participate in blood vessel branching such as guidance molecules (netrins, semaphorins, and slits).

11.3.4.2 Signaling in the Tracheobronchial Tree

In developing lungs, branching is bounded by a volume of mesenchymal tissue. The bronchial tree indeed develops by branching of the airway epithelium into a surrounding mesenchyme. Reciprocal feedbacks between the airway epithelium and surrounding mesenchyme control the normal sequence and three-dimensional pattern of branching events [1392].

The branched tree can be generated by a genetically tractable, hierarchical, and modular program that contains 3 local modes of branching used in 3 different orders, i.e., at different times: domain branching at the branching onset and planar and orthogonal bifurcations [1392]. These branching modes may arise from the combination of a few elements, i.e., a master branch generator and 3 genetically encoded slaves that govern the pattern of branching events [1392]: (1) a domain specifier; (2) a genetically programmed periodicity generator that determines when side branches arise; (3) a bifurcator that defines branching loci, i.e., their occurrence time; and (4) a rotator that directs the bifurcation angle w.r.t. the stem plane. These branching modules are characterized by their coupling and repetitive occurrence.

A primary conduit constructs branching domains that serve as scaffolds. Branching arises at different positions directed by the periodicity generator to give rise to 2 coplanar branches, one of which undergoes domain branching and the other bifurcates again in a planar manner. Planar bifurcation, i.e., splitting of a branch tip into 2 branches, generates most ternary and some quaternary branches. Orthogonal bifurcation involves planar bifurcation with rotation of the bifurcation plane to give rise to higher-order generations.

Several types of sequences arise according to branching modes [1392]. In sequence `1`, a founder branch formed by domain branching uses immediately and permanently the orthogonal branching mode. In sequence `2`, a founder branch formed by domain branching utilizes domain branching for some daughter tubes that then employ permanently the orthogonal branching mode. The founder forms other daughter ducts by planar bifurcation that themselves undergo planar bifurcation at their tips to form domain branches along their length. These domain branches switch permanently to orthogonal bifurcation. Other sequences can be observed.

However, branching pattern can vary [1392]. Branches can originate off an unusual parent branch. A branch can miss and daughter branches spring directly from the grandparent conduit. In addition, branching network size and architecture differ among species.

Domain Branching

In mice, branches form in 4 domains: lateral, dorsal, medial, and ventral [1392]. The first secondary branch from the left primary bronchus lineage buds off the lateral sprout of the founder branch. Over the next 2 days, additional branches sprout distal to the first secondary branch and create a row of lateral secondary branches. Then, another row begins to form along the dorsal surface of the left primary bronchus. As this domain develops, a third row starts to sprout from the medial surface of the left primary bronchus. Afterward, a fourth domain eventually arises from the ventral surface. This ventral domain often consists of a single distal branch. Secondary branches off the distal portion of the right primary branch also arise by domain branching, beginning with a row of lateral branches. The branch spacing in each row is similar to the left pattern, but neither the row-initiation positions nor branch number in each row are the same.[7]

In domain branching, the main secondary daughter branches form in domains at different positions around the wall circumference of the parent branch, like rows of bristles on a bottle brush [1392]. However, each row is characterized by an independently patterned domain. A patterning controller of domain branching that directs the branching situation (either proximal or distal) is constituted of a periodicity generator that regulates the sequence of branching within each domain. A circumferential patterning controller specifies the positions of domains and the order with which domains are used.

Planar and Orthogonal Bifurcations

Many ternary branches are constructed by domain branching. Nonetheless, some ternary and higher-order generations arise from a different mode of tip expansion and bifurcation. Planar bifurcation corresponds to branching with similar orientation w.r.t. the founder branch, such as that of certain ternary and quaternary branches [1392].

On the other hand, some ternary and most higher-order generations use a third branching mode, the orthogonal bifurcation. Branches bifurcate as in planar bifurcation, but between each round of branching with an approximately 90-degree rotation of the bifurcation plane. The daughter tubes are thus arranged in a rosette [1392].

---

[7]The first right dorsal branch is proximal to the first right lateral branch, whereas the first left dorsal branch is distal to the first left lateral branch.

Genetic Control and Epithelial–Mesenchymal Feedback Loops

Lung development relies on epithelial–mesenchymal feedback loops that use gremlin-1, bone morphogenetic protein BMP4, fibroblast growth factor, as well as sonic Hedgehog and Wnt morphogens, and retinoic acid. Many transcription factors, such as zinc finger protein, multitype ZFPM2,[8] GATA4, and NKx2.1, are implicated in pulmonary branching [1394]. Components of the extracellular matrix, such as fibronectin and laminin, are also involved [1401].

FGF

Fibroblast growth factor secreted from mesenchymal cells near to epithelial tip cells, especially FGF10 that targets predominantly FGFR2b receptor on emerging branch epithelial cells, promotes epithelial budding, outgrowth, and branching. Signaling launched by FGF10 provokes local sprouty-2 production at the branch tips [1392]. Sprouty-2 hinders the mitogen-activated protein kinase modules and limits the proliferation and migration of tip cells, hence prohibiting further branch extension, but initiating new branching. Sprouty-2 can then be associated with the periodicity generator [1392]. An elevated activity of the periodicity generator reduces interbranch length and provokes a shift from bifurcation to trifurcation.

The signaling pathways mediated by hypoxia-inducible factor and vascular endothelial growth factor cooperate with the FGF–FGFR–sprouty-2 pathway to match the developing capillary network to the growing airway tree, thus optimizing gas-exchange capacity of lungs [1401]. Moreover, FGF upregulates expression of the Bmp2 and Bmp4 genes in tip cells. Agent FGF10 may also act as a chemoattractant [1394].

BMP

Bone morphogenetic protein BMP4 as well as sonic Hedgehog signaling by tip cells exert an autocrine regulation that restrict FGF-mediated branching. On the other hand, BMP4 synthesized by mesenchymal cells enhances branching (paracrine regulation) [1394]. Factor BMP4 favors proliferation and survival of lung epithelial tip cells.

[8]This modulator of GATA4 transcription factor in cardiomyocytes is also called friend of GATA protein FOG2.

Sonic Hedgehog

Sonic Hedgehog regulates the progression rather than initiation of airway branching [1394]. It inhibits near-tip mesenchymal FGF10 synthesis, hence new branching in non-allocated loci. Therefore, sonic Hedgehog contributes to the structural reproducibility.

Wnt

Wnt morphogen uses its canonical pathway to reinforce FGFR2b expression in epithelial cells [1394]. On the other hand, non-canonical Wnt5 signaling precludes FGF10 synthesis.

11.4 Therapeutic Repair

Cell transplantation, gene delivery, and molecule administration are studied to treat various disorders. In addition, therapeutic strategies are developed either to promote revascularization of ischemic tissues and regenerate organs or inhibit angiogenesis in some diseases.

Certain organs and tissues, such as liver and skin, regenerate, but most organs lack regenerative capacity, hence being unable to repair aging, diseased, and injured tissues. Heart has a limited intrinsic regenerative capacity. Somatic cells normally cannot regenerate, except when adequately treated. Tissue stem cells generally have limited capacity.

Various types of differentiated cell can be generated from embryonic (ESC) or induced pluripotent stem (iPSC) cells. Moreover, these cell types can be used to model diseases and optimize therapies. Regenerative medicine aims not only at stopping progression of degenerative diseases of more or less known pathogenesis, but also at reversing their course.

In addition to therapies based on stem cells and virus-injected genes, some strategies rely on locally prolonged administration of growth factors. Intramyocardial injection of self-assembling peptide nanofibers combined with VEGF can create a microenvironment that can recruit myofibroblasts and cardiomyocyte-like cells [1402]. Sustained VEGF release (14 d) improves postinfarct neovascularization in rats.

11.4.1 Stem Cells and Regenerative Therapy

Tissue repair and regeneration involve cell interaction over a long time scale to produce correct cell type, number, location, and function. A successful cell-based

Table 11.4. In the heart, transplantation of stem cells (including epicardial progenitor cells) is aimed at repairing an infarcted myocardial region upon trilineage differentiation (cardiac and smooth myocytes and endothelial cells) to recover the cardiac function, especially contractility, and myocardial blood flow (Source: [1404]). Implanted stem cells can stimulate the recruitment, survival, and differentiation of host stem cells. The former can liberate growth factors (e.g., angiopoietin-1 and insulin-like [IGF1] and vascular endothelial [VEGF] growth factors), chemokines (e.g., CXCL12), and antifibrotic mediators. Timing, delivery method (intravenous, transarterial, as well as intracoronary, transvenous injection into coronary veins, and intramyocardial, either epicardial or catheter-based transendocardial), and amount contribute to the clinical outcome.

Cell differentiation and migration	Paracrine signaling
Tissue regeneration	Protection against apoptosis
Adequate scar	Homing of exogenous stem cells
Angiogenesis	Migration of endogenous stem cells
Restitution of stem cell niches	

therapy relies on long-term stem cell engraftment and survival as well as cooperation between transplanted and host cells (Table 11.4). Performance of cell therapy indeed relies on paracrine signaling launched by transplanted cells to support stem cell homing, cell survival, angiogenesis, tissue development, and organ functioning.

Regeneration potential relies on tissue-resident stem cells. However, the number of stem cells in tissues is less than 1 for 10^4 resident cells. Multiple stem cell populations have been discovered in various organs in adults (e.g., blood, brain, myocardium, skeletal muscle, and skin). In particular, hematopoietic stem cells (HSC) can proliferate and differentiate to produce lymphoid and myeloid cell types. Bone marrow-derived stem cells (bmSC) can differentiate into various unipotent progenitors and then specialized cell types (e.g., cardiomyocytes).

Cells of certain tissues, such as skin, blood, and gut epithelia, have a great turnover in normal conditions and are permanently replaced. These tissues can thus be restored to their original state after injury. On the other hand, other tissues such as the central nervous system can be repaired, but not completely restored.

Stem cells modulate tissue formation, maintenance, and repair. They most often stay in relative quiescence, and reside in specialized microenvironments (niches). Stem cells both: (1) self-renew and generate additional stem cells and (2) differentiate into various progenitor cells.[9]

Stem cell can thus be used in therapies that require repair, replacement, and regeneration to stop and reverse damages caused by degenerative diseases or by

[9] Transcription factors Nanog (for Tir Na Nog that means land of the ever young), octamer-binding protein Oct4, and sex-determining region Y-box Sox2, coordinately contribute to stem cell pluripotency and self-renewal of embryonic stem cells, activating genes that maintain pluripotency and repressing those for differentiation. Moreover, Nanog, Oct4, and Sox2 have an autoregulatory loop to allow rapid responses to environmental stimuli [1403].

perfusion defects such as heart failure arising from ischemic and non-ischemic cardiomyopathies.

Pluripotent stem cells such as embryonic stem cells (ESC) can differentiate into all cell types. Ethical, legal, and biological issues limit their use.

Reprogrammation of adult somatic cells into induced pluripotent stem cells or induced specialized cells such as cardiomyocytes yields an alternative. However, further investigations are required. Adult stem cells are most often only able to form their tissue of origin. In any case, therapy benefits depend on an efficient engraftment of administered stem cells, proper differentiation, and interaction with host cells.

Adult multipotent precursors offer alternatives to embryonic stem cells if they have the same potential. Stem cell function can be enhanced by suitable drugs and growth factors that promote cell growth, differentiation, survival, and homing.

Bone marrow accomodates various types of stem and precursor cells. They yield a pro-angiogenic microenvironment. Bone marrow-derived mesenchymal stem cells (bmMSC) can differentiate into multinucleated myotubes and exhibit cardiomyocyte-specific gene expression [1404].

Umbilical cord blood contains a wide variety of stem cells, such as hematopoietic, mesenchymal, and unrestricted somatic stem cells, that can generate numerous cell types.

Reprogramming of fibroblasts, B lymphocytes, hepatocytes, and gastric epithelial cells to induced pluripotent cells was carried out using retrovirus-mediated transduction of a defined set of transcription factors in mice (octamer-binding transcription factor Oct4, SRY-related HMG box Sox2, Krüppel-like factor KLF4, and MyC). Functionally redundant KLF2 and KLF5 can replace KLF4 in fibroblast reprogramming [1405]. In addition, nuclear receptor ERRβ (NR3b2) that forms a complex with Nanog and Oct4 and targets many genes involved in self-renewal and pluripotency in stem cells participates with Oct4 and Sox2 to conversion of mouse embryonic fibroblasts to induced pluripotent cells. Nuclear receptor NR3b2 can replace KLF4 in the presence of Oct4 and Sox2 and in the absence of MyC [1405].

Embryonic or induced pluripotent stem cells are most often suitable to generate given cell lineages, but have little or no capacity to commit into other cell lineages. Multipotent cardiovascular progenitors can enter cardiovascular cell lineages to give birth to distinct vascular cells. They must not only differentiate into multiple cell types (arterial, capillary, and venous endothelial and smooth muscle cells, nodal cells, and atrial and ventricular myocytes) according to types of stimulated pathways, but also keep the adequate proportion of involved cell types as well as tissue architecture. *Cardiovascular master stem cell* is a common multipotent progenitor that is characterized by cardiac signature gene NKx2-5.

Growth factor-mobilized peripheral blood mononuclear cells represent a suitable source of autologous progenitor cells when proper differentiation can be achieved. Activators of early cardiac lineage-restricted gene NKx2-5 include small sulfonyl hydrazones that can trigger expression of cardiac markers (e.g., transcription factor NKx2-5, myocardin, troponin-I, and sarcomeric α-tropomyosin)[10] in various embryonic and adult stem and progenitor cells such as mobilized peripheral blood mononuclear cells [1406].

Except left ventriculomyocytes, most cardiac cells (vascular smooth and cardiac myocytes and nodal and endothelial cells) are derived from multipotent progenitors that express gene Islet-1 (Isl1) [1407]. A small quantity of cells that express stem cell markers such as stem cell factor receptor (SCFR), stem cell antigen SCA1, and multidrug resistance protein MDR1, have been identified in the human heart [1408]. However, SCFR+ cells can be scouting cells of the immune system.

11.4.1.1 Cardiac Stem Cells

Cardiac stem cells divide either symmetrically or asymmetrically.[11] Myocardium contains stem cell niches [1409]. Cardiac stem cells are smaller then mature cardiomyocytes. Cardiac stem cells and supporting cells are connected by gap and adherens junctions. Integrins of primitive and committed cells within the niche attach these cells to extracellular matrix proteins (fibronectin and laminin) that transduce mechanical signals for differentiation regulation.

Stem cells may regenerate the myocardium, especially after myocardial infarction and in ischemic cardiomyopathy, because the heart is a weakly regenerative organ. Progenitor cells secrete paracrine factors and may contribute to vasculogenesis and tissue repair and remodeling [1410].

In the heart, resident stem cells can lead to endothelial and smooth muscle cells as well as cardiomyocytes [1411]. The heart may contain various populations of resident myocardial progenitors with different expression modes that may differentiate into cardiomyocytes and endothelial cells.[12]

Although cardiogenic progenitor cells maintain myocardial turnover, they do not lead to heart regeneration. Consequently, injured adult hearts form scars without cardiomyocyte proliferation. Appropriate differentiation of stem cell and induction of the division cycle of resident cardiomyocytes can be stimulated with suitable factors. Cytokine treatment (e.g., colony-stimulating factor CSF3) improves infarction repair, but can also increase mortality [1412]. Cell delivery can be done either by intracoronary infusion or direct intramyocardial injection during grafting or using endovascular procedures. Periostin activates plasmalemmal α_V-, α_1-, α_3-, and α_5-integrins, and induces re-entry of cardiomyocytes into the cell cycle [1413].

---

[10] These markers are supposed to be better than markers of late cardiogenic differentiation without an important role in cardiac fate, such as α-myosin heavy chain and atrial natriuretic factor.

[11] Asymmetric division predominates. Replicating cardiac stem cell gives birth to one daughter cardiac stem cell and one daughter committed cell (cardiomyocyte, endothelial cell, or smooth myocyte).

[12] In vitro, these observations may correspond to culture artifacts.

11.4.1.2 Endothelial Progenitor Cells

Circulating endothelial progenitor cells (EPC) promote neovascularization; hence, they are also called circulating angiogenic cells. They are reduced in number. They can become dysfunctional in chronic diseases. Circulating progenitors contribute only sligthly to cardiomyocyte repopulation. Endothelial progenitors can induce myogenesis in vitro, and angiogenesis for O_2 supply. A time window exists for optimal cell therapy. Infusion of EPCs promotes neovascularization after ischemia. Peptidase *cathepsin-L* expressed in EPCs (but not strongly in endothelial cells) degrades the matrix, thereby allowing invasion by EPCs [1414].

11.4.1.3 Skeletal Myoblasts

Skeletal myoblasts are quiescent stem cells with proliferative capacity in vitro and ischemic tolerance in vivo. Moreover, the risk of aberrant differentiation is limited [1415]. Unfortunately, implanted myoblasts do not necessarily differentiate into excitable cardiomyocytes. In particular, they are unable to couple electromechanically with cardiomyocytes. Skeletal myocytes actually do not express adhesion and junction proteins required for electromechanical coupling.

11.4.1.4 Bone Marrow Cells and Hematopoietic Stem Cells

Transplantation of bone marrow cells possibly enriched with hematopoietic stem cells can generate coronary endothelium to a greater extent than myocardium [1412]. Bone marrow cells also enhance infarct healing. Works on bone marrow-derived mononuclear cells fail to demonstrate efficiency.

Adult bone marrow cells retain dedifferentiation capacity. They can enter into cardiomyocyte and vascular lineages. Bone marrow cells engraft, survive, and grow within the myocardium and form junctional complexes based on connexin-43 and N-cadherin with resident cardiomyocytes [1416]. Implanted hematopoietic stem cells differentiate into blood cells, but not significantly into cardiomyocytes.

11.4.1.5 Mesenchymal Stem Cells

Mesenchymal stem cells (MSC) of the stromal region (non-hematopoietic compartment) of the bone marrow produce growth factors and cytokines. They colonize sites of injury. Implanted MSCs decrease pathological wall remodeling. Bone marrow mesenchymal stem cells can differentiate into cardiomyocytes and vascular cells [1417], but they can have limited survival after transplantation in the myocardium [1418]. Furthermore, cell therapy may be ineffective or even hazardous in certain clinical settings and specific subgroups of patients [1419, 1420].

Bone marrow-derived mesenchymal stem cells that overexpress the Pkb1 gene (that encode PKB1 kinase) are better for cell therapy of acute myocardial infarction. Secretion of paracrine cytoprotective factors from stem cells exerts an effect on cardiomyocytes exposed to hypoxia [1421].

A mixed ester of hyaluronan with butyric and retinoic acid acts as a cardio- and vasculogenic agent in human mesenchymal stem cells isolated particularly from fetal membranes of term placenta. Hyaluronan with butyric and retinoic acids primes stem cell differentiation into endothelial and cardiac cells [1422].

11.4.1.6 Embryonic Stem Cells

Embryonic stem cells differentiate into all cell types, especially cardiomyocytes organized in myofibers when stimulated by suitable cardiac promoters, but they can produce teratomas. Moreover, they can form ectopic pacemakers.

Human embryonic stem cells can differentiate to cardiomyocytes using activin-A and BMP4 factors. However, a mixed cell population with cardiomyocytes, endothelial cells, and fibroblasts must be cultured in a rotating orbital shaker to create human, functional, prevascularized, cardiac tissue patches [1423]. Preformed microvessels can anastomose with the host coronary circulation.

In summary, transdifferentiation has stimulated multiple clinical trials. Cardiac progenitors from the cardiac population and cardiomyogenic cells derived from adult bone marrow stromal cells are able to differentiate into cardiomyocytes when interacting with cardiomyocytes. Regenerative cardiology trials most often use bone marrow cells and mesenchymal stem cells, because cardiac stem cells do not generate significant repair. However, the results of stem cell injection in hearts after heart attack at most show slight, transient benefit. Cell treatment optimization hence remains challenging, taking into account disease type, patient history, associated therapy for the selection of the cell type, associated growth factors, delivery route, and administration timing.

11.4.2 Gene Therapy

Genes encode transcripts for the synthesis of proteins required for cell structures and functions. Gene therapy is the delivery of genetic materials into cells using viral and non-viral gene carriers, or vectors. Gene transfer technology is aimed at correcting defective genes responsible for lesions and thus repairing diseased tissues. Normal genes are inserted within the genome to replace abnormal genes.

Current vectors are genetically modified, safe viruses (e.g., adenoviruses, lentiviruses, retroviruses, and adeno-associated virus)[13] that unload the therapeutic human gene into target cells. Gene therapy vectors using adeno-associated virus can lodge without incorporating the genome (extrachromosomal residence) of host cells. In particular, recombinant human adenoviruses and adeno-associated virus efficiently deliver and express their genomes in both dividing and quiescent cells. Adeno-associated virus can cause a long-lasting gene expression with respect to adenovirus, because of its minimal immunogenicity. Major drawbacks comprise oncogenic transformation and inflammation with elimination of infected cells.

Non-viral gene transfer uses plasmids, polymers, and liposomes,[14] which are less expensive and safer than viral vectors, but less efficient. Manufactured non-viral vectors with low toxicity and immunogenicity can efficiently enter into cells, but yield low gene expression.

Genetic transfer to restore a normal cardiac function struggles against cardiomyocyte apoptosis and targets angiogenesis using growth factors and cytokines ($VEGFa_{121}$, $VEGFa_{165}$, VEGFc, FGF4, IGF1, CSF3, and erythropoietin),[15] nervous control[16] (e.g., C-terminus of β-adrenergic receptors and adenylate cyclase AC6),[17] and Ca^{2+} handling during excitation-contraction coupling (systole) and relaxation

[13]Adeno-associated virus is a non-enveloped species of the Dependovirus genus (synonym: adeno-associated virus set) of the Parvoviridae family that contains a single-stranded DNA in an icosahedral capsid (proteic shell). This small virus infects humans without causing disease, but only a very mild immune response.

[14]Liposome is an artificial lipid capsule with an aqueous core with therapeutic DNA that is able to cross the plasma membrane.

[15]Receptor CSF3R resides on cardiomyocytes and endothelial cells. Factor CSF3 represses apoptosis, increases NO production, and supports angiogenesis, as it enhances proliferation and migration of endothelial cells. Both CSF3 and erythropoietin mobilize hematopoietic cells and endothelial progenitors from the bone marrow. They activate JaK2 and its anti-apoptotic and angiogenic signaling via STAT3, STAT5, the PI3K–PKB axis, and MAPK module. In addition, CSF3 promotes early myocardial healing via TGFβ1 and prevents later excessive fibrosis, as it hampers angiotensin-2 receptor AT_1 and TNFα. The Epo receptor is also synthesized in both ventriculomyocytes and endothelial cells. Erythropoietin, which is liberated by cardiomyocytes, causes secretion of sonic Hedgehog from cardiomyocytes; it can then launches angiogenesis by stimulating cardiomyocytes to release VEGF and angiopoietin-1. Erythropoietin prevents apoptosis of endothelial cells and cardiomyocytes, as it activates PKC and PI3K–PKB pathway. In addition, Epo fosters cardiac mitochondrion genesis.

[16]In the heart, β1-adrenergic receptors are the predominant subtype.

[17]Cardiac failure is characterized by a reduction of plasmalemmal density of β1-adrenoceptors and by an uncoupling of these receptors from G proteins (desensitization). The C-terminus of β-adrenergic receptor, a GRK2 inhibitor, impedes recruitment of G-protein-coupled receptor kinase GRK2, which phosphorylates (desensitizes) β-adrenoceptors and promotes binding to β-arrestins, thereby supporting cAMP signaling. Adenylate cyclase-6 increases Ca^{2+} handling via PKA, PI3K–PKB, and ATF3 pathways.

(diastole; i.e., (sarco(endo)plasmic reticulum Ca^{2+} ATPase SERCA2a, phospholamban, and S100 Ca^{2+}-binding protein S100a1) [1424].[18]

A combination of 3 transcription factors, GATA4, myocyte enhancer factor MEF2c, and T-box factor Tbx5, reprograms fibroblasts into functional cardiomyocytes in vitro. Transfer of the Tbx5 gene using viral vector can convert cardiac fibroblasts from the infarct region into cardiomyocyte-like cells [1426].

Sonic Hedgehog[19] (Shh; Vol. 3 – Chap. 10. Morphogen Receptors) that regulates heart development during embryogenesis, can preserve, after intramyocardial gene transfer in adult animal hearts, the left ventricular function after both acute and chronic myocardial ischemia by enhanced neovascularization, as well as reduced fibrosis and apoptosis [1427].

11.4.3 Tissue Engineering

Tissue engineering[20] is the construction of living tissue. Bioreactors are devices used for the growth of tissues in an artificial environment that mimicks the physiological conditions.

In vivo biomechanical[21] and chemicophysical conditions that mimic the natural environment (mechanical loading such as flow in particular) are created for in vitro cell conditioning and construction of blood vessel, heart valves, etc., with features similar to the native tissue ones. Mechanical loading with a conditioning protocol (cycle frequency and magnitude range) of tissue in formation in bioreactors

---

[18]The amplitude and frequency of Ca^{2+} transients are regulated by the phosphorylation of Ca^{2+} regulators, which depends on the balance between the activity of kinases, such as PKA and CamK2, and phosphatases, such as PP1 and PP2. Protein kinase-A, an amplifier of β-adrenergic receptor signaling, phosphorylates (activates) $Ca_V1.2$ channels and ryanodine receptors as well as (inactivates) phospholamban, hence raising SERCA activity, in addition to (activates) $PP1_{r1a}$, $PP1_{r1b}$, and $PP1_{r2}$, converting them in potent inhibitors of protein phosphatase-1. Cardiac-specific Ca^{2+} sensor S100a1 is preferentially produced and abounds in the heart; it is the predominant subtype among other S100 isoforms, such as S100a4, S100a6, and S100b. It lodges mainly in ventriculomyocytes. It localizes to the sarcoplasmic reticulum, sarcomere, and mitochondria. It interacts in a Ca^{2+}-dependent manner with the cardiac ryanodine receptor RyR2 (augmented cytosolic Ca^{2+} transients), the SERCA2a–phospholamban complex (increased Ca^{2+} resequestration), titin, and mitochondrial F_1 ATPase (enhanced ATP generation) [1425]. Its nitric oxide-dependent S-glutathionylation increases its Ca^{2+} affinity (nanomolar range). Therefore, it is an inotropic and lusitropic enhancer.

[19]Sonic Hedgehog is a member of the Hedgehog family of secreted glycoproteins. It acts via Patched-1–Smoothened receptor complex to upregulate the transcription of zinc finger Gli transcription factor.

[20]Tissue engineering is a multidisciplinary topic that involves, in particular, cell and molecular biologists, biomaterial scientists, and imaging specialists to develop materials that can replace or support diseased or damaged tissues and restore or improve tissue functions.

[21]Because mechanical forces regulate the physiological functioning of the cells, these influences must be reproduced in cell and tissue engineering.

increases synthesis of extracellular matrix constituents much more than in the absence of dynamical loading.

Smooth myocytes cultured in a three-dimensional collagen-1 matrix, stimulated by PDGF and TGFβ and cyclic mechanical loading respond accordingly to the combination of chemical and mechanical stimuli [1428]. The produced effects depend on the nature of the extracellular matrix.

In vitro investigations on cultured cells can simplify the reality because important, more or less unknown, factors can be neglected, in particular blood supply. Signaling associated with local and regional needs as well as the state of the whole organism is neglected.

By analogy with ecology, histological *niches* better define the cell status, living site, functioning, and interactions with the environment. Bioreactors must then replicate the niche. For example, endothelial cells constitute a niche component of adult neural stem cells.[22] Osteoblasts lining the inner bone surface are associated with hematopoietic stem cells.[23]

A scaffold made of natural and synthetic materials that acts as a temporary extracellular matrix, with a nourishing supply bathed with growth factors is seeded by cells that thrive and adapt to produce engineered tissue to be implanted in the body. Synthetic biomaterials are thus developed for use as three-dimensional extracellular environments that mimic the regulatory characteristics of natural extracellular matrix and ECM-bound growth factors. Support biomaterials include self-assembling fibrillar networks and other protein polymers that present bioactive ligands and respond to cell signals. Such cues can lead to remodeling and developmental processes involved in tissue-specific differentiation with suitable structure–function relationships. Besides, bone marrow cultured in growth matrix can be used to yield angiogenesis. Moreover, mesoscale tissular modules can be assembled and associated with vascular supplies to form vascularized engineered tissues with significant viable cell densities [1431].

Tissue production depends on interactions between the growing tissue and its environment. The dynamic environment within bioreactors, in particular stimulation by mechanical stresses under control to avoid cell damage significantly affect tissue growth and development.

---

[22]In the hippocampus, adult neural stem cells receive cues not only from astrocytes [1429], but also from endothelial cells [1430]. Endothelial cells, but not vascular smooth myocytes, release factors that stimulate the self-renewal of neural stem cells, inhibit their differentiation, and enhance neuron production, activating the Notch pathway. The signaling between adult stem and support cells involves the same molecular pathways as during embryogenesis, such as the Notch, TGFβ, BMP, and Wnt axes.

[23]Osteoblasts have plasmalemmal Jagged-1 ligand that activates Notch receptor of hematopoietic stem cells. Stimulated Notch pathway leads to hematopoietic stem cell self-renewal.

11.4.3.1 Heart Tissue Engineering

Engineered heart tissue can form a thick myocardium after implantation on myocardial infarcts in immune-suppressed rats [1432]. When evaluated 28 days later, engineered heart tissue shows neither undelayed electrical coupling to the native myocardium, arrhythmia, nor dilation. Cardiomyocytes seeded onto mechanically conditioned, perfused, porous scaffolds can conduct action potentials and beat synchronously, but the scar hinders a proper integration of implanted cells.

Cell sheet technology applied to a monolayer of adipose tissue-derived, pluripotent, mesenchymal (non-hematopoietic) stem cells allows the repair of scarred myocardium after myocardial infarction in rat hearts [1433]. Engrafted mesenchymal stem cells prevent wall thinning in the scar area and improve the cardiac function.

Composite matrices can be used for heart valve engineering. Enzymatically decellularized porcine aortic valves that are impregnated with biodegradable polymer by a stepwise exchange provide a scaffold for heart valve tissue engineering with complete endothelialization [1434]. Ovine cell seeding of decellularized porcine heart valves (with almost complete preservation of the extracellular matrix) is improved under pulsatile flow conditions [1435].

Preliminary numerical experiments have been carried out to study the content and orientation of collagen fibers[24] of the loaded aortic valve [1437]. In an isotropic cusp, the fiber orientation is driven by the principal strain directions. The remodeling depends on the transvalvular pressure, fiber stiffness, initial fiber direction (close to the principal strain directions in this work), and loading conditions.

11.4.3.2 Vascular Tissue Engineering

Sheet-based tissue engineering, using fibroblasts and endothelial cells, without smooth myocytes and exogenous biomaterials, can form cylindrical blood vessels, characterized by a composite multilayered wall with vasa vasorum [1438].

Arterial stenoses and lumen blockage by clots locally or emboli remotely can be treated by bypass grafting surgery. Autologous grafts (internal mammary artery, saphenous vein) are used in most patients. Synthetic grafts can also be utilized. To reduce the failure rate of synthetic (polytetrafluoroethylene) grafts due to thrombosis and scar within graft lumen, biocompatible, durable biosynthetic grafts have been made. Synthetic tubes are coated by an adhesion matrix (fibronectin) and endothelial cells on its interior surface. Endothelial cells are genetically changed to overexpress

[24]The cardiac extracellular matrix is mainly composed of collagenous fibers. The fiber network is involved in the distribution of forces generated in the heart. Studies on collagen synthesis by cardiac fibroblasts show that the ratio of collagen-3 to collagen-1 increases in mechanically stretched cells [1436]. Collagen-3 mRNA level increases in response to cyclic mechanical stretch in 12 h, whereas collagen-1 mRNA concentration does not change.

fibulin-5 in order to improve their adhesion on matrix and avoid flow-induced detachment [1439, 1440]. Because of their decayed capacity of proliferation, these modified endothelial cells are also genetically altered to overexpress vascular endothelial growth factor. A coating of adhesion matrix on the tube exterior surface allows seeding of smooth myocytes, which are also genetically altered to overexpress adhesion factors and growth factors.

Mesenchymal stem cells can also be used for the construction of vascular grafts. Strain influences the differentiation (particularly into smooth myocytes) and proliferation of mesenchymal stem cells, mechanotransduction depending on the orientation of cells with respect to the strain axis [1441].[25] The implicated signaling pathway can be affected by the activity of the cytoskeleton and focal adhesions, both subjected to applied loading.

A better understanding of mechanotransduction processes can improve the tissue engineering setup. Numerical simulations, once validated, are important because they: (1) provide the fields of imposed stresses and (2) can optimize the bioreactor design.

11.4.3.3 Lung Tissue Engineering

Lung design is characterized by an efficient structure–function relationship. The pulmonary architecture enables ventilation, perfusion, matching of these 2 functions, and gas exchange across an optimal gas transfer surface area. The anticoagulant and -immunogenic air–blood interface is immersed in a deformable pump. Because lung tissue has a limited regeneration capacity in vivo, lung transplantation is used in some circumstances, but is limited by a shortage of donor organs. In the case of bone marrow, successful transplantation relies on tissue devitalization followed by reintroduction of a regenerating cell population. After complete decellularization of isolated adult rat lungs, the acellular extracellular matrix that preserves the hierarchical branching structure of its vasculature and airways, i.e., that keeps scaffolds of vasculature, airways, and alveoli, is reseeded with lung epithelial and endothelial cells of neonatal lungs and cultured in a bioreactor that retains the developmental environment of fetal lung [1442, 1443]. The resulting tissue contains alveoli, microvessels, and small airways with the appropriate cell types and rheology. After sevarl days, the construct can be perfused with blood and ventilated with physiological pressures.

---

[25]Cyclic uniaxial strain on elastomeric membranes coated by mesenchymal stem cells upregulates smooth myocyte markers, such as α-actin, caldesmon, calponin, and myosin, until the cells align perpendicularly to the strain axis. Microgrooves created in the elastomeric cell culture membrane in the loading direction keep mesenchymal stem cells aligned parallel to the strain axis and change the cell behavior, with a production increase in calponin-1 and decrease in cartilage matrix markers. Microgrooves normal to the loading direction lead to attenuation in matrix remodeling and cell signaling.

11.5 Inflammation

The immune system responds to aggression by inflammation, a defensive event cascade. Inflammation includes: (1) effector and cell delivery to the involved region; (2) limitation of the damage spreading by a physical barrier; and (3) angiogenesis and tissue repair.

The major inflammatory events comprise: (1) an immediate local increase in blood supply with vasodilation upstream and vasoconstriction downstream from the affected region; (2) a decrease in blood velocity for cell adhesion; (3) an increase in local capillary permeability, resulting in plasma leakage in affected tissues and creating interstitial space congestion (the exudated fluid has a higher protein content than the fluid that is normally filtrated by the capillary wall); (4) an increase in expression of endothelial adhesion molecules[26] for cell adhesion and extravasation; and (5) an influx of phagocytic cells into damaged tissues.[27] Later, lymphocytes enter the inflammation site.

Leukocyte chemotaxis under molecular guidance is controlled by external and internal signals that activate signal transduction cascades and regulate the cytoskeleton dynamics (Vol. 2 – Chap. 6. Cell Motility). Several signaling pathways are involved in sensing chemoattractant gradient and amplification, among which is the phosphatidylinositol 3-kinase axis.

In addition to migration of leukocytes during inflammation, cell subpopulations continuously patrol organ microvasculature at rest to search for invading particles. Subpopulations of monocytes permanently scout the skin, mesentery, and brain microcirculation. Upon infection, they are the very first cells to enter tissue. Intravascular CD1d-restricted invariant natural killer T cells patrol within liver sinusoids at a speed of 10 to 20 μm/min [1444]. Vascular patrolling CSF1R+, Ly6g− (in Mus musculus) monocytes are also able to extravasate [1445].

Vascular inflammation is detected by augmented circulating interleukin-6 that is produced by activated inflammatory cells within the vessel wall. Synthesis of IL6, in turn, causes the expression of hepatic acute-phase reactants, such as C-reactive protein, γ-fibrinogen, and angiotensinogen [1446].

11.5.1 Extravasation

Alarm signals generated at inflammatory foci reach the vascular lumen to attract diverse leukocyte subsets that decipher and integrate these signals. Soluble factors, i.e., cytokines, chemokines, chemoattractants, and adhesion molecules as well as adhesion and chemotactic receptors in both endothelial cells and leukocytes, orchestrate various steps of leukocyte recruitment during inflammation. Cytokines

[26]Endothelial cells, activated by released IL1 and TNF, express VCAM, ICAM1, E-selectin, and L-selectin. Mastocytes release histamine and prostaglandin, macrophages TNFα and IL1.

[27]Neutrophils, followed by macrophages, are the first leukocyte type to migrate.

and chemokines locally released by an inflammatory focus reach the surrounding endothelium, which becomes activated.

The repertoire of adhesion molecules expressed by different populations of leukocytes as well as chemoattractants and their receptors determines their recruitment. These molecules follow a spatiotemporal organization at the wetted (luminal) surface of the vascular endothelium and leukocyte membrane or create chemical gradients in damaged tissue.

Margination allows circulating leukocytes to move away from the blood stream. Capture, or tethering, represents the first contact of a leukocyte with the activated endothelium. During the first steps — tethering and rolling, labile contacts between leukocytes and activated endothelial cells depend on selectins and their ligands. Once leukocytes are captured, they transiently adhere to the endothelium and begin to roll.[28] Rolling allows leukocytes to search for cues such as glycosaminoglycan-immobilized chemokines on the endothelial surface [1445]. Trapped chemokines activate leukocyte chemokine receptors to express leukocyte integrins that bind endothelial ligands for firm adhesion.

During the transition from rolling to firm adhesion and crawling, leukocytes undergo a morphological change from round to polarized promigratory shape. Endothelial cells emit docking microvilli around adhered leukocytes to prevent their detachment. Leukocyte crawl on the endothelium to find a suitable place for transmigration. Crawling includes cell polarization with a leading edge (pseudopod) and a trailing side (uropod), associated with cytoskeletal rearrangement within cell protrusion and retraction. The contractile force that pulls the uropod results from stress fibers with linked myosin-2 nanomotor. Afterward, leukocytes migrate across endothelium if a chemoattractant is present (chemotactic transmigration).

Mechanotaxis, i.e., response to hemodynamical forces, influences crawling. In the absence of flow, neutrophils crawl in all directions. Once shear is applied, neutrophils move perpendicularly to exerted shear [1444].

After transendothelial migration between adjacent endothelial cells (paracellular passage) or across them (transcellular diapedesis), leukocytes traverse the endothelial basement membrane. They then continue to travel through the extracellular matrix guided by a chemotactic gradient.

11.5.2 Molecular Basis of Extravasation

When an inflammatory zone occurs, surrounding endothelial cells become activated by cytokines and chemokines. They increase their content of adhesion molecules on their luminal surface. Extravasation during inflammation occurs preferentially at postcapillary venules.

---

[28]During rolling, bonds are formed at the leading edge of rolling cells and broken at the trailing edge.

11.5.2.1 Pro-Inflammatory Cytokines Involved in Extravasation

Leukocyte extravasation is influenced by: (1) pro-inflammatory cytokines, such as interleukins IL1α, IL1β, and IL6, and TNFα, among others, and (2) chemoattractants for myeloid cells, such as ^{N}formyl methionyl leucyl phenylalanine (fMLP), complement component C5a, leukotriene LTb4, and platelet-activating factor, produced in the inflammatory site [1445].

The secretion of a selective set of chemokines and upregulated expression of endothelial adhesion molecules caused by specific cytokines direct the subpopulations of leukocytes that are recruited during different phases of inflammation. In response to pro-inflammatory stimuli, such as TNFα, IL1β, Ifnγ, and thrombin, endothelial cells synthesize different chemokines, such as CCL2, CCL5, CXCL1, CXCL8, CXCL10, or CXCL12 [1445].

Endothelial cells produce chemokines attached to glycosaminoglycans and Duffy receptor for chemokines, especially in apical endothelial microvilli [1445]. These endothelium-immobilized chemokines promote the transition from selectin-mediated tethering and rolling of leukocytes to integrin-mediated firm attachment and spreading, as they activate their cognate Gi/o-coupled chemokine receptors to express leukocyte integrins. In addition, neutrophils secrete chemoattractants to recruit other leukocyte types in later stages of inflammation.

During inflammation, myeloid cell subpopulations, in which several chemokine receptors are upregulated, migrate from the bone marrow to inflammation loci and uptake antigens. The innate immune response that involves monocytes, macrophages, and dendritic cells, occurs in the inflamed tissue. Migration of immature dendritic cells to inflammation zones depends on the expression of chemokine receptors such as CCR2, CCR5, and CCR6 [1445]. After antigen uptake, the density of these chemokine receptors on mature dendritic cells decays, whereas CCR7 expression rises.

Conventional dendritic cells can then move via lymphatics to draining lymph nodes. In lymph nodes, mature dendritic cells present antigens to specific naive T cells. Naive T lymphocytes differentiate into effector (T_{H1}, T_{H2}, T_{H9}, T_{H17}, T_{H22}, and follicular T cells) and regulatory T lymphocytes (Table 11.5). Effector CD4+ T cells, T_{Reg}, cytotoxic CD8+ T cells, and NK cells upregulate chemokine receptors to also migrate to inflammation sites and mount a specialized immune response. In addition, Ly6c^{high}[29] Ly6g+[30] monocytes migrate to participate in mucosal immunity [1445].

[29] Lymphocyte antigen-6 complex, locus C in rodents.

[30] Lymphocyte antigen-6 complex, locus G in Mus musculus. Lymphocyte antigen-6 complex, loci Ly6g5c, Ly6g6c, and Ly6g6d exist in humans, in addition to Ly6d and Ly6h.

Table 11.5. Chemokine receptors and other products of subpopulations of T lymphocytes (Source: [1445]; Chap. 3; CTL: cytotoxic T lymphocyte; NK: natural killer cell; T_{FH}: follicular helper T lymphocyte; T_H: helper T lymphocyte; T_{Reg}: regulatory T lymphocyte; Ifn: interferon; IL: interleukin; TNF: tumor-necrosis factor). T_{H2} cells express chemoattractant receptor homologous molecule expressed in T_{H2} cells (CRTH2; a.k.a. prostaglandin-D2 receptor [DP_2], G-protein-coupled receptor GPR44, and CD294). The synthesis of T_{H2} cytokines is enhanced by CCL2 and CXCL4 chemokines. T_{FH} cells localize to the follicular regions of B cells owing to chemokine receptor CXCR5 and its ligand CXCL13 secreted by follicular stromal cells. T_{Reg} cells modulate the immune response to maintain immune tolerance. They reside mainly in the bone marrow owing to CXCR4 chemokine receptor. During inflammation, they migrate using CCR2, CCR4 to CCR6, and CXCR3 chemokine receptors. CD8+ cytotoxic T cells (CTL) migrate to tissues where CCL3 and CCL4 are synthesized. NK cells upregulate CCR7 to migrate in response to CCL9 and/or CCL21 chemokine.

T lymphocyte type	Chemokine receptors	Adhesion molecules	Synthesized cytokines and chemokines
T_{H1}	CCR5/7, CXCR3/4/6 CX3CR1	ICAM1, PSGL1	CCL5 IL1/2, 2Ifnγ, TNFSF1/2
T_{H2}	CCR4/7/8 CXCR4		IL4/5/10/13
T_{H9}			IL9/10
T_{H17}	CCR2/4 (at rest), CXCR4/6 (at rest) CCR6 (activated)		CCL20 IL17/21/22/26
T_{H22}	CCR4/6/10		IL22
T_{FH}	CXCR5		IL21
T_{Reg}	CXCR4 (at rest), CCR2/4/5/6/7, CXCR3 (migration)		IL10, TNFSF4/5
CD8+ CTL	CCR5		
NK	CCR7		

11.5.2.2 Adhesion Molecules Involved in Extravasation

Numerous adhesion molecules are implicated in transmigration, such as: (1) selectins (E-, L-, and P-selectins);[31] (2) cadherins such as vascular endothelial (VE)-cadherin, or cadherin-5; (3) immunoglobulin-like adhesion molecules, such as platelet–endothelial cell adhesion molecule PECAM1, intercellular adhesion

[31] E-selectin is also called endothelial-leukocyte adhesion molecule ELAM1 and leukocyte-endothelial cell adhesion molecule LECAM2; L-selectin lymphocyte adhesion molecule-1 and sulfated form of sialyl Lewis-X; P-selectin granule membrane protein GMP140 and platelet activation-dependent granule to external membrane protein (PADGEM).

Table 11.6. Adhesion molecules involved in leukocyte migration across blood vessel wall during inflammation (Source: [1447, 1448]; GAGCAM: glycosaminoglycan cell adhesion molecule, or glycosylation-dependent CAM [GlyCAM1]); ICAM: intercellular adhesion molecule; MAdCAM: mucosal vascular addressin cell adhesion molecule; PECAM: platelet endothelial cell adhesion molecule; VCAM: vascular cell adhesion molecule). 3-fucosyl ^{N}acetyl lactosamine (FAL; a.k.a. Lewis X or CD15) is a glucid adhesion molecule expressed on glycoproteins and -lipids as well as proteoglycans that is synthezised by fucosyltransferases FuT4 and FuT9. It mediates neutrophil chemotaxis and phagocytosis. The random contact is followed by rolling to a stop involving selectins. Sticking and extravasation require integrins.

Extravasation step	Adhesion molecules
Rolling	Selectins (CD62E/L/P), FAL (CD15), PSGL1 (CD162) GAGCAM1, neutrophil addressins
Stopping	$\alpha_4\beta_1$-, β_2-, $\alpha_4\beta_7$-, $\alpha_L\beta_2$-integrins ICAM1, VCAM1, MadCAM1
Transmigration	β_2-integrins ICAM1, VCAM1, PECAM1

molecules (e.g., ICAM1 and ICAM2), and vascular cell adhesion molecule VCAM1; (4) integrins, such as $\alpha_4\beta_1$-, $\alpha_4\beta_7$-, $\alpha_L\beta_2$-, and $\alpha_M\beta_2$-integrin; as well as (5) integrin-associated protein, or leukocyte surface antigen CD47 (Table 11.6; Vols. 1 – Chap. 7. Plasma Membrane and 2 – Chap. 6. Cell Motility). Leukocyte integrins connect to their endothelial ligands that are organized in tetraspanin-enriched microdomains [1445]. These cellular structures mediate the contact between leukocytes and endothelial cells during extravasation.

Endothelial adhesion receptors of activated endothelial cells localize to adhesion platforms that are enriched in tetraspanins in apical microvilli. Tetraspanins interact simultaneously with other tetraspanins and other types of transmembrane receptors. These adhesion platforms contain selectins and members of the Ig superfamily (VCAM1, ICAM1, ICAM2, PECAM1, and JAMs) [1445]. These functional units that coalesce around adherent leukocytes are connected to the actin cytoskeleton and contain α-actinin, ezrin, moesin, paxillin, talin, vinculin, and vasoactive stimulatory phosphoprotein. They are regulated by the Rho–RoCK pathway and PIP_2 second messengers. These docking structures, or transmigratory cups, stabilize firm adhesion of leukocyte to endothelium, hence preventing leukocyte detachment.

During leukocyte displacement over the endothelium, L-selectin, epican (CD44), ICAM1, and ICAM3, among others redistribute to the leukocyte rear pole. On the oher hand, integrins, such as $\alpha_L\beta_2$- and $\alpha_M\beta_2$-integrins, and chemokine receptors localize to the leading edge and endothelial contact area.[32]

Leukocytes probe for sites suitable for diapedesis by constructing protrusions. When they find an appropriate site for migration, filopodia evolve into an invasive pseudopod to cross the endothelium. During paracellular transmigration, ICAM1

[32]Leukocyte crawling relies on $\alpha_L\beta_2$-integrin in lymphocytes, $\alpha_L\beta_2$- $\alpha_M\beta_2$-integrins in monocytes, and $\alpha_M\beta_2$-integrin in neutrophils [1445].

and JAMa interact with $\alpha_L\beta_2$-integrin and form a ring-shaped cluster around the leukocyte [1445]. Transcellular migration requires the supply of additional membrane by the vacuolovesicular organelles to build a transcellular pore that lenghtens toward the basal membrane. Both ICAM1 and caveolin-1 translocate toward the transcellular pore. Membrane fusion is regulated by Ca^{2+} and SNARE-containing complexes [1445]. Intermediate filament constituent vimentin contributes to the transcellular pore.

Selectins

The selectin family of transmembrane adhesion receptors mediates leukocyte tethering and rolling. Selectin-P on endothelial cells is the primary adhesion molecule for cell capture and rolling onto vascular endothelium. The main leukocyte ligand for P-selectin is selectin-P ligand (SelPLg; a.k.a. P-selectin glycoprotein ligand-1 [PSGL1] and CD162).[33]

Selectin-L also contributes significantly to leukocyte capture. Selectin-L and -E also take part in rolling. Selectin-L is much less efficient than P-selectin in cell rolling. Nonetheless, it is necessary to capture leukocytes and initiate their rolling. Selectin-E may be responsible for slow rolling interactions and possibly the initiation of firm adhesion.

Selectin-E and -P interact with L-selectin and PSGL1 expressed by leukocytes (Table 11.7). In addition, E-selectin targets other specific ligands, such as E-selectin ligand ESL1 (a.k.a. Golgi body sialoglycoprotein GlG1 and cysteine-rich fibroblast growth factor receptor CFR1), epican (a.k.a. extracellular matrix receptor ECMR3, GP90 lymphocyte homing–adhesion receptor, heparan sulfate proteoglycan receptor, and CD44).

Integrins

β_2 Integrins participate in leukocyte deceleration and arrest. Neutrophils express small amounts of other integrins such as $\alpha_4\beta_1$-integrin that binds to endothelial VCAM1 molecule. Integrin-$\alpha_L\beta_2$ is the most important integrin in firm leukocyte adhesion, whereas $\alpha_M\beta_2$-integrin seems to be important in neutrophil activation and phagocytosis. Both $\alpha_L\beta_2$- and $\alpha_M\beta_2$-integrins can bind to ICAM1 and ICAM2 molecules.

Neutrophils and some lymphocyte subpopulations use preferentially $\alpha_L\beta_2$-integrin for firm adhesion. On the other hand, monocytes and lymphoblasts target predominantly $\alpha_4\beta_1$-integrin [1445].

---

[33] P-selectin glycoprotein ligand-1 on all lymphocytes, monocytes, eosinophils, and neutrophils allows these cells to bind to endothelial P-selectin and, then, to roll along the endothelium.

Table 11.7. Adhesion molecules of leukocyte binding to endothelium and their interactions and functions (Source: [1448]; HCAM: homing cell adhesion molecule; I: integrin; TEM: transendothelial migration). High endothelial venules (HEV) are specialized postcapillary venous swellings with cuboidal endothelial cells in all secondary lymphoid organs, except the spleen, that enable circulating leukocytes to enter lymph nodes in any organ, tonsils and adenoids in the pharynx, Peyer's patches in the small intestine, and small aggregates of lymphoid tissue in the stomach and large intestine, once they have crossed high endothelial venules. Sialyl Lewis antigen X (sLe^x), a carbohydrate E-selectin ligand, is constitutively expressed on granulocytes and monocytes; it mediates their extravasation. It is also expressed by activated T (e.g., T_{H1}, but not T_{H2} cells) and B lymphocytes.

	Neutrophil Receptor	Endothelial cell Receptor	Role
Selectin interactions	L-selectin, sLe^x	E/P-selectins GAGCAM, neutrophil addressins	Rolling, binding to HEV
Integrin–IgCAM interactions	$\alpha_L\beta_2$-I $\alpha_4\beta_1$-I	ICAM1/2, VCAM1	Adhesion; homing
IgCAM interactions	PECAM1 HCAM	PECAM1	Adhesion potentiation; TEM; binding to connective tissue components

Immunoglobulin-like Adhesion Molecules

Several integrin ligands, such as vascular cell (VCAM1) and intercellular (ICAM1–ICAM2) adhesion molecules, are involved in firm adhesion, crawling, and paracellular and transcellular migration. Other types of Ig-like adhesion molecules, such as platelet–endothelial cell (PECAM1), junctional (JAM), and endothelial cell-selective adhesion molecule (ESAM), T-cell surface glycoprotein-E2 (or CD99), among others, act during paracellular transmigration [1445] (Table 11.8).

Receptor VCAM1 is the endothelial ligand for the $\alpha_4\beta_1$- and $\alpha_4\beta_7$-integrins. Both ICAM1 and ICAM2 tether $\alpha_L\beta_2$- and $\alpha_M\beta_2$-integrins. Mucosal vascular addressin cell adhesion molecule (MAdCAM) is a major ligand of $\alpha_4\beta_7$-integrin and L-selectin. Adhesion molecules PECAM1 and JAMs can interact with themselves as well as with integrins [1445].

Table 11.8. Endothelial cell junctional molecules in leukocyte transmigration (Source: [1449]). Substance CD99 expressed on all leukocytes is also called T-cell surface glycoprotein-E2. Molecule CD99-like-2 protein (CD99L2) is synthesized in leukocytes and endothelial cells. It blocks neutrophil transmigration through the perivascular basement membrane,

Type	Leukocyte type specificity	Stimulus specificity	Extravasation stage
CD99			Transcellular route
CD99L2	Neutrophils		Basement membrane crossing
ESAM	Neutrophils		
ICAM2		IL1β	
JAMa		IL1β	Transcellular route
PECAM1	Monocytes, neutrophils, NK cells, some lymphocyte types	IL1β	Paracellular and transcellular route Basement membrane crossing

11.5.2.3 Signaling during Extravasation

Rolling

The first step of leukocyte attachment for extravasation is the rolling of leukocytes over the endothelium owing to selectins. Lymphocytes, eosinophils, neutrophils, monocytes, and mastocytes utilize selectins as rolling receptors [1450].

Expression of E-selectin results from cytokines, such as TNFα, interleukins IL1β and IL10, oncostatin-M, and lipopolysaccharide. Within 3 to 4 h after stimulation, E-selectin reaches its maximal level at the plasma membrane [1450]. On the other hand, interleukin-4 prevents E-selectin production via STAT6 factor. Glucocorticoids, transforming growth factor-β, and cAMP can counteract cytokine-induced expression of E-selectin [1450].

Interleukin-4 and oncostatin-M trigger P-selectin synthesis, but not TNFα, IL1β, and LPS [1450]. L-selectin is constitutively expressed on myeloid cells and a large subset of lymphocytes. Lymphocyte and neutrophil activation provokes rapid (within minutes) proteolytic cleavage of L-selectin [1450]. Various chemoattractants and activators, such as C5a, fMLP, leukotriene-B4, IL8, TNF, and granulocyte–macrophage colony-stimulating factor (gmCSF or CSF2), but not gCSF (CSF3), mCSF (CSF1), IL1, and interferon-γ, cause this proteolytic shedding.

Leukocyte clustering of E-selectin is associated with the linkage to the actin cytoskeleton, as this clustering causes the recruitment of actin-binding proteins, such as α-actinin, filamin, focal adhesion kinase, paxillin, and vinculin [1451]. Clustering of P-selectins is also associated with changes in the actin cytoskeleton.

Selectins and their ligands, i.e., carbohydrate structures on glycoproteins (P-selectin ligands PSGL1 and CD24; L-selectin ligands GlyCAM1, MAdCAM1, and CD34; E-selectin ligands ESL1 and PSGL1) and -lipids (sialyl Lewis^x-carrying glycolipids) and proteoglycans participate in the activation of leukocyte integrins [1450].

L-selectin can prime signaling cascades that involve lymphocyte-specific protein Tyr kinase (LCK), GRB2 adaptor, SOS GEF, Ras and Rac2 GTPases, MAPK kinases, and O_2^- [1450]. E-selectin can stimulate focal adhesion kinase. P-selectin sensitizes cells for PAF-triggered secretion of CCL2 and TNFα as well as CCL5-stimulated secretion of CCL2 and IL8 (CXCL8) [1450].

Firm Leukocyte Adhesion on Endothelium

Stimulated guanine nucleotide-exchange factors by clustered integrin ligands such as ICAM1 activates a subset of Rho GTPases. Monomeric RhoA GTPase promotes additional clustering of ICAM1 and VCAM1 [1452]. Adhesion-induced clustering of ICAM1 or VCAM1 on the surface of endothelial cells launches intracellular signaling. Subsequently activated GTPase RhoA provokes bundling of existing ^{F}actin filaments into stress fibers. Small GTPases RhoG and Rac1 mediate membrane ruffling and the formation of the apical docking structure. The DOCK–ELMO complex allows crosstalk between RhoG and Rac1 GTPases.

Crosslinking of ICAM1 molecules causes the production of inositol phosphate and phosphorylation of phospholipase-Cγ [1452]. In addition, ICAM1 clustering rapidly triggers calcium influx. Calcium-dependent activation of protein kinase-C regulates the phosphorylation of Src kinase and its substrate cortactin. Clustering of ICAM1 molecules also provokes RhoA-dependent phosphorylation of cytoskeletal and regulatory proteins, such as paxillin, FAK, and BCAR1 (or P130CAS), as well as interaction of CRK adaptor with paxillin and BCAR1 adaptor.

Crawling

Crawling of leukocytes on the wetted surface of the vascular endothelium is governed by β2-integrins. The Rho/Rac guanine nucleotide-exchange factor Vav1 participates in the control of location and activity of β2-integrins [1444]. Mechanoreceptors activate signaling mediators such as Vav1 protein.

Small GTPase RhoG that is activated upon ICAM1 clustering mediates the formation of docking structures, actin-rich membrane protrusions that are formed by endothelial cells around adhered leukocytes upon adhesion of leukocytes to ICAM1 and/or VCAM1 [1452]. Small GTPase Rac1 is also activated upon ICAM1 clustering. Activated RhoG and Rac1 generate dorsal and lateral membrane ruffles, respectively. Besides, RhoG can activate Rac1 via the DOCK1–ElMo complex and CRK adaptor. Tetraspanins (Tspan24, Tspan28, and Tspan29) are involved in ICAM1 and VCAM1 functioning and clustering.

Transendothelial Migration

Signals generated by interactions between leukocytes and endothelial cells as well as antibody-induced clustering of adhesion receptors cooperate to disassemble inter-endothelial junctions to raise endothelial paracellular permeability. Involved molecules include small GTPases Rac1, RhoA, and RhoG, second messengers calcium and reactive oxygen species, FAK2 and Src kinases, and cytosolic PTPn1 as well as receptor PTPRb and PTPRm protein Tyr phosphatases [1452].

Transcellular Migration

Molecule ICAM1 is involved in the transcellular migration of lymphocytes, during which it remains bound to the leukocyte. Internalization of ICAM1 depends on PKC and Src kinases as well as RhoA GTPase, in addition to caveolin-1-mediated endocytosis that is regulated by filamin-A [1452].

Homophilic interaction of leukocyte PECAM1 with endothelial PECAM1 is needed for leukocyte extravasation. Molecule PECAM1 induces the recruitment of PTPn6 and PTPn11 as well as SHIP phosphatases. PECAM1 may contribute to both para- and transcellular routes [1449]. The functional role of PECAM1 may be governed by differences in endothelial cell phenotype in different vascular beds.

Paracellular Migration

Between-endothelial cell adhesion is primarily mediated by adherens and tight junctions. Adherens and tight junctions are intermingled and regulate each other. Junctional (JAM) and endothelial cell-selective (ESAM) adhesion molecule of tight junctions modulate paracellular mode of leukocyte transendothelial migration [1452].

Cadherin-5 is a major constituent of adherens junctions that are stabilized by PTPRb phosphatase. Leukocyte adhesion to endothelial cells triggers signaling from ICAM1 and VCAM1 that dissociates cadherin-5 from PTPRb and subsequent phosphorylation by Src and FAK2 kinases for junction disassembly [1452].

Reactive oxygen species produced by the Rac1–NADPH oxidase pathway also participate in cadherin-5 and catenin phosphorylation. Activity of RhoA and calcium influx provoke stress fiber contraction that increases the endothelial permeability.

In addition, a disintegrin and metallopeptidase ADAM10 cleaves and generates soluble cadherin-5 ectodomain fragment as well as soluble JAMa that facilitates T-cell and antagonizes neutrophil diapedesis [1453, 1454]. However, JAMa is cleaved predominantly by ADAM17 and, to a lesser extent, by ADAM10 peptidase.

11.5.3 Crossing of the Vascular Basement Membrane

After transendothelial migration, leukocytes cross the vascular basement membrane. The venular basement membrane that has low content of matrix protein is a preferential site for neutrophil and monocyte migration under a CCL2 chemotactic gradient [1445]. Leukocyte migration across the perivascular basement membrane involves α_6-integrins and PECAM-1 [1445].

Each leukocyte subtype uses a different mechanism to cross the vascular basement membrane. Neutrophils enlarge regions of low-protein content and degrade laminin [1445]. Monocytes can use pre-existing smaller permeable regions of the basement membrane due to their higher deformability.

11.5.4 Interstitial Migration of Leukocytes

Migratory behavior differs between neutrophils and monocytes. Monocytes begin their interstitial migration later and are slower than neutrophils [1444]. Neutrophil migration in conjonctive tissue involves β_1-integrin.

Molecular guidance for selective recruitment into certain tissues depends on some environmental factors. Vitamin-A and -D3 influence lymphocyte homing to small intestine and epidermis [1445]. Dendritic cells intervene in the transformation of vitamin precursors into active derivatives such as retinoic acid.

11.5.5 Molecular Mechanisms of Leukocyte Chemotaxis

Once leukocytes have crossed the vascular endothelium, they integrate and prioritize multiple combinatorial chemotactic signals, using $PI3K_{c1\gamma}$ kinase as well as PTen and SHIP1 phosphatase, among others.

11.5.5.1 Phosphatidylinositol 3-Kinase

Chemotactic gradient signal received by specific G-protein-coupled receptors on the surface of leukocytes is transduced and amplified within the cell. Phosphatidylinositol 3-kinase is activated in the cell region that faces the chemoattractant [1444]. It then phosphorylates phosphatidylinositol (4,5)-bisphosphate to produce phosphatidylinositol trisphosphate. PIP_3-binding proteins then accumulate in the pseudopods of migrating cells. Simultaneously, enzymes that degrade PIP_3 are active on the cell sides and uropod, thereby neutralizing PI3K activity in these regions. Phosphatase PTen dephosphorylates PIP_3 into $PI(4,5)P_2$. Src homology-2 domain-containing inositol 5-phosphatase SHIP1 also regulates PIP_3 levels. Enzymes $PI3K_{c1\gamma}$, PTen, and SHIP1 cooperatively confine PIP_3 to the leading edge for full polarization and motility.

Isotypes PI3Kγ and -δ operate in neutrophil migration from blood circulation in response to brief chemokine stimulation [1444]. After longer stimulation, migration depends on PI3Kδ.

Different classes of chemoattractants use different signaling pathways. Kinase PI3K accelerates the initial response to bacterial chemoattractant ^{N}formyl methionyl leucyl phenylalanine (fMLP). Afterward, an alternative pathway replaces the PI3K axis. Neutrophil migration launched by chemoattractants fMLP and C5a can indeed rely on P38MAPK [1444]. On the other hand, chemoattractants IL8 and LTb4 trigger PI3K-dependent migration.

11.5.5.2 Phospholipase-A2

The family of phospholipase-A2 is divided into 4 groups: cytosolic (cPLA2), calcium-independent (iPLA2), secreted (sPLA2), and a group that includes platelet-activating factor acetylhydrolase and related PLA2s. Both cPLA2α and iPLA2β use different lipid mediators, arachidonic acid and lipopolysaccharide, respectively, to provoke migration of monocytes subjected to CCL2 chemokine [1444].

11.5.5.3 Guanylate Cyclase

During chemotaxis, cyclic guanosine monophosphate and soluble guanylate cyclase operate in the pseudopod and uropod, respectively [1444].

11.5.5.4 Monomeric Guanosine Triphosphatases

During chemotaxis, activated small guanosine triphosphatase Rac organizes membrane protrusions in the direction of migration. In neutrophils, Rac activation is primarily mediated by DOCK2 guanine nucleotide-exchange factor [1455]. Phosphatidylinositol (3,4,5)-trisphosphate supports rapid recruitment of DOCK2 to the plasma membrane. Subsequent accumulation of DOCK2 at the leading edge requires synthesis by phospholipase-D of phosphatidic acid. Phosphatidic acid recruits DOCK2 to the pseudopod to initiate chemotaxis.

Monomeric Rap GTPase participates in the regulation of the integrin inside–out signaling pathway. Activated Rap1 redistributes integrins and promotes their clustering. Several guanine nucleotide-exchange factors activate Rap GTPases. Activator RasGRP2 that binds both Ca^{2+} and diacylglycerol (DAG) contributes to integrin-based adhesion regulated by chemokines via Rap GTPases [1444].

11.5.5.5 Integrins

Migrating cells adhere to the crossed matrix. The role of integrins vary according to the traversed medium and chemoattractant type. Neutrophil chemotaxis depends on β_2-integrins. In response to TNFα, neutrophil extravasation relies on $\alpha_L\beta_2$-integrin, whereas $\alpha_M\beta_2$-integrin slows down neutrophil extravasation [1444]. On the other hand, $\alpha_M\beta_2$-integrin is the predominant integrin involved in chemotaxis through synovial and dermal fibroblast barriers [1444].

Integrin-α_4 is expressed at low levels on human neutrophils. It may repress neutrophil chemotaxis [1444]. It impedes B cells from leaving the spleen marginal zone. On the other hand, it enables the recruitment of T cells and eosinophils in inflammation sites. Integrins regulate chemotaxis owing to their dual roles as both cell adhesion molecules and signal transducers.

11.5.6 Inflammatory Mediators

Inflammation involves mediators released by damaged tissue, generated by plasma enzymes, and produced by leukocytes. Activated Toll-like receptors, cytokines, as well as inflammation messengers such as angiotensin-2 activate nuclear factor-κB factor. Transcription factor NFκB acts as an integrator that controls initial steps of inflammation — leukocyte extravasation and chemotaxis — using canonical and non-canonical pathways, as well as monocyte activation owing to interleukin-6 and signal transducer and activator of transcription STAT3 [1446].

11.5.6.1 Toll-like Receptors

Inflammation is a process during which immunocytes are recruited to sites of infection and injury. Sensors of infectious agents and transducers of inflammation include pattern-recognition receptors such as Toll-like receptors that bind to pathogen-associated molecules.

Toll-like receptors on antigen-presenting cells, especially macrophages and dendritic cells, stimulated by pathogen-associated molecules, such as lipopolysaccharide (LPS; ligand for TLR4)[34] and nucleic acids (ligands for TLR3, TLR7, TLR8 and TLR9), rapidly activate transcription factors, such as interferon-regulatory

---

[34]Lipopolysaccharides of bacterial walls activate CD14–TLR4 complexes on the surface of lymphocytes, macrophages, and dendritic cells. They then cause the secretion of inflammatory cytokines, chemokines, and enzymes that mobilize immune effector cells and prime the adaptive immune system. Multiple post-transcriptional regulations attenuate the production of pro-inflammatory proteins to avoid excessive response by restricting mRNA stability and/or translation [1456].

factors for type-I interferons (Ifnα, Ifnβ, Ifnκ, and Ifnω)[35] as well as members of the families of nuclear factor-κB and AP1 Activator protein.

Excessive TLR-mediated inflammation is avoided after TLR4-mediated NFκB1 activation by a negative feedback based on tripartite-motif protein TRIM30α.[36] Agent TRIM30α binds to MAP3K7 in the TLR–NFκB pathway.[37] Protein TRIM30α associates with and degrades MAP3K7-interacting proteins MAP3K7IP2[38] and MAP3K7IP3,[39] hence reducing production of NFκB-dependent cytokines TNF and IL6 and impeding secondary stimulation of immunocytes [1457].

Chemokine-binding protein CCBP2,[40] also called CCR9 (GPR28) and CCR10 (GPR2), is required in inflammation [1458]. In case of CCBP2 deficiency, an excessive concentration of residual chemokines induces an inflammatory pathology. Chronic inflammation, characterized by numerous CD4+ T lymphocytes and macrophages, can lead to fibrosis and granuloma.

Toll-like receptors can interact with caveolae. For example, TLR4 connects to caveolin-1 due to heme oxygenase-1 and carbon monoxide to reduce the production of tumor-necrosis factor-α and interleukin-6 in macrophages [878].

11.5.6.2 Cytokines and Chemokines

Recruitment of circulating leukocytes with adhesion, chemotaxis, and cellular activation stages requires cell adhesion molecules, chemokines (CCL2 and gmCSF), and matrix degradation. Chemokines, such as CCL2 and CXCL8 (interleukin-8), attract circulating leukocytes to the inflammation site. Once they have entered the

[35]Receptors TLR3 and TLR4 activate IRF3; TLR7, TLR8, and TLR9 stimulate IRF7. Nine members of the interferon-regulatory factor family bind to interferon-stimulated response elements. Expression of several IRFs is induced by interferons and cytokines. Interferon-regulatory factors are also regulated by post-translational modification. Phosphorylation of IRF3 and IRF7 by kinases TBK1 and IKKϵ leads to the formation of dimers and nuclear translocation to regulate the production of interferon-I in response to TLR activation. Interferon produced by stimulated TLRs triggers an autocrine loop when it binds to cognate receptor (heterodimer made of interferon-α and -β (and -ω) receptor α [IfnaR1] and β [IfnaR2] chains) and activates the interferon-stimulated gene factor ISGF3 complex (heterotrimer of IRF9, STAT1 and STAT2). Trimer ISGF3 also binds to interferon-stimulated response elements. In addition, IRF1 regulates the synthesis of inducible nitric oxide synthase in macrophages.

[36]Members of the TRIM family regulate innate immunity. Many TRIM proteins have an antiviral function, such as TRIM5α, TRIM25, TRIM28, etc. Synthesis of TRIM30α depends on NFκB factor.

[37]Enzyme MAP3K7 is recruited to TRAF6 via MAP3K7-interacting protein MAP3K7IP2 and MAP3K7IP3 for activation. The TRAF6–MAP3K7–MAP3K7IP2/3 complex recruits and activates the IKK complex (MAP3K7 phosphorylates IKKβ) to liberate NFκB from IκB inhibitor.

[38]A.k.a. TGFβ-activated kinase (TAK)-binding protein TAB2.

[39]A.k.a. TAB3.

[40]A.k.a. D6 chemokine receptor.

inflammation site, phagocytes release inflammatory cytokines, such as interleukin-1 and -6 as well as tumor-necrosis factor-α.

Cytokines IL1 and TNFα not only stimulate adhesion molecules, but also activate macrophages and initiate production of growth factors required for healing and inflammatory mediators. These cytokines induce the production of acute-phase proteins by the liver, such as C-reactive protein and mannan-binding lectin, and increase hematopoiesis.

In vascular inflammation, various pro-inflammatory factors activate resident vascular smooth myocytes, fibroblasts, and endothelial cells via the NFκB transcription factor, such as TNFSF1, TNFSF5, angiotensin-2, and oxidized LDLs. Circulating monocytes enter the vessel wall from the arterial lumen or vaso vasorum. Monocyte activation is controlled by local IL6 production. Interleukin-6 is a pleiotropic glycoprotein secreted by diverse vascular cell types, such as macrophages, lymphocytes, fibroblasts, and endothelial and smooth muscle cells. Production of IL6 is upregulated by angiotensin-2, cytokines, and reactive oxygen species [1446]. Interleukin-6 signals via the IL6 receptor and Ras–ERK and JaK–STAT pathways.[41]

11.5.6.3 Tumor-Necrosis Factor

Tumor-necrosis factor binds to 2 distinct receptors: (1) TNFR1, i.e., receptor subunits $P55_{TNFR1}$ and $P60_{TNFR1}$ that are constitutively expressed by most cell types and (2) TNFR2, i.e., receptor subunits $P75_{TNFR2}$ and $P80_{TNFR2}$, the expression of which is induced mainly in immune and endothelial cells.

Both receptors can cooperate to transduce signals, as they activate MAPK enzymes, NFκB, and AP1 transcriptional activator. Tumor-necrosis factor not only quickly activates inflammation genes via NFκB and MAPKs, but also induces delayed responses.

High-affinity TNFR1 lacks intrinsic kinase activity. Binding of TNFα to TNFR1 provokes TNFR1 trimerization and recruitment of adaptors and signaling mediators, such as TNFR-associated factors TRAF2, TRAF5, and TRAF6, receptor-interacting protein kinase (RIPK), mitogen-activated protein kinase kinase kinases MAP3K3 and MAP3K7, that build an inducible submembranous complex.

---

[41] In normal conditions, vascular endothelial and smooth muscle cells do not express high IL6R levels. Trans-signaling by soluble IL6Rα that results from ectodomain shedding by adamlysin ADAM10 and ADAM17 raises expression of IL6R in vascular endothelial and smooth muscle cells. Enzyme JaK1 phosphorylates STAT1 and STAT3 and causes homo- and heterodimerization and nuclear translocation for gene transcription. Activity of STAT factor is also modulated by acetylation. Once it resides in the nucleus, STAT3 associates with P300 and CBP coactivators. Activated nuclear STAT3 controls the expression of genes for tissue remodeling (collagen and fibronectin), cell growth (cyclin-D1 and MyC factor), in addition to mediators (angiotensinogen, C-reactive protein, and γ-fibrinogen) [1446].

Activated circulating and tissue-resident monocytes and macrophages are the main producers of TNF monokine in innate immune responses and in chronic inflammatory diseases. This cytokine generates an autocrine loop characterized by low, sustained production of interferon-β. Whereas TLRs prime massive, rapid, transient Ifnβ production via interferon-regulatory factors IRF3 or IRF7, TNF-mediated Ifnβ acts via IRF1. Interferon-β-mediated autocrine loop sustains the expression of inflammation genes and causes delayed expression of interferon-response genes that encode transcription factors STAT1 and IRF7 [1459]. This autocrine regulation of gene expression depends on IRF1 and synergy between produced interferon and activated NFκB factor. The latter enhance macrophage response to cytokines and Toll-like receptors. Monokine TNF thus activates a feedback loop, but avoids the toxicity of high interferon production induced by stimulated Toll-like receptors.

Tumor-necrosis factor signals via the canonical NFκB pathway that activates the *IKK signalosome* (Sect. 11.5.6.4).[42] It modulates cellular responses via the MAPK and NFκB signaling pathways to activate gene transcription. Another regulatory mechanism of TNF-dependent MAPK signaling exists [1460].

Non-receptor protein Tyr phosphatase PTPn2 is involved in hematopoiesis and impedes inflammation. Phosphatase PTPn2 selectively regulates TNF–MAPK signaling, suppressing the activation of MAPK enzymes by tumor-necrosis factor, without acting on the NFκB pathway. It interacts with adaptor TRAF2 and inactivates SRC family kinase, modulating TNF-mediated inflammation. Besides, TNF is a potent inducer of interleukin-6. Deficiency of PTPn2 leads to: (1) enhanced signaling by ERK1 and ERK2, but not P38MAPK, and (2) higher IL6 levels. Kinases of the SRC family are involved in the elevated TNF–MAPK signaling observed in PTPn2-deficient cells. Phosphatase PTPn2 selectively regulates the TNF–MAPK pathway by inhibiting SRC family kinases. In addition, the control by PTPn2 of the activation of extracellular signal-regulated protein kinases ERK1 and ERK2 induced by SRC family kinases in hematopoietic cells appears to be specific to cell type.

11.5.6.4 Nuclear Factor-κB

Nuclear factor-κB factors operate in immune responses and inflammation. Category-1 NFκB heterodimers, i.e., NFκB1 ($P105_{NF\kappa B}$ precursor monomer processed to generate $P50_{NF\kappa B}$ dimer) and NFκB2 ($P100_{NF\kappa B}$ precursor monomer processed to generate $P52_{NF\kappa B}$ dimer), can associate with transcriptionally active category-2 NFκB proteins, i.e., Rel, RelA ($P65_{NF\kappa B}$ or NFκB3), or RelB; Vol. 4 – Chap. 10. Other Major Signaling Mediators). Subunits Rel, RelA, and RelB have a transactivating domain (TAD) to trigger transcription in synergy with recruited transcriptional coactivators.

---

[42]This kinase complex contains 2 kinases IKKα and IKKβ and the regulatory IKKγ subunit.

NFκB Activation and Nuclear Translocation

In resting cells, NFκB is sequestered in the cytoplasm in a complex with one of the several inhibitors of NFκB (in particular IκBα, IκBβ, and IκBϵ). Factor NFκB responds to many agents, such as vasoactive peptides, oxidized LDLs, activated TNFRSF5 receptor, inflammatory cytokines such as tumor-necrosis factor-α, advanced glycation end products, and lipopolysaccharide.

Activators of NFκB trigger IκB phosphorylation, hence relieving their inhibition via ubiquitination and proteasomal degradation. Liberated NFκB then translocates to the nucleus and binds DNA to synthesize inflammatory cytokines, such as TNFα and interleukin-6.

The NFκB pathway is composed of linked, distinct signaling modules that control NFκB translocation and activation. The NFκB translocation module that results from phosphorylation-induced IκBα degradation is necessary, but may be not sufficient for gene activation. An NFκB activation module can indeed be required for gene activation such as during angiotensin-2 stimulation in vascular smooth myocytes [1446]. The relative contribution of transactivation and translocation modules depends on cell types.

Reactive oxygen species act as second messengers to cause RelA phosphorylation (Ser276) by protein kinase-A. The catalytic subunit of protein kinase-A is activated by TNFα. Activation of NFκB follows phosphorylation by PKA of RelA subunit. In addition, phosphorylation at Ser276 can result from the PI3K pathway as well as, during virus infection, ribosomal protein S6 kinase RSK5.[43] Phosphorylation of RelA (Ser276) reduces interactions between N- and C-termini, thereby allowing RelA to complex with histone acetyltransferases, polyubiquitin ligases, and adaptors of the P300–CBP family of transcriptional coactivators. Subsequently, RelA is acetylated and links to chromatin as well as the dimer formed by cyclin-dependent kinase CDK9 and cyclin-T1 that acts as a transcriptional elongation complex [1446]. This complex then phosphorylates RNA polymerase-2 (Ser2) for gene transcription.

Factor NFκB provokes the expression of molecules involved in leukocyte recruitment, such as intercellular (ICAM1) and vascular (VCAM1) cell adhesion molecules, chemokines (e.g., CCL2), and growth factors (e.g., granulocyte–macrophage colony-stimulating factor or CSF2).

In addition, NFκB activates the synthesis of negative feedback regulators that terminate its action, such as IκB inhibitor and TNFαIP3 deubiquitinase. The latter removes Lys_{63} linker of ubiquitin chains from IKK-regulating adaptors to stop NFκB signaling.

[43] A.k.a. mitogen- and stress-activated protein kinase MSK1.

Canonical NFκB Pathway

The canonical NFκB pathway is activated by monokines, such as TNFα and interleukin-1, as well as angiotensin-2. The NFκB canonical pathway is associated with the *IKK signalosome* composed of 2 homologous protein Ser/Thr kinases — IKKα and IKKβ — and regulatory IKKγ subunit. The latter binds ubiquitinated adaptors, recruits the IκBα inhibitor, and liberates NFκB for nuclear translocation. In the cytosol, the active IKK signalosome phosphorylates IκBα (Ser32 and Ser36) that then undergoes proteolysis. The canonical pathway controls the nuclear translocation primarily of NFκB1–RelA heterodimer.

Activated B-cell receptors of B lymphocytes provoke the formation of a plasmalemmal signalosome that contains scaffold BCL10, paracaspase scaffold mucosa-associated lymphoid tissue lymphoma translocation gene product MALT1, IκB kinase, and the IκB–NFκB (Rel) complex (*BCL10–MALT1–NFκB signalosome*) [1461]. Kinase IκB phosphorylates IκB inhibitor that is then degraded, hence allowing NFκB nuclear import.

The activation of the IκB kinase complex requires phosphorylation of catalytic subunits IKKα and IKKβ and additional ubiquitination of regulatory subunit IKKγ. Plasmalemmal scaffold CARMA1 regulates IKKγ ubiquitination, whereas activated MAP3K7 phosphorylates IKKα and IKKβ when T-cell receptor is stimulated [1462].

Angiotensin-2 also activates the canonical pathway via its high-affinity receptor AT_{1A} and the CARMA3–BCL10–MALT1 complex (*CBM signalosome*; Sect. 11.5.6.7) [1446] . The canonical pathway triggered by angiotensin-2 thus differs from that primed by TNFα.

Non-Canonical NFκB Pathway

The non-canonical pathway is activated by diverse signals, such as TNFSF3, platelet-derived TNFSF5, TNFSF13b, and viral RNA. The non-canonical pathway activates the P100 precursor into its mature DNA-binding NFκB2 form via phosphorylation by IKKα, itself phosphorylated (stimulated) by NFκB-inducing kinase (NIK or MAP3K14).

In unstimulated cells, MAP3K14 binds a ubiquitin ligase complex made of TRAF2, TRAF3, and cIAP for constitutive degradation. Activators induce TRAF3 degradation, thereby stabilizing MAP3K14 and causing its oligomerization and auto-activation. Hence, the non-canonical pathway frees RelA–NFκB2 and RelB–NFκB1 complexes. The latter selectively controls genes that produce CCL19, CXCL12, and TNFSF13b that are involved in lymph node organogenesis and immunity [1446].

Table 11.9. Nuclear receptors that operate transcription regulators in immunocytes (Source: [1463]; AP1: activator protein-1; CoREST: corepressor for RE1 silencing transcription factor; FOxP3: forkhead box-P3; GR: glucocorticoid receptor; GRIP: glucocorticoid receptor-interacting protein; IRF: interferon-regulatory factor; LXR: liver X receptor; NCoA: nuclear receptor coactivator; NCoR: nuclear receptor corepressor; NuRR: nuclear receptor-related factor; PPAR: peroxisome proliferator-activated receptor; RAR: retinoic acid receptor; RelA: component of nuclear factor-κB; ROR: RAR-related orphan receptor; T_{H17}: helper 17 T lymphocyte; TLR: Toll-like receptor).

Type	Immunocyte	Partners	Effect
GR	Macrophages	RelA, AP1, IRF3;	Inhibition of some TLR4/9 target genes
	T lymphocytes	NCoA2 (GRIP1)	Inhibition of T_{H17}-cell differentiation
LXR	Macrophages	NCoR1–NCoR2	Inhibition of some TLR4 target genes
PPARγ	Macrophages	NCoR1–NCoR2	Inhibition of some TLR4 target genes
	T lymphocytes		Inhibition of T_{H17}-cell differentiation
			Inhibition of RORγ
NuRR1	Microglia	CoREST	Inhibition of RelA-dependent TLR4 target genes
			T_{H17}-cell differentiation
RAR	T lymphocytes		Inhibition of RORγ2
			FOxP3 synthesis

11.5.6.5 Nuclear Receptors

Members of the nuclear receptor superfamily of agonist-activated transcription factors (Vol. 3 – Chap. 6. Receptors) contain: (1) an N-terminus that serves in transcriptional activation; (2) central DNA-binding domain; and (3) C-terminus with a ligand-binding domain that mediates ligand-regulated transcriptional activation or repression and frequently receptor homo- or heterodimerization.

Nuclear receptors are mainly aimed at regulating the initiation, magnitude, and duration of gene transcription. Many nuclear receptors are widespread, but can have cell-specific functions via collaborations with transcription factors. In particular, nuclear receptors that act as activators and repressors control the beginning, propagation, and resolution of inflammation (Table 11.9). Furthermore, nuclear receptors regulate the differentiation and function of helper T-cell subpopulations [1463].

Classical steroid hormone receptors bind to gluco- (cortisol) and mineralocorticoids (aldosterone), estrogens, progesterone, and androgens. These homo- and heterodimeric activators bind to hormone response elements.

Orphan nuclear receptors, for which agonists have not yet been identified, constitute a second family of the nuclear receptor superfamily. They bind to DNA as monomers or homo- or heterodimers with retinoid X receptors (NR2b).

Adopted orphan nuclear receptors are receptors for which ligands have been identified afterward. Nuclear receptors for retinoic acid, fatty acid metabolites, and oxysterols, form a third family.

Nuclear receptors for non-steroids, such as peroxisome proliferator-activated (NR1c1–NR1c3) and liver X (NR1h2–NR1h3) receptors usually build heterodimers with retinoid X receptors (NR2b1–NR2b3) and bind constitutively to their specific hormone response elements, liganded or not. Both PPAR–RXR and LXR–RXR heterodimers act as transcriptional repressors in the absence of ligand. They actually interact with corepressor complexes.

Nuclear receptors impede pro-inflammatory programs of gene expression in macrophages, microglial cells [1463]. They indeed operate in the turnover or recruitment of corepressors and coactivators of gene expression. *Transrepression* refers to indirect tethering of nuclear receptors with target genes rather than direct, sequence-specific DNA connection.

The nuclear receptor corepressors NCoR1[44] and NCoR2[45] complex to maintain a basal repression of a subset of genes activated by Toll-like receptors and other pro-inflammatory pathways [1463].

The NCoR corepressor complex interact with transcriptional repressors, such as C-promoter binding factor CBF1,[46] myoblast determination protein (MyoD), and signal transducer and activator of transcription STAT5 [1463].[47] The NCoR corepressor complex must be cleared for gene activation.[48]

Peroxisome proliferator-activated receptor PPARγ (NR1c3) and liver X receptors prevent the removal of NCoR1–NCoR2 complexes in response to inflammatory signals [1463]. Liganded PPARγ attaches to small ubiquitin-related modifier SUMo1 using Ub conjugase UbC9 and SUMo ligase protein inhibitor of activated STAT PIAS1 [1463]. Sumoylated PPARγ targets the NCoR complex and prevents its ubiquitination that leads to NCoR clearance from gene promoters. In addition, liganded LXRα and LXRβ (NR1h3 and NR1h2) are sumoylated with SUMo2 or SUMo3 using UbC9 and HDAC4 histone deacetylase. Sumoylated LXRs also

---

[44] A.k.a. T_3 receptor-associating cofactor or thyroid hormone and retinoic acid receptor-associated corepressor TRAC1.

[45] A.k.a. silencing mediator of retinoic acid and thyroid hormone receptor [SMRT] as well as thyroid hormone and retinoic acid receptor-associated corepressor TRAC1 (!).

[46] A.k.a. recombination signal-binding protein for immunoglobulin-κ J region (RBPJ).

[47] Nuclear receptor corepressor NCoR1 can be recruited to inflammatory response genes by non-phosphorylated Jun and NCoR2 by translocated ETS leukemia repressor (TEL) and P50 subunit of NFκB. Factor Jun can be phosphorylated by IKKε to liberate the promoters of NFκB-binding genes.

[48] Components of the NCoR–HDAC3 complex transducin-β-like protein TBL1 (X-linked [TBL1X] or Y-linked [TBL1Y]) and/or TBL1-related protein TBLR1 (in fact, TBL1XR1) operate as adaptors that, once phosphorylated by activator kinases such as calmodulin-dependent protein kinase CamK2, mediate the recruitment of the ubiquitin conjugase homolog UbCH5 that ubiquitinates the NCoR1–HDAC3 and/or NCoR2–HDAC3 complexes for removal from gene promoters by the 19S proteasome subunit.

interact with the NCoR complex and impede the recruitment of UbCH5 and 19S proteasome.

The glucocorticoid receptor (GR or NR3c1) prevents the interaction of nuclear factor-κB with coactivators [1463]. It connects to DNA via AP1 (e.g. Jun–Fos complex) or NFκB (transrepression), thereby hampering their linkage to transcription factors, such as interferon-regulatory factor IRF3 that acts as a coactivator and positive transcription elongation factor PTEFb, and the assembly of coactivator complexes. In addition, the glucocorticoid receptor recruits nuclear receptor coactivator NCoA2[49] to repress inflammation. Coregulator NCoA2 indeed is converted from an activator to a repressor of transcription upon activation of the NCoA2-specific repression domain by interaction with GR–NFκB or GR–AP1 complex. Glucocorticoid receptor may tether to AP1 via thyroid hormone receptor-interacting protein TRIP6 that is required for glucocorticoid receptor transrepression. Moreover, the glucocorticoid receptor and nuclear receptor-related regulator NuRR1 (NR4a2) permit the recruitment of the corepressor complexes to activator protein AP1 and NFκB factors bound to target genes.[50]

11.5.6.6 Vasoactive Amines Histamine and Serotonin

Released histamine is a vasodilator that increases capillary permeability. Serotonin that is a monoamine neurotransmitter also acts as a growth factor that circulates in blood to target cells with $5HT_{2A}$ and $5HT_{2B}$ receptors. 5-Hydroxytryptamine indeed actuates action of endothelial nitric oxide synthase and stimulates phosphorylation of extracellular signal-regulated kinases ERK1 and ERK2 in endothelial cells [1464].

11.5.6.7 Angiotensin-2

Vasoactive peptide angiotensin-2 promotes vascular inflammation via activation of transcription factors, such as nuclear factor-κB, that provoke synthesis of inflammatory mediators, such as adhesion molecules and chemotaxins. Angiotensin-2-induced vascular inflammation triggers mainly the production of interleukin-6 that controls monocyte activation [1446].

Stimulated AT_{1A} activates phospholipase-Cβ that generates inositol trisphosphate and diacylglycerol to increase cytosolic calcium content and activate typical

---

[49] A.k.a. class-E basic helix–loop–helix protein bHLHe75 and glucocorticoid receptor-interacting protein GRIP1.

[50] Factor NuRR1 can also bind to NFκB component RelA on inflammatory gene promoters, when RelA inhibitory site is phosphorylated by glycogen synthase kinase GSK3 [1463]. The RelA–NuRR1 dimer recruits corepressor for RE1-silencing transcription factor (CoREST) complex that provokes RelA removal, hence restoring a repressed state to inflammatory genes (NuRR1–CoREST transrepression).

Table 11.10. Phosphoinositide 3-kinases in inflammation.

PI3K isoform	Stimulus	Effect
$PI3K_{c1\delta}$	Ig receptors	Mastocyte degranulation
$PI3K_{c1\gamma}$	Chemokines, complement fragments, bacterial products, etc.	Leukocyte migration
	GPCR, Gβγ	Mastocyte activation

protein kinase-C isoforms. Moreoevr, RhoA GTPase is activated by GDP–GTP exchange factors, isoprenylated by geranylgeranyl pyrophosphate, and translocates to the plasma membrane to reorganize the cytoskeleton and produce reactive oxygen species.

In hepatocytes, activated PKC phosphorylates CARD and MAGuK domain-containing protein CARMA3, a tissue-specific member of the MAGUK superfamily. Scaffold $CARMA3^P$ complexes with B-cell lymphoma protein BCL10 and mucosa-associated lymphoid tissue lymphoma translocation MALT1 to form the *CBM signalosome*. Activated MALT1 causes Lys63-linked ubiquitination of IKK and cleaves the IKK inhibitor TNFαIP3 that deubiquitinates TRAF6, RIPK1, and IKKγ. Therefore, in hepatocytes, where NFκB is sequestered in the cytoplasm, both the translocation and transactivation modules using the CBM signalosome and the RhoA–MAP3K14 axis are involved in angiotensin-2 signaling [1446].

In vascular smooth myocytes, inactive NFκB is constitutively nuclear. The predominant signaling arises from the activation of the NFκB transactivation module. Angiotensin-2 rapidly provokes NFκB activation, as RelA is phosphorylated (Ser536) by MAP3K14 using the RhoA-NIK pathway [1446].

11.5.6.8 Lipid Signaling

Lipid signaling contributes to inflammation and its deregulation to metabolic and degenerative diseases and cancer. Inflammation is characterized by excess activation of PI3K (Table 11.10) and SphK pathways [1465]. Excess circulating fatty acids cause long-term inflammation via Toll-like receptor TLR4, and production of reactive oxygen species and activation of inflammatory kinase pathways (PKC, IKKβ).

11.5.6.9 Leukotrienes and Prostaglandins

The plasmalemmal, polyunsaturated fatty acid arachidonic acid forms potent lipid mediators leukotrienes that are vasoconstrictors and prostaglandins (PG) that are either vasoconstrictors or vasodilators (Chaps. 8 and 9). Leukotrienes and

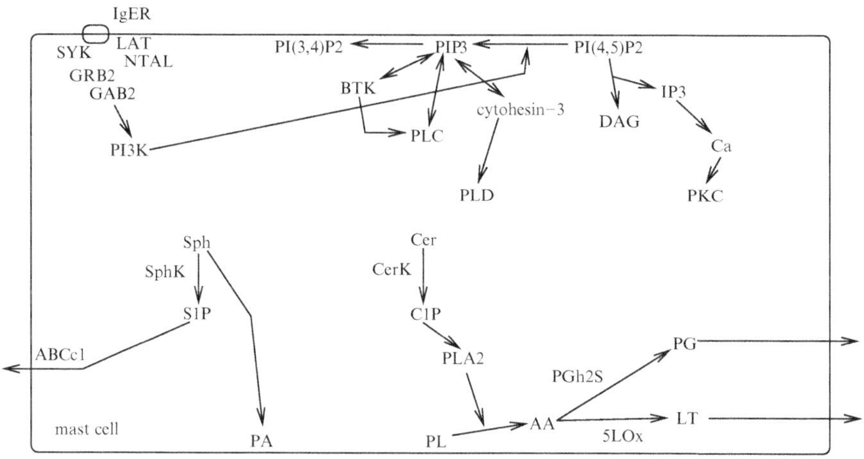

Fig. 11.2 Antigen receptor and lipid signaling in inflammation and allergy (Source: [1465]). IgE–antigen complex binding to IgE receptors (IgER) primes IgER phosphorylation by membrane-bound protein tyrosine kinases (i.e., Fyn). Subsequently, Syk, which is recruited to phosphorylated IgE receptor, mediates the phosphorylation of adaptors, such as linker for activation of T cells (LAT), non-T-cell activation linker (NTAL), and GRB2-associated binder (GAB2), which together favors the recruitment of class-1A phosphoinositide 3-kinases. At the plasma membrane, PI3K generates phosphatidylinositol (3,4,5)-trisphosphate (PIP_3) from PIP_2 lipid. PI3Kδ relays signals from not only the IgE receptor, but also the B-cell receptor and the T-cell receptor. Overstimulation of immunocytes at low antigen concentrations is hindered by the SH2-domain-containing inositol 5-phosphatase-1 (SHIP1). Agent PIP_3 recruits proteins, such as Bruton tyrosine kinase (BTK), phospholipase-Cγ (PLC), and cytohesin-3 (an ArfGEF). Activated BTK phosphorylates PLCγ, which cleaves PIP_2 into diacylglycerol (DAG) and IP_3. Messenger IP_3 releases Ca^{2+} from cellular stores. Phospholipase-D (PLD) and sphingosine kinase (SphK) are activated when PIP_3 concentration remains high. Activated SphKs generate phosphatidic acid (PA) and S1P mediators. In addition, SphK reinforces Ca^{2+} release that causes activation of mastocytes, hence stimulating PKC, ceramide kinase (CerK), phospholipase-A2 (PLA2), 5-lipoxygenase (5LOx) and prostaglandin-H2 synthase (PGh2S), and triggering degranulation of histamine-containing granules and production of prostaglandins (PG) and leukotrienes (LT) from arachidonic acid (AA) by PLA2, PGh2S, and 5LOx. Eicosanoids drive inflammation. Prostaglandins, leukotrienes, and sphingosine 1-phosphate (S1P) exit the cell via cognate carriers, such as ATP-binding cassette transporter ABCC1 for S1P, and act on G-protein-coupled receptors of neighboring cells that relay signals to effectors, such as PLCβ and PI3Kγ enzymes. Eicosanoids and S1P act with chemokines, cytokines, and histamine to promote constriction of downstream arteries and bronchi, dilation of upstream arteries, and to increase vascular permeability and attract leukocytes. Ceramide 1-phosphate (C1P) is produced by CerK from ceramide (Cer).

prostaglandins promote leukocyte chemotaxis. In particular, they are chemoattractants for neutrophils.

Eicosanoids — prostaglandins and leukotrienes — act via G-protein-coupled receptors and intervene in inflammation and allergy (Fig. 11.2).

Prostanoids derive from unsaturated ω6-fatty acids, such as arachidonic acid. They produce inflammatory mediators. On the other hand, unsaturated ω3-fatty acids are a source for anti-inflammatory and homeostatic prostanoids.

Table 11.11. Leukotriene receptors (Source: [815]). These receptors have a tissue-specific expression pattern. They are classified into receptors for LTb4 (BLT_1 and BLT_2) and for cysteinyl leukotrienes (LTc4, LTd4, and LTe4; $CysLT_1$, $CysLT_2$, and LTe4-specific $CysLT_E$). Receptor BLT_1 lodges mainly in leukocytes. BLT_2 Receptor is a low-affinity receptor or LTb4 and 12-hydroxy-heptadecadienoic acid on leukocytes and other cell types. The CysLT receptors reside on airway smooth myocytes, B lymphocytes, eosinophils, and monocytes (CysLT1), mastocytes and macrophages (CysLT1 and CysLT2), and endothelial cells (mainly CysLT2).

Type	Effect
BLT_1	Cell adhesion and chemotaxis (high affinity); vasoconstriction
BLT_2	Leukocyte chemotaxis (low affinity)
$CysLT_1$	Monocyte migration; bronchoconstriction
$CysLT_2$	Vascular permeability
$CysLT_E$	Eosinophil recruitment, vascular permeability

Table 11.12. Leukotriene function (Source: [815]). Leukotriene-B4 can bind to the nuclear receptor PPARα. Leukotrienes typically act as autacoids near their origin, via capillaries or interstitial space.

Type	Effect
5oxoETE	Chemoattractant (granulocytes)
5HETE	Weak activator of neutrophils and eosinophils; pulmonary vasoconstriction
LTb4	Neutrophil chemotaxis and adhesion
CysLTs	Postcapillary venule permeability; vasoconstriction; activation and chemotaxis of eosinophils and monocytes; bronchoconstriction; mucus secretion
Lipoxins	Reduction of neutrophil chemotaxis and extravasation; monocyte attraction; stimulation of macrophage phagocytosis; attenuation of edema

Leukotriene synthesis involves 5-lipoxygenase (5LOx), 5LOx-activating protein, LTa4 hydrolase, and LTc4 synthase [815]. Messenger LTb4 targets BLT_1 and BLT_2 receptors. Leukotrienes LTc4, LTd4, and LTe4 that are designated as cysteinyl leukotrienes (cysLT) bind to $CysLT_1$, $CysLT_2$, and LTe4-specific $CysLT_E$ receptors (Table 11.11). Leukotrienes possess more or less potent biological functions (Table 11.12).

Prostaglandins are synthesized by cyclooxygenases. Neutrophil recruitment to inflammatory sites by interleukins IL17 and IL23 is enhanced by prostaglandin PGe2, but inhibited by IL12 and Ifnγ [1466]. Interleukin-17 is a chemoattractant for neutrophils owing to CXC chemokines. Prostaglandin PGe2 raises IL17 production induced by IL23 as it simultaneously impedes IL12 and Ifnγ synthesis.

11.5.6.10 Platelet-Activating Factor

Phospholipid platelet-activating factor provokes platelet aggregation and attracts neutrophils. It is produced in response to stimuli by basophils, neutrophils, platelets, and endothelial cells. It also causes bronchoconstriction and vasodilation.

Basophils sensitized with IgE degranulate release histamine and platelet-activating factor that causes aggregation of platelets and release of their histamine content [1467].

11.5.6.11 TRAF2 and Sphingosine 1-Phosphate

Tumor-necrosis factor receptor (TNFR)-associated factor TRAF2 participates in signaling by the classical inflammatory cytokine TNFα. Sphingosine 1-phosphate is a cofactor for TRAF2 ubiquitin ligase. It binds to TRAF2 and promotes polyubiquitination of receptor-interacting protein kinase RIPK1 by TRAF2 downstream from TNFR and Toll-like receptors [1468]. Kinase RIPK1 then activates the IκB kinase complex and consequently nuclear factor-κB.

11.5.6.12 Complement System

Acute-phase proteins trigger formation of the membrane attack complex and release of complement components, such as fragment C3b of complement component-3.[51] Phagocytosis also leads to the production of inflammation mediators, like NO, peroxides, and oxygen radicals, which are toxic to invading microorganisms. Fibroblasts form connective tissue.

11.5.6.13 Inflammasome and Caspases

Inflammasome, a cytosolic protein complex, is made of precursors of pro -inflammatory caspases. These caspases cleave the precursor of interleukin-1β. Active IL1β causes a potent inflammatory response. Inflammasome contains NALP1,

[51] The complement system is an important component of innate immunity. Activation of the complement system produces derived products that contribute to pathogen elimination; but inappropriate stimulation of the complement system leads to inflammatory diseases. Complement component-3 (C3), once activated, forms fragments. The C3b fragment cements the assembly of convertases for C3 and C5 activation. C3 cleavage and activation of C3 allows binding with various molecules. The proteolysis of C3b gives birth to other fragments, such as C3c.

NALP2, and NALP3[52] that links to adaptor *apoptosis-associated speck-like protein* (ASC). ASC recruits pro-inflammatory caspase precursors, which are activated by NALP3 [1469, 1470].

In fact, 3 types of inflammasomes have been identified [1471]: (1) NALP1 inflammasome that is composed of ASC, caspase-1, and caspase-5, in addition to NALP1;[53] (2) NALP2 and NALP3 inflammasomes that contain, in addition to NALP2 or NALP3, Cardinal,[54] ASC, and caspase-1; and (3) NLRC4 inflammasome made of NLR family, CARD domain-containing protein NLRC4, or caspase-1.[55] The NALP and NLRC4 proteins are members of the nucleotide-binding oligomerization domain (NOD)-containing protein-like receptor (NOD-like; NLR) family of cytoplasmic proteins. They serve as pathogen-recognition receptors.

Vessel walls and heart express fewer types of TLRs and NLRs than immunocytes and other cell types of defense tissues, such as blood, lymph nodes, thymus, and trachea. Therefore, a model of 3-tier tissues according to synthesis of inflammasomes has been proposed [1472]. Tier-1 tissues, such as blood, brain, thymus, and placenta, constitutively express inflammasomes. Tier-2 tissues have inflammasomes in nearly ready status that requires upregulation of one component to assemble functional inflammasomes. Tier-3 tissues, such as heart and bone marrow, need upregulation of at least 2 components.

11.5.7 Angiogenesis and Inflammation

Vascular endothelial cells are hubs in the body's defense, as they are involved in 2 linked processes: angiogenesis and inflammation. Angiogenesis and inflammation are indeed often coupled. Many inflammatory stimuli can activate the angiogenic program of endothelial cells that destabilizes cell junctions (tight and adherens junctions) and elevates vessel permeability.

Increased capillary permeability in inflammatory sites that can result from action of VEGF and NO, among other mediators, causes leukocyte exit and deposition of a fibrin-rich matrix. Subsequently, matrix peptidases permit displacement of endothelial cells.

---

[52]NALP3 is also named cryopyrin. NALP stands for NACHT (NB-ARC [Apaf-1/R gene/CED4] nucleotide-binding domain), LRR (leucine-rich repeat), and pyrin domain-containing proteins.

[53]Protein NALP1 is also named NLR family, pyrin domain-containing protein NLRP1; NB and CARD domain-containing protein NAC; caspase activation and recruitment domain-containing protein CARD7; and Death effector filament-forming Ced-4-like apoptosis protein (DEFCAP)

[54]Cardinal is also termed caspase recruitment domain-containing CARD8 and tumor-upregulated CARD-containing antagonist of caspase-9 (TUCAN).

[55]A.k.a. interleukin-1βconverting enzyme (ICE) peptidase-activating factor (IPAF); CARD, LRR, and NACHT motif-containing protein (CLAN); and CARD domain-containing protein CARD12.

11.5.7.1 Role of the Extracellular Matrix

Many factors associated with the extracellular matrix control angiogenesis [1185]. Matrix-sequestered growth factors support angiogenesis. Mechanical forces generated by endothelial cells and mainly myofibroblasts influence angiogenesis as well as non-angiogenic translocation of the vasculature. Matricryptic sites are created by peptidases on matrix constituents (fibronectin, collagens, vitronectin, and osteopontin). They bind to cell integrins, hence priming signaling to moving endothelial cells. Released matrikines[56] positively or negatively regulate angiogenesis (Sect. 10.5.2).

Matrix remodeling and vascular regression contribute to the inflammation-driven angiogenesis as well as tissue repair. In the absence of vascular regression, positive feedback between vessels and inflammatory infiltrate sustain the new vasculature and exacerbate inflammation [1185]. Macrophages participate in pericyte apoptosis and capillary regression. Peptidases MMP1 and MMP10 intervene in capillary regression. Plasminogen and plasma kallikrein may also accelerate the process.

11.5.7.2 Angiopoietins

Among common regulators of angiogenesis and inflammation, angiopoietin-1 and -2 (Ang1–Ang2) hinder and enhance inflammation, respectively [1286]. The Ang–TIE2 pathway regulates vessel maturation and quiescence to stabilize newly formed blood vessels.[57] Angiopoietin-1 is secreted by pericytes and activates TIE2 receptor protein Tyr kinase (paracrine TIE2 activation). Angiopoietin-2 antagonizes the anti-inflammatory effect of angiopoietin-1, as it enhances leukocyte extravasation. Angiopoietin-2 released by endothelial cells is an antagonist for TIE2 (autocrine TIE2 regulation) on vascular endothelial cells that destabilizes mature blood vessels.[58] Angiopoietin-2 also acts as an agonist of TIE2 on lymphatic endothelial cells. The quiescent endothelium is not affected by local perfusion of tumor-necrosis factor-α, whereas tumor vasculature is disrupted [1287].

---

[56] Peptides generated by proteolytic cleavage of constituents of the extracellular matrix that regulate positively or negatively cell activities, similarly to cytokines or other types of regulatory peptides.

[57] Stimulated TIE2 activates the PI3K–PKB pathway. Stimulated TIE2 interacts with ABIN2 adaptor and blocks NFκB implicated in the synthesis of tissue factor as well as ICAM1 and VCAM1 adhesion molecules expressed by adjoining endothelial cells.

[58] Angiopoietin-2 sensitizes endothelium for TNFα-mediated inflammation and upregulates adhesion molecules ICAM1 and VCAM1 on endothelial cells.

11.5.7.3 Carboxyalkylpyrrole Protein Adducts

Inflammation leads to the release of oxidized phospholipids that activate Toll-like receptor TLR2 on endothelial cells and provoke angiogenesis (Chap. 10). During tissue repair, leukocytes contribute to the production of carboxyalkylpyrrole protein adducts (CAP), such as ω-2-carboxyethyl pyrrole (CEP) [1329] (Sect. 10.6.15). These adducts, CEP in particular, promote angiogenesis.

11.5.8 Inflammatory Diseases

The migration patterns of activated immunocytes depend on the stimulus features. Innate bone marrow-derived immunocytes react to inflammatory signals, whereas naive thymus-derived T lymphocytes poorly respond to such cues. Neutrophils are often twinned to T_{H1} cells in inflammatory infiltrates.[59] Eosinophils participate in inflammation that implicates T_{H2} cells.[60] Monocytes that differentiate to give birth to macrophages or dendritic cells, are often involved in inflammatory lesions.

Inflamed endothelium undergoes changes in blood molecule permeability and flowing cell adhesiveness. Chemoattractants trigger leukocyte adhesion to endothelial cells and guide leukocyte migration and positioning. They mainly act via G-protein-coupled receptors. Homing receptors are involved in the cell migration.[61] Except T and B lymphoblasts, which express adhesive integrins, circulating leukocytes have inactive integrins. Leukocyte activation leads to active integrin, which is able to bind to specific endothelial ligands. Integrin activation by endothelial chemokines occurs very quickly (a few milliseconds).

Combinations of chemokines and GPCRs (Gi–RhoA and Rap1 pathways mainly) activate endothelium adhesion under shear. L-selectin, expressed on most circulating leukocytes, initiates leukocyte extravasation in venular endothelium in inflammatory sites. Selectin-P and -E are inducibly produced by stimulated endothelial cells to attract neutrophils, eosinophils, monocytes, natural killer cells, and T and B lymphocytes. Moreover, L-selectin on circulating leukocytes can also bind surface molecules (such as P-selectin glycoprotein ligand-1) on wall-adherent leukocytes, enhancing the extravasation rate.

---

[59] T_{H1}-based inflammation is characterized by tissue infiltration of interferon-γ-secreting, CD4+ and CD8+ T lymphocytes and activated macrophages.

[60] Allergic inflammation is characterized by infiltration of T_{H2} cells. Eosinophils and mastocytes secrete interleukin-4, -5, and -13.

[61] Chemokines CCR4 and CCR8 may be involved in the displacement of T_{H2} cells and CXCR3, CXCR6, and CCR5 in the T_{H1}-cell recruitment. Eosinophil uses CCR3 at sites of allergic inflammation, and certain T_{H2} cells and monocytes CCR2 at sites of T_{H1} inflammation. Tissue-specific homing receptors, such as CCR10 for the skin and CCR9 for the gut explain the specificity of leukocyte recruitment.

Diapedesis requires transient disassembly of endothelial junctions. Monomeric Rac, Rap1 and RhoA GTPases, focal adhesion kinases, protein kinase-C, and phosphatidylinositol 3-kinase promote leukocyte extravasation. Extravasation depends on basement membrane composition: leukocytes cross vascular basement membranes that contain laminin-8.[62] The interstitial matrices remodel during inflammation with an increased production of fibronectin, tenascin, collagen, and sulfated proteoglycans. Peptidases and glycosaminoglycan-degrading enzymes are upregulated shortly after transendothelial migration by inflammatory cells to penetrate the extracellular matrix.

The strong neutrophil inflammatory response can damage host tissues.[63] Any anti-inflammation therapy must target the infiltrated immunocytes translated in effectors of inflammatory diseases without too much disturbance of the migration and function of leukocytes required in immunity. Effective therapies target specific immunocytes and focus on particular chemical reactions involved in chemotaxis and cell motion.[64] Recruitment of T_{H1} cells in T_{H1}-based inflammation disorders, such as atherosclerosis, can be targeted.

11.6 Healing

Repair of damaged tissues involves blood cells and molecules to seal the vessel wall gap off, clean the wound, and reconstruct tissues. Various intra- and intercellular mechanisms are activated and coordinated to restore homeostasis and tissue integrity. Blood coagulation, immune system, and inflammation are primed. Wound repair usually occurs with a similar sequence of events after exposure to a destructive agent whatever tissue (e.g., myocardial infarction and skin wound) and cause.

Delayed wound healing occurs in patients with diabetes or after radiation exposure. Excessive healing that manifests by deposition of large amounts of extracellular matrix and alterations in local vascularization and cell proliferation leads to hypertrophic and keloid scars. In some circumstances, healing leads to fibrosis. Fibrotic healing prevents tissue regeneration.

---

[62]Leukocytes cannot migrate through a basement membrane that contains laminin-10.

[63]Inflammatory diseases can target any body tissue, but they mainly damage target organs or tissues, such as the skin, intestine, central nervous system, skeleton articulations, respiratory conduits, and arteries.

[64]Blockade of neutrophil chemokine receptors CXCR1 and CXCR2 hinders the neutrophil flux.

11.6.1 Plasma Enzymatic Systems

Four interconnected plasma enzymatic systems produce mediators: (1) kinins, (2) complement, (3) coagulation factors, and (4) components of the fibrinolytic system. Tissue damage activates clotting factor-XII, which triggers the kinin and clotting cascades.

Bradykinin causes vasodilation and increased permeability. Cell-bound bradykinin also promotes histamine production. The fibrinolytic system leads to plasmin synthesis. Plasmin degrades the fibrin clot and activates the complement system.

11.6.2 Involved Cells

Many cell types are involved: (1) platelets (Table 11.13); (2) immunocytes (dendritic cells, lymphocytes, monocytes, and neutrophils); (3) vascular cells such as endothelial cells, and (4) tissue cells, such as fibroblasts and keratinocytes in the case of skin wound.[65] Nucleated cells change their gene expression to proliferate, differentiate, and migrate. Repair begins with a scar that mainly contains fibroblasts and extracellular matrix predominantly made of collagens. Tissue remodeling and regeneration is ensured by stem cells.[66]

---

[65]Wound re-epithelialization involves many growth factors: fibroblast growth factors FGF7 and FGF10 that bind to FGFR2-3b receptor; hepatocyte growth factor that targets HGFR receptor; and transforming growth factor-α and heparin-binding epidermal growth factor-like growth factor that associates with epidermal growth factor receptor [1473]. Activated receptors trigger signaling via signal transducer and activator of transcription STAT3 and AP1 proteins. Whereas TGFβ impedes wound re-epithelialization via transcriptional regulator SMAD3, activin that also signals via SMAD3, promotes keratinocyte proliferation in the wound site. In addition, acetylcholine and its receptors synthesized by keratinocytes in an autocrine loop either stimulate or hamper migration via muscarinic acetylcholine muscarinic M_3 and M_4 receptors, respectively. Keratinocytes also produce catecholamines and their receptors that inhibit re-epithelialization in an autocrine manner. Polyunsaturated fatty acids and derivatives that activate peroxisome-proliferator-activated receptors PPARα and PPARβ in keratinocytes after skin injury also regulate re-epithelialization. Factor PPARβ is also upregulated by pro-inflammatory cytokines via AP1 proteins and the stress-activated protein-kinase cascade. Factor PPARβ enhances cell survival by upregulating the expression of integrin-linked kinase (ILK) and 3-phosphoinositide-dependent protein kinase PDK1 that phosphorylate (activate) anti-apoptotic PKB.

[66]Epidermis permanently and rapidly renews itself owing to epidermal stem cells. The upper region of hair follicle below the sebaceous glands contains multipotent progenitor cells.

Table 11.13. Released content of platelet, in particular from α and dense granules at repair sites (Source: [1377]; CXCL4: chemokine CXC motif-containing ligand-4, also termed platelet factor-4; HETE: hydroxyeicosatetranoic acid; IGFBP: insulin-like growth factor-binding protein).

Substance	Effect
α Granules	
Platelet-derived growth factor	Matrix deposition
Transforming growth factor-α	Epithelialization
Transforming growth factor-β	Matrix deposition
CXCL4	Activation of growth factors
IGFBP3	Matrix deposition
β-Thromboglobulin	Activation of growth factors
Dense granules	
Adenosine diphosphate	Platelet aggregation
Calcium	Platelet aggregation
Serotonin	Vasoconstriction
Cytosol	
von Willebrand factor	Mediator of platelet adhesion
Fibronectin	Platelet aggregation
Fibrinogen	Platelet aggregation
Thrombospondin	Platelet aggregation
Coagulation factor V	Hemostasis
Platelet-activating factor	Platelet activation
Thromboxane-A2	Vasoconstriction
12-HETE	Vasoconstriction
Other	
Vascular endothelial growth factor	Angiogenesis

11.6.3 Myofibroblasts

Myofibroblasts, or mesenchyme-like interstitial cells, are contractile cells that participate in wound healing, fibrosis, and tumors, as well as vascular diseases with vessel wall remodeling, such as arterial restenosis and aneurysms. Myofibroblast markers include α smooth muscle actin, or α-actin-2, intermediate filament vimentin, and desmin.

Myofibroblasts can develop under concomitant mechanical and biochemical stimuli from [1474]: (1) fibroblasts and fibrocytes, the latter being defined as CD34+ (hematopoietic progenitor cell antigen), CD45+ (PTPRc), and collagen-1+ mesenchymal progenitors; (2) smooth myocytes;[67] (3) pericytes (4) endothelial and epithelial cells after reversible epit(endo)thelial–mesenchymal transition; (5)

[67]Activated myofibroblasts and smooth myocytes share markers, such as αsmA, transgelin (a.k.a. smooth muscle protein SM22 or SM22α), caldesmon, vinculin, and calponin$_H$.

Table 11.14. Signaling from TGFβ in myofibroblast differentiation (Source: [1474]; Abl: Abelson protein Tyr kinase; MAPK: mitogen-activated protein kinase; PAK: P21-activated kinase; PI3K, phosphatidylinositol 3-kinase; rSMAD: receptor-regulated SMAD; TOR: target of rapamycin). Homodimer TGFβ1 binds to TβR1–TβR2 complex that auto- and transphosphorylates. It targets myocardin during myofibroblast differentiation. Alternative TGFβ1 pathway triggers the release of matrix metallopeptidases MMP2, MMP9, and MMP13, as well as plasminogen activator, thereby matrix disruption.

Pathway	Effect
TGFβ–TβR1/2–rSMAD–SMAD4	Matrix production; myofibroblast differentiation
TGFβ–TβR1/2–PI3K–PKB–TOR; PI3K–PAK2–Abl1	Fibroblast proliferation
RhoA/Ras/MAP3K7–ERK1/2/JNK/P38MAPK	Matrix remodeling
Non-canonical ALK1–SMAD1/5/8	Matrix degradation

stellate cells (e.g., hepatic Ito and pancreatic stellate cells); (6) stromal progenitor cells; and (7) circulating mesenchymal stem and precursor cells.

Three features discriminate myofibroblasts from fibroblasts [1474]: (1) bundles of contractile microfilaments; (2) abundant cell–matrix attachment sites; and (3) intercellular adherens and gap junctions.

Myofibroblasts produce and remodel the extracellular matrix. They produce alternatively spliced fibronectin isoform with extradomain A (EDA [fibronectin^{EDA}]). In addition to fibronectin^{EDA}, they synthesize collagen-1 and -3. They can generate matrix tension. Moreover, they secrete angiogenic and pro-inflammatory factors.

Differentiation of myofibroblasts occurs during wound healing. Numerous agents trigger myofibroblast differentiation, especially transforming growth factor-β (Table 11.14) [1474]. Fibronectin^{EDA} cooperates with TGF-β to activate α-actin-2 and form α-actin-2-based stress fibers, whereas protomyofibroblasts contain stress fibers composed of only β- and γ-actins.

Myofibroblast recruitment and altered TGFβ1 signaling are involved in the early stage of thoracic aortic aneurysms [1474]. Production of TGFβ1 in myofibroblasts followed by secretion and autocrine control upregulates the expression of the focal adhesion molecule TGFβ1-induced transcript TGFβ1I1[68] that reduces the proliferation of pathogenic myofibroblasts [1474].

Transforming growth factor-β1 effect depends on its concentration. At low concentration, it stimulates proliferation and migration of fibroblasts and smooth myocytes. At high concentration, it precludes cell proliferation. Local TGFβ1 overexpression can stabilize abdominal aortic aneurysms [1474].

Angiotensin-2 can prime myofibroblast differentiation via nicotinamide adenine dinucleotide phosphate oxidase that generates reactive oxygen species and activation of P38MAPK and JNK [1474]. In addition, angiotensin-2 enhances

[68] A.k.a. 55-kDa androgen receptor-associated protein ARA55 and hydrogen peroxide-inducible clone-5 protein HIC5.

myofibroblast migration from the adventitia to the neointima in arterial restenosis after stent implantation via increased density of AT_1 and VEGFR1 receptor. Angiotensin-2 causes ROS generation via NADPH oxidase that stimulates myofibroblast migration and differentiation via AT_1 receptors and P38MAPK and JNK kinases [1474]. Endothelin-1 can also trigger myofibroblast differentiation via ET_A receptor.

Myofibroblasts disappear at the end of tissue repair. They undergo apoptosis or dedifferentiate in the absence of growth factors, whereas the extracellular matrix is stabilized.

11.6.4 Healing Phases

The wound healing process has 3 distinct, overlapping phases: inflammatory (phase 1), proliferative (phase 2), and maturational (phase 3) phases.

The inflammatory phase is characterized by hemostasis and inflammation. Blood flow is regionally regulated. Exposed collagen activates the clotting cascade and initiates the inflammatory phase. Damaged cells release vasoconstrictors, thromboxane-A2 and PGf2α, and wounded vessels immediately constrict to reduce hemorrhage. Hemostatic plug that will be transformed into a stable clot occurs.

Platelets, the first responding cells, release growth factors and chemokines, such as platelet-derived growth factor[69] and transforming growth factor,[70] which initiate healing (Table 11.15). Other involved growth factors include insulin-like (IGF), epidermal (EGF),[71] and fibroblast (FGF) growth factors.[72] Fibronectin, fibrinogen, histamine, serotonin, and von Willenbrand factor are also recruited. These mediators control bleeding through clot formation. Platelet degranulation activates the complement cascade, specifically C5a, a chemoattractant for macrophages, neutrophils, and mastocytes, which also increases the wall permeability of nearby vessels and stimulates LTc4 and LTd4 leukotrienes. Capillary vasodilation is later induced by local histamine release, and inflammation cells migrate to the wound-healing locus. The early infiltrated neutrophils are attracted by proteolytic degradation products of fibrin, complement components C3a and C5a, CXCL4 chemokine, PDGF, scavenger cell debris, and foreign bodies.[73] In addition, neutrophils produce elastase

---

[69]Factor PDGF is chemotactic for fibroblasts, neutrophils, and macrophages. Once attracted to the wound site, PDGF then activates macrophages and induces fibroblast proliferation.

[70]Transforming growth factor is chemotactic for monocytes, lymphocytes, and fibroblasts. It regulates matrix proteins, such as collagen, proteoglycans, fibronectin, and matrix-degrading peptidases, and their inhibitors.

[71]EGF, produced by macrophages, induces epidermal regeneration.

[72]Isotypes FGF1 and FGF2 are involved in wound healing. Isoform FGF2 stimulates angiogenesis.

[73]Local decontamination deals with bacterium destruction. Bacteria are phagocytosed and contained within phagosomes, where they are killed by reactive oxygen species, such as superoxide, hydroxyl free radicals, and hydrogen peroxide.

Table 11.15. Growth factors in healing site (Source: [1377]; CTGF: connective tissue growth factor; EGF: epidermal growth factor; FGF: fibroblast growth factor; IGF: insulin-like growth factor; IL: interleukin; KGF: keratinocyte growth factor; PDGF: platelet-derived growth factor; SmC: somatostatin-C; TGF: transforming growth factor; TNF: tumor-necrosis factor; VEGF, vascular endothelial cell growth factor; EC: endothelial cell; FB: fibroblast; KC: keratinocyte; Mφ: macrophage; Nφ: neutrophil; P: platelet; VSMC: vascular smooth muscle cell).

Type	Cell source	Target cells	Effects
CTGF	FB, EC	FB	TGFβ1 effector
EGF	KC, Mφ, P	FB, KC	Proliferation, chemotaxis
FGF1/2	EC, FB, Mφ	Fb, KC	Proliferation, angiogenesis
FGF4		EC	Chemotaxis
FGF7	FB	KC	Proliferation, chemotaxis
IGF1	Fb, Mφ	EC, FB	Proliferation, collagen synthesis
IL1α/β	Mφ, Nφ	FB, Mφ	Proliferation, collagenase synthesis, chemotaxis
PDGF	EC, FB, Mφ, P, VSMC	EC, FB, VSMC, Mφ, Nφ	Proliferation, chemotaxis, matrix production
TGFβ1/2	FB, KC, Mφ, P	Inflammatory cells, FB, KC	Proliferation, chemotaxis, Matrix production
TGFβ3	Mφ	FB	
TGFα	KC, Mφ, P	FB, KC	Proliferation,
TNFα	Nφ	FB, KC, Mφ,	GF expression
VEGF	Mφ, KC	EC	Angiogenesis

and collagenase. The inflammatory stage (duration $\sim$48 h) is thus characterized by a a fibrin clot and the presence of platelets and neutrophils.

The proliferative phase follows. Cells present in the wound are other types of leukocytes, particularly monocytes and macrophages. Attracted monocytes differentiate into macrophages. Macrophages kill bacteria that have been opsonized by binding specific antibody and complement fixation products from serum exudate. Macrophages also phagocytose cell debris. Enzymes and cytokines are secreted by macrophages: collagenases, interleukins, and tumor-necrosis factor, all of which stimulating fibroblasts and promote angiogenesis, and transforming growth factor. Lymphocytes participate in the late inflammatory infiltration. Fibroblasts, which have migrated from the vicinity under chemotaxis, produce ground matrix and then collagen to build a new extracellular matrix.

Angiogenesis appears to eliminate hypoxic environment mainly from capillary sprouts and endothelial cell migration, accessorily via bone marrow-derived endothelial progenitor cells, owing primarily to vascular endothelial VEGFa and fibroblast FGF2 growth factors.

Cell migration and signal release lead to tissue reconstruction. Tissue remodeling is characterized by a balance among synthesis, deposition, and degradation. The connective tissue is able to receive water (edema), minerals, molecules, and cells. The edema region is invaded by various substances and cells to become an inflammatory granuloma. Granulation tissue is formed, composed of a matrix of fibrin, fibronectin, glycosaminoglycans with proliferating endothelial cells, and fibroblasts mixed with macrophages and lymphocytes. Epithelialization occurs early in wound repair.

Angiogenesis stimulated by TNFα leads to formation of new capillaries that deliver nutrients to the healing region. Fibroblasts differentiate. Wound extracellular matrix, composed of collagen and elastin fibrils interspersed with glycosaminoglycans, polysaccharides, and proteoglycans, fulfills a structural and regulatory role in cell behavior by contact or as a reservoir of growth factors. Fibroblasts transform into myofibroblasts with contractile features.

During the final maturational phase, the tissue contracts. Collagen turnover is increased. Remodeling of the collagen fibers with degradation by matrix metallopeptidase manufactured by macrophages and fibroblasts among others, and with synthesis stimulated by cytokines, such as TNFα and IL1, is associated with reorientation of collagen fibers.

Tissue injury creates an environment that releases growth factors and neurohormones to recruit stromal-like cells for the repair process. In response to cytokines, stromal-like cells can migrate with hematopoietic stem cells and endothelial precursor cells and then differentiate.

Neuropeptide substance-P is able to induce migration of human mesenchymal stem cells from the bone marrow stroma, to increase expression of several matrix metallopeptidases for cell mobilization, and to stimulate proliferation of bone marrow stromal mesenchymal stem cells via extracellular signal-related kinases ERK1 and ERK2. Substance-P also primes translocation of β-catenin to the nucleus, so that the TCF/LEF–βCtn pathway involved in stem cell self-renewal may help to repopulate the bone marrow stroma. CD29+ (β_1-integrin) stromal-like stem cells are able to circulate in the blood stream owing to, at least partly, substance-P [1475].

11.6.5 Modeling of Healing

Simple models of wound healing are based on conservation equations based on cell and matrix density and concentration of stimulators and inhibitors in cell division [1476]. The simplest model consists of 2 conservation equations in the case of a thin epithelium made only of connected cells and in the absence of convective mass transfer:

$$\begin{aligned}\text{rate of change of cell density} &= \text{cell migration}\\ &\quad + \text{cell proliferation}\\ &\quad - \text{cell death},\end{aligned} \tag{11.1}$$

$$\begin{aligned}\text{rate of change of activator concentration} &= \text{activator diffusion}\\ &\quad + \text{activator synthesis}\\ &\quad - \text{activator degradation}.\end{aligned} \tag{11.2}$$

$$\begin{aligned}\text{rate of change of inhibitor concentration} &= \text{inhibitor diffusion}\\ &\quad + \text{inhibitor synthesis}\\ &\quad - \text{inhibitor degradation}.\end{aligned} \tag{11.3}$$

The change rate in cell density (n) depends on the cell migration flux and production and loss rates. The death rate is supposed to be proportional to the cell density. The controlled cell proliferation takes into account the cell density (n_0) and regulator concentration (c_0) in the unwounded state. The proliferation model states that [1476]:

$$\text{cell proliferation} = \mathsf{P}n\left(2 - \frac{n}{n_0}\right) \tag{11.4}$$

($\mathsf{P} = \mathsf{P}(c)$: chemically controlled cell division rate)

The time gradient of substance concentration (c) depends on the chemical diffusion flux and its production and decay rates. The role of blood flow in the transport of cells and compounds is ignored. The temporal decay of an active compound is currently assumed to be governed by fisrt-order kinetics modeled by a rate constant $-\tau c$. The production (p) of the regulator (subscript r), assuming that $\mathsf{p}(n_0) = \tau c_0$ and introducing a parameter α, which is related to the maximal rate of regulator production, can be given by [1476]:

$$\mathsf{p}_r(n) = \tau c_0 \frac{n}{n_0}\left(\frac{n_0^2 + \alpha^2}{n^2 + \alpha^2}\right) \quad \text{for an activator}, \tag{11.5}$$

$$= \tau c_0 \frac{n}{n_0} \quad \text{for an inhibitor}. \tag{11.6}$$

In the presence of either an activator or an inhibitor, equation 11.1 can be written as [1476]:

$$\frac{\partial n}{\partial t} = \mathcal{D}_c \nabla^2 n + \mathsf{P} n\left(2 - \frac{n}{n_0}\right) - \mathsf{d} n,$$
$$\frac{\partial c}{\partial t} = \mathcal{D}_r \nabla^2 c + \mathcal{P}_r(n) - \tau c. \quad (11.7)$$

This model does not take into account wound contraction in the final healing stage. A model of the deformation of epithelial sheets was then developed [1476]. Wounded epithelium is assumed to be an isotropic, viscoelastic medium. The mechanical behavior of a cell sheet depends on the state of the actin cytoskeleton, hence on the intracellular calcium concentration. Two force types are considered: intra- and extracellular elastic restoring and active traction forces [1476]:

$$\mathbf{C} = \rho_a(\kappa_1 \mathbf{D} + \kappa_2 \nabla \cdot \mathbf{u} \mathbf{I}) + \rho_a T \mathbf{I}, \quad (11.8)$$

where $\mathbf{C}$ is the stress tensor, ρ_a the density of actin filaments, $\mathbf{D}$: deformation rate tensor, T traction, $\kappa_{1/2}$ elastic parameters, $\mathbf{u}$ displacement, and $\mathbf{I}$ the unit tensor.

When a larger region of a tissue is considered such as in dermal wound healing, cell–matrix interactions must be incorporated in the model. Matrix structure and density as well as its turnover and reaction influence cell proliferation, differentiation, and migration. Conversely, cells determine matrix composition, turnover, and deformation.

The change rate of matrix density depends on its production and degradation rates and on its deformation rate. Cell–matrix interaction is introduced by a force balance in which the body forces are in equilibrium with the divergence of the traction stress exerted by the cells on the extracellular matrix and the divergence of the resistive stress of the extracellular matrix that is composed of an elastoplastic and a viscous component.

11.7 Respiratory Epithelium Repair

In adult lungs, epithelial cell populations include basal, secretory, ciliated, neuroendocrine, and alveolar cells. Under normal conditions, cell turnover in the lung is relatively low. Nevertheless, when the respiratory epithelium is damaged, stem or progenitor cell populations are activated to replace lost cells.

Optimal airway healing and repair is controlled by both bronchial epithelial and endothelial cells of the perfusing bronchial vasculature [1477].[74] Both cell types work synergistically and produce multiple mediators, although at different levels [1477]:[75] (1) prostaglandins such as PGe2, a potent bronchodilator that stabilizes extracellular matrix in airway walls and augments vascular permeability; (2) growth factors such as TGFβ1 and TGFβ2 as well as FGF2 that stimulates the proliferation of vascular endothelial and bronchial smooth muscle cells; (3) cytokines such as CSF2; (4) chemokines such as CCL2 and CXCL8; and (5) soluble adhesion molecules $VCAM1^S$ and $ICAM1^S$. An ordered cellular architecture is dispensable to achieve these effects.

Notch signaling is involved in the specification of the ciliated lineage in airways. Transcription factor FoxJ1 expressed in ciliated cells of adult lungs, is required in late stages of ciliogenesis. Factor FoxJ1 can then serve as a marker of precursor cells already committed to the ciliated lineage. FoxJ1+ ciliated cells transiently change their morphology in response to lung injury, but neither proliferate nor transdifferentiate during repair [1393].

Nevertheless, ciliated cells can function as progenitors in response to different types of damages and according to proximal or distal injury location, i.e., trachea and primary bronchi or distal bronchi, bronchioles, and bronchoalveolar duct junctions. Ciliated cells may transdifferentiate into mucus-secreting cells in murine models of asthma and viral infection [1393].

11.8 Vascular Tissue Remodeling

The wall of the heart and the blood vessels bear stresses applied by the flowing blood. The wall reacts, and then strengthens in high-stress regions. The short-term wall adaptation can indeed lead to long-term remodeling when abnormal stress magnitude and oscillation amplitude during the cardiac cycle are sensed. An integrative model incorporates behaviors at various length and time scales in order to efficiently describe structure–function relationships of physiological systems.

---

[74] Airway epithelial and vascular endothelial cells cultured separately or together secrete different patterns of repair mediators. Furthermore, their release levels depend on whether these cells are embedded into a matrix or not [1477]. Matrix-embedded endothelial cells express lower amounts of MHC class-2 molecules, chemokines, and adhesion molecules than endothelial cells cultured as a sheet. Matrix-embedded endothelial cell synthesizes smaller quantities of chemokines CCL2 and CXCL8 and soluble adhesion molecules sICAM1 and sVCAM1 in cocultures with bronchial epithelial cells than alone.

[75] Matrix-embedded bronchial epithelial cells secrete higher amounts of PGe2, gmCSF, and TGFβ1 than vascular endothelial cells [1477]. Conversely, matrix-embedded bronchial epithelial cells produce lower levels of FGF2 than vascular endothelial cells.

11.8.1 Myocardium Remodeling

Acute myocardial infarction leads to necrosis of cardiomyocytes and other cells. The inflammatory response triggers the migration of platelets, neutrophils, macrophages, monocytes, and other inflammatory mediators to the infarction site, where matrix metallopeptidases (MMP) degrade components of the extracellular matrix. Regions of fibrosis are formed as collagen is deposited. The fibrotic areas of infarctions do not have the contractile function, without regional myocyte regeneration, appropriate matrix formation, and angiogenesis.

The heart contains very rare cardioblasts (with a decreasing number with age) for self-regeneration, especially after a heart attack [1478]. These cardioblasts are susceptible to divide and mature after birth. Unlike stem cells, cardiac progenitor cells undergo a finite number of divisions and become fully specialized cardiomyocytes.

Cardiomyocyte proliferation from cardiomyocytes, resident stem cells, endothelial cells, fibroblasts, or migrated hematopoietic stem cells in areas adjacent to the infarcted zone can be a regeneration source.

Adequate input of growth factors is necessary to stimulate the myocardial regeneration with needed angiogenesis and matrix formation to avoid maladaptive remodeling of the myocardium.

Embryonic stem cells lead to 2 types of cardiovascular progenitors that express VEGFR2 [546]. The first type of VEGFR2+ cells act as *hemangioblasts*, giving birth to hematopoietic and endothelial cells. The second type gives rise to cardiomyocytes and endothelial and smooth muscle cells. The pathways (Wnt, TGFβ1, BMPs, and their antagonists, FGFs, etc.), which regulate cardiogenesis, may be used for stem cell-based cardiac repair.

Granulocyte colony-stimulating factor (CSF3) participates in heart regeneration after myocardial infarction. It activates the JaK–STAT pathway in cardiomyocytes, phosphorylating STAT3 and JaK2 [1479]. It has an anti-apoptotic effect, with increased levels of the anti-apoptotic BCL2 and BCLxL proteins. However, CSF3 stimulates angiogenesis, but not the proliferation of cardiomyocytes.

Lipopolysaccharides and postischemic reperfusion activate myocardial P38MAPK and nuclear factor-κB that lead to TNF production by cardiomyocytes [1480]. Tumor-necrosis factor depresses myocardial function by NO- and sphingosine-dependent mechanisms. It can also cause cardiomyocyte apoptosis.

Tumor-necrosis factors and lipopolysaccharides induce distinct biological responses, although they use the same signaling pathway with the effector nuclear factor-κB, via TNF receptor-1 and Toll-like receptor-4, respectively. Positive and negative feedbacks determine the timing of transcription factor activity and control the expression of genes mediated by the same transcription factor [1481, 1482]. TNF-dependent activation of IKK reaches its peak between 5 and 15 minutes and then shows oscillatory behavior if the stimulus is continued, whereas LPS-mediated signaling induces a slower biphasic IKK response (a small initial followed by a larger increase and slow attenuation).

Cardiac hypertrophy is induced by sustained pressure overload. Multiple hypertrophy signaling pathways are triggered by pressure: PP3–NFAT and PI3K–PKB axes and ERK1 and ERK2 kinases. The calcium-dependent PP3 phosphatase influences the growth and gene expression of the myocardium. Calcineurin (PP3) binds to *calsarcins*.[76] Calsarcin-1 impedes the functions of PP3 [1483]. The absence of calsarcin-1 enhances the cardiac growth response to pressure overload and favors hypertrophy.

Vasoactive substances such as angiotensin-2 are secreted by the overloaded myocardium. Angiotensin-2 induces superoxide anion production in cardiomyocytes and heart hypertrophy. Cardiac hypertrophy is mediated by myocardial oxidative stress, Rac1 GTPase, and NADPH oxidase [1484].

Cardiac hypertrophy caused by chronic hypertension is associated with fibrosis, such as in myocardial infarction. Fibrosis due to hypertension involves reciprocal interactions between stimulatory and inhibitory factors, resulting in increased deposition of collagen-1 and -3 within the adventitia of coronary arteries (perivascular fibrosis), which extends progressively.

11.8.2 Vessel Wall Remodeling

Blood vessels are subjected to mechanical forces that are implicated in vascular development, adaptation, and genesis of vascular diseases. Vessel caliber changes depend on a combination of wall wetted surface and intramural stresses. Blood pressure changes quickly induce tissue remodeling. Hypoperfusion reduces lumen caliber and media mass, whereas hyperperfusion increases these two quantities. Chronic increases in arterial blood flow lead to vessel enlargement and reduction of mechanical stress to physiological values.

Arteries enlarge in response to increased blood flow and wall shear stresses. Gaps in the internal elastic lamina have been observed in arteries exposed to high blood flow, whereas EC proliferation assures a continuous lining [1485]. Artery enlargement and adaptive remodeling is associated with FGF2 expression which acts on endothelial and smooth muscle cells [1486]. Conversely, upregulation of endothelin-1 and downregulation of nitric oxide can initiate and mediate the remodeling after blood flow reduction.

Step pressures induce a bore increase followed by a reactive SMC contraction (quick adaptation). Delayed slow structural changes are characterized by SMC proliferation. The time required by internal vessel radius variations after flow rate changes in large arteries is few days to weeks [1487]. Wall remodeling is characterized by a rapid first stage of geometrical changes. Besides, hypertension attenuates endothelial-to-endothelial or smooth muscle-to-endothelial

[76]The calsarcin family contains proteins of the sarcomeric Z disc (Sect. 5.3.4).

communications in skeletal muscle resistance arteries [1488]. Wall distensibility is more progressively restored. The mechanical changes occur during the second stage of wall remodeling.

Under sustained hypertension, the wall thickness of segments of the descending thoracic aorta increases in hypertensive dogs with respect to normotensive dogs, with a concurrent prestress reduction and an axial stiffening increase [1489]. Wall thickness of rat thoracic aortas correlates significantly with systolic blood pressure [1490]. The wall elastic moduli of hypertensive rats becomes equal to the normal values after a relatively long time.

Rapid changes in stiffness of the basilar artery occurs during hypertension, with increased collagen content in the media and adventitia and in smooth myocytes in the media, within 2 weeks after the initiation of hypertension, using an aortic coarctation model [1491].

Hypoxia-induced pulmonary hypertension leads to remodeling of pulmonary vessels and right ventricular hypertrophy. Pulmonary vessel remodeling is characterized by wall thickening, collagen deposition, and infiltration by pulmonary macrophages.

Vascular smooth myocytes participate, in response to hypertension, in geometrical and rheological adaptation, characterized not only by SMC proliferation but also migration, with myogenic adjustment. Vascular smooth myocytes migrate from the media into the intima, proliferate, synthesize extracellular matrix, and form the neointima. Hypertension induces change in vasomotor tone and wall circumferential stress [1492].

Flow-dependent remodeling is more pronounced in muscular arteries than in elastic arteries. These regional differences may be associated with the local flow pattern and the quantity of smooth myocytes and matrix components.

The arterial wall layers non-uniformly thicken during induced hypertension [1493]. The inner wall layers thicken more in the acute phase of hypertension, whereas the outer layers are thicker than the inner strata when the vessel is subjected to long-term (up to 56 days) hypertension. The medial collagen content quickly increases during the acute hypertension phase and slowly later. The elastin level slightly and steadily increases. Vascular smooth muscle tone rapidly rises and nearly returns to control levels.

Hypertensive remodeling is characterized by wall accumulation of collagen-1, -3, and -4 to counteract wall distention, but it yields wall stiffness. However, released matrix metallopeptidases, especially MMP9 [1494], lead to wall degradation with subsequent enhanced wall distensibility.

Nitric oxide is a predominant mediator of vasodilation and remodeling in response to increased wall shear stresses. NO release from endothelial cells exposed to excessive shear triggers a set of events, including growth factor induction and MMP activation, which together contribute to restructuring of the vessel wall [1495]. Moreover, degradation by MMP can overcome concomitant matrix synthesis. Although the elastic content remains normal, fragmentations in medial elastic laminae induce vessel tortuosity.

Platelet-derived growth factor primes phosphorylation of signaling proteins, such as the PDGF receptor and phospholipase-Cγ1, via intracellular production of H_2O_2 in particular. Factor PDGF is especially involved in smooth myocyte proliferation and migration during vascular remodeling.

The *peroxiredoxin-2* (Prx2), a peroxidase that eliminates endogenous H_2O_2 in response to growth factor signaling, is a negative regulator of PDGF signaling, especially during vascular remodeling [1496]. Neointimal layers of injured carotid arteries of mice lacking Prx2 are thicker than those of wild-type mice. Prx2 is involved in PDGF-dependent neointimal thickening, atherosclerosis, and other lesions of blood vessels.

Urokinase (uPA), involved in smooth myocyte behavior (proliferation, adhesion, and migration)[77] participates in wall remodeling via the uPA receptor (uPAR). Activated uPAR associates with PDGFRβ receptor,[78] hence leading to PDGF-independent PDGFRβ activation of its protein Tyr kinase activity [1497].

The expression of $P2X_4$ receptors in the vessel wall strongly increases after balloon injury, with a much higher density in the media, a very high density in the enlarged neointima, and a ten-time density increase in the endothelium [1498]. Activated $P2X_4$ receptors can thus be involved in neointimal proliferation. Examinations of blood vessels used for coronary artery grafts show that levels of $P2X_4$ are much lower in arterial vessels than vein samples. Therefore, artery grafts are more suitable than venous grafts to prevent restenosis and atheroma. $P2Y_2$ Expression also increases in balloon-injured endothelium and neointima, but without discernible differences between different examined vessel types.

In a dynamical adaptation model of the vessel wall under hypertension, the time evolution of the vessel geometry (external χ_e and internal χ_i perimeters, wall thickness h, vessel length L) has been described, assuming a thick-walled tube made of non-linear elastic incompressible material [1499]. The geometrical quantities are expressed as functions of the respective stresses in normal (superscript n) and hypertensive (superscript h) states. For example, the remodeling rate equation of the wall thickness is given by:

$$dh/dt = (1/\tau_h)[(\check{c}^h_\theta(t) - \check{c}^h_\theta)/\check{c}^n_\theta], \qquad (11.9)$$

where τ_h is the corresponding time constant and $\check{c}_\theta$ the wall-averaged circumferential stress. In particular, the model predicts that the vessel caliber increases up to a maximum and, then, decreases to a limit value 1 month later. The mechanical properties vary during the remodeling. The constitutive equations must then include

[77]Migration of smooth myocytes involves activation of TyK2 kinase, phosphatidylinositol 3-kinase, and the RoCK pathway.

[78]The 2 receptors PDGFRα and PDGFRβ are expressed in smooth myocytes, but PDGFRβ expression is stronger. Receptor uPAR interacts also with integrins, LDL receptor-related proteins, G-protein-coupled receptor formyl peptide receptor (FPRL1, or lipoxin A4 receptor LXa4R), FGF1 receptor.

a growth–stress relationship in addition to the usual stress–strain relationship. Bending experiments of excised vessel samples have been proposed to determine the rheology constants of the wall layers during the layer structure-dependent remodeling [1500]. However, in vitro data may be very far from in vivo values.

11.9 Growth Modeling

In planar cultures on matrices rich in collagen-4 and laminin, the endothelial cells form clusters and pull on the matrix, generating tension lines that can extend between the cell aggregates. The matrix eventually condenses along the tension lines, along which the cells elongate and migrate, building cellular rods. The rate of change in cell density is equal to the balance between the convection and the strain-dependent motion. The inertia being negligible, the forces implicated in the vasculogenesis model include the traction exerted by the cells on the extracellular matrix, cell anchoring forces, and recoil forces of the matrix [1501]. A mathematical model has been carried out using a finite difference scheme to study the role of the mechanical and chemical forces in blood vessel formation, and to simulate the formation of vascular networks in a plane [1502]. The numerical model assumes: (1) traction forces exerted by the cells onto the extracellular matrix, (2) a linear viscoelastic matrix, and (3) chemotaxis. The equation set is composed of: (1) a convection–diffusion–reaction equation that describes the cell density, (2) a conservation equation for extracellular matrix density, (3) a traction–displacement equation associated with extracellular matrix organization, which contains a non-linear term due to cell traction saturation at high cell densities, leading to two scalar equations for the force and displacement vector components (2D problem), and (4) diffusion–reaction equation for the chemotactic molecule concentration. Spontaneous formation of networks can be explained via a purely mechanical interaction between cells and the extracellular matrix.

Chemotaxis alone is not sufficient to generate tissue formation, with cell proliferation and ECM formation. During vessel sprouting, mechanical forces can help in the formation of well-defined vascular structures. The modeling can take into account intercellular interactions and feeding [1503]. A model, which is based on the Navier-Stokes equations in steady state and a simple mechanistic tissue response, can predict bifurcation formation and microvessel separation in a porous cellular medium [1504]. The tissue is remodeled according to the tangential shear stress; the convection is approximated by simple non-diffusive heuristics at each remodeling step.

Whereas angiogenesis cannot be explained by parabolic models, numerical simulations based on hyperbolic models of chemotaxis mimic migration of endothelial cells on Matrigel and the formation of networks that lead to vasculature [1505, 1506].

Table 11.16. Values of coefficients of cutaneous wound healing and angiogenesis model (Source: [1507]; ρ_t, ρ_s: endothelial cell density of tip and stalk of capillary sprout; ρ_f, ρ_l: density of fibroblasts and leukocytes). Approximative densities of endothelial cells, macrophages, and fibroblasts are equal to $10\times10\times1$ μm (107 cells/cm^3), $10\times10\times10$ μm (65 cells/0.07 mm^3), and $100\times100\times10$ μm (104 cells/cm^3).

ρ_{t_0}	10^3 g /cm^3
ρ_{s_0}	10^4 g /cm^3
ρ_{ic_0}	10^3 g /cm^3
ρ_{f_0}	10^3 g /cm^3

Injury leads to quick healing associated with angiogenesis. Wound environment characterized by moderate hypoxia, lactate accumulation, slight pH lowering, and hypercapnia primes cell migration and proliferation. A system of non-linear partial differential equations describing space and time interactions between coagulation proteins, endothelial cells of capillary tips and sprouts (density ρ_t and ρ_s; Table 11.16), other recruited cells such as fibroblasts (density ρ_f) and inflammatory leukocytes (neutrophils and macrophages; density ρ_l), chemoattractants (concentration c_{gf}), oxygen (concentration c_{O_2}), and the extracellular matrix (density ρ_{ecm}) has been used to model cutaneous wound healing and angiogenesis as a function of oxygen availability [1507]. Capillary tip flux is assumed to be

$$J_t = -\mathcal{D}_t \nabla \rho_t - \mathcal{C}_t(\rho_t, \rho_{ecm}) \nabla c_{gf} \tag{11.10}$$

($\mathcal{D}_t$: diffusion coefficient; $\mathcal{C}_t$ chemotactic coefficient). Capillary sprouts bear diffusion and drag caused by tip flux (tip velocity $v_t = J_t/\rho_t$) and move with velocity $v_s = g_s(\rho_t)v_t$ (assumption: $g_s(\rho_t) = A\rho_t/(\rho_{t_0})$. Fibroblast and leukocyte fluxes are supposed to be given by

$$J_f = -\mathcal{D}_f \nabla \rho_f - \mathcal{C}_f(\rho_f) \nabla c_{O_2} \tag{11.11}$$

and

$$J_l = -\mathcal{D}_l \nabla \rho_l + \mathcal{C}_l(\rho_l) \nabla c_{O_2}. \tag{11.12}$$

An effective growth factor is assumed to combine all involved growth factors.

When tissues remodel, zero-stress state of the vessel is modified. Wall remodeling implies changes in rheological properties, and consequently, if the elastic behavior is expressed in terms of a strain–energy function (SEF), the material constants in the function must be updated. Moreover, the constitutive equation must include not only stress history but also material history due to wall restructuring.

The vessel lumen subjected to sustained increments in blood flow enlarges. The arterial wall thickens in response to sustained increases in blood pressure. Walls remodel, mainly to restore the stress field toward their homeostatic values. Flow-induced changes involved in long-term vascular tissue growth and remodeling have been studied using the continuum approach and motion decomposition [1508, 1509].

A homogenized, constrained mixture theory is used to develop a 3D constitutive law that takes into account the 3 primary load-bearing constituents (smooth myocyte, collagen, and elastin) with time-varying mass fractions due to the turnover of cells and extracellular matrix fibers during the wall remodeling under a varying stress field. The turnover of constituent i is described by its total mass evolution, introducing two evolution functions for production and degradation rates. In addition, axial extension quickly increases the length of a carotid artery and the rate of turnover of cells and matrix, the turnover rates correlating with the stress magnitude. Numerical simulations show that moderate (15%) increases in axial extension generate much greater axial stress than circumferential stress augmentation induced by marked (50%) rise in blood pressure [1510]. A 2D constrained mixture model based on different constitutive relations shows that the turnover of cells and matrix in altered configurations is effective in restoring nearly normal wall mechanics.

Chapter 12
Airway Surface Liquid and Respiratory Mucus

Mucosa is a lining involved in absorption and secretion of mucus, especially, among other materials. Mucosal surface comprises a wet epithelium, a structural barrier, covered by a protective mucous barrier. The mucus layer in airways is thin and mobile.

Human conducting airways are mostly lined with a pseudostratified, secretory, and ciliated epithelium that comprises 3 major cell types — ciliated, secretory, and basal cells — with submucosal glands and cartilaginous elements in large bronchi (Vol. 1 – Chap. 3. Cells of the Ventilatory Tract).

The lung resists to damages caused by the body's outside environment following a permanent exposure to pathogens and more or less toxic chemical pollutants in inhaled air. Airway mucus represents the first line of defense. Inhaled particles (allergens, dust, microorganisms, and cellular debris) trapped in the mucus gel are removed from the respiratory tract by the mucociliary clearance. In fact, airway mucus traps inhaled toxins and transfers them out of the airways not only using mucus transport that results from ciliary beats, but also cough. Mucus cleans airways, as it flows from either the tracheobronchial tree or upper airways toward the pharynx, where it is swallowed (or expectorated).

The airway surface liquid consists of 2 components: (1) a gel-like mucus layer that traps inhaled particles and transports them out of airways by a cilia-generated motion and (2) a periciliary liquid layer that supports ciliary beating, hydration of airway mucus, and lubrication of the wetted epithelial surface. Cilia of respiratory epithelia beat in a coordinated manner within a periciliary fluid underneath a layer of mucus gel and transiently impact the mucus layer during the beating cycle, thereby avoiding an oscillatory motion. The luminal gel comprises water and heavily glycosylated mucins.

Pulmonary epithelial cells provide a mechanical barrier to microbial entry and participate in innate defense. They sense the presence of microbes and then augment their barrier function to resist to microbial penetration, signal to leukocytes, release

M. Thiriet, *Tissue Functioning and Remodeling in the Circulatory and Ventilatory Systems*, Biomathematical and Biomechanical Modeling of the Circulatory and Ventilatory Systems 5, DOI 10.1007/978-1-4614-5966-8_12,

of antimicrobial peptides, and kill pathogens [1511].[1] In the nearly sterile distal respiratory tract of healthy lungs, innate immune epithelial function is reduced due to small constitutive stimulation, when the mucociliary clearance functions effectively.

Secretory cells include Clara, goblet, and serous cells (Sect. 12.3.1). Various endo- and exogenous factors modify the secretory cell phenotype. Therefore, the overall term "secretory cell" is more appropriate to designate non-ciliated cells that liberate materials in lumens of the respiratory tract. Secretory cells release mucins as well as antimicrobial (defensins, lysozyme, and immunoglobulin-A), immunomodulatory (secretoglobins and cytokines), and protective (stable, secretory trefoil proteins and heregulin) molecules, either constitutively and inducibly, that can be incorporated into mucus [1511]. In large airways (internal caliber > 2 mm), submucosal glands located between airway smooth muscle and cartilage and connected to the airway lumen by a duct (ciliated in its downstram segment) contribute to the secretion of mucins and liquid.

Mucus is a viscoelastic fluid (thickness 2–5 μm) secreted by the respiratory epithelium. It not only traps inhaled particles (allergens, carcinogens, dust, microorganisms, and inflammatory debris) that come into contact with it to clear them from airways, but also protects the mucosa of the tracheobronchial tree from dehydration.

Ciliary motions that are associated with mucus propulsion in the human respiratory tract were described in the nineteenth century by Purkinje and Valentin (1834) and Sharpey (1835). The human respiratory tract and its related cavities and conduits (nose, sinuses, Eustachian tubes, middle ear, pharynx, trachea, and bronchi

---

[1]Neither resident nor recruited leukocytes are required for innate resistance of the pulmonary epithelium, although resident leukocytes amplify sensing of pathogen-associated molecular patterns and signal to adjoining epithelial cells and recruited neutrophil enable microbial clearance [1511]. Respiratory epithelial cells respond to stimulated Toll-like receptors (class-1 transmembrane innate immune receptors); they then produce pro-inflammatory and antimicrobial mediators. They also use cytosolic NOD-like receptors especially against Streptococcus pneumoniae, Pseudomonas aeruginosa, Moraxella catarrhalis, Chlamydophila pneumoniae, and Mycobacterium tuberculosis [1511]. RIG-I-like receptors are other types of cytosolic pattern recognition receptors involved in sensing of viruses. Additional innate receptors include the glycosphingolipid lactosylceramide on the apical surface of lung epithelial cells, which detects fungal β-glucans, and class-A scavenger receptors (e.g., ScaRa1 and ScaRa2), which participate in lung defense. Moreover, epithelial cells sense the release of constituents from injured neighboring cells (ATP in large quantities, adenosine, urate, and alarmins [e.g., HMGB1, S100a8, S100a9, and cathelicidin]), activated extracellular proteins (e.g., kallikrein–kinin and complement cascades), degraded extracellular components (hyaluronan), and inflammatory signals secreted from leukocytes (cytokines and eicosanoids). Airway epithelial cells express receptors for the C3a and C5a anaphylatoxins [1511]. In response to infection or injury, epithelial cells produce pro-inflammatory cytokines and chemokines as well as small cationic bacteriostatic and bactericidal peptides, such as β-defensins (constitutive β-defensin-1 and inducile ubeta-defensin-2, -3, and -4) and cathelicidins, and Large antimicrobial proteins, such as lysozyme and lactoferrin, in addition to surfactant collectins and lipocalin-2 [1511].

down to terminal bronchioles) have mucosal surfaces that exhibit mucus flow toward the esophagus. Mucus is more or less continuously secreted, shed, and recycled, discarded, or degraded.

Mucus delivery and motion are investigated in the framework of flow and transport of gases and inhaled particles in the respiratory tract under several alternative modes of ventilation together with aerosol transfer and surfactant dynamics, especially in the terminal airways, where it yields alveolar stability and increased lung compliance.

12.1 Double-Layered Airway Surface Fluid

The liquid layer at the luminal surface of the respiratory epithelium has 2 strata: (1) a deep, more aqueous, glycoprotein-free, lubricating region that bathes cilia (interciliary space ~200 nm) — the *epithelial lining fluid* — that is also called the *periciliary liquid* (thickness ~7 μm) and (2) a superficial, viscous domain, the gel or *mucus layer*.

Membrane-bound mucins contribute to the physical properties of liquid near the cell wetted surface. They confer the features of an anchored gel rather than a simple gel on top of a liquid. Mucin-4 is densely expressed on cilia and yields a brush-like configuration. Tethered mucins form an endoluminal brush [1512]. This brush layer establishes a mesh that prevents mucins of the mucus gel layer as well as deposited inhaled particles to penetrate into the periciliary space and causes mucus to form a distinct layer. Moreover, it provides lubrication through bound water.

In the current gel-on-liquid mucus clearance model, a mucus gel is propelled on top of a periciliary fluid layer over the respiratory epithelium that hence baths cilia. In the gel-on-brush model, the periciliary layer is occupied by membrane-spanning mucins and mucopolysaccharides tethered to the airway wall wetted surface (cilia, microvilli, and epithelial surface; mesh structure 20–40 nm) [1513]. The relatively high concentration of membrane-tethered mucins in the endoluminal brush layer produces intermolecular repulsion within this layer, which stabilizes the periciliary liquid against compression by an osmotically active mucus layer. The relative osmotic moduli of the mucus (~200 Pa) and periciliary brush (>300 Pa) layers explain that the mucus layer acts as a reservoir for water in the periciliary liquid, thereby stabilizing the mucus clearance in healthy airways [1513]. In airway diseases, when the airway surface is sufficiently dehydrated (the partial osmotic modulus of the mucus layer exceeds the minimal modulus of the periciliary brush), the mucus layer compresses the periciliary brush and cilia, slowing down and eventually stopping mucus clearance [1513]. Immobile mucus favors chronic inflammation and bacterial infections. Increase in the partial osmotic modulus of the mucus layer results from either a decrease in the amount of water solvent, as in mucovicidosis, or an increase in content of secreted mucins as in chronic obstructive pulmonary diseases.

The mucus layer is needed for particle transport. Cilia beat depends mostly on stimulation. The epithelial lining fluid is more stable than the overlying mucus layer, the latter being replaced every 10 to 20 mn. Each layer has a corresponding glandular apparatus: mucoid and serous secretions for the superficial and deep layer, respectively. Cilia freely beat in deep sol layer, but their tips contact the overlying gel layer. The mucus layer entraps all deposited particles that are transported by coordinated ciliary movements. The superficial gel layer and deep fluid layer are separated by surfactant.

Furthermore, airway surface liquid can be decomposed into 3 compartments when the glycocalyx coating of the apical cell surface is added to the epithelial lining fluid and the overlying mucus layer. Carbohydrate-rich glycocalyx contains intrinsic glycolipids, glycoproteins, and proteoglycans as well as adsorbed glycoproteins and proteoglycans.

The height of the airway surface liquid layer varies from about 3 μm over goblet or non-ciliated brush cells to 7 μm over ciliated cells (layer height is only estimated due to measurement disturbance) [1514].

The lubricating layer contains tethered mucins and other molecules such as cell surface glycolipids, glycoproteins, and proteoglycans, whereas the mucus layer consists of high-molecular-weight mucin dimers and trimers (length 0.5–20 μm) that interact with crosslinkers and innate defense proteins.

12.1.1 Control of the Quantity and Composition of the Airway Surface Fluid

Mechanism of mucus transport thus depends on interaction between the near-wall environment and overlying mucus, regulation of mucus hydration, and mucus adhesion. Electrolyte transport through and between airway epithelial cells controls the quantity and composition of the airway surface liquid.

Water transport across the apical plasma membrane of respiratory epithelial cells, i.e., the hydration status of airway surface liquid, indeed relies on ion fluxes, mainly chloride ion export through both cystic fibrosis transmembrane conductance regulator (CFTR) and calcium-activated chloride channels (CaCC) as well as sodium import through epithelial Na^+ channel (ENaC; Fig. 12.1).

Water flux across the wetted surface of the respiratory epithelium is controlled not only by motions of Cl^- across the apical membrane, but also K^+ flux across the basolateral membrane. Both Cl^- and K^+ conductances of airway epithelial cells are influenced by calcium ions. Conversely, features of the extracellular fluid play a role in calcium signaling [1515].

Two parallel routes — the trans- and paracellular paths — are used for the transepithelial electrolyte transport that determine the quantity and composition of the airway surface liquid and gel. The transcellular route depends on distinct populations of active and passive ion channels, transporters, and pumps in apical and basolateral membranes. Two main transcellular mechanisms of active ion transport comprise ENaC-dependent Na^+ absorption and CFTR-dependent anion

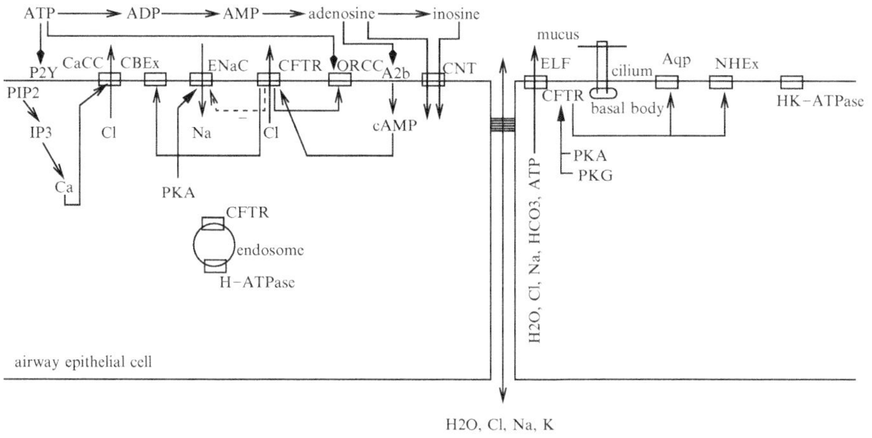

Fig. 12.1 Ion transport across the apical plasma membrane of airway epithelial cells (Source: [1514]). The liquid layer at the luminal surface (airway surface liquid) of the respiratory epithelium has 2 strata: epithelial lining fluid (ELF) and mucus layer. The respiratory cilium beats in epithelial lining fluid, but its tip contact overlying mucus. Aquaporins and ion channels that regulate water and ion transport are involved in production and maintenance of epithelial lining fluid as well as mucus homeostasis. Chloride ions exit through cystic fibrosis transmembrane conductance regulator (CFTR) are stimulated by adenosine binding to its receptor A_{2B} or calcium ions that excite Ca^{2+}-activated Cl^- channels (CaCC) after Ca^{2+} influx upon stimulation of nucleotide receptor $P2Y_2$ by ATP messenger. Channel CFTR also transports transport ATP, sodium, bicarbonate ions, and water. Channel CFTR is activated by protein kinase-A (PKA), cAMP, and ATP binding. In addition, CFTR is phosphorylated by protein kinases PKG1a and PKG2. Sodium ions are imported through epithelial Na^+ channel (ENaC). Chloride channel CFTR inhibits ENaC channel. It also regulates Cl^-–HCO_3^- exchanger (CBEx) and outwardly rectifying Cl^- channels (ORCC). Airway surface liquid pH is regulated by H^+–K^+ ATPase and Cl^-–HCO_3^- exchanger. Channel CFTR can activate aquaporin (Aqp). It also targets regulator NHERF of Na^+–H^+ exchanger NHE3 (NHEx). Channel CFTR located in membrane of endosomal vesicles can counterbalance the activity of H^+ ATPase. Adenosine and inosine are removed from airway surface by concentrative nucleoside transporters (CNT).

(Cl^- and HCO_3^-) secretion. On the other hand, the passive paracellular ion transfer depends on adhesion plaques between apposed epithelial cells that confer the permeability and ion selectivity of the paracellular path. In polarized respiratory epithelia, the paracellular pathway is limited by tight junctions and scattered adherens junctions. The latter localize in the lateral intercellular spaces, where adjacent epithelial cells interdigitate. Circumferential tight junctions form the functional and structural border that separates apical and basolateral compartments. Like many other epithelia, respiratory epithelia produce multiple claudins, the primary determinants of paracellular ion conductance.[2] Rapid regulation of transcellular

[2]Different cell types of the respiratory epithelium synthesize the same claudin types that form homo- and hetero-oligomers, both on a given cell and with those of its neighbors. Aldosterone increases claudin-4 phosphorylation and paracellular Cl^- permeability.

ion transport occurs in airway epithelia. Several hormones, neurotransmitters, and other agents regulate ion transport through the cell. On the other hand, paracellular ion transfer relies on dilation and disruption of tight junctions that result from the contraction of the actomyosin ring of the cortical cytoskeleton. Like the transcellular ion transport, the paracellular ion transfer can change quickly [1516]. Histamine that can be released by mastocytes binds to its H_1 receptor and provokes a Ca^{2+} influx, hence promoting cell contraction, and a rapid and transient increase in the paracellular Na^+ conductance, in addition to a smaller elevation in Cl^- conductance.

Airway surface liquid hydration is also controlled by extracellular nucleotides, such as adenosine triphosphate, uridine triphosphate, and UDP, and nucleosides such as adenosine. These substances bind to cognate receptors. Nucleotides ATP and UTP target nucleotides (purinergic) $P2Y_2$ receptors; UDP binds to receptor $P2Y_6$; adenosine interacts with A_{2B} receptor to stimulate Cl^- export into the airway lumen (Vol. 3 – Chap. 7. G-Protein-Coupled Receptors). Moreover, activated $P2Y_2$ inhibits Na^+ absorption by airway epithelial cells. Nucleotides receptors thus participate in the regulation of airway surface liquid volume. In normal conditions, the respiratory epithelium adjusts Na^+ and Cl^- fluxes so that stretched cilia are almost completely bathed in the lubricating epithelial lining fluid. In the airway lumen, released ATP is converted into adenosine that binds to receptor A_{2B} (Fig. 12.1).

Cyclic motion of the airway wall that is associated with successive lung inflation and deflation regulates homeostasis of airway surface liquid via nucleotides and nucleosides [1514]. Aerodynamical stress applied to the airway wall activates $P2Y_2$ receptors to increase luminal ATP level. Oscillatory motion in vitro that mimics tidal volume breathing in vivo favors airway surface liquid production, at least partly, because of ATP-stimulated nucleotide receptor-mediated ENaC inhibition and CaCC activation.

Airway surface liquid pH influences activity of ion channels, mucus adhesion to membranes, and attachment of bacteria and viruses to mucus. The glycocalyx pH is supposed to be influenced by negative charges of sugar moieties and membrane phospholipid head groups that can attract H^+ ion. In addition, glycoaminoglycans and glycoproteins are sialiated and sulfated, thereby accumulating cations on their surfaces. In healthy airways, airway surface liquid pH is regulated by H^+–K^+ ATPase and CFTR-mediated $HCO3^-$ flux through Cl^-–HCO_3^- exchanger.

12.2 Mucociliary Clearance

Mucociliary clearance is a defense mechanism that prevents accumulation of inhaled contaminants within the respiratory tract. Mucociliary clearance that comprises cough results from continuous flow of airway surface liquid over surfaces of airway lumens. Usually, airway surface liquid is cleared by cilium-mediated mucus transport. Normal daily volume of respiratory secretion that arrives at the larynx is estimated to be about 10 ml, as a part of its water content is reabsorbed via ion transport.

In healthy non-smoking adults, tracheal mucociliary clearance rate can be assessed by a radioisotope technique (70–85 μm/s; i.e., from 4.3 ± 1.1 mm/mn [1518] to 5.1 ± 2.9 [1517]). However, mucociliary clearance rates of 100 to 300 μm/s have been measured in the human trachea, i.e., nearly an order of magnitude greater [1519]. In the latter study, intersubject variations of mucus transport rate is not correlated to variations in ciliary beat frequency of nasal cells of examined subjects (12.1 ± 1.7 Hz) measured in vitro.

12.2.1 Factors of the Mucociliary Clearance

Four main factors determine efficiency of basic mucociliary clearance: (1) number of cilia that synchronously beat; (2) cilium beating amplitude and frequency (i.e., velocity of cilium tips); (3) thickness of serous fluid and mucus layers produced by surface goblet cells and submucosal glands; and (4) composition and rheology of serous fluid and mucus (i.e., ion transport and associated epithelial water flux as well as macromolecule content and its degree of entanglement and density of protein crosslinking).

Efficient mucociliary transport requires appropriate mucus composition for optimal flow, i.e., appropriate epithelial water and ion transport and mucin secretion, adequate numbers of functioning ciliated cells and their coordinated motion for mucus propulsion, suitable airway surface liquid depth, as well as respiratory epithelium integrity over long distances. Efficient transfer of momentum between cilia and mucus requires that cilia contact mucus during forward stroke, but minimally interact with it during return. When epithelial lining fluid is either too deep or too shallow, the mucociliary clearance rate decreases. Moreover, ciliary activity is adjusted in response to varying environmental conditions (Sect. 12.6.2).

Hyaluronan, or hyaluronic acid, a glycosaminoglycan lubricant at the airway epithelium surface, enhances the transport of airway mucus by cilia and by cough as well as protects the airway epithelial barrier; the lower the hyaluronan molecular weight, the stronger the effect [1520]. Low-molecular-weight hyaluronan (40 kDa) increases the expression of tight junction proteins such as ZO1 as well as the gap junction function.

12.2.2 Cough Clearance

Air flow–mucus interaction becomes important in clearing respiratory mucus as a lung disease develops. Moreover, various lung diseases are characterized by mucus hypersecretion and impaired airway clearance. Hence, excess mucus must be eliminated by coughing that propels mucus layer parts.

Efficiency of cough clearance depends on mucus rheology and surface properties at both air–mucus and epithelial lining fluid–mucus interfaces. Strong mucus elasticity that favors recoil after cough impedes mucus propulsion toward the larynx. High adhesivity precludes cough clearance. In addition, mucus sheared by strong expiratory air flow during coughing is prevented from sedimenting, as it rapidly recovers its rheological features and does not undergo tearing.

12.2.3 Mucociliary Clearance and Infection

The mucociliary clearance is a major defense process of the airway epithelium against inhaled pathogens and noxious entrapped micro- and nanoparticles that are removed with the mucus layer from the respiratory tract. As an innate airway defense mechanism, the mucociliary clearance affects drug delivery to airways. Drug absorption depends on duration during which drug is retained, i.e., the rate of mucus transport that is usually slowed down in airway pathologies. Engineered nanoparticles must be able to cross viscoelastic and adhesive mucus barrier, limit interactions with mucus constituents, and avoid adhesion and rapid clearance to be delivered in target cells.

Among normal volunteers exposed to rhinovirus or influenza-B virus, some remain uninfected, whereas others experience subclinical or symptomatic infection. Most symptomatic infected volunteers have prolonged nasal clearance, and less than 50% of their epithelium ciliated, but without significant reduction in ciliary beat frequency [1521].

12.2.4 Alterations in Mucociliary Clearance

Altered mucus composition causes abnormal mucociliary transport. Dysfunctional ion channels, such as CFTR and ENaC, that generate depletion of airway surface liquid produce mucus adhesion that favors formation of mucus plaque and plug, thereby an attenuated clearance of inhaled pathogens. Defective chloride transport across mucous membranes prevents normal hydration of mucus, leading to accumulation of highly viscous mucus such as that observed in cystic fibrosis. Mucus-generated clogging in airways, inflammation, and infection cause structural changes in lungs and sinuses. Mucus with elevated viscosity reduces air flow and fails to adequately clear germs.

Loss of water and salt in the airway surface liquid induces collapse of both the lubricating and mucus layers and mucus adhesion to cell wetted surface. On the other hand, addition of water and electrolytes to the airway surface liquid swells the mucus layer, but maintains apposition of both strata.

Ciliary activity that depends on appropriately hydrated airway surface liquid is a dominant factor that govern mucus clearance. In fact, individuals with mucoviscidosis, hence depleted airway surface liquid, quickly develop severe airway infection, whereas asthma that is associated with mucin hypersecretion is not necessarily associated with infection.

Primary ciliary dyskinesia causes impaired coordination of ciliary beating. Primary ciliary dyskinesia (Kartagener's, or immotile cilia syndrome) causes impaired coordination of ciliary beating. This disease causes a chronic cough, recurrent infections of the lung (bronchitis and pneumonia), and bronchiectasis, as well as chronic sinusitis and otitis.

Leukocyte migration through mucus during mucosal invasion by pathogens is a component of the mucosal immune response. Neutrophils can rapidly travel through mucus [1522]. In the absence of stimulation, random motility of neutrophils and lymphocytes is similar in collagen gels that have a structure comparable to mucus, whereas monocytes remain immotile. Motility of these cells is primed or enhanced upon suitable excitation. Neutrophils are able to cross a gel layer of thickness less than 500 μm.

Immunoglobulins are small enough to diffuse through the mucus mesh. Nevertheless, IgM diffusion is slowed by mucus, as immunoglobulins fabricate low-affinity bonds with mucins [1523]. Nonetheless, IgA, IgG, and IgM diffuse only slightly less in mucus than in water ($0.7 < \mathcal{D}(\text{mucus})/\mathcal{D}(\text{water}) < 1.0$; $\mathcal{D}$: diffusion coefficient) [1524].

Small viruses (size 38–55 nm) are able to diffuse through mucus as rapidly as in saline because mesh spacing between mucin fibers is large enough (20–200 nm; estimated hydrodynamic pore size for cervical mucus ~100 nm [1523, 1524]). However, herpes simplex virus (180 nm) can make low-affinity bonds with mucins that slow its running rate.

Mucus composition can be altered during infection. Moreover, after respiratory infections, accumulation of neutrophils and cellular debris raises mucus viscosity and impedes lung clearance and sinus drainage. Impaired mucociliary transport is often associated with viral or bacterial respiratory infections. Viral infections can reduce ciliary beating frequency as well as number of cilia and ciliated cells. Among viruses, influenza viruses destroy ciliated respiratory epithelia. Inhaled toxins from cigarette smoking and pollution depress ciliary function.

Purulent sputum from bronchitic patients characterized by a higher viscosity and a lower elastic recoil is transported at a slower rate than mucus of low viscosity and high elastic recoil [1525, 1526]. In dogs, the higher the mucus elastic modulus, the lower the clearance rate [1527]. The mucociliary clearance rate is halved by a 10-fold elevation in mucus elastic modulus. In frog palate, the mucociliary clearance rate is maximal for a mucus storage modulus of 0.16 Pa and sharply decreases above and below this value [1528]. In human subjects, optimal mucus rheology for mucociliary transport is given by a viscosity of 12 Pa·s [1529] and an elastic modulus of 0.4 to 0.8 Pa [1530].

Chronic bacterial infection of a compartment of the respiratory tract that is often associated with tumors or foreign bodies usually leads to obstruction. However, airway obstruction is generally caused by mucus plugs that follow mucus adhesion and formation of mucus plaques.

When the mucus layer is too thick and clearance by the cilia is hindered, clearance by coughing takes over. In healthy subjects, coughing is less efficient than by mucociliary clearance, but not in patients with airway diseases.

12.3 Mucus Delivery

Mucus is secreted by secretory cells of the respiratory epithelium. Mucus is released by exocytosis in respiratory conduits down to terminal bronchioles. The mucus extends over the tops of the cilia in a layer of thickness of about 5 μm. Mucus has also been observed to exist as droplets, flakes, and plaques in airways. Droplets (size 4 μm) can aggregate into flakes (size 10–70 μ), and then plaques, but does not necessarily form a complete mucus blanket (mucus rafts). Nasal mucus has a thickness greater than 200 μm.

Mucus is secreted in healthy upper airways primarily by submucosal glands. It contains defense molecules mixed with mucins, electrolytes, and water. Mucus traps pathogens, impedes their proliferation using molecular (e.g., lysozyme) and cellular (e.g., neutrophils and macrophages) defenses, and sweeps them out of airways using ciliary beats.

12.3.1 Secretory Cells

Mucous and serous secretions form the protective airway surface fluid. Airway mucus is secreted by mucous cells, goblet cells of the respiratory epithelium and mucous cells of the submucosal glands. In large airways, the major part of respiratory mucus is produced by submucosal glands, whereas epithelial goblet cells yield a smaller contribution.

The secretory cell population is usually defined by 3 cell types according to their microscopical features — non-ciliated bronchiolar Clara, goblet, and serous cells. Serous and goblet cells reside principally in trachea and bronchi. Secretory cells of airways are classically identified according to morphological characteristics, such as the density and type of secretory granules and the proportion of smooth to rough endoplasmic reticulum. However, secretory cells are characterized by a great phenotype (structural, molecular, and functional) changeability.

Secretoglobins (ScGb1a1–ScGb1a2, ScGb1c1, ScGb1d1–ScGb1d2, and ScGb1d4, ScGb2a1, and ScGb3a1–ScGb3a2) that are expressed in secretory epithelia can serve as cell markers. Secretoglobin-1A1 (ScGb1a1), or Clara cell secretory protein (CCSP),[3] is synthesized in cells that also produce mucins. Secretaglobin-3A1 is, like mucins, expressed selectively in secretory cells of large bronchi in humans [1532]. On the other hand, secretaglobin-3A2 is an early molecular marker for bronchiolar Clara cells.

[3]A.k.a. uteroglobin (UGb), Clara cell phospholipid-binding protein (CCPBP), Clara cells 10-kDa secretory protein CC10, and urinary protein-1 (UP1). Secretoglobin-1A1 is the founding member of the secretaglobin family. Secretory cells of the conducting airway epithelium produce different members of the secretoglobin family, such as ScGb3a1 and ScGb3a2 in neonatal human lungs [1532].

12.3.1.1 Goblet Cell

Airway goblet cells are cylindrical-shaped, glandular, simple columnar epithelial cells. They use both apocrine (decapitation secretion, the apical portion of the secretory cell of the gland pinching off and entering the lumen) and merocrine (exocytosis, i.e., membrane-bound vesicle-mediated secretion through the plasma membrane in the lumen of an epithelial-walled duct) processes for secretion. They quickly (<100 ms) secrete mucins that covers airway lumen surface with a protective mucus that is exposed to more or less polluted inhaled air. Numerous goblet cells (density $3–5 \times 10^4$ /mm^3) reside in the respiratory epithelium. Serous and Clara cells in small airways can transform into goblet cells.

Their apical secretory vesicles are discharged into the airway lumen to form a mucus layer over the epithelial surface. Peripheral granules in goblet cells produce the continuous baseline mucus, whereas core granules are secreted in response to stimuli [1531]. Goblet cells contain sialoglycoproteins or a mixture of sialo- and sulfoglycoproteins.

Secretory cells undergo metaplasia, i.e., a change in phenotype.[4] Interleukin-13 causes goblet cell metaplasia via production and subsequent activation of epidermal growth factor receptor. Interleukin-13 provokes transcrition of the SCGB1A1 gene in epithelial non-ciliated cells, more precisely, in non-granulated secretory and goblet cells [1533]. This change in phenotype does not depend upon proliferation; it is restricted to ScGb1a1+ cells of the tracheobronchial tree, but not ScGb1a1+ cells of bronchioles [1534].

Secretion of oligomeric mucins from airway goblet cells is regulated primarily by ATP and UTP nucleotides that activate Gq-coupled $P2Y_2$ receptors, thereby triggering the PLC–IP_3–Ca^{2+} and PLC–DAG–PKC intracellular cascades [1535]. These pathways coordinate actin cytoskeleton remodeling, thus enabling interaction of mucin secretory granules with apical membrane exocytic docking sites [1535].

Goblet and ciliated cells are indeed characterized by an apical distribution of β- and γ-actin. In goblet cells, actin forms a cortical sheet between granules and the apical plasma membrane that ruptures under ATP and UTP stimulation. Myristoylated alanine-rich C kinase substrate (MARCKS), a PKC-activated actin–plasma membrane-tethering protein, is phosphorylated upon nucleotide stimulation and translocates to the core cytosol [1535]. In addition, scinderin, or adseverin, a Ca^{2+}-activated actin filament severing and capping protein in human airway cells, once it is activated by Ca^{2+} influx, can disassemble actin filaments at sites of exocytosis and permit mucin secretion. Calcium ion contributes to the regulation of the final steps of exocytosis in goblet cells, as it interacts with rabphilin homolog, Uncoordinated-13 homolog Unc13b (or MUnc13),[5] and synaptotagmin.

---

[4]In Greek, μ ετα) means after, later; πλασια): formation. Metaplasia is the reversible replacement of a given differentiated cell type by another mature differentiated cell type.

[5]The MUNC13 family includes 4 known isoforms (MUnc13-1–MUnc13-4). subtype MUnc13-2 has 2 splice variants. The brain variant bMunc13-2 is similar to MUnc13-3, as it lacks the

12.3.1.2 Submucosal Glands

In healthy humans, submucosal glands (thickness 0.15 mm for airway wall thickness 0.64 mm [mean gland–wall ratio range 0.14–0.36] [1536]) are estimated to provide most of the upper airway mucus. About half of the amount of intracellular mucins is stored in submucosal glands. Submucosal glands secrete about 40 times more mucus than secretory epithelial cells.

The body of the gland is located between the spiral bands of smooth muscle and cartilage plates. Tracheobronchial submucosal glands are tubular branching structures. Each gland contains multiple tubules that merge in a collecting duct.

Submucosal glands are only found in cartilagous-walled airways (upper aiways, trachea, and large bronchi) between the epithelium and cartilage. They abound down to about generation 10 (airway lumen caliber 1–2 mm) [1537]. Density of submucosal glands decays with increasing distance from the pharynx, whereas that of brush and intermediate cells rises.

In submucosal glands after transplantion in human patients with diseases other than cystic fibrosis, the secretion rate per gland upon stimulation by vasoactive intestinal peptide[6] in tracheal and bronchial glands was 1.0 ± 0.2 and 1.1 ± 0.4 nl/mn, respectively [1538]. The volume of secretion depends on the Na^+–K^+–$2Cl^-$ cotransporter and on HCO_3^- ion.

Submucosal glands are lined with 2 cell types, proximal mucous and distal serous cells. Mucous and serous cells constitute approximately 60 and 40% of the gland volume, respectively [1538]. Mucous and serous cells secrete a viscoelastic gel and a lubricating fluid (water and electrolytes) that bathes epithelial cilia, respectively.

At their distal ends (acini), serous cells secrete proteoglycans and numerous antimicrobial, anti-inflammatory, and anti-oxidant proteins [1538]. The serous fluid actually contains lysozyme, which breaks down cell walls and lactoferrin, which limits iron availability to bacteria, immunoglobulin-A, in addition to transferrin, albumin, Muc7, serum leukocyte protease inhibitor, and surfactant protein-A [1531, 1538]. Serous fluid washes proximal mucous cells that secrete mucins. Mucous cells contain acidic glycoproteins, whereas serous ceils produce mainly neutral glycoproteins.

Cystic fibrosis transmembrane regulator localizes primarily to serous cells. Elevation of cytosolic Ca^{2+} level is supposed to activate secretion of fluid, ions (Na^+, Cl^-, and HCO_3^-), mucins, and other proteins from both serous and mucous cells and increase in cytosolic cAMP level to stimulate serous cells and mucin, but not fluid secretion, from mucous cells [1538].

C2A domain. The ubiquitous variant ubMUnc13-2 is similar to MUnc13-1, as it contains a C2A domain. Isoforms MUnc13-1, MUnc13-2, and MUnc13-3 complex individually with Rab3 and Rab3-interacting protein (RIM) GTPases in a trimer required for exocytosis of neuronal synaptic vesicles and endocrine secretory granules. Protein Unc13 was identified from gene mutations in Caenorhabditis elegans that cause a severe uncoordinated phenotype, i.e., paralysis.

[6]VIP+ neurons localize around submucosal glands, in addition to smooth muscle layer, lamina propria, and walls of pulmonary and bronchial arteries (Sect. 12.3.1.4). Moreover, VIP lodges also in mastocytes.

Each gland with serous acini and tubules followed by mucous tubules is connected to the airway lumen (gland opening density $\sim$1/mm^2) by an upstream non-ciliated collecting duct and a downstream narrower ciliated duct that propels secretions outward.

The ciliated duct may have an absorptive function, as secretions from bronchial submucosal and nasal glands contain significantly less ions than the bath [1538]. In addition, they may acidify secretions using H^+–K^+ ATPase.

Collecting duct (length $\sim$1 mm) runs obliquely to the wall lumen. Its caliber rises along its length. Duct-lining non-ciliated cells may control ion and water content of gland secretion, as they are surrounded by abundant blood capillaries in opposition to the ciliated duct to which it connects, which is covered by a duct-lining, ciliated, mucus-secreting epithelium.

Both cholinergic and adrenergic innervations of submucosal glands exist. However, responsiveness (affinity) of serous and mucous cells to α- and β-adrenergic transmitters differs. In addition, serous cells receive a greater number of both nerve inputs (overall ratio of cholinergic to adrenergic innervation 9:1).

12.3.1.3 Clara Cells

Mucins are secreted not only by goblet cells and submucosal glands, but also Clara cells. Clara cells are the most abundant secretory cell type in distal airways of the human lung.

Clara cell produces high levels of cytochrome-P450. It secretes the lipoprotein secretoglobin-1A1, or Clara cell secretory protein (CCSP). Whereas Muc5ac localizes to the surface epithelium, Muc5b lodges not only in submucosal glands, but also in Clara cells.

Clara cells can differentiate into mucous cells such as goblet cells [1539, 1540] (metaplasia). However, the molecular and functional features of the Clara cell are retained in mucous cells [1534]. Clara cells synthesize the regulated exocytic marker Rab3d GTPase.[7]

Upon antigen and aerosolized ATP exposure, Clara cells rapidly secrete mucin. Extracellular purine and pyrimidine nucleotide triphosphates (adenosine [ATP] and uridine [UTP] 5′-triphosphate) stimulate mucin liberation by airway epithelial cells [1541]. The order of potency of nucleosides and nucleotides in stimulating mucin secretion in cultures of human tracheobronchial epithelial cells is the following:

$$\text{ATP} \sim \text{UTP} > \text{ADP} \sim \text{UDP} \gg \text{adenosine}.$$

[7] Among the 4 Rab3 GTPases (Rab3a–Rab3d), Rab3a and Rab3c localize mainly to neurons (synaptic vesicles); Rab3b to adherens junction of colonic epithelium; Rab3d to secretory granules of exocrine pancreatic and salivary epithelial cells, mastocytes, and several types of endocrine cells.

Agonists of nucleotide receptors are potent mucin secretagogues. Mucin is released in response to the G-protein–PLC–PKC pathway. However, only UTP can increase mucin gene (MUC5AC and MUC5B) transcription using the G-protein–MAPK axis. Therefore, in differentiated airway epithelial cell cultures, ATP causes mucin secretion and UTP both mucin synthesis and secretion.

The exocytic priming protein MUnc13-2, or Uncoordinated-13 homolog Unc13b, is required in the regulated, baseline secretory pathway of Muc5b in Clara cells [1542]. On the other hand, MUnc13-4, or Unc13d, a Rab27 effector, triggers ATP and UTP nucleotide-regulated release of accumulated mucins in secretory granules.

12.3.1.4 Nervous Control

The lungs and pleura are innervated by the mixed, anterior and posterior, pulmonary plexi that contain vagal (parasympathetic) and sympathetic nerves. The pulmonary parasympathetic and sympathetic innervations originate from thoracic plexi. The pulmonary nerve plexus lies behind each hilum, receiving fibers from both vagi and thoracic T1 to T4 ganglia of the sympathetic trunk. Each vagus contains sensory afferents from airways and bronchoconstrictor and secretomotor efferents.

The main pulmonary nerves enter the lungs at their respective hila and then accompany bronchi and associated pulmonary arteries and veins. Each lobar bronchus has 4 to 5 nerves around it. Peribronchial nerves emit thick and thin fibers that enter the bronchial wall [1543]. Intrabronchial nerves give rise to smaller branches and bronchial ganglia.

Most airway mucus is produced by submucosal glands in response to neural signals [1544]. Various secretory chemical messengers synthesized in the neuroendocrine system enable short-range (neurotransmitters and paracrine regulators) and long-range (circulating hormones) communications.

Lung Innervation

The lung is supplied by motor nerves of the sympathetic and parasympathetic compartments of the autonomic nervous system and by sensory nerves that originate primarily from sensory ganglia of the vagus nerve, but also arise from dorsal root ganglia (Table 12.1) [1544].

Pulmonary plexi are located anterior and posterior to the lung roots, close to lung hila, which are constituted by stem bronchi, pulmonary arteries and veins, bronchial arteries and veins, and lymphatic vessels with bronchial lymph glands enclosed by a reflection of the pleura.

Pulmonary plexi innervate smooth muscles of airways and blood vessels, and glands of the bronchial tree. Inner peribronchial nerve plexi reside in the lamina propria under the respiratory epithelium. They feed the bronchial smooth muscle and glands. Large bronchi possess peribronchial and subepithelial nervous plexi with intraepithelial nerve terminals. In bronchioles, the inner and outer nerve plexi merge to form a peribronchiolar plexus.

Table 12.1. Sensory and preganglionic and postganglionic sympathetic and parasympathetic nerves and peptidergic innervation of the respiratory tract (Source: [1544]; CGRP: calcitonin gene-related peptide; NAd: noradrenaline; NPY: neuropeptide-Y; VIP: vasoactive intestinal peptide). Sensory nerves, which have cell bodies mainly in the vagal ganglia, especially the inferior ganglion of vagus nerve (or nodose ganglion), project axons into the pulmonary blood vessels, bronchi, cartilage, and other structures. The parasympathetic preganglionic neurons (vagus nerve) originate from the brain stem (primarily the dorsal motor nucleus and nucleus ambiguus); the parasympathetic postganglionic neurons innervate the intrapulmonary ganglia in the vascular adventitia and airway submucosa. Neurons of airway ganglia produce acetylcholine (ACh), VIP, substance-P (SP), galanin, enkephalins, and nitric oxide (NO). Preganglionic sympathetic fibers of the lungs originate from the intermediolateral horn of spinal cord segments (T1–T5 column), postganglionic neurons from ganglia of the sympathetic chain (superior and stellate ganglia). VIP+ and NPY+ neurons of sympathetic chain ganglia project to airways.

Nerve type	Released messengers Targets
Parasympathetic nerves (vagus nerve)	
Sensory nerves	CGRP, SP
Preganglionic nerves	ACh
Postganglionic nerves	ACh, NO, SP, VIP Respiratory epithelium, airway smooth muscle, glands, bronchial arteries
Sympathetic nerves	
Preganglionic nerves	ACh
Postganglionic nerves	NAd, NPY, VIP Respiratory epithelium, airway smooth muscle, glands, bronchial arteries

Postganglionic parasympathetic efferent nerves cause bronchoconstriction, vasodilation; and bronchial gland secretion. Parasympathetic afferent nerves are sensory fibers of the respiratory epithelium (stretch sensor) that ascend via the vagus nerves. Stretch receptors lodge in smooth muscles of bronchial walls, most being slowly adapting (firing with sustained stimulation), but others are rapidly adapting (transiently firing). Juxtapulmonary (J) receptors localize to alveolar and bronchial walls are connected to unmyelinated (C-fiber) or myelinated afferent nerves of vagus nerves. Postganglionic, adrenergic, sympathetic efferent nerves are bronchodilators, vasoconstrictors, and inhibitors of the glandular secretion.

The parasympathetic nervous system uses not only acetylcholine as a neurotransmitter, but also peptides. Stimulated preganglionic nerves (cell bodies in the medulla oblongata and nucleus ambiguus) release acetylcholine in the ganglion that targets nicotinic receptors of postganglionic neurons. The latter then releases acetylcholine that activate muscarinic receptors of target cells.

Acetylcholine is also the principal neurotransmitter in all autonomic ganglia. In the autonomic nervous system, acetylcholine is actually released by all pre- and post-ganglionic parasympathetic neurons, all preganglionic sympathetic neurons, and some postganglionic sympathetic fibers.

The sympathetic innervation consists of cell bodies in the lateral horn of the spinal cord (intermediolateral cell columns from T1 to L3). These cell bodies belong to general visceral efferent (GVE) neurons and preganglionic neurons. The sympathetic trunks (a paired bundle of nerve fibers laterally along the spine column from the base of the skull to the coccyx) generate the paravertebral ganglia, or ganglia of the sympathetic trunk. Prevertebral ganglia (collateral or preaortic ganglia) localize between the sympathetic chain and supplied abdominal and pelvic organs. These ganglia give rise to postganglionic neurons. In fact, cardiopulmonary, thoracic splanchnic, sympathetic nerves that originate from the paravertebral ganglia contain noradrenergic postganglionic efferent and visceral afferent (sensory) fibers (general visceral afferent [GVA] neurons). Noradrenaline released from postganglionic sympathetic neurons participates to the control of airway behavior.

Neuropeptides

Neuropeptides contribute to regulate the body's functions, in particular the body's respiration, as: (1) para- and autocrine messengers; (2) neurohormones, which are produced by central neurons and reach their target cells of the nervous system via the blood circulation; (3) circulating hormones, which target remote cells; and (4) neurotransmitters and neuromodulators, which are released from nerve endings to act locally on nearby cell. A given neuron can manufactures one or more peptide hormones together with one or more neurotransmitters.

Numerous neuropeptides can be identified in the lung or influence its function. These neuropeptides are synthesized in neurons and released from nerve terminals. They encompass vasoactive intestinal peptide, bombesin, gastrin-releasing peptide (GRP), calcitonin gene-related peptide (CGRP), which are produced by neuroendocrine cells, in addition to serotonin [1544]. Calcitonin, endothelins, and angiotensin-converting enzyme, which forms angiotensin and degrades bradykinin, localize mainly to the endothelium.

In humans, the VIP/secretin family of brain–gut endocrine peptides include 10 members. In addition to vasoactive intestinal peptide and secretin, a hormone that controls water homeostasis throughout the body, this family encompasses adenylate cyclase-activating polypeptide AdCyAP1, or pituitary adenylate cyclase-activating peptide (PACAP), which may act as a sensory neuropeptide,[8] PACAP-related peptide (PRP), peptide histidine methionine (PHM), glucagon,

[8]Two AdCyAP1 isoforms exist: $AdCyAP1_{38}$ and truncated $AdCyAP1_{27}$ (C-terminally truncated $AdCyAP1^{\Delta^{CT}}$).

glucagon-like peptides GLP1 and GLP2,[9] glucose-dependent insulinotropic peptide (GIP), and growth hormone-releasing hormone (GHRH).[10] Vasoactive intestinal peptide is coproduced with peptide histidine methionine. Vasoactive intestinal peptide modulates cholinergic transmission in airways. In addition, the coexistence of VIP and substance-P in cholinergic nerves of airways yields a balance between the bronchoconstrictory, pro-inflammatory substance-P and the bronchiorelaxant, anti-inflammatory VIP [1544].[11] Moreover, VIP and NO cooperate to relaxing airway and vascular smooth myocytes. Nitric oxide acts as a cotransmitter of non-adrenergic non-cholinergic (NANC) relaxation. Vasoactive intestinal peptide, directly or by interacting with other transmitters such as acetylcholine, stimulates airway secretion and ciliary mobility.

Pituitary adenylate cyclase-activating polypeptide AdCyAP1 can be observed in nerves around tracheal and bronchial smooth muscle, glands, and blood vessels [1544]. It coexists with the neuropeptides calcitonin gene-related peptide and substance-P.

Tachykinin+ nerves feed airways. Tachykinins are synthesized in the neuronal bodies of a subset of sensory neurons in dorsal root ganglia. Tachykinin+ nerves distribute to the tracheobronchial epithelium, lamina propria, and smooth muscle, as well as to alveolar walls, systemic and pulmonary blood vessels, and airway ganglia [1544]. Inflammation is triggered by tachykinin release from nerve endings. Stimuli of tachykinin liberation in airways include bradykinin[12] and various prostanoids.

The tachykinin family includes substance-P, neurokinin-A (substance-K) and -B (neuromedin-K), and neuropeptide-K (neurokinin-K) and -γ. The TAC1 and TAC3 genes encode the tachykinin precursors preprotachykinin-A and -B. The TAC1 gene (or TAC2) generates multiple splice variants: substance-P, neurokinin-A, neuropeptide-K, and neuropeptide-γ. The TAC3 gene encodes neurokinin-B. The primary afferent neurons that produce tachykinins and innervate airways express the TAC1 gene [1544].

In airways, tachykinins coexist with calcitonin gene-related peptide, a potent vasodilator. Tachykinins pertain to the most powerful spasmogens on airway smooth myocytes. Substance-P and neurokinin-A target preferentially NK_1 and NK_2 receptors (Table 12.2).

In human isolated airways, neurokinin-A is more potent than substance-P in causing long-lasting contraction [1544]. On the other hand, tachykinins provoke endothelium-dependent vasodilation of isolated precontracted pulmonary arteries

[9]Glucagon results from the cleavage of proglucagon in pancreatic islet β cells. In intestinal L cells, proglucagon is split into glicentin, GLP1, and GLP2 peptides.

[10]A.k.a. growth hormone-releasing factor (GRF or GHRF), somatoliberin, and somatocrinin.

[11]Vasoactive intestinal peptide represses the broncho- and vasoconstrictor effect of histamine, prostaglandin-F2α, leukotriene-C4 and -D4, neurokinin-A and -B, and endothelin. It reduces platelet-activating factor (PAF)-induced edema.

[12]Tachykinins are characterized by a fast onset of action, unlike slower acting bradykinins.

Table 12.2. Tachykinins receptors in cells of the respiratory tract (Source: [1544]).

Target cell	Tachykinin receptor types
Postganglionic nerves	NK_1, NK_2 (between-species variability)
Tracheobronchial ganglia	NK_1, NK_2
Airway epithelial cell	NK_1
Alveolar macrophage	NK_2
Airway smooth myocyte	NK_2
Seromucous glands	NK_1
Pulmonary endothelial cell	NK_1
Pulmonary smooth myocyte	NK_2
Postcapillary venule	NK_1

as well as in perfused tracheal arteries of anesthetized dog, possibly preceded by transient vasoconstriction. They can thus have a mixed effects on the pulmonary and bronchial circulations. Tachykinins also raise airway mucosa microvascular permeability. Last, but not least, these powerful secretagogues stimulate ciliary beating in synergy with mucus secretion.

12.3.2 Mucus Release

Mucins are stored in a dehydrated form in secretory granules. Exocytosis is executed in 3 main steps: (1) motion of mucin granules to the apical plasma membrane; (2) fusion with the plasma membrane; and (3) opening onto the airway surface. Each calcium counterion within the granule is exchanged for 2 sodium ions in the extracellular space [1545]. Rapid secretion enables the formation of concentrated mucus that is resistant to dilution once the mucin network is formed.

Step 1 is regulated by *myristoylated alanine-rich C-kinase substrate* (MARCKS) [1546]. Activated protein kinase-C phosphorylates MARCKS, causing MARCKS translocation from the plasma membrane to the cytoplasm. Agent MARCKS is then dephosphorylated by protein phosphatase PP2 that is activated by cGMP-dependent protein kinase PKG, itself stimulated by the NO–cGMP pathway. A cooperative interaction between protein kinases PKC and PKG is then required. Dephosphorylated cytoplasmic MARCKS associates with the contractile cytoskeleton (Vol. 1 – Chap. 6. Cell Cytoskeleton). Step 2 probably implicates a coordinated interaction between soluble ^{N}ethylmaleimide-sensitive factor receptors (SNARE), MUnc13-4,[13] and small GTPase Rab.

[13]MUnc13-4 is a protein expressed in lungs by goblet cells and by alveolar cells 2 [1547].

12.3.3 Regulation of Mucus Secretion

Submucosal glands continuously secrete polymeric mucins at a low level. They can be further stimulated by adrenergic, cholinergic, and non-adrenergic, non-cholinergic nervous signals (Sect. 12.3.1.4). Therefore, 2 secretion mechanisms exist. The *basal secretion* at low level is unregulated. It is carried out by continuous movement of secretory granules driven by the cytoskeleton. The *regulated secretion* corresponds to granule exocytosis in response to extracellular stimuli to increase mucus secretion. Mucus secretion is elicited by irritations, such as dust and smoke.

The secretion of polymeric mucins is regulated separately from mucin production [1545]. Many mediators trigger mucin secretion, such as cholinergic agonists (also called parasympathometic agents), lipid mediators, oxidants, cytokines, neuropeptides, neutrophil elastase, ATP and UTP, etc. On the other hand, interleukin-4, -9, and -13 provoke mucin synthesis.

Muscarinic M_3 receptors are detected on airway smooth myocytes as well as submucosal glands, whereas M_l and M_2 receptors are located in parasympathic ganglia and cholinergic nerves in lungs, respectively. Receptor M_l is also observed in submucosal glands.

In humans, *adrenergic nerves* of airways could stimulate mucus secretion via α- and β-adrenoreceptors. β-Agonists stimulate ciliary beating in a dose-dependent fashion. A third component of the autonomic nervous system of lungs, the *non-adrenergic, non-cholinergic nervous component* represents the single inhibitory nervous mechanism in humans. Vasoactive intestinal peptide in conjunction with peptide histidine methamine serves as neurotransmitter. It not only increases fluid transport across the epithelium and mucin secretion, but also can inhibit resting secretion of mucus. It raises cAMP concentration in submucosal glands, respiratory epithelium, and airway smooth muscle.

Substance P and calcitonin gene-related peptide are colocalized in afferent nerves of human airways. They contract airway smooth myocytes, elevate mucus secretion, and can provoke mucosal edema. Histamine, leukotrienes, and prostaglandins PGa2 and PGf2 also augment airway mucus secretion.

The most important secretagogue in the respiratory epithelium is ATP that binds to $P2Y_2$ receptors [1545]. Other messengers, such as acetylcholine and histamine, may directly activate receptors on airway epithelial cells or cause airway smooth myocyte contraction, thereby leading to ATP release.

Airway surface liquid hydration, hence effective mucin dispersion into airway surface liquid volume, requires electrolytes and water secretion onto the wetted respiratory surface. Therefore, mucin secretion rate must match ion and water transfer fluxes.

The volume of the airway surface liquid layer depends on the amount of Na^+ and Cl^- in the airway lumen, hence of Na^+ absorption by epithelial Na^+ channel and Cl^- secretion by the Ca^{2+}-activated Cl^- channel and cystic fibrosis transmembrane conductance regulator [1548].[14] Airway surface liquid is not only isotonic with plasma [1548]. In addition, mucus has a similar ion concentration to that of the periciliary liquid.

Oscillatory-like tidal air flow exerts quasi-cyclic shear and stretch with given magnitude and frequency on mural cells. Shear stress may activate signaling pathways in epithelial cells. In particular, time change in high-magnitude mechanical stress raises ATP release into the extracellular medium [1548]. The shear frequency influences more time-varying ion motions than the shear magnitude to maintain a constant overall volume of airway surface liquid, some amount possibly being swept away duting expiration.

Locally, the volume of the periciliary fluid layer is controlled by the concentrations of adenyl and uridyl nucleotides and their metabolite adenosine. Adenine nucleotides (AMP, ADP, and ATP) are released from ciliated cells that sense mechanical stress during ventilation via dedicated carriers such as pannexons and vesicular liberation as well as coexocytosis with uridyl nucleotides from secretory cells. Hydrolytic and transphosphorylating ectoenzymes then convert triphosphate nucleotides into ADP, AMP, and adenosine, UDP, UMP, and uridine.

Mechanical stresses stimulate cosecretion of nucleotides (ATP and UTP) and mucin from goblet cells [1549]. Pannexin-1 (but not Panx3) is synthesized in normal human airway epithelial cells [1550]. Apical pannexons of airway epithelial cells contribute to ATP release. Pannexons open at resting membrane potentials in response to long-range intercellular calcium wave initiated by mechanical stress [1551].[15] In addition, Panx1 colocalized with P2Y receptors open in response to extracellular ATP via intracellular Ca^{2+} ion [1552].

Adenosine triphosphate and its metabolite adenosine regulate secretion via activation of Gq-coupled $P2Y_2$ and Gs-coupled A_{2B} receptors on the apical membrane of ciliated cells, in addition to $P2X_4$ channel and $P2Y_2$ receptors on goblet cells, causing changes in intracellular second messengers that promote chloride release and inhibit sodium absorption, thereby favoring water flux into the airway lumen.

Activated A_{2B} receptor launches cAMP-dependent activation of CFTR Cl^- channel and stimulates ciliary beat frequency [1553]. The adenosine–cAMP axis regulates CFTR positively and ENaC negatively [1548].

Liganded $P2Y_2$ receptor prevents Na^+ absorption as well as CFTR-dependent Cl^- secretion, in addition to precluding ciliary beating and mucin secretion. On the other hand, ATP stimulates Ca^{2+}-mediated CL^- secretion through CaCC channels [1548].

---

[14]The coordinated action of ion channels, pumps, and transporters, maintains secretion by highly water-permeable airway epithelia. These leaky epithelia are capable of both absorbing Na^+ and secreting Cl^- ion, thereby controlling the airway surface liquid (ASL) volume and generating an efficient mucus clearance. When the airway epithelium secretes more salts into the airway surface liquid, water follows passively; the ASL volume rapidly increases to maintain isotonicity with plasma; and vice versa.

[15]Long-range communication using intercellular calcium wave results from the flux of IP_3 through connexons (gap junctions) as well as an extracellular release of ATP messenger through pannexons.

Goblet cells possess subapical granules similar to Muc5ac granules that undergo a Ca^{2+}-regulated exocytosis [1554]. These secretory granules contain ATP and ^{UDP}glucose. Stimulated nucleotide $P2Y_2$ receptor increases cytosolic Ca^{2+} concentration, hence coupled secretion of mucins and nucleotides, the latter creating a paracrine signal that enables mucin hydration fitting to mucin release.

Concomitant nucleotide release by exocytosis and calcium-regulated mucin secretion are elevated in goblet cell metaplasia [1555]. Increased nucleotide liberation from goblet cell and resultant adenosine accumulation yield a compensatory mechanisms to hydrate mucins by paracrine control of ion and water secretion by ciliated cells.

Irritants such as cigarette smoke cause a change in mucus glycoprotein type within goblet cells, from an acid-to-neutral ratio of about 50:50 to 80:20 [1531]. The number of goblet cells increases at the expense of ciliated cells. Prolonged exposure to irritants leads to an increase in size of submucosal glands and, subsequently, a reduction in airway diameter. Mucus hypersecretion overwhelms the mucociliary apparatus and mucociliary clearance is attenuated. A significant part of mucus removal from lungs results from coughing.

12.4 Mucus Composition

Mucus is a heterogeneous fluid aimed at protecting airways against flow stresses and chemical damages, entrapping and eliminating inhaled particulate matter and microorganisms. Suitable quality and quantity of airway surface liquid are necessary for cilium activity. Respiratory mucus is approximately composed of 90 to 98% water, 2 to 3% proteins, glycoproteins, and proteoglycans, and 2 to 3% lipids, carbohydrates, and minerals in addition to DNA (0.02%) and other debris of shed epithelial cells as well as cells such as neutrophils that can influence mucus rheology. Mucus constituents are interdependent. Active ion transport across the epithelial wetted surface modulates airway surface liquid hydration.

Mucus forms a barrier over epithelial cells in the conductive airways that is composed of highly branched, negatively charged glycoproteins intertwined with other mucosal constituents (water, salts, and other macromolecules, as well as cellular debris and entrapped particles). Entanglements between mucosal constituents are stabilized by covalent and non-covalent links, such as hydrogen bonding and electrostatic and hydrophobic interactions. Crosslinking rises when mucus hydration decays.

Mucus glycoproteins are produced from glycopeptide subunit with a linear peptide backbone and linear or branched carbohydrate side chains (2–14 oligosaccharides).[16] Subunits can be linked by disulfide bonds or a link protein.

---

[16]Hydroxy amino acids serine and threonine that are the most abundant amino acids in the peptide backbone form the ^{O}glycosidic link with ^{N}acetylgalactosamine, the initial sugar in the

12.4.1 Mucin

Mucus is composed of *mucins*, heavily glycosylated proteins that are encoded by MUC genes. Glycoproteic mucins have a composition that is dominated by carbohydrates (up to 80% of the molecular weight). Mucins correspond to less than 2% of the wet weight of the mucus. Mucin size and concentration influence gel formation.

In the respiratory tract and its annexes, mucin gene expression varies according to the compartment. The middle ear expresses mainly Muc5b and the Eustachian tube produces equal amounts of Muc5b and Muc5ac, whereas the tracheobronchial epithelium synthesizes primarily Muc5ac.

Mucins are characterized by a variable number of tandem repeats — mucin domain — that constitutes the main component of mucus structure [1512].

Among 13 identified mucin genes, only 4 (MUC2, MUC5AC, MUC5B, and MUC6) encode the cysteine-rich motifs of mucin C- and N-termini that are required for oligomerization. Three of these 4 genes (MUC2, MUC5AC, and MUC5B) are expressed in airway cells. The 2 predominant mucins in airway mucus — Muc5ac and Muc5b — have different chemical and physical properties (charge density, buoyancy, etc.). Mucins of airway mucus constitute a mixture of heterogeneous glycoproteins that are polydisperse (mass 2–40 MDa; size 0.5–10 μm).

Among mucins, Muc5ac and Muc5b form homotypic polymers, i.e., Muc5ac monomers bind only to Muc5ac and Muc5b monomers to Muc5b. These polymers that form the mucus gel are mainly structured as long single chains rather than branches. Entangled, bundled, and possibly crosslinked mucins constitute glycosylated proteic fibers (caliber 3–10 nm) that are suspended in a solution of electrolytes and other molecules.

Mucins can be divided into 2 subsets according to their location relative to the wetted epithelial surface. Two major forms of mucin comprise monomers that are primarily, but not exclusively, located at the cell surface and oligomers that are secreted [1556].

Secreted, gel-forming mucins are entirely extracellular. The majority of mucins produced in the lung belong to the gel-forming subset. The most notable members are Muc5ac and Muc5b (~ 90% of the mucin content of sputum) [1512].

Transmembrane (or membrane-tethered) mucins, such as Muc1, Muc4, and Muc16 (~ 10% of the mucin content), have a single membrane-spanning domain, a short cytoplasmic tail, and an extensive extracellular segment.[17] Other members of the mucin family (Muc1, Muc4, Muc7, Muc11, and Muc13) also expressed by

---

side chain. Four monosaccharides constitute side chains: galactose, fucose, ^{N}acetylglucosamine, and ^{N}acetylneuraminic acid (or sialic acid). Carbohydrates can have sulfate residues and sialic acids that confer a negative charge to the molecule.

[17]Other tethered mucins (Muc3a, Muc3b, Muc12, Muc13, Muc15, Muc17, and Muc20) are either largely uncharacterized, or are not elaborated in the lung [1512]. Mucin-3 is produced in lung tumors; Muc13 and Muc15 in the respiratory tract.

airway wall cells, except Muc7, are primarily, but not exclusively, large transmembrane glycoproteins that can be released in airway surface liquid after shedding from the cell surface. In addition, splice variants of Muc1 and Muc4 are secreted [1556].

Apical mucins (Muc1, Muc4, and Muc16) tethered to the plasma membrane are involved in shielding the airway epithelium against pathogens as well as in cellular signaling [1512]. On the wetted surface, cilium Muc4 density prevents mucus penetration into the periciliary fluid and provides lubrication through bound water. In addition, mucin-1 on the cell surface and microvilli of both ciliated and secretory cells has a cytoplasmic tail capable of intracellular signaling. It modulates airway defense and inflammation. Mucin-16, the largest mucin, is expressed by both ciliated and secretory cells.

Mucin-1 extends at least 100 nm from the cell surface; Muc4 at least 300 nm [1512]. Both the MUC1 and MUC4 genes encode heterodimers that are post-translationally processed. In the lung, Muc1 localizes mainly around microvilli, whereas Muc4 and Muc16 are expressed on the surface of cilia [1512]. Functions of Muc1, Muc4, and Muc16, like those of extracellular mucins, comprise hydration, lubrication, protection from proteases, and defense against pathogens.

Mucus mesh is composed primarily of entangled mucins and other mucus constituents with reversible linkages rather than crosslinked polymers, as mucus sample can reach complete dissolution in saline and recover its viscous and elastic properties within seconds, when it bears shearing, whereas gels characterized by covalently crosslinked constituents do not dissolve and irreversibly tear upon shear [1557]. Nevertheless, low-affinity non-covalent bonds and stronger disulfide links exist between mucin fibers as well as other mucus constituents.

Mucins are enriched in serine and threonine that are sites for attachment to ^{N}acetylgalactosamine. Other mucin-linked oligosaccharides comprise ^{N} acetylglucosamine, galactose, fucose, and sialic acids [1556]. Oligosaccharide diversity enhances the probability that bacteria bind to mucus for removal by mucociliary clearance.

Mucin Muc5ac is predominantly produced by goblet cells, whereas Muc5b is mainly synthesized by mucous cells of submucosal glands [1556]. Production of Muc5ac can also occur in glandular mucous cells, whereas synthesis of Muc5b and Muc2 is detected in goblet cells. Sites of formation of various Muc5b glycoform types remain unknown.

Mucins can have both anti- and pro-adhesive capacities. The extended structure of mucins creates steric hindrance, in addition to act as an electrostatic barrier. However, Muc1 and Muc16 provide binding sites for various types of adhesion molecules (e.g., ICAM1 and selectins) [1512].

Between-mucin interactions (e.g., Muc1–ICAM1 connection) can bridge different cell types. These associations rapidly elevate cytosolic calcium concentration in Muc1+ cells. The cell response involves Src, PI3K, and PLC enzymes [1512].

The cytoplasmic tail of mucin-1 influences multiple signaling axes, especially upon phosphorylation by Src, HER receptors, GSK3β, and PKCδ [1512]. Other binding partners encompass catenin-δ1, β-catenin and its controller the tumor suppressor adenomatous polyposis coli (APC), GRB2 adaptor, P53 transcription factor, ERK1 and ERK2 kinases, estrogen receptor-α (NR3a1), and HSP70 and HSP90 chaperones.

Mucus contains bacteriostatic components, such as lysozyme to break down pathogens and lactoferrin to limit iron availability to bacteria [1558]. Secretory immunoglobulin-A produced by plasmocytes in mucus prevents bacterial adhesion by binding to the bacterial surface within the mucus.

The mucus in goblet cell granules is condensed. Negative polyanionic charges within the secretory granules are neutralized by calcium ions. When the granule content is released across a membrane pore into the airway lumen, calcium ions diffuse and polyanionic charges repulse, leading to mucus hydration. Owing to its composition, condensed intragranular glycoproteic mucins expand up to several hundred-fold once released in the airway lumen.

Once secreted, the mucin gel expands almost instantaneously. It is able to increase in volume 500-fold in about 20 ms. Mucins then mix with proteins, lipids and glycoconjugates in order to form a dilute aqueous solution, the mucus.

Mucins increase mucus viscosity. Protein Muc5b has especially a strong tendency to polymerize and form a gel. Mucins also function in agglutination and opsonization of foreign particles and pathogens, thereby participating in the innate immune system.

12.4.2 Other Proteic Constituents of Mucus

Mucus glycoproteins are important determinants of mucus viscoelasticity. Proteoglycans are a major source of mucus hydration. Mucus proteins have a protective effect against H^+ penetration into tissues. Buffer capacity is accompanied by pH-dependency of mucus viscosity.

Some airway surface liquid components contribute to airway protection, as they target pathogens (e.g., secretory IgA, proline-rich proteins, defensins, lysozyme, and transferrin). Glycoprotein GP340[18] that has an antibacterial role complexes with Muc5b in airways [1556]. It directly binds to bacteria or associates with bacterial-binding collectin surfactant protein-D. Other mucus components modulate the organization and hence the properties of the mucus gel. Addition of proteins that can interact with mucins, such as immunoglobulin-A and -M or lysozyme, can increase mucus viscoelasticity [1557].

[18] A.k.a. Deleted in malignant brain tumor protein DMBT1, hensin, muclin, salivary agglutinin, and pulmonary surfactant protein-D-binding protein.

Table 12.3. Ion composition of airway surface liquid in human xenograft model of fetal airways developed in severe combined immunodeficiency mice (Source: [1560]).

Ion	Concentration (mmol/l)
Cl^-	65
Na^+	64
SO_4^{2-}	5
K^+	2
Mg^{2+}	2
Ca^{2+}	1

12.4.3 Lipids

Lipids represent another important part of mucus molecules (1–2%). Most of the lipid content is associated with mucin glycoproteins. Phosphatidylethanolamine, sphingomyelins, and lysophosphatidylcholine raise mucus viscosity, whereas phosphatidylglycerol reduces it [1557].

12.4.4 Surfactant

Constitutive surfactants that are constituted of proteins and phospholipids have several functions, as they regulate surface tension, control Eustachian tube opening pressure, modify mucus rheology, mitigate oxygen free radical damage, and decrease inflammation [1559].

In addition, surfactant components enable sliding of deep gel phase over superficial sol phase of the liquid interface between flowing air and respiratory duct wall. Surfactant lowers adhesion at the gel–sol interface so that cilium tips can better penetrate the gel layer and more effectively propel this layer. Surfactant favors high ciliary beat frequency, hence mucus speed. Surfactant treatment in infants with respiratory distress syndrome elevates mucociliary transport rate.

12.4.5 Minerals

Various salts with mono-, di- and trivalent ions account for up to 1% of mucus mass (Table 12.3). Change in ionic content leads to mucus shrinkage or swelling. High concentration of multivalent cations, such as calcium and magnesium, can facilitate reversible crosslinking between mucin monomers.

12.5 Mucus Properties

Biophysical properties of airway secretions include rheology, *cohesivity* that results from molecular attraction and depends on polymer length and density, and surface properties.

Rheological properties deal with deformation and energy storage when a stress is applied and release of stored energy when the stress is removed. Restitution of stored energy explains mucus clearability. The respiratory mucus is characterized by a viscoelastic behavior. Mucus actually possess *pseudoplasticity*, *elastothixotropy*, *spinnability* (capacity to be drawn out into long threads [Spinnbarkeit]), and *adhesiveness* [1529].

Surface interactions of the airway surface liquid with the respiratory epithelium on the one hand and airflow on the other comprise *wettability* (interfacial spreading capacity measured by the contact angle), adhesiveness (interfacial tension quantified by adhesivity), and *tenacity* (product of adhesive work and cohesivity). Contact and rheological properties determine mucus transport function. The latter also depends on mucosal *permeability* to particles.

12.5.1 Mucus Permeability

Airway surface liquid operates as a protective barrier at mucosal surfaces that traps and removes foreign particles such as bacteria and toxins (size $\sim$ 1 mm) as well as viruses and environmental small particles (size 100–500 nm), but allows flux of nutrients, selected gases, ions, antibodies, and many proteins as well as cells of the mucosal immune system. It is impermeable to large particles, the permeability threshold depending on strength of association of mucin fibers that determines the typical mucin mesh size. Selective permeability and dynamic viscoelastic behavior of mucus are also controlled by mucin environmental condition (lipids, salts, cells, etc.). Mucus behavior is governed by viscous drag of interstitial fluids moving within the mucin mesh and elastic recoil from mucin fibers.

Mucus is characterized by heterogeneity of the pore network of the viscoelastic matrix made of highly branched and strongly negatively charged mucus glycoproteins that are entangled with other mucosal constituents.

Diffusion of amine-modified and carboxylated polystyrene particles[19] (size 100–500 nm) has been investigated in sputum of mucoviscidosis patients by particle tracking with good spatial and temporal resolution (5 nm and 33 ms) [1561]. Small (size $<$ 200 nm), neutrally charged particles travel more quickly inside mucus than large, charged particles. Particle transport rate depends on the particle size range, as

[19]Amine-modified and carboxylated polystyrene particles have neutral and negative surface charges at physiological pH, respectively.

small particles undergo more diffusive transport,[20] whereas large particles interact with a porous, charged, viscoelastic network.

When a particle set has a large particle size range, the average particle transport rate is influenced by transport rates of large particles. Assessed ensemble-average diffusivities of 100-, 200-, and 500-nm particles in cystic fibrotic sputum range from about 1.5×10^{-2} to 3.8×10^{-4} $\mu m^2/s$ [1561].[21]

12.5.2 Mucus Contact Properties

Wettability and adhesiveness govern contact properties with the respiratory epithelium. These 2 properties must be high enough in order to avoid sedimentation down to the distal airways in lower lung lobes, and low enough for mucus mobilization by air flow during coughing. Surface-active phospholipids, such as phosphatidylcholine and phosphatidylglycerol, improve mucus wettability [1562].

12.5.3 Mucus Rheology

Biological materials are commonly viscoelastic. When they are subjected to a small-amplitude, sinusoidal load, and display a linear response, they can be characterized by 2 frequency-dependent properties: (1) *elastic or storage modulus*, a measure of material stiffness or tendency for mucus non-Newtonian gel (with a viscosity that varies non-linearly with shear rate) to recover its original configuration following load-induced deformation, and (2) *viscous or loss modulus*, i.e., extent to which the gel resists the tendency to flow that is also a measure of energy dissipation. Additional rheological quantities can also describe biological materials. *Creep* describes strain as a function of time when a constant stress is applied. It quantifies the tendency to deform permanently. *Spinnability* measures the capacity of fluids to be shaped into threads.

At the macroscopic scale that yields averaged measures of rheological features, mucus is a viscoelastic, thixotropic, non-Newtonian gel, as its viscosity varies non-linearly with shear rate (strong and weak resistance to deformation at low and high shear rate, respectively). Mucus is indeed characterized by a shear thinning behavior. Viscosity of human respiratory mucus of 12 to 15 Pa·s with a relaxation time of about 40 s and elastic modulus of 1 Pa is supposed to represent an optimal rheological profile for mucociliary clearance [1557]. The rheological properties

---

[20] Small particles (100–200 nm) can move almost linearly with time and more freely than larger particles.

[21] In water, diffusion coefficients of 100-, 200-, and 500-nm particles are equal to 4.5, 2.3, and 0.9 $\mu m^2/s$, respectively.

Table 12.4. Mucus rigidity measured by magnetic rheometry at low rotation speed (1 rad/s) and clearance indices (Source: [1565]).

	Rigidity factor (log) index	Mucociliary clearability index	Cough clearability	Secretion rate (mg/min)
Baseline	2.25±0.60	0.9±0.1	1.3±0.6	0.12±0.12
Acetylcholine	1.58±0.40	1.0±0.1	1.8±0.6	0.63±0.49
Tasuldine	1.80±0.64	1.0±0.1	1.7±0.7	0.11±0.12

of respiratory mucus vary according to the site along the airway network. Mucus heterogeneous microscopic domains contain entangled fibers and spaces filled by a low-viscosity fluid. For a given sampling site, mucus rheology depends not only on loading strength and application rate, but also constituent crosslinking and entanglement.

Airway surface liquid is subjected to low and high stresses that correspond to cilium beat and cough, respectively. An optimal rheology of the airway surface liquid is required for the mucociliary clearance, as its deformation features determine efficiency of its coupling to cilia. Mucus with a low or high storage modulus (i.e., small or large number of crosslinks) yields reduced transport speed. When mucus becomes too thick (e.g., in severe bronchitis or cystic fibrosis), mucociliary clearance decays and bacterial growth is favored.

Elasticity is the capacity to store a recoil force and restore energy, whereas viscosity that measures resistance to deformation is associated with energy loss. Elasticity of a mucus gel depends on density of crosslinks and time scale over which crosslinks rearrange in response to applied stress. Mucus viscosity decays with heightening stress or rate of strain.

Dynamic viscosity of normal mucus is $\sim$ 1 Pa·s.[22] Current values of the dynamic storage (elastic) modulus range 1 to 2 N/m^2 for efficient transport. The crosslinking degree of mucus gel influences the storage modulus.

Glycosphingolipids increase mucus viscosity. Mucus viscosity is correlated with sialomucin concentration. In patients with cystic fibrosis, cholesterol, glycosphingolipids, and sphingomyelin are positively correlated to the viscosity of expectorated airway secretions, whereas phosphoglycerol is negatively correlated [1564]. Certain bronchosecretolytic agents decrease and increase the sialomucin and sulfomucin content, respectively, thereby attenuating mucus viscosity [1565] (Table 12.4).

[22] In normal, anesthetized dogs, mean values of elasticity and viscosity of tracheal mucus at 1 rad/s are about 30 Pa and 7.5 Pa·s , respectively, whereas steady-state viscosity is approximately 3×10^3 Pa·s [1563]. Mean viscosity as well as water content (98%) of airway liquid in normal (0.6 ± 0.5 Pa.s) and cystic fibrosis (0.2 ± 0.1 Pa·s) fetal tracheal xenografts is similar, although previous literature data report high viscosity values in cystic fibrosis (from 1 Pa·s to 2 kPa·s) [1560]. Hyperviscosity and dehydration in cystic fibrosis airway secretions result from inflammation and infection.

Nanoscale mucus rheology assessed by amplitude and time scale of geometrically averaged ensemble mean-square displacements of fluorescent particles[23] can change without influencing macrorheological properties measured by torque required to apply a small-amplitude oscillatory stress at specified frequencies using a cone-and-plate rheometer [1566]. Brownian displacements of non-interacting beads (coated with a non-mucoadhesive surface) is more or less precluded according to bead size. Rheological parameters vary according to trapped particle size.[24]

Many factors contribute to mucus rheology, such as mucus glycoprotein type, hydration degree, and crosslink density [1567]. The latter is influenced by pH and ion content as well as the presence of possible inflammatory mediators and enzymes. Mucolytics reduce viscosity by disrupting polymer network in airway surface liquid. Classical mucolytics work by severing of disulfide bonds, binding of calcium, depolymerizing mucopolysaccharides and liquefying proteins. Some peptide mucolytics degrade DNA and actin.

Microrheology of human mucus deals with behavior at length scale of pores in mucus mesh and macromolecules ($\mathcal{O}$[1 nm]) or nanoparticles ($\mathcal{O}$[10–100 nm]). Microviscosity at length scales higher than average interfiber spacing is associated with a transport rate that can be assessed by particle tracking. Diffusion at these length scales in mucus appears to be similar as that in water [1566]. Nanopolymers with a characteristic length of few to several hundreds of nanometers are able to cross mucus with only 4- to 6-fold decrease in effective diffusivity with respect to water. Conversely, heterogeneity of mucus mesh that results from increased content in nanoparticles entrapped in mucus could create larger pores for mass transfer [1566]

---

[23]The bulk elastic modulus of concentrated isotropic solutions of entangled polymers is described by the storage modulus. The storage modulus for semiflexible, flexible, and stiff, crosslinked polymer networks is estimated by [1566]:

$$G'(\omega) \sim \kappa^2 \xi^{-2} L_c^{-3} / k_B T$$

(xi: mesh size; L_c chain length; $k_B T$: thermal energy; k_B:Boltzmann constant; T: absolute temperature; and κ is related to the persistence length of the chain $Lp \approx \kappa / k_B T$), $G'(\omega) \sim \kappa^2 \xi^{-5} / k_B T$, and $G'(\omega) \sim \xi^{-3} k_B T$, respectively. The viscoelastic spectrum $G(s) = 2k_B T / 3\pi R s < \Delta r^2(s) >$ (R: particle radius; s: Laplace frequency; and $< \Delta r^2(s) >$: unilateral Laplace transform of time-averaged, mean-square displacement $\Delta r^2(\tau)) >=< [x(t+\tau) - x(t)]^2 + [y(t+\tau) - y(t)]^2$ ($x(t), y(t)$: nanoparticle coordinates at a given time t; and τ time scale). The Fourier transform equivalent of $G(s)$ is the complex shear modulus $G^*(\omega)$ [1566]. The complex shear modulus of mucus that bears oscillatory deformations of small amplitude (1% strain) yields frequency-dependent (dynamic), storage (elastic; $G'(\omega)$), and loss (viscous; $G''(\omega)$) moduli, which are in-phase and out-of-phase components with respect to loading, respectively.

[24]Effective storage modulus of cervicovaginal mucus at the lowest loading frequency is equal to 1.0 and 4.3 mPa when mucus traps non-interacting 200- and 500-nm beads and 400 to 154,000 mPa with 1-mm beads [1566]. In addition, loss modulus is higher than storage modulus for mucus that traps 100- to 500-nm particles, and conversely for 1-mm beads. At high frequency, beads sufficiently small relative to mesh spacing experience the viscous drag of interstitial fluid, whereas at low frequency (large time scale), elastic behavior contributes to mucin mesh hindrance.

12.6 Mucus Motility

Once delivered in the airway lumen, mucus is driven by cilia of the respiratory epithelium that beat in a coordinated manner within the intraciliary fluid layer (epithelial lining fluid) beneath a viscoelastic (more or less continuous) mucus film.

Ciliated cells of the respiratory mucosa line the respiratory tract from the nose and nasal sinuses down to terminal bronchioles. Ciliated columnar cells outnumber the other cell types. The percentage of ciliated cells in human airway epithelium decreases from approximately 80% to 50% in the tracheal epithelium to 20% in the bronchiolar compartment. Ciliated cells are very vulnerable to injury by infection and pollutant exposure. Interleukins IL4 and IL13 increase the population of goblet cells and decrease that of ciliated cells.

Each cell has 200 to 300 cilia at its apical surface that beat in unison, propelling mucus and entrapped foreign material toward the oropharynx. Some substances such as cigarette smoke are ciliotoxic.

12.6.1 Respiratory Cilium

Respiratory cylindrical cilia have a caliber of 0.10 to 0.25 μm, a height of 0.5 to 7 μm, with a mean separation between cilia of 2 to 5 μm [1568]. The cilium originates in a corpuscule beneath the plasma membrane. Between cilia, cytoplasmic projections (microvilli) can be observed. The distal, narrowed end of cilia is associated with the glycocalyx. Microtubule capping structures at tips of respiratory cilia that are firmly attached to specific peripheral A-tubule and central pair microtubules restrict microtubule sliding.

The cilium is covered by a ciliary membrane that is continuous with the plasma membrane, as it is a specialized extension of the cell membrane. An array of microtubules (Vol. 1 – Chap. 6. Cell Cytoskeleton), the *axoneme*, runs longitudinally inside the cilium, with a characteristic arrangement of 9 peripheral doublets of microtubules and 2 central single microtubules (Fig. 12.2). Axonemal microtubules end in the basal body within the cell cortex.

Each outer doublet microtubule consists of tubule-A and -B. Each tubule-A supports inner and outer dynein arms and a radial spoke that has a head close to one of the 2 single microtubules of the central sheath complex. In addition, 2 sets of links — the interdoublet linkers — join adjacent microtubular doublets. Nexin and radial spoke linkages serve as circumferential and radial connections. Two inner and outer dynein arms that yield axonemal ATPases exist on each outer microtubular pair implicated in sliding of microtubular pairs relative to one another. Dynein undergoes cyclical shape changes in the presence of its preferred substrate ATP^{Mg} to propel adjacent microtubules tipward. Accessory structural elements (radial spokes, interdoublet linkers) are supposed to provide shear resistance to microtubular sliding. Tektins are proteins associated with microtubules.

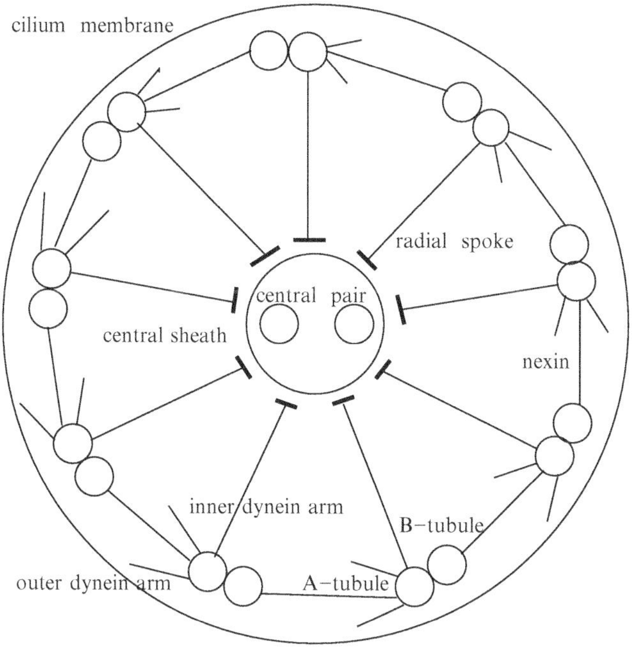

Fig. 12.2 Structure of the respiratory cilium. The cytoskeletal structure of the cilium is the axoneme. The building block of the axonene is the microtubule. Each axoneme is composed of several microtubules aligned in parallel. It is actually constituted of an array of 9 peripheral doublets of microtubules (A- and B-tubules) and 2 central single microtubules. Each A-tubule possesses: (1) inner and outer dynein arms that cause microtubules to slide against each other and cilium to bend back and forth; (2) nexin linkage to adjacent microtubular doublet; and (3) radial spoke with head close to one of the 2 single microtubules of the central sheath complex. Energy source ATP allows active sliding of axonemal doublet microtubules.

Both plasma and ciliary membranes control the ciliary activity by regulating the ionic environment and ATP concentration for normal enzymatic function of the axoneme. Regulators modify the spatiotemporal pattern of the dynein–microtubule interactions within the axoneme.

Bardet-Biedl Syndrome Proteins

Mutations in genes that generate Bardet-Biedl syndrome affect primary, sensory cilia as well as motile cilia of the respiratory tract.[25] The BBS genes are expressed

[25]The Bardet-Biedl syndrome is a rare, pleiotropic, autosomal recessive disorder with diverse manifestations (obesity, hypertension, retinopathy, polydactyly, hypogenitalism, renal abnormalities, and developmental delay) and increased incidence of asthma. It results from mutations in any of BBS genes (BBS1–BBS12).

in human airway epithelia [1569]. The BBS proteins operate in cilium formation and function. They localize to the neighborhood of primary cilia, especially the basal body, ciliary axoneme, and pericentriolar region. The proteic complex — the BBSome — that contains 7 BBS proteins lodges in centrioles and primary cilia. Proteins BBS2 and BBS4 reside in structures associated with motile cilia. Although BBS proteins are not required for ciliogenesis, their loss causes structural and functional defects in cilia, such as bulges filled with vesicles near the tips of cilia as well as reduced ciliary beat frequency, possibly due to an impaired intraflagellar transport complex.

Inositol Pentakisphosphate 2-Kinase

Inositol (1,3,4,5,6)-pentakisphosphate (IP_5) 2-kinase (IPPK; a.k.a. IPK1) that generates inositol hexakisphosphate (IP_6) intervenes in ciliary beating and length maintenance [1570].[26] Enzyme IPPK abounds in centrosomes and basal bodies, where it colocalizes with BBS4 and BBS8 proteins. It influences microtubule-based transport. Inositol polyphosphates modulate protein activity and may contribute to the intraflagellar transport complex. They may act as cofactors for a component of intraflagellar transport complex, cilium, or basal body, thereby allowing for its optimal activity.

12.6.2 Cilium Beats

Cilia from the set of ciliated cells form a coordinated beating cluster. Cilia synchronously beat, thereby induces rythmic waves, which propels the mucus layer with trapped particles and germs. The epithelial lining fluid thickness is slightly less than the length of an extended cilium. The cilium tips thus can impact the superficial mucus layer. The mucus is propelled by momentum transfer from the less viscous liquid film. At the airway branching, ciliary motions impose a particle path lateral from the apex.

12.6.2.1 Cilium Beating Stages

The beat cycle of respiratory cilia is composed of 3 phases [1515]: rest, recovery, and effective stroke. In functionally intact frog palate epithelium, a resting state

[26]Lipid-anchored phosphatidylinositol (4,5)-bisphosphate is the starting molecule for soluble inositol polyphosphate production. Phospholipase-C produces diacylglycerol and soluble inositol (1,4,5)-triphosphate. The latter is sequentially phosphorylated to produce inositol tetrakisphosphate, pentakisphosphate, and hexakisphosphate.

Table 12.5. Duration and percentage of beat cycle of respiratory cilium beat stages assuming a beat frequency of 20 Hz (Source: [1531]). Estimated speed of a cilium tip during effective stroke is 1 mm/s.

Stage	Duration (ms)	Percentage of beat cycle
Effective stroke	10	20
Rest	13	26
Recovery stroke	27	54

exists during which cilia do not beat [1571].[27] At rest, the cilium lies parallel to the cell surface pointing in the direction of mucus flow. The resting period separates effective stroke that is mostly planar with an arc of about 110 degrees and recovery stroke during which cilia sweep backward remaining close to the epithelium and immersed in the periciliary fluid (without contact with the mucus layer).

Two beating stages result from the sliding of axonemal microtubules and permit unidirectional propulsion of the mucus layer (Table 12.5): (1) a preparatory phase with a backward motion and, after maximizing its height to interact with the mucus gel, (2) a propulsion phase that provide impulse to the mucus. During effective stroke, the bending cilium region passes progressively from the base to the tip to become fully extended. During recovery stroke, the cilium curls backward in the same plane. Cilia are more rigid in the forward power stroke and more flexible in recovery stroke in preparation for the next effective stroke. The ciliary motion speed differs between the 2 stages, with a quick rising phase and a slow sagging phase. Elevation in ciliary beat frequency can be achieved by shortening a single or, most often, several phases.

12.6.2.2 Ciliary Beat Frequency

Ex vivo, cilia of the nose, trachea, and large bronchi beat at a similar frequency (10–20 Hz). The ciliary beat frequency of samples of human respiratory epithelium ranges from 9.1 to 16.8 Hz [1572]. In patients examined by bronchoscopy, the ciliary beat frequency is similar in nasal, tracheal, and lower lobe bronchial epithelia (14 Hz), whereas it is slower in peripheral bronchi (10 Hz) [1573]. Ciliary beat frequency in human nose in vivo is equal to 11.5 ± 1.6 Hz [1574] and that of maxillary sinus mucosa in vitro to 16.6 ± 3.0 Hz [1575]. Ciliary beat frequency hence slightly differs between in vivo and in vitro condition.

[27]Contact with a particle (size 50–75 μm) stimulates ciliary beat. Only cilia surrounding the particle in a patch about 10 times larger are beating at one time. Beat frequency is the highest in the patch center, close to the particle, and decays toward patch edge. The moving patch of beating cilia corresponds to a mucus plaque that is produced upon particle impact and surrounds the particle to be removed. Contact on mucus-depleted epithelium produces a lower ciliary beat frequency.

12.6.2.3 Ciliary Beat Speed

The nasal mucus is displaced toward the pharynx at a speed of 1 to 3 cm/mn. Because the density of ciliated cells in the tracheobronchial tree decreases distally (from a proportion of about 22% in the trachea to 3% in generation 3), mucus speed ranges from 4.9 mm/mn in the trachea to 1.7 mm/mn in generation 3.

12.6.2.4 Ciliary Beat Regulators

Ciliary beat is regulated by intracellular second messengers, such as cAMP and cGMP as well as Ca^{2+} and H^{+} ions [1576]. Messenger cAMP activates protein kinase-A localized to the axoneme. Phosphorylation of PKA substrates increase ciliary beat frequency. Messenger cGMP also raises ciliary beat frequency using protein kinase-G and possibly PKA kinase. On the other hand, protein kinase-C lowers ciliary beating. Intracellular pH may affect ciliary beat frequency via dynein arms.

Increase in intracellular concentration of calcium ions that heightens the ciliary beat frequency results from Ca^{2+} influx from both the extracellular medium and intracellular stores. Reversal of beat direction is triggered by quick calcium influx through voltage-sensitive calcium channels distributed over the cilium length [1577].

Deformations of the cell surface induce a rapid, transient increase in ciliary beat frequency. Mechanical stimuli of ciliated or non-ciliated cells of the respiratory epithelium immediately increase the Ca^{2+} level via stretch-activated calcium channels [1515]. Mechanotransduction in cultured ciliated cells of the respiratory epithelium is lost when extracellular Ca^{2+} is removed.

In addition, mechanical stimuli that excite some respiratory epithelial cells activate adjacent cells via messenger transmission through gap junctions [1578]. Calcium waves primed by intra- and intercellular messenger inositol trisphosphate in cells of the respiratory epithelium indeed coordinate a cooperative increase in ciliary beat frequency by releasing calcium ions from stores.

Duration of effective, recovery, and rest phases of the ciliary beat cycle depends on dual regulation of intracellular concentration of cAMP and calcium ions [1515]. Mucus load yields a local control of the ciliary beat frequency, whereas neurohormonal stimuli regulate ciliary activity throughout the respiratory tract.

Ciliary beat frequency rises (duration of the 3 ciliary beat phases decays) when temperature elevates and intracellular concentrations of cAMP and Ca^{2+} heighten [1515]. Regulators Ca^{2+} and cAMP can regulate ciliary beat frequency by influencing ATP availability to or its use rate by axonemes. Calcium ions act in conjunction with calmodulin at some axonemal sites [1578]. Excitation by cAMP of ciliary beat frequency follows phosphorylation of axonemal proteins by cAMP-dependent protein kinase PKA. The latter is a component of ciliary axoneme.

Transient increases in cytosolic Ca^{2+} concentration that augment the respiratory cilium activity in cultures of rabbit oviductal ciliated cells can be caused by micromolar concentrations of ATP and prostaglandin-F2α [1579]. Ciliostimulator ATP acts via nucleotide receptors that are coupled to plasmalemmal calcium channels, whereas PGf2α releases Ca^{2+} from its intracellular store.

Adenosine reduces intracellular cAMP content and ciliary beat frequency in cultured rabbit tracheal epithelium [1580]. Adenosine A_2-receptor agonist acts only slightly on ciliary activity, whereas adenosine A_1-receptor agonist attenuates the ciliary beat frequency in a dose-dependent fashion. Adenosine A_1-receptor inhibits adenylate cyclase and impairs respiratory ciliary activity. Conversely, adenosine receptor antagonist 8-phenyltheophylline reverses the effect of adenosine on cAMP level and ciliary beat frequency. On the other hand, adenosine-induced activation of apical receptor A_{2B} primes PLC–Ca^{2+} signaling that stimulates basolateral Ca^{2+}-activated K^+ channels and contributes substantially to anion secretion [1581]. Hence, apical adenosine activates: (1) contralateral K^+ channels via the PLC–Ca^{2+} axis for transepithelial anion secretion in coordination with (2) ipsilateral cystic fibrosis transmembrane conductance regulators via the cAMP–PKA axis. It then causes secretion of both chloride and bicarbonate.

Neurokinin-A augments the ciliary beat frequency via tachykinin receptors in cultured rabbit tracheal epithelium [1582]. Substance-P does not significantly affect the airway ciliary activity. However, neutral endopeptidase inhibitor potentiates substance-P response, but not that of neurokinin-A. Other peptidase inhibitors do not alter tachykinin-induced increase in ciliary beat frequency.[28]

Angiotensin-2 elevates the ciliary beat frequency [1583]. Ciliostimulator angiotensin-2 targets its cognate receptors to provoke a prostaglandin action without changing intracellular cAMP level.

Other regulators, such as cGMP, protein kinases PKG and PKC, calmodulin, phospholipase-C, and nitric oxide influence ciliary beating (Table 12.6). Human ciliary beat disappears in vitro at a temperature of 4 °C. Above this threshold, protein kinase-C and Ca^{2+}–calmodulin-dependent kinase CamK2 regulate nasal ciliary beat frequency in response to rising temperature [1584].

Exhaled nitric oxide is produced mainly in paranasal sinuses and nasal mucosa. Nasal nitric oxide can serve as an anti-infectious agent in the nose and associated sinuses and contributes to the mucociliary clearance. Nitric oxide production by inducible nitric oxide synthase is enhanced in nasal epithelium of patients with allergic rhinitis. Inducible nitric oxide synthase (NOS2) is upregulated in respiratory epithelium excited by tumor-necrosis factor-α and lipopolysaccharides. Baseline ciliary activity depends on NO production [1585]. In addition, at very low concentration (1–10 mmol), alcohol stimulates ciliary motility by rapidly and sequentially activating the NOS3–NO–GC–cGMP–PKG axis and soluble adenylate cyclase signaling (sAC–cAMP–PKA pathway) [1586]. However, long-term, heavy alcohol consumption disturbs mucociliary clearance and nitric oxide signaling in ciliated epithelial cells.

---

[28]Therefore, angiotensin-converting enzyme, aminopeptidases, and serine peptidases do not modulate ciliary activity in response to tachykinins [1582].

Table 12.6. Factors that stimulate or inhibit mucociliary clearance.

Factor class	Increase	Decrease
Biochemical factors	Adenosine cAMP–PKA signaling PLC–Ca^{2+} axis Nitric oxide Adrenergic agonists Cholinergic agonists Histamine ATP Amiloride Hyperosmolar solution	Leukotriene-C4 Atropine
Environmental factors	Acute exposure to sulfur dioxide, cigarette smoke, ozone	Chronic exposure to sulfur dioxide, cigarette smoke, ozone, NO_2
Pathological factors	Pseudohypoaldosteronism Chronic cough	Cystic fibrosis Primary ciliary dyskinesia Chronic bronchitis Asthma Respiratory infections

Transport speed depends on air flow and airway bore, as well as mucus layer thickness, rheology, and production rate [1587, 1588]. The air flow rate required for upward transport in vertical tubes of a quasi-uniform liquid layer of rheological properties comparable to human sputum is in the Reynolds number range of 142 to 1132 and 708 to 2830 whether the tube internal diameter is equal to 5 mm or 1 cm, depending on the liquid types. The mean mucus layer thickness range from 0.2 to 0.5 mm and from 0.8 to 1.4 mm whether the internal diameter is equal to 5 mm or 1 cm. The mean mucus layer thickness diminishes with increasing flow rate and decreasing mucus viscosity. The advancing speed of the leading edge of mucus layer in vertical tube ranges from 1.1 to 3.1 cm/min with a mucus feed rate of 0.5 ml/min at air flow rates of 0.33 to 1.17 l/s. The mucus speed increases almost proportionally with elevating mucus feed rate. The mucus speed rises with increasing air flow rates and decreases with increasing mucus viscosity. The transport speed in a horizontal tube is 5 to 60% faster than in a vertical tube.

Propulsion of mucus is assisted by coughing to expel large inhaled materials and chloride pumps to osmotically hydrate mucosal secretions. When the respiratory mucus interacts with air flow during coughing, micron-sized droplets form and are expelled to the environment. The droplets can hence possibly participate in propagation of respiratory infections.

12.6.3 Ciliogenesis

The respiratory epithelium is often restored after biological, chemical, or physical damage. Ciliogenesis is complete within 14 days in vivo. Motility and morphology of cilia during ciliogenesis can be studied in human respiratory epithelial cells in floating culture. On day 7, cells exhibit numerous microvilli, but no cilia. On day 8, cilium length was approximately 2 μm, whereas on days 12 to 14, it was 5 to 7 μm.

On day 8 of culture (rarely before), cells with numerous short moving cilia can be observed (mean ciliary beat frequency 17.8 ± 2.5 Hz) [1589]. Mean ciliary beat frequency that begins on day 8 of culture (17.8 ± 2.5 Hz) reaches its highest value on day 10 (19.6 ± 3.9 Hz) to decay afterward (17.4 ± 2.5 Hz on day 14).

Ciliary beat amplitude detected on day 8 rises on day 9 and then continues. Immature short cilia have a rigid beat pattern. As cilia lenghten, flexible cilium forward and backward motions gradually synchronize. Ciliary beat direction is random on day 8, but beat coordination for a given cell and for cell population is established after days 12 and 14, respectively.

Therefore, ciliogenesis is complete within 14 days. At early stage of ciliogenesis, ciliary beat frequency and amplitude is high and low, respectively. High ciliary beat frequency does not reflect activity of mature cilia.

During differentiation of tracheal epithelial cells, hundreds of centrioles migrate to the apical surface of the cell, dock with the membrane to form basal bodies, and act as templates for the outgrowth of the outer microtubule doublets of the ciliary axoneme [1590].

Centrioles involved in ciliogenesis are generated by 2 processes [1591]. Early stages of formation of both primary cilia and ciliated border are similar. Rudimentary (primary) cilia arise from centrioles in differentiating pulmonary cells of all types throughout the fetal period. Annular procentrioles lengthen into cylinders to give rise to daughter centrioles. When mature, newly formed centrioles align in rows beneath the apical plasma membrane, where each centriole produces satellites from its sides, a root from its base, and a cilium from its apex. In developing cilia of the ciliated border, outer ciliary tubules rapidly reach the tips of the elongating shafts, and a central pair of tubules is formed, whereas in primary cilia, tubule development lags behind shaft elongation and only outer ciliary tubules appear.

Hedgehog signaling is mediated by protein Ser/Thr kinase Fused for assembly of the central microtubule pair of motile cilia that regulates ciliary motility [1590].[29] Fused is strongly expressed in the respiratory epithelium. Fused interacts with kinesin family member KIF27 (but not KIF7) to link Fused to structural components of the central pair apparatus. Expression of Fused and KIF27 is upregulated during tracheal epithelial cell differentiation.

[29] Mice lacking functional Fused have defects or loss in central pairs, hence impaired cilium motion and mucociliary transport (reduced forward power stroke amplitude, uncoordinated ciliary beating, etc.).

Table 12.7. Simulation domain of the airway surface liquid constituted by a periciliary fluid and a mucus layer (Source: [1594]).

Subdomain	Features
Ciliated epithelium	Rigid, straight, smooth interface
Periciliary liquid	Cell-surface-anchored mucins immersed in water Newtonian fluid Constant physical properties (mass density, dynamic viscosity) Layer thickness 5 μm
Mucus layer	Mucin mesh Non-Newtonian gel Variable physical properties (mass density, dynamic viscosity) Layer thickness 2.5 μm
Air	Convective transport Oscillatory flow (frequency 0.20–0.25 Hz) Incompressible, Newtonian fluid

12.7 Mucus Transport Modeling

Numerical investigations have been carried out on interactions between mucus motion and air flow at the microscopic scale [1592]. Mucus displacement is driven by respiratory epithelium cilia, but not markedly by air stream.

A quasi-steady Stokes flow of fluid with spatially variable viscosity has been simulated over a moving boundary [1593]. A second-order elliptic partial differential equation is non-linearly coupled to a first-order hyperbolic equation that describes the cilium beating, which takes into account the actin polymerization via a model of elastic beams of variable modulus.

The simulation domain of the airway surface liquid is defined in Table 12.7. In a preliminary study, 2 cilia states are only considered: a rest (without contact with the mucus layer) and an erected state, partly penetrating the mucus layer. In the former, the cilium lies parallel to the airway wall, immersed in the periciliary fluid, pointing in the direction of mucus flow. A cilium is modeled by a succession of constant radius spheres separated by a constant distance. When the cilium is straight, all the sphere centers are aligned. A random process is used to provide cilium positions at rest. A stiffening mechanism based on angle between 3 successive spheres along the cilium models the cilium beating cycle. Model parameters and their values are listed in Table 12.8. In the preliminary work, mucus is assumed to be Newtonian (constant dynamic viscosity $\mu = 1$) and the air–mucus interface is flat.

From a macroscopic point of view, mucus is a thixotropic gel that behaves as: (1) an elastic solid at low shear and (2) a viscous liquid at high shear. The dynamical

Table 12.8. Airway surface liquid model parameters (Source: [1594]).

Quantity	Symbol	Value
Cilium height	H_c	7 μm
Cilium radius	R_c	0.3 μm
Number of cilia per cell	N_c	4 or 16
Number of spheres per cilium	N_s	25
Angle between wall and cilium	α	110 degrees
Distance between-cilium center	d_c	$4R_c$
Length of epithelium cell	L_c	$d_c\sqrt{N_c}$
Domain edge length	L	12 μm
Number of x_i-space steps	N_{x_i}	32
Propulsion duration	T_p	17 ms
Recovery duration	T_p	34 ms
Cilia oscillation frequency	f	19.6 Hz

viscoelastic properties of mucus are constantly regulated to ensure an efficient clearance and to maintain sufficient adhesive strength to be retained on the epithelial surface despite external shearing forces.

12.7.1 Governing Equations

Fluid flow equations are given by laws of mass and momentum conservation and a convection equation for the dynamic viscosity (μ):

$$\begin{aligned} \partial_t \rho + \nabla \cdot (\rho \mathbf{v}) &= 0, \\ \partial_t (\rho \mathbf{v}) + \nabla \cdot (\rho \mathbf{v} \otimes \mathbf{v}) - \nabla \cdot \mathbf{C} &= \mathbf{b}, \\ \partial_t \mu + \mathbf{v} \cdot \nabla \mu &= 0, \end{aligned} \tag{12.1}$$

where $\mathbf{v}$ is the flow velocity vector, $\mathbf{C}$ the stress tensor, $\mathbf{b}$ the body forces, the third equation arising from the transport of the mucin mass fraction, and

$$\rho \mathbf{v} \otimes \mathbf{v} = \rho \mathbf{v}(\nabla \cdot \mathbf{v}) + \rho(\mathbf{v} \cdot \nabla)\mathbf{v} + (\mathbf{v} \cdot \nabla \rho)\mathbf{v} \tag{12.2}$$

The cilium is represented by a moving, piecewise articulated rode with a given velocity [1594].

Chapter 13
Surfactant

> *" ...; omne oleo tranquillari, et ob id urinantes ore spargere, quoniam mitiget naturam asperam lucemque deportet;...*
> *[...; the whole sea is made still with oil and therefore the divers under the water scatter it with their mouths, because it allaies the rough nature thereof, and carries a light with it;...] "*
>
> (Caius Plinius Secundus [Pliny the Elder; 23–79], Naturalis Historiae, Book II)

Surfactant is an interfacial material of respiratory conduits and its annexes. It modulates the surface tension and innate immune defense of the lung. The alveolar surface film stretches as the lung expands, raising the surface tension, then molecules are packed as the lung deflates, lowering the surface tension.

The substances that lower the surface tension in small amounts are called surface-active agents or surfactants. They reduce the energy required for the surface production.

13.1 Surface Tension

The free surface of a liquid at rest is not necessarily flat and horizontal, especially in the neighborhood of a vertical solid wall. The surface effects are indeed not always negligible with respect to the surface (pressure) and body (gravity) forces. The free surface of a liquid behaves like an elastic membrane bearing a uniform force per unit length.

The surface tension is the force acting on the surface of a liquid that tends to minimize the surface area.[1] The smaller the surface area of any interface between a

[1]The surface tension can also be named superficial tension or interfacial tension whether the interface separates a liquid and gas or 2 liquids.

M. Thiriet, *Tissue Functioning and Remodeling in the Circulatory and Ventilatory Systems*, Biomathematical and Biomechanical Modeling of the Circulatory and Ventilatory Systems 5,
DOI 10.1007/978-1-4614-5966-8_13,

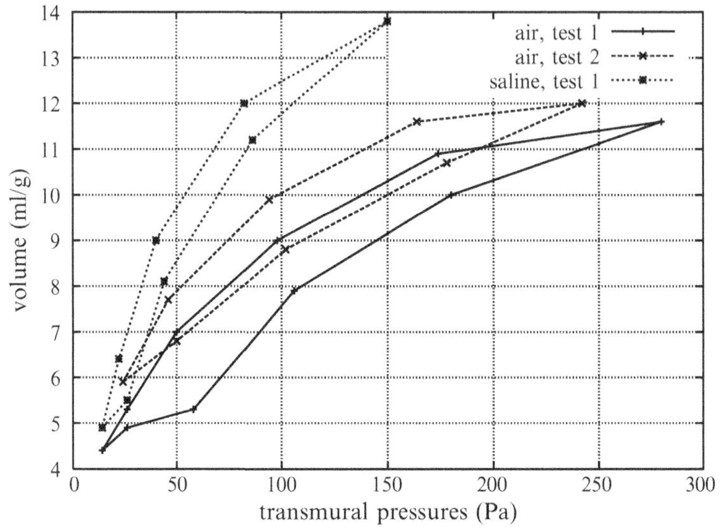

Fig. 13.1 Relationships between volume (ml per g of tissue) and pressure (Pa) in air-filled and saline-filled lungs (Source: [1595]). In the case of air-filled lung, the effect of loading history is demonstrated with the upper shift of the loop during the second cyclic loading with respect to the first. When the lung is filled with saline, only the first cyclic loading is displayed, owing to a quasi-superimposition of the loops of the following successive cyclic loadings with the first. The loop is displaced leftward and upward with respect to curves in air-filled lung.

liquid and another medium, the greater the stability of the interface, and the lower the energy involved in the formation of the interface.

Surface tension is the manifestation of intermolecular forces in the interface, the surface shrinking to the smallest possible area. The surface tension is the force of attraction between adjacent molecules of a liquid. Therefore, this molecular cohesive force in the surface film tends to contract the surface.

The pressure–volume curves obtained in air-filled lungs and after filling lungs with saline show that: (1) surface tension forces are greater during inflation than deflation and (2) a large portion of transpulmonary pressure can be attributed to surface tension (Fig. 13.1).[2]

The superficial tension changes with the radius of curvature of the alveolar surface. The *Laplace law* states that the transmural pressure increases as the

[2]Lung elasticity can be assessed by administration of saline solution in isolated lungs that removes the contribution of the superficial tension. For a given lung volume, a greater pressure is required to inflate the lung filled with air than lung filled with a saline solution. Pressure–volume loops of liquid-filled lungs are displaced leftward and upward (low transpulmonary pressure and high compliance) and display less hysteresis. Lung inflation with air measures both tissue and surface tension components. Furthermore, there is a much smaller difference between the air- and liquid-filled curves on deflation than on inflation.

curvature radius decreases, when the surface tension is constant.[3] In a bialveolar model with 2 communicating alveoli of different curvature radii, the pressure of the small alveolus is higher than the pressure of the large one. Such a bialveolar model is not synchronized. The force exerted on a very thin curved shell is equal to twice the tension divided by the surface curvature radius. For a surface tension of 5 N/m^2 (physiological magnitude) and a radius of 50 μm, the force is equal to 200 kPa. The superficial tension is then minimized in proportion to the alveolar volume by the secretion of a wetting agent.

The interface forces govern adhesion, friction, wettability (wetting or non-wetting)[4] of solids by liquids (Fig. 13.2), capillarity of liquids in small-bore tubes[5] and the curvature of gas–liquid or solid–liquid interfaces.

In the liquid domain, the time and space-averaged attraction force (resultant) exerted on any liquid molecule by its neighbors is equal to zero, even though the molecules diffuse and undergo random collisions. But at the interface of the liquid with the surrounding medium, no molecule can balance the effects of the interior liquid molecules (symmetry rupture). Cohesion forces are hence responsible for

---

[3]The Laplace law indicates that the pressure within a spherical structure with surface tension associated with a liquid–gas interface is inversely proportional to the radius of the sphere:

$$p = 2T/R$$

(T: surface tension, R: sphere radius). At a constant surface tension, small hollow spheres generate greater pressures than large spheres. Small spherical tanks connected in parallel to large spherical reservoirs can therefore empty into large ones.

[4]The criterion for the wetting and non-wetting of solids by liquids is the value of the contact angle between the solid and the liquid. A liquid wets a solid when the contact angle ranges between 0 and 90 degrees (concave meniscus in capillarity experiments). When the contact angle is greater than 90 degrees, the liquid does not wet the solid (convex meniscus in capillarity experiments). Three interfaces exist when a liquid droplet contacts a solid or a liquid rises in a capillary tube: (1) the gas–solid (GS), (2) liquid–solid (LS), and (3) gas–liquid (GL) interface. Each interface is associated with a surface tension $T_{GS/LS/GL}$. The contact angle α can be calculated from the following formula:

$$\cos\alpha = (T_{GS} - T_{LS})/T_{GL}.$$

Partial or total wetting can be characterized by a spreading coefficient. When it is positive, the energy of the dry surface is greater than that of the wetted surface, and conversely when the spreading coefficient is negative. The droplet volume change can modify the contact angle without disturbing the equilibrium between the liquid droplet and the more or less rough solid wall. However, when the contact angle is greater than the expansion limit or smaller than the retraction limit, the droplet configuration changes.

[5]The motion of liquid in capillary tubes (rise or fall whether the liquid wets or not the solid wall of the tube of small radius ($R < 5$ mm) plunged into the liquid bath, according to the magnitude of capillarity forces, pressure and opposing gravity) is characterized by a meniscus (downward or upward meniscus whether the liquid wets or not the solid wall) at a height given by the *Jurin law*:

$$H = 2T_s \cos\alpha/(R\rho g).$$

Water barometers must then be based on tubes of inner diameter equal to 8 mm at least in order to avoid disturbed measurements.

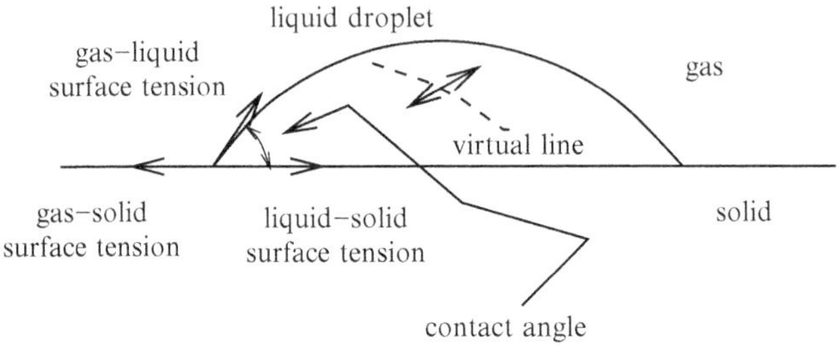

Fig. 13.2 Forces acting on a liquid droplet in a gas wetting a solid. Small droplets (size of $\mathcal{O}[1\text{ mm}]$) are spherical whereas larger ones are ellipsoidal under the effect of gravity.

surface phenomena.[6] The surface molecules then bear a net attraction toward the liquid body. The centrally directed forces minimize both the free energy[7] and the surface area. The surface tension expresses cohesion internal forces that hinder the liquid molecules from occupying the available space.

Two fluids in contact, such as a liquid and its vapor, are separated by an interface. The interface properties differ from those of 2 uniform phases. The interface is modeled by a zero-thickness membrane undergoing an uniform tension. Let us define 2 surface regions separated by a virtual line on the surface (Fig. 13.2). In any point of this line, the length unit ds is subjected to a force from one of the 2 regions, normal to the line and tangent to the surface at the selected point, of magnitude:

$$df = T_s ds, \tag{13.1}$$

where T_s is the surface tension ($M.T^{-2}$; N/m).[8] The surface tension decays when the temperature rises and becomes zero at the critical temperature. A surface energy (W) corresponds to the surface force on the surface area (A): $W = T_s A$.

At the beginning of the nineteenth century (1805–1806), T. Young and P.S. de Laplace found the equation that describes the relationship between the curvature

---

[6]The cohesion forces are more important in liquids than in gases, with smaller molecule concentration. The higher the cohesion forces of a liquid, the stronger the surface tension, and the lower the liquid wetting.

[7]The total free energy of a system composed of 2 uniform fluids of densities ρ_1 and ρ_2, of volumes V_1 and V_2, and of specific free energy e_1 and e_2, separated by an interface of area A is given by:

$$e_{tot} = \rho_1 V_1 e_1 + \rho_2 V_2 e_2 + T_s A,$$

where $T_s A$ represents the surface energy, T_s being the surface tension that can hence be interpreted as free energy per unit area of the interface. The interface stretching work done by the surface tension in small reversible isothermal changes in the fluid system is equal to the gain in total free energy.

[8]The surface tension between air and water is equal to 73.10^{-3} N/m at 20°C (293 K) and to 68.10^{-3} N/m at 50°C (323 K).

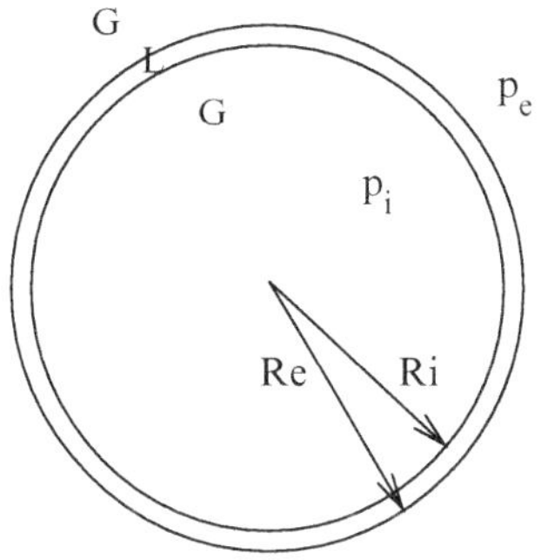

Fig. 13.3 Soap bubble in air. $p_i - p_e = p_i - p_L + p_L - p_e \sim 2(2T_s/R)$. A spherical bubble provides the smallest surface containing a given volume. The higher the surface area, the greater the associated energy. The equilibrium is reached when the energy, hence the area, is minimal

radius of a surface and the pressure. At the interface, a pressure difference exists between the concave and the convex sides of the surface. When a stretched zero-thickness membrane is crossed, an increment of pressure is observed from the convex side to the concave side that is given by the Laplace law. The excess pressure p on the concave side over the convex side is indeed expressed by the principal radii of curvature of the surface R_{c1} and R_{c2}:

$$\Delta p = T_s(1/R_{c1} + 1/R_{c2}). \tag{13.2}$$

Using this principle, the vapor pressure p is greater for small bubbles in a liquid:

$$\ln p/p_0 = 2T_s/(\rho R_c)(M/(R_g T)), \tag{13.3}$$

where p_0 is the vapor pressure over a liquid plane surface (infinite curvature radius) and M the molecular weight.

Any hollow bubble is characterized by 2 interfaces, an inner (convex with respect to the liquid domain) between the liquid and the internal gas (curvature radius R_i), and an outer (concave with respect to the liquid domain) between the liquid and the external gas (curvature radius $R_e \sim R_i \equiv R$). The Laplace law ($p = 2T_s/R$) applied to both interfaces then leads to: $p_i - p_e = 4T_s/R$ (Fig. 13.3).

In Eustachian tubes, surfactant slightly lowers surface tension and reduces opening pressure for better aeration and middle ear drainage. Eustachian tube surfactant is a much weaker surface tension lowering agent than pulmonary surfactant. Eustachian tube surfactant primarily acts as an anti-adhesive agent.

Alveolus surface tension is associated with neonatal respiratory distress syndrome [1596]. A greater pressure is required to inflate lungs filled with air than with aqueous solution. Surfactant effects were studied in 1950s by various teams of investigators [1597–1601]. The surface tension hampers transudation from capillaries to alveoli through the alveolocapillary membrane. Furthermore, the absence of a lung lining substance can cause neonatal atelectasis [1602]. Surface tension varies with lung compression and expansion, i.e., with changes in the surface area of the lung parenchyma. Surface tension rises with lung expansion to prevent the lung from overexpanding and helps lungs to deflate. Conversely, surface tension falls with lung deflation, maintains the alveoli in an inflation state, and helps them to expand again rather than to collapse. The surfactant, thereby, prevents atelectasis. Surfactant deficiency in respiratory distress syndrome was demonstrated from lung extracts of babies in 1959 [1603]. Surface tension in compressed lung extracts of babies with respiratory distress syndrome remains high. Surface tension increases in expended lung extracts, but to much greater values compared to normal lung extracts.

13.2 Alveolar Surfactant

Alveolar epithelium produces a liquid film over its surface in contact with air, generating a fluid–air interface. Surfactant is made in type-2 alveolar cells. The functional surfactant adsorbs very rapidly into the liquid interface to form a tensoactive monolayer. After secretion, the surfactant causes less densely packed arrangements in the structure of the air–liquid interface.

Surfactant is a material that contains phospholipids, predominantly dipalmitoyl lecithin, and apoproteins. It is included in a fluid layer (thickness ~70 nm). The surfactant reduces surface tension throughout the major part of the lung parenchyma, thereby contributing to lung compliance. It also stabilizes the alveoli.

13.2.1 Formation of the Alveolar Surface Film

Surfactant is formed relatively late in fetal life. In utero, the fetal lung secretes fluid into the alveolar spaces to promote lung development. Shortly before birth, the fetal lung switches from fluid secretion to resorption to clear alveoli of excess fluid and establish a proper air–blood barrier for gas exchange. After birth, gas exchange depends on adequate regulation of the amount of fluid in the very thin liquid layer (thickness ~0.2 mm) that lines the alveolar epithelium. Premature infants born without adequate amounts experience respiratory distress and can die in the absence of clearance of the excess alveolar fluid and restoration of alveolar fluid homeostasis to reinstate normal gas exchange. The half-life of the surfactant is equal to about 18 h.

The formation of the alveolar surface film encompasses 3 successive steps: (1) the transport stage during which surfactant aggregates move from the subphase toward the interface; (2) the attachment stage during which surfactant bilayers associate with the pre-existing monolayer; and (3) the transfer stage during which the structure of the surfactant-associated phospholipid bilayers in the aqueous subphase open in order to enable surfactant transfer to the interfacial monolayer.

Once secreted, surfactant forms *tubular myelin*, a lung-specific lipid transport molecule for insertion into the air–liquid interface. Tubular myelin, an ordered array of tubules, is formed from secreted lamellar bodies.[9] Tubular myelin is a precursor of the interfacial film.

Phase 1 is characterized by unpacking of lamellar bodies, their transformation into tubular myelin, and the transfer of molecules toward the interface. Phase 1 is affected by the local physicochemical conditions (hydration and clcium levels, pH, etc.).

13.2.2 Alveolar Liquid Film Homeostasis

Alveolar epithelium (99% of the lung internal surface area) produces a liquid film and maintains its homeostasis for effective gas exchange. The fluid balance results from osmotic gradients created by active solute transport. In lungs, the alveolar epithelium maintains air spaces free of excess liquid via Na^+ influx through highly selective, apical epithelial Na^+ channels in response to an electrochemical gradient created by basolateral Na^+–K^+ ATPases. The resulting osmotic gradient leads to fluid reabsorption from alveolar to interstitial spaces.

13.2.2.1 Ion Carriers of Type-1 Pneumocyte

Type-1 and -2 alveolar epithelial cells, or pneumocyte-1 and -2, that cover the alveolar wall possess many types of ion channels. Alveolar epithelial type-1 cells (length$\sim$50–100 μm) have epithelial sodium, cyclic nucleotide-gated, and potassium channels, as well as cAMP-stimulated cystic fibrosis transmembrane conductance regulators [1604]. Potassium channels prevent cell depolarization during Cl^- efflux and Na^+ influx. Chloride uptake across CFTR contributes to fluid absorption. In addition, pneumocyte-1 expresses aquaporin-5.

Pneumocyte-1 has the highest known osmotic water permeability of any mammalian cell type. The function of ENaC channel can be modulated by catecholamines. Alveolar type-1 cells indeed possess dopamine type-1 and -2 receptors that can raise ENaC activity [1604]. Moreover, they also contain γ-aminobutyric acid receptor, a ligand-gated chloride channel.

Both type-1 and -2 pneumocytes have subunit A1 of Na^+–K^+ ATPase, but only type-1 pneumocyte possesses subunit A2 [1604]. Using this difference with associated distinct antagonist sensitivity, it can be demonstrated that type-1 pneumocyte is responsible for at least 60% of the total alveolar fluid clearance in rats.

[9] Surfactant components are packed by type-2 pneumocytes in lamellar bodies, although certain components such as SPa can also be secreted separately from lamellar bodies.

13.2.2.2 Ion Carriers of Type-2 Pneumocyte

Alveolar type-2 cells (size~10 μm) cover 2 to 5% of the alveolar surface area. They produce, secrete, and recycle alveolar surfactant. They also participate in active ion transport, as they have Na^+–K^+ ATPases, K^+ and epithelial Na^+ channels, and cystic fibrosis transmembrane conductance regulators [1604]. However, they contain fewer Na^+ channels than alveolar type-1 cells, at least in rodents.

13.2.2.3 Adenosine Receptors and Ion Transport

Adenosine operates via G-protein-coupled receptors to activate or inhibit adenylate cyclases. Messenger cAMP regulates active Na^+ transport in alveolar epithelium. Adenosine hence regulates alveolar ion transport via elevated activity of ENaC and CFTR and airspace fluid content (para- and autocrine regulation of the alveolar fluid clearance). At least in rodents, type-1 and -2 alveolar epithelial cells possess all 4 types of adenosine receptors (A_1, A_{2A}, A_{2B}, and A_3) [1605]. Receptors A_1 and A_3 inhibit adenylate cyclase and signal via inositol 3-phosphate and phospholipase-C.

Type-2 adenosine receptor activates adenylate cyclase via $G\alpha_s$ subunit. Receptor A_{2A} is expressed 5- to 9-fold more than the 3 other types in alveolar cells.

At low doses, adenosine increases alveolar fluid clearance owing to A_{2A} and A_3 receptors. At physiological doses (micromolar), adenosine reduces alveolar fluid clearance via A_1 receptor that causes Cl^- efflux through CFTR, which counterbalances active Na^+ absorption.

13.2.3 Secretion and Role of Alveolar Surfactant

The surfactant that covers the alveolar wetted surface is a phospholipid, mainly lecithin and dipalmitoyl lecithin. Lecithin consists in a long chain of fatty acids bound to glycerol and to electrically charged phosphate and choline groups that are attracted by polar water molecules. Pure oxygen and atmospheric pollutants affect the alveolar surface.

The surfactant production is controlled by the vagus nerve. A decline in surfactant activity indeed results from a cut in vagus nerve. Therefore, the release of surfactant depends on the local blood flow to match ventilation to perfusion.

The surfactant is secreted by the type-2 granular pneumocytes of the alveolar wall. It induces variable surface tension according to the lung volume, i.e., the degree of alveolar expansion. It contributes to the alveolar stability, preserving the alveoli from full collapse during expiration and hyperinflation during inspiration.

Alveoli expand more easily due to the presence of surfactant. This surface-active agent also varies the surface tension with changes in the alveolus surface area. The surfactant layer becomes thinner in inflated alveoli, thereby increasing the surface tension. This augmentation is balanced by the alveolus size elevation. Conversely, rising concentration of surfactant balances the effect of reduction in alveolar size during exhalation.

The surfactant is thus very important because: (1) it promotes alveolar stability, alveoli of varying radii coexisting without ventilation between them, and (2) it reduces the transpulmonary pressure. Furthermore, the surfactant hinders fluid exudation from the alveolar septa within the alveolar space. The surface tension, the tissue osmotic pressure difference in the alveolar septum, and the capillary blood pressure work to move fluid out of the alveolar capillaries. The capillary osmotic pressure opposes this force combination. A favorable balance can be maintained by a reduction in the surface tension.

The absence of surfactant can induce atelectasis (lung collapse). A reduction in surfactant can occur during prolonged inhalation of 100% oxygen. Premature newborns often have a reduction in or lack of surfactant (respiratory distress syndrome).

13.3 Surfactant Composition

Surfactant is a mixture of approximately 90% lipids, mostly phospholipids, and 10% proteins, mostly surfactant proteins (SPa–SPd). Constituents mainly responsible for the surface tension are phospholipids, organized in a thick monomolecular layer at the air–liquid interface of the alveoli, and contiguous with the underlying layer. The most abundant surfactant phospholipid is dipalmitoyl phosphatidylcholine (DPPC) [1606, 1607].

Lung stability at low surface tension requires that the material of the monolayer must sustain surface tension (up to 30 mN/m [1608]) over a long time. A monolayer of dipalmitoyl phosphatidylcholine can bear such a loading, but the surfactant monolayer must undergo DPPC enrichment during breathing. This adaptive process is done via exclusion of non-DPPC molecules and insertion of new DPPCs. However, surfactant replacement therapy requires both lipids and proteins [1609].

13.3.1 Surfactant Proteins

Surfactant proteins are either hydrophobic or -philic. They participate in: (1) interfacial layer formation; (2) layer stabilization during deflation; (3) reformation of surface film on expansion; and (4) monolayer turnover and recycling [1610]. Cyclic compression and traction of the liquid interface indeed induces a continuous loss of material from the interface, leading to cellular uptake and reprocessing of surfactant lipids and proteins, followed by very quick transfer to the liquid film.

The hydrophobic proteins constitute about 40% of the total protein of lamellar bodies[10] isolated from pig lungs and about 13% of the total protein of surfactant isolated from bronchopulmonary lavage fluid [1612].

Several types of surfactant proteins exist (SPa–SPd) [1613]. Surfactant proteins are detected not only in pulmonary acini, but also the digestive, urinary, and reproductive tracts, as well as synovial and pericardial fluids, various organs, such as spleen, thymus, and pancreas as well as middle ear and Eustachian tubes. Surfactant composition depends on tissue type.[11]

Hydrophobic surfactant proteins SPb and SPc help release of dipalmitoyl phosphatidylcholine. They promote rapid transfer of lamellar bodies from the subphase to the interface. Proteins SPb and SPc are also involved in protein respreading upon expansion. They bind preferentially to anionic phospholipids.

Hydrophilic surfactant proteins SPa and SPd that form trimers are members of a family of collagenous carbohydrate binding proteins, the *collectins* that are involved in the immune system. Collectins are a subset of calcium-dependent C-type lectins that bind carbohydrates. Both SPa and SPd interact with viruses, bacteria, fungi, and allergens. They are thus present in extrapulmonary tissues. Proteins SPa and SPd bind to alveolar macrophages.

Hexadecameric SPa is the most abundant surfactant protein. Protein SPa is the product of 2 distinct genes on chromosome 10q22-q23: SPa1 and SPa2. It is involved in the structural organization of surfactant.[12] Protein SPa that is necessary for tubular myelin production is implicated in spacing of common and tubular myelins.

Functions of SPa include surface tension lowering in combination with phospholipids in the form of tubular myelin, maintenance of surfactant protein pool homeostasis, and defense against pathogens. Protein SPa increases phagocytosis of cells infected by Streptococcus pneumoniae, Haemophilus influenza, respiratory syncytial virus (Pneumovirus), and herpes virus (Simplexvirus), among other pathogens [1559]. It also modulates uptake and secretion of phospholipids by isolated type-2 alveolar cells in vitro, hence balancing surfactant release and clearance. Moreover, SPa improves adsorption of compounds made of phospholipids and hydrophobic proteins.

Dimeric SPb is encoded by the spb gene that belongs to chromosome 2. Protein SPb is a member of a family of saposins that are sphingolipid activators.[13] It

[10]Lamellar bodies are granules that contain SPa to SPc and lipids [1611].

[11]The ratio of phosphatidylcholine to sphyngomyelin varies from 2:1 in Eustachian tube surfactant to 67:1 in lung surfactant [1559]. In particular, dipalmitoyl phosphatidylcholine concentration is much higher in lung surfactant than in that of Eustachian tube surfactant.

[12]Protein SPa neck domain is involved in protein trimerization. Its globular C-terminus acts in lipid binding as well as formation and stabilization of curved membranes. Protein SPa exists in open or closed forms according to the medium composition. Calcium ions produce the closed structure of 6 trimers. The N-terminus of SPa protein is required for oligomerization as well as binding and aggregation of phospholipids [1614]. Its collagen-like domain is involved in SPa stability and oligomerization. It contributes to SPa shape and dimension.

[13]Saposins activate several lysosomal hydrolases involved in sphingolipid metabolism.

is synthesized in type-2 alveolar epithelial and Clara cells. A preproprotein is synthesized and then cleaved to form a proprotein, itself splitted to produce the mature protein. Protein SPb fluidizes lipid monolayers. It is a tubular myelin component. It enhances lipid insertion into the air–liquid interface and respreading during inflation, hence adjusting the surface tension. It is the most involved surfactant protein in surface tension reduction. Protein SPb prevents lipid packing, thereby airspace collapse. It is dissolved in the liquid-expanded rather than in the liquid-condensed phase of artificial monolayers [1615].[14] Protein-rich domains form protrusions in compressed monolayers in order to cope with low surface tension during exhalation, proteins being squeezed out of the monolayer. It then reduces the size of the compressed monolayer.

Monomeric SPc is encoded by gene spc that belongs to chromosome 8. SPc Protein is expressed in type-2 alveolar epithelial cells. Once it is synthesized, SPc proprotein is cleaved to form the mature protein. Protein SPc is the lightest and smallest surfactant protein. It is located only in the respiratory tract. This transmembrane protein is highly hydrophobic and contributes to the organization and maintenance of the surface active monolayer that attenuates surface tension. Protein SPc enhances phospholipid adsorption rate to maintain a low surface tension during inhalation and intervenes in the molecular ordering of the phospholipid interfacial layer. It can also act on the liquid film stability.

Dodecameric SPd is the largest surfactant protein.[15] It regulates surfactant composition, modulating surfactant lipid turnover. SPd Protein binds to phosphatidylinositol and glucosylceramide. It has also immunomodulatory function and operates in mucociliary transport. Protein SPd has been identified in different mucosal secretions, such as those of parotid, gall bladder, pancreas, digestive tract, urinary conduits, and Eustachian tube [1559].

13.3.2 Surfactant Lipids

Surfactant proteins lower surface tension, as they combine with phospholipids to form tubular myelin, recruit phospholipids and facilitate their spread on air–fluid interfaces, and organize and maintain the surface active monolayer.

[14]The artificial monolayer is composed of dipalmitoylphosphatidylcholine and dipalmitoylphosphatidylglycerol in a molar ratio of 4:1 with 0.2 mol% SPb.

[15]Protein SPd is equivalent to conglutinin in blood circulation.

13.3.3 Calcium

Calcium heightens phospholipid–protein aggregation, thereby reducing surfactant viscosity [1559]. Calcium is required for the formation of tubular myelin by SPa, as well as for agglutination, opsonization, and ligand binding of surfactant proteins.

13.4 Immune Function of Surfactant

Surfactant proteins belong to a class of C-type lectins, the collectins. Collectins have 2 domains: (1) a carbohydrate-binding site at the surface of the mucosal layer that can bind to various foreign substances (e.g., bacteria, viruses, fungi, molds, and other allergens) and (2) a collagenous region underneath the mucous layer that serves as a ligand for the collectin receptor on phagocytes and regulates C1q-independent activation of the complement cascade.

Surfactant can activate macrophages. On the other hand, it hinders neutrophil function [1559]. In addition, it operates as a prophagocytic and anti-inflammatory material. Surfactant proteins have 2 immune functions: agglutination and opsonization of foreign particles and pathogens.

Surfactant proteins stimulate, but surfactant lipids inhibit production of oxidants, cytokines, and collagens and collagenases, NFκB activity, and fibroblast proliferation. Decreased synthesis of pro-inflammatory cytokines and blocking of complement activation associates with a shift from neutrophil activity to macrophage phagocytosis.

13.5 Governing Equations and Surfactant Transport Modeling

Molecules in a fluid experience a mutual attraction. When the attractive force is overcome by thermal agitation, molecules enter in the gazeaous phase. At a free surface such as an air–water (gas–liquid) interface, water molecules in the bulk fluid are surrounded by attractive molecules; at the interface, they are attracted by a reduced number of molecules. As the surface has an energetic cost, fluid minimizes its area. The surface tension (T_s; with unit of force per unit length) is a measure of energy loss per unit free surface area. Similarly, interfacial tension at an immiscible liquid–liquid interface relies on different attractive intermolecular forces in the 2 phases in contact.

13.5.1 Interfacial Tension and Marangoni Stress

The surface tension is approximately equal to [1616]:

$$T_s \sim U/(2L^2), \tag{13.4}$$

where U is the total cohesive energy per molecule and L the characteristic size of the molecule. The surface tension increases when the intermolecular attraction rises and molecular size lowers.

When the surface tension is uniform at the interface, the net surface tension is nul and the fluid remains static. However, surface tension gradients generate a net interfacial force that can drive flow and distort surface. Flows within thin viscous layers due to Marangoni stresses, film thickness, and material concentration can be derived from lubrication theory. Surface tension acts as a boundary condition in the governing equations of fluid dynamics.

The normal stress at the free surface is in equilibrium with the curvature force associated with the surface tension:

$$\mathbf{n} \cdot \mathbf{T} \cdot \mathbf{n} = T_s(\nabla \cdot \mathbf{n}), \tag{13.5}$$

where $\mathbf{n}$ is the unit outward normal to the surface and

$$\mathbf{T} = -p\mathbf{I} + \mu(\nabla \mathbf{v} + [\nabla \mathbf{v}]^T) \tag{13.6}$$

the stress tensor ($\mathbf{v}$: velocity vector).

The tangential component of the hydrodynamic stress at the surface balance the tangential Marangoni stress associated with surface tension gradient that results itself from gradients in chemical composition at the interface, in the absence of temperature changes:

$$\mathbf{n} \cdot \mathbf{T} \cdot \mathbf{t} = \nabla \mathbf{T_S} \cdot \mathbf{t}, \tag{13.7}$$

where $\mathbf{t}$ is the unit tangent vector to the interface.

13.5.2 Dimensionless Governing Parameters

The motion of a fluid (density ρ and viscosity μ) with a free surface at which is exerted a surface tension (T_s) under a gravitation field (gravitational acceleration $\mathbf{g}$), characterized by characteristic length ($L^\star$) and velocity ($V^\star$), can be defined by a set of 6 physical quantities (ρ, μ, $L^\star$, $V^\star$, g, and T_s). This set gives rise to a set of dimensionless parameters (Table 13.1) [1616].

When the *Marangoni stress* has a characteristic magnitude $\Delta T_s/L$, the dimensionless parameter $L^\star \Delta T_s/(LT_s)$ characterizes the relative effects of the Marangoni stress to the curvature stress.

Table 13.1. Flow dimensionless parameters in the presence of a surface tension.

Governing parameter	Symbol	Expression	Meaning (ratio)
Bond number	Bo	$\rho g L^{\star 2}/T_s$	gravity/centrifugal force
Eötvös number	Eo	$\Delta\rho g L^{\star 2}/T_s$	buoyancy force/surface tension
Capillary number	Ca	$\mu U^\star/T_s$	friction force/centrifugal force
Froude number	Fr	$V^{\star 2}/(gL^\star)$	inertia force/gravity
Marangoni number	Ma	$-\frac{1}{\mu\mathcal{D}_T}\frac{dT_s}{dT}L^{\star 2}\Delta T$	surface tension/friction
Reynolds number	Re	$\rho L^\star V^\star/\mu$	inertia force/friction force
Weber number	We	$\rho L^\star V^{\star 2}/T_s$	inertia force/centrifugal force

The *Bond number* is the most common index of the relative magnitude of surface tension compared to gravity force. A low Bond number (< 1) indicates that the surface tension dominates.

The *capillary number* represents the relative effect of viscous forces vs. surface or interfacial tension acting across an interface between a liquid and a gas or between 2 immiscible liquids.

The *capillary length* ($\ell_c = (T_s/[\rho g])^{1/2}$) prescribes the maximal drop size.

The *Froude number* is the ratio of inertia forces to gravitational force.

The *Marangoni number* is proportional to the ratio of thermal gradient-induced surface tension to viscous forces ($\mathcal{D}_T$: thermal diffusivity). It is used in thermocapillary convection along an interface between 2 fluids due to surface tension gradient created by a temperature gradient.

The *Weber number* is a measure of the relative magnitude of inertia forces with respect to its surface tension.

13.5.3 Surfactant and Marangoni Flows

In general, the air-liquid interface cannot induce Marangoni flow since the air is not viscous enough with respect to the liquid [1617].

Surfactant has a molecular structure often characterized by a hydrophylic head and hydrophobic tail that enables it to reside at interfaces. Insoluble surfactants remain on the surface, whereas surfactants soluble in the suspending fluid diffuse. Surfactant reduces the surface tension. Therefore, gradients in surfactant concentration cause gradients in surface tension. The dependence of surface tension on surfactant concentration ($T_s(c_s)$) requires the coupling of the flow field and surfactant field.

Surfactants affect the normal stress balance, as they reduce T_s, but also the tangential stress balance, as they generate Marangoni stresses. Tangential surface stresses can only be balanced by viscous stresses associated with fluid motion.

Surfactant brings an effective elasticity to the interface. In a fluid motion associated with a radial surface motion, the presence of surfactant provokes the redistribution of surfactant. Whatever the evolution of the surfactant concentration, either surfactant accumulation or drops, the resulting Marangoni stress aims at suppressing the surface motion, resisting it through an effective surface elasticity. When surfactant accumulates, Marangoni stresses again resist to the wall movement. Surfactant is aimed at avoiding interface collapse and rupture.

Mass transfer of surfactant in thin, viscous layers such as airway surface liquid at constant temperature along an interface between 2 fluids depends on the surface tension gradient caused by a concentration gradient. Surfactant creates a type of Marangoni flow.

The transient spreading of an insoluble surfactant on an interface causes motion in the direction of higher surface tension, i.e., lower surfactant concentration [1618]. When the surface diffusion and gravity are negligible, a large change in film height and surface tension over a very short distance can appear. The film thickens and thins downstream with possible rupture.

The evolving concentration of surfactant on a free surface (c_s) is governed by the *surfactant surface transport equation*:

$$\partial_t c_s + \nabla \cdot (c_s \mathbf{v}_s) + c_s(\nabla_s \cdot \mathbf{v}_s)(\mathbf{v} \cdot \mathbf{n}) = \mathcal{D}_s \nabla_s c_s + J(c_s, C_s), \tag{13.8}$$

where $\mathbf{v}_s \equiv (\mathbf{I} - \mathbf{nn}) \cdot \mathbf{v}$ is the surface velocity, $\nabla_s \equiv (\mathbf{I} - \mathbf{nn}) \cdot \nabla$ ($\nabla_s = \nabla - \mathbf{n}\frac{\partial}{\partial n}$) the surface gradient operator ($\mathbf{I}$: identity tensor), $\mathcal{D}_s$ the surface diffusivity of the surfactant, J flux of surfactant from the source to the interface associated with adsorption onto or desorption from the surface that depends on both the surface surfactant concentration and the bulk concentration (C_s).

Other governing equations include the Navier-Stokes equations (mass and momentum conservation) and conservation of surfactant species in the bulk phase (*surfactant bulk transport equation*):

$$\partial_t C_s + \nabla \cdot (C_s \mathbf{v}) = \mathcal{D} \nabla^2 C_s, \tag{13.9}$$

where $\mathcal{D}$ is the bulk diffusivity.

In surfactant replacement therapy, several factors influence the transport: the bolus volume, its injection rate, dose, instillation site, physical properties (viscosity, density, and surface tension) of the injected fluid, gravity, ventilation mode, clearance of instilled surfactant, and treatment history. Surfactant delivery comprises 4 stages of instilled bolus transport, starting from a liquid plug that possibly occludes the large airways, progression with a trailing film to a deposited layer along large airway walls that moves under the influence of gravity, spreading in smaller airways due to Marangoni flows, and arriving in alveoli, where surfactant lipids and proteins can be removed. Alveolar type-2 cells can ensure surfactant clearance and recycle a fraction upon secretion. A surfactant amount is taken up by alveolar macrophages, in adition to bronchial Clara cells. The transport of liquid and insoluble surfactant through airways from a source located at the tracheal end was computed [1619]. Its transit time is strongly influenced by the concentration of pre-existing surfactant and airway geometrical properties.

The effect of in-plane wall oscillations on surfactant transport in a thin fluid film on a membrane to mimic airway wall motion was studied in a 3-bifurcation airway tree model [1618]. Walls covered by a pre-established liquid film thickness and a steady-state surfactant concentration (background surfactant transport) were subjected to cyclic stretches with variable amplitude and frequency. The surface tension is related to the surfactant concentration by a linear equation of state. The flux of soluble surfactant between from the bulk (thin fluid film) and interface is computed using a linear sorption interfacial kinetics equation. Surfactant transport is enhanced at large stretching cycle amplitudes. The amplitude being constant, high stretching cycle frequency improves and attenuates surfactant transport at small and large amplitudes, respectively.

Chapter 14
Conclusion

Volume 5 aims at providing the fundamental knowledge of the respiratory and vascular tissues that convey and regulate air and blood flows. These serial flows supply oxygen to the body's tissues and successively remove carbon dioxide produced by the body's cells. The right cardiac pump drives blood into pulmonary circulation. The pulmonary perfusion must locally match the ventilation. Lung territories oxygenate blood that then returns to the left cardiac pump. The left venticle expels blood into the systemic arterial bed to irrigate all the body's tissues, including the heart wall and lung parenchyma.

To raise the surface area of gas exchange in lungs between alveolar air and capillary blood, the design of the tracheobronchial tree relies on branchings of short parent ducts to irrigate and drain all the gas exchange units inside the confined volume of lung parenchyma within the thoracic cage.

Quasi-periodic ventilation is caused by the motion of the thoracic wall, i.e., by cycles of contraction and relaxation of respiratory mucles. The ventilatory pump that limits the external borders of lung parenchymas with their bronchial trees changes the intrathoracic pressure, and hence the volume of deformable bronchioles and alveoli. The end pressure within lung acini enables successively inhalation and exhalation of air from the atmosphere. Rib cage displacements are controlled by the central nervous system.

In blood circulation, the cardiac pump is a component of the blood circuitry. Blood circulates in two serial compartments: pulmonary and systemic circulation. Two closely apposed left and right pumps, each made of two chambers to tackle pressures in the venous and arterial compartments, synchronously expel blood in their respective circuits.

The heart generates starting–stopping-like flows. Arterial unsteady blood flow becomes continuous during the cardiac cycle, but can become bidirectional or even reverse during diastole according to the rheology of the arterial wall.

The vasculature is characterized by a complicated architecture with variable geometry both in space and time. Moreover, the vessel geometry changes over short

M. Thiriet, *Tissue Functioning and Remodeling in the Circulatory and Ventilatory Systems*, Biomathematical and Biomechanical Modeling of the Circulatory and Ventilatory Systems 5, DOI 10.1007/978-1-4614-5966-8_14,

distances. The vasculature is composed of curved blood vessels with successive branchings to irrigate the body's tissues.

Adaptation of blood flow to the body's activity and tissue needs also relies on the central nervous system. In addition, local mechanisms regulate blood flows, using three main mechanisms from large arteries to small arterioles: (1) mechanotransduction-driven postload control; (2) autoregulation; and (3) functional (reactive) hyperemia.

Both air and blood streams are three-dimensional and developing, as they are conveyed in entrance length (Vol. 7 – Chaps. 2. Hemodynamics and 3. Air Transport). Moreover, airways and blood vessels are deformable. Changes in transmural pressure (the pressure difference between the pressure at the wetted surface of the lumen applied by the moving fluid on the pipe wall and the pressure at the external wall side that depends on the activity of neighboring tissues and organs) can also affect the shape of the vessel cross-section, especially when it becomes negative. More generally, the change in cross-section shape can result from taper, possible prints of adjacent organs with more or less progressive constriction and enlargment, and adaptation to branching (transition zone). These changes can induce local three-dimensional blood motions displayed by virtual transverse currents, even if the vessel is considered locally straight.

The magnitude and direction of mechanical stresses applied by air and blood flows on the wetted surface of respiratory epithelia and vascular endothelia as well as within the conduit wall vary during the ventilatory and cardiac cycles. Inside conduit lumens, local changes in the direction of stress components are caused by flow separations and reversal. Flow separation is set by an adverse pressure gradient when inertia forces and blood vorticity are high enough, especially in branching vascular segments. Due to the time-dependent feature of flows, the flow separation region spreads over a variable length during the physiological cycle and can move. The location and variable size of the flow separation region depends on the flow distribution between the branches, which can vary during the physiological cycle. Flow reversal occurs during the diastole of the left ventricle in elastic arteries, such as the aorta, and most of the muscular arteries, such as brachial and femoral arteries (but not in the carotid arteries). Flow reversal can be observed either in a region near the wall, more or less wide with respect to the position of the local center of vessel curvature, or in the entire lumen.

During normal tidal breathing, neurohumoral variations are slight. Breathing influences blood circulation, especially the venous return, by changes in intrathoracic pressure. Because lungs constantly tend to shrink away from the costal pleura, the pressure in the pleural cavity is smaller than that of the atmosphere (negative pressure). Lungs also exert a traction on the pericardium, on arteries in which the blood pressure is high, especially in those of the systemic circulation, and on venae cavae in which the blood pressure is low, thereby deforming intrathoracic veins much more than the intrathoracic arteries. During inspiration, the intrathoracic pressure lowers, thereby reducing right and left auricular preload and right ventricular postload (heightening vascular bore and attenuating flow resistance).[1] The cardiac chambers may then receive an amount of blood during inspiration slightly greater than that during expiration. However, the aid given to the right cardiac pump by a low intrathoracic pressure may not significantly influence the left cardiac pump, which works against a much higher outlet pressure range.

In addition, the natural respiratory sinus arrhythmia defines a variation in cardiac frequency during a breathing cycle. The heart beat rate increases during inspiration and decreases during expiration. The cardiac frequency is controlled by the nucleus ambiguus among the centers in the medulla oblongata. The nucleus ambiguus increases parasympathetic nervous system inputs to the heart via the cardioinhibitory vagus nerve. During expiration, neurons in the nucleus ambiguus are activated and the cardiac frequency slows. Conversely, inspiration triggers inhibitory signals to the nucleus ambiguus and consequently the vagus nerve remains in a resting state. In summary, the heart beats more slowly during expiration than inspiration.

Stepwise inspiration and expiration provoke an increase in cardiac frequency followed by a rapid decrease [1620]. The fastest and slowest heart rates are obtained in response to inspiration with respect to expiration. Changes in breathing frequency (from 0.05 to 0.2 Hz) also cause cardiac frequency variations, especially in a respiration frequency range of 0.1 to 0.12 Hz for which heart rate variation is maximal. Heart rate variability with deep breathing serves as a measure of parasympathetic function.

Many factors influence heart rate variability [1621]: (1) age, as the variability decreases with advancing age; (2) body position, as the variability is maximal when the patient is lying supine; (3) respiration frequency, as the variability is maximal when the breathing rate equals 0.08 to 0.1 Hz; and (4) respiration amplitude, as the variability is maximal when the tidal volume is approximately 1.2 l for an average adult.

Both breathing and blood circulation adapt to the body's needs. Blood contents of oxygen and carbon dioxide primarily control the breathing and cardiac frequencies. Both the body's ventilation and blood convection are controlled by the autonomic nervous system.

In the central nervous system, four main centers regulate respiration (inspiratory, expiratory, pneumotaxic, and apneustic centers; Table 14.1). Inspiration results from a sudden, ramped increase in motor discharge to the inspiratory muscles. The inspiratory center (dorsal group of respiratory neurons) formed by neurons of the nucleus of tractus solitarius situated in the upper region of the medulla oblongata controls the diaphragm and intercostal muscles responsible for inspiration. These inspiratory muscles are connected to the nervous centers via spinal nerves. Contraction of the diaphragm is triggered by 2 phrenic nerves that emanate from cervical spinal nerves 3 to 5. Eleven pairs of intercostal nerves, the anterior branches (rami anteriores) of the thoracic T1 to T11 spinal nerves regulate the activity of intercostal muscles. Before the end of inhalation, motor discharge declines. Exhalation is usually passive, except at high ventilation rates. At rest, expiration

[1] On the other hand, when arterial blood pressure rises, left ventricular postload elevates.

Table 14.1. Respiratory centers of the central nervous system.

Brain region	Respiratory centers	Function
Pons	Pneumotaxic center	Inhibition of inspiratory center (limitation of inspiration duration)
	Apneustic center	Inspiration activation
Medulla oblongata	Inspiratory center	Inspiration Activation of diaphragm and external intercostal muscles
	Expiratory center	Deep and/or forced expiration Activation of abdominal muscles and internal intercostal muscles

results from the relaxation of the diaphragm and intercostal muscles and lung elastic recoil. A feedback control is ensured by central and peripheral chemoreceptors and mechanoreceptors.

When breathing increases, such as during exercise and coughing that helps to remove large foreign bodies from the respiratory tract as well as smaller particles entrapped in the mucus coating the respiratory epithelium, the expiratory center primes the contraction of abdominal muscles and 11 internal intercostals (or intercostales interni).[2] The expiratory center (ventral group of respiratory neurons) formed by neurons of the nuclei ambiguus and retro-ambiguus is located anteriorly and laterally with respect to the inspiratory center.

The nervous centers change the amplitude and rhythm of respiration upon reception of feedback signals sent by sensory neurons. The Hering-Breuer reflex that uses the vagus nerve prevents lung overinflation (Table 14.2). When lung airways expand during inhalation, the vagus nerve conveys information to the medulla oblongata that, in response, temporarily inhibits the inspiratory center, thereby priming expiration. Once bronchi and bronchioles return to resting size, the vagus nerve ceases to send inhibitory signals, and the inspiratory center reinitiates inspiration.

The pneumotaxic center (or pontine respiratory group) located in the pons provides cyclic inhibitory impulses to the inspiratory center, thereby decreasing the tidal volume. This center, which controls the breathing amplitude and respiratory frequency, is only active during elevated breathing. The apneustic center, also situated in the pons, activates inspiration at rest. The apneustic center activity is canceled out by the pneumotaxic center when the breathing rate increases. Conversely, the apneustic center delays arrest signals from the pneumotaxic center.

[2]External intercostals (or intercostales externi) work in unison during deep inspiration. These thick muscles are directed obliquely downward and laterally. Internal intercostals are also directed obliquely, but in a direction opposite to that of external intercostals. Internal intercostal muscles counteract the action of external intercostal muscles.

Table 14.2. Hering-Breuer reflex.

Inspiratory center	$\oplus\longrightarrow$	Phrenic and intercostal nerves,
	$\oplus\longrightarrow$	inspiratory muscles,
	$\oplus\longrightarrow$	lung inflation
Pulmonary stretch receptors	$\oplus\longrightarrow$	vagus nerve,
	$\ominus\longrightarrow$	inspiratory center

Chemicals such as oxygen, carbon dioxide, and hydrogen ion influence the body's respiration. Chemical concentrations are monitored by chemoreceptors located centrally in the medulla oblongata and peripherally at the bifurcation of the carotid arteries and aortic arch. The peripheral chemoreceptors augment afferent inputs to the respiratory centers in the brainstem when perfused with hypoxemic, hypercapnic, and acidotic arterial blood. The central chemoreceptors are sensitive to brain hypercapnia and acidosis.

Other influence factors of the body's respiration include voluntary control, i.e., the cerebral cortex, emotions driven by the hypothalamus and limbic system, body temperature, and mechanically and chemically gated receptors.

Baroreceptors are pressure sensors of the carotid sinuses and aortic arch. Proprioceptors localize to skeletal muscles, tendons, and joints that sense body movements. During exercise, these receptors transmit signals to the respiratory centers that increase breathing frequency and amplitude.

Cutaneous thermoreceptors can also send signals to the cerebral cortex, which, in turn, activates respiratory centers, causing hyperventilation. Pain receptors operates similarly to provoke hyperventilation.

In addition, juxtacapillary (J) receptors in alveolar walls are innervated by fibers of the vagus nerve. Irritant receptors are sensory terminals of myelinated afferent fibers that reside in bronchial and bronchiolar walls. Irritant receptors respond to smoke and dust; they are stimulated by histamine, serotonine, and prostaglandins, as well as inflation, deflation, and acute congestion of blood capillaries [1622]. Their stimulation causes prolonged inspiratory duration and excitatory responses, such as cough and gasp. Juxtacapillary receptors, or unmyelinated C-fiber endings, reside in alveolar walls, close to capillaries of pulmonary circulation (hence their name), as well as in bronchial walls, near capillaries of the bronchial (systemic) circulation. They are stimulated by hyperinflation, lung injury, and edema (especially alveolar J receptors), and by exposure to chemicals, such as histamine and serotonine (especially bronchial J receptors) [1622]. Their acute and strong stimulation can cause apnea and a fall in systemic vascular resistance. Afferent, unmyelinated C fibers of the somatic sensory system innervate both pulmonary capillary bed and airways, thereby yielding vagal inputs with different properties from these two neighboring structures.

The human body uses two major ventilatory control systems to maintain the content of oxygen, carbon dioxide, and protons: the chemical and behavioral control of breathing. Whereas chemical control of breathing relies on arterial and brain partial pressure in O_2 and CO_2 and pH sensed by peripheral and central chemoreceptors, the behavioral control of breathing is based on cues received from different sites (e.g., cortical and subcortical regions and hypothalamus, as well as proprio- and nociceptors). The behavioral control of breathing regulates breathing either by direct control of respiratory motoneurons (corticospinal control of respiration) or by modulation of respiratory centers in the brainstem via the reticular activating system (or extrathalamic control modulatory system). In addition, the behavioral control can modulate the chemical control to optimize breathing to the body's needs.

The heart is capable of beating independently of any nervous or hormonal influences, hence is endowed with intrinsic automaticity. Yet, its activity is influenced by nervous signals as well as by regulatory chemicals. Like other muscle or nerve cells, cardiomyocytes are excitable cells that synchronously contract at a given rate when their plasma membrane depolarizes upon the arrival of electrochemical impulses — action potentials — that are initiated in and run through the nodal tissue.

Like breathing, nervous control of the heart relies on parasympathetic fibers that travel in vagus nerves and sympathetic nerves. The vagus nerve operates as a cardiac inhibitor, whereas sympathetic nerves are cardiac exciters. The sympathetic nervous system supply to the heart leaves the spinal cord at the first 4 thoracic (T1–T4) vertebra.

The sympathetic nervous system causes bronchodilation and vasoconstriction, thus diverting blood flow away from some vascular compartments to others, but it promotes vasodilation for the coronary arteries, and increases the cardiac frequency and myocardium contractility. On the other hand, the parasympathetic nervous system provokes bronchoconstriction and vasodilation.

At the effector organs, sympathetic postganglionic neurons release noradrenaline (or norepinephrine), along with other cotransmitters such as ATP, among others. Noradrenaline binds to adrenergic receptors (Table 14.3). Acetylcholine is the neurotransmitter for both preganglionic sympathetic and parasympathetic neurons as well as for postganglionic parasympathetic neurons. Acetylcholine targets muscarinic receptors on the plasma membrane of effector cells (Table 14.4). At the adrenal medulla, an endocrine gland, presynaptic neurons release acetylcholine that binds to nicotinic receptors, i.e., ligand-gated ion channels. Stimulated adrenal medulla releases adrenaline (epinephrine) into the blood stream. Adrenaline targets also adrenergic receptors, thereby enhancing sympathetic activity.

Any modeling and simulation tests explore a biological event at a given time. In the case of physiological flows, numerical simulations use computational domains based on the three-dimensional reconstruction of organs of interest from a medical imaging data set to get a patient-specific geometry, as organ configuration varies strongly between human subjects. However, medical imaging, an element of personnalized medicine, not only captures data at a given moment, but also yields a model of real anatomy.

Therefore, a wide gap still exists between competences, on the one hand, in modeling and numerical analysis, development of appropriate simulation algorithms, softwares, and coupling platforms for solving multiscale, multiphysics problems related to complex behavior and non-linear dynamics that incorporate

Table 14.3. Receptors of the sympathetic postganglionic neurons (Vol. 3 – Chap. 7. G-Protein-Coupled Receptors).

Type	Effect
	Cardiovascular apparatus
α1a GPCR	Positive inotropy
	Vasoconstriction
α1b GPCR	Positive inotropy
	Vasoconstriction
α1d GPCR	Vasoconstriction
α2a GPCR	Sympathetic output reduction
α2b GPCR	Vasoconstriction
α2c GPCR	Venoconstriction
β1 GPCR	Positive bathmotropy, chronotropy, dromotropy, inotropy, lusitropy
β2 GPCR	Positive inotropy, lusitropy
β3 GPCR	Negative inotropy
	Respiratory apparatus
α1 GPCR	Bronchoconstriction (minor contribution)
β2 GPCR	Bronchodilation (major contribution)
	Adrenal medulla
Nicotinic receptor (ion channel)	Secretion of adrenaline and noradrenaline (preganglionic neuron)
	Kidney (juxtaglomerular apparatus)
β1 GPCR	Secretion of renin
	Liver
α1 GPCR	Glycogenolysis, gluconeogenesis
β2 GPCR	Glycogenolysis, gluconeogenesis
	Adipose tissue
β1/3 GPCR	Increased lipolysis

phenomenological models of nano- and microscale events and, on the other hand, virtual physiology that results from cell signaling with its spatiotemporal fine-tuned regulation.

However, grant programs have been launched to promote and support research networks that propose innovative, interdisciplinary approaches with a four-stage philosophy: conception, development, implementation, and optimization. Research programs are commonly based on multiscale approaches (from genes, proteins and other molecule types, cells, tissues, organs, and organisms), i.e., at different biological scales involved in structure–function relations.

Biological organization at the nanoscopic scale relies on intermolecular interactions, especially those participating in gene transcription regulation and signal transduction. Cell signaling pathways determine cell fate, thereby governing structure–function relations at the tissue, organ, and body scales.

Table 14.4. Receptors of the parasympathetic postganglionic neurons (Vol. 3 – Chap. 7. G-Protein-Coupled Receptors).

Type	Effect
	Cardiovascular apparatus
Muscarinic M_2 GPCR	Decreased cardiac output
	Attenuated cardiac frequency (negative chronotropy)
	Reduced contractility (negative inotropy)
	Decreased conduction (negative dromotropy)
	Vascular smooth muscle relaxation
Muscarinic M_3 GPCR	Vascular smooth muscle contraction
Muscarinic M_5 GPCR	Vasodilation of cerebral arteries
	Respiratory apparatus
Muscarinic M_3 GPCR	Airway smooth muscle contraction

These collaborative programs aims at boosting, on the one hand, health knowledge acquisition, formalization, representation, and communication, as well as, on the other, training to diagnostic and therapeutic procedures, identification of relevant biomarkers to foster early diagnosis of chronic diseases and prevention, development of drug nanovectors, optimization of treatment planning, elaboratation of augmented reality-based image-guided interventions, optimal design of artificial organs, and remote medical decision and gesture.

Strategies usually combine data mining, advanced data analysis techniques, non-invasive imaging, image and signal processing, parameter identification, cell and tissue characterization, biochemical and biophysical modeling, high-performance computing, verification and experimental validation, and production of coupling platforms and computer-assisted medicine softwares.

References

Chap. 1. Blood

1. Harvey W (1628) Exercitatio anatomica de motus cordis et sanguinis in animalibus [An anatomical study on the motion of the heart and blood in animals]. London
2. Snabre P, Mills P (1996) Rheology of weakly flocculated suspensions of rigid particles I. Journal de Physique III 6:1811–1834
3. Kolb M, Jullien R (1984) Chemically limited versus diffusion limited aggregation. Journal de Physique – Lettres 45:977–981
4. Willer CJ, Sanna S, Jackson AU, Scuteri A, Bonnycastle LL, Clarke R, Heath SC, Timpson NT, Najjar SS, Stringham HM, Strait J, Duren WL, Maschio A, Busonero F, Mulas A, Albai G, Swift AJ, Morken MA, Narisu N, Bennett D, Parish S, Shen H, Galan P, Meneton P, Hercberg S, Zelenika D, Chen WM, Li Y, Scott LJ, Scheet PA, Sundvall J, Watanabe RM, Nagaraja R, Ebrahim S, Lawlor DA, Ben-Shlomo Y, Davey-Smith G, Shuldiner AR, Collins R, Bergman RN, Uda M, Tuomilehto J, Cao A, Collins FS, Lakatta E, Lathrop GM, Boehnke M, Schlessinger D, Mohlke KL, Abecasis GR (2008) Newly identified loci that influence lipid concentrations and risk of coronary artery disease. Nature – Genetics 40:161–169
5. Kathiresan S, Melander O, Guiducci C, Surti A, Burtt NP, Rieder MJ, Cooper GM, Roos C, Voight BF, Havulinna AS, Wahlstrand B, Hedner T, Corella D, Tai ES, Ordovas JM, Berglund G, Vartiainen E, Jousilahti P, Hedblad B, Taskinen MR, Newton-Che C, Salomaa V, Peltonen L, Groop L, Altshuler DM, Orho-Melander M (2008) Six new loci associated with blood low-density lipoprotein cholesterol, high-density lipoprotein cholesterol or triglycerides in humans. Nature – Genetics 40:189–197
6. Brown MS, Goldstein JL (1993) A receptor-mediated pathway for cholesterol homeostasis. In: Nobel lectures in physiology or medecine 1981–1990. World Scientific Publishing, Singapore
7. Mahley RW (1988) Apolipoprotein E: cholesterol transport protein with expanding role in cell biology. Science 240:622–630
8. Linsel-Nitschke P, Tall AR (2005) HDL as a target in the treatment of atherosclerotic cardiovascular disease. Nature Reviews – Drug Discovery 4:193–205
9. Brunham LR, Kruit JK, IqbalJ, Fievet C, Timmins JM, Pape TD, Coburn BA, Bissada N, Staels B, Groen AK, Hussain MM, Parks JS, Kuipers F, Hayden MR (2006) Intestinal ABCA1 directly contributes to HDL biogenesis in vivo. Journal of Clinical Investigation 116: 1052–1062
10. Brown MS, Goldstein JL (1983) Lipoprotein metabolism in the macrophage: implications for cholesterol deposition in atherosclerosis. Annual Review of Biochemistry 52:223–261

M. Thiriet, *Tissue Functioning and Remodeling in the Circulatory and Ventilatory Systems*, Biomathematical and Biomechanical Modeling of the Circulatory and Ventilatory Systems 5, DOI 10.1007/978-1-4614-5966-8,

11. Forte TM, Bielicki JK, Goth-Goldstein R, Selmek J, McCall MR (1995) Recruitment of cell phospholipids and cholesterol by apolipoproteins A-II and A-I: formation of nascent apolipoprotein-specific HDL that differ in size, phospholipid composition, and reactivity with LCAT. Journal of Lipid Research 36:148–57
12. Duverger N, Ghalim N, Ailhaud G, Steinmetz A, Fruchart JC, Castro G (1993) Characterization of apoA-IV-containing lipoprotein particles isolated from human plasma and interstitial fluid. Arteriosclerosis and Thrombosis 13:126-132
13. Sattler K, Levkau B (2009) Sphingosine-1-phosphate as a mediator of high-density lipoprotein effects in cardiovascular protection. Cardiovascular Research 82:201–211
14. Gautier T, Masson D, Pais de Barros J-P, Athias A, Gambert P, Aunis D, Metz-Boutigue M-H, Lagrost L (2000) Human apolipoprotein C-I accounts for the ability of plasma high density lipoproteins to inhibit the cholesteryl ester transfer protein activity. Biological Chemistry 275:37504–37509
15. Wolfrum C, Poy MN, Stoffel M (2005) Apolipoprotein M is required for preβ-HDL formation and cholesterol efflux to HDL and protects against atherosclerosis. Nature – Medicine 11:418–422
16. Esteve E, Ricart W, Fernandez-Real JM (2005) Dyslipidemia and inflammation: an evolutionary conserved mechanism. Clinical Nutrition 24:16–31
17. Khovidhunkit W, Kim MS, Memon RA, Shigenaga JK, Moser AH, Feingold KR, Grunfeld C (2004) Effects of infection and inflammation on lipid and lipoprotein metabolism: mechanisms and consequences to the host. Journal of Lipid Research 45:1169–1196
18. Kontush A, Chapman MJ (2006) Antiatherogenic small, dense HDL guardian angel of the arterial wall? Nature – Clinical Practice Cardiovascular Medicine 3:144–153
19. Greene DJ, Skeggs JW, Morton RE (2001) Elevated triglyceride content diminishes the capacity of high density lipoprotein to deliver cholesteryl esters via the scavenger receptor class B type I. Journal of Biological Chemistry 276:4804–48011
20. Malle E, Steinmetz A, Raynes JG (1993) Serum amyloid A: an acute phase protein and apolipoprotein. Atherosclerosis 102:131–146
21. Cai L, de Beer MC, de Beer FC, van der Westhuyzen DR (2005) Serum amyloid A is a ligand for scavenger receptor class B type I and inhibits high density lipoprotein binding and selective lipid uptake. Journal of Biological Chemistry 280:2954–2961
22. Houten SM, Watanabe M, Auwerx J (2006) Endocrine functions of bile acids. EMBO Journal 25:1419–1425
23. Huang W, Ma K, Zhang J, Qatanani M, Cuvillier J, Liu J, Dong B, Huang X, Moore DD (2006) Nuclear receptor-dependent bile acid signaling is required for normal liver regeneration. Science 312:233–236
24. Bohlius J, Weingart O, Trelle S, Engert A (2006) Cancer-related anemia and recombinant human erythropoietin: an updated overview. Nature – Clinical Practice – Oncology 3:152–164
25. Giarratana MC, Kobari L, Lapillonne H, Chalmers D, Kiger L, Cynober T, Marden MC, Wajcman H, Douay L (2004) Ex vivo generation of fully mature human red blood cells from hematopoietic stem cells. Nature – Biotechnology 23:69–74
26. Koder RL, Anderson JLR, Solomon LA, Reddyl KS, Moser CC, Dutton PL (2009) Design and engineering of an O_2 transport protein. Nature 458:305–309
27. Gupta BB, Jaffrin MY, Ding LH, Dohi T (1986) Membrane plasma separation through small-area, hollow-fiber filters. Artificial Organs 10:45–51
28. Ding LH, Gupta BB, Jaffrin MY (1986) Augmentation de filtration en plamaphrse par dbit sanguin puls [Filtration enhancement in plasmapheresis by pulsatile blood flow] Journal de biophysique et de biomcanique 10:71–77
29. Belhadj M, Cancs E, Gerbeau J-F, Mikelić A (2004) Homogeneization approach to filtration through a fibrous medium. INRIA Research Report RR-5277
30. Allaire G (1991) Homogenization of the Navier–Stokes equations in open set perforated with tiny holes I: Abstract framework, a volume distribution of holes. Archive for Rational Mechanics and Analysis 113:209–259

31. Allaire G (1991) Homogenization of the Navier-Stokes equations in open set perforated with tiny holes II: Non-critical size of the holes for a volume distribution and a surface distribution of holes. Archive for Rational Mechanics and Analysis 113:261–298
32. Allaire G (1991) Homogenization of the Navier–Stokes equations with a slip boundary condition. Communications on Pure and Applied Mathematics 44:605–641
33. Allaire G (1991) Continuity of the Darcy's law in the low-volume fraction limit. Annali della Scuola Normale Superiore di Pisa 18:475–499
34. Allaire G (1992) Homogeneization and two-scale convergence. SIAM Journal on Mathematical Analysis 23:1482–1518
35. Briane M (1993) Three models of nonperiodic fibrous materials obtained by homogenization. RAIRO Modlisation Mathmatique et Analyse Numrique 27:759–775
36. Briane M (1994) Homogenization of a non-periodic material. Journal de Mathmatiques Pures et Appliques 73:47–66

Chap. 2. Hematopoiesis

37. Lajtha LG (1979) Stem cell concepts. Differentiation 14:23–34
38. Szabo E, Rampalli S, Risueño RM, Schnerch A, Mitchell R, Fiebig-Comyn A, Levadoux-Martin M, Bhatia M (2010) Direct conversion of human fibroblasts to multilineage blood progenitors. Nature 468:521–526
39. Ogawa M (1993) Differentiation and proliferation of hematopoietic stem cells. Blood 81:2844-2853
40. Li L, Clevers H (2010) Coexistence of quiescent and active adult stem cells in mammals. Science 327:542–545
41. Beerman I, Bhattacharya D, Zandi S, Sigvardsson M, Weissman IL, Bryder D, Rossi DJ (2010) Functionally distinct hematopoietic stem cells modulate hematopoietic lineage potential during aging by a mechanism of clonal expansion. Proceedings of the National Academy of Sciences of the United States of America 107:5465–5470
42. Nakada D, Saunders TL, SJ Morrison (2010) Lkb1 regulates cell cycle and energy metabolism in haematopoietic stem cells. Nature 468:653–658
43. Gurumurthy S, Xie SZ, Alagesan B, Kim J, Yusuf RZ, Saez B, Tzatsos A, Ozsolak F, Milos P, Ferrari F, Park PJ, OS Shirihai, Scadden DT, Bardeesy N (2010) The Lkb1 metabolic sensor maintains haematopoietic stem cell survival. Nature 468:659–666
44. Gan B, Hu J, Jiang S, Liu Y, Sahin E, Zhuang L, Fletcher-Sananikone E, Colla S, Wang YA, Chin L, DePinho RA (2010) Lkb1 regulates quiescence and metabolic homeostasis of haematopoietic stem cells. Nature 468:701–704
45. Taniguchi Ishikawa E, Gonzalez-Nieto D, Ghiaur G, Dunn SK, Ficker AM, Murali B, Madhu M, Gutstein DE, Fishman GI, Barrio LC, Cancelas JA (2012) Connexin-43 prevents hematopoietic stem cell senescence through transfer of reactive oxygen species to bone marrow stromal cells. Proceedings of the National Academy of Sciences of the United States of America 109:9071–9076
46. Sirin O, Lukov GL, Mao R, Conneely OM, Goodell MA (2010) The orphan nuclear receptor Nurr1 restricts the proliferation of haematopoietic stem cells. Nature – Cell Biology 12:1213–1219
47. Dingli D, Pacheco JM (2006) Allometric scaling of the active hematopoietic stem cell pool across mammals. PLoS ONE 1: e2
48. Kiel MJ, Yilmaz OH, Iwashita T, Yilmaz OH, Terhorst C, Morrison SJ (2005) SLAM family receptors distinguish hematopoietic stem and progenitor cells and reveal endothelial niches for stem cells. Cell 121:1109–1121
49. Eilken HM, Nishikawa SI, Schroeder T (2009) Continuous single-cell imaging of blood generation from haemogenic endothelium. Nature 457:896–900

50. Lancrin C, Sroczynska P, Stephenson C, Allen T, Kouskoff V, Lacaud G (2009) The haemangioblast generates haematopoietic cells through a haemogenic endothelium stage. Nature 457:892–895
51. Kissa K, Herbomel P (2010) Blood stem cells emerge from aortic endothelium by a novel type of cell transition. Nature 464:112–115
52. Boisset JC, van Cappellen W, Andrieu-Soler C, Galjart N, Dzierzak E, Robin C (2010) In vivo imaging of haematopoietic cells emerging from the mouse aortic endothelium. Nature 464:116–120
53. Bertrand JY, Chi NC, Santoso B, Teng S, Stainier DYR, Traver D (2010) Haematopoietic stem cells derive directly from aortic endothelium during development. Nature 464:108–111
54. Semerad CL, Mercer EM, Inlay MA, Weissman IL, Murre C (2009) E2A proteins maintain the hematopoietic stem cell pool and promote the maturation of myelolymphoid and myeloerythroid progenitors. Proceedings of the National Academy of Sciences of the United States of America 106:1930–1935
55. Ceredig R, Rolink AG, Brown G (2009) Models of haematopoiesis: seeing the wood for the trees. Nature Reviews – Immunology 9:293–300
56. Wilson A, Trumpp A (2006) Bone-marrow haematopoietic-stem-cell niches. Nature Reviews – Immunology 6:93–106
57. Kiel MJ, Morrison SJ (2008) Uncertainty in the niches that maintain haematopoietic stem cells. Nature Reviews – Immunology 8:290–301
58. North TE, Goessling W, Walkley CR, Lengerke C, Kopani KR, Lord AM, Weber GJ, Bowman TV, Jang IH, Grosser T, FitzGerald GA, Daley GQ, Orkin SH, Zon LI (2007) Prostaglandin E2 regulates vertebrate haematopoietic stem cell homeostasis. Nature 447:1007–1011
59. Naveiras O, Nardi V, Wenzel PL, Hauschka PV, Fahey F, Daley GQ (2009) Bone-marrow adipocytes as negative regulators of the haematopoietic microenvironment. Nature 460:259–263
60. Adams GB, Scadden DT (2006) The hematopoietic stem cell in its place. Nature – Immunology 7:333–337
61. Adams GB, Chabner KT, Alley IR, Olson DP, Szczepiorkowski ZM, Poznansky MC, Kos CH, Pollak MR, Brown EM, Scadden DT (2006) Stem cell engraftment at the endosteal niche is specified by the calcium-sensing receptor. Nature 439:599–603
62. Jones DL, Wagers AJ (2008) No place like home: anatomy and function of the stem cell niche. Nature Reviews – Molecular Cell Biology 9:11–21
63. Kollet O, Dar A, Shivtiel S, Kalinkovich A, Lapid K, Sztainberg Y, Tesio M, Samstein RM, Goichberg P, Spiegel A, Elson A, Lapidot T (2006) Osteoclasts degrade endosteal components and promote mobilization of hematopoietic progenitor cells. Nature – Medicine 12:657–664
64. Kobayashi H, Butler JM, O'Donnell R, Kobayashi M, Ding BS, Bonner B, Chiu VK, Nolan DJ, Shido K, Benjamin L, Rafii S (2010) Angiocrine factors from Akt-activated endothelial cells balance self-renewal and differentiation of haematopoietic stem cells. Nature – Cell Biology 12:1046–1056
65. Furness SG, McNagny K (2006) Beyond mere markers: functions for CD34 family of sialomucins in hematopoiesis. Immunologic Research 34:13–32
66. Nielsen JS, McNagny KM (2008) Novel functions of the CD34 family. Journal of Cell Science 121:3683–3692
67. Miccio A, Wang Y, Hong W, Gregory GD, Wang H, Yu X, Choi JK, Shelat S, Tong W, Poncz M, Blobel GA (2010) NuRD mediates activating and repressive functions of GATA-1 and FOG-1 during blood development. EMBO Journal 29:442–456
68. Hueber SD, Weiller GF, Djordjevic MA, Frickey T (2010) Improving Hox protein classification across the major model organisms. PLoS ONE 5:e10820
69. Li Z (2008) Intrinsic and extrinsic control of haematopoietic stem-cell self-renewal. Nature 453:306–313
70. Kee BL (2009) E and ID proteins branch out. Nature Reviews – Immunology 9:175–184

71. Lcuyer E, Herblot S, Saint-Denis M, Martin R, Begley CG, Porcher C, Orkin SH, Hoang T (2002) The SCL complex regulates c-kit expression in hematopoietic cells through functional interaction with Sp1. Blood 100:2430–2440
72. Huang S, Brandt SJ (2000) mSin3A regulates murine erythroleukemia cell differentiation through association with the TAL1 (or SCL) transcription factor. Molecular and Cellular Biology 20:2248–2259
73. Chen MJ, Yokomizo T, Zeigler BM, Dzierzak E, Speck NA (2009) Runx1 is required for the endothelial to haematopoietic cell transition but not thereafter. Nature 457:887–891
74. Adamo L, Naveiras O, Wenzel PL, McKinney-Freeman S, Mack PJ, Gracia-Sancho J, Suchy-Dicey A, Yoshimoto M, Lensch W, Yoder MC, García-Cardeña G, Daley GQ (2009) Biomechanical forces promote embryonic haematopoiesis. Nature 459:1131–1135
75. Chang HH, Hemberg M, Barahona M, Ingber DE, Huang S (2008) Transcriptome-wide noise controls lineage choice in mammalian progenitor cells. Nature 453:544–547
76. Stehling-Sun S, Dade J, Nutt SL, DeKoter RP, Camargo FD (2009) Regulation of lymphoid versus myeloid fate 'choice' by the transcription factor Mef2c. Nature – Immunology 10:289–296
77. Thierfelder WE, van Deursen JM, Yamamoto K, Tripp RA, Sarawar SR, Carson RT, Sangster MY, Vignali DA, Doherty PC, Grosveld GC, Ihle JN (1996) Requirement for Stat4 in interleukin-12-mediated responses of natural killer and T cells. Nature 382:171–174
78. Kaplan MH, Schindler U, Smiley ST, Grusby MJ (1996) Stat6 is required for mediating responses to IL-4 and for development of Th2 cells. Immunity 4:313–319
79. Nakajima K, Yamanaka Y, Nakae K, Kojima H, Ichiba M, Kiuchi N, Kitaoka T, Fukada T, Hibi M, Hirano T (1996) A central role for Stat3 in IL-6-induced regulation of growth and differentiation in M1 leukemia cells. EMBO Journal 15:3651–3658
80. Gupta R, Hong D, Iborra F, Sarno S, Enver T (2007) Nov (CCN3) functions as a regulator of human hematopoietic stem or progenitor cells. Science 316:590–593
81. Yamazaki S, Iwama A, Takayanagi SI, Morita Y, Eto K, Ema H, Nakauchi H (2006) Cytokine signals modulated via lipid rafts mimic niche signals and induce hibernation in hematopoietic stem cells. The EMBO Journal 25:3515–3523
82. Zhu J, Zhang Y, Joe GJ, Pompetti R, Emerson SG (2005) NF-Ya activates multiple hematopoietic stem cell (HSC) regulatory genes and promotes HSC self-renewal. Proceedings of the National Academy of Sciences of the United States of America 102:11728–11733
83. Gtz R, Wiese S, Takayama S, Camarero GC, Rossoll W, Schweizer U, Troppmair J, Jablonka S, Holtmann B, Reed JC, Rapp UR, Sendtner M (2005) Bag1 is essential for differentiation and survival of hematopoietic and neuronal cells. Nature – Neuroscience 8:1169–1178
84. Adams GB, Alley IR, Chung UI, Chabner KT, Jeanson NT, Celso CL, Marsters ES, Chen M, Weinstein LS, Lin CP, Kronenberg HM, Scadden DT (2009) Haematopoietic stem cells depend on Galphas-mediated signalling to engraft bone marrow. Nature 459:103–107
85. Kurzrock R (2003) Hematopoietic growth factors. In: Bast RC, Kufe DW, Pollock RE, Weichselbaum RR, Holland JF, Frei E (eds) Cancer Medicine. BC Decker, Hamilton, Ontario, Canada
86. Duncan AW, Rattis FM, DiMascio LN, Congdon KL, Pazianos G, Zhao C, Yoon K, Cook JM, Willert K, Gaiano N, Reya T (2005) Integration of Notch and Wnt signaling in hematopoietic stem cell maintenance. Nature – Immunology 6:314–322
87. Avalos BR, Parker JM, Ware DA, Hunter MG, Sibert KA, Druker BJ (1997) Dissociation of the Jak kinase pathway from G-CSF receptor signaling in neutrophils. Experimental Hematology 25:160–168
88. Irandoust MI, Aarts LHJ, Roovers O, Gits J, Erkeland SJ, Touw IP (2007) Suppressor of cytokine signaling 3 controls lysosomal routing of G-CSF receptor. EMBO Journal 26:1782–1793
89. Kishimoto S, Nakamura S, Hattori H, Nakamura SI, Oonuma F, Kanatani Y, Tanaka Y, Mori Y, Harada Y, Tagawa M, Ishihara M (2009) Human stem cell factor (SCF) is a heparin-binding cytokine. Journal of Biochemistry 145:275–278

90. Himburg HA, Muramoto GG, Daher P, Meadows SK, Russell JL, Doan P, Chi JT, Salter AB, Lento WE, Reya T, Chao NJ, Chute JP (2010) Pleiotrophin regulates the expansion and regeneration of hematopoietic stem cells. Nature – Medicine 16:475–482
91. Scott CS, Richards SJ, Roberts BE (1990) Patterns of membrane TcRαβ and TcRγδ chain expression by normal blood CD4+CD8−, CD4−CD8+, CD4−CD8dim+ and CD4−CD8− lymphocytes. Immunology 70:351–356
92. Ohta M, Greenberger JS, Anklesaria P, Bassols A, Massague J (1987) Two forms of transforming growth factor-β distinguished by multipotential haematopoietic progenitor cells. Nature 329:539–541
93. Kosaki K, Ando J, Korenaga R, Kurokawa T, Kamiya A (1998) Fluid shear stress increases the production of granulocyte-macrophage colony-stimulating factor by endothelial cells via mRNA stabilization. Circulation Research 82:794–802
94. Lal A, Pan Y, Navarro F, Dykxhoorn DM, Moreau L, Meire E, Bentwich Z, Lieberman J, Chowdhury D (2009) MiR-24-mediated downregulation of H2AX suppresses DNA repair in terminally differentiated blood cells. Nature – Structural and Molecular Biology 16:492–498
95. Sakata-Yanagimoto M, Nakagami-Yamaguchi E, Saito T, Kumano K, YasutomoK, Ogawa S, Kurokawa M, Chiba S (2008) Coordinated regulation of transcription factors through Notch2 is an important mediator of mast cell fate. Proceedings of the National Academy of Sciences of the United States of America 105:7839–7844
96. Chang KH, Chan-Ling T, McFarland EL, AfzalA, Pan H, Baxter LC, Shaw LC, Caballero S, Sengupta N, Calzi SL, Sullivan SM, Grant MB (2007) IGF binding protein-3 regulates hematopoietic stem cell and endothelial precursor cell function during vascular development. Proceedings of the National Academy of Sciences of the United States of America 104:10595–10600

Chap. 3. Blood Cells

97. Hickey MJ, Kubes P (2009) Intravascular immunity: the host–pathogen encounter in blood vessels. Nature Reviews – Immunology 9:364–375
98. Ozaki K, Leonard WJ (2002) Cytokine and cytokine receptor pleiotropy and redundancy. Journal of Biological Chemistry 277:29355–29358
99. Shaw AS, Filbert EL (2009) Scaffold proteins and immune-cell signalling. Nature Reviews – Immunology 9:47–56
100. Matza D, Badou A, Kobayashi KS, Goldsmith-Pestana K, Masuda Y, Komuro A, McMahon-Pratt D, Marchesi VT, Flavell RA (2008) A scaffold protein, AHNAK1, is required for calcium signaling during T cell activation. Immunity 28:64–74
101. Cokelet GR, Meiselman HJ (1968) Rheological comparison of hemoglobin solutions and erythrocyte suspensions. Science 162:275–277
102. Yoshida H, Kawane K, Koike M, Mori Y, Uchiyama Y, Nagata S (2005) Phosphatidylserine-dependent engulfment by macrophages of nuclei from erythroid precursor cells. Nature 437:754–758
103. Zhang L, Flygare J, Wong P, Lim B, Lodish HF (2011) MiR-191 regulates mouse erythroblast enucleation by down-regulating Riok3 and Mxi1. Genes and Development 25:119–124
104. Higgins JM, Mahadevan L (2010) Physiological and pathological population dynamics of circulating human red blood cells. Proceedings of the National Academy of Sciences of the United States of America 107:20587–20592
105. Betz T, Lenz M, Joanny JF, Sykes C (2009) ATP-dependent mechanics of red blood cells. Proceedings of the National Academy of Sciences of the United States of America 106:15320–15325

106. Sprague RS, Ellsworth ML, Stephenson AH, Lonigro AJ (2001) Participation of cAMP in a signal-transduction pathway relating erythrocyte deformation to ATP release. American Journal of Physiology – Cell Physiology 281:C1158–C1164
107. Carroll J, Raththagala M, Subasinghe W, Baguzis S, D'amico Oblak T, Root P, Spence D (2006) An altered oxidant defense system in red blood cells affects their ability to release nitric oxide-stimulating ATP. Molecular Biosystems 2:305–311
108. Carroll JS, Ku CJ, Karunarathne W, Spence DM (2007) Red blood cell stimulation of platelet nitric oxide production indicated by quantitative monitoring of the communication between cells in the bloodstream. Analytical Chemistry 79:5133–5138
109. Li J, Lykotrafitis G, Dao M, Suresh S (2007) Cytoskeletal dynamics of human erythrocyte. Proceedings of the National Academy of Sciences of the United States of America 104:4937–4942
110. Li J, Dao M, Lim CT, Suresh S (2005) Spectrin-level modeling of the cytoskeleton and optical tweezers stretching of the erythrocyte. Biophysical Journal 88:3707–3719
111. Lang PA, Kempe DS, Myssina S, Tanneur V, Birka C, Laufer S, Lang F, Wieder T, S M Huber SM (2005) PGE2 in the regulation of programmed erythrocyte death. Cell Death and Differentiation 12:415–428
112. Bruce LJ, Robinson HC, Guizouarn H, Borgese F, Harrison P, King MJ, Goede JS, Coles SE, Gore DM, Lutz HU, Ficarella R, Layton DM, Iolascon A, Ellory JC, Stewart GW (2005) Monovalent cation leaks in human red cells caused by single amino-acid substitutions in the transport domain of the band 3 chloride-bicarbonate exchanger, AE1. Nature – Genetics 37:1258–1263
113. McWhirter JL, Noguchi H, Gompper G (2009) Flow-induced clustering and alignment of vesicles and red blood cells in microcapillaries. Proceedings of the National Academy of Sciences of the United States of America 106:6039-6043
114. Delaunay J, Boivin P (1990) Le squelette du globule rouge [The skeleton of the red blood cell]. La Recherche 223:845–852
115. Michaely P, Tomchick DR, Machius M, Anderson RGW (2002) Crystal structure of a 12 ANK repeat stack from human ankyrinR. EMBO Journal 21:6387–6396
116. Salomao M, Zhang X, Yang Y, Lee S, Hartwig JH, Chasis JA, Mohandas N, An X (2008) Protein 4.1R-dependent multiprotein complex: new insights into the structural organization of the red blood cell membrane. Proceedings of the National Academy of Sciences of the United States of America 105:8026–8031
117. Park Y, Best CA, Auth T, Gov NS, Safran SA, Popescu G, Suresh S, Feld MS (2010) Metabolic remodeling of the human red blood cell membrane. Proceedings of the National Academy of Sciences of the United States of America 107:1289–1294
118. Locovei S, Bao L, Dahl G (2006) Pannexin 1 in erythrocytes: Function without a gap. Proceedings of the National Academy of Sciences of the United States of America 103:7655–7659
119. Wan J, Ristenpart WD, Stone HA (2008) Dynamics of shear-induced ATP release from red blood cells. Proceedings of the National Academy of Sciences of the United States of America 105:16432–16437
120. Karunarathne W, Ku CJ, Spence DM (2009) The dual nature of extracellular ATP as a concentration-dependent platelet $P2X_1$ agonist and antagonist. Integrative Biology 1:655–663
121. Oury C, Toth-Zsamboki E, Vermylen J, Hoylaerts MF (2002) $P2X_1$-mediated activation of extracellular signal-regulated kinase 2 contributes to platelet secretion and aggregation induced by collagen. Blood 100:2499–2505
122. Medawala W, McCahill P, Giebink A, Meyer J, Ku CJ, Spence DM (2009) A molecular level understanding of zinc activation of C-peptide and its effects on cellular communication in the bloodstream. Review of Diabetic Studies 6:148–158
123. Keller SR, Skalak R (1982) Motion of a tank-treading ellipsoidal particle in a shear-flow. Journal of Fluid Mechanics 120:27–47

124. Skotheim JM, Secomb TW (2007) Red blood cells and other nonspherical capsules in shear flow: oscillatory dynamics and the tank-treading-to-tumbling transition. Physical Review Letters 98:078301
125. Abkarian M, Faivre M, Viallat A (2007) Swinging of red blood cells under shear flow. Physical Review Letters 98:188302
126. Forsyth AM, Wan J, Owrutsky PD, Abkarian M, Stone HA (2011) Multiscale approach to link red blood cell dynamics, shear viscosity, and ATP release. Proceedings of the National Academy of Sciences of the United States of America 108:10986–10991
127. Chasan B, Geisse NA, Pedatella K, Wooster DG, Teintze M, Carattino MD, Goldmann WH, Cantiello HF (2002) Evidence for direct interaction between actin and the cystic fibrosis transmembrane conductance regulator. European Biophysics Journal 30:617–624
128. Lewis IA, Campanella ME, Markley JL, Low PS (2009) Role of band 3 in regulating metabolic flux of red blood cells. Proceedings of the National Academy of Sciences of the United States of America 106:18515–18520
129. Pittman RN (2010) Erythrocytes: surveyors as well as purveyors of oxygen? American Journal of Physiology – Heart and Circulatory Physiology 298:H1637–H1638
130. Ellis CG, Goldman D, Hanson M, Stephenson AH, Milkovich S, Benlamri A, Ellsworth ML, Sprague RS (2010) Defects in oxygen supply to skeletal muscle of prediabetic ZDF rats. American Journal of Physiology – Heart and Circulatory Physiology 298:H1661–1670
131. Chaudhuri A, Nielsen S, Elkjaer ML, Zbrzezna V, Fang F, Pogo AO (1997) Detection of Duffy antigen in the plasma membranes and caveolae of vascular endothelial and epithelial cells of nonerythroid organs. Blood 89:701–712
132. Hadley TJ, Lu ZH, Wasniowska K, Martin AW, Peiper SC, Hesselgesser J, Horuk R (1994) Postcapillary venule endothelial cells in kidney express a multispecific chemokine receptor that is structurally and functionally identical to the erythroid isoform, which is the Duffy blood group antigen. Journal of Clinical Investigation 94:985–991
133. Chapel-Fernandes S, Callebaut I, Halverson GR, Reid ME, Bailly P, Chiaroni J (2009) Dombrock genotyping in a native Congolese cohort reveals two novel alleles. Transfusion 49:1661–1671
134. Perutz MF (1964) The hemoglobin molecule. Scientific American 211:64–76
135. Hu RG, Wang H, Xia Z, Varshavsky A (2008) The N-end rule pathway is a sensor of heme. Proceedings of the National Academy of Sciences of the United States of America 105:76-81
136. Yin L, Wu N, Curtin JC, Qatanani M, Szwergold NR, Reid RA, Waitt GM, Parks DJ, Pearce KH, Wisely GB, Lazar MA (2007) Rev-erbα, a heme sensor that coordinates metabolic and circadian pathways. Science 318:1786–1789
137. Brasier G, Tikellis C, Xuereb L, Craigie J, Casley D, Kovacs CS, Fudge NJ, Kalnins R, Cooper ME, Wookey PJ (2004) Novel hexad repeats conserved in a putative transporter with restricted expression in cell types associated with growth, calcium exchange and homeostasis. Experimental Cell Research 293:31–42
138. Meyer E, Ricketts C, Morgan NV, Morris MR, Pasha S, Tee LJ, Rahman F, Bazin A, Bessires B, Dchelotte P, Yacoubi MT, Al-Adnani M, Marton T, Tannahill D, Trembath RC, Fallet-Bianco C, Cox P, Williams D, Maher ER (2010) Mutations in FLVCR2 are associated with proliferative vasculopathy and hydranencephaly-hydrocephaly syndrome (Fowler syndrome). American Journal of Human Genetics 86:471–478
139. Keel SB, Doty RT, Yang Z, Quigley JG, Chen J, Knoblaugh S, Kingsley PD, De Domenico I, Vaughn MB, Kaplan J, Palis J, Abkowitz JL (2008) A heme export protein is required for red blood cell differentiation and iron homeostasis. Science 319:825–828
140. Yang Z, Philips JD, Doty RT, Giraudi P, Ostrow JD, Tiribelli C, Smith A, Abkowitz JL (2010) Kinetics and specificity of feline leukemia virus subgroup C receptor (FLVCR) export function and its dependence on hemopexin. Journal of Biological Chemistry 285:28874–28882

141. Chen W, Paradkar PN, Li L, Pierce EL, Langer NB, Takahashi-Makise N, Hyde BB, Shirihai OS, Ward DM, Kaplan J, Paw BH (2009) Abcb10 physically interacts with mitoferrin-1 (Slc25a37) to enhance its stability and function in the erythroid mitochondria. Proceedings of the National Academy of Sciences of the United States of America 106:16263–16268
142. Chen W, Dailey HA, Paw BH (2010) Ferrochelatase forms an oligomeric complex with mitoferrin-1 and Abcb10 for erythroid heme biosynthesis. Blood 116:628–630
143. Higgs DR, Vickers MA, Wilkie AO, Pretorius IM, Jarman AP, Weatherall DJ (1989) A review of the molecular genetics of the human α-globin gene cluster. Blood 73:1081-1104
144. Snyder SH, Ferris CD (2000) Novel neurotransmitters and their neuropsychiatric relevance. American Journal of Psychiatry 157:1738–1751
145. Hon YY, Sun H, Dejam A, Gladwin MT (2010) Characterization of erythrocytic uptake and release and disposition pathways of nitrite, nitrate, methemoglobin, and iron-nitrosyl hemoglobin in the human circulation. Drug Metabolism and Disposition 38:1707–1713
146. Palmer LA, Doctor A, Gaston B (2008) SNO-hemoglobin and hypoxic vasodilation. Nature – Medicine 14:1009
147. Palmer LA, Doctor A, Chhabra P, Sheram ML, Laubach VE, Karlinsey MZ, Forbes MS, Macdonald T, Gaston B (2007) S-nitrosothiols signal hypoxia-mimetic vascular pathology. Journal of Clinical Investigation 117:2592–2601
148. Stamler JS, Singel DJ, Piantadosi CA (2008) SNO-hemoglobin and hypoxic vasodilation. Nature – Medicine 14:1008–1009
149. Patel R, Townes T (2008) SNO-hemoglobin and hypoxic vasodilation. Nature – Medicine 14:1009–1010
150. Shi H, Bencze KZ, Stemmler TL, Philpott CC (2008) A cytosolic iron chaperone that delivers iron to ferritin. Science 320:1207–1210
151. De Domenico I, McVey Ward D, Kaplan J (2008) Regulation of iron acquisition and storage: consequences for iron-linked disorders. Nature Reviews – Molecular Cell Biology 9:72-81
152. De Domenico I, McVey Ward D, Kaplan J (2009) Serum ferritin regulates blood vessel formation: A role beyond iron storage Proceedings of the National Academy of Sciences of the United States of America 106:1683–1684
153. Meynard D, Kautz L, Darnaud V, Canonne-Hergaux F, Coppin H, Roth MP (2009) Lack of the bone morphogenetic protein BMP6 induces massive iron overload. Nature – Genetics 41:478–481
154. Andriopoulos B, CorradiniE, Xia Y, Faasse SA, Chen S, Grgurevic L, Knutson MD, Pietrangelo A, Vukicevic S, Lin HY, Babitt JL (2009) BMP6 is a key endogenous regulator of hepcidin expression and iron metabolism. Nature – Genetics 41:482–487
155. Mason KD, Carpinelli MR, Fletcher JI, Collinge JE, Hilton AA, Ellis S, Kelly PN, Ekert PG, Metcalf D, Roberts AW, Huang DC, Kile BT (2007) Programmed anuclear cell death delimits platelet life span. Cell 128:1173–1186
156. Junt T, Schulze H, Chen Z, Massberg S, Goerge T, Krueger A, Wagner DD, Graf T, Italiano JE, Shivdasani RA, von Andrian UH (2007) Dynamic visualization of thrombopoiesis within bone marrow. Science 317:1767–1770
157. Landry P, Plante I, Ouellet DL, Perron MP, Rousseau G, Provost P (2009) Existence of a microRNA pathway in anucleate platelets. Nature – Structural and Molecular Biology 16:961–966
158. Cramer EM, Berger G, Berndt MC (1994) Platelet alpha-granule and plasma membrane share two new components: CD9 and PECAM-1. Blood 84:1722–1730
159. Lorenz HP, Longaker MT (2004) Wounds: biology, pathology, and management. http://recon.stanford.edu/Articles/LorenzWH.pdf
160. Borel J-P, Maquart F-X (1991) La cicatrisation [The healing]. La Recherche 236:1174–1181
161. Boilard E, Nigrovic PA, Larabee K, Watts GF, Coblyn JS, Weinblatt ME, Massarotti EM, Remold-O'Donnell E, Farndale RW, Ware J, Lee DM (2010) Platelets amplify inflammation in arthritis via collagen-dependent microparticle production. Science 327:580–583

162. Heijnen HF, Van Lier M, Waaijenborg S, Ohno-Iwashita Y, Waheed AA, Inomata M, Gorter G, Mbius W, Akkerman JW, Slot JW (2003) Concentration of rafts in platelet filopodia correlates with recruitment of c-Src and CD63 to these domains. Journal of Thrombosis and Haemostasis 1:1161–1173
163. McNicol A, Robson CA (1997) Thrombin receptor-activating peptide releases arachidonic acid from human platelets: a comparison with thrombin and trypsin. Journal of Pharmacology and Experimental Therapeutics 281:861–867
164. Michelson AD, Benoit SE, Furman MI, Barnard MR, Nurden P, Nurden AT (1996) The platelet surface expression of glycoprotein V is regulated by two independent mechanisms: proteolysis and a reversible cytoskeletal-mediated redistribution to the surface-connected canalicular system. Blood 87:1396–1408
165. Moraes LA, Swales KE, Wray JA, Damazo A, Gibbins JM, Warner TD, Bishop-Bailey D (2007) Nongenomic signaling of the retinoid X receptor through binding and inhibiting Gq in human platelets. Blood 109:3741-3744
166. O'Brien JJ, Ray DM, Spinelli SL, Blumberg N, Taubman MB, Francis CW, Wittlin SD, Phipps RP (2007) The platelet as a therapeutic target for treating vascular diseases and the role of eicosanoid and synthetic PPARγ ligands. Prostaglandins and Other Lipid Mediators 82:68-76
167. Moraes LA, Spyridon M, Kaiser WJ, Jones CI, Sage T, Atherton RE, Gibbins JM (2010) Non-genomic effects of PPARγ ligands: inhibition of GPVI-stimulated platelet activation. Journal of Thrombosis and Haemostasis 8:577–587
168. Gambaryan S, Kobsar A, Rukoyatkina N, Herterich S, Geiger J, Smolenski A, Lohmann SM, Walter U (2010) Thrombin and collagen induce a feedback inhibitory signaling pathway in platelets involving dissociation of the catalytic subunit of protein kinase A from an NFκB–IκB complex. Journal of Biological Chemistry 285:18352–18363
169. Malaver E, Romaniuk MA, D'Atri LP, Pozner RG, Negrotto S, Benzadn R, Schattner M (2009) NF-kappaB inhibitors impair platelet activation responses. Journal of Thrombosis and Haemostasis 7:1333–1343
170. Weyrich AS, Dixon DA, Pabla R, Elstad MR, McIntyre TM, Prescott SM, Zimmerman GA (1998) Signal-dependent translation of a regulatory protein, Bcl-3, in activated human platelets. Proceedings of the National Academy of Sciences of the United States of America 95:5556–5561
171. Spinelli SL, Maggirwar SB, Blumberg N, Phipps RP (2010) Nuclear emancipation: a platelet tour de force. Science Signaling 3:pe37
172. Steele BM, Harper MT, Macaulay IC, Morrell CN, Perez-Tamayo A, Foy M, Habas R, Poole AW, Fitzgerald DJ, Maguire PB (2009) Canonical Wnt signaling negatively regulates platelet function. Proceedings of the National Academy of Sciences of the United States of America 106:19836–19841
173. Min B (2008) Basophils: what they "can do" versus what they "actually do". Nature – Immunology 9:1333–1339
174. Denzel A, Maus UA, Gomez MR, Moll C, Niedermeier M, Winter C, Maus R, Hollingshead S, Briles DE, Kunz-Schughart LA, Talke Y, Mack M (2008) Basophils enhance immunological memory responses. Nature – Immunology 9:733–742
175. Lucey DR, Clerici M, Shearer GM (1996) Type 1 and type 2 cytokine dysregulation in human infectious, neoplastic, and inflammatory diseases. Clinical Microbiology Reviews 9:532–562
176. Ibelgaufts H (2010) Cytokines and Cells Online Pathfinder Encyclopaedia (www.copewithcytokines.de/cope.cgi)
177. Yousefi S, Gold JA, Andina N, Lee JJ, Kelly AM, Kozlowski E, Schmid I, Straumann A, Reichenbach J, Gleich GJ, Simon HU (2008) Catapult-like release of mitochondrial DNA by eosinophils contributes to antibacterial defense. Nature – Medicine 14:949–953
178. De Santo C, Arscott R, Booth S, Karydis I, Jones M, Asher R, Salio M, Middleton M, Cerundolo V (2010) Invariant NKT cells modulate the suppressive activity of IL-10-secreting neutrophils differentiated with serum amyloid A. Nature – Immunology 11:1039–1046

179. McDonald B, Pittman K, Menezes GB, Hirota SA, Slaba I, Waterhouse CC, Beck PL, Muruve DA, Kubes P (2010) Intravascular danger signals guide neutrophils to sites of sterile inflammation. Science 330:362–366
180. Schroder K, Zhou R, Tschopp J (2010) The NLRP3 inflammasome: a sensor for metabolic danger? Science 327:296–300
181. Syeda F, Tullis E, Slutsky AS, Zhang H (2008) Human neutrophil peptides upregulate expression of COX-2 and endothelin-1 by inducing oxidative stress. American Journal of Physiology – Heart and Circulatory Physiology 294:H2769–H2774
182. Afonso PV, Janka-Junttila M, Lee YJ, McCann CP, Oliver CM, Aamer KA, Losert W, Cicerone MT, Parent CA (2012) LTB_4 is a signal-relay molecule during neutrophil chemotaxis. Developmental Cell 22:1079–1091
183. Heit B, Robbins SM, Downey CM, Guan Z, Colarusso P, Miller BJ, Jirik FR, Kubes P (2008) PTEN functions to 'prioritize' chemotactic cues and prevent "distraction" in migrating neutrophils. Nature – Immunology 9:743–752
184. McEver RP (2010) Rolling back neutrophil adhesion. Nature – Immunology 11:282–284
185. Deban L, Russo RC, Sironi M, Moalli F, Scanziani M, Zambelli V, Cuccovillo I, Bastone A, Gobbi M, Valentino S, Doni A, Garlanda C, Danese S, Salvatori G, Sassano M, Evangelista V, Rossi B, Zenaro E, Constantin G, Laudanna C, Bottazzi B, Mantovani A (2010) Regulation of leukocyte recruitment by the long pentraxin PTX3. Nature – Immunology 11:328–334
186. Choi EY, Chavakis E, Czabanka MA, Langer HF, Fraemohs L, Economopoulou M, Kundu RK, Orlandi A, Zheng YY, Prieto DA, Ballantyne CM, Constant SL, Aird WC, Papayannopoulou T, Gahmberg CG, Udey MC, Vajkoczy P, Quertermous T, Dimmeler S, Weber C, Chavakis T (2008) Del-1, an endogenous leukocyte-endothelial adhesion inhibitor, limits inflammatory cell recruitment. Science 322:1101–1104
187. Chavakis E, Hain A, Orlandi A, Carmona G, Quertermous T, Chavakis T, Dimmeler S (2008) Del-1, a new β_2-integrin ligand, which inhibits adhesion and homing of progenitor cells. Circulation 118:S502
188. Koyama S, Sato E, Masubuchi T, Takamizawa A, Kubo K, Nagai S, Izumi T (1998) Alveolar type II-like cells release G-CSF as neutrophil chemotactic activity. American Journal of Physiology – Lung Cellular and Molecular Physiology 275:L687–L693
189. Weathington NM, van Houwelingen AH, Noerager BD, Jackson PL, Kraneveld AD, Galin FS, Folkerts G, Nijkamp FP, Blalock JE (2006) A novel peptide CXCR ligand derived from extracellular matrix degradation during airway inflammation. Nature – Medicine 12:317–323
190. Moon TC, St Laurent CD, Morris KE, Marcet C, Yoshimura T, Sekar Y, Befus AD (2010) Advances in mast cell biology: new understanding of heterogeneity and function. Mucosal Immunology 3:111–128
191. Galli SJ, Grimbaldeston M, Tsai M (2008) Immunomodulatory mast cells: negative, as well as positive, regulators of immunity. Nature Reviews – Immunology 8:478–486
192. Taghon T, Yui MA, Rothenberg EV (2007) Mast cell lineage diversion of T lineage precursors by the essential T cell transcription factor GATA-3. Nature – Immunology 8:845–855
193. Metz M, Maurer M (2007) Mast cells – key effector cells in immune responses. Trends in Immunology 28:234–241
194. Henz BM (2003) What is the physiological function of mast cells? Viewpoint 6. Experimental Dermatology 12:900–902
195. Vliagoftis H, Forsythe P, Befus D (2003) What is the physiological function of mast cells? Commentary 10. Experimental Dermatology 12:906–907
196. Metz M, Siebenhaar F, Maurer M (2008) Mast cell functions in the innate skin immune system. Immunobiology 213:251–260
197. Theoharides TC (2003) What is the physiological function of mast cells? Viewpoint 2. Experimental Dermatology 12:891–894

198. Kovanen PT (2003) What is the physiological function of mast cells? Commentary 7. Experimental Dermatology 12:902–903
199. Mackins CJ, Kano S, Seyedi N, Schfer U, Reid AC, Machida T, Silver RB, Levi R (2006) Cardiac mast cell-derived renin promotes local angiotensin formation, norepinephrine release, and arrhythmias in ischemia/reperfusion. Journal of Clinical Investigation 116:1063–1070
200. Levick SP, Meléndez GC, Plante E, McLarty JL, Brower GL, Janicki JS (2011) Cardiac mast cells: the centrepiece in adverse myocardial remodelling. Cardiovascular Research 89:12-9
201. Veerappan A, Reid AC, Estephan R, O'Connor N, Thadani-Mulero M, Salazar-Rodriguez M, Levi R, Silver RB (2008) Mast cell renin and a local renin-angiotensin system in the airway: role in bronchoconstriction. Proceedings of the National Academy of Sciences of the United States of America 105:1315–1320
202. Lo WL, Donermeyer DL, Allen PM (2012) A voltage-gated sodium channel is essential for the positive selection of CD4(+) T cells. Nature – Immunology 13:880–887
203. Alarcón B, van Santen HM (2010) Two receptors, two kinases, and T cell lineage determination. Science Signaling 3:pe11
204. Saini M, Sinclair C, Marshall D, Tolaini M, Sakaguchi S, Seddon B (2010) Regulation of Zap70 expression during thymocyte development enables temporal separation of CD4 and CD8 repertoire selection at different signaling thresholds. Science Signaling 3:ra23
205. Iwasaki A, Medzhitov R (2010) Regulation of adaptive immunity by the innate immune system. Science 327:291–295
206. Bonneville M, O'Brien RL, Born WK (2010) $\gamma\delta$ T Cell effector functions: a blend of innate programming and acquired plasticity. Nature Reviews – Immunology 10:467–478
207. Bousso P, Albert ML (2010) Signal 0 for guided priming of CTLs: NKT cells do it too. Nature – Immunology 11:284–286
208. Bogeski I, Kummerow C, Al-Ansary D, Schwarz EC, Koehler R, Kozai D, Takahashi N, Peinelt C, Griesemer D, Bozem M, Mori Y, Hoth M, Niemeyer BA (2010) Differential redox regulation of ORAI ion channels: a mechanism to tune cellular calcium signaling. Science Signaling 3:ra24
209. Sauer K, Cooke MP (2010) Regulation of immune cell development through soluble inositol-1,3,4,5-tetrakisphosphate. Nature Reviews – Immunology 10:257–271
210. Ugolini S, Vivier E (2009) Immunology: natural killer cells remember. Nature 457:544–545
211. Sun JC, Beilke JN, Lanier LL (2009) Adaptive immune features of natural killer cells. Nature 457:557–561
212. Shinoda K, Tokoyoda K, Hanazawa A, Hayashizaki K, Zehentmeier S, Hosokawa H, Iwamura C, Koseki H, Tumes DJ, Radbruch A, Nakayama T (2012) Type II membrane protein CD69 regulates the formation of resting T-helper memory. Proceedings of the National Academy of Sciences of the United States of America 109:7409–7414
213. Schoenberger SP (2012) CD69 guides CD4+ T cells to the seat of memory. Proceedings of the National Academy of Sciences of the United States of America 109:8358–8359
214. Paust S, Gill HS, Wang BZ, Flynn MP, Moseman EA, Senman B, Szczepanik M, Telenti A, Askenase PW, Compans RW, von Andrian UH (2010) Critical role for the chemokine receptor CXCR6 in NK cell-mediated antigen-specific memory of haptens and viruses. Nature – Immunology 11:1127–1135
215. Di Santo JP (2010) A guardian of T cell fate. Science 329:44–45
216. Li P, Burke S, Wang J, Chen X, Ortiz M, Lee SC, Lu D, Campos L, Goulding D, Ng BL, Dougan G, Huntly B, Gottgens B, Jenkins NA, Copeland NG, Colucci F, Liu P (2010) Reprogramming of T cells to natural killer-like cells upon Bcl11b deletion. Science 329:85–89
217. Li L, Leid M, Rothenberg EV (2010) An early T cell lineage commitment checkpoint dependent on the transcription factor Bcl11b. Science 329:89–93
218. Ikawa T, Hirose S, Masuda K, Kakugawa K, Satoh R, Shibano-Satoh A, Kominami R, Katsura Y, Kawamoto H (2010) An essential developmental checkpoint for production of the T cell lineage. Science 329:93–96

219. Bi S, Earl LA, Jacobs L, Baum LG (2008) Structural features of galectin-9 and galectin-1 that determine distinct T cell death pathways. Journal of Biological Chemistry
220. Turley SJ, Fletcher AL, Elpek KG (2010) The stromal and haematopoietic antigen-presenting cells that reside in secondary lymphoid organs. Nature Reviews – Immunology 10:813–825
221. Huse M, Quann EJ, Davis MM (2008) Shouts, whispers and the kiss of death: directional secretion in T cells. Nature – Immunology 9:1105–1111
222. Guo F, Hildeman D, Tripathi P, Velu CS, Grimes HL, Zheng Y (2010) Coordination of IL-7 receptor and T-cell receptor signaling by cell-division cycle 42 in T-cell homeostasis. Proceedings of the National Academy of Sciences of the United States of America 107:18505–18510
223. Ciofani M, Zúñiga-Pflücker JC (2010) Determining $\gamma\delta$ versus $\alpha\beta$ T cell development. Nature Reviews – Immunology 10:657–663
224. Johnson BJ, Costelloe EO, Fitzpatrick DR, Haanen JB, Schumacher TN, Brown LE, Kelso A (2003) Single-cell perforin and granzyme expression reveals the anatomical localization of effector CD8+ T cells in influenza virus-infected mice. Proceedings of the National Academy of Sciences of the United States of America 100:2657–2662
225. Dong C, Martinez GJ (2010) T cells: the usual subsets (poster). Nature Reviews – Immunology (http://www.nature.com/reviews/posters/Tcellsubsets)
226. Bettelli E, Korn T, Oukka M, Kuchroo VK (2008) Induction and effector functions of TH17 cells. Nature 453:1051–1057
227. Muroi S, Naoe Y, Miyamoto C, Akiyama K, Ikawa T, Masuda K, Kawamoto H, Taniuchi I (2008) Cascading suppression of transcriptional silencers by ThPOK seals helper T cell fate. Nature – Immunology 9:1113–1121
228. Wang L, Wildt KF, Zhu J, Zhang X, Feigenbaum L, Tessarollo L, Paul WE, Fowlkes BJ, Bosselut R (2008) Distinct functions for the transcription factors GATA-3 and ThPOK during intrathymic differentiation of CD4+ T cells. Nature – Immunology 9:1122–1130
229. Egawa T, Littman DR (2008) ThPOK acts late in specification of the helper T cell lineage and suppresses Runx-mediated commitment to the cytotoxic T cell lineage. Nature – Immunology 9:1131–1139
230. Bonneville M (2008) $\gamma\delta$ T Cell selection: is anyone useless? Immunity 29:3–5
231. Jensen KDC, Su X, Shin S, Li L, Youssef S, Yamasaki S, Steinman L, Saito T, Locksley RM, Davis MM, Baumgarth N, Chien Y (2008) Thymic selection determines $\gamma\delta$ T cell effector fate: antigen-naive cells make interleukin-17 and antigen-experienced cells make interferon γ. Immunity 29:90–100
232. Godfrey DI, Rossjohn J, McCluskey J (2010) Fighting infection with your MAITs. Nature – Immunology 11:693–695
233. Le Bourhis L, Martin E, Péguillet I, Guihot A, Froux N, Coré M, Lévy E, Dusseaux M, Meyssonnier V, Premel V, Ngo C, Riteau B, Duban L, Robert D, Huang S, Rottman M, Soudais C, Lantz O (2010) Antimicrobial activity of mucosal-associated invariant T cells. Nature – Immunology 11:701–708
234. Harrington LE, Janowski KM, Oliver JR, Zajac AJ, Weaver CT (2008) Memory CD4 T cells emerge from effector T-cell progenitors. Nature 452:356–360
235. Woodland DI, Kohlmeier JE (2009) Migration, maintenance and recall of memory T cells in peripheral tissues. Nature Reviews – Immunology 9:153–161
236. Araki K, Turner AP, Shaffer VO, Gangappa S, Keller SA, Bachmann MF, Larsen CP, Ahmed R (2009) mTOR regulates memory CD8 T-cell differentiation. Nature 460:108-112
237. Mold JE, Venkatasubrahmanyam S, Burt TD, Michaëlsson J, Rivera JM, Galkina SA, Weinberg K, Stoddart CA, McCune JM (2010) Fetal and adult hematopoietic stem cells give rise to distinct T cell4 lineages in humans. Science 330:1695–1699
238. Rubtsov YP, Niec RE, Josefowicz S, Li L, Darce J, Mathis D, Benoist C, Rudensky AY (2010) Stability of the regulatory T cell lineage in vivo. Science 329:1667–1671
239. Zheng Y, Chaudhry A, Kas A, deRoos P, Kim JM, Chu TT, Corcoran L, Treuting P, Klein U, Rudensky AY (2009) Regulatory T-cell suppressor program co-opts transcription factor IRF4 to control T_H2 responses. Nature 458:351–356

240. Feuerer M, Hill JA, Kretschmer K, von Boehmer H, Mathis D, Benoist C (2010) Genomic definition of multiple ex vivo regulatory T cell subphenotypes. Proceedings of the National Academy of Sciences of the United States of America 107:5919–5924
241. Feuerer M, Hill JA, Mathis D, Benoist C (2009) $Foxp3^+$ regulatory T cells: differentiation, specification, subphenotypes. Nature – Immunology 10:689–695
242. Zanin-Zhorov A, Ding Y, Kumari S, Attur M, Hippen KL, Brown M, Blazar BR, Abramson SB, Lafaille JJ, Dustin ML (2010) Protein kinase C-θ mediates negative feedback on regulatory T cell function. Science 328:372–376
243. Vignali DAA, Collison LW, Workman CJ (2008) How regulatory T cells work. Nature Reviews – Immunology 8:523–532
244. Sakaguchi S, Miyara M, Costantino CM, Hafler DA (2010) FOXP3+ regulatory T cells in the human immune system. Nature Reviews – Immunology 10:490–500
245. Shevach E, Davidson T (2010) Regulatory T cells (poster nri1001_treg_poster.pdf), Ryan G (copyed.), Bradbrook S (design). Nature Reviews – Immunology (Macmillan Publishers Ltd. http://www.nature.com/nri/posters/tregcells)
246. Zou W (2006) Regulatory T cells, tumour immunity and immunotherapy. Nature Reviews – Immunology 6:295–307
247. Miyara M, Yoshioka Y, Kitoh A, Shima T, Wing K, Niwa A, Parizot C, Taflin C, Heike T, Valeyre D, Mathian A, Nakahata T, Yamaguchi T, Nomura T, Ono M, Amoura Z, Gorochov G, Sakaguchi S (2009) Functional delineation and differentiation dynamics of human CD4+ T cells expressing the FoxP3 transcription factor. Immunity 30:899–911
248. Liu B, Tahk S, Yee KM, Fan G, Shuai K (2010) The ligase PIAS1 restricts natural regulatory T cell differentiation by epigenetic repression. Science 330:521–525
249. Liu B, Yang Y, Chernishof V, Loo RR, Jang H, Tahk S, Yang R, Mink S, Shultz D, Bellone CJ, Loo JA, Shuai K (2007) Proinflammatory stimuli induce IKKalpha-mediated phosphorylation of PIAS1 to restrict inflammation and immunity. Cell 129:903–914
250. Lund JM, Hsing L, Pham TT, Rudensky AY (2008) Coordination of early protective immunity to viral infection by regulatory T cells. Science 320:1220–1224
251. O'Shea JJ, Paul WE (2010) Mechanisms underlying lineage commitment and plasticity of helper $CD4^+$ T cells. Science 327:1098–1102
252. Paul WE, Zhu J (2010) How are TH2-type immune responses initiated and amplified? Nature Reviews – Immunology 10:225–235
253. Larch M, Akdis A, Valenta R (2006) Immunological mechanisms of allergen-specific immunotherapy. Nature Reviews – Immunology 6:761–771
254. Neill DR, McKenzie ANJ (2010) TH9: the latest addition to the expanding repertoire of IL-25 targets. Immunology and Cell Biology 88:502–504
255. Ghoreschi K, Laurence A, Yang XP, Tato CM, McGeachy MJ, KonkelJE, Ramos HL, Wei L, Davidson TS, Bouladoux N, Grainger JR, Chen Q, Kanno Y, Watford WT, Sun HW, Eberl G, Shevach EM, Belkaid Y, Cua DJ, Chen WJ, O'Shea JJ (2010) Generation of pathogenic TH17 cells in the absence of TGF-β signalling. Nature 467:967–971
256. Schraml BU, Hildner K, Ise W, Lee WL, Smith WA, Solomon B, Sahota G, Sim J, Mukasa R, Cemerski S, Hatton RD, Stormo GD, Weaver CT, Russell JH, Murphy TL, Murphy KM (2009) The AP-1 transcription factor Batf controls T_H17 differentiation. Nature 460:405–409
257. Zou W, Restifo NP (2010) TH17 cells in tumour immunity and immunotherapy. Nature Reviews – Immunology 10:248–256
258. Duhen T, Geiger R, Jarrossay D, Lanzavecchia A, Sallusto F (2009) Production of interleukin 22 but not interleukin 17 by a subset of human skin-homing memory T cells. Nature – Immunology 10:857–863
259. Trifari S, Kaplan CD, Tran EH, Crellin NK, Spits H (2009) Identification of a human helper T cell population that has abundant production of interleukin 22 and is distinct from TH-17, TH1 and TH2 cells. Nature – Immunology 10:864–871

260. Demotte N, Stroobant V, Courtoy P, van der Smissen P, Colau D, Luescher I, Hivroz C, Nicaise J, Squifflet J, Mourad M (2008) Restoring the association of the T cell receptor with CD8 reverses anergy in human tumor-infiltrating lymphocytes. Immunity 28:414–424
261. Goetz JG, Joshi B, Lajoie P, Strugnell SS, Scudamore T, Kojic LD, Nabi IR (2008) Concerted regulation of focal adhesion dynamics by galectin-3 and tyrosine-phosphorylated caveolin-1. Journal of Cell Biology 180:1261–1275
262. Buzza MS, Bird PI (2006) Extracellular granzymes: current perspectives. Biological Chemistry 387:827–837
263. Semmling V, Lukacs-Kornek V, Thaiss CA, Quast T, Hochheiser K, Panzer U, Rossjohn J, Perlmutter P, Cao J, Godfrey DI, Savage PB, Knolle PA, Kolanus W, Frster I, Kurts C (2010) Alternative cross-priming through CCL17-CCR4-mediated attraction of CTLs toward NKT cell-licensed DCs. Nature – Immunology 11:313–320
264. Nieda M, Okai M, Tazbirkova A, Lin H, Yamaura A, Ide K, Abraham R, Juji T, Macfarlane DJ, Nicol AJ (2004) Therapeutic activation of $V_{\alpha 24}+V_{\beta 11}+$ NKT cells in human subjects results in highly coordinated secondary activation of acquired and innate immunity. Blood 103:383–389
265. Zajonc DM, Wilson IA (2007) Architecture of CD1 proteins. Current Topics in Microbiology and Immunology 314:27–50
266. Sköld M, Behar SM (2005) The role of group 1 and group 2 CD1-restricted T cells in microbial immunity. Microbes and Infection 7:544–551
267. Song L, Asgharzadeh S, Salo J, Engell K, Wu HW, Sposto R, Ara T, Silverman AM, DeClerck YA, Seeger RC, Metelitsa LS (2009) $V_{\alpha 24}$-invariant NKT cells mediate antitumor activity via killing of tumor-associated macrophages. Journal of Clinical Investigation 119:1524–1536
268. Kovalovsky D, Uche OU, Eladad S, Hobbs RM, Yi W, Alonzo E, Chua K, Eidson M, Kim HJ, Im JS, Pandolfi PP, Sant'Angelo DB (2008) The BTB-zinc finger transcriptional regulator PLZF controls the development of invariant natural killer T cell effector functions. Nature – Immunology 9:1055–1064
269. Hu T, Simmons A, Yuan J, Bender TP, Alberola J (2010) The transcription factor c-Myb primes $CD4^+CD8^+$ immature thymocytes for selection into the iNKT lineage. Nature – Immunology 11:435–441
270. Kamijuku H, Nagata Y, Jiang X, Ichinohe T, Tashiro T, Mori K, Taniguchi M, Hase K, Ohno H, Shimaoka T, Yonehara S, Odagiri T, Tashiro M, Sata T, Hasegawa H, Seino KI (2008) Mechanism of NKT cell activation by intranasal coadministration of alpha-galactosylceramide, which can induce cross-protection against influenza viruses. Mucosal Immunology 1:208–218
271. Cerutti A (2008) The regulation of IgA class switching. Nature Reviews – Immunology 8:421–434
272. Spits H, Di Santo JP (2011) The expanding family of innate lymphoid cells: regulators and effectors of immunity and tissue remodeling. Nature – Immunology 12:21–27
273. Veldhoen M, Withers DR (2010) Innate lymphoid cell relations. Science 330:594–595
274. Sawa S, Cherrier M, Lochner M, Satoh-Takayama N, Fehling HJ, Langa F, Di Santo JP, Eberl G (2010) Lineage relationship analysis of $ROR\gamma t^+$ innate lymphoid cells. Science 330:665–669
275. Neill DR, Wong SH, Bellosi A, Flynn RJ, Daly M, Langford TK, Bucks C, Kane CM, Fallon PG, Pannell R, Jolin HE, McKenzie ANJ (2010) Nuocytes represent a new innate effector leukocyte that mediates type-2 immunity. Nature 464:1367–1370
276. Vivier E, Raulet DH, Moretta A, Caligiuri MA, Zitvogel L, Lanier LL, Yokoyama WM, Ugolini S (2011) Innate or adaptive immunity? The example of natural killer cells. Science 331:44–49
277. Shi FD, van Kaer L (2006) Reciprocal regulation between natural killer cells and autoreactive T cells. Nature Reviews – Immunology 6:751–760
278. Lanier LL (2008) Up on the tightrope: natural killer cell activation and inhibition. Nature – Immunology 9:495–502

279. Raulet DH, Guerra N (2009) Oncogenic stress sensed by the immune system: role of natural killer cell receptors. Nature Reviews – Immunology 9:568–580
280. Cruz-Munoz ME, Veillette A (2010) Do NK cells always need a license to kill? Nature – Immunology 11:279–280
281. Orr MT, Murphy WJ, Lanier LL (2010) "Unlicensed" natural killer cells dominate the response to cytomegalovirus infection. Nature – Immunology 11:321–327
282. Vivier E, Tomasello E, Baratin M, Walzer T, Ugolini S (2008) Functions of natural killer cells. Nature – Immunology 9:503–510
283. Rachitskaya AV, Hansen AM, Horai R, Li Z, Villasmil R, Luger D, Nussenblatt RB, Caspi RR (2008) Cutting edge: NKT cells constitutively express IL-23 receptor and RORγt and rapidly produce IL-17 upon receptor ligation in an IL-6-independent fashion. Journal of Immunology 180:5167–5171
284. Martin F, Kearney JF (2000) B-cell subsets and the mature preimmune repertoire. Marginal zone and B1 B cells as part of a "natural immune memory". Immunological Reviews 175:70–79
285. Tarlinton DM, McLean M, Nossal GJV (1995) B1 and B2 cells differ in their potential to switch immunoglobulin isotype. European Journal of Immunology 25:3388–3393
286. Khalil AM, Cambier JC, Shlomchik MJ (2012) B cell receptor signal transduction in the GC is short-circuited by high phosphatase activity. Science 336:1178–1181
287. Ehrenstein MR, Notley CA (2010) The importance of natural IgM: scavenger, protector and regulator. Nature Reviews – Immunology 10:778–786
288. Pierce SK, Liu W (2010) The tipping points in the initiation of B cell signalling: how small changes make big differences. Nature Reviews – Immunology 10:767–777
289. Vinuesa CG, Tangye SG, Moser B, Mackay CR (2005) Follicular B helper T cells in antibody responses and autoimmunity. Nature Reviews – Immunology 5:853–865
290. Lefebvre C, Rajbhandari P, Alvarez MJ, Bandaru P, Lim WK, Sato M, Wang K, Sumazin P, Kustagi M, Bisikirska BC, Basso K, Beltrao P, Krogan N, Gautier J, Dalla-Favera R, Califano A (2010) A human B-cell interactome identifies MYB and FOXM1 as master regulators of proliferation in germinal centers. Molecular Systems Biology 6:377
291. Lund FE, Randall TD (2010) Effector and regulatory B cells: modulators of $CD4^+$ T cell immunity. Nature Reviews – Immunology 10:236–247
292. Radbruch A, Muehlinghaus G, Luger EO, Inamine A, Smith KGC, Drner T, Hiepe F (2006) Competence and competition: the challenge of becoming a long-lived plasma cell. Nature Reviews – Immunology 6:741–750
293. Doulatov S, Notta F, Eppert K, Nguyen LT, Ohashi PS, Dick JE (2010) Revised map of the human progenitor hierarchy shows the origin of macrophages and dendritic cells in early lymphoid development. Nature – Immunology 11:585–593
294. Saha P, Geissmann F (2011) Multifunctional monocytes. Toward a functional characterization of blood monocytes. Immunology and Cell Biology 89:2–4
295. Slaney CY, Toker A, La Flamme A, Bäckström BT, Harper JL (2011) Naïve blood monocytes suppress T-cell function. A possible mechanism for protection from autoimmunity. Immunology and Cell Biology 89:7–13
296. Geissmann F, Manz MG, Jung S, Sieweke MH, Merad M, Ley K (2010) Development of monocytes, macrophages, and dendritic cells. Science 327:656–661
297. Otero K, Turnbull IR, Poliani PL, Vermi W, Cerutti E, Aoshi T, Tassi I, Takai T, Stanley SL, Miller M, Shaw AS, Colonna M (2009) Macrophage colony-stimulating factor induces the proliferation and survival of macrophages via a pathway involving DAP12 and β-catenin. Nature Immunology 10:734–743
298. Tickenbrock L, Schwble J, Strey A, Sargin B, Hehn S, Baas M, Choudhary C, Gerke V, Berdel WE, Mller-Tidow C, Serve H (2006) Wnt signaling regulates transendothelial migration of monocytes. Journal of Leukocyte Biology 79:1306–1313
299. Mosser DM, Edwards JP (2008) Exploring the full spectrum of macrophage activation. Nature Reviews – Immunology 8:958–969

300. Gordon S, Taylor PR (2005) Monocyte and macrophage heterogeneity. Nature Reviews – Immunology 5:953–964
301. Auffray C, Fogg D, Garfa M, Elain G, Join-Lambert O, Kayal S, Sarnacki S, Cumano A, Lauvau G, Geissmann F (2007) Monitoring of blood vessels and tissues by a population of monocytes with patrolling behavior. Science 317:666–670
302. Liu H, Shi B, Huang CC, Eksarko P, Pope RM (2008) Transcriptional diversity during monocyte to macrophage differentiation. Immunology Letters 117:70–80
303. Montaner LJ, da Silva RP, Sun J, Sutterwala S, Hollinshead M, Vaux D, Gordon S (1999) Type 1 and type 2 cytokine regulation of macrophage endocytosis: differential activation by IL-4/IL-13 as opposed to IFN-γ or IL-10. Journal of Immunology 162:4606–4613
304. Pollard JW (2009) Trophic macrophages in development and disease. Nature Reviews – Immunology 9:259–270
305. Kohyama M, Ise W, Edelson BT, Wilker PR, Hildner K, Mejia C, Frazier WA, Murphy TL, Murphy KM (2009) Role for Spi-C in the development of red pulp macrophages and splenic iron homeostasis. Nature 457:318–321
306. Snelgrove RJ, Goulding J, Didierlaurent AM, Lyonga D, Vekaria S, Edwards L, Gwyer E, Sedgwick JD, Barclay AN, Hussell T (2008) A critical function for CD200 in lung immune homeostasis and the severity of influenza infection. Nature – Immunology 9:1074–1083
307. Strauch UG, Mueller RC, Li XY, Cernadas M, Higgins JM, Binion DG, Parker CM (2001) Integrin α_E (CD103) β_7 mediates adhesion to intestinal microvascular endothelial cell lines via an E-cadherin-independent interaction. Journal of Immunology 166:3506–3514
308. Norris PC, Dennis EA (2012) Omega-3 fatty acids cause dramatic changes in TLR4 and purinergic eicosanoid signaling. Proceedings of the National Academy of Sciences of the United States of America 109:8517–8522
309. Karin M, Greten FR (2005) NF-κB: linking inflammation and immunity to cancer development and progression. Nature Reviews – Immunology 5:749–759
310. Takamatsu H, Takegahara N, Nakagawa Y, Tomura M, Taniguchi M, Friedel RH, Rayburn H, Tessier-Lavigne M, Yoshida Y, Okuno T, Mizui M, Kang S, Nojima S, Tsujimura T, Nakatsuji Y, Katayama I, Toyofuku T, Kikutani H, Kumanogoh A (2010) Semaphorins guide the entry of dendritic cells into the lymphatics by activating myosin II. Nature – Immunology 11:594–600
311. Geissmann F (2007) The origin of dendritic cells. Nature – Immunology 8:558–560
312. Asagiri M, Hirai T, Kunigami T, Kamano S, Gober HJ, Okamoto K, Nishikawa K, Latz E, Golenbock DT, Aoki K, Ohya K, Imai Y, Morishita Y, Miyazono K, Kato S, Saftig P, Takayanagi H (2008) Cathepsin K-dependent toll-like receptor 9 signaling revealed in experimental arthritis. Science 319:624–627
313. Wing K, Onishi Y, Prieto-Martin P, Yamaguchi T, Miyara M, Fehervari Z, Nomura T, Sakaguchi S (2008) CTLA-4 control over Foxp3+ regulatory T cell function. Science 322:271–275
314. Sasai M, Linehan MM, Iwasaki A (2010) Bifurcation of Toll-like receptor 9 signaling by adaptor protein 3. Science 329:1530–1534
315. Proietto AI, van Dommelen S, Zhou P, Rizzitelli A, D'Amico A, Steptoe RJ, Naik SH, Lahoud MH, Liu Y, Zheng P, Shortman K, Wu L (2008) Dendritic cells in the thymus contribute to T-regulatory cell induction. Proceedings of the National Academy of Sciences of the United States of America 105:19869–19874
316. Gabrilovich DI, Nagaraj S (2009) Myeloid-derived suppressor cells as regulators of the immune system. Nature Reviews – Immunology 9:162–174
317. Rose S, Misharin A, Perlman H (2012) A novel Ly6C/Ly6G-based strategy to analyze the mouse splenic myeloid compartment. Cytometry Part A 81(4):343–350

Chap. 4. Lymph Drainage

318. Edwards JR, Williams K, Kindblom LG, Meis-Kindblom JM, Hogendoorn PC, Hughes D, Forsyth RG, Jackson D, Athanasou NA (2008) Lymphatics and bone. Human Pathology 39:49–55
319. Cupedo T, Mebius RE (2005) Cellular interactions in lymph node development. Journal of Immunology 174:21–25
320. Venugopal AM, Quick CM, Laine GA, Stewart RH (2009) Optimal postnodal lymphatic network structure that maximizes active propulsion of lymph. American Journal of Physiology – Heart and Circulatory Physiology 296:H303–H309
321. Davis MJ, Davis AM, Ku CW, Gashev AA (2009) Myogenic constriction and dilation of isolated lymphatic vessels. American Journal of Physiology – Heart and Circulatory Physiology 296:H293–H302
322. Drayton DL, Liao S, Mounzer RH, Ruddle NH (2006) Lymphoid organ development: from ontogeny to neogenesis. Nature Immunology 7:344–353
323. Scandella E, Bolinger B, Lattmann E, Miller S, Favre S, Littman DR, Finke D, Luther SA, Junt T, Ludewig B (2008) Restoration of lymphoid organ integrity through the interaction of lymphoid tissue-inducer cells with stroma of the T cell zone. Nature Immunology 9:667–675
324. Foo SY, Phipps S (2010) Regulation of inducible BALT formation and contribution to immunity and pathology. Mucosal Immunology 3:537–544
325. Azeredo da Silveira S, Kikuchi S, Fossati-Jimack L, Moll T, Saito T, Verbeek JS, Botto M, Walport MJ, Carroll M, Izui S (2002) Complement activation selectively potentiates the pathogenicity of the IgG2b and IgG3 isotypes of a high affinity anti-erythrocyte autoantibody. Journal of Experimental Medicine 195:665–672
326. Csencsits KL, Jutila MA, Pascual DW (1999) Nasal-associated lymphoid tissue: phenotypic and functional evidence for the primary role of peripheral node addressin in naive lymphocyte adhesion to high endothelial venules in a mucosal site. Journal of Immunology 163:1382–1389
327. Cherrier M, Sawa S, Eberl G (2012) Notch, Id2, and RORγt sequentially orchestrate the fetal development of lymphoid tissue inducer cells. Journal of Experimental Medicine 209:729–740

Chap. 5. Cardiomyocytes

328. ter Keurs HEDJ (2012) The interaction of Ca^{2+} with sarcomeric proteins: role in function and dysfunction of the heart. American Journal of Physiology – Heart and Circulatory Physiology 302:H38–H50
329. Young AA, Legrice IJ, Young MA, Smaill BH (1998) Extended confocal microscopy of myocardial laminae and collagen network. Journal of Microscopy 192:139–150
330. Algranati D, Kassab GS, Lanir Y (2010) Mechanisms of myocardium-coronary vessel interaction. American Journal of Physiology – Heart and Circulatory Physiology 298:H861–H873
331. Robinson TF, Factor SM, Sonnenblick EH (1986) The heart as a suction pump. Scientific American 6:62-69
332. Borg TK, Caulfield JB (1981) The collagen matrix of the heart. Federation Proceedings 40:2037–2041
333. Severs NJ (2000) The cardiac muscle cell. Bioessays 22:188–199
334. Palatinus JA, O'Quinn MP, Barker RJ, Harris BS, Jourdan J, Gourdie RG (2011) ZO-1 determines adherens and gap junction localization at intercalated disks. American Journal of Physiology – Heart and Circulatory Physiology 300:H583–H594

335. Delmar M (2012) Connexin43 regulates sodium current; ankyrin-G modulates gap junctions: the intercalated disc exchanger. Cardiovascular Research 93:220–222
336. d'Alch P (1973) Pourquoi le cœur bat [Why the heart beats]. La Recherche 33:327–336
337. Heinzel FR, Bito V, Volders PG, Antoons G, Mubagwa K, Sipido KR (2002) Spatial and temporal inhomogeneities during Ca^{2+} release from the sarcoplasmic reticulum in pig ventricular myocytes. Circulation Research 91:1023–1030
338. Al-Qusairi L, Weiss N, Toussaint A, Berbey C, Messaddeq N, Kretz C, Sanoudou D, Beggs AH, Allard B, Mandel JL, Laporte J, Jacquemond V, Buj-Bello A (2009) T-tubule disorganization and defective excitation-contraction coupling in muscle fibers lacking myotubularin lipid phosphatase. Proceedings of the National Academy of Sciences of the United States of America 106:18763–18768
339. Maguy A, Hebert TE, Nattel S (2006) Involvement of lipid rafts and caveolae in cardiac ion channel function. Cardiovascular Research 69:798–807
340. Fujita T, Toya Y, Iwatsubo K, Onda T, Kimura K, Umemura S, Ishikawa Y (2001) Accumulation of molecules involved in α1-adrenergic signal within caveolae: caveolin expression and the development of cardiac hypertrophy. Cardiovascular Research 51:709–716
341. Chopra N, Yang T, Asghari P, Moore ED, Huke S, Akin B, Cattolica RA, Perez CF, Hlaing T, Knollmann-Ritschel BE, Jones LR, Pessah IN, Allen PD, Franzini-Armstrong C, Knollmann BC (2009) Ablation of triadin causes loss of cardiac Ca^{2+} release units, impaired excitation-contraction coupling, and cardiac arrhythmias. Proceedings of the National Academy of Sciences of the United States of America 106:7636–7641
342. Ramay HR, Liu OZ, Sobie EA (2011) Recovery of cardiac calcium release is controlled by sarcoplasmic reticulum refilling and ryanodine receptor sensitivity. Cardiovascular Research 91:598–605
343. Palmer JW, Tandler B, Hoppel CL (1977) Biochemical properties of subsarcolemmal and interfibrillar mitochondria isolated from rat cardiac muscle. Journal of Biological Chemistry 252:8731–8739
344. Schgger H (2002) Respiratory chain supercomplexes of mitochondria and bacteria. Biochimica et Biophysica Acta 1555:154–159
345. Finck BN, Kelly DP (2006) PGC-1 coactivators: inducible regulators of energy metabolism in health and disease. Journal of Clinical Investigation 116:615–622
346. Arany Z, He H, Lin J, Hoyer K, Handschin C, Toka O, Ahmad F, Matsui T, Chin S, Wu PH, Rybkin II, Shelton JM, Manieri M, Cinti S, Schoen FJ, Bassel-Duby R, Rosenzweig A, Ingwall JS, Spiegelman BM (2005) Transcriptional coactivator PGC-1 alpha controls the energy state and contractile function of cardiac muscle. Cell Metabolism 1:259–271
347. Miura T, Tanno M (2012) The mPTP and its regulatory proteins: final common targets of signalling pathways for protection against necrosis. Cardiovascular Research 94:181–189
348. Rottlaender D, Boengler K, Wolny M, Schwaiger A, Motloch LJ, Ovize M, Schulz R, Heusch G, Hoppe UC (2012) Glycogen synthase kinase 3β transfers cytoprotective signaling through connexin 43 onto mitochondrial ATP-sensitive K^+ channels. Proceedings of the National Academy of Sciences of the United States of America 109:E242–E251
349. Ishikawa S, Kuno A, Tanno M, Miki T, Kouzu H, Itoh T, Sato T, Sunaga D, Murase H, Miura T (2012) Role of connexin-43 in protective PI3K-Akt-GSK-3β signaling in cardiomyocytes. American Journal of Physiology – Heart and Circulatory Physiology 302:H2536–H2544
350. Hendgen-Cotta UB, Merx MW, Shiva S, Schmitz J, Becher S, Klare JP, Steinhoff HJ, Goedecke A, Schrader J, Gladwin MT, Kelm M, Rassaf T (2008) Nitrite reductase activity of myoglobin regulates respiration and cellular viability in myocardial ischemia-reperfusion injury. Proceedings of the National Academy of Sciences of the United States of America 105:10256–10261
351. Kaasik A, Kuum M, Joubert F, Wilding J, Ventura-Clapier R, Veksler V (2010) Mitochondria as a source of mechanical signals in cardiomyocytes. Cardiovascular Research 87:83–91
352. Gucek M, Murphy E (2010) What can we learn about cardioprotection from the cardiac mitochondrial proteome? Cardiovascular Research 88:211–218

353. Cadenas S, Aragonés J, Landázuri MO (2010) Mitochondrial reprogramming through cardiac oxygen sensors in ischaemic heart disease. Cardiovascular Research 88:219–228
354. Vaniotis G, Allen BG, Hébert TE (2011) Nuclear GPCRs in cardiomyocytes: an insider's view of β-adrenergic receptor signaling. American Journal of Physiology – Heart and Circulatory Physiology 301:H1754–H1764
355. Xia J, Weaver A, Gerrard DE, Yao G (2006) Monitoring sarcomere structure changes in whole muscle using diffuse light reflectance. Journal of Biomedical Optics 11:40504–40504
356. Llewellyn ME, Barretto RPJ, Delp SL, Schnitzer MJ (2008) Minimally invasive high-speed imaging of sarcomere contractile dynamics in mice and humans. Nature 454:784-788
357. Vandekerckhove J, Bugaisky G, Buckingham M (1986) Simultaneous expression of skeletal muscle and heart actin proteins in various striated muscle tissues and cells. A quantitative determination of the two actin isoforms. Journal of Biological Chemistry 261:1838–1843
358. Zou P, Pinotsis N, Lange S, Song YH, Popov A, Mavridis I, Mayans OM, Gaute M, Wilmanns M (2006) Palindromic assembly of the giant muscle protein titin in the sarcomeric Z-disk. Nature 439:229–233
359. Leach IH, Tsang ML, Church RJ, Lowe J (1994) α-B crystallin in the normal human myocardium and cardiac conducting system. Journal of Pathology 173:255–260
360. Whittaker R, Glassy MS, Gude N, Sussman MA, Gottlieb RA, Glembotski CC (2009) Kinetics of the translocation and phosphorylation of αB-crystallin in mouse heart mitochondria during ex vivo ischemia. American Journal of Physiology – Heart and Circulatory Physiology 296:H1633–H1642
361. Bagshaw CR (1993) Muscle contraction. Chapman & Hall, London
362. Pope B, Hoh JF, Weeds A (1980) The ATPase activities of rat cardiac myosin isoenzymes. FEBS Letters 118:205–208
363. Day SM, Westfall MV, Fomicheva EV, Yasuda S, La Cross NC, D'Alecy LG, Ingwall JS, Metzger JM (2006) Histidine button engineered into cardiac troponin I protects the ischemic and failing heart. Nature – Medicine 12:181–189
364. de Tombe PP (2003) Cardiac myofilaments: mechanics and regulation. Journal of Biomechanics 36:721–730
365. Rayment I, Holden HM, Whittaker M, Yohn CB, Lorenz M, Holmes KC, Milligan RA (1993) Structure of the actin-myosin complex and its implications for muscle contraction. Science 261:58–65
366. Ishijima A, Doi T, Sakurada K, Yanagida T (1991) Sub-piconewton force fluctuations of actomyosin in vitro. Nature 352:301–306
367. Arts T, Reneman RS (1980) Measurement of deformation of canine epicardium in vivo during cardiac cycle. American Journal of Physiology – Heart and Circulatory Physiology 239:432–437
368. Whitten AE, Jeffries CM, Harris SP, Trewhella J (2008) Cardiac myosin-binding protein C decorates F-actin: implications for cardiac function. Proceedings of the National Academy of Sciences of the United States of America 105:18360-18365
369. Previs MJ, Beck Previs S, Gulick J, Robbins J, Warshaw DM (2012) Molecular mechanics of cardiac myosin-binding protein C in native thick filaments. Science 337:1215–1218
370. Sadayappan S, Osinska H, Klevitsky R, Lorenz JN, Sargent M, Molkentin JD, Seidman CE, Seidman JG, Robbins J (2006) Cardiac myosin binding protein c phosphorylation is cardioprotective. Proceedings of the National Academy of Sciences of the United States of America 103:16918–16923
371. Blankenfeldt W, Thoma NH, Wray JS, Gautel M, Schlichting I (2006) Crystal structures of human cardiac beta-myosin II S2-Delta provide insight into the functional role of the S2 subfragment. Proceedings of the National Academy of Sciences of the United States of America 103:17713–17717
372. Kunst G, Kress KR, Gruen M, Uttenweiler D, Gautel M, Fink RH (2000) Myosin binding protein C, a phosphorylation-dependent force regulator in muscle that controls the attachment of myosin heads by its interaction with myosin S2. Circulation Research 86:51–58

373. Frank D, Kuhn C, Katus HA, Frey N (2006) The sarcomeric Z-disc: a nodal point in signalling and disease. Journal of Molecular Medicine 84:446-468
374. Park JB, Kim JH, Kim Y, Ha SH, Yoo JS, Du G, Frohman MA, Suh PG, Ryu SH (2000) Cardiac phospholipase D2 localizes to sarcolemmal membranes and is inhibited by alpha-actinin in an ADP-ribosylation factor-reversible manner. Journal of Biological Chemistry 275:21295–21301
375. Freeman JL, Pitcher JA, Li X, Bennett V, Lefkowitz RJ (2000) alpha-Actinin is a potent regulator of G protein-coupled receptor kinase activity and substrate specificity in vitro. FEBS Letters 473:280–284
376. Puchner EM, Alexandrovich A, Kho AL, Hensen U, Schafer LV, Brandmeier B, Grater F, Grubmuller H, Gaub HE, Gautel M (2008) Mechanoenzymatics of titin kinase. Proceedings of the National Academy of Sciences of the United States of America 105:13385–13390
377. LeWinter MM, Wu Y, Labeit S, Granzier H (2007) Cardiac titin: structure, functions and role in disease. Clinica Chimica Acta 375:1–9
378. Yamasaki R, Wu Y, McNabb M, Greaser M, Labeit S, Granzier H (2002) Protein kinase A phosphorylates titin's cardiac-specific N2B domain and reduces passive tension in rat cardiac myocytes. Circulation Research 90:1181–1188
379. Borisov AB, Raeker MO, Russell MW (2008) Developmental expression and differential cellular localization of obscurin and obscurin-associated kinase in cardiac muscle cells. Journal of Cellular Biochemistry 103:1621–1635
380. Sauer F, Vahokoski J, Song YH, Wilmanns H (2010) Molecular basis of the head-to-tail assembly of giant muscle proteins obscurin-like 1 and titin. EMBO Reports 11:534–540
381. Ford-Speelman DL, Roche JA, Bowman AL, Bloch RJ (2009) The rho-guanine nucleotide exchange factor domain of obscurin activates rhoA signaling in skeletal muscle. Molecular Biology of the Cell 20:3905–3917
382. Perry NA, Shriver M, Mameza MG, Grabias B, Balzer E, Kontrogianni-Konstantopoulos A (2012) Loss of giant obscurins promotes breast epithelial cell survival through apoptotic resistance. FASEB Journal
383. Geisler SB, Robinson D, Hauringa M, Raeker MO, Borisov AB, Westfall MV, Russell MW (2007) Obscurin-like 1, OBSL1, is a novel cytoskeletal protein related to obscurin. Genomics 89:521–531
384. Yamazaki T, Wälchli S, Fujita T, Ryser S, Hoshijima M, Schlegel W, Kuroda S, Maturana AD (2010) Splice variants of Enigma homolog, differentially expressed during heart development, promote or prevent hypertrophy. Cardiovascular Research 86:374–382
385. Camarata T, Bimber B, Kulisz A, Chew TL, Yeung J, Simon HG (2006) LMP4 regulates Tbx5 protein subcellular localization and activity. Journal of Cell Biology 174:339–348
386. Fimia GM, De Cesare D, Sassone-Corsi P (2000) A family of LIM-only transcriptional coactivators: tissue-specific expression and selective activation of CREB and CREM. Molecular and Cellular Biology 20:8613–8622
387. Xia Y, McMillin JB, Lewis A, Moore M, Zhu WG, Williams RS, Kellems RE (2000) Electrical stimulation of neonatal cardiac myocytes activates the NFAT3 and GATA4 pathways and up-regulates the adenylosuccinate synthetase 1 gene. Journal of Biological Chemistry 275:1855–1863
388. Wu X, Sun Z, Foskett A, Trzeciakowski JP, Meininger GA, Muthuchamy M (2010) Cardiomyocyte contractile status is associated with differences in fibronectin and integrin interactions. American Journal of Physiology – Heart and Circulatory Physiology 298:H2071–H2081
389. Brancaccio M, Fratta L, Notte A, Hirsch E, Poulet R, Guazzone S, De Acetis M, Vecchione C, Marino G, Altruda F, Silengo L, Tarone G, Lembo G (2003) Melusin, a muscle-specific integrin β1-interacting protein, is required to prevent cardiac failure in response to chronic pressure overload. Nature – Medicine 9:68–75
390. Chen J, Levin LR, Buck J (2012) Role of soluble adenylyl cyclase in the heart. American Journal of Physiology – Heart and Circulatory Physiology 302:H538–H543

391. Clusin WT (2008) Mechanisms of calcium transient and action potential alternans in cardiac cells and tissues. American Journal of Physiology – Heart and Circulatory Physiology 294:H1–H10
392. Woodcock EA, Matkovich SJ (2005) Ins(1,4,5)P_3 receptors and inositol phosphates in the heart – Evolutionary artefacts or active signal transducers? Pharmacology and Therapeutics 107:240–251
393. Drago I, De Stefani D, Rizzuto R, Pozzan T (2012) Mitochondrial Ca^{2+} uptake contributes to buffering cytoplasmic Ca^{2+} peaks in cardiomyocytes. Proceedings of the National Academy of Sciences of the United States of America 109:12986–12991
394. Prosser BL, Ward CW, Lederer WJ (2011) X-ROS signaling: rapid mechano-chemo transduction in heart. Science 333:1440–1445
395. Despa S, Lingrel JB, Bers DM (2012) Na^+/K^+-ATPase α2-isoform preferentially modulates Ca^{2+} transients and sarcoplasmic reticulum Ca^{2+} release in cardiac myocytes. Cardiovascular Research 95:480-486
396. Nivala M, Qu Z (2012) Calcium alternans in a couplon network model of ventricular myocytes: role of sarcoplasmic reticulum load. American Journal of Physiology – Heart and Circulatory Physiology 303:H341–H352
397. Rovetti R, Cui X, Garfinkel A, Weiss JN, Qu Z (2010) Spark-induced sparks as a mechanism of intracellular calcium alternans in cardiac myocytes. Circulation Research 106:1582–1591
398. Restrepo JG, Karma A (2009) Spatiotemporal intracellular calcium dynamics during cardiac alternans. Chaos 19:037115
399. Klaiber M, Dankworth B, Kruse M, Hartmann M, Nikolaev VO, Yang RB, Völker K, Gaβner B, Oberwinkler H, Feil R, Freichel M, Groschner K, Skryabin BV, Frantz S, Birnbaumer L, Pongs O, Kuhn M (2011) A cardiac pathway of cyclic GMP-independent signaling of guanylate cyclase A, the receptor for atrial natriuretic peptide. Proceedings of the National Academy of Sciences of the United States of America 108:18500–18505
400. Herring N, Zaman JA, Paterson DJ (2001) Natriuretic peptides like NO facilitate cardiac vagal neurotransmission and bradycardia via a cGMP pathway. American Journal of Physiology – Heart and Circulatory Physiology 281:H2318–H2327
401. Chen W, Gaβner B, Bröner S, Nikolaev VO, Schlegel N, Waschke J, Steinbronn N, Strasser R, Kuhn M (2012) Atrial natriuretic peptide enhances microvascular albumin permeability by the caveolae-mediated transcellular pathway. Cardiovascular Research 93:141–151
402. Das BB, Solinger R (2009) Role of natriuretic peptide family in cardiovascular medicine. Cardiovascular and Hematological Agents in Medicinal Chemistry 7:29–42
403. Forssmann W, Meyer M, Forssmann K (2001) The renal urodilatin system: clinical implications. Cardiovascular Research 51:450–462
404. Cheung BMY, Kumana CR (1998) Natriuretic peptides: relevance in cardiovascular disease. Journal of the American Medical Association 280:1983–1984
405. Gallagher PE, Ferrario CM, Tallant EA (2008) Regulation of ACE2 in cardiac myocytes and fibroblasts. American Journal of Physiology – Heart and Circulatory Physiology 295: H2373–H2379
406. Babitt JL, Zhang Y, Samad TA, Xia Y, Tang J, Campagna JA, Schneyer AL, Woolf CJ, Lin H (2005) Repulsive guidance molecule (RGMa), a DRAGON homologue, is a bone morphogenetic protein co-receptor. Journal of Biological Chemistry 280:29820–29827
407. Babitt JL, Huang FW, Wrighting DM, Xia Y, Sidis Y, Samad TA, Campagna JA, Chung RT, Schneyer AL, Woolf CJ, Andrews NC, Lin HY (2006) Bone morphogenetic protein signaling by hemojuvelin regulates hepcidin expression. Nature – Genetics 38:531–539
408. Sheikh F, Raskin A, Chu PH, Lange S, Domenighetti AA, Zheng M, Liang X, Zhang T, Yajima T, Gu Y, Dalton ND, Mahata SK, Dorn GW, Heller-Brown J, Peterson KL, Omens JH, McCulloch AD, Chen J(2008) An FHL1-containing complex within the cardiomyocyte sarcomere mediates hypertrophic biomechanical stress responses in mice. Journal of Clinical Investigation 118:3870–3880
409. DiMichele LA, Doherty JT, Rojas M, Beggs HE, Reichardt LF, Mack CP, Taylor JM (2006) Myocyte-restricted focal adhesion kinase deletion attenuates pressure overload-induced hypertrophy. Circulation Research 99:636–645

410. Scimia MC, Hurtado C, Ray S, Metzler S, Wei K, Wang J, Woods CE, Purcell NH, Catalucci D, Akasaka T, Bueno OF, Vlasuk GP, Kaliman P, Bodmer R, Smith LH, Ashley E, Mercola M, Brown JH, Ruiz-Lozano P (2012) APJ acts as a dual receptor in cardiac hypertrophy. Nature 488:394–398
411. Heineke J, Molkentin JD (2006) Regulation of cardiac hypertrophy by intracellular signalling pathways. Nature Reviews – Molecular Cell Biology 7:589–600
412. Fisch S, Gray S, Heymans S, Haldar SM, Wang B, Pfister O, Cui L, Kumar A, Lin Z, Sen-Banerjee S, Das H, Petersen CA, Mende U, Burleigh BA, Zhu Y, Pinto Y, Liao R, Jain MK (2007) Kruppel-like factor 15 is a regulator of cardiomyocyte hypertrophy. Proceedings of the National Academy of Sciences of the United States of America 104:7074–7079
413. Colella M, Grisan F, Robert V, Turner JD, Thomas AP, Pozzan T (2008) Ca^{2+} oscillation frequency decoding in cardiac cell hypertrophy: Role of calcineurin/NFAT as Ca^{2+} signal integrators. Proceedings of the National Academy of Sciences of the United States of America 105:2859–2864
414. Onohara N, Nishida M, Inoue R, Kobayashi H, Sumimoto H, Sato Y, Mori Y, Nagao T, Kurose H (2006) TRPC3 and TRPC6 are essential for angiotensin II-induced cardiac hypertrophy. EMBO Journal 25:5305–5316
415. Gul R, Kim SY, Park KH, Kim BJ, Kim SJ, Im MJ, Kim UH (2008) A novel signaling pathway of ADP-ribosyl cyclase activation by angiotensin II in adult rat cardiomyocytes. American Journal of Physiology – Heart and Circulatory Physiology 295:H77–H88
416. Hoshijima M, Knoll R, Pashmforoush M, Chien KR (2006) Reversal of calcium cycling defects in advanced heart failure; toward molecular therapy. Journal of the American College of Cardiology 48:A15–A23
417. Spector NL, Yarden Y, Smith B, Lyass L, Trusk P, Pry K, Hill JE, Xia W, Seger R, Bacus SS (2007) Activation of AMP-activated protein kinase by human EGF receptor 2/EGF receptor tyrosine kinase inhibitor protects cardiac cells. Proceedings of the National Academy of Sciences of the United States of America 104:10607–10612
418. McMullen JR, Amirahmadi F, Woodcock EA, Schinke-Braun M, Bouwman RD, Hewitt KA, Mollica JP, Zhang L, Zhang Y, Shioi T, Buerger A, Izumo S, Jay PY, Jennings GL (2007) Protective effects of exercise and phosphoinositide 3-kinase(p110α) signaling in dilated and hypertrophic cardiomyopathy. Proceedings of the National Academy of Sciences of the United States of America 104:612–617
419. Morissette MR, Cook SA, Foo SY, McKoy G, Ashida N, Novikov M, Scherrer-Crosbie M, Li L, Matsui T, Brooks G, Rosenzweig A (2006) Myostatin regulates cardiomyocyte growth through modulation of Akt signaling. Circulation Research 99:15–24
420. Russell MA, Lund LM, Haber R, McKeegan K, Cianciola N, Bond M (2006) The intermediate filament protein, synemin, is an AKAP in the heart. Archives of Biochemistry and Biophysics 456:204–215
421. Scholten A, Poh MK, van Veen TA, van Breukelen B, Vos MA, Heck AJ (2006) Analysis of the cGMP/cAMP interactome using a chemical proteomics approach in mammalian heart tissue validates sphingosine kinase type 1-interacting protein as a genuine and highly abundant AKAP. Journal of Proteome Research 5:1435–1447
422. Appert-Collin A, Cotecchia S, Nenniger-Tosato M, Pedrazzini T, Diviani D (2007) The A-kinase anchoring protein (AKAP)-Lbc-signaling complex mediates alpha1 adrenergic receptor-induced cardiomyocyte hypertrophy. Proceedings of the National Academy of Sciences of the United States of America 104:10140–10145
423. He Q, Harding P, LaPointe MC (2010) PKA, Rap1, ERK1/2, and p90RSK mediate PGE2 and EP4 signaling in neonatal ventricular myocytes. American Journal of Physiology – Heart and Circulatory Physiology 298:H136–H143
424. Merkel MJ, Liu L, Cao Z, Packwood WH, Young J, Alkayed NJ, Van Winkle DM (2010) Inhibition of soluble epoxide hydrolase preserves cardiomyocytes: role of STAT3 signaling. American Journal of Physiology – Heart and Circulatory Physiology 298:H679–H687

425. Xu D, Li N, He Y, Timofeyev V, Lu L, Tsai HJ, Kim IH, Tuteja D, Mateo RKP, Singapuri A, Davis BB, Low R, Hammock BD, Chiamvimonvat N (2006) Prevention and reversal of cardiac hypertrophy by soluble epoxide hydrolase inhibitors. Proceedings of the National Academy of Sciences of the United States of America 103:18733–18738
426. Fukuzawa J, Booz GW, Hunt RA, Shimizu N, Karoor V, Baker KM, Dostal DE (2000) Cardiotrophin-1 increases angiotensinogen mRNA in rat cardiac myocytes through STAT3. An autocrine loop for hypertrophy. Hypertension 35:1191–1196
427. Shimano M, Ouchi N, Nakamura K, van Wijk B, Ohashi K, Asaumi Y, Higuchi A, Pimentel DR, Sam F, Murohara T, van den Hoff MJ, Walsh K (2011) Cardiac myocyte follistatin-like 1 functions to attenuate hypertrophy following pressure overload. Proceedings of the National Academy of Sciences of the United States of America 108:E899–E906
428. Shimano M, Ouchi N, Nakamura K, Oshima Y, Higuchi A, Pimentel DR, Panse KD, Lara-Pezzi E, Lee SJ, Sam F, Walsh K (2011) Cardiac myocyte-specific ablation of follistatin-like 3 attenuates stress-induced myocardial hypertrophy. Journal of Biological Chemistry 286:9840–9848
429. Streeter DD Jr (1979) Gross morphology and fiber geometry of the heart. In: Beme RM, Sperelakis N (eds) The heart. Handbook of physiology. The cardiovascular system. American Physiology Society, Bethesda, MD
430. Geerts L, Bovendeerd P, Nicolay K, Arts T (2002) Characterization of the normal cardiac myofiber field in goat measured with MR-diffusion tensor imaging, American Journal of Physiology – Heart and Circulatory Physiology 283:139–145
431. Jouk PS, Usson Y, Michalowicz G, Parazza F (1995) Mapping of the orientation of myocardial cells by means of polarized light and confocal scanning laser microscopy. Microscopy Research and Technique 30:480–490
432. Mourad A, Biard L, Caillerie D, Jouk PS, Raoult A, Szafran N, Usson Y (2001) Geometrical modelling of the fibre organization in the human left ventricle. In: Katila T, Magnin IE, Clarysse P, Montagnat J, Nenonen J (eds) Functional imaging and modeling of the heart. Springer, Berlin
433. Chadwick RS (1982) Mechanics of the left ventricle. Biophysical Journal 39:279–288
434. van Campen DH, Huyghe JM, Bovendeerd PHM, Arts, T (1994) Biomechanics of the heart muscle, European Journal of Mechanics A – Solids 13:19–41
435. Rijcken J, Bovendeerd PHM, Schoofs AJG, van Campen DH, Arts T (1999) Optimization of cardiac fiber orientation for homogeneous strain during ejection. Annals of Biomedical Engineering 27:289–297
436. Arts T, Bovendeerd P, Delhaas T, Prinzen F (2003) Modeling the relation between cardiac pump function and myofiber mechanics. Journal of Biomechanics 36:731–736
437. Savadjiev P, Strijkers GJ, Bakermans AJ, Piuze E, Zucker SW, Siddiqi K (2012) Heart wall myofibers are arranged in minimal surfaces to optimize organ function. Proceedings of the National Academy of Sciences of the United States of America 109:9248–9253
438. Hume JR, Duan D, Collier ML, Yamazaki J, Horowitz B (2000) Anion transport in heart. Physiological Reviews 80:31–81
439. Goldin AL (2001) Resurgence of sodium channel research. Annual Review of Physiology 63:871–894
440. Camacho JA, Hensellek S, Rougier JS, Blechschmidt S, Abriel H, Benndorf K, Zimmer T (2006) Modulation of $Na_V1.5$ channel function by an alternatively spliced sequence in the DII/DIII linker region. Journal of Biological Chemistry 281:9498–9506
441. Lei M, Zhang H, Grace AA, Huang CL (2007) SCN5A and sinoatrial node pacemaker function. Cardiovascular Research 74:356–365
442. Hallaq H, Yang Z, Viswanathan PC, Fukuda K, Shen W, Wang DW, Wells KS, Zhou J, Yi J, Murray KT (2006) Quantitation of protein kinase A-mediated trafficking of cardiac sodium channels in living cells. Cardiovascular Research 72:250–261
443. Aiba T, Hesketh GG, Liu T, Carlisle R, Villa-Abrille MC, O'Rourke B, Akar FG, Tomaselli GF (2010) Na^+ channel regulation by Ca^{2+}/calmodulin and Ca^{2+}/calmodulin-dependent protein kinase II in guinea-pig ventricular myocytes. Cardiovascular Research 85:454–463

444. Wagner S, Dybkova N, Rasenack EC, Jacobshagen C, Fabritz L, Kirchhof P, Maier SK, Zhang T, Hasenfuss G, Brown JH, Bers DM, Maier LS (2006) Ca^{2+}/calmodulin-dependent protein kinase II regulates cardiac Na^{+} channels. Journal of Clinical Investigation 116:3127–3138
445. Abriel H (2007) Roles and regulation of the cardiac sodium channel $Na_V1.5$: Recent insights from experimental studies. Cardiovascular Research 76:381–389
446. Levi AJ, Spitzer KW, Kohmoto O, Bridge JHB (1994) Depolarization-induced Ca entry vai Na–Ca exchange triggers SR release in guinea pig cardiac myocytes. American Journal of Physiology – Heart and Circulatory Physiology 266:H1422–H1433
447. Song J, Zhang XQ, Wang J, Cheskis E, Chan TO, Feldman AM, Tucker AL, Cheung JY (2008) Regulation of cardiac myocyte contractility by phospholemman: Na^{+}/Ca^{2+} exchange versus Na^{+}-K^{+}-ATPase. American Journal of Physiology – Heart and Circulatory Physiology 295:H1615–H1625
448. Wang J, Gao E, Song J, Zhang XQ, Li J, Koch WJ, Tucker AL, Philipson KD, Chan TO, Feldman AM, Cheung JY (2010) Phospholemman and β-adrenergic stimulation in the heart. American Journal of Physiology – Heart and Circulatory Physiology 298:H807–H815
449. Van Petegem F, Chatelain FC, Minor DL (2005) Insights into voltage-gated calcium channel regulation from the structure of the CaV1.2 IQ domain–Ca^{2+}/calmodulin complex. Nature – Structural and Molecular Biology 12:1108–1115
450. Tuluc P, Kern G, Obermair GJ, Flucher BE (2007) Computer modeling of siRNA knockdown effects indicates an essential role of the Ca^{2+} channel $\alpha2\delta$-1 subunit in cardiac excitation–contraction coupling. Proceedings of the National Academy of Sciences of the United States of America 104:11091–11096
451. Zuhlke RD, Pitt GS, Deisseroth K, Tsien RW, Reuter H (1999) Calmodulin supports both inactivation and facilitation of L-type calcium channels. Nature 399:159–162
452. Sipido KR, Callewaert G, Carmeliet E (1995) Inhibition and rapid recovery of Ca^{2+} current during Ca^{2+} release from sarcoplasmic reticulum in guinea pig ventricular myocytes. Circulation Research 76:102–109
453. Dixon RE, Yuan C, Cheng EP, Navedo MF, Santana LF (2012) Ca^{2+} signaling amplification by oligomerization of L-type $Ca_V1.2$ channels. Proceedings of the National Academy of Sciences of the United States of America 109:1749–1754
454. Balijepalli RC, Foell JD, Hall DD, Hell JW, Kamp TJ (2006) Localization of cardiac L-type Ca^{2+} channels to a caveolar macromolecular signaling complex is required for β2-adrenergic regulation. Proceedings of the National Academy of Sciences of the United States of America 103:7500–7505
455. Zhou Z, January CT (1998) Both T- and L-type Ca^{2+} channels can contribute to excitation-contraction coupling in cardiac Purkinje cells. Biophysical Journal 74:1830–1839
456. Ferrier GR, Howlett SE (2001) Cardiac excitation-contraction coupling: role of membrane potential in regulation of contraction. American Journal Physiology – Heart and Circulatory Physiology 280:H1928–H1944
457. Hobai IA, Howarth FC, Pabbathi VK, Dalton GR, Hancox JC, Zhu JQ, Howlett SE, Ferrier GR, and Levi AJ (1997) Voltage-activated Ca release in rabbit, rat and guinea-pig cardiac myocytes, and modulation by internal cAMP. Pflgers Archiv 435:164–173
458. Kiriazis H, Kranias EG (2000) Genetically engineered models with alterations in cardiac membrane calcium-handling proteins. Annual Review of Physiology 62:321–351
459. Franzini-Armstrong C, Protasi F, Ramesh V (1999) Shape, size, and distribution of Ca^{2+} release units and couplons in skeletal and cardiac muscles. Biophysical Journal 77:1528–1539
460. Marx SO, Gaburjakova J, Gaburjakova M, Henrikson C, Ondrias K, Marks AR (2001) Coupled gating between cardiac calcium release channels (ryanodine receptors). Circulation Research 88:1151–1158
461. Soeller C, Crossman D, Gilbert R, Cannell MB (2007) Analysis of ryanodine receptor clusters in rat and human cardiac myocytes. Proceedings of the National Academy of Sciences of the United States of America 104:14958–14963

462. Lehnart SE, Wehrens XH, Reiken S, Warrier S, Belevych AE, Harvey RD, Richter W, Jin SL, Conti M, Marks AR (2005) Phosphodiesterase 4D deficiency in the ryanodine-receptor complex promotes heart failure and arrhythmias. Cell 123:25–35
463. Kasneci A, Kemeny-Suss NM, Komarova SV, Chalifour LE (2009) Egr-1 negatively regulates calsequestrin expression and calcium dynamics in ventricular cells. Cardiovascular Research 81:695–702
464. Bers DM (2002) Cardiac excitation-contraction coupling. Nature 415:198–205
465. Pahlavan S, Oberhofer M, Sauer B, Ruppenthal S, Tian Q, Scholz A, Kaestner L, Lipp P (2012) $G\alpha q$ and $G\alpha 11$ contribute to the maintenance of cellular electrophysiology and Ca^{2+} handling in ventricular cardiomyocytes. Cardiovascular Research 95:48–58
466. Snyders DJ (1999) Structure and function of cardiac potassium channels. Cardiovascular Research 42:377–390
467. Zhou J, Kodirov S, Murata M, Buckett PD, Nerbonne JM, Koren G (2003) Regional upregulation of $K_V2.1$-encoded current, $I_{K,slow2}$, in Kv1DN mice is abolished by crossbreeding with Kv2DN mice. American Journal of Physiology – Heart and Circulatory Physiology 284:H491–H500
468. Baumgarten CM (2004) Cardiac bioelectricity. Encyclopedia of biomaterials and biomedical engineering, Marcel Dekker, New York
469. Nerbonne JM, Guo W (2002) Heterogeneous expression of voltage-gated potassium channels in the heart: roles in normal excitation and arrhythmias. Journal of Cardiovascular Electrophysiology 13:406-409
470. Loewen ME, Wang Z, Eldstrom J, Dehghani Zadeh A, Khurana A, Steele DF, Fedida D (2009) Shared requirement for dynein function and intact microtubule cytoskeleton for normal surface expression of cardiac potassium channels. American Journal of Physiology – Heart and Circulatory Physiology 296:H71–H83
471. Nerbonne JM, http://www.j-circ.or.jp/english/sessions/reports/64th-ss/nerbonne.htm
472. Xu H, Guo W, Nerbonne JM (1999) Four kinetically distinct depolarization-activated K^+ currents in adult mouse ventricular myocytes. Journal of General Physiology 113:661–678
473. Brahmajothi MV, Morales MJ, Liu S, Rasmusson RL, Campbell DL, Strauss HC (1996) In situ hybridization reveals extensive diversity of K^+ channel mRNA in isolated ferret cardiac myocytes. Circulation Research 78:1083–1089
474. Brunner M, Guo W, Mitchell GF, Buckett PD, Nerbonne JM, Koren G (2001) Characterization of mice with a combined suppression of I_{to} and $I_{K,slow}$. American Journal of Physiology – Heart and Circulatory Physiology 281:H1201–H1209
475. Costantini DL, Arruda EP, Agarwal P, Kim KH, Zhu Y, Zhu W, Lebel M, Cheng CW, Park CY, Pierce SA, Guerchicoff A, Pollevick GD, Chan TY, Kabir MG, Cheng SH, Husain M, Antzelevitch C, Srivastava D, Gross GJ, Hui CC, Backx PH, Bruneau BG (2005) The homeodomain transcription factor Irx5 establishes the mouse cardiac ventricular repolarization gradient. Cell 123:347–358
476. He W, Jia Y, Takimoto K (2009) Interaction between transcription factors Iroquois proteins 4 and 5 controls cardiac potassium channel Kv4.2 gene transcription. Cardiovascular Research 81:64–71
477. Zhang YH, Wu W, Sun HY, Deng XL, Cheng LC, Li X, Tse HF, Lau CP, Li GR (2012) Modulation of human cardiac transient outward potassium current by EGFR tyrosine kinase and Src-family kinases. Cardiovascular Research 93:424–433
478. Kurachi Y, Ishii M (2004) Cell signal control of the G protein-gated potassium channel and its subcellular localization. Journal of Physiology 554.2:285–294
479. Ishii M, Inanobe A, Kurachi Y (2002) PIP3 inhibition of RGS protein and its reversal by Ca^{2+}/calmodulin mediate voltage-dependent control of the G protein cycle in a cardiac K^+ channel. Proceedings of the National Academy of Sciences 99:4325–4330
480. Jovanovi S, Du Q, Crawford RM, Budas GR, Stagljar I, Jovanovi A (2005) Glyceraldehyde 3-phosphate dehydrogenase serves as an accessory protein of the cardiac sarcolemmal K-ATP channel. EMBO Reports 6:848–852

481. Li RA, Leppo M, Miki T, Seino S, Marban E (2000) Molecular basis of electrocardiographic ST-segment elevation. Circulation Research 87:837–839

Chap. 6. Heart Wall

482. Forouhar AS, Liebling M, Hickerson A, Nasiraei-Moghaddam A, Tsai HJ, Hove JR, Fraser SE, Dickinson ME, Gharib M (2006) The embryonic vertebrate heart tube is a dynamic suction pump. Science 312:751–753
483. Brand T (2003) Heart development: molecular insights into cardiac specification and early morphogenesis. Developmental Biology 258:1–19
484. Red-Horse K, Ueno H, Weissman IL, Krasnow MA (2010) Coronary arteries form by developmental reprogramming of venous cells. Nature 464:549–553
485. Armstrong EJ, Bischoff J (2004) Heart valve development: endothelial cell signaling and differentiation. Circulation Research 95:459–470
486. Brand T (2010) Exciting news: catecholamines in induction and regionalization of the heart. Cardiovascular Research 88:1–2
487. Bu L, Jiang X, Martin-Puig S, Caron L, Zhu S, Shao Y, Roberts DJ, Huang PL, Domian IJ, Chien KR (2009) Human ISL1 heart progenitors generate diverse multipotent cardiovascular cell lineages. Nature 460:113–117
488. Gupta V, Poss KD (2012) Clonally dominant cardiomyocytes direct heart morphogenesis. Nature 484:479–484
489. Eisenberg LM, Markwald RR (1995) Molecular regulation of atrioventricular valvuloseptal morphogenesis. Circulation Research 77:1–6
490. Katsuragi N, Morishita R, Nakamura N, Ochiai T, Taniyama Y, Hasegawa Y, Kawashima K, Kaneda Y, Ogihara T, Sugimura K (2004) Periostin as a novel factor responsible for ventricular dilation. Circulation 110:1806–1813
491. Kim J, Wu Q, Zhang Y, Wiens KM, Huang Y, Rubin N, Shimada H, Handin RI, Chao MY, Tuan TL, Starnes VA, Lien CL (2010) PDGF signaling is required for epicardial function and blood vessel formation in regenerating zebrafish hearts. Proceedings of the National Academy of Sciences of the United States of America 107:17206–17210
492. Kamei M, Saunders WB, Bayless KJ, Dye L, Davis GE, Weinstein BM (2006) Endothelial tubes assemble from intracellular vacuoles in vivo. Nature 442:453–456
493. Bhattacharya S, Macdonald ST, Farthing CR (2006) Molecular mechanisms controlling the coupled development of myocardium and coronary vasculature. Clinical Science (London) 111:35–46
494. Saito D, Takasel Y, Murai H, Takahashi Y (2012) The dorsal aorta initiates a molecular cascade that instructs sympatho-adrenal specification. Science 336:1578–1581
495. Mabe AM, Hoover DB (2009) Structural and functional cardiac cholinergic deficits in adult neurturin knockout mice. Cardiovascular Research 82:93–99
496. Sanicola M, Hession C, Worley D, Carmillo P, Ehrenfels C, Walus L, Robinson S, Jaworski G, Wei H, Tizard R, Whitty A, Pepinsky RB, Cate RL (1997) Glial cell line-derived neurotrophic factor-dependent RET activation can be mediated by two different cell-surface accessory proteins. Proceedings of the National Academy of Sciences of the United States of America 94:6238–6243
497. Johnson EN, Lee YM, Sander TL, Rabkin E, Schoen FJ, Kaushal S, Bischoff J (2003) NFATc1 mediates vascular endothelial growth factor-induced proliferation of human pulmonary valve endothelial cells. Journal of Biological Chemistry 278:1686–1692
498. Nakajima Y, Yamagishi T, Hokari S, Nakamura H (2000) Mechanisms involved in valvuloseptal endocardial cushion formation in early cardiogenesis: roles of transforming growth factor (TGF)-β and bone morphogenetic protein (BMP). Anatomical Record 258:119–127

499. Yoshioka M, Yuasa S, Matsumura K, Kimura K, Shiomi T, Kimura N, Shukunami C, Okada Y, Mukai M, Shin H, Yozu R, Sata M, Ogawa S, Hiraki Y, Fukuda K (2006) Chondromodulin-I maintains cardiac valvular function by preventing angiogenesis. Nature – Medicine 12:1151–1159
500. Murry CE, Lee RT (2009) Turnover after the fallout. Science 324:47–48
501. Olivetti G, Cigola E, Maestri R, Corradi D, Lagastra C, Gambert SR, Anversa P (1996) Aging, cardiac hypertrophy and ischemic cardiomyopathy do not affect the proportion of mononucleated and multinucleated myocytes in the human heart. Journal of Molecular and Cellular Cardiology 28:1463–1477
502. Bergmann O, Bhardwaj RD, Bernard S, Zdunek S, Barnab-Heider F, Walsh S, Zupicich J, Alkass K, Buchholz BA, Druid H, Jovinge S, Frisn J (2009) Evidence for cardiomyocyte renewal in humans. Science 324:98–102
503. Huang J, Min Lu M, Cheng L, Yuan LJ, Zhu X, Stout AL, Chen M, Li J, Parmacek MS (2009) Myocardin is required for cardiomyocyte survival and maintenance of heart function. Proceedings of the National Academy of Sciences of the United States of America 106:18734–18739
504. Taurin S, Sandbo N, Yau DM, Sethakorn N, Kach J, Dulin NO (2009) Phosphorylation of myocardin by extracellular signal-regulated kinase. Journal of Biological Chemistry 284:33789–33794
505. Muraski JA, Rota M, Misao Y, Fransioli J, Cottage C, Gude N, Esposito G, Delucchi F, Arcarese M, Alvarez R, Siddiqi S, Emmanuel GN, Wu W, Fischer K, Martindale JJ, Glembotski CC, Leri A, Kajstura J, Magnuson N, Berns A, Beretta RM, Houser SR, Schaefer EM, Anversa P, Sussman MA (2007) Pim-1 regulates cardiomyocyte survival downstream of Akt. Nature – Medicine 13:1467–1475
506. Lage K, Møllgd̊ K, Greenway S, Wakimoto H, Gorham JM, Workman CT, Bendsen E, Hansen NT, Rigina O, Roque FS, Wiese C, Christoffels VM, Roberts AE, Smoot LB, Pu WT, Donahoe PK, Tommerup N, Brunak S, Seidman CE, Seidman JG, Larsen LA (2010) Dissecting spatio-temporal protein networks driving human heart development and related disorders. Molecular Systems Biology 6:381
507. Ferdous A, Caprioli A, Iacovino M, Martin CM, Morris J, Richardson JA, Latif S, Hammer RE, Harvey RP, Olson EN, Kyba M, Garry DJ (2009) Nkx2-5 transactivates the Ets-related protein 71 gene and specifies an endothelial/endocardial fate in the developing embryo. Proceedings of the National Academy of Sciences of the United States of America 106:814–819
508. Monzen K, Ito Y, Naito AT, Kasai H, Hiroi Y, Hayashi D, Shiojima I, Yamazaki T, Miyazono K, Asashima M, Nagai R, Komuro I (2008) A crucial role of a high mobility group protein HMGA2 in cardiogenesis. Nature – Cell Biology 10:567–574
509. Bushdid PB, Osinska H, Waclaw RR, Molkentin JD, Yutzey KE (2003) NFATc3 and NFATc4 are required for cardiac development and mitochondrial function. Circulation Research 92:1305–1313
510. van Amerongen MJ, Diehl F, Novoyatleva T, Patra C, Engel FB (2010) E2F4 is required for cardiomyocyte proliferation. Cardiovascular Research 86:92–102
511. López-Sánchez C, Bártulos O, Martnez-Campos E, Gañán C, Valenciano AI, García-Martnez V, De Pablo F, Hernández-Sánchez C (2010) Tyrosine hydroxylase is expressed during early heart development and is required for cardiac chamber formation. Cardiovascular Research 88:111–120
512. Ilagan R, Abu-Issa R, Brown D, Yang YP, Jiao K, Schwartz RJ, Klingensmith J, Meyers EN (2006) Fgf8 is required for anterior heart field development. Development 133:2435–2445
513. Park EJ, Ogden LA, Talbot A, Evans S, Cai CL, Black BL, Frank DU, Moon AM (2006) Required, tissue-specific roles for Fgf8 in outflow tract formation and remodeling. Development 133:2419–2433
514. Wadugu B, Kühn B (2012) The role of neuregulin/ErbB2/ErbB4 signaling in the heart with special focus on effects on cardiomyocyte proliferation. American Journal of Physiology – Heart and Circulatory Physiology 302:H2139–H2147

515. Elenius K, Choi CJ, Paul S, Santiestevan E, Nishi E, Klagsbrun M (1999) Characterization of a naturally occurring ErbB4 isoform that does not bind or activate phosphatidyl inositol 3-kinase. Oncogene 18:2607–2615
516. Novoyatleva T, Diehl F, van Amerongen MJ, Patra C, Ferrazzi F, Bellazzi R, Engel FB (2010) TWEAK is a positive regulator of cardiomyocyte proliferation. Cardiovascular Research 85:681–690
517. Nagy II, Railo A, Rapila R, Hast T, Sormunen R, Tavi P, Räsänen J, Vainio SJ (2009) Wnt-11 signalling controls ventricular myocardium development by patterning N-cadherin and β-catenin expression. Circulation Research 85:100–109
518. Schwarz-Romond T, Asbrand C, Bakkers J, Kühl M, Schaeffer HJ, Huelsken J, Behrens J, Hammerschmidt M, Birchmeier W (2002) The ankyrin repeat protein Diversin recruits Casein kinase Iε to the beta-catenin degradation complex and acts in both canonical Wnt and Wnt/JNK signaling. Genes and Development 16:2073–2084
519. Moeller H, Jenny A, Schaeffer HJ, Schwarz-Romond T, Mlodzik M, Hammerschmidt M, Birchmeier W (2006) Diversin regulates heart formation and gastrulation movements in development. Proceedings of the National Academy of Sciences of the United States of America 103:15900–15905
520. Debril MB, Gelman L, Fayard E, Annicotte JS, Rocchi S, Auwerx J (2004) Transcription factors and nuclear receptors interact with the SWI/SNF complex through the BAF60c subunit. Journal of Biological Chemistry 279:16677–16686
521. Liang FS, Crabtree GR (2009) Developmental biology: the early heart remodelled. Nature 459:654–655
522. Takeuchi JK, Bruneau BG (2009) Directed transdifferentiation of mouse mesoderm to heart tissue by defined factors. Nature 459:708–711
523. Hang CT, Yang J, Han P, Cheng HL, Shang C, Ashley E, Zhou B, Chang CP (2010) Chromatin regulation by Brg1 underlies heart muscle development and disease. Nature 466:62–67
524. Stankunas K, Hang CT, Tsun ZY, Chen H, Lee NV, Wu JI, Shang C, Bayle JH, Shou W, Iruela-Arispe ML, Chang CP (2008) Endocardial Brg1 represses ADAMTS1 to maintain the microenvironment for myocardial morphogenesis. Developmental Cell 14:298–311
525. Sasse P, Malan D, Fleischmann M, Roell W, Gustafsson E, Bostani T, Fan Y, Kolbe T, Breitbach M, Addicks K, Welz A, Brem G, Hescheler J, Aszodi A, Costell M, Bloch W, Fleischmann BK (2008) Perlecan is critical for heart stability. Cardiovascular Research 80:435–444
526. Yang B, Lu Y, Wang Z (2008) Control of cardiac excitability by microRNAs. Cardiovascular Research 79:571–580
527. Yang B, Lin H, Xiao J, Lu Y, Luo X, Li B, Zhang Y, Xu C, Bai Y, Wang H, Chen G, Wang Z (2007) The muscle-specific microRNA miR-1 regulates cardiac arrhythmogenic potential by targeting GJA1 and KCNJ2. Nature – Medicine 13:486–491
528. Zhao Y, Ransom JF, Li A, Vedantham V, von Drehle M, Muth AN, Tsuchihashi T, McManus MT, Schwartz RJ, Srivastava D (2007) Dysregulation of cardiogenesis, cardiac conduction, and cell cycle in mice lacking miRNA-1-2. Cell 129:303–317
529. Care A, Catalucci D, Felicetti F, Bonci D, Addario A, Gallo P, Bang ML, Segnalini P, Gu Y, Dalton ND, Elia L, Latronico MV, Hoydal M, Autore C, Russo MA, Dorn GW, Ellingsen O, Ruiz-Lozano P, Peterson KL, Croce CM, Peschle C, Condorelli G (2007) MicroRNA-133 controls cardiac hypertrophy. Nature – Medicine 13:613–618
530. van Rooij E, Sutherland LB, Qi X, Richardson JA, Hill J, Olson EN (2007) Control of stress-dependent cardiac growth and gene expression by a MicroRNA. Science 316:575–579
531. Thum T, Catalucci D, Bauersachs J (2008) MicroRNAs: novel regulators in cardiac development and disease. Cardiovascular Research 79:562–570
532. Morton SU, Scherz PJ, Cordes KR, Ivey KN, Stainier DY, Srivastava D (2008) MicroRNA-138 modulates cardiac patterning during embryonic development. Proceedings of the National Academy of Sciences of the United States of America 105:17830–17835

533. Thum T, Gross C, Fiedler J, Fischer T, Kissler S, Bussen M, Galuppo P, Just S, Rottbauer W, Frantz S, Castoldi M, Soutschek J, Koteliansky V, Rosenwald A, Basson MA, Licht JD, Pena JT, Rouhanifard SH, Muckenthaler MU, Tuschl T, Martin GR, Bauersachs J, Engelhardt S (2008) MicroRNA-21 contributes to myocardial disease by stimulating MAP kinase signalling in fibroblasts. Nature 456:980–984
534. Roy S, Khanna S, Hussain SR, Biswas S, Azad A, Rink C, Gnyawali S, Shilo S, Nuovo GJ, Sen CK (2009) MicroRNA expression in response to murine myocardial infarction: miR-21 regulates fibroblast metalloprotease-2 via phosphatase and tensin homologue. Cardiovascular Research 82:21–29
535. Lunde BM, Moore C, Varani G (2007) RNA-binding proteins: modular design for efficient function. Nature Reviews – Molecular Cell Biology 8:479–490
536. Glisovic T, Bachorik JL, Yong J, Dreyfuss G (2008) RNA-binding proteins and post-transcriptional gene regulation. FEBS Letters 582:1977–1986
537. Dreyfuss G, Kim VN, Kataoka N (2002) Messenger-RNA-binding proteins and the messages they carry. Nature Reviews – Molecular Cell Biology 3:195–205
538. Lobbardi R, Lambert G, Zhao J, Geisler R, Kim HR, Rosa FM (2011) Fine-tuning of Hh signaling by the RNA-binding protein Quaking to control muscle development. Development 138:1783–1794
539. Beauchamp P, Nassif C, Hillock S, van der Giessen K, von Roretz C, Jasmin BJ, Gallouzi IE (2010) The cleavage of HuR interferes with its transportin-2-mediated nuclear import and promotes muscle fiber formation. Cell Death and Differentiation 17:1588–1599
540. Gerber WV, Yatskievych TA, Antin PB, Correia KM, Conlon RA, Krieg PA (1999) The RNA-binding protein gene, hermes, is expressed at high levels in the developing heart. Mechanisms of Development 80:77–86
541. Guo W, Schafer S, Greaser ML, Radke MH, Liss M, Govindarajan T, Maatz H, Schulz H, Li S, Parrish AM, Dauksaite V, Vakeel P, Klaassen S, Gerull B, Thierfelder L, Regitz-Zagrosek V, Hacker TA, Saupe KW, Dec GW, Ellinor PT, MacRae CA, Spallek B, Fischer R, Perrot A, Ozcelik C, Saar K, Hubner N, Gotthardt M (2012) RBM20, a gene for hereditary cardiomyopathy, regulates titin splicing. Nature – Medicine 18:766–773
542. Poon KL, Tan KT, Wei YY, Ng CP, Colman A, Korzh V, Xu XQ (2012) RNA-binding protein RBM24 is required for sarcomere assembly and heart contractility. Cardiovascular Research 94:418–427
543. Buckingham M, Meilhac S, Zaffran S (2005) Building the mammalian heart from two sources of myocardial cells. Nature Reviews – Genetics 6:826–837
544. Gaetani R, Ledda M, Barile L, Chimenti I, De Carlo F, Forte E, Ionta V, Giuliani L, D'Emilia E, Frati G, Miraldi F, Pozzi D, Messina E, Grimaldi S, Giacomello A, Lisi A (2009) Differentiation of human adult cardiac stem cells exposed to extremely low-frequency electromagnetic fields. Cardiovascular Research 82:411–420
545. Moretti A, Caron L, Nakano A, Lam JT, Bernshausen A, Chen Y, Qyang Y, Bu L, Sasaki M, Martin-Puig S, Sun Y, Evans SM, Laugwitz KL, Chien KR (2006) Multipotent embryonic isl1+ progenitor cells lead to cardiac, smooth muscle, and endothelial cell diversification. Cell 127:1151–1165
546. Kattman SJ, Huber TL, Keller GM (2006) Multipotent flk-1+ cardiovascular progenitor cells give rise to the cardiomyocyte, endothelial, and vascular smooth muscle lineages. Developmental Cell 11:723–732
547. Wu SM, Fujiwara Y, Cibulsky SM, Clapham DE, LienCL, Schultheiss TM, Orkin SH (2006) Developmental origin of a bipotential myocardial and smooth muscle cell precursor in the mmmalian heart. Cell 127:1137–1150
548. Kwon C, Qian L, Cheng P, Nigam V, Arnold J, Srivastava D (2009) A regulatory pathway involving Notch1/β-catenin/Isl1 determines cardiac progenitor cell fate. Nature – Cell Biology 11:951–957
549. Lian X, Hsiao C, Wilson G, Zhu K, Hazeltine LB, Azarin SM, Raval KK, Zhang J, Kamp TJ, Palecek SP (2012) Robust cardiomyocyte differentiation from human pluripotent stem cells via temporal modulation of canonical Wnt signaling. Proceedings of the National Academy of Sciences of the United States of America 109:E1848–E1857

550. Tang JM, Wang JN, Zhang L, Zheng F, Yang JY, Kong X, Guo LY, Chen L, Huang YZ, Wan Y, Chen SY (2011) VEGF/SDF-1 promotes cardiac stem cell mobilization and myocardial repair in the infarcted heart. Cardiovascular Research 91:402–411
551. Bearzi C, Leri A, Lo Monaco F, Rota M, Gonzalez A, Hosoda T, Pepe M, Qanud K, Ojaimi C, Bardelli S, D'Amario D, D'Alessandro DA, Michler RE, Dimmeler S, Zeiher AM, Urbanek K, Hintze TH, Kajstura J, Anversa P (2009) Identification of a coronary vascular progenitor cell in the human heart. Proceedings of the National Academy of Sciences of the United States of America 106:15885–15890
552. Boni A, Urbanek K, Nascimbene A, Hosoda T, Zheng H, Delucchi F, Amano K, Gonzalez A, Vitale S, Ojaimi C, Rizzi R, Bolli R, Yutzey KE, Rota M, Kajstura J, Anversa P, Leri A (2008) Notch1 regulates the fate of cardiac progenitor cells. Proceedings of the National Academy of Sciences of the United States of America 105:15529–15534
553. Chen VC, Stull R, Joo D, Cheng X, Keller G (2008) Notch signaling respecifies the hemangioblast to a cardiac fate. Nature – Biotechnology 26:1169–1178
554. Cai CL, Martin JC, Sun Y, Cui L, Wang L, Ouyang K, Yang L, Bu L, Liang X, Zhang X, Stallcup WB, Denton CP, McCulloch A, Chen J, Evans SM (2008) A myocardial lineage derives from Tbx18 epicardial cells. Nature 454:104–108
555. Zhou B, Ma Q, Rajagopal S, Wu SM, Domian I, Rivera-Feliciano J, Jiang D, von Gise A, Ikeda S, Chien KR, Pu WT (2008) Epicardial progenitors contribute to the cardiomyocyte lineage in the developing heart. Nature 454:109–113
556. Beltrami AP, Barlucchi L, Torella D, Baker M, Limana F, Chimenti S, Kasahara H, Rota M, Musso E, Urbanek K, Leri A, Kajstura J, Nadal-Ginard B, Anversa P (2003) Adult cardiac stem cells are multipotent and support myocardial regeneration. Cell 114:763–776
557. Tallini YN, Greene KS, Craven M, Spealman A, Breitbach M, Smith J, Fisher PJ, Steffey M, Hesse M, Doran RM, Woods A, Singh B, Yen A, Fleischmann BK, Kotlikoff MI (2009) c-Kit Expression identifies cardiovascular precursors in the neonatal heart. Proceedings of the National Academy of Sciences of the United States of America 106:1808–1813
558. Tevosian SG, Deconinck AE, Tanaka M, Schinke M, Litovsky SH, Izumo S, Fujiwara Y, Orkin SH (2000) FOG-2, a cofactor for GATA transcription factors, is essential for heart morphogenesis and development of coronary vessels from epicardium. Cell 101:729–739
559. Smart N, Risebro CA, Melville AAD, Moses K, Schwartz RJ, Chien KR, Riley PR (2007) Thymosin bold beta4 induces adult epicardial progenitor mobilization and neovascularization. Nature 445:177–182
560. Liu Y, Asakura M, Inoue H, Nakamura T, Sano M, Niu Z, Chen M, Schwartz RJ, Schneider MD (2007) Sox17 is essential for the specification of cardiac mesoderm in embryonic stem cells. Proceedings of the National Academy of Sciences of the United States of America 104:3859–3864
561. David R, Brenner C, Stieber J, Schwarz F, Brunner S, Vollmer M, Mentele E, Mller-Hcker J, Kitajima S, Lickert H, Rupp R, Franz WM (2008) MesP1 drives vertebrate cardiovascular differentiation through Dkk-1-mediated blockade of Wnt-signalling. Nature – Cell Biology 10:338–345
562. Kwon C, Arnold J, Hsiao EC, Taketo MM, Conklin BR, Srivastava C (2007) Canonical Wnt signaling is a positive regulator of mammalian cardiac progenitors. Proceedings of the National Academy of Sciences of the United States of America 104:10894–10899
563. Naito AT, Shiojima I, Akazawa H, Hidaka K, Morisaki T, Kikuchi A, Komuro I (2006) Developmental stage-specific biphasic roles of Wnt/beta-catenin signaling in cardiomyogenesis and hematopoiesis. Proceedings of the National Academy of Sciences of the United States of America 103:19812–19817
564. Zamora M, Mnner J, Ruiz-Lozano P (2007) Epicardium-derived progenitor cells require β-catenin for coronary artery formation. Proceedings of the National Academy of Sciences of the United States of America 104:18109–18114
565. Chang DF, Belaguli NS, Chang J, Schwartz RJ (2007) LIM-only protein, CRP2, switched on smooth muscle gene activity in adult cardiac myocytes. Proceedings of the National Academy of Sciences of the United States of America 104:157–162

566. Parker KK, Ingber DE (2007) Extracellular matrix, mechanotransduction and structural hierarchies in heart tissue engineering. Philosophical Transactions of the Royal Society – London – B Biological Sciences 362:1267–1279
567. Weber KT (2000) Fibrosis and hypertensive heart disease. Current Opinion in Cardiology 15:264–272
568. Ongstad EL, Gourdie RG (2012) Myocyte–fibroblast electrical coupling: the basis of a stable relationship? Cardiovascular Research 93:215–217
569. He K, Shi X, Zhang X, Dang S, Ma X, Liu F, Xu M, Lv Z, Han D, Fang X, Zhang Y (2011) Long-distance intercellular connectivity between cardiomyocytes and cardiofibroblasts mediated by membrane nanotubes. Cardiovascular Research 92:39–47
570. Nguyen TP, Xie Y, Garfinkel A, Qu Z, Weiss JN (2012) Arrhythmogenic consequences of myofibroblast-myocyte coupling. Cardiovascular Research 93:242–251
571. Glenn DJ, Rahmutula D, Nishimoto M, Liang F, Gardner DG (2009) Atrial natriuretic peptide suppresses endothelin gene expression and proliferation in cardiac fibroblasts through a GATA4-dependent mechanism. Cardiovascular Research 84:209–217
572. Jankowski M (2009) GATA4, a new regulator of cardiac fibroblasts, is sensitive to natriuretic peptides. Cardiovascular Research 84:176–177
573. Heineke J, Auger-Messier M, Xu J, Oka T, Sargent MA, York A, Klevitsky R, Vaikunth S, Duncan SA, Aronow BJ, Robbins J, Crombleholme TM, Molkentin JD (2007) Cardiomyocyte GATA4 functions as a stress-responsive regulator of angiogenesis in the murine heart. Journal of Clinical Investigation 117:3198–3210
574. Ieda M, Tsuchihashi T, Ivey KN, Ross RS, Hong TT, Shaw RM, Srivastava D (2009) Cardiac fibroblasts regulate myocardial proliferation through β_1 integrin signaling. Developmental Cell 16:233–244
575. Li L, Fan D, Wang C, Wang JY, Cui XB, Wu D, Zhou Y, Wu LL (2011) Angiotensin II increases periostin expression via Ras/p38 MAPK/CREB and ERK1/2/TGF-β1 pathways in cardiac fibroblasts. Cardiovascular Research 91:80–89
576. LM Hinescu ME, Popescu LM (2005) Interstitial Cajal-like cells (ICLC) in human atrial myocardium. Journal of Cellular and Molecular Medicine 9:972–975
577. Gherghiceanu M, Hinescu ME, Andrei F, Mandache E, Macarie CE, Faussone-Pellegrini MS, Popescu LM (2008) Interstitial Cajal-like cells (ICLC) in myocardial sleeves of human pulmonary veins. Journal of Cellular and Molecular Medicine 12:1777–1781
578. Gherghiceanu M, Manole CG, Popescu LM (2010) Telocytes in endocardium: electron microscope evidence. Journal of Cellular and Molecular Medicine 14:2330–2334
579. Popescu LM, Gherghiceanu M (2010) Cardiomyocyte precursors and telocytes in epicardial stem cells niche. Journal of Cellular and Molecular Medicine (http://telocytes.com/Telocyte/Telocytes_in_the_heart_files/Cardiomyocyteprecursorsandtelocytesinepicardialstemcellsniche-Electronmicroscopeimages.pdf)
580. Iacobellis G, Corradi D, Sharma AM (2005) Epicardial adipose tissue: anatomic, biomolecular and clinical relationships with the heart. Nature – Clinical Practice – Cardiovascular Medicine 2:536–543.
581. Kadowaki T, Yamauchi T (2005) Adiponectin and adiponectin receptors. Endocrine Reviews 26:439–451
582. Cui XB, Wang C, Li L, Fan D, Zhou Y, Wu D, Cui QH, Fu FY, Wu LL (2012) Insulin decreases myocardial adiponectin receptor 1 expression via PI3K/Akt and FoxO1 pathway. Cardiovascular Research 93:69–78
583. Shibata R, Sato K, Pimentel DR, Takemura Y, Kihara S, Ohashi K, Funahashi T, Ouchi N, Walsch K (2005) Adiponectin protects against myocardial ischemia-reperfusion injury through AMPK- and COX-2-dependent mechanisms. Nature – Medicine 11:1096–1103
584. Karmazyn M, Purdham DM, Rajapurohitam V, Zeidan A (2008) Signalling mechanisms underlying the metabolic and other effects of adipokines on the heart. Cardiovascular Research 79:279–286

585. Barrenechea GR, Le Tallec P, Valentin F (2002) New wall laws for the unsteady incompressible Navier–Stokes equations on rough domains. Mathematical Modelling and Numerical Analysis 36:177–203
586. Sacks MS, Smith DB, Hiester ED (1998) The aortic valve microstructure: effects of transvalvular pressure. Journal of Biomedical Materials Research 41:131–141.
587. Mendelson K, Schoen F (2006) Heart valve tissue engineering: concepts, approaches, progress, and challenges. Annals of Biomedical Engineering 34:1799–1819
588. Peskin CS, McQueen DM (1994) Mechanical equilibrium determines the fractal fiber architecture of aortic heart valve leaflets. American Journal of Physiology – Heart and Circulatory Physiology 266:319–328
589. Peskin CS (1989) Fiber architecture of the left ventricular wall. an asymptotic analysis. Communications on Pure and Applied Mathematics 42:79–113
590. Bukauskas FF, Kreuzberg MM, Rackauskas M, Bukauskiene A, Bennett MLV, Verselis VK, Willecke K (2006) Properties of mouse connexin 30.2 and human connexin 31.9 hemichannels: Implications for atrioventricular conduction in the heart. Proceedings of the National Academy of Sciences of the United States of America 103:9726–9731
591. Gros D, Thveniau-Ruissy M, Bernard M, Calmels T, Kober F, Söhl G, Willecke K, Nargeot J, Jongsma HJ, Mangoni ME (2009) Connexin 30 is expressed in the mouse sino-atrial node and modulates heart rate. Cardiovascular Research 85:45–55
592. Fedorov VV, Glukhov AV, Chang R (2012) Conduction barriers and pathways of the sinoatrial pacemaker complex: their role in normal rhythm and atrial arrhythmias. American Journal of Physiology – Heart and Circulatory Physiology 302:H1773–H1783
593. Robinson RB (2011) The long and short of calcium-dependent automaticity in the sinoatrial node. American Journal of Physiology – Heart and Circulatory Physiology 300:H31–32
594. Choi BR, Salama G (1998) Optical mapping of atrioventricular node reveals a conduction barrier between atrial and nodal cells. American Journal of Physiology – Heart and Circulatory Physiology 275:H1905–H1909
595. Stuyvers BD, Dun W, Matkovich S, Sorrentino V, Boyden PA, ter Keurs HE (2005) Ca^{2+} sparks and waves in canine Purkinje cells: a triple layered system of Ca2+ activation. Circulation Research 97:35–43
596. Nonidez JF (1943) The structure and innervation of the conductive system of the heart of the dog and rhesus monkey, as seen with a silver impregnation technique. American Heart Journal 26:577–597
597. Grueter CE, van Rooij E, Johnson BA, DeLeon SM, Sutherland LB, Qi X, Gautron L, Elmquist JK, Bassel-Duby R, Olson EN (2012) A cardiac microRNA governs systemic energy homeostasis by regulation of MED13. Cell 149:671–683
598. Zhang M, Wang Y, Jiang M, Zankov DP, Chowdhury S, Kasirajan V, Tseng GN (2011) KCNE2 protein is more abundant in ventricles than in atria and can accelerate hERG protein degradation in a phosphorylation-dependent manner. American Journal of Physiology – Heart and Circulatory Physiology 302:H910–H922
599. Rozanski GJ (2012) Physiological remodelling of potassium channels in the heart. Cardiovascular Research 93:218–219
600. Yang KC, Jay PY, McMullen JR, Nerbonne JM (2012) Enhanced cardiac PI3Kα signalling mitigates arrhythmogenic electrical remodelling in pathological hypertrophy and heart failure. Cardiovascular Research 93:252–262
601. Irisawa H, Brown HF, Giles W (1993) Cardiac pacemaker in the sinoatrial node, Physiological Reviews 73:197–227
602. Demir SS, Clark JW, Murphey CR, Giles WR (1994) A mathematical model of a rabbit sinoatrial node cell. American Journal of Physiology 266:C832–C852.
603. Courtemanche M, Ramirez RJ, Nattel S (1998) Ionic mechanisms underlying human atrial action potential properties: insights from a mathematical model. American Journal of Physiology – Heart and Circulatory Physiology 275:H301–H321.

604. Djabella K, Sorine M (2006) A differential model of controlled cardiac pacemaker cell, SIAM Conference on the Life Sciences, Raleigh, USA (http://hal.inria.fr/docs/00/06/46/22/PDF/Djabella_Sorine.pdf)
605. Garcia-Frigola C, Shi Y, Evans SM (2003) Expression of the hyperpolarization-activated cyclic nucleotide-gated cation channel HCN4 during mouse heart development. Gene Expression Patterns 3:777–783
606. Harzheim D, Pfeiffer KH, Fabritz L, Kremmer E, Buch T, Waisman A, Kirchhof P, Kaupp UB, Seifert R (2008) Cardiac pacemaker function of HCN4 channels in mice is confined to embryonic development and requires cyclic AMP. EMBO Journal 27:692–703
607. Wu Y, Gao Z, Chen B, Koval OM, Singh MV, Guan X, Hund TJ, Kutschke W, Sarma S, Grumbach IM, Wehrens XH, Mohler PJ, Song LS, Anderson ME (2009) Calmodulin kinase II is required for fight or flight sinoatrial node physiology. Proceedings of the National Academy of Sciences of the United States of America 106:5972–5977
608. Tingley WG, Pawlikowska L, Zaroff JG, Kim T, Nguyen T, Young SG, Vranizan K, Kwok PY, Whooley MA, Conklin BR (2007) Gene-trapped mouse embryonic stem cell-derived cardiac myocytes and human genetics implicate AKAP10 in heart rhythm regulation. Proceedings of the National Academy of Sciences of the United States of America 104:8461-8466
609. Vinogradova TM, Maltsev VA, Bogdanov KY, Lyashkov AE, Lakatta EG (2005) Rhythmic Ca^{2+} oscillations drive sinoatrial nodal cell pacemaker function to make the heart tick. Annals of the New York Academy of Sciences 1047:138–156
610. Maltsev VA, Lakatta EG (2010) A novel quantitative explanation for the autonomic modulation of cardiac pacemaker cell automaticity via a dynamic system of sarcolemmal and intracellular proteins. American Journal of Physiology – Heart and Circulatory Physiology 298:H2010–H2023
611. Lyashkov AE, Vinogradova TM, Zahanich I, Li Y, Younes A, Nuss HB, Spurgeon HA, Maltsev VA, Lakatta EG (2009) Cholinergic receptor signaling modulates spontaneous firing of sinoatrial nodal cells via integrated effects on PKA-dependent Ca^{2+} cycling and $_{KACh}$. American Journal of Physiology – Heart and Circulatory Physiology 297:H949–H959
612. ten Tusscher KHWJ, Noble D, Noble PJ, Panfilov AV (2004) A model for human ventricular tissue. American Journal of Physiology – Heart and Circulatory Physiology 286:H1573–H1589
613. Beeler GW, Reuter H (1977) Reconstruction of the action potential of ventricular myocardial fibres. Journal of Physiology 268:177-210
614. Luo CH, Rudy Y (1994) A dynamic model of the cardiac ventricular action potential. I. Simulations of ionic currents and concentration changes. Circulation Research 74:1071–1096
615. Luo CH, Rudy Y (1994) A dynamic model of the cardiac ventricular action potential. II. Afterdepolarizations, triggered activity, and potentiation. Circulation Research 74:1097–1113
616. Noble D, Varghese A, Kohl P, Noble P (1998) Improved guinea-pig ventricular cell model incorporating a diadic space, IKr and IKs, and length- and tension-dependent processes. Canadian Journal of Cardiology 14:123–134
617. Priebe L, Beuckelmann DJ (1998) Simulation study of cellular electric properties in heart failure. Circulation Research 82:1206–1223
618. Sipido KR (1998) Efficiency of L-type Ca^{2+} current compared with reverse mode Na/Ca exchange or T-type Ca^{2+} current as trigger for Ca^{2+} release from the sarcoplasmic reticulum. Annals of the New York Academy of Sciences 853:357–360
619. Strom M, Wan X, Poelzing S, Ficker E, Rosenbaum DS (2010) Gap junction heterogeneity as mechanism for electrophysiologically distinct properties across the ventricular wall. American Journal of Physiology – Heart and Circulatory Physiology 298:H787–H794
620. Casini S, Verkerk AO, van Borren MMGJ, van Ginneken ACG, Veldkamp MW, de Bakker JMT, Tan HL (2009) Intracellular calcium modulation of voltage-gated sodium channels in ventricular myocytes. Cardiovascular Research 81:72–81

621. Wang LJ, Sobie EA (2008) Mathematical model of the neonatal mouse ventricular action potential. American Journal of Physiology – Heart and Circulatory Physiology 294:H2565–H2575
622. Malmivuo J, Plonsey R (2006) Bioelectromagnetism; Principles and Applications of Bioelectric and Biomagnetic Fields, University of Technology, Finland and Department of Biomedical Engineering, Duke University, Durham, NC, http://butler.cc.tut.fi/ malmivuo/bem/bembook, Ragnar Granit Institute, Tampere
623. Lipsius SL, Huser J, Blatter LA (2001) Intracellular Ca^{2+} release sparks atrial pacemaker activity. News in Physiological Sciences 16:101–106
624. Liu DW, Antzelevitch C (1995) Characteristics of the delayed rectifier current (IKr and IKs) in canine ventricular epicardial, midmyocardial and endocardial myocytes: a weaker IKs contributes to the longer action potential of the M cell. Circulation Research 76:351–365
625. Laurita KR, Katra R, Wible B, Wan X, Koo MH (2003) Transmural heterogeneity of calcium handling in canine. Circulation Research 92:668–675
626. Joseph T, Coirault C, Lecarpentier Y (2000) Species-dependent changes in mechano-energetics of isolated cardiac muscle during hypoxia. Basic Research in Cardiology 95:378–384
627. Fedida D, Giles WR (1991) Regional variations in action potentials and transient outward current in myocytes isolated from rabbit left ventricle. Journal of Physiology (London) 442:191–209
628. Sah R, Ramirez RJ, Backx PH (2002) Modulation of Ca^{2+} release in cardiac myocytes by changes in repolarization rate: role of phase-1 action potential repolarization in excitation-contraction coupling. Circulation Research 90:165–173
629. Drouin E, Charpentier F, Gauthier C, Chevallier JC, Laurent K, Michaud JL, Le Marec H (1993) Evidence for the presence of M cells in the human ventricle. Pacing and Clinical Electrophysiology (PACE) 16:876
630. Antzelevitch C (2001) Heterogeneity of cellular repolarization in LQTS: the role of M cells. European Heart Journal Supplements 3:K2–K16
631. Zygmunt AC, Goodrow RJ, Antzelevitch C (2000) INa-Ca contributes to electrical heterogeneity within the canine ventricle. American Journal of Physiology – Heart and Circulatory Physiology 278:H1671–H1678
632. Zygmunt AC, Eddlestone GT, Thomas GP, Nesterenko VV, Antzelevitch C (2001) Larger late sodium conductance in M cells contributes to electrical heterogeneity in canine ventricle. American Journal of Physiology – Heart and Circulatory Physiology 281:H689–H697
633. Cordeiro JM, Greene L, Heilmann C, Antzelevitch D, Antzelevitch C (2004) Transmural heterogeneity of calcium activity and mechanical function in the canine left ventricle. American Journal of Physiology – Heart and Circulatory Physiology 286:H1471–H1479
634. Janssen PM (2010) Myocardial contraction-relaxation coupling. American Journal of Physiology – Heart and Circulatory Physiology 299:H1741–H1749
635. McCain ML, Lee H, Aratyn-Schaus Y, Kléber AG, Parker KK (2012) Cooperative coupling of cell-matrix and cell-cell adhesions in cardiac muscle. Proceedings of the National Academy of Sciences of the United States of America 109:9881–9886
636. Bers DM (2000) Calcium fluxes involved in control of cardiac myocyte contraction. Circulation Research 87:275–281
637. Babu GJ, Bhupathy P, Timofeyev V, Petrashevskaya NN, Reiser PJ, Chiamvimonvat N, Periasamy M (2007) Ablation of sarcolipin enhances sarcoplasmic reticulum calcium transport and atrial contractility. Proceedings of the National Academy of Sciences of the United States of America 104:17867–17872
638. Tuncay E, Bilginoglu A, Sozmen NN, Zeydanli EN, Ugur M, Vassort G, Turan B (2011) Intracellular free zinc during cardiac excitation-contraction cycle: calcium and redox dependencies. Cardiovascular Research 89:634–642
639. Cazorla O, Lucas A, Poirier F, Lacampagne A, Lezoualc'h F (2009) The cAMP binding protein Epac regulates cardiac myofilament function. Proceedings of the National Academy of Sciences of the United States of America 106:14144–14149

640. Marenholz I, Heizmann CW, Fritz G (2004) S100 proteins in mouse and man: from evolution to function and pathology. Biochemical and Biophysical Research Communications 322:1111–1122
641. Ambartsumian N, Klingelhofer J, Grigorian M, Christensen C, Kriajevska M, Tulchinsky E, Georgiev G, Berezin V, Bock E, Rygaard J, Cao R, Cao Y, Lukanidin E (2001) The metastasis-associated Mts1(S100A4) protein could act as an angiogenic factor. Oncogene 20:4685–4695
642. Ehlermann P, Remppis A, Guddat O, Weimann J, Schnabel PA, Motsch J, Heizmann CW, Katus HA (2000) Right ventricular upregulation of the Ca^{2+} binding protein S100A1 in chronic pulmonary hypertension. Biochimica et Biophysica Acta 1500:249–255
643. Remppis A, Greten T, Schafer BW, Hunziker P, Erne P, Katus HA, Heizmann CW (1996) Altered expression of the Ca^{2+}-binding protein S100A1 in human cardiomyopathy. Biochimica et Biophysica Acta 1313:253–257
644. Most P, Pleger ST, Volkers M, Heidt B, Boerries M, Weichenhan D, Loffler E, Janssen PM, Eckhart AD, Martini J, Williams ML, Katus HA, Remppis A, Koch WJ (2004) Cardiac adenoviral S100A1 gene delivery rescues failing myocardium. Journal of Clinical Investigation 114:1550–1563
645. Hulme JT, Westenbroek RE, Scheuer T, Catterall WA (2006) Phosphorylation of serine 1928 in the distal C-terminal domain of cardiac CaV1.2 channels during beta1-adrenergic regulation. Proceedings of the National Academy of Sciences of the United States of America 103:16574–16579
646. Hulme JT, Yarov-Yarovoy V, Lin TWC, Scheuer T, Catterall WA (2006) Autoinhibitory control of the CaV1.2 channel by its proteolytically processed distal C-terminal domain. Journal of Physiology 576:87–102
647. Lygren B, Carlson CR, Santamaria K, Lissandron V, McSorley T, Litzenberg J, Lorenz D, Wiesner B, Rosenthal W, Zaccolo M, Taskn K, Klussmann E (2007) AKAP complex regulates Ca^{2+} re-uptake into heart sarcoplasmic reticulum. EMBO reports 8:1061–1067
648. Aprigliano O, Rybin VO, Pak E, Robinson RB, Steinberg SF (1997) β1- and β2-adrenergic receptors exhibit differing susceptibility to muscarinic accentuated antagonism. American Journal of Physiology – Heart and Circulatory Physiology 272:2726–2735
649. Saucerman JJ, Zhang J, Martin JC, Peng LX, Stenbit AE, Tsien RY, McCulloch AD (2006) Systems analysis of PKA-mediated phosphorylation gradients in live cardiac myocytes. Proceedings of the National Academy of Sciences of the United States of America 103:12923–12928
650. Zaugg M, Schaub MC (2004) Cellular mechanisms in sympatho-modulation of the heart. British Journal of Anaesthesia 93:34–52
651. Potter EK, Smith-White MA (2004) Galanin modulates cholinergic neurotransmission in the heart. Neuropeptides 39:345–348

Chap. 7. Vessel Wall

652. Yashiro K, Shiratori H, Hamada H (2007) Haemodynamics determined by a genetic programme govern asymmetric development of the aortic arch. Nature 450:285–288
653. Mironneau J (1976) Relationship between contraction and transmembrane ionic currents in voltage-clamped uterine smooth muscle. In: Bülbring E, Shuba MF (eds) Physiology of the smooth muscle. Raven Press, New York
654. Schmid F, Sommer G, Rappolt M, Schulze-Bauer CA, Regitnig P, Holzapfel GA, Laggner P, Amenitsch H (2005) In situ tensile testing of human aortas by time-resolved small-angle X-ray scattering. Journal of Synchrotron Radiation 12:727–733
655. Milnor WR (1982) Hemodynamics. Williams and Wilkins, Baltimore
656. Fung YC (1981) Biomechanics. Springer, New York

657. Passman JN, Dong XR, Wu SP, Maguire CT, Hogan KA, Bautch VL, Majesky MW (2008) A sonic hedgehog signaling domain in the arterial adventitia supports resident Sca1+ smooth muscle progenitor cells. Proceedings of the National Academy of Sciences of the United States of America 105:9349–9354
658. Young PP, Modur V, Teleron AA, Ladenson JH (2005) Enrichment of genes in the aortic intima that are associated with stratified epithelium: implications of underlying biomechanical and barrier properties of the arterial intima. Circulation 111:2382–2390
659. Pyle AL, Li B, Maupin AB, Guzman RJ, Crimmins DL, Olson S, Atkinson JB, Young PP (2010) Biomechanical stress induces novel arterial intima-enriched genes: implications for vascular adaptation to stress. Cardiovascular Pathology 19:e13–e20
660. Miyake M, Ito N, Nagahata S, Takeuchi Y, Fukui Y (1996) Morphological study of the human maxillofacial venous vasculature: examination of venous valves using the corrosion resin cast technique. Anatomical Record 244:126–132
661. Chauvin M, Shah DC, Hassaguerre M, Marcellin L, Brechenmacher C (2000) The anatomic basis of connections between the coronary sinus musculature and the left atrium in humans. Circulation 101:647–652
662. Gandhi RH, Irizarry E, Nackman GB, Halpern VJ, Mulcare RJ, Tilson MD (1993) Analysis of the connective tissue matrix and proteolytic activity of primary varicose veins. Journal of Vascular Surgery 18:814–820
663. Sansilvestri-Morel P, Rupin A, Jaisson S, Fabiani JN, Verbeuren TJ, Vanhoutte PM (2002) Synthesis of collagen is dysregulated in cultured fibroblasts derived from skin of subjects with varicose veins as it is in venous smooth muscle cells. Circulation 106:479–483
664. Sansilvestri-Morel P, Rupin A, Badier-Commander C, Kern P, Fabiani JN, Verbeuren TJ, Vanhoutte PM (2001) Imbalance in the synthesis of collagen type I and collagen type III in smooth muscle cells derived from human varicose veins. Journal of Vascular Research 38:560–568
665. Van Cleef JF (1993) A vein has a preferential axis of flattening. Journal of Dermatologic Surgery and Oncology 19:468–470
666. Hossler FE, West RF (1988) Venous valve anatomy and morphometry: studies on the duckling using vascular corrosion casting. American Journal of Anatomy 181:425-432
667. Abbott NJ, Ronnback L, Hansson E (2006) Astrocyte–endothelial interactions at the blood–brain barrier. Nature Reviews – Neuroscience 7:41–53
668. Coomber BL, Stewart PA (1985) Morphometric analysis of CNS microvascular endothelium. Microvascular Research 30:99–115
669. Oldendorf WH, Cornford ME, Brown WJ (1977) The large apparent work capability of the blood-brain barrier: a study of the mitochondrial content of capillary endothelial cells in brain and other tissues of the rat. Annals of Neurology 1:409–417
670. Daneman R, Zhou L, Kebede AA, Barres BA (2010) Pericytes are required for blood–brain barrier integrity during embryogenesis. Nature 468:562–566
671. Armulik A, Genové G, Mäe M, Nisancioglu MH, Wallgard E, Niaudet C, He L, Norlin J, Lindblom P, Strittmatter K, Johansson BR, C Betsholtz (2010) Pericytes regulate the blood–brain barrier. Nature 468:557–561
672. Betz AL, Firth JA, Goldstein GW (1980) Polarity of the blood-brain barrier: distribution of enzymes between the luminal and antiluminal membranes of brain capillary endothelial cells. Brain Research 192:17–28
673. Farrell CL, Pardridge WM (1991) Blood-brain barrier glucose transporter is asymmetrically distributed on brain capillary endothelial lumenal and ablumenal membranes: an electron microscopic immunogold study. Proceedings of the National Academy of Sciences of the United States of America 88:5779–5783
674. BangsowT, Baumann E, Bangsow C, Jaeger MH, Pelzer B, Gruhn P, Wolf S, von Melchner H, Stanimirovi DB (2008) The epithelial membrane protein 1 is a novel tight junction protein of the blood–brain barrier. Journal of Cerebral Blood Flow and Metabolism 28::1249–1260
675. Takano T, Tian GF, Peng W, Lou N, Libionka W, Han X, Nedergaard M (2005) Astrocyte-mediated control of cerebral blood flow. Nature – Neuroscience 9:260–267

676. Deane R, LaRue B, Sagare AP, Castellino FJ, Zhong Z, Zlokovic BV (2009) Endothelial protein C receptor-assisted transport of activated protein C across the mouse blood–brain barrier. Journal of Cerebral Blood Flow and Metabolism 29:25–33
677. Spencer BJ, Verma IM (2007) Targeted delivery of proteins across the blood–brain barrier. Proceedings of the National Academy of Sciences of the United States of America 104:7594–7599
678. Daneman R, Zhou L, Kebede AA, Barres BA (2010) Pericytes are required for blood–brain barrier integrity during embryogenesis. Nature 468:562–566
679. Spector R, Johanson CE (1989) The mammalian choroid plexus. Scientific American 261:68–74
680. Burnstock G (1999) Release of vasoactive substances from endothelial cells by shear stress and purinergic mechanosensory transduction. Journal of Anatomy 194:335–342
681. Honma Y, Araki T, Gianino S, Bruce A, Heuckeroth R, Johnson E, Milbrandt J (2002) Artemin is a vascular-derived neurotropic factor for developing sympathetic neurons. Neuron 35:267–282
682. Shepro D, Morel NM (1993) Pericyte physiology. The FASEB Journal 7:1031–1038
683. Peppiatt CM, Howarth C, Mobbs P, Attwell D (2006) Bidirectional control of CNS capillary diameter by pericytes. Nature 443:700–704
684. Cohn HI, Harris DM, Pesant S, Pfeiffer M, Zhou RH, Koch WJ, Dorn GW 2nd, Eckhart AD (2008) Inhibition of vascular smooth muscle G protein-coupled receptor kinase 2 enhances alpha1D-adrenergic receptor constriction. American Journal of Physiology – Heart and Circulatory Physiology 295:H1695-H1704
685. Lazarowski ER, Boucher RC, Harden TK (2003) Mechanisms of release of nucleotides and integration of their action as P2X- and P2Y-receptor activating molecules. Molecular Pharmacology 64:785–795
686. Lohman AW, Billaud M, Isakson BE (2012) Mechanisms of ATP release and signalling in the blood vessel wall. Cardiovascular Research 95:269–280
687. Ellsworth ML (2000) The red blood cell as an oxygen sensor: what is the evidence? Acta Physiologica Scandinavica 168:551–559
688. González-Alonso J (2012) ATP as a mediator of erythrocyte-dependent regulation of skeletal muscle blood flow and oxygen delivery in humans. Journal of Physiology
689. Milner P, Bodin P, Loesch A, Burnstock G (1992) Increased shear stress leads to differential release of endothelin and ATP from isolated endothelial cells from 4- and 12-month-old male rabbit aorta. Journal of Vascular Research 29:420–425
690. Tateishi J, Faber JE (1995) ATP-sensitive K^+ channels mediate α2D-adrenergic receptor contraction of arteriolar smooth muscle and reversal of contraction by hypoxia. Circulation Research 76:53–63
691. Wettschureck N, Offermanns S (2005) Mammalian G proteins and their cell type specific functions. Physiological Reviews 85:1159–1204
692. Lazarowski ER (2012) Vesicular and conductive mechanisms of nucleotide release. Purinergic Signalling 8:359–373
693. Sawada K, Echigo N, Juge N, Miyaji T, Otsuka M, Omote H, Yamamoto A, Moriyama Y (2008) Identification of a vesicular nucleotide transporter. Proceedings of the National Academy of Sciences of the United States of America 105:5683–5686
694. Miyaji T, Sawada K, Omote H, Moriyama Y (2011) Divalent cation transport by vesicular nucleotide transporter. Journal of Biological Chemistry 286:42881–42887
695. Bodin P, Burnstock G (2001) Evidence that release of adenosine triphosphate from endothelial cells during increased shear stress is vesicular. Journal of Cardiovascular Pharmacology 38:900–908
696. Peti-Peterdi J (2006) Calcium wave of tubuloglomerular feedback. American Journal of Physiology – Renal Physiology 291:F473–F480
697. Toma I, Bansal E, Meer EJ, Kang JJ, Vargas SL, Peti-Peterdi J (2008) Connexin 40 and ATP-dependent intercellular calcium wave in renal glomerular endothelial cells. American Journal of Physiology – Regulatory, Integrative and Comparative Physiology 294:R1769–R1776

698. Billaud M, Lohman AW, Straub AC, Looft-Wilson R, Johnstone SR, Araj CA, Best AK, Chekeni FB, Ravichandran KS, Penuela S, Laird DW, Isakson BE (2011) Pannexin1 regulates α1-adrenergic receptor-mediated vasoconstriction. Circulation Research 109:80–85
699. Gao Q, Wolin MS (2008) Effects of hypoxia on relationship between cytosolic and mitochondrial NAD(P)H redox and superoxide generation in coronary arterial smooth muscle. American Journal of Physiology – Heart and Circulatory Physiology 295:H978–H989
700. Pakala R, Willerson JT, Benedict CR (1997) Effect of serotonin, thromboxane A2, and specific receptor antagonists on vascular smooth muscle cell proliferation. Circulation 96.2280–2286
701. Heusch G, Post H, Michel MC, Kelm M, Schulz R (2000) Endogenous nitric oxide and myocardial adaptation to ischemia. Circulation Research 87:146–152
702. Laude K, Favre J, Thuillez C, Richard V (2003) NO produced by endothelial NO synthase is a mediator of delayed preconditioning-induced endothelial protection. American Journal of Physiology – Heart and Circulatory Physiology 284:2053–2060
703. Lévy BI, Michel JB, Salzmann J-L, Azizi M, Poitevin P, Safar M, Camilleri J-P. Effects of chronic inhibition of converting enzyme on mechanical and structural properties of arteries in rat renovascular hypertension. Circulation Research 63:227–239
704. Sarzani R, Brecher P, Chobanian AV (1989) Growth factor expression in aorta of normotensive and hypertensive rats. The Journal of Clinical Investigation 83:1404–1408

Chap. 8. Smooth Muscle Cells

705. Rovner AS, Freyzon Y, Trybus KM (1997) An insert in the motor domain determines the functional properties of expressed smooth muscle myosin isoforms. Journal of Muscle Research and Cell Motility 18:103–110
706. Mani RS, McCubbin WD, Kay, CM (1992) Calcium-dependent regulation of caldesmon by an 11-kDa smooth muscle calcium-binding protein, caltropin. Biochemistry 31:11896–11901
707. Ishikawa R, Kagami O, Hayashi C, Kohama K (1992) Characterization of smooth muscle caldesmon as a microtubule-associated protein. Cell Motility and the Cytoskeleton 23:244–251
708. Franklin MT, Wang CL, Adam LP (1997) Stretch-dependent activation and desensitization of mitogen-activated protein kinase in carotid arteries. American Journal of Physiology – Cell Physiology 273:1819–1827
709. Winder SJ, Walsh MP (1990) Smooth muscle calponin. Inhibition of actomyosin MgATPase and regulation by phosphorylation. Journal of Biological Chemistry 265:10148–10155
710. Dos Remedios CG, Chhabra D, Kekic M, Dedova IV, Tsubakihara M, Berry DA, Noseworthy NJ (2003) Encyclopaedia of Actin-Binding Proteins (and Drugs), www.bms.ed.ac.uk/research/others/smaciver/Encyclop/encycloABP.htm, also Actin-binding proteins: regulation of cytoskeletal microfilaments. Physiological Reviews 83:433–473
711. Ebong EE, Macaluso FP, Spray DC, Tarbell JM (2011) Imaging the endothelial glycocalyx in vitro by rapid freezing/freeze substitution transmission electron microscopy. Arteriosclerosis, Thrombosis, and Vascular Biology 31:1908–1915
712. Shi ZD, Wang H, Tarbell JM (2011) Heparan sulfate proteoglycans mediate interstitial flow mechanotransduction regulating MMP-13 expression and cell motility via FAK-ERK in 3D collagen. PLoS One 6:e15956
713. Hemric ME, Chalovich JM (1988) Effect of caldesmon on the ATPase activity and the binding of smooth and skeletal myosin subfragments to actin. Journal of Biological Chemistry 263:1878–1885

714. Wang Z, Yang ZQ (2000) Casein kinase II phosphorylation of caldesmon downregulates myosin-caldsemon interaction. Biochemistry 39:11114–11120
715. Katayama T, Watanabe M, Tanaka H, Hino M, Miyakawa T, Ohki T, Ye LH, Xie C, Yoshiyama S, Nakamura A, Ishikawa R, Tanokura M, Oiwa K, Kohama K (2010) Stimulatory effects of arachidonic acid on myosin ATPase activity and contraction of smooth muscle via myosin motor domain. American Journal of Physiology – Heart and Circulatory Physiology 298:H505–H514
716. Desch M, Sigl K, Hieke B, Salb K, Kees F, Bernhard D, Jochim A, Spiessberger B, Hcherl K, Feil R, Feil S, Lukowski R, Wegener JW, Hofmann F, Schlossmann J (2010) IRAG determines nitric oxide- and atrial natriuretic peptide-mediated smooth muscle relaxation. Cardiovascular Research 86:496–505
717. Narayanan D, Adebiyi A, Jaggar JH (2012) Inositol trisphosphate receptors in smooth muscle cells. American Journal of Physiology – Heart and Circulatory Physiology 302:H2190–H2210
718. Zhang W, Gunst SJ (2008) Interactions of airway smooth muscle cells with their tissue matrix: implications for contraction. Proceedings of the American Thoracic Society 5:32–39
719. Halayko AJ, Salari H, MA X, Stephens NL (1996) Markers of airway smooth muscle cell phenotype. American Journal of Physiology – Lung Cellular and Molecular Physiology 270:L1040–L1051
720. Liu QH, Zheng YM, Korde AS, Yadav VR, Rathore R, Wess J, Wang YX (2009) Membrane depolarization causes a direct activation of G protein-coupled receptors leading to local Ca^{2+} release in smooth muscle. Proceedings of the National Academy of Sciences of the United States of America 106:11418–11423
721. Einstein R, Jordan H, Zhou W, Brenner M, Moses EG, Liggett SB (2008) Alternative splicing of the G protein-coupled receptor superfamily in human airway smooth muscle diversifies the complement of receptors. Proceedings of the National Academy of Sciences of the United States of America 105:5230–5235
722. Billington CK, Penn RB (2003) Signaling and regulation of G protein-coupled receptors in airway smooth muscle. Respiratory Research 4:2
723. Ibitayo AI, Sladick J, Tuteja S, Louis-Jacques O, Yamada H, Groblewski G, Welsh M, Bitar KN (1999) HSP27 in signal transduction and association with contractile proteins in smooth muscle cells. American Journal of Physiology – Gastrointestinal and Liver Physiology 277:G445–G454
724. An SS, Fabry B, Mellema M, Bursac P, Gerthoffer WT, Kayyali US, Gaestel M, Shore SA, Fredberg JJ (2004) Role of heat shock protein 27 in cytoskeletal remodeling of the airway smooth muscle cell. Journal of Applied Physiology 96:1701–1713
725. Goldie RG, Paterson JW, Lulich KM (1990) Adrenoceptors in airway smooth muscle. Pharmacology and Therapeutics 48:295–322
726. Kassel KM, Wyatt TA, Panettieri RA, Toews ML (2008) Inhibition of human airway smooth muscle cell proliferation by β2-adrenergic receptors and cAMP is PKA independent: evidence for EPAC involvement. American Journal of Physiology – Lung Cellular and Molecular Physiology 294:L131–L138
727. Eglen RM, Reddy H, Watson N, Challiss RA (1994) Muscarinic acetylcholine receptor subtypes in smooth muscle. Trends in Pharmacological Sciences 15:114–119
728. Murthy KS, Makhlouf GM (1997) Differential coupling of muscarinic M_2 and M_3 receptors to adenylyl cyclases V/VI in smooth muscle. Concurrent M_2-mediated inhibition via $G\alpha i3$ and M_3-mediated stimulation via $G\beta\gamma q$. Journal of Biological Chemistry 272:21317–21324
729. Puddicombe SM, Polosa R, Richter A, Krishna MT, Howarth PH, Holgate ST, Davies DE (2000) Involvement of the epidermal growth factor receptor in epithelial repair in asthma. FASEB Journal 14:1362–1374
730. James A (2005) Airway remodeling in asthma. Current Opinion in Pulmonary Medicine 11:1–6

731. Knox AJ, Corbett L, Stocks J, Holland E, Zhu YM, Pang L (2001) Human airway smooth muscle cells secrete vascular endothelial growth factor: up-regulation by bradykinin via a protein kinase C and prostanoid-dependent mechanism. FASEB Journal 15:2480–2488
732. Noveral JP, Bhala A, Hintz RL, Grunstein MM, Cohen P (1994) Insulin-like growth factor axis in airway smooth muscle cells. American Journal of Physiology – Lung Cellular and Molecular Physiology 267:L761–L765
733. Boström H, Gritli-Linde A, Betsholtz C (2002) PDGF-A/PDGF alpha-receptor signaling is required for lung growth and the formation of alveoli but not for early lung branching morphogenesis. Developmental Dynamics 223:155–162
734. Bischof RJ, Bourke JE, Hirst SJ, Meeusen EN, Snibson KJ, Van Der Velden J (2009) Measurement and impact of remodeling in the lung: airway neovascularization in asthma. Proceedings of the American Thoracic Society 6:673–677
735. Barnes PJ, Shapiro SD, Pauwels RA (2003) Chronic obstructive pulmonary disease: molecular and cellular mechanisms. European Respiratory Journal 22:672–688
736. Saunders R, Sutcliffe A, Woodman L, Kaur D, Siddiqui S, Okayama Y, Wardlaw A, Bradding P, Brightling C (2008) The airway smooth muscle CCR3/CCL11 axis is inhibited by mast cells. Allergy 63:1148–1155
737. Gizycki MJ, Adelroth E, Rogers AV, O'Byrne PM, Jeffery PK (1997) Myofibroblast involvement in the allergen-induced late response in mild atopic asthma. American Journal of Respiratory Cell and Molecular Biology 16:664–673
738. Zhang S, Smartt H, Holgate ST, Roche WR (1999) Growth factors secreted by bronchial epithelial cells control myofibroblast proliferation: an in vitro co-culture model of airway remodeling in asthma. Laboratory Investigation 79:395–405
739. Chung KF (2000) Airway smooth muscle cells: contributing to and regulating airway mucosal inflammation? European Respiratory Journal 15:961–968
740. Ammit AJ, Hastie AT, Edsall LC, Hoffman RK, Amrani Y, Krymskaya VP, Kane SA, Peters SP, Penn RB, Spiegel S, Panettieri RA (2001) Sphingosine 1-phosphate modulates human airway smooth muscle cell functions that promote inflammation and airway remodeling in asthma. FASEB Journal 15:1212–1214
741. Trepat X, Deng L, An SS, Navajas D, Tschumperlin DJ, Gerthoffer WT, Butler JP, Fredberg JJ (2007) Universal physical responses to stretch in the living cell. Nature 447:592–595
742. Seow CY (2005) Myosin filament assembly in an ever-changing myofilament lattice of smooth muscle. American Journal of Physiology – Cell Physiology 289:C1363–C1368
743. Krishnan R, Trepat X, Nguyen TT, Lenormand G, Oliver M, Fredberg JJ (2008) Airway smooth muscle and bronchospasm: fluctuating, fluidizing, freezing. Respiratory Physiology and Neurobiology 163:17–24
744. Krishnan R, Park CY, Lin YC, Mead J, Jaspers RT, Trepat X, Lenormand G, Tambe D, Smolensky AV, Knoll AH, Butler JP, Fredberg JJ (2009) Reinforcement versus fluidization in cytoskeletal mechanoresponsiveness. PLoS ONE 4:e5486
745. Lacolley P, Regnault V, Nicoletti A, Li Z, Michel JB (2012) The vascular smooth muscle cell in arterial pathology: a cell that can take on multiple roles. Cardiovascular Research 95:194–204
746. Hao H, Ropraz P, Verin V, Camenzind E, Geinoz A, Pepper MS, Gabbiani G, Bochaton-Piallat M-L (2002) Heterogeneity of smooth muscle cell populations cultured from pig coronary artery. Arteriosclerosis, Thrombosis, and Vascular Biology 22:1093–1099
747. Michel JB, Li Z, Lacolley P (2012) Smooth muscle cells and vascular diseases. Cardiovascular Research 95:135–137
748. Spin JM, Maegdefessel L, Tsao PS (2012) Vascular smooth muscle cell phenotypic plasticity: focus on chromatin remodelling. Cardiovascular Research 95:147–155
749. Demasi M, Laurindo FRM (2012) Physiological and pathological role of the ubiquitin2013proteasome system in the vascular smooth muscle cell. Cardiovascular Research 95:183–193
750. Liu WF (2012) Mechanical regulation of cellular phenotype: implications for vascular tissue regeneration. Cardiovascular Research 95:215–222

751. Chang L, Noseda M, Higginson M, Ly M, Patenaude A, Fuller M, Kyle AH, Minchinton AI, Puri MC, Dumont DJ, Karsan A (2012) Differentiation of vascular smooth muscle cells from local precursors during embryonic and adult arteriogenesis requires Notch signaling. Proceedings of the National Academy of Sciences of the United States of America 109:6993–6998
752. Fouillade C, Monet-Leprêtre M, Baron-Menguy C, Joutel A (2012) Notch signalling in smooth muscle cells during development and disease. Cardiovascular Research 95:138–146
753. Mill C, George SJ (2012) Wnt signalling in smooth muscle cells and its role in cardiovascular disorders. Cardiovascular Research 95:233–240
754. Sun SG, Zheng B, Han M, Fang XM, Li HX, SB M (2011) MiR-146a and Krüppel-like factor 4 form a feedback loop to participate in vascular smooth muscle cell proliferation. EMBO Reports 12:56–62
755. Valente AJ, Yoshida T, Murthy SN, Sakamuri SS, Katsuyama M, Clark RA, Delafontaine P, Chandrasekar B (2012) Angiotensin II enhances AT1-Nox1 binding and stimulates arterial smooth muscle cell migration and proliferation through AT1, Nox1, and interleukin-18. American Journal of Physiology – Heart and Circulatory Physiology 303:H282–H296
756. Pentz ES, Cordaillat M, Carretero OA, Tucker AE, Sequeira-Lopez ML, Gomez RA (2012) Histone acetyl transferases CBP and p300 are necessary for maintenance of renin cell identity and transformation of smooth muscle cells to the renin phenotype. American Journal of Physiology – Heart and Circulatory Physiology 302:H2545–H2552
757. Zheng T, Li W, Altura BT, Shah NC, BM Altura (2011) Sphingolipids regulate $[Mg^{2+}]_o$ uptake and $[Mg^{2+}]_i$ content in vascular smooth muscle cells: potential mechanisms and importance to membrane transport of Mg^{2+}. American Journal of Physiology – Heart and Circulatory Physiology 300:H486–H492
758. Fatagati V, Murphy RA (1984) Actin and tropomyosin variants in smooth muscles: dependence on tissue type. The Journal of Biological Chemistry 259:14383–14388
759. Rovner AS, Thompson MM, Murphy RA (1986) Two different heavy chains are found in smooth muscle myosin. American Journal of Physiology – Cell Physiology 250:861–870
760. Amer MS, Li J, O'Regan DJ, Steele DS, Porter KE, Sivaprasadarao A, Beech DJ (2009) Translocon closure to Ca^{2+} leak in proliferating vascular smooth muscle cells. American Journal of Physiology – Heart and Circulatory Physiology 296:H910–H916
761. Klabunde RE (2004) Cardiovascular Physiology Concepts. Lippincott Williams and Wilkins, Philadelphia (http://cvphysiology.com/)
762. O'Donnell ME, Owen NE (1994) Regulation of ion pumps and carriers in vascular smooth muscle. Physiological Reviews 74:683–722
763. Duling BR, Gore RW, Dacey RG Jr, Damon DN (1981) Methods for isolation, cannulation, and in vitro study of single microvessels. American Journal of Physiology – Heart and Circulatory Physiology 241:108–116
764. Davis MJ, Kuo L, Chilian WM, Muller JM (1995) Isolated, perfused microvessels. In: Barker JH, Anderson GL, Menger MD (eds) Clinically applied microcirculation research. CRC Press, Boca Raton, FL
765. Johansson B, Mellander S (1975) Static and dynamic components in the vascular myogenic response to passive changes in length as revealed by electrical and mechanical recordings from the rat portal vein. Circulation Research 36:76–83
766. Nakayama K (1982) Calcium-dependent contractile activation of cerebral artery produced by quick stretch. American Journal of Physiology – Heart and Circulatory Physiology 242:760–768
767. Dhein S (1998) Gap junction channels in the cardiovascular system: pharmacological and physiological modulation. Trends in Pharmacological Science 19:229–241
768. Evans AM, Osipenko ON, Gurney AM (1996) Properties of a novel K+ current that is active at resting potential in rabbit pulmonary artery smooth muscle cells. Journal of Physiology 496:407–420

769. Nelson MT, Patlak JB, Worley JF, Standen NB (1990) Calcium channels, potassium channels, and voltage dependence of arterial smooth muscle tone. American Journal of Physiology – Cell Physiology 259:3–18
770. Dought JM, Miller AL, Langton PD (1998) Non-specificity of chloride channel blockers in rat cerebral arteries: block of the L-type calcium channel. Journal of Physiology 507:433-439
771. Davis MJ, Hill MA (1999) Signaling mechanisms underlying the vascular myogenic response. Physiological Reviews 79:387–423
772. Knock GA, Shaifta Y, Snetkov VA, Vowles B, Drndarski S, Ward JP, Aaronson PI (2008) Interaction between src family kinases and rho-kinase in agonist-induced Ca^{2+}-sensitization of rat pulmonary artery. Cardiovascular Research 77:570–579
773. Nakao F, Kobayashi S, Mogami K, Mizukami Y, Shirao S, Miwa S, Todoroki-Ikeda N, Ito M, Matsuzaki M (2002) Involvement of Src family protein tyrosine kinases in Ca^{2+} sensitization of coronary artery contraction mediated by a sphingosylphosphorylcholine-Rho-kinase pathway. Circulation Research 91:953–960
774. Surks HK, Mochizuki N, Kasai Y, Georgescu SP, Tang KM, Ito M, Lincoln TM, Mendelsohn ME (1999) Regulation of myosin phosphatase by a specific interaction with cGMP-dependent protein kinase Iα. Science 286:1583–1587
775. Chen X, Pavlish K, Benoit JN (2008) Myosin phosphorylation triggers actin polymerization in vascular smooth muscle. American Journal of Physiology – Heart and Circulatory Physiology 295:H2172–H2177
776. Miralles F, Posern G, Zaromytidou AI, Treisman R (2003) Actin dynamics control SRF activity by regulation of its coactivator MAL. Cell 113:329–342
777. Cordes KR, Sheehy NT, White MP, Berry EC, Morton SU, Muth AN, Lee TH, Miano JM, Ivey KN, Srivastava D (2009) miR-145 and miR-143 regulate smooth muscle cell fate and plasticity. Nature 460:705–710
778. Ansari HR, Teng B, Nadeem A, Roush KP, Martin KH, Schnermann J, Mustafa SJ (2009) A_1 adenosine receptor-mediated PKC and p42/p44 MAPK signaling in mouse coronary artery smooth muscle cells. American Journal of Physiology – Heart and Circulatory Physiology 297:H1032–H1039
779. Zacchigna L, Vecchione C, Notte A, Cordenonsi M, Dupont S, Maretto S, Cifelli G, Ferrari A, Maffei A, Fabbro C, Braghetta P, Marino G, Selvetella G, Aretini A, Colonnese C, Bettarini U, Russo G, Soligo S, Adorno M, Bonaldo P, Volpin D, Piccolo S, Lembo G, Bressan GM (2006) Emilin1 links TGF-beta maturation to blood pressure homeostasis. Cell 124:929–942
780. Osanai T, Tomita H, Kushibiki M, Yamada M, Tanaka M, Ashitate T, Echizen T, Katoh C, Magota K, Okumura K (2009) Coupling factor 6 enhances Src-mediated responsiveness to angiotensin II in resistance arterioles and cells. Cardiovascular Research 81:780–787
781. Lo IC, Lin TM, Chou LH, Liu SL, Wu LW, Shi GY, Wu HL, Jiang MJ (2009) Ets-1 mediates platelet-derived growth factor-BB-induced thrombomodulin expression in human vascular smooth muscle cells. Cardiovascular Research 81:771–779
782. Greco CM, Camera M, Facchinetti L, Brambilla M, Pellegrino S, Gelmi ML, Tremoli E, Corsini A, Ferri N (2012) Chemotactic effect of prorenin on human aortic smooth muscle cells: a novel function of the (pro)renin receptor. Cardiovascular Research 95:366–374
783. Snyder GD, Oberley-Deegan RE, Goss KL, Romig-Martin SA, Stoll LL, Snyder JM, Weintraub NL (2008) Surfactant protein D is expressed and modulates inflammatory responses in human coronary artery smooth muscle cells. American Journal of Physiology – Heart and Circulatory Physiology 294:H2053–H2059
784. Gerthoffer WT (2007) Mechanisms of vascular smooth muscle cell migration. Circulation Research 100:607–621
785. Vuori K, Hirai H, Aizawa S, Ruoslahti E (1996) Introduction of p130cas signaling complex formation upon integrin-mediated cell adhesion: a role for Src family kinases. Molecular and Cellular Biology 16:2606–2613

786. Bayliss WM (1902) On the local reactions of the arterial wall to changes of internal pressure. Journal of Physiology 28:220–231
787. Broegger T, Jacobsen JC, Secher Dam V, Boedtkjer DM, Kold-Petersen H, Pedersen FS, Aalkjaer C, Matchkov VV (2011) Bestrophin is important for the rhythmic but not the tonic contraction in rat mesenteric small arteries. Cardiovascular Research 91:685–693
788. Johnson PC (1981) The myogenic response. In: Bohr DF, Somlyo AP, Sparks HV Jr (eds) Vascular smooth muscle. Handbook of physiology. The cardiovascular system. American Physiological Society, Bethesda, MD
789. Harder DR, Gilbert R, Lombard JH (1987) Vascular muscle cell depolarization and activation in renal arteries on elevation of transmural pressure. American Journal of Physiology – Renal Physiology 253:778–781
790. Knot HJ, Nelson MT (1995) Regulation of membrane potential and diameter by voltage-dependent K+ channels in rabbit myogenic cerebral arteries. American Journal of Physiology – Heart and Circulatory Physiology 269:348–355
791. Kirber MT, Walsh JV, Singer JJ (1988) Stretch-activated ion channels in smooth muscle: a mechanism for the initiation of stretch-induced contraction. Pflgers Archives 412:339–345
792. Davis MJ, Donovitz JA, Hood JD (1992) Stretch-activated single-channel and whole-cell currents in vascular smooth muscle cells. American Journal of Physiology – Cell Physiology 262:1083–1088
793. Davis MJ, Meininger GA, Zawieja DC (1992) Stretch-induced increases in intracellular calcium in isolated vascular smooth muscle cells. American Journal of Physiology – Heart and Circulatory Physiology 263:1292–1299
794. Zou AP, Fleming JT, Falck JR, Jacobs ER, Gebremedhin D, Harder DR, R. J. Roman RJ (1996) 20-HETE is an endogenous inhibitor of the large-conductance Ca(2+)-activated K+ channel in renal arterioles. American Journal of Physiology – Regulatory, Integrative and Comparative Physiology 270:228–237
795. Harder DR, Narayanan J, Gebremedhin D (2011) Pressure-induced myogenic tone and role of 20-HETE in mediating autoregulation of cerebral blood flow. American Journal of Physiology – Heart and Circulatory Physiology 300:H1557–H1565
796. Kauffenstein G, Laher I, Matrougui K, Guérineau NC, Henrion D (2012) Emerging role of G protein-coupled receptors in microvascular myogenic tone. Cardiovascular Research 95:223–232
797. Goldschmidt-Clermont PJ, Machesky LM, Baldassare JJ, Pollard TD (1990) The actin-binding protein profilin binds to PIP2 and inhibits its hydrolysis by phospholipase C. Science 247:1575–1578
798. Sheng M, McFadden G, Greenberg ME (1990) Membrane depolarization and calcium induce c-fos transcription via phosphorylation of transcription factor CREB. Neuron 4:571–582
799. Nollert MU, Eskin SG, McIntire LV (1990) Shear stress increases inositol trisphosphate levels in human endothelial cells. Biochemical and Biophysical Research Communications 170:281–287
800. Meininger GA, Davis MJ (1992) Cellular mechanisms involved in the vascular myogenic response. American Journal of Physiology – Heart and Circulatory Physiology 263:647–659
801. Miller FJ, Dellsperger KC, Guttermann DD (1997) Myogenic constriction of human coronary arterioles. American Journal of Physiology – Heart and Circulatory Physiology 273:H257–H264
802. Karibe A, Watanabe J, Horiguchi S, Takeuchi M, Suzuki S, Funakoshi M, Katoh H, Keitoku M, Satoh S, Shirato K (1997) Role of cytosolic Ca2+ and protein kinase C in developing myogenic contraction in isolated rat small arteries. American Journal of Physiology – Heart and Circulatory Physiology 272:1165–1172
803. Wiersbitzky M, Mills I, Sumpio BE, Gerwitz H (1994) Chronic cyclic strain reduces adenylate cyclase activity and stimulatory G protein subunit levels in coronary smooth muscle cells. Experimental Cell Research 210:52–55

804. Mills I, Letsou G, Rabban J, Sumpio B, Gewirtz H (1990) Mechanosensitive adenylate cyclase activity in coronary vascular smooth muscle cells. Biochemical and Biophysical Research Communications 171:143–147
805. Gudi S, Nolan JP, Frangos JA (1998) Modulation of GTPase activity of G proteins by fluid shear stress and phospholipid composition. Proceedings of the National Academy of Sciences of the United States of America 95:2515–2519
806. Davies PF, Mundel T, Barbee KA (1995) A mechanism for heterogeneous endothelial responses to flow in vivo and in vitro. Journal of Biomechanics 28:1553–1560
807. Burridge K, Fath K, Kelly T, Nuckolls G, Turner C (1988) Focal adhesions: transmembrane junctions between the extracellular matrix and the cytoskeleton. Annual Review of Cell Biology 4:1487–525
808. Chicurel ME, Chen CS, Ingber DE (1998) Cellular control lies in the balance of forces. Current Opinion in Cell Biology 10:232–239
809. Frangos JA, Eskin SG, McIntire LV, Ives CL (1985) Flow effects on prostacyclin production by cultured human endothelial cells. Science 227:1477–1479
810. Topper JN, Gimbrone MA (1999) Blood flow and vascular gene expression: fluid shear stress as a modulator of endothelial phenotype. Molecular Medicine Today 5:40–46
811. Khachigian LM, Resnick N, Gimbrone MA, Collins T (1995) Nuclear factor-kB interacts functionally with the platelet-derived growth facor B-chain shear-stress-response-element in vascular endothelial cells exposed to fluid shear stress. Journal of Clinical Investigation 96:1169–1175
812. Maguire JJ, Davenport AP (2005) Regulation of vascular reactivity by established and emerging GPCRs. Trends in Pharmacological Sciences 26:448–454
813. King MW (2005) The medical biochemistry page. http://web.indstate.edu/thcme/mwking/
814. Belton O, Fitzgerald DJ (2003) Cyclooxygenase isoforms and atherosclerosis. Expert Reviews in Molecular Medicine 5:1–18
815. Poeckel D, Funk CD (2010) The 5-lipoxygenase/leukotriene pathway in preclinical models of cardiovascular disease. Cardiovascular Research 86:243–253
816. Frazziano G, Champion HC, Pagano PJ (2012) NADPH oxidase-derived ROS and the regulation of pulmonary vessel tone. American Journal of Physiology – Heart and Circulatory Physiology 302:H2166–H2177
817. Nelson CP, Rainbow RD, Brignell JL, Perry MD, Willets JM, Davies NW, Standen NB, Challiss RA (2011) Principal role of adenylyl cyclase 6 in K^+ channel regulation and vasodilator signalling in vascular smooth muscle cells. Cardiovascular Research 91:694–702
818. Fellner SK, Arendshorst WJ (2010) Complex interactions of NO/cGMP/PKG systems on Ca^{2+} signaling in afferent arteriolar vascular smooth muscle. American Journal of Physiology – Heart and Circulatory Physiology 298:H144–H151
819. Rosenfeld CR, Liu XT, Despain K (2009) Pregnancy modifies the large conductance Ca^{2+}-activated K^+ channel and cGMP-dependent signaling pathway in uterine vascular smooth muscle. American Journal of Physiology – Heart and Circulatory Physiology 296:H1878-H1887
820. Cao Z, Bell JB, Mohanty JG, Nagababu E, Rifkind JM (2009) Nitrite enhances RBC hypoxic ATP synthesis and the release of ATP into the vasculature: a new mechanism for nitrite-induced vasodilation. American Journal of Physiology – Heart and Circulatory Physiology 297:H1494–H1503
821. Mombouli JV, Vanhoutte PM (1997) Endothelium-derived hyperpolarizing factor(s): updating the unknown. Trends in Pharmacological Sciences 18:252–256
822. Fleming I (2001) Cytochrome P450 and vascular homeostasis. Circulation Research 89:753–762
823. Bauersachs J, Popp R, Hecker M, Sauer E, Fleming I, Busse R (1996) Nitric oxide attenuates the release of endothelium-derived hyperpolarizing factor. Circulation 94:3341–3347
824. Jacobs ER, Zeldin DC (2001) The lung HETEs (and EETs) up. American Journal of Physiology – Heart and Circulatory Physiology 280:H1–H10

825. Chawengsub Y, Gauthier KM, Campbell WB (2009) Role of arachidonic acid lipoxygenase metabolites in the regulation of vascular tone. American Journal of Physiology – Heart and Circulatory Physiology 297:H495–H507
826. Hinson JP, Kapas S, Smith DM (2000) Adrenomedullin, a multifunctional regulatory peptide. Endocrine Reviews 21:138–167.
827. Dettmann ES, Vysniauskiene I, Wu R, Flammer J, Haefliger IO (2003) Adrenomedullin-induced endothelium-dependent relaxation in porcine ciliary arteries. Investigative Ophthalmology and Visual Science 44:3961–3966
828. Burnstock G (2006) Vessel tone and remodeling Nature Medicine 12:16–17
829. Milner P, Kirkpatrick KA, Ralevic V, Toothill V, Pearson J, Burnstock G (1990) Endothelial cells cultured from human umbilical vein release ATP, substance P and acetylcholine in response to increased flow. Proceedings of the Royal Society of London – Series B: Biological Sciences 241:245–248
830. Yamamoto K, Sokabe T, Matsumoto T, Yoshimura K, Shibata M, Ohura N, Fukuda T, Sato T, Sekine K, Kato S, Isshiki M, Fujita T, Kobayashi M, Kawamura K, Kamiya A, Ando J (2005) Impaired flow-dependent control of vascular tone and remodeling in P2X4-deficient mice. Nature – Medicine 12:133–137
831. Alexander SPH, Mathie A, Peters JA (2007) Guide to receptors and channels. British Journal of Pharmacology 150:S1–S168
832. Granata F, Potenza RL, Fiori A, Strom R, Caronti B, Molinari P, Donsante S, Citro G, Iacovelli L, De Blasi A, Ngomba RT, Palladini G, Passarelli F (2003) Expression of OP4 (ORL1, NOP1) receptors in vascular endothelium. European Journal of Pharmacology 482:17–23
833. Hou M, Uddman R, Tajti J, Edvinsson L (2003) Nociceptin immunoreactivity and receptor mRNA in the human trigeminal ganglion. Brain Research 964:179–186
834. Katusic ZS (1996) Superoxide anion and endothelial regulation of arterial tone. Free Radical Biology and Medicine 20:443–448
835. Jankowski V, Tlle M, Vanholder R, Schnfelder G, van der Giet M, Henning L, Schlter H, Paul M, Zidek W, Jankowski J (2005) Uridine adenosine tetraphosphate: a novel endothelium-derived vasoconstrictive factor. Nature – Medicine 11:223–227
836. Gui Y, Walsh MP, Jankowski V, Jankowski J, Zheng XL (2008) Up4A stimulates endothelium-independent contraction of isolated rat pulmonary artery. American Journal of Physiology – Lung Cellular and Molecular Physiology 294:L733–L738
837. Luykenaar KD, El-Rahman RA, Walsh MP, Welsh DG (2009) Rho-kinase-mediated suppression of KDR current in cerebral arteries requires an intact actin cytoskeleton. American Journal of Physiology – Heart and Circulatory Physiology 296:H917–H926
838. Harrington LS, Mitchell JA (2004) Novel role for P2X receptor activation in endothelium-dependent vasodilation. British Journal of Pharmacology 143:611–617
839. Wiemer G, Scholkens BA, Becker RH, Busse R (1991) Ramiprilat enhances endothelial autacoid formation by inhibiting breakdown of endothelium-derived bradykinin. Hypertension 18:558–563
840. Mitchell JD, Maguire JJ, Kuc RE, Davenport AP (2009) Expression and vasoconstrictor function of anorexigenic peptides neuromedin U-25 and S in the human cardiovascular system. Cardiovascular Research 81:353–361

Chap. 9. Endothelium

841. Galley HF, Webster NR (2004) Physiology of the endothelium. British Journal of Anaesthesia 93:105–113
842. Lewis JD, Destito G, Zijlstra A, Gonzalez MJ, Quigley JP, Manchester M, Stuhlmann H (2006) Viral nanoparticles as tools for intravital vascular imaging. Nature – Medicine 12:354–360

843. Reichlin T, Wild A, Durrenberger M, Daniels AU, Aebi U, Hunziker PR, Stolz M (2005) Investigating native coronary artery endothelium in situ and in cell culture by scanning force microscopy. Journal of Structural Biology 152:52–63
844. Fadini GP, Avogaro A (2010) Cell-based methods for ex vivo evaluation of human endothelial biology. Cardiovascular Research 87:12–21
845. Urao N, Okigaki M, Yamada H, Aadachi Y, Matsuno K, Matsui A, Matsunaga S, Tateishi K, Nomura T, Takahashi T, Tatsumi T, Matsubara H (2006) Erythropoietin-mobilized endothelial progenitors enhance reendothelialization via Akt-endothelial nitric oxide synthase activation and prevent neointimal hyperplasia. Cardiovascular Research 98:1405–1413
846. Michaux G, Abbitt KB, Collinson LM, Haberichter SL, Norman KE, Cutler DF (2006) The physiological function of von Willebrand's factor depends on its tubular storage in endothelial Weibel-Palade bodies. Developmental Cell 10:223–232
847. Tarbell JM, Lever MJ, Caro CG (1988) The effect of varying albumin concentration of the hydraulic conductivity of the rabbit common carotid artery. Microvascular Research 35:204–220
848. Beaucourt J, Biben T, Misbah C (2004) Optimal lift force on vesicles near a compressible substrate. Europhysics Letters 67:676–682
849. Wang L, Fuster M, Sriramarao P, Esko JD (2005) Endothelial heparan sulfate deficiency impairs L-selectin- and chemokine-mediated neutrophil trafficking during inflammatory responses. Nature – Immunology 6:902–910
850. Mokady AJ, Mestel AJ, Winlove CP (1999) Flow through a charged biopolymer layer. Journal of Fluid Mechanics 383:353–378
851. Ramiro-Diaz J, Barajas-Espinosa A, Chi-Ahumada E, Perez-Aguilar S, Torres-Tirado D, Castillo-Hernandez J, Knabb M, de Rosa AB, Rubio R (2010) Luminal endothelial lectins with affinity for N-acetylglucosamine determine flow-induced cardiac and vascular paracrine-dependent responses. American Journal of Physiology – Heart and Circulatory Physiology 299:H743–H751
852. Liu Z, Tan JL, Cohen DM, Yang MT, Sniadecki NJ, Ruiz SA, Nelson CM, Chen CS (2010) Mechanical tugging force regulates the size of cell-cell junctions. Proceedings of the National Academy of Sciences of the United States of America 107:9944–9949
853. Nyqvist D, Giampietro C, Dejan E (2008) Deciphering the functional role of endothelial junctions by using in vivo models. EMBO Reports 9:742–747
854. Mehta D, Malik AB (2006) Signaling mechanisms regulating endothelial permeability. Physiological Reviews 86:279-367
855. Taddei A, Giampietro C, Conti A, Orsenigo F, Breviario F, Pirazzoli V, Potente M, Daly C, Dimmeler S, Dejana E (2008) Endothelial adherens junctions control tight junctions by VE-cadherin-mediated upregulation of claudin-5. Nature – Cell Biology 10:923–934
856. Sandow SL (2004) Factors, fiction and endothelium-derived hyperpolarizing factor. Clinical and Experimental Pharmacology and Physiology 31:563–570
857. Chadha PS, Liu L, Rikard-Bell M, Senadheera S, Howitt L, Bertrand RL, Grayson TH, Murphy TV, Sandow SL (2011) Endothelium-dependent vasodilation in human mesenteric artery is primarily mediated by myoendothelial gap junctions intermediate conductance calcium-activated K^+ channel and nitric oxide. Journal of Pharmacology and Experimental Therapeutics 336:701–708
858. Aaronson PI (2012) Does TRPC3 macrodominate the myoendothelial gap junction microdomain? Cardiovascular Research 95:399–400
859. Edwards G, Dora KA, Gardener MJ, Garland CJ, Weston AH (1998) K^+ is an endothelium-derived hyperpolarizing factor in rat arteries. Nature 396:269–272
860. Ledoux J, Taylor MS, Bonev AD, Hannah RM, Solodushko V, Shui B, Tallini Y, Kotlikoff MI, Nelson MT (2008) Functional architecture of inositol 1,4,5-trisphosphate signaling in restricted spaces of myoendothelial projections. Proceedings of the National Academy of Sciences of the United States of America 105:9627–9632
861. Senadheera S, Kim Y, Grayson TH, Toemoe S, Kochukov MY, Abramowitz J, Housley GD, Bertrand RL, Chadha PS, Bertrand PP, Murphy TV, Tare M, Birnbaumer L, Marrelli

SP, Sandow SL (2012) Transient receptor potential canonical type 3 channels facilitate endothelium-derived hyperpolarization-mediated resistance artery vasodilator activity. Cardiovascular Research 95:439–447
862. Zhao D, Qin L, Bourbon PM, James L, Dvorak HF, Zeng H (2011) Orphan nuclear transcription factor TR3/Nur77 regulates microvessel permeability by targeting endothelial nitric oxide synthase and destabilizing endothelial junctions. Proceedings of the National Academy of Sciences of the United States of America 108:12066–12071
863. Girard JP, Springer TA (1995) High endothelial venules (HEVs): specialized endothelium for lymphocyte migration. Immunology Today 16:449–457
864. Chen Q, Fisher DT, Clancy KA, Gauguet JMM, Wang WC, Unger E, Rose-John S, von Andrian UH, Baumann H, Evans SS (2006) Fever-range thermal stress promotes lymphocyte trafficking across high endothelial venules via an interleukin 6 trans-signaling mechanism. Nature – Immunology 7:1299–1308
865. Kanda H, Newton R, Klein R, Morita Y, Gunn MD, Rosen SD (2008) Autotaxin, an ectoenzyme that produces lysophosphatidic acid, promotes the entry of lymphocytes into secondary lymphoid organs. Nature – Immunology 9:415–423
866. Backhed F, Crawford PA, O'Donnell D, Gordon JI (2007) Postnatal lymphatic partitioning from the blood vasculature in the small intestine requires fasting-induced adipose factor. Proceedings of the National Academy of Sciences of the United States of America 104:606–611
867. Ioannidou S, Deinhardt K, Miotla J, Bradley J, Cheung E, Samuelsson S, Ng YS, Shima DT (2006) An in vitro assay reveals a role for the diaphragm protein PV-1 in endothelial fenestra morphogenesis. Proceedings of the National Academy of Sciences of the United States of America 103:16770–16775
868. Hagensen MK, Vanhoutte PM, Bentzon JF (2012) Arterial endothelial cells: still the craftsmen of regenerated endothelium. Cardiovascular Research 95:281–289
869. Asahara T, Murohara T, Sullivan A, Silver M, van der Zee R, Li T, Witzenbichler B, Schatteman G, and Isner JM (1997) Isolation of putative progenitor endothelial cells for angiogenesis. Science 275:964–967
870. Leone AM, Valgimigli M, Giannico MB, Zaccone V, Perfetti M, D'Amario D, Rebuzzi AG, Crea F (2009) From bone marrow to the arterial wall: the ongoing tale of endothelial progenitor cells. European Heart Journal 30:890–899
871. Asahara T, Kawamoto A (2004) Endothelial progenitor cells for postnatal vasculogenesis. American Journal of Physiology – Cell Physiology 287:C572–579
872. Purhonen S, Palm J, Rossi D, Kaskenp N, Rajantie I, Yl-Herttuala S, Alitalo K, Weissman IL, Salven P (2008) Bone marrow-derived circulating endothelial precursors do not contribute to vascular endothelium and are not needed for tumor growth. Proceedings of the National Academy of Sciences of the United States of America 105:6620–6625
873. Ladhoff J, Fleischer B, Hara Y, Volk HD, Seifert M (2010) Immune privilege of endothelial cells differentiated from endothelial progenitor cells. Cardiovascular Research 88:121–129
874. Pi X, Wu Y, Ferguson JE 3rd, Portbury AL, Patterson C (2009) SDF-1α stimulates JNK3 activity via eNOS-dependent nitrosylation of MKP7 to enhance endothelial migration. Proceedings of the National Academy of Sciences of the United States of America 106:5675–5680
875. Traub O, Berk BC (1998) Laminar shear stress: mechanisms by which endothelial cells transduce an atheroprotective force. Arteriosclerosis, Thrombosis, and Vascular Biology 18:677–685
876. Dhanasekaran A, Bodiga S, Gruenloh S, Gao Y, Dunn L, Falck JR, Buonaccorsi JN, Medhora M, Jacobs ER (2009) 20-HETE increases survival and decreases apoptosis in pulmonary arteries and pulmonary artery endothelial cells. American Journal of Physiology – Heart and Circulatory Physiology 296:H777–H786
877. Guo AM, Scicli G, Sheng J, Falck JC, Edwards PA, Scicli AG (2009) 20-HETE can act as a nonhypoxic regulator of HIF-1α in human microvascular endothelial cells. American Journal of Physiology – Heart and Circulatory Physiology 297:H602–H613

878. Chidlow JH, Sessa WC (2010) Caveolae, caveolins, and cavins: complex control of cellular signalling and inflammation. Cardiovascular Research 86:219–225
879. Swales KE, Moore R, Truss NJ, Tucker A, Warner TD, Negishi M, Bishop-Bailey D (2012) Pregnane X receptor regulates drug metabolism and transport in the vasculature and protects from oxidative stress. Cardiovascular Research 93:674–681
880. Bretschneider E, Uzonyi B, Weber AA, Fischer JW, Pape R, Ltzer K, Schrr K (2007) Human vascular smooth muscle cells express functionally active endothelial cell protein C receptor. Circulation Research 100:255–262
881. Schulte G, Shenoy SK (2011) β-Arrestin and dishevelled coordinate biased signaling. Proceedings of the National Academy of Sciences of the United States of America 108:19839–19840
882. Soh UJK Trejo JA (2011) Activated protein C promotes protease-activated receptor-1 cytoprotective signaling through β-arrestin and dishevelled-2 scaffolds. Proceedings of the National Academy of Sciences of the United States of America 108:19849-19850
883. Svensson KJ, Kucharzewska P, Christianson HC, Sköld S, Löfstedt T, Johansson MC, Mörgelin M, Bengzon J, Ruf W, Belting M (2011) Hypoxia triggers a proangiogenic pathway involving cancer cell microvesicles and PAR-2-mediated heparin-binding EGF signaling in endothelial cells. Proceedings of the National Academy of Sciences of the United States of America 108:13147–13152
884. Hatakeyama S, Sugihara K, Nakayama J, Akama TO, Wong SM, Kawashima H, Zhang J, Smith DF, Ohyama C, Fukuda M, Fukuda MN (2009) Identification of mRNA splicing factors as the endothelial receptor for carbohydrate-dependent lung colonization of cancer cells. Proceedings of the National Academy of Sciences of the United States of America 106:3095-3100
885. Elfenbein A, Lanahan A, Zhou TX, Yamasaki A, Tkachenko E, Matsuda M, Simons M (2012) Syndecan 4 regulates FGFR1 signaling in endothelial cells by directing macropinocytosis. Science Signaling 5:ra36
886. Herbert SP, Stainier DYR (2011) Molecular control of endothelial cell behaviour during blood vessel morphogenesis. Nature Reviews – Molecular Cell Biology 12:551–564
887. Rudini N, Felici A, Giampietro C, Lampugnani MG, Corada M, Swirsding K, Garr M, Liebner S, Letarte M, ten Dijke P, Dejana E (2008) VE-cadherin is a critical endothelial regulator of TGF-β signalling. EMBO Journal 27:993–1004
888. Medici D, Shore EM, Lounev VY, Kaplan FS, Kalluri R, Olsen BR (2010) Conversion of vascular endothelial cells into multipotent stem-like cells. Nature – Medicine 16:1400–1406
889. Zhou J, Lee PL, Tsai CS, Lee CI, Yang TL, Chuang HS, Lin WW, Lin TE, Lim SH, Wei SY, Chen YL, Chien S, Chiu JJ (2012) Force-specific activation of Smad1/5 regulates vascular endothelial cell cycle progression in response to disturbed flow. Proceedings of the National Academy of Sciences of the United States of America 109:7770–7775
890. Meuchel LW, Thompson MA, Cassivi SD, Pabelick CM, Prakash YS (2011) Neurotrophins induce nitric oxide generation in human pulmonary artery endothelial cells. Cardiovascular Research 91:668–676
891. Zhu L, Carretero OA, Xu J, Wang L, Harding P, Rhaleb NE, Yang JJ, Sumners C, Yang XP (2012) Angiotensin II type 2 receptor-stimulated activation of plasma prekallikrein and bradykinin release: role of SHP-1. American Journal of Physiology – Heart and Circulatory Physiology 302:H2553–H2559
892. Jones CA, London NR, Chen H, Park KW, Sauvaget D, Stockton RA, Wythe JD, Suh W, Larrieu-Lahargue F, Mukouyama YS, Lindblom P, Seth P, Frias A, Nishiya N, Ginsberg MH, Gerhardt H, Zhang K, Li DY (2008) Robo4 stabilizes the vascular network by inhibiting pathologic angiogenesis and endothelial hyperpermeability. Nature – Medicine 14:448–453
893. Lundström A, Gallio M, Englund C, Steneberg P, Hemphälä J, Aspenström P, Keleman K, Falileeva L, Dickson BJ, Samakovlis C (2004) Vilse, a conserved Rac/Cdc42 GAP mediating Robo repulsion in tracheal cells and axons. Genes and Development 18:2161–2171
894. Hsieh MY, Chen WY, Jiang MJ, Cheng BC, Huang TY, Chang MS (2006) Interleukin-20 promotes angiogenesis in a direct and indirect manner. Genes and Immunity 7:234–242

895. Nilius B, Droogmans G (2001) Ion channels and their functional role in vascular endothelium, Physiological Reviews 81:1415–1459
896. Yao X, Garland CJ (2005) Recent developments in vascular endothelial cell transient receptor potential channels. Circulation Research 97:853–863
897. Bishara NB, Ding H (2010) Glucose enhances expression of TRPC1 and calcium entry in endothelial cells. American Journal of Physiology – Heart and Circulatory Physiology 298:H171–H178
898. Wlfle SE, Schmidt VJ, Hoyer J, Khler R, de Wit C (2009) Prominent role of KCa3.1 in endothelium-derived hyperpolarizing factor-type dilations and conducted responses in the microcirculation in vivo. Cardiovascular Research 82:476–483
899. Helmlinger G, Berk BC, Nerem RM (1995) The calcium responses of endothelial cell monolayers subjected to pulsatile and steady laminar flow differ. American Journal of Physiology – Cell Physiology 269:C367–C375
900. Qi YX, Jiang J, Jiang XH, Wang XD, Ji SY, Han Y, Long DK, Shen BR, Yan ZQ, Chien S, Jiang ZL (2011) PDGF-BB and TGF-β1 on cross-talk between endothelial and smooth muscle cells in vascular remodeling induced by low shear stress. Proceedings of the National Academy of Sciences of the United States of America 108:1908–1913
901. Parnavelas JG, Kelly W, Burnstock G (1985) Ultrastructural localization of choline acetyltransferase in vascular endothelial cells in rat brain. Nature 316:724–725
902. Zhou Y, Varadharaj S, Zhao X, Parinandi N, Flavahan NA, Zweier JL (2005) Acetylcholine causes endothelium-dependent contraction of mouse arteries. American Journal of Physiology – Heart and Circulatory Physiology 289:H1027–H1032
903. Tousoulis D, Davies G, Crake T [correspondence]; Lerman A, Hasdai D, Holmes D [response] (1998) Acetylcholine and endothelial function. Circulation 98:1587–1590
904. Milner P, Ralevic V, Hopwood AM,Feher E, Lincoln J, Kirkpatrick A, Burnstock G (1989) Ultrastructural localisation of substance P and choline acetyltransferase in endothelial cells of rat coronary artery and release of substance P and acetylcholine during hypoxia. Experientia 45:121–125
905. Loesch A, Burnstock G (1988) Ultrastructural localisation of serotonin and substance P in vascular endothelial cells of rat femoral and mesenteric arteries. Anatomy and Embryology 178:137–142
906. Caron K, Hagaman J, Nishikimi T, Kim HS, Smithies O (2007) Adrenomedullin gene expression differences in mice do not affect blood pressure but modulate hypertension-induced pathology in males. Proceedings of the National Academy of Sciences of the United States of America 104:3420–3425
907. Chakrabarti S, Lekontseva O, Peters A, Davidge ST (2010) 17-β-Estradiol induces protein S-nitrosylation in the endothelium. Cardiovascular Research 85:796–805
908. Li L, Hisamoto K, Kim KH, Haynes MP, Bauer PM, Sanjay A, Collinge M, Baron R, Sessa WC, Bender JR (2007) Variant estrogen receptor c-Src molecular interdependence and c-Src structural requirements for endothelial NO synthase activation. Proceedings of the National Academy of Sciences of the United States of America 104:16468–16473
909. Oberleithner H, Callies C, Kusche-Vihrog K, Schillers H, Shahin V, Riethmller C, Macgregor GA, de Wardener HE (2009) Potassium softens vascular endothelium and increases nitric oxide release. Proceedings of the National Academy of Sciences of the United States of America 106:2829–2834
910. Tanizawa K (2011) Production of H_2S by 3-mercaptopyruvate sulphurtransferase. Journal of Biochemistry 149:357–359
911. Leffler CW, Parfenova H, Jaggar JH (2011) Carbon monoxide as an endogenous vascular modulator. American Journal of Physiology – Heart and Circulatory Physiology 1:H1–H11
912. Gilchrist M, Shore AC, Benjamin N (2011) Inorganic nitrate and nitrite and control of blood pressure. Cardiovascular Research 89:492–498
913. Patel RP, Hogg N, Kim-Shapiro DB (2011) The potential role of the red blood cell in nitrite-dependent regulation of blood flow. Cardiovascular Research 89:507–515

914. Shibuya N, Mikami Y, Kimura Y, Nagahara N, Kimura H (2009) Vascular endothelium expresses 3-mercaptopyruvate sulfurtransferase and produces hydrogen sulfide. Journal of Biochemistry 146:623–626
915. Liu L, Shi GP (2012) CD31: Beyond a marker for endothelial cells. Cardiovascular Research 94:3–5
916. Fornasa G, Clement M, Groyer E, Gaston AT, Khallou-Laschet J, Morvan M, Guedj K, Kaveri SV, Tedgui A, Michel JB, Nicoletti A, Caligiuri G (2012) A CD31-derived peptide prevents angiotensin II-induced atherosclerosis progression and aneurysm formation. Cardiovascular Research 94:30–37
917. Ikeda K, Nakano R, Uraoka M, Nakagawa Y, Koide M, Katsume A, Minamino K, Yamada E, Yamada H, Quertermous T, Matsubara H (2009) Identification of ARIA regulating endothelial apoptosis and angiogenesis by modulating proteasomal degradation of cIAP-1 and cIAP-2. Proceedings of the National Academy of Sciences of the United States of America 106:8227–8232
918. Zeng L, Zampetaki A, Margariti A, Pepe AE, Alam S, Martin D, Xiao Q, Wang W, Jin ZG, Cockerill G, Mori K, Li YS, Hu Y, Chien S, Xu Q (2009) Sustained activation of XBP1 splicing leads to endothelial apoptosis and atherosclerosis development in response to disturbed flow. Proceedings of the National Academy of Sciences of the United States of America 106:8326–8331
919. Bachar AR, Scheffer L, Schroeder AS, Nakamura HK, Cobb LJ, Oh YK, Lerman LO, Pagano RE, Cohen P, Lerman A (2010) Humanin is expressed in human vascular walls and has a cytoprotective effect against oxidized LDL-induced oxidative stress. Cardiovascular Research 88:360-366
920. Horowitz A (2010) Angiomotin. UCSD-Nature Molecule Pages, UCSD-Nature Signaling Gateway (www.signaling-gateway.org)
921. Lelivre E, Hinek A, Lupu F, Buquet C, Soncin F, Mattot V (2008) VE-statin/egfl7 regulates vascular elastogenesis by interacting with lysyl oxidases. EMBO Journal 27:1658–1670
922. Tomkowicz B, Rybinski K, Foley B, Ebel W, Kline B, Routhier E, Sass P, Nicolaides NC, Grasso L, Zhou Y (2007) Interaction of endosialin/TEM1 with extracellular matrix proteins mediates cell adhesion and migration. Proceedings of the National Academy of Sciences of the United States of America 104:17965–17970
923. Jones RA, Kotsakis P, Johnson TS, Chau DYS, Ali S, Melino G, Griffin M (2006) Matrix changes induced by transglutaminase 2 lead to inhibition of angiogenesis and tumor growth. Cell Death and Differentiation 13:1442–1453
924. Parish CR (2006) The role of heparan sulphate in inflammation. Nature Reviews – Immunology 6:633–643
925. Ettenson DS, Koo EW, Januzzi JL, Edelman ER (2000) Endothelial heparan sulfate is necessary but not sufficient for control of vascular smooth muscle cell growth. Journal of Cellular Physiology 184:93–100
926. Lok J, Sardi SP, Guo S, Besancon E, Ha DM, Rosell A, Kim WJ, Corfas G, Lo EH (2009) Neuregulin-1 signaling in brain endothelial cells. Journal of Cerebral Blood Flow and Metabolism 29:39–43
927. Lee DY, Lee CI, Lin TE, Lim SH, Zhou J, Tseng YC, Chien S, Chiu JJ (2012) Role of histone deacetylases in transcription factor regulation and cell cycle modulation in endothelial cells in response to disturbed flow. Proceedings of the National Academy of Sciences of the United States of America 109:1967–1972
928. Ibelgaufts H (2010) Cytokines and Cells Online Pathfinder Encyclopaedia (www.copewithcytokines.de/cope.cgi)
929. Kobayashi H, Butler JM, O'Donnell R, Kobayashi M, Ding BS, Bonner B, Chiu VK, Nolan DJ, Shido K, Benjamin L, Rafii S (2010) Angiocrine factors from Akt-activated endothelial cells balance self-renewal and differentiation of haematopoietic stem cells. Nature – Cell Biology 12:1046–1056

930. Muehlich S, Schneider N, Hinkmann F, Garlichs CD, Goppelt-Struebe M (2004) Induction of connective tissue growth factor (CTGF) in human endothelial cells by lysophosphatidic acid, sphingosine-1-phosphate, and platelets. Atherosclerosis 175:261–268
931. Lampugnani MG, Orsenigo F, Gagliani MC, Tacchetti C, Dejana E (2006) Vascular endothelial cadherin controls VEGFR-2 internalization and signaling from intracellular compartments. Journal of Cell Biology 174:593–604
932. Autschbach F, Palou E, Mechtersheimer G, Rohr C, Pirotto F, Gassler N, Otto HF, Schraven B, Gaya A (1999) Expression of the membrane protein tyrosine phosphatase CD148 in human tissues. Tissue Antigens 54:485–498
933. Urbich C, Kuehbacher A, Dimmeler S (2008) Role of microRNAs in vascular diseases, inflammation, and angiogenesis. Cardiovascular Research 79:581–588
934. Fang Y, Shi C, Manduchi E, Civelek M, Davies PF (2010) MicroRNA-10a regulation of proinflammatory phenotype in athero-susceptible endothelium in vivo and in vitro. Proceedings of the National Academy of Sciences of the United States of America 107:13450–13455
935. Kaur S, Castellone MD, Bedell VM, Konar M, Gutkind JS, Ramchandran R (2006) Robo4 signaling in endothelial cells implies attraction guidance mechanisms. Journal of Biological Chemistry 281:11347–11356
936. Wang B, Xiao Y, Ding BB, Zhang N, Yuan X, Gui L, Qian KX, Duan S, Chen Z, Rao Y, Geng JG. (2003) Induction of tumor angiogenesis by Slit-Robo signaling and inhibition of cancer growth by blocking Robo activity. Cancer Cell 4:19–29
937. Moehler TM, Sauer S, Witzel M, Andrulis M, Garcia-Vallejo JJ, Grobholz R, Willhauck-Fleckenstein M, Greiner A, Goldschmidt H, Schwartz-Albiez R (2008) Involvement of α1-2-fucosyltransferase I (FUT1) and surface-expressed Lewis(y) (CD174) in first endothelial cell-cell contacts during angiogenesis. Journal of Cellular Physiology 215:27–36
938. Tarbell JM (2010) Shear stress and the endothelial transport barrier. Cardiovascular Research 87:320–330
939. Florian JA, Kosky JR, Ainslie K, Pang Z, Dull RO, Tarbell JM (2003) Heparan sulfate proteoglycan is a mechanosensor on endothelial cells. Circulation Research 93:e136—e142
940. DeMaio L, Chang YS, Gardner TW, Tarbell JM, Antonetti DA (2001) Shear stress regulates occludin content and phosphorylation. American Journal of Physiology – Heart and Circulatory Physiology 281:105–113
941. Collins NT, Cummins PM, Colgan OC, Ferguson G, Birney YA, Murphy RP, Meade G, Cahill PA (2006) Cyclic strain-mediated regulation of vascular endothelial occludin and ZO-1: influence on intercellular tight junction assembly and function. Arteriosclerosis, Thrombosis, and Vascular Biology 26:62–68
942. Schnittler HJ, Püschel B, Drenckhahn D (1997) Role of cadherins and plakoglobin in interendothelial adhesion under resting conditions and shear stress, American Journal of Physiology – Heart and Circulatory Physiology 273:H2396–H2405
943. Yao Y, Rabodzey A, Dewey CF (2010) Glycocalyx modulates the motility and proliferative response of vascular endothelium to fluid shear stress. American Journal of Physiology – Heart and Circulatory Physiology 293:H1023–H1030
944. Rizzo V, Morton C, DePaola N, Schnitzer JE, Davies PF (2003) Recruitment of endothelial caveolae into mechanotransduction pathways by flow conditioning in vitro. American Journal of Physiology – Heart and Circulatory Physiology 285:H1720–H1729
945. Walker DC, MacKenzie A, Hosford S (1994) The structure of the tricellular region of endothelial tight junctions of pulmonary capillaries analyzed by freeze-fracture. Microvascular Research 48:259–281
946. Burns AR, Walker DC, Brown ES, Thurmon LT, Bowden RA, Keese CR, Simon SI, Entman ML, Smith CW (1997) Neutrophil transendothelial migration is independent of tight junctions and occurs preferentially at tricellular corners. Journal of Immunology 159:2893–2903.
947. Lin SJ, Jan KM, Weinbaum S, Chien S (1989) Transendothelial transport of low density lipoprotein in association with cell mitosis in rat aorta. Arteriosclerosis 9:230–236

948. Tsay R, Weinbaum S, Pfeffer R (1989) A new model for capillary filtration based on recent electron microscopic studies of endothelial junctions. Chemical Engineering Communications 82:67–102
949. Dvorak AM, Feng D (2001) The vesiculo-vacuolar organelle (VVO): a new endothelial cell permeability organelle. Journal of Histochemistry and Cytochemistry 49:419–432
950. Bendayan M (2002) Morphological and cytochemical aspects of capillary permeability. Microscopy Research and Technique 57:327–349
951. Casanello P, Sobrevia L (2002) Intrauterine growth retardation is associated with reduced activity and expression of the cationic amino acid transport systems y+/hCAT-1 and y+/hCAT-2B and lower activity of nitric oxide synthase in human umbilical vein endothelial cells. Circulation Research 91:127–134
952. Knott RM, Robertson M, Muckersie E, Forrester JV (1996) Regulation of glucose transporters (GLUT-1 and GLUT-3) in human retinal endothelial cells. Biochemical Journal 318:313–317
953. Ethier CR, Zhang J (2005) Mass transfer to vascular endothelial cells. Workshop on mini-invasive procedures in medicine and surgery: Mathematical and Numerical Challenges, Montreal, May 23-27
954. Baluk P, McDonald DM (1994) The β2-adrenergic receptor agonist formoterol reduces microvascular leakage by inhibiting endothelial gap formation. American Journal of Physiology – Lung Cellular and Molecular Physiology 266:461–468
955. Martinez-Rico C, Pincet F, Thiery JP, Dufour S (2010) Integrins stimulate E-cadherin-mediated intercellular adhesion by regulating Src-kinase activation and actomyosin contractility. Journal of Cell Science 123:712–722
956. Mucha DR, Myers CL, Schaeffer RC (2003) Endothelial contraction and monolayer hyperpermeability are regulated by Src kinase. American Journal of Physiology – Heart and Circulatory Physiology 284:H994–H1002
957. Suzuki-Inoue K, Hughes CE, Inoue O, Kaneko M, Cuyun-Lira O, Takafuta T, Watson SP, Ozaki Y (2007) Involvement of Src kinases and PLCγ2 in clot retraction. Thrombosis Research 120:251–258
958. Schlegel N, Waschke J (2010) Vasodilator-stimulated phosphoprotein (VASP): crucial for activation of Rac1 in endothelial barrier maintenance. Cardiovascular Research 87:1–3
959. Mehta D, Ahmmed GU, Paria BC, Holinstat M, Voyno-Yasenetskaya T, Tiruppathi C, Minshall RD, Malik AB (2003) RhoA interaction with inositol 1,4,5-trisphosphate receptor and transient receptor potential channel-1 regulates Ca2+ entry. Role in signaling increased endothelial permeability. Journal of Biological Chemistry 278:33492–33500
960. Ahmmed GU, Mehta D, Vogel S, Holinstat M, Paria BC, Tiruppathi C, Malik AB (2004) Protein kinase Cα phosphorylates the TRPC1 channel and regulates store-operated Ca2+ entry in endothelial cells. Journal of Biological Chemistry 279:20941–20949
961. Bearer EL, Prakash JM, Manchester RD, Allen PG (2000) VASP protects actin filaments from gelsolin: An in vitro study with implications for platelet actin reorganizations. Cell Motility and the Cytoskeleton 47:351–364
962. Bubb MR, Yarmola EG, Gibson BG, Southwick FS (2003) Depolymerization of actin filaments by profilin. Effects of profilin on capping protein function. Journal of Biological Chemistry 278:24629–24635
963. Comerford KM, Lawrence DW, Synnestvedt K, Levi BP, Colgan SP (2002) Role of vasodilator-stimulated phosphoprotein in PKA-induced changes in endothelial junctional permeability. FASEB Journal 16:583–585
964. Schulz E, Tsilimingas N, Rinze R, Reiter B, Wendt M, Oelze M, Woelken-Weckmller S, Walter U, Reichenspurner H, Meinertz T, Mnzel T (2002) Functional and biochemical analysis of endothelial (dys)function and NO/cGMP signaling in human blood vessels with and without nitroglycerin pretreatment. Circulation 105:1170–1175
965. Fryer BH, Wang C, Vedantam S, Zhou GL, Jin S, Fletcher L, Simon MC, Field J (2006) cGMP-dependent protein kinase phosphorylates p21-activated kinase (Pak) 1, inhibiting Pak/Nck binding and stimulating Pak/vasodilator-stimulated phosphoprotein association. Journal of Biological Chemistry 281:11487–11495

966. Schnitzer JE, Sung A, Horvat R, Bravo J (1992) Preferential interaction of albumin-binding proteins, gp30 and gp18, with conformationally modified albumins. Presence in many cells and tissues with a possible role in catabolism. Journal of Biological Chemistry 267:24544–24553
967. Patterson CE, Lum H, Schaphorst KL, Verin AD, Garcia JG (2000) Regulation of endothelial barrier function by the cAMP-dependent protein kinase. Endothelium 7:287–308
968. Schaphorst KL, Jacobs KN, Verin AD, Garcia JGN (2001) Sphingosine 1-phosphate increases F-actin/adherins junction linkage and enhances barrier protection. American Journal of Respiratory and Critical Care Medicine 163:A615
969. Aslam M, Härtel FV, Arshad M, Gündüz D, Abdallah Y, Sauer H, Piper HM, Noll T (2010) cAMP/PKA antagonizes thrombin-induced inactivation of endothelial myosin light chain phosphatase: role of CPI-17. Cardiovascular Research 87:375–384
970. Christoffersen C, Obinata H, Kumaraswamy SB, Galvani S, Ahnström J, Sevvana M, Egerer-Sieber C, Muller YA, Hla T, Nielsen LB, Dahlbäck B (2011) Endothelium-protective sphingosine-1-phosphate provided by HDL-associated apolipoprotein M. Proceedings of the National Academy of Sciences of the United States of America 108:9613–9618
971. Salmon AH, Neal CR, Sage LM, Glass CA, Harper SJ, Bates DO (2009) Angiopoietin-1 alters microvascular permeability coefficients in vivo via modification of endothelial glycocalyx. Cardiovascular Research 83:24–33
972. Snchez FA, Rana R, Kim DD, Iwahashi T, Zheng R, Lal BK, Gordon DM, Meininger CJ, Durn WN (2009) Internalization of eNOS and NO delivery to subcellular targets determine agonist-induced hyperpermeability. Proceedings of the National Academy of Sciences of the United States of America 16:6849–6853
973. Kusaba T, Okigaki M, Matui A, Murakami M, Ishikawa K, Kimura T, Sonomura K, Adachi Y, Shibuya M, Shirayama T, Tanda S, Hatta T, Sasaki S, Mori Y, Matsubara H (2010) Klotho is associated with VEGF receptor-2 and the transient receptor potential canonical-1 Ca^{2+} channel to maintain endothelial integrity. Proceedings of the National Academy of Sciences of the United States of America 107:19308–19313
974. Knezevic N, Tauseef M, Thennes T, Mehta D (2009) The G protein $\beta\gamma$ subunit mediates reannealing of adherens junctions to reverse endothelial permeability increase by thrombin. Journal of Experimental Medicine 206:2761–2777
975. Jin BY, Lin AJ, Golan DE, Michel T (2012) MARCKS protein mediates hydrogen peroxide regulation of endothelial permeability. Proceedings of the National Academy of Sciences of the United States of America 109:14864–14869
976. Pearson MJ, Lipowsky HH (2000) Influence of erythrocyte aggregation on leukocyte margination in postcapillary venules of rat mesentery. American Journal of Physiology – Heart and Circulatory Physiology 279:1460–1471
977. Mills P, Snabre P (1995) Rheology and structure of concentrated suspensions of hard spheres. Shear induced particle migration. Journal de Physique II 5:1597–1608
978. Zhao Y, Chien S, Weinbaum S (2000) Dynamic contact forces on leukocyte microvilli and their penetration of the endothelial glycocalyx. Biophysical Journal 80:1124–1140
979. Schenkel AR, Mamdouh Z, Muller WA (2004) Locomotion of monocytes on endothelium is a critical step during extravasation. Nature – Immunology 5:393–400
980. Nourshargh S, Hordijk PL, Sixt M (2010) Breaching multiple barriers: leukocyte motility through venular walls and the interstitium. Nature Reviews – Molecular Cell Biology 11:366–378
981. HarrisTA, Yamakuchi M, Ferlito M, Mendell JT, Lowenstein CJ (2008) MicroRNA-126 regulates endothelial expression of vascular cell adhesion molecule 1. Proceedings of the National Academy of Sciences of the United States of America 105:1516–1521
982. Wu Y, Zhan L, Ai Y, Hannigan M, Gaestel M, Huang CK, Madri JA (2007) MAPKAPK2-mediated LSP1 phosphorylation and FMLP-induced neutrophil polarization. Biochemical and Biophysical Research Communications 358:170–175

983. Moser M, Bauer M, Schmid S, Ruppert R, Schmidt S, Sixt M, Wang HV, Sperandio M, Fssler R (2009) Kindlin-3 is required for β2 integrin-mediated leukocyte adhesion to endothelial cells. Nature – Medicine 15:300–305
984. Phan UT, Waldron TT, Springer TA (2006) Remodeling of the lectin–EGF-like domain interface in P- and L-selectin increases adhesiveness and shear resistance under hydrodynamic force. Nature – Immunology 7:883–889
985. Stolen CM, Marttila-Ichihara F, Koskinen K, Yegutkin GG, Turja R, Bono P, Skurnik M, Hanninen A, Jalkanen S, Salmi M (2005) Absence of the endothelial oxidase AOC3 leads to abnormal leukocyte traffic in vivo. Immunity 22:105–115
986. Luster AD, Alon R, von Andrian UH (2005) Immune cell migration in inflammation: present and future therapeutic targets. Nature – Immunology 6:1182–1190
987. Pero RS, Borchers MT, Spicher K, Ochkur SI, Sikora L, Rao SP, Abdala-Valencia H, O'Neill KR, Shen H, McGarry MP, Lee NA, Cook-Mills JM, Sriramarao P, Simon MI, Birnbaumer L, Leeb JJ (2007) Gαi2-mediated signaling events in the endothelium are involved in controlling leukocyte extravasation. Proceedings of the National Academy of Sciences of the United States of America 104:4371–4376
988. Shulman Z, Pasvolsky R, Woolf E, Grabovsky V, Feigelson SW, Erez N, Fukui Y, Alon R (2006) DOCK2 regulates chemokine-triggered lateral lymphocyte motility but not transendothelial migration. Blood 108:2150–2158
989. Carman CV, Springer TA (2004) A transmigratory cup in leukocyte diapedesis both through individual vascular endothelial cells and between them. Journal of Cell Biology 167:377–388
990. Nieminen M, Henttinen T, Merinen M, Marttila-Ichihara F, Eriksson JE, Jalkanen S (2006) Vimentin function in lymphocyte adhesion and transcellular migration. Nature – Cell Biology 8, 156–162
991. Salmi M, Jalkanen S (2005) Cell-surface enzymes in control of leukocyte trafficking. Nature Reviews – Immunology 5:760–771
992. Domogatskaya A, Rodin S, Boutaud A, Tryggvason K (2008) Laminin-511 but not -332, -111, or -411 enables mouse embryonic stem cell self-renewal in vitro. Stem Cells 26:2800–2809
993. Renkawitz J, Sixt M (2010) Mechanisms of force generation and force transmission during interstitial leukocyte migration. EMBO Reports 11:744–750
994. Uruno T, Zhang P, Liu J, Hao JJ, Zhan X (2003) Haematopoietic lineage cell-specific protein 1 (HS1) promotes actin-related protein (Arp) 2/3 complex-mediated actin polymerization. Biochemical Journal 371:485–493
995. de Hostos EL (1999) The coronin family of actin-associated proteins. Trends in Cell Biology 9:345–350
996. Tooley AJ, Gilden J, Jacobelli J, Beemiller P, Trimble WS, Kinoshita M, Krummel MF (2009) Amoeboid T lymphocytes require the septin cytoskeleton for cortical integrity and persistent motility. Nature – Cell Biology 11:17–26
997. Medzhitov R (2008) Origin and physiological roles of inflammation. Nature 454:428–435
998. Pober JS, Sessa WC (2007) Evolving functions of endothelial cells in inflammation. Nature Reviews – Immunology 7:803–815
999. Choi EY, Chavakis E, Czabanka MA, Langer HF, Fraemohs L, Economopoulou M, Kundu RK, Orlandi A, Zheng YY, Prieto DA, Ballantyne CM, Constant SL, Aird WC, Papayannopoulou T, Gahmberg CG, Udey MC, Vajkoczy P, Quertermous T, Dimmeler S, Weber C, Chavakis T (2008) Del-1, an endogenous leukocyte-endothelial adhesion inhibitor, limits inflammatory cell recruitment. Science 322:1101–1104
1000. Rhee WJ, Ni CW, Zheng Z, Chang K, Jo H, Bao G (2010) HuR regulates the expression of stress-sensitive genes and mediates inflammatory response in human umbilical vein endothelial cells. Proceedings of the National Academy of Sciences of the United States of America 107:6858-6863
1001. Esmon C (2005) Do-all receptor takes on coagulation, inflammation. Nature – Medicine 11:475–477

1002. Abeyama K, Stern DM, Ito Y, Kawahara K, Yoshimoto Y, Tanaka M, Uchimura T, Ida N, Yamazaki Y, Yamada S, Yamamoto Y, Yamamoto H, Iino S, Taniguchi N, Maruyama I (2005) The N-terminal domain of thrombomodulin sequesters high-mobility group-B1 protein, a novel antiinflammatory mechanism. Journal of Clinical Investigation 115:1267–1274
1003. Bae JS, Yang L, Rezaie AR (2007) Receptors of the protein C activation and activated protein C signaling pathways are colocalized in lipid rafts of endothelial cells. Proceedings of the National Academy of Sciences of the United States of America 104:2867–2872
1004. Novotny WF, Girard TJ, Miletich JP, Broze GJ Jr (1988) Platelets secrete a coagulation inhibitor functionally and antigenically similar to the lipoprotein associated coagulation inhibitor. Blood 72:2020–2025
1005. Morrison SA, Jesty J (1984) Tissue factor-dependent activation of tritium labeled factor IX and factor X in human plasma. Blood 63:1338–1347
1006. Girard TJ, Warren LA, Novotny WF, Likert KM, Brown SG, Miletich PJ, Broze GJ (1989) Functional significance of the Kunitz-type inhibitory domains coagulation inhibitor. Nature 338:518–520
1007. Kastrup CJ, Runyon MK, Shen F, Ismagilov RF (2006) Modular chemical mechanism predicts spatiotemporal dynamics of initiation in the complex network of hemostasis. Proceedings of the National Academy of Sciences of the United States of America 103:15747–15752
1008. Wohner N (2008) Role of cellular elements in thrombus formation and dissolution. Cardiovascular and Hematological Agents in Medicinal Chemistry 6:224–228
1009. Bergh N, Ulfhammer E, Glise K, Jern S, Karlsson L (2009) Influence of TNF-α and biomechanical stress on endothelial anti- and prothrombotic genes. Biochemical and Biophysical Research Communications 385:314–318
1010. Kannemeier C, Shibamiya A, Nakazawa F, Trusheim H, Ruppert C, Markart P, Song Y, Tzima E, Kennerknecht E, Niepmann M, von Bruehl ML, Sedding B, Massberg S, Gunther A, Engelmann B, Preissner KT (2007) Extracellular RNA constitutes a natural procoagulant cofactor in blood coagulation. Proceedings of the National Academy of Sciences of the United States of America 104:6388–6393
1011. Schneider SW, Nuschele S, Wixforth A, Gorzelanny C, Alexander-Katz A, Netz RR, Schneider MF (2007) Shear-induced unfolding triggers adhesion of von Willebrand factor fibers. Proceedings of the National Academy of Sciences of the United States of America 104:7899–7903
1012. Tolmachova T, brink M, Futter CE, Authi KS, Seabra MC (2007) Rab27b regulates number and secretion of platelet dense granules. Proceedings of the National Academy of Sciences of the United States of America 104: 5872–5877
1013. Liu W, Jawerth LM, Sparks EA, Falvo MR, Hantgan RR, Superfine R, Lord ST, Guthold M (2006) Fibrin fibers have extraordinary extensibility and elasticity. Science 313:634
1014. Collet JP, Shuman H, Ledger RE, Lee S, Weisel JW (2005) The elasticity of an individual fibrin fiber in a clot. Proceedings of the National Academy of Sciences of the United States of America 102:9133–9137
1015. Storm C, Pastore JJ, MacKintosh FC, Lubensky TC, Janmey PA (2005) Nonlinear elasticity in biological gels. Nature 435:191–194
1016. Nesbitt WS, Westein E, Tovar-Lopez FJ, Tolouei E, Mitchell A, Fu J, Carberry J, Fouras A, Jackson SP (2009) A shear gradient-dependent platelet aggregation mechanism drives thrombus formation. Nature – Medicine 15:665–673
1017. Ono A, Westein E, Hsiao S, Nesbitt WS, Hamilton JR, Schoenwaelder SM, Jackson SP (2008) Identification of a fibrin-independent platelet contractile mechanism regulating primary hemostasis and thrombus growth. Blood 112:90–99
1018. Moser M, Nieswandt B, Ussar S, Pozgajova M, Fssler R (2008) Kindlin-3 is essential for integrin activation and platelet aggregation. Nature – Medicine 14:325–330

1019. Kulkarni S, Dopheide SM, Yap CL, Ravanat C, Freund M, Mangin P, Heel KA, Street A, Harper IS, Lanza F, Shaun P, Jackson SP (2000) A revised model of platelet aggregation. Journal of Clinical Investigation 105:783–791
1020. Bergmeier W, Piffath CL, Goerge T, Cifuni SM, Ruggeri ZM, Ware J, Wagner DD (2006) The role of platelet adhesion receptor GPIbαfar exceeds that of its main ligand, von Willebrand factor, in arterial thrombosis. Proceedings of the National Academy of Sciences of the United States of America 103: 16900–16905
1021. Kim JS, Zhang CZ, Zhang XH, Springer TA (2010) A mechanically stabilized receptor2013ligand flex-bond important in the vasculature. Nature 466:992–995
1022. Maas C, Meijers JC, Marquart JA, Bakhtiari K, Weeterings C, de Groot PG, Urbanus RT (2010) Activated factor V is a cofactor for the activation of factor XI by thrombin in plasma. Proceedings of the National Academy of Sciences of the United States of America 107:9083–9087
1023. Carlsson K, Freskgård PO, Persson E, Carlsson U, Svensson M (2003) Probing the interface between factor Xa and tissue factor in the quaternary complex tissue factor-factor VIIa-factor Xa-tissue factor pathway inhibitor. European Journal of Biochemistry 270:2576–2582
1024. Bogdanov VY, Balasubramanian V, Hathcock J, Vele O, Lieb M, Nemerson Y (2003) Alternatively spliced human tissue factor: a circulating, soluble, thrombogenic protein. Nature – Medicine 9:458–462
1025. Wikimedia Foundation. Coagulation. http://www.free-definition.com/Clotting-factors.html
1026. Kim SH, Park SJ, Lee JW, Jin JY (2002) Platelet-agglutinating protein p37 from a patient with thrombotic thrombocytopenic purpura has characteristics similar to prethrombin 2. Journal of Biochemistry and Molecular Biology 35:524–531
1027. Strickland DK, Kounnas MZ (1997) Mechanisms of cellular uptake of thrombin-antithrombin II complexes. Role of the low-density lipoprotein receptor-related protein as a serpin-enzyme complex receptor. Trends in Cardiovascular Medicine 7:9–16
1028. Lam CK, Yoo T, Hiner B, Liu Z, Grutzendler J (2010) Embolus extravasation is an alternative mechanism for cerebral microvascular recanalization. Nature 465:478–482
1029. Lin Z, Kumar A, SenBanerjee S, Staniszewski K, Parmar K, Vaughan DE, Gimbrone MA Jr, Balasubramanian V, Garca-Cardea G, Jain MK (2005) Kruppel-like factor 2 (KLF2) regulates endothelial thrombotic function. Circulation Research 96:e48–e57
1030. Miller MW, Basra S, Kulp DW, Billings PC, Choi S, Beavers MP, McCarty OJ, Zou Z, Kahn ML, Bennett JS, Degrado WF (2009) Small-molecule inhibitors of integrin $\alpha_2\beta_1$ that prevent pathological thrombus formation via an allosteric mechanism. Proceedings of the National Academy of Sciences of the United States of America 106:719–724
1031. Jackson SP, Schoenwaelder SM, Goncalves I Nesbitt WS, Yap CL, Wright CE, Kenche V, Anderson KE, Dopheide SM, Yuan Y, Sturgeon SA, Prabaharan H, Thompson PE, Smith GD, Shepherd PR, Daniele N, Kulkarni S, Abbott B, Saylik D, Jones C, Lu L, Giuliano S, Hughan SC, Angus JA, Robertson AD, Salem HH (2005) PI 3-kinase p110: a new target for antithrombotic therapy. Nature Medicine 11:507–514
1032. Yamasaki M, Li W, Johnson DJD, Huntington JA (2008) Crystal structure of a stable dimer reveals the molecular basis of serpin polymerization. Nature 455:1255–1258
1033. Angiolillo DJ, Capodanno D, Goto S (2010) Platelet thrombin receptor antagonism and atherothrombosis. European Heart Journal 31:17–28
1034. Kawano H, Tsuji H, Nishimura H, Kimura S, Yano S, Ukimura N, Kunieda Y, Yoshizumi M, Sugano T, Nakagawa K, Masuda H, Sawada S, Nakagawa M (2001) Serotonin induces the expression of tissue factor and plasminogen activator inhibitor-1 in cultured rat aortic endothelial cells. Blood 97:1697–1702
1035. Paul BZ, Jin J, Kunapuli SP (1999) Molecular mechanism of thromboxane A2-induced platelet aggregation. Essential role for $P2T_{AC}$ and α_{2a} receptors. Journal of Biological Chemistry 274:29108–29114
1036. Smith CC, Wilson AP, Prichard BN, Betteridge DJ (1986) Stimulus-induced release of endogenous catecholamines from human washed platelets. Clinical Science (London) 70:495–500

1037. Kuharsky AL, Fogelson AL (2001) Surface-mediated control of blood coagulation: the role of binding site densities and platelet deposition. Biophysical Journal 80:1050–1074
1038. Fogelson AL, Guy RD (2004) Platelet-wall interactions in continuum models of platelet thrombosis: formulation and numerical solution. Mathematical Medicine and Biology 21:293–334
1039. Guy RD, Fogelson AL (2002) Probabilistic modeling of platelet aggregation: effects of activation time and receptor occupancy. Journal of Theoretical Biology 219:33–53
1040. Anand M, Rajagopal K, Rajagopal KR (2003) A model incorporating some of the mechanical and biochemical factors underlying clot formation and dissolution in flowing blood. Journal of Theoretical Medicine 5:183–218
1041. Anand M, Rajagopal K, Rajagopal KR (2005) A model for the formation and lysis of blood clots. Pathophysiology of Haemostasis and Thrombosis 34:109–120
1042. Mackness MI, Durrington PN, Mackness B (2000) How high-density lipoprotein protects against the effects of lipid peroxidation. Current Opinion in Lipidology 11:383–388
1043. Calabresi L, Gomaraschi M, Franceschini G (2003) Endothelial protection by high-density lipoproteins: from bench to bedside. Arteriosclerosis, Thrombosis, and Vascular Biology 23:1724-1731
1044. Vogel V, Sheetz M (2006) Local force and geometry sensing regulate cell functions. Nature Reviews – Molecular Cell Biology 7:265–275
1045. Davies PF, Robotewskyj A, Griem ML (1994) Quantitative studies of endothelial cell adhesion. Directional remodeling of focal adhesion sites in response to flow forces. Journal of Clinical Investigation 93:2031–2038
1046. Chuong CJ, Fung YC (1986) On residual stresses in arteries. Journal of Biomechanical Engineering 108:189–192
1047. Wang Y, Liu H, McKenzie G, Witting PK, Stasch JP, Hahn M, Changsirivathanathamrong D, Wu BJ, Ball HJ, Thomas SR, Kapoor V, Celermajer DS, Mellor AL, Keaney JF Jr, Hunt NH, Stocker R (2010) Kynurenine is an endothelium-derived relaxing factor produced during inflammation. Nature – Medicine 16:279–285
1048. Topper JN, Cai J, Falb D, Gimbrone MA (1996) Identification of vascular endothelial genes differentially responsive to fluid mechanical stimuli: Cyclooxygenase-2, manganese superoxide dismutase, and endothelial cell nitric oxide synthase are selectively up-regulated by steady laminar shear stress. Proceedings of the National Academy of Sciences of the United States of America 93:10417–10422
1049. Pohl U, Holtz J, Busse R, Bassenge E (1986) Crucial role of endothelium in the vasodilator response to increased flow in vivo. Hypertension 8:37–44
1050. Fung YC, Liu SQ (1993) Elementary mechanics of the endothelium of blood vessels, Journal of Biomechanical Engineering 115:1–12
1051. Olesen SP, Clapham DE, Davies PF (1988) Haemodynamic shear stress activates a K+ current in vascular endothelial cells. Nature 331:168–170
1052. Morris CE, Sigurdson WJ (1989) Stretch-inactivated ion channels coexist with stretch-activated ion channels in neuronal membranes. Science 243:807–809
1053. Lehoux S, Esposito B, Merval R, Tedgui A (2005) Differential regulation of vascular focal adhesion kinase by steady stretch andpulsatility. Circulation 111:643–649
1054. Spofford CM, Chilian WM (2003) Mechanotransduction via the elastin–laminin receptor (ELR) in resistance arteries. Journal of Biomechanics 36:645–652
1055. Hur SS, Del Alamo JC, Park JS, Li YS, Nguyen HA, Teng D, Wang KC, Flores L, Alonso-Latorre B, Lasheras JC, Chien S (2012) Roles of cell confluency and fluid shear in 3-dimensional intracellular forces in endothelial cells. Proceedings of the National Academy of Sciences of the United States of America 109:11110–11115
1056. Andersson M, Karlsson L, Svensson PA, Ulfhammer E, Ekman M, Jernas M, Carlsson LM, Jern S (2005) Differential global gene expression response patterns of human endothelium exposed to shear stress and intraluminal pressure. Journal of Vascular Research 42:441–452

1057. Resnick N, Yahav H, Khachigian LM, Collins T, Anderson KR, Dewey FC, Gimbrone MA (1997) Endothelial gene regulation by laminar shear stress. Advances in Experimental Medicine and Biology 430:155–164
1058. Lehoux S, Tedgui A (2003) Cellular mechanics and gene expression in blood vessels. Journal of Biomechanics 36:631–643
1059. Tzima E, Irani-Tehrani M, Kiosses WB, Dejana E, Schultz DA, Engelhardt B, Cao G, DeLisser H, Schwartz MA (2005) A mechanosensory complex that mediates the endothelial cell response to fluid shear stress. Nature 437:426–431
1060. Yamawaki H, Pan S, Lee RT, Berk BC (2005) Fluid shear stress inhibits vascular inflammation by decreasing thioredoxin-interacting protein in endothelial cells. Journal of Clinical Investigation 115:733–738
1061. Kohler R, Grundig A, Brakemeier S, Rothermund L, Distler A, Kreutz R, Hoyer J (2001) Regulation of pressure-activated channel in intact vascular endothelium of stroke-prone spontaneously hypertensive rats. American Journal of Hypertension 14:716–721
1062. Cai H (2006) Potassium channel as endothelial mechanosensor and modulator of NAD(P)H oxidase: new mechanism for flow-mediated vasoprotection. American Journal of Physiology – Cell Physiology 290:C35–C36
1063. Gudi SRP, Huver IV, Taliana AP, Boss GR, Frangos JA (1997) Fluid flow-induced ras activation is mediated by Gaq in human vascular endothelial cells. FASEB Journal 11:A223.
1064. Park H, Go YM, St John PL, Maland MC, Lisanti MP, Abrahamson DR, Jo H (1998) Plasma membrane cholesterol is a key molecule in shear stress-dependent activation of extracellular signal-regulated kinase. Journal of Biological Chemistry 273:32304–32311
1065. Resnick N, Yahav H, Schubert S, Wolfovitz E, Shay A (2000) Signalling pathways in vascular endothelium activated by shear stress: relevance to atherosclerosis. Current Opinion in Lipidology 11:167–177
1066. Chen KD, Li YS, Kim M, Li S, Yuan S, Chien S, Shyy JY (1999) Mechanotransduction in response to shear stress. Roles of receptor tyrosine kinases, integrins, and Shc. Journal of Biological Chemistry 274:18393–18400
1067. Park H, Go YM, Darji R, Choi JW, Lisanti MP, Maland MC, Jo H (2000) Caveolin-1 regulates shear stress-dependent activation of extracellular signal-regulated kinase. American Journal of Physiology – Heart and Circulatory Physiology 278:1285–1293
1068. Ohno M, Cooke JP, Dzau VJ, Gibbons GH (1995) Fluid shear stress induces endothelial transforming growth factor beta-1 transcription and production. Modulation by potassium channel blockade. Journal of Clinical Investigation 95:1363–1369
1069. Traub O, Ishida T, Ishida M, Tupper JC, Berk BC (1999) Shear stress-mediated extracellular signal-regulated kinase activation is regulated by sodium in endothelial cells. Potential role for a voltage-dependent sodium channel. Journal of Biological Chemistry 274:20144–20150
1070. Barakat AI, Leaver EV, Pappone PA, Davies PF (1999) A flow-activated chloride-selective membrane current in vascular endothelial cells. Circulation Research 85:820–828
1071. Hoger JH, Ilyin VI, Forsyth S, Hoger A (2002) Shear stress regulates the endothelial Kir2.1 ion channel. Proceedings of the National Academy of Sciences of the United States of America 99:7780–7785
1072. Loot AE, Popp R, Fisslthaler B, Vriens J, Nilius B, Fleming I (2008) Role of cytochrome P450-dependent transient receptor potential V4 activation in flow-induced vasodilatation. Cardiovascular Research 80:445–452
1073. Conway DE, Sakurai Y, Weiss D, Vega JD, Taylor WR, Jo H, Eskin SG, Marcus CB, McIntire LV (2009) Expression of CYP1A1 and CYP1B1 in human endothelial cells: regulation by fluid shear stress. Cardiovascular Research 81:669–677
1074. Boittin FX, Gribi F, Serir K, Bny JL (2008) Ca^{2+}-independent PLA2 controls endothelial store-operated Ca^{2+} entry and vascular tone in intact aorta. American Journal of Physiology – Heart and Circulatory Physiology 295:H2466–H2474
1075. Fujiwara K, Masuda M, Osawa M, Kano Y, Katoh K (2001) Is PECAM-1 a mechanoresponsive molecule? Cell Structure and Function 26:11–17

1076. Carmeliet P, Lampugnani MG, Moons L, Breviario F, Compernolle V, Bono F, Balconi G, Spagnuolo R, Oostuyse B, Dewerchin M, Zanetti A, Angellilo A, Mattot V, Nuyens D, Lutgens E, Clotman F, de Ruiter MC, Gittenberger-de Groot A, Poelmann R, Lupu F, Herbert JM, Collen D, Dejana E (1999) Targeted deficiency or cytosolic truncation of the VE-cadherin gene in mice impairs VEGF-mediated endothelial survival and angiogenesis. Cell 98:147–157
1077. Ukropec JA, Hollinger MK, Woolkalis MJ (2002) Regulation of VE-cadherin linkage to the cytoskeleton in endothelial cells exposed to fluid shear stress. Experimental Cell Research 273:240–247
1078. Urbich C, Dernbach E, Reissner A, Vasa M, Zeiher AM, Dimmeler S (2002) Shear stress-induced endothelial cell migration involves integrin signaling via the fibronectin receptor subunits alpha(5) and beta(1). Arteriosclerosis, Thrombosis, and Vascular Biology 22:69–75
1079. Tergaonkar V, Correa RG, Ikawa M, Verma IM (2005) Distinct roles of IκB proteins in regulating constitutive NF-κB activity. Nature – Cell Biology 7:921–923
1080. Bhullar IS, Li YS, Miao H, Zandi E, Kim M, Shyy JY, Chien S (1998) Fluid shear stress activation of IkappaB kinase is integrin-dependent. Journal of Biological Chemistry 273:30544–30549
1081. Ganguli A, Persson L, Palmer IR, Evans I, Yang L, Smallwood R, Black R, Qwarnstrom EE (2005) Distinct NF-κB regulation by shear stress through Ras-dependent IκBα oscillations. Circulation Research 96:626–634
1082. Helmke BP, Davies PF (2002) The cytoskeleton under external fluid mechanical forces: hemodynamic forces acting on the endothelium. Annals of Biomedical Engineering 30:284–296
1083. Ma W, Sun Y, Han D, Chu W, Lin D, Chen D (2006) Cytoskeletal response of microvessel endothelial cells to an applied stress force at the submicrometer scale studied by atomic force microscopy. Microscopy Research and Technique 69:784–793
1084. Dewey CF, Bussolari SR, Gimbrone MA, Davies PF (1981) The dynamic response of vascular endothelial cells to fluid shear stress. Journal of Biomechanical Engineering 103:177–185
1085. Barbee KA, Davies PF, Lal R (1994) Shear stress-induced reorganization of the surface topography of living endothelial cells imaged by atomic force microscopy. Circulation Research 74:163–171
1086. Nollert MU, Diamond SL, McIntire LV (1991) Hydrodynamic shear stress and mass transport modulation of endothelial cell metabolism. Biotechnology and Bioengineering 38:588–602
1087. Theret DP, Levesque MJ, Sato M, Nerem RM, Wheeler LT (1988) The application of a homogeneous half-space model in the analysis of endothelial cell micropipette measurements. Journal of Biomechanical Engineering 1988, 110:190–199
1088. Sato M, Theret DP, Wheeler LT, Ohshima N, Nerem RM (1990) Application of the micropipette technique to the measurement of cultured porcine aortic endothelial cells viscoelastic properties. Journal of Biomechanical Engineering 112:263–268
1089. Davies PF, Tripathi SC (1993) Mechanical stress mechnisms and the cell. An endothelial paradigm. Circulation Research 72:239–245
1090. Naili S, Ribreau C (1999) Wall shear stress in collapsed tubes. European Physical Journal – Applied Physics 5:95–100
1091. Thiriet M, Naili S, Ribreau C (2003) Entry length and wall shear stress in uniformly collapsed veins. Computer Modeling in Engineering & Sciences 4:473–488
1092. Franke, RP, Grfe M, Schnittler H, Seiffge, D, Mittermayer, D, Drenckhahn, D (1984) Induction of human vascular endothelial stress fibers by fluid shear stress. Nature 307:648–649
1093. De Paola N, Dewey CF, Davies PF, Gimbrone, MA (1992) Vascular endothelium responds to fluid shear stress gradients. Arteriosclerosis, Thrombosis, and Vascular Biology 12:1254–1259

1094. Barbee KA, Davies PF, Ratneshwar L (1994) Shear stress-induced reorganization of the surface topography of living endothelial cells imaged by atomic force microscopy. Circulation Research 74:163–171
1095. Sun RJ, Muller S. Wang X, Lucius, M, Sotolz, JF (2000) Effect of shear stress on the regulation of von Willebrand factor of human endothelial cells. In: Ribreau C, Berthaud Y, Moreau M-R, Ratier L, Renaudaux JP, Thiriet M, Wendling S (eds) Mcanotransduction [Mechanotransduction] 2000. Tec & Doc, Paris
1096. Haond C, Ribreau C, Boutherin-Falson O, Finet, M (1999) Laminar flow through a model of collapsed veins. Morphometric response of endothelial vascular cells to a longitudinal shear stress non uniform cross-wise. European Physical Journal – Applied Physics 8:87–96
1097. Barbee KA, Mundel T, Lal R, Davies PF (1995) Subcellular distribution of shear stress at the surface of flow-aligned and nonaligned endothelial monolayers. American Journal of Physiology – Heart and Circulatory Physiology 268:1765–1772
1098. Dardik A, Chen L, Frattini J, Asada H, Aziz F, Kudo FA, Sumpio BE (2005) Differential effects of orbital and laminar shear stress on endothelial cells. Journal of Vascular Surgery 41:869–880
1099. Bao X, Lu C, Frangos JA (2001) Mechanism of temporal gradients in shear-induced ERK1/2 activation and proliferation in endothelial cells. American Journal of Physiology – Heart and Circulatory Physiology 281:22–29
1100. White CR, Haidekker MA, Stevens HY, Frangos JA (2004) Extracellular signal-regulated kinase activation and endothelin-1 production in human endothelial cells exposed to vibration. Journal of Physiology 555:565–572
1101. Wu CC, Li YS, Haga JH, Kaunas R, Chiu JJ, Su FC, Usami S, Chien S (2007) Directional shear flow and Rho activation prevent the endothelial cell apoptosis induced by micropatterned anisotropic geometry. Proceedings of the National Academy of Sciences of the United States of America 104:1254–1259
1102. Furchgott RF, Zawadzki JV (1980) The obligatory role of the endothelium in the relaxation of arterial smooth muscle by acetylcholine. Nature 288:373–376
1103. Vanhoutte PM (1997) Endothelial dysfunction and atherosclerosis. European Heart Journal 18:E19–E29
1104. Perrier E, Fournet-Bourguignon MP, Royere E, Molez S, Reure H, Lesage L, Gosgnach W, Frapart Y, Boucher JL, Villeneuve N, Vilaine JP (2009) Effect of uncoupling endothelial nitric oxide synthase on calcium homeostasis in aged porcine endothelial cells. Cardiovascular Research 82:133–142
1105. Pohl U, Busse R (1989) Hypoxia stimulates release of endothelium-derived relaxant factor. American Journal of Physiology – Heart and Circulatory Physiology 256:1595–1600
1106. Busse R, Mulsch A, Fleming I, Hecker M (1993) Mechanisms of nitric oxide release from the vascular endothelium. Circulation 87:V18–V25
1107. Garcia-Cardena G, Martasek P, Masters BS, Skidd PM, Couet J, Li S, Lisanti MP, Sessa W C (1997) Dissecting the interaction between nitric oxide synthase (NOS) and caveolin. Functional significance of the NOS caveolin binding domain in vivo. Journal of Biological Chemistry 272:25437–25440
1108. Palacios-Callender M, Hollis V, Mitchison M, Frakich N, Unitt D, Moncada S (2007) Cytochrome c oxidase regulates endogenous nitric oxide availability in respiring cells: a possible explanation for hypoxic vasodilation. Proceedings of the National Academy of Sciences of the United States of America 104:18508–18513
1109. Sugiyama T, Michel T (2010) Thiol-metabolizing proteins and endothelial redox state: differential modulation of eNOS and biopterin pathways. American Journal of Physiology – Heart and Circulatory Physiology 298:H194–H201
1110. Ivashchenko CY, Bradley BT, Ao Z, Leiper J, Vallance P, Johns DG (2010) Regulation of the ADMA-DDAH system in endothelial cells: a novel mechanism for the sterol response element binding proteins, SREBP1c and -2. American Journal of Physiology – Heart and Circulatory Physiology 298:H251–H258

1111. Schwarz G, Callewaert G, Droogmans G, Nilius B (1992) Shear stress-induced calcium transients in endothelial cells from human umbilical cord veins. Journal of Physiology 458:527–538
1112. Shen J, Luscinskas FW, Connolly A, Dewey CF Jr, Gimbrone MA Jr (1992) Fluid shear stress modulates cytosolic free calcium in vascular endothelial cells. American Journal of Physiology – Cell Physiology 262:384–390
1113. Chen Z, Peng IC, Cui X, Li YS, Chien S, Shyy JY (2010) Shear stress, SIRT1, and vascular homeostasis. Proceedings of the National Academy of Sciences of the United States of America 107:10268–10273
1114. Inoue A, Yanagisawa M, Kimura S, Kasuya Y, Miyauchi T, Goto K, Masaki T (1989) The human endothelin family: three structurally and pharmacologically distinct isopeptides predicted by three separate genes. Proceedings of the National Academy of Sciences of the United States of America 86:2863–2867
1115. Bhm F, Pernow J (2007) The importance of endothelin-1 for vascular dysfunction in cardiovascular disease. Cardiovascular Research 76:8–18
1116. Rizvi MA, Katwa L, Spadone DP, Myers PR (1996) The effects of endothelin-1 on collagen type I and type III synthesis in cultured porcine coronary artery vascular smooth muscle cells. Journal of Molecular and Cellular Cardiology 28:243–252
1117. Kedzierski RM, Grayburn PA, Kisanuki YY, Williams CS, Hammer RE, Richardson JA, Schneider MD, Yanagisawa M (2003) Cardiomyocyte-specific endothelin A receptor knockout mice have normal cardiac function and an unaltered hypertrophic response to angiotensin II and isoproterenol. Molecular and Cellular Biology 23:8226–8232
1118. Pieske B, Beyermann B, Breu V, Loffler BM, Schlotthauer K, Maier LS, Schmidt-Schweda S, Just H, Hasenfuss G (1999) Functional effects of endothelin and regulation of endothelin receptors in isolated human nonfailing and failing myocardium. Circulation 99:1802–1809
1119. Watanabe T, Kusumoto K, Kitayoshi T, Shimamoto N (1989) Positive inotropic and vasoconstrictive effects of endothelin-1 in in vivo and in vitro experiments: characteristics and the role of L-type calcium channels. Journal of Cardiovascular Pharmacology 13:S108–S111
1120. Ito H, Hirata Y, Hiroe M, Tsujino M, Adachi S, Takamoto T, Nitta M, Taniguchi K, Marumo F (1991) Endothelin-1 induces hypertrophy with enhanced expression of muscle-specific genes in cultured neonatal rat cardiomyocytes. Circulation Research 69:209–215
1121. Kuchan MJ, Frangos JA (1993) Shear stress regulates endothelin-1 release via protein kinase C and cGMP in cultured endothelial cells. American Journal of Physiology – Heart and Circulatory Physiology 264:150–156
1122. Morawietz H, Talanow R, Szibor M, Rueckschloss U, Schubert A, Bartling B, Darmer D, Holtz J (2000) Regulation of the endothelin system by shear stress in human endothelial cells. Journal of Physiology 525:761–770
1123. Ramzy D, Rao V, Tumiati LC, Xu N, Sheshgiri R, Miriuka S, Delgado DH, Ross HJ (2006) Elevated endothelin-1 levels impair nitric oxide homeostasis through a PKC-dependent pathway. Circulation 114:I319–I326
1124. Sonkusare SK, Bonev AD, Ledoux J, Liedtke W, Kotlikoff MI, Heppner TJ, Hill-Eubanks DC, Nelson MT (2012) Elementary Ca^{2+} signals through endothelial TRPV4 channels regulate vascular function. Science 336:597–601
1125. Dekker RJ, van Soest S, Fontijn RD, Salamanca S, de Groot PG, VanBavel E, Pannekoek H, Horrevoets AJ (2002) Prolonged fluid shear stress induces a distinct set of endothelial cell genes, most specifically lung Kruppel-like factor (KLF2). Blood 100:1689–1698
1126. Dekker RJ, van Thienen JV, Rohlena J, de Jager SC, Elderkamp YW, Seppen J, de Vries CJ, Biessen EA, van Berkel TJ, Pannekoek H, Horrevoets AJ (2005) Endothelial KLF2 links local arterial shear stress levels to the expression of vascular tone-regulating genes. American Journal of Pathology 167:609–618
1127. Mayr M, Hu Y, Hainaut P, Xu Q (2002) Mechanical stress-induced DNA damage and rac-p38MAPK signal pathways mediate p53-dependent apoptosis in vascular smooth muscle cells. FASEB Journal 16:1423–1425

1128. Qin X, Wang X, Wang Y, Tang Z, Cui Q, Xi J, Li YS, Chien S, Wang N (2010) MicroRNA-19a mediates the suppressive effect of laminar flow on cyclin D1 expression in human umbilical vein endothelial cells. Proceedings of the National Academy of Sciences of the United States of America 107:3240-3244
1129. Wang KC, Garmire LX, Young A, Nguyen P, Trinh A, Subramaniam S, Wang N, Shyy JY, Li YS, Chien S (2010) Role of microRNA-23b in flow-regulation of Rb phosphorylation and endothelial cell growth. Proceedings of the National Academy of Sciences of the United States of America 107:3234–3239
1130. Tzima E, Kiosses WB, del Pozo MA, Schwartz MA (2003) Localized cdc42 activation, detected using a novel assay, mediates microtubule organizing center positioning in endothelial cells in response to fluid shear stress. Journal of Biological Chemistry 278:31020–31023.
1131. Dimmeler S, Hermann C, Galle J, Zeiher AM (1999) Upregulation of superoxide dismutase and nitric oxide synthase mediates the apoptosis-suppressive effects of shear stress on endothelial cells. Arteriosclerosis, Thrombosis, and Vascular Biology 19:656–664
1132. Takeshita S, Inoue N, Ueyama T, Kawashima S, Yokoyama M (2000) Shear stress enhances glutathione peroxidase expression in endothelial cells. Biochemical and Biophysical Research Communications 273:66–71
1133. Park HS, Yu JW, Cho JH, Kim MS, Huh SH, Ryoo K, Choi EJ (2004) Inhibition of apoptosis signal-regulating kinase 1 by nitric oxide through a thiol redox mechanism. Journal of Biological Chemistry 279:7584–7590
1134. Levonen AL, Patel RP, Brookes P, Go YM, Jo H, Parthasarathy S, Anderson PG, Darley-Usmar VM (2001) Mechanisms of cell signaling by nitric oxide and peroxynitrite: from mitochondria to MAP kinases. Antioxidants & Redox Signaling 3:215–229
1135. Dardik A, Yamashita A, Aziz F, Asada H, Sumpio BE (2005) Shear stress-stimulated endothelial cells induce smooth muscle cell chemotaxis via platelet-derived growth factor-BB and interleukin-1alpha. Journal of Vascular Surgery 41:321–331
1136. Tzima E, Reader JS, Irani-Tehrani M, Ewalt KL, Schwartz MA, Schimmel P (2003) Biologically active fragment of a human tRNA synthetase inhibits fluid shear stress-activated responses of endothelial cells. Proceedings of the National Academy of Sciences of the United States of America 100:14903–14907
1137. Tzima E, Reader JS, Irani-Tehrani M, Ewalt KL, Schwartz MA, Schimmel P (2005) VE-cadherin links tRNA synthetase cytokine to anti-angiogenic function. Journal of Biological Chemistry 280:2405–2408
1138. Bonini MG, Stadler K, de Oliveira Silva S, Corbett J, Dore M, Petranka J, Fernandes DC, Tanaka LY, Duma D, Laurindo FRM, Mason RP (2008) Constitutive nitric oxide synthase activation is a significant route for nitroglycerin-mediated vasodilation. Proceedings of the National Academy of Sciences of the United States of America 105:8569–8574
1139. Chan CK, Vanhoutte PM (2011) Secretoneurin facilitates endothelium-dependent relaxations in porcine coronary arteries. American Journal of Physiology – Heart and Circulatory Physiology 300:H1159–H1165
1140. Castier Y, Brandes RP, Leseche G, Tedgui A, Lehoux S (2005) p47phox-dependent NADPH oxidase regulates flow-induced vascular remodeling. Circulation Research 97:533–540
1141. Jin BY, Sartoretto JL, Gladyshev VN, Michel T (2009) Endothelial nitric oxide synthase negatively regulates hydrogen peroxide-stimulated AMP-activated protein kinase in endothelial cells. Proceedings of the National Academy of Sciences of the United States of America 106:17343–17348
1142. Capettini LS, Cortes SF, Gomes MA, Silva GA, Pesquero JL, Lopes MJ, Teixeira MM, Lemos VS (2008) Neuronal nitric oxide synthase-derived hydrogen peroxide is a major endothelium-dependent relaxing factor. American Journal of Physiology – Heart and Circulatory Physiology 295:H2503–H2511
1143. Snetkov VA, Smirnov SV, Kua J, Aaronson PI, Ward JP, Knock GA (2011) Superoxide differentially controls pulmonary and systemic vascular tone through multiple signalling pathways. Cardiovascular Research 89:214–224

1144. Faraci FM (2006) Reactive oxygen species: influence on cerebral vascular tone. Journal of Applied Physiology 100:739–743
1145. Wei EP, Kontos HA, Beckman JS (1996) Mechanisms of cerebral vasodilation by superoxide, hydrogen peroxide, and peroxynitrite. American Journal of Physiology – Heart and Circulatory Physiology 271:H1262–H1266
1146. Xie H, Ray PE, Short BL (2005) NF-κB activation plays a role in superoxide-mediated cerebral endothelial dysfunction after hypoxia/reoxygenation. Stroke 36:1047–1052
1147. Szabó C, Ischiropoulos H, Radi R (2007) Peroxynitrite: biochemistry, pathophysiology and development of therapeutics. Nature Reviews – Drug Discovery 6:662–680
1148. Janssen LJ (2001) Isoprostanes: an overview and putative roles in pulmonary pathophysiology. American Journal of Physiology – Lung Cellular and Molecular Physiology 280:L1067-L1082
1149. Knock GA, Snetkov VA, Shaifta Y, Drndarski S, Ward JP, Aaronson PI (2008) Role of src-family kinases in hypoxic vasoconstriction of rat pulmonary artery. Cardiovascular Research 80:453–462
1150. Ellsworth ML, Forrester T, Ellis CG, Dietrich HH (1995) The erythrocyte as a regulator of vascular tone. American Journal of Physiology – Heart and Circulatory Physiology 269:H2155–H2161
1151. Bergfeld GR, Forrester T (1992) Release of ATP from human erythrocytes in response to a brief period of hypoxia and hypercapnia. Cardiovascular Research 26:40–47
1152. Light DB, Capes TL, Gronau RT, Adler MR (1999) Extracellular ATP stimulates volume decrease in Necturus red blood cells. American Journal of Physiology – Cell Physiology 277:C480–C491
1153. Moehlenbrock MJ, Price AK, Martin RS (2006) Use of microchip-based hydrodynamic focusing to measure the deformation-induced release of ATP from erythrocytes. Analyst 131:930–937
1154. Sprague RS, Ellsworth ML, Stephenson AH, Kleinhenz ME, Lonigro AJ (1998) Deformation-induced ATP release from red blood cells requires CFTR activity. American Journal of Physiology – Heart and Circulatory Physiology 275:H1726–H1732
1155. Sprague RS, Stephenson AH, Ellsworth ML, Keller C, Lonigro AJ (2001) Impaired release of ATP from red blood cells of humans with primary pulmonary hypertension. Experimental Biology and Medicine 226:434–439
1156. Sprague RS, Stephenson AH, Bowles EA, Stumpf MS, Lonigro AJ (2006) Reduced expression of G(i) in erythrocytes of humans with type 2 diabetes is associated with impairment of both cAMP generation and ATP release. Diabetes 55:3588–3593
1157. Morikawa T, Kajimura M, Nakamura T, Hishiki T, Nakanishi T, Yukutake Y, Nagahata Y, Ishikawa M, Hattori K, Takenouchi T, Takahashi T, Ishii I, Matsubara K, Kabe Y, Uchiyama S, Nagata E, Gadalla MM, Snyder SH, Suematsu M (2012) Hypoxic regulation of the cerebral microcirculation is mediated by a carbon monoxide-sensitive hydrogen sulfide pathway. Proceedings of the National Academy of Sciences of the United States of America 109:1293–1298
1158. Pearce WJ (1995) Mechanisms of hypoxic cerebral vasodilatation. Pharmacology and Therapeutics 65:75–91
1159. Beaudin AE, Brugniaux JV, Vöhringer M, Flewitt J, Green JD, Friedrich MG, Poulin MJ (2011) Cerebral and myocardial blood flow responses to hypercapnia and hypoxia in humans. American Journal of Physiology – Heart and Circulatory Physiology 301:H1678–H1686
1160. Faraci FM, Breese KR, Heistad DD (1994) Cerebral vasodilation during hypercapnia. Role of glibenclamide-sensitive potassium channels and nitric oxide. Stroke 25:1679–1683
1161. Ogoh S, Ainslie PN, Miyamoto T (2009) Onset responses of ventilation and cerebral blood flow to hypercapnia in humans: rest and exercise. Journal of Applied Physiology 106:880–886
1162. Zhang P, Huang G, Shi X (2011) Cerebral vasoreactivity during hypercapnia is reset by augmented sympathetic influence. Journal of Applied Physiology 110:352–358

1163. Gokina NI, Kuzina OY, Vance AM (2010) Augmented EDHF signaling in rat uteroplacental vasculature during late pregnancy. American Journal of Physiology – Heart and Circulatory Physiology 299:H1642–H1652

Chap. 10. Vasculature Growth

1164. Koivisto L, Larjava K, Hakkinen L, Uitto VJ, Heino J, Larjava H (1999) Different integrins mediate cell spreading, haptotaxis and lateral migration of HaCaT keratinocytes on fibronectin. Cell Adhesion and Communication 7:245–257
1165. DuBois-Reymond E (1843) Vorlufiger Abriss einer Untersuchung ber den sogenannten Froschstrom und ber die elektromotorischen Fische [Preliminary outline of an investigation of the so-called frog current and of the electromotive fish]. Annalen der Physik und Chemie 58:1–30
1166. Zhao M, Song B, Pu J, Wada T, Reid B, Tai G, Wang F, Guo A, Walczysko P, Gu Y, Sasaki T, Suzuki A, Forrester JV, Bourne HR, Devreotes PN, McCaig CD, Penninger JM (2006) Electrical signals control wound healing through phosphatidylinositol-3-OH kinase-big gamma and PTEN. Nature 442:457–460
1167. Coultas L, Chawengsaksophak K, Rossant J (2005) Endothelial cells and VEGF in vascular development. Nature 438:937–945
1168. Kokovay E, Li L, Cunningham LA (2006) Angiogenic recruitment of pericytes from bone marrow after stroke. Journal of Cerebral Blood Flow & Metabolism 26:545–555
1169. Huu SN, Oster M, Uzan S, Chareyre F, Aractingi S, Khosrotehrani K (2007) Maternal neoangiogenesis during pregnancy partly derives from fetal endothelial progenitor cells. Proceedings of the National Academy of Sciences of the United States of America 104:1871–1876
1170. Bousquenaud M, Schwartz C, Léonard F, Rolland-Turner M, Wagner D, Devaux Y (2012) Monocyte chemotactic protein 3 is a homing factor for circulating angiogenic cells. Cardiovascular Research 94:519–525
1171. Gao D, Nolan DJ, Mellick AS, Bambino K, McDonnell K, Mittal V (2008) Endothelial progenitor cells control the angiogenic switch in mouse lung metastasis. Science 319:195–198
1172. Bailey AS, Willenbring H, Jiang S, Anderson DA, Schroeder DA, Wong MH, Grompe M, Fleming WH (2006) Myeloid lineage progenitors give rise to vascular endothelium. Proceedings of the National Academy of Sciences of the United States of America 103: 13156-13161
1173. Szczerba D, Szkely G (2005) A computational model of micro-vascular growth, In: Lecture Notes in Computer Science, Vol. 3516, 17–24, Springer, Berlin
1174. De Palma M, Venneri MA, Galli R, Sergi LS, Politi LS, Sampaolesi M, Naldini L (2005) Tie2 identifies a hematopoietic lineage of proangiogenic monocytes required for tumor vessel formation and a mesenchymal population of pericyte progenitors. Cancer Cell 8:211–226
1175. Jakobsson L, Franco CA, Bentley K, Collins RT, Ponsioen B, Aspalter IM, Rosewell I, Busse M, Thurston G, Medvinsky A, Schulte-Merker S, Gerhardt H (2010) Endothelial cells dynamically compete for the tip cell position during angiogenic sprouting. Nature – Cell Biology 12:943–953
1176. Kurz H, Burri PH, Djonov VG (2003) Angiogenesis and vascular remodeling by intussusception: From form to function. News in Physiological Sciences 18:65–70
1177. Apodaca G (2010) Opening ahead: early steps in lumen formation revealed. Nature – Cell Biology 12:1026–1028

1178. Bryant DM, Datta A, Rodríguez-Fraticelli AE, Peränen J, Martn-Belmonte F, Mostov KE (2010) A molecular network for de novo generation of the apical surface and lumen. Nature – Cell Biology 12:1035–1045
1179. Hattula K, Furuhjelm J, Arffman A, Pernen J (2002) A Rab8-specific GDP/GTP exchange factor is involved in actin remodeling and polarized membrane transport. Molecular Biology of the Cell 13:3268–3280
1180. Salazar MA, Kwiatkowski AV, Pellegrini L, Cestra G, Butler MH, Rossman KL, Serna DM, Sondek J, Gertler FB, De Camilli P (2003) Tuba, a novel protein containing bin/amphiphysin/Rvs and Dbl homology domains, links dynamin to regulation of the actin cytoskeleton. Journal of Biological Chemistry 278:49031–49043.
1181. van Royen N, Piek JJ, Buschmann I, Hoefer I, Voskuil M, Schaper W (2001) Stimulation of arteriogenesis: a new concept for the treatment of arterial occlusive disease. Cardiovascular Research 49:543–553
1182. Murray CD (1926) The physiological principle of minimum work I: The vascular system and the cost of blood volume. Proceedings of the National Academy of Sciences of the United States of America 12:207–214
1183. le Noble F, Moyon D, Pardanaud L, Yuan L, Djonov V, Matthijsen R, Breant C, Fleury V, Eichmann A (2004) Flow regulates arterial-venous differentiation in the chick embryo yolk sac. Development 131:361–375
1184. Adams RH, Alitalo K (2007) Molecular regulation of angiogenesis and lymphangiogenesis. Nature Reviews – Molecular Cell Biology 8:464–478
1185. Arroyo AG, Iruela-Arispe ML (2010) Extracellular matrix, inflammation, and the angiogenic response. Cardiovascular Research 86:226–235
1186. Armulik A, Abramsson A, Betsholtz C (2005) Endothelial/pericyte interactions. Circulation Research 97:512–523
1187. Kupprion C, Motamed K, Sage EH (1998) SPARC (BM-40, osteonectin) inhibits the mitogenic effect of vascular endothelial growth factor on microvascular endothelial cells. Journal of Biological Chemistry 273:29635–29640
1188. Mundlos S, Schwahn B, Reichert T, Zabel B (1992) Distribution of osteonectin mRNA and protein during human embryonic and fetal development. Journal of Histochemistry and Cytochemistry 40:283–291
1189. Chlenski A, Guerrero LJ, Peddinti R, Spitz JA, Leonhardt PT, Yang Q, Tian Y, Salwen HR, Cohn SL (2010) Anti-angiogenic SPARC peptides inhibit progression of neuroblastoma tumors. Molecular Cancer 9:138
1190. Iozzo RV (2005) Basement membrane proteoglycans: from cellar to ceiling. Nature Reviews – Molecular Cell Biology 6:646–656
1191. Sunshine SB, Dallabrida SM, Durand E, Ismail NS, Bazinet L, Birsner AE, Sohn R, Ikeda S, Pu WT, Kulke MH, Javaherian K, Zurakowski D, Folkman JM, Rupnick M (2012) Endostatin lowers blood pressure via nitric oxide and prevents hypertension associated with VEGF inhibition. Proceedings of the National Academy of Sciences of the United States of America 109:11306–11311
1192. Kitazume S, Imamaki R, Ogawa K, Komi Y, Futakawa S, Kojima S, Hashimoto Y, Marth JD, Paulson JC, Taniguchi N (2010) α-2,6-sialic acid on platelet endothelial cell adhesion molecule (PECAM) regulates its homophilic interactions and downstream antiapoptotic signaling. Journal of Biological Chemistry 285:6515–6521
1193. Francis SE, Goh KL, Hodivala-Dilke K, Bader BL, Stark M, Davidson D, Hynes RO (2002) Central roles of $\alpha5\beta1$ integrin and fibronectin in vascular development in mouse embryos and embryoid bodies. Arteriosclerosis, Thrombosis, and Vascular Biology 22:927–933
1194. Huang S, Ingber DE (1999) The structural and mechanical complexity of cell-growth control. Nature – Cell Biology 1:E131–E138
1195. Silvestre J-S, Thry C, Hamard G, Boddaert J, Aguilar B, Delcayre A, Houbron C, Tamarat R, Blanc-Brude O, Heeneman S, Clergue M, Duriez M, Merval R, Lvy B, Tedgui A, Amigorena S, Mallat Z (2005) Lactadherin promotes VEGF-dependent neovascularization. Nature – Medicine 11:499–506

1196. Lorger M, Krueger JS, O'Neal M, Staflin K, Felding-Habermann B (2009) Activation of tumor cell integrin $\alpha_V\beta_3$ controls angiogenesis and metastatic growth in the brain. Proceedings of the National Academy of Sciences of the United States of America 106:10666–10671
1197. Jones FS, Jones PL (2000) The tenascin family of ECM glycoproteins: structure, function, and regulation during embryonic development and tissue remodeling. Developmental Dynamics 218:235–259.
1198. Chung CY, Murphy-Ullrich JE, Erickson HP (1996) Mitogenesis, cell migration, and loss of focal adhesions induced by tenascin-C interacting with its cell surface receptor, annexin II. Molecular Biology of the Cell 7:883–892
1199. Twal WO, Czirok A, Hegedus B, Knaak C, Chintalapudi MR, Okagawa H, Sugi Y, Argraves WS (2001) Fibulin-1 suppression of fibronectin-regulated cell adhesion and motility. Journal of Cell Science 114:4587–4598
1200. Hunzelmann N, Nischt R, Brenneisen P, Eickert A, Krieg T (2001) Increased deposition of fibulin-2 in solar elastosis and its colocalization with elastic fibres. British Journal of Dermatology 145:217–222
1201. Hopf M, Göhring W, Mann K, Timpl R (2001) Mapping of binding sites for nidogens, fibulin-2, fibronectin and heparin to different IG modules of perlecan. Journal of Molecular Biology 311:529–541
1202. Seeliger H, Camaj P, Ischenko I, Kleespies A, De Toni EN, Thieme SE, Blum H, Assmann G, Jauch KW, Bruns CJ (2009) EFEMP1 expression promotes in vivo tumor growth in human pancreatic adenocarcinoma. Molecular Cancer Research 7:189–198
1203. Chen Q, Zhang T, Roshetsky JF, Ouyang Z, Essers J, Fan C, Wang Q, Hinek A, Plow EF, Dicorleto PE (2009) Fibulin-4 regulates expression of the tropoelastin gene and consequent elastic-fibre formation by human fibroblasts. Biochemical Journal 423:79–89
1204. Nakamura T, Ruiz-Lozano P, Lindner V, Yabe D, Taniwaki M, Furukawa Y, Kobuke K, Tashiro K, Lu Z, Andon NL, Schaub R, Matsumori A, Sasayama S, Chien KR, Honjo T (1999) DANCE, a novel secreted RGD protein expressed in developing, atherosclerotic, and balloon-injured arteries. Journal of Biological Chemistry 274:22476–22483
1205. Conway RE, Petrovic N, Li Z, Heston W, Wu D, Shapiro LH (2006) Prostate-specific membrane antigen regulates angiogenesis by modulating integrin signal transduction. Molecular and Cellular Biology 26:5310–5324
1206. Teodoro JG, Parker AE, Zhu X, Green MR (2006) p53-mediated inhibition of angiogenesis through up-regulation of a collagen prolyl hydroxylase. Science 313:968–971
1207. Klagsbrun M, Eichmann A (2005) A role for axon guidance receptors and ligands in blood vessel development and tumor angiogenesis. Cytokine Growth Factor Reviews 16:535–548
1208. Weinstein BM (2005) Vessels and nerves: marching to the same tune. Cell 120:299–302
1209. Serini G, Valdembri D, Zanivan S, Morterra G, Burkhardt C, Caccavari F, Zammataro L, Primo L, Tamagnone L, Logan M, Tessier-Lavigne M, Taniguchi M, Puschel AW, Bussolino F (2003) Class 3 semaphorins control vascular morphogenesis by inhibiting integrin function. Nature 424:391–397
1210. Kawasaki T, Kitsukawa T, Bekku Y, Matsuda Y, Sanbo M, Yagi T, Fujisawa H (1999) A requirement for neuropilin-1 in embryonic vessel formation. Development 126:4895–4902
1211. Gu C, Rodriguez ER, Reimert DV, Shu T, Fritzsch B, Richards LJ, Kolodkin AL, Ginty DD (2003) Neuropilin-1 conveys semaphorin and VEGF signaling during neural and cardiovascular development. Developmental Cell 5:45–57
1212. Gu C, Yoshida Y, Livet J, Reimert DV, Mann F, Merte J, Henderson CE, Jessell TM, Kolodkin AL, Ginty DD (2005) Semaphorin 3E and plexin-D1 control vascular pattern independently of neuropilins. Science 307:265–268
1213. Toyofuku T, Yabuki M, Kamei J, Kamei M, Makino N, Kumanogoh A, Hori M (2007) Semaphorin-4A, an activator for T-cell-mediated immunity, suppresses angiogenesis via Plexin-D1. EMBO Journal 26:1373–1384
1214. Conrotto P, Valdembri D, Corso S, Serini G, Tamagnone L, Comoglio PM, Bussolino F, Giordano S (2005) Sema4D induces angiogenesis through Met recruitment by Plexin B1. Blood 105:4321–4329

1215. Cirulli V, Yebra M (2007) Netrins: beyond the brain. Nature Reviews – Molecular Cell Biology 8, 296-306
1216. Pandey A, Shao H, Marks RM, Polverini PJ, Dixit VM (1995) Role of B61, the ligand for the Eck receptor tyrosine kinase, in TNF-alpha-induced angiogenesis. Science 268:567–569
1217. Georgakopoulos A, Litterst C, Ghersi E, Baki L, Xu C, Serban G, Robakis NK (2006) Metalloproteinase/Presenilin1 processing of ephrinB regulates EphB-induced Src phosphorylation and signaling. EMBO Journal 25:1242–1252
1218. Erber R, Eichelsbacher U, Powajbo V, Korn T, Djonov V, Lin J, Hammes HP, Grobholz R, Ullrich A, Vajkoczy P (2006) EphB4 controls blood vascular morphogenesis during postnatal angiogenesis. EMBO Journal 25:628–641
1219. Park KW, Morrison CM, Sorensen LK, Jones CA, Rao Y, Chien CB, Wu JY, Urness LD, Li DY (2003) Robo4 is a vascular-specific receptor that inhibits endothelial migration. Developmental Biology 61:251–267
1220. Fischer A, Schumacher N, Maier M, Sendtner M, Gessler M (2004) The Notch target genes Hey1 and Hey2 are required for embryonic vascular development. Genes and Development 18:901–911
1221. Winnik S, Klinkert M, Kurz H, Zoeller C, Heinke J, Wu Y, Bode C, Patterson C, Moser M (2009) HoxB5 induces endothelial sprouting in vitro and modifies intussusceptive angiogenesis in vivo involving angiopoietin-2. Cardiovascular Research 83:558–565
1222. Vasudevan A, Long JE, Crandall JE, Rubenstein JLR, Bhide PG (2008) Compartment-specific transcription factors orchestrate angiogenesis gradients in the embryonic brain. Nature – Neuroscience 11:429–439
1223. Roukens MG, Alloul-Ramdhani M, Baan B, Kobayashi K, Peterson-Maduro J, van Dam H, Schulte-Merker S, Baker DA (2010) Control of endothelial sprouting by a Tel-CtBP complex. Nature – Cell Biology 12:933–942
1224. Tabruyn SP, Griffioen AW (2007) A new role for NF-kappaB in angiogenesis inhibition. Cell Death and Differentiation 14:1393–1397
1225. Scoditti E, Massaro M, Carluccio MA, Distante A, Storelli C, De Caterina R (2010) PPARγ agonists inhibit angiogenesis by suppressing PKCα- and CREB-mediated COX-2 expression in the human endothelium. Cardiovascular Research 86:302–310
1226. Chinsomboon J, Ruas J, Gupta RK, Thom R, Shoag J, Rowe GC, Sawada N, Raghuram S, Arany Z (2009) The transcriptional coactivator PGC-1α mediates exercise-induced angiogenesis in skeletal muscle. Proceedings of the National Academy of Sciences of the United States of America 106:21401–21406
1227. Ray LB (2009) Physiology: marathon mouse mechanism. Science Signaling 2:ec403
1228. Boon RA, Urbich C, Fischer A, Fontijn RD, Seeger FH, Koyanagi M, Horrevoets AJG, Dimmeler S (2011) Krüppel-like factor 2 improves neovascularization capacity of aged proangiogenic cells. European Heart Journal 32:371–377
1229. Rey S, Semenza GL (2010) Hypoxia-inducible factor-1-dependent mechanisms of vascularization and vascular remodelling. Cardiovascular Research 86:236–242
1230. Martin M, Potente M, Janssens V, Vertommen D, Twizere JC, Rider MH, Goris J, Dimmeler S, Kettmann R, Dequiedt F (2008) Protein phosphatase 2A controls the activity of histone deacetylase 7 during T cell apoptosis and angiogenesis. Proceedings of the National Academy of Sciences of the United States of America 105:4727–4732
1231. Economopoulou M, Langer HF, Celeste A, Orlova VV, Choi EY, Ma M, Vassilopoulos A, Callen E, Deng C, Bassing CH, Boehm M, Nussenzweig A, Chavakis T (2009) Histone H2AX is integral to hypoxia-driven neovascularization. Nature – Medicine 15:553–558
1232. Kaikkonen MU, Lam MTY, Glass CK (2011) Non-coding RNAs as regulators of gene expression and epigenetics. Cardiovascular Research 90:430–440
1233. Suarez Y, Fernandez-Hernando C, Yu J, Gerber SA, Harrison KD, Pober JS, Iruela-Arispe L, Merkenschlager M, Sessa WC (2008) Dicer-dependent endothelial microRNAs are necessary for postnatal angiogenesis. Proceedings of the National Academy of Sciences of the United States of America 105:14082-14087

1234. Nicoli S, Standley C, Walker P, Hurlstone A, Fogarty KE, Lawson ND (2010) MicroRNA-mediated integration of haemodynamics and Vegf signalling during angiogenesis. Nature 464:1196–1200
1235. Halberg N, Alarcón C, Tavazoie SF (2012) MicroRNA regulation of cancer–endothelial interactions: Vesicular microRNAs on the move... EMBO Journal 31:3509–3510
1236. Anand S, Majeti BK, Acevedo LM, Murphy EA, Mukthavaram R, Scheppke L, Huang M, Shields DJ, Lindquist JN, Lapinski PE, King PD, Weis SM, Cheresh DA (2010) MicroRNA-132-mediated loss of p120RasGAP activates the endothelium to facilitate pathological angiogenesis. Nature – Medicine 16:909–914
1237. Zhou Q, Gallagher R, Ufret-Vincenty R, Li X, Olson EN, Wang S (2011) Regulation of angiogenesis and choroidal neovascularization by members of microRNA-23 27 24 clusters. Proceedings of the National Academy of Sciences of the United States of America 108:8287–8292
1238. Zhuang G, Wu X, Jiang Z, Kasman I, Yao J, Guan Y, Oeh J, Modrusan Z, Bais C, Sampath D, Ferrara N (2012) Tumour-secreted miR-9 promotes endothelial cell migration and angiogenesis by activating the JAK-STAT pathway. EMBO Journal 31:3513–3523
1239. Turing AM (1952) The chemical basis of morphogenesis. Philosophicals Transactions of the Royal Society of London, Series B, Biological Sciences 237:37–72
1240. Hellström M, Phng LK, Hofmann JJ, Wallgard E, Coultas L, Lindblom P, Alva J, Nilsson AK, Karlsson L, Gaiano N, Yoon K, Rossant J, Iruela-Arispe L, Kalén M, Gerhardt H, Betsholtz C (2007) Dll4 signalling through Notch1 regulates formation of tip cells during angiogenesis. Nature 445:776–780
1241. Siekmann AF, Lawson ND (2007) Notch signalling limits angiogenic cell behaviour in developing zebrafish arteries. Nature 445:781–784
1242. Noguera-Troise I, Daly C, Papadopoulos NJ, Coetzee S, Boland P, Gale NW, Lin HC, Yancopoulos GD, Thurston G (2006) Blockade of Dll4 inhibits tumour growth by promoting non-productive angiogenesis. Nature 444:1032–1037
1243. Ridgway J, Zhang G, Wu Y, Stawicki S, Liang WC, Chanthery Y, Kowalski J, Watts RJ, Callahan C, Kasman I, Singh M, Chien M, Tan C, Hongo JAS, de Sauvage F, Plowman G, Yan M (2006) Inhibition of Dll4 signalling inhibits tumour growth by deregulating angiogenesis. Nature 444:1083–1087
1244. Benedito R, Roca C, Sörensen I, Adams S, Gossler A, Fruttiger M, Adams RH (2009) The notch ligands Dll4 and Jagged1 have opposing effects on angiogenesis. Cell 137:1124–1135
1245. Wüstehube J, Bartol A, Liebler SS, Brütsch R, Zhu Y, Felbor U, Sure U, Augustin HG, Fischer A (2010) Cerebral cavernous malformation protein CCM1 inhibits sprouting angiogenesis by activating DELTA-NOTCH signaling. Proceedings of the National Academy of Sciences of the United States of America 107:12640–12645
1246. Benedito R, Rocha SF, Woeste M, Zamykal M, Radtke F, Casanovas O, Duarte A, Pytowski B, Adams RH (2012) Notch-dependent VEGFR3 upregulation allows angiogenesis without VEGF–VEGFR2 signalling. Nature 484:110–114
1247. Rodríguez P, Higueras MA, Gonzlez-Rajal A, Alfranca A, Fierro-Fernndez M, Garca-Fernndez RA, Ruiz-Hidalgo MJ, Monsalve M, Rodríguez-Pascual F, Redondo JM, de la Pompa JL, Laborda J, Lamas S (2012) The non-canonical NOTCH ligand DLK1 exhibits a novel vascular role as a strong inhibitor of angiogenesis. Cardiovascular Research 93:232–241
1248. Zhao N, Liu H, Lilly B (2012) Reciprocal regulation of syndecan-2 and Notch signaling in vascular smooth muscle cells. Journal of Biological Chemistry 287:16111–16120
1249. Ju R, Cirone P, Lin S, Griesbach H, Slusarski DC, Crews CM (2010) Activation of the planar cell polarity formin DAAM1 leads to inhibition of endothelial cell proliferation, migration, and angiogenesis. Proceedings of the National Academy of Sciences of the United States of America 107:6906–6911
1250. Stefater JA, Lewkowich I, Rao S, Mariggi G, Carpenter AC, Burr AR, Fan J, Ajima R, Molkentin JD, Williams BO, Wills-Karp M, Pollard JW, Yamaguchi T, Ferrara N, Gerhardt H, Lang RA (2011) Regulation of angiogenesis by a non-canonical Wnt–Flt1 pathway in myeloid cells. Nature 474:511–515

1251. Coultas L, Nieuwenhuis E, Anderson GA, Cabezas J, Nagy A, Henkelman RM, Hui CC, Rossant J (2010) Hedgehog regulates distinct vascular patterning events through VEGF-dependent and -independent mechanisms. Blood 116:653–660
1252. Tedgui A, Lévy B (1994) Biologie de la paroi artrielle [Biology of the arterial wall]. Masson, Paris
1253. Wahlberg E (2005) Angiogenesis and arteriogenesis – the basics. svs.vascularweb.org/_CONTRIBUTION_PAGES/Research/Basic_Articles/Angiogenesis_Wahlberg.html
1254. Angiogenesis: an introduction. http://www.med.unibs.it/~airc/
1255. Ferrari G, Pintucci G, Seghezzi G, Hyman K, Galloway AC, Mignatti P (2006) VEGF, a prosurvival factor, acts in concert with TGF-beta1 to induce endothelial cell apoptosis. Proceedings of the National Academy of Sciences of the United States of America 103:17260–17265
1256. Ferrara N, Davis-Smyth T (1997) The biology of vascular endothelial growth factor. Endocrine Reviews 18:4–25
1257. www.sigmaaldrich.com/
1258. Storkebaum E, Carmeliet P (2004) VEGF: a critical player in neurodegeneration. Journal of Clinical Investigation 113:14–18
1259. Witmer AN, van Blijswijk BC, van Noorden CJ, Vrensen GF, Schlingemann RO (2004) In vivo angiogenic phenotype of endothelial cells and pericytes induced by vascular endothelial growth factor-A. Journal of Histochemistry and Cytochemistry 52:39–52
1260. Dominguez MG, Hughes VC, Pan L, Simmons M, Daly C, Anderson K, Noguera-Troise I, Murphy AJ, Valenzuela DM, Davis S, Thurston G, Yancopoulos GD, Gale NW (2007) Vascular endothelial tyrosine phosphatase (VE-PTP)-null mice undergo vasculogenesis but die embryonically because of defects in angiogenesis. Proceedings of the National Academy of Sciences of the United States of America 104:3243–3248
1261. Hiratsuka S, Nakamura K, Iwai S, Murakami M, Itoh T, Kijima H, Shipley JM, Senior RM, Shibuya M (2002) MMP9 induction by vascular endothelial growth factor receptor-1 is involved in lung-specific metastasis. Cancer Cell 2:289–300
1262. LeCouter J, Moritz DR, Li B, Phillips GL, Liang XH, Gerber HP, Hillan KJ, Ferrara N (2003) Angiogenesis-independent endothelial protection of liver: role of VEGFR-1. Science 299:890–893
1263. Wang S, Li X, Parra M, Verdin E, Bassel-Duby R, Olson EN (2008) Control of endothelial cell proliferation and migration by VEGF signaling to histone deacetylase 7. Proceedings of the National Academy of Sciences of the United States of America 105:7738–7743
1264. Zeng H, Zhao D, Yang S, Datta K, Mukhopadhyay D (2003) Heterotrimeric $G\alpha_q/G\alpha_{11}$ proteins function upstream of vascular endothelial growth factor (VEGF) receptor-2 (KDR) phosphorylation in vascular permeability factor/VEGF signaling. The Journal of Biological Chemistry 278:20738–20745
1265. Rousseau S, Houle F, Huot J (2000) Integrating the VEGF signals leading to actin-based motility in vascular endothelial cells. Trends in Cardiovascular Medicine 10:321–327
1266. Landry J, Huot J (1999) Regulation of actin dynamics by stress-activated protein kinase 2 (SAPK2)-dependent phosphorylation of heat-shock protein of 27 kDa (Hsp27). Biochemical Society Symposium 64:79–89
1267. Kobayashi M, Nishita M, Mishima T, Ohashi K, Mizuno K (2006) MAPKAPK-2-mediated LIM-kinase activation is critical for VEGF-induced actin remodeling and cell migration. EMBO Journal 25:713–726
1268. Dawe HR, Minamide LS, Bamburg JR, Cramer LP (2003) ADF/cofilin controls cell polarity during fibroblast migration. Current Biology 13:252–257
1269. Endo M, Ohashi K, Sasaki Y, Goshima Y, Niwa R, Uemura T, Mizuno K (2003) Control of growth cone motility and morphology by LIM kinase and Slingshot via phosphorylation and dephosphorylation of cofilin. Journal of Neuroscience 23:2527–2537
1270. Del Valle-Pérez B, Martínez VG, Lacasa-Salavert C, Figueras A, Shapiro SS, Takafuta T, Casanovas O, Capellà G, Ventura F, Viñals F (2010) Filamin B plays a key role in vascular endothelial growth factor-induced endothelial cell motility through its interaction with Rac-1 and Vav-2. Journal of Biological Chemistry 285:10748–10760

1271. Gavard J, Gutkind JS (2006) VEGF controls endothelial-cell permeability by promoting the bold beta-arrestin-dependent endocytosis of VE-cadherin. Nature – Cell Biology 8:1223–1234
1272. Powis G, Kirkpatrick L (2004) Hypoxia inducible factor-1 as a cancer drug target. Molecular Cancer Therapeutics 3:647–654
1273. Schmidt D, Textor B, Pein OT, Licht AH, Andrecht S, Sator-Schmitt M, Fusenig NE, Angel P, Schorpp-Kistner M (2007) Critical role for NF-κB-induced JunB in VEGF regulation and tumor angiogenesis. EMBO Journal 26:710–719
1274. Licht AH, Pein OT, Florin L, Hartenstein B, Reuter H, Arnold B, Lichter P, Angel P, Schorpp-Kistner MJ (2006) JunB is required for endothelial cell morphogenesis by regulating core-binding factor β. Journal of Cell Biology 175:981–991
1275. Ray PS, Fox PL (2007) A post-transcriptional pathway represses monocyte VEGF-A expression and angiogenic activity. EMBO Journal 26:3360–3372
1276. Fox PL, Mukhopadhyay R, Ray PS, Arif A (2008) The GAIT (interferon-γ-activated inhibitor of translation) system defines an auto-regulatory, negative-feedback circuit that controls inflammatory gene expression. FASEB Journal 22:601.4
1277. Fischer C, Jonckx B, Mazzone M, Zacchigna S, Loges S, Pattarini L, Chorianopoulos E, Liesenborghs L, Koch M, De Mol M, Autiero M, Wyns S, Plaisance S, Moons L, van Rooijen N, Giacca M, Stassen JM, Dewerchin M, Collen D, Carmeliet P (2007) Anti-PlGF inhibits growth of VEGF(R)-inhibitor-resistant tumors without affecting healthy vessels. Cell 131:463–475
1278. Greenberg JI, Shields DJ, Barillas SG, Acevedo LM, Murphy E, Huang J, Scheppke L, Stockmann C, Johnson RS, Angle N, Cheresh DA (2008) A role for VEGF as a negative regulator of pericyte function and vessel maturation. Nature 456:809–813
1279. Dong J, Grunstein J, Tejada M, Peale F, Frantz G, Liang WC, Bai W, Yu L, Kowalski J, Liang X, Fuh G, Gerber HP, Ferrara N (2004) VEGF-null cells require PDGFR alpha signaling-mediated stromal fibroblast recruitment for tumorigenesis. EMBO Journal 23:2800–2810
1280. Hou X, Kumar A, Lee C, Wang B, Arjunan P, Dong L, Maminishkis A, Tang Z, Li Y, Zhang F, Zhang SZ, Wardega P, Chakrabarty S, Liu B, Wu Z, Colosi P, Fariss RN, Lennartsson J, Nussenblatt R, Gutkind JS, Cao Y, Li X (2010) PDGF-CC blockade inhibits pathological angiogenesis by acting on multiple cellular and molecular targets. Proceedings of the National Academy of Sciences of the United States of America 107:12216–12221
1281. Song S, Ewald AJ, Stallcup W, Werb Z, Bergers G (2005) PDGFRβ+ perivascular progenitor cells in tumours regulate pericyte differentiation and vascular survival. Nature – Cell Biology 7:870–879
1282. Shojaei F, Wu X, Zhong C, Yu L, Liang XH, Yao J, Blanchard D, Bais C, Peale FV, van Bruggen N, Ho C, Ross J, Tan M, Carano RAD, Meng YG, Ferrara N (2007) Bv8 regulates myeloid-cell-dependent tumour angiogenesis. Nature 450:825–831
1283. Weckbach LT, Groesser L, Borgolte J, Pagel JI, Pogoda F, Schymeinsky J, Mller-Hcker J, Shakibaei M, Muramatsu T, Deindl E, Walzog B (2012) Midkine acts as proangiogenic cytokine in hypoxia-induced angiogenesis. American Journal of Physiology – Heart and Circulatory Physiology 303:H429–H438
1284. Ge G, Fernndez CA, Moses MA, Greenspan DS (2007) Bone morphogenetic protein 1 processes prolactin to a 17-kDa antiangiogenic factor. Proceedings of the National Academy of Sciences of the United States of America 104:10010–10015
1285. Tritsaris K, Myren M, Ditlev SB, Hbschmann MV, van der Blom I, Hansen AJ, Olsen UB, Cao R, Zhang J, Jia T, Wahlberg E, Dissing S, Cao Y (2007) IL-20 is an arteriogenic cytokine that remodels collateral networks and improves functions of ischemic hind limbs. Proceedings of the National Academy of Sciences of the United States of America 104:15364–15369
1286. Imhof BA, Aurrand-Lions M (2006) Angiogenesis and inflammation face off. Nature – Medicine 12:171–172

1287. Fiedler U, Reiss Y, Scharpfenecker M, Grunow V, Koidl S, Thurston G, Gale NW, Witzenrath M, Rosseau S, Suttorp N, Sobke A, Herrmann M, Preissner KT, Vajkoczy P, Augustin HG (2006) Angiopoietin-2 sensitizes endothelial cells to TNF-alpha and has a crucial role in the induction of inflammation. Nature – Medicine 12:235–239
1288. Suri C, Jones PF, Patan S, Bartunkova S, Maisonpierre PC, Davis S, Sato TN (1996) Yancopoulos G.D. Requisite role of angiopoietin-1, a ligand for the TIE2 receptor, during embryonic angiogenesis. Cell 87:1171–1180
1289. Augustin HG, Koh GY, Thurston G, Alitalo K (2009) Control of vascular morphogenesis and homeostasis through the angiopoietin–Tie system. Nature Reviews – Molecular Cell Biology 10:165–177
1290. Gale NW, Thurston G, Hackett SF, Renard R, Wang Q, McClain J, Martin C, Witte C, Witte MH, Jackson D, Suri C, Campochiaro PA, Wiegand SJ, Yancopoulos GD. Angiopoietin-2 is required for postnatal angiogenesis and lymphatic patterning, and only the latter role is rescued by Angiopoietin-1. Developmental Cell 3:411–423
1291. Li Z, Huang H, Boland P, Dominguez MG, Burfeind P, Lai KM, Lin HC, Gale NW, Daly C, Auerbach W, Valenzuela D, Yancopoulos GD, Thurston G (2009) Embryonic stem cell tumor model reveals role of vascular endothelial receptor tyrosine phosphatase in regulating Tie2 pathway in tumor angiogenesis. Proceedings of the National Academy of Sciences of the United States of America 106:22399–22404
1292. Kidoya H, Ueno M, Yamada Y, Mochizuki N, Nakata M, Yano T, Fujii R, Takakura N (2008) Spatial and temporal role of the apelin/APJ system in the caliber size regulation of blood vessels during angiogenesis. EMBO Journal 27:522–534
1293. Kidoya H, Takakura N (2012) Biology of the apelin–APJ axis in vascular formation. Journal of Biochemistry 152:125–131
1294. Ribatti D, Nico B, Spinazzi R, Vacca A, Nussdorfer GG (2005) The role of adrenomedullin in angiogenesis. Peptides 26:1670–1675
1295. Smith RS Jr, Gao L, Bledsoe G, Chao L, Chao J (2009) Intermedin is a new angiogenic growth factor. American Journal of Physiology – Heart and Circulatory Physiology 297:H1040–H1047
1296. English D, Welch Z, Kovala AT, Harvey K, Volpert OV, Brindley DN, Garcia JGN (2000) Sphingosine 1-phosphate released from platelets during clotting accounts for the potent endothelial cell chemotactic activity of blood serum and provides a novel link between hemostasis and angiogenesis. FASEB Journal 14:2255–2265
1297. Watanabe K, Hasegawa Y, Yamashita H, Shimizu K, Ding Y, Abe M, Ohta H, Imagawa K, Hojo K, Maki H, Sonoda H, Sato Y (2004) Vasohibin as an endothelium-derived negative feedback regulator of angiogenesis. Journal of Clinical Investigation 114:898–907
1298. Naito H, Kidoya H, Sato Y, Takakura N (2009) Induction and expression of anti-angiogenic vasohibins in the hematopoietic stem/progenitor cell population. Journal of Biochemistry 145:653–659
1299. Shimizu K, Watanabe K, Yamashita H, Abe M, Yoshimatsu H, Ohta H, Sonoda H, Sato Y (2005) Gene regulation of a novel angiogenesis inhibitor, vasohibin, in endothelial cells. Biochemical and Biophysical Research Communications 327:700–706
1300. Shibuya T, Watanabe K, Yamashita H, Shimizu K, Miyashita H, Abe M, Moriya T, Ohta H, Sonoda H, Shimosegawa T, Tabayashi K, Sato Y (2006) Isolation and characterization of vasohibin-2 as a homologue of VEGF-inducible endothelium-derived angiogenesis inhibitor vasohibin. Arteriosclerosis, Thrombosis, and Vascular Biology 26:1051–1057
1301. Sonoda H, Ohta H, Watanabe K, Yamashita H, Kimura H, Sato Y (2006) Multiple processing forms and their biological activities of a novel angiogenesis inhibitor vasohibin. Biochemical and Biophysical Research Communications 342:640–646
1302. Kimura H, Miyashita H, Suzuki Y, Kobayashi M, Watanabe K, Sonoda H, Ohta H, Fujiwara T, Shimosegawa T, Sato Y (2009) Distinctive localization and opposed roles of vasohibin-1 and vasohibin-2 in the regulation of angiogenesis. Blood 113:4810–4818
1303. Sato Y (2010) The vasohibin family. Pharmaceuticals 3:433–440

1304. Kuhnert F, Mancuso MR, Shamloo A, Wang HT, Choksi V, Florek M, Su H, Fruttiger M, Young WL, Heilshorn SC, Kuo CJ (2010) Essential regulation of CNS angiogenesis by the orphan G-protein-coupled receptor GPR124. Science 330:985–989
1305. Mammoto A, Connor KM, Mammoto T, Yung CW, Huh D, Aderman CM, Mostoslavsky G, Smith LEH, Ingber DE (2009) A mechanosensitive transcriptional mechanism that controls angiogenesis. Nature 457:1103–1108
1306. Wieland T, Mittmann C (2003) Regulators of G-protein signalling: multifunctional proteins with impact on signalling in the cardiovascular system. Pharmacology & Therapeutics 97:95–115
1307. Albig AR, Schiemann WP (2005) Identification and characterization of regulator of G protein signaling 4 (RGS4) as a novel inhibitor of tubulogenesis: RGS4 inhibits mitogen-activated protein kinases and vascular endothelial growth factor signaling. Molecular Biology of the Cell 16:609–625
1308. Hamzah J, Jugold M, Kiessling F, Rigby P, Manzur M, Marti HH, Rabie T, Kaden S, Grne HJ, Hmmerling GJ, Arnold B, Ganss R (2008) Vascular normalization in Rgs5-deficient tumours promotes immune destruction. Nature 453:410–414
1309. Im E, Kazlauskas A (2006) Regulating angiogenesis at the level of PtdIns-4,5-P2. EMBO Journal 25:2075–2082
1310. Graupera M, Guillermet-Guibert J, Foukas LC, Phng LK, Cain RJ, Salpekar A, Pearce W, Meek S, Millan J, Cutillas PR, Smith AJH, Ridley AJ, Ruhrberg C, Gerhardt H, Vanhaesebroeck B (2008) Angiogenesis selectively requires the p110alpha isoform of PI3K to control endothelial cell migration. Nature 453:662–666
1311. Gambardella L, Hemberger M, Hughes B, Zudaire E, Andrews S, Vermeren S (2010) PI3K Signaling through the dual GTPase-activating protein ARAP3 is essential for developmental angiogenesis. Science Signaling 3:ra76
1312. Chen J, Somanath PR, Razorenova O, Chen WS, Hay N, Bornstein P, Byzova TV (2005) Akt1 regulates pathological angiogenesis, vascular maturation and permeability in vivo. Nature – Medicine 11:1188–1196
1313. Shiojima I, Sato K, Izumiya Y, Schiekofer S, Ito M, Liao R, Colucci WS, Walsh K (2005) Disruption of coordinated cardiac hypertrophy and angiogenesis contributes to the transition to heart failure. The Journal of Clinical Investigation 115:2059–2064
1314. Sun JF, Phung T, Shiojima I, Felske T, Upalakalin JN, Feng D, Kornaga T, Dor T, Dvorak AM, Walsh K, Benjamin LE (2005) Microvascular patterning is controlled by fine-tuning the Akt signal. Proceedings of the National Academy of Sciences of the United States of America 102:128–133
1315. Nagoshi T, Matsui T, Aoyama T, Leri A, Anversa P, Li L, Ogawa W, del Monte F, Gwathmey JK, Grazette L, Hemmings B, Kass DA, Champion HC, Rosenzweig A (2005) PI3K rescues the detrimental effects of chronic Akt activation in the heart during ischemia/reperfusion injury. The Journal of Clinical Investigation 115:2059–2064
1316. Bornstein P, Agah A, Kyriakides TR (2004) The role of thrombospondins 1 and 2 in the regulation of cell-matrix interactions, collagen fibril formation, and the response to injury. The International Journal of Biochemistry & Cell Biology 36:1115–1125
1317. Lee NV, Sato M, Annis DS, Loo JA, Wu L, Mosher DF, Iruela-Arispe ML (2006) ADAMTS1 mediates the release of antiangiogenic polypeptides from TSP1 and 2. EMBO Journal 25:5270–5283
1318. Kitamura T, Asai N, Enomoto A, Maeda K, Kato T, Ishida M, Jiang P, Watanabe T, Usukura J, Kondo T, Costantini F, Murohara T, Takahashi M (2008) Regulation of VEGF-mediated angiogenesis by the Akt/PKB substrate Girdin. Nature – Cell Biology 10:329–337
1319. Alfranca A, Lpez-Oliva JM, Gens L, Lpez-Maderuelo D, Mirones I, Salvado D, Quesada AJ, Arroyo AG, Redondo JM (2008) PGE2 induces angiogenesis via MT1-MMP-mediated activation of the TGFbeta/Alk5 signaling pathway. Blood 112:1120–1128
1320. van den Berg YW, van den Hengel LG, Myers HR, Ayachi O, Jordanova E, Ruf W, Spek CA, Reitsma PH, Bogdanov VY, Versteeg HH (2009) Alternatively spliced tissue factor induces angiogenesis through integrin ligation. Proceedings of the National Academy of Sciences of the United States of America 106:19497–19502

1321. Yao YY, Yin H, Shen B, Smith RS, Liu Y, Gao L, Chao L, Chao J (2008) Tissue kallikrein promotes neovascularization and improves cardiac function by the Akt-glycogen synthase kinase-3β pathway. Cardiovascular Research 80:354–364
1322. Emanueli C, Salis MB, Van Linthout S, Meloni M, Desortes E, Silvestre JS, Clergue M, Figueroa CD, Gadau S, Condorelli G, Madeddu P (2004) Akt/protein kinase B and endothelial nitric oxide synthase mediate muscular neovascularization induced by tissue kallikrein gene transfer. Circulation 110:1638–1644
1323. Coffman LG, Parsonage D, D'Agostino R Jr, Torti FM, Torti SV (2009) Regulatory effects of ferritin on angiogenesis. Proceedings of the National Academy of Sciences of the United States of America 106:570–575
1324. Khan MM, Liu Y, Khan ME, Gilman ML, Khan ST, Bromberg ME, Colman RW (2010) Upregulation of tissue factor in monocytes by cleaved high molecular weight kininogen is dependent on TNFα and IL-1β. American Journal of Physiology – Heart and Circulatory Physiology 298:H652–H658
1325. Wang R (2012) Shared signaling pathways among gasotransmitters. Proceedings of the National Academy of Sciences of the United States of America 109:8801–8802
1326. Tang G, Wu L, Liang W, Wang R (2005) Direct stimulation of K_{ATP} channels by exogenous and endogenous hydrogen sulfide in vascular smooth muscle cells. Molecular Pharmacology 68:1757–1764
1327. Papapetropoulos A, Pyriochou A, Altaany Z, Yang G, Marazioti A, Zhou Z, Jeschke MG, Branski LK, Herndon DN, Wang R, Szabó C (2009) Hydrogen sulfide is an endogenous stimulator of angiogenesis. Proceedings of the National Academy of Sciences of the United States of America 106:21972–21977
1328. Coletta C, Papapetropoulos A, Erdelyi K, Olah G, Módis K, Panopoulos P, Asimakopoulou A, Gerö D, Sharina I, Martin E, Szabo C (2012) Hydrogen sulfide and nitric oxide are mutually dependent in the regulation of angiogenesis and endothelium-dependent vasorelaxation. Proceedings of the National Academy of Sciences of the United States of America 109:9161–9166
1329. West XZ, Malinin NL, Merkulova AA, Tischenko M, Kerr BA, Borden EC, Podrez EA, Salomon RG, Byzova TV (2010) Oxidative stress induces angiogenesis by activating TLR2 with novel endogenous ligands. Nature 467:972–976
1330. Zhang M, Brewer AC, Schröder K, Santos CX, Grieve DJ, Wang M, Anilkumar N, Yu B, Dong X, Walker SJ, Brandes RP, Shah AM (2010) NADPH oxidase-4 mediates protection against chronic load-induced stress in mouse hearts by enhancing angiogenesis. Proceedings of the National Academy of Sciences of the United States of America 107:18121–18126
1331. Kilarski WW, Samolov B, Petersson L, Kvanta A, Gerwins P (2009) Biomechanical regulation of blood vessel growth during tissue vascularization. Nature – Medicine 15:657–664
1332. Yung YC, Chae J, Buehler MJ, Hunter CP, Mooney DJ (2009) Cyclic tensile strain triggers a sequence of autocrine and paracrine signaling to regulate angiogenic sprouting in human vascular cells. Proceedings of the National Academy of Sciences of the United States of America 106:15279–15284
1333. Song JW, Munn LL (2011) Fluid forces control endothelial sprouting. Proceedings of the National Academy of Sciences of the United States of America 108:15342–15347
1334. Rivron NC, Vrij EJ, Rouwkema J, Le Gac S, van den Berg A, Truckenmller RK, van Blitterswijk CA (2012) Tissue deformation spatially modulates VEGF signaling and angiogenesis. Proceedings of the National Academy of Sciences of the United States of America 109:6886–6891
1335. Boerckel JD, Uhrig BA, Willett NJ, Huebsch N, Guldberg RE (2011) Mechanical regulation of vascular growth and tissue regeneration in vivo. Proceedings of the National Academy of Sciences of the United States of America 108:E674–E680
1336. Charlesworth PJS, Harris AL (2006) Mechanisms of disease: angiogenesis in urologic malignancies. Nature – Clinical Practice – Urology 3:157–169

1337. Murohara T, Ikeda H, Duan J, Shintani S, Sasaki KI, Eguchi H, Onitsuka I, Matsui K, Imaizumi T (2000) Transplanted cord blood-derived endothelial precursor cells augment postnatal neovascularization. Journal of Clinical Investigation 105:1527–1536
1338. Madeddu P, Emanueli C, Pelosi E, Salis MB, Cerio AM, Bonanno G, Patti M, Stassi G, Condorelli G, Peschle C (2004) Transplantation of low dose CD34+KDR+ cells promotes vascular and muscular regeneration in ischemic limbs. FASEB Journal 18:1737–1739
1339. Bir SC, Xiong Y, Kevil CG, Luo J (2012) Emerging role of PKA/eNOS pathway in therapeutic angiogenesis for ischaemic tissue diseases. Cardiovascular Research 95:7–18
1340. Ambati BK, Nozaki M, Singh N, Takeda A, Jani PD, Suthar T, Albuquerque RJC, Richter E, Sakurai E, Newcomb MT, Kleinman ME, Caldwell RB, Lin Q, Ogura Y, Orecchia A, Samuelson DA, Agnew DW, St Leger J, Green WR, Mahasreshti PJ, Curiel DT, Kwan D, Marsh H, Ikeda I, Leiper LJ, Collinson JM, Bogdanovich S, Khurana TS, Shibuya M, Baldwin ME, Ferrara N, Gerber HP, De Falco S, Witta J, Baffi JZ, Raisler BJ, Ambati J (2006) Corneal avascularity is due to soluble VEGF receptor-1. Nature 443:993–997
1341. Karagiannis ED, Popel AS (2008) A systematic methodology for proteome-wide identification of peptides inhibiting the proliferation and migration of endothelial cells. Proceedings of the National Academy of Sciences of the United States of America 105:13775–13780
1342. Singhn AJ, Meyer RD, Navruzbekov G, Shelke R, Duan L, Band H, Leeman SE, Rahimi N (2007) A critical role for the E3-ligase activity of c-Cbl in VEGFR-2-mediated PLCγ1 activation and angiogenesis. Proceedings of the National Academy of Sciences of the United States of America 104:5413–5418
1343. Streit M, Velasco P, Brown LF, Skobe M, Richard L, Riccardi L, Lawler J, Detmar M (1999) Overexpression of thrombospondin-1 decreases angiogenesis and inhibits the growth of human cutaneous squamous cell carcinomas. American Journal of Pathology 155:441–452
1344. Hamzah J, Jugold M, Kiessling F, Rigby P, Manzur M, Marti HH, Rabie T, Kaden S, Grne HJ, Hmmerling GJ, Arnold B, Ganss R (2008) Vascular normalization in Rgs5-deficient tumours promotes immune destruction. Nature 453:410-414
1345. Seaman S, Stevens J, Yang MY, Logsdon D, Graff-Cherry C, St. Croix B (2007) Genes that distinguish physiological and pathological angiogenesis. Cancer Cell 11:539–554
1346. Ardi VC, Kupriyanova TA, Deryugina EI, Quigley JP (2007) Human neutrophils uniquely release TIMP-free MMP-9 to provide a potent catalytic stimulator of angiogenesis. Proceedings of the National Academy of Sciences of the United States of America 104:20262–20267
1347. Lee DY, Deng Z, Wang CH, Yang BB (2007) MicroRNA-378 promotes cell survival, tumor growth, and angiogenesis by targeting SuFu and Fus-1 expression. Proceedings of the National Academy of Sciences of the United States of America 104:20350–20355
1348. Gomez DE, Alonso DF, Yoshiji H, Thorgeirsson UP (1997) Tissue inhibitors of metalloproteinases: structure, regulation and biological functions. European Journal of Cell Biology 74:111–122
1349. Stromblad S, Becker JC, Yebra M, Brooks PC, Cheresh DA (1996) Suppression of p53 activity and p21WAF1/CIP1 expression by vascular cell integrin alphaVbeta3 during angiogenesis. Journal of Clinical Investigation 98:426–433
1350. Mizukami Y, Jo WS, Duerr EM, Gala M, Li J, Zhang X, Zimmer MA, Iliopoulos O, Zukerberg LR, Kohgo Y, Lynch MP, Rueda BR, Chung DC (2005) Induction of interleukin-8 preserves the angiogenic response in HIF-1-deficient colon cancer cells. Nature Medicine 11:992–997
1351. Vincent L, Kermani P, Young LM, Cheng J, Zhang F, Shido K, Lam G, Bompais-Vincent H, Zhu Z, Hicklin DJ, Bohlen P, Chaplin DJ, May C, Rafii S (2005) Combretastatin A4 phosphate induces rapid regression of tumor neovessels and growth through interference with vascular endothelial-cadherin signaling. The Journal of Clinical Investigation 115:2992–3006

1352. Thijssen VLJL, Postel R, Brandwijk RJMGE, Dings RPM, Nesmelova I, Satijn S, Verhofstad N, Nakabeppu Y, Baum LG, Bakkers J, Mayo KH, Poirier F, Griffioen AW (2006) Galectin-1 is essential in tumor angiogenesis and is a target for antiangiogenesis therapy. Proceedings of the National Academy of Sciences of the United States of America 103:15975–15980
1353. Bernardi R, Guernah I, Jin D, Grisendi S, Alimonti A, Teruya-Feldstein J, Cordon-Cardo C, Simon MC, Rafii S, Pandolfi PP (2006) PML inhibits HIF-1alpha translation and neoangiogenesis through repression of mTOR. Nature 442:779–785
1354. Mancuso MR, Davis R, Norberg SM, O'Brien S, Sennino B, Nakahara T, Yao VJ, Inai T, Brooks P, Freimark B, Shalinsky DR, Hu-Lowe DD, McDonald DM (2006) Rapid vascular regrowth in tumors after reversal of VEGF inhibition. Journal of Clinical Investigation 116:2610–2621
1355. Juarez JC, Manuia M, Burnett M, Betancourt O, Boivin B, Shaw DE, Tonks NK, Mazar AP, Doate F (2008) Superoxide dismutase 1 (SOD1) is essential for H2O2-mediated oxidation and inactivation of phosphatases in growth factor signaling. Proceedings of the National Academy of Sciences of the United States of America 105:7147–7152
1356. Kleinman ME, Yamada K, Takeda A, Chandrasekaran V, Nozaki M, Baffi JZ, Albuquerque RJC, Yamasaki S, Itaya M, Pan Y, Appukuttan B, Gibbs D, Yang Z, Karik K, Ambati BK, Wilgus TA, DiPietro LA, Sakurai E, Zhang K, Smith JR, Taylor EW, Ambati J (2008) Sequence- and target-independent angiogenesis suppression by siRNA via TLR3. Nature 452:591–597
1357. Goel S, Fukumura D, Jain RK (2012) Normalization of the tumor vasculature through oncogenic inhibition: An emerging paradigm in tumor biology. Proceedings of the National Academy of Sciences of the United States of America 109:E1214
1358. Bottos A, Martini M, Di Nicolantonio F, Comunanza V, Maione F, Minassi A, Appendino G, Bussolino F, Bardelli A (2012) Targeting oncogenic serine/threonine-protein kinase BRAF in cancer cells inhibits angiogenesis and abrogates hypoxia. Proceedings of the National Academy of Sciences of the United States of America 109:E353–E359
1359. Oliver G, Harvey N (2002) A stepwise model of the development of lymphatic vasculature. Annals of the New York Academy of Sciences 979:159–165
1360. Gale NW, Prevo R, Espinosa J, Ferguson DJ, Dominguez MG, Yancopoulos GD, Thurston G, Jackson DG (2007) Normal lymphatic development and function in mice deficient for the lymphatic hyaluronan receptor LYVE-1. Molecular and Cellular Biology 27:595–604
1361. Saharinen P, Petrova TV (2004) Molecular regulation of lymphangiogenesis. Annals of the New York Academy of Sciences 1014:76–87
1362. Franois M, Caprini A, Hosking B, Orsenigo F, Wilhelm D, Browne C, Paavonen K, Karnezis T, Shayan R, Downes M, Davidson T, Tutt D, Cheah KSE, Stacker SA, Muscat GEO, Achen MG, Dejana E, Koopman P (2008) Sox18 induces development of the lymphatic vasculature in mice. Nature 456:643–647
1363. Watabe T (2012) Roles of transcriptional network during the formation of lymphatic vessels. Journal of Biochemistry 152:213–220
1364. Dagenais SL, Hartsough RL, Erickson RP, Witte MH, Butler MG, Glover TW (2004) Foxc2 is expressed in developing lymphatic vessels and other tissues associated with lymphedema-distichiasis syndrome. Gene Expression Patterns 4:611–619
1365. Petrova TV, Karpanen T, Norrmén C, Mellor R, Tamakoshi T, Finegold D, Ferrell R, Kerjaschki D, Mortimer P, Ylä-Herttuala S, Miura N, Alitalo K (2004) Defective valves and abnormal mural cell recruitment underlie lymphatic vascular failure in lymphedema distichiasis. Nature – Medicine 10:974–981
1366. Tammela T, Zarkada G, Wallgard E, Murtomäk A, Suchting S, Wirzenius M, Waltari M, Hellström M, Schomber T, Peltonen R, Freitas C, Duarte A, Isoniemi H, Laakkonen P, Christofori G, Ylä-Herttuala S, Shibuya M, Pytowski B, Eichmann A, Betsholtz C, Alitalo K (2008) Blocking VEGFR3 suppresses angiogenic sprouting and vascular network formation. Nature 454:656–660

1367. Alitalo K, Tammela T, Petrova TV (2005) Lymphangiogenesis in development and human disease. Nature 438:946–953
1368. Böhmer R, Neuhaus B, Bühren S, Zhang D, Stehling M, Böck B, Kiefer F (2010) Regulation of developmental lymphangiogenesis by Syk+ leukocytes. Developmental Cell 18:437–449
1369. Tammela T, Petrova TV, Alitalo K (2005) Molecular lymphangiogenesis: new players. Trends in Cell Biology 15:434–441
1370. Gale NW, Thurston G, Hackett SF, Renard R, Wang Q, McClain J, Martin C, Witte C, Witte MH, Jackson D, Suri C, Campochiaro PA, Wiegand SJ, Yancopoulos GD (2002) Angiopoietin-2 is required for postnatal angiogenesis and lymphatic patterning, and only the latter role is rescued by Angiopoietin-1. Developmental Cell 3:411–423
1371. Mäkinen T, Adams RH, Bailey J, Lu Q, Ziemiecki A, Alitalo K, Klein R, Wilkinson GA (2005) PDZ interaction site in ephrinB2 is required for the remodeling of lymphatic vasculature. Genes and Development 19:397–410
1372. Foo SS, Turner CJ, Adams S, Compagni A, Aubyn D, Kogata N, Lindblom P, Shani M, Zicha D, Adams RH (2006) Ephrin-B2 controls cell motility and adhesion during blood-vessel-wall assembly. Cell 124:161–173
1373. Qutub AA, Gabhann FM, Karagiannis ED, Vempati P, Popel AS (2009) Multiscale models of angiogenesis. IEEE Engineering in Medicine and Biology Magazine 28:14–31
1374. Schneider M, Reichold J, Weber B, Székely G, Hirsch S (2012) Tissue metabolism driven arterial tree generation. Medical Image Analysis (dx.doi.org/10.1016/j.media.2012.04.009)
1375. d'Onofrio A, Gandolfi A (2009) A family of models of angiogenesis and anti-angiogenesis anti-cancer therapy. Mathematical Medicine and Biology 26:63-95

Chap. 11. Tissue Development, Repair, and Remodeling

1376. Greenhalgh DG (1998) The role of apoptosis in wound healing. International Journal of Biochemistry and Cell Biology 30:1019–1030
1377. Lorenz HP, Longaker MT (2003) Wounds: biology, pathology, and Management. Stanford University Medical Center (http://recon.st.edu/Articles/LorenzWH.pdf)
1378. Zernecke A, Bidzhekov K, Noels H, Shagdarsuren E, Gan L, Denecke B, Hristov M, Kppel T, Jahantigh MN, Lutgens E, Wang S, Olson EN, Schober A, Weber C (2009) Delivery of microRNA-126 by apoptotic bodies induces CXCL12-dependent vascular protection. Science Signaling 2:ra81
1379. Steinberg MS (1970) Does differential adhesion govern self-assembly processes in histogenesis? Equilibrium configurations and the emergence of a hierarchy among populations of embryonic cells. Journal of Experimental Zoology 173:395–433
1380. Steinberg MS (2007) Differential adhesion in morphogenesis: a modern view. Current Opinion in Genetics and Development 17:281–286
1381. Harris AK (1976) Is cell sorting caused by differences in the work of intercellular adhesion? A critique of the Steinberg hypothesis. Journal of Theoretical Biology 61:267–285
1382. Ninomiya H, Winklbauer R (2008) Epithelial coating controls mesenchymal shape change through tissue-positioning effects and reduction of surface-minimizing tension. Nature – Cell Biology 10:61–69
1383. Krieg M, Arboleda-Estudillo Y, Puech PH, Kfer J, Graner F, Mller DJ, Heisenberg CP (2008) Tensile forces govern germ-layer organization in zebrafish. Nature – Cell Biology 10:429–436
1384. Hoffman LM, Carpenter MK (2005) Characterization and culture of human embryonic stem cells. Nature – Biotechnology 23:699–708
1385. Knox SM, Lombaert IMA, Reed X, Vitale-Cross L, Gutkind JS, Hoffman MP (2010) Parasympathetic innervation maintains epithelial progenitor cells during salivary organogenesis. Science 329:1645–1647

1386. Zile MH (1998) Vitamin A and embryonic development: an overview. Journal of Nutrition 128:455S–458S
1387. Niederreither K, Doll P (2008) Retinoic acid in development: towards an integrated view. Nature Reviews – Genetics 9:541–553
1388. Amaral PP, Dinger ME, Mercer TR, Mattick JS (2008) The eukaryotic genome as an RNA machine. Science 319:1787–1789
1389. Losick R, Desplan C (2008) Stochasticity and cell fate. Science 320:65–68
1390. Brandenberger R, Schmidt A, Linton J, Wang D, Backus C, Denda S, Mller U, Reichardt LF (2001) Identification and characterization of a novel extracellular matrix protein nephronectin that is associated with integrin alpha8beta1 in the embryonic kidney. Journal of Cell Biology 154:447–458
1391. Hobert O (2008) Gene regulation by transcription factors and microRNAs. Science 319:1785–1786
1392. Metzger RJ, Klein OD, Martin GR, Krasnow MA (2008) The branching programme of mouse lung development. Nature 453:745–750
1393. Rawlins EL, Ostrowski LE, Randell SH, Hogan BLM (2007) Lung development and repair: contribution of the ciliated lineage. Proceedings of the National Academy of Sciences of the United States of America 104:410–417
1394. Affolter M, Zeller R, Caussinus E (2009) Tissue remodelling through branching morphogenesis. Nature Reviews – Molecular Cell Biology 10:831–842
1395. Cirulli V, Yebra M (2007) Netrins: beyond the brain. Nature Reviews – Molecular Cell Biology 8:296–306
1396. De Paepe ME (2005) Lung growth and development. In: Churg AM, Myers JL, Tazelaar HD, Wright JL (eds), Thurlbeck's Pathology of the Lung, third edition, Thieme, New York
1397. Bellusci S (2008) Lung stem cells in the balance. Nature – Genetics 40:822–823
1398. Zhang Y, Goss AM, Cohen ED, Kadzik R, Lepore JJ, Muthukumaraswamy K, Yang J, DeMayo FJ, Whitsett JA, Parmacek MS, Morrisey EE (2008) A Gata6–Wnt pathway required for epithelial stem cell development and airway regeneration. Nature – Genetics 40:862–870
1399. Ventura JJ, Tenbaum S, Perdiguero E, Huth M, Guerra C, Barbacid M, Pasparakis M, Nebreda AR (2007) P38α MAP kinase is essential in lung stem and progenitor cell proliferation and differentiation. Nature – Genetics 39:750–758
1400. Kalin TV, Wang IC, Meliton L, Zhang Y, Wert SE, Ren X, Snyder J, Bell SM, Graf L Jr, Whitsett JA, Kalinichenko VV (2008) Forkhead Box m1 transcription factor is required for perinatal lung function. Proceedings of the National Academy of Sciences of the United States of America 105:19330–19335
1401. Warburton D (2008) Developmental biology: order in the lung. Nature 453:733–735
1402. Lin YD, Luo CY, Hu YN, Yeh ML, Hsueh YC, Chang MY, Tsai DC, Wang JN, Tang MJ, Wei EI, Springer ML, Hsieh PC (2012) Instructive nanofiber scaffolds with VEGF create a microenvironment for arteriogenesis and cardiac repair. Science Translational Medicine 4:146ra109
1403. Boyer LA, Lee TI, Cole MF, Johnstone SE, Levine SS, Zucker JP, Guenther MG, Kumar RM, Murray HL, Jenner RG, Gifford DK, Melton DA, Jaenisch R, Young RA (2005) Core transcriptional regulatory circuitry in human embryonic stem cells. Cell 122:947–956
1404. Karantalis V, Balkan W, Schulman IH, Hatzistergos KE, Hare JM (2012) Cell-based therapy for prevention and reversal of myocardial remodeling. American Journal of Physiology – Heart and Circulatory Physiology 303:H256–H270
1405. Feng B, Jiang J, Kraus P, Ng JH, Heng JCD, Chan YS, Yaw LP, Zhang W, Loh YH, Han J, Vega VB, Cacheux-Rataboul V, Lim B, Lufkin T, Ng HH (2009) Reprogramming of fibroblasts into induced pluripotent stem cells with orphan nuclear receptor Esrrb. Nature – Cell Biology 11:197–203

1406. Sadek H, Hannack B, Choe E, Wang J, Latif S, Garry MG, Garry DJ, Longgood J, Frantz DE, Olson EN, Hsieh J, Schneider JW (2008) Cardiogenic small molecules that enhance myocardial repair by stem cells. Proceedings of the National Academy of Sciences of the United States of America 105:6063–6068
1407. Chien KR (2008) Regenerative medicine and human models of human disease. Nature 453:302–305
1408. Passier R, van Laake LW, Mummery CL (2008) Stem-cell-based therapy and lessons from the heart. Nature 453:322–329
1409. Urbanek K, Cesselli D, Rota M, Nascimbene A, de Angelis A, Hosoda T, Bearzi C, Boni A, Bolli R, Kajstura J, Anversa P, Leri A (2006) Stem cell niches in the adult mouse heart. Proceedings of the National Academy of Sciences of the United States of America 103:9226-9231
1410. Caplice NM, Gersh BJ, Alegria JR (2005) Cell therapy for cardiovascular disease: what cells, what diseases and for whom? Nature – Clinical Practice – Cardiovascular Medicine 2:37–43
1411. Beltrami AP, Barlucchi L, Torella D, Baker M, Limana F, Chimenti S, Kasahara H, Rota M, Musso E, Urbanek K, Leri A, Kajstura J, Nadal-Ginard B, Anversa P (2003) Adult cardiac stem cells are multipotent and support myocardial regeneration. Cell 114:763–776
1412. Laflamme MA, Murry CE (2005) Regenerating the heart. Nature – Biotechnology 23:845–856.
1413. Kuhn B, Del Monte F, Hajjar RJ, Chang YS, Lebeche D, Arab S, Keating MT (2007) Periostin induces proliferation of differentiated cardiomyocytes and promotes cardiac repair. Nature – Medicine 13:962–969
1414. Urbich C, Heeschen C, Aicher A, Sasaki K, Bruhl T, Farhadi MR, Vajkoczy P, Hofmann WK, Peters C, Pennacchio LA, Abolmaali ND, Chavakis E, Reinhecke T, Zeiher AM, Dimmeler S (2005) Cathepsin L is required for endothelial progenitor cell-induced neo-vascularization. Nature – Medicine 11:206–213
1415. Menasche P (2004) Cellular transplantation: hurdles remaining before widespread clinical use. Current Opinion in Cardiology 19:154–161
1416. Rota M, Kajstura J, Hosoda T, Bearzi C, Vitale S, Esposito G, Iaffaldano G, Padin-Iruegas ME, Gonzalez A, Rizzi R, Small N, Muraski J, Alvarez R, Chen X, Urbanek K, Bolli R, Houser SR, Leri A, Sussman MA, Anversa P (2007) Bone marrow cells adopt the cardiomyogenic fate in vivo. Proceedings of the National Academy of Sciences of the United States of America 104:17783–17788
1417. Jiang Y, Jahagirdar BN, Reinhardt RL, Schwartz RE, Keene CD, Ortiz-Gonzalez XR, Reyes M, Lenvik T, Lund T, Blackstad M, Du J, Aldrich S, Lisberg A, Low WC, Largaespada DA, Verfaillie CM (2002) Pluripotency of mesenchymal stem cells derived from adult marrow. Nature 418:41–49
1418. Toma C, Pittenger MF, Cahill KS, Byrne BJ, Kessler PD (2002) Human mesenchymal stem cells differentiate to a cardiomyocyte phenotype in the adult murine heart, Circulation 105:93–98
1419. Hill JM, Zalos G, Halcox JPJ, Schenke WH, Waclawiw MA, Quyyumi AA, Finkel T (2003) Circulating endothelial progenitor cells, vascular function, and cardiovascular risk. New England Journal of Medicine 348:593–600
1420. Heeschen C, Lehmann R, Honold J, Assmus B, Aicher A, Walter DH, Martin H, Zeiher AM, S. Dimmeler S (2004) Profoundly reduced neovascularization capacity of bone marrow mononuclear cells derived from patients with chronic ischemic heart disease. Circulation 109:1615–1622
1421. Gnecchi M, He H, Liang OD, Melo LG, Morello F, Mu H, Noiseux N, Zhang L, Pratt RE, Ingwall JS, Dzau VJ (2005) Paracrine action accounts for marked protection of ischemic heart by Akt-modified mesenchymal stem cells. Nature – Medicine 11:367–368
1422. Ventura C, Cantoni S, Bianchi F, Lionetti V, Cavallini C, Scarlata I, Foroni L, Maioli M, Bonsi L, Alviano F, Fossati V, Bagnara GP, Pasquinelli G, Recchia FA, Perbellini A (2007)

Hyaluronan mixed esters of butyric and retinoic Acid drive cardiac and endothelial fate in term placenta human mesenchymal stem cells and enhance cardiac repair in infarcted rat hearts. Journal of Biological Chemistry 282:14243–14252

1423. Stevens KR, Kreutziger KL, Dupras SK, Korte FS, Regnier M, Muskheli V, Nourse MB, Bendixen K, Reinecke H, Murry CE (2009) Physiological function and transplantation of scaffold-free and vascularized human cardiac muscle tissue. Proceedings of the National Academy of Sciences of the United States of America 106:16568–16573
1424. Nagai T, Komuro I (2012) Gene and cytokine therapy for heart failure: molecular mechanisms in the improvement of cardiac function. American Journal of Physiology – Heart and Circulatory Physiology 303:H501–H512
1425. Most P, Remppis A, Pleger ST, Katus HA, Koch WJ (2007) S100A1: a novel inotropic regulator of cardiac performance. Transition from molecular physiology to pathophysiological relevance. American Journal of Physiology – Regulatory, Integrative and Comparative Physiology 293:R568–R577
1426. Inagawa K, Miyamoto K, Yamakawa H, Muraoka N, Sadahiro T, Umei T, Wada R, Katsumata Y, Kaneda R, Nakade K, Kurihara C, Obata Y, Miyake K, Fukuda K, Ieda M (2012) Induction of cardiomyocyte-like cells in infarct hearts by gene transfer of Gata4, Mef2c, and Tbx5. Circulation Research
1427. Kusano KF, Pola R, Murayama T, Curry C, Kawamoto A, Iwakura A, Shintani S, Ii M, Asai J, Tkebuchava T, Thorne T, Takenaka H, Aikawa R, Goukassian D, von Samson P, Hamada H, Yoon YS, Silver M, Eaton E, Ma H, Heyd L, Kearney M, Munger W, Porter JA, Kishore R, Losordo DW (2005) Sonic hedgehog myocardial gene therapy: tissue repair through transient reconstitution of embryonic signaling. Nature – Medicine 11:1197–1204
1428. Stegemann JP, Nerem RM (2003) Phenotype modulation in vascular tissue engineering using biochemical and mechanical stimulation. Annals of Biomedical Engineering 31:391–402
1429. Song H, Stevens CF, Gage FH (2002) Astroglia induce neurogenesis from adult neural stem cells. Nature 417:39–44
1430. Shen Q, Goderie SK, Jin L, Karanth N, Sun Y, Abramova N, Vincent P, Pumiglia K, Temple S (2004) Endothelial cells stimulate self-renewal and expand neurogenesis of neural stem cells. Science 304:1338–1340.
1431. McGuigan AP, Sefton MV (2006) Vascularized organoid engineered by modular assembly enables blood perfusion. Proceedings of the National Academy of Sciences of the United States of America 103:11461–11466
1432. Zimmermann WH, Melnychenko I, Wasmeier G, Didie M, Naito H, Nixdorff U, Hess A, Budinsky L, Brune K, Michaelis B, Dhein S, Schwoerer A, Ehmke H, Eschenhagen T (2006) Engineered heart tissue grafts improve systolic and diastolic function in infarcted rat hearts. Nature – Medicine 12:452–458
1433. Miyahara Y, Nagaya N, Kataoka M, Yanagawa B, Tanaka K, Hao H, Ishino K, Ishida H, Shimizu T, Kangawa K, Sano S, Okano T, Kitamura S, Mori H (2006) Monolayered mesenchymal stem cells repair scarred myocardium after myocardial infarction. Nature – Medicine 12:459–465
1434. Stamm C, Khosravi A, Grabow N, Schmohl K, Treckmann N, Drechsel A, Nan M, Schmitz KP, Haubold A, Steinhoff G (2004) Biomatrix/polymer composite material for heart valve tissue engineering. Annals of Thoracic Surgery 78:2084–2093
1435. Schenke-Layland K, Opitz F, Gross M, Doring C, Halbhuber KJ, Schirrmeister F, Wahlers T, Stock UA (2003) Complete dynamic repopulation of decellularized heart valves by application of defined physical signals – an in vitro study. Cardiovascular Research 60:497–509
1436. Carver W, Nagpal ML, Nachtigal M, Borg TK, Terracio L (1991) Collagen expression in mechanically stimulated cardiac fibroblasts. Circulation Research 69:116–122
1437. Driessen NJ, Boerboom RA, Huyghe JM, Bouten CV, Baaijens FP (2003) Computational analyses of mechanically induced collagen fiber remodeling in the aortic heart valve. Journal of Biomechanical Engineering 125:549–557

1438. L'Heureux N, Dusserre N, Konig G, Victor B, Keire P, Wight TN, Chronos NAF, Kyles AE, Gregory CR, Hoyt G, Robbins RC, McAllister TN (2006) Human tissue-engineered blood vessels for adult arterial revascularization. Nature – Medicine 12:361–365
1439. Staudacher DL, Flugelman MY (2006) Cell and gene therapies in cardiovascular disease with secial focus on the no option patient. Current Gene Therapy 6:609–623
1440. Zarbiv G, Preis M, Ben-Yosef Y, Flugelman MY (2007) Engineering blood vessels by gene and cell therapy. Expert Opinion on Biological Therapy 7:1183–1191
1441. Kurpinski K, Chu J, Hashi C, Li S (2006) Anisotropic mechanosensing by mesenchymal stem cells. Proceedings of the National Academy of Sciences of the United States of America 103:16095–16100
1442. Petersen TH, Calle EA, Zhao L, Lee EJ, Gui L, Raredon MB, Gavrilov K, Yi T, Zhuang ZW, Breuer C, Herzog E, Niklason LE (2010) Tissue-engineered lungs for in vivo implantation. Science 329:538–541
1443. Ott HC, Clippinger B, Conrad C, Schuetz C, Pomerantseva I, Ikonomou L, Kotton D, Vacanti JP (2010) Regeneration and orthotopic transplantation of a bioartificial lung. Nature – Medicine 16:927–933
1444. Wong CHY, Heit B, Kubes P (2010) Molecular regulators of leucocyte chemotaxis during inflammation. Cardiovascular Research 86:183–191
1445. Barreiro O, Martín P, González-Amaro R, Sánchez-Madrid F (2010) Molecular cues guiding inflammatory responses. Cardiovascular Research 86:174–182
1446. Brasier AR (2010) The nuclear factor-κB–interleukin-6 signalling pathway mediating vascular inflammation. Cardiovascular Research 86:211–218
1447. Pestronk A (2005) Adhesion molecules (http://www.neuro.wustl.edu/neuromuscular/lab/adhesion.htm)
1448. Štvrtinová V, Jakubovký J, Hulín I (1995) Inflammation and fever, Academic Electronic Press (http://nic.sav.sk/logos/books/scientific/node37.html)
1449. Woodfin A, Voisin MB, Nourshargh S (2007) PECAM-1: a multi-functional molecule in inflammation and vascular biology. Arteriosclerosis, Thrombosis, and Vascular Biology 27:2514.
1450. Vestweber D, Blanks JE (1999) Mechanisms that regulate the function of the selectins and their ligands. Physiological Reviews 79:181–213
1451. Yoshida M, Westlin WF, Wang N, Ingber DE, Rosenzweig A, Resnick N, Gimbrone MA (1996) Leukocyte adhesion to vascular endothelium induces E-selectin linkage to the actin cytoskeleton. Journal of Cell Biology 133:445–455
1452. Fernandez-Borja M, van Buul JD, Hordijk PL (2010) The regulation of leucocyte transendothelial migration by endothelial signalling events. Cardiovascular Research 86:202–210
1453. Schulz B, Pruessmeyer J, Maretzky T, Ludwig A, Blobel CP, Saftig P, Reiss K (2008) ADAM10 regulates endothelial permeability and T-Cell transmigration by proteolysis of vascular endothelial cadherin. Circulation Research 102:1192–1201
1454. Koenen RR, Pruessmeyer J, Soehnlein O, Fraemohs L, Zernecke A, Schwarz N, Reiss K, Sarabi A, Lindbom L, Hackeng TM, Weber C, Ludwig A (2009) Regulated release and functional modulation of junctional adhesion molecule A by disintegrin metalloproteinases. Blood 113:4799–4809
1455. Nishikimi A, Fukuhara H, Su W, Hongu T, Takasuga S, Mihara H, Cao Q, Sanematsu F, Kanai M, Hasegawa H, Tanaka Y, Shibasaki M, Kanaho Y, Sasaki T, Frohman MA, Fukui Y (2009) Sequential regulation of DOCK2 dynamics by two phospholipids during neutrophil chemotaxis. Science 324:384–387
1456. Anderson P, Kedersha N (2007) On again, off again: the SRC-3 transcriptional coactivator moonlights as a translational corepressor. Molecular Cell 25:765–778
1457. Shi M, Deng W, Bi E, Mao K, Ji Y, Lin G, Wu X, Tao Z, Li Z, Cai X, Sun S, Xiang C, Sun B (2008) TRIM30α negatively regulates TLR-mediated NF-κB activation by targeting TAB2 and TAB3 for degradation. Nature – Immunology 9:369–377

1458. Jamieson T, Cook DN, Nibbs JB, Rot A, Nixon C, McLean P, Alcami A, Lira SA, Wiekowski M, Graham GJ (2005) The chemokine receptor D6 limits the inflammatory response in vivo. Nature – Immunology 6:403–411
1459. Yarilina A, Park-Min KH, Antoniv T, Hu X, Ivashkiv LB (2008) TNF activates an IRF1-dependent autocrine loop leading to sustained expression of chemokines and STAT1-dependent type I interferon-response genes. Nature – Immunology 9:378–387
1460. van Vliet C, Bukczynska PE, Puryer MA, Sadek CM, Shields BJ, Tremblay ML, Tiganis T (2005) Selective regulation of tumor necrosis factor-induced Erk signaling by Src family kinases and the T cell protein tyrosine phosphatase. Nature – Immunology 6:253–260
1461. Ferch U, Meyer zum Bschenfelde C, Gewies A, Wegener E, Rauser S, Peschel C, Krappmann D, Ruland J (2007) MALT1 directs B cell receptor-induced canonical nuclear factor-κB signaling selectively to the c-Rel subunit. Nature – Immunology 8:984–991
1462. Shambharka PB, Blonska M, Pappu BP, Li H, You Y, Sakurai H, Darnay BG, Hara H, Penninger J, Lin X (2007) Phosphorylation and ubiquitination of the IκB kinase complex by two distinct signaling pathways. EMBO Journal 26:1794–1805
1463. Glass CK, Saijo K (2010) Nuclear receptor transrepression pathways that regulate inflammation in macrophages and T cells. Nature Reviews – Immunology 10:365–376
1464. McDuffie JE, Motley ED, Limbird LE, Maleque MA (2000). 5-hydroxytryptamine stimulates phosphorylation of p44/p42 mitogen-activated protein kinase activation in bovine aortic endothelial cell cultures. Journal of Cardiovascular Pharmacology 35:398–402
1465. Wymann MP, Schneiter R (2008) Lipid signalling in disease. Nature Reviews – Molecular Cell Biology 9:162–176
1466. Lemos HP, Grespan R, Vieira SM, Cunha TM, Verri WA, Fernandes KS, Souto FO, McInnes IB, Ferreira SH, Liew FY, Cunha FQ (2009) Prostaglandin mediates IL-23/IL-17-induced neutrophil migration in inflammation by inhibiting IL-12 and IFNγ production. Proceedings of the National Academy of Sciences of the United States of America 106:5954–5959
1467. Benveniste J, Henson PM, Cochrane CG (1972) Leukocyte-dependent histamine release from rabbit platelets. The role of IgE, basophils, and a platelet-activating factor. Journal of Experimental Medicine 136:1356–1377
1468. Napolitano G, Karin M (2010) Sphingolipids: the oil on the TRAFire that promotes inflammation. Science Signaling 3:pe34
1469. Agostini L, Martinon F, Burns K, McDermott MF, Hawkins PN, Tschopp J (2004) NALP3 forms an IL-1beta-processing inflammasome with increased activity in Muckle-Wells autoinflammatory disorder. Immunity 20:319–325
1470. Kanneganti TD, Ozoren N, Body-Malapel M, Amer A, Park JH, Franchi L, Whitfield J, Barchet W, Colonna M, Vandenabeele P, Bertin J, Coyle A, Grant EP, Akira S, Nunez G (2006) Bacterial RNA and small antiviral compounds activate caspase-1 through cryopyrin/Nalp3. Nature 440:4337–4342
1471. Martinon F, Tschopp J (2005) NLRs join TLRs as innate sensors of pathogens. Trends in Immunology 26:447–454
1472. Yin Y, Yan Y, Jiang X, Mai J, Chen NC, Wang H, Yang XF (2009) Inflammasomes are differentially expressed in cardiovascular and other tissues. International Journal of Immunopathology and Pharmacology 22:311–322
1473. Gurtner GC, Werner S, Barrandon Y, Longaker MT (2008) Wound repair and regeneration. Nature 453:314–321
1474. Forte A, Della Corte A, De Feo M, Cerasuolo F, Cipollaro M (2010) Role of myofibroblasts in vascular remodelling: focus on restenosis and aneurysm. Cardiovascular Research 88:395–405
1475. Hong HS, Lee J, Lee EA, Kwon YS, Lee E, Ahn W, Jiang MH, Kim JC, Son Y (2009) A new role of substance P as an injury-inducible messenger for mobilization of CD29+ stromal-like cells. Nature – Medicine 15:425–435
1476. Murray JD (2002) Mathematical Biology. Volume II: Spatial Models and Biomedical Applications. Springer, New York

1477. Zani BG, Kojima K, Vacanti CA, Edelman ER (2008) Tissue-engineered endothelial and epithelial implants differentially and synergistically regulate airway repair. Proceedings of the National Academy of Sciences of the United States of America 105:7046–7051
1478. Laugwitz KL, Moretti A, Lam J, Gruber P, Chen Y, Woodard S, Lin LZ, Cai CL, Lu MM, Reth M, Platoshyn O, Yuan JX, Evans S, Chien KR (2005) Postnatal isl1+ cardioblasts enter fully differentiated cardiomyocyte lineages. Nature 433:647–653
1479. Harada M, Qin Y, Takano H, Minamino T, Zou Y Toko H, Ohtsuka M, Matsuura K, Sano M, Nishi J, Iwanaga K, Akazawa H, Kunieda T, Zhu W, Hasegawa H, Kunisada K, Nagai T, Nakaya H, Yamauchi-Takihara K, Komuro I (2005) G-CSF prevents cardiac remodeling after myocardial infarction by activating the Jak-Stat pathway in cardiomyocytes. Nature – Medicine 11:305–311
1480. Meldrum DR, Wang M, Tsai BM, Kher A, Pitcher JM, Brown JW, Meldrum KK (2005) Intracellular signaling mechanisms of sex hormones in acute myocardial inflammation and injury. Frontiers in Bioscience 10:1835–1867
1481. Werner SL, Barken D, Hoffmann A (2005) Stimulus specificity of gene expression programs determined by temporal control of IKK activity. Science 309:1857–1861
1482. Covert MW, Leung TH, Gaston JE, Baltimore D (2005) Achieving stability of lipopolysaccharide-induced NF-kappaB activation. Science 309:1854–1857
1483. Frey N, Barrientos T, Shelton JM, Frank D, Rtten H, Gehring D, Kuhn C, Lutz M, Rothermel B, Bassel-Duby R, Richardson JA, Katus HA, Hill JA, Olson EN (2004) Mice lacking calsarcin-1 are sensitized to calcineurin signaling and show accelerated cardiomyopathy in response to pathological biomechanical stress. Nature – Medicine 10:1336–1343
1484. Satoh M, Ogita H, Takeshita K, Mukai Y, Kwiatkowski DJ, Liao JK (2006) Requirement of Rac1 in the development of cardiac hypertrophy. Proceedings of the National Academy of Sciences of the United States of America 103:7432-7437
1485. Masuda H, Zhuang YJ, Singh TM, Kawamura K, Murakami M, Zarins CK, Glagov S (1999) Adaptive remodeling of internal elastic lamina and endothelial lining during flow-induced arterial enlargement. Arteriosclerosis, Thrombosis, and Vascular Biology 19:2298–2307
1486. Singh TM, Abe KY, Sasaki T, Zhuang YJ, Masuda H, Zarins CK (1998) Basic fibroblast growth factor expression precedes flow-induced arterial enlargement. Journal of Surgical Research 77:165–173
1487. Langille BL (1993) Remodeling of developing and mature arteries: endothelium, smooth muscle, and matrix. Journal of Cardiovascular Pharmacology 21:S11–S117
1488. Kurjiaka DT, Bender SB, Nye DD, Wiehler WB, Welsh DG (2005) Hypertension attenuates cell-to-cell communication in hamster retractor muscle feed arteries. American Journal of Physiology – Heart and Circulatory Physiology 288:H861–H870
1489. Vaishnav RN, Vossoughi J, Patel DJ, Cothran LN, Coleman BR, Ison-Franklin EL (1990) Effect of hypertension on elasticity and geometry of aortic tissue from dogs. Journal of Biomechanical Engineering 112:70–74
1490. Matsumoto T, Hayashi K (1994) Mechanical and dimensional adaptation of rat aorta to hypertension. Journal of Biomechanical Engineering 116:278–283
1491. Hu JJ, Fossum TW, Miller MW, Xu H, Liu JC, Humphrey JD (2007) Biomechanics of the porcine basilar artery in hypertension. Annals of Biomedical Engineering 35:19–29
1492. Fridez P, Rachev A, Meister J-J, Hayashi K, Stergiopulos N (2001) Model of geometrical and vascular smooth muscle tone adaptation on carotid artery subjected to step change in pressure. American Journal of Physiology – Heart and Circulatory Physiology 280:2752–2760
1493. Fridez P, Zulliger M, Bobard F, Montorzi G, Miyazaki H, Hayashi K, Stergiopulos N (2003) Geometrical, functional, and histomorphometric adaptation of rat carotid artery in induced hypertension. Journal of Biomechanics 36:671-680
1494. Lehoux S, Lemarie CA, Esposito B, Lijnen HR, Tedgui A (2004) Pressure-induced matrix metalloproteinase-9 contributes to early hypertensive remodeling. Circulation 109:1041–1047

1495. Tronc F, Mallat Z, Lehoux S, Wassef M, Esposito B, Tedgui A (2000) Role of matrix metalloproteinases in blood flow-induced arterial enlargement: interaction with NO. Arteriosclerosis, Thrombosis and Vascular Biology 20:E120–E126
1496. Choi MH, Lee IK, Kim GW, Kim BU, Han YH, Yu DY, Park HS, Kim KY, Lee JS, Choi C, Bae YS, Lee BI, Rhee SG, Kang SW (2005) Regulation of PDGF signalling and vascular remodelling by peroxiredoxin II. Nature 435:347–353
1497. Kiyan J, Kiyan R, Haller H, Dumler I (2005) Urokinase-induced signaling in human vascular smooth muscle cells is mediated by PDGFR-β. EMBO Journal 24:1787–1797
1498. Ray FR, Huang W, Slater M, Barden JA (2002) Purinergic receptor distribution in endothelial cells in blood vessels: a basis for selection of coronary artery grafts. Atherosclerosis 162:55–61
1499. Rachev A, Stergiopulos N, Meister J-J, A model for geometric and mechanical adaptation of arteries to sustained hypertension. Journal of Biomechanical Engineering 120:9–17
1500. Fung YC, Liu SQ, Zhou JB (1993) Remodeling of the constitutive equation while a blood vessel remodels itself under stress. Journal of Biomechanical Engineering 115:453–459
1501. Murray J, Osher G, Harris A (1983) A mechanical model for mesenchymal morphogenesis. Journal of Mathematical Biology 17:125–129
1502. Manoussaki D (2003) A mechanochemical model of angiogenesis and vasculogenesis. Mathematical Modelling and Numerical Analysis 37:581–599
1503. Byrne HM, Chaplain MA (1996) Modelling the role of cell-cell adhesion in the growth and development of carcinomas. Mathematical and Computer Modelling 24:1–17
1504. Szczerba D, Szkely G (2005) Computational model of flow-tissue interactions in intussusceptive angiogenesis. Journal of Theoretical Biology 234:87–97
1505. Filbet F, Laurencot P, Perthame B (2004) Derivation of hyperbolic models for chemosensitive movement. Journal of Mathematical Biology 50:189–207.
1506. Ambrosi D, Gamba A, Serini G (2004) Cell directional persistence and chemotaxis in vascular morphogenesis. Bulletin of Mathematical Biology 66:1851–1873.
1507. Schugart RC, Friedman A, Zhao R, Sen CK (2008) Wound angiogenesis as a function of tissue oxygen tension: A mathematical model. Proceedings of the National Academy of Sciences of the United States of America 105:2628–2633
1508. Humphrey JD, Rajagopal KR (2002) A constrained mixture model for growth and remodeling of soft tissues. Mathematical Models and Methods in Applied Sciences 12:407–430
1509. Humphrey JD (2003) Continuum biomechanics of soft biological tissues, Proceedings of the Royal Society of London – Series A: Mathematical, Physical and Engineering Sciences 459:3–46
1510. Gleason RL, Humphrey JD (2005) Effects of a sustained extension on arterial growth and remodeling: a theoretical study. Journal of Biomechanics 38:1255–1261

Chap. 12. Airway Surface Liquid and Respiratory Mucus

1511. Evans SE, Xu Y, Tuvim MJ, Dickey BF (2010) Inducible innate resistance of lung epithelium to infection. Annual Review of Physiology 72:413–435
1512. Hattrup CL, Gendler SJ (2008) Structure and function of the cell surface (tethered) mucins. Annual Review of Physiology 70:431–457
1513. Button B, Cai LH, Ehre C, Kesimer M, Hill DB, Sheehan JK, Boucher RC, Rubinstein M (2012) A periciliary brush promotes the lung health by separating the mucus layer from airway epithelia. Science 337:937–941
1514. Randell SH, Boucher RC (2006) Effective mucus clearance is essential for respiratory health. American Journal of Respiratory Cell and Molecular Biology 35:20–28
1515. Sanderson MJ, Lansley AB, Dirksen ER (1992) Regulation of ciliary beat frequency in respiratory tract cells. Chest 101:69S–71S

1516. Flynn AN, Itani OA, Moninger TO, Welsh MJ (2009) Acute regulation of tight junction ion selectivity in human airway epithelia. Proceedings of the National Academy of Sciences of the United States of America 106:3591–3596
1517. Yeates DB, Pitt BR, Spektor DM, Karron GA, Albert RE (1981) Coordination of mucociliary transport in the human trachea and intrapulmonary airways. Journal of Applied Physiology – Respiratory, Environmental and Exercise Physiology 51:1057–1064
1518. Katz I, Zwas T, Baum GL, Aharonson E, Belfer B (1987) Ciliary beat frequency and mucociliary clearance. What is the relationship? Chest 92:491–493
1519. Wanner A (1981) Alteration of tracheal mucociliary transport in airway disease: effect of pharmacologic agents. Chest 80:867-870
1520. Zahm JM, Milliot M, Bresin A, Coraux C, Birembaut P (2011) The effect of hyaluronan on airway mucus transport and airway epithelial barrier integrity: potential application to the cytoprotection of airway tissue. Matrix Biology 30:389–395
1521. Wilson R, Alton E, Rutman A, Higgins P, Al Nakib W, Geddes DM, Tyrrell DA, Cole PJ (1987) Upper respiratory tract viral infection and mucociliary clearance. European Journal of Respiratory Diseases 70:272–279
1522. Parkhurst MR, Saltzman WM (1994) Leukocytes migrate through three-dimensional gels of midcycle cervical mucus. Cellular Immunology 156:77–94
1523. Olmsted SS, Padgett JL, Yudin AI, Whaley KJ, Moench TR, Cone RA (2001) Diffusion of macromolecules and virus-like particles in human cervical mucus. Biophysical Journal 81:1930–1937
1524. Saltzman WM, Radomsky ML, Whaley KJ, Cone RA (1994) Antibody diffusion in human cervical mucus. Biophysical Journal 66:508–515
1525. Dulfano MJ, Adler KB (1975) Physical properties of sputum VII. Rheologic properties and mucociliary transport. American Review of Respiratory Diseases 112:341–347
1526. Puchelle E, Girard F, Zahm JM (1976) Rh´ologie des sécrétions bronchitiques et transport muco-cilaire. Rheology of bronchial secretions and mucociliary transport. Bulletin Européen de Physiopathologie Respiratoire 12:771–779
1527. Giordano AM, Holsclaw D, Litt M (1978) Mucus rheology and mucociliary clearance: normal physiologic state. American Review of Respiratory Diseases 118:245–250
1528. Gelman RA, Meyer FA (1979) Mucociliary transference rate and mucus viscoelasticity. Dependence on dynamic storage and loss modulus. American Review of Respiratory Diseases 120:553–557
1529. Puchelle E, Zahm JM, Quemada D (1987) Rheological properties controlling mucociliary frequency and respiratory mucus transport. Biorheology 24:557–563
1530. Puchelle E, Zahm JM, Girard F, Bertrand A, Polu JM, Aug F, Sadoul P (1980) Mucociliary transport in vitro and in vivo. European Journal of Respiratory Diseases 61:254–264
1531. Marriott C (1990) Mucus and mucociliary clearance in the respiratory tract. Advanced Drug Delivery Reviews 5:5–19
1532. Reynolds SD, Reynolds PR, Pryhuber GS, Finder JD, Stripp BR (2002) Secretoglobins SCGB3A1 and SCGB3A2 define secretory cell subsets in mouse and human airways. American Journal of Respiratory and Critical Care Medicine 166:1498–1509
1533. Kim S, Shim JJ, Burgel PR, Ueki IF, Dao-Pick T, Tam DC, Nadel JA (2002) IL-13-induced Clara cell secretory protein expression in airway epithelium: role of EGFR signaling pathway. American Journal of Physiology – Lung Cellular and Molecular Physiology 283:L67–L75
1534. Evans CM, Williams OW, Tuvim MJ, Nigam R, Mixides GP, Blackburn MR, DeMayo FJ, Burns AR, Smith C, Reynolds SD, Stripp BR, Dickey BF (2004) Mucin is produced by clara cells in the proximal airways of antigen-challenged mice.
1535. Ehre C, Rossi AH, Abdullah LH, De Pestel K, Hill S, Olsen JC, Davis CW (2005) Barrier role of actin filaments in regulated mucin secretion from airway goblet cells. American Journal of Physiology – Cell Physiology 288:C46–C56
1536. Reid L (1960) Measurement of the bronchial mucous gland layer: a diagnostic yardstick in chronic bronchitis. Thorax 15:132–141

1537. Tos M (1966) Development of the tracheal glands in man. Number, density, structure, shape, and distribution of mucous glands elucidated by quantitative studies of whole mounts. Acta Pathologica and Microbiologica Scandinavica 68:1–130
1538. Wine JJ, Joo NS (2004) Submucosal glands and airway defense. Proceedings of the American Thoracic Society 1:47–53
1539. Reader JR, Tepper JS, Schelegle ES, Aldrich MC, Putney LF, Pfeiffer JW, Hyde DM (2003) Pathogenesis of mucous cell metaplasia in a murine asthma model. American Journal of Pathology 162:2069–2078
1540. Hayashi T, Ishii A, Nakai S, Hasegawa K (2004) Ultrastructure of goblet-cell metaplasia from Clara cell in the allergic asthmatic airway inflammation in a mouse model of asthma in vivo. Virchows Archiv 444:66–73
1541. Chen Y, Zhao YH, Wu R (2001) Differential regulation of airway mucin gene expression and mucin secretion by extracellular nucleotide triphosphates. American Journal of Respiratory Cell and Molecular Biology 25:409–417
1542. Zhu Y, Ehre C, Abdullah LH, Sheehan JK, Roy M, Evans CM, Dickey BF, Davis CW (2008) Munc13-2−/− baseline secretion defect reveals source of oligomeric mucins in mouse airways. Journal of Physiology 586:1977–1992
1543. Spencer H, Leof D (1964) The innervation of the human lung. Journal of Anatomy 98:599–609
1544. Maggi CA, Giachetti A, Dey RD, Said SI (1995) Neuropeptides as regulators of airway function: vasoactive intestinal peptide and the tachykinins. Physiological Reviews 75:277–322
1545. Fahy JV, Dickey BF (2010) Airway mucus function and dysfunction. New England Journal of Medicine 363:2233–2247
1546. Li Y, Martin LD, Spizz G, Adler KB (2001) MARCKS protein is a key molecule regulating mucin secretion by human airway epithelial cells in vitro. Journal of Biological Chemistry 276:40982–40990
1547. Koch H, Hofmann K, Brose N (2000) Definition of Munc13-homology-domains and characterization of a novel ubiquitously expressed Munc13 isoform. Biochemical Journal 349:247–253
1548. Tarran R, Button B, Boucher RC (2006) Regulation of normal and cystic fibrosis airway surface liquid volume by phasic shear stress. Annual Review of Physiology 68:543–561
1549. Davis CW, Lazarowski ER (2008) Coupling of airway ciliary activity and mucin secretion to mechanical stresses by purinergic signaling. Respiratory Physiology and Neurobiology 163:208–213
1550. Ransford GA, Fregien N, Qiu F, Dahl G, Conner GE, Salathe M (2009) Pannexin 1 contributes to ATP release in airway epithelia. American Journal of Respiratory Cell and Molecular Biology 41:525–534
1551. Bao L, Locovei S, Dahl G (2004) Pannexin membrane channels are mechanosensitive conduits for ATP. FEBS Letters 572:65–68
1552. Locovei S, Wang J, Dahl G (2006) Activation of pannexin 1 channels by ATP through P2Y receptors and by cytoplasmic calcium. FEBS Letters 580:239–244
1553. Lazarowski ER, Boucher RC (2009) Purinergic receptors in airway epithelia. Current Opinion in Pharmacology 9:262–267
1554. Kreda SM, Okada SF, van Heusden CA, O'Neal W, Gabriel S, Abdullah L, Davis CW, Boucher RC, Lazarowski ER (2007) Coordinated release of nucleotides and mucin from human airway epithelial Calu-3 cells. Journal of Physiology 584:245–259
1555. Okada SF, Zhang L, Kreda SM, Abdullah LH, Davis CW, Pickles RJ, Lazarowski ER, Boucher RC (2011) Coupled nucleotide and mucin hypersecretion from goblet-cell metaplastic human airway epithelium. American Journal of Respiratory Cell and Molecular Biology 45:253–260
1556. Thornton DJ, Sheehan JK (2004) From mucins to mucus: toward a more coherent understanding of this essential barrier. Proceedings of the American Thoracic Society 1: 54–61

1557. Lai SK, Wang Y-Y, Wirtz D, Hanes J (2009) Micro- and macrorheology of mucus. Advanced Drug Delivery Reviews 61:86–100
1558. Sleigh MA, Blake JR, Liron N (1988) The propulsion of mucus by cilia. American Review of Respiratory Diseases 137:726–741
1559. McGuire JF (2002) Surfactant in the middle ear and eustachian tube: a review. International Journal of Pediatric Otorhinolaryngology 66:1–15
1560. Baconnais S, Tirouvanziam R, Zahm JM, de Bentzmann S, Pault B, Balossier G, Puchelle E (1999) Ion composition and rheology of airway liquid from cystic fibrosis fetal tracheal xenografts. American Journal of Respiratory Cell and Molecular Biology 20:605–611
1561. Dawson M, Wirtz D, Hanes J (2003) Enhanced viscoelasticity of human cystic fibrotic sputum correlates with increasing microheterogeneity in particle transport. Journal of Biological Chemistry 278: 50393–50401
1562. Girod S, Zahm JM, Plotkowski C, Beck G, Puchelle E (1992) Role of the physiochemical properties of mucus in the protection of the respiratory epithelium. European Respiratory Journal 5:477–487
1563. King M, Macklem PT (1977) Rheological properties of microliter quantities of normal mucus. Journal of Applied Physiology 42:797–802
1564. Galabert C, Jacquot J, Zahm JM, Puchelle E (1987) Relationships between the lipid content and the rheological properties of airway secretions in cystic fibrosis. Clinica Chimica Acta 164:139–149
1565. App EM, Tomkiewicz RP, Hahn HL, Engler H, Vergin H, King M (1997) The effect of Tasuldine, a bronchosecretolytic agent, on mucus rheology and clearability and the interaction with acetylcholine in ferrets. Pulmonary Pharmacology and Therapeutics 10:271–276
1566. Lai SK, Wang Y-Y, Cone R, Wirtz D, Hanes J (2009) Altering mucus rheology to "solidify" human mucus at the nanoscale. PLoS One 4: e4294
1567. King M (2006) Physiology of mucus clearance. Paediatric Respiratory Reviews 7:S212–S214
1568. Foliguet B, Puchelle E (1986) Apical structure of human respiratory cilia. Bulletin Européen de Physio-Pathologie Respiratoire 22:43–47
1569. Shah AS, Farmen SL, Moninger TO, Businga TR, Andrews MP, Bugge K, Searby CC, Nishimura D, Brogden KA, Kline JN, Sheffield VC, Welsh MJ (2008) Loss of Bardet-Biedl syndrome proteins alters the morphology and function of motile cilia in airway epithelia. Proceedings of the National Academy of Sciences of the United States of America 105:3380–3385
1570. Sarmah B, Winfrey VP, Olson GE, Appel B, Wente SR (2007) A role for the inositol kinase Ipk1 in ciliary beating and length maintenance. Proceedings of the National Academy of Sciences of the United States of America 104:19843–19848
1571. Spungin B, Silberberg A (1984) Stimulation of mucus secretion, ciliary activity and transport in frog palate epithelium. American Journal of Physiology – Cell Physiology 247:C299–C308
1572. Yager J, Chen TM, Dulfano MJ (1978) Frequency of ciliary beats of human respiratory epithelium. Chest 73:627–633
1573. Rutland J, Griffin WM, Cole PJ (1982) Human ciliary beat frequency in epithelium from intrathoracic and extrathoracic airways. American Review of Respiratory Diseases 125:100–105
1574. Lindberg S, Runer T (1994) Method for in vivo measurement of mucociliary activity in the human nose. Annals of Otology, Rhinology and Laryngology 103:558–566
1575. Nuutinen J, Toskala ER, Saano V, Susanna J (1993) Ciliary beating frequency in chronic sinusitis. Archives of Otolaryngology – Head and Neck Surgery 119:645–647
1576. Salathe M (2007) Regulation of mammalian ciliary beating. Annual Review of Physiology 69:401–422
1577. Tamm SL, Terasaki M (1994) Visualization of calcium transients controlling orientation of ciliary beat. Journal of Cell Biology 125:1127–1135

1578. Sanderson MJ, Dirksen ER (1986) Mechanosensitivity of cultured ciliated cells from the mammalian respiratory tract: implications for the regulation of mucociliary transport. Proceedings of the National Academy of Sciences of the United States of America 83:7302–7306
1579. Villaln M, Hinds TR, Verdugo P (1989) Stimulus response coupling in mammalian ciliated cells. Biophysical Journal 56:1255–1258
1580. Tamaoki J, Kondo M, Takizawa T (1989) Adenosine-mediated cAMP-dependent inhibition of ciliary activity in rabbit tracheal epithelium. American Review of Respiratory Diseases 139:441–445
1581. Wang D, Sun Y, Zhang W, Huang P (2008) Apical adenosine regulates basolateral Ca^{2+}-activated potassium channels in human airway Calu-3 epithelial cells. American Journal of Physiology – Cell Physiology 294:C1443–C1453
1582. Kondo M, Tamaoki J, Takizawa T (1990) Neutral endopeptidase inhibitor potentiates the tachykinin-induced increase in ciliary beat frequency in rabbit trachea. American Review of Respiratory Diseases 142:403–406
1583. Kobayashi K, Tamaoki J, Sakai N, Kanemura T, Horii S, Takizawa T (1990) Angiotensin II stimulates airway ciliary motility in rabbit cultured tracheal epithelium. Acta Physiologica Scandinavica 138:497–502
1584. Mwimbi XK, Muimo R, Green MW, Mehta A (2003) Making human nasal cilia beat in the cold: a real time assay for cell signalling. Cellular Signalling 15:395–402
1585. Alberty J, August C, Stoll W, Rudack C (2004) The effect of endogenous nitric oxide on cholinergic ciliary stimulation of human nasal mucosa. Laryngoscope 114:1642-1647
1586. Sisson JH, Pavlik JA, Wyatt TA (2009) Alcohol stimulates ciliary motility of isolated airway axonemes through a nitric oxide, cyclase, and cyclic nucleotide-dependent kinase mechanism. Alcoholism, Clinical and Experimental Research 33:610–616
1587. Kim CS, Rodriguez CR, Eldridge MA, Sackner MA (1986) Criteria for mucus transport in the airways by two-phase gas-liquid flow mechanism. Journal of Applied Physiology 60:901–907
1588. Kim CS, Greene MA, Sankaran S, Sackner MA (1986) Mucus transport in the airways by two-phase gas-liquid flow mechanism: continuous flow model. Journal of Applied Physiology 60:908–917
1589. Yoshitsugu M, Hanamure Y, Furuta S, Deguchi K, Ueno K, Rautiainen M (1994) Ciliary motility and surface morphology of cultured human respiratory epithelial cells during ciliogenesis. Biology of the Cell 82:211–216
1590. Wilson CW, Nguyen CT, Chen MH, Yang JH, Gacayan R, Huang J, Chen JN, Chuang PT (2009) Fused has evolved divergent roles in vertebrate Hedgehog signalling and motile ciliogenesis. Nature 459:98–102
1591. Sorokin SP (1968) Reconstruction of centriole formation and ciliogenesis in mammalian lungs. Journal of Cell Science 3:207–230
1592. Enault S, Lombardi D, Poncet P, Thiriet M (2010) Mucus dynamics subject to air and wall motion, ESAIM: Proceedings 30:125–141
1593. Chatelin R, Poncet P, Didier A, Murris-Espin M, Anne-Archard D, Thiriet M (2012) Mucus and ciliated cells of the human lung: splitting strategies for particle methods and 3D Stokes flows. International Union of Theoretical and Applied Mechanics (IUTAM) Symposium on Particle Methods in Fluid Mechanics, October 15–17, 2012, Lyngby, Denmark
1594. Chatelin R, Poncet P, Tokman M (2010) Computational aspects of mucus propulsion by ciliated epithelium. Proceedings of the 2nd European Conference on Microfluidics - Microfluidics 2010 - Toulouse, December 8-10, 2010 (paper μFLU10-107)

Chap. 13. Surfactant

1595. Hoppin FG, Lee GC, Dawson SV (1975) Properties of lung parenchyma in distortion. Journal of Applied Physiology 39:742–751
1596. von Neergaard K (1929) Neue Auffassungen über einen Grundbegriff der Atemmechanik. Zeitschrift für die Gesamte Experimentelle Medizin 66:373–394
1597. Pattle RE (1955) Properties, function and origin of the alveolar lining layer. Nature 175:1125–1126
1598. Radford EP (1963) Mechanical stability of the lung. Determination by surface active agents. Archives of Environmental Health 6:128–133
1599. Clements JA (1956) Dependence of pressure-volume characteristics of lungs on intrinsic surface active material. American Journal of Physiology 187:592
1600. Clements JA (1957) Surface tension of lung extracts. Proceedings of the Society of Experimental Biology and Medicine 95:170-172
1601. Clements JA, Brown ES, Johnson RP (1958) Pulmonary surface tension and the mucus lining of the lungs: some theoretical considerations. Journal of Applied Physiology 12:262–268
1602. Pattle RE (1958) Properties, function, and origin of the alveolar lining layer. Proceedings of the Royal Society London – Series B Biological Sciences 148:217–240
1603. Avery ME, Mead J (1959) Surface properties in relation to atelectasis and hyaline membrane disease. American Journal of Diseases of Childhood 97,517–523
1604. Johnson MD (2007) Ion transport in alveolar type I cells. Molecular BioSystems 3:178–186
1605. Factor P, Mutlu GM, Chen L, Mohameed J, Akhmedov AT, Meng FJ, Jilling T, Lewis ER, Johnson MD, Xu A, Kass D, Martino JM, Bellmeyer A, Albazi JS, Emala C, Lee HT, Dobbs LG, Matalon S (2007) Adenosine regulation of alveolar fluid clearance. Proceedings of the National Academy of Sciences of the United States of America 104:4083–4088
1606. Klaus MH, Clements JA, Havel RJ (1961) Composition of surface-active material isolated from beef lung. Proceedings of the National Academy of Sciences of the United States of America 47:1858–1859
1607. Gluck L, Motoyama EK, Smits HL, Kulovich MV (1967) The biochemical development of surface activity in mammalian lung. I. The surface-active phospholipids; the separation and distribution of surface-active lecithin in the lung of the developing rabbit fetus. Pediatric Research 1:237–246
1608. Schürch S, Goerke J, Clements JA (1978) Direct determination of volume- and time-dependence of alveolar surface tension in excised lungs. Proceedings of the National Academy of Sciences of the United States of America 75:3417–3421
1609. Fujiwara T, Maeta H, Chida S, Morita T, Watabe Y, Abe, T (1980) Artificial surfactant therapy in hyaline-membrane disease. Lancet 1:55–59
1610. Perez-Gil J, Keough KM (1998) Interfacial properties of surfactant proteins. Biochimica et Biophysica Acta 1408:203–217
1611. Oosterlaken-Dijksterhuis MA, van Eijk M, van Buel BL, van Golde LMG, Haagsman HP (1991) Surfactant protein composition of lamellar bodies isolated from rat lung. Biochemical Journal 274:115–119
1612. Petrache I, Natarajan V, Zhen L, Medler TR, Richter AT, Cho C, Hubbard WC, Berdyshev EV, Tuder RM (2005) Ceramide upregulation causes pulmonary cell apoptosis and emphysema-like disease in mice. Nature Medicine 11:491–498
1613. Haagsman HP, Diemel RV (2001) Surfactant-associated proteins: functions and structural variation. Comparative Biochemistry and Physiology. Part A, Molecular & Integrative Physiology 129:91–108
1614. Palaniyar N, Ikegami M, Korfhagen T, Whitsett J, McCormack FX (2001) Domains of surfactant protein A that affect protein oligomerization, lipid structure and surface tension. Comparative Biochemistry and Physiology. Part A, Molecular & Integrative Physiology 129:109–127

1615. Krol S, Ross M, Sieber M, Kunneke S, Galla HJ, Janshoff A (2000) Formation of three-dimensional protein-lipid aggregates in monolayer films induced by surfactant protein B. Biophysical Journal 79:904–918
1616. Bush JMW (2011) Surface tension. In Ben Amar M, Goriely A, Müller MM, Cagliandolo LF (eds) New Trends in the Physisc and Mechanics of Biological Systems, Oxford University Press, New York
1617. Tadmor R (2009) Marangoni flow revisited. Journal of Colloid and Interface Science 332:451-454
1618. Halpern D, Bull JL, Grotberg JB (2004) The effect of airway wall motion on surfactant delivery. ASME Journal of Biomechanical Engineering 126:410–419
1619. Halpern D, Jensen OE, Grotberg JB (1998) A theoretical study of surfactant and liquid delivery into the lung. Journal of Applied Physiology 85:333–352

Chap. 14. Conclusion

1620. Mehlsen J, Pagh K, Nielsen JS, Sestoft L, Nielsen SL (1987) Heart rate response to breathing: dependency upon breathing pattern. Clinical Physiology 7:115–124
1621. Shields RW (2009) Heart rate variability with deep breathing as a clinical test of cardiovagal function. Cleveland Clinic Journal of Medicine 76:S37-S40
1622. Rhoades R, Bell DR (2009) Medical Physiology: Principles for Clinical Medicine, 3rd ed., Wolters Kluwer/Lippincott Williams & Wilkins, Philadelphia

Notation Rules: Aliases and Symbols

1623. Hoffmann R, Valencia A (2004) A gene network for navigating the literature. Nature – Genetics 36:664 (Information Hyperlinked over Proteins (www.ihop-net.org/)

Notation Rules: Aliases and Symbols

Aliases that designate different types of molecules as well as those that do not carry an obvious meaning should be eliminated; they are thus not used in the present text. For example, P35 is an alias for annexin-A1, brain syntaxin-1A, ficolin-2, interleukin-12A, the cyclin-H assembly factor Ménage à trois homolog-1, regulatory subunit-1 of cyclin-dependent kinase CDK5, and uroplakin-3B, among others. It is substituted by AnxA1, Stx1a, Fcn2, IL12a, MAT1, $CDK5_{r1}$, and UPk3b aliases, respectively. Protein P39 corresponds to the subunit D1 of the lysosomal V-type H^+ ATPase (ATP6v0d1), Jun transcription factor, a component of the Activator protein AP1, and regulatory subunit-2 of cyclin-dependent kinase CDK5 ($CDK5_{r2}$). Extracellular signal-regulated protein kinases ERK1 and ERK2, members of the mitogen-activated protein kinase (MAPK) module (last tier), are also abbreviated P44 and P42 (also P40 and P41). However, both P42 and P44 correspond to the 26S protease regulatory AAA ATPase subunit (PSMC6). Alias P42 is also utilized for cyclin-dependent kinase CDK20, cyclin-dependent kinase-like protein CDKL1, and 43-kDa NuP43 nucleoporin. Alias P44 can also refer to interferon-induced protein IFI44 (or microtubule-associated protein MTAP44) and androgen receptor cofactor P44 (a.k.a. methylosome protein MeP50 and WD repeat-containing protein WDR77). In the present text, P38 members (P38α–P38δ) of the mitogen-activated protein kinase modules (i.e., MAPK11–MAPK14)[1] are designated as P38MAPKs to avoid confusion with other molecules, the alias of which is also P38.[2]

---

[1]Protein P38α is also known as MAPK14, cytokine suppressive anti-inflammatory drug (CSAID)-binding protein CSBP, CSBP1, or CSBP2, and stress-activated protein kinase SAPK2a; P38β as MAPK11 and SAPK2b; P38γ as MAPK12, ERK6, and SAPK3; P38δ as MAPK13 and SAPK4.

[2]Alias P38 is used for: (1) mitogen-activated protein kinase MAPK1, extracellular signal-regulated kinase ERK2, as well as P40, P41, and P42; (2) adaptor CRK (chicken tumor virus regulator of kinase, or v-crk sarcoma virus CT10 oncogene homolog); (3) growth factor receptor-binding protein GRB2-related adaptor protein GRAP2 (a.k.a. GRID, GADS, GRB2L, GRF40, GRPL, and Mona); (4) ubiquitin ligase RING finger protein RNF19a, or dorfin; (5) 38-kDa DNA polymerase-δ-interacting protein PolδIP2 (a.k.a. polymerase [DNA-directed] PDIP38 and PolD4); (6) activator of 90-kDa heat shock protein ATPase homolog AHSA1; and (7) aminoacyl

M. Thiriet, *Tissue Functioning and Remodeling in the Circulatory and Ventilatory Systems*, Biomathematical and Biomechanical Modeling of the Circulatory and Ventilatory Systems 5, DOI 10.1007/978-1-4614-5966-8,

Aliases for Molecules

Aliases include all written variants, i.e., any abbreviation[3] such as acronyms. An *acronym* corresponds to a word made from the initial letters or syllables of nouns that are pronounceable as a word. Acronyms are generally written with all letters in upper case. Yet, some acronyms are treated as words and written in lower case (e.g., laser [originally LASER] is an acronym for light amplification by stimulated emission of radiation, sonar [originally SONAR] for sound navigation and ranging). A substance's name can derive from its chemical name (e.g., amphetamine: α-methylphenethylamine).

Acronyms can give rise to molecule names by adding a scientific suffix such as "-in", a common ending of molecule nouns (e.g., sirtuin, a portmanteau, that comes from the alias SIRT, which stands for silent information regulator-2 [two]). Other scientific prefixes and suffixes can be frequently detected throughout the present text. Their meaning is given in appendix Notations – Prefixes and Suffixes, particularly for readers from Asia. Many prefixes are used to specify position, configuration and behavior, quantity, direction and motion, structure, timing, frequency, and speed.

A *portmanteau* is a word that combines initials and some inner letters of at least 2 words (e.g., calmodulin stands for calcium modulated protein; caspase for cysteine-dependent aspartate-specific protease; chanzyme for ion channel and enzyme; chemokine for chemoattractant cytokine;[4] emilin for elastin microfibril interfacer; endorphins and endomorphins for endogenous morphines; ephrin for erythropoietin-producing hepatocyte (EPH) receptor kinase interactor; granzyme for granule enzyme; moesin for membrane-organizing extension spike protein; porin for pore-forming protein; restin for Reed-Steinberg cell-expressed intermediate filament-associated protein, an alias for cytoplasmic linker protein CLiP1 (or CLiP170); serpin for serine protease inhibitor; siglec for sialic acid-binding Ig-like lectin; transceptor for transporter-related receptor; and Prompt for promoter upstream transcript).[5]

---

tRNA synthase complex-interacting multifunctional protein AIMP2, or tRNA synthase complex component JTV1 [1623].

[3]In general, abbreviations exclude the initials of short function words, such as "and", "or", "of", or "to". However, they are sometimes included in acronyms to make them pronounceable (e.g., radar [originally RADAR] for radio detection and ranging). These letters are often written in lower case. In addition, both cardinal (size, molecular weight, etc.) and ordinal (isoform discovery order) numbers in names are represented by digits.

[4]Cytokines are peptidic, proteic, or glycoproteic regulators that are secreted by cells of the immune system. These immunomodulating agents serve as auto- or paracrine signals.

[5]The upper case initial P in Prompt is used to avoid confusion with command-line interpreter prompt or prompt book to direct precise timing of actions on theater stage.

Initialisms are abbreviations that are formed from initial letters of a single long noun or several nouns and, instead of being pronounced like an ordinary word, are read letter-by-letter (e.g., DNA stands deoxyribonucleic acid).

Some abbreviations can give rise to alphabetisms that are written as new words (e.g., Rho-associated, coiled-coil-containing protein kinase [RoCK] that is also called Rho kinase). In biochemistry, multiple-letter abbreviations can also be formed from a single word that can be long (e.g., Cam stands for calmodulin, which is itself a portmanteau word, Trx for thioredoxin, etc.) as well as short (e.g., Ttn for titin, etc.). In addition, single-letter symbols of amino acids are often used to define a molecule alias (e.g., tyrosine can be abbreviated as Tyr or Y, hence SYK stands for spleen tyrosine kinase).

Aliases use, in general, capital letters and can include hyphens and dots. Yet, as a given protein can represent a proto-oncogene[6] encoded by a gene that can give rise to an oncogene (tumor promoter) after gain- or loss-of-function mutations,[7] the same acronym represents 3 different entities.[8]

Besides, a given abbreviation can designate distinct molecules without necessarily erroneous consequence in a given context (e.g., PAR: poly^{ADP}ribose or protease-activated receptor and GCK: germinal center kinases or glucokinase; in the latter case, the glucokinase abbreviation should be written as GcK or, better, GK).

In addition, a large number of aliases that designate a single molecule results from the fact that molecules have been discovered independently several times with possibly updated functions. Some biochemists uppercase the name of a

[6]In 1911, P. Rous isolated a virus that was capable of generating tumors of connective tissue (sarcomas) in chicken. Proteins were afterward identified, the activity of which, when uncontrolled, can provoke cancer, hence the name oncogene given to genes that encode these proteins. Most of these proteins are enzymes, more precisely kinases. The first oncogene was isolated from the avian Rous virus by D. Stéhelin and called Src (from sarcoma). This investigator demonstrated that the abnormal functioning of the Src protein resulted from mutation of a normal gene, or proto-oncogene, which is involved in cell division.

[7]Loss-of-function mutations cause complete or partial loss of function of gene products that operate as tumor suppressors, whereas gain-of-function mutations generate gene products with new or abnormal function that can then act as oncogenes. Typical tumor-inducing agents are enzymes, mostly regulatory kinases and small guanosine triphosphatases, that favor proliferation of cells, which do normally need to be activated to exert their activities. Once their genes are mutated, these enzymes become constitutively active. Other oncogenes include growth factors (a.k.a. mitogens) and transcription factors. Mutations can also disturb signaling axis regulation, thereby raising protein expression. Last, but not least, chromosomal translocation can also provoke the expression of a constitutively active hybrid protein.

[8]Like Latin-derived shortened expressions – as well as foreign words – that are currently written in italics, genes can be italicized. However, this usage is not required in scientific textbooks published by Springer. Italic characters are then used to highlight words within a text to easily target them. Proteins are currently romanized (ordinary print), but with a capital initial. Nevertheless, names (not aliases) of chemical species are entirely lowercased in most – if not all – scientific articles, except to avoid confusion with a usual word (e.g., hedgehog animal vs. Hedgehog protein and raptor [bird of prey] vs. Raptor molecule).

given molecule, whereas others lowercase (e.g., cell division cycle guanosine triphosphatase of the Rho family CDC42 or Cdc42, adaptor growth factor receptor-bound protein GRB2 or Grb2, chicken tumor virus regulator of kinase CRK or Crk, guanine nucleotide-exchange factor Son of sevenless SOS or Sos, etc.). Acronyms are then not always capitalized. Printing style of aliases should not only avoid confusion, but also help one in remembering alias meaning.

In the present textbook, choice of lower and upper case letters in molecule aliases is dictated by the following criteria.

(1) An upper case letter is used for initials of words that constitute molecule nouns (e.g., receptor tyrosine kinase RTK). An alias of any compound takes into account added atoms or molecules (e.g., PI: phosphoinositide and PIP: phosphoinositide phosphate) as well as their number (e.g., PIP_2: phosphatidylinositol bisphosphate, DAG: diacylglycerol, and PDE: [cyclic nucleotide] phosphodiesterases).

(2) A lower case letter is used when a single letter denotes a subfamily or an isoform when it is preceded by a capital letter (e.g., PTPRe: protein tyrosine phosphatase receptor-like type-E). Nevertheless, an upper case letter is used in an alias after a single or several lower case letters to distinguish the isoform type (e.g., RhoA isoform and DNA-repair protein RecA for recombination protein-A), but OSM stands for oncostatin-M, not osmole Osm[9] to optimize molecule identification.

These criteria enable to use differently written aliases with the same sequence of letters for distinct molecules (e.g., CLIP for corticotropin-like intermediate peptide, CLiP: cytoplasmic CAP-Gly domain-containing linker protein, and iCliP: intramembrane-cleaving protease).

As the exception proves the rule, current aliases, such as PKA and PLA that designate protein kinase-A and phospholipase-A, respectively, have been kept. Preceded by only 2 upper case letters, a lower case letter that should be used to specify an isoform can bring confusion with acronyms of other protein types (e.g., phospholamban alias PLb).

Nouns (e.g., hormone-like fibroblast growth factor [hFGF] and urokinase-type plasminogen activator [uPA]) or adjectives (e.g., intracellular FGF isoform [iFGF]) that categorize a subtype of a given molecule correspond to a lower case letter to emphasize the molecule species. Hence, an upper case letter with a commonly used hyphen (e.g., I[R]-SMAD that stands for inhibitory [receptor-regulated] SMAD; V-ATPase for vacuolar adenosine triphosphatase; MT1-MMP for membrane type-1 matrix metalloproteinase; and T[V]-SNARE for target [vesicle-associated] soluble ^{N}ethylmaleimide-sensitive factor-attachment protein receptor) is then replaced by a lower case letter (e.g., i[r]SMAD, vATPase, mt1MMP, and t[v]SNARE), as is usual for RNA subtypes (mRNA, rRNA, snRNA, and tRNA for messenger, ribosomal, small nuclear, and transfer RNA, respectively). Similarly, membrane-bound and secreted forms of receptors and coreceptors that can derive from alternative mRNA

---

[9]Osmole: the amount of osmotically active particles that exerts an osmotic pressure of 1 atm when dissolved in 22.4 l of solvent at 0 °C.

splicing are defined by a lower case letter (e.g., sFGFR for secreted extracellular FGFR form and sFRP for soluble Frizzled-related protein), as well as eukaryotic translation elongation (eEF) and initiation (eIF) factors.

(3) Although `l`, `r`, and `t` can stand for molecule-like, -related, and -type, respectively, when a chemical is related to another one, in general, upper case letters are used for the sake of homogenity and to clearly distinguish between the letter L and numeral 1 (e.g., KLF: Krüppel-like factor, CTK: C-terminal Src kinase (CSK)-type kinase, and SLA: Src-like adaptor).

(4) An upper case letter is most often used for initials of adjectives contained in the molecule name (e.g., AIP: actin-interacting protein; BAX: BCL2-associated X protein; HIF: hypoxia-inducible factor; KHC: kinesin heavy chain; LAB: linker of activated B lymphocytes; MAPK: mitogen-activated protein kinase; and SNAP: soluble N-ethylmaleimide-sensitive factor-attachment protein).

(5) Lower case letters are used when alias letters do not correspond to initials (e.g., Fox – not fox –: forkhead box), except for portmanteau words that are entirely written in minuscules (e.g., gadkin: γ1-`ad`aptin and `kin`esin `in`teractor).

This rule applies, whether alias letters do correspond to successive noun letters (e.g., Par: partitioning defective protein and Pax: paxillin, as well as BrK: breast tumor kinase and ChK: checkpoint kinase, whereas CHK denotes C-terminal Src kinase [CSK]-homologous kinase) or not (e.g., Fz: Frizzled and HhIP: Hedgehog-interacting protein),[10] except for composite chemical species (e.g., DAG: diacylglycerol). However, some current usages have been kept for short aliases of chemical species name (e.g., Rho for Ras homolog rather than RHo).

In any case, molecule (super)family (class) aliases as well as those of their members are written in capital letters, such as the IGSF (IGSF*i*: member *i*; immunoglobulin), KIF (KIF*i*; kinesin), SLC (SLC*i*; solute carrier), TNFSF (TNFSF*i*; tumor-necrosis factor), and TNFRSF (TNFRSF*i*; tumor-necrosis factor receptor) superfamily.

Gene names are also written with majuscules when the corresponding protein name contains at least one minuscule, otherwise only the gene name initial is written with an upper case letter that is then followed by lower case letters.

To highlight its function, substrate aliases (e.g., ARF GTPases) contained in a molecule alias are partly written with lower case letters (e.g., ArfRP, ArfGEF, ArfGAP stand for ARF-related protein, ARF guanine nucleotide-exchange factor, and ARF GTPase-activating protein, respectively).

Last, but not least, heavy and pedantic designation of protein isoforms based on roman numerals has been avoided and replaced by usual arabic numerals (e.g., angiotensin-2 rather than angiotensin-II), except for coagulation (or clotting)

---

[10]The Hedgehog gene was originally identified in the fruit fly Drosophila melanogaster. It encodes a protein involved in the determination of segmental polarity and intercellular signaling during morphogenesis. Homologous gene and protein exist in various vertebrate species. The name of the mammal hedgehog comes from hecg and hegge (dense row of shrubs or low trees), as it resides in hedgerows, and hogg and hogge, due to its pig-like, long projecting nose (snout). The word Hedgehog hence is considered as a seamless whole.

factors. Moreover, the character I can mean either letter I or number 1 without obvious discrimination at first glance (e.g., GAPI that stands for Ras GTPase-activating protein GAP1, but can be used to designate a growth-associated protein inhibitor).

Unnecessary hyphenation in aliases of substances (between an upper case letter, which can define the molecule function, and the chemical alias, or between it and assigned isotype number) has been avoided. In any case, the Notation section serves not only to define aliases, but also, in some instances, as disambiguation pages.

A space rather than hyphen is used in: (1) structural components at the picoscale (e.g., P loop), nanoscale (e.g., G protein [G standing for guanine nucleotide-binding]), microscale (e.g., H zone, M line, A band, I band, and Z disc of the sarcomere and T tubule of the cardiomyocyte); (2) process stages (e.g., M phase of the cell division cycle); and (3) cell types (e.g., B and T lymphocytes). When these terms are used as adjectives, an hypen is then employed (e.g., P-loop Cys–X_5–Arg (CX_5R) motif, G-protein-coupled receptor, Z-disc ligand, M-phase enzyme, and T-cell activation).

In terms incorporating a Greek letter, similarly, a space is used in: (1) structural components (e.g., α and β chains and subunits); (2) cellular organelles (e.g., α granule); and (3) cell types (e.g., pancreatic β cell). On the other hand, terms are hyphenated when they refer to (1) structural shape (e.g., α-helix and α(β)-sheet) and (2) molecule subtype (e.g., α-actinin, β-glycan, and γ-secretase).

Symbols for Physical Variables

Unlike substances aliases, symbols for physical quantities are most often represented by a single letter of the Latin or Greek alphabet (i: current; J: flux; L: length; m: mass; p: pressure; P: power; T: temperature; t: time; u: displacement; v: velocity; x: space; λ: wavelength; μ: dynamic viscosity; ρ: mass density; etc.). These symbols are specified using sub- and superscripts (c_p and c_v: heat capacity at constant pressure and volume, respectively; $\mathcal{D}_T$: thermal diffusivity; G_h: hydraulic conductivity; G_T: thermal conductivity; α_k: kinetic energy coefficient; α_m: momentum coefficient; etc.).

A physical quantity associated with a given point in space at a given time can be: (1) a scalar uniquely defined by its magnitude; (2) a vector characterized by a magnitude, a support, and a direction represented by an oriented line segment defined by a unit vector; and (3) a tensor specified by a magnitude and a few directions. To ensure a straightforward meaning of symbols used for scalar, vectorial, and tensorial quantities, bold face upper (**T**) and lower (**v**) case letters are used to denote a tensor and a vector, respectively, whereas both roman (plain, upright)-style upper and lower case letters designate a scalar.

List of Currently Used Prefixes and Suffixes

Prefixes (localization)

"ab-" (Latin) and "apo-" (Greek: απο): away from or off (abluminal: endothelial edge opposite to wetted surface; apolipoproteins: lipid carriers that cause egress [also ingress] from cells; aponeurosis (απονευρωσις; νευρον: sinew, tendon) muscle sheath that limits radial motion and enhances axial contraction; and apoptosis: separation ["-ptosis": fall (πτωσις): as leaves fall away from a tree], a type of programmed cell death)
"acr-" (variant "acro-" [ακρος]): top or apex
"ad-" (adfecto: to reach; adfio: to blow toward; adfluo: to flow toward): toward (ad- becomes "ac-" before c, k, or q; "af-" before f [afferent]; "ag-" before g [agglutination]; "al-" before l; "ap-" before p [approximation]; "as-" before s; and "at-" before t)
"cis-", "juxta-", and "para-" (παρα): near, beside, or alongside
"contra-": opposite side; "ipsi-" (ipse): same side; "latero-": side;
"ecto-" (εκτος), "exo-" (εξο), and "extra-": outside, outer, external, or beyond (exogenous chemicals produced by an external source, or xenobiotics ["xeno-": foreigner])
"endo-" (ενδον) and "intra-": inside (endogenous substances synthesized by the body's cells; endomembranes at organelle surfaces within the cell)
"ep-" (variant "eph-", or "epi-" [επι]): upon (epigenetics refers to the inheritance ("-genetic": ability to procreate [γεννητικος]) of variations in gene expression beyond ("epi-": on, upon, above, close to, beside, near, toward, against, among, beyond, and also) change in the DNA sequence.
"front-" and "pre-": anterior or in front of
"post-": behind
"infra-" and "sub-": under or below
"super-" and "supra-": above
"inter-": between or among
"peri-" (περι): around
"tele-" (τελε): remote
"trans-": across

Prefixes (composition)

"an-" and "aniso-" (ανισος): unequal, uneven, heterogeneous
"iso-" (ισος): equal, alike (isomer [μερος: part, portion]
"mono-" (μονος) and "uni-" (unicus): single
"oligo-" (ολιγος): few, little, small
"multi-" (multus), "pluri-" (plus, plures), and "poly-" (πολυς): many, much
"ultra-": in excess.

Prefixes (quantity)

"demi-" (dimidius) and "hemi-" (ημι): half
"sesqui-": one and a half (half more)

M. Thiriet, *Tissue Functioning and Remodeling in the Circulatory and Ventilatory Systems*, Biomathematical and Biomechanical Modeling of the Circulatory and Ventilatory Systems 5, DOI 10.1007/978-1-4614-5966-8,

"di-" or "dis-" (δυο; δις) as well as "bi-" or "bis-": 2, twice
"tri" (τρεις, τρι-; tres, tria): 3
"tetra-" (τετρα), "quadri-" (variant: "quadr-" and "quadru-"): 4
"penta-" (πεντας; pentas), "quinqu-", and "quint-": 5
"hexa-" (εξ) and "sexa-": 6
"hepta-" (επτα): 7
"octa-" (οκτα): 8
"nona-" (εννεα): 9 (ninth part)
"deca-" (δεκα): 10
"quadra-" (quadragenarius): 40 (elements)
"quinqua-" (quinquagenarius): 50
"sexa-" (sexagenarius [sex: 6]: 60
"septua-" (septuagenarius [septem: 7]): 70
"nona-" (nonagenarius): 90

Prefixes (motion and direction)

"af-": toward the center (single master object); e.g., nerve and vascular afferents (ferre: to carry) to brain and heart, respectively, rather than toward any slave, supplied tissue from the set of the body's organs; also affector, i.e., chemical messenger that brings a signal to the cell considered as the object of interest, this exploration focus being virtually excised from the organism with its central command system, except received signals
"ef-" (effero: to take away): from the center (efferent; effector, i.e., chemical transmitter recruited by the previous mediator of a signaling cascade at a given locus to possibly translocate to another subcellular compartment)
"antero-" (anterior): before, in front of, facing, or forward
"retro-": behind or backward
"tropo-" (τροπος): duct direction; (tropa: rotation; celestial revolution); e.g., tropomyosin (μυς, musculus: muscle; μυο-: refers to muscle [μυοτρωτος: injured at a muscle])

Prefixes (structure and size)

"macro-" (μακρος): large, long, or big
"mega-" (μεγας): great, large
"meso-" (μεσος): middle
"micro-" (μικρος): small
"nano-" (νανος): dwarf, tiny
"homo-" (ομο-): same (ομολογος: agreeing, corroborating; variant: "homeo-" [homeostasis])
"hetero-" (ετερο-): other

Prefixes (timing)

"ana-" (ανα): culminating (anaphase of the cell division cycle), up, above (ανοδος: a way up, anode [positive electrode; οδος; way, path, road, track])
"ante-": before
"circa-": approximately, around (circadian: approximately one day)
"infra-": below, shorter (infradian: rhythm with lower frequency than that of circadian rhythm, not smaller period)
"inter-": among, between, during
"meta-" (μετα): after, beyond, behind, later; in the middle of (metaphase of the cell division cycle); as well as connected to, but with a change of state (metabolism) and about (metadata)
"post-": after
"pre-": earlier
"pro-" (προ): preceding, first, before (prophase of the cell division cycle)
"telo-" (τελος): end, completion
"ultra-": beyond, longer (ultradian: period smaller than that of 24–28-hour cycle, i.e., frequency greater than that of the circadian rhythm)

Prefixes (functioning modality)

"auto-" (αυτος): same, self
"brady-" (βραδυς): slow (decelerate)
"tachy-" (ταχος): rapid (accelerate)
"amphi-" (αμφι): both (amphiphilic substances are both hydrophilic and lipophilic; amphisomes are generated by both autophagosomes and endosomes)

"ana-" : upward (anabolism) or against (anaphylaxis)
"cata-" (κατα): downward (catabolism, cathode [negative electrode; οδος; way, path, road, track])
"anti-" (αντι): against
"pro-": favoring
"co-" (coaccedo: add itself to): together
"contra-": adverse, against, beside, next to, opposite
"de-": remove, reduce, separation after association (Latin de; e.g., deoxy-)
"dys-" (δυς): abnormal (δυσαης): ill-blowing)
"equi-" (æque): equal or alike
"hem-" or "hemat-" (αιμα: blood): related to blood
"hypo-" (υπο): under, beneath, and low
"hyper-" (υπερ): above, beyond, and large
"per-": through (e.g., percutaneous) and during (e.g., peroperative)
"pseudo-" (ψευδο): pretended, false
"re-"; again

Scientific suffixes

"-ase": enzyme (synthase, lipase, etc.)
"-ate": salt of a base
"-cyte" (κυτος): cell (erythro- [ερυθρος: red], leuko- [λευκος: light, bright, clear, white], thrombo- [θρομβος: lump, clot], adipo- [adeps: fat; adipalis, adipatus, adipeus, adipinus: fatty], fibro- [fibra: fiber, filament], myo- [μυς: muscle, mouse, mussel], myocardiocyte [κραδια: heart; cardiacus: related to heart, stomach; to have heart trouble, stomach trouble], etc.);
"-crine" (κρινω): to decide, to separate, and to secrete (e.g., endocrine regulator) (ευκρινεω: keep in order)
"-elle": small (organelle in a cell [like an organ in a body])
"-ium", "-ion", "-isk", and "-iscus": little ("-ium": tissue interface and envelope, such as endothelium and pericardium)
"-phil" (φιλια): attracted (αφιλια: want of friends)
"-phob" (φοβια): repulsed (υδροφοβια, hydrophobia [Latin]: horror of water)
"-phore" (φερω): carrier (αμφερω: to bring up)
"-yl" denotes a radical (molecules with unpaired electrons)
"-ploid" (πλοω): double, fold (diploid, twofold; διπλοω: to double; διαπλοω: unfold)
"-emia": in relation to flow (ανεμια: flatulence; ευηνεμια: fair wind), particularly blood condition
"-genesis" (γενεσις): cause, generation, life source, origin, productive force
"-iasis": for diseased condition
"-itis": inflammation
"-lemma" (λεμμα: skin): sheath
"-ole" and "-ule": small (arteriole and venule; variant "-ula" [blastula] and "-ulum")
"-plasma" (πλασμα): anything molded (plasma: creature generated from silt of earth)
"-plasia" (πλασια): formation, molding
"-podium" (ποδος: foot; podium [Latin]: small knoll, small protuberance): protrusion
"-poiesis" (ποιεω): production
"-soma" (σωμα): body
"-sclerosis" (σκλημα): hardness, induration
"-stasis" (στασις): stabilization (αποκαταστασις: restoration; ανυποστασις: migration)
"-stomosis" (στομα: mouth): equipped with an outlet
"-taxy/tactic" (ταχυ: rapid; τακτικος: to maneuver): related to motion (also prefix, i.e., ταχυκινησις: quick motion; ταχυνω: to accelerate; and ταχυπνοια: short breath; not [δια]ταξις: disposition, arrangement)
"-trophy/trophic" (τροφις: well fed): related to growth
"-oma": tumor of
"-pathy" (παθος, παθεια): disease of
"-tomy" (τομια) and "-ectomy": surgical removal (απλοτομια: simple incision; φαrhουγγοτομια: laryngotomy)

ASAP: artery-specific antigenic protein
ASAP: ArfGAP with SH3, ankyrin repeat, PH domains
ASIC: acid-sensing ion channel
ASK: apoptosis signal-regulating kinase
aSMC: airway smooth muscle cell
ASP: actin-severing protein
AT: antithrombin
ATAA: ascending thoracic aortic aneurysm
ATF: activating transcription factor
AtG: autophagy-related gene product
ATMK: ataxia telangiectasia mutated kinase
ATn: angiotensin
ATng: angiotensinogen
AtOx: anti-oxidant protein (metallochaperone)
ATP: adenosine triphosphate
ATPase: adenosine triphosphatase
ATR ($AT_{1/2}$): angiotensin receptor
ATRK: ataxia telangiectasia and Rad3-related kinase
AVN: atrioventricular node
AVV: atrioventricular valves
AW: analysis window

B

B: Biot-Finger strain tensor
B: bulk modulus
$\mathcal{B}$: bilinear form
b: minor semi-axis
b: body force
$\hat{\mathbf{b}}$: unit binormal
b: birth rate
B lymphocyte (B cell): bone marrow lymphocyte
BACE: β-amyloid precursor protein-converting enzyme
BAD: BCL2 antagonist of cell death
BAF: barrier-to-autointegration factor
BAG: BCL2-associated athanogene (chaperone regulator)
BAI: brain-specific angiogenesis inhibitor (adhesion-GPCR)
BAIAP: brain-specific angiogenesis inhibitor-1-associated protein (insulin receptor substrate)
BAK: BCL2-antagonist–killer
(i)BALT: (inducible) bronchus-associated lymphoid tissue
BAMBI: BMP and activin membrane-bound inhibitor homolog
BAnk: B-cell scaffold with ankyrin repeats
Barkor: beclin-1-associated autophagy-related key regulator
BAT: brown adipose tissue
BATF: basic leucine zipper ATF-like transcription factor (B-cell-activating transcription factor)
BAX: BCL2-associated X protein
BBB: blood–brain barrier
BBS: Bardet-Biedl syndrome protein
BBSome: BBS coat complex (transport of membrane proteins into cilium)
BC: boundary condition
bCAM: basal cell adhesion molecule (Lutheran blood group glycoprotein)
BCAP: B-cell adaptor for phosphatidylinositol 3-kinase
BCAR: Breast cancer anti-estrogen resistance docking protein
BCL: B-cell lymphoma (leukemia) protein
BCLxL: B-cell lymphoma extra-large protein
BCR: B-cell receptor
BCR: breakpoint cluster region protein (GAP and GEF)
Bdk: bradykinin
BDNF: brain-derived neurotrophic factor
Be: Bejan number
Becn, beclin: BCL2-interacting protein
BEM: boundary element method
Best: bestrophin
bFGF: basic fibroblast growth factor (FGF2)
BFUe: burst-forming unit-erythroid
BFUmeg: burst-forming unit-megakaryocyte
BGT: betaine–GABA transporter
BH_4: tetrahydrobiopterin (enzyme cofactor)
BID: BH3-interacting domain death agonist
BIG: brefeldin-A-inhibited GEFs for ARFs
BIK: BCL2-interacting killer
BIM: BH3-containing protein BCL2-like 11 (BCL2L11)
BK: high-conductance, Ca^{2+}-activated, voltage-gated K^+ channel
BLK: B-lymphoid tyrosine kinase
Blm: Bloom syndrome, RecQ DNA helicase-like protein
BLnk: B-cell linker protein
BLOC: biogenesis of lysosome-related organelles
BM: basement membrane
BMAL: brain and muscle ARNT-like protein (gene Bmal)
BMAT: bone-marrow adipose tissue
BMF: BCL2 modifying factor
BMP: bone morphogenetic protein (TGFβ superfamily)

BMPR: bone morphogenetic protein receptor
BNIP: BCL2/adenovirus E1B 19-kDa protein-interacting protein
BNP: B-type natriuretic peptide
BMX: bone marrow Tyr kinase gene in chromosome-X product
Bo: Boltzmann constant
BOC: brother of CDO
BOK: BCL2-related ovarian killer
BORG: binder of Rho GTPase
Br: Brinkman number
BRAG: brefeldin-resistant ArfGEF
BrCa: breast cancer-associated (susceptibility) protein (tumor suppressor; DNA-damage repair; a.k.a. FancD1)
BrD: bromodomain-containing protein
BrK: breast tumor kinase
BrSK: brain-selective kinase
BSEP: bile salt export pump
BTF: basic transcription factor
BTK: Bruton Tyr kinase
BUB: budding uninhibited by benzimidazoles

C

C: stress tensor
C: compliance
C: heat capacity
C: chronotropy
Cx: type-x chemokine C (γ)
C_D: drag coefficient
C_f: friction coefficient
C_L: lift coefficient
C_p: pressure coefficient
c: stress vector
c_τ: shear
c_w: wall shear stress
c: concentration
$c(p)$: wave speed
c_p: isobar heat capacity
c_v: isochor heat capacity
C1P: ceramide 1-phosphate
C-terminus: carboxy (carboxyl group COOH)-terminus
C/EBP: CCAAT/enhancer-binding protein
CA: computed angiography
CAi: carbonic anhydrase isoform i
Ca: calcium
Ca_V: voltage-gated Ca^{2+} channel
$Ca_V1.x$: L-type high-voltage-gated Ca^{2+} channel
$Ca_V2.x$: P/Q/R-type Ca^{2+} channel
$Ca_V3.x$: T-type low-voltage-gated Ca^{2+} channel
CAAT: cationic amino acid transporter
CABG: coronary artery bypass grafting
Cables: CDK5 and Abl enzyme substrate
CAK: CDK-activating kinase (pseudokinase)
Cam: calmodulin (calcium-modulated protein)
CamK: calmodulin-dependent kinase
cAMP: cyclic adenosine monophosphate
CAP: adenylate cyclase-associated protein
CAP: carboxyalkylpyrrole protein adduct
CAP: chromosome-associated protein (BrD4)
CAPN: calpain gene
CaPON: carboxy-terminal PDZ ligand of NOS1 (NOS1AP)
CAR: constitutive androstane receptor (NR1i3)
CaR: calcium-sensing receptor
CARD: caspase activation and recruitment domain-containing protein
CARMA: CARD and membrane-associated guanylate kinase-like (MAGuK) domain-containing protein
CARP: cell division cycle and apoptosis regulatory protein
CAS: cellular apoptosis susceptibility protein
CAS: CRK-associated substrate (or P130CAS and BCAR1)
CAs: cadherin-associated protein
CASK: calcium–calmodulin-dependent serine kinase (pseudokinase)
CASL: CRK-associated substrate-related protein (CAS2)
CASP: cytohesin-associated scaffold protein
caspase: cysteine-dependent aspartate-specific peptidase
Cav: caveolin
CBF: coronary blood flow
CBF: core-binding factor
CBL: Casitas B-lineage lymphoma adaptor and Ub ligase
CBLb: CBL-related adaptor
CBP: cap-binding protein
CBP: CREB-binding protein
CBP: C-terminal Src kinase-binding protein
CBS: cystathionine β-synthase (H_2S production)
CCD: cortical collecting duct
CCDC: coiled-coil domain-containing protein
CCE: capacitative Ca^{2+} entry channel (SOC channel)
CCICR: calcium channel-induced Ca^{2+} release
CCK4: colon carcinoma kinase 4 (PTK7)
CCL: chemokine CC-motif ligand

CCN: CyR61, CTGF, and NOv (CCN1–CCN3) family
Ccn: cyclin
Ccnx–CDK*i*: type-x cyclin–type-*i* cyclin-dependent kinase dimer
CCPg: cell cycle progression protein
CCS: copper chaperone for superoxide dismutase
CCT: chaperonin containing T-complex protein
CCx: type-x chemokine CC (β)
CCR: chemokine CC motif receptor
CD: cluster determinant protein (cluster of differentiation)
CDase: ceramidase
CDC: cell division cycle protein
cDC: classical dendritic cell
CDH: CDC20 homolog
Cdh: cadherin
CDK: cyclin-dependent kinase
Cdm: caldesmon
CDO: cell adhesion molecule-related/downregulated by oncogenes
CE (CsE): cholesteryl esters
CEC: circulating endothelial cell
CELSR: cadherin, EGF-like, LAG-like, and seven-pass receptor
CenP: centromere protein
CEP: carboxyethylpyrrole
CeP: centrosomal protein
CEPC: circulating endothelial progenitor cell
Cer: ceramide
CerK: ceramide kinase
CerT: ceramide transfer protein
CETP: cholesterol ester transfer protein
CFD: computational fluid dynamics
CFLAR: caspase-8 and FADD-like apoptosis regulator
CFTR: cystic fibrosis transmembrane conductance regulator
CFU: colony-forming unit
CFUb: CFU-basophil (basophil-committed stem cells)
CFUc: CFU in culture (granulocyte precursors, i.e., CFUgm)
CFUe: CFU-erythroid
CFUeo: CFU-eosinophil
CFUg: CFU-granulocyte
CFUgm: CFU-granulocyte–macrophage
CFUgemm: CFU-granulocyte–erythroid–macrophage–megakaryocyte
CFUm: CFU-macrophage
CFUmeg: CFU-megakaryocyte
CFUs: CFU-spleen (pluripotent stem cells)
CG: chromogranin
cGK: cGMP-dependent protein kinase (protein kinase G)
cGMP: cyclic guanosine monophosphate
CGN: cis-Golgi network
CGRP: calcitonin gene-related peptide
chanzyme: ion channel and enzyme
chemokine: chemoattractant cytokine
CHIP: C-terminus heat shock cognate-70-interacting protein
ChK: checkpoint kinase
CHK: CSK homologous kinase
Chn: chimerin (GAP)
CHOP: CCAAT/enhancer-binding protein homologous protein
CHREBP: carbohydrate-responsive element-binding protein
ChT: choline transporter
CI: cardiac index
CICR: calcium-induced calcium release
Cin: chronophin
CIP: CDC42-interacting protein
CIP2a: cancerous inhibitor of protein phosphatase-2A
CIPC: CLOCK-interacting protein, circadian
CIS: cytokine-inducible SH2-containing protein
CITED: CBP/P300-interacting transactivator with glutamic (E) and aspartic acid (D)-rich C-terminus-containing protein
CK: creatine kinase
CK: casein kinase
CKI: cyclin-dependent kinase inhibitor
CLAsP: CLiP-associated protein (microtubule binder)
ClASP: clathrin-associated sorting protein
CLC: cardiotrophin-like cytokine
ClC: voltage-gated chloride channel
ClCa: calcium-activated chloride channel
ClIC: chloride intracellular channel
CLINT: clathrin-interacting protein located in the trans-Golgi network
CLIP: corticotropin-like intermediate peptide
CLiP: cytoplasmic CAP-Gly domain-containing linker protein
iCliP: intramembrane-cleaving peptidase (that clips)
CLK: CDC-like kinase
ClNS: Cl^- channel nucleotide-sensitive
CLOCK: circadian locomotor output cycles kaput
CLP: common lymphoid progenitor
CLS: ciliary localization signal
Cmi: chylomicron

CMLP: common myeloid–lymphoid progenitor
CMP: common myeloid progenitor
CMC: cardiomyocyte
Col: collagen
CoLec: collectin
ColF: collagen fiber
CORM: carbon monoxide (CO)-releasing molecule
CNG: cyclic nucleotide-gated channel
CnK: connector enhancer of kinase suppressor of Ras
CNS: central nervous system
CNT: connecting tubule
CNTi: concentrative nucleoside transporter (SLC28ai)
CNTF: ciliary neurotrophic factor
CntnAP: contactin-associated protein
CO: cardiac output
CoA: coenzyme-A
CoBl: Cordon-bleu homolog (actin nucleator)
COLD: chronic obstructive lung disease
COOL: Cloned out of library (RhoGEF6/7)
coSMAD: common (mediator) SMAD (SMAD4)
COx: cyclooxygenase (prostaglandin endoperoxide synthase)
COx17: cytochrome-C oxidase copper chaperone
CoP: coat protein
CoP: constitutive photomorphogenic protein (Ub ligase)
COPD: chronic obstructive pulmonary disease
COUPTF: chicken ovalbumin upstream promoter transcription factor (NR2f1/2)
CP4H: collagen prolyl 4-hydroxylase
CPC: chromosomal passenger complex
CpG: cytidine^{P}–guanosine oligodeoxynucleotide (motif)
cPKC: conventional protein kinase C
Cpx: complexin
CR: complement component receptor
Cr: creatine
cRABP: cellular retinoic acid-binding protein
cRBP: cellular retinol-binding protein
CRAC: Ca^{2+} release-activated Ca^{2+} channel
CRACR: CRAC regulator
Crb: Crumbs homolog polarity complex
CRE: cAMP-responsive element
CREB: cAMP-responsive element-binding protein
CRF: corticotropin-releasing factor (family)
CRH: corticotropin-releasing hormone
CRIB: CDC42/Rac interactive-binding protein
CRIK: citron Rho-interacting, Ser/Thr kinase (STK21)
CRK: chicken tumor virus CT10 regulator of kinase
CRKL: CRK avian sarcoma virus CT10 homolog-like
CRL4: cullin-4A RING ubiquitin ligase
CRLR: calcitonin receptor-like receptor
CRP: C-reactive protein
Crt: calreticulin
CRTC: CREB-regulated transcription coactivator
Cry: cryptochrome
Cs: cholesterol
CSBP: cytokine-suppressive anti-inflammatory drug-binding protein
CSE: cystathionine γ-lyase (H_2S production)
CSF: cerebrospinal fluid
CSF: colony-stimulating factor
CSF1: macrophage colony-stimulating factor (mCSF)
CSF2: granulocyte–macrophage colony-stimulating factors (gmCSF and sargramostim)
CSF3: granulocyte colony-stimulating factors (gCSF and filgrastim)
CSK: C-terminal Src kinase
Csk: cytoskeleton
Csq: calsequestrin
CSS: candidate sphingomyelin synthase
CT: cardiotrophin
CT: computed tomography
CTBP: C-terminal-binding protein
CTen: C-terminal tensin-like protein
CTF: C-terminal fragment
CTGF: connective tissue growth factor
CTL: cytotoxic T lymphocyte
CTLA: cytotoxic T-lymphocyte-associated protein
Ctn: catenin
CTr: copper transporter
CtR: calcitonin receptor
CTRC: CREB-regulated transcription coactivator
Cul: cullin
CUT: cryptic unstable transcript
CVI: chronic venous insufficiency
CVLM: caudal ventrolateral medulla
CVP: central venous pressure
CVS: cardiovascular system
Cx: connexin
CXCL*i*: type-*i* CXC (C-X-C motif; α) chemokine ligand

CXCRi: type-i CXC (C-X-C motif; α) chemokine receptor
CX3CLi: type-i CX3C (δ) chemokine ligand
CX3CRi: type-i CX3C (δ) chemokine receptor
cyCK: cytosolic creatine kinase
Cyld: cylindromatosis tumor suppressor protein (deubiquitinase USPL2)
CyP: member of the cytochrome-P450 superfamily
C3G: Crk SH3-binding GEF

D

D: dromotropy
D: vessel distensibility
$\mathcal{D}$: diffusion coefficient
$\mathcal{D}_T$: thermal diffusivity
D: deformation rate tensor
d: displacement vector
D: flexural rigidity
D: demobilization function (from proliferation to quiescence)
d: death, decay, degradation rate
d: duration
Dab: Disabled homolog
DAD: delayed afterdepolarization
DAG: diacylglycerol
DAPC: dystrophin-associated protein complex
DAPK: death-associated protein kinase
DARC: Duffy antigen receptor for chemokine
DAT: dopamine active transporter
DAX: dosage-sensitive sex reversal, adrenal hypoplasia critical region on chromosome X (NR0b1)
DBC: deleted in breast cancer protein
DBF: dumbbell formation kinase (in Saccharomyces cerevisiae; e.g., DBF2)
DBP: albumin D-element binding protein (PAR/b–ZIP family)
DC: dendritic cell
DCA: directional coronary atherectomy
DCAF: DDB1- and Cul4-associated factor
DCC: deleted in colorectal carcinoma (netrin receptor)
DCT: distal convoluted tubule
Dctn: dynactin
DDAH: dimethylarginine dimethylaminohydrolase
DDB: damage-specific DNA-binding protein
DDEF: development and differentiation-enhancing factor (ArfGAP)
DDR: discoidin domain receptor
De: Dean number
Deb: Deborah number
DEC: differentially expressed in chondrocytes (DEC1 and DEC2 are a.k.a bHLHe40 and bHLHe41, bHLHb2 and bHLHb3, or HRT2 and HRT1)
DEC: deleted in esophageal cancer
DEG: delayed-early gene
deoxyHb: deoxyhemoglobin (deoxygenated hemoglobin)
DETC: dendritic epidermal $\gamma\delta$ T cell
DH: Dbl homology
DHA: docosahexaenoic acid
DHET: dihydroxyeicosatrienoic acid
DHh: desert Hedgehog
Dia: Diaphanous
DICOM: digital imaging and communication for medicine
DICR: depolarization-induced Ca^{2+} release
DISC: death-inducing signaling complex
Dkk: Dickkopf
DLg: Disc large homolog
DLL: Delta-like (Notch) ligand
DLx: distal-less homeobox protein
DM: double minute
DMM: DNA methylation modulator
DMPK: myotonic dystrophy-associated protein kinase
DMT: divalent metal transporter
DN1: double-negative-1 cell
DN2: double-negative-2 cell
DN3: double-negative-3 cell
DNA: deoxyribonucleic acid
DNAPK: DNA-dependent protein kinase
DoC2: double C2-like domain-containing protein
DOCK: dedicator of cytokinesis (GEF)
DOK: downstream of Tyr kinase docking protein
DOR: δ-opioid receptor
DPG: diphosphoglyceric acid
DRAM: damage-regulated modulator of autophagy
DRF: Diaphanous-related formin (for GTPase-triggered actin rearrangement)
DRG: dorsal root ganglion
Drl: Derailed
Dsc: desmocollin
Dsg: desmoglein
Dsh: Disheveled (Wnt-signaling mediator)
DSK: dual-specificity kinase
dsRNA: double-stranded RNA
Dst: dystonin

DUb: deubiquitinase
DuOx: dual oxidase
DUS: Doppler ultrasound
DuSP: dual-specificity phosphatase
DV: dead space volume
Dvl: Disheveled (cytoplasmic phosphoprotein; other alias Dsh)
DVT: deep-vein thrombosis
dynactin: dynein activator
DYRK: dual-specificity Tyr (Y) phosphorylation-regulated kinase

E

E: strain tensor
E: electric field
E: elastic modulus
E: elastance
$\mathcal{E}$: energy
$\{\hat{\mathbf{e}}_i\}_{i=1}^3$: basis
e: strain vector
e: specific free energy
E-box: enhancer box sequence of DNA
E2: ubiquitin conjugase
E3: ubiquitin ligase
EAAT: excitatory amino acid (glutamatee–aspartate) transporter
EAD: early afterdepolarization
EAR: V-erbA-related nuclear receptor (NR2f6)
EB: end-binding protein
EBCT: electron beam CT
EBF: early B-cell factor
EC: endothelial cell
Ec: Eckert number
ECA: external carotid artery
ECF: extracellular fluid
ECG: electrocardiogram
ECM: extracellular matrix
ED1L: EGF-like repeat- and discoidin-1-like domain-containing protein
EDGR: endothelial differentiation gene receptor
EDHF: endothelial-derived hyperpolarizing factor
EDIL: EGF-like repeats and discoidin-1 (I)-like domain-containing protein
EDV: end-diastolic volume
EEA: early endosomal antigen
eEF: eukaryotic translation elongation factor
EEL: external elastic lamina
EET: epoxyeicosatrienoic acid
EFA6: exchange factor for ARF6 (ArfGEF)
EF-Tu: elongation factor Tu
EGF: epidermal growth factor
EGFL: EGF-like domain-containing protein
EGFR: epidermal growth factor receptor
EGR: early growth response transcription factor
EHD: C-terminal EGFR substrate-15 homology domain-containing protein
eIF: eukaryotic translation initiation factor
EL: endothelial lipase
ELAM: endothelial–leukocyte adhesion molecules
ELCA: excimer laser coronary angioplasty
ELk: ETS-like transcription factor (ternary complex factor [TCF] subfamily)
ElMo: engulfment and cell motility adaptor
Eln: elastin
ElnF: elastin fiber
ELP: early lymphoid progenitor
EMI: early mitotic inhibitor
EMR: EGF-like module-containing, mucin-like, hormone receptor-like protein
EMT: epithelial–mesenchymal transition
ENA–VASP: Enabled homolog and vasoactive (vasodilator)-stimulated phosphoprotein family
ENaC: epithelial Na^+ channel
EnaH: Enabled homolog
endo-siRNA: endogenous small interfering RNA
ENPP: ectonucleotide pyrophosphatase–phosphodiesterase
Ens: endosulfine
ENT: equilibrative nucleoside transporter
ENTPD: ectonucleoside triphosphate diphosphohydrolase
EPAC: exchange protein activated by cAMP
EPAS: endothelial PAS domain protein
EPC: endothelial progenitor cell
EPCR: endothelial protein-C receptor
EPDC: epicardial-derived cell
Epgn: epigen (EGF superfamily member)
EPH: erythropoietin-producing hepatocyte receptor kinase or pseudokinase (EPHa10 and EPHb6)
ephrin: EPH receptor interactor
Epo: erythropoietin
EPS: epidermal growth factor receptor pathway substrate
ER: endoplasmic reticulum
ERx: type-x estrogen receptor (NR3a1/2)

eRas: embryonic stem cell-expressed Ras (or hRas2)
ErbB: erythroblastoma viral gene product B (HER)
ERE: estrogen response element (DNA sequence)
Ereg: epiregulin (EGF superfamily member)
eRF: eukaryotic release factor
ERGIC: endoplasmic reticulum–Golgi intermediate compartment
ERK: extracellular signal-regulated protein kinase
ERK1/2: usually refers to ERK1 and ERK2
ERM: ezrin–radixin–moesin
ERMES: endoplasmic reticulum–mitochondrion encounter structure
ERP: effective refractory period
ERR: estrogen-related receptor (NR3b1–NR3b3)
ESCRT: endosomal sorting complex required for transport
ESL: E-selectin ligand
ESRP: epithelial splicing regulatory protein
ESV: end-systolic volume
ET: endothelin
ETP: early thymocyte progenitor
ETR ($ET_{A/B}$): endothelin receptor
ETS: E-twenty six (transcription factor; erythroblastosis virus E26 proto-oncogene product homolog)
ETV: ETS-related translocation variant
EVAR: endovascular aneurysm repair
Exo: exocyst subunit
Ext: exostosin (glycosyltransferase)

F

F: transformation gradient tensor
F: function fraction of proliferating cells
F: erythrocytic rouleau fragmentation rate
f: surface force
$\hat{\mathbf{f}}$: fiber direction unit vector
f: binding frequency
f_C: cardiac frequency
f_R: breathing frequency
f: friction shape factor
f_v: head loss per unit length
f_i: molar fraction of gas component i
FA: fatty acid
FABP: fatty acid-binding protein
FABP: filamentous actin-binding protein
FACAP: F-actin complex-associated protein
FAD: flavine adenine dinucleotide
FADD: Fas receptor-associated death domain
FAK: focal adhesion kinase
Fanc: Fanconi anemia protein
FAN: Fanconi anemia-associated nuclease
FAPP: phosphatidylinositol four-phosphate adaptor protein
Fas: death receptor (TNFRSF6a)
FasL: death ligand (TNFSF6)
FAST: Forkhead activin signal transducer
FB: fibroblast
Fbln (Fibl): fibulin
Fbn: fibrillin
FBS: F-box, Sec7 protein (ArfGEF)
FBx: F-box only protein (ArfGEF)
FC: fibrocyte
FCHO: FCH domain only protein
FcαR: Fc receptor of IgA
FcγR: Fc receptor of IgG
FcϵR: Fc receptor of IgE
FCP: TF2F-associating C-terminal domain phosphatase
FDM: finite difference method
FEM: finite element method
FERM: four point-1, ezrin–radixin–moesin domain
FeR: FeS-related Tyr kinase
FeS: feline sarcoma kinase
FFA: free fatty acid
FGF: fibroblast growth factor
FGFR: fibroblast growth factor receptor
FGR: viral feline Gardner-Rasheed sarcoma oncogene homolog kinase
FHL: four-and-a-half LIM-only protein
FHoD: formin homology domain-containing protein (FmnL)
FIH: factor inhibiting HIF1α (asparaginyl hydroxylase)
FIP: family of Rab11-interacting protein
FIP: focal adhesion kinase family-interacting protein
FIT: Fat-inducing transcript
FKBP: FK506-binding protein
FlIP: flice-inhibitory protein
FLK: fetal liver kinase
fMLP: N-formyl methionyl-leucyl-phenylalanine
FN: fibronectin
Fn: fibrin
Fng: fibrinogen
Fos: Finkel Biskis Jinkins murine osteosarcoma virus sarcoma proto-oncogene product

Fox: forkhead box transcription factor
Fpn: ferroportin
FR: flow ratio
FRK: Fyn-related kinase
FrmD: FERM domain-containing adaptor
FRNK: FAK-related non-kinase
FRS: fibroblast growth factor receptor substrate
FSH: follicle-stimulating hormone
FSI: fluid–structure interaction
FVM: finite volume method
FXR: farnesoid X receptor (NR1h4)
Fz: Frizzled (Wnt GPCR)

G

G: Green-Lagrange strain tensor
G: shear modulus
G': storage modulus
G'': loss modulus
$\mathcal{G}$: Gibbs function
G: conductance
G_p: pressure gradient
$\mathcal{G}_b$: perfusion conductivity
$\mathcal{G}_e$: electrical conductivity
$\mathcal{G}_h$: hydraulic conductivity
$\mathcal{G}_T$: thermal conductivity
g: gravity acceleration
g: physical quantity
g: gravity
g: detachment frequency
g: free enthalpy
G protein: guanine nucleotide-binding protein ($G\alpha\beta\gamma$ trimer)
$G\alpha$: α subunit (signaling mediator) of G protein
$G\alpha_i$ (Gi): inhibitory $G\alpha$ subunit
$G\alpha_s$ (Gs): stimulatory $G\alpha$ subunit
$G\alpha_t$ (Gt): transducin, $G\alpha$ subunit of rhodopsin
Gs_{XL}: extra-large Gs protein
$G\alpha_{i/o}$ (Gi/o): $G\alpha$ subunit class
$G\alpha_{q/11}$ (Gq/11): $G\alpha$ subunit class
$G\alpha_{12/13}$ (G12/13): $G\alpha$ subunit class
$G\beta\gamma$: dimeric subunit (signaling effector) of G protein
G_{gust}: gustducin, G protein α subunit (Gi/o) of taste receptor
G_{olf}: G protein α subunit (Gs) of olfactory receptor
GAB: GRB2-associated binder
GABA: γ-aminobutyric acid
$GABA_A$: GABA ionotropic receptor (Cl^- channel)
$GABA_B$: GABA metabotropic receptor (GPCR)
GABARAP: $GABA_A$ receptor-associated protein
GaBP: globular actin-binding protein
GADD: growth arrest and DNA-damage-induced protein
gadkin: γ1-adaptin and kinesin interactor
GAG: glycosaminoglycan
GAK: cyclin G-associated kinase
Gal: galanin
GAP: GTPase-activating protein
GAPDH: glyceraldehyde 3-phosphate dehydrogenase
GARP: Golgi-associated retrograde protein complex
GAS: growth arrest-specific non-coding, single-stranded RNA
GAT: γ-aminobutyric acid transporter
GATA: DNA sequence GATA-binding transcrition factor
GBF: Golgi-associated brefeldin-A-resistant guanine nucleotide-exchange factor
GBP: guanylate-binding protein
GCAP: guanylate cyclase-activating protein
GCC: Golgi coiled-coil domain-containing protein
GCK: germinal center kinase
GCKR: GCK-related kinase
GCNF: germ cell nuclear factor (NR6a1)
GCN2: general control non-derepressible 2 (pseudokinase)
gCSF: granulocyte colony-stimulating factor (CSF3)
GD: disialoganglioside
GDP: guanosine diphosphate
GDF: growth differentiation factor
GDF: (Rab)GDI displacement (dissociation) factor
GDI: guanine nucleotide-dissociation inhibitor
GDNF: glial cell line-derived neurotrophic factor
GEF: guanine nucleotide (GDP-to-GTP)-exchange factor
GF: growth factor
GFAP: glial fibrillary acidic protein (intermediate filament)
GFL: GDNF family of ligands
GFP: geodesic front propagation
GFR: growth factor receptor
$GFR\alpha i$: type-i GDNF family receptor-α
GGA: Golgi-localized γ-adaptin ear-containing Arf-binding protein
Ggust: (G protein) $G\alpha$ subunit gustducin

GH: growth hormone
GHR: growth hormone receptor
GHRH: growth hormone-releasing hormone
GIP: GPCR-interacting protein
GIRK: Gβγ-regulated inwardly rectifying K^+ channel
GIT: GPCR kinase-interacting protein
GKAP: G-kinase-anchoring protein
GKAP: glucokinase-associated phosphatase (DuSP12)
GKAP: guanylate kinase-associated protein
GLK: GCK-like kinase
GluK: ionotropic glutamate receptor (kainate type)
GluN: ionotropic glutamate receptor (NMDA type)
GluR: ionotropic glutamate receptor (AMPA type)
GluT: glucose transporter
GlyCAM: glycosylation-dependent cell adhesion molecule
GlyR: glycine receptor (channel)
GlyT: glycine transporter
GM: monosialoganglioside
gmCSF: granulocyte–monocyte colony-stimulating factor (CSF2)
GMP: granulocyte–monocyte progenitor
GMP: guanosine monophosphate
GnRH: gonadotropin-releasing hormone
GP: glycoprotein
Gpc: glypican
GPI: glycosyl-phosphatidylinositol anchor
gpiAP: GPI-anchored protein
GPCR: G-protein-coupled receptor
GPx: glutathione peroxidase
GQ: quadrisialoganglioside
GR: glucocorticoid receptor (NR3c1)
Gr: Graetz number
GRAP: GRB2-related adaptor protein (or GAds)
GRB: growth factor receptor-bound protein
GRC: growth factor-regulated, Ca^{2+}-permeable, cation channel (TRPV2)
GRE: glucocorticoid response element (DNA sequence)
GRK: G-protein-coupled receptor kinase
GRP: G-protein-coupled receptor phosphatase
GSH: reduced form of glutathione
GSK: glycogen synthase kinase
GSSG: oxidized form of glutathione (glutathione disulfide)
GT: trisialoganglioside
GTF: general transcription factor
GTP: guanosine triphosphate
GTPase: guanosine triphosphatase
GuCy: guanylate cyclase (CyG)
GWAS: genome-wide association study

H

H: height
$\mathcal{H}$: history function
$\mathtt{H}$: dissipation
$\mathtt{h}$: head loss
h: thickness
h: specific enthalpy
h_{m}: mass transfer coefficient
h_{T}: heat transfer coefficient
HA: hyaluronic acid
HAD: haloacid dehalogenase
HAP: huntingtin-associated protein
HAT: histone acetyltransferase
HAAT: heterodimeric amino acid transporter
HAND: heart and neural crest derivatives expressed protein
Hb: hemoglobin
Hb^{SNO}: ^{S}nitrosohemoglobin
HBEGF: heparin-binding EGF-like growth factor
HCK: hematopoietic cell kinase
HCLS: hematopoietic lineage cell-specific Lyn substrate protein
HCN: hyperpolarization-activated, cyclic nucleotide-gated K^+ channel
HCNP: hippocampal cholinergic neurostimulatory peptide
HCT: helical CT
HDAC: histone deacetylase complex
HDL: high-density lipoprotein
HDL–C: HDL–cholesterol
HDL–CE: HDL–cholesteryl ester
HDM: human double minute (Ub ligase)
HEET: hydroxyepoxyeicosatrienoic acid
hemin: heme oxygenase-1 inducer
HERG: human ether-a-go-go related gene
HER: human epidermal growth factor receptor (HER3: pseudokinase)
HES: Hairy enhancer of split
HETE: hydroxyeicosatetraenoic acid
HETEE: HETE ethanolamide
HEV: high endothelial venule
HGF: hepatocyte growth factor
HGFA: hepatocyte growth factor activator (serine peptidase)
HGFR: hepatocyte growth factor receptor

HGS: HGF-regulated Tyr kinase substrate (HRS)
HhIP: Hedgehog-interacting protein
HIF: hypoxia-inducible factor
HIP: huntingtin-interacting protein
HIP1R: HIP1-related protein
His: histamine
Hjv: hemojuvelin
HK: hexokinase
HL: hepatic lipase
HMG: high-mobility group protein
HMGB: high-mobility group box protein
HMGCoAR: hydroxy methyl glutaryl coenzyme-A reductase
HMT: histone methyl transferase
HMWK: high-molecular-weight kininogen
HNF: hepatocyte nuclear factor (NR2a1/2)
HNP: human neutrophil peptide
hnRNP: heterogeneous nuclear ribonucleoprotein
HODE: hydroxyoctadecadienoic acid
HOP: HSP70–HSP90 complex-organizing protein
HoPS: homotypic fusion and vacuole protein sorting complex
HotAIR: HOX antisense intergenic RNA (large intergenic non-coding RNA)
HOx: heme oxygenase
Hox: homeobox DNA sequence (encodes homeodomain-containing morphogens)
HpCa: hippocalcin
HPETE: hydroperoxyeicosatetraenoic acid
HPETEE: HPETE ethanolamide
HPK: hematopoietic progenitor kinase (MAP4K)
hpRNA: long hairpin RNA
hRas: Harvey Ras
HRE: hormone response element (DNA sequence)
HRM: hypoxia-regulated microRNA
hRNP: heterogeneous ribonucleoprotein
HRS: hepatocyte growth factor-regulated Tyr kinase substrate
HRT: Hairy and enhancer of Split-related transcription factor
HS: heparan sulfate
HSC: hematopoietic stem cell
HSC: heat shock cognate
HSER: heat stable enterotoxin receptor (guanylate cyclase 2C)
HSP: heat shock protein (chaperone)
HSPG: heparan sulfate proteoglycan
Ht: hematocrit
HTR: high temperature requirement endopeptidase
HUNK: hormonally up-regulated Neu-associated kinase

I

I: identity tensor
i: current
I: inotropy
IAP: inhibitor of apoptosis protein
IBABP: intestinal bile acid-binding protein
IC: isovolumetric contraction
ICA: internal carotid artery
ICAM: intercellular adhesion molecule (IgCAM member)
IgCAM: immunoglobulin-like cell adhesion molecule
ICF: intracellular fluid
ICliP: intramembrane-cleaving peptidase
ID: inhibitor of DNA binding
IDL: intermediate-density lipoprotein
IDmiR: immediately downregulated microRNA
IDOL: inducible degrader of LDL receptor (Ub ligase)
IEG: immediate-early gene
IEL: internal elastic lamina
IEL: intra-epithelial lymphocyte
IfIH: interferon-induced with helicase-C domain-containing protein
Ifn: interferon
IfnAR: interferon-$\alpha/\beta/\omega$ receptor
IFT: intraflagellar transport complex
Ig: immunoglobulin
IGF: insulin-like growth factor
IGFBP: IGF-binding protein
IgHC: immunoglobulin heavy chain
IgLC: immunoglobulin light chain
iGluR: ionotropic glutamate receptor
IH: intimal hyperplasia
IHh: indian Hedgehog
IK: intermediate-conductance Ca^{2+}-activated K^+ channel
IκB: inhibitor of NFκB
IKK: IκB kinase
IL: interleukin
ILC: innate lymphoid cell
iLBP: intracellular lipid-binding protein
ILK: integrin-linked (pseudo)kinase
ILKAP: integrin-linked kinase-associated Ser/Thr phosphatase-2C

IMM: inner mitochondrial membrane
IMP: Impedes mitogenic signal propagation
INADl: inactivation no after-potential D-protein
InCenP: inner centromere protein
InF: inverted formin
InsIG: insulin-induced gene product (ER anchor)
InsL: insulin-like peptide
InsR (IR): insulin receptor
InsRR: insulin receptor-related receptor
IP: inositol phosphate
IP_3: inositol (1,4,5)-trisphosphate
IP_3R: IP_3 receptor (IP_3-sensitive Ca^{2+}-release channel)
IP_4: inositol (1,3,4,5)-tetrakisphosphate
IP_5: inositol pentakisphosphate
IP_6: inositol hexakisphosphate
IPCEF: interaction protein for cytohesin exchange factor
IPOD: (perivacuolar) insoluble protein deposit
IPP: inositol polyphosphate phosphatase
IPP: ILK–PINCH–parvin complex
iPSC: induced pluripotent stem cell
IQGAP: IQ motif-containing GTPase-activating protein (IQ: first 2 amino acids of the motif: isoleucine [I; commonly] and glutamine [Q; invariably]).
IR: isovolumetric relaxation
IRAK: IL1 receptor-associated kinase (IRAK2: pseudokinase)
IRE: irreversible electroporation
IRES: internal ribosome entry site
IRF: interferon-regulatory protein (transcription factor)
IRFF: interferon-regulatory factor family
IRP: iron regulatory protein
IRS: insulin receptor substrate
ISA: intracranial saccular aneurysm
ISG: interferon-stimulated gene product
iSMAD: inhibitory SMAD (SMAD6 or SMAD7)
ITAM: immunoreceptor tyrosine-based activation motif
Itch: Itchy homolog (Ub ligase)
Itg: integrin
ITIM: immunoreceptor tyrosine-based inhibitory motif
ITK: interleukin-2-inducible T-cell kinase
ITPK: inositol trisphosphate kinase
IVC: inferior vena cava
IVP: initial value problem
IVUS: intravascular ultrasound

J

J: flux
J_m: cell surface current density
JAM: junctional adhesion molecule
JaK: Janus (pseudo)kinase
JIP: JNK-interacting protein (MAPK8IP1 and -2)
JMy: junction-mediating and regulatory protein
JNK: Jun N-terminal kinase (MAPK8–MAPK10)
JNKBP: JNK-binding protein
JNKK: JNK kinase
JSAP: JNK/SAPK-associated protein
Jun: avian sarcoma virus-17 proto-oncogene product (Japanese juunana: seventeen [17]; TF)
JUNQ: juxtanuclear quality-control compartment

K

$\mathbf{K}$: conductivity tensor
K: bending stiffness
K: reflection coefficient
K_d: dissociation constant (index of ligand–target affinity: ([L][T])/[C]; [L], [T], [C]: molar concentrations of the ligand, target, and created complex, respectively)
$\mathrm{K_M}$: Michaelis constant (chemical reaction kinetics)
K_m: material compressibility
k: cross-section ellipticity
k_{ATP}: myosin ATPase rate
k_B: Boltzmann constant (1.38×10^{-23} J/K)
k_c: spring stiffness
k_m: mass-transfer coefficient
k_P: Planck constant
K_R: resistance coefficient
KaP: karyopherin
K_{ATP}: ATP-sensitive K^+ channel
$K_{Ca}1.x$: BK channel
$K_{Ca}2/3/4.x$: SK channel
$K_{Ca}5.x$: IK channel
K_{IR}: inwardly rectifying K^+ channel
K_V: voltage-gated K^+ channel
KAP: kinesin (KIF)-associated protein

Kap: karyopherin
KAT: lysine (K) acetyltransferase
KCC: K^+–Cl^- cotransporter
KChAP: K^+ channel-associated protein
KChIP: K_V channel-interacting protein
KDELR: KDEL (Lys–Asp–Glu–Leu) endoplasmic reticulum retention receptor
KDR: kinase insert domain receptor
KHC: kinesin heavy chain
KIF: kinesin family
KIR: killer cell immunoglobulin-like receptor
KIT: cellular kinase in tyrosine (SCFR)
Kk: kallikrein
KLC: kinesin light chain
KLF: Krüppel-like factor
KLR: killer cell lectin-like receptor
Kn: Knudsen number
KOR: κ-opioid receptor
kRas: Kirsten Ras
Krt: keratin
KSR: kinase suppressor of Ras (adaptor; pseudokinase)

L

L: velocity gradient tensor
L: inertance
L: length
L_e: entry length
LA: left atrium
LAB: linker of activated B lymphocyte
LAd: LCK-associated adaptor
LAMTOR: late endosomal and lysosomal adaptor, MAPK and TOR activator
LANP: long-acting natriuretic peptide
LAP: leucine-rich repeat and PDZ domain-containing protein (4-member family)
LAP: latency-associated peptide (4 isoforms LAP1–LAP4)
LAP: nuclear lamina-associated polypeptide
LAR: leukocyte common antigen-related receptor (PTPRF)
LAT: linker of activated T lymphocytes
LaTS: large tumor suppressor
LAX: linker of activated X cells (both B and T cells)
LBR: lamin-B receptor
LCA: left coronary artery
LCAT: lysolecithin cholesterol acyltransferase
LCC: left coronary cusp
LCK: leukocyte-specific cytosolic (non-receptor) Tyr kinase
LCP: lymphocyte cytosolic protein (adaptor SLP76)
LDL: low-density lipoprotein
LDLR: low-density lipoprotein receptor
LDV: laser Doppler velocimetry
Le: entry length
LEF: lymphoid enhancer-binding transcription factor
LGalS: lectin, galactoside-binding, soluble cell adhesion molecule
LGIC: ligand-gated ion channel
LGL: lethal giant larva protein
LH: luteinizing hormone
LIF: leukemia-inhibitory factor
LIFR: leukemia-inhibitory factor receptor
LIMA: LIM domain and actin-binding protein
LIME: LCK-interacting molecule
LIMK: Lin1, Isl1, and Mec3 motif-containing kinase
LIMS: LIM and senescent cell antigen-like-containing domain protein
LiNC: linker of nucleoskeleton and cytoskeleton
lincRNA: large intergenic non-coding RNA
LipC: hepatic lipase
LipD: lipoprotein lipase
LipE: hormone-sensitive lipase
LipG: endothelial lipase
LipH: lipase-H
liprin: LAR PTP-interacting protein
LIR: leukocyte immunoglobulin-like receptor
LIS: lissencephaly protein
LKB: liver kinase-B
LKLF: lung Krüppel-like factor
LLTC: large latent TGFβ complex
LMan: lectin, mannose-binding
LMO: LIM domain-only-7 protein
Lmod: leiomodin (actin nucleator)
LMPP: lymphoid-primed multipotent progenitor
LMR: laser myocardial revascularization
Lam: laminin
LOx: lipoxygenase
LP: lipoprotein
LPA: lysophosphatidic acid
LPase: lipoprotein lipase
lpDC: lamina propria dendritic cell
Lphn: latrophilin (adhesion-GPCR)
LPL: lysophospholipid
LPLase: lysophospholipase
LPP: lipid phosphate phosphatase
LPR: lipid phosphatase-related protein

LPS: lipopolysaccharide
LQTS: long-QT syndrome
LRAT: lecithin–retinol acyltransferase
LRH: liver receptor homolog (NR5a2)
LRO: lysosome-related organelle
LRP: LDL receptor-related protein
LRRTM: leucine-rich repeat-containing transmembrane protein
LSK: Lin−, SCA1+, KIT+ cell
LST: lethal with Sec-thirteen
LSV: long saphenous vein
LT (Lkt): leukotriene
LTBP: latent TGFβ-binding protein
LTCC: L-type Ca^{2+} channel (Ca_V1)
LTK: leukocyte tyrosine kinase
LUbAC: linear ubiquitin chain assembly complex
LV: left ventricle
LVAD: left ventricular assist device
LX: lipoxin
LXR: liver X receptor (NR1h2/3)
LyVE: lymphatic vessel endothelial hyaluronan receptor

M

M: molar mass
M: metabolic rate
$\mathcal{M}$: moment
m: mass
Ma: Mach number
MACF: microtubule-actin crosslinking factor
mAChR: acetylcholine muscarinic receptor (metabotropic; GPCR)
MAD: mothers against decapentaplegic homolog
MAD: mitotic arrest-deficient protein
MAdCAM: mucosal vascular addressin cell adhesion molecule
MAF: musculoaponeurotic fibrosarcoma oncogene homolog (TF)
MAGI: membrane-associated guanylate kinase-related protein with inverted domain organization
MAGP: microfibril-associated glycoprotein
MAGuK: membrane-associated guanylate kinase
MAIT: mucosal-associated invariant T lymphocyte
MALT: mucosa-associated lymphoid tissue
MALT1: mucosa-associated lymphoid tissue lymphoma translocation peptidase
MAO: monoamine oxidase
MAP: microtubule-associated protein
MAP1LC3: microtubule-associated protein-1 light chain-3 (LC3)
mAP: mean arterial pressure
MAPK: mitogen-activated protein kinase
MAP2K: MAPK kinase
MAP3K: MA2KP kinase
MAP3K7IP: MAP3K7-interacting protein
MAPKAPK: MAPK-activated protein kinase
MARCKS: myristoylated alanine-rich C kinase substrate
MaRCo: macrophage receptor with collagenous structure (ScaRa2)
MARK: microtubule affinity-regulating kinase
MASTL: microtubule-associated Ser/Thr kinase-like protein
MAT: ménage à trois
MATK: megakaryocyte-associated Tyr kinase
MAVS: mitochondrial antiviral signaling protein
MAX: MyC-associated factor-X (bHLHd4–bHLHd8)
MBP: myosin-binding protein
MBP: myeloid–B-cell progenitor
MBTPS*i*: membrane-bound transcription factor peptidase site *i*
MCAK: mitotic centromere-associated kinesin
MCAM: melanoma cell adhesion molecule
MCD: medullary collecting duct
MCL1: BCL2-related myeloid cell leukemia sequence protein-1
MCLC: stretch-gated Mid1-related chloride channel
MCM: minichromosome maintenance protein
MCP: monocyte chemoattractant protein
mCSF: macrophage colony-stimulating factor (CSF1)
MCT: monocarboxylate–proton cotransporter
mDC: myeloid dendritic cell
MDM: mitochondrial distribution and morphology protein
MDR: multiple drug resistance (ABC transporter)
MEF: myocyte enhancer factor
megCSF: megakaryocyte colony-stimulating factor
MEJ: myoendothelial junction
MELK: maternal embryonic leucine zipper kinase
MEP: megakaryocyte erythroid progenitor
MEP: myeloid–erythroid progenitor
MET: mesenchymal–epithelial transition factor (proto-oncogene; HGFR)

METC: mitochondrial electron transport chain
metHb: methemoglobin
MGIC: mechanogated ion channel
mGluR: metabotropic glutamate receptor
MGP: matrix Gla protein
MHC: major histocompatibility complex
MHC: myosin heavy chain
MyHC or MYH: myosin heavy chain gene
miCK: mitochondrial creatine kinase
MiCU: mitochondrial calcium uptake protein
Mid: midline
MinK: misshapen-like kinase
miR: microRNA
miRNP: microribonucleoprotein
MiRP: MinK-related peptide
MIRR: multichain immune-recognition receptor
MIS: Müllerian-inhibiting substance
MIS: mini-invasive surgery
MIS: mitochondrial intermembrane space
MIST: mastocyte immunoreceptor signal transducer
MIT: mini-invasive therapy
MiV: mitral valve
MIZ: Myc-interacting zinc finger protein
MJD: Machado-Joseph disease protein domain-containing peptidase (DUb)
MKL: megakaryoblastic leukemia-1 fusion coactivator
MKnK: MAPK-interacting protein Ser/Thr kinase (MnK)
MKP: mitogen-activated protein kinase phosphatase
MLC: myosin light chain
MLCK: myosin light-chain kinase
MLCP: myosin light-chain phosphatase
MLK: mixed lineage kinase
MLKL: mixed lineage kinase-like pseudokinase
MLL: mixed lineage [myeloid–lymphoid] leukemia factor
MLLT: mixed lineage leukemia translocated protein
MLP: muscle LIM protein
mmCK: myofibrillar creatine kinase
MME: membrane metalloendopeptidase
MMM: maintenance of mitochondrial morphology protein
MMP: matrix metallopeptidase
MO: mouse protein
Mo: monocyte
MOMP: mitochondrial outer membrane permeabilization
MOR: μ-opioid receptor
MP: MAPK partner
MPF: mitosis (maturation)-promoting factor (CcnB–CDK1 complex)
MPG: N-methylpurine (N-methyladenine)-DNA glycosylase
MPO: median preoptic nucleus
Mpo: myeloperoxidase
MP_P: membrane protein, palmitoylated
MPP: multipotent progenitor
MR: mineralocorticoid receptor (NR3c2)
mRas: muscle Ras (or rRas3)
MRCK: myotonic dystrophy kinase-related CDC42-binding kinase
MRI: (nuclear) magnetic resonance imaging
mRNA: messenger RNA
mRNP: messenger ribonucleoprotein
MRTF: myocardin-related transcription factor
MSC: mesenchymal stem cell
MSH: melanocyte-stimulating hormone
MSIC: mechanosensitive ion channel
MSSCT: multi-slice spiral CT
MST: mammalian sterile-twenty-like kinase
MSt1R: macrophage-stimulating-1 factor receptor (RON)
MT: metallothionein
MTM: myotubularin (myotubular myopathy-associated gene product)
mtMMP: membrane-type MMP (mt*i*MMP: type-*i* mtMMP)
MTMR: myotubularin-related phosphatase
MTOC: microtubule organizing center
MTP: myeloid–T-cell progenitor
MTP: microsomal triglyceride transfer protein
MuRF: muscle-specific RING finger (Ub ligase)
MuSK: muscle-specific kinase
MVB: multivesicular body
MVE: multivesicular endosome (MVB)
MVO2: myocardial oxygen consumption
MWSS: maximal wall shear stress
MyB: myeloblastosis viral oncogene homolog (TF)
MyC: myelocytomatosis viral oncogene homolog (TF)
MyD88: myeloid differentiation primary response gene product-88
MyHC: myosin heavy chain
MyLC or MYL: myosin light-chain gene
MyPT: myosin phosphatase targeting subunit
MyT: myelin transcription factor

N

N: sarcomere number
$\hat{\mathbf{n}}$: unit normal vector
n: mole number
n: PAM density with elongation x
$\mathtt{n}$: myosin head density
$\mathcal{N}_A$: Avogadro number
N-terminus: amino (amine group NH_2)-terminus
NAADP: nicotinic acid adenine dinucleotide phosphate
nAChR: acetylcholine nicotinic receptor (ionotropic; LGIC)
NAD: nicotine adenine dinucleotide
NADPH: reduced form of nicotinamide adenine dinucleotide phosphate
NAd: noradrenaline
NAF: nutrient-deprivation autophagy factor
NALT: nasal-associated lymphoid tissue
NAmPT: nicotinamide phosphoribosyltransferase
Nanog: ever young (Gaelic)
NAP: NCK-associated protein (NCKAP)
NAT: nucleobase–ascorbate transporter
NAT1: noradrenaline transporter
Na_V: voltage-gated Na^+ channel
NBC: Na^+–HCO_3^- cotransporters
NCC: non-coronary cusp
NCC: Na^+–Cl^- cotransporter
Ncdn: neurochondrin
NCK: non-catalytic region of Tyr kinase adaptor
NCoA: nuclear receptor coactivator
NCoR: nuclear receptor corepressor
NCR: natural cytotoxicity-triggering receptor
ncRNA: non-coding RNA
NCS: neuronal calcium sensor
NCKX: Na^+–Ca^{2+}–K^+ exchanger
NCLX: Na^+–Ca^{2+}–Li^+ exchanger
NCX: Na^+–Ca^{2+} exchanger
NDCBE: Na^+-dependent Cl^-–HCO_3^- exchanger
NecL: nectin-like molecule
NEDD: neural precursor cell expressed, developmentally downregulated
NDFIP: NEDD4 family-interacting protein
NeK: never in mitosis gene-A (NIMA)-related kinase
NES: nuclear export signal
NESK: NIK-like embryo-specific kinase
nesprin: nuclear envelope spectrin repeat protein
NeuroD: neurogenic differentiation protein
NF: neurofilament protein (intermediate filament)
NF: neurofibromin (RasGAP)
NFAT: nuclear factor of activated T cells
NFe2: erythroid-derived nuclear factor-2
NFH: neurofilament, heavy polypeptide
NFκB: nuclear factor κ light chain enhancer of activated B cells
NFL: neurofilament, light polypeptide
NFM: neurofilament, medium polypeptide
NGAL: neutrophil gelatinase-associated lipocalin
NGF: nerve growth factor
Ngn: neogenin (netrin receptor)
NHA: Na^+–H^+ antiporter
NHE: Na^+–H^+ exchanger
NHERF: NHE regulatory factor
NHR: nuclear hormone receptor
NIc: nucleoporin-interacting protein
NIK: NFκB-inducing kinase
NIK: NCK-interacting kinase
NIP: neointimal proliferation
NK: natural killer cell
NKCC: Na^+–Ka^+–$2Cl^-$ cotransporter
NKG: NK receptor group
NKT: natural killer T cell
NKx2: NK2 transcription factor-related homeobox protein
NLR: NOD-like receptor (nucleotide-binding oligomerization domain, Leu-rich repeat-containing)
NLS: nuclear localization signal
NMDAR: ^{N}methyl ^{D}aspartate receptor
NmU: neuromedin-U
NO: nitric oxide (nitrogen monoxide)
NOD: nucleotide-binding oligomerization domain
NonO: non-POU domain-containing octamer-binding protein
NOR: neuron-derived orphan receptor (NR4a3)
NOS: nitric oxide synthase
NOS1: neuronal NOS
NOS1AP: NOS1 adaptor protein
NOS2: inducible NOS
NOS3: endothelial NOS
NOx: NAD(P)H oxidase
Noxa: damage (Latin)
NPAS: neuronal PAS domain-containing transcription factor
NPC: nuclear-pore complex
NPC: Niemann-Pick disease type-C protein
NPC1L: Niemann-Pick protein-C1-like
nPKC: novel protein kinase C

NPY: neuropeptide Y
NR: nuclear receptor
NRAP: nebulin-related actinin-binding protein
nRas: neuroblastoma Ras
NRBP: nuclear receptor-binding protein
NRF: nuclear factor erythroid-derived-2 (NF-E2)-related factor
NRF1: nuclear respiratory factor-1
Nrg: neuregulin (EGF superfamily member)
Nrgn: neuroligin
Nrp: neuropilin (VEGF-binding molecule; VEGFR coreceptor)
NRPTP: non-receptor protein Tyr phosphatase
NRSTK: non-receptor Ser/Thr kinase
NRTK: non-receptor protein Tyr kinase
NRx: nucleoredoxin
Nrxn: neurexin
NSC: non-selective cation channel
NSCLC: non-small-cell lung cancer
NSF: N-ethylmaleimide-sensitive factor
NSLTP: non-specific lipid-transfer protein
NST: nucleus of the solitary tract
NT: neurotrophin
NT5E: ecto-5′-nucleotidase
NTCP: sodium–taurocholate cotransporter polypeptide
NTF: N-terminal fragment
NTP: nucleoside triphosphate
NTPase: nucleoside triphosphate hydrolase superfamily member
NTRK: neurotrophic tyrosine receptor kinase (TRK)
NTRKR: neurotrophic protein Tyr receptor kinase-related protein ($ROR_{(RTK)}$)
NTS: nucleus tractus solitarius
Nu: Nusselt number
NuAK: nuclear AMPK-related kinase
NuP: nucleoporin (nuclear-pore complex protein)
NuRD: nucleosome remodeling and histone deacetylase
NuRR: nuclear receptor-related factor (NR4a2)
nWASP: neuronal WASP

O

$^{O}Glc^{N}Ac$: β^{N}acetyl ^{D}glucosamine
OCRL: oculocerebrorenal syndrome of Lowe phosphatase
Oct: octamer-binding transcription factor
ODE: ordinary differential equation
OGA: $^{O}Glc^{N}Acase$ (β^{N}acetylglucosaminidase)
OMCD: outer medullary collecting duct
OMM: outer mitochondrial membrane
ORC: origin recognition complex
ORF: open reading frame
ORP: OSBP-related protein
OSA: obstructive sleep apnea
OSBP: oxysterol-binding protein
OSI: oscillatory shear index
OSM: oncostatin M
OSMR: oncostatin M receptor
OSR (OxSR): oxidative stress-responsive kinase
OTK: off-track (pseudo)kinase
OTU: ovarian tumor superfamily peptidase (deubiquitinase)
OTUB: otubain (Ub thioesterase of the OTU superfamily)
OVLT: organum vasculosum lamina terminalis
oxyHb: oxyhemoglobin (oxygenated hemoglobin)

P

$\mathcal{P}$: permeability
P: power
P: cell division rate
p: production rate
p: pressure
$p_{\mathtt{i}}$: partial pressure of gas component $\mathtt{i}$
PA: phosphatidic acid
PAAT: proton–amino acid transporter
PACS: phosphofurin acidic cluster sorting protein
PAF: platelet-activating factor
PAFAH: platelet-activating factor acetylhydrolase
PAG: phosphoprotein associated with glycosphingolipid-enriched microdomains
PAH: polycyclic aromatic hydrocarbon
PAH: pulmonary arterial hypertension
PAI: plasminogen activator inhibitor
PAK: P21-activated kinase
PALR: promoter-associated long RNA
PALS: protein associated with Lin-7
PAMP: pathogen-associated molecular pattern
PAMP: proadrenomedullin peptide
PAR: poly^{ADP}ribose
PAR: promoter-associated, non-coding RNA
PAR*i*: type-*i* peptidase-activated receptor
Par: partitioning defective protein
PARG: polyADPribosyl glycosidase

PARP: polyADPribose polymerase
PASR: promoter-associated short RNA
PATJ: protein (PALS1) associated to tight junctions
Pax: paxillin
Pax*i*: paired box protein-*i* (transcription regulator)
PBC: pre-Bötzinger complex (ventilation frequency)
PBIP: Polo box-interacting protein
PC: phosphatidylcholine
PC: polycystin
PC: protein C
PCMRV: phase-contrast MR velocimetry
PCr: phosphocreatine
PCT: proximal convoluted tubule
PCTP: phosphatidylcholine-transfer protein
PD: pharmacodynamics
pDC: plasmacytoid dendritic cell
PdCD: programmed cell death protein
PdCD6IP: PdCD 6-interacting protein
PdCD1Lg: programmed cell death-1 ligand
PDE: phosphodiesterase
PDE: partial differential equation
PDGF: platelet-derived growth factor
PDGFR: platelet-derived growth factor receptor
PDI: protein disulfide isomerase
PDK: phosphoinositide-dependent kinase
PDP: pyruvate dehydrogenase phosphatase
Pe: Péclet number
PE: phosphatidylethanolamine
PE: pulmonary embolism
PEBP: phosphatidylethanolamine-binding protein
PECAM: platelet–endothelial cell adhesion molecule
PEDF: pigment epithelium-derived factor (serpin F1)
PEn2: presenilin enhancer-2
PEO: proepicardial organ
Per: Period homolog
PERK: protein kinase-like endoplasmic reticulum kinase
PERP: P53 apoptosis effector related to peripheral myelin protein PMP22
PET: positron emission tomography
Pex: peroxin
PF: platelet factor
PFK: phosphofructokinase
pFRG: parafacial respiratory group
PG: prostaglandin
PGC: PPARγ coactivator
pGC: particulate guanylate cyclase
PGEA: prostaglandin ethanolamide
PGF: paracrine growth factor
PGG: prostaglandin glycerol ester
PGi2: prostacyclin
PGP: permeability glycoprotein
PGx: type-x (D, E, F, H, I) prostaglandin
PGxS: type-x prostaglandin synthase
PH: pleckstrin homology domain
PHD: prolyl hydroxylase
PhK: phosphorylase kinase
PHLPP: PH domain and Leu-rich repeat protein phosphatase
PI: phosphoinositide (phosphorylated phosphatidylinositol)
PI(4)P: phosphatidylinositol 4-phosphate
PI(*i*)P*i*K: phosphatidylinositol *i*-phosphate *i*-kinase
PI(i,j)P_2: phosphatidylinositol (i,j)-bisphosphate (PIP_2)
PI(3,4,5)P_3: phosphatidylinositol (3,4,5)-trisphosphate (PIP_3)
PI3K: phosphatidylinositol 3-kinase
PI3KAP: PI3K adaptor protein
PI*i*K: phosphatidylinositol *i*-kinase
PIAS: protein inhibitor of activated STAT (SUMo ligase)
PIC: pre-initiation complex
PICK: protein that interacts with C-kinase
PIDD: P53-induced protein with a death domain
PIKE: phosphoinositide 3-kinase enhancer (GTPase; ArfGAP)
PIKK: phosphatidylinositol 3-kinase-related kinase (pseudokinase)
PIM: provirus insertion of Molony murine leukemia virus gene product
PIN: peptidyl prolyl isomerase interacting with NIMA
PINCH: particularly interesting new Cys–His protein (or LIMS1)
PInK: PTen-induced kinase
PIP: phosphoinositide monophosphate
PIP*i*K: phosphatidylinositol phosphate *i*-kinase
PIP_2: phosphatidylinositol bisphosphate
PIP_3: phosphatidylinositol triphosphate
PIPP: proline-rich inositol polyphosphate 5-phosphatase
PIR: paired immunoglobulin-like receptor
piRNA: P-element-induced wimpy testis-interacting (PIWI) RNA
PIRT: phosphoinositide-interacting regulator of TRP channels
PITP: phosphatidylinositol-transfer protein

Pitx: pituitary (or paired-like) homeobox transcription factor
PIV: particle image velocimetry
PIX: P21-activated kinase (PAK)-interacting exchange factor (RhoGEF6/7)
PK: pharmacokinetics
PK: protein kinase
PKA: protein kinase A
PKB: protein kinase B
PKC: protein kinase C
PKD: protein kinase D
PKG: protein kinase G
PKL: paxillin kinase linker
PKM: pyruvate kinase muscle isozyme
PKMYT (MYT): membrane-associated Tyr/ Thr protein kinase
PKN: protein kinase novel
Pkp: plakophilin
PL: phospholipase
PLA2: phospholipase A2
PLC: phospholipase C
PLD: phospholipase D
PLb: phospholamban
PLd: phospholipid
PlGF: placental growth factor
PLK: Polo-like kinase
PLTP: phospholipid transfer protein
Plxn: plexin
PMCA: plasma membrane Ca^{2+} ATPase
PML: promyelocytic leukemia protein
PMR: percutaneous (laser) myocardial revascularization
PMRT: protein arginine methyltransferase
Pn: plasmin
Png: plasminogen
PNS: peripheral nervous system
PoG: proteoglycan
PoM: pore membrane protein
Pon: paraoxonase
POPx: partner of PIX
POSH: scaffold plenty of SH3 domains
POT: Protection of telomeres (single-stranded telomeric DNA-binding protein)
PP: protein phosphatase
PP3: protein phosphatase 3 (PP2b or calcineurin)
PPAR: peroxisome proliferator-activated receptor (NR1c1–3)
PPG: photoplethysmography
PPId: peptidyl prolyl isomerase-D
PPIP: monopyrophosphorylated inositol phosphate
$(PP)_2IP$: bispyrophosphorylated inositol phosphate
PPK: PIP kinase
PPM: protein phosphatase (magnesium-dependent)
PPR: pathogen-recognition receptor
PPRE: PPAR response element (DNA sequence)
PPTC: protein phosphatase T-cell activation (TAPP2c)
PR: progesterone receptor (NR3c3)
Pr: Prandtl number
PRC: protein regulator of cytokinesis
PRC: Polycomb repressive complex
Prdx: peroxiredoxin
pre-cDC: pre-classical dendritic cell
pre-miR: precursor microRNA
preBotC: pre-Bötzinger complex
preKk: prekallikrein
PREx: PIP_3-dependent Rac exchanger (RacGEF)
PRG: plasticity-related gene product
PRH: prolactin-releasing hormone
pri-miR: primary microRNA
PRL: phosphatase of regenerating liver
Prl: prolactin
PrlR: prolactin receptor
PRMT: protein arginine (R) N-methyltransferase
Prompt: promoter upstream transcript
Protor: protein observed with Rictor
PROX: prospero homeobox gene
Prox: PROX gene product (transcription factor)
PrP: processing protein
PRPK: P53-related protein kinase
PRR: pattern recognition receptor
PRR: prorenin and renin receptor
PS: presenilin
PS: protein S
PSC: pluripotent stem cell
PSD: postsynaptic density adaptor
PsD: postsynaptic density
PSEF: pseudo-strain energy function
PSer: phosphatidylserine
PSGL: P-selectin glycoprotein ligand
PSKh: protein serine kinase H
Psm: proteasome subunit
PSTPIP: Pro–Ser–Thr phosphatase-interacting protein
PTA: plasma thromboplastin antecedent
Ptc: Patched receptor (Hedgehog signaling)
PTCA: percutaneous transluminal coronary angioplasty
PtcH: Patched Hedgehog receptor
PTCRA: PTC rotational burr atherectomy
PtdCho (PC): phosphatidylcholine

PtdEtn (PE): phosphatidylethanolamine
PtdSer (PS): phosphatidylserine
PtdIns (PI): phosphatidylinositol
PTen: phosphatase and tensin homolog deleted on chromosome ten (phosphatidylinositol 3-phosphatase)
PTFE: polytetrafluoroethylene
PTH: parathyroid hormone
PTHRP: parathyroid hormone-related protein
PTK: protein Tyr kinase
PTK7: pseudokinase (RTK)
PTP: protein Tyr phosphatase
PTPn*i*: protein Tyr phosphatase non-receptor type *i*
PTPR: protein Tyr phosphatase receptor
PTRF: RNA polymerase-1 and transcript release factor
PUFA: polyunsaturated fatty acid
PUMA: P53-upregulated modulator of apoptosis
PuV: pulmonary valve
PVF: PDGF- and VEGF-related factor
PVNH: paraventricular nucleus of hypothalamus
PVR: pulmonary vascular resistance
PWS: pulse wave speed
Px: pannexin
PXR: pregnane X receptor (NR1i2)
PYK: proline-rich tyrosine kinase
P2X: purinergic ligand-gated channel
P53AIP: P53-regulated apoptosis-inducing protein
p75NtR: pan-neurotrophin receptor

Q

$\mathtt{Q}$: material quantity
Q: thermal energy
Q_e: electric current density
Q_T: thermal energy (heat)
q_T: transfer rate of thermal energy (power)
q_{met}: metabolic heat source
q: flow rate

R

$\mathtt{R}$: resistance
$\mathcal{R}$: local reaction term
R_h: hydraulic radius
R_g: gas constant
R_R: respiratory quotient
R: recruitment function (from quiescence to proliferation)
r: cell renewal rate
r: radial coordinate
RA: right atrium
RAAS: renin–angiotensin–aldosterone system
Rab: Ras from brain
Rab11FIP: Rab11 family-interacting protein
Rac: Ras-related C3-botulinum toxin substrate
RACC: receptor-activated cation channel
RACK: receptor for activated C-kinase
RAD: recombination protein-A (RecA)-homolog DNA-repair protein
Rad: radiation sensitivity protein
Rag: Ras-related GTP-binding protein
Ral: Ras-related protein
RAlBP: retinaldehyde-binding protein
RalGDS: Ral guanine nucleotide-dissociation stimulator
RAMP: (calcitonin receptor-like) receptor-activity-modifying protein
Ran: Ras-related nuclear protein
RANTES: regulated upon activation, normal T-cell expressed, and secreted product (CCL5)
RAP: receptor-associated protein
Rap: Ras-proximate (Ras-related) protein
Raptor: regulatory-associated protein of TOR
RAR: retinoic acid receptor (NR1b2/3)
Ras: rat sarcoma viral oncogene homolog (small GTPase)
RasA: Ras p21 protein activator
rasiRNA: repeat-associated small interfering RNA (PIWI)
RASSF: Ras interaction/interference protein RIN1, afadin, and Ras association domain-containing protein family member
RB: retinoblastoma protein
RBC: red blood capsule (cell, or erythrocyte [without nucleus])
RBP: retinoid-binding protein
RC: ryanodine calcium channel (RyR)
RCA: right coronary artery
RCan: regulator of calcineurin
RCC: right coronary cusp
RCC: regulator of chromosome condensation
Re: Reynolds number
REDD: regulated in development and DNA-damage response gene product

Rel: reticuloendotheliosis proto-oncogene product (TF; member of NFκB)
REP: Rab escort protein
ReR: renin receptor (PRR)
restin: Reed-Steinberg cell-expressed intermediate filament-associated protein (CLiP1)
ReT: rearranged during transfection (receptor Tyr kinase)
RevRE: reverse (Rev)-ErbA (NR1d1/2) response element (DNA sequence)
RFA: radiofrequency ablation
RGL: Ral guanine nucleotide-dissociation stimulator-like protein (GEF)
RGS: regulator of G-protein signaling
RHEB: Ras homolog enriched in brain
RHS: equation right-hand side
Rho: Ras homologous
RIAM: Rap1-GTP-interacting adaptor
RIBP: RLK- and ITK-binding protein
RICH: RhoGAP interacting with CIP4 homolog
RICK: receptor for inactive C-kinase
Rictor: rapamycin-insensitive companion of TOR
RIF: Rho in filopodium
RIN: Ras-like protein expressed in neurons (GTPase)
RIn: Ras and Rab interactor (RabGEF)
RIP: regulated intramembrane proteolysis
RIPK: receptor-interacting protein kinase
RISC: RNA-induced silencing complex
RIT: Ras-like protein expressed in many tissues
RKIP: Raf kinase inhibitor protein
RlBP: retinaldehyde-binding protein
RLC: RISC-loading complex
RLK: resting lymphocyte kinase (TXK)
RNA: ribonucleic acid
RNABP: RNA-binding protein
RNase: ribonuclease
RnBP: renin-binding protein
RNF2: RING finger protein-2 (Ub ligase)
RNP: ribonucleoprotein
Robo: roundabout
ROC: receptor-operated channel
RoCK: Rho-associated, coiled-coil-containing protein kinase
ROI: region of interest
ROMK: renal outer medullary potassium channel
ROR: RAR-related orphan receptor (NR1f1–NR1f3)
$ROR_{(RTK)}$: receptor Tyr kinase-like orphan receptor
ROS: reactive oxygen species
Ros: ros UR2 sarcoma virus proto-oncogene product (RTK)
RPIP: Rap2-interacting protein
RPS6: ribosomal protein S6
RPTP: receptor protein Tyr phosphatase
rRas: related Ras
rRNA: ribosomal RNA
RSA: respiratory sinus arrhythmia
RSE: rapid systolic ejection
RSK: P90 ribosomal S6 kinase (^{P90}RSK)
RSKL: ribosomal protein S6 kinase-like (pseudokinase)
rSMAD: receptor-regulated SMAD (SMAD1–SMAD3, SMAD5, and SMAD9)
RSMCS: robot-supported medical and surgical system
RSpo: R-spondin
RSTK: receptor Ser/Thr kinase
RTK: receptor protein Tyr kinase
RTN: retrotrapezoid nucleus
Rubicon: RUN domain and Cys-rich domain-containing, beclin-1-interacting protein
Runx: Runt-related transcription factor
RV: right ventricle
RVF: rapid ventricular filling
RVLM: rostral ventrolateral medulla
RVMM: rostral ventromedial medulla
RXR: retinoid X receptor (NR2b1–NR2b3)
RYK: receptor-like Tyr (Y) kinase (pseudokinase)
RyR: ryanodine receptor (ryanodine-sensitive Ca^{2+}-release channel)

S

S: Cauchy-Green deformation tensor
s: entropy
s: sarcomere length
s: evolution speed
SAA: serum amyloid A
$SAC_{Cl(K)}$: stretch-activated Cl^- (K^+)-selective channel
SAc: suppressor of actin domain-containing 5-phosphatase
sAC: soluble adenylate cyclase
$SACC_{NS}$: stretch-activated cation non-selective channel
SACM1L: suppressor of actin mutation-1-like

SAH: subarachnoid hemorrhage
SAIC: stretch-activated ion channel
SAN: sinoatrial node
SAP: SLAM-associated protein
SAP: stress-activated protein
SAP*i*: synapse-associated protein *i*
SAPK: stress-activated protein kinase (MAPK)
SAR: secretion-associated and Ras-related protein
SBE: SMAD-binding element
SBF: SET-binding factor
Sc: Schmidt number
SCA: stem cell antigen
SCAMP: secretory carrier membrane protein
SCAP: SREBP cleavage-activating protein (SREBP escort)
SCAR: suppressor of cAMP receptor (WAVe)
ScaR: scavenger receptor
SCF: SKP1–Cul1–F-box Ub-ligase complex
SCF: stem cell factor
SCFR: stem cell factor receptor (KIT)
Scgb: secretoglobin
SCLC: small-cell lung cancer
scLC: squamous-cell lung cancer (NSCLC subtype)
SCN: suprachiasmatic nucleus
SCO: synthesis of cytochrome-C oxidase
SCP (CTDSP): small C-terminal domain (CTD)-containing phosphatase
Scp: stresscopin (urocortin 3)
Scrib: Scribble polarity protein
Sdc: syndecan
SDF: stromal cell-derived factor
SDPR: serum deprivation protein response
SE: systolic ejection
SEF: strain-energy function
SEF: similar expression to FGF genes (inhibitor of RTK signaling)
SEK: SAPK/ERK kinase
Sema: semaphorin (Sema, Ig, transmembrane, and short cytoplasmic domain)
SERCA: sarco(endo)plasmic reticulum calcium ATPase
serpin: serine peptidase inhibitor
SerT: serotonin transporter
SF: steroidogenic factor (NR5a1)
SFK: SRC-family kinase
SFO: subfornical organ
SFPQ: splicing factor proline and glutamine-rich
sFRP: secreted Frizzled-related protein
SftP (SP): surfactant protein
sGC: soluble guanylate cyclase
SGK: serum- and glucocorticoid-regulated kinase
SGlT: Na^+–glucose cotransporter (SLC5a)
Sgo: shugoshin (Japanese: guardian spirit)
SH: Src homology domain
Sh: Sherwood number
SH3P: Src homology-3 domain-containing adaptor protein
SHAnk: SH3 and multiple ankyrin repeat domain-containing protein
SHAX: SNF7 (VSP32) homolog associated with ALIX
SHB: Src homology-2 domain-containing adaptor
SHC: Src-homologous and collagen-like substrate
SHC: Src homology-2 domain-containing transforming protein
SHh: sonic Hedgehog
SHIP: SH-containing inositol phosphatase
SHP: SH-containing protein Tyr phosphatase (PTPn6/11)
SHP: small heterodimer partner (NR0b2)
shRNA: small (short) hairpin RNA
SIAH: Seven in absentia homolog (Ub ligase)
siglec: sialic acid-binding Ig-like lectin
SIK: salt-inducible kinase
SIn: stress-activated protein kinase-interacting protein
SIP: steroid receptor coactivator-interacting protein
siRNA: small interfering RNA
SiRP: signal-regulatory protein
SIRT: sirtuin (silent information regulator-2 [two]; histone deacetylase)
SIT: SHP2-interacting transmembrane adaptor
SK: small conductance Ca^{2+}-activated K^+ channel
SK*i*: sphingosine kinase-*i*
SKIP: sphingosine kinase-1-interacting protein
SKIP: skeletal muscle and kidney-enriched inositol phosphatase
SKP: S-phase kinase-associated protein
SLA: Src-like adaptor
SLAM: signaling lymphocytic activation molecule
SLAMF: SLAM family member
SLAP: Src-like adaptor protein
SLC: solute carrier class member
SLCO: solute carrier organic anion class transporter
SLK: Ste20-like kinase
Sln: sarcolipin
SLPI: secretory leukocyte peptidase inhibitor

SLTC: small latent TGFβ complex
SM: sphingomyelin
SMA: smooth muscle actin
SMAD: small (son of, similar to) mothers against decapentaplegia homolog
SMAP: Small ArfGAP protein, stromal membrane-associated GTPase-activating protein
SMase: sphingomyelinase
SMC: smooth muscle cell
Smo: Smoothened
SMPD: sphingomyelin phosphodiesterase
SMRT: silencing mediator of retinoic acid and thyroid hormone receptor (NCoR2)
SMS: sphingomyelin synthase
SMURF: SMAD ubiquitination regulatory factor
SNAAT: sodium-coupled neutral amino acid transporter
SNAP: soluble N-ethylmaleimide-sensitive factor-attachment protein
SnAP: synaptosomal-associated protein
SNARE: SNAP receptor
SNF7: sucrose non-fermenting (VPS32)
SNIP: SMAD nuclear-interacting protein
snoRNA: small nucleolar RNA
snoRNP: small nucleolar ribonucleoprotein
SNP: single-nucleotide polymorphism
snRNA: small nuclear RNA
snRNP: small nuclear ribonucleoprotein
SNx: sorting nexin
SOC: store-operated Ca^{2+} channel
SOCE: store-operated Ca^{2+} entry
SOCS: suppressor of cytokine signaling protein
SOD: superoxide dismutase
SorbS: sorbin and SH3 domain-containing adaptor
SOS: Son of sevenless (GEF)
Sost: sclerostin
SostDC: sclerostin domain-containing protein
SOX: sex-determining region Y (SRY)-box gene
Sox: SOX gene product (transcription factor)
SP1: specificity protein (transcription factor)
SPARC: secreted protein acidic and rich in cysteine
SPC: sphingosylphosphorylcholine
SPCA: secretory pathway Ca^{2+} ATPase
SPECT: single-photon emission CT
Sph: sphingosine
SphK: sphingosine kinase
SPI: spleen focus-forming virus (SFFV) proviral integration proto-oncogene product (transcription factor)
SPInt: serine peptidase inhibitor
SPN: supernormal period
SPP: sphingosine phosphate phosphatase
SpRED: Sprouty-related protein with an EVH1 domain
SPURT: secretory protein in upper respiratory tract
SQTS: short-QT syndrome
SR: sarcoplasmic reticulum
SR: Arg/Ser domain-containing protein (alternative splicing)
SRA: steroid receptor RNA activator
SRC: steroid receptor coactivator
Src: sarcoma-associated (Schmidt-Ruppin A2 viral oncogene homolog) kinase
SREBP: sterol regulatory element-binding protein
SRF: serum response factor
SRM/SMRS: Src-related kinase lacking regulatory and myristylation sites
SRP: stresscopin-related peptide (urocortin 2)
SRPK: splicing factor RS domain-containing protein kinase
SRY: sex-determining region Y
SSAC: shear stress-activated channel
SSE: slow systolic ejection
Ssh: slingshot homolog phosphatase
SSI: STAT-induced STAT inhibitor
ssRNA: single-stranded RNA
Sst: somatostatin
SSV: short saphenous vein
St: Strouhal number
STAM: signal-transducing adaptor molecule
STAMBP: STAM-binding protein (Ub isopeptidase)
StAR: steroidogenic acute regulatory protein
StART: StAR-related lipid transfer protein
STAT: signal transducer and activator of transduction
STEAP: six transmembrane epithelial antigen of the prostate
STICK: substrate that interacts with C-kinase
StIM: stromal interaction molecule
STK: protein Ser/Thr kinase
STK1: stem cell protein Tyr kinase receptor
STLK: Ser/Thr kinase-like (pseudo)kinase
Sto: Stokes number
StRAd: STe20-related adaptor
STRAP: Ser/Thr kinase receptor-associated protein
StRAP: stress-responsive activator of P300

Stx: syntaxin ($SNARE^Q$)
SUMo: small ubiquitin-related modifier
SUn: Sad1 and Unc84 homology protein
SUR: sulfonylurea receptor
SUT: stable unannotated transcript
SV: stroke volume
SVC: superior vena cava
SVCT: sodium-dependent vitamin-C transporter
SVF: slow ventricular filling
SVP: synaptic vesicle precursor
SVR: systemic vascular resistance
SW: stroke work
SwAP70: 70-kDa switch-associated protein (RacGEF)
SYK: spleen tyrosine kinase
Synj: synaptojanin
Syp: synaptophysin
Syt: synaptotagmin
S1P: sphingosine 1-phosphate
S6K: P70 ribosomal S6 kinase (^{P70}RSK)

T

T: extrastress tensor
T: transition rate from a cell cycle phase to the next
T: temperature
T_s: surface tension
T lymphocyte (T cell): thymic lymphocyte
T_C: cytotoxic T lymphocyte (CD8+ effector T cell; CTL)
T_{C1}: type-1 cytotoxic T lymphocyte
T_{C2}: type-2 cytotoxic T lymphocyte
T_{CM}: central memory T lymphocyte
T_{Conv}: conventional T lymphocyte
T_{Eff}: effector T lymphocyte
T_{EM}: effector memory T lymphocyte
T_{FH}: follicular helper T lymphocyte
T_H: helper T lymphocyte (CD4+ effector T cell)
T_{Hi}: type-i helper T lymphocyte ($i = 1/2/9/17/22$)
T_{H3}: TGFβ-secreting T_{Reg} lymphocyte
T_L: lung transfer capacity (alveolocapillary membrane)
T_{R1}: type-1, IL10-secreting, regulatory T lymphocyte
T_{Reg}: regulatory T lymphocyte
aT_{Reg}: CD45RA−, FoxP3^{hi}, activated T_{Reg} cell
iT_{Reg}: inducible T_{Reg} lymphocyte
nT_{Reg}: naturally occurring (natural) T_{Reg} lymphocyte
rT_{Reg}: CD45RA+, FoxP3^{low}, resting T_{Reg} cell
$\hat{\mathbf{t}}$: unit tangent
t: time
TβRi: type-i TGFβ receptor
TAA: thoracic aortic aneurysm
TAB: TAK1-binding protein
TACE: tumor-necrosis factor-α-converting enzyme (ADAM17)
TACE: transarterial chemoembolization
TAF: TBP associated factor
TAK: TGFβ-activated kinase (MAP3K7)
TALK: TWIK-related alkaline pH-activated K^+ channel
TANK: TRAF family member-associated NFκB activator
TASK: TWIK-related acid-sensitive K^+ channel
TASR: terminus-associated short RNA
TAP: transporter associated with antigen processing (ABC transporter)
Taz: taffazin
TBC1D: Tre2 (or USP6), BUB2, CDC16 domain-containg RabGAP
TBCK: tubulin-binding cofactor kinase (pseudokinase)
TBK: TANK-binding kinase
TBP: TATA box-binding protein (subclass-4F transcription factor)
TBx: T-box transcription factor
TC: thrombocyte (platelet)
TCA: tricarboxylic acid cycle
TCF: T-cell factor
TCF: ternary complex factor
TcFi: type-i transcription factor
TCP: T-complex protein
TCR: T-cell receptor
TEA: transluminal extraction atherectomy
TEC: Tyr kinase expressed in hepatocellular carcinoma
TEF: thyrotroph embryonic factor (PAR/b–ZIP family)
TEK: Tyr endothelial kinase
TEM: transendothelial migration
Ten: tenascin
TF: transcription factor
Tf: transferrin
TFPI: tissue factor pathway inhibitor
TfR: transferrin receptor
TG: triglyceride (triacylglycerol)
TGF: transforming growth factor
TGFBR: TGFβ receptor gene

TGFβRAP: TGFβ receptor-associated protein
TGN: trans-Golgi network
THET: trihydroxyeicosatrienoic acid
THIK: tandem pore-domain halothane-inhibited K^+ channel
THR: thyroid hormone receptor (NR1a1/2)
TIAM: T-lymphoma invasion and metastasis-inducing protein (RacGEF)
TICE: transintestinal cholesterol efflux
TIE: Tyr kinase with Ig and EGF homology domains (angiopoietin receptor)
TIEG: TGFβ-inducible early gene product
TIF: transcription intermediary factor (kinase and Ub. ligase)
TIGAR: TP53-inducible glycolysis and apoptosis regulator
TIM: T-cell immunoglobulin and mucin domain-containing protein
Tim: timeless homolog
TIMM: translocase of inner mitochondrial membrane
TIMP: tissue inhibitor of metallopeptidase
TIRAP: Toll–IL1R domain-containing adaptor protein
tiRNA: transcription initiation RNA
TJ: tight junction
TKR: Tyr kinase receptor
TLC: total lung capacity
TLR: Toll-like receptor
TLT: TREM-like transcript
TLX: tailless receptor (NR2e1)
TM: thrombomodulin
TM*i*: transmembrane segment-*i* of membrane protein
TMC: twisting magnetocytometry
TMePAI: transmembrane prostate androgen-induced protein
TMy: tropomyosin
Tnn (TN): troponin
Tn: thrombin
TNF: tumor-necrosis factor
TNFαIP: tumor-necrosis factor-α-induced protein
TNFR: tumor-necrosis factor receptor
TNFRSF: tumor-necrosis factor receptor superfamily member
TNFSF: tumor-necrosis factor superfamily member
TNK: Tyr kinase inhitor of NFκB
Tns: tensin
TOR: target of rapamycin
TORC: target of rapamycin complex
TORC: transducer of regulated CREB activity (a.k.a. CRTC)
TP: thromboxane-A2 Gq/11-coupled receptor
TP53I: tumor protein P53-inducible protein
tPA: tissue plasminogen activator
Tpo: thrombopoietin
TpM: tropomyosin
TPPP: tubulin polymerization-promoting protein
TPST: tyrosylprotein sulftotransferase
TR: testicular receptor (NR2c1/2)
TRAAK: TWIK-related arachidonic acid-stimulated K^+ channel
TRADD: tumor-necrosis factor receptor-associated death domain adaptor
TRAF: tumor-necrosis factor receptor-associated factor
TRAM: TRIF-related adaptor molecule
transceptor: transporter-related receptor
TRAP: TNF receptor-associated protein (HSP75)
TraPP: transport protein particle
TRAT: T-cell receptor-associated transmembrane adaptor
Trb: Tribbles homolog (pseudokinase)
TRE: TPA-response element (AP1/CREB-binding site on promoters)
TRE: trapped in endoderm
TREK: TWIK-related K^+ channel
TREM: triggering receptor expressed on myeloid cells
TRESK: TWIK-related spinal cord K^+ channel
TRF: TBP-related factor
TRF: double-stranded telomeric DNA-binding repeat-binding factor
TRH: thyrotropin-releasing hormone
TRIF: Toll–IL1R domain-containing adaptor inducing Ifnβ
TRIM: T-cell receptor-interacting molecule
TRIP: TGFβ receptor-interacting protein (eIF3S2)
TRK: tropomyosin receptor kinase (NTRK)
tRNA: transfer RNA
TRP: transient receptor potential channel
TRPA: ankyrin-like transient receptor potential channel
TRPC: canonical transient receptor potential channel
TRPM: melastatin-related transient receptor potential channel
TRPML: mucolipin-related transient receptor potential channel
TRPN: no mechanoreceptor potential C
TRPP: polycystin-related transient receptor potential channel

TRPV: vanilloid transient receptor potential channel
TrrAP: transactivation (transformation)/transcription domain-associated protein (pseudokinase)
TrV: tricuspid valve
TRx: thioredoxin
TRxIP: thioredoxin-interacting protein
TSC: tuberous sclerosis complex
TSH: thyroid-stimulating hormone
TSLP: thymic stromal lymphopoietin
Tsp: thrombospondin
Tspan: tetraspanin
TsPO: translocator protein of the outer mitochondrial membrane
tSNARE: target SNARE
tsRNA: tRNA-derived small RNA
tssaRNA: transcription start site-associated RNA
TTbK: Tau-tubulin kinase
TTK: dual-specificity Thr/Tyr kinase
Ttn: titin (pseudokinase)
TUT: terminal uridine transferase
TWIK: tandem of P domains in a weak inwardly rectifying K^+ channel
TxA2: thromboxane A2 (thromboxane)
TxB2: thromboxane B2 (thromboxane metabolite)
TXK: Tyr kinase mutated in X-linked agammaglobulinemia
TxaS: thromboxane-A synthase
TyK: tyrosine kinase
T_3: tri-iodothyronine
T_4: thyroxine
$^+$TP: plus-end-tracking proteins

U

U: right stretch tensor
u: displacement vector
u: electrochemical command
u: specific internal energy
Ub: ubiquitin
UbC: ubiquitin conjugase
UbE2: E2 ubiquitin conjugase
UbE3: E3 ubiquitin ligase
UbL: ubiquitin-like protein
UCH: ubiquitin C-terminal hydrolase (DUb)
Ucn: urocortin
UCP: uncoupling protein
UDP: uridine diphosphate-glucose
UK: urokinase
ULK: uncoordinated-51-like kinase (pseudokinase)
Unc: uncoordinated receptor
uPA: urokinase-type plasminogen activator (urokinase)
uPAR: uPA receptor
uPARAP: uPAR-associated protein (CLec13e)
UPR: unfolded protein response
UPS: ubiquitin–proteasome system
UP4A: uridine adenosine tetraphosphate
Uro: urodilatin
US: ultrasound
USC: unipotential stem cell
USF: upstream stimulatory factor
USI: ultrasound imaging
USP: ubiquitin-specific peptidase (deubiquitinase)
UTP: uridine triphosphate
UTR: untranslated region
UVRAG: ultraviolet wave resistance-associated gene product

V

V: left stretch tensor
V: volume
V_q: cross-sectional average velocity
V_s: specific volume
v: velocity vector
v: recovery variable
V1(2)R: type-1(2) vomeronasal receptor
$V_{1A/1B/2}$: type-1A/1B/2 arginine vasopressin receptor
VAAC: volume-activated anion channel
$VAC_{Cl(K)}$: volume-activated Cl^- (K^+)-selective channel
VACamKL: vesicle-associated CamK-like (pseudokinase)
$VACC_{NS}$: volume-activated cation non-selective channel
VAChT: vesicular acetylcholine transporter
VAIC: volume-activated ion channel
VAMP: vesicle-associated membrane protein (synaptobrevin)
VanGL: Van Gogh (Strabismus)-like protein
VAP: VAMP-associated protein
VASP: vasoactive stimulatory phosphoprotein
VAT: vesicular amine transporter
vATPase: vesicular-type H^+ ATPase
VAV: ventriculoarterial valve
Vav: GEF named from Hebrew sixth letter

VC: vital capacity
VCAM: vascular cell adhesion molecule
VCt: vasoconstriction
VDAC: voltage-dependent anion channel (porin)
VDACL: plasmalemmal, volume- and voltage-dependent, ATP-conductive, large-conductance, anion channel
VDCC: voltage-dependent calcium channel
VDP: vesicle docking protein
VDt: vasodilation
VEGF: vascular endothelial growth factor
VEGFR: vascular endothelial growth factor receptor
VF: ventricular fibrillation
VF: ventricular filling
VGAT: vesicular GABA transporter
VGC: voltage-gated channel
VgL: Vestigial-like protein
VGluT: vesicular glutamate transporter
VHL: von Hippel-Lindau Ub ligase
VIP: vasoactive intestinal peptide
VLDL: very-low-density lipoprotein
VLDLR: very-low-density lipoprotein receptor
VMAT: vesicular monoamine transporter
VN: vitronectin
VPO: vascular peroxidase
VPS: vacuolar protein sorting-associated kinase
VR: venous return
VRAC: volume-regulated anion channel
VRC: ventral respiratory column
VRK: vaccinia-related kinase
VS: vasostatin
vSMC: vascular smooth myocyte
vSNARE: vesicular SNAP receptor (SNARE)
VSOR: volume-sensitive outwardly rectifying anion channel
VSP: voltage-sensing phosphatase
VVO: vesiculo-vacuolar organelle
vWF: von Willebrand factor

W

$\mathbf{W}$: vorticity tensor
$\mathcal{W}$: strain energy density
W: work, deformation energy
$\mathtt{w}$: weight
$\mathbf{w}$: grid velocity
WASH: WASP and SCAR homolog
WASP: Wiskott-Aldrich syndrome protein
WAT: white adipose tissue
WAVe: WASP-family verprolin homolog
WBC: white blood cell
WDR: WD repeat-containing protein
Wee: small (Scottish)
WHAMM: WASP homolog associated with actin, membranes, and microtubules
WIP: WASP-interacting protein
WIPF: WASP-interacting protein family protein
WIPI: WD repeat domain-containing phosphoinositide-interacting protein
WNK: with no K (Lys) kinase
Wnt: wingless-type
WPWS: Wolff-Parkinson-White syndrome
WNRRTK: Wnt and neurotrophin receptor-related receptor Tyr kinase ($ROR_{(RTK)}$)
WSB: WD-repeat and SOCS box-containing protein (Ub ligase)
WSS: wall shear stress
WSSTG: WSS transverse gradient
WWTR: WW domain-containing transcription regulator

X

$\mathcal{X}$: trajectory
$\mathtt{X}$: reactance
$\mathbf{X}$: Lagrangian position vector
$\mathbf{x}$: position vector
$\{x, y, z\}$: Cartesian coordinates
XBE: X-factor-binding element
XBP: X-box-binding protein (transcription factor)
XIAP: X-linked inhibitor of apoptosis protein (Ub ligase)

Y

$\mathtt{Y}$: admittance coefficient
YAP: Yes-associated protein
YBP: Y-box-binding protein (transcription factor)
YY: yin yang (transcriptional repressor)

Z

$\mathtt{Z}$: impedance
ZAP70: ζ-associated protein 70

ZBTB: zinc finger and BTB (Broad complex, Tramtrack, and bric-à-brac) domain-containing transcription factor
ZnF: zinc finger protein
ZO: zonula occludens

Miscellaneous

2-5A: 5′-triphosphorylated, (2′,5′)-phosphodiester-linked oligoadenylate
2AG: 2-arachidonyl glycerol
3DR: three-dimensional reconstruction
3BP2: Abl Src homology-3 domain-binding adaptor
4eBP1: inhibitory eIF4e-binding protein
5HT: serotonin
7TMR: 7-transmembrane receptor (GPCR)

Complementary Lists of Notations

Greek Symbols

α: volumic fraction
α: convergence/divergence angle
α: attenuation coefficient
α_k: kinetic energy coefficient
α_m: momentum coefficient
β: inclination angle
$\{\beta_i\}_1^2$: myocyte parameters
β_T: coefficient of thermal expansion
Γ: domain boundary
Γ_L: local reflection coefficient
Γ_G: global reflection coefficient
γ: heat capacity ratio
γ: activation factor
γ_G: amplitude ratio (modulation rate) of G
$\dot{\gamma}$: shear rate
δ: boundary layer thickness
ϵ_T: emissivity (thermal energy radiation)
ϵ_e: electric permittivity
ε: strain
ε: dimensionless small quantity
ζ: singular head loss coefficient
ζ: transmural coordinate
$\{\zeta_j\}_1^3$: local coordinate
η: azimuthal spheroidal coordinate
θ: circumferential polar coordinate
θ: $(\hat{\mathbf{e}}_x, \hat{\mathbf{t}})$ angle
κ: wall curvature
κ_c: curvature ratio
κ_d: drag reflection coefficient
κ_h: hindrance coefficient
κ_o: osmotic reflection coefficient
κ_s: size ratio
$\{\kappa_k\}_{k=1}^9$: tube law coefficients
κ_e: correction factor
Λ: head loss coefficient
λ_L: Lamé coefficient
λ: stretch ratio
λ: wavelength
λ_A: area ratio
λ_a: acceleration ratio
λ_L: length ratio
λ_q: flow rate ratio
λ_t: time ratio
λ_v: velocity ratio
μ: dynamic viscosity
μ_L: Lamé coefficient
ν: kinematic viscosity
ν_P: Poisson ratio
Π: osmotic pressure
ρ: mass density
τ: time constant
Φ: potential
$\phi(t)$: creep function
φ: phase
χ: Lagrangian label
χ_i: molar fraction of species i
χ_i: wetted perimeter
$\psi(t)$: relaxation function
Ψ: porosity
ω: angular frequency
Ω: computational domain

Dual Notations

Bφ: basophil
Eφ: eosinophil
Lφ: lymphocyte
Mφ: macrophage
aaMφ: alternatively activated macrophage

M. Thiriet, *Tissue Functioning and Remodeling in the Circulatory and Ventilatory Systems*, Biomathematical and Biomechanical Modeling of the Circulatory and Ventilatory Systems 5,
DOI 10.1007/978-1-4614-5966-8,

caMφ: classically activated macrophage
Nφ: neutrophil
Σc: sympathetic
pΣc: parasympathetic

Subscripts

$_A$: alveolar, atrial
$_{Ao}$: aortic
$_a$: arterial
$_{app}$: apparent
$_{atm}$: atmospheric
$_b$: blood
$_\mathbf{c}$: contractile
$_c$: center
$_c$: point-contact
$_D$: Darcy (filtration)
$_d$: diastolic
$_{dyn}$: dynamic
$_E$: expiration, Eulerian
$_e$: external
$_e$: extremum
$_{eff}$: effective
$_f$: fluid
$_g$: grid
$_I$: inspiration
$_i$: internal
$_{inc}$: incremental
$_L$: Lagrangian
$_l$: limit
$_\ell$: line-contact
$_M$: macroscopic
$_m$: mean
$_{max}$: maximum
$_m$: muscular, mouth
$_{met}$: metabolic
$_\mu$: microscopic
$_P$: pulmonary
$_\mathbf{p}$: parallel
$_p$: particle
$_q$: quasi-ovalization
$_r$: radial
$_{rel}$: relative
$_S$: systemic
$_s$: solute
$_\mathbf{s}$: serial
$_s$: systolic
$_t$: stream division
$_T$: total
$_t$: turbulence
$_t$: time derivative of order 1
$_{tt}$: time derivative of order 2
$_{tis}$: tissue
$_V$: ventricular
$_v$: venous
$_w$: wall
$_w$: water (solvent)
$_\Gamma$: boundary
$_\theta$: azimuthal
$_+$: positive command
$_-$: negative command
$_*$: at interface
$_0$: reference state ($\cdot_0$: unstressed or low shear rate)
$_\infty$: high shear rate

Superscripts

^{a}: active state
^{e}: elastic
^{f}: fluid
^{h}: hypertensive
^{n}: normotensive
^{p}: passive state
^{P}: power
^{s}: solid
^{T}: transpose
^{v}: viscoelastic
*: scale
*: complex variable
$\cdot'$: first component of complex elastic and shear moduli
$\cdot''$: second component of complex elastic and shear moduli
$^\natural$: static, stationary, steady variable

Mathematical Notations

T: bold face capital letter means tensor
v: bold face lower case letter means vector
S, s: upper or lower case letter means scalar
$\Delta\bullet$: difference
$\delta\bullet$: increment
$d\bullet/dt$: time gradient
∂_t: first-order time partial derivative
∂_{tt}: second-order time partial derivative
∂_i: first-order space partial derivative with respect to spatial coordinate x_i
∇: gradient operator
$\nabla\mathbf{u}$: displacement gradient tensor
$\nabla\mathbf{v}$: velocity gradient tensor
$\nabla\cdot$: divergence operator

∇^2: Laplace operator
$|\ |_+$: positive part
$|\ |_-$: negative part
$\dot{\bullet}$: time derivative
$\bar{\bullet}$: time mean
$\breve{\bullet}$: space averaged
$\langle\bullet\rangle$: ensemble averaged
$\tilde{\bullet}$: dimensionless
$\bullet^+$: normalized ($\in [0,1]$)
$\hat{\bullet}$: peak value
$\bullet_\sim$: modulation amplitude
det($\bullet$): determinant
cof($\bullet$): cofactor
tr($\bullet$): trace

Cranial Nerves

I: olfactory nerve (sensory)
II: optic nerve (sensory)
III: oculomotor nerve (mainly motor)
IV: trochlear nerve (mainly motor)
V: trigeminal nerve (sensory and motor)
VI: abducens nerve (mainly motor)
VII: facial nerve (sensory and motor)
VIII: vestibulocochlear (auditory-vestibular) nerve (mainly sensory)
IX: glossopharyngeal nerve (sensory and motor)
X: vagus nerve (sensory and motor)
XI: cranial accessory nerve (mainly motor)
XII: hypoglossal nerve (mainly motor)

Chemical Notations

[X]: concentration of X species
X (x): upper and lower case letters correspond to gene and corresponding protein or conversely (i.e., Fes, FES, and fes designate protein, a proto-oncogene product that acts as a kinase, and corresponding gene and oncogene product, respectively)
$^\bullet$: radical (unpaired electron[s])
$^{\Delta NT}$: truncated form without the N-terminal domain
$^{\Delta CT}$: truncated form without the C-terminal domain
^{C}X: cardiac-specific isoform of X molecule
$^{D(L)}$X: D (L)-stereoisomer of amino acids and carbohydrates (chirality prefixes for dextro- [dexter: right] and levorotation [lævus: left]), i.e., dextro(levo)rotatory enantiomer
^{G}X: globular form of X molecule
$^{F(G)}$actin: polymeric, filamentous (monomeric, globular) actin
$_{C}$X: carboxy (carboxyl group COOH [C])-terminal cleaved part of X molecule
$_{c}$X: cytosolic molecule
$_{L,Ac}$X: lysosomal, acidic X molecule (e.g., sphingomyelinase)
$_{m}$X: membrane-bound molecule
$_{N}$X: amino (amine group NH_2 [N])-terminal cleaved part of X molecule
$_{S}$X: secreted form of X molecule
$_{S,Ac}$X: secreted, acidic molecule X (e.g., sphingomyelinase)
$_{t}$X: truncated isoform
X_i: type-i isoform of the receptor of ligand X (i: integer)
XRi: receptor isoform i of ligand X (i: integer)
X+: molecule X expressed (X-positive)
X^+: cation; also intermediate product X of oxidation (loss of electron) from a reductant (or reducer) by an oxidant (electron acceptor that removes electrons from a reductant)
X−: molecule X absent (X-negative)
X^-: anion; also intermediate product X of reduction (gain of electron) from an oxidant (or oxidizer) by a reductant (electron donor that transfers electrons to an oxidant)
X^A: activator form of molecule X
X^a: active form of molecule X
X^{ECD}: soluble fragment correponding to the ectodomain of molecule X after extracellular proteolytic cleavage and shedding (possible extracellular messenger or sequestrator)
$X^{(ER)}$: endoplasmic reticulum type of molecule X
small GTPase$^{GTP(GDP)}$: active (inactive) form of small (monomeric), regulatory guanosine triphosphatase
$X^{GTP(GDP)}$: GTP (GDP)-loaded molecule X
X^{high}: molecule X produced at high levels
X^{ICD}: soluble fragment correponding to intracellular domain of molecule X after intracellular proteolytic cleavage (possible messenger and/or transcription factor; e.g., Notch^{ICD}: intracellular Notch fragment)

X^{low}: molecule X produced at low levels
X^{MT}: mitochondrial type of molecule X
X^{PM}: plasmalemmal type of molecule X
X^{R}: repressor form of molecule X
X^{S}: soluble form
X_{alt}: alternative splice variant
X_{FL}: full-length protein X
$X_{h(l,m)MW}$: high (low, mid)-molecular-weight isotype
$X_{L(S)}$: long (short) isoform (splice variants)
X_c: catalytic subunit
X_i: number of molecule or atom (i: integer, often 2 or 3)
$(X_1–X_2)_i$: oligomer made of i complexes constituted of molecules X_1 and X_2 (e.g., histones)
a, c, nX: atypical, conventional, novel molecule X (e.g., PKC)
al, ac, nX: alkaline, acidic, neutral molecule X (e.g., sphingomyelinase)
asX: alternatively spliced molecule X (e.g., asTF)
cX: cellular, cytosolic, constitutive (e.g., cNOS), or cyclic (e.g., cAMP and cGMP) X molecule
caX: cardiomyocyte isoform (e.g., caMLCK)
dX: deoxyX
eX: endothelial isoform (e.g., eNOS and eMLCK)
hX: human form (ortholog); heart type (e.g., hFABP); hormone-like isoform (FGF)
iX: inhibitory mediator (e.g., iSMAD) or intracellular (e.g., iFGF) or inducible (e.g., iNOS) isoform
kX: renal type (kidney) X molecule
ksX: kidney-specific isoform of X molecule
lX: lysosomal X molecule
lpX: lipoprotein-associated X molecule (e.g., lpPLA2)
mX: mammalian species or membrane-associated X molecule (e.g., mTGFβ)
mtX: mitochondrial type of X molecule
nX: neutral X; neuronal type (e.g., nWASP)
oxX: oxidized X molecule (e.g., oxLDL)
plX: plasmalemmal type of X molecule
rX: receptor-associated mediator or receptor-like enzyme; also regulatory type of molecular species (e.g., rSMAD)
skX: skeletal myocyte isoform (e.g., skMLCK)
smcX: smooth muscle cell isoform (e.g., smcMLCK)
tX: target type of X (e.g., tSNARE); tissue type (e.g., tPA)
tmX: transmembrane type of X
vX: vesicle-associated (e.g., vSNARE) or vacuolar (e.g., vATPase) type of X
GPX: glycoprotein (X: molecule abbreviation or assigned numeral)
Xx: (x: single letter) splice variants
X1: human form (ortholog)
Xi: isoform type i (paralog or splice variant; i: integer)
Xi/j: (i,j: integers) refers to either both isoforms (i.e., Xi and Xj, such as ERK1/2) or heterodimer (i.e., Xi–Xj, such as ARP2/3)
X1/X2: molecular homologs or commonly used aliases (e.g., contactin-1/F3)
PI(i)P, PI(i,j)P_2, PI(i,j,k)P_3: i,j,k (integers): position(s) of phosphorylated OH groups of the inositol ring of phosphatidylinositol mono-, bis-, and trisphosphates

Post-Translational Modification

X^{M}: methylated molecule X
X^{P}: phosphorylated molecule X
pAA: phosphorylated amino acid (pSer, pThr, and pTyr)
X^{SNO}: ^{S}nitrosylated molecule X
X^{U}: ubiquitinated protein X
X_A: acetylated molecule X
acX: acetylated molecule X (e.g., acLDL)
X_M: myristoylated molecule X
X_P: palmitoylated molecule X

Amino Acids

Ala (A): alanine
Arg (R): arginine
Asn (N): asparagine
Asp (D): aspartic acid
Asp^{COO^-}: aspartate
CysH (C): cysteine
Cys: cystine
Gln (Q): glutamine
Glu (E): glutamic acid
Glu^{COO^-}: glutamate
Gly (G): glycine
His (H): histidine

Iso, Ile (I): isoleucine
Leu (L): leucine
Lys (K): lysine
Met (M): methionine
Orn: ornithine (not encoded by DNA, but use in the urea cycle)
Phe (F): phenylalanine
Pro (P): proline
Ser (S): serine
Thr (T): threonine
Trp (W): tryptophan
Tyr (Y): tyrosine
Val (V): valine

Ions

Asp^-: aspartate (carboxylate anion of aspartic acid)
ADP^{3-}: ADP anion
ATP^{4-}: ATP anion
Ca^{2+}: calcium cation
Cl^-: chloride anion
Co^{2+}: cobalt cation
Cu^+: copper monovalent cation
Cu^{2+}: copper divalent cation
Fe^{2+}: ferrous iron cation
Fe^{3+}: ferric iron cation
Glu^-: glutamate (carboxylate anion of glutamic acid)
H^+: hydrogen cation (proton)
H_3O^+: hydronium (oxonium or hydroxonium) cation
HCO_3^-: bicarbonate anion
HPO_4^{2-}: hydrogen phosphate anion
K^+: potassium cation
Mg^{2+}: magnesium cation
$^{Mg}ATP^{2-}$: ATP anion
Mn^{2+}: manganese cation
Na^+: sodium cation
Ni^{2+}: nickel cation (common oxidation state)
OH^-: hydroxide anion
PO_4^{3-}: phosphate anion
SO_4^{2-}: sulfate anion
Zn^{2+}: zinc cation (common oxidation state)

Inhaled and Signaling Gas

CO: carbon monoxide (or carbonic oxide; signaling gas and pollutant)
CO_2: carbon dioxide (cell waste)
H_2S: hydrogen sulfide (signaling gas)
He: helium (inert monatomic gas)
N_2: nitrogen (inert diatomic gas)
NO: nitric oxide (or nitrogen monoxide; signaling gas and pollutant)
NO_2: nitrogen dioxide (air pollutant)
O_2: oxygen (cell energy producer)
SO_2: sulfur dioxide (air pollutant)

Nitric Oxide Derivatives

$NO^\bullet$: free radical form
NO^+: nitrosyl or nitrosonium cation
NO^-: nitroxyl or hyponitrite anion (inodilator)
HNO: protonated nitroxyl anion
HNO_2: nitrous acid
NO_2^-: nitrite anion
NO_3^-: nitrate anion

Reactive Oxygen and Nitrogen Species

H_2O_2: hydrogen peroxide
HOCl: hypochlorous acid
N_2O_3: dinitrogen trioxide
$NO_2^\bullet$: nitrogen dioxide
$^1O_2$: singlet oxygen
O_2^-: superoxide ($O_2^{\bullet -}$)
$O{=}C(O^\bullet)O^-$: carbonate radical
$OH^\bullet$: hydroxyl radical (hydroxide ion neutral form)
$ONOO^-$: peroxynitrite
$RO^\bullet$: alkoxyl
$RO_2^\bullet$: peroxyl

Moieties (R denotes an organic group)

R: alkyl group (with only carbon and hydrogen atoms linked exclusively by single bonds)
$R{-}CH_3$: methyl group (with 3 forms: methanide anion [CH_3^-], methylium cation [CH_3^+], and methyl radical [$CH_3^\bullet$])
R–CHO: aldehyde group
R–CN: nitrile group
R–CO: acyl group
R–CO–R: carbonyl group
$R{-}COO^-$: carboxylate group

R–COOH: carboxyl group
R–NC: isonitrile group
R–NCO: isocyanate group
R–NH_2: amine group
R–NO: nitroso group
R–NO_2: nitro group
R–O: alkoxy group
R=O: oxo group
R–OCN: cyanate group
R–OH: hydroxyl group
R–ONO: nitrosooxy group
R–ONO_2: nitrate group
R–OO–R: peroxy group
R–OOH: hydroperoxy group
R–S–R: sulfide group
R–SH: thiol (or sulfhydryl) moiety
R–SN: sulfenyl-amide moiety
R–SNO: nitrosothiol (or thionitrite) moiety
R–SO: sulfinyl
R–SO–R: sulfoxide group
R–SO_2: sulfonyl group
R–SO_2H: sulfinic acid (sulfinyl moiety)
R–SO_2N: sulfonyl-amide moiety
R–SO_3H: sulfonic acid (sulfonyl moiety)
R–SOH: sulfenic acid (sulfenyl moiety)
R–SON: sulfinyl-amide moiety
R–SS–R: disulfide group

Time Units

d: day
h: hour
mn: minute
s: second
wk: week

SI-Based and Non-SI Units of Quantity

mmHg: millimeter of mercury (133.322 Pa [$\sim$ 0.1333 kPa])
mmol, nmol, μmol: milli-, nano-, micromoles
(amount of a chemical species, one mole containing about $6.02214078 \times 10^{23}$ molecules)
mosm: milliosmole
(osm: number of moles of a osmotically active chemical compound)
kDa: kiloDalton
(Da: atomic or molecular mass unit)
ppm: parts per million
l: liter

Index

M. Thiriet, *Tissue Functioning and Remodeling in the Circulatory and Ventilatory Systems*, Biomathematical and Biomechanical Modeling of the Circulatory and Ventilatory Systems 5, DOI 10.1007/978-1-4614-5966-8, © Springer Science+Business Media New York 2013

B

C

H

I

J

K

L

M

R

GPSR Compliance

The European Union's (EU) General Product Safety Regulation (GPSR) is a set of rules that requires consumer products to be safe and our obligations to ensure this.

If you have any concerns about our products, you can contact us on ProductSafety@springernature.com

In case Publisher is established outside the EU, the EU authorized representative is:

Springer Nature Customer Service Center GmbH
Europaplatz 3
69115 Heidelberg, Germany

Batch number: 10584682

Printed by Printforce, the Netherlands